FUNDAMENTALS OF THE THEORY OF OPERATOR ALGEBRAS

VOLUME II

Advanced Theory

This is a volume in
PURE AND APPLIED MATHEMATICS

A Series of Monographs and Textbooks

Editors: SAMUEL EILENBERG AND HYMAN BASS

A list of recent titles in this series is available from the publisher upon request.

FUNDAMENTALS OF THE THEORY OF OPERATOR ALGEBRAS

VOLUME II

Advanced Theory

Richard V. Kadison

Department of Mathematics
University of Pennsylvania
Philadelphia, Pennsylvania

John R. Ringrose

School of Mathematics
University of Newcastle
Newcastle upon Tyne, England

1986

ACADEMIC PRESS, INC.
Harcourt Brace Jovanovich, Publishers
Orlando San Diego New York Austin
Boston London Sydney Tokyo Toronto

ACADEMIC PRESS, INC.
Orlando, Florida 32887

United Kingdom Edition published by
ACADEMIC PRESS INC. (LONDON) LTD.
24–28 Oval Road, London NW1 7DX

Library of Congress Cataloging in Publication Data
(Revised for vol. 2)

Kadison, Richard V., Date
Fundamentals of the theory of operator algebras.
(Pure and applied mathematics ; 100)
Includes bibliographies and indexes.
Contents: v. 1. Elementary theory – v. 2.
Advanced theory.
1. Operator algebras. I. Ringrose, John R.
II. Title. III. Series: Pure and applied mathematics
(Academic Press) ; 100.
QA3.P8 vol. 100 510 s [512'.55] 82-1376
[QA326]
ISBN 0–12–393302–1 (v. 2 : alk. paper)

PRINTED IN THE UNITED STATES OF AMERICA

86 87 88 89 9 8 7 6 5 4 3 2 1

CONTENTS

Chapter 9. **Algebra and Commutant**

Chapter 10. **Special Representation of C^*-Algebras**

Chapter 11. **Tensor Products**

PREFACE

Most of the comments in the preface appearing at the beginning of Volume I are fully applicable to this second volume. This is particularly so for the statement of our primary goal: to *teach* the subject rather than be encyclopaedic. Some of those comments refer to possible styles of reading and using Volume I. The reader who has studied the first volume following the plan that avoids all the material on unbounded operators can continue in this volume, deferring Lemma 6.1.10, Theorem 6.1.11, and Theorem 7.2.1′ with its associated discussion to a later reading. This program will take the reader to Section 9.2, where Tomita's modular theory is developed. At that point, an important individual decision should be made: Is it time to retrieve the unbounded operator theory or shall the first reading proceed without it? The reader can continue without that material through all sections of Chapters 9 (other than Section 9.2), 10, 11, and 12 (ignoring Subsection 11.2, *Tensor products of unbounded operators*, which provides an alternative approach to the commutant formula for tensor products of von Neumann algebras). However, avoiding Section 9.2 makes a large segment of the post-1970 literature of von Neumann algebras unavailable. Depending on the purposes of the study of these volumes, that might not be a workable restriction. Very little of Chapter 13 is accessible without the results of Section 9.2, but Chapter 14 can be read completely.

Another shortened path through this volume can be arranged by omitting some of the alternative approaches to results obtained in one way. For example, the first subsection of Section 9.2 may be read and the last two omitted on the first reading. The last subsection of Section 11.2 may also be omitted. It is not recommended that Section 7.3 be omitted on the first reading although it does deal primarily with an alternative approach to the theory of normal states. Too many of the results and techniques appearing in that section reappear in the later chapters. Of course, all omissions affect the exercises and groups of exercises that can be undertaken.

As noted in the preface appearing in Volume I, certain exercises (and groups of exercises) "constitute small (guided) research projects." Samples of this are: the Banach–Orliz theorem developed in Exercises 1.9.26 and 1.9.34; the theory of compact operators developed in Exercises 2.8.20–2.8.29, 3.5.17, and 3.5.18; the theory of $\beta(\mathbb{N})$ developed in Exercises 3.5.5, 3.5.6, and 5.7.14–5.7.21. There are many other such instances. To a much greater extent, this process was used in the design of exercises for the present volume; results on diagonalizing abelian, self-adjoint families of matrices over a von Neumann algebra are developed in Exercises 6.9.14–6.9.35; the algebra of unbounded operators affiliated with a finite von Neumann algebra is constructed in Exercises 6.9.53–6.9.55, 8.7.32–8.7.35, and 8.7.60. The representation-independent characterizations of von Neumann algebras appear in Exercises 7.6.35–7.6.45 and 10.5.85–10.5.87. The Friedrichs extension of a positive symmetric operator affiliated with a von Neumann algebra is described in Exercises 7.6.52–7.6.55, and this topic is needed in the development of the theory of the positive dual and self-dual cones associated with von Neumann algebras that appears in Exercises 9.5.51–9.6.65. A detailed analysis of the intersection with the center of various closures of the convex hull of the unitary conjugates of an operator in a von Neumann algebra is found in Exercises 8.7.4–8.7.22, and the relation of these results to the theory of conditional expectations in von Neumann algebras is the substance of the next seven exercises; this analysis is also applied to the development of the theory of (bounded) derivations of von Neumann algebras occurring in Exercises 8.7.51–8.7.55 and 10.5.76–10.5.79. Portions of the theory of representations of the canonical anticommutation relations appear in Exercises 10.5.88–10.5.90, 12.5.39, and 12.5.40. This list could continue much further; there are more than 1100 exercise tasks apportioned among 450 exercises in this volume. The index provides a usable map of the topical relation of exercises through key-word references.

Each exercise has been designed, by arrangement in parts and with suitable hints, to be realistically capable of solution by the techniques and skills that will have been acquired in a careful study of the chapters preceding the exercise. However, full solutions to all the exercises in a topic grouping may require serious devotion and time. Such groupings provide material for special seminars, either in association with a standard course or by themselves. Seminars of that type are an invaluable "hands-on" experience for active students of the subject.

Aside from the potential for working seminars that the exercises supply, a fast-paced, one-semester course could cover Chapters 6–9. The second semester might cover the remaining chapters of this volume. A more leisurely pace might spread Chapters 6–10 over a one-year course, with an expansive treatment of modular theory (Section 9.2) and a careful review (study) of the unbounded operator theory developed in Sections 2.7 and 5.6 of Volume I. Chapters 11–14 could be dealt with in seminars or in an additional semester course. In addition to these course possibilities, both volumes have been written with the possibility of self-study very much in mind.

The list of references and the index in this volume contain those of Volume I. Again, the reference list is relatively short, for the reasons mentioned in the preface in Volume I. A special comment must be made about the lack of references in the exercise sections. Many of the exercises (especially the topic groupings) are drawn from the literature of the subject. In designing the exercises (parts, hints, and formulation), complete, model solutions have been constructed. These solutions streamline, simplify, and unify the literature on the topic in almost all cases; on occasion, new results are included. References to the literature in the exercise sets could misdirect more than inform the reader. It seems expedient to defer references for the exercises to volumes containing the exercises and model solutions; a significant number of references pertain directly to the solutions. We hope that the benefits from the more sensible references in later volumes will outweigh the present lack; our own publications have been one source of topic groupings subject to this policy.

Again, individual purposes should play a dominant role in the proportion of effort the reader places on the text proper and on the exercises. In any case, a good working procedure might be to include a careful scanning of the exercise sets with a reading of the text even if the decision has been made not to devote significant time to solving exercises.

CONTENTS OF VOLUME I

Chapter 4. **Elementary C^*-Algebra Theory**

Chapter 5. **Elementary von Neumann Algebra Theory**

Bibliography

CHAPTER 6

COMPARISON THEORY OF PROJECTIONS

We take up the detailed study of von Neumann algebras in this chapter. The principal tool for this study is the technique of "comparison" of the projections in a von Neumann algebra $\mathscr{R}$ relative to $\mathscr{R}$. By these means we develop a notion of "equivalence" of projections in $\mathscr{R}$ (meaning, loosely, "of the same size relative to $\mathscr{R}$"). Associated with this equivalence, we have a partial ordering of (the equivalence classes of) projections in $\mathscr{R}$—with corresponding notions of "finite" and "infinite" projections relative to $\mathscr{R}$. In these terms, we can separate von Neumann algebras into broad types (algebraically non-isomorphic) and show that each such algebra is a direct sum of algebras of the various types (the so-called "type decomposition" of von Neumann algebras). The simplest of the types ("Type I von Neumann algebras") is analyzed and examples of some of the other types are studied.

6.1. Polar decomposition and equivalence

In the discussion following Lemma 2.4.8, we observed that each bounded operator T on a Hilbert space $\mathscr{H}$ can be expressed as $H + iK$, with H and K self-adjoint operators. We referred to H and K as the "real" and "imaginary" parts of T—noting the analogy between this representation of T and the corresponding representation of a complex number in terms of its real and imaginary parts.

If we pursue the analogy between representations (decompositions) of complex numbers and those of linear operators, we are led to consider the possibility of a "polar decomposition" of operators analogous to the decomposition of a complex number as the product of a positive number (its modulus) and a number of modulus 1.

With the function calculus for self-adjoint operators at our disposal, there is no problem in producing a "polar decomposition" for an invertible operator T. As modulus, both $(T^*T)^{1/2}$ and $(TT^*)^{1/2}$ suggest themselves. At first guess, we might expect the number of modulus 1 in the polar decomposition of a complex number to correspond to a unitary operator in the case of an operator. The non-commutativity of the operator situation introduces a complicating factor. Shall we multiply the modulus of T on the left or right by the unitary operator (if it is, indeed, to be a unitary operator); and which of $(T^*T)^{1/2}$, $(TT^*)^{1/2}$ shall we use as modulus? A small amount of experimentation shows that writing $T = U(T^*T)^{1/2}$ (somewhat hopefully), and, then, "solving" for U as $T(T^*T)^{-1/2}$ produces a unitary operator U (while $T(TT^*)^{-1/2}$ will not, in general, be unitary—nor would $(T^*T)^{-1/2}T$). The computation involved in this is

$$\langle T(T^*T)^{-1/2}x, T(T^*T)^{-1/2}x\rangle = \langle (T^*T)^{-1/2}T^*T(T^*T)^{-1/2}x, x\rangle = \langle x, x\rangle.$$

If WH is another "polar decomposition" of T (with W unitary and H positive), then $H = W^*T$ so that $H^2 = H^*H = T^*WW^*T = T^*T$. As $H \geq 0$, and the positive square root of a positive operator is unique (see Theorem 4.2.6), $H = (T^*T)^{1/2}$ and $W = T(T^*T)^{-1/2} = U$. Of course, $T^* = (T^*T)^{1/2}U^*$, while T^* has its own polar decomposition, $T^* = V^*(T^{**}T^*)^{1/2} = V^*(TT^*)^{1/2}$. Thus $T = (TT^*)^{1/2}V$; and this last equality provides a "polar decomposition" for T with the positive operator factor appearing on the left. This, incidentally, redresses the balance between the two candidates for "modulus" of T. Combining $T = U(T^*T)^{1/2}$ and $T^* = (T^*T)^{1/2}U^*$, we have $TT^* = U(T^*T)U^*$ (so that TT^* and T^*T are unitarily equivalent, when T is invertible). Since $U(T^*T)^{1/2}U^*$ is a positive square root of $U(T^*T)U^*$, $(TT^*)^{1/2} = U(T^*T)^{1/2}U^*$. But $V^*(TT^*)^{1/2} = T^* = (T^*T)^{1/2}U^*$, so that $UV^*(TT^*)^{1/2} = U(T^*T)^{1/2}U^* = (TT^*)^{1/2}$; and $V = U$. Thus the same unitary operator appears in the "right" and "left" polar decomposition of T.

For the polar decomposition of the general bounded operator, we must replace the unitary operators of the preceding discussion by operators that map one (closed) subspace of a Hilbert space isometrically onto another and annihilate the orthogonal complement of the first subspace. Such operators are called *partial isometries*. The first subspace is called the *initial space* of the partial isometry, and the second subspace (its range) is called its *final space*. The projections with these subspaces as ranges are called the *initial* and *final projections*, respectively, of the partial isometry.

6.1.1. PROPOSITION. *The operator V acting on the Hilbert space $\mathscr{H}$ is a partial isometry if and only if V^*V is a projection E. In this case, E is the initial projection of V, VV^* is the final projection F of V, and V^* is a partial isometry with initial projection F and final projection E.*

Proof. Suppose, first, that V is a partial isometry with initial projection E. Then $\|Vx\| = \|VEx + V(I - E)x\| = \|VEx\| = \|Ex\| \leq \|x\|$; so that $\|V\| \leq 1$. If x is a unit vector in the range of E, then $1 = \langle x, x\rangle = \langle Vx, Vx\rangle =$

$\langle V^*Vx, x\rangle$. From Proposition 2.1.3 (the "Cauchy–Schwarz equality"), $V^*Vx = x$. If y is in the range of $I - E$, $V^*Vy = V^*(0) = 0$. Thus $V^*V = E$.

Suppose, now, that V^*V is a projection E. Then for each x in the range of E, $\langle x, x\rangle = \langle V^*Vx, x\rangle = \langle Vx, Vx\rangle$; while, for y orthogonal to the range of E, $0 = \langle V^*Vy, y\rangle = \langle Vy, Vy\rangle$. Thus V is isometric on $E(\mathscr{H})$ and 0 on $(I - E)(\mathscr{H})$. It follows that V is a partial isometry with initial projection E. In addition, $V = VE = VV^*V$, and $VV^*VV^* = VEV^* = VV^*$. Thus VV^* is a projection F and $FV = V$. Consequently $F(\mathscr{H})$ contains $V(\mathscr{H})$. But $F(\mathscr{H}) = VV^*(\mathscr{H}) \subseteq V(\mathscr{H})$. Hence F is the final projection of V. As $VV^* = (V^*)^*V^* = F$, we conclude, from the foregoing, that V^* is a partial isometry with initial projection F and final projection E. ∎

6.1.2. THEOREM (Polar decomposition). *If T is a bounded operator on the Hilbert space $\mathscr{H}$, there is a partial isometry V with initial space the closure* $\mathrm{r}(T^*)$ *of the range of T^* and final space* $\mathrm{r}(T)$ *such that $T = V(T^*T)^{1/2} = (TT^*)^{1/2}V$. If $T = WH$ with H positive and W a partial isometry whose initial space is* $\mathrm{r}(H)$, *then $H = (T^*T)^{1/2}$ and $W = V$. If neither T nor T^* annihilates a non-zero vector, then V is a unitary operator.*

Proof. Recall from Proposition 2.5.13 that $\mathrm{r}(T^*) = \mathrm{r}(T^*T)$ so that $\mathrm{r}(T^*) = \mathrm{r}((T^*T)^{1/2})$. Since

$$\langle (T^*T)^{1/2}x, (T^*T)^{1/2}x\rangle = \langle T^*Tx, x\rangle = \langle Tx, Tx\rangle,$$

there is a partial isometry V with initial space $\mathrm{r}(T^*)$ and final space $\mathrm{r}(T)$ such that $T = V(T^*T)^{1/2}$. Thus $T^* = (T^*T)^{1/2}V^*$ and $TT^* = VT^*TV^*$. Now

$$[V(T^*T)^{1/2}V^*]^2 = VT^*TV^* = TT^*,$$

so that $V(T^*T)^{1/2}V^* = (TT^*)^{1/2}$. Hence

$$T = V(T^*T)^{1/2} = V(T^*T)^{1/2}V^*V = (TT^*)^{1/2}V.$$

(Note, for this, that $V^*V = R((T^*T)^{1/2})$, from Proposition 6.1.1, so that $(T^*T)^{1/2} = V^*V(T^*T)^{1/2} = (T^*T)^{1/2}V^*V$.)

With W and H as described, $W^*WH = H$, so that $T^*T = HW^*WH = H^2$. Hence $H = (T^*T)^{1/2}$ and $W = V$.

If T and T^* have (0) as null space, their ranges are dense in $\mathscr{H}$. Hence V is a unitary operator, in this case. ∎

Note that $(T^*T)^{1/2}$ and $(TT^*)^{1/2}$ are contained in each C^*-algebra containing T. However, V may not lie in such an algebra. If T is a positive operator, V is $R(T)$. With $\mathfrak{A}$ the algebra of multiplications by continuous functions on $L_2([0, 1])$ (relative to Lebesgue measure) and H multiplication by a positive function that vanishes on $[0, \frac{1}{2}]$, $R(H)$ is a projection different from 0 and I. Since $\mathfrak{A}$ contains no projections other than 0 and I, the polar decomposition of H cannot be effected in $\mathfrak{A}$. If T is invertible, $(T^*T)^{1/2}$ and $U\ (=T(T^*T)^{-1/2})$ lie in each C^*-algebra containing T. The critical information concerning the possibility of polar decomposition within a C^*-algebra is found in the proposition that follows.

6.1.3. PROPOSITION. *If T lies in a von Neumann algebra $\mathcal{R}$ and UH is the polar decomposition of T, then U and H are in $\mathcal{R}$.*

Proof. As noted, $H = (T^*T)^{1/2} \in \mathcal{R}$, since $\mathcal{R}$ is, in particular, a C^*-algebra containing T. If $T' \in \mathcal{R}'$, $T'UHx = T'Tx = TT'x$; while $UT'Hx = UHT'x = TT'x$. Thus UT' and $T'U$ agree on the range of H. Since T' commutes with H, both the range of H and its orthogonal complement are stable under T'. As U is 0 on this complement, both UT' and $T'U$ are 0 there. Thus $UT' = T'U$ and $U \in \mathcal{R}'' = \mathcal{R}$. ■

If T is normal, $(T^*T)^{1/2} = (TT^*)^{1/2}(=H)$. Thus $UH = T = HU$ (from Theorem 6.1.2). Conversely, from uniqueness of the polar decomposition ("left" and "right"), if $UH = HU$, $(T^*T)^{1/2} = (TT^*)^{1/2}$ and $T^*T = TT^*$.

To compare the dimensions of the ranges of two projections E and F acting on a Hilbert space, we compare the cardinality of orthonormal bases for each of these subspaces. Another (equivalent) technique for comparing the dimensions of the ranges of E and F to see if they are the same would be to seek a partial isometry with one as initial projection and the other as final projection. If E and F lie in a von Neumann algebra $\mathcal{R}$ and we insist that our partial isometry lie in $\mathcal{R}$, we are demanding a stricter comparison of E and F—a comparison relative to $\mathcal{R}$. The structure of $\mathcal{R}$ would seem to exert an important influence on the possibility of comparison; and, consequently, the structure this comparison process imposes on the projections of $\mathcal{R}$ will reflect the structure of $\mathcal{R}$.

Elaborating this idea leads to the Murray–von Neumann comparison theory of projections in a factor and its extension to a comparison theory of projections in a von Neumann algebra.

6.1.4. DEFINITION. Two projections E and F are said to be equivalent relative to a von Neumann algebra $\mathcal{R}$ (written, $E \sim F(\mathcal{R})$) when $V^*V = E$ and $VV^* = F$ for some V in $\mathcal{R}$. ■

In view of Proposition 6.1.1, the operator V in $\mathcal{R}$ is a partial isometry with initial projection E and final projection F. Since $E = V^*V$ and $F = VV^*$, both E and F are in $\mathcal{R}$. Most often, the von Neumann algebra $\mathcal{R}$ relative to which the equivalence of E and F is being asserted will be clearly indicated by the context. In this case we say that E is equivalent to F and write $E \sim F$.

In the proposition that follows, we show that the relation $\sim$ defined on the projections of $\mathscr{R}$ is an equivalence relation.

6.1.5. PROPOSITION. *If projections E, F, G in a von Neumann algebra $\mathscr{R}$ satisfy $E \sim F$ and $F \sim G$, then $F \sim E$, $E \sim G$, and $E \sim E$.*

Proof. Since $E \sim F$ and $F \sim G$, there are partial isometries V and W in $\mathscr{R}$ such that $V^*V = E$, $VV^* = F$, $W^*W = F$, and $WW^* = G$. Thus $F = (V^*)^*V^*$ and $E = V^*(V^*)^*$; so that $F \sim E$. As $E = E^*E = EE^*$, E is a partial isometry with initial and final projection E; and $E \sim E$. Finally,

$$(WV)^*WV = V^*W^*WV = V^*FV = V^*V = E;$$

while

$$WV(WV)^* = WVV^*W^* = WFW^* = WW^* = G.$$

Thus $E \sim G$. ∎

Employing the polar decomposition, we establish a result that provides the main technique for proving equivalence of projections in a von Neumann algebra.

6.1.6. PROPOSITION. *If $\mathscr{R}$ is a von Neumann algebra and $T \in \mathscr{R}$, then $R(T) \sim R(T^*)$.*

Proof. From Theorem 6.1.2 and Proposition 6.1.3, $T = V(T^*T)^{1/2}$, $V \in \mathscr{R}$, and V is a partial isometry with initial projection $R(T^*)$ and final projection $R(T)$. Thus $R(T) \sim R(T^*)$. ∎

6.1.7. THEOREM (Kaplansky formula). *If E and F are projections in a von Neumann algebra $\mathscr{R}$, then $(E \vee F - F) \sim (E - E \wedge F)$.*

Proof. We note that $E \vee F - F$ is the range projection of $(I - F)E$, while $E - E \wedge F$ is the range projection of $E(I - F)$ $(=[(I - F)E]^*)$. Once this has been established, the Kaplansky formula follows from Proposition 6.1.6. From Proposition 2.5.14, $R(E(I - F)) = E - E \wedge F$; and $R((I - F)E) = I - F - (I - F) \wedge (I - E) = E \vee F - F$ (since $I - (I - F) \wedge (I - E) = E \vee F$). ∎

6.1.8. PROPOSITION. *Two projections E and F in a von Neumann algebra $\mathscr{R}$ have non-zero equivalent subprojections if and only if $C_E C_F \neq 0$.*

Proof. If $C_E C_F = 0$, $E_0 \leq E$, $F_0 \leq F$, and $E_0 \sim F_0$, there is a partial isometry V in $\mathscr{R}$ such that $V^*V = E_0$ and $VV^* = F_0$. Since $F_0 \leq F \leq C_F$ and $E_0 \leq E \leq C_E$, $V = F_0VE_0 = F_0C_FVC_EE_0 = F_0VE_0C_FC_E = 0$. Thus $0 = E_0 = F_0$.

If $C_E C_F \neq 0$, then $[\mathscr{R}E(\mathscr{H})] \wedge [\mathscr{R}F(\mathscr{H})] \neq (0)$, from Proposition 5.5.2 (and Proposition 2.5.3). Thus there are operators A, B in $\mathscr{R}$ and vectors x, y such that $0 \neq \langle AEx, BFy \rangle = \langle FB^*AEx, y \rangle$, and $FTE \neq 0$, where $T = B^*A \in \mathscr{R}$. It follows that $R(FTE)$ and $R(ET^*F)$ are non-zero projections in $\mathscr{R}$. From Proposition 6.1.6, they are equivalent. Of course $R(FTE) \leq F$ and $R(ET^*F) \leq E$. ∎

6.1.9. Corollary. *Each pair of non-zero projections in a factor have equivalent non-zero subprojections.*

Proof. If $E \neq 0 \neq F$, then $C_E \neq 0 \neq C_F$. Since the only non-zero central projection in a factor is I, $C_E = I = C_F$. Thus $C_E C_F = I \neq 0$ and Proposition 6.1.8 applies. It follows that E and F have equivalent non-zero subprojections. ∎

An extension of the polar decomposition to the case of a closed densely defined linear transformation from one Hilbert space to another forms the basis for the developments in Section 9.2. We describe this extension before passing to a detailed study of the partial ordering of (the equivalence classes of) projections associated with our equivalence relation. The following simple lemma will prove useful to us.

6.1.10. Lemma. *If A and C are densely defined preclosed operators and B is a bounded operator such that $A = BC$, then $A^* = C^*B^*$.*

Proof. If $y \in \mathscr{D}(A^*)$, then, for each x in $\mathscr{D}(A)$ $(=\mathscr{D}(C))$,

$$\langle x, A^*y \rangle = \langle Ax, y \rangle = \langle BCx, y \rangle = \langle Cx, B^*y \rangle;$$

so that $B^*y \in \mathscr{D}(C^*)$ and $C^*B^*y = A^*y$. If $y \in \mathscr{D}(C^*B^*)$, then $B^*y \in \mathscr{D}(C^*)$ and, for each x in $\mathscr{D}(C)$ $(=\mathscr{D}(A))$,

$$\langle x, C^*B^*y \rangle = \langle Cx, B^*y \rangle = \langle BCx, y \rangle = \langle Ax, y \rangle;$$

so that $y \in \mathscr{D}(A^*)$ and $A^*y = C^*B^*y$. ∎

6.1.11. Theorem. *If T is a closed densely defined linear transformation from one Hilbert space to another, there is a partial isometry V with initial space the closure of the range of $(T^*T)^{1/2}$ and final space the closure of the range of T such that $T = V(T^*T)^{1/2} = (TT^*)^{1/2}V$. Restricted to the closures of the ranges of T^* and T, respectively, T^*T and TT^* are unitarily equivalent (and V implements this equivalence). If $T = WH$, where H is a positive operator and W is a partial isometry with initial space the closure of the range of H, then $H = (T^*T)^{1/2}$ and $W = V$. If $\mathscr{R}$ is a von Neumann algebra, $T \,\eta\, \mathscr{R}$ if and only if $V \in \mathscr{R}$ and $(T^*T)^{1/2} \,\eta\, \mathscr{R}$.*

Proof. From Theorem 2.7.8(v), T^*T is self-adjoint. If $x \in \mathscr{D}(T^*T)$, then $x \in \mathscr{D}(T)$, $Tx \in \mathscr{D}(T^*)$, and

$$0 \leq \langle Tx, Tx\rangle = \langle T^*Tx, x\rangle.$$

Thus T^*T is positive and has a (unique) positive square root $(T^*T)^{1/2}$. (See Proposition 5.6.21 and Remark 5.6.32.) From Remark 2.7.7, $\mathscr{D}(T^*T)$ is a core for $(T^*T)^{1/2}$ and for T. Thus $(T^*T)^{1/2}$ and T map $\mathscr{D}(T^*T)$ onto dense subsets of their ranges. Defining $V_0(T^*T)^{1/2}x$ to be Tx, for x in $\mathscr{D}(T^*T)$, V_0 extends to a partial isometry V with initial space the closure of the range of $(T^*T)^{1/2}$ and final space the closure of the range of T, since

$$\langle (T^*T)^{1/2}x, (T^*T)^{1/2}x\rangle = \langle T^*Tx, x\rangle = \langle Tx, Tx\rangle.$$

Moreover, $Tx = V(T^*T)^{1/2}x$ for each x in $\mathscr{D}(T^*T)$.

With x in $\mathscr{D}(V(T^*T)^{1/2})$, choose x_n in $\mathscr{D}(T^*T)$ such that $x_n \to x$ and $(T^*T)^{1/2}x_n \to (T^*T)^{1/2}x$. Then $Tx_n = V(T^*T)^{1/2}x_n \to V(T^*T)^{1/2}x$. Since T is closed, $x \in \mathscr{D}(T)$ and $Tx = V(T^*T)^{1/2}x$. Thus $V(T^*T)^{1/2} \subseteq T$.

Conversely, if $x \in \mathscr{D}(T)$ and x_n is chosen in $\mathscr{D}(T^*T)$ such that $x_n \to x$ and $Tx_n \to Tx$, then $(T^*T)^{1/2}x_n = V^*V(T^*T)^{1/2}x_n = V^*Tx_n \to V^*Tx$. Since $(T^*T)^{1/2}$ is closed, $x \in \mathscr{D}((T^*T)^{1/2})$. It follows that $T = V(T^*T)^{1/2}$.

From Lemma 6.1.10, $T^* = (T^*T)^{1/2}V^*$, so that $TT^* = VT^*TV^*$. Thus the restriction of TT^* to the closure of the range of T is unitarily equivalent to the restriction of T^*T to the closure of the range of T^*, and V implements this equivalence. It follows that $(TT^*)^{1/2} = V(T^*T)^{1/2}V^*$; so that

$$T = V(T^*T)^{1/2} = V(T^*T)^{1/2}V^*V = (TT^*)^{1/2}V.$$

If $T = WH$ with H positive and W a partial isometry with initial space the closure of the range of H, then, from Lemma 6.1.10, $T^* = HW^*$ and $T^*T = H^2$. From Remark 5.6.32, $H = (T^*T)^{1/2}$, so that $W = V$.

Let $\mathscr{R}$ be a von Neumann algebra and U be a unitary operator in $\mathscr{R}'$. Then $UVU^*U(T^*T)^{1/2}U^*$ is the polar decomposition of UTU^*. From uniqueness of the polar decomposition, $T = UTU^*$ if and only if $V = UVU^*$ and $(T^*T)^{1/2} = U(T^*T)^{1/2}U^*$. Thus $T \,\eta\, \mathscr{R}$ if and only if $V \in \mathscr{R}$ and $(T^*T)^{1/2} \,\eta\, \mathscr{R}$. ■

Bibliography: [56]

6.2. Ordering

The equivalence relation on projections in a von Neumann algebra introduced in Definition 6.1.4 indicates an extension of the usual ordering of projections.

6.2.1. Definition. If E and F are projections in a von Neumann algebra $\mathscr{R}$, we say that E is *weaker* than F (and write $E \precsim F$) when E is equivalent to a subprojection of F. ■

We shall establish that $\precsim$ is a partial ordering on the (equivalence classes of) projections in $\mathscr{R}$. In the sequel, free use will be made of all the notational variations and terminology that are associated with an ordering. For example, in the circumstances of Definition 6.2.1, we say that F is *stronger* than E and write $F \succsim E$ (as well as $E \precsim F$). It is worth emphasizing that $E \prec F$ (E is *strictly weaker* than F) is the same as $E \precsim F$ and E is not equivalent to F (written $E \nsim F$).

6.2.2. Proposition. *If $\{E_a\}$ and $\{F_a\}$ are orthogonal families of projections in a von Neumann algebra $\mathscr{R}$ such that $E_a \precsim F_a$ for all a, then $\sum E_a \precsim \sum F_a$. If $E_a \sim F_a$ for all a, then $\sum E_a \sim \sum F_a$.*

Proof. Suppose $E_a \sim F_a$ for all a; and suppose V_a is a partial isometry in $\mathscr{R}$ with initial projection E_a and final projection F_a. Defining V to be V_a on the range of E_a, for each a, and to be 0 on the range of $I - E$, where $E = \sum E_a$ and $F = \sum F_a$, V extends (linearly) to a partial isometry with initial projection E and final projection F. To see that V is in $\mathscr{R}$, note that V coincides with $V_{a_1} + \cdots + V_{a_n}$ on the range of $E_{a_1} + \cdots + E_{a_n} + I - E$. Since $V_{a_1} + \cdots + V_{a_n}$ is a partial isometry in $\mathscr{R}$, the set of such operators has the uniform bound 1. At the same time, the ranges of the projections $E_{a_1} + \cdots + E_{a_n} + I - E$ span the Hilbert space. From the discussion preceding Remark 2.5.9, V is in the strong-operator closure of $\mathscr{R}$; and V is in $\mathscr{R}$.

If $E_a \precsim F_a$ for all a, then $E_a \sim G_a \le F_a$ for all a. Thus $E \sim G \le F$, where $G = \sum G_a$, from what we have proved to this point. Hence $E \precsim F$. ■

6.2.3. Proposition. *If E and F are projections in a von Neumann algebra $\mathscr{R}$ and $E \sim F$, then $PE \sim PF$ for each central projection P in $\mathscr{R}$. If $E \precsim F$, then $PE \precsim PF$.*

Proof. If V is a partial isometry in $\mathscr{R}$ such that $V^*V = E$ and $VV^* = F$, then $(PV)^*PV = PE$ and $PV(PV)^* = PF$. Thus $PE \sim PF$. If $E \sim F_0 \le F$, then $PE \sim PF_0 \le PF$; so that $PE \precsim PF$. ■

The argument used to prove the proposition that follows is a Hilbert-space version of the standard (set-theoretic) proof of the Cantor–Bernstein theorem.

6.2.4. Proposition. *If E and F are projections in a von Neumann algebra $\mathscr{R}$ such that $E \precsim F$ and $F \precsim E$, then $E \sim F$.*

Proof. Let V and W be partial isometries in $\mathscr{R}$ such that $V^*V = E(=E_0)$, $VV^* = F_1 \leq F\ (=F_0)$, $W^*W = F\ (=F_0)$, and $WW^* = E_1 \leq E$. Since V maps the range of E isometrically onto that of F_1, V maps the range of E_1 isometrically onto that of a subprojection F_2 of F_1—algebraically, $(VE_1)^*VE_1 = E_1$ and $VE_1(VE_1)^* = F_2$. Similarly W maps the range of F_1 onto that of a subprojection E_2 of E_1. Moreover, $V(E - E_1)$ is a partial isometry in $\mathscr{R}$ with initial projection $E - E_1$ and final projection $F_1 - F_2$.

Continuing in this way, by taking successive images under V and W, we construct two sequences $\{E_n\}$, $\{F_n\}$ of projections in $\mathscr{R}$ such that $E = E_0 \geq E_1 \geq E_2 \geq E_3 \geq \cdots$, $F = F_0 \geq F_1 \geq F_2 \geq \cdots$, V maps the range of E_n isometrically onto that of F_{n+1}, and W maps the range of F_n isometrically onto that of E_{n+1}. Thus V maps the range of E_∞ onto that of F_∞, where $E_\infty = \bigwedge_n E_n$ and $F_\infty = \bigwedge_n F_n$. In addition $E_n - E_{n+1} \sim F_{n+1} - F_{n+2}$ and $F_n - F_{n+1} \sim E_{n+1} - E_{n+2}$ for $n = 0, 1, 2, \ldots,$ since $V(E_n - E_{n+1})$ and $W(F_n - F_{n+1})$ are partial isometries in $\mathscr{R}$ with initial projections $E_n - E_{n+1}$, $F_n - F_{n+1}$ and final projections $F_{n+1} - F_{n+2}$ and $E_{n+1} - E_{n+2}$, respectively. From Proposition 6.2.2,

$$\sum_{n=0}^{\infty} (E_{2n} - E_{2n+1}) \sim \sum_{n=0}^{\infty} (F_{2n+1} - F_{2n+2})$$

and

$$\sum_{n=0}^{\infty} (E_{2n+1} - E_{2n+2}) \sim \sum_{n=0}^{\infty} (F_{2n} - F_{2n+1}).$$

Thus,

$$\begin{aligned} E &= E_\infty + \sum_{n=0}^{\infty} (E_{2n} - E_{2n+1}) + \sum_{n=0}^{\infty} (E_{2n+1} - E_{2n+2}) \\ &\sim F_\infty + \sum_{n=0}^{\infty} (F_{2n+1} - F_{2n+2}) + \sum_{n=0}^{\infty} (F_{2n} - F_{2n+1}) = F. \quad \blacksquare \end{aligned}$$

6.2.5. Proposition. *If E, F, and G are projections in a von Neumann algebra $\mathscr{R}$ and $E \precsim F$, $F \precsim G$, then $E \precsim G$.*

Proof. Suppose V and W are partial isometries in $\mathscr{R}$ such that $V^*V = E$, $VV^* = F_0 \leq F$, $W^*W = F$, and $WW^* = G_0 \leq G$. In this case, WF_0W^* is a subprojection G_1 of G_0 in $\mathscr{R}$; and WV is a partial isometry in $\mathscr{R}$ with initial projection E and final projection G_1. Thus $E \sim G_1 \leq G$; and $E \precsim G$. $\blacksquare$

Propositions 6.2.4 and 6.2.5 tell us that $\precsim$ is a partial ordering of the classes of equivalent projections. In the usual loose way, we speak of this relation as a partial ordering on the projections.

Corollary 6.1.9 and Proposition 6.2.2 can be combined with a maximality ("measure-theoretic exhaustion") argument to show that the equivalence classes of projections in a factor are totally ordered. While this result is subsumed in the comparison theorem (Theorem 6.2.7—the more general argument using Proposition 6.1.8 in place of Corollary 6.1.9), it is instructive to see the argument for the case of factors.

6.2.6. PROPOSITION. *If E and F are projections in a factor $\mathcal{M}$, either $E \precsim F$ or $F \precsim E$.*

Proof. Let $\mathcal{F}$ be the family of sets $\{\langle E_a, F_a\rangle\}$ of ordered pairs $\langle E_a, F_a\rangle$ of equivalent subprojections E_a, F_a of E, F such that $\{E_a\}$ and $\{F_a\}$ are orthogonal families. Since $\{\langle 0, 0\rangle\} \in \mathcal{F}$, $\mathcal{F}$ is non-empty. If we partially order $\mathcal{F}$ by inclusion, the union of a totally ordered subset is an upper bound for it. From Zorn's lemma, $\mathcal{F}$ has a maximal element $\{\langle E_a, F_a\rangle\}$. If $E - \sum E_a$ and $F - \sum F_a$ are non-zero, they have equivalent non-zero subprojections E_0 and F_0, from Corollary 6.1.9. Adjoining $\langle E_0, F_0\rangle$ to $\{\langle E_a, F_a\rangle\}$ contradicts the maximality of $\{\langle E_a, F_a\rangle\}$. Thus one of $E - \sum E_a$ and $F - \sum F_a$ is 0 (possibly both are). From Proposition 6.2.2, $\sum E_a \sim \sum F_a$. Thus either E is equivalent to a subprojection of F or F is equivalent to a subprojection of E. That is, either $E \precsim F$ or $F \precsim E$. ∎

The basic ingredient of "total comparability" in factors is found in Corollary 6.1.9, where we learn that non-zero projections E and F have equivalent non-zero subprojections. This is established by showing that $FTE \neq 0$ for some T in the factor. Roughly speaking, T "compares" a piece of E with a piece of F. While *equivalence* calls for "comparison" by partial isometries, the polar decomposition of T supplies this. Actually the transition from comparison by an operator T in $\mathcal{M}$ to comparison by the partial isometry appearing in its polar decomposition is the essence of Proposition 6.1.6; and it is most expedient to use this proposition on FTE. Finally, the fact that FTE is non-zero for some T in $\mathcal{M}$ reduces to the observation that the only projections in $\mathcal{M}$ whose ranges are stable under $\mathcal{M}$ are 0 and I. As T takes on various values in $\mathcal{M}$, the range of TE sweeps out a set of vectors that spans the range of C_E (see Proposition 5.5.2). With E non-zero and $\mathcal{M}$ a factor, $C_E = I$, and F cannot annihilate this set unless $F = 0$. In a von Neumann algebra $\mathcal{R}$, if $C_E \neq I$, it is precisely C_E that can "block" comparison of F and E (as indicated in Proposition 6.1.8). Formalizing these ideas, we arrive at the comparison theorem, a generalization of Proposition 6.2.6.

From the point of view of the comparison theory of projections, it is very useful to treat a von Neumann algebra as a direct sum of factors (although this is not generally valid). For most purposes, three factors will suffice. The sum of the unit projections of the factors in which an operator A of the algebra has a non-zero component is C_A. If E and F are two projections in the algebra, we compare their component projections in each factor. Summing the unit projections of the factors in which E and F have equivalent components yields the central projection Q of the comparison theorem. The sum of the unit projections of the factors in which the component of E is strictly weaker than that of F yields the central projection P of that theorem.

6.2.7. THEOREM (Comparison). *If E and F are projections in a von Neumann algebra $\mathcal{R}$, there are unique orthogonal central projections P and Q maximal with respect to the properties $QE \sim QF$, and, if P_0 is a non-zero central subprojection of P, then $P_0E \prec P_0F$. If R_0 is a non-zero central subprojection of $I - P - Q$, then $R_0F \prec R_0E$.*

Proof. We begin by describing the structure of the proof. A maximality argument allows us to locate the central projections Q and P. Replacing $\mathcal{R}$ by $\mathcal{R}(I - P - Q)$, E by $(I - P - Q)E$, and F by $(I - P - Q)F$, we may now assume that $R_0E \precsim R_0F$ for a central projection only if $R_0 = 0$. Under this assumption and applying that argument with the roles of E and F reversed, we construct a maximal central projection R_1 with the property asserted in the statement of the theorem for $I - P - Q$. We must show that $R_1 = I$ (or, in terms of the initial notation, that $R_1 = I - P - Q$). If this is not the case, replacing $\mathcal{R}$, E, and F, again, by $\mathcal{R}(I - R_1)$, $(I - R_1)E$, and $(I - R_1)F$, we may assume that if $R_0E \precsim R_0F$ or $R_0F \precsim R_0E$, for a central projection, then $R_0 = 0$. The last stage of the proof consists of proving that this situation leads to a contradiction. That is accomplished by means of the argument of Proposition 6.2.6 (the factor version of the present theorem)—with Proposition 6.1.8 in place of Corollary 6.1.9. We proceed to the details.

Let $\{Q_a\}$ be an orthogonal family of central projections in $\mathcal{R}$ maximal with respect to the property that $Q_aE \sim Q_aF$ for each a. From Proposition 6.2.2, $\sum Q_aE = QE \sim QF = \sum Q_aF$, where $Q = \sum Q_a$. By maximality of $\{Q_a\}$, if $P_0E \sim P_0F$ for a central subprojection P_0 of $I - Q$, then $P_0 = 0$. If Q_0 is a central projection such that $Q_0E \sim Q_0F$, then $(Q_0 - Q_0Q)E \sim (Q_0 - Q_0Q)F$, from Proposition 6.2.3. Since $Q_0 - Q_0Q$ is a subprojection P_0 of $I - Q$, $Q_0 - Q_0Q = 0$, that is, $Q_0 \le Q$. The uniqueness of Q follows.

Replacing $\mathcal{R}$, E, and F by $\mathcal{R}(I - Q)$, $(I - Q)E$, and $(I - Q)F$, we may assume that $P_0 = 0$ if P_0 is a central projection such that $P_0E \sim P_0F$. Let $\{P_a\}$ be a maximal orthogonal family of central projections such that $P_aE \prec P_aF$, and let P be its union—provided there is at least one such P_a. (Note that P_a must be non-zero if $P_aE \prec P_aF$.) Otherwise, let P be 0. From Proposition 6.2.3, $P_0P_aE \precsim P_0P_aF$ for each central projection P_0. Thus, from Proposition 6.2.2, if $P_0 \le P$,

$$\sum P_0P_aE = P_0PE = P_0E \precsim \sum P_0P_aF = P_0F.$$

Under our present assumption, if $P_0E \sim P_0F$, then $P_0 = 0$. Thus $P_0E \prec P_0F$ if $0 \neq P_0 \leq P$.

Again, if P_0 is a central projection such that $P_0E \prec P_0F$, then

$$(P_0 - P_0P)E \precsim (P_0 - P_0P)F;$$

and $P_0 - P_0P$ is orthogonal to each P_a. By maximality of $\{P_a\}$,

$$(P_0 - P_0P)E \prec (P_0 - P_0P)F$$

does not hold. Thus $(P_0 - P_0P)E \sim (P_0 - P_0P)F$. With the present assumption on $\mathscr{R}$, $P_0 - P_0P = 0$. Hence $P_0 \leq P$; and P is unique.

From the preceding, we note that $R_0 = 0$ if R_0 is a central projection in $\mathscr{R}(I - P - Q)$ such that $R_0E \precsim R_0F$ (using our original notation). Replace $\mathscr{R}$, E, and F by $\mathscr{R}(I - P - Q)$, $(I - P - Q)E$, and $(I - P - Q)F$. Applying what we have proved with the roles of E and F reversed, there is a maximal central projection R_1 with the property asserted in the statement for $I - P - Q$. If either $R_0E \precsim R_0F$ or $R_0F \precsim R_0E$ for a central subprojection R_0 of $I - R_1$, then $R_0 = 0$. If $R_1 = I$ (that is, in the preceding notation, if $R_1 = I - P - Q$), then the proof is complete. Assuming that $I - R_1 \neq 0$, replace $\mathscr{R}$, E, and F by $\mathscr{R}(I - R_1)$, $(I - R_1)E$, and $(I - R_1)F$. If either $R_0E \precsim R_0F$ or $R_0F \precsim R_0E$, for a central projection R_0, then $R_0 = 0$. This situation will lead us to a contradiction.

Let $\{\langle E_a, F_a\rangle\}$ be a family of ordered pairs of equivalent subprojections of E and F maximal with respect to the property that $\{E_a\}$ and $\{F_a\}$ are orthogonal families. Then $\sum E_a \sim \sum F_a$, from Proposition 6.2.2; and $R_0 \sum E_a \sim R_0 \sum F_a$, for each central projection R_0, from Proposition 6.2.3. Thus if $R_0(E - \sum E_a) = 0$, $R_0E \sim R_0 \sum F_a \leq R_0F$; and $R_0 = 0$. It follows that $E - \sum E_a$ and, similarly, $F - \sum F_a$ have I as central carriers. From Proposition 6.1.8, $E - \sum E_a$ and $F - \sum F_a$ have equivalent non-zero subprojections E_0 and F_0. Adjoining $\langle E_0, F_0\rangle$ to $\{\langle E_a, F_a\rangle\}$ contradicts its maximality. ∎

6.2.8. PROPOSITION. *If E and F are equivalent projections in a von Neumann algebra $\mathscr{R}$ acting on a Hilbert space $\mathscr{H}$, then $C_E = C_F$.*

Proof. By assumption, $V^*V = E$ and $VV^* = F$, for some partial isometry V in $\mathscr{R}$. Thus $E = V^*FV$. From Proposition 5.5.2, $[\mathscr{R}E(\mathscr{H})]$ and $[\mathscr{R}F(\mathscr{H})]$ are the ranges of C_E and C_F, respectively. Now

$$[\mathscr{R}E(\mathscr{H})] = [\mathscr{R}V^*FV(\mathscr{H})] \subseteq [\mathscr{R}F(\mathscr{H})].$$

Similarly $[\mathscr{R}F(\mathscr{H})] \subseteq [\mathscr{R}E(\mathscr{H})]$. Thus $[\mathscr{R}E(\mathscr{H})] = [\mathscr{R}F(\mathscr{H})]$, and $C_E = C_F$. ∎

6.2.9. PROPOSITION. *If E is a cyclic projection in the von Neumann algebra $\mathscr{R}$, acting on the Hilbert space $\mathscr{H}$, and $F \precsim E$, then F is cyclic in $\mathscr{R}$.*

Proof. Since $F \sim E_0 \le E$, and, from Proposition 5.5.9, E_0 is cyclic, we may assume $F \sim E$. Let V be a partial isometry in $\mathscr{R}$ such that $V^*V = E$ and $VV^* = F$; and let x be a generating vector for E under $\mathscr{R}'$. Then

$$[\mathscr{R}'Vx] = [V\mathscr{R}'x] = V[\mathscr{R}'x] = VE(\mathscr{H}) = F(\mathscr{H}),$$

where the equality $[V\mathscr{R}'x] = V[\mathscr{R}'x]$ follows from the fact that V is isometric on $[\mathscr{R}'x]$. ■

Bibliography: [56]

6.3. Finite and infinite projections

We noted and used the analogy with set theory in proving that $E \sim F$ when $E \precsim F$ and $F \precsim E$ (Proposition 6.2.4). This analogy suggests extending the concepts of finite and infinite to the projection-class ordering.

6.3.1. DEFINITION. A projection E in a von Neumann algebra $\mathscr{R}$ is said to be *infinite* relative to $\mathscr{R}$ when $E \sim E_0 < E$ for some projection E_0 in $\mathscr{R}$. Otherwise, E is said to be *finite* relative to $\mathscr{R}$. If E is infinite and PE is either 0 or infinite, for each central projection P, E is said to be *properly infinite*. We say that $\mathscr{R}$ is a *finite* or *properly infinite* von Neumann algebra when I is, respectively, finite or properly infinite. ■

We have avoided the use of the terminology "purely infinite von Neumann algebra" since it appears in the literature with two distinct senses: one to mean what we have defined above as a "properly infinite von Neumann algebra" and the other to mean what will later be called a "type III von Neumann algebra" (see Definition 6.5.1).

6.3.2. PROPOSITION. *If E is a finite projection in the von Neumann algebra $\mathscr{R}$, each subprojection of E is finite. Each minimal projection in $\mathscr{R}$ is finite; and* 0 *is finite. If $E \sim F$ and E is finite, then F is finite.*

Proof. If E_0 in $\mathscr{R}$ is a subprojection of E and V is a partial isometry in $\mathscr{R}$ with initial projection E_0 and final projection E_1, with $E_1 \le E_0$, then $E - E_0 + V$ is a partial isometry in $\mathscr{R}$ with initial projection E and final projection $E - E_0 + E_1$ $(\le E)$. As E is finite (in $\mathscr{R}$), $E = E - E_0 + E_1$; and $E_0 = E_1$. Thus E_0 is finite.

If $E \sim F$, E is finite, $F_0 \leq F$, and $F_0 \sim F$, then there are partial isometries V and W in $\mathcal{R}$ with initial projections E and F and final projections F and F_0, respectively. In this case, V^*WV is a partial isometry with initial projection E and final projection V^*F_0V. As $V^*F_0V \leq E$ and E is finite, $V^*F_0V = E$. Thus $VV^*F_0VV^* = FF_0F = F_0 = VEV^* = F$; and F is finite. It follows that a projection equivalent to a finite or infinite projection is, respectively, finite or infinite.

Since 0 has no proper subprojection, it is finite. If G is a minimal projection in $\mathcal{R}$, its only proper subprojection in $\mathcal{R}$ is 0; and only 0 is equivalent to 0 (for, if $V^*V = 0$, then $V = 0$, and $VV^* = 0$). Thus G is finite. ■

In the lemma that follows, we prove the analogue, for infinite projections, of the possibility of "halving" an infinite set, with each of the halves in one-to-one correspondence with the original set.

6.3.3. Lemma (Halving). *If E is a properly infinite projection in a von Neumann algebra $\mathcal{R}$, there is a projection F in $\mathcal{R}$ such that $F \leq E$ and $F \sim E - F \sim E$.*

Proof. Since E is infinite, $E \sim E_1 < E$. If $V \in \mathcal{R}$, $V^*V = E$, and $VV^* = E_1$, then $(E_2 =)VE_1V^* < E_1$ and $E - E_1 \sim E_1 - E_2$, Continuing in this way ($VE_2V^* = E_3 < E_2$ and $E_1 - E_2 \sim E_2 - E_3$), we construct a countably infinite orthogonal family $\{E_n - E_{n+1}\}$ of equivalent non-zero subprojections of E. This family is contained in a maximal such family $(F_a)_{a \in \mathbb{A}}$. By maximality, we cannot have $F_a \precsim E - \sum F_a (= E_0)$, for then $F_a \sim F_0 \leq E_0$, and adjoining F_0 to $\{F_a\}$ contradicts its maximality. From the comparison theorem (Theorem 6.2.7), there is a non-zero central projection P such that $PE_0 \prec PF_a$. Since $\mathbb{A}$ is an infinite set, there is a subset $\mathbb{A}_0$ of $\mathbb{A}$ such that if $a_0 \in \mathbb{A}_0$, $\mathbb{A} \backslash \mathbb{A}_0 \ (= \mathbb{A}_1)$, $\mathbb{A}_0$, and $\mathbb{A}_0 \backslash \{a_0\}$ can each be put into one-to-one correspondence with $\mathbb{A}$. From Proposition 6.2.3, $PF_a \sim PF_{a'}$ for a and a' in $\mathbb{A}$. From Proposition 6.2.2,

$$PE = \sum_{a \in \mathbb{A}} PF_a + PE_0 \precsim \sum_{a \in \mathbb{A}_0 \backslash \{a_0\}} PF_a + PF_{a_0} = \sum_{a \in \mathbb{A}_0} PF_a \sim \sum_{a \in \mathbb{A}_1} PF_a$$
$$\leq \sum_{a \in \mathbb{A}_1} PF_a + PE_0 \leq PE.$$

With G equal to $\sum_{a \in \mathbb{A}_1} PF_a + PE_0$, we have

$$PE \sim G \sim PE - G = \sum_{a \in \mathbb{A}_0} PF_a.$$

To this point, we have proved that if E is a properly infinite projection in $\mathcal{R}$, there is a non-zero central projection P in $\mathcal{R}$ such that PE can be "halved"—that is, there is a subprojection G of PE in $\mathcal{R}$ such that $G \sim PE - G \sim PE \neq 0$.

Let $\{Q_a\}$ be a maximal orthogonal family of non-zero central subprojections of C_E such that each Q_aE can be halved; and let G_a be a subprojection of Q_aE in $\mathscr{R}$ such that $G_a \sim Q_aE - G_a \sim Q_aE$. If $C_E - \sum Q_a$ is not 0 then

$$(C_E - \sum Q_a)E$$

is properly infinite; and, from what we have proved, there is a non-zero central subprojection Q_0 of $C_E - \sum Q_a$ such that Q_0E can be halved. Adjoining Q_0 to $\{Q_a\}$ contradicts its maximality. Thus $C_E = \sum Q_a$. Letting F be $\sum G_a$, from Proposition 6.2.2 we have,

$$F \sim \sum (Q_aE - G_a) = E - F \sim \sum Q_aE = E,$$

so that E can be halved. ■

6.3.4. THEOREM. *If E is a properly infinite projection in the von Neumann algebra $\mathscr{R}$, F is a countably decomposable projection in $\mathscr{R}$ (in particular, if F is cyclic in $\mathscr{R}$), and $C_F \leq C_E$, then $F \precsim E$.*

Proof. If $0 = C_FC_E (= C_F)$, then $F = 0$; and $F \precsim E$. If $F \neq 0$, $C_EC_F \neq 0$; and E, F have equivalent non-zero subprojections E_0, F_0, respectively. By use of the halving lemma, we construct a countably infinite orthogonal family $\{E_n\}$ of subprojections of E with sum E such that each $E_n \sim E$. (Halve E as $E_1 + F_1$; then halve F_1 as $E_2 + F_2$, and so on. Now replace E_1 by $E - \sum_{n=2}^{\infty} E_n$.) Let $\{F_n\}$ be a maximal orthogonal family of subprojections of F with each F_n equivalent to F_0. (As a consequence of the countable decomposability of F, the family $\{F_n\}$ can be indexed by integers.) Since $\{F_n\}$ is maximal, the relation $F_0 \precsim F - \sum F_n$ cannot hold (for, otherwise, a "copy" of F_0 in $F - \sum F_n$ adjoined to $\{F_n\}$ would contradict that maximality). From the comparison theorem (Theorem 6.2.7), there is a central projection P such that $P(F - \sum F_n) \prec PF_0$ (and, in particular, $PF \neq 0$). Since $F_n \sim F_0 \sim E_0 \precsim E_{n+1}$, $P(F - \sum F_n) \precsim PE_1$ and $PF_n \precsim PE_{n+1}$, from Proposition 6.2.3. Thus

$$0 \neq PF = P(F - \sum F_n) + \sum PF_n \precsim \sum PE_n = PE,$$

from Proposition 6.2.2. We have proved that there is a central projection P such that $0 \neq PF \precsim PE$, under the given hypotheses.

Let $\{P_a\}$ be a maximal orthogonal family of non-zero central subprojections of C_F such that $P_aF \precsim P_aE$. If $C_F - \sum P_a \neq 0$, from what we have just proved, there is a non-zero central subprojection P_0 of $C_F - \sum P_a$ such that $P_0(C_F - \sum P_a)F = P_0F \precsim P_0E$. Adjoining P_0 to $\{P_a\}$ contradicts its maximality. Thus $C_F = \sum P_a$. From Proposition 6.2.2,

$$\sum P_aF = C_FF = F \precsim \sum P_aE \leq E. \quad ■$$

6.3.5. COROLLARY. *Two properly infinite, countably decomposable projections in a von Neumann algebra are equivalent if and only if they have the same central carrier. Two infinite projections in a factor on a separable space are equivalent.*

For the final assertion of this corollary, note that, in a factor, an infinite projection is properly infinite.

6.3.6. LEMMA. *If $\{P_a\}$ is a family of central projections in a von Neumann algebra $\mathscr{R}$ and E is a projection in $\mathscr{R}$ such that P_aE is finite for each a, then PE is finite, where $P = \bigvee_a P_a$.*

Proof. If $F \sim PE$ and $F < PE$, then $0 \neq PE - F \leq P$. If $(PE - F)P_a = 0$ for all a, then $PE - F$ annihilates the range of $P(= \bigvee P_a)$; and we would have $0 = (PE - F)P = PE - F$, contrary to assumption. Thus $(PE - F)P_a \neq 0$ for some a; and $P_aF < P_aPE = P_aE$. From Proposition 6.2.3, $P_aF \sim P_aPE = P_aE$; so that P_aE is infinite in $\mathscr{R}$—contrary to hypothesis. Thus PE is finite. ■

6.3.7. PROPOSITION. *If E is an infinite projection in the von Neumann algebra $\mathscr{R}$, there is a (unique) central projection P in $\mathscr{R}$ such that $P \leq C_E$, PE is properly infinite, and $(I - P)E$ is finite. If E is properly infinite and $F \sim E$, then F is properly infinite.*

Proof. Let $\{Q_a\}$ be a maximal orthogonal family of central projections in $\mathscr{R}$ such that Q_aE is finite for each a. Then, from Lemma 6.3.6, QE is finite, where $Q = \sum Q_a$. Moreover, PE is properly infinite, by maximality of $\{Q_a\}$, where $P = I - Q$; and P is unique.

If E is properly infinite, $F \sim E$, and P is a central projection such that $PF \neq 0$, then $PF \sim PE \neq 0$, from Proposition 6.2.3. Since E is properly infinite, PE is infinite; and, from Proposition 6.3.2, PF is infinite. Thus F is properly infinite. ■

6.3.8. THEOREM. *If E and F are finite projections in the von Neumann algebra $\mathscr{R}$, then $E \vee F$ is finite in $\mathscr{R}$.*

Proof. Since $E \vee F - F \sim E - E \wedge F$ (from the Kaplansky formula—Theorem 6.1.7) and $E - E \wedge F$ is finite (see Proposition 6.3.2), $E \vee F - F$ is finite (again, from Proposition 6.3.2). As $E \vee F = F + E \vee F - F$, it will suffice to show that the sum (union) of two orthogonal finite projections in $\mathscr{R}$ is finite.

We assume that $EF = 0$. Suppose $E + F$ is infinite. From Proposition 6.3.7, there is a central projection P such that $P(E + F)$ is properly infinite.

From Proposition 6.3.2, PE and PF are finite. We may assume, therefore, that $E + F$ is properly infinite. From the halving lemma (Lemma 6.3.3), there is a subprojection G of $E + F$ such that $G \sim E + F - G\,(=G') \sim E + F$. From the comparison theorem (Theorem 6.2.7), there is a central projection Q such that $Q(G \wedge E) \precsim Q(G' \wedge F)$ and $(I - Q)(G' \wedge F) \precsim (I - Q)(G \wedge E)$. One at least, of $Q(E + F)$ and $(I - Q)(E + F)$ is not zero. If, say, $Q(E + F) \neq 0$, then $Q(E + F)$ is infinite; while QE and QF are finite and orthogonal. Moreover, $QG \sim QG' \sim Q(E + F)$; and

$$Q(G \wedge E) = Q \wedge (G \wedge E) = (Q \wedge G) \wedge (Q \wedge E) = QG \wedge QE \precsim QG' \wedge QF.$$

If $(I - Q)(E + F) \neq 0$ and $Q(E + F) = 0$, we reverse the roles of E and F and of G and G'.

We may assume, thus, that $G \wedge E \precsim G' \wedge F$. Then, from Proposition 6.2.2, $G = G - G \wedge E + G \wedge E \precsim E \vee G - E + G' \wedge F \leq F$, since $G - G \wedge E \sim E \vee G - E$ and $E \vee G - E$ and $G' \wedge F$ are orthogonal subprojections of F. To see this last, note that a vector in the range of $G' \wedge F$ is orthogonal to both the range of G and of E—hence, to the range of $E \vee G$; while $E \vee G \leq E + F$, so that $E \vee G - E \leq F$. From Proposition 6.3.2, it follows that G is finite (for $G \precsim F$ and F is finite). But $G \sim E + F$ and $E + F$ was assumed to be infinite. Thus $E \vee F$ is finite. ■

It is instructive to review the preceding argument assuming that $\mathscr{R}$ is a factor. In this case, all reference to Q disappears; and we can pass from the introduction of G to the argument of the last paragraph of the proof.

6.3.9. Lemma. *If $\{E_a\}_{a \in \mathbb{A}}$ and $\{F_b\}_{b \in \mathbb{B}}$ are infinite, orthogonal families of non-zero projections in a von Neumann algebra $\mathscr{R}$, each E_a is cyclic, and the unions E and F of $\{E_a\}$ and $\{F_b\}$ are such that $F \precsim E$; then $\aleph' \leq \aleph$, where $\aleph$ and $\aleph'$ are the cardinal numbers of $\mathbb{A}$ and $\mathbb{B}$, respectively. If $F \sim E$ and each F_b is cyclic, then $\aleph = \aleph'$.*

Proof. Let V be a partial isometry in $\mathscr{R}$ such that $V^*V = F$ and $VV^* = E_0 \leq E$. Then $\{VF_bV^*\}$ is an orthogonal family of non-zero projections with union E_0 and cardinal number $\aleph'$. We may assume that $F \leq E$.

If x_a is a unit generating vector for E_a under $\mathscr{R}'$ and $F_b x_a = 0$, then $(0) = [\mathscr{R}'F_b x_a] = [F_b \mathscr{R}' x_a]$. With the assumption $F \leq E$, $F_b x_a \neq 0$ for some a in $\mathbb{A}$; otherwise F_b annihilates the range of each E_a, hence of E, and $F_b = 0$ contrary to hypothesis. Thus, if $\mathscr{S}_a = \{b : b \in \mathbb{B},\ F_b x_a \neq 0\}$, $\mathbb{B} = \bigcup_{a \in \mathbb{A}} \mathscr{S}_a$. Now, $\sum_{b \in \mathbb{B}} \|F_b x_a\|^2 \leq \|x_a\|^2 = 1$; so that $F_b x_a \neq 0$ for at most a countable number of elements b of $\mathbb{B}$, and $\mathscr{S}_a$ has cardinal number at most $\aleph_0$. As $\mathbb{B} = \bigcup_{a \in \mathbb{A}} \mathscr{S}_a$, $\aleph' \leq \aleph \cdot \aleph_0 = \aleph$. ■

6.3.10. PROPOSITION. *If G is a finite projection in a von Neumann algebra $\mathcal{R}$ and C_G is countably decomposable relative to the center $\mathcal{C}$ of $\mathcal{R}$, then G is the sum of a countable number of cyclic projections in $\mathcal{R}$. In particular, G is countably decomposable.*

Proof. Let G_0 be a non-zero cyclic subprojection of G. Let $\{G_a\}$ be an orthogonal family of subprojections of G maximal with respect to the property that each $G_a \sim G_0$. The family $\{G_a\}$ is finite; for otherwise it could be put into one-to-one correspondence with a proper subfamily. Using Proposition 6.2.2, G would be equivalent to a proper subprojection—contradicting the hypothesis that G is finite.

If $G_0 \sim G_1 \leq G - \sum G_a$, adjoining G_1 to $\{G_a\}$ contradicts the maximality of $\{G_a\}$. Thus $G_0 \not\precsim G - \sum G_a$. Using the comparison theorem (Theorem 6.2.7), there is a (non-zero) central subprojection P of C_G such that

$$P(G - \sum G_a) \prec PG_0.$$

From Proposition 6.2.9, $P(G - \sum G_a)$ and each PG_a are cyclic. Thus PG is the sum of a finite number of cyclic projections.

Let $\{P_b\}$ be an orthogonal family of non-zero central subprojections of C_G maximal with respect to the property that $P_b G$ is the sum of a finite number of cyclic projections. If $P_0 = C_G - \sum P_b$ and $P_0 \neq 0$, from the preceding, P_0 contains a non-zero central subprojection P such that PG is the sum of a finite number of cyclic projections. By maximality of $\{P_b\}$, $P_0 = 0$. Since C_G is countably decomposable in $\mathcal{C}$, $\{P_b\}$ is a countable family. As $G = \sum P_b G$, each $P_b G$ is the sum of a finite number of cyclic projections, and $\{P_b\}$ is a countable family; G is the sum of a countable number of cyclic projections.

From Proposition 5.5.19, G is countably decomposable in $\mathcal{R}$. ∎

6.3.11. THEOREM (Generalized invariance of dimension). *If G is a finite projection in a von Neumann algebra $\mathcal{R}$, $\{E_a\}_{a \in \mathbb{A}}$ and $\{F_b\}_{b \in \mathbb{B}}$ are orthogonal families of subprojections of a projection E in $\mathcal{R}$ maximal with respect to the property that $E_a \sim F_b \sim G$ for each a and b; then $\mathbb{A}$ and $\mathbb{B}$ have the same cardinal number.*

Proof. If $G \not\precsim E$, E has no subprojection equivalent to G; and both $\mathbb{A}$ and $\mathbb{B}$ are empty. On the other hand, if $G \precsim E$, then, by maximality, both $\mathbb{A}$ and $\mathbb{B}$ have at least one element. Assuming this, we can replace G (our "test" projection) by any E_a and, so, assume that $G \leq E$. Since two projections in $E\mathcal{R}E$ are equivalent (relative to $E\mathcal{R}E$) if and only if they are equivalent in $\mathcal{R}$, we may work in $E\mathcal{R}E$ and assume that $E = I$. Since $C_{E_a} = C_{F_b} = C_G$ (from Proposition 6.2.8), $\{E_a\}$ and $\{F_b\}$ are maximal families in $\mathcal{R}C_G$. Again, working in $\mathcal{R}C_G$, we may assume that $C_G = C_{E_a} = C_{F_b} = I$.

From Proposition 6.3.7, there is a central projection Q such that Q is finite and $I - Q$ is either 0 or properly infinite.

(i) Suppose $Q \neq 0$. We show that $\mathbb{A}$ (and $\mathbb{B}$) are finite. Suppose the contrary. Then the orthogonal family $\{QE_a\}_{a \in \mathbb{A}}$ of non-zero, equivalent subprojections of Q can be put into one-to-one correspondence with a proper subfamily. Using Proposition 6.2.2, Q is, then, equivalent to a proper subprojection of itself—contradicting the fact that Q is finite. Thus $\mathbb{A}$ (and $\mathbb{B}$) are finite.

Let $\bar{E}$ be $\sum_{a \in \mathbb{A}} E_a$. We show that $(I - Q)G \precsim (I - Q)(I - \bar{E})$. Suppose the contrary. In particular, then, $I - Q \neq 0$ and $I - Q$ is properly infinite. From Theorem 6.3.8, $(I - Q)\bar{E}$ is finite; since $E_a \sim G$, G is finite, $\bar{E} = \sum_{a \in \mathbb{A}} E_a$, and $\mathbb{A}$ is a finite set. If P is a non-zero central subprojection of $I - Q$, $P(=P(I - Q))$ is infinite, $P\bar{E}$ is finite, and $P = P(I - \bar{E}) + P\bar{E}$. Thus, from Theorem 6.3.8, $P(I - \bar{E})(=P(I - Q)(I - \bar{E}))$ is infinite; and $(I - Q)(I - \bar{E})$ is properly infinite. We are assuming that $(I - Q)G \not\precsim (I - Q)(I - \bar{E})$ (to reach a contradiction). From the comparison theorem, there is a (non-zero) central subprojection P of $I - Q$ such that $P(I - Q)(I - \bar{E}) \prec PG$. But PG is finite and $P(I - Q)(I - \bar{E})$ $(=P(I - \bar{E}))$ is infinite, contradicting Proposition 6.3.2. Thus $(I - Q)G \precsim (I - Q)(I - \bar{E})$, and $(I - Q)G \sim E_1 \leq (I - Q)(I - \bar{E})$.

We note, next, that $(QE_a)_{a \in \mathbb{A}}$ is maximal (as an orthogonal family of subprojections of Q) with respect to the property that each $QE_a \sim QG$. If this is not the case, there is a subprojection E_0 of $Q(I - \bar{E})$ equivalent to QG. From Proposition 6.2.2, $G = QG + (I - Q)G \sim E_0 + E_1 \leq Q(I - \bar{E}) + (I - Q)(I - \bar{E}) = I - \bar{E}$. Adjoining $E_0 + E_1$ to $\{E_a\}$ contradicts the maximality of $\{E_a\}$, so that $\{QE_a\}_{a \in \mathbb{A}}$ is maximal.

It follows that $QG \not\precsim Q(I - \bar{E})$; and there is a central subprojection P of Q such that $P(I - \bar{E}) \prec PG$. If there are fewer elements in $\mathbb{A}$ than in $\mathbb{B}$, $\{PE_a, P(I - \bar{E})\}$ can be put in one-to-one correspondence with a subset of $\{PF_b\}$. As $PE_a \sim PF_b$, $P(I - \bar{E}) \prec PG \sim PF_b$, and $P = P\bar{E} + P(I - \bar{E})$, from Proposition 6.2.2, P is equivalent to a proper subprojection of itself. But P is a subprojection of Q and Q is finite. Thus $\mathbb{A}$ cannot have fewer elements than $\mathbb{B}$; and, symmetrically, $\mathbb{B}$ cannot have fewer elements than $\mathbb{A}$.

(ii) We assume, now, that $Q = 0$. Thus I is properly infinite. We proved in (i), with $I - Q$ in place of I, that $G \precsim I - \bar{E}$, when $\mathbb{A}$ is a finite set, But then $\{E_a\}$ is not maximal. Thus $\mathbb{A}$ and $\mathbb{B}$ are infinite sets. Let $\aleph$ and $\aleph'$ be their (respective) cardinal numbers.

Since $\{E_a\}$ is maximal, there is a central projection P such that $P(I - \bar{E}) \prec PG$. Employing Proposition 6.2.2, $P \sim P\bar{E}$, since $PE_a \sim PG$ for all a and $\{PE_a\}$ is an infinite orthogonal family. Thus $P\bar{F} \precsim P\bar{E}$, where $\bar{F} = \sum_{b \in \mathbb{B}} F_b$. If P_0 is a non-zero central subprojection of P cyclic in the center of $\mathscr{R}$, then

$P_0 G$ is the sum of a countable number n of cyclic projections in $\mathscr{R}$, from Proposition 6.3.10. As $P_0 G \sim P_0 E_a$, each $P_0 E_a$ is the sum of n orthogonal cyclic projections (from Proposition 6.2.9). The family of all these cyclic projections in all $P_0 E_a$ has sum $P_0 \bar{E}$ and cardinal number $\aleph$ $(=n \cdot \aleph)$. As $P_0 \bar{F} \precsim P_0 \bar{E}$ and $\{P_0 F_b\}$ has cardinal number $\aleph'$, from Lemma 6.3.9, $\aleph' \le \aleph$. Symmetrically $\aleph \le \aleph'$; and $\aleph = \aleph'$. ■

If $\mathscr{R}$ is $\mathscr{B}(\mathscr{H})$ in the preceding theorem, and G is a minimal projection (see Proposition 6.3.2), then $\{E_a\}$ and $\{F_b\}$ are families of minimal projections with sum I. They correspond to orthonormal bases for $\mathscr{H}$. The fact that these families have the same cardinal number establishes, again, that the dimension of a Hilbert space is an invariant, independent of the orthonormal basis used in calculating it (see Theorem 2.2.10). The theorem draws its name from this application.

6.3.12. PROPOSITION. *If E is a properly infinite projection in a von Neumann algebra $\mathscr{R}$ and G is a finite projection in $\mathscr{R}$ such that $C_G = C_E$, then E is the sum of a family $\{G_a\}$ of projections equivalent to G if either of the following conditions is satisfied:*

(i) *$\mathscr{R}$ is a factor;*
(ii) *E is countably decomposable in $\mathscr{R}$.*

Proof. Let $\{E_a\}$ be an orthogonal family of subprojections of E maximal with respect to the property that each $E_a \sim G$. As in the proof of Theorem 6.3.11, there is some (non-zero) central subprojection P of C_G $(=C_E = C_{E_a})$ such that $P(E - \sum E_a) \prec PG$ and $\sum PE_a \sim PE$. Let V be a partial isometry in $\mathscr{R}$ with initial projection $\sum PE_a$ and final projection PE. Then $\{VPE_a V^*\}$ is an orthogonal family of projections with sum PE, each equivalent to PG.

If $\mathscr{R}$ is a factor, $P = I$. Taking $VE_a V^*$ for G_a, our assertion follows.

If E is countably decomposable, $\{PE_a\}$ is a countably infinite family. We relabel it $\{PE_n\}$. Let $\{P_b\}$ be an orthogonal family of non-zero central subprojections of C_G maximal with respect to the property that $P_b E$ is the sum of a countable family $\{G_{bn}\}$ of projections equivalent to $P_b G$. If $C_G - \sum P_b \neq 0$, then, from the foregoing, there is a non-zero central subprojection P_0 of $C_G - \sum P_b$ such that $P_0 E$ is the sum of a countable family of projections equivalent to $P_0 G$. Adjoining P_0 to $\{P_b\}$ contradicts the maximality of $\{P_b\}$. Thus $C_G = \sum P_b$. From Proposition 6.2.2, $\sum_b G_{bn}$ $(=G_n) \sim \sum_b P_b G = C_G G = G$. Moreover, $\sum G_n = E$. ■

Conditions (i) and (ii) are curiously different restrictions. One is a cardinality restriction on E, and the other is a restriction on the center of $\mathscr{R}$. With E properly infinite, it will take an infinite number of copies of G to sum to E over each non-zero central portion PE of E. As P varies over an orthogonal family, the cardinal number of copies necessary may vary. Condition (i) says that P cannot vary. Condition (ii) says that as P varies only $\aleph_0$ copies of G are needed to sum to PE.

Bibliography: [47, 48, 56]

6.4. Abelian projections

As noted in Example 5.1.7, the projections in $\mathscr{B}(\mathscr{H})$ with one-dimensional ranges—the minimal projections—occupy a special position in the theory. At the same time, $\mathscr{B}(\mathscr{H})$ is a factor; and we have been viewing the general von Neumann algebra (in a heuristic way) as a direct sum of factors. One important process in the development of the subject involves the translation of each structural aspect of factors into a form suitable for application in a general von Neumann algebra. The *abelian projections* provide such a "translation" of minimal projections. In terms of direct sums of factors, the abelian projections are those whose component projections in each factor are either 0 or a minimal projection. With this (heuristic) description in mind, the statements of the results of this section become transparent. For working purposes, we must devise a general ("global") characterization of abelian projections.

6.4.1. Definition. A projection E in a von Neumann algebra $\mathscr{R}$ is said to be an *abelian projection* in $\mathscr{R}$ when $E\mathscr{R}E$ is abelian. ∎

6.4.2. Proposition. *Each subprojection of an abelian projection in a von Neumann algebra $\mathscr{R}$ is the product of the abelian projection and a central projection. A projection in $\mathscr{R}$ is abelian if and only if it is minimal in the class of projections in $\mathscr{R}$ with the same central carrier. Each abelian projection in $\mathscr{R}$ is finite. If $\mathscr{C}$ is the center of $\mathscr{R}$ and E is an abelian projection in $\mathscr{R}$, then $E\mathscr{R}E = \mathscr{C}E$.*

Proof. Suppose E is abelian in $\mathscr{R}$. From Proposition 5.5.6, $\mathscr{C}E$ is the center of $E\mathscr{R}E$. Since $E\mathscr{R}E$ is abelian, $\mathscr{C}E = E\mathscr{R}E$. If F is a subprojection in $\mathscr{R}$ of E, then $F = EFE \in E\mathscr{R}E = \mathscr{C}E$. Thus $F = C_0E = C_0C_EE = CE$, where $C\ (=C_0C_E) \in \mathscr{C}$ and $CC_E = C$. Now $F = CE = F^2 = C^2E$, so that

$$(C - C^2)E = 0.$$

Since $C - C^2 \in \mathscr{C} \subseteq \mathscr{R}'$, $0 = (C - C^2)C_E = C - C^2$, from Theorem 5.5.4. Thus C is a (normal) idempotent in $\mathscr{C}$, so that C is a central projection. It follows that each subprojection in $\mathscr{R}$ of an abelian projection E has the form PE with P a central projection. If $C_F = C_E$ and $F = PE$, then $C_E = C_F \le P$;

and $PE = E = F$. Thus E is minimal in the class of projections in $\mathscr{R}$ with central carrier C_E, and, in particular, E is finite.

Suppose, now, that E is minimal in the class of projections in $\mathscr{R}$ with central carrier C_E. If G is a subprojection in $\mathscr{R}$ of E, then $G \leq C_G E$. If $G < C_G E$, then $G + (I - C_G)E < E$ and $G + (I - C_G)E$ has central carrier C_E—contradicting our present assumption. Thus $G = C_G E$. It follows that each projection in $E\mathscr{R}E$ is in $\mathscr{C}E$. Since $E\mathscr{R}E$ is a von Neumann algebra (Corollary 5.5.7), it is generated by its projections (Theorem 5.2.2); and $E\mathscr{R}E \subseteq \mathscr{C}E$. Thus E is abelian. ∎

It follows that each non-zero abelian projection in a factor is a minimal projection in that factor.

6.4.3. PROPOSITION. *A projection E is a minimal projection in a von Neumann algebra $\mathscr{R}$ acting on a Hilbert space $\mathscr{H}$, if and only if $E\mathscr{R}E$ consists of scalar multiples of E. If E is a minimal projection in $\mathscr{R}$, E is an abelian projection in $\mathscr{R}$, C_E is a minimal projection in the center $\mathscr{C}$ of $\mathscr{R}$, $\mathscr{R}C_E$ is a factor, and $\mathscr{R}'E = \mathscr{B}(E(\mathscr{H}))$.*

Proof. If E is a projection in $\mathscr{R}$ and $E\mathscr{R}E$ consists of scalar multiples of E, then, if F is a subprojection in $\mathscr{R}$ of E, $F = EFE \in E\mathscr{R}E$; so that $F = aE$. As F is a projection, $F = E$ or $F = 0$; and E is a minimal projection in $\mathscr{R}$.

If E is a minimal projection in $\mathscr{R}$, then each projection in $E\mathscr{R}E$ is either E or 0. Since $E\mathscr{R}E$ is a von Neumann algebra, it is generated by its projections. Thus $E\mathscr{R}E$ consists of scalar multiples of E. From Proposition 5.5.6, $E\mathscr{R}E$ is the commutant of $\mathscr{R}'E$ in $\mathscr{B}(E(\mathscr{H}))$. Thus $\mathscr{R}'E = \mathscr{B}(E(\mathscr{H}))$. From Proposition 5.5.5, $\mathscr{R}'C_E$ is * isomorphic to $\mathscr{R}'E$, so that $\mathscr{R}'C_E$ is a factor. From Proposition 5.5.6, again, $\mathscr{R}'C_E$ has $\mathscr{R}C_E$ as commutant on $\mathscr{B}(C_E(\mathscr{H}))$. Thus $\mathscr{R}C_E$ is a factor. Each central subprojection P of C_E is a central projection in $\mathscr{R}C_E$. Since $\mathscr{R}C_E$ is a factor, P is either 0 or C_E. Hence C_E is a minimal projection in $\mathscr{C}$. As $E\mathscr{R}E$ consists of scalar multiples of E, it is abelian, so that E is an abelian projection in $\mathscr{R}$. ∎

The preceding argument establishes that a projection Q in the center $\mathscr{C}$ of a von Neumann algebra $\mathscr{R}$ is a minimal projection in $\mathscr{C}$ if $\mathscr{R}Q$ is a factor. Conversely, if Q is a minimal projection in $\mathscr{C}$, $\mathscr{R}Q$ is a factor since each central projection in $\mathscr{R}Q$ is a subprojection of Q in $\mathscr{C}$ (hence, equal either to 0 or Q).

6.4.4. PROPOSITION. *If E is a minimal projection in a von Neumann algebra $\mathscr{R}$ acting on a Hilbert space $\mathscr{H}$ and x is a unit vector in $E(\mathscr{H})$, then $[\mathscr{R}x]$ is the range of a minimal projection E' in $\mathscr{R}'$ and EE' is the one-dimensional projection G in $\mathscr{B}(\mathscr{H})$ with x in its range.*

Proof. Since E is a minimal projection in $\mathscr{R}$, from Proposition 6.4.3, $\mathscr{R}'E = \mathscr{B}(E(\mathscr{H}))$. From Proposition 5.5.5, the mapping $T'C_E \to T'E$ is a * isomorphism of $\mathscr{R}'C_E$ onto $\mathscr{R}'E$. Since G is a minimal projection in $\mathscr{R}'E$ $(=\mathscr{B}(E(\mathscr{H})))$, $G = G'E$, where G' is a minimal projection in $\mathscr{R}'C_E$. As $G'x = x$, $[\mathscr{R}x] \subseteq G'(\mathscr{H})$. Thus $E' \le G'$ and $E' \in \mathscr{R}'C_E$. It follows that $E' = G'$. Thus E' is a minimal projection in $\mathscr{R}'C_E$, hence in $\mathscr{R}'$; and $E'E = G$. ■

6.4.5. Proposition. *If $\{E_a\}_{a \in \mathbb{A}}$ is a family of abelian projections in a von Neumann algebra $\mathscr{R}$ and $\{C_{E_a}\}$ is an orthogonal family, then $\sum E_a$ is an abelian projection in $\mathscr{R}$.*

Proof. Let E be $\sum_{a \in \mathbb{A}} E_a$. Then $EAE = \sum_{a, a' \in \mathbb{A}} E_a A E_{a'} = \sum_{a \in \mathbb{A}} E_a A E_a$, since $E_a A E_{a'} = E_a C_{E_a} A C_{E_{a'}} E_{a'} = E_a A E_{a'} C_{E_a} C_{E_{a'}} = 0$, if $a \ne a'$, where the infinite sums are understood in the sense of strong-operator convergence over the net of finite subsets of $\mathbb{A}$. Now,

$$E_a A E_a E_{a'} B E_{a'} = E_{a'} B E_{a'} E_a A E_a,$$

for all a, a' in $\mathbb{A}$ and A, B in $\mathscr{R}$, since both products are 0 if $a \ne a'$, and $E_a \mathscr{R} E_a$ is abelian. Thus EAE and EBE commute, and E is an abelian projection in $\mathscr{R}$. ■

6.4.6. Proposition. *If E and F are projections in a von Neumann algebra $\mathscr{R}$, E is abelian, and*

(i) *$E \sim F$, then F is abelian;*
(ii) *$C_E \le C_F$, then $E \precsim F$;*
(iii) *$C_E = C_F$ and F is abelian, then $E \sim F$.*

Proof. (i) If V, A, and B are in $\mathscr{R}$ and $V^*V = E$, $VV^* = F$, then

$$\begin{aligned} FAFBF &= VV^*AVV^*BVV^* = VEV^*AVEV^*BVEV^* \\ &= VEV^*BVEV^*AVEV^* = FBFAF, \end{aligned}$$

since $E\mathscr{R}E$ is abelian. Thus $F\mathscr{R}F$ is abelian; and F is an abelian projection in $\mathscr{R}$.

(ii) If $E \not\precsim F$, then, from the comparison theorem, there is some (nonzero) central subprojection P of C_E such that $PF \prec PE$. In this case, $0 \ne PF \sim E_1 < PE$, contradicting Proposition 6.4.2, since PE is an abelian projection and $C_{E_1} = C_{PF} = P = C_{PE}$. Thus $E \precsim F$.

(iii) If $C_E = C_F$ and F is abelian, then $E \sim F$, since $E \precsim F$ and $F \precsim E$, from (ii). ■

6.4.7. Corollary. *If E and F are projections in a factor $\mathscr{M}$ and E is minimal, then $E \precsim F$ if $F \ne 0$. If F is minimal as well, $E \sim F$. If $E \sim F$, then F is minimal.*

6.4.8. PROPOSITION. *If E is an abelian projection in a von Neumann algebra $\mathscr{R}$ and F is a projection in $\mathscr{R}$ such that $F \leq C_E$, then F is a sum of abelian projections.*

Proof. If $F = 0$, there is nothing to prove. Assume $F \neq 0$. As $C_F \leq C_E$, $C_F E$ is a non-zero abelian projection with central carrier C_F. From Proposition 6.4.6, $C_F E \sim F_1 \leq F$, and F_1 is a (non-zero) abelian projection. Let $\{F_a\}$ be a maximal orthogonal family of non-zero abelian subprojections of F. Since $F - \sum F_a$ is a subprojection of C_E, it is either 0 or dominates a non-zero abelian projection. By maximality of $\{F_a\}$, $0 = F - \sum F_a$, and F is a sum of abelian projections. ■

Bibliography: [47, 48]

6.5. Type decomposition

With the results of the preceding sections at our disposal, several ways of distinguishing among von Neumann algebras, algebraically, suggest themselves. Have they abelian projections? Have they non-zero finite projections? Have they infinite projections? The various combinations of affirmative and negative replies to these questions lead us to the "type" description of von Neumann algebras.

6.5.1. DEFINITION. A von Neumann algebra $\mathscr{R}$ is said to be of type I if it has an abelian projection with central carrier I—of type I_n if I is the sum of n equivalent abelian projections. If $\mathscr{R}$ has no non-zero abelian projections but has a finite projection with central carrier I, then $\mathscr{R}$ is said to be of type II—of type II_1 if I is finite—of type II_∞ if I is properly infinite. If $\mathscr{R}$ has no non-zero finite projections, $\mathscr{R}$ is said to be of type III. ■

A number of simple facts, related to the above definition, follow easily from the techniques developed in the earlier sections of this chapter. If a von Neumann algebra $\mathscr{R}$ is of type I (or I_n, or II, or II_1, or II_∞, or III) the same is true of $\mathscr{R}P$ for each non-zero central projection P in $\mathscr{R}$. If $\mathscr{R}$ is of type I_n, it is also of type I; and, of course, if $\mathscr{R}$ is of type either II_1 or II_∞, it is also of type II. However, a von Neumann algebra cannot be of more than one of the types I, II, III, nor can it be of both types II_1 and II_∞. Slightly less obvious is the fact (established at the end of the proof of the following theorem) that if a von Neumann algebra is of both types I_m and I_n, then $m = n$.

6.5.2. THEOREM (Type decomposition). *If $\mathscr{R}$ is a von Neumann algebra acting on a Hilbert space $\mathscr{H}$, there are (mutually orthogonal) central projections P_n, n not exceeding* $\dim \mathscr{H}$, *P_{c_1}, P_{c_∞}, and P_∞, with sum I, maximal with respect* *to the properties that* $\mathscr{R}P_n$ *is of type* I_n *or* $P_n = 0$, $\mathscr{R}P_{c_1}$ *is of type* II_1 *or* $P_{c_1} = 0$, $\mathscr{R}P_{c_\infty}$ *is of type* II_∞ *or* $P_{c_\infty} = 0$, *and* $\mathscr{R}P_\infty$ *is of type* III *or* $P_\infty = 0$.

Proof. Let $\{E_a\}$ be a family of abelian projections maximal with respect to the property that $\{C_{E_a}\}$ is orthogonal. From Propositions 6.4.5 and 5.5.3, $\sum E_a$ is an abelian projection with central carrier $\sum C_{E_a}$ $(=P_d)$. Thus, either $P_d = 0$ or $\mathscr{R}P_d$, acting on $P_d(\mathscr{H})$, is a type I von Neumann algebra. By maximality of $\{E_a\}$, $I - P_d$ has no non-zero abelian subprojections.

From Proposition 6.3.7, there is a central subprojection P_{c_1} of $I - P_d$ such that P_{c_1} is finite and $I - P_d - P_{c_1}$ is either 0 or properly infinite. Since $\mathscr{R}P_{c_1}$ has no non-zero abelian projections and P_{c_1} is finite, either $P_{c_1} = 0$ or $\mathscr{R}P_{c_1}$ is a type II_1 von Neumann algebra.

Let $\{G_c\}$ be a family of finite subprojections of $I - P_d - P_{c_1}$ maximal with respect to the property that $\{C_{G_c}\}$ is an orthogonal family. From Lemma 6.3.6 and Proposition 5.5.3, $\sum G_c$ is a finite projection with central carrier $\sum C_{G_c}$ $(=P_{c_\infty})$. By maximality of $\{G_c\}$, $I - P_d - P_{c_1} - P_{c_\infty}$ $(=P_\infty)$ has no non-zero finite subprojections in $\mathscr{R}$, so that either $P_\infty = 0$ or $\mathscr{R}P_\infty$ is a von Neumann algebra of type III. As $I - P_d - P_{c_1}$ is either 0 or properly infinite, P_{c_∞} is either 0 or properly infinite. If $P_{c_\infty} \neq 0$, since P_{c_∞} is the central carrier of the finite projection $\sum G_c$ and P_{c_∞} has no non-zero abelian subprojections, $\mathscr{R}P_{c_\infty}$ is of type II_∞.

It remains to show that P_d is the sum of a family $\{P_n\}$ of central projections such that P_n is the sum of n equivalent abelian projections. Let $\{Q_a\}$ be an orthogonal family of central subprojections of P_d each of which is the sum of n equivalent abelian projections E_{aj}, $j = 1, 2, \ldots, n$, where n is some cardinal number not exceeding $\dim \mathscr{H}$. In addition, let $\{Q_a\}$ be maximal with respect to the property of being orthogonal. Since $E_{a1} \sim E_{a2} \cdots$ and $\sum_j E_{aj} = Q_a$, $C_{E_{a1}} = C_{E_{a2}} = \cdots = Q_a$, from Proposition 5.5.3. Thus $\sum_a E_{aj}$ $(=E_j)$ is an abelian projection with central carrier $\sum_a Q_a (=P_n)$, from Proposition 6.4.5. It follows, from Proposition 6.4.6(iii), that $E_1 \sim E_2 \sim \cdots$. As $P_n = \sum_{j=1}^n E_j$, either P_n is 0 or $\mathscr{R}P_n$ is of type I_n.

Since 0 is an abelian projection and is the sum of n equivalent abelian projections, the preceding discussion envisages, as it must, the possibility that each Q_a is 0. In this case $P_n = 0$. The essence of the argument appears now. We show that $P_d = \sum_{n \le \dim \mathscr{H}} P_n$. If $0 \neq P_d - \bigvee P_n$ $(=P)$, then, for each n, P will fail to have a single non-zero central subprojection that is the sum of n equivalent abelian projections, by maximality of the family $\{Q_a\}$ used to define P_n. Since P_d is C_E for some abelian projection E, P is the central carrier of the abelian projection PE $(=F)$. Let $\{F_b\}$ be a maximal orthogonal family of subprojections of P equivalent to F. By maximality of $\{F_b\}$, $F \not\precsim P - \sum F_b$; and, from the comparison theorem, there is a non-zero central subprojection P_0 of P such that $P_0(P - \sum F_b) = P_0 - \sum P_0 F_b \sim F_0 < P_0 F$. From

Proposition 6.4.2, $F_0 = C_{F_0}P_0F = C_{F_0}F$; and F_0 is abelian. If $F_0 = 0$, then $P_0 = \sum P_0F_b$, and $\{P_0F_b\}$ is a family of equivalent abelian projections, from Propositions 6.2.3 and 6.4.6(i) (since $F_b \sim F$, for all b). If $F_0 \neq 0$, then $P_0 - \sum P_0F_b = C_{F_0}(P_0 - \sum P_0F_b) = C_{F_0} - \sum C_{F_0}F_b \sim F_0$, from Proposition 6.2.8; and from Proposition 6.4.6(i), C_{F_0} is the sum $C_{F_0} - \sum C_{F_0}F_b + \sum C_{F_0}F_b$ of equivalent abelian projections. In any event, if $P \neq 0$, P has a non-zero central subprojection that is the sum of equivalent abelian subprojections—contradicting the maximality of $\{Q_a\}$ (for some n not exceeding $\dim \mathscr{H}$). Thus $P = 0$, and $P_d = \bigvee P_n$.

We note that $P_nP_m = 0$ if $n \neq m$; so that $\bigvee P_n = \sum P_n = P_d$. From the preceding discussion, if $0 \neq P = P_nP_m$, then P is the sum both of n and of m equivalent abelian projections. Each of these abelian projections is finite, from Proposition 6.4.2, so that Theorem 6.3.11 applies, and $n = m$. ■

Type I von Neumann algebras are sometimes called *discrete* von Neumann algebras to indicate the fact that the identity can be decomposed as a sum of central projections (the P_n of Theorem 6.5.2) each of which is the "discrete" sum of projections minimal with the given central projection as central carrier. The type II von Neumann algebras, by contrast, are described as *continuous*. (In some of the literature, this description is applied to von Neumann algebras of type III as well.) This terminology is the basis for the notation $P_d, P_{c_1}, P_{c_\infty}$. In case $P_\infty = 0$, the von Neumann algebra is sometimes referred to as *semifinite*.

6.5.3. Corollary. *A factor $\mathscr{M}$ is either of type* I_n, *or* II_1, *or* II_∞, *or* III. *It is of type* I *if it has a minimal projection—of type* I_n *if I is the sum of n minimal projections. If $\mathscr{M}$ has no minimal projections but has a non-zero finite projection, it is of type* II*—of type* II_1 *if I is finite—of type* II_∞ *if I is infinite. If $\mathscr{M}$ has no non-zero finite projections, it is of type* III.

6.5.4. Remark. Note that factors of type I_n (n finite) and type II_1 are finite von Neumann algebras, while factors of the other types are properly infinite von Neumann algebras (see Definition 6.3.1). We shall note in Theorem 6.6.1 that factors of type I_n are * isomorphic to $\mathscr{B}(\mathscr{H})$, where $\mathscr{H}$ has dimension n, so that such factors have finite linear dimension when n is finite. On the other hand, the fact that a factor of type II_1 has no minimal projections allows us to choose an infinite orthogonal family of non-zero projections in it. It follows that each factor of type II_1 has infinite linear dimension. ■

The corollary that follows extends, to general von Neumann algebras and abelian projections, the fact that each projection in $\mathscr{B}(\mathscr{H})$ is a sum of (minimal) projections with one-dimensional ranges.

6.5.5. COROLLARY. *If E and E_0 are projections in a von Neumann algebra $\mathscr{R}$, $C_E = C_{E_0}$, and E_0 is abelian in $\mathscr{R}$, then there is a family $\{Q_j\}$ of central projections in $\mathscr{R}$ with sum C_E such that Q_jE is the sum of j equivalent abelian projections. If $\mathscr{R}C_{E_0}$ is of type I_n, then $1 \le j \le n$. In particular, if E_0 is a minimal projection in $\mathscr{R}$, then E is the sum of j equivalent minimal projections in $\mathscr{R}$.*

Proof. From Proposition 6.4.8, E is a sum of abelian projections in $\mathscr{R}$—each of which is abelian in the von Neumann algebra $E\mathscr{R}E$ $(= \mathscr{R}_0)$. Arguing as in the first paragraph of the proof of Theorem 6.5.2, we conclude that $\mathscr{R}_0$ is of type I. Applying Theorem 6.5.2 to $\mathscr{R}_0$, let $\tilde{P}_j$ be the central projection in $\mathscr{R}_0$ such that either $\mathscr{R}_0\tilde{P}_j$ is of type I_j or $\tilde{P}_j = 0$. Now $\tilde{P}_j$ is a (central) projection in $\mathscr{R}_0'$, and $\mathscr{R}_0' = \mathscr{R}'E$, from Corollary 5.5.7. From Proposition 5.5.5, the mapping $T'E \to T'C_E$ is a * isomorphism of $\mathscr{R}'E$ onto $\mathscr{R}'C_E$. Let Q_j be the image of $\tilde{P}_j$ under this mapping. Then Q_j is a central projection in $\mathscr{R}'C_E$, hence in $\mathscr{R}'$ and in $\mathscr{R}$. Moreover, $Q_jE = \tilde{P}_j$ (for, in general, $T'C_EE = T'E$). By construction, either $\tilde{P}_j$ is 0 or $Q_jE = \tilde{P}_j = F_1 + \cdots + F_j$, where $F_1, \ldots, F_j$ are equivalent (non-zero) abelian projections in $E\mathscr{R}E$. As each abelian projection in $E\mathscr{R}E$ is abelian in $\mathscr{R}$ and equivalence in $E\mathscr{R}E$ persists in $\mathscr{R}$, $\{Q_j\}$ serves as the required family of central projections in $\mathscr{R}$. (Note for this that $\sum_j \tilde{P}_j = E$, so that $\sum_j Q_j = C_E$.)

If $\mathscr{R}C_E(= \mathscr{R}C_{E_0})$ is of type I_n, then $C_E = E_1 + \cdots + E_n$, where $E_1, \ldots, E_n$ are equivalent abelian projections in $\mathscr{R}C_E$. From Proposition 6.4.6(iii), $Q_jE_1 \sim F_1$. Generalized invariance of dimension (Theorem 6.3.11) applied to Q_j and to the families $\{Q_jE_1, \ldots, Q_jE_n\}$, $\{F_1, \ldots, F_j, \ldots, F_m\}$ (this latter family being an extension of $\{F_1, \ldots, F_j\}$ to a maximal orthogonal family of subprojections of Q_j equivalent to F_1) tells us that $n = m$, so that $j \le n$. (Recall, for this, that each abelian projection is finite, from Proposition 6.4.2.)

If E_0 is minimal in $\mathscr{R}$, then $\mathscr{R}C_{E_0}(= \mathscr{R}C_E)$ is a factor of type I_n, for some n, from Proposition 6.4.3 and Corollary 6.5.3. From the preceding (or from Proposition 6.4.8), E is a sum of (non-zero) abelian projections in $\mathscr{R}C_E$. Since $\mathscr{R}C_E$ is a factor, each of these abelian projections is minimal in $\mathscr{R}C_E$, hence in $\mathscr{R}$. ■

We proved, in Lemma 6.3.3, that a properly infinite projection can be "halved." This result can be extended to assert divisibility into *any number* of equivalent subprojections. Keeping in mind the situation of projections in $\mathscr{B}(\mathscr{H})$, we should not expect to be able to halve a projection whose range has dimension 5. More generally, if $\mathscr{R}$ is a von Neumann algebra of type I_n with n odd, the identity in $\mathscr{R}$ cannot be halved. The possibility of finer and finer subdivision of projections combined with the fact that we can pass to strong-operator limits suggests that (finite) divisibility can always be effected in a von Neumann algebra with no central portion of type I. This assertion is the substance of the lemma that follows.

6.5.6. LEMMA. *If E is a projection in a von Neumann algebra with no central portion of type* I *(equivalently, with no non-zero abelian projections), for each positive integer n, there are n equivalent (orthogonal) projections with sum E.*

Proof. Since E is not abelian, there is a proper subprojection F of E in $\mathscr{R}$ such that $C_F = C_E$ (from Proposition 6.4.2). With Q the central carrier of $E - F$ in $\mathscr{R}$, it follows that QF has central carrier Q. From Proposition 6.1.8, QF and $E - F$ have non-zero equivalent subprojections F_1 and F_2 (and, of course $F_1F_2 = 0$). Now F_1 is not abelian in $\mathscr{R}$, so that, from what we have just established, F_1 has two equivalent orthogonal non-zero subprojections M_1 and M_2. The equivalence of F_1 and F_2 provides us with two equivalent orthogonal non-zero subprojections, N_1 and N_2, of F_2, where $M_1 \sim M_2 \sim N_1 \sim N_2$. Continuing in this way we produce 2^n (and, hence, n) equivalent orthogonal non-zero subprojections of E.

Let $\mathscr{S}$ be the family of sets $\{\mathscr{F}_1, \ldots, \mathscr{F}_n\}$ of n elements, where each $\mathscr{F}_j$ is an orthogonal family $\{E_j^{(a)}\}_{a \in \mathbb{A}}$ of subprojections of E, each family is indexed by $\mathbb{A}$, $E_1^{(a)} \sim \cdots \sim E_n^{(a)}$ for each a in $\mathbb{A}$, and $\bigcup_{j=1}^n \mathscr{F}_j$ is an orthogonal family. We define a partial ordering $\leq$ of $\mathscr{S}$ so that $\{\mathscr{F}_1, \ldots, \mathscr{F}_n\} \leq \{\mathscr{F}'_1, \ldots, \mathscr{F}'_n\}$ precisely when the indexing of the sets of the second family extends the indexing of those of the first family (which entails, in particular, that $\mathscr{F}_j \subseteq \mathscr{F}'_j$ for all j). Let $\{\mathscr{F}_1, \ldots, \mathscr{F}_n\}$ be a maximal element of $\mathscr{S}$ relative to this ordering; and let E_j be the union of the projections in $\mathscr{F}_j$. From Proposition 6.2.2 $\{E_1, \ldots, E_n\}$ is an equivalent orthogonal family of subprojections of E. If $E - \sum_{j=1}^n E_j \neq 0$, then it is not abelian in $\mathscr{R}$. From the preceding paragraph, we can find an orthogonal family $\{F_1, \ldots, F_n\}$ of equivalent non-zero subprojections of $E - \sum_{j=1}^n E_j$, in that case. Adjoining F_j to $\mathscr{F}_j$, we construct a set in $\mathscr{S}$ properly larger than $\{\mathscr{F}_1, \ldots, \mathscr{F}_n\}$ (relative to the given partial ordering on $\mathscr{S}$). This would contradict the maximality of $\{\mathscr{F}_1, \ldots, \mathscr{F}_n\}$ so that $E = \sum_{j=1}^n E_j$. ■

Bibliography: [47, 48, 56]

6.6. Type I algebras

The results on abelian and minimal projections, obtained in Section 6.4, provide easy access to the algebraic structure of type I von Neumann algebras, in general, and type I factors, in particular.

6.6.1. THEOREM. *If $\mathscr{M}$ is a type* I_n *factor, then $\mathscr{M}$ is * isomorphic to $\mathscr{B}(\mathscr{H})$, where $\mathscr{H}$ has dimension n.*

Proof. According to Corollary 6.5.3, I is the sum of n minimal projections in $\mathscr{M}$. If E is a minimal projection in $\mathscr{M}$ and x is a unit vector in its range, $[\mathscr{M}x]\ (=\mathscr{H})$ is the range of a minimal projection E' in $\mathscr{M}'$ and $\mathscr{M}E'$ is $\mathscr{B}(\mathscr{H})$ from Propositions 6.4.3 and 6.4.4. Since $\mathscr{M}$ is a factor, $\mathscr{M}'$ is a factor and $C_{E'} = I$. From Proposition 5.5.5, $\mathscr{M}E'\ (=\mathscr{B}(\mathscr{H}))$ is * isomorphic to $\mathscr{M}(=\mathscr{M}C_{E'})$. As n minimal projections in $\mathscr{M}$ have sum I, there are n one-dimensional projections (corresponding to them) in $\mathscr{B}(\mathscr{H})$ that have sum I. Thus $\mathscr{H}$ is n-dimensional. ■

In Section 2.6, *Matrix representations*, we discussed the concept of $n \times n$ matrices $[T_{a,b}]_{a,b \in \mathbb{B}}$ whose entries are bounded operators $T_{a,b}$ on a Hilbert space $\mathscr{H}$. We noted that these matrices acted (in the usual matrix fashion) as linear operators on the direct sum $\sum \oplus \mathscr{H}_b (=\tilde{\mathscr{H}})$ of n copies $\mathscr{H}_b$ of $\mathscr{H}$—the n-fold direct sum of $\mathscr{H}$ with itself—and that some of the operators so obtained were bounded. Each operator in $\mathscr{B}(\tilde{\mathscr{H}})$ arises in this way—that is, has a matrix representation. If n is finite, we saw that each such matrix corresponds to a bounded operator, but that this is not the case if n is infinite.

Suppose now that $\mathscr{R}$ is a von Neumann algebra acting on $\mathscr{H}$. If $n \otimes \mathscr{R}$ is the set of operators in $\mathscr{B}(\tilde{\mathscr{H}})$ whose matrix representations have each entry in $\mathscr{R}$ and $\mathscr{R} \otimes I_n$ is the set of operators in $n \otimes \mathscr{R}$ with all diagonal entries equal to the same operator in $\mathscr{R}$ and all other entries 0, we shall note that both sets are von Neumann algebras. If φ is a * isomorphism of $\mathscr{R}$ with a von Neumann algebra $\mathscr{T}$ acting on a Hilbert space $\mathscr{K}$, then $n \otimes \varphi$ and $\varphi \otimes I_n$ are * isomorphisms of $n \otimes \mathscr{R}$ with $n \otimes \mathscr{T}$ and $\mathscr{R} \otimes I_n$ with $\mathscr{T} \otimes I_n$, respectively, where $(n \otimes \varphi)([T_{a,b}]) = [\varphi(T_{a,b})]$ for an operator in $n \otimes \mathscr{R}$ and $\varphi \otimes I_n$ is the restriction of $n \otimes \varphi$ to $\mathscr{R} \otimes I_n$.

6.6.2. LEMMA. *If $\mathscr{R}$ is a von Neumann algebra acting on a Hilbert space $\mathscr{H}$ and $\tilde{\mathscr{H}}$ is the n-fold direct sum of $\mathscr{H}$ with itself, then $n \otimes \mathscr{R}$ and $\mathscr{R} \otimes I_n$ acting on $\tilde{\mathscr{H}}$ are von Neumann algebras, $\mathscr{R} \otimes I_n$ is * isomorphic to $\mathscr{R}$, and $(\mathscr{R}' \otimes I_n)' = n \otimes \mathscr{R}$. If $\mathscr{T}$ is a von Neumann algebra acting on a Hilbert space $\mathscr{K}$ and φ is a * isomorphism of $\mathscr{R}$ onto $\mathscr{T}$, then $n \otimes \varphi$ is a * isomorphism of $n \otimes \mathscr{R}$ onto $n \otimes \mathscr{T}$.*

Proof. The rules of (infinite) matrix multiplication, established in Section 2.6 for the matrix representations of operators in $\mathscr{B}(\tilde{\mathscr{H}})$, make it apparent that

(i) $(\mathscr{R}' \otimes I_n)' = n \otimes \mathscr{R}$;
(ii) $(n \otimes \mathscr{R}')' = \mathscr{R} \otimes I_n$;
(iii) $\mathscr{R} \otimes I_n$ is * isomorphic to $\mathscr{R}$.

It follows from (i) that $n \otimes \mathscr{R}$ acting on $\mathscr{H}$ is a von Neumann algebra, and from (ii) that $\mathscr{R} \otimes I_n$ is a von Neumann algebra.

It is equally apparent, from the discussion in Section 2.6, that $n \otimes \varphi$ is a * isomorphism of $n \otimes \mathscr{R}$ onto $n \otimes \mathscr{T}$ when n is finite—so that $n \otimes \varphi$ is an isometry, from Theorem 4.1.8(iii), in this case. If n is infinite, this comment applies to the C^*-(von Neumann) algebra of matrices whose entries outside a given finite diagonal block are 0. From Proposition 2.6.13, $[\varphi(T_{a,b})]$ is a bounded operator on $\tilde{\mathscr{K}}$, the n-fold direct sum of $\mathscr{K}$ with itself, if $[T_{a,b}] \in n \otimes \mathscr{R}$, since each finite diagonal block of $[\varphi(T_{a,b})]$ has the same bound as that of the corresponding finite diagonal block in $[T_{a,b}]$. Thus $[\varphi(T_{a,b})]$ is in $n \otimes \mathscr{T}$ and has the same bound as $[T_{a,b}]$, so that $n \otimes \varphi$ is an isometric (adjoint-preserving) linear mapping of $n \otimes \mathscr{R}$ onto $n \otimes \mathscr{T}$ that carries the unit of $n \otimes \mathscr{R}$ onto that of $n \otimes \mathscr{T}$. Since, for a self-adjoint operator H, $0 \le H \le 2I$ if and only if $\|I - H\| \le 1$, $n \otimes \varphi$ maps an element in $n \otimes \mathscr{R}$ onto a positive element in $n \otimes \mathscr{T}$ if and only if that element is positive. (That is, $n \otimes \varphi$ is an *order isomorphism* of $n \otimes \mathscr{R}$ onto $n \otimes \mathscr{T}$.)

To conclude that $(n \otimes \varphi)(AB) = (n \otimes \varphi)(A)(n \otimes \varphi)(B)$ for each A and B in $n \otimes \mathscr{R}$—and, hence, that $n \otimes \varphi$ is a * isomorphism—we encounter the question of whether $\varphi(\sum_{b \in \mathbb{B}} T_{a,b}S_{b,c})$ and $\sum_{b \in \mathbb{B}} \varphi(T_{a,b})\varphi(S_{b,c})$ are the same, where $A = [T_{a,b}]$ and $B = [S_{a,b}]$. Both sums are strong-operator convergent. The fact that φ is a * isomorphism does, indeed, imply this equality, for we shall prove (Corollary 7.1.16) that such mappings are strong-operator continuous on bounded sets. At this occasion, an easy *ad hoc* argument will allow us to draw the desired conclusion.

If $\sum_{b \in \mathbb{B}} T_{a,b}S_{b,c}$ has only a finite number of non-zero terms, then $\varphi(\sum_b T_{a,b}S_{b,c}) = \sum_b \varphi(T_{a,b})\varphi(S_{b,c})$, since φ is an isomorphism. Thus

$$(n \otimes \varphi)(AB) = (n \otimes \varphi)(A)(n \otimes \varphi)(B),$$

if either A has only a finite number of columns with non-zero entries or B has at most a finite number of rows with non-zero entries. If $E_{\mathbb{B}_0}$ is the projection in $n \otimes \mathscr{R}$ with matrix $[R_{a,b}]$, where $R_{b,b} = I$ if b is in the finite subset $\mathbb{B}_0$ of $\mathbb{B}$ and $R_{a,b} = 0$ for all other entries, then $\{E_{\mathbb{B}_0}\}$ and $\{(n \otimes \varphi)(E_{\mathbb{B}_0})\}$ are strong-operator convergent to I over the net of finite subsets of $\mathbb{B}$. Now $\{A^*E_{\mathbb{B}_0}A\}$ is a monotone increasing net with least upper bound A^*A. Since $n \otimes \varphi$ is an order isomorphism and

$$\begin{aligned}(n \otimes \varphi)(A^*E_{\mathbb{B}_0}A) &= (n \otimes \varphi)(A^*)(n \otimes \varphi)(E_{\mathbb{B}_0}A)\\ &= (n \otimes \varphi)(A^*)(n \otimes \varphi)(E_{\mathbb{B}_0})(n \otimes \varphi)(A),\end{aligned}$$

$\{(n \otimes \varphi)(A^*E_{\mathbb{B}_0}A)\}$ is a monotone increasing net with least upper bound $(n \otimes \varphi)(A^*A)$ and $(n \otimes \varphi)(A^*)(n \otimes \varphi)(A)$. Thus

$$(n \otimes \varphi)(A^*A) = (n \otimes \varphi)(A^*)(n \otimes \varphi)(A).$$

Substituting, successively, $H + iK$, H, K, and $H + K$ for A, where H and K are self-adjoint operators in $n \otimes \mathscr{R}$, and combining, we conclude that

$(n \otimes \varphi)(HK) = (n \otimes \varphi)(H)(n \otimes \varphi)(K)$. Replacing A and B by their decomposition as a sum of their real and imaginary parts, we have $(n \otimes \varphi)(AB) = (n \otimes \varphi)(A)(n \otimes \varphi)(B)$. ∎

As with direct sums, it is possible and useful to recognize a given von Neumann algebra $\mathscr{R}$ as being (* isomorphic to) an algebra of matrices with entries in another von Neumann algebra from the "internal" structure of $\mathscr{R}$. For this purpose, we single out a system of $n \times n$ *matrix units* in $\mathscr{R}$. Such a system is a family $(E_{a,b})_{a,b \in \mathbb{B}}$ of operators in $\mathscr{R}$ such that $E_{a,b}E_{c,d} = 0$ if $b \neq c$ and $E_{a,c}E_{c,d} = E_{a,d}$, where $\mathbb{B}$ has cardinal number n and $\sum_{b \in \mathbb{B}} E_{b,b}$ is strong-operator convergent to I. If, in addition, $E_{a,b}^* = E_{b,a}$, we say that the system is *self-adjoint*. In any case, $E_{b,b}^2 = E_{b,b}$. When the system is self-adjoint, $\{E_{b,b}\}$ is an orthogonal family of projections with sum I; and each $E_{a,b}$ is a partial isometry (with initial projection $E_{b,b}$ and final projection $E_{a,a}$). In matrix terms, $E_{a,b}$ corresponds to the matrix with all entries 0 except in position (a, b)—where the entry is I.

6.6.3. LEMMA. *If $(E_{a,b})_{a,b \in \mathbb{B}}$ is a self-adjoint system of $n \times n$ matrix units for a von Neumann algebra $\mathscr{R}$ acting on a Hilbert space $\mathscr{H}$ and $\mathscr{T}$ is the subalgebra of $\mathscr{R}$ consisting of those elements commuting with $\{E_{a,b}\}$, then, for each T in $\mathscr{R}$, $\sum_{c \in \mathbb{B}} E_{c,a}TE_{b,c}$ is strong-operator convergent to an element $T_{a,b}$ of $\mathscr{T}$, $[T_{a,b}]_{a,b \in \mathbb{B}}\ (=\varphi(T)) \in n \otimes \mathscr{T}$, and φ is a * isomorphism of $\mathscr{R}$ onto $n \otimes \mathscr{T}$. Moreover, $E_{a,a}\mathscr{R}E_{a,a}$ is * isomorphic to $\mathscr{T}$ for each a.*

Proof. Note, first, that the commutant of $\{E_{a,b}\}$ is a von Neumann algebra, since $\{E_{a,b}\}$ is a self-adjoint family, and that $\mathscr{T}$ is the intersection of $\mathscr{R}$ with this commutant. Thus $\mathscr{T}$ is a von Neumann algebra. If $\mathbb{B}_0$ is a finite subset of $\mathbb{B}$ and x is a vector in $\mathscr{H}$,

$$\begin{aligned}\Big\|\sum_{c \in \mathbb{B}_0} E_{c,a}TE_{b,c}x\Big\|^2 &= \sum_{c \in \mathbb{B}_0} \|E_{c,a}TE_{b,c}x\|^2 \\ &\leq \|T\|^2 \sum_{c \in \mathbb{B}_0} \|E_{b,c}x\|^2 = \|T\|^2 \sum_{c \in \mathbb{B}_0} \|E_{b,c}E_{c,c}x\|^2 \\ &\leq \|T\|^2 \sum_{c \in \mathbb{B}_0} \|E_{c,c}x\|^2 \leq \|T\|^2\|x\|^2.\end{aligned}$$

It follows that the net $\{\sum_{c \in \mathbb{B}_0} E_{c,a}TE_{b,c}\}$ of finite partial sums of

$$\sum_{c \in \mathbb{B}} E_{c,a}TE_{b,c}$$

has $\|T\|$ as uniform bound and is Cauchy convergent in the strong-operator topology. From Proposition 2.5.11, it is strong-operator convergent to an operator $T_{a,b}$ in $\mathscr{R}$. Since multiplication by an operator is strong-operator continuous on $\mathscr{B}(\mathscr{H})$, $E_{c,d}T_{a,b} = E_{c,a}TE_{b,d} = T_{a,b}E_{c,d}$. Thus $T_{a,b} \in \mathscr{T}$.

If $\mathbb{B}_0$ is a finite subset of $\mathbb{B}$ with n_0 elements, $E_{\mathbb{B}_0} = \sum_{b\in\mathbb{B}_0} E_{b,b}$, and $\varphi_{\mathbb{B}_0}(E_{\mathbb{B}_0} T E_{\mathbb{B}_0}) = [T_{a,b}]_{a,b\in\mathbb{B}_0}$, then $\varphi_{\mathbb{B}_0}$ is a * isomorphism of $E_{\mathbb{B}_0}\mathscr{R}E_{\mathbb{B}_0}$ onto $n_0 \otimes \mathscr{T}$. To see this, note that $E_{\mathbb{B}_0} T E_{\mathbb{B}_0} = \sum_{a,b\in\mathbb{B}_0} E_{a,a} T E_{b,b}$ and $E_{\mathbb{B}_0}\mathscr{R}E_{\mathbb{B}_0}$ is a von Neumann algebra acting on $E_{\mathbb{B}_0}(\mathscr{H})$. If $E_{\mathbb{B}_0} T E_{\mathbb{B}_0} = 0$, then $0 = E_{c,a}E_{a,a}E_{\mathbb{B}_0} T E_{\mathbb{B}_0} E_{b,b} E_{b,c} = E_{c,a} T E_{b,c}$, for all a and b in $\mathbb{B}_0$ and c in $\mathbb{B}$. Thus $0 = \sum_{c\in\mathbb{B}} E_{c,a} T E_{b,c} = T_{a,b}$, for all a and b in $\mathbb{B}_0$; and $\varphi_{\mathbb{B}_0}$ is well defined. Linearity of $\varphi_{\mathbb{B}_0}$ is evident. Moreover, $T^*_{a,b} = \sum_{c\in\mathbb{B}} E_{c,a} T^* E_{b,c} = \sum_{c\in\mathbb{B}} (E_{c,b} T E_{a,c})^*$. Since $\sum_{c\in\mathbb{B}} E_{c,b} T E_{a,c}$ converges to $T_{b,a}$ in the weak- as well as in the strong-operator topology, and the adjoint operation is weak-operator continuous on $\mathscr{B}(\mathscr{H})$, $(T^*)_{a,b} = (T_{b,a})^*$. Thus $\varphi_{\mathbb{B}_0}$ preserves adjoints. Now, $(TE_{\mathbb{B}_0}S)_{a,b} = \sum_{c\in\mathbb{B}} E_{c,a} T(\sum_{d\in\mathbb{B}_0} E_{d,d}) S E_{b,c}$; while the entry in the a, b position of the product of $[T_{a,b}]_{a,b\in\mathbb{B}_0}$ and $[S_{a,b}]_{a,b\in\mathbb{B}_0}$ is

$$\sum_{d\in\mathbb{B}_0} T_{a,d} S_{d,b} = \sum_{d\in\mathbb{B}_0} \left(\sum_{c\in\mathbb{B}} E_{c,a} T E_{d,c}\right)\left(\sum_{c\in\mathbb{B}} E_{c,d} S E_{b,c}\right)$$

$$= \sum_{d\in\mathbb{B}_0} \sum_{c\in\mathbb{B}} E_{c,a} T E_{d,d} S E_{b,c} = \sum_{c\in\mathbb{B}} E_{c,a} T\left(\sum_{d\in\mathbb{B}_0} E_{d,d}\right) S E_{b,c}.$$

Thus $(TE_{\mathbb{B}_0}S)_{a,b}$ is the entry in the a, b position of this product; and

$$\varphi_{\mathbb{B}_0}(E_{\mathbb{B}_0} T E_{\mathbb{B}_0} S E_{\mathbb{B}_0}) = \varphi_{\mathbb{B}_0}(E_{\mathbb{B}_0} T E_{\mathbb{B}_0}) \varphi_{\mathbb{B}_0}(E_{\mathbb{B}_0} S E_{\mathbb{B}_0}).$$

Finally, if $A \in \mathscr{T}$ and $a, b \in \mathbb{B}_0$, then $\sum_{c\in\mathbb{B}} E_{c,a}(AE_{a,b})E_{b,c} = \sum_{c\in\mathbb{B}} AE_{c,c} = A$, so that $\varphi_{\mathbb{B}_0}(AE_{a,b})$ has A at the a, b position and 0 at all other positions. Thus $\varphi_{\mathbb{B}_0}$ is a * homomorphism of $E_{\mathbb{B}_0}\mathscr{R}E_{\mathbb{B}_0}$ *onto* $n_0 \otimes \mathscr{T}$. If $T_{a,b} = 0$ when a, $b \in \mathbb{B}_0$, then $0 = T_{a,b} = \sum_{c\in\mathbb{B}} E_{c,a} T E_{b,c} = E_{a,a}(\sum_{c\in\mathbb{B}} E_{c,a} T E_{b,c}) E_{a,b} = E_{a,a} T E_{b,b}$ and $0 = \sum_{a,b\in\mathbb{B}_0} E_{a,a} T E_{b,b} = E_{\mathbb{B}_0} T E_{\mathbb{B}_0}$. Thus $\varphi_{\mathbb{B}_0}$ is a * isomorphism. From Theorem 4.1.8(iii), $[T_{a,b}]_{a,b\in\mathbb{B}_0}$ has the same bound as $E_{\mathbb{B}_0} T E_{\mathbb{B}_0}$. Since $\|E_{\mathbb{B}_0} T E_{\mathbb{B}_0}\|$ is monotone increasing to $\|T\|$, over the net of finite subsets of $\mathbb{B}$, $[T_{a,b}]_{a,b\in\mathbb{B}} \in n \otimes \mathscr{T}$, and has the same bound as T. Thus φ is a linear, isometric, adjoint-preserving mapping of $\mathscr{R}$ into $n \otimes \mathscr{T}$. The restriction of φ to $E_{\mathbb{B}_0}\mathscr{R}E_{\mathbb{B}_0}$ is a * isomorphism onto the subalgebra of $n \otimes \mathscr{T}$ consisting of those matrices with 0 in the a, b position unless a and b are both in $\mathbb{B}_0$. Changing notation slightly, let $\varphi_{\mathbb{B}_0}$ now denote this restriction of φ, and $\mathscr{T}_{\mathbb{B}_0}$ denote the subalgebra of $n \otimes \mathscr{T}$. Of course, $\mathscr{T}_{\mathbb{B}_0}$ is a von Neumann algebra * isomorphic to $n_0 \otimes \mathscr{T}$.

If x and y are vectors in $\mathscr{H}$ and $\varepsilon > 0$, there is a finite set $\mathbb{B}_0(\subseteq \mathbb{B})$ such that

$$\sum_{c\notin\mathbb{B}_0} \|E_{c,c}x\| \, \|E_{c,c}y\| \le \left(\sum_{c\notin\mathbb{B}_0} \|E_{c,c}x\|^2\right)^{1/2} \left(\sum_{c\notin\mathbb{B}_0} \|E_{c,c}y\|^2\right)^{1/2}$$

$$= \|(I - E_{\mathbb{B}_0})x\| \, \|(I - E_{\mathbb{B}_0})y\| < \varepsilon.$$

Thus, with T and S in the unit ball of $\mathscr{R}$,

$$|\langle (T_{a,b} - S_{a,b})x, y\rangle| = \left| \sum_{c\in\mathbb{B}} \langle (E_{c,a}(T - S)E_{b,c})x, y\rangle \right|$$
$$\leq \sum_{c\in\mathbb{B}_0} |\langle (T - S)E_{b,c}x, E_{a,c}y\rangle| + 2\sum_{c\notin\mathbb{B}_0} \|E_{c,c}x\|\,\|E_{c,c}y\|.$$

It follows that the mapping assigning $T_{a,b}$ to T is weak-operator continuous on the unit ball of $\mathscr{R}$.

Let $\tilde{\mathscr{H}}\,(=\sum_{b\in\mathbb{B}} \oplus \mathscr{H}_b)$ be the n-fold direct sum of $\mathscr{H}$ with itself, and let F_b be the projection of $\tilde{\mathscr{H}}$ on $\mathscr{H}_b$. Since φ is isometric and $\bigcup_{b\in\mathbb{B}} F_b(\tilde{\mathscr{H}})$ spans an everywhere-dense linear subspace of $\tilde{\mathscr{H}}$, weak-operator continuity of φ on $(\mathscr{R})_1$ will be established if we show that the mapping

$$T \to \langle \varphi(T)F_b x, F_a y\rangle$$

is weak-operator continuous on $(\mathscr{R})_1$, for all a, b in $\mathbb{B}$ and x, y in $\tilde{\mathscr{H}}$. This follows from the preceding paragraph, since $\langle \varphi(T)F_b x, F_a y\rangle = \langle T_{a,b}x_0, y_0\rangle$, for some x_0, y_0 in $\mathscr{H}$. (See the discussion following Theorem 5.1.2.) Now $\varphi(\mathscr{R})$ contains $\mathscr{T}_{\mathbb{B}_0}$, for each finite subset $\mathbb{B}_0$ of $\mathbb{B}$, and the union of these is weak-operator dense in $n \otimes \mathscr{T}$. We conclude, with the aid of the Kaplansky density theorem, that $\varphi(\mathscr{R}) = n \otimes \mathscr{T}$.

It follows that φ is a * isomorphism, from the fact that each $\varphi_{\mathbb{B}_0}$ is and the fact that φ is weak-operator continuous on the unit ball of $\mathscr{R}$. Note that $\varphi(E_{a,b})$ is the matrix with I as the (a, b) entry and 0 as all other entries. In particular $\varphi(E_{a,a})$ is the projection whose matrix has I as the (a, a) entry and 0 as all other entries. Of course, $\varphi(E_{a,a})\varphi(\mathscr{R})\varphi(E_{a,a})$ $(=\varphi(E_{a,a}\mathscr{R}E_{a,a}))$ is isomorphic to $\mathscr{T}$, so that $E_{a,a}\mathscr{R}E_{a,a}$ is isomorphic to $\mathscr{T}$. ∎

6.6.4. Lemma. *If $(E_b)_{b\in\mathbb{B}}$ is a family of equivalent projections with sum I in a von Neumann algebra $\mathscr{R}$, E_{a,b_0} is a partial isometry in $\mathscr{R}$ with initial projection E_{b_0} and final projection E_a, and $E_{b_0,b} = E_{b_0}$, then $(E_{a,b_0}E_{b,b_0}^*)_{a,b\in\mathbb{B}}$ is a self-adjoint system of matrix units for $\mathscr{R}$.*

Proof. Note that $E_{a,b_0}E_{b_0,b_0}^* = E_{a,b_0}E_{b_0} = E_{a,b_0}$. Let $E_{a,b}$ be $E_{a,b_0}E_{b,b_0}^*$. Then $E_{a,b}^* = E_{b,b_0}E_{a,b_0}^* = E_{b,a}$ and $E_{a,a} = E_{a,b_0}E_{a,b_0}^* = E_a$. Thus $\sum_{a\in\mathbb{B}} E_{a,a} = I$. Note, too, that

$$E_{a,b}E_{c,d} = E_{a,b_0}E_{b_0,b}E_{c,b_0}E_{b_0,d} = E_{a,b_0}E_{b_0,b}E_bE_cE_{c,b_0}E_{b_0,d} = 0,$$

unless $b = c$. Finally,

$$E_{a,c}E_{c,b} = E_{a,b_0}E_{c,b_0}^*E_{c,b_0}E_{b,b_0}^* = E_{a,b_0}E_{b_0}E_{b,b_0}^* = E_{a,b_0}E_{b,b_0}^* = E_{a,b}.$$

Thus $(E_{a,b})_{a,b\in\mathbb{B}}$ is a self-adjoint system of matrix units for $\mathscr{R}$. ∎

By representing the operators in $\mathscr{B}(\mathscr{H})$ as matrices of scalars relative to an orthonormal basis for $\mathscr{H}$ (see Section 2.6, *Matrix representations*), Theorem 6.6.1 can be reformulated to state that a factor of type I_n is * isomorphic to $n \otimes \mathbb{C}$. Remarking that $\mathbb{C}$ is * isomorphic to the center of $\mathscr{M}$, our factor, we arrive at a statement that applies, as well, to a von Neumann algebra of type I_n.

6.6.5. THEOREM *If $\mathscr{R}$ is a von Neumann algebra of type I_n with center $\mathscr{C}$ and E is an abelian projection with central carrier I, then $\mathscr{R}$ is * isomorphic to both $n \otimes \mathscr{C}$ and $n \otimes (E\mathscr{R}E)$.*

Proof. Since $\mathscr{R}$ is of type I_n, there is a family $(E_b)_{b \in \mathbb{B}}$ of equivalent abelian projections in $\mathscr{R}$ such that $\mathbb{B}$ has cardinal number n and $\sum E_b = I$. Choose b_0 in $\mathbb{B}$, let E_{b_0, b_0} be E_{b_0}, and let E_{a, b_0} be a partial isometry in $\mathscr{R}$ with initial projection E_{b_0} and final projection E_a for each a different from b_0. Let $E_{a,b}$ be $E_{a, b_0} E^*_{b, b_0}$. From Lemma 6.6.4, $(E_{a,b})_{a, b \in \mathbb{B}}$ is a self-adjoint system of $n \times n$ matrix units for $\mathscr{R}$. From Lemma 6.6.3, $\mathscr{R}$ is isomorphic to $n \otimes \mathscr{T}$, where $\mathscr{T}$ is the algebra of elements in $\mathscr{R}$ commuting with $\{E_{a,b}\}$.

We show that $\mathscr{T} = \mathscr{C}$. Clearly $\mathscr{C} \subseteq \mathscr{T}$. Suppose $A \in \mathscr{T}$. For each B in $\mathscr{R}$, $B = \sum_{a, b \in \mathbb{B}} E_a B E_b$. It will suffice to show that $AE_a BE_b = E_a BE_b A$. Now,

$$\begin{aligned} AE_a BE_b &= AE_{a,b} E_b E_{b,a} BE_b = E_{a,b} E_b AE_b E_{b,a} BE_b \\ &= E_{a,b} E_b E_{b,a} BE_b E_b AE_b = E_a BE_b A, \end{aligned}$$

since $E_b \mathscr{R} E_b$ is abelian and A commutes with $\{E_{a,b}\}$.

Since $C_E = I$, the mapping $A' \to A'E$ of $\mathscr{R}'$ onto $\mathscr{R}'E$ is a * isomorphism, from Proposition 5.5.5. Thus $\mathscr{C}$ and $\mathscr{C}E$ are * isomorphic. Since E is abelian in $\mathscr{R}$, $E\mathscr{R}E = \mathscr{C}E$. From Lemma 6.6.2, $n \otimes \mathscr{C}$ and $n \otimes (E\mathscr{R}E)$ are * isomorphic. ■

If $\mathfrak{A}$ is a finite-dimensional C^*-algebra, its center $\mathscr{C}$ is * isomorphic to a finite-dimensional $C(X)$. Thus $\mathscr{C}$ is the linear span of projections Q minimal in $\mathscr{C}$. If $\mathfrak{A}$ acts on $\mathscr{H}$, $\mathfrak{A}$ being finite-dimensional is weak-operator closed. Thus $\mathfrak{A}Q$ is a (finite-dimensional) factor (from the comments following Proposition 6.4.3). These considerations yield the result that follows.

6.6.6. PROPOSITION. *Each finite-dimensional C^*-algebra is a finite direct sum of factors of (finite) type* I.

Bibliography: [47, 48, 56]

6.7. Examples

In Section 6.5, we separated the family of von Neumann algebras into subfamilies designated as types I_n, II_1, II_∞, and III. It was remarked (see Corollary 6.5.3) that each factor lies in one of these subfamilies. In Theorem 6.6.1, we note the existence of factors of type I_n (namely, $\mathscr{B}(\mathscr{H})$, with $\mathscr{H}$ an n-dimensional Hilbert space). The present section is devoted to examples that establish the existence of factors of types II_1 and II_∞. Factors of type III will be constructed both in Section 8.6 and in Section 12.3.

The examples of factors of type II_1 we construct in this section are associated with (countable) discrete groups. They constitute one possible extension of the concept of (complex) group algebra of a finite group to the case of infinite discrete groups.

Throughout this section, G will be a discrete group with unit element e, and $\mathscr{H}$ will be the Hilbert space $l_2(G)$ (see Example 2.1.12). For each pair of elements (square-summable, complex-valued functions on G) x, y in $\mathscr{H}$ define $(x * y)(g_0)$ to be $\sum_{g \in G} x(g_0 g^{-1})y(g)$. If $x'(g) = \overline{x(g_0 g^{-1})}$, then both x' and y lie in $l_2(G)$, so that the infinite sum defining $(x * y)(g_0)$ converges, and

$$|(x * y)(g_0)| = |\langle y, x' \rangle| \leq \|y\| \, \|x'\| = \|x\| \, \|y\|.$$

From this, the mapping $y \to (x * y)(g_0)$ is a bounded linear functional on $\mathscr{H}$, when $x \in \mathscr{H}$. In addition, $(x * y)(g_0) = \sum_{g \in G} x(g^{-1})y(gg_0)$, for, replacing g by gg_0 in $\sum_{g \in G} x(g_0 g^{-1})y(g)$ and observing that $g \to gg_0$ is a one-to-one mapping of G onto G, we have the desired equality. The function $x * y$ on G is the *convolution* of x and y. It is bounded, and is therefore an element of the Banach space $l_\infty(G)$, but need not lie in $l_2(G)$. We define linear mappings L_x and R_x, from $\mathscr{H}$ to $l_\infty(G)$, by $L_x(y) = x * y$ and $R_x(y) = y * x$. We shall see that, for certain x in $\mathscr{H}$, the mappings L_x and R_x have ranges contained in $l_2(G)$ $(= \mathscr{H})$, and can therefore be regarded as linear mappings from $\mathscr{H}$ into itself. With g in G, we denote by x_g the function on G that takes the value 1 at g and 0 at other elements of G. Then $\{x_g\}$ is an orthonormal basis for $\mathscr{H}$. Simple calculation shows that $(x_g * x)(g_0) = x(g^{-1}g_0)$ and $(x * x_g)(g_0) = x(g_0 g^{-1})$, from which it follows that $x_g * x$ and $x * x_g$ are in $\mathscr{H}$ whenever $x \in \mathscr{H}$ and $g \in G$. Hence, $x * y \in \mathscr{H}$ when x and y are in $\mathscr{H}$ and either x or y is non-zero at only finitely many points of G (and is thus a finite linear combination of elements x_g).

6.7.1. LEMMA. *If $T \in \mathscr{B}(\mathscr{H})$, $x \in \mathscr{H}$, and $\langle Tx_g, x_h \rangle = \langle x * x_g, x_h \rangle$ for all g and h in G, then $T = L_x$.*

Proof. Since $x * x_g \in \mathscr{H}$, the inner product $\langle x * x_g, x_h \rangle$ is defined; moreover,

$$(x * x_g)(h) = \langle x * x_g, x_h \rangle = \langle Tx_g, x_h \rangle$$

for all g and h in G. Since the two mappings

$$y \to (x * y)(h) \qquad \text{and} \qquad y \to \langle Ty, x_h \rangle$$

are continuous linear functionals on $\mathscr{H}$, and take the same values at each member of the orthonormal basis $\{x_g\}$, it follows that $(x * y)(h) = \langle Ty, x_h \rangle = (Ty)(h)$ for all h in G and y in $\mathscr{H}$. Hence $L_x y = x * y = Ty$, and $L_x = T$. ∎

If we knew at the outset that L_x is a bounded linear operator acting on $\mathscr{H}$, the equation $\langle Tx_g, x_h \rangle = \langle L_x x_g, x_h \rangle$ would imply, by linearity and continuity, that $\langle Ty, z \rangle = \langle L_x y, z \rangle$ for all y and z in $\mathscr{H}$, whence $T = L_x$. The force of the preceding lemma is to allow us to identify L_x as a bounded operator, under the given conditions.

6.7.2. THEOREM. *If x and y in $l_2(G)$ are such that L_x and L_y are bounded operators on $l_2(G)$, then*

(i) $L_x + L_y = L_{x+y}$, $aL_x = L_{ax}$, $L_x L_y = L_{x*y}$, $L_x^* = L_{x^*}$ *where* $x^*(g) = \overline{x(g^{-1})}$, $L_{x_e} = I$, *and* $x = y$ *if* $L_x = L_y$;

(ii) *the sets* $\{L_x : x \in \mathscr{H}, L_x \in \mathscr{B}(\mathscr{H})\}$ $(= \mathscr{L}_G)$ *and*

$$\{R_x : x \in \mathscr{H}, R_x \in \mathscr{B}(\mathscr{H})\} \ (= \mathscr{R}_G)$$

are von Neumann algebras such that $\mathscr{L}'_G = \mathscr{R}_G$;

(iii) $\{L_{x_g} : g \in G\}$ *generates* $\mathscr{L}_G$ *and* $\{R_{x_g} : g \in G\}$ *generates* $\mathscr{R}_G$ *as von Neumann algebras; and* L_{x_g}, R_{x_g} *are unitary operators.*

Proof. With z in $\mathscr{H}$ $(= l_2(G))$ and g_0 in G,

$$\begin{aligned}(L_{ax+y}(z))(g_0) &= ((ax + y) * z)(g_0) = \sum_g (ax + y)(g_0 g^{-1}) z(g) \\ &= a \sum_g x(g_0 g^{-1}) z(g) + \sum_g y(g_0 g^{-1}) z(g) \\ &= a(x * z)(g_0) + (y * z)(g_0) = (aL_x z + L_y z)(g_0).\end{aligned}$$

Thus $L_{ax+y} = aL_x + L_y$.

Since

$$\langle L_x^* x_g, x_h \rangle = \langle x_g, L_x x_h \rangle = \overline{(x * x_h)(g)} = \overline{x(gh^{-1})},$$

while

$$\langle L_{x^*} x_g, x_h \rangle = (x^* * x_g)(h) = x^*(hg^{-1}) = \overline{x(gh^{-1})},$$

L_{x^*} is bounded and $L_{x^*} = L_x^*$, from the preceding lemma.

From the definition of L_x, $x * (y * x_g) = L_x L_y x_g \in l_2(G)$. Replacing g by e in this observation, we have that $x * y \in l_2(G)$, and L_{x*y} is defined. Note that

$$\langle L_{x*y} x_g, x_h \rangle = ((x * y) * x_g)(h) = \sum_f (x * y)(hf^{-1}) x_g(f) = (x * y)(hg^{-1}),$$

while

$$\begin{aligned}\langle L_x L_y x_g, x_h \rangle &= (x * (y * x_g))(h) = \sum_f x(hf^{-1})(y * x_g)(f)\\ &= \sum_f x(hf^{-1}) y(fg^{-1}) = \sum_f x(hg^{-1}(fg^{-1})^{-1}) y(fg^{-1})\\ &= (x * y)(hg^{-1}).\end{aligned}$$

Thus, from the preceding lemma, L_{x*y} is in $\mathscr{B}(\mathscr{H})$ and $L_{x*y} = L_x L_y$.

Since $\langle L_{x_e} x_g, x_h \rangle = (x_e * x_g)(h) = x_e(hg^{-1}) = \langle x_g, x_h \rangle$, it follows that $L_{x_e} = I$.

If $L_x = L_y$, then

$$\langle L_x x_e, x_g \rangle = (x * x_e)(g) = x(g) = \langle L_y x_e, x_g \rangle = y(g)$$

for all g in G. Thus $x = y$.

From the foregoing, we conclude that $\mathscr{L}_G$ and $\mathscr{R}_G$ are * algebras of operators on $\mathscr{H}$. If $L_x \in \mathscr{L}_G$ and $R_y \in \mathscr{R}_G$, then

$$\begin{aligned}(L_x R_y x_g)(h) &= (x * (x_g * y))(h) = \sum_f x(hf^{-1})(x_g * y)(f)\\ &= \sum_f x(hf^{-1}) y(g^{-1}f) = \sum_f x(h(gf)^{-1}) y(f)\\ &= \sum_f (x * x_g)(hf^{-1}) y(f) = ((x * x_g) * y)(h) = (R_y L_x x_g)(h).\end{aligned}$$

Thus $L_x R_y x_g = R_y L_x x_g$ for each g in G, and $L_x R_y = R_y L_x$. It follows that each of $\mathscr{L}_G$ and $\mathscr{R}_G$ is contained in the other's commutant.

If $T \in \mathscr{R}_G'$ and $x = Tx_e$, we show that $T = L_x$. Note that

$$\begin{aligned}\langle x * x_g, x_h \rangle &= \langle (Tx_e) * x_g, x_h \rangle = \langle R_{x_g}(Tx_e), x_h \rangle = \langle T(R_{x_g} x_e), x_h \rangle\\ &= \langle Tx_g, x_h \rangle.\end{aligned}$$

Thus L_x is a bounded operator and $T = L_x \in \mathscr{L}_G$, from the preceding lemma. It follows that $\mathscr{R}_G'$ is contained in $\mathscr{L}_G$. Thus $\mathscr{L}_G = \mathscr{R}_G'$, and, symmetrically, $\mathscr{R}_G = \mathscr{L}_G'$.

Having established that $T \in \mathscr{L}_G$ when T commutes with all R_{x_g}, we see that $\mathscr{R}'$ is contained in $\mathscr{L}_G$, where $\mathscr{R}$ is the von Neumann algebra generated by $\{R_{x_g} : g \in G\}$. Of course $\mathscr{R} \subseteq \mathscr{R}_G$, so that $\mathscr{L}_G = \mathscr{R}_G' \subseteq \mathscr{R}'$. Thus $\mathscr{L}_G = \mathscr{R}' = \mathscr{R}_G'$, and $\mathscr{R} = \mathscr{R}_G$. Symmetrically, the von Neumann algebra generated by $\{L_{x_g} : g \in G\}$ is $\mathscr{L}_G$. Since $(L_{x_g} x)(h) = x(g^{-1}h)$, L_{x_g} is isometric on $\mathscr{H}$ and has $L_{x_{g^{-1}}}$ as its inverse. Thus L_{x_g}, and, similarly, R_{x_g}, are unitary operators. ∎

6.7.3. REMARK. As $R_{x_g}x_e = L_{x_g}x_e = x_g$ and $\{x_g\}$ spans $\mathscr{H}$, x_e is a generating vector for $\mathscr{L}_G$ and $\mathscr{R}_G$. For each g in G, x_g, the transform of x_e under the unitary operators L_{x_g} (in $\mathscr{L}_G$) and R_{x_g} (in $\mathscr{R}_G$), is a generating vector for both $\mathscr{L}_G$ and $\mathscr{R}_G$. We noted, in the proof of Theorem 6.7.2, that

$$\langle L_x L_y x_g, x_h\rangle = (x * y)(hg^{-1}).$$

Thus

$$\begin{aligned}\langle L_x L_y x_g, x_g\rangle &= (x * y)(e) = \sum_h x(h^{-1})y(h) = \sum_h y(h^{-1})x(h) = (y * x)(e)\\ &= \langle L_y L_x x_g, x_g\rangle.\end{aligned}$$

The vectors with the properties of x_g are *trace vectors*. We treat the trace function in detail in Chapter VIII, although trace vectors will be used in Section 7.2.

Using the existence of trace vectors just noted, we prove, in the proposition that follows, that the von Neumann algebras $\mathscr{L}_G$ and $\mathscr{R}_G$ are finite. In the theorem following that proposition we show that, for those groups in which the conjugacy class of each element other than the identity is infinite (*i.c.c. groups*) $\mathscr{L}_G$ and $\mathscr{R}_G$ are factors (necessarily of type II_1, as we shall note).

In Theorem 6.7.8 we describe two i.c.c. groups (Π and $\mathscr{F}_2$) for which $\mathscr{L}_\Pi$ and $\mathscr{L}_{\mathscr{F}_2}$ are not * isomorphic. Ultimately, the fact that * isomorphisms preserve the trace (coupled with certain properties of the trace in $\mathscr{L}_\Pi$ and $\mathscr{L}_{\mathscr{F}_2}$) will be used to establish that these factors are not * isomorphic. The uniqueness of the trace in finite factors (Theorem 8.2.8) is the key to this.

It follows from the uniqueness of the dimension function (Theorem 8.4.3) that $\langle Ex_g, x_g\rangle$ is the (normalized) dimension of E (relative to $\mathscr{L}_G$) when G is an i.c.c. group. Since each operator in $\mathscr{L}_G$ is a norm limit of finite, linear combinations of projections in $\mathscr{L}_G$, each state (trace) τ on $\mathscr{L}_G$ that restricts to the dimension function on projections has the property $\tau(A) = \langle Ax_g, x_g\rangle$ for all A in $\mathscr{L}_G$. (We have just noted, in effect, the "uniqueness of the trace" in a type II_1 factor—see Theorem 8.2.8). In particular, $\langle Ax_g, x_g\rangle = \langle Ax_e, x_e\rangle$ for all A in $\mathscr{L}_G$ and all g in G, although this equality can be deduced from the observation that $x_g = R_{x_g}x_e$ and the fact that R_{x_g} is a unitary operator in $\mathscr{L}_G'$ (Theorem 6.7.2(ii) and (iii)). Thus if φ is a * isomorphism of $\mathscr{L}_G$ onto $\mathscr{L}_{\mathscr{F}}$, where G and $\mathscr{F}$ are i.c.c. groups, then $\langle Ax_g, x_g\rangle = \langle\varphi(A)x_f, x_f\rangle$ for all A in $\mathscr{L}_G$, g in G, and f in $\mathscr{F}$, since E and $\varphi(E)$ have the same (normalized) dimension for each projection E in $\mathscr{L}_G$ (again from uniqueness of the dimension function). ∎

6.7.4. PROPOSITION. *The von Neumann algebras $\mathscr{L}_G$ and $\mathscr{R}_G$ are finite.*

Proof. If the projection E in $\mathscr{L}_G$ is equivalent to I and V is a partial isometry in $\mathscr{L}_G$ such that $V^*V = E$ and $VV^* = I$, then $\langle Ex_e, x_e\rangle = \langle V^*Vx_e, x_e\rangle = \langle VV^*x_e, x_e\rangle = \langle x_e, x_e\rangle$, so that $\langle (I - E)x_e, x_e\rangle = 0$, and

$(I - E)x_e = 0$ (from the preceding remark). Since x_e is generating for $\mathscr{R}_G$, it is separating for $\mathscr{L}_G$ and $I - E = 0$. It follows that $\mathscr{L}_G$, and, similarly, $\mathscr{R}_G$, are finite von Neumann algebras. ■

6.7.5. THEOREM. *If G is a group with unit e and the conjugacy class (g) of each element g different from e is infinite, then $\mathscr{L}_G$ and $\mathscr{R}_G$ are factors of type* II_1 *when $G \neq (e)$.*

Proof. If $\{g_1, \ldots, g_n\}$ is a set of n distinct group elements, then

$$a_1 L_{x_{g_1}} + \cdots + a_n L_{x_{g_n}} = L_{a_1 x_{g_1} + \cdots + a_n x_{g_n}} = 0$$

only if $a_1 x_{g_1} + \cdots + a_n x_{g_n} = 0$, from Theorem 6.7.2(i). But

$$(a_1 x_{g_1} + \cdots + a_n x_{g_n})(g_j) = a_j,$$

so that $\{L_{x_g} : g \in G\}$ is a linearly independent set in $\mathscr{L}_G$. Thus $\mathscr{L}_G$ and, similarly, $\mathscr{R}_G$ have infinite linear dimension.

If $L_x \in \mathscr{L}_G$ and L_x commutes with L_{x_g}, then $x * x_g = x_g * x$; and

$$x(gg_0g^{-1}) = (x * x_g)(gg_0) = (x_g * x)(gg_0) = x(g_0).$$

If L_x is in the center of $\mathscr{L}_G$, then x is constant on (g_0) for all g_0 in G. Since $x \in l_2(G)$, $x = ax_e$, in this case; $L_x = aI$, and $\mathscr{L}_G$ is a factor of type II_1 (as is $\mathscr{R}_G$)—see Remark 6.5.4. ■

We turn, now, to specific examples of groups satisfying the infinite-conjugacy-class condition (*i.c.c. groups*).

6.7.6. EXAMPLE. Let $\mathscr{F}_n$ be the free (non-abelian) group on n generators ($n \geq 2$, and an infinity of generators is permissible). Loosely speaking, the elements of $\mathscr{F}_n$ are "words" formed from the n generators $a_1, a_2, \ldots, a_n$. A word is a juxtaposition of the symbols $a_1, a_2, \ldots, a_n, a_1^{-1}, \ldots, a_n^{-1}$ in any order with any number of repetitions (or deletions). Two words are equivalent if a sequence of insertions or deletions of combinations $a_j a_j^{-1}$ or $a_j^{-1} a_j$ applied to each of them leads, after a finite number of steps to identical words. So, for example, $a_3 a_2^{-1} a_2 a_1$ and $a_1 a_1^{-1} a_3 a_1$ are equivalent words in $\mathscr{F}_3$ (both equivalent to $a_3 a_1$). A word is in *reduced* form if it contains no combination $a_j a_j^{-1}$ or $a_j^{-1} a_j$. Each equivalence class contains a (unique) word in reduced form. The elements of $\mathscr{F}_n$ are the equivalence classes of words together with the class e of the "null" word. The product of two classes is the class of the word formed by juxtaposing a representative of each of the classes. As usual with such constructs, we use the relaxed convention of treating words (representatives) as elements (classes) of $\mathscr{F}_n$.

If g is a word in reduced form, distinct from e, say, $g = a_j^{\pm 1} \cdots a_k^{\pm 1}$, where j may be k, then $a_h^r g a_h^{-r}$ $(= g_r)$ is in reduced form, where $h \neq j$ and r is an integer of sign opposite to that of the exponent in the last occurrence of a_k (in g)—the obvious meaning being assigned to a_h^r. Moreover, $g_r \neq g_m$ for $r \neq m$, so that $\{g_r\}$ constitutes an infinite set of conjugates of g, and $\mathscr{F}_n$ is an i.c.c. group. It follows from Theorem 6.7.5 that $\mathscr{L}_{\mathscr{F}_n}$ and its commutant $\mathscr{R}_{\mathscr{F}_n}$ are factors of type II_1. ■

6.7.7. EXAMPLE. Let Π be the group of permutations of the integers that leave fixed all but a finite set of integers (the set may vary with the permutation). The restriction to "finite set" keeps the group countable. It also provides us with an example of a "locally finite" group—each finite set of group elements generates a finite subgroup, so that the group is an ascending union of finite subgroups. To see this, note that $\Pi = \bigcup \Pi_n$, where Π_n is the subgroup of Π consisting of those elements that leave fixed all integers of absolute value exceeding n. Suppose g is in Π_n and g moves j to k, where $j \neq k$. For each positive integer m, let g_m be the permutation ("transposition") that interchanges j and $n + m$ and leaves all other integers fixed. Then $g_m g g_m^{-1}$ $(= g_m g g_m)$ moves $n + m$ to k. Thus $\{g_m g g_m^{-1} : m = 1, 2, \ldots\}$ is an infinite set of conjugates of g, and Π is an i.c.c. group. From Theorem 6.7.5, $\mathscr{L}_\Pi$ and its commutant $\mathscr{R}_\Pi$ are factors of type II_1. ■

Further examples can be constructed by using the (simple) observation that a direct product of i.c.c. groups is an i.c.c. group. The two examples that we have at hand provide us with the first clue that the division of the factors into types I_n, II_1, II_∞, and III does not complete the determination of (algebraic) isomorphism classes of factors.

6.7.8. THEOREM. *The factors $\mathscr{L}_\Pi$ and $\mathscr{L}_{\mathscr{F}_2}$ are not * isomorphic.*

Proof. Let e and e' be the group units of $\mathscr{F}_2$ and Π, respectively. Let a_1 and a_2 be the two generators of $\mathscr{F}_2$. The fact that $\mathscr{L}_\Pi$ and $\mathscr{L}_{\mathscr{F}_2}$ are not * isomorphic is established with the aid of the last comments in Remark 6.7.3. We show that for each unitary operator U in $\mathscr{L}_{\mathscr{F}_2}$ such that $\langle Ux_e, x_e \rangle = 0$, one of $\|(UL_{x_{a_j}} - L_{x_{a_j}}U)x_e\|$ $(= t_j)$, $j = 1, 2$, exceeds $\frac{1}{25}$; while, for each pair of operators A_1, A_2 in $\mathscr{L}_\Pi$ and positive ε, there is a unitary operator V in $\mathscr{L}_\Pi$ such that $\langle Vx_{e'}, x_{e'} \rangle = 0$ and $\|(VA_j - A_jV)x_{e'}\| < \varepsilon$, for $j = 1, 2$. (Note, in this connection, that $\|Tx_e\|^2 = \langle T^*Tx_e, x_e \rangle$.)

Since $L_{x_{a_j}}$ is unitary, we have, with L_x for U,

$$t_j^2 = \|(L_{x_{a_j}}^* L_x L_{x_{a_j}} - L_x)x_e\|^2 = \sum_{g \in \mathscr{F}_2} |(x_{a_j^{-1}} * x * x_{a_j} - x)(g)|^2$$

$$= \sum_{g \in \mathscr{F}_2} |x(a_j g a_j^{-1}) - x(g)|^2 \geq \sum_{g \in S} |(y_j - x)(g)|^2 = \|y_j - x\|_S^2,$$

for each subset S of $\mathscr{F}_2$, where $y_j(g) = x(a_j g a_j^{-1})$ and $\| \ \|_S$ denotes the norm in $l_2(S)$. Writing $\mu(S)$ for $\sum_{g \in S} |x(g)|^2$ $(= \|x\|_S^2)$, we have

$$\begin{aligned} |\mu(a_j S a_j^{-1}) - \mu(S)| &= \big| \|y_j\|_S^2 - \|x\|_S^2 \big| \\ &= \big| \|y_j\|_S - \|x\|_S \big| (\|y_j\|_S + \|x\|_S) \le 2t_j, \end{aligned}$$

since

$$\|x\|_S \le \|x\|, \qquad \|y_j\|_S \le \|x\|, \qquad \mu(\mathscr{F}_2) = \|x\|^2 = \|L_x x_e\|^2 = 1,$$

and

$$\|y_j\|_S^2 = \sum_{g \in S} |x(a_j g a_j^{-1})|^2 = \sum_{g \in a_j S a_j^{-1}} |x(g)|^2 = \mu(a_j S a_j^{-1}).$$

Substituting $a_j^{-1} S a_j$ for S, $|\mu(S) - \mu(a_j^{-1} S a_j)| \le 2t_j$. Choosing for S all words in $\mathscr{F}_2$ in reduced form that begin with $a_1^{\pm 1}$, observe that $S, a_2 S a_2^{-1}$, and $a_2^{-1} S a_2$ are disjoint and that $S \cup a_1^{-1} S a_1 = \mathscr{F}_2 \backslash \{e\}$. Thus at least one of $\mu(S)$, $\mu(a_2 S a_2^{-1})$, or $\mu(a_2^{-1} S a_2)$ does not exceed $\frac{1}{3}$; while one of $\mu(S)$ or $\mu(a_1^{-1} S a_1)$ is not less than $\frac{1}{2}$, since $x(e) = \langle L_x x_e, x_e \rangle = 0$ so that $1 = \mu(\mathscr{F}_2) = \mu(\mathscr{F}_2 \backslash \{e\})$. In any event, $\mu(S) \le \frac{1}{3} + 2t_2$ and $\frac{1}{2} - 2t_1 \le \mu(S)$, so that $\frac{1}{12} \le t_1 + t_2$ and one of t_1 or t_2 exceeds $\frac{1}{25}$.

Suppose, now, that $A_1(= L_{x_1})$ and $A_2(= L_{x_2})$ are two elements of $\mathscr{L}_\Pi$. Let $g_1, g_2, \ldots$ be an enumeration of the elements of Π. Choose n so large that $\sum_{k>n} |x_j(g_k)|^2 < \varepsilon^2, j = 1, 2$, for a preassigned positive ε. Let $y_j(g_k)$ be $x_j(g_k)$ if $k \le n$ and 0 if $k > n$. Writing B_j for L_{y_j}, we have

$$\|(A_j - B_j)x_{e'}\|^2 = \sum_{k>n} |x_j(g_k)|^2 < \varepsilon^2.$$

Let m be an integer such that $g_1, \ldots, g_n$ leave fixed all integers not in $[-m, m]$. Let g be the transposition (in Π) that interchanges $m + 1$ and $m + 2$. Then $g g_k = g_k g$ for $k = 1, \ldots, n$, so that $L_{x_g} B_j = B_j L_{x_g}$ for $j = 1, 2$. Now

$$\|L_{x_g}(A_j - B_j)x_{e'}\| = \|(A_j - B_j)x_{e'}\| < \varepsilon$$

and

$$\|(A_j - B_j)L_{x_g} x_{e'}\| = \|(A_j - B_j)x_g\| = \|(A_j - B_j)x_{e'}\| < \varepsilon,$$

so that $\|(L_{x_g} A_j - A_j L_{x_g})x_{e'}\| < 2\varepsilon$ for $j = 1, 2$. The argument is completed by noting, in addition, that $\langle L_{x_g} x_{e'}, x_{e'} \rangle = \langle x_g, x_{e'} \rangle = 0$. ■

6.7.9. REMARK. The argument of Theorem 6.7.8 can be broadened, with little effort, to show that $\mathscr{L}_{\mathscr{F}_n}$ and $\mathscr{L}_\Pi$ are not * isomorphic for $n = 2, 3, \ldots$; and, more generally, that $\mathscr{L}_G$ and $\mathscr{L}_\Pi$ are not * isomorphic when G is the "free product" of two groups, one of order at least 2 and the other of order at least 3. (The free product is built on words formed from the elements of both groups allowing cancellations dictated only by the multiplication tables of each of the groups.) ■

We turn, now, to the construction of factors of type II_∞. Having specific factors of type II_1 at hand, the matrix techniques of the preceding section (especially Lemma 6.6.2) provide us easy access to factors of type II_∞. Indeed, if $\mathscr{R}$ is a factor of type II_1, we shall see that $n \otimes \mathscr{R}$ is a factor of type II_1 when n is finite and is a factor of type II_∞ when n is infinite. We shall also note (using Lemma 6.6.3) that each factor of type II_∞ has the form $n \otimes \mathscr{R}$ with n infinite and $\mathscr{R}$ a factor of type II_1. We state the theorem that details this information in a form that encompasses the more general von Neumann algebra situation.

6.7.10. THEOREM. *If $\mathscr{R}$ is a von Neumann algebra (factor) of type* II, *then $n \otimes \mathscr{R}$ is a von Neumann algebra (factor) of type* II. *If $\mathscr{R}$ is of type II_1, $n \otimes \mathscr{R}$ is of type II_1 if n is finite—and of type II_∞ if n is infinite. If $\mathscr{T}$ is a countably decomposable von Neumann algebra (or if $\mathscr{T}$ is a factor) of type II_∞, then $\mathscr{T} \cong n \otimes \mathscr{R}$, where $\mathscr{R}$ is a countably decomposable von Neumann algebra of type II_1 and n is $\aleph_0$ (or $\mathscr{R}$ is a factor of type II_1 and n is infinite).*

Proof. Suppose $\mathscr{R}$ is a von Neumann algebra of type II. Let F be a finite projection in $\mathscr{R}$ with central carrier I. From Lemma 6.6.2, $n \otimes \mathscr{R}$ has center $\mathscr{C} \otimes I_n$, where $\mathscr{R}$ has center $\mathscr{C}$. Thus F_a has central carrier I relative to $n \otimes \mathscr{R}$, where F_a is the $n \times n$ matrix whose only non-zero entry is F in the (a, a) position. Now F_a is finite in $n \otimes \mathscr{R}$; for if V_a is a partial isometry in $n \otimes \mathscr{R}$ with initial projection F_a and final projection G_a in $n \otimes \mathscr{R}$, where G_a is the matrix whose only non-zero entry is a projection G, less than F, in the (a, a) position, then $V_a = F_a V_a F_a$, so that V_a has its only non-zero entry a partial isometry V in $\mathscr{R}$ at position (a, a). In this case, V has initial projection F and final projection G, contradicting the choice of F finite in $\mathscr{R}$. It follows that $n \otimes \mathscr{R}$ has no central portion of type III.

Let $\mathbb{A}$ be the indexing set (with cardinality n) for the rows and columns of the $n \times n$ matrices with entries in $\mathscr{R}$ and let $E_{a,b}$ be the matrix whose only non-zero entry is I in row a and column b (that is, in the (a, b) position). Then $\{E_{a,b}\}_{a,b\in\mathbb{A}}$ is a self-adjoint system of $n \times n$ matrix units for $n \otimes \mathscr{R}$, $\sum_{a\in\mathbb{A}} E_{a,a} = I$, $E_{a,a}(n \otimes \mathscr{R})E_{a,a} \cong \mathscr{R}$ for each a in $\mathbb{A}$, and $E_{a,a} \sim E_{b,b}$ for all a and b in $\mathbb{A}$. From this (or, as above) each $E_{a,a}$ has central carrier I in $n \otimes \mathscr{R}$. If E is an abelian projection in $n \otimes \mathscr{R}$, then $E \precsim E_{a,a}$, from Proposition 6.4.6(ii). Thus $E \sim E_0 \le E_{a,a}$, and E_0 is an abelian projection in $n \otimes \mathscr{R}$ and in $E_{a,a}(n \otimes \mathscr{R})E_{a,a}$, from Proposition 6.4.6(i). If E were not 0, $\mathscr{R}$ would contain a non-zero abelian projection, contradicting the assumption that $\mathscr{R}$ is of type II. Thus $E = 0$ and $n \otimes \mathscr{R}$ is of type II. If $\mathscr{R}$ is a factor, then $\mathscr{R}'$ is a factor (as noted in the comments following Theorem 5.3.1). From Lemma 6.6.2, $\mathscr{R}' \otimes I_n$ is a factor, so that $(\mathscr{R}' \otimes I_n)'(= n \otimes \mathscr{R})$ is a factor.

If $\mathscr{R}$ is of type II_1, $E_{a,a}$ is finite in $n \otimes \mathscr{R}$ for each a in $\mathbb{A}$ (by an argument similar to that of the first paragraph). If n is finite, $n \otimes \mathscr{R}$ is finite, since

$I = \sum_{a \in \mathbb{A}} E_{a,a}$. (See Theorem 6.3.8.) When n is infinite, $n \otimes \mathcal{R}$ is of type II_∞, since I is the sum of an infinite family $\{E_{a,a}\}$ of (orthogonal) equivalent projections (so that I is properly infinite).

Suppose, now, that $\mathcal{T}$ is a countably decomposable von Neumann algebra of type II_∞. Let F be a finite projection in $\mathcal{T}$ with central carrier I. From Proposition 6.3.12, there is an orthogonal family $\{E_n\}$ (countable, since $\mathcal{T}$ is countably decomposable) of projections in $\mathcal{T}$, each equivalent to F, having sum I. From Lemma 6.6.4, there is a self-adjoint system $\{E_{j,k}\}_{j,k=1,2,\ldots}$ of $\aleph_0 \times \aleph_0$ matrix units for $\mathcal{T}$ such that $E_{n,n} = E_n$ for $n = 1, 2, \ldots$. Lemma 6.6.3 applies and $\mathcal{T} \cong \aleph_0 \otimes \mathcal{R}$, where $\mathcal{R} = E_{1,1}\mathcal{T}E_{1,1}$. Now $E_{1,1}\mathcal{T}E_{1,1}$ is of type II_1, since $E_{1,1}$ is a finite projection in $\mathcal{T}$ and $\mathcal{T}$ (and, hence, $E_{1,1}\mathcal{T}E_{1,1}$) has no non-zero abelian projections. If $\mathcal{T}$ is a factor, Proposition 6.3.12 applies without the restriction of countable-decomposability. In this case, $\mathcal{R}$ is a factor since $E_{1,1}\mathcal{T}E_{1,1}$ is a factor (from Proposition 5.5.6). ■

Bibliography: [56, 58]

6.8. Ideals

Using the techniques of spectral theory and comparison of projections, it is possible, now, to say much about the nature of ideals in a von Neumann algebra. A key result states that ideals in a von Neumann algebra have many projections.

6.8.1. LEMMA. *If $\mathcal{I}$ is an ideal (left, right, or two-sided) in a von Neumann algebra $\mathcal{R}$ and A is a self-adjoint operator in $\mathcal{I}$, then $E_\lambda \in \mathcal{I}$ when $\lambda < 0$ and $I - E_\lambda \in \mathcal{I}$ when $\lambda > 0$, where $\{E_\lambda\}$ is the resolution of the identity for A.*

Proof. From Theorem 5.2.2, $E_\lambda \in \mathcal{A}$, where $\mathcal{A}$ is the (abelian) von Neumann algebra generated by A and I. From Theorem 5.2.1, $\mathcal{A}$ is isomorphic to $C(X)$, where X is an extremely disconnected compact Hausdorff space. Let f be the function representing A in $C(X)$. Then e_λ, the characteristic function of X_λ, the complement of the closure of the set of points in X at which the values of f exceed λ, represents E_λ (see Theorem 5.2.2).

Suppose $0 < \lambda$. Let g be the function defined as 0 on X_λ and $1/f$ on $X \backslash X_\lambda$. (Note that $\lambda \le f(p)$ if $p \in X \backslash X_\lambda$.) Since X_λ is a clopen set, g is continuous. Moreover, $gf = fg = 1 - e_\lambda$. If B in $\mathcal{A}$ corresponds to g, then $BA = AB = I - E_\lambda$, so that $I - E_\lambda \in \mathcal{I}$ when $0 < \lambda$.

If $\lambda < 0$, let h be 0 on $X \backslash X_\lambda$ and $1/f$ on X_λ. Again, h is continuous and $hf = fh = e_\lambda$. If C in $\mathcal{A}$ corresponds to h, then $CA = AC = E_\lambda$, so that $E_\lambda \in \mathcal{I}$ if $\lambda < 0$. ■

6.8.2. REMARK. If $\mathscr{I}$ is a left (right) ideal in a von Neumann algebra $\mathscr{R}$, E is a projection in $\mathscr{I}$, and V is a partial isometry in $\mathscr{R}$ with initial (final) projection E, then $V \in \mathscr{I}$; for $V = VE$ (or $V = EV$). If $\mathscr{I}$ is a two-sided ideal in $\mathscr{R}$ and $F \precsim E$, then $F \in \mathscr{I}$; for $F \sim E_0 \leq E$. Since $E_0 = E_0 E$, $E_0 \in \mathscr{I}$. If $V \in \mathscr{R}$ and $V^*V = F$, $VV^* = E_0$, then $V \in \mathscr{I}$, since V has E_0 as final projection, so that $F(= V^*V) \in \mathscr{I}$. ■

6.8.3. THEOREM. *The set $\mathscr{I}$ of operators with range projection finite in a von Neumann algebra $\mathscr{R}$ is a two-sided ideal in $\mathscr{R}$. Each non-zero, two-sided ideal in a factor $\mathscr{M}$ contains this ideal.*

Proof. Suppose A and B are in $\mathscr{I}$. From Theorem 6.3.8, $R(A) \vee R(B)$ is finite. Since $R(A + B) \leq R(A) \vee R(B)$, $R(A + B)$ is finite, from Proposition 6.3.2, and $A + B \in \mathscr{I}$. If $a \neq 0$, $R(aA) = R(A)$, so that $aA \in \mathscr{I}$. From Proposition 6.1.6, $R(A) \sim R(A^*)$, so that $A^* \in \mathscr{I}$. If $T \in \mathscr{R}$, $R(TA) \sim R(A^*T^*) \leq R(A^*)$, so that $TA \in \mathscr{I}$. As $AT = (T^*A^*)^*$, $AT \in \mathscr{I}$. Thus $\mathscr{I}$ is a two-sided ideal in $\mathscr{R}$.

Suppose $\mathscr{K}$ is a non-zero, two-sided ideal in the factor $\mathscr{M}$ and F is a finite projection in $\mathscr{M}$. If T is a non-zero operator in $\mathscr{K}$, then T^*T is a non-zero positive operator in $\mathscr{K}$. If $\{E_\lambda\}$ is the resolution of the identity for T^*T, then $I - E_\lambda \in \mathscr{K}$ for each positive λ, from Lemma 6.8.1. Since $0 < T^*T$, $I - E_\lambda \neq 0$ for some positive λ. Thus $\mathscr{K}$ contains a non-zero projection E. A maximal orthogonal family of subprojections of F equivalent to E can have at most a finite number of members, $E_1, \ldots, E_n$. By maximality of $\{E_j\}$ and comparability of projections in a factor $F - \sum E_j \prec E$. From Remark 6.8.2, $F - \sum E_j$ and each E_j are in $\mathscr{K}$. Thus $F \in \mathscr{K}$. If A is an operator in $\mathscr{M}$ with finite range projection, then $R(A) \in \mathscr{K}$. Since $A = R(A)A$, $A \in \mathscr{K}$. ■

It follows from the preceding theorem that if I is finite relative to the factor $\mathscr{M}$, then I lies in each non-zero, two-sided ideal in $\mathscr{M}$, so that $\mathscr{M}$ has no *proper* two-sided ideals—that is, $\mathscr{M}$ is *simple*.

6.8.4. COROLLARY. *Each finite factor is simple.*

6.8.5. COROLLARY. *Each countably decomposable type* III *factor is simple.*

Proof. From Lemma 6.8.1, a non-zero, two-sided ideal $\mathscr{I}$ contains a non-zero projection E. If F is another projection in the factor, $F \precsim E$, from Theorem 6.3.4, since E is infinite, $C_E = I$, and F is countably decomposable. From Remark 6.8.2, $F \in \mathscr{I}$. With T in the factor, $R(T) \in \mathscr{I}$, so that $T = R(T)T \in \mathscr{I}$. ■

The proof of Corollary 6.8.5 establishes, at the same time, the following lemma.

6.8.6. LEMMA. *No proper, two-sided ideal in a countably decomposable factor contains an infinite projection.*

6.8.7. THEOREM. *If $\mathscr{M}$ is a countably decomposable factor of type I_∞ or II_∞, the norm closure $\mathscr{F}$ of the two-sided ideal $\mathscr{I}$ of operators with finite range projection is the only proper, norm-closed, two-sided ideal in $\mathscr{M}$.*

Proof. If $\mathscr{K}$ is a norm-closed, non-zero, two-sided ideal in $\mathscr{M}$, then $\mathscr{I} \subseteq \mathscr{K}$, from Theorem 6.8.3. Thus $\mathscr{F} \subseteq \mathscr{K}$. Suppose $\mathscr{F} \neq \mathscr{K}$, and T is an operator in $\mathscr{K}$ not in $\mathscr{F}$. Since $T = V(T^*T)^{1/2}$ (polar decomposition), $(T^*T)^{1/2} \notin \mathscr{F}$. Passing to the function representation of the C^*-algebra generated by T^*T and I (on the spectrum of $(T^*T)^{1/2}$) and applying the Stone–Weierstrass theorem to the square-root function restricted to the spectrum of $(T^*T)^{1/2}$ and noting that this function vanishes at 0, we see that $(T^*T)^{1/2}$ is a norm limit of polynomials without constant term in T^*T. Thus $T^*T \notin \mathscr{F}$. If $\{E_\lambda\}$ is the resolution of the identity for T^*T, from Lemma 6.8.1, $I - E_\lambda \in \mathscr{K}$ for each positive λ. Thus $E_\lambda - E_\mu \in \mathscr{K}$ for all positive λ and μ. From Theorem 5.2.2, T^*T is a norm limit of finite linear combinations of projections of the form $E_\lambda - E_\mu, 0 < \mu < \lambda$. Thus at least one such projection E is not in $\mathscr{F}$, and E is infinite. It follows from Lemma 6.8.6 that $\mathscr{K} = \mathscr{M}$. ■

6.8.8. THEOREM. *If $\mathscr{K}$ is a weak-operator-closed left (or right) ideal in the von Neumann algebra $\mathscr{R}$, then $\mathscr{K} = \mathscr{R}E$ (or $\mathscr{K} = E\mathscr{R}$) for some projection E in $\mathscr{R}$. If $\mathscr{K}$ is a two-sided ideal, E is a central projection in $\mathscr{R}$.*

Proof. If H is a positive operator in $\mathscr{K}$ with resolution of the identity $\{E_\lambda\}$, then, from Lemma 6.8.1, $I - E_\lambda \in \mathscr{K}$ for each positive λ. Since $E_0 = \bigwedge_{\lambda > 0} E_\lambda$, from Theorem 5.2.2 and the definition of a resolution of the identity, $I - E_\lambda \to I - E_0$ as $\lambda \to 0^+$ in the strong-operator topology. (Therefore $I - E_0 \in \mathscr{K}$.) From Theorem 5.2.2, $HE_0 \leq 0$. Since $H \geq 0$ and H commutes with E_0, $HE_0 \geq 0$. Thus $HE_0 = 0$ and $R(H) \leq I - E_0$. It follows that $R(H) \in \mathscr{K}$ for each positive H in $\mathscr{K}$. Assuming $\mathscr{K}$ is a left ideal, $R(T^*T) = R(T^*) \in \mathscr{K}$ if $T \in \mathscr{K}$. The union of a finite family of projections in $\mathscr{K}$ is in $\mathscr{K}$, since that union is the range projection of their sum. Since $\mathscr{K}$ is weak-operator closed, $\bigvee_{T \in \mathscr{K}} R(T^*)\ (= E) \in \mathscr{K}$. As $ET^* = T^*$, $TE = T$ for each T in $\mathscr{K}$; and $\mathscr{K} = \mathscr{R}E$.

If $\mathscr{K}$ is a right ideal, $\mathscr{K}^*$ is a left ideal. Applying the foregoing, $\mathscr{K}^* = \mathscr{R}F$ and $\mathscr{K} = F\mathscr{R}$, for some projection F in $\mathscr{R}$. If $\mathscr{K}$ is a two-sided ideal, $\mathscr{K} = F\mathscr{R} = \mathscr{R}E$, so that $E = FE = F$. It follows that $E\mathscr{R}E = \mathscr{R}E$; and the range of E is invariant under $\mathscr{R}$. Since $\mathscr{R}$ is a self-adjoint family, E commutes with all the operators in $\mathscr{R}$ (see Section 2.6, *Subspaces*); and E is a central projection in $\mathscr{R}$. ■

It follows from the preceding theorem that each weak-operator-closed, two-sided ideal $\mathscr{K}$ in a von Neumann algebra $\mathscr{R}$ is a self-adjoint family. This holds, as well, for each norm-closed, two-sided ideal $\mathscr{K}$ in a C^*-algebra (as we noted in Corollary 4.2.10). In a von Neumann algebra more is true.

6.8.9. PROPOSITION. *Each two-sided ideal $\mathscr{K}$ in a von Neumann algebra $\mathscr{R}$ is self-adjoint.*

Proof. With T in $\mathscr{K}$ and $V(T^*T)^{1/2}$ its polar decomposition, $(T^*T)^{1/2} = V^*V(T^*T)^{1/2} \in \mathscr{K}$, since $V \in \mathscr{R}$. Thus $(T^*T)^{1/2}V^* = T^* \in \mathscr{K}$, and $\mathscr{K}$ is self-adjoint. ■

6.8.10. REMARK. For factors of type I_∞ and II_∞, Theorem 6.8.3 tells us that each non-zero, two-sided ideal contains the ideal of operators with finite range projection. Since I is the strong-operator limit of finite projections in such a factor (see Proposition 6.3.12), each non-zero ideal is weak-operator dense. ■

6.8.11. PROPOSITION. *The family $\mathscr{I}_c$ of operators with countably decomposable range in a von Neumann algebra $\mathscr{R}$ is a norm-closed, two-sided ideal that is weak-operator dense in $\mathscr{R}$. If $\mathscr{R}$ is a factor of type* III, *each non-zero, two-sided ideal in $\mathscr{R}$ contains $\mathscr{I}_c$.*

Proof. The argument of the first paragraph of Theorem 6.8.3 applies, noting that $R(A) \vee R(B)$ is countably decomposable, by virtue of Proposition 5.5.19 (in place of $R(A) \vee R(B)$ is finite, from Theorem 6.3.8); and $\mathscr{I}_c$ is a two-sided ideal in $\mathscr{R}$. If $\|A - A_n\| \to 0$, with A_n in $\mathscr{I}_c$, then $R(A) \le \bigvee_n R(A_n)$; so that $A \in \mathscr{I}_c$. Thus $\mathscr{I}_c$ is a norm-closed, two-sided ideal in $\mathscr{R}$. Since I is the sum of cyclic projections in $\mathscr{R}$ (each being countably decomposable, by Proposition 5.5.15), $\mathscr{I}_c$ is weak-operator dense in $\mathscr{R}$.

Suppose, now, that $\mathscr{R}$ is a factor of type III and that $\mathscr{K}$ is a non-zero, two-sided ideal in $\mathscr{R}$. From Lemma 6.8.1, $\mathscr{K}$ contains a non-zero projection E. From Theorem 6.3.4, $F \precsim E$ for each countably decomposable projection F in $\mathscr{R}$ (for $C_F = C_E = I$ and E is infinite, under the assumption that $\mathscr{R}$ is a factor of type III). Thus $\mathscr{K}$ contains $R(A)$ for each A in $\mathscr{I}_c$, from Remark 6.8.2; and $\mathscr{K}$ contains $A(=R(A)A)$. Thus $\mathscr{I}_c \subseteq \mathscr{K}$. ■

With the aid of "universal representation" techniques (see Proposition 10.1.5), we shall describe left and right, norm-closed ideals in a C^*-algebra in a manner analogous to the description given in Theorem 6.8.8 for weak-operator-closed ideals in a von Neumann algebra.

Bibliography: [28]

6.9. Exercises

6.9.1. Suppose E and F are projections in a von Neumann algebra $\mathscr{R}$, E is properly infinite, $C_E = I$, and $E \precsim F$. Show that F is properly infinite and $C_F = I$.

6.9.2. Let $\mathscr{R}$ be a von Neumann algebra and E and F be projections in $\mathscr{R}$ such that $F \precsim E$ and E is countably decomposable. Show that F is countably decomposable.

6.9.3. Let E, F, M, and N be projections in a von Neumann algebra $\mathscr{R}$ such that $E \precsim M$ and $F \precsim N$.

(i) Suppose $MN = 0$. Show that $E \vee F \precsim M + N$.
(ii) Is it true that $E \vee F \precsim M \vee N$? Proof—counterexample?

6.9.4. Let E be a properly infinite projection in a von Neumann algebra $\mathscr{R}$. Let F be a projection in $\mathscr{R}$ such that $F \precsim E$. Show that $E \sim E \vee F$.

6.9.5. Let E be a finite projection and F be a properly infinite projection in a von Neumann algebra $\mathscr{R}$. Suppose $C_E \leq C_F$. Show that $QE \prec QF$ for each non-zero central subprojection Q of C_F.

6.9.6. Let E and F be equivalent projections in a finite von Neumann algebra $\mathscr{R}$. Show that $I - E \sim I - F$.

6.9.7. Let E and F be equivalent finite projections in a von Neumann algebra $\mathscr{R}$. Show that $I - E \sim I - F$.

6.9.8. Let E_1, E_2, F_1, and F_2 be finite projections in a von Neumann algebra $\mathscr{R}$ such that $0 = E_1E_2 = F_1F_2$, $E_1 \sim F_1$, and $E_1 + E_2 \sim F_1 + F_2$. Show that $E_2 \sim F_2$.

6.9.9. Let E, F, M, and N be projections in a von Neumann algebra $\mathscr{R}$. Suppose that M and N are finite, $M \sim N$, $0 = ME = NF$, and $E \prec F$. Show that $E + M \prec F + N$.

6.9.10. Let E and F be projections in a von Neumann algebra $\mathscr{R}$. Suppose $E \sim F$ and V is a partial isometry in $\mathscr{R}$ with initial projection E and final projection F. Show that

(i) there is a unitary operator U in $\mathscr{R}$ such that $UE = V$ if E is finite;
(ii) there is an isometry W in $\mathscr{R}$ (that is, $W^*W = I$) such that $WE = V$ if $I - E$ is countably decomposable, $I - F$ is properly infinite, and $C_{I-E} \leq C_{I-F}$.

6.9.11. Let $\mathscr{R}$ be a von Neumann algebra. Show that the following two statements are equivalent:

(i) $\mathscr{R}$ is finite;
(ii) for each pair of equivalent projections E and F in $\mathscr{R}$, there is a unitary operator U in $\mathscr{R}$ such that $UEU^* = F$.

What are some of the consequences of defining "equivalence" of projections E and F in $\mathscr{R}$ to be "unitary equivalence" ($UEU^* = F$ for some unitary operator U in $\mathscr{R}$)?

6.9.12. Show that, if $\mathscr{R}$ is a semifinite von Neumann algebra, there is an orthogonal family $\{Q_a\}$ of non-zero central projections in $\mathscr{R}$ such that $\sum Q_a = I$ and each Q_a is the sum of an orthogonal family of mutually equivalent finite projections in $\mathscr{R}$. [*Hint*. Look at the proof of Proposition 6.3.12.]

6.9.13. Let E be a properly infinite projection in a von Neumann algebra $\mathscr{R}$ and F be a cyclic projection in $\mathscr{R}$ such that $C_E \leq C_F$. Show that there is an orthogonal family $\{Q_a\}$ of central subprojections of C_E with sum C_E such that, for each a, Q_aE is the sum of an orthogonal family of projections each equivalent to Q_aF.

6.9.14. Let $\mathscr{R}$ be a von Neumann algebra and $\{E_1, \ldots, E_n\}$, $\{F_1, \ldots, F_n\}$ be two finite sets of projections in $\mathscr{R}$ such that $E_1 \sim \cdots \sim E_n$, $F_1 \sim \cdots \sim F_n$, $\sum_{j=1}^n E_j = I$, and $\sum_{j=1}^n F_j = I$. Show that $E_j \sim F_j$.

6.9.15. Let E be a non-zero projection in a von Neumann algebra $\mathscr{R}$ acting on a Hilbert space $\mathscr{H}$. Show that, in the von Neumann algebra $E\mathscr{R}E$ acting on $E(\mathscr{H})$,

(i) G is an abelian projection if and only if G is a subprojection of E abelian in $\mathscr{R}$;
(ii) each central projection has the form PE with P a central projection in $\mathscr{R}$;

(iii) G is a finite projection if and only if G is finite in $\mathscr{R}$ and G is a subprojection of E;

(iv) G is properly infinite if and only if G is a subprojection of E properly infinite in $\mathscr{R}$.

6.9.16. Let $\mathscr{R}$ be a von Neumann algebra acting on a Hilbert space $\mathscr{H}$, and let E be a non-zero projection in $\mathscr{R}$. Show that $E\mathscr{R}E$ acting on $E(\mathscr{H})$

(i) is finite if $\mathscr{R}$ is finite;
(ii) is of type I if $\mathscr{R}$ is of type I;
(iii) is of type I_∞ if $\mathscr{R}$ is of type I_∞ and E is properly infinite;
(iv) is of type II_1 if $\mathscr{R}$ is of type II_1;
(v) is of type II_∞ if $\mathscr{R}$ is of type II_∞ and E is properly infinite;
(vi) is of type III if $\mathscr{R}$ is of type III.

6.9.17. Let E be a projection in a maximal abelian (self-adjoint) subalgebra $\mathscr{A}$ of a von Neumann algebra $\mathscr{R}$. Suppose E is minimal in the family of projections in $\mathscr{A}$ with central carrier C_E (relative to $\mathscr{R}$). Show that E is abelian in $\mathscr{R}$.

6.9.18. Let $\mathscr{A}$ be a maximal abelian (self-adjoint) subalgebra of a von Neumann algebra $\mathscr{R}$. Suppose $\mathscr{A} \neq \mathscr{R}$. Show that $\mathscr{A}$ contains two orthogonal non-zero projections E and F such that $C_E = C_F$ and $E \precsim F$.

6.9.19. Let $\mathscr{R}$ be a von Neumann algebra with no abelian central summands, and let $\mathscr{A}$ be a maximal abelian (self-adjoint) subalgebra of $\mathscr{R}$. Show that $\mathscr{A}$ contains a projection E such that $C_E = C_{I-E} = I$ and $E \precsim I - E$.

6.9.20. Let $\mathscr{R}$ be a countably decomposable von Neumann algebra and $\mathscr{A}$ a maximal abelian (self-adjoint) subalgebra of $\mathscr{R}$ that contains no non-zero projection finite in $\mathscr{R}$. Show that $\mathscr{A}$ contains n orthogonal projections with sum I equivalent in $\mathscr{R}$ for each positive integer n.

6.9.21. Let $\mathscr{R}$ be a von Neumann algebra of type I with no infinite central summand, and let P_n be a central projection in $\mathscr{R}$ such that $\mathscr{R}P_n$ is of type I_n. Let $\mathscr{A}$ be a maximal abelian subalgebra of $\mathscr{R}$. Show that

(i) some non-zero subprojection of P_n in $\mathscr{A}$ is abelian in $\mathscr{R}$;
(ii) $\mathscr{A}$ contains an abelian projection with central carrier I.

6.9.22. Let E_1 be an abelian projection with central carrier I in a von Neumann algebra $\mathscr{R}$ of type I_n with n finite. Show that

(i) there is a set of n orthogonal equivalent projections with sum I in $\mathscr{R}$ containing E_1 (so that each is abelian in $\mathscr{R}$);
(ii) $(I - E_1)\mathscr{R}(I - E_1)$ is of type I_{n-1}.

6.9.23. Let $\mathscr{R}$ be a von Neumann algebra of type I_n with n finite, and let $\mathscr{A}$ be a maximal abelian subalgebra of $\mathscr{R}$. Show that

(i) there are n (orthogonal, equivalent) projections in $\mathscr{A}$ with sum I each abelian in $\mathscr{R}$ with central carrier I relative to $\mathscr{R}$;
(ii) $\mathscr{A}$ is generated algebraically by the n abelian projections of (i) and the center of $\mathscr{R}$;
(iii) $\mathscr{A}$ contains p orthogonal projections with sum I equivalent in $\mathscr{R}$ if $n = pq$ (with p and q positive integers).

6.9.24. Let $\mathscr{R}$ be a countably decomposable von Neumann algebra of type I_∞, and let $\mathscr{A}$ be a maximal abelian subalgebra of $\mathscr{R}$ in which I is the union of projections in $\mathscr{A}$ finite in $\mathscr{R}$. Show that

(i) $\mathscr{A}$ has a projection finite in $\mathscr{R}$ with central carrier I;
(ii) $\mathscr{A}$ has a projection abelian in $\mathscr{R}$ with central carrier I;
(iii) some non-zero central projection Q in $\mathscr{R}$ is the sum of projections in $\mathscr{A}$ abelian in $\mathscr{R}$ with central carriers Q in $\mathscr{R}$;
(iv) there is a countable family of orthogonal projections in $\mathscr{A}$ with sum I each abelian with central carrier I;
(v) there are n orthogonal projections in $\mathscr{A}$ with sum I equivalent in $\mathscr{R}$ for each positive integer n.

6.9.25. Let $\mathscr{R}$ be a countably decomposable von Neumann algebra of type I_∞. Show that each maximal abelian subalgebra $\mathscr{A}$ of $\mathscr{R}$ contains n orthogonal projections with sum I equivalent in $\mathscr{R}$ for each positive integer n. [*Hint.* Consider the union E_0 of all projections in $\mathscr{A}$ finite in $\mathscr{R}$ and apply the results of Exercises 6.9.20 and 6.9.24 as well as Proposition 6.3.7.]

6.9.26. Let $\mathscr{R}$ be a finite von Neumann algebra and E, E_0, F, F_0 be projections in $\mathscr{R}$ such that $E_0 \le E$, $F_0 \le F$, $E_0 \sim F_0$, and $E \precsim F$.

(i) Show that $E - E_0 \precsim F - F_0$.
(ii) Suppose $E \sim F$. Show that $E - E_0 \sim F - F_0$.

6.9.27. Let $\mathscr{R}$ be a von Neumann algebra of type II_1. Let $\mathscr{A}$ be a maximal abelian subalgebra of $\mathscr{R}$ and E be a projection in $\mathscr{A}$.

(i) Show that there is a sequence $\{E_n\}$ of projections in $\mathscr{A}$ such that $E_0 = E$, $C_{E_n} = C_E$, $E_n \leq E_{n-1}$, and $E_n \precsim E_{n-1} - E_n$ for n in $\{1, 2, \ldots\}$.

(ii) Suppose F is a projection in $\mathscr{R}$ such that $C_E C_F \neq 0$. Show that there is a non-zero projection G in $\mathscr{A}$ such that $G \leq E$ and $G \precsim F$.

(iii) Suppose F is a projection in $\mathscr{R}$ such that $F \precsim E$. Show that some subprojection E_1 of E in $\mathscr{A}$ is equivalent to F.

(iv) Show that $\mathscr{A}$ contains n orthogonal equivalent projections with sum I for each positive integer n.

6.9.28. Let $\mathscr{R}$ be a countably decomposable von Neumann algebra and $\mathscr{A}$ a maximal abelian subalgebra of it. Suppose $\mathscr{R}$ has no central summand of type I. Show that each non-zero projection in $\mathscr{A}$ contains n non-zero orthogonal projections in $\mathscr{A}$ equivalent in $\mathscr{R}$.

6.9.29. Let $\mathscr{R}$ be a countably decomposable von Neumann algebra with no central summand of type I, and let n be a positive integer. Show that each maximal abelian subalgebra of $\mathscr{R}$ contains n orthogonal projections with sum I equivalent in $\mathscr{R}$.

6.9.30. Let $\mathscr{R}$ be a von Neumann algebra with center $\mathscr{C}$. Show that $n \otimes \mathscr{R}$ has center $\mathscr{C} \otimes I_n$.

6.9.31. Let $\mathscr{R}$ be a von Neumann algebra and n be a positive integer. Suppose P is a central projection in $n \otimes \mathscr{R}$ such that $(n \otimes \mathscr{R})P$ is of type I_m. Show that m is divisible by n.

6.9.32. Let $\mathscr{R}$ be a von Neumann algebra and n be a finite cardinal. Show that

(i) $n \otimes \mathscr{R}$ is finite if $\mathscr{R}$ is finite;
(ii) $n \otimes \mathscr{R}$ is properly infinite if $\mathscr{R}$ is properly infinite;
(iii) $n \otimes \mathscr{R}$ is countably decomposable if $\mathscr{R}$ is countably decomposable.

6.9.33. Let $\mathscr{R}$ be a von Neumann algebra and n be a cardinal. Show that

(i) $\mathscr{R}$ is finite if $n \otimes \mathscr{R}$ is finite;
(ii) $\mathscr{R}$ is properly infinite if n is finite and $n \otimes \mathscr{R}$ is properly infinite;
(iii) $\mathscr{R}$ is countably decomposable if $n \otimes \mathscr{R}$ is countably decomposable.

6.9.34. Let $\mathscr{R}$ be a countably decomposable von Neumann algebra and n a positive integer. Show that each maximal abelian subalgebra of $n \otimes \mathscr{R}$ contains n (orthogonal) equivalent projections with sum I.

6.9.35. Let $\mathscr{S}$ be an abelian self-adjoint subset of $n \otimes \mathscr{R}$, where n is a positive integer and $\mathscr{R}$ is a countably decomposable von Neumann algebra. Show that there is a unitary operator U in $n \otimes \mathscr{R}$ such that UAU^{-1} has all its non-zero entries on the diagonal for each A in $\mathscr{S}$.

6.9.36. With the notation and assumptions of Exercise 5.7.40(iii), show that

(i) the center $\mathscr{C}$ of $\mathfrak{A}'$ is generated as a finite-dimensional linear space by its minimal projections;
(ii) $\mathfrak{A}'P$ is a factor of type I_n with n finite and $\mathfrak{A}P$ has linear dimension not less than n^2 when P is a minimal projection in $\mathscr{C}$.

6.9.37. Let $\mathscr{R}$ be a von Neumann algebra with center $\mathscr{C}$. Show that $\mathscr{R}$ is of type I and $\mathscr{C}$ is the weak-operator-closed linear span of its minimal projections (we say $\mathscr{C}$ is *totally atomic*) if and only if $\mathscr{R}$ is the weak-operator-closed linear span of its minimal projections.

6.9.38. Let $\mathfrak{A}$ be a C^*-algebra acting on a Hilbert space $\mathscr{H}$. Suppose $\mathfrak{A}'$ is a factor of type I_n with n finite and $\mathfrak{A}$ has linear dimension at least n^2. Show that there is a vector x_0 in $\mathscr{H}$ such that $\mathscr{H} = \{Ax_0 : A \in \mathfrak{A}\}$.

6.9.39. Let G be a (discrete) group and F be the set of elements in G whose conjugacy classes are finite. Show that the center $\mathscr{C}$ of $\mathscr{L}_G$ (and $\mathscr{R}_G$) is precisely the set of elements L_x in $\mathscr{L}_G$ such that x is constant on the conjugacy class (g_0) for each g_0 in F and $x(g) = 0$ if $g \notin F$.

6.9.40. Let G be a (discrete) group and $\mathscr{F}$ be the net of finite subsets of G partially ordered by inclusion. For x in $l_2(G)$ and $\mathbb{F}$ in $\mathscr{F}$, define $x_{\mathbb{F}}(g)$ to be $x(g)$ if $g \in \mathbb{F}$ and 0 if $g \notin \mathbb{F}$.

(i) Show that $\lim_{\mathbb{F}} L_{x_{\mathbb{F}}} x_g = L_x x_g$ in $l_2(G)$ for each g in G.
(ii) Suppose $L_x \in \mathscr{L}_G$ and y in $l_2(G)$ is such that $\lim_{\mathbb{F}} L_{x_{\mathbb{F}}} y$ exists. Show that this limit is $L_x y$.

6.9.41. Let G_0 be a subgroup of the (discrete) group G. Let $\mathscr{R}$ be $\mathscr{L}_G$ acting on $l_2(G)$, and let $\mathscr{R}_0$ be $\{L_x \in \mathscr{R} : x(g) = 0, g \notin G_0\}$. Show that

(i) $\mathscr{R}_0$ is a von Neumann subalgebra of $\mathscr{R}$;
(ii) $\mathscr{R}_0$ is the weak-operator closure of the linear span of $\{L_{x_g} : g \in G_0\}$;
(iii) $\mathscr{R}_0$ is $*$ isomorphic to $\mathscr{L}_{G_0}$ acting on $l_2(G_0)$.

6.9.42. With the notation of Example 6.7.6, let $\mathscr{A}_j$ be the von Neumann subalgebra of $\mathscr{L}_{\mathscr{F}_n}$ generated by $L_{x_{a_j}}$. Show that $\mathscr{A}_j$ is a maximal abelian subalgebra of $\mathscr{L}_{\mathscr{F}_n}$.

6.9.43. Let G be a (discrete) group and α be an automorphism of G.

(i) Show that the mapping $L_x \to L_{x \circ \alpha}$ is a * automorphism of $\mathscr{L}_G$.

(ii) Let G be $\mathscr{F}_2$ and α be the automorphism of G that interchanges the generators a_1 and a_2. Show that therc is no unitary operator U in $\mathscr{L}_{\mathscr{F}_2}$ such that $UL_xU^{-1} = L_{x \circ \alpha}$ for each L_x in $\mathscr{L}_{\mathscr{F}_2}$. (We say that the automorphism $L_x \to L_{x \circ \alpha}$ is *outer* in this case.)

6.9.44. Let $\mathscr{F}_\infty$ be the free group on a countable number of generators $a_1, a_2, \ldots$. Let G_n be the subgroup generated by $a_1, a_{n+1}, a_{n+2}, \ldots$. Let $\mathscr{R}_n$ be the von Neumann subalgebra of $\mathscr{L}_{\mathscr{F}_\infty}$ generated by $\{L_g : g \in G_n\}$. Show that

(i) each $\mathscr{R}_n$ is a factor * isomorphic to $\mathscr{L}_{\mathscr{F}_\infty}$;

(ii) $\bigcap_{n=1}^{\infty} \mathscr{R}_n = \mathscr{A}_1$, where $\mathscr{A}_1$ is the (abelian) von Neumann subalgebra of $\mathscr{L}_{\mathscr{F}_\infty}$ generated by L_{a_1}, and that $\mathscr{A}_1$ is a maximal abelian subalgebra of $\mathscr{L}_{\mathscr{F}_\infty}$.

6.9.45. (i) Show that $G_1 \oplus G_2$ is an i.c.c. group if and only if G_1 and G_2 are i.c.c. groups.

(ii) Let G be $\mathscr{F}_2 \oplus \Pi$. Show that $\mathscr{L}_G$ is not * isomorphic to $\mathscr{L}_{\mathscr{F}_2}$.

6.9.46. Let $\mathscr{I}^=$ be the norm closure of a two-sided ideal $\mathscr{I}$ in a von Neumann algebra $\mathscr{R}$. Show that each projection in $\mathscr{I}^=$ lies in $\mathscr{I}$.

6.9.47. Suppose that $\mathscr{I}$ and $\mathscr{J}$ are two-sided ideals in a von Neumann algebra $\mathscr{R}$, and $\mathscr{J}$ is norm closed. Prove that

(i) $\mathscr{I} \subseteq \mathscr{J}$ if each projection in $\mathscr{I}$ lies in $\mathscr{J}$;

(ii) $\mathscr{I} \subseteq \mathscr{J}$ if, for each projection E in $\mathscr{I}$, there is a projection F in $\mathscr{J}$ such that $E \precsim F$.

6.9.48. Suppose that $\mathscr{R}$ is a factor.

(i) Show that if $\mathscr{I}$ and $\mathscr{J}$ are norm-closed, two-sided ideals in $\mathscr{R}$, then either $\mathscr{I} \subseteq \mathscr{J}$ or $\mathscr{J} \subseteq \mathscr{I}$. [*Hint*. Use the result of Exercise 6.9.47(ii).]

(ii) Prove that there is a norm-closed, two-sided ideal $\mathscr{I}$ in $\mathscr{R}$ such that $\mathscr{I} \neq \mathscr{R}$ and $\mathscr{I}$ contains every proper two-sided ideal in $\mathscr{R}$. [$\mathscr{I}$ is $\{0\}$ in some cases; see Corollaries 6.8.4 and 6.8.5.]

6.9.49. Let $\mathscr{R}$ be a von Neumann algebra.

(i) Suppose that $\mathscr{I}$ is a two-sided ideal in $\mathscr{R}$ and $\mathscr{P}_0$ is the set of all projections in $\mathscr{I}$. Prove that

(a) $E \vee F \in \mathscr{P}_0$ if $E, F \in \mathscr{P}_0$;
(b) if E and F are projections in $\mathscr{R}$ and $E \precsim F \in \mathscr{P}_0$, then $E \in \mathscr{P}_0$.

(ii) Suppose that a family $\mathscr{P}_0$ of projections in $\mathscr{R}$ satisfies the conditions (a) and (b) set out in (i). Let $\mathscr{I}_0$ be the set of all operators in $\mathscr{R}$ with range projection in $\mathscr{P}_0$. Show that $\mathscr{I}_0$ is a two-sided ideal in $\mathscr{R}$, and $\mathscr{P}_0$ is the set of all projections in $\mathscr{I}_0$. Prove also that if $\mathscr{J}$ is a two-sided ideal in $\mathscr{R}$, then $\mathscr{P}_0$ is the set of all projections in $\mathscr{J}$ if and only if $\mathscr{I}_0 \subseteq \mathscr{J} \subseteq \mathscr{I}_0^=$.

6.9.50. Suppose that E, F, G are projections in a von Neumann algebra $\mathscr{R}$, such that G is properly infinite and $PE \prec PG$, $PF \prec PG$ whenever P is a central projection in $\mathscr{R}$ for which $PG \neq 0$. Show that $PE \vee PF \prec PG$ for each such projection P.

6.9.51. Let $\mathscr{R}$ be a properly infinite von Neumann algebra with center $\mathscr{C}$, and let $\mathscr{P}_0$ be the set of all projections E in $\mathscr{R}$ such that $PE \prec P$ for each non-zero projection P in $\mathscr{C}$. From the result of Exercise 6.9.50, with $G = I$, note that $E \vee F \in \mathscr{P}_0$ whenever $E, F \in \mathscr{P}_0$. Prove that

(i) the set $\mathscr{I}$ of all operators in $\mathscr{R}$ with range projection in $\mathscr{P}_0$ is a two-sided ideal in $\mathscr{R}$;
(ii) the norm-closed, two-sided ideal $\mathscr{I}^=$ in $\mathscr{R}$ satisfies $\mathscr{I}^= \cap \mathscr{C} = \{0\}$;
(iii) if $\mathscr{J}$ is a two-sided ideal in $\mathscr{R}$, and $\mathscr{J} \cap \mathscr{C} = \{0\}$, then $\mathscr{J} \subseteq \mathscr{I}^=$.

Interpret these results in the case in which $\mathscr{R}$ is an infinite factor.

6.9.52. Let $\mathscr{R}$ be a von Neumann algebra. Suppose $T \,\eta\, \mathscr{R}$. Show that

(i) $R(T)$ and $N(T)$ are in $\mathscr{R}$ (see Exercise 2.8.45);
(ii) $R(T^*) = R(T^*T) = R((T^*T)^{1/2})$;
(iii) $R(T)$ and $R(T^*)$ are equivalent in $\mathscr{R}$.

6.9.53. Let S be a symmetric operator affiliated with a finite von Neumann algebra $\mathscr{R}$. Show that S is self-adjoint. [*Hint*. Use Proposition 2.7.10, especially (i) and (ii).]

6.9.54. Let A and B be operators affiliated with a finite von Neumann algebra $\mathscr{R}$, and let VH be the polar decomposition of B. Suppose $A \subseteq B$. Show that

(i) V^*A is a symmetric operator affiliated with $\mathscr{R}$;
(ii) $A = B$.

6.9.55. Let E be a projection in a finite von Neumann algebra $\mathscr{R}$ acting on a Hilbert space $\mathscr{H}$. With T in $\mathscr{R}$, let F be the projection with range $\{x : Tx \in E(\mathscr{H})\}$. Show that $F \in \mathscr{R}$ and that $E \precsim F$.

CHAPTER 7

NORMAL STATES AND UNITARY EQUIVALENCE OF VON NEUMANN ALGEBRAS

In a broad sense the results of this chapter are devoted to the study of the way in which the action of a von Neumann algebra on a Hilbert space is governed by its algebraic structure. The basic result of this part of the theory is the *unitary implementation theorem* (Theorem 7.2.9), which states that a * isomorphism between von Neumann algebras is implemented by a unitary transformation when each possesses a separating and generating vector. Thus, under the normalization of a generating and separating vector, the spatial action is *completely* determined by the algebraic structure. With the aid of this result and the techniques of tensoring von Neumann algebras and restricting them to invariant subspaces, the problem of spatial action can be reduced to one of algebraic structure.

The main tool in this process is a theory of *normal states* on von Neumann algebras. The states in question have special weak-operator continuity properties, which may be described in terms of *order* continuity. Thus * isomorphisms preserve these states. The key result in this area is Theorem 7.2.3, which tells us that, if the von Neumann algebra has a separating vector, each normal state is a vector state. Combining this with the fact that * isomorphisms preserve normal states provides the algebraic structure with a firm grip on the spatial action.

7.1. Completely additive states

The goal of this section is to characterize states that are convex combinations of vector states of a von Neumann algebra in terms of their continuity properties (Theorem 7.1.12).

7.1.1. Definition. A completely additive state of a von Neumann algebra $\mathscr{R}$ is a state ω of $\mathscr{R}$ such that $\omega(\sum_a E_a) = \sum_a \omega(E_a)$ for each orthogonal family $\{E_a\}$ of projections in $\mathscr{R}$. A carrier (support) for ω is a projection E such that $I - E = \sum_a E_a$ and $\{E_a\}$ is an orthogonal family of projections in $\mathscr{R}$ maximal with respect to the property that $\omega(E_a) = 0$ for all a. ■

With the notation of the preceding definition,

$$\omega(I - E) = \omega\left(\sum_a E_a\right) = \sum_a \omega(E_a) = 0.$$

Since $0 \le I - E$ and ω is a state, $0 = \omega(T(I - E)) = \omega((I - E)T)$ for each T in $\mathscr{R}$. (See Proposition 4.5.1.) If $A > 0, A \in \mathscr{R}$, and $EAE = A$, then $\omega(A) > 0$. Indeed, $A \ge \lambda E_0$ for some non-zero spectral projection E_0 of A and some positive λ (use Theorem 5.2.2(iv) for this). Thus $E_0 \le E$; and $0 < \omega(E_0) \le \lambda^{-1}\omega(A)$ by maximality of $\{E_a\}$. At this point, a special argument could be given to show that there is only one carrier for ω; but this fact is not needed and will become clear as we proceed.

7.1.2. Lemma. *If ω is a completely additive state of the von Neumann algebra $\mathscr{R}$, there is a countable orthogonal family of projections $\{E_n\}$ in $\mathscr{R}$ and vectors $\{x_n\}$ such that $\omega(E_n A E_n) \le \langle Ax_n, x_n\rangle$ for each positive A in $\mathscr{R}$ and $n = 1, 2, \ldots$, and $\omega(\sum_n E_n) = \sum_n \omega(E_n) = 1$.*

Proof. Let F be a carrier for ω. Suppose that we can find $\{E_n\}$ and $\{x_n'\}$ with the stated properties, but for $F\mathscr{R}F$ in place of $\mathscr{R}$ and $\omega|F\mathscr{R}F$ in place of ω. Then, with x_n equal to Fx_n' and A a positive operator in $\mathscr{R}$, $\omega(E_n A E_n) = \omega(E_n FAF E_n) \le \langle FAFx_n', x_n'\rangle = \langle Ax_n, x_n\rangle$. Thus we may assume that $F = I$ and that $\omega(A) > 0$ for each positive A in $\mathscr{R}$.

Let us note that if $\omega(G) \le \langle Gx, x\rangle$ for each subprojection G in $\mathscr{R}$ of a projection E in $\mathscr{R}$, then $\omega(EAE) \le \langle AEx, Ex\rangle$ for each positive A in $\mathscr{R}$. In this case, we can approximate EAE as closely as we wish, in norm, by linear combinations $\sum_{j=1}^n \lambda_j F_j$, where $\lambda_j \ge 0$ and $\{F_j\}$ is a family of mutually orthogonal spectral projections (for EAE) in $\mathscr{R}$ contained in E (see Theorem 5.2.2(v)). Thus $\sum_{j=1}^n \lambda_j \omega(F_j) \le \sum_{j=1}^n \lambda_j \langle F_j x, x\rangle = \langle(\sum_{j=1}^n \lambda_j F_j)x, x\rangle$, so that $\omega(EAE) \le \omega_x(EAE)$. (Recall the notation ω_x and the discussion preceding Proposition 4.3.1.)

Suppose that E is a projection in $\mathscr{R}$ and x is a vector such that $\|Ex\| = 1$. We show that E has a non-zero subprojection E_0 in $\mathscr{R}$ such that $\omega(E_0 A E_0) \le \langle AE_0 x, E_0 x\rangle$ for each positive A in $\mathscr{R}$. From the preceding paragraph, either we can choose E as E_0 or there is some subprojection F_0 of E in $\mathscr{R}$ such that $\omega(F_0) > \langle F_0 x, x\rangle$. In the latter case, let $\{F_b\}$ be a maximal orthogonal family of such projections. Then

$$\omega\left(\sum_b F_b\right) = \sum_b \omega(F_b) > \sum_b \langle F_b x, x\rangle = \left\langle\left(\sum_b F_b\right)x, x\right\rangle,$$

while $\omega(E) \le 1 = \langle Ex, x\rangle$. Thus $E \ne \sum_b F_b$. Let E_0 be $E - \sum_b F_b$. By maximality of $\{F_b\}$, $\omega(G) \le \langle Gx, x\rangle$ for each subprojection G of E_0 in $\mathscr{R}$. Again, from the argument of the preceding paragraph, $\omega(E_0 A E_0) \le \langle AE_0 x, E_0 x\rangle$, for each positive A in $\mathscr{R}$.

Let $\{E_a\}$ be an orthogonal family of projections in $\mathscr{R}$ maximal with respect to the property that there are vectors $\{x_a\}$ such that $\omega(E_a A E_a) \leq \langle Ax_a, x_a \rangle$ for each positive A in $\mathscr{R}$. Since $\sum_a \omega(E_a) = \omega(\sum_a E_a) \leq 1$ and $\omega(E_a) > 0$ unless $E_a = 0$, $\{E_a\}$ is a countable family. We relabel it $\{E_n\}$. If $0 \neq I - \sum_n E_n \,(=E)$, there is a unit vector x in the range of E; and, from our earlier argument, there is a subprojection E_0 of E in $\mathscr{R}$ such that $\omega(E_0 A E_0) \leq \langle AE_0 x, E_0 x \rangle$ for each positive A in $\mathscr{R}$, and $E_0 \neq 0$. But E_0 is orthogonal to each E_n, contradicting the maximality of $\{E_n\}$. Thus $I = \sum_n E_n$; and $\sum_n \omega(E_n) = 1$. ■

The technical lemma that follows is stated in a more general form than we need at present. It asserts, in effect, that weak- and strong-operator continuity are the same for linear mappings from a von Neumann algebra. The proof relies on noting that it suffices to check the closure properties of the inverse image of a convex set under the linear mapping, that this inverse is convex, and that such sets have the same weak- and strong-operator closures (Theorem 5.1.2).

7.1.3. Lemma. *If $\mathfrak{A}$ is a C*-algebra acting on the Hilbert space $\mathscr{H}$ and η is a linear mapping of $\mathfrak{A}$ into $\mathscr{B}(\mathscr{K})$ that is continuous at 0 on $(\mathfrak{A})_r^+$, the set of positive operators in $(\mathfrak{A})_r$, the ball of radius r with center 0 in $\mathfrak{A}$, from $\mathfrak{A}$ in the strong-operator topology to $\mathscr{B}(\mathscr{K})$ in the weak-operator topology, then η is continuous on $(\mathfrak{A})_s$ from $\mathfrak{A}$ in the weak-operator topology to $\mathscr{B}(\mathscr{K})$ in the weak-operator topology.*

Proof. Since $A \to tA$ and $B \to tB$ are strong- and weak-operator continuous mappings on $\mathscr{B}(\mathscr{H})$ and $\mathscr{B}(\mathscr{K})$, $A \to tA \to \eta(tA) = t\eta(A) \to n(A)$ has the same continuity on $(\mathfrak{A})_{t^{-1}r}$ as η has on $(\mathfrak{A})_r$ for each positive scalar t. Thus we have the same hypothesis for $(\mathfrak{A})_1$ as for $(\mathfrak{A})_r$ and can draw the conclusion we wish for all $(\mathfrak{A})_s$ if we establish the stated continuity for a single $(\mathfrak{A})_s$.

The mappings $A \to A^+$ and $A \to A^-$ carry $(\mathfrak{A}_h)_1$, the set of self-adjoint operators in the unit ball of $\mathfrak{A}$, into $(\mathfrak{A})_1^+$ and are strong-operator continuous at 0 on $\mathfrak{A}_h$, the set of self-adjoint operators in $\mathfrak{A}$, since $\|Ax\|^2 = \|A^+x\|^2 + \|A^-x\|^2$. (Recall that $A = A^+ - A^-$ and $A^+A^- = 0$, where A^+ and A^- are the "positive" and "negative" parts of A—see Proposition 4.2.3.) Thus $A = A^+ - A^- \to \eta(A^+) - \eta(A^-) = \eta(A)$ is continuous on $(\mathfrak{A}_h)_1$ at 0 from $\mathfrak{A}$ in the strong-operator topology to $\mathscr{B}(\mathscr{K})$ in the weak-operator topology. Thus η has the same continuity on $(\mathfrak{A}_h)_2$ at 0; and

$$A \to A - A_0 \to \eta(A - A_0) = \eta(A) - \eta(A_0) \to \eta(A)$$

is continuous on $(\mathfrak{A}_h)_1$ at A_0 from $\mathfrak{A}$ in its strong-operator topology to $\mathscr{B}(\mathscr{K})$ in its weak-operator topology.

We describe a subbase for the weak-operator open sets in $\mathscr{B}(\mathscr{K})$ such that each set has a convex complement. The inverse image under η of such a complement intersects $(\mathfrak{A}_h)_1$ in a strong-operator closed convex set. From Theorem 5.1.2, this inverse image is weak-operator closed, and its complement, the inverse image of the set of the subbase, is weak-operator open in $(\mathfrak{A}_h)_1$. It will follow that η is continuous on $(\mathfrak{A}_h)_1$ in the weak-operator topology to $\mathscr{B}(\mathscr{K})$ in its weak-operator topology. Now $A \to (A + A^*)/2$ and $A \to (A - A^*)/2i$ are weak-operator continuous mappings of $(\mathfrak{A})_1$ into $(\mathfrak{A}_h)_1$, so that

$$A = \frac{1}{2}(A + A^*) + i\left(\frac{1}{2i}(A - A^*)\right) \to \eta\left(\frac{1}{2}(A + A^*)\right) + \eta\left[i\left(\frac{1}{2i}(A - A^*)\right)\right]$$
$$= \eta(A)$$

is continuous on $(\mathfrak{A})_1$ in the weak-operator topology to $\mathscr{B}(\mathscr{K})$ in the weak-operator topology.

It remains to describe a subbase of open sets each of which has convex complement for the weak-operator topology on $\mathscr{B}(\mathscr{K})$. The sets

$$\{B : \mathrm{Re}\langle(B - B_0)x, y\rangle < a\}, \qquad \{B : \mathrm{Re}\langle(B - B_0)x, y\rangle > a\},$$
$$\{B : \mathrm{Im}\langle(B - B_0)x, y\rangle < a\}, \qquad \{B : \mathrm{Im}\langle(B - B_0)x, y\rangle > a\}$$

for real a and x, y in $\mathscr{K}$ will satisfy the requirements. The complement of the first type of set, for example, is the inverse image of the half-space

$$\{z : \mathrm{Re}\, z \geq \mathrm{Re}\langle B_0 x, y\rangle + a\}$$

under the linear functional $B \to \langle Bx, y\rangle$. The intersection of four of these sets, one of each type, consists of operators B for which $|\langle(B - B_0)x, y\rangle| < a\sqrt{2}$. These sets are weak-operator open. ■

Of course the identity mapping of $\mathscr{B}(\mathscr{H})$ onto $\mathscr{B}(\mathscr{H})$ is continuous on $(\mathscr{B}(\mathscr{H}))_1$ in its weak-operator topology (hence, in its strong-operator topology) to $\mathscr{B}(\mathscr{H})$ in its weak-operator topology without being continuous from $(\mathscr{B}(\mathscr{H}))_1$ with its weak-operator topology to $\mathscr{B}(\mathscr{H})$ with its strong-operator topology (for this would imply that the strong- and weak-operator topologies coincide on $(\mathscr{B}(\mathscr{H}))_1$).

7.1.4. LEMMA. *Each completely additive state of a von Neumann algebra is weak-operator continuous on the unit ball of that algebra.*

Proof. Let ω be a completely additive state of the von Neumann algebra $\mathscr{R}$, and let $\{E_n\}$ and $\{x_n\}$ be chosen as in Lemma 7.1.2. If $\varepsilon > 0$, choose N such that $\omega(\sum_{n>N} E_n) < \varepsilon^2/4$. Let A be an operator in the unit ball of $\mathscr{R}$ such that $\|Ax_n\| < \varepsilon/2N$, $n = 1, \ldots, N$. Then, writing H for A^*A,

$$\begin{aligned}|\omega(A)|^2 \le \omega(H) &= \omega\left(H\sum_{n=1}^{N} E_n\right) + \omega\left(H\sum_{n>N} E_n\right)\\ &\le \sum_{n=1}^{N} \omega(H^{1/2}H^{1/2}E_n) + \omega(H^2)^{1/2}\omega\left(\sum_{n>N} E_n\right)^{1/2}\\ &\le \sum_{n=1}^{N} \omega(H)^{1/2}\omega(E_nHE_n)^{1/2} + \frac{\varepsilon}{2}\\ &\le \sum_{n=1}^{N} \|Ax_n\| + \frac{\varepsilon}{2} \le \varepsilon.\end{aligned}$$

Thus ω is strong-operator continuous at 0 on $(\mathscr{R})_1$. From the preceding lemma, ω is weak-operator continuous on $(\mathscr{R})_1$. ∎

With ω a state of $\mathscr{R}$, let ω_B be the positive linear functional on $\mathscr{R}$ defined by the equality $\omega_B(A) = \omega(B^*AB)$ for A and B in $\mathscr{R}$. If $\omega(B^*B) = 1$, then ω_B is a state of $\mathscr{R}$. Since the mapping $A \to B^*AB$ is a weak-operator continuous mapping of $(\mathscr{R})_1$ into $(\mathscr{R})_{\|B\|^2}$, Lemma 7.1.4 yields the following result.

7.1.5. COROLLARY. *If ω is a completely additive state of the von Neumann algebra $\mathscr{R}$, ω_B is weak-operator continuous on $(\mathscr{R})_1$, so that ω_B is a completely additive state of $\mathscr{R}$ when $\omega(B^*B) = 1$.*

In the lemma that follows, we shall make use of the direct sum of all representations of the von Neumann algebra $\mathscr{R}$ engendered by the completely additive states of $\mathscr{R}$. We encountered a similar process in Remark 4.5.8, where the resulting representation was denoted by Φ and called the *universal* representation of the algebra. Its more detailed properties will be studied in Section 10.1. The completely additive states will be identified, presently, with the *normal* states (see Definition 7.1.11), after which the more usual terminology, *normal*, will be retained to refer to them. By analogy with the earlier discussion of the universal representation, we denote the direct sum of those representations of a von Neumann algebra engendered (through the GNS construction) by its completely additive states by Φ_{n} and refer to it as the *universal normal representation* of the algebra.

We note that Φ_{n} is faithful (and, hence, by Theorem 4.1.8(iii), isometric). With x a unit vector in the Hilbert space $\mathscr{H}$ on which $\mathscr{R}$ acts and ω the completely additive state, $\omega_x|\mathscr{R}$, of $\mathscr{R}$, the representation π_ω of $\mathscr{R}$, obtained from the GNS construction, is unitarily equivalent to the representation

$A \to AE'$ of $\mathscr{R}$ as $\mathscr{R}E'$ acting on $E'(\mathscr{H})$, where E' is the projection with range $[\mathscr{R}x]$ (from Corollary 4.5.4); and the mapping $Ax \to \pi_\omega(A)x_\omega$ extends to a unitary transformation of $E'(\mathscr{H})$ onto $\mathscr{H}_\omega$ that implements π_ω. Now π_ω is a direct summand of Φ_{n}, by construction. Thus, if $\Phi_{\mathrm{n}}(A) = 0$ for some A in $\mathscr{R}$, $AE' = 0$ for each cyclic projection E' in $\mathscr{R}'$, and $A = 0$. From this same discussion we see that, if ω and ω' are distinct completely additive states of $\mathscr{R}$, x_ω and $x_{\omega'}$ appear as unit vectors in orthogonal subspaces invariant under $\Phi_{\mathrm{n}}(\mathscr{R})$—corresponding to $\mathscr{H}_\omega$ and $\mathscr{H}_{\omega'}$, respectively. Thus, for each A in $\mathscr{R}$, $\omega(A) = \langle \Phi_{\mathrm{n}}(A)x_\omega, x_\omega \rangle$, $\omega'(A) = \langle \Phi_{\mathrm{n}}(A)x_{\omega'}, x_{\omega'} \rangle$, and

$$\langle \Phi_{\mathrm{n}}(A)x_\omega, \Phi_{\mathrm{n}}(B)x_{\omega'} \rangle = 0$$

for all A and B in $\mathscr{R}$.

7.1.6. LEMMA. *If $\mathscr{R}$ is a von Neumann algebra and Φ_{n} is its universal normal representation, then $\Phi_{\mathrm{n}}(\mathscr{R})$ is weak-operator closed.*

Proof. If Φ_{n} represents $\mathscr{R}$ on $\mathscr{H}_{\mathrm{n}}$, ω is a completely additive state of $\mathscr{R}$, and x is a unit vector in $\mathscr{H}_{\mathrm{n}}$ such that $\omega(A) = \langle \Phi_{\mathrm{n}}(A)x, x \rangle$ for all A in $\mathscr{R}$, then $\omega_B(A) = \langle \Phi_{\mathrm{n}}(A)\Phi_{\mathrm{n}}(B)x, \Phi_{\mathrm{n}}(B)x \rangle$. Thus $A \to \langle \Phi_{\mathrm{n}}(A)\Phi_{\mathrm{n}}(B)x, \Phi_{\mathrm{n}}(B)x \rangle$ is weak-operator continuous on $(\mathscr{R})_1$. With $\{\omega_1, \ldots, \omega_n\}$ a finite set of completely additive states of $\mathscr{R}$, as noted above, we can choose vectors $x_1, \ldots, x_n$ in $\mathscr{H}_{\mathrm{n}}$ such that $\omega_j(A) = \langle \Phi_{\mathrm{n}}(A)x_j, x_j \rangle$ for all A in $\mathscr{R}$ and $j = 1, \ldots, n$ and such that $\langle \Phi_{\mathrm{n}}(A)x_j, \Phi_{\mathrm{n}}(B)x_k \rangle = 0$ when $j \neq k$ for all A and B in $\mathscr{R}$. If we allow $\{\omega_1, \ldots, \omega_n\}$ to vary, the set $\{\sum_{j=1}^n \Phi_{\mathrm{n}}(B_j)x_j : B_j \text{ in } \mathscr{R}\}$ is dense in $\mathscr{H}_{\mathrm{n}}$. Moreover,

$$A \to \left\langle \Phi_{\mathrm{n}}(A) \sum_{j=1}^n \Phi_{\mathrm{n}}(B_j)x_j, \sum_{j=1}^n \Phi_{\mathrm{n}}(B_j)x_j \right\rangle = \sum_{j=1}^n \langle \Phi_{\mathrm{n}}(A)\Phi_{\mathrm{n}}(B_j)x_j, \Phi_{\mathrm{n}}(B_j)x_j \rangle$$

is weak-operator continuous on $(\mathscr{R})_1$. For each z in $\mathscr{H}_{\mathrm{n}}$, $A \to \langle \Phi_{\mathrm{n}}(A)z, z \rangle$ is a uniform limit of weak-operator continuous functions on $(\mathscr{R})_1$ and is itself weak-operator continuous on $(\mathscr{R})_1$. Thus Φ_{n} is weak-operator continuous on $(\mathscr{R})_1$ to $(\Phi_{\mathrm{n}}(\mathscr{R}))_1$ in its weak-operator topology. But $(\mathscr{R})_1$ is weak-operator compact, by Theorem 5.1.3, so that $(\Phi_{\mathrm{n}}(\mathscr{R}))_1$ is weak-operator compact (hence, closed). From the Kaplansky density theorem (Theorem 5.3.5), $(\Phi_{\mathrm{n}}(\mathscr{R})^-)_1 = (\Phi_{\mathrm{n}}(\mathscr{R}))_1^-$ $(=(\Phi_{\mathrm{n}}(\mathscr{R}))_1)$, so that $\Phi_{\mathrm{n}}(\mathscr{R}) = \Phi_{\mathrm{n}}(\mathscr{R})^-$. ■

7.1.7. COROLLARY. *If ω is a completely additive state of the von Neumann algebra $\mathscr{R}$ and φ is the representation it engenders, then $\varphi(\mathscr{R})^- = \varphi(\mathscr{R})$.*

Proof. If E' is the projection in $\Phi_{\mathrm{n}}(\mathscr{R})'$ with range $[\Phi_{\mathrm{n}}(\mathscr{R})x]$, where $\omega(A) = \langle \Phi_{\mathrm{n}}(A)x, x \rangle$ for each A in $\mathscr{R}$, then φ is unitarily equivalent to $A \to \Phi_{\mathrm{n}}(A)E'$, from Proposition 4.5.3, and $\Phi_{\mathrm{n}}(\mathscr{R})E'$ is weak-operator closed (Proposition 5.5.6). ■

7.1.8. THEOREM. *If ω is a completely additive state of a von Neumann algebra $\mathscr{R}$ acting on a Hilbert space $\mathscr{H}$, there is a countable set of vectors $\{x_n\}$ in $\mathscr{H}$ such that $\sum_{m=1}^{\infty} \|x_m\|^2 = 1$ and $\omega = \sum_{m=1}^{\infty} \omega_{x_m}|\mathscr{R}$, in the sense of norm convergence.*

Proof. Let $\{E'_a\}$ be a family of cyclic projections in $\mathscr{R}'$ maximal with respect to the property that their central carriers are orthogonal. By maximality, $\sum_a C_{E'_a} = I$. Let ω_a be the vector state of $\mathscr{R}$ corresponding to a unit generating vector for E'_a, and let x_a be a unit vector in $\mathscr{H}_{\mathrm{n}}$ such that $\omega_a = \omega_{x_a} \circ \Phi_{\mathrm{n}}$. Then the representation $A \to \Phi_{\mathrm{n}}(A)F'_a$, where F'_a is the projection with range $[\Phi_{\mathrm{n}}(\mathscr{R})x_a]$, and $A \to AE'_a$ are unitarily equivalent representations of $\mathscr{R}$ (see Proposition 4.5.3). As the states ω_a are distinct (since $\omega_a(C_{E'_b})$ is 1 when $a = b$ and 0 otherwise), the x_a may be chosen (by construction of the universal normal representation Φ_{n}) such that $\{F'_a\}$ is an orthogonal family. It follows, then, that $A \to AE'$ and $A \to \Phi_{\mathrm{n}}(A)F'$ are unitarily equivalent representations of $\mathscr{R}$, where $E' = \sum E'_a$ and $F' = \sum F'_a$. Let U be a unitary transformation of $F'(\mathscr{H}_{\mathrm{n}})$ onto $E'(\mathscr{H})$ such that $U^{-1}AE'U = \Phi_{\mathrm{n}}(A)F'$ for all A in $\mathscr{R}$.

From Proposition 5.5.3, $C_{E'} = I$; and, from Proposition 5.5.5, the mapping $A \to AE'$ is a * isomorphism of $\mathscr{R}$ onto $\mathscr{R}E'$. Thus $A \to \Phi_{\mathrm{n}}(A)F'$ is a * isomorphism, as is $\Phi_{\mathrm{n}}(A) \to \Phi_{\mathrm{n}}(A)F'$. From Lemma 7.1.6, $\Phi_{\mathrm{n}}(\mathscr{R})$ is a von Neumann algebra. As $(I - C_{F'})F' = 0$, $C_{F'} = I$.

Let $\{G'_b\}$ be an orthogonal family of projections in $\Phi_{\mathrm{n}}(\mathscr{R})'$ maximal with respect to the property that each $G'_b \precsim F'$. If $I - \sum G'_b \neq 0$, then, from Proposition 6.1.8, F' and $I - \sum G'_b$ have non-zero equivalent subprojections (since $C_{F'} = I$), contradicting the maximality of $\{G'_b\}$. Thus $\sum G'_b = I$. Let V'_b be a partial isometry in $\Phi_{\mathrm{n}}(\mathscr{R})'$ with initial projection G'_b and final projection a subprojection of F'.

Since ω is a completely additive state of $\mathscr{R}$, there is a unit vector y in $\mathscr{H}_{\mathrm{n}}$ such that $\omega = \omega_y \circ \Phi_{\mathrm{n}}$. All but a countable number of vectors $V'_bG'_by$ are non-zero, since $\sum_b \|G'_by\|^2 = \|y\|^2 = 1$. Denote these by $y_1, y_2, \ldots$; and note that $y_m \in F'(\mathscr{H}_{\mathrm{n}})$. If $x_m = Uy_m$, then

$$\begin{aligned}\sum_m \langle Ax_m, x_m\rangle &= \sum_m \langle AE'Uy_m, Uy_m\rangle = \sum_m \langle \Phi_{\mathrm{n}}(A)y_m, y_m\rangle \\ &= \sum_b \langle \Phi_{\mathrm{n}}(A)V'_bG'_by, V'_bG'_by\rangle = \sum_b \langle \Phi_{\mathrm{n}}(A)y, G'_by\rangle \\ &= \langle \Phi_{\mathrm{n}}(A)y, y\rangle = \omega(A).\end{aligned}$$

Since $\omega(I) = 1$, $\sum_m \|x_m\|^2 = 1$. ■

7.1.9. THEOREM. *If ω is a completely additive state of the von Neumann algebra $\mathscr{R}$ acting on the Hilbert space $\mathscr{H}$, there is an orthogonal family $\{y_n\}$ of vectors in $\mathscr{H}$ such that $\omega = \sum_n \omega_{y_n}|\mathscr{R}$ and $\sum_n \|y_n\|^2 = 1$.*

Proof. From Theorem 7.1.8, there is a family $\{x_m\}$ of vectors in $\mathscr{H}$ such that $\sum_m \omega_{x_m}|\mathscr{R} = \omega$ and $\sum_m \|x_m\|^2 = 1$. If $x'_m = x_m/\|x_m\|$ and E_m is the one-dimensional projection in $\mathscr{B}(\mathscr{H})$ with x'_m in its range, then $\sum_m \|x_m\|^2 E_m$ converges in norm to a positive operator H. If E is a projection, the mapping $x \to \langle E_m x, x\rangle = |\langle x, x'_m\rangle|^2$ is continuous on the unit ball $(E(\mathscr{H}))_1$ of $E(\mathscr{H})$ in its weak topology. The same is true for the mapping

$$x \to \left\langle \left(\sum_{j=1}^{m} \|x_j\|^2 E_j \right) x, x \right\rangle$$

and for the uniform limit $x \to \langle Hx, x\rangle$ of these mappings. Since $(E(\mathscr{H}))_1$ is compact in the weak topology (see Corollary 2.3.4), there is a vector x_0 in $(E(\mathscr{H}))_1$ at which the mapping $x \to \langle Hx, x\rangle\ (= \|H^{1/2}Ex\|^2)$ attains its maximum. Thus $\langle Hx_0, x_0\rangle = \|EHE\|$. Assuming that $HE \neq 0$ and $HE = EH$, $H_0^2 \leq H_0$, where $H_0 = HE/\|HE\|$; and

$$1 = \langle H_0 x_0, x_0\rangle^2 \leq \|H_0 x_0\|^2 = \langle H_0^2 x_0, x_0\rangle \leq \langle H_0 x_0, x_0\rangle = 1.$$

Thus $\langle H_0 x_0, x_0\rangle = \|H_0 x_0\|$; and, from Corollary 2.1.4, x_0 is an eigenvector for H_0 (and for H). If $HE = 0$, then each unit vector in $(E(\mathscr{H}))_1$ is an eigenvector for H. In any event, if E is a non-zero projection commuting with H (equivalently, with range invariant under H), then H has a unit eigenvector in $(E(\mathscr{H}))_1$. With $\{y'_a\}$ a maximal orthogonal family of unit eigenvectors for H, the existence of an eigenvector in each subspace invariant under H, combined with this maximality, guarantees that $\{y'_a\}$ is a complete orthonormal basis for $\mathscr{H}$. Suppose $Hy'_a = \lambda_a^2 y'_a$ and $\lambda_a \geq 0$. Then, from Theorem 2.2.9(ii),

$$(1) \qquad \begin{aligned} \langle Ax'_m, x'_m\rangle &= \langle x'_m, A^*x'_m\rangle = \sum_a \langle x'_m, y'_a\rangle\overline{\langle A^*x'_m, y'_a\rangle} \\ &= \sum_a \langle Ay'_a, x'_m\rangle\langle x'_m, y'_a\rangle = \sum_a \langle E_m A y'_a, y'_a\rangle. \end{aligned}$$

Now

$$\begin{aligned} &\sum_m \|x_m\|^2 \sum_a |\langle E_m A y'_a, y'_a\rangle| \\ &\quad = \sum_m \|x_m\|^2 \sum_a |\langle Ay'_a, x'_m\rangle\langle x'_m, y'_a\rangle| \\ &\quad \leq \sum_m \|x_m\|^2 \left(\sum_a |\langle A^*x'_m, y'_a\rangle|^2 \right)^{1/2} \left(\sum_a |\langle x'_m, y'_a\rangle|^2 \right)^{1/2} \\ &\quad = \sum_m \|x_m\|^2 \|A^*x'_m\|\ \|x'_m\| \leq \|A\|, \end{aligned}$$

which justifies the interchange of the order of summation in

$$\begin{aligned}\omega(A) &= \sum_m \langle Ax_m, x_m\rangle = \sum_m \|x_m\|^2 \sum_a \langle E_m A y'_a, y'_a\rangle \\ &= \sum_a \sum_m \langle \|x_m\|^2 E_m A y'_a, y'_a\rangle = \sum_a \langle HAy'_a, y'_a\rangle = \sum_a \langle \lambda_a^2 A y'_a, y'_a\rangle\end{aligned}$$

(where (1) is used for the second equality). Writing y_a for $\lambda_a y'_a$, we conclude that $\omega(A) = \sum_a \langle Ay_a, y_a\rangle$. Thus $1 = \omega(I) = \sum_a \|y_a\|^2$, and there are at most a countable number of y_a different from 0. Relabeling these as $y_1, y_2, \ldots$, we have that $\omega = \sum_n \omega_{y_n}|\mathscr{R}$, $\sum_n \|y_n\|^2 = 1$, and $\{y_n\}$ is an orthogonal family. ■

7.1.10. REMARK. In the preceding proof, we use the vectors $\{x_m\}$ to construct the operator H as an absolutely convergent sum of positive multiples of one-dimensional projections relative to the operator norm. In the course of the argument, we find an orthonormal basis of unit eigenvectors $\{y'_a\}$ for H. (Compare Exercises 22(iv), 23, and 29, of Section 2.8.) We prove that $\omega(A) = \sum_a \langle HAy'_a, y'_a\rangle$. If the von Neumann algebra $\mathscr{R}$ is $\mathscr{B}(\mathscr{H})$, then the basis $\{y'_a\}$ provides a frame in which operators can be represented as matrices (see Section 2.6, *Matrix representations*). Since $1 = \omega(I) = \sum_a \langle Hy'_a, y'_a\rangle$, the (positive) diagonal entries of H have (absolutely) convergent sum. By analogy with the case of finite matrix algebras, we may refer to this sum as "the trace of H" and denote it by tr H. Of course, not all positive operators in $\mathscr{B}(\mathscr{H})$ have a finite trace (for example, I does not). Those that do are referred to as "trace class operators" (see Remark 8.5.6). Our argument shows that $\omega(A) = \operatorname{tr}(HA)$ for each A in $\mathscr{B}(\mathscr{H})$. In the context of applications to physics (in particular, statistical mechanics) this representation of a completely additive (normal) state in terms of the trace tr and the "density matrix" H is the usual one. Theorem 7.1.9 establishes that each such state of a von Neumann algebra $\mathscr{R}$ is the restriction of a state of $\mathscr{B}(\mathscr{H})$ representable in terms of the trace and a positive trace class operator with trace 1. Chapter VIII is devoted to the study of the concept of *trace* extended to von Neumann algebras. ■

7.1.11. DEFINITION. A state ω of a von Neumann algebra $\mathscr{R}$ is said to be *normal* when $\omega(H_a) \to \omega(H)$ for each monotone increasing net of operators $\{H_a\}$ in $\mathscr{R}$ with least upper bound H. ■

7.1.12. THEOREM. *The following conditions on a state ω of a von Neumann algebra $\mathscr{R}$ acting on a Hilbert space $\mathscr{H}$ are equivalent:*

(a) $\omega = \sum_{n=1}^{\infty} \omega_{y_n}|\mathscr{R}$ *and* $\sum_n \|y_n\|^2 = 1$, *with* $\{y_n\}$ *an orthogonal family of vectors in* $\mathscr{H}$.

(b) $\omega = \sum_{n=1}^{\infty} \omega_{x_n}|\mathscr{R}$ *and* $\sum_n \|x_n\|^2 = 1$.

(c) *ω is weak-operator continuous on the unit ball of $\mathscr{R}$.*
(d) *ω is strong-operator continuous on the unit ball of $\mathscr{R}$.*
(e) *ω is normal on $\mathscr{R}$.*
(f) *ω is a completely additive state of $\mathscr{R}$.*
(g) ω is ultraweakly cont. on $\mathscr{R}$ (see p. 484)

Proof. The implication (a) → (b) is immediate. Since $\sum_n \|x_n\|^2 = 1$, $\{\sum_{n=1}^m \omega_{x_n}|\mathscr{R}\}$ converges in norm to ω, assuming (b). As each $\sum_{n=1}^m \omega_{x_n}|\mathscr{R}$ is weak-operator continuous on $(\mathscr{R})_1$, and ω is the uniform limit of $\{\sum_{n=1}^m \omega_{x_n}|(\mathscr{R})_1\}$, ω is weak-operator continuous on $(\mathscr{R})_1$ (that is, (b) → (c)). The implication (c) → (d) is a consequence of the fact that the weak-operator topology on $\mathscr{B}(\mathscr{H})$ is weaker (coarser) than the strong-operator topology. If $\{H_a\}$ is a monotone increasing net with least upper bound H, the same is true for the (bounded) cofinal subnet of $\{H_a\}$ consisting of those H_a for which $a \geq a_0$. Since the convergence involved is strong-operator convergence, $\omega(H_a) \to \omega(H)$, under assumption (d). Thus (d) → (e). If $E = \sum_a E_a$ for an orthogonal family $\{E_a\}$ of projections in $\mathscr{R}$, then the monotone increasing net of sums of finite subfamilies of $\{E_a\}$ has E as least upper bound. Assuming (e), we have from this that $\omega(E) = \sum_a \omega(E_a)$. Thus ω is completely additive. Finally, (f) → (a) is precisely the assertion of Theorem 7.1.9. ∎

Using the equivalence of (c) and (e) in the preceding theorem, we have the following result:

7.1.13. Corollary. *The family of normal states of a von Neumann algebra is norm closed.*

7.1.14. Lemma. *If φ is a representation of the C^*-algebra $\mathfrak{A}$ acting on the Hilbert space $\mathscr{H}$ into $\mathscr{B}(\mathscr{K})$ and φ is continuous at 0 from $(\mathfrak{A})_1^+$ into $\mathscr{B}(\mathscr{K})$, each in its weak-operator topology, then φ is continuous from $(\mathfrak{A})_1$ to $\mathscr{B}(\mathscr{K})$, each in its strong-operator topology.*

Proof. Since $|\langle A^*Ax, y\rangle| \leq \|Ax\| \cdot \|Ay\|$, the mapping $A \to A^*A$ is continuous at 0 on $\mathfrak{A}$ (and, hence, on $(\mathfrak{A})_2$) in its strong-operator topology to $\mathfrak{A}$ in its weak-operator topology. At the same time, $\langle B^*Bx, x\rangle = \|Bx\|^2$, so that B is near 0 in the strong-operator topology on $\mathscr{B}(\mathscr{K})$ if B^*B is sufficiently close to 0 in the weak-operator topology on $\mathscr{B}(\mathscr{K})$. Thus,

$$A \to A^*A \to \varphi(A^*A) = \varphi(A)^*\varphi(A)$$

is continuous at 0 on $(\mathfrak{A})_2$ in its strong-operator topology to $\mathscr{B}(\mathscr{K})$ in its weak-operator topology; and $A \to \varphi(A)$ is continuous at 0 from $(\mathfrak{A})_2$ to $\mathscr{B}(\mathscr{K})$, each in its strong-operator topology. It follows that $A \to \varphi(A)$ is continuous from $(\mathfrak{A})_1$ to $\mathscr{B}(\mathscr{K})$, each in its strong-operator topology. ∎

We gather various portions of arguments together in the following useful proposition.

7.1.15. PROPOSITION. *If φ is a representation of a von Neumann algebra $\mathscr{R}$ such that $\omega_x \circ \varphi$ is a normal state of $\mathscr{R}$ for each vector x in a family $\mathscr{F}$ of vectors whose transforms under $\varphi(\mathscr{R})$ span a dense linear submanifold $\mathscr{D}$ of $\mathscr{H}$, the representation space of φ, then φ is continuous from the unit ball $(\mathscr{R})_1$ of $\mathscr{R}$ to $\mathscr{B}(\mathscr{H})$, both in their weak-operator topologies or both in their strong-operator topologies; and $\varphi(\mathscr{R})$ is a von Neumann algebra.*

Proof. Note, first, that if $\omega_x \circ \varphi$ and $\omega_y \circ \varphi$ are weak-operator continuous on $(\mathscr{R})_1$, then $A \to \langle \varphi(A)x, y \rangle$ is weak-operator continuous on $(\mathscr{R})_1$; for, applying the Cauchy–Schwarz inequality to the inner product that assigns $\langle \varphi(A)u, v \rangle$ to u, v in $\mathscr{H}$, where A is a positive operator in $\mathscr{R}$, we have

$$|\langle \varphi(A)x, y \rangle| \leq |\langle \varphi(A)x, x \rangle|^{1/2} |\langle \varphi(A)y, y \rangle|^{1/2}.$$

Thus $A \to \langle \varphi(A)x, y \rangle$ is weak-operator continuous on $(\mathscr{R})_1^+$ at 0 and Lemma 7.1.3 applies to give the stated continuity of $A \to \langle \varphi(A)x, y \rangle$.

If $u = \varphi(B)v$, with v in $\mathscr{F}$, then $(\omega_u \circ \varphi)(A) = \langle \varphi(A)\varphi(B)v, \varphi(B)v \rangle = \langle \varphi(B^*AB)v, v \rangle = (\omega_v \circ \varphi)_B(A)$, so that $\omega_u \circ \varphi$ is a normal positive linear functional on $\mathscr{R}$. (See Corollary 7.1.5 and the discussion preceding it.) Combining this observation with that of the preceding paragraph, we have that $\omega_x \circ \varphi$ is a normal state of $\mathscr{R}$ for each unit vector x in $\mathscr{D}$. If u is a unit vector in $\mathscr{H}$ and $\{z_n\}$ is a sequence of unit vectors in $\mathscr{D}$ tending to u, then $\{\omega_{z_n} \circ \varphi\}$ is a sequence of normal states of $\mathscr{R}$ tending in norm to $\omega_u \circ \varphi$. From Corollary 7.1.13, $\omega_u \circ \varphi$ is a normal state of $\mathscr{R}$ (hence, weak-operator continuous on $(\mathscr{R})_1$). Thus φ is weak-operator continuous on $(\mathscr{R})_1$. Applying Lemma 7.1.14, φ is continuous from $(\mathscr{R})_1$ to $\mathscr{B}(\mathscr{H})$ where both are endowed with their strong-operator topologies.

Since φ maps $\mathscr{R}$ onto the C^*-algebra $\varphi(\mathscr{R})$ (see Theorem 4.1.9), φ is an open mapping (see Theorem 1.8.4). Thus $\varphi((\mathscr{R})_1)$ contains $(\varphi(\mathscr{R}))_r$ for some positive r (see Proposition 1.8.1). As $(\mathscr{R})_1$ is weak-operator compact (see Theorem 5.1.3) and φ is weak-operator continuous on $(\mathscr{R})_1$, $\varphi((\mathscr{R})_1)$ is weak-operator compact (and closed). Thus

$$([\varphi(\mathscr{R})]^-)_r = [(\varphi(\mathscr{R}))_r]^- \subseteq (\varphi(\mathscr{R}))_1 \subseteq \varphi(\mathscr{R}),$$

where the first equality results from the Kaplansky density theorem. Hence $[\varphi(\mathscr{R})]^- \subseteq \varphi(\mathscr{R})$; and $\varphi(\mathscr{R})$ is a von Neumann algebra. ■

In the circumstances of the preceding theorem, we refer to φ as a *normal* representation of $\mathscr{R}$. As an important and easy corollary of the preceding proposition, we have that * isomorphisms between von Neumann algebras are weak- and strong-operator homeomorphisms on bounded sets.

7.1.16. COROLLARY. *If φ is a * isomorphism of one von Neumann algebra $\mathscr{R}_1$ onto another $\mathscr{R}_2$, then φ is a homeomorphism of the unit ball $(\mathscr{R}_1)_1$ of $\mathscr{R}_1$*

onto that of $\mathscr{R}_2$ when both $\mathscr{R}_1$ and $\mathscr{R}_2$ are endowed with either their weak- or their strong-operator topologies.

Proof. From Theorem 4.1.8(iii), φ and φ^{-1} are isometries. Hence φ maps $(\mathscr{R}_1)_1$ onto $(\mathscr{R}_2)_1$. Moreover, $\varphi(A) \geq 0$ if and only if $A \geq 0$, so that φ preserves least upper bounds and $\varphi(\sum_a E_a) = \sum_a \varphi(E_a)$ for each orthogonal family $\{E_a\}$ of projections in $\mathscr{R}$. From Theorem 7.1.12, $\omega_x \circ \varphi$ is normal, for each unit vector x in the Hilbert space on which $\mathscr{R}_2$ acts. Proposition 7.1.15 applies (to both φ and φ^{-1}) to yield the assertion of this corollary. ∎

Bibliography: [18, 19, 24]

7.2. Vector states and unitary implementation

The goal of this section is Theorem 7.2.3, in the presence of a separating vector, each normal state is a vector state. With its aid, we prove the unitary implementation theorem (Theorem 7.2.9). Further use of it results in the comparison theorem for pairs of jointly generated cyclic projections in a von Neumann algebra and its commutant (Theorem 7.2.12).

7.2.1. THEOREM. *If $\mathscr{R}$ is a von Neumann algebra acting on a Hilbert space $\mathscr{H}$ and x_0 is a unit generating vector for $\mathscr{R}$, then, for each vector z_0 in $\mathscr{H}$, there are operators T and S in $\mathscr{R}$ and a vector y_0 orthogonal to the null space of T such that $Ty_0 = x_0$ and $Sy_0 = z_0$.*

Proof. Assume $\|z_0\| = 1$. Since x_0 is generating for $\mathscr{R}$, there are operators T_n in $\mathscr{R}$ such that $\sum_{n=0}^{\infty} T_n x_0 = z_0$ and $\|T_n x_0\| \leq 4^{-n}$. For this, choose T_0 in $\mathscr{R}$ so that $\|z_0 - T_0 x_0\| \leq 4^{-1}$ and $\|T_0 x_0\| \leq \|z_0\| = 1$. Then choose T_1 in $\mathscr{R}$ so that $\|z_0 - T_0 x_0 - T_1 x_0\| \leq 4^{-2}$ and $\|T_1 x_0\| \leq \|z_0 - T_0 x_0\| \leq 4^{-1}$. Continuing in this way produces the sequence of operators T_n in $\mathscr{R}$ with the stated properties.

If $H_n^2 = I + \sum_{k=0}^{n} 4^k T_k^* T_k$, then $\{H_n^2\}$ is a monotone increasing sequence of positive invertible operators in $\mathscr{R}$; and, from Proposition 4.2.8(iii), $\{H_n^{-2}\}$ is a monotone decreasing sequence of positive invertible operators in $\mathscr{R}$. From Lemma 5.1.4, $\{H_n^{-2}\}$ has a greatest lower bound, which is its strong-operator limit. Since $\{H_n^{-2}\}$ is a bounded set of positive operators, $\{H_n^{-1}\}$ $(= \{[H_n^{-2}]^{1/2}\})$ has a strong-operator limit T in $\mathscr{R}$, from Proposition 5.3.2. Note that

$$\|H_n x_0\|^2 = \langle H_n^2 x_0, x_0\rangle = \left\langle \left(I + \sum_{k=0}^{n} 4^k T_k^* T_k\right) x_0, x_0 \right\rangle$$

$$= \langle x_0, x_0\rangle + \sum_{k=0}^{n} 4^k \|T_k x_0\|^2 \leq 1 + \sum_{k=0}^{n} 4^{-k} < 3.$$

Since the ball of radius 3 in $\mathscr{H}$ is weakly compact, there is a y_0 in this ball, each of whose weak neighborhoods contains an infinite number of terms of $\{H_n x_0\}$. We show that $Ty_0 = x_0$. If $\varepsilon > 0$ and x is a vector in $\mathscr{H}$, there is an n such that

$$\|(H_n^{-1} - T)x\| < \frac{\varepsilon}{6}, \qquad |\langle H_n x_0 - y_0, Tx\rangle| < \frac{\varepsilon}{2}.$$

Thus

$$\begin{aligned}|\langle H_n x_0, Tx\rangle - \langle x_0, x\rangle| &= |\langle H_n x_0, Tx\rangle - \langle H_n x_0, H_n^{-1}x\rangle| \\ &< \frac{\varepsilon}{6}\|H_n x_0\| < \frac{\varepsilon}{2};\end{aligned}$$

so that

$$|\langle y_0, Tx\rangle - \langle x_0, x\rangle| = |\langle Ty_0 - x_0, x\rangle| < \varepsilon,$$

and $Ty_0 = x_0$ as asserted.

With m larger than n,

$$0 \le H_m^{-1}4^n T_n^* T_n H_m^{-1} \le H_m^{-1}\left(\sum_{k=0}^{m} 4^k T_k^* T_k\right)H_m^{-1} \le I.$$

Now $H_m^{-1}4^n T_n^* T_n H_m^{-1}$ is strong-operator convergent to $4^n T T_n^* T_n T$ as m tends to ∞, so that $0 \le 4^n T T_n^* T_n T \le I$, and $\|T_n T\| \le 2^{-n}$. Thus $\sum_{n=0}^{\infty} T_n T$ converges in norm to an operator S in $\mathscr{R}$. Since both T and S have the same effect on y_0 as they do on its component orthogonal to the null space of T, we may assume y_0 is orthogonal to that null space. As $Ty_0 = x_0$,

$$Sy_0 = \sum_{n=0}^{\infty} T_n T y_0 = \sum_{n=0}^{\infty} T_n x_0 = z_0. \quad ■$$

In the preceding theorem, we prove that we can "reach" an arbitrary vector z_0 from a generating vector x_0 for $\mathscr{R}$, through an intermediate vector y_0, by means of two bounded operators S and T ($Ty_0 = x_0$ and $Sy_0 = z_0$). Proceeding formally, $ST^{-1}x_0 = Sy_0 = z_0$, so that one "operator," ST^{-1}, transforms x_0 onto z_0. Of course, T may not have an inverse. From our proof, T is a (bounded) self-adjoint operator (in $\mathscr{R}$) and $N(T)y_0 = 0$. From Proposition 2.5.13, $N(T) = I - R(T)$. Thus, if $T_0(Tx + y) = x$, where $N(T)x = 0$ and $N(T)y = y$, T_0 is a well-defined, densely defined, linear transformation on $\mathscr{H}$; and $T_0 x_0 = y_0$. (In a sense, T_0 is "inverse" to T.) We note that T_0 is self-adjoint. With x and y as above, if $N(T)u = 0$ and $N(T)v = v$, then $0 = \langle x, v\rangle = \langle y, u\rangle$ and

$$\begin{aligned}\langle T_0(Tx + y), Tu + v\rangle &= \langle x, Tu\rangle + \langle x, v\rangle = \langle Tx + y, u\rangle \\ &= \langle Tx + y, T_0(Tu + v)\rangle.\end{aligned}$$

Thus $T_0 \subseteq T_0^*$. With x, y, u, v as above, suppose that $u + v \in \mathcal{D}(T_0^*)$ and that $T_0^*(u + v) = z + w$, where $N(T)z = 0$ and $N(T)w = w$. Then

$$\begin{aligned}\langle x, u\rangle = \langle T_0(Tx + y), u + v\rangle &= \langle Tx + y, z + w\rangle \\ &= \langle Tx, z\rangle + \langle y, w\rangle,\end{aligned}$$

so that $\langle x, u - Tz\rangle = \langle y, w\rangle$. We may choose y to be w and x to be $t(u - Tz)$ for any scalar t, since this last equality is valid for each x orthogonal to the null space of T and each y in this null space. Thus $t\|u - Tz\|^2 = \|w\|^2$ for each scalar t; and $0 = w = u - Tz$. It follows that $u + v = Tz + v \in \mathcal{D}(T_0)$ and $T_0^*(u + v) = z = T_0(Tz + v)$. Hence $T_0^* \subseteq T_0$; and T_0 is self-adjoint.

If U' is a unitary operator in $\mathcal{R}'$, then $U'(Tx + y) = TU'x + U'y$; and $N(T)U'x = U'N(T)x = 0$ while $N(T)U'y = U'N(T)y = U'y$. Hence U' maps $\mathcal{D}(T_0)$ onto (since the same is true for U'^*) itself. At the same time,

$$U'T_0(Tx + y) = U'x = T_0(TU'x + U'y) = T_0U'(Tx + y).$$

Thus $T_0 \,\eta\, \mathcal{R}$.

From the preceding discussion, we may reformulate Theorem 7.2.1 in terms of unbounded operators as follows.

7.2.1′. THEOREM. *If $\mathcal{R}$ is a von Neumann algebra acting on a Hilbert space $\mathcal{H}$ and x_0 is a unit generating vector for $\mathcal{R}$, then, for each vector z_0 in $\mathcal{H}$, there are operators B in $\mathcal{R}$ and T_0 self-adjoint and affiliated with $\mathcal{R}$ such that $BT_0x_0 = z_0$.*

7.2.2. PROPOSITION. *If $\mathcal{R}$ is a von Neumann algebra acting on $\mathcal{H}$, x_0 is a separating vector for $\mathcal{R}$, and F' is the projection (in $\mathcal{R}'$) with range $[\mathcal{R}x_0]$, then $E' \precsim F'$ for each cyclic projection E' in $\mathcal{R}'$.*

Proof. Suppose $[\mathcal{R}z_0]$ is the range of E'. From Proposition 5.5.11, x_0 is generating for $\mathcal{R}'$. Thus, from Theorem 7.2.1, there are operators T' and S' in $\mathcal{R}'$ and a vector y_0, orthogonal to the null space of T', such that $T'y_0 = x_0$ and $S'y_0 = z_0$. If M' is the projection in $\mathcal{R}'$ with range $[\mathcal{R}y_0]$, then $E' \precsim M'$; for $[\mathcal{R}z_0] = [S'\mathcal{R}y_0] = R(S'M')(\mathcal{H})$. But $R(S'M') \sim R(M'S'^*) \le M'$. Similarly, $F' \precsim M'$, and $F' \sim R(M'T'^*)$. Since y_0 is orthogonal to the null space of T', y_0 is in $[T'^*(\mathcal{H})]$. If $\{y_n\}$ is a sequence of vectors in $\mathcal{H}$ such that $T'^*y_n \to y_0$, then $M'T'^*y_n \to M'y_0 = y_0$. Thus $y_0 \in [M'T'^*(\mathcal{H})]$ and $[\mathcal{R}y_0] \subseteq [M'T'^*(\mathcal{H})]$. It follows that $M' \le R(M'T'^*) \sim F'$; and $E' \precsim F'$. ■

7.2.3. THEOREM. *If ω is a normal state of the von Neumann algebra $\mathcal{R}$ acting on the Hilbert space $\mathcal{H}_0$ and x_0 is a separating vector for $\mathcal{R}$, there is a vector y_0 in $\mathcal{H}_0$ such that $\omega = \omega_{y_0}|\mathcal{R}$.*

Proof. If Φ_n is the universal normal representation of $\mathscr{R}$ on $\mathscr{H}_n$ (see the remarks preceding Lemma 7.1.6), there are vectors x and y in $\mathscr{H}_n$ such that $\omega(A) = \langle\Phi_n(A)y, y\rangle$ and $\omega_{x_0}(A) = \langle\Phi_n(A)x, x\rangle$ for each A in $\mathscr{R}$. The mapping $Ax_0 \to \Phi_n(A)x$ extends to a unitary transformation U of $[\mathscr{R}x_0]$ onto $[\Phi_n(\mathscr{R})x]$ such that $UAG'U^* = \Phi_n(A)F'$ for all A in $\mathscr{R}$, where F' is the projection (in $\Phi_n(\mathscr{R})'$) with range $[\Phi_n(\mathscr{R})x]$ and G' (in $\mathscr{R}'$) has range $[\mathscr{R}x_0]$. Since x_0 is separating for $\mathscr{R}$, x is separating for $\Phi_n(\mathscr{R})$. (Recall that Φ_n is a * isomorphism and that $\Phi_n(\mathscr{R})$ is weak-operator closed in $\mathscr{B}(\mathscr{H}_n)$). Thus, from Proposition 7.2.2, $E' \precsim F'$, where E' is the projection with range $[\Phi_n(\mathscr{R})y]$. If V' is a partial isometry in $\Phi_n(\mathscr{R})'$ such that $V'^*V' = E'$ and $V'V'^* \leq F'$, then $\omega = \omega_{y_0}|\mathscr{R}$, where $y_0 = U^*V'y$. ■

7.2.4. Definition. The *support* (or, *carrier*) of a normal state of a von Neumann algebra is the orthogonal complement of the union of all projections annihilated by the state. ■

7.2.5. Remark. In Definition 7.1.1 we introduced the *carrier* of a completely additive state, but were not in a position to prove (easily) that it is unique (and coincides with the *support* described in Definition 7.2.4) with the desired properties. With the results at hand, there is no longer any difficulty in doing this. If E is the support of the normal state ω on the von Neumann algebra $\mathscr{R}$ and A is a positive operator in $\mathscr{R}$ such that $\omega(A) = 0$, then

$$0 \leq \omega(A^n) = \omega(A^{n-1/2}A^{1/2}) \leq \omega(A^{2n-1})^{1/2}\omega(A)^{1/2} = 0$$

(using Proposition 4.3.1, the Cauchy–Schwarz inequality for the state ω). If f is a continuous real-valued function on $\mathrm{sp}(A)$ and $f(0) = 0$, then f is a norm limit of polynomials without constant term. Thus $\omega(f(A)) = 0$. In particular, $\omega(A^{1/n}) = 0$. From Lemma 5.1.5, $\{A^{1/n}\}$ is a monotone increasing sequence with least upper bound $R(A)$ (we may assume that $\|A\| \leq 1$); so that $\omega(A^{1/n}) = \omega(R(A)) = 0$. Conversely, if $\omega(R(A)) = 0$, then

$$0 \leq \omega(A) = \omega(R(A)A) \leq \omega(R(A))^{1/2}\omega(A^2)^{1/2} = 0.$$

Thus two normal states of $\mathscr{R}$ have the same support if and only if they annihilate the same positive operators. If M and N are projections in $\mathscr{R}$ such that $\omega(M) = \omega(N) = 0$, then $\omega(M + N) = 0$; and, from our discussion, $\omega(R(M + N)) = 0$. In Proposition 2.5.14 we note that $R(M + N) = M \vee N$, so that $\omega(M \vee N) = 0$. Since ω is normal, it annihilates the union $I - E$ of all projections on which it vanishes. ■

7.2.6. Remark. The support of $\omega_z|\mathscr{R}$ has range $[\mathscr{R}'z]$. If $\omega_z(F) = 0$ for some projection F in $\mathscr{R}$, then $\|Fz\|^2 = 0$, and $[F\mathscr{R}'z] = (0)$. Thus $F \leq I - E$, where E is the projection with range $[\mathscr{R}'z]$. Of course $\omega_z(I - E) = 0$; and E is the support of $\omega_z|\mathscr{R}$. ■

7.2.7. PROPOSITION. *A normal state ω of a von Neumann algebra $\mathscr{R}$ acting on the Hilbert space $\mathscr{H}$ is a vector state of $\mathscr{R}$ if and only if its support is a cyclic projection in $\mathscr{R}$.*

Proof. Let E be the support of ω. If $\omega = \omega_y|\mathscr{R}$, then E has range $[\mathscr{R}'y]$, from Remark 7.2.6.

Suppose E has range $[\mathscr{R}'z]$. Then z is separating for $E\mathscr{R}E$ acting on $E(\mathscr{H})$; and $\omega|E\mathscr{R}E$ is a vector state $\omega_y|E\mathscr{R}E$ (from Theorem 7.2.3). Since $\omega(I - E) = 0$ (from Remark 7.2.5),

$$\omega(A) = \omega(EAE) = \langle EAEy, y\rangle = \langle Ay, y\rangle = \omega_y(A),$$

for each A in $\mathscr{R}$. ■

7.2.8. LEMMA. *If ω is a normal state of the von Neumann algebra $\mathscr{R}$ acting on the Hilbert space $\mathscr{H}$ and z is a unit generating vector for $\mathscr{R}$ such that ω and $\omega_z|\mathscr{R}$ have the same support, then $\omega = \omega_x|\mathscr{R}$ with x a unit generating vector for $\mathscr{R}$.*

Proof. From Remark 7.2.6, the support of $\omega_z|\mathscr{R}$ (and ω) has range $[\mathscr{R}'z]$. From Proposition 7.2.7, $\omega = \omega_y|\mathscr{R}$ for some vector y in $\mathscr{H}$. By assumption, $[\mathscr{R}'z] = [\mathscr{R}'y]$. From Proposition 5.5.13, E', the projection (in $\mathscr{R}'$) with range $[\mathscr{R}y]$, has the same central carrier as the projection with range $[\mathscr{R}z]$ ($= \mathscr{H}$). Thus $C_{E'} = I$ and the mapping $A \to AE'$ of $\mathscr{R}$ onto $\mathscr{R}E'$ is a * isomorphism (from Proposition 5.5.5). This * isomorphism carries $\omega_z|\mathscr{R}$ onto a normal state ω_0 of $\mathscr{R}E'$ (and $\omega_0(AE') = \langle Az, z\rangle$ for each A in $\mathscr{R}$). Now $\langle Fz, z\rangle = 0$ if and only if $\langle Fy, y\rangle = \langle FE'y, y\rangle = 0$. Thus ω_0 and $\omega_y|\mathscr{R}E'$ have the same support; and, from Proposition 7.2.7 (and Remark 7.2.6), $\omega_0 = \omega_v|\mathscr{R}E'$ for some unit vector v in $[\mathscr{R}y]$. Since $\langle Av, v\rangle = \langle Az, z\rangle$ for each A in $\mathscr{R}$, the mapping $Az \to Av$ of $\mathscr{R}z$ onto $\mathscr{R}v$ extends to a (partial) isometry V' in $\mathscr{R}'$ of $\mathscr{H}$ onto $[\mathscr{R}v]$. If F' is the projection (in $\mathscr{R}'$) with range $[\mathscr{R}v]$, it follows that $F' = V'V'^* \sim I$. As $[\mathscr{R}v] \subseteq [\mathscr{R}y]$, $F' \leq E'$ and $E' \sim I$. If W' is a partial isometry in $\mathscr{R}'$ with initial space $[\mathscr{R}y]$ and final space $\mathscr{H}$, then $[\mathscr{R}W'y] = W'[\mathscr{R}y] = \mathscr{H}$. If $x = W'y$, then $\langle Ax, x\rangle = \langle AW'y, W'y\rangle = \langle Ay, y\rangle = \omega(A)$ for each A in $\mathscr{R}$. Thus $\omega = \omega_x|\mathscr{R}$, and $[\mathscr{R}x] = \mathscr{H}$. ■

7.2.9. THEOREM (Unitary implementation). *If φ is a * isomorphism of the von Neumann algebra $\mathscr{R}_1$, with unit separating and generating vector x, onto the von Neumann algebra $\mathscr{R}_2$, with unit separating and generating vector y, then there is a unitary transformation U of the Hilbert space $\mathscr{H}_1$, on which $\mathscr{R}_1$ acts, onto the Hilbert space $\mathscr{H}_2$, on which $\mathscr{R}_2$ acts, such that $\varphi(A) = UAU^{-1}$ for each A in $\mathscr{R}_1$.*

Proof. Let ω be the state $\omega_y \circ \varphi$ of $\mathscr{R}_1$. Since φ is a * isomorphism of $\mathscr{R}_1$ onto $\mathscr{R}_2$, ω is normal (as in Corollary 7.1.16). Since x is a separating vector for $\mathscr{R}_1$, $\omega = \omega_z|\mathscr{R}_1$ for some unit vector z in $\mathscr{H}_1$. As y is separating for $\mathscr{R}_2$ and x is separating for $\mathscr{R}_1$, ω and $\omega_x|\mathscr{R}_1$ have the same support I in $\mathscr{R}_1$. From Lemma 7.2.8, we can choose z so that $[\mathscr{R}_1 z] = [\mathscr{R}_1 x] = \mathscr{H}_1$. Thus φ^{-1} is a cyclic representation of $\mathscr{R}_2$ as $\mathscr{R}_1$ acting on $\mathscr{H}_1$ with generating vector z such that $\omega_z \circ \varphi^{-1} = \omega_y|\mathscr{R}_2$; and y is a generating vector for $\mathscr{R}_2$. From Proposition 4.5.3 the mapping $Az \to \varphi(A)y$ extends to a unitary transformation U of $\mathscr{H}_1$ onto $\mathscr{H}_2$ such that $\varphi(A) = UAU^{-1}$ for each A in $\mathscr{R}_1$. ■

7.2.10. REMARK. The preceding theorem, in conjunction with the techniques of restricting a von Neumann algebra $\mathscr{R}$ to a projection in $\mathscr{R}$ or in $\mathscr{R}'$ (see Section 5.5, *Some constructions*) and the technique of forming matrix algebras of various orders over $\mathscr{R}$ (see Section 6.6) can be used to reduce the question of when two von Neumann algebras on a Hilbert space are unitarily equivalent, to a question of algebraic isomorphism. A number of results that can be obtained in this way are set out in Exercises 9.6.22 to 9.6.36. In the special cases of a countably decomposable, maximal abelian algebra and of a von Neumann algebra that, together with its commutant, is properly infinite and countably decomposable, the unitary implementation theorem applies as it stands; for, in each of these cases, there is a separating and generating vector. Corollary 5.5.17 establishes this in the case of a maximal abelian algebra. In the properly infinite case, we note the existence of a separating and generating vector in Proposition 9.1.6. ■

7.2.11. LEMMA. *If ω is a normal state of the von Neumann algebra $\mathscr{R}$ acting on the Hilbert space $\mathscr{H}$, and ω has the same support as $\omega_z|\mathscr{R}$, then there is a unit vector x such that $[\mathscr{R}x] = [\mathscr{R}z]$ and $\omega = \omega_x|\mathscr{R}$.*

Proof. Let E' be the projection (in $\mathscr{R}'$) with range $[\mathscr{R}z]$; and define ω_0 on $\mathscr{R}E'$ by $\omega_0(AE') = \omega(A)$. Note that ω_0 is well defined, for if $AE' = 0$, then $\langle A^*Az, z\rangle = 0$; so that $\omega(A^*A)$, and, hence, $\omega(A)$ are 0. Since $C_{E'}$ is the central carrier of the projection E (in $\mathscr{R}$) with range $[\mathscr{R}'z]$ and E is the support of $\omega_z|\mathscr{R}$ and ω, we have that $\omega(A) = \omega(AC_{E'}) = \omega_0(AE')$. As the mapping $AC_{E'} \to AE'$ of $\mathscr{R}C_{E'}$ onto $\mathscr{R}E'$ is an isomorphism, ω_0 is a normal state of $\mathscr{R}E'$. Moreover, ω_0 has the same support as $\omega_z|\mathscr{R}E'$ for $0 = \langle FE'z, z\rangle = \langle Fz, z\rangle$ if and only if $\omega(F) = \omega_0(FE') = 0$. As z is cyclic for $E'(\mathscr{H})$ under $\mathscr{R}E'$, Lemma 7.2.8 applies, and there is a unit vector x in $E'(\mathscr{H})$ such that $\omega(A) = \omega_0(AE') = \langle AE'x, x\rangle = \langle Ax, x\rangle$, for all A in $\mathscr{R}$, and $[\mathscr{R}E'x] = [\mathscr{R}x] = E'(\mathscr{H}) = [\mathscr{R}z]$. ■

We use the notation $\mathscr{M} \sim \mathscr{N}$ (and the associated notation $\precsim, \succ$) to mean $E \sim F$. where $\mathscr{M}$ and $\mathscr{N}$ are the ranges of E and F.

7.2.12. Theorem. *Suppose $\mathscr{R}$ is a von Neumann algebra acting on the Hilbert space $\mathscr{H}$ and x, y are vectors in $\mathscr{H}$. Then $[\mathscr{R}'x] \precsim_{\succ} [\mathscr{R}'y]$ if and only if $[\mathscr{R}x] \precsim_{\succ} [\mathscr{R}y]$.*

Proof. Suppose we have proved that if $[\mathscr{R}'x] \sim [\mathscr{R}'y]$, then $[\mathscr{R}x] \sim [\mathscr{R}y]$. In this case, if we are given that $[\mathscr{R}'u] \prec [\mathscr{R}'v]$, then $[\mathscr{R}'u] \sim [\mathscr{R}'v_0] < [\mathscr{R}'v]$, where $v_0 = Fv$ and F is the projection (in $\mathscr{R}$) with range $[\mathscr{R}'v_0]$. By assumption, then, $[\mathscr{R}u] \sim [\mathscr{R}v_0] = [\mathscr{R}Fv] \leq [\mathscr{R}v]$; so that $[\mathscr{R}u] \precsim [\mathscr{R}v]$. On the other hand, if $[\mathscr{R}u] \sim [\mathscr{R}v]$, then, by symmetry, $[\mathscr{R}'u] \sim [\mathscr{R}'v]$, contrary to assumption ($[\mathscr{R}'u] \prec [\mathscr{R}'v]$). Thus $[\mathscr{R}u] \prec [\mathscr{R}v]$.

It remains to establish that if $[\mathscr{R}'x] \sim [\mathscr{R}'y]$, then $[\mathscr{R}x] \sim [\mathscr{R}y]$. Suppose V is a partial isometry in $\mathscr{R}$ with initial space $[\mathscr{R}'x]$ and final space $[\mathscr{R}'y]$. Then $[\mathscr{R}'Vx] = V[\mathscr{R}'x] = [\mathscr{R}'y]$; and $[\mathscr{R}Vx] \subseteq [\mathscr{R}x] = [\mathscr{R}V^*Vx] \subseteq [\mathscr{R}Vx]$. Thus $[\mathscr{R}Vx] = [\mathscr{R}x]$. Replacing x by Vx, we may assume that $[\mathscr{R}'x] = [\mathscr{R}'y]$. In this case, $\omega_x|\mathscr{R}$ and $\omega_y|\mathscr{R}$ have the same support. From the preceding lemma, we can find z in $[\mathscr{R}y]$ such that $\omega_z|\mathscr{R} = \omega_x|\mathscr{R}$ with $[\mathscr{R}z] = [\mathscr{R}y]$. The mapping $Ax \to Az$ (A in $\mathscr{R}$) extends to a partial isometry in $\mathscr{R}'$ with initial space $[\mathscr{R}x]$ and final space $[\mathscr{R}y]$. Thus $[\mathscr{R}x] \sim [\mathscr{R}y]$ (in $\mathscr{R}'$). ∎

We recall, from Remark 6.7.3, that a vector x_0 is said to be a trace vector for a von Neumann algebra $\mathscr{R}$ when $\langle ABx_0, x_0 \rangle = \langle BAx_0, x_0 \rangle$ for all A and B in $\mathscr{R}$.

7.2.13. Lemma. *If x_0 is a generating unit trace vector for the von Neumann algebra $\mathscr{R}$ acting on $\mathscr{H}$ and A' is a self-adjoint operator in $\mathscr{R}'$, there are self-adjoint operators A_n in $\mathscr{R}$ such that $A_n x_0 \to A'x_0$.*

Proof. Since x_0 is generating for $\mathscr{R}$, there are operators B_n in $\mathscr{R}$ such that $B_n x_0 \to A'x_0$. If B is in $\mathscr{R}$, then $\langle B_n^* x_0, Bx_0 \rangle = \langle x_0, B_n B x_0 \rangle = \langle x_0, BB_n x_0 \rangle \to \langle x_0, BA'x_0 \rangle = \langle A'x_0, Bx_0 \rangle$. Moreover,

$$\|(B_n^* - B_m^*)x_0\|^2 = \langle (B_n - B_m)(B_n^* - B_m^*)x_0, x_0 \rangle$$
$$= \langle (B_n^* - B_m^*)(B_n - B_m)x_0, x_0 \rangle = \|(B_n - B_m)x_0\|^2 \to 0$$

as $n, m \to \infty$. Thus $\{B_n^* x_0\}$ converges to some vector y. From the first computation, $\langle y, Bx_0 \rangle = \langle A'x_0, Bx_0 \rangle$ for each B in $\mathscr{R}$. As $[\mathscr{R}x_0] = \mathscr{H}$, $B_n^* x_0 \to y = A'x_0$. Hence $A_n x_0 \to A'x_0$ where $A_n = \frac{1}{2}(B_n + B_n^*)$. ∎

7.2.14. LEMMA. *If x_0 is a generating trace vector for a von Neumann algebra $\mathscr{R}$, then x_0 is a generating trace vector for $\mathscr{R}'$.*

Proof. With A' and B' self-adjoint operators in $\mathscr{R}'$, from Lemma 7.2.13, we can find sequences $\{A_n\}$ and $\{B_n\}$ of self-adjoint operators in $\mathscr{R}$ such that $A_n x_0 \to A'x_0$ and $B_n x_0 \to B'x_0$. Then $\langle A_n x_0, B_n x_0\rangle \to \langle A'x_0, B'x_0\rangle$ and $\langle B_n x_0, A_n x_0\rangle \to \langle B'x_0, A'x_0\rangle$. But

$$\langle A_n x_0, B_n x_0\rangle = \langle B_n A_n x_0, x_0\rangle = \langle A_n B_n x_0, x_0\rangle = \langle B_n x_0, A_n x_0\rangle.$$

Thus

$$\langle B'A'x_0, x_0\rangle = \langle A'x_0, B'x_0\rangle = \langle B'x_0, A'x_0\rangle = \langle A'B'x_0, x_0\rangle.$$

The same now holds for all A', B' in $\mathscr{R}'$; and x_0 is a trace vector for $\mathscr{R}'$.

We note that x_0 is separating for $\mathscr{R}$ and, hence, generating for $\mathscr{R}'$. Suppose $Ax_0 = 0$ for some A in $\mathscr{R}$. Then, with B and C in $\mathscr{R}$,

$$\langle ABx_0, Cx_0\rangle = \langle C^*ABx_0, x_0\rangle = \langle BC^*Ax_0, x_0\rangle = 0,$$

since x_0 is a trace vector for $\mathscr{R}$. As x_0 is generating for $\mathscr{R}$, $A = 0$. ∎

In the theorem that follows, we show that a generating trace vector for $\mathscr{R}$ gives rise to a * anti-isomorphism between $\mathscr{R}$ and $\mathscr{R}'$ (that is, to an adjoint-preserving, *product-reversing*, linear isomorphism between $\mathscr{R}$ and $\mathscr{R}'$). This theorem is the precursor to the fundamental theorem of Tomita, which is the keystone of the modular theory described in Section 9.2. In that theorem, a * anti-isomorphism is established in the presence of a joint generating vector x_0 for $\mathscr{R}$ and $\mathscr{R}'$ (that need not be a trace vector). In this case, $\mathscr{R}$ and $\mathscr{R}'$ may not be finite; and the argument is considerably more involved.

7.2.15. THEOREM. *If the von Neumann algebra $\mathscr{R}$ acting on the Hilbert space $\mathscr{H}$ has a generating trace vector x_0, then $\mathscr{R}'$ is finite. For each A in $\mathscr{R}$, there is a unique A' in $\mathscr{R}'$ such that $Ax_0 = A'x_0$. The mapping $A \to A'$ is a * anti-isomorphism of $\mathscr{R}$ onto $\mathscr{R}'$.*

Proof. From the preceding lemma, x_0 is a generating trace vector for $\mathscr{R}'$. If $I \sim E'$ (modulo $\mathscr{R}'$), there is a (partial isometry) V' in $\mathscr{R}'$ such that $I = V'^*V'$, $V'V'^* = E'$. Since $\langle x_0, x_0\rangle = \langle V'^*V'x_0, x_0\rangle = \langle V'V'^*x_0, x_0\rangle = \langle E'x_0, x_0\rangle$, we have $\langle (I - E')x_0, x_0\rangle = 0$. As x_0 is generating for $\mathscr{R}$, it is separating for $\mathscr{R}'$, and $I - E' = 0$. Thus $\mathscr{R}'$ (and, likewise, $\mathscr{R}$) is finite.

Since x_0 is a trace vector for $\mathscr{R}$, if U is a unitary operator in $\mathscr{R}$, $\omega_{Ux_0}|\mathscr{R} = \omega_{x_0}|\mathscr{R}$. Thus the mapping $Tx_0 \to TUx_0$ (T in $\mathscr{R}$) extends to a unitary operator U' in $\mathscr{R}'$. (Recall, for this, that $[\mathscr{R}x_0] = [\mathscr{R}Ux_0] = \mathscr{H}$.) Of course, $Ux_0 = U'x_0$. Since each operator A in $\mathscr{R}$ is a linear combination of at most four unitary operators in $\mathscr{R}$ (see Theorem 4.1.7), there is an operator A'

in $\mathscr{R}'$ such that $Ax_0 = A'x_0$. As x_0 is separating for $\mathscr{R}'$, there is just one such A'. Now, $(aA + B)x_0 = aA'x_0 + B'x_0 = (aA' + B')x_0$; so that $A \to A'$ is linear. Moreover, $ABx_0 = AB'x_0 = B'Ax_0 = B'A'x_0$; whence $(AB)' = B'A'$. By symmetry $A \to A'$ maps $\mathscr{R}$ onto $\mathscr{R}'$; so that it is an anti-isomorphism of $\mathscr{R}$ onto $\mathscr{R}'$. Noting, again, that with H and K self-adjoint operators in $\mathscr{R}$,

$$\langle Hx_0, Kx_0\rangle = \langle KHx_0, x_0\rangle = \langle HKx_0, x_0\rangle = \langle Kx_0, Hx_0\rangle = \overline{\langle Hx_0, Kx_0\rangle},$$

we see that $\langle Hx_0, Kx_0\rangle$ is real. Thus if A is a self-adjoint operator in $\mathscr{R}$ and B is an arbitrary operator in $\mathscr{R}$, $\langle A'Bx_0, Bx_0\rangle = \langle A'x_0, B^*Bx_0\rangle = \langle Ax_0, B^*Bx_0\rangle$, which is real. Thus A' is self-adjoint, and $A \to A'$ is a * anti-isomorphism of $\mathscr{R}$ onto $\mathscr{R}'$. ■

7.2.16. COROLLARY. *If $\mathscr{A}$ is an abelian von Neumann algebra acting on the Hilbert space $\mathscr{H}$ and x_0 is a generating vector for $\mathscr{A}$, then $\mathscr{A} = \mathscr{A}'$ (that is, $\mathscr{A}$ is maximal abelian).*

Proof. Since $\mathscr{A}$ is abelian, $\mathscr{A} \subseteq \mathscr{A}'$; and x_0 is a generating vector for $\mathscr{A}'$. Thus x_0 is a separating vector for $\mathscr{A}$; and x_0 is a trace vector for $\mathscr{A}$. From the preceding theorem $\mathscr{A}$ and $\mathscr{A}'$ are * anti-isomorphic. Thus $\mathscr{A}'$ is abelian; and $\mathscr{A}' \subseteq \mathscr{A}'' = \mathscr{A}$. Hence $\mathscr{A} = \mathscr{A}'$. ■

Bibliography: [20, 31, 32, 43, 56]

7.3. A second approach to normal states

With the aid of the Sakai–Radon–Nikodým theorem (Theorem 7.3.6), we shall establish the key result (Theorem 7.2.3) of the preceding section once again. For this purpose, we shall need a decomposition of bounded linear functionals on a C^*-algebra analogous to the polar decomposition of operators on a Hilbert space (Theorem 7.3.2). In this connection, the additive decomposition of such functionals (Theorem 4.3.6 and Corollary 4.3.7) should be recalled. The "invertibility" of the polar decomposition of functionals is crucial to our discussion and results from an application of the precise determination of the extreme points of the unit ball of a C^*-algebra. We begin our treatment with that determination.

7.3.1. THEOREM. *The set of extreme points of the unit ball $(\mathfrak{A})_1$ of a C^*-algebra $\mathfrak{A}$ consists precisely of those partially isometric operators V in $\mathfrak{A}$ such that $(I - F)\mathfrak{A}(I - E) = (0)$, where $E = V^*V$ and $F = VV^*$.*

Proof. Suppose, first, that V is an extreme point of $(\mathfrak{A})_1$, and let $\mathfrak{A}(V^*V)$ be the C^*-subalgebra of $\mathfrak{A}$ generated by V^*V. Then the spectrum $\text{sp}(V^*V)$

of V^*V is contained in $[0, 1]$. If t is a point of $(0, 1)$ and h is a continuous function, small on a small neighborhood of t, vanishing outside that neighborhood, and non-zero at t, then $\|V^*V(I \pm h(V^*V))^2\| \le 1$. Thus

$$\|V(I \pm h(V^*V))\| \le 1;$$

and $V = \frac{1}{2}[V(I + h(V^*V)) + V(I - h(V^*V))]$. Since V is extreme on $(\mathfrak{A})_1$, $V = V + Vh(V^*V)$ and $0 = Vh(V^*V) = V^*Vh(V^*V)$. Since $h(t) \ne 0$, $t \notin \mathrm{sp}(V^*V)$. Thus the spectrum of V^*V contains, at most, 0 and 1; so that V^*V is a projection E in $\mathfrak{A}$. It follows that VV^* is a projection F in $\mathfrak{A}$, from Proposition 6.1.1.

If A is an operator in the unit ball of $(I - F)\mathfrak{A}(I - E)$ and z is a unit vector, then $z = x + y$ where $y = Ez$, $x = (I - E)z$; and

$$\begin{aligned}\|(V \pm A)z\|^2 = \|Vy \pm Ax\|^2 &= \|FVy \pm (I - F)Ax\|^2\\ &= \|Vy\|^2 + \|Ax\|^2 \le 1.\end{aligned}$$

Thus $V \pm A \in (\mathfrak{A})_1$. As $V = \frac{1}{2}(V + A + V - A)$, $V = V + A$ and $A = 0$. Hence $(I - F)\mathfrak{A}(I - E) = (0)$.

Suppose U is a partially isometric operator in $\mathfrak{A}$ with initial projection E and final projection F such that $(I - F)\mathfrak{A}(I - E) = (0)$. If $U = \frac{1}{2}(A + B)$ with A, B in $(\mathfrak{A})_1$, then $1 = \langle Ux, Ux\rangle = \frac{1}{2}[\langle Ax, Ux\rangle + \langle Bx, Ux\rangle]$, if x is a unit vector in the range of E. As $\langle Ax, Ux\rangle$ and $\langle Bx, Ux\rangle$ lie in the unit disk in the complex numbers, and 1 is an extreme point of that disk, $1 = \langle Ax, Ux\rangle = \langle Bx, Ux\rangle$. From the limit case of the Cauchy–Schwarz inequality, $Ax = Bx = Ux$. Thus $AE = BE = U$; and both A and B map the range of E isometrically onto that of F. We note that since both A and B have norm not exceeding 1, this last implies that $FA(I - E) = FB(I - E) = 0$. Otherwise, say, $FA(I - E) \ne 0$. There is a y in the range of $I - E$ such that $FAy = z$ for some unit vector z in the range of F. There is a unit vector x in the range of E such that $Ax = z$. Then $1 = \|z\| = \|FAy\| \le \|y\| = t^{-1}$; and $x \cos\theta + yt \sin\theta$ is a unit vector u for θ in $[0, 2\pi]$. But

$$\|FAu\| = |\cos\theta + t\sin\theta| = (1 + t^2)^{1/2} > 1,$$

when $\tan\theta = t$.

Having noted that $FA(I - E) = FB(I - E) = 0$, and given

$$0 = (I - F)A(I - E) = (I - F)B(I - E),$$

we have $A(I - E) = B(I - E) = 0$; so that $A = AE = U = BE = B$. Thus U is an extreme point of $(\mathfrak{A})_1$. ∎

7.3.2. THEOREM (Polar decomposition of linear functionals). *If $\mathscr{R}$ is a von Neumann algebra on $\mathscr{H}$ and ρ is a linear functional weak-operator continuous on the unit ball $(\mathscr{R})_1$ of $\mathscr{R}$, then there is a partial isometry U in $\mathscr{R}$,* *extreme on* $(\mathscr{R})_1$, *such that* ω *is a positive normal linear functional on* $\mathscr{R}$, *where* $\omega(A) = \rho(UA)$ *for each* A *in* $\mathscr{R}$. *Moreover* $\omega(U^*A) = \rho(A)$.

Proof. Since ρ is weak-operator continuous on $(\mathscr{R})_1$, and $(\mathscr{R})_1$ is weak-operator compact, there is some U in $(\mathscr{R})_1$ such that $|\rho(U)| = \|\rho\|$. We may assume that $\|\rho\| = 1$; and, multiplying U by a suitable scalar of modulus 1, we may assume that $\rho(U) = 1$. The subset of $(\mathscr{R})_1$ at which ρ takes the value 1 is a compact non-null face of $(\mathscr{R})_1$, and thus its extreme points are extreme on $(\mathscr{R})_1$. We may assume that U is an extreme point of $(\mathscr{R})_1$; so that U is a partial isometry with initial projection E and final projection F such that $(I - F)\mathscr{R}(I - E) = (0)$. If $P = I - C_{I-E}$, then, since $I - E \leq C_{I-E}$, $P \leq E$. Since $C_{I-F}C_{I-E} = 0$ and $I - F \leq C_{I-F}$, $I - P = C_{I-E} \leq I - C_{I-F} \leq F$; so that $U^*UP = P$ and $UU^*(I - P) = I - P$. Thus $UP + I - P (= V)$ and $U^*(I - P) + P (= W)$ are isometries ($V^*V = I$ and $W^*W = I$) in $\mathscr{R}$. Note that $W^*V = U$.

We may assume that $\mathscr{R}$ acting on $\mathscr{H}$ is the universal normal representation of $\mathscr{R}$. Since the functional which has the value $\rho(W^*AV)$ at A assumes its norm, 1, at I, it is a (normal) state of $\mathscr{R}$ (by Theorem 4.3.2); and there is a unit vector z in $\mathscr{H}$ such that $\rho(W^*AV) = \langle Az, z\rangle$ for all A in $\mathscr{R}$. Then $\rho(A) = \rho(W^*WAV^*V) = \langle AV^*z, W^*z\rangle$ for all A in $\mathscr{R}$. As

$$1 = \rho(U) = \langle UV^*z, W^*z\rangle,$$

and $\|V^*z\| \leq 1$, $\|W^*z\| \leq 1$, from the Cauchy–Schwarz inequality, $Uu = v$, where $u = V^*z$ and $v = W^*z$. Let $\omega(A)$ be $\rho(UA)$. Since ω takes its norm, 1, at I, ω is a (normal) state of $\mathscr{R}$. Since $FU = U$, we have that $Fv = v$. Hence $\omega(U^*A) = \rho(UU^*A) = \rho(FA) = \langle FAu, v\rangle = \langle Au, v\rangle = \rho(A)$, for all A in $\mathscr{R}$. ■

7.3.3. COROLLARY. *If* ρ *is a norm* 1, *linear functional on the von Neumann algebra* $\mathscr{R}$, ρ *is weak-operator continuous on* $(\mathscr{R})_1$, *and* $\mathscr{R}$ *acting on the Hilbert space* $\mathscr{H}$ *has a separating vector (or is the universal normal representation of* $\mathscr{R}$), *then there are unit vectors* u, v *in* $\mathscr{H}$ *such that* $\rho = \omega_{u,v}|\mathscr{R}$.

Proof. In the last paragraph of the proof of Theorem 7.3.2, we show that $\rho(A) = \langle Au, v\rangle$ $(= \omega_{u,v}(A))$ for all A in $\mathscr{R}$, where $\|u\| = \|v\| = 1$. We assume that $\mathscr{R}$ acting on $\mathscr{H}$ is the universal normal representation of $\mathscr{R}$, but make use of this assumption to assert only that the normal state assigning $\rho(W^*AV)$ to A in $\mathscr{R}$ is a vector state. We could conclude this from the assumption that $\mathscr{R}$ has a separating vector by applying Theorem 7.2.3. ■

One of our goals in this section is an independent proof of Theorem 7.2.3. We shall not make use of the "separating vector" part of Corollary 7.3.3 in this section.

7.3.4. LEMMA. *If ρ is a state of the C^*-algebra $\mathfrak{A}$ and A in $\mathfrak{A}$ is such that $B \to \rho(BA)$ is hermitian, then $|\rho(AH)| \le \|A\|\rho(H)$ for each positive H in $\mathfrak{A}$.*

Proof. Since $B \to \rho(BA)$ is hermitian, $\rho(B^*A) = \overline{\rho(BA)} = \rho(A^*B^*)$ for each B in $\mathfrak{A}$. Thus $\rho(BA^{2n}) = \rho(A^{*n}BA^n)$. With H a positive operator in $\mathfrak{A}$,

$$|\rho(HA^n)| = |\rho(H^{1/2}H^{1/2}A^n)| \le \rho(A^{*n}HA^n)^{1/2}\rho(H)^{1/2} = \rho(HA^{2n})^{1/2}\rho(H)^{1/2};$$

so that

$$\begin{aligned}|\rho(HA)| &\le \rho(HA^4)^{1/4}\rho(H)^{1/2}\rho(H)^{1/4} \le \cdots \le \rho(HA^{2^n})^{2^{-n}}\rho(H)^{1/2+\cdots+1/2^n}\\ &\le (\|\rho\|\,\|H\|)^{2^{-n}}\|A\|\rho(H)^{1-1/2^n}\end{aligned}$$

and

$$|\rho(HA)| \le \|A\|\rho(H). \quad ■$$

Note that if $B \to \rho(AB)$ is hermitian, then

$$|\rho(AH)| = |\rho(HA^*)| \le \|A^*\|\rho(H) = \|A\|\rho(H),$$

by applying the preceding lemma to the hermitian functional $B \to \rho(BA^*)$.

7.3.5. PROPOSITION. *If $\mathfrak{A}$ is a self-adjoint algebra of operators (containing I) on the Hilbert space $\mathscr{H}$ and ω is a positive linear functional on $\mathfrak{A}$ such that $\omega \le \omega_x|\mathfrak{A}$ for some vector x in $\mathscr{H}$, then there is a positive operator H' in the unit ball $(\mathfrak{A}')_1$ of $\mathfrak{A}'$ such that $\omega(A) = \omega_x(H'A) = \omega_{H'^{1/2}x}(A)$ for all A in $\mathfrak{A}$.*

Proof. With $\varphi(Ax, Bx)$ defined as $\omega(B^*A)$,

$$|\varphi(Ax, Bx)|^2 = |\omega(B^*A)|^2 \le \omega(A^*A)\omega(B^*B) \le \|Ax\|^2\|Bx\|^2.$$

Thus φ is a well-defined positive conjugate-bilinear functional on $\mathfrak{A}x$, bounded by 1. It follows that φ has a unique extension to $[\mathfrak{A}x]$; and that there is a positive operator H' on $[\mathfrak{A}x]$ such that $\|H'\| \le 1$ and $\langle H'Ax, Bx\rangle = \varphi(Ax, Bx) = \omega(B^*A)$ (see Theorem 2.4.1 and the discussion preceding Proposition 2.4.6). Thus $\omega(A) = \langle H'Ax, x\rangle = \omega_x(H'A)$ for all A in $\mathfrak{A}$. Extending H' by defining it to be 0 on $\mathscr{H} \ominus [\mathfrak{A}x]$ leaves it positive, of norm not exceeding 1, and leaves the foregoing equalities unaltered. Since

$$\langle H'ABx, Cx\rangle = \omega(C^*AB) = \langle H'Bx, A^*Cx\rangle = \langle AH'Bx, Cx\rangle$$

for all A, B, C in $\mathfrak{A}$; $H'A - AH'$ is 0 on $[\mathfrak{A}x]$. As H' is 0 on $\mathscr{H} \ominus [\mathfrak{A}x]$, $H'A - AH' = 0$ for all A in $\mathfrak{A}$; and $H' \in (\mathfrak{A}')_1$. Thus

$$\omega(A) = \langle H'Ax, x\rangle = \langle AH'^{1/2}x, H'^{1/2}x\rangle = \omega_{H'^{1/2}x}(A)$$

for all A in $\mathfrak{A}$. ■

7.3.6. THEOREM (Sakai–Radon–Nikodým). *If ω and ω_0 are normal positive linear functionals on a von Neumann algebra $\mathscr{R}$ and $\omega_0 \le \omega$, then there is a positive operator H_0 in the unit ball of $\mathscr{R}$ such that $\omega_0(A) = \omega(H_0 A H_0)$, for all A in $\mathscr{R}$.*

Proof. We may assume that $\mathscr{R}$ acting on the Hilbert space $\mathscr{H}$ is the universal normal representation of $\mathscr{R}$. In this case, $\omega = \omega_x|\mathscr{R}$ for some vector x in $\mathscr{H}$. The support E of $\omega_x|\mathscr{R}$ has range $[\mathscr{R}'x]$; and, by hypothesis, the support of ω_0 is dominated by E. By considering $E\mathscr{R}E$ acting on $[\mathscr{R}'x]$ in place of $\mathscr{R}$ acting on $\mathscr{H}$, we may assume that $\omega = \omega_x|\mathscr{R}$ and $[\mathscr{R}'x] = \mathscr{H}$. From Proposition 7.3.5, there is a positive operator H_0' in $(\mathscr{R}')_1$ such that $\omega_0 = \omega_{H_0'x}|\mathscr{R}$. From Theorem 7.3.2 (applied to $\omega_{x, H_0'x}|\mathscr{R}'$), there is a partial isometry V' in $\mathscr{R}'$ such that $\omega_{x, V'^*H_0'x}|\mathscr{R}'$ is positive and $\omega_{x, H_0'x}|\mathscr{R}' = \omega_{x, V'V'^*H_0'x}|\mathscr{R}'$. Since $[\mathscr{R}'x] = \mathscr{H}$, $H_0'x = V'V'^*H_0'x$. From Lemma 7.3.4, $\omega_{x, V'^*H_0'x}(H') \le \|H_0' V'\|\omega_x(H') \le \omega_x(H')$ for each positive operator H' in $\mathscr{R}'$. From Proposition 7.3.5, there is a positive operator $H_0^{1/2}$ in $(\mathscr{R})_1$ such that $\omega_{x, V'^*H_0'x}|\mathscr{R}' = \omega_{H_0^{1/2}x}|\mathscr{R}'$. Hence $\langle A'x, V'^*H_0'x\rangle = \langle A'x, H_0 x\rangle$ for all A' in $\mathscr{R}'$. It follows that $V'^*H_0'x = H_0 x$. Since $H_0'x = V'V'^*H_0'x$, $\omega_{H_0x}|\mathscr{R} = \omega_{H_0'x}|\mathscr{R} = \omega_0$. ■

7.3.7. LEMMA. *If $\mathscr{R}$ is a von Neumann algebra acting on the Hilbert space $\mathscr{H}$, H is a positive operator in $\mathscr{R}$, N is the projection on the null space of H, and x_0 is a vector in $\mathscr{H}$ such that $Nx_0 = 0$, then $x_0 \in [\mathscr{R}Hx_0]$.*

Proof. By Proposition 2.5.13 we have

$$x_0 = Mx_0 + Nx_0 = Mx_0,$$

where $M = R(H)$. We may assume that $0 \le H \le I$, so that $x_0 = \lim H^{1/n}x_0$ by Lemma 5.1.5. Since $H^{1/n}$ is the norm limit of a sequence of polynomials (without constant term) in H, we have $H^{1/n}x_0 \in [\mathscr{R}Hx_0]$ and thus $x_0 \in [\mathscr{R}Hx_0]$. ■

7.3.8. THEOREM. *If ω is a normal state of the von Neumann algebra $\mathscr{R}$ acting on the Hilbert space $\mathscr{H}$ and z is a unit separating vector for $\mathscr{R}$, then $\omega = \omega_x|\mathscr{R}$, for some unit vector x in $\mathscr{H}$.*

Proof. If $\omega_0 = \omega + \omega_z|\mathscr{R}$, then ω_0 engenders a faithful representation φ of $\mathscr{R}$ onto a von Neumann algebra $\varphi(\mathscr{R})$ acting on a Hilbert space $\mathscr{K}$. Let y_0 be a generating vector for $\varphi(\mathscr{R})$ in $\mathscr{K}$ such that $\omega_0(A) = \langle\varphi(A)y_0, y_0\rangle$, for all A in $\mathscr{R}$. As $\omega \le \omega_0$ and $\omega_z \le \omega_0$, there is a unit vector x_0 in $\mathscr{K}$ such that $\omega(A) = \langle\varphi(A)x_0, x_0\rangle$, and an operator $\varphi(H)$ in $\varphi(\mathscr{R})$ such that $\omega_z(A) = \langle\varphi(A)\varphi(H)y_0, \varphi(H)y_0\rangle$, for all A in $\mathscr{R}$, and $0 \le H \le I$. If N is the projection on the null space of H, $\varphi(N)$ is the projection on the null space of $\varphi(H)$; and $0 = \langle\varphi(N)\varphi(H)y_0, \varphi(H)y_0\rangle = \omega_z(N)$. Thus N and $\varphi(N)$ are 0. From the preceding lemma, $[\varphi(\mathscr{R})\varphi(H)y_0]$ contains y_0 and coincides with $\mathscr{K}$. The mapping $Az \to \varphi(A)\varphi(H)y_0$ extends to a unitary transformation U of $[\mathscr{R}z]$ onto $\mathscr{K}$ such that $UAE'U^{-1} = \varphi(A)$, for all A in $\mathscr{R}$ where E' is the projection with range $[\mathscr{R}z]$. If x is $U^{-1}x_0$, then x is a unit vector, and $\langle Ax, x\rangle = \langle UAU^{-1}x_0, x_0\rangle = \langle\varphi(A)x_0, x_0\rangle = \omega(A)$, for all A in $\mathscr{R}$. ∎

7.3.9. REMARK. In the proof of Theorem 7.2.15, we established that, for each A in $\mathscr{R}$, there is a unique A' in $\mathscr{R}'$ such that $Ax_0 = A'x_0$, where x_0 is a generating trace vector for $\mathscr{R}$ (and, hence, for $\mathscr{R}'$). This was achieved by appealing to the decomposition of operators in $\mathscr{R}$ as linear combinations of unitary operators in $\mathscr{R}$. Proposition 7.3.5 and Theorem 7.3.6 each provide another route to this conclusion.

With H and A positive operators in $\mathscr{R}$,

$$\begin{aligned} 0 \le \omega_{Ax_0}(H) &= \langle HAx_0, Ax_0\rangle = \langle AHAx_0, x_0\rangle = \langle HA^2x_0, x_0\rangle \\ &= \langle H^{1/2}A^2H^{1/2}x_0, x_0\rangle \le \|A^2\|\langle Hx_0, x_0\rangle. \end{aligned}$$

From Proposition 7.3.5, there is a positive operator A' in $\mathscr{R}'$ such that $\omega_{A'x_0}|\mathscr{R} = \omega_{Ax_0}|\mathscr{R}$. It follows that, for each T in $\mathscr{R}$,

$$\begin{aligned} \langle TA'x_0, A'x_0\rangle &= \langle Tx_0, A'^2x_0\rangle = \langle TAx_0, Ax_0\rangle \\ &= \langle A^2Tx_0, x_0\rangle = \langle Tx_0, A^2x_0\rangle. \end{aligned}$$

Since x_0 is a generating vector for $\mathscr{R}$, $A'^2x_0 = A^2x_0$. Decomposing a self-adjoint operator as the difference of two positive operators, and each operator as the sum of a self-adjoint and $\sqrt{-1}$ multiplied by a self-adjoint operator, we have the desired mapping ($A \to A'$ of Theorem 7.2.15). In this approach the mapping is seen to be adjoint preserving.

If somewhat less emphasis on the use of a trace vector is desired, the Sakai–Radon–Nikodým theorem can be used. With H' positive in $\mathscr{R}'$ and A positive in $\mathscr{R}$, $\omega_{Ax_0}(H') = \langle H'A^2x_0, x_0\rangle \le \|A^2\|\langle H'x_0, x_0\rangle$, so that there is a positive operator A' in $\mathscr{R}'$ such that $\omega_{Ax_0}|\mathscr{R}' = \omega_{A'x_0}|\mathscr{R}'$. Thus

$$\langle T'x_0, A^2x_0\rangle = \langle T'A'x_0, A'x_0\rangle = \langle T'x_0, A'^2x_0\rangle$$

for all T' in $\mathscr{R}'$. As x_0 is generating for $\mathscr{R}'$, $A^2x_0 = A'^2x_0$, and we proceed as before. ∎

Making use, once again, of the trace function τ of Section 8.2, we prove, in the next result, that strong-operator limits of cyclic projections are, themselves, cyclic—and from this that norm limits of vector states are vector states. Needless to say, these results will not be used in Section 8.2.

7.3.10. PROPOSITION. *If the sequence* $\{E_n\}$ *of cyclic projections in a von Neumann algebra* $\mathscr{R}$ *acting on a Hilbert space* $\mathscr{H}$ *converge in the strong-operator topology to* E, *then* E *is a cyclic projection in* $\mathscr{R}$.

Proof. Since $\{E_n\}$ converges to E in the weak as well as the strong-operator topology, E is self-adjoint. Since $\{E_n^2\} = \{E_n\}$ converges to E^2, it follows that E is self-adjoint and idempotent. Thus E is a projection in $\mathscr{R}$.

If F_n is the range projection of EE_n, then $F_n \leq E$, F_n is cyclic in $\mathscr{R}$, and F_n tends to E in the strong-operator topology. To see this, note that, with x_n a generating vector for E_n,

$$[\mathscr{R}'Ex_n] = [E\mathscr{R}'x_n] = [EE_n(\mathscr{H})];$$

so that Ex_n is a generating vector for F_n. Moreover, with x a unit vector in $E(\mathscr{H})$, $\langle F_n x, x\rangle \geq \langle EE_n Ex, x\rangle = \langle EE_n x, x\rangle \to \langle x, x\rangle = 1$; the first inequality resulting from the facts that $EE_n E$ has range in $F_n(\mathscr{H})$ and $EE_n E \leq I$. Thus $F_n x \to x$.

Restricting attention to $E\mathscr{R}E$ and $\mathscr{R}'E$ acting on $E(\mathscr{H})$, we may assume that $E = I$. In this case, I is countably decomposable, from Proposition 5.5.19. It follows from Proposition 5.5.18 that there is a central projection cyclic in $\mathscr{R}$ whose orthogonal complement P is cyclic in $\mathscr{R}'$. If we can show that P is cyclic in $\mathscr{R}$, as well, then, under the present assumptions, Proposition 5.5.10 yields that I is cyclic in $\mathscr{R}$.

Restricting attention to $\mathscr{R}P$ and $\mathscr{R}'P$ acting on $P(\mathscr{H})$, we may assume that $\mathscr{R}$ has a generating vector x_0. From Proposition 7.2.2, $E_n \precsim F$ for all n, where F is the projection with range $[\mathscr{R}'x_0]$. There is a central projection Q, such that either Q is 0 and F is finite or FQ is properly infinite in $\mathscr{R}$ and $F(I - Q)$ is finite in $\mathscr{R}$, from Proposition 6.3.7. It follows from Theorem 6.3.4 that $Q \precsim FQ$; since I, and, hence, Q are countably decomposable in $\mathscr{R}$ and $C_{FQ} = Q$ (see Proposition 5.5.13, from which $C_F = I$). It follows that Q is cyclic (see Proposition 6.2.9).

Restricting attention to $\mathscr{R}(I - Q)$ and $\mathscr{R}'(I - Q)$ acting on $(I - Q)(\mathscr{H})$, we may assume that F is finite. If I is not cyclic, then $F \neq I$, so F is contained in a properly larger finite projection G. As before, the range projections of GE_n are cyclic, contained in G, and converge to G in the strong-operator topology. Restricting attention to $G\mathscr{R}G$ and $\mathscr{R}'G$ acting on $G(\mathscr{H})$, we may assume that I is a finite projection in $\mathscr{R}$ and $F < I$. With x a unit vector in $\mathscr{H}$, from Theorem 8.2.8, $\langle x, x\rangle \geq \langle \tau(F)x, x\rangle \geq \langle \tau(E_n)x, x\rangle \to \langle x, x\rangle = 1$; since $E_n \precsim F \leq I$, so that $\tau(F) = I = F$, a contradiction. Hence I and, after various identifications, E are cyclic. ■

We shall see, in Exercise 7.6.11, that "sequence" rather than "net" is essential to the preceding result.

7.3.11. THEOREM. *If $\mathscr{R}$ is a von Neumann algebra acting on the Hilbert space $\mathscr{H}$ and ω is a linear functional on $\mathscr{R}$ such that $\|\omega - \omega_{x_n}|\mathscr{R}\| \to_n 0$, for some sequence $\{x_n\}$ of vectors in $\mathscr{H}$, then there is a vector x in $\mathscr{H}$ such that $\omega = \omega_x|\mathscr{R}$.*

Proof. As a norm limit of positive normal functionals, ω is positive, and, from Corollary 7.1.13, ω is normal. With E_n the projection (in $\mathscr{R}$) on $[\mathscr{R}'x_n]$ and E the support of ω, let F_n be the range projection of EE_n. Since $[\mathscr{R}'Ex_n] = [EE_n(\mathscr{H})]$, F_n is a cyclic projection in $\mathscr{R}$. We prove that E is the strong-operator limit of $\{F_n\}$. From Proposition 7.3.10, we can, then, conclude that E is cyclic, and, from Proposition 7.2.7, that $\omega = \omega_x|\mathscr{R}$ for some vector x in $\mathscr{H}$.

Let a_n be $\|\omega - \omega_{x_n}|\mathscr{R}\|$. Then

$$\begin{aligned}\omega(E - F_n) &\le a_n + \langle (E - F_n)x_n, x_n\rangle = a_n + \|(E - F_n)x_n\|^2\\ &= a_n + \|EE_nx_n - F_nx_n\|^2 = a_n + \|F_n(EE_nx_n - x_n)\|^2\\ &\le a_n + \|EE_nx_n - x_n\|^2 = a_n + \langle (I - E)x_n, x_n\rangle\\ &\le a_n + a_n + \omega(I - E) = 2a_n \to_n 0.\end{aligned}$$

From Theorem 7.1.12(b), $\omega = \sum_{j=1}^{\infty} \omega_{y_j}|\mathscr{R}$. If $M_n = E - F_n$, then $M_n y_j \to_n 0$ for each j, since $\|M_n y_j\|^2 \le \omega(M_n)$. Thus $M_n A' y_j = A' M_n y_j \to_n 0$, when $A' \in \mathscr{R}'$; and $M_n G_j \to_n 0$, in the strong-operator topology, where G_j (with range $[\mathscr{R}'y_j]$) is the support of $\omega_{y_j}|\mathscr{R}$. Since $E = \bigvee_j G_j$, $M_n E$ $(= M_n) \to 0$, in the strong-operator topology. ■

7.3.12. PROPOSITION. *If $\mathscr{R}$ is a von Neumann algebra acting on the Hilbert space $\mathscr{H}$ and $\omega_{x,y}|\mathscr{R}$ is a state of $\mathscr{R}$, then there is a unit vector z in $\mathscr{H}$ such that $\omega_z|\mathscr{R} = \omega_{x,y}|\mathscr{R}$.*

Proof. If H is a positive operator in $\mathscr{R}$, then

$$0 \le \omega_{x-y}(H) = \langle Hx, x\rangle + \langle Hy, y\rangle - \langle Hx, y\rangle - \langle Hy, x\rangle.$$

As $\omega_{x,y}|\mathscr{R}$ is a state, $\langle Hx, y\rangle = \langle y, Hx\rangle = \langle Hy, x\rangle$; and $2\langle Hx, y\rangle \le \langle Hx, x\rangle + \langle Hy, y\rangle$. Thus $4\omega_{x,y}(H) \le \omega_{x+y}(H)$. From Proposition 7.3.5, $\omega_{x,y}|\mathscr{R} = \omega_z|\mathscr{R}$ for some unit vector z. ■

7.3.13. THEOREM. (Linear Radon–Nikodým). *If ω is a positive normal linear functional on the von Neumann algebra $\mathscr{R}$, $\rho_0 \in \mathscr{R}^{\sharp}$, and $0 \le \rho_0 \le \omega$, there is a K_0 in $(\mathscr{R}^+)_1$ such that $\rho_0 = \omega_{K_0}$, where $\omega_{K_0}(A) = \frac{1}{2}\omega(K_0A + AK_0)$.*

Proof. Since $\omega_{aK+a'K'} = a\omega_K + a'\omega_{K'}$ and the mapping $K \to \omega_K$ is continuous from $(\mathscr{R}^+)_1$ in its weak-operator topology to $\mathscr{R}^{\sharp}$ in its weak* topology, the set $\{\omega_K : K \in (\mathscr{R}^+)_1\}$ $(=\mathscr{K})$ is a weak* compact convex subset of $\mathscr{R}^{\#}$. If $\rho_0 \notin \mathscr{K}$, from Corollary 1.2.12 and Proposition 1.3.5, there is a B in $\mathscr{R}$ and a real a such that

$$\omega_K(A) = \text{Re } \omega_K(B) \leq a < \text{Re } \rho_0(B) = \rho_0(A)$$

for all K in $(\mathscr{R}^+)_1$, where $A = \frac{1}{2}(B + B^*)$. If $A^+ - A^-$ is the decomposition of A into its positive and negative parts and E is the range projection of A^+, then $E \in (\mathscr{R}^+)_1$, so that

$$\rho_0(A^+) \leq \omega(A^+) = \omega_E(A) \leq a < \rho_0(A) \leq \rho_0(A^+).$$

We conclude, from this contradiction, that $\rho_0 \in \mathscr{K}$, that is, $\rho_0 = \omega_{K_0}$ for some K_0 in $(\mathscr{R}^+)_1$. ■

Bibliography: [40, 79, 80]

7.4. The predual

Using the analysis of normal states effected in the preceding sections, we can obtain useful information about the Banach-space structure of von Neumann algebras. In Theorem 5.1.3 we noted that the unit ball of a von Neumann algebra is compact in the weak-operator topology. Comparing this fact with the Alaoglu–Bourbaki theorem (see Theorem 1.6.5(i)), we observe that a von Neumann algebra exhibits one of the key features of the norm (continuous) dual of a Banach space, where the weak-operator topology plays the role of the weak* topology. Carrying this analogy further, we recall that the weak-operator topology is the weak topology induced by the linear span of vector states. This leads us to the following definition.

7.4.1. Definition. If $\mathscr{R}$ is a von Neumann algebra acting on the Hilbert space $\mathscr{H}$, we denote by $\mathscr{R}_{\#}$ the linear space of linear functionals on $\mathscr{R}$ that are weak-operator continuous on the unit ball of $\mathscr{R}$ (the "normal" linear functionals on $\mathscr{R}$). We refer to $\mathscr{R}_{\#}$ as the *predual* of $\mathscr{R}$. ■

If ρ is an element of $\mathscr{R}_{\#}$, then the image of $(\mathscr{R})_1$, the unit ball of $\mathscr{R}$, under ρ is a compact subset of the complex numbers. It follows that ρ is bounded and that $\mathscr{R}_{\#}$ is a subspace of $\mathscr{R}^{\#}$, the norm dual of $\mathscr{R}$. As in the proof of Lemma 7.1.6, a norm limit of elements of $\mathscr{R}_{\#}$ is weak-operator continuous on $(\mathscr{R})_1$. Thus $\mathscr{R}_{\#}$ is a norm-closed subspace of $\mathscr{R}^{\#}$. Provided with its norm, $\mathscr{R}_{\#}$ is a Banach space.

As defined, $\mathscr{R}_{\#}$ seems to depend on the representation (that is, action) of $\mathscr{R}$ as a von Neumann algebra on $\mathscr{H}$ (through the weak-operator topology). In point of fact, $\mathscr{R}_{\#}$ does not vary with the Hilbert space on which $\mathscr{R}$ acts as a von Neumann algebra. Put in more precise form, what is indicated by this comment is that a * isomorphism φ of one von Neumann algebra $\mathscr{R}$ onto another $\mathscr{T}$, induces a linear isomorphism (in fact, isometry) of $\mathscr{T}_\sharp$ onto $\mathscr{R}_\sharp$. This is apparent on comparing Definition 7.4.1 with Corollary 7.1.16 (and noting that φ is an isometry). In this sense $\mathscr{R}_\sharp$ and the weak-operator topology on $(\mathscr{R})_1$ are independent of the (faithful) representation in which $\mathscr{R}$ appears. For many purposes, then, we may choose a (faithful) representation to suit our convenience in proving (or studying) certain properties of $\mathscr{R}$. Even though $\mathscr{R}$ is given as acting on a Hilbert space $\mathscr{H}$, we often indicate the change of representation, without introducing notation for a * isomorphism, by saying "we may assume that $\mathscr{R}$ acting on $\mathscr{H}$ is the such-and-such (for example, universal normal) representation."

7.4.2. THEOREM. *If $\mathscr{R}$ is a von Neumann algebra acting on the Hilbert space $\mathscr{H}$ and $\mathscr{R}_\sharp$ is the space of linear functionals that are weak-operator continuous on $(\mathscr{R})_1$, the unit ball of $\mathscr{R}$, then $\mathscr{R}$, acting as linear functionals on $\mathscr{R}_\sharp$, is the norm-dual space of $\mathscr{R}_\sharp$; and the weak* topology on $(\mathscr{R})_1$ coincides with the weak-operator topology on $(\mathscr{R})_1$.*

Proof. From the discussion preceding this theorem, we may assume that $\mathscr{R}$ acting on $\mathscr{H}$ is the universal normal representation of $\mathscr{R}$. Since, from Theorem 2.4.1, $\|A\| = \sup\{\langle Ax, y\rangle : \|x\| \le 1,\ \|y\| \le 1,\ x,\ y \text{ in } \mathscr{H}\}$ and $\|\omega_{x,y}|\mathscr{R}\| \le 1$ when $\|x\| \le 1$, $\|y\| \le 1$, the imbedding of $\mathscr{R}$ into the dual of $\mathscr{R}_\sharp$ is linear and isometric. It remains to show that this imbedding is onto the dual space.

If φ is a bounded linear functional on $\mathscr{R}_\sharp$, defining $\varphi'(x, y)$ to be $\varphi(\omega_{x,y}|\mathscr{R})$, we see that φ' is a bounded, conjugate-bilinear functional on $\mathscr{H}$. From Theorem 2.4.1, there is an operator A on $\mathscr{B}(\mathscr{H})$ such that $\langle Ax, y\rangle = \varphi'(x, y) = \varphi(\omega_{x,y}|\mathscr{R})$ for all x, y in $\mathscr{H}$. With T' a self-adjoint operator in $\mathscr{R}'$, $\omega_{T'x,y}|\mathscr{R} = \omega_{x,T'y}|\mathscr{R}$; so that $\langle AT'x, y\rangle = \langle T'Ax, y\rangle$ for all x, y in $\mathscr{H}$. Thus $A \in \mathscr{R}''$ $(= \mathscr{R})$. Since $\omega_{x,y}(A) = \langle Ax, y\rangle = \varphi'(x, y) = \varphi(\omega_{x,y}|\mathscr{R})$, the image of A in the dual of $\mathscr{R}_\sharp$ coincides with φ, by Corollary 7.3.3, and the weak* topology induced on $(\mathscr{R})_1$ by $\mathscr{R}_\sharp$ is the weak-operator topology on $(\mathscr{R})_1$. ∎

Exercise 7.6.45 is a converse to Theorem 7.4.2. In Exercise 7.6.9, we note that continuity in the weak-operator topology on the unit ball, $(\mathscr{R})_1$, of a von Neumann algebra $\mathscr{R}$ does not entail continuity on $\mathscr{R}$ in that topology. Since continuity on $(\mathscr{R})_1$ is critical to our study of normal states, we are led to ask whether some topology on $\mathscr{R}$ is "compatible" with that continuity.

7.4.3. DEFINITION. The *ultraweak topology* on a von Neumann algebra $\mathscr{R}$ acting on a Hilbert space $\mathscr{H}$ is the weakest topology relative to which all functionals of the form $\sum_{n=1}^{\infty} \omega_{x_n,y_n}|\mathscr{R}$, with $\sum (\|x_n\|^2 + \|y_n\|^2) < \infty$, are continuous. ∎

7.4.4. REMARK. From Theorem 1.3.1, each functional ω on $\mathscr{R}$ continuous in the ultraweak topology has the form $\sum_{n=1}^{\infty} \omega_{x_n, y_n}|\mathscr{R}$, with

$$\sum (\|x_n\|^2 + \|y_n\|^2) < \infty,$$

since the set of such functionals is a linear subspace of the dual of $\mathscr{R}$. Note, too, that

$$\begin{aligned}|\sum \omega_{x_n, y_n}(A)| = |\sum \langle Ax_n, y_n\rangle| &\leq \sum \|A\| \, \|x_n\| \, \|y_n\| \\ &\leq \|A\|(\sum \|x_n\|^2)^{1/2}(\sum \|y_n\|^2)^{1/2};\end{aligned}$$

so that, under the assumption that $\sum (\|x_n\|^2 + \|y_n\|^2) < \infty$, $\sum_{n=1}^{\infty} \omega_{x_n, y_n}(A)$ converges (absolutely) and $\sum \omega_{x_n, y_n}|\mathscr{R}$ is in the norm dual of $\mathscr{R}$. From this same computation, ω is the norm limit of the sequence of finite partial sums of $\sum_{n=1}^{\infty} \omega_{x_n, y_n}|\mathscr{R}$, and $\omega \in \mathscr{R}_\sharp$ (as noted in Proposition 7.4.5, which follows). Of course each functional $\sum_{n=1}^{\infty} \omega_{x_n, y_n}|\mathscr{R}$, that is ultraweakly continuous, has an extension, $\sum_{n=1}^{\infty} \omega_{x_n, y_n}|\mathscr{B}(\mathscr{H})$, to $\mathscr{B}(\mathscr{H})$, continuous in the ultraweak topology on $\mathscr{B}(\mathscr{H})$; so that the ultraweak topology on $\mathscr{R}$ is the relative topology induced by the ultraweak topology on $\mathscr{B}(\mathscr{H})$. Since each functional weak-operator continuous on $(\mathscr{R})_1$ has the form $\omega_{u,v}|\mathscr{R}$ (from Corollary 7.3.3) when $\mathscr{R}$ is taken in its universal normal representation, the ultraweak topology coincides with the weak-operator topology in this representation. From the discussion following Definition 7.4.1, the ultraweak and weak-operator topologies on $(\mathscr{R})_1$ are, therefore, the same for all (faithful) representations of $\mathscr{R}$ (as a von Neumann algebra). Again, since $\omega_{u,v}|\mathscr{R}$ is a finite linear combination of vector states of $\mathscr{R}$ (by polarization), the ultraweak topology on $\mathscr{R}$ is induced (in each normal faithful representation of $\mathscr{R}$) by the normal *states* of $\mathscr{R}$. If φ is a * isomorphism of $\mathscr{R}$ onto another von Neumann algebra $\mathscr{T}$ and ω is a normal state of $\mathscr{T}$, then $\omega \circ \varphi$ is a normal state of $\mathscr{R}$ (as in the proof of Corollary 7.1.16). Thus φ is a homeomorphism when each of $\mathscr{R}$ and $\mathscr{T}$ are provided with its ultraweak topology. ∎

7.4.5. PROPOSITION. *If $\mathscr{R}$ is a von Neumann algebra acting on a Hilbert space $\mathscr{H}$ and ω is a linear functional on $\mathscr{R}$, then ω is ultraweakly continuous on $\mathscr{R}$ if and only if ω is weak-operator continuous on the unit ball $(\mathscr{R})_1$ of $\mathscr{R}$.*

Proof. If ω is ultraweakly continuous on $\mathscr{R}$, from the preceding remark, $\omega = \sum_{n=1}^{\infty} \omega_{x_n, y_n}|\mathscr{R}$, where $\sum_{n=1}^{\infty}(\|x_n\|^2 + \|y_n\|^2) < \infty$. Thus $\|\omega - \omega_k\| \to 0$ as $k \to \infty$, where $\omega_k = \sum_{n=1}^{k} \omega_{x_n, y_n}|\mathscr{R}$. As each ω_k is weak-operator continuous on $(\mathscr{R})_1$, the same is true of ω.

If ρ is weak-operator continuous on $(\mathscr{R})_1$, from Theorem 7.3.2, there is a partial isometry U in $\mathscr{R}$ such that $\rho(A) = \omega(UA)$ for all A in $\mathscr{R}$, where ω is positive and normal. From Theorem 7.1.12(b) and (c), $\omega = \sum \omega_{x_n}|\mathscr{R}$, where $\sum \|x_n\|^2 < \infty$. Thus $\rho = \sum \omega_{x_n, y_n}|\mathscr{R}$ where $y_n = U^*x_n$; so that

$$\sum (\|x_n\|^2 + \|y_n\|^2) < \infty;$$

and ρ is ultraweakly continuous. ∎

Applied to ultraweakly continuous states, Proposition 7.4.5 constitutes an additional condition to Theorem 7.1.12. The terms "normal" and "ultraweakly continuous" are now equally applicable to functionals on $\mathscr{R}$.

If we define $\rho^*(A)$ to be $\overline{\rho(A^*)}$, as in Section 4.3, when ρ is a normal linear functional on the von Neumann algebra $\mathscr{R}$, then ρ^* is normal; and the decomposition of ρ as $\rho_1 + i\rho_2$ yields hermitian functionals ρ_1 and ρ_2 in $\mathscr{R}_\sharp$. Applying Theorem 4.3.6 to a normal hermitian functional ρ on $\mathscr{R}$, we have that $\rho = \rho^+ - \rho^-$, where ρ^+ and ρ^- are positive. With the aid of more information about extreme points of subsets of $(\mathscr{R})_1$ (compare Theorem 7.3.1), we shall show that ρ^+ and ρ^- are normal (see Theorem 7.4.7).

7.4.6. PROPOSITION. *If $\mathfrak{A}$ is a C^*-algebra, E is an extreme point of the set $(\mathfrak{A}^+)_1$ of positive elements in the unit ball $(\mathfrak{A})_1$ of $\mathfrak{A}$ if and only if E is a projection in $\mathfrak{A}$. With $(\mathfrak{A}_{\mathrm{h}})_1$ the self-adjoint elements in the unit ball of $\mathfrak{A}$, A is an extreme point of $(\mathfrak{A}_{\mathrm{h}})_1$ if and only if $A = 2E - I$ with E a projection in $\mathfrak{A}$.*

Proof. If E is extreme on $(\mathfrak{A}^+)_1$, then E is extreme on $(\mathfrak{A}(E)^+)_1$ where $\mathfrak{A}(E)$ is the commutative C^*-subalgebra of $\mathfrak{A}$ generated by E. Passing to the algebra of functions representing $\mathfrak{A}(E)$ (see Theorem 4.4.3), let f be the function representing E (so that $0 \le f \le 1$). If f takes a value different from 0 or 1 at some point, there is a non-zero function h in the algebra such that $0 \le f - h \le 1$ and $0 \le f + h \le 1$. If H_1 and H_2 in $\mathfrak{A}(E)$ correspond to $f - h$ and $f + h$, respectively, then $H_1 \ne H_2$, H_1 and H_2 are in $(\mathfrak{A}^+)_1$ and $E = \frac{1}{2}(H_1 + H_2)$, contradicting the assumption that E is an extreme point of $(\mathfrak{A}^+)_1$. Thus f has range in $\{0, 1\}$ and E is a projection in $\mathfrak{A}$.

If E is a projection in $\mathfrak{A}$, and $E = \frac{1}{2}(H_1 + H_2)$ with H_1 and H_2 in $(\mathfrak{A}^+)_1$, then $\langle H_1 x, x\rangle = 0$ when $\langle Ex, x\rangle = 0$. Thus $H_1 x = 0$ (similarly, $H_2 x = 0$) when $Ex = 0$. With x a unit vector in the range of E,

$$1 = \langle Ex, x\rangle = \tfrac{1}{2}[\langle H_1 x, x\rangle + \langle H_2 x, x\rangle].$$

Since $0 \le \langle H_1 x, x\rangle \le 1$ and $0 \le \langle H_2 x, x\rangle \le 1$, we conclude that $\langle H_1 x, x\rangle = \langle H_2 x, x\rangle = 1$. From the Cauchy–Schwarz inequality $H_1 x = H_2 x = x$. Thus $H_1 = H_2 = E$; and E is an extreme point of $(\mathfrak{A}^+)_1$.

If A is an extreme point of $(\mathfrak{A}_{\mathrm{h}})_1$ and $A = A^+ - A^-$, where A^+ and A^- are in $(\mathfrak{A}^+)_1$ and $A^+A^- = 0$ (see Proposition 4.2.3); then A^+ is an extreme point of $(\mathfrak{A}^+)_1$. Indeed, if $A^+ = \frac{1}{2}(H_1 + H_2)$ with H_1 and H_2 in $(\mathfrak{A}^+)_1$, then $A = \frac{1}{2}[H_1 - A^- + H_2 - A^-]$, where $H_1 - A^-$ and $H_2 - A^-$ are in

$(\mathfrak{A}_h)_1$ (since $0 \le H_1 \le 2A^+$, $0 \le H_2 \le 2A^+$, $H_1A^- = H_2A^- = 0$, so that $\|H_1 - A^-\| = \max\{\|H_1\|, \|A^-\|\} \le 1$). As A is an extreme point of $(\mathfrak{A}_h)_1$, $H_1 - A^- = H_2 - A^-$ and $H_1 = H_2$. Thus A^+ and, similarly, A^- are (mutually) orthogonal projections E and F in $\mathfrak{A}$. If $I - E - F\ (= G) \ne 0$, then $A = \frac{1}{2}[E - F + G + E - F - G]$ and $E - F + G \ne E - F - G$, which contradicts the assumption that A is an extreme point of $(\mathfrak{A}_h)_1$. Thus $A = 2E - I$. From Theorem 7.3.1, $2E - I$ is extreme on $(\mathfrak{A})_1$ hence on $(\mathfrak{A}_h)_1$. ■

7.4.7. THEOREM. *If ρ is a normal hermitian functional on the von Neumann algebra $\mathscr{R}$, then ρ attains its norm at an element $2E - I$ of the unit ball $(\mathscr{R})_1$ of $\mathscr{R}$, where E is a projection in $\mathscr{R}$. Moreover, $\rho = \rho^+ - \rho^-$ where $\rho^+(A) = \rho(EAE)$ and $\rho^-(A) = -\rho((I - E)A(I - E))$. The linear functionals ρ^+ and ρ^- are positive and normal, and $\|\rho\| = \|\rho^+\| + \|\rho^-\|$.*

Proof. Since ρ is normal and $(\mathscr{R})_1$ is weak-operator compact, $A \to |\rho(A)|$ assumes its maximum $\|\rho\|$ at some point A of $(\mathscr{R})_1$. Multiplying by an appropriate scalar of modulus 1, we may assume that $\rho(A) = \|\rho\|$. Thus $\rho(A^*) = \overline{\rho(A)} = \|\rho\|$, and $\rho(\frac{1}{2}[A + A^*]) = \|\rho\|$. Since ρ takes its maximum on $(\mathscr{R}_h)_1$ (on a face), there is an extreme point $2E - I$ of $(\mathscr{R}_h)_1$ at which ρ takes the value $\|\rho\|$, where E is a projection in $\mathscr{R}$ (see the preceding proposition).

From Theorem 4.3.6, we know that $\rho = \rho^+ - \rho^-$ with ρ^+ and ρ^- positive linear functions on $\mathscr{R}$ such that $\rho(2E - I) = \|\rho\| = \rho^+(2E - I) - \rho^-(2E - I) = \|\rho^+\| + \|\rho^-\| = \rho^+(I) + \rho^-(I)$. Thus $\rho^+(E) - \rho^-(E) = \rho^+(I)$, and $\rho^+(I - E) = -\rho^-(E)$. Since $I - E \ge 0$, $E \ge 0$, and both ρ^+ and ρ^- are positive functionals; $\rho^+(I - E) = \rho^-(E) = 0$. From Proposition 4.5.1, $0 = \rho^+(A(I - E)) = \rho^+((I - E)A) = \rho^-(AE) = \rho^-(EA)$ for all A in $\mathscr{R}$. Thus $\rho(EAE) = \rho^+(EAE) = \rho^+(A)$ and $\rho((I - E)A(I - E)) = -\rho^-(A)$ for all A in $\mathscr{R}$. It follows that ρ^+ and ρ^- are normal on $\mathscr{R}$. ■

Bibliography: [18, 19]

7.5. Normal weights on von Neumann algebras

Each regular Borel measure on a compact Hausdorff space X gives rise to a state of $C(X)$, namely, integration relative to the measure. Conversely, each state of $C(X)$ gives rise to a regular Borel measure on X (see Remark 1.7.6). Theorem 4.4.3 identifies the abelian C^*-algebras with the algebras of continuous functions on compact Hausdorff spaces. This result gives substance to the view that non-commutative C^*-algebras are "non-commutative continuous function algebras." In this view, a state of a non-commutative

C^*-algebra may be thought of as a "non-commutative integral" and corresponding, in an implicit sense, to a "non-commutative measure." In Example 5.1.6, we noted that the algebra of multiplications by essentially bounded measurable functions is an abelian von Neumann algebra. Conversely, each abelian von Neumann algebra is isomorphic to the multiplication algebra of some measure space. (This is proved in Section 9.4 for the case in which the underlying Hilbert space is separable.) Thus if non-commutative C^*-algebras are viewed as non-commutative continuous function algebras, then non-commutative von Neumann algebras are to be viewed as non-commutative algebras of multiplication by essentially bounded measurable functions. The projections in the von Neumann algebra are (multiplication by) the characteristic functions of (implicit) measurable sets. A normal state on the algebra corresponds to integration relative to the (finite) non-commutative measure obtained by restricting the state to the projections in the algebra.

In the case of ("commutative") measure spaces, it is interesting and technically useful to deal with infinite measures on the spaces and the associated integrals. For arbitrary von Neumann algebras, the appropriate extension is a functional defined on positive elements in the algebra (a "weight") taking real non-negative values or $+\infty$ with suitable additivity and continuity properties. In Chapter VIII the "extended trace" on a factor of type II_∞ will provide us with a natural example of such an extended functional. Nonetheless, weights will play, for us, only a peripheral role. For this reason, the discussion of weights we present in this section avoids the deeper technical aspects of the theory. We take, as our starting point, a definition of (normal) weight that incorporates conclusions that could be drawn from more basic assumptions.

7.5.1. Definition. A *weight* on a C^*-algebra $\mathfrak{A}$ is a mapping ρ from the set $\mathfrak{A}^+$ of positive elements in $\mathfrak{A}$ into $[0, \infty]$ such that

$$\rho(H + K) = \rho(H) + \rho(K) \qquad (H, K \in \mathfrak{A}^+),$$

$$\rho(aH) = a \cdot \rho(H) \qquad (a \geq 0,\ H \in \mathfrak{A}^+).$$

We adopt the notation

$$\mathcal{N}_\rho = \{A : A \in \mathfrak{A},\ \rho(A^*A) < \infty\},$$

$$N_\rho = \{A : A \in \mathfrak{A},\ \rho(A^*A) = 0\},$$

$$F_\rho = \{A : A \in \mathfrak{A}^+,\ \rho(A) < \infty\},$$

$$\mathcal{M}_\rho = \text{linear span of } F_\rho.$$

When $N_\rho = (0)$, we say that ρ is *faithful.* If $\mathfrak{A}$ is a von Neumann algebra $\mathscr{R}$ acting on the Hilbert space $\mathscr{H}$, we say that ρ is *semi-finite* when $\mathscr{M}_\rho$ is weak-operator dense in $\mathscr{R}$. We say that ρ is *normal* when there is a family $\{\rho_a : a \in \mathbb{A}\}$ of positive normal *functionals* ρ_a on $\mathscr{R}$ such that $\rho(H) = \sum_{a \in \mathbb{A}} \rho_a(H)$ for each H in $\mathscr{R}^+$. ■

Concerning algebraic manipulations with ∞, we use the following conventions: $\infty + \infty = \infty$; $\infty + a = a + \infty = \infty$; $a \cdot \infty = \infty$; and $0 \cdot \infty = 0$ $(a \in (0, \infty))$. Our primary concern is with normal faithful weights on von Neumann algebras. By virtue of Theorem 7.1.12, such weights are sums of vector states. Much of the ensuing discussion is valid for general weights on C^*-algebras. We phrase that portion of the discussion in those terms.

If ρ is a state on $\mathfrak{A}$, then $\rho|\mathfrak{A}^+$ is a weight on $\mathfrak{A}$. For such a weight, $\mathscr{N}_\rho = \mathfrak{A}$ and $N_\rho = \mathscr{L}_\rho$, the left-kernel of ρ (see Proposition 4.5.1). It is clear, for a general weight ρ, that $H_0 \in F_\rho$ if $0 \le H_0 \le H$ and $H \in F_\rho$ (for, then, $H = H - H_0 + H_0$, so that $\rho(H_0) + \rho(H - H_0) = \rho(H) < \infty$). For these more general weights ρ, the following lemma gives important information about $\mathscr{N}_\rho$, N_ρ, and $\mathscr{M}_\rho$.

7.5.2. Lemma. *If $\mathfrak{A}$ is a C^*-algebra and ρ is a weight on $\mathfrak{A}$, then*

(i) *$\mathscr{N}_\rho$ and N_ρ are left ideals in $\mathfrak{A}$, $\mathscr{M}_\rho$ is the linear span, $\mathscr{N}_\rho^*\mathscr{N}_\rho$, of $\{A^*B : A, B \in \mathscr{N}_\rho\}$, and $\mathscr{M}_\rho \subseteq \mathscr{N}_\rho \cap \mathscr{N}_\rho^*$;*

(ii) *$\rho|F_\rho$ extends, uniquely, to a positive hermitian linear functional on $\mathscr{M}_\rho$; and $\mathscr{M}_\rho^+ = F_\rho$.*

Proof. (i) With A and B in $\mathfrak{A}$,

$$(A + B)^*(A + B) + (A - B)^*(A - B) = 2(A^*A + B^*B);$$

so that

$$(A + B)^*(A + B) \le 2(A^*A + B^*B)$$

and

$$\rho[(A + B)^*(A + B)] \le 2\rho(A^*A + B^*B).$$

Thus, sums of elements in $\mathscr{N}_\rho$ and N_ρ are, again, in $\mathscr{N}_\rho$ and N_ρ. Moreover,

$$A^*B^*BA \le \|B\|^2 A^*A;$$

so that

$$\rho(A^*B^*BA) \le \|B\|^2 \rho(A^*A),$$

and both $\mathscr{N}_\rho$ and N_ρ are left ideals in $\mathfrak{A}$.

When $H \in F_\rho$, we have $H^{1/2} \in \mathcal{N}_\rho$; so that $H = (H^{1/2})^* H^{1/2} \in \mathcal{N}_\rho^* \mathcal{N}_\rho$. Thus

$$\text{linear span of } F_\rho = \mathcal{M}_\rho \subseteq \mathcal{N}_\rho^* \mathcal{N}_\rho.$$

If A and B are in $\mathcal{N}_\rho$, a form of the polarization identity (see 2.1(7)) yields

$$A^*B = \tfrac{1}{4}[(B + A)^*(B + A) - (B - A)^*(B - A) + i(B + iA)^*(B + iA) - i(B - iA)^*(B - iA)]$$

so that $A^*B \in \mathcal{M}_\rho$. Thus $\mathcal{N}_\rho^*\mathcal{N}_\rho \subseteq \mathcal{M}_\rho$; and $\mathcal{N}_\rho^*\mathcal{N}_\rho = \mathcal{M}_\rho$. At the same time, $A^*B \in \mathcal{N}_\rho \cap \mathcal{N}_\rho^*$, since $\mathcal{N}_\rho$ is a left ideal and $\mathcal{N}_\rho^*$ is a right ideal. Hence $\mathcal{M}_\rho = \mathcal{N}_\rho^*\mathcal{N}_\rho \subseteq \mathcal{N}_\rho \cap \mathcal{N}_\rho^*$.

(ii) For each A in $\mathcal{M}_\rho$,

$$A = A_1 - A_2 + iA_3 - iA_4$$

where A_1, A_2, A_3, A_4 are in F_ρ. If we also have

$$A = B_1 - B_2 + iB_3 - iB_4,$$

with B_1, B_2, B_3, B_4 in F_ρ, then $A_1 - A_2 = B_1 - B_2$ and $A_3 - A_4 = B_3 - B_4$. Thus

$$\rho(A_1) + \rho(B_2) = \rho(B_1) + \rho(A_2), \qquad \rho(A_3) + \rho(B_4) = \rho(B_3) + \rho(A_4);$$

and

$$\rho(A_1) - \rho(A_2) + i\rho(A_3) - i\rho(A_4) = \rho(B_1) - \rho(B_2) + i\rho(B_3) - i\rho(B_4).$$

Letting $\rho(A)$ be $\rho(A_1) - \rho(A_2) + i\rho(A_3) - i\rho(A_4)$ defines (unambiguously) a hermitian linear functional (we denote again by ρ) on $\mathcal{M}_\rho$. If A, above, is in $\mathcal{M}_\rho^+$, then $0 \le A = A_1 - A_2 \le A_1$ with A_1 in F_ρ, whence $A \in F_\rho$ and $\mathcal{M}_\rho^+ = F_\rho$. Thus ρ is positive on $\mathcal{M}_\rho$. ■

From the fact that $\mathcal{M}_\rho = \mathcal{N}_\rho^*\mathcal{N}_\rho$, we have that $\mathcal{M}_\rho$ is a self-adjoint subalgebra of $\mathfrak{A}$. Thus, when $\mathfrak{A}$ is a von Neumann algebra $\mathcal{R}$ and ρ is semi-finite, the unit ball in $\mathcal{M}_\rho$ is both weak- and strong-operator dense in the unit ball of $\mathcal{R}$. Since $\mathcal{M}_\rho$ is contained in $\mathcal{N}_\rho \cap \mathcal{N}_\rho^*$, the same is true of $\mathcal{N}_\rho \cap \mathcal{N}_\rho^*$, $\mathcal{N}_\rho$ and $\mathcal{N}_\rho^*$. If ρ is normal, then $\rho = \sum_{a \in \mathbb{A}} \rho_a$ with $\{\rho_a\}$ a family of positive normal functionals on $\mathcal{R}$. If

$$A = A_1 - A_2 + iA_3 - iA_4 \qquad (A_1, A_2, A_3, A_4 \in F_\rho),$$

then

$$\rho(A) = \sum \rho_a(A_1) - \sum \rho_a(A_2) + i \sum \rho_a(A_3) - i \sum \rho_a(A_4) = \sum \rho_a(A)$$

(in the sense of the unordered summation over the net of finite subsets of $\mathbb{A}$ directed by inclusion described at the end of Section 1.2).

Despite the fact that ρ need not take finite values on $\mathcal{N}_\rho$ (that is, $\mathcal{N}_\rho \nsubseteq \mathcal{M}_\rho$, in general) and that $\mathcal{N}_\rho^* \neq \mathcal{N}_\rho$ (in general), the mapping

$$(A, B) \to \rho(B^*A) : \mathcal{N}_\rho \times \mathcal{N}_\rho \to \mathbb{C}$$

defines a (positive) inner product on $\mathcal{N}_\rho$ (though not *definite*, in general); since $B^*A \in \mathcal{N}_\rho^*\mathcal{N}_\rho = \mathcal{M}_\rho$. The Cauchy–Schwarz inequality is available, so that $\rho(B^*A) = 0$ when at least one of A or B is in N_ρ and both are in $\mathcal{N}_\rho$. (It is not true, generally, as in the case of a state, that $\rho(A) = 0$ for A in N_ρ even though A may lie in F_ρ—see Exercise 7.6.48). It follows, now, that the formula

$$\langle A + N_\rho, B + N_\rho \rangle = \rho(B^*A) \qquad (A, B \in \mathcal{N}_\rho)$$

defines a definite inner product on the (linear) quotient space $\mathcal{N}_\rho/N_\rho$. If we denote the completion of $\mathcal{N}_\rho/N_\rho$ relative to the norm associated with this inner product by $\mathscr{H}_\rho$ and the mapping

$$B + N_\rho \to AB + N_\rho : \mathcal{N}_\rho/N_\rho \to \mathcal{N}_\rho/N_\rho$$

by $\pi_\rho(A)'$, then the argument of the GNS construction (see Theorem 4.5.2) yields the fact that $\pi_\rho(A)'$ has a (unique) bounded extension $\pi_\rho(A)$ to $\mathscr{H}_\rho$ and that π_ρ is a representation of $\mathfrak{A}$ on $\mathscr{H}_\rho$.

If ρ is a faithful weight on the von Neumann algebra $\mathscr{R}$, $N_\rho = (0)$. If, in addition, ρ is semi-finite, then, as noted, $\mathcal{N}_\rho \cap \mathcal{N}_\rho^*$ and $\mathcal{N}_\rho$ are weak-operator dense in $\mathscr{R}$. If $\pi_\rho(A) = 0$, in this case, then $AB \in N_\rho = (0)$ for all B in $\mathcal{N}_\rho$. Since left multiplication by A is weak-operator continuous and $\mathcal{N}_\rho$ is dense in $\mathscr{R}$, we have $0 = A \cdot I = A$, and π_ρ is faithful.

If $B \in \mathcal{N}_\rho$, then $A \to \rho_a(B^*AB)$ $(= \rho_a'(A))$ is a positive normal functional on $\mathscr{R}$, when $\{\rho_a : a \in \mathbb{A}\}$ is a family of such functionals on $\mathscr{R}$ such that $\rho = \sum \rho_a$, and

$$\langle \pi_\rho(A)(B + N_\rho), B + N_\rho \rangle = \rho(B^*AB) = \sum \rho_a'(A),$$

for each A in $\mathscr{R}$. Choosing A to be I, we have, in particular, that $\sum \|\rho_a'\|$ converges. Thus, given a positive ε, there is a finite subset $\mathbb{A}_0$ of $\mathbb{A}$ such that $\sum_{a \notin \mathbb{A}_0} \|\rho_a'\| < \varepsilon/2$. Since $\sum_{a \in \mathbb{A}_0} \rho_a'$ is normal, we can choose a weak-operator open neighborhood of 0 in $(\mathscr{R})_1$ such that, for A in this neighborhood $|\sum_{a \in \mathbb{A}_0} \rho_a'(A)| < \varepsilon/2$. Thus, for such A, $|\sum_{a \in \mathbb{A}} \rho_a'(A)| < \varepsilon$. It follows from Lemma 7.1.3 and Theorem 7.1.12 that $A \to \langle \pi_\rho(A)(B + N_\rho), B + N_\rho \rangle$ is normal on $\mathscr{R}$ for each B in $\mathcal{N}_\rho$. Since $\{B + N_\rho : B \in \mathcal{N}_\rho\}$ is dense in $\mathscr{H}_\rho$, the representation π_ρ is normal and $\pi_\rho(\mathscr{R})$ is a von Neumann algebra (see Proposition 7.1.15).

We summarize the preceding discussion in the following theorem.

7.5.3. THEOREM. *If ρ is a weight on a C*-algebra $\mathfrak{A}$, then*

$$\langle A + N_\rho, B + N_\rho \rangle = \rho(B^*A) \qquad (A, B \in \mathcal{N}_\rho) \tag{1}$$

defines a definite inner product on the (linear) quotient space $\mathcal{N}_\rho/N_\rho$; and

$$\pi_\rho(A)'(B + N_\rho) = AB + N_\rho \qquad (A \in \mathfrak{A}, B \in \mathcal{N}_\rho)$$

defines a bounded linear operator $\pi_\rho(A)'$ on $\mathcal{N}_\rho/N_\rho$. If $\pi_\rho(A)$ is the (unique) bounded extension of $\pi_\rho(A)'$ to the completion $\mathcal{H}_\rho$ of $\mathcal{N}_\rho/N_\rho$ relative to the norm defined by the inner product (1), *then π_ρ is a representation of $\mathfrak{A}$ on $\mathcal{H}_\rho$. If ρ is a normal weight on a von Neumann algebra $\mathcal{R}$, then π_ρ is a normal representation of $\mathcal{R}$ and $\pi_\rho(\mathcal{R})$ is a von Neumann algebra. If ρ is faithful and semi-finite, as well, then π_ρ is faithful.*

Bibliography: [69]

7.6. Exercises

7.6.1. Let $\mathcal{R}$ be a von Neumann algebra acting on a Hilbert space $\mathcal{H}$, and let ω be a normal state of $\mathcal{R}$. Show that

(i) the support of ω has range $[\mathcal{R}'x_j : j = 1, 2, \ldots]$, where $\omega = \sum_{j=1}^{\infty} \omega_{x_j} | \mathcal{R}$;
(ii) the support of ω is the union of the projections E_j, $j \in \mathbb{N}$, where E_j has range $[\mathcal{R}'x_j]$.

7.6.2. Let ω be a normal positive linear functional on a von Neumann algebra $\mathcal{R}$ acting on a Hilbert space $\mathcal{H}$. Let E be the support of ω and F' be a projection in $\mathcal{R}'$ such that $E \le C_{F'}$. Show that

(i) there is a normal positive linear functional ω_0 on $\mathcal{R}F'$ such that $\omega_0(AF') = \omega(A)$ for each A in $\mathcal{R}$;
(ii) the support of ω_0 is EF'.

7.6.3. Adopt the notation of Exercise 7.6.2.

(i) Must EF' be cyclic in $\mathcal{R}F'$ under $F'\mathcal{R}'F'$ when E is cyclic in $\mathcal{R}$ under $\mathcal{R}'$? Proof? Counterexample?
(ii) Must ω_0 be a vector state of $\mathcal{R}F'$ when ω is a vector state of $\mathcal{R}$? Proof? Counterexample?

7.6.4. Let $\mathscr{A}$ be an abelian von Neumann algebra acting on a Hilbert space $\mathscr{H}$, and let ω be a normal state of $\mathscr{A}$. Show that

(i) ω is a vector state of $\mathscr{A}$;
(ii) the weak-operator topology and the ultraweak topology coincide on $\mathscr{A}$.

7.6.5. (i) Suppose $(S, \mathscr{S}, m)$ is a σ-finite measure space, $\{g_j\}$ is a sequence of functions in $L_1 (= L_1(S, m))$, and $\sum_{j=1}^{\infty} \|g_j\|_1 < \infty$. Show that the series $\sum_{j=1}^{\infty} g_j$ converges almost everywhere to a function g in L_1, and

$$\sum_{j=1}^{\infty} \int f g_j \, dm = \int f g \, dm$$

for each f in L_∞ $(= L_\infty(S, m))$.
(ii) How is the result of (i) related to Exercise 7.6.4?

7.6.6. Show that a normal state of a von Neumann algebra is faithful (see Exercise 4.6.15) if and only if its support is I.

7.6.7. Suppose that, for $j = 1, 2$, ρ_j is a faithful normal state of a von Neumann algebra $\mathscr{R}_j$, and $\mathfrak{A}_j$ is a weak-operator dense self-adjoint subalgebra of $\mathscr{R}_j$. Let φ be a * isomorphism from $\mathfrak{A}_1$ onto $\mathfrak{A}_2$, such that $\rho_1(A) = \rho_2(\varphi(A))$ for each A in $\mathfrak{A}_1$. Show that φ extends to a * isomorphism ψ from $\mathscr{R}_1$ onto $\mathscr{R}_2$, such that $\rho_1(R) = \rho_2(\psi(R))$ for each R in $\mathscr{R}_1$. [*Hint*. Consider the representation φ_j of $\mathscr{R}_j$ engendered by ρ_j, and use Proposition 7.1.15.]

7.6.8. (i) Find an example of a factor (acting on a separable Hilbert space) for which there is no separating vector.
(ii) Construct a faithful normal state for the factor exhibited in (i).
(iii) Conclude that a normal state of a factor need not be a vector state.

7.6.9. Let $\{e_n : n \in \mathbb{N}\}$ be an orthonormal basis for the separable Hilbert space $\mathscr{H}$. Define $\omega(A)$ to be $\sum_{n=1}^{\infty} 2^{-n}\langle Ae_n, e_n\rangle$ for each A in $\mathscr{B}(\mathscr{H})$. Show that

(i) ω is a faithful normal state of $\mathscr{B}(\mathscr{H})$;
(ii) ω is not weak-operator continuous on $\mathscr{B}(\mathscr{H})$;
(iii) the weak-operator topology is strictly coarser than the ultraweak topology on $\mathscr{B}(\mathscr{H})$.

7.6.10. Let $\mathscr{R}$ be a von Neumann algebra acting on a Hilbert space $\mathscr{H}$, $\{Q_a\}_{a \in \mathbb{A}}$ be an orthogonal family of central projections in $\mathscr{R}$ with sum I, and

ω be a normal state of $\mathscr{R}$. Suppose that $\omega \mid \mathscr{R}Q_a = \sum_{j=1}^{n_a} \omega_{x_{ja}} \mid \mathscr{R}Q_a$, where $n_a \le m$, for all a in $\mathbb{A}$. Show that $\omega = \sum_{j=1}^{m} \omega_{x_j} \mid \mathscr{R}$.

7.6.11. Find an example of a von Neumann algebra $\mathscr{R}$ acting on a Hilbert space $\mathscr{H}$ and a net of cyclic projections in $\mathscr{R}$ converging in the strong-operator topology to a projection in $\mathscr{R}$ that is not cyclic.

7.6.12. Let $\mathscr{R}$ be a countably decomposable von Neumann algebra acting on a Hilbert space $\mathscr{H}$. Suppose that E is the strong-operator limit of a net of cyclic projections in $\mathscr{R}$. Show that E is a cyclic projection in $\mathscr{R}$.

7.6.13. Show that a projection in a von Neumann algebra is the support of a normal state if and only if it is countably decomposable.

7.6.14. Show that each normal state of a von Neumann algebra $\mathscr{R}$ is a vector state of $\mathscr{R}$ if and only if each countably decomposable projection in $\mathscr{R}$ is cyclic.

7.6.15. Suppose that $\mathfrak{A}$ is a C^*-algebra acting on a Hilbert space $\mathscr{H}$. Let δ be a derivation of $\mathfrak{A}$, and recall that δ is norm continuous (Exercise 4.6.65(v)). By using Lemma 7.1.3, show that δ is weak-operator continuous on bounded subsets of $\mathfrak{A}$. Deduce that δ extends (uniquely) to a derivation $\bar{\delta}$ of $\mathfrak{A}^-$, and that $\|\bar{\delta}\| = \|\delta\|$.

7.6.16. Let η be a linear mapping of a C^*-algebra $\mathfrak{A}$ into a C^*-algebra $\mathscr{B}$. Suppose $\eta(I) = I$ and $\|\eta(T)\| = \|T\|$ for each normal element T in $\mathfrak{A}$. Show that

(i) $\eta(A)$ is self-adjoint if A is a self-adjoint element of $\mathfrak{A}$ [*Hint*. Consider $\rho \circ \eta$ for an appropriate state ρ of $\mathscr{B}$.];
(ii) $\eta(B^*) = \eta(B)^*$ for each B in $\mathfrak{A}$.

7.6.17. Let η be a linear isometry of one C^*-algebra $\mathfrak{A}$ onto another C^*-algebra $\mathscr{B}$.

(i) Show that U is a partial isometry in $\mathscr{B}$, where $U = \eta(I)$.
(ii) Let $\eta_0(T)$ be $U^*\eta(T)E$ for each T in $\mathfrak{A}$, where $E = U^*U$. Show that $\|\eta_0(A)\| = \|A\|$ for each normal element A in $\mathfrak{A}$. [*Hint*. If $1 \in \mathrm{sp}(A)$, there is a state ρ of $\mathscr{B}$ such that $\rho([U + \eta(A)]^*[U + \eta(A)]) = 4$.]
(iii) Show that η_0 is hermitian. [*Hint*. Use Exercise 7.6.16.]
(iv) Show that U is a unitary operator in $\mathscr{B}$. [*Hint*. Consider T in $\mathfrak{A}$ such that $\eta(T) = I - E$.]
(v) Show that $\eta(V)$ is unitary when V is a unitary element of $\mathfrak{A}$.

7.6.18. With the notation and assumptions of Exercise 7.6.17, define $\eta'(T)$ to be $U^*\eta(T)$ for each T in $\mathfrak{A}$. Show that

(i) $\eta'(H^2) = \eta'(H)^2$ for each self-adjoint H in $\mathfrak{A}$ [*Hint*. Consider $\eta'(\exp itH)$ in series form and use (v) of Exercise 7.6.17.];

(ii) $\eta'(A^2) = \eta'(A)^2$ for each A in $\mathfrak{A}$;

(iii) $\eta = U\eta'$ and $\eta'(AB + BA) = \eta'(A)\eta'(B) + \eta'(B)\eta'(A)$ for all A and B in $\mathfrak{A}$. (We call η' a Jordan * isomorphism of $\mathfrak{A}$ onto $\mathscr{B}$. Compare Exercise 10.5.28.)

7.6.19. Let ω and ω' be normal positive linear functionals on a von Neumann algebra $\mathscr{R}$. Suppose $\omega - \omega' = \omega_1 - \omega_2$, where ω_1 and ω_2 are normal positive linear functionals on $\mathscr{R}$ with orthogonal supports M and N, respectively. Let G be a projection in $\mathscr{R}$ such that $\omega(GTG) = \omega(T)$ and $\omega'(GTG) = \omega'(T)$ for each T in $\mathscr{R}$. Show that

(i) ω_1 and ω_2 are the *unique* positive linear functionals on $\mathscr{R}$ that are normal, have orthogonal supports, and whose difference is $\omega - \omega'$;

(ii) $M + N \leq G$.

7.6.20. Let $\mathscr{R}$ be a von Neumann algebra with center $\mathscr{C}$ acting on a Hilbert space $\mathscr{H}$, and let x be a vector in $\mathscr{H}$. Show that

(i) the range of the support of $\omega_x | \mathscr{C}$ is $[AA'x : A \in \mathscr{R}, A' \in \mathscr{R}']$;

(ii) the support of $\omega_x | \mathscr{C}$ is C_E where E is the support of $\omega_x | \mathscr{R}$.

7.6.21. Let $\mathscr{R}$ be a von Neumann algebra acting on a Hilbert space $\mathscr{H}$ and $\{x_n\}$ be a sequence of unit vectors in $\mathscr{H}$ tending in norm to x_0. Let E_n be the support of $\omega_{x_n} | \mathscr{R}$ and P_n be the support of $\omega_{x_n} | \mathscr{C}$, where $\mathscr{C}$ is the center of $\mathscr{R}$. Show that, in the strong-operator topology,

(i) $\{E_n E_0\}$ tends to E_0;

(ii) $\{P_n P_0\}$ tends to P_0.

7.6.22. Let $\mathscr{R}$ be a von Neumann algebra acting on a Hilbert space $\mathscr{H}$, $\{x_n\}$ be a sequence of unit vectors in $\mathscr{H}$ tending (in norm) to x_0, and E_n be the support of $\omega_{x_n} | \mathscr{R}$. Does $\{E_n\}$ tend to E_0 in the strong-operator topology? Proof? Counterexample?

7.6.23. Let $\mathscr{R}$ be a von Neumann algebra acting on a Hilbert space $\mathscr{H}$, and let x and y be vectors in $\mathscr{H}$ such that $\omega_x | \mathscr{R} = \omega_y | \mathscr{R}$. Show that

(i) the mapping $Ax \to Ay$ extends to an isometry of $[\mathscr{R}x]$ onto $[\mathscr{R}y]$;

(ii) there is a unique partial isometry V' in $\mathscr{R}'$ with initial space $[\mathscr{R}x]$ and final space $[\mathscr{R}y]$ such that $V'Ax = Ay$ for each A in $\mathscr{R}$.

7.6.24. Let ω and ω' be normal positive linear functionals on a von Neumann algebra $\mathscr{R}$ acting on a Hilbert space $\mathscr{H}$. Suppose that the union of the support projections of ω and ω' is a cyclic projection G in $\mathscr{R}$. Show that

(i) there are normal positive linear functionals ω_1 and ω_2 on $\mathscr{R}$ with orthogonal supports contained in G such that

$$\omega - \omega' = \omega_1 - \omega_2, \qquad \|\omega - \omega'\| = \|\omega_1\| + \|\omega_2\|$$

and that there is a vector z in $\mathscr{H}$ such that $\omega_z | \mathscr{R} = \omega + \omega_2 = \omega' + \omega_1$;

(ii) there are operators H' and K' in $\mathscr{R}'$ such that $0 \le H' \le I$, $0 \le K' \le I$, $\omega = \omega_{H'z} | \mathscr{R}$, and $\omega' = \omega_{K'z} | \mathscr{R}$;

(iii) $\|z - H'z\|^2 \le \|\omega_2\|$, $\|z - K'z\|^2 \le \|\omega_1\|$;

(iv) there are vectors x and x' in $\mathscr{H}$ such that $\omega_x | \mathscr{R} = \omega$, $\omega_{x'} | \mathscr{R} = \omega'$, and

$$\|x - x'\| \le [2\|\omega - \omega'\|]^{1/2}.$$

7.6.25. Let ω be a normal positive linear functional on a von Neumann algebra $\mathscr{R}$ acting on a Hilbert space $\mathscr{H}$, and let v be a vector in $\mathscr{H}$. Suppose that the union of the supports of ω and $\omega_v | \mathscr{R}$ is cyclic in $\mathscr{R}$. Show that

(i) there are vectors x and x' in $\mathscr{H}$ such that

$$\omega_x | \mathscr{R} = \omega, \qquad \omega_{x'} | \mathscr{R} = \omega_v | \mathscr{R}, \qquad \|x - x'\| \le [2\|\omega - \omega_v | \mathscr{R}\|]^{1/2};$$

(ii) there is a partial isometry V' in $\mathscr{R}'$ with initial space $[\mathscr{R}x']$ and final space $[\mathscr{R}v]$ such that $V'x' = v$ and that if W' is a partial isometry in $\mathscr{R}'$ whose initial space contains $[\mathscr{R}x]$ and $[\mathscr{R}x']$ (in particular if W' is an isometry) and W' agrees on $[\mathscr{R}x']$ with V', then

$$(*) \qquad \omega = \omega_u | \mathscr{R}, \qquad \|u - v\| \le [2\|\omega - \omega_v | \mathscr{R}\|]^{1/2},$$

where $u = W'x$.

7.6.26. With the notation and assumptions of Exercise 7.6.25, let E' be the projection with range $[\mathscr{R}v]$ and P be the central projection in $\mathscr{R}$ (whose existence and uniqueness is guaranteed by Proposition 6.3.7) such that $P \le C_{E'}$, PE' is either properly infinite or 0 (in which case P is 0), and $(I - P)E'$ is finite. Show that there is an isometry U' in $\mathscr{R}'$ such that $U' | [\mathscr{R}x'] = V' | [\mathscr{R}x']$ and conclude that there is a vector u satisfying (*) if either of the following two conditions holds:

(i) $PE' = 0$ [*Hint*. Use Exercise 6.9.10.];

(ii) $\mathscr{R}'P$ is countably decomposable and both PE' and $P(I - E')$ are properly infinite with central carrier P.

7.6.27. With the assumptions of Exercise 7.6.25 and the notation of Exercise 7.6.26, show that there is a vector u satisfying

$$\omega_u \mid \mathscr{R} = \omega, \qquad \|u - v\| \leq 2\,\|\omega - \omega_v \mid \mathscr{R}\|^{1/2}$$

when $\mathscr{R}'$ is countably decomposable. [*Hint*. In case PE' is properly infinite and $P(I - E')$ is not, express PE' as a countably infinite sum of projections in $\mathscr{R}'$ equivalent to PE' and use these projections to replace v by a close approximant v' such that PG' and $P(I - G')$ are properly infinite, where G' has range $[\mathscr{R}v']$. Remember to prove that the union of the supports of ω and $\omega_{v'} \mid \mathscr{R}$ is cyclic in applying Exercise 7.6.26(ii).]

7.6.28. Let ω be a normal positive linear functional on a von Neumann algebra $\mathscr{R}$ acting on a Hilbert space $\mathscr{H}$, and let v be a vector in $\mathscr{H}$. Suppose that the union of the supports of ω and $\omega_v \mid \mathscr{R}$ is cyclic in $\mathscr{R}$. Show that there is a vector u in $\mathscr{H}$ such that

$$(**) \qquad \omega_u \mid \mathscr{R} = \omega, \qquad \|u - v\| \leq 2\,\|\omega - \omega_v \mid \mathscr{R}\|^{1/2}.$$

[*Hint*. Reduce to the case where $\mathscr{R}'$ is countably decomposable and apply Exercise 7.6.27.]

7.6.29. Let ρ be an ultraweakly continuous linear functional on a von Neumann algebra $\mathscr{R}$ acting on a Hilbert space $\mathscr{H}$. Let U_0 be *any* partial isometry in $\mathscr{R}$ such that $|\rho(U_0)| = \|\rho\|$, and let U be aU_0, where $|a| = 1$ and $a\rho(U_0) = |\rho(U_0)|$.

(i) Show that ω is a normal positive linear functional on $\mathscr{R}$, where $\omega(A) = \rho(UA)$ for each A in $\mathscr{R}$. [*Hint*. Use Theorem 4.3.2.]

(ii) Let F be UU^*. Show that for each A in $\mathscr{R}$,

$$\rho(A) = \rho(FA) = \omega(U^*A).$$

[*Hint*. Suppose the contrary and choose A in $\mathscr{R}$ such that $\|(I - F)A\| = 1$ and $\rho((I - F)A) > 0$. Let θ satisfy $0 < \theta < \pi/2$ and $\tan\theta = \rho((I - F)A)/\|\rho\|$, and consider $\rho(U\cos\theta + (I - F)A\sin\theta)$.]

7.6.30. Let $\mathscr{R}$ be a von Neumann algebra and ρ be an ultraweakly continuous linear functional on $\mathscr{R}$. Show that

(i) there are a normal positive linear functional ω_0 on $\mathscr{R}$ and a partial isometry U_0 in $\mathscr{R}$ such that $U_0U_0^*$ $(= E_0)$ is the support of ω_0 and

$$\rho(A) = \omega_0(U_0A), \qquad \omega_0(A) = \rho(U_0^*A) \qquad (A \in \mathscr{R})$$

[*Hint*. Use Theorem 7.3.2.];

(ii) $U_0 = U_1$ and $\omega_0 = \omega_1$ when U_1 is a partial isometry in $\mathscr{R}$ and ω_1 is a normal positive linear functional on $\mathscr{R}$ with support $U_1 U_1^*$ $(= E_1)$ such that $\rho(A) = \omega_1(U_1 A)$. [*Hint*. Show that $\|\omega_1\| = \|\rho\|$ and reduce to the case where $\|\rho\| = 1$. Prove that $1 = \omega_1(E_1) = \omega_0(U_0 U_1^*) = \langle U_1^* y, U_0^* y \rangle$, where $\mathscr{R}$ acting on $\mathscr{H}$ is the universal normal representation of $\mathscr{R}$ and y is a (unit) vector in $\mathscr{H}$ such that $\omega_0 = \omega_y | \mathscr{R}$. Use the equality clause of the Cauchy–Schwarz inequality to conclude that $U_0 U_1^* y = y$. Deduce that $E_0 = E_1$ and $U_0 U_1^* = E_0$.]

7.6.31. Let E be a projection in a von Neumann algebra, and let P be the union of all projections in $\mathscr{R}$ equivalent to E.

(i) Show that $P = C_E$.
(ii) Show that the support of $\omega_{x_0} | \mathscr{R}$ is a central projection when x_0 is a trace vector for $\mathscr{R}$.

7.6.32. Let x_0 be a trace vector for a von Neumann algebra $\mathscr{R}$ acting on a Hilbert space $\mathscr{H}$. Suppose that $\omega_{x_0} | \mathscr{C}$ has support I, where $\mathscr{C}$ is the center of $\mathscr{R}$. Show that x_0 is separating for $\mathscr{R}$.

7.6.33. Let $\mathfrak{A}$ be a C^*-algebra acting on a Hilbert space $\mathscr{H}$, and let $\{\omega_{x_n} | \mathfrak{A}\}$ be a sequence of vector states of $\mathfrak{A}$ tending in norm to ρ. Show that ρ is a vector state of $\mathfrak{A}$.

7.6.34. Let $\{\rho_n\}$ be a sequence of states of a C^*-algebra $\mathfrak{A}$ converging in norm to a state ρ_0. Let π_n be the GNS representation corresponding to ρ_n. Suppose that $\pi_1, \pi_2, \ldots$ are equivalent to a single representation π of $\mathfrak{A}$ on a Hilbert space $\mathscr{H}$.

(i) Show that π_0 is equivalent to the representation $A \to \pi(A)E'$ of $\mathfrak{A}$ on $E'(\mathscr{H})$ for some projection E' in $\pi(\mathfrak{A})'$.
(ii) Find an example in which π_0 is not equivalent to the representation π.

7.6.35. Let $\mathfrak{A}$ be a C^*-algebra acting on a Hilbert space $\mathscr{H}$. Suppose that each increasing net of operators in $\mathfrak{A}$ that is bounded above has its strong-operator limit in $\mathfrak{A}$. Show that

(i) each decreasing net of operators in $\mathfrak{A}$ that is bounded below has its strong-operator limit in $\mathfrak{A}$;
(ii) the range projection of each operator in $\mathfrak{A}$ lies in $\mathfrak{A}$;
(iii) the union and intersection of each finite set of projections in $\mathfrak{A}$ lie in $\mathfrak{A}$;

(iv) the union and intersection of an arbitrary set of projections in $\mathfrak{A}$ lie in $\mathfrak{A}$;

(v) $E \in \mathfrak{A}$, where E is a cyclic projection in $\mathfrak{A}^-$ with generating vector x, provided that for each vector y in $(I - E)(\mathscr{H})$ there is a self-adjoint A_y in $\mathfrak{A}$ such that $A_y x = x$ and $A_y y = 0$;

(vi) $\mathfrak{A}^- = \mathfrak{A}$ if each cyclic projection in $\mathfrak{A}^-$ lies in $\mathfrak{A}$.

7.6.36. Let $\mathfrak{A}$ and $\mathscr{H}$ be as in Exercise 7.6.35. Suppose E is a cyclic projection in $\mathfrak{A}^-$ and x is a unit generating vector for $E(\mathscr{H})$ under $\mathfrak{A}'$. With y a unit vector in $(I - E)(\mathscr{H})$, show that

(i) there is a sequence $\{A_n\}$ in $(\mathfrak{A}_h)_1$ such that $A_n x \to x$, $A_n y \to 0$, $\|(A_n - A_{n-1})^+ x\| < 2^{1-n}$, and $\|(A_n - A_{n-1})^+ y\| < 2^{1-n}$, where $A_0 = 0$;

(ii) $\{T_n\}$ is a bounded monotone decreasing sequence of positive elements of $\mathfrak{A}$, where $T_n = (I + \sum_{k=1}^{n} (A_k - A_{k-1})^+)^{-1}$, and $T^{1/2}(\sum_{k=1}^{n} (A_k - A_{k-1})^+) T^{1/2} \le I$ for each n, where T is the strong-operator limit of $\{T_n\}$ (in $\mathfrak{A}$);

(iii) for each j in $\{1, \ldots, n\}$, $\{T^{1/2}(\sum_{k=j}^{n} (A_k - A_{k-1})^+) T^{1/2}\}$ is monotone increasing with n, bounded above by I, and if C_j is its strong-operator limit, then $0 \le C_j \le I$, $\{C_j\}$ is decreasing,

$$T^{1/2}\left(\sum_{k=1}^{n} (A_k - A_{k-1})^+\right) T^{1/2} + C_{n+1} = C_1$$

and

$$T^{1/2} A_n T^{1/2} + C_{n+1} = T^{1/2}\left(\sum_{k=1}^{n} - (A_k - A_{k-1})^-\right) T^{1/2} + C_1;$$

(iv) $\{T^{1/2} A_n T^{1/2} + C_{n+1}\}$ is monotone decreasing and bounded and $T^{1/2} A T^{1/2} \in \mathfrak{A}$, where A is a weak-operator limit point of $\{A_n\}$;

(v) $R(T) \in \mathfrak{A}$, $R(T)x = x$, $R(T)y = y$;

(vi) each maximal abelian (self-adjoint) subalgebra of $\mathfrak{A}$ is weak-operator closed. [*Hint*. Note that $AT \in \mathfrak{A}$ if $\mathfrak{A}$ is abelian, and apply (i)–(v) and Exercise 7.6.35 to the subalgebra.]

7.6.37. With the notation and assumptions of Exercise 7.6.36, show that

(i) MAN lies in $\mathfrak{A}$, where M and N are spectral projections for T corresponding to bounded intervals with positive left endpoints;

(ii) $M_m AF$ and FAM_m are in $\mathfrak{A}$, where $F = R(T)$ and $\{M_m\}$ is a sequence of spectral projections for T corresponding to bounded intervals with positive left endpoints such that $\sum_m M_m = F$ [*Hint*. Consider $(M_m A M_n + M_n)(M_n A M_m + M_n)$ and suitable monotone limits.];

(iii) $FAFAF \in \mathfrak{A}$;
(iv) $FAFAFx = x$ and $FAFAFy = 0$;
(v) $\mathfrak{A} = \mathfrak{A}^{-}$.

7.6.38. Let $\mathfrak{A}$ be a C^*-algebra with the property that each bounded monotone-increasing net in $\mathfrak{A}$ has a least upper bound in $\mathfrak{A}$ and suppose $\mathfrak{A}$ has a separating family of states (if each state of the family is 0 at some positive A in $\mathfrak{A}$, then A is 0) whose limits on such nets are their values at the least upper bounds. We call such states of $\mathfrak{A}$ *normal* and refer to a C^*-algebra satisfying these conditions as a *W^*-algebra*. Show that a C^*-algebra is * isomorphic to a von Neumann algebra if and only if it is a W^*-algebra. [*Hint*. Observe that $A \to T^*AT$ is an order isomorphism of $\mathfrak{A}_h$ onto itself for every invertible T in $\mathfrak{A}$.]

7.6.39. Let $\mathfrak{A}$ be a C^*-algebra acting on a Hilbert space. Suppose that each bounded increasing *sequence* in $\mathfrak{A}$ has its strong-operator limit in $\mathfrak{A}$ and that each orthogonal family of non-zero projections in $\mathfrak{A}$ is countable. Show that $\mathfrak{A} = \mathfrak{A}^{-}$. [*Hint*. Note that the only use of nets (as opposed to sequences) in Exercises 7.6.35–37 is to establish that arbitrary unions of projections in $\mathfrak{A}$ lie in $\mathfrak{A}$ and prove this under the present assumptions.]

7.6.40. Show that a C^*-algebra $\mathfrak{A}$ is * isomorphic to a countably decomposable von Neumann algebra $\mathscr{R}$ if and only if each bounded increasing sequence in $\mathfrak{A}$ has a least upper bound in $\mathfrak{A}$, there is a separating family of (normal) states of $\mathfrak{A}$ whose limits on such a sequence are their values at the least upper bound, and each orthogonal family of non-zero projections in $\mathfrak{A}$ is countable.

7.6.41. Let $\mathfrak{A}$ be a C^*-algebra and $\mathfrak{A}_\#$ be a Banach space such that $\mathfrak{A}$ is (isometrically isomorphic to) the (norm) dual space of $\mathfrak{A}_\#$.

(i) Show that an element A in $\mathfrak{A}$ is a self-adjoint element in the ball $(\mathfrak{A})_r$ of radius r in $\mathfrak{A}$ with center 0 if and only if $\|A + inI\|^2 \leq r^2 + n^2$ for each integer n.

(ii) Show that the set of self-adjoint elements in $(\mathfrak{A})_r$ is weak * closed in $\mathfrak{A}$.

(iii) Show that the set $(\mathfrak{A}^+)_r$ of positive elements in $(\mathfrak{A})_r$ is weak * closed in $\mathfrak{A}$.

7.6.42. Adopt the notation of Exercise 7.6.41, and let $\mathscr{F}$ be the family of subsets of $\mathfrak{A}$ whose intersection with every $(\mathfrak{A})_n$ is weak * closed, where n

is a positive integer. With $\mathscr{S}$ a subset of $\mathfrak{A}_\sharp$ and a a positive number, denote by $\mathfrak{A}(\mathscr{S}, a)$ and $\mathfrak{A}^0(\mathscr{S}, a)$, respectively, the subsets

$$\{A \in \mathfrak{A} : |\eta(A)| \leq a, \eta \in \mathscr{S}\}, \qquad \{A \in \mathfrak{A} : |\eta(A)| < a, \eta \in \mathscr{S}\};$$

let $\mathfrak{A}_n(\mathscr{S}, a)$ and $\mathfrak{A}_n^0(\mathscr{S}, a)$ denote the sets $\mathfrak{A}(\mathscr{S}, a) \cap (\mathfrak{A})_n$ and $\mathfrak{A}^0(\mathscr{S}, a) \cap (\mathfrak{A})_n$, respectively.

(i) Show that $\mathscr{F}$ is the family of closed sets of a topology (the "$\mathscr{F}$-topology") for $\mathfrak{A}$.

(ii) Let $\mathscr{O}$ be an $\mathscr{F}$-open subset of $\mathfrak{A}$. Show that for each A in $\mathfrak{A}$, $A + \mathscr{O}$ is an $\mathscr{F}$-open set. Conclude that the mapping $B \to A + B$ of $\mathfrak{A}$ onto itself is an $\mathscr{F}$-homeomorphism.

(iii) Let $\{\eta_j\}$ be a sequence of elements of $\mathfrak{A}_\sharp$ tending to 0 in norm. Show that $\mathfrak{A}^0(\{\eta_j\}, a)$ is $\mathscr{F}$-open for each positive a.

(iv) Let $\mathscr{O}$ be an $\mathscr{F}$-open set containing 0 and n be a positive integer. Show that there is a finite subset $\mathscr{S}$ of $\mathfrak{A}_\sharp$ such that $\mathfrak{A}_n(\mathscr{S}, 1)$ is contained in $\mathscr{O} \cap (\mathfrak{A})_n$, and that there is a finite subset $\mathscr{T}$ of $(\mathfrak{A}_\sharp)_{1/n}$ such that

$$\mathfrak{A}_{n+1}(\mathscr{S} \cup \mathscr{T}, 1) \subseteq \mathscr{O} \cap (\mathfrak{A})_{n+1}.$$

[*Hint*. Assume the contrary and establish the finite intersection property for the family

$$\{\mathfrak{A}_{n+1}(\mathscr{S} \cup \mathscr{T}, 1) \cap (\mathfrak{A} \setminus \mathscr{O}) : \mathscr{T} \text{ a finite subset of } (\mathfrak{A}_\sharp)_{1/n}\};$$

show that an element of the intersection of this family lies in $(\mathfrak{A})_n$ and deduce a contradiction.]

(v) Let $\mathscr{O}$ be an $\mathscr{F}$-open set containing 0. Construct a sequence $\{\mathscr{S}_n\}$ of finite sets $\mathscr{S}_n$ such that $\mathscr{S}_{n+1} \subseteq (\mathfrak{A}_\sharp)_{1/n}$ for n in $\{1, 2, \ldots\}$ and such that for n in $\mathbb{N}$,

$$\mathfrak{A}_n(\mathscr{S}_1 \cup \cdots \cup \mathscr{S}_n, 1) \subseteq \mathscr{O} \cap (\mathfrak{A})_n.$$

(vi) For a given $\mathscr{F}$-open set $\mathscr{O}$ containing 0, construct a sequence $\{\eta_j\}$ in $\mathfrak{A}_\sharp$ tending to 0 in norm such that $\mathfrak{A}(\{\eta_j\}, 1) \subseteq \mathscr{O}$.

(vii) Given an $\mathscr{F}$-open set $\mathscr{O}$ in $\mathfrak{A}$ containing 0, find an $\mathscr{F}$-open set $\mathscr{O}_0$ containing 0 such that $\mathscr{O}_0 + \mathscr{O}_0 \subseteq \mathscr{O}$ and conclude (using (ii)) that addition is $\mathscr{F}$-continuous on $\mathfrak{A}$.

(viii) Show that the mapping

$$(a, A) \to aA : \mathbb{C} \times \mathfrak{A} \to \mathfrak{A}$$

is $\mathscr{F}$-continuous. Conclude (using (vi) and (vii)) that $\mathscr{F}$ provides $\mathfrak{A}$ with a locally convex linear topological structure.

7.6.43. With the notation and terminology of Exercise 7.6.42:

(i) identify the $\mathcal{F}$-continuous linear functionals on $\mathfrak{A}$ [*Hint*. Use Exercise 1.9.15.];
(ii) show that a convex subset of $\mathfrak{A}$ is weak * closed if and only if it lies in $\mathcal{F}$;
(iii) conclude that the sets of self-adjoint and positive elements in $\mathfrak{A}$ are weak * closed.

7.6.44. With the notation and assumptions of Exercise 7.6.42, let $\mathfrak{A}_\#^{\mathrm{h}}$ and $\mathfrak{A}_{\mathrm{h}}$ be the real-linear spaces of hermitian elements in $\mathfrak{A}_\#$ and $\mathfrak{A}$, respectively.

(i) Suppose $T \in \mathfrak{A} \backslash \mathfrak{A}_{\mathrm{h}}$. Show that there is an η in $\mathfrak{A}_\#^{\mathrm{h}}$ such that $\operatorname{Im} \eta(T) \neq 0$. [*Hint*. Use Exercise 7.6.43(iii) and a Hahn–Banach separation theorem.]
(ii) Suppose A is a non-zero element of $\mathfrak{A}_{\mathrm{h}}$. Show that there is an η in $\mathfrak{A}_\#^{\mathrm{h}}$ such that $\eta(A) \neq 0$ and that $\mathfrak{A}_\#^{\mathrm{h}}$ separates $\mathfrak{A}$.
(iii) Show that $\mathfrak{A}_\#^{\mathrm{h}} + i\mathfrak{A}_\#^{\mathrm{h}} = \mathfrak{A}_\#$. [*Hint*. Show that $\mathfrak{A}_\#^{\mathrm{h}}$ is norm closed and use Exercise 1.9.5.]
(iv) Suppose $A \in \mathfrak{A}_{\mathrm{h}} \backslash \mathfrak{A}^+$. Show that there is a state η of $\mathfrak{A}$ in $\mathfrak{A}_\#$ such that $\eta(A) < 0$.
(v) Deduce that, with A and B in $\mathfrak{A}_{\mathrm{h}}$, $A \leq B$ if and only if $\eta(A) \leq \eta(B)$ for each state η of $\mathfrak{A}$ in $\mathfrak{A}_\#$.

7.6.45. With the notation and assumptions of Exercise 7.6.41, show that

(i) each monotone increasing net in $\mathfrak{A}$ with an upper bound has a least upper bound in $\mathfrak{A}$;
(ii) $\mathfrak{A}$ is a W^*-algebra in the sense of Exercise 7.6.38.

7.6.46. Let $\mathcal{R}$ be a von Neumann algebra. Prove that

(i) $\mathcal{R}$ has a faithful normal semi-finite weight;
(ii) $\mathcal{R}$ has a faithful normal state if and only if $\mathcal{R}$ is countably decomposable.

7.6.47. Find an example of a normal semi-finite weight ρ on a von Neumann algebra $\mathfrak{R}$ and an operator A in $\mathcal{R}$ such that

(i) $A \in N_\rho$ but $A^* \notin N_\rho$;
(ii) ρ is faithful, $A \in \mathcal{N}_\rho$ but $A^* \notin \mathcal{N}_\rho$.

7.6.48. Let $\mathcal{M}$ be the subspace of $C([0, 1])$ consisting of those f such that f/ι is bounded on $(0, 1]$, where $\iota(\lambda) = \lambda$ for each λ in $[0, 1]$. Show that

(i) $\mathcal{M}$ is a proper ideal in $C([0, 1])$ and if $g_1 \leq g \leq g_2$ with g_1, g_2 in $\mathcal{M}$, then $g \in \mathcal{M}$;

(ii) the real-linear space $\mathcal{M}_{\mathrm{h}}$ of real-valued functions in $\mathcal{M}$ is an archimedian partially ordered vector space with order unit ι (see pp. 212, 213, 297);

(iii) ρ_0 is a positive linear functional on $\mathcal{M}_0$, where $\rho_0(b\iota - f\iota^2) = b$ and

$$\mathcal{M}_0 = \{b\iota - f\iota^2 : b \in \mathbb{R}, f \in C([0, 1], \mathbb{R})\};$$

(iv) ρ_0 extends to a positive linear functional ρ_1 on $\mathcal{M}_{\mathrm{h}}$;

(v) ρ is a weight on $C([0, 1])$, where $\rho(f) = \rho_1(f)$ for each positive f in $\mathcal{M}_{\mathrm{h}}$ and $\rho(g) = +\infty$ for each positive g in $C([0, 1]) \setminus \mathcal{M}_{\mathrm{h}}$;

(vi) $\iota \in F_\rho \subseteq \mathcal{M}_\rho = \mathcal{M}$, $\rho(\iota^2) = 0$, $\iota \in N_\rho$, and $\rho(\iota) = 1$ (with the notation of Section 7.5).

7.6.49. Let $\mathcal{R}$ be a von Neumann algebra acting on a Hilbert space $\mathcal{H}$ and x_0 be a generating trace vector for $\mathcal{R}$. For A in $\mathcal{R}$, let $J_0 A x_0$ be $A^* x_0$. Show that

(i) J_0 extends to an isometric conjugate-linear mapping J of $\mathcal{H}$ onto $\mathcal{H}$ such that $J^2 = I$ (J is an *involution*);

(ii) $JA^*J = A'(\in \mathcal{R}')$ for each A in $\mathcal{R}$, where $A \to A'$ is the * anti-isomorphism described in Theorem 7.2.15.

7.6.50. Let $\{e_n : n \in \mathbb{N}\}$ be an orthonormal basis for a (separable) Hilbert space $\mathcal{H}_0$, and let $\mathcal{H}$ be the (Hilbert space) countable direct sum of $\mathcal{H}_0$ with itself. Suppose T is the operator on $\mathcal{H}$ with domain consisting of all vectors $\{x_k\}$ such that $\sum_{k=1}^{\infty} k^2 \|x_k\|^2 < \infty$ and $T(\{x_k\}) = \{kx_k\}$, B is the operator on $\mathcal{H}$ that maps $\{y_k\}$ to $\{\sum_{k=1}^{\infty} k^{-1} y_k, 0, 0, \ldots\}$, x_0 is $\{2^{-1}e_1, 2^{-2}e_2, \ldots\}$, and z_0 is $\{\sum_{k=1}^{\infty} 2^{-k} e_k, 0, 0, \ldots\}$. Show that

(i) $T = T^*$ and $T \,\eta\, (\mathcal{B}(\mathcal{H}_0) \otimes I)'(= \mathcal{R})$;

(ii) $B \in \mathcal{R}$;

(iii) $BTx_0 = z_0$;

(iv) $Sx_0 \neq z_0$ if $S \,\eta\, \mathcal{R}$; (But see Exercise 8.7.60(vi).)

(v) x_0 is generating and separating for $\mathcal{R}$.

7.6.51. Let $\mathcal{S}_0$ be the set of faithful normal states of a von Neumann algebra $\mathcal{R}$ acting on a separable Hilbert space $\mathcal{H}$. Show that

(i) $\mathcal{S}_0$ satisfies some one (and hence *all*) of the conditions (i), (ii), (iii), and (iv) of Theorem 4.3.9, where $\mathcal{M}$ of that theorem is replaced by $\mathcal{R}$;

(ii) H is self-adjoint if $\omega(H)$ is real for each ω in $\mathcal{S}_0$.

7.6.52. Let A_0 be a closed operator acting on a Hilbert space $\mathscr{H}$, and suppose $\langle A_0 x, x\rangle \geq 0$ for each x in $\mathscr{D}(A_0)$.

(i) Show that A_0 and $A_0 + I$ are closed symmetric operators on $\mathscr{H}$. [*Hint*. Use Proposition 2.1.7.]

(ii) Suppose A is a positive self-adjoint extension of A_0. Note that $A + I$ is a positive self-adjoint extension of $A_0 + I$ and that $A + I$ is a one-to-one linear transformation with range $\mathscr{H}$. Show that the inverse B of $A + I$ is in $(\mathscr{B}(\mathscr{H}))^+$.

(iii) With B as in (ii), y in $\mathscr{H}$, and x in $\mathscr{D}(A_0)$ $(= \mathscr{D}(A_0 + I))$, show that

$$(*) \qquad \langle x, y\rangle = \langle (A_0 + I)x, By\rangle$$

and that $By \in \mathscr{D}(A_0^*)$.

7.6.53. Let A_0 be a closed operator acting on a Hilbert space $\mathscr{H}$, and suppose $\langle A_0 x, x\rangle \geq 0$ for each x in $\mathscr{D}(A_0)$. Define $\langle u, v\rangle'$ for each pair of vectors u, v in $\mathscr{D}(A_0)$ to be $\langle (A_0 + I)u, v\rangle$ and let $\mathscr{D}'$ be the completion of $\mathscr{D}(A_0)$ relative to the definite inner product $(u, v) \to \langle u, v\rangle'$ on $\mathscr{D}(A_0)$.

(i) Show that the "identity" mapping of $\mathscr{D}(A_0)$ onto itself has a (unique) bounded extension ι mapping $\mathscr{D}'$ into $\mathscr{H}$, ι is one-to-one, and $\|\iota\| \leq 1$. [*Hint*. Choose x_n in $\mathscr{D}(A_0)$ tending to z' in $\mathscr{D}'$. If $\iota(z') = 0$, show that $\|x_n\| \to 0$ and $\langle z', x_m\rangle' = 0$ for each m.]

(ii) With y in $\mathscr{H}$, show that $x \to \langle x, y\rangle$ $(x \in \mathscr{D}(A_0))$ extends to a bounded linear functional on $\mathscr{D}'$ of norm not exceeding $\|y\|$.

(iii) Show that there is a vector By in $\mathscr{D}(A_0^*)$ satisfying $(*)$ of Exercise 7.6.52(iii). [*Hint*. Find z' in $\mathscr{D}'$ such that $\langle x, y\rangle = \langle x, z'\rangle'$ for all x in $\mathscr{D}(A_0)$ and let By be $\iota(z')$.]

(iv) Show that $B \in (\mathscr{B}(\mathscr{H}))_1^+$. [*Hint*. Use (i) and (ii) to show that $\|B\| \leq 1$. Use the relation $\langle x, y\rangle = \langle x, \iota^{-1}(By)\rangle'$ for each y in $\mathscr{H}$ and each x in $\mathscr{D}(A_0)$ (implicit in the hint of (iii)) to show that $\langle By, y\rangle = \langle \iota^{-1}(By), \iota^{-1}(By)\rangle'$.]

(v) Show that B is a one-to-one mapping and that its inverse A_1 is a self-adjoint extension of $A_0 + I$. [*Hint*. With y not 0, choose x in $\mathscr{D}(A_0)$ such that $\langle x, y\rangle \neq 0$. Use the equality $\langle x, y\rangle = \langle x, \iota^{-1}(By)\rangle'$. Apply the discussion following Theorem 7.2.1. With x and u in $\mathscr{D}(A_0)$, show that $\langle u, x\rangle' = \langle u, \iota^{-1}(B(A_0 + I)x)\rangle'$ and conclude that $x = B(A_0 + I)x$.]

(vi) Show that $A_1 - I (= A)$ is a positive self-adjoint extension of A_0, and $\mathscr{D}(A) \subseteq \iota(\mathscr{D}')$.

7.6.54. With the notation of Exercise 7.6.53, show that A is the *unique* positive self-adjoint extension of A_0 whose domain is contained in $\iota(\mathscr{D}')$. (This extension is known as the *Friedrichs extension* of A_0.) [*Hint*. Use

Exercise 7.6.52(ii) and (iii) and an argument of the type indicated in the hint to Exercise 7.6.53(i) to show that ι is one-to-one.]

7.6.55. Let $\mathscr{R}$ be a von Neumann algebra acting on a Hilbert space $\mathscr{H}$ and A_0 be a symmetric operator affiliated with $\mathscr{R}$. Suppose $\langle A_0x, x\rangle \geq 0$ for each x in $\mathscr{D}(A_0)$. Show that the Friedrichs extension of A_0 (see Exercise 7.6.54) is affiliated with $\mathscr{R}$.

CHAPTER 8

THE TRACE

This chapter is concerned with the (center-valued) trace defined on a finite von Neumann algebra, the dimension function obtained by restricting the trace to the set of projections in the algebra, tracial weights on factors, and another construction for producing various types of factors. After a number of illustrative examples, the main result, dealing with the existence and properties of the trace, is proved in Section 8.2. An alternative approach to the theory, which takes as its starting point the Dixmier approximation theorem, is described in outline, after a proof of that theorem, in Section 8.3. Section 8.4 is concerned with the existence and uniqueness of the dimension function, and its connection with the comparison theory of projections. Criteria for determining the type of a factor, from information about its tracial weights, are derived in Section 8.5. They are used in Section 8.6, where a measure-theoretic construction is described and shown to produce examples of factors of all types I_n, II_1, II_∞, and III.

8.1. Traces

Let $\mathscr{R}$ be a von Neumann algebra with center $\mathscr{C}$ and unitary group $\mathscr{U}$. By a *center-valued trace* on $\mathscr{R}$ we mean a linear mapping $\tau : \mathscr{R} \to \mathscr{C}$ such that

(i) $\tau(AB) = \tau(BA)$ for each A, B in $\mathscr{R}$;
(ii) $\tau(C) = C$ for each C in $\mathscr{C}$;
(iii) $\tau(A) > 0$ if $A \in \mathscr{R}$ and $A > 0$.

If such a mapping τ exists, then $\mathscr{R}$ is finite. Indeed, if E and F are projections in $\mathscr{R}$ such that $E \sim F \leq E$, there is a partial isometry V from E to F in $\mathscr{R}$, and properties (i) and (iii) of τ yield

$$\tau(E - F) = \tau(V^*V) - \tau(VV^*) = 0, \qquad E - F = 0.$$

In Theorem 8.2.8 we show conversely that, if $\mathscr{R}$ is finite, there is a unique center-valued trace τ on $\mathscr{R}$. Moreover, τ has the additional properties

(iv) $\tau(CA) = C\tau(A)$ for each A in $\mathscr{R}$ and C in $\mathscr{C}$;
(v) $\|\tau(A)\| \leq \|A\|$ for each A in $\mathscr{R}$;
(vi) τ is ultraweakly continuous.

A bounded linear functional ρ on $\mathscr{R}$ is said to be *central* if $\rho(AB) = \rho(BA)$ for each A and B in $\mathscr{R}$. By a *numerical trace* on $\mathscr{R}$, we mean a positive central linear functional. A *tracial state* of $\mathscr{R}$ is a numerical trace ρ such that $\rho(I) = 1$. In accordance with our terminology for weights (including states), we say that a numerical trace ρ is *faithful* if $\rho(A) > 0$ whenever $A \in \mathscr{R}$ and $A > 0$. The existence of a faithful numerical trace ρ on $\mathscr{R}$ implies that $\mathscr{R}$ is finite; indeed, the argument that follows the statement of conditions (i), (ii), and (iii) above can be applied, with ρ in place of τ. When $\mathscr{R}$ is a factor, $\mathscr{C} = \{aI : a \in \mathbb{C}\}$, so a center-valued trace necessarily has the form $\tau(A) = \rho(A)I$, where ρ is a faithful tracial state.

Our main purpose in the present section is to give some examples of von Neumann algebras in which it is possible to write down a simple formula for the center-valued trace. Before doing this, we show that condition (i) above can be written in three other forms; subsequently, the four variants will be used interchangeably, frequently without comment.

8.1.1. PROPOSITION. *If $\mathscr{R}$ is a von Neumann algebra with unitary group $\mathscr{U}$, $\mathfrak{X}$ is a Banach space and $\eta: \mathscr{R} \to \mathfrak{X}$ is a bounded linear mapping, then the following four conditions are equivalent*:

(i) $\eta(AB) = \eta(BA)$ *for each A and B in* $\mathscr{R}$.
(ii) $\eta(AA^*) = \eta(A^*A)$ *for each A in* $\mathscr{R}$.
(iii) $\eta(E) = \eta(F)$ *whenever E and F are equivalent projections in* $\mathscr{R}$.
(iv) $\eta(UAU^*) = \eta(A)$ *for each A in* $\mathscr{R}$ *and U in* $\mathscr{U}$.

Proof. It is evident that (i) implies (ii), and that (ii) implies (iii). If (iv) is satisfied then, for each B in $\mathscr{R}$ and U in $\mathscr{U}$,

$$\eta(UB) = \eta(U(BU)U^*) = \eta(BU).$$

Since each A in $\mathscr{R}$ is a linear combination of unitary elements, it follows that $\eta(AB) = \eta(BA)$. Hence (iv) implies (i), and it now suffices to show that (iii) implies (iv).

Suppose that (iii) is satisfied. With U in $\mathscr{U}$, and E a projection in $\mathscr{R}$, $UEU^* \sim E$ and so $\eta(UEU^*) = \eta(E)$. Since $\mathscr{R}$ is the norm-closed linear hull of its projections, it follows from the boundedness and linearity of η that $\eta(UAU^*) = \eta(A)$ for each A in $\mathscr{R}$. Thus (iii) implies (iv). ∎

8.1.2. EXAMPLE. Suppose that $\mathscr{H}$ is a finite-dimensional Hilbert space, $\mathscr{R} = \mathscr{B}(\mathscr{H})$ and $\{e_1, \ldots, e_n\}$ is an orthonormal basis of $\mathscr{H}$. We shall show that the most general numerical trace ρ on $\mathscr{R}$ is given by

$$(1) \qquad \rho(A) = c \sum_{j=1}^{n} \langle Ae_j, e_j \rangle,$$

where c is a non-negative scalar. Moreover, the right-hand side of equation (1) is independent of the choice of the orthonormal basis.

With E_j the one-dimensional projection whose range contains e_j, $\{E_1, \ldots, E_n\}$ is an orthogonal family of minimal projections in $\mathscr{R}$, with sum I; and

$$E_j A E_j = \langle Ae_j, e_j \rangle E_j \qquad (A \in \mathscr{B}(\mathscr{H}),\ \ j = 1, \ldots, n).$$

A numerical trace ρ on $\mathscr{R}$ takes the same value at all minimal projections in $\mathscr{R}$, since any two such projections are equivalent. With c (≥ 0) this common value, we have

$$\begin{aligned}\rho(A) = \sum_{j=1}^{n} \rho(E_j A) &= \sum_{j=1}^{n} \rho(E_j A E_j)\\ &= \sum_{j=1}^{n} \langle Ae_j, e_j \rangle \rho(E_j) = c \sum_{j=1}^{n} \langle Ae_j, e_j \rangle.\end{aligned}$$

Conversely, suppose that $c \geq 0$ and $\rho\colon \mathscr{R} \to \mathbb{C}$ is defined by (1). Then ρ is a positive linear functional on $\mathscr{R}$, and is a numerical trace since, for each A in $\mathscr{R}$,

$$\begin{aligned}\rho(A^*A) &= c \sum_{j=1}^{n} \langle A^*Ae_j, e_j \rangle = c \sum_{j=1}^{n} \|Ae_j\|^2\\ &= c \sum_{j=1}^{n} \sum_{k=1}^{n} |\langle Ae_j, e_k \rangle|^2\\ &= c \sum_{k=1}^{n} \sum_{j=1}^{n} |\langle A^*e_k, e_j \rangle|^2 = \rho(AA^*).\end{aligned}$$

We have now shown that the numerical traces on $\mathscr{R}$ are precisely the linear functionals ρ given by equation (1), and that c (≥ 0) is the value taken by ρ at minimal projections. This implies that the right-hand side of (1) is independent of the choice of the orthonormal basis $\{e_1, \ldots, e_n\}$. Note also that, if A has matrix $[a_{jk}]$ with respect to this basis, then

$$\rho(A) = c \sum_{j=1}^{n} \langle Ae_j, e_j \rangle = c \sum_{j=1}^{n} a_{jj} = c\,\mathrm{tr}([a_{jk}]),$$

where tr denotes the elementary trace for $n \times n$ complex matrices.

The unique tracial state ρ_0 on $\mathscr{R}$ is obtained by taking $c = n^{-1}$ in (1). Since $\mathscr{R}$ is a factor, it follows that there is a unique center-valued trace, given by

$$\tau(A) = \rho_0(A)I = \left(\frac{1}{n} \sum_{j=1}^{n} \langle Ae_j, e_j \rangle\right) I.$$

If, at the outset, it were known that there is at least one tracial state ρ of $\mathscr{B}(\mathscr{H})$, the first three paragraphs above would already suffice to show that ρ is unique and is given by (1) (with $c = 1/n$), the right-hand side of (1) being independent of the choice of the orthonormal basis $e_1, \ldots, e_n$. The existence of a tracial state was proved, above, by computations that verify that the linear functional ρ, defined in (1), has the requisite properties. There is merit in an alternative proof, which uses the fact that the unitary group $\mathscr{U}$ of $\mathscr{B}(\mathscr{H})$ is compact (since $\mathscr{H}$ is finite dimensional), and so has a translation-invariant (Haar) measure μ such that $\mu(\mathscr{U}) = 1$. When $A \in \mathscr{B}(\mathscr{H})$, we can define an element $\tau_0(A)$ of $\mathscr{B}(\mathscr{H})$ by

$$\tau_0(A) = \int_{\mathscr{U}} U^*AU\, d\mu(U);$$

the integral exists, as the limit in norm of approximating Riemann sums, since the mapping $U \rightarrow U^*AU$ is norm continuous on $\mathscr{U}$. (By considering these Riemann sums, one sees that $\tau_0(A)$ lies in the norm-closed convex hull of the set $\{U^*AU : U \in \mathscr{U}\}$.) Invariance of μ under left translation implies that $\tau_0(V^*AV) = \tau_0(A)$, when $V \in \mathscr{U}$; and by Proposition 8.1.1, $\tau_0(AB) = \tau_0(BA)$ for all A, B in $\mathscr{B}(\mathscr{H})$. From the right invariance of μ, $V^*\tau_0(A)V = \tau_0(A)$; so $\tau_0(A)$ commutes with each V in $\mathscr{U}$, and hence lies in the center of $\mathscr{B}(\mathscr{H})$ (the scalars). It is easily verified that τ_0 is linear, $\tau_0(A) > 0$ when $A > 0$, and $\tau_0(C) = C$ when C lies in the center of $\mathscr{B}(\mathscr{H})$. Accordingly, τ_0 is a center-valued trace on $\mathscr{B}(\mathscr{H})$, and $\tau_0(A)$ has the form $\rho_0(A)I$, where ρ_0 is a faithful tracial state of $\mathscr{B}(\mathscr{H})$.

Of course, this integration technique is not available when $\mathscr{H}$ is infinite dimensional, since the unitary group is then not compact; and indeed, $\mathscr{B}(\mathscr{H})$ is of type I_∞, and admits no tracial state, in that case. ■

8.1.3. Example. Suppose that n is a positive integer, and $\mathscr{R}$ is a type I_n von Neumann algebra with center $\mathscr{C}$. Thus $\mathscr{R}$ is isomorphic to the algebra of all $n \times n$ matrices with entries in $\mathscr{C}$, by Theorem 6.6.5. Although we shall not use this matrix representation explicitly, it motivates the calculations that follow. The end result amounts to the assertion that a matrix $[C_{jk}]$, with each C_{jk} in $\mathscr{C}$, has trace equal to the diagonal matrix with $n^{-1}\sum C_{jj}$ in each diagonal position. We show also that the trace of an element A of $\mathscr{R}$ is a convex combination of operators of the form UAU^*, with U unitary in $\mathscr{R}$.

Let $\{E_{jk} : j, k = 1, \ldots, n\}$ be a self-adjoint system of matrix units in $\mathscr{R}$, with each E_{jj} an abelian projection and $\sum E_{jj} = I$. With A in $\mathscr{R}$, by Proposition 6.4.2, $E_{jj}AE_{jj} \in E_{jj}\mathscr{R}E_{jj} = \mathscr{C}E_{jj}$, so we can choose C_j in $\mathscr{C}$ such that

(2) $$E_{jj}AE_{jj} = C_jE_{jj} \qquad (j = 1, \ldots, n).$$

Since

$$C_j = \sum_{k=1}^{n} C_j E_{kk} = \sum_{k=1}^{n} C_j E_{kj} E_{jj} E_{jk} = \sum_{k=1}^{n} E_{kj} C_j E_{jj} E_{jk},$$

it follows from (2) that

$$C_j = \sum_{k=1}^{n} E_{kj} A E_{jk}. \tag{3}$$

(If φ is the isomorphism from $\mathcal{R}$ onto the matrix algebra $n \otimes \mathcal{C}$, described in Theorem 6.6.5 and Lemma 6.6.3, then C_j is the entry occurring at the (j, j) position of the matrix $\varphi(A)$.)

If τ is a center-valued trace on $\mathcal{R}$, then

$$\begin{aligned}\tau(C_j E_{jj}) = \tau(C_j E_{jk} E_{kj}) &= \tau(E_{kj} C_j E_{jk}) \\ &= \tau(C_j E_{kj} E_{jk}) = \tau(C_j E_{kk})\end{aligned}$$

for all j and k. Hence

$$C_j = \tau(C_j) = \sum_{k=1}^{n} \tau(C_j E_{kk}) = n\tau(C_j E_{jj}), \qquad \tau(C_j E_{jj}) = \frac{1}{n} C_j.$$

From (2)

$$\begin{aligned}\tau(A) = \sum_{j=1}^{n} \tau(E_{jj} A) &= \sum_{j=1}^{n} \tau(E_{jj} A E_{jj}) \\ &= \sum_{j=1}^{n} \tau(C_j E_{jj}) = \frac{1}{n} \sum_{j=1}^{n} C_j.\end{aligned}$$

This, with (3), gives

$$\tau(A) = \frac{1}{n} \sum_{j=1}^{n} \sum_{k=1}^{n} E_{kj} A E_{jk}. \tag{4}$$

We prove next that the linear mapping $\tau: \mathcal{R} \to \mathcal{R}$, defined by (4), does indeed have the properties required of a center-valued trace. (In understanding all the subsequent computations, it may help to keep in mind the representation of $\mathcal{R}$ as $n \times n$ matrices over $\mathcal{C}$.) It follows from (3) that τ maps $\mathcal{R}$ into $\mathcal{C}$. When $C \in \mathcal{C}$,

$$\begin{aligned}\tau(C) &= \frac{1}{n} \sum_{j=1}^{n} \sum_{k=1}^{n} E_{kj} C E_{jk} \\ &= \frac{1}{n} \sum_{j=1}^{n} \sum_{k=1}^{n} C E_{kk} = \frac{1}{n} \sum_{j=1}^{n} C = C.\end{aligned}$$

If $A \in \mathscr{R}$ and $A > 0$, then $A^{1/2} \neq 0$ and therefore $A^{1/2}E_{mm} \neq 0$ for some m. Thus

$$\tau(A) = \frac{1}{n}\sum_{j=1}^{n}\sum_{k=1}^{n}(A^{1/2}E_{jk})^*(A^{1/2}E_{jk}) \geq (A^{1/2}E_{mm})^*(A^{1/2}E_{mm}) > 0.$$

Finally, if $A, B \in \mathscr{R}$, then, since $E_{kk}\mathscr{R}E_{kk}$ is abelian,

$$\begin{aligned}\tau(AB) &= \frac{1}{n}\sum_{j}\sum_{k}E_{kj}ABE_{jk} = \frac{1}{n}\sum_{j}\sum_{k}\sum_{m}E_{kj}AE_{mm}BE_{jk}\\ &= \frac{1}{n}\sum_{j}\sum_{k}\sum_{m}(E_{kj}AE_{mk})(E_{km}BE_{jk})\\ &= \frac{1}{n}\sum_{m}\sum_{k}\sum_{j}(E_{km}BE_{jk})(E_{kj}AE_{mk})\\ &= \frac{1}{n}\sum_{m}\sum_{k}\sum_{j}E_{km}BE_{jj}AE_{mk}\\ &= \frac{1}{n}\sum_{m}\sum_{k}E_{km}BAE_{mk} = \tau(BA).\end{aligned}$$

We have now shown that $\mathscr{R}$ has a unique center-valued trace, defined by (4). We next derive another expression, in which the trace is exhibited in a different form. For this, let $S(n)$ denote the symmetric group of all permutations of the set $\{1, 2, \ldots, n\}$, and let F be the class of all the 2^n mappings from $\{1, 2, \ldots, n\}$ into $\{1, -1\}$. With f in F and π in $S(n)$, the equation

$$U(f, \pi) = \sum_{j=1}^{n} f(j)E_{\pi(j)j}$$

defines a unitary element $U(f, \pi)$ of $\mathscr{R}$. With j and k in $\{1, 2, \ldots, n\}$,

$$\sum_{f\in F} f(j)f(k) = \begin{cases} 0 & (j \neq k) \\ 2^n & (j = k). \end{cases}$$

It follows that, for each A in $\mathscr{R}$ and π in $S(n)$,

$$\begin{aligned}\sum_{f\in F} U(f, \pi)AU(f, \pi)^* &= \sum_{j=1}^{n}\sum_{k=1}^{n}\sum_{f\in F} f(j)f(k)E_{\pi(j)j}AE_{k\pi(k)}\\ &= 2^n\sum_{j=1}^{n} E_{\pi(j)j}AE_{j\pi(j)}.\end{aligned}$$

For each j and k in $\{1, 2, \ldots, n\}$, there are $(n-1)!$ elements π in $S(n)$ for which $\pi(j) = k$, so

$$\sum_{\pi \in S(n)} \sum_{f \in F} U(f, \pi) A U(f, \pi)^* = 2^n (n-1)! \sum_{j=1}^{n} \sum_{k=1}^{n} E_{kj} A E_{jk}$$
$$= 2^n n! \, \tau(A)$$

and

$$\tau(A) = \frac{1}{2^n n!} \sum_{\pi \in S(n)} \sum_{f \in F} U(f, \pi) A U(f, \pi)^*. \tag{5}$$

From (5), one sees that $\tau(A)$ lies in the convex hull $\text{co}_{\mathscr{R}}(A)$ of the set of all operators of the form UAU^*, with U unitary in $\mathscr{R}$. In Section 8.3, we study $\text{co}_{\mathscr{R}}(A)$ and its norm closure $\text{co}_{\mathscr{R}}(A)^=$ when A is an element of an arbitrary von Neumann algebra $\mathscr{R}$. ■

8.1.4. Example. We exhibit the trace in certain factors of type II_1.

In Section 6.7, we constructed the "left convolution" von Neumann algebra $\mathscr{L}_G$ associated with a discrete group G, together with its commutant, the "right convolution" algebra $\mathscr{R}_G$. We showed (Theorem 6.7.5) that $\mathscr{L}_G$ and $\mathscr{R}_G$ are factors of type II_1, when G is an i.c.c. group. In so doing, we proved also (Remark 6.7.3) that the unit vectors x_g in $l_2(G)$ (where $x_g(h)$ is 0 or 1 according as $h \neq g$ or $h = g$) are separating for $\mathscr{L}_G$ (and $\mathscr{R}_G$), and are trace vectors for $\mathscr{L}_G$ (and $\mathscr{R}_G$) in the sense that

$$\langle ABx_g, x_g \rangle = \langle BAx_g, x_g \rangle \qquad (A, B \in \mathscr{L}_G).$$

From these properties, it follows that $\omega_{x_g} | \mathscr{L}_G$ is a faithful tracial state of $\mathscr{L}_G$, and that the equation $\tau(A) = \langle Ax_g, x_g \rangle I$ defines a center-valued trace τ on $\mathscr{L}_G$ when G is an i.c.c. group.

These elementary considerations prove (in the present examples of factors of type II_1) the *existence*, but not the *uniqueness*, of the trace; for the latter, we appeal to the general theory developed in the next section. ■

8.2. The trace in finite algebras

Throughout this section, $\mathscr{R}$ is a von Neumann algebra with center $\mathscr{C}$ and unitary group $\mathscr{U}$. Our main objective is to prove the existence and properties of the center-valued trace when $\mathscr{R}$ is finite.

We say that a projection E in $\mathscr{R}$ is *monic* if $E \neq 0$ and there exist a positive integer k and projections $E_1, \ldots, E_k$ in $\mathscr{R}$ and Q in $\mathscr{C}$ such that

$$E_1 \sim E_2 \sim \cdots \sim E_k \sim E, \qquad E_1 + E_2 + \cdots + E_k = Q.$$

Since each E_j has central carrier C_E, it is easily verified that $Q = C_E$. It is apparent that, if $E \sim F$ and E is monic, then so is F.

8.2.1. PROPOSITION. *Each non-zero projection E in a finite von Neumann algebra $\mathscr{R}$ is the sum of an orthogonal family of monic projections in $\mathscr{R}$.*

Proof. Since $\mathscr{R}$ is finite it is a direct sum of algebras of types I_n $(n = 1, 2, 3, \ldots)$ and II_1, from Theorem 6.5.2. It therefore suffices to consider only the two cases in which $\mathscr{R}$ is either type I_n or type II_1. If $\mathscr{R}$ is type I_n, E is a (finite) sum of abelian projections by Corollary 6.5.5. Moreover, each abelian projection F in $\mathscr{R}$ is monic, since there are n projections in $\mathscr{R}$, each equivalent to F, whose sum is C_F. We assume henceforth that $\mathscr{R}$ is type II_1.

Let $\{E_a\}$ be a maximal orthogonal family of monic subprojections of E in $\mathscr{R}$, and let $F = E - \sum E_a$. The maximality assumption implies that F has no monic subprojection in $\mathscr{R}$, and we want to show that $F = 0$.

It now suffices to prove that each non-zero projection F in a type II_1 von Neumann algebra $\mathscr{R}$ has a monic subprojection in $\mathscr{R}$. Let $\{F_1, \ldots, F_n\}$ be a maximal orthogonal family of projections in $\mathscr{R}$ such that $F_j \sim F$ for each j (such a family is necessarily finite, since $\mathscr{R}$ is finite). As $\mathscr{R}$ is of type II_1, by Lemma 6.5.6, we can find projections $M_0, \ldots, M_n$ in $\mathscr{R}$ such that

$$M_0 \sim M_1 \sim \cdots \sim M_n, \qquad \sum_{j=0}^{n} M_j = C_F.$$

Now $PM_0 \precsim PF$ for some central projection P such that $0 < P \le C_F$, and PM_0 is equivalent to the desired monic subprojection of F; for otherwise, from the comparison theorem, $F \prec M_0$ and F has $n + 1$ orthogonal equivalent copies in $\mathscr{R}$—contradicting Theorem 6.3.11. ■

In what follows, we shall make use of linear mappings between operator algebras that map positive elements onto positive elements. We refer to these as *positive linear mappings*. (More generally, we use this terminology for such mappings between partially ordered vector spaces.) A linear mapping η between operator algebras that preserves adjoints $(\eta(A^*) = \eta(A)^*)$ is said to be *hermitian*.

8.2.2. LEMMA. *Each positive linear mapping η from $\mathscr{R}$ into $\mathscr{C}$ is bounded, with $\|\eta\| = \|\eta(I)\|$.*

Proof. From the function representation of the commutative C^*-algebra $\mathscr{C}$, it is apparent that

$$\|C\| = \sup\{|\rho(C)| : \rho \text{ a pure state of } \mathscr{C}\}$$

for each C in $\mathscr{C}$. With ρ a pure state of $\mathscr{C}$, $\rho \circ \eta$ is a positive linear functional on $\mathscr{R}$, and is therefore bounded, with $\|\rho \circ \eta\| = \rho(\eta(I)) \le \|\eta(I)\|$. Hence

$$|\rho(\eta(A))| \le \|\rho \circ \eta\| \, \|A\| \le \|\eta(I)\| \, \|A\| \qquad (A \in \mathscr{R});$$

and by taking the supremum as ρ varies, we have $\|\eta(A)\| \le \|\eta(I)\| \, \|A\|$. This shows that η is bounded, with $\|\eta\| \le \|\eta(I)\|$, and the reverse inequality is apparent. ■

By a *center state* on $\mathscr{R}$ we mean a positive linear mapping $\eta: \mathscr{R} \to \mathscr{C}$ such that $\eta(C) = C$ and $\eta(CA) = C\eta(A)$ for each C in $\mathscr{C}$ and A in $\mathscr{R}$. From Lemma 8.2.2, such a mapping η is necessarily bounded, with $\|\eta\| = 1$.

8.2.3. LEMMA. *$\mathscr{R}$ has an ultraweakly continuous center state.*

Proof. Let $\{P_a\}$ be an orthogonal family of cyclic projections in $\mathscr{C}$, with sum I, and choose a vector x_a so that P_a has range $[\mathscr{C}'x_a]$. If E_a is the projection in $\mathscr{C}'$ with range $[\mathscr{C}x_a]$, then x_a is a cyclic vector for the abelian von Neumann algebra $\mathscr{C}E_a$, and so $E_a\mathscr{C}'E_a$ $(= (\mathscr{C}E_a)')$ is abelian by Corollary 7.2.16. Since E_a is abelian and has central carrier P_a (relative to $\mathscr{C}'$), the projection $E = \sum E_a$ in $\mathscr{C}'$ has central carrier $\sum P_a$ $(= I)$, and is abelian by Proposition 6.4.5. Thus $E\mathscr{C}'E$ is abelian, and

$$E\mathscr{R}E \subseteq E\mathscr{C}'E \subseteq (E\mathscr{C}'E)' = \mathscr{C}E.$$

The mapping $C \to CE$ is a * isomorphism φ from $\mathscr{C}$ onto $\mathscr{C}E$, from Proposition 5.5.5, and is therefore ultraweakly bicontinuous, by the final assertion of Remark 7.4.4. It is now easily verified that η, defined by $\eta(A) = \varphi^{-1}(EAE)$ $(A \in \mathscr{R})$, is an ultraweakly continuous center state on $\mathscr{R}$. ■

8.2.4. LEMMA. *If $\mathscr{R}$ is finite, η is a positive linear mapping from $\mathscr{R}$ into another von Neumann algebra $\mathscr{S}$ and $\varepsilon > 0$, the following three conditions are equivalent:*

(i) $\eta(UAU^*) \le (1 + \varepsilon)\eta(A) \qquad (A \in \mathscr{R}^+, U \in \mathscr{U})$.

(ii) $\eta(AA^*) \le (1 + \varepsilon)\eta(A^*A) \qquad (A \in \mathscr{R})$.

(iii) $\eta(E) \le (1 + \varepsilon)\eta(F)$ *whenever E and F are equivalent projections in $\mathscr{R}$.*

Proof. Suppose that (i) is satisfied. Each A in $\mathscr{R}$ has a polar decomposition VH, with H in $\mathscr{R}^+$ and V a partial isometry in $\mathscr{R}$. The initial and final projections of V, E and F, satisfy $E \sim F$, $H = EH$. Since $\mathscr{R}$ is finite, $I - E \sim I - F$ (otherwise, the comparison theorem yields a contradiction), so there is a partial isometry W in $\mathscr{R}$, from $I - E$ to $I - F$. With $U = V + W$, we have $U \in \mathscr{U}$ and $A = UH$. From (i),

$$\eta(AA^*) = \eta(UH^2U^*) \le (1 + \varepsilon)\eta(H^2) = (1 + \varepsilon)\eta(A^*A).$$

This proves that (i) implies (ii), and it is evident that (ii) implies (iii).

Finally, suppose that (iii) is satisfied. Given U in $\mathscr{U}$ and a projection E in $\mathscr{R}$, $UEU^* \sim E$, whence $\eta(UEU^*) \le (1 + \varepsilon)\eta(E)$ by (iii). Since each A in $\mathscr{R}^+$

is a norm limit of positive linear combinations of projections in $\mathscr{R}$, while positivity of η implies that η is bounded, it follows that $\eta(UAU^*) \leq (1 + \varepsilon)\eta(A)$. Thus (iii) implies (i). ■

We have now assembled the various pieces of information needed to show the existence of the center-valued trace in a finite von Neumann algebra $\mathscr{R}$. The proof is given as a series of lemmas in which, starting from an ultraweakly continuous center state on $\mathscr{R}$, we construct a sequence of such states that satisfy an approximate form of the properties required of a trace. It turns out that this sequence converges in norm, and its limit is the trace.

8.2.5. LEMMA. *If $\mathscr{R}$ is finite, P is a non-zero projection in $\mathscr{C}$, η is an ultraweakly continuous center state on $\mathscr{R}$, and $\varepsilon > 0$, then there is a projection G in $\mathscr{R}$ such that $G \leq P$, $\eta(G) > 0$, and*

$$\eta(AA^*) \leq (1 + \varepsilon)\eta(A^*A)$$

for each A in $G\mathscr{R}G$.

Proof. With x a unit vector in the range of P, the restriction $\omega_x \circ \eta \,|\, \mathscr{R}P$ is a positive normal functional on $\mathscr{R}P$, and $(\omega_x \circ \eta)(P) = \omega_x(P) = 1$. The support of $\omega_x \circ \eta \,|\, \mathscr{R}P$ is a projection E in $\mathscr{R}$, with $0 < E \leq P$. If F is a projection in $\mathscr{R}$ such that $0 < F \leq E$, then $(\omega_x \circ \eta)(F) > 0$ and thus $\eta(F) > 0$.

Let $\{(M_a', M_a''): a \in \mathbb{A}\}$ be a family of pairs of subprojections of E in $\mathscr{R}$, maximal subject to the conditions that each of the families $\{M_a'\}$, $\{M_a''\}$ is pairwise orthogonal, $M_a' \sim M_a''$ and $\eta(M_a') < \eta(M_a'')$ $(a \in \mathbb{A})$. Thus $\sum M_a'$ and $\sum M_a''$ are equivalent subprojections of E in $\mathscr{R}$ and (unless the index set $\mathbb{A}$ is vacuous) ultraweak continuity of η entails

$$\eta(\textstyle\sum M_a') < \eta(\textstyle\sum M_a'') \leq \eta(E).$$

If $E' = E - \sum M_a'$ and $E'' = E - \sum M_a''$, then $E' \neq 0$ and $E' \sim E''$ since $\mathscr{R}$ is finite (otherwise, the comparison theorem yields a contradiction). Moreover, $\eta(M') \geq \eta(M'')$ whenever $M' \sim M''$, $M' \leq E'$, and $M'' \leq E''$; for otherwise, the range projection Q of $[\eta(M'') - \eta(M')]^+$ satisfies

$$Q \in \mathscr{C}, \qquad \eta(QM'') = Q\eta(M'') > Q\eta(M') = \eta(QM'),$$

and the pair (QM', QM'') can be added to the family $\{(M_a', M_a'')\}$, contradicting the assumed maximality.

Let μ (≤ 1) be the smallest real number such that, whenever $M' \sim M''$, $M' \leq E'$, and $M'' \leq E''$, we have $\eta(M'') \leq \mu\eta(M')$. Then $\mu\eta(E') \geq \eta(E'')$, and $\eta(E'') > 0$ since $0 < E'' \leq E$; so $\mu > 0$. Since $0 < \mu(1 + \varepsilon)^{-1} < \mu$, there are equivalent projections F' $(\leq E')$ and F'' $(\leq E'')$ in $\mathscr{R}$ such that $(1+\varepsilon)\eta(F'') \nleq \mu\eta(F')$. With Q the range projection of $[(1+\varepsilon)\eta(F'') - \mu\eta(F')]^+$, $Q \in \mathscr{C}$ and

$$(1+\varepsilon)\eta(QF'') = (1+\varepsilon)Q\eta(F'') > \mu Q\eta(F') = \mu\eta(QF').$$

Upon replacing F' and F'' by QF' and QF'', respectively, we have

$$F' \leq E', \qquad F'' \leq E'', \qquad F' \sim F'', \qquad (1+\varepsilon)\eta(F'') > \mu\eta(F').$$

Let $\{(N_b', N_b''): b \in \mathbb{B}\}$ be a family of pairs of projections in $\mathscr{R}$, maximal subject to the conditions that each of the families $\{N_b'\}, \{N_b''\}$ is pairwise orthogonal, $N_b' \leq F'$, $N_b'' \leq F''$, $N_b' \sim N_b''$, and $(1+\varepsilon)\eta(N_b'') < \mu\eta(N_b')$, for all b in $\mathbb{B}$. Then $\sum N_b'$ and $\sum N_b''$ are equivalent projections in $\mathscr{R}$, dominated by F' and F'', respectively. With $G' = F' - \sum N_b'$ and $G'' = F'' - \sum N_b''$, the finiteness of $\mathscr{R}$ entails $G' \sim G''$.

If $G' = 0$, then $G'' = 0$, so $F' = \sum N_b'$, $F'' = \sum N_b''$. The ultraweak continuity of η then gives

$$(1+\varepsilon)\eta(F'') = \sum (1+\varepsilon)\eta(N_b'') < \sum \mu\eta(N_b') = \mu\eta(F'),$$

contradicting our choice of F', F''. Thus $G' \neq 0 \neq G''$.

Suppose that N' $(\leq G')$ and N'' $(\leq G'')$ are equivalent projections in $\mathscr{R}$. Then $N' \leq E'$, $N'' \leq E''$, and, by the definition of μ, $\eta(N'') \leq \mu\eta(N')$. If $(1+\varepsilon)\eta(N'') \ngeq \mu\eta(N')$, then we have that the range projection Q of $[\mu\eta(N') - (1+\varepsilon)\eta(N'')]^+$ satisfies $Q \in \mathscr{C}$,

$$\mu\eta(QN') = \mu Q\eta(N') > (1+\varepsilon)Q\eta(N'') = (1+\varepsilon)\eta(QN'').$$

Then (QN', QN'') can be added to the family $\{(N_b', N_b'')\}$, contradicting the assumed maximality. Thus

$$\eta(N'') \leq \mu\eta(N') \leq (1+\varepsilon)\eta(N'').$$

If N_1'' and N_2'' are equivalent subprojections of G'' in $\mathscr{R}$, then $N_2'' \leq G'' \sim G'$, so there is a projection N' in $\mathscr{R}$ such that $N_1'' \sim N_2'' \sim N' \leq G'$. From the preceding paragraph,

$$\eta(N_1'') \leq \mu\eta(N') \leq (1+\varepsilon)\eta(N_2'').$$

It follows that the positive linear mapping $\eta\,|\,G''\mathscr{R}G''$, from $G''\mathscr{R}G''$ into $\mathscr{C}$, satisfies condition (iii) of Lemma 8.2.4, and therefore satisfies also condition (ii) of that lemma. Moreover, $\eta(G'') > 0$ since $0 < G'' \leq E$. This proves the required result, with $G = G''$. ∎

8.2.6. Lemma. *If $\mathscr{R}$ is finite, P is a non-zero projection in $\mathscr{C}$, and $\varepsilon > 0$, there is an ultraweakly continuous positive linear mapping $\tau: \mathscr{R} \to \mathscr{C}$ and a projection Q in $\mathscr{C}$ such that $0 < Q \leq P$, and*

$$\tau(AA^*) \leq (1 + \varepsilon)\tau(A^*A), \qquad \tau(CA) = C\tau(A), \qquad \tau(CQ) = CQ$$

whenever $A \in \mathscr{R}$ and $C \in \mathscr{C}$.

Proof. Let η be an ultraweakly continuous center state on $\mathscr{R}$ (Lemma 8.2.3), and choose a projection G in $\mathscr{R}$ such that $G \leq P$, $\eta(G) > 0$, and

$$\eta(AA^*) \leq (1 + \varepsilon)\eta(A^*A)$$

whenever $A \in G\mathscr{R}G$ (Lemma 8.2.5). Since G is the sum of an orthogonal family of monic projections, the above conditions are still satisfied when G is replaced by a suitable monic subprojection. We may therefore suppose that G is monic, and choose projections $G_1, \ldots, G_n$ in $\mathscr{R}$ and P_0 in $\mathscr{C}$ so that

$$G \sim G_1 \sim G_2 \sim \cdots \sim G_n, \qquad G_1 + G_2 + \cdots + G_n = P_0.$$

For $j = 1, \ldots, n$, let V_j be a partial isometry in $\mathscr{R}$, from G_j to G. We define an ultraweakly continuous positive linear mapping $\tau_0: \mathscr{R} \to \mathscr{C}$ by

$$\tau_0(A) = \sum_{j=1}^{n} \eta(V_j A V_j^*),$$

and note that

$$\tau_0(A) = \tau_0(P_0 A), \qquad \tau_0(CA) = C\tau_0(A)$$

for each A in $\mathscr{R}$ and C in $\mathscr{C}$. Since $V_j \mathscr{R} V_k^* \subseteq G\mathscr{R}G$, we have

$$\begin{aligned}
\tau_0(AA^*) = \tau_0(P_0 AA^*) = \tau_0(AP_0A^*) &= \sum_{k=1}^{n} \tau_0(AG_k A^*) \\
&= \sum_{k=1}^{n}\sum_{j=1}^{n} \eta(V_j A V_k^* V_k A^* V_j^*) \\
&= \sum_{k=1}^{n}\sum_{j=1}^{n} \eta((V_j A V_k^*)(V_j A V_k^*)^*) \\
&\leq (1 + \varepsilon) \sum_{j=1}^{n}\sum_{k=1}^{n} \eta((V_j A V_k^*)^*(V_j A V_k^*)) \\
&= (1 + \varepsilon) \sum_{j=1}^{n}\sum_{k=1}^{n} \eta(V_k A^* V_j^* V_j A V_k^*) = (1 + \varepsilon)\tau_0(A^*A)
\end{aligned}$$

for each A in $\mathscr{R}$. Also

$$\tau_0(P) = \sum_{j=1}^{n} \eta(V_j P V_j^*) = \sum_{j=1}^{n} \eta(G) > 0.$$

We can choose a positive real number a and a spectral projection Q of $\tau_0(P)$ ($\in \mathscr{C}^+$) so that

$$\tau_0(P)Q \geq aQ > 0.$$

Then $Q \in \mathscr{C}$, and $Q \leq P$ since $\tau_0(P) = P\tau_0(I)$. As an element of $\mathscr{C}Q$, $\tau_0(P)Q$ is invertible, so there is a positive operator C_0 in $\mathscr{C}Q$ such that $\tau_0(P)C_0 = Q$. With C in $\mathscr{C}Q$

$$\tau_0(C)C_0 = \tau_0(CP)C_0 = C\tau_0(P)C_0 = CQ = C.$$

The conclusions of the lemma are now satisfied if we define $\tau(A)$ to be $\tau_0(A)C_0$ for each A in $\mathscr{R}$. ■

8.2.7. Lemma. *If $\mathscr{R}$ is finite and $\varepsilon > 0$, there is an ultraweakly continuous center state τ on $\mathscr{R}$ such that*

$$\tau(AA^*) \leq (1 + \varepsilon)\tau(A^*A)$$

for each A in $\mathscr{R}$.

Proof. Let $\{Q_b : b \in \mathbb{B}\}$ be a maximal orthogonal family of non-zero projections in $\mathscr{C}$ with the following property: for each b in $\mathbb{B}$, there is an ultraweakly continuous positive linear mapping $\tau_b : \mathscr{R} \to \mathscr{C}$ such that

$$\text{(1)} \quad \tau_b(AA^*) \leq (1 + \varepsilon)\tau_b(A^*A), \qquad \tau_b(CA) = C\tau_b(A), \qquad \tau_b(CQ_b) = CQ_b,$$

whenever $A \in \mathscr{R}$ and $C \in \mathscr{C}$. If $\sum Q_b \neq I$, it follows from Lemma 8.2.6 that $I - \sum Q_b$ has a non-zero subprojection Q that can be added to the family $\{Q_b\}$, contradicting the maximality assumption. Thus $\sum Q_b = I$.

With A in $\mathscr{R}$ and b in $\mathbb{B}$,

$$\tau_b(Q_b A) = Q_b \tau_b(A) \in \mathscr{C}Q_b,$$

so $\tau_b | \mathscr{R}Q_b$ maps $\mathscr{R}Q_b$ into its center $\mathscr{C}Q_b$. From Lemma 8.2.2, $\|\tau_b | \mathscr{R}Q_b\| = \|\tau_b(Q_b)\| = \|Q_b\| = 1$, so $\|\tau_b(Q_b A)\| \leq \|Q_b A\| \leq \|A\|$ $(A \in \mathscr{R})$. We can now define a positive linear mapping $\tau : \mathscr{R} \to \mathscr{C}$ by

$$\tau(A) = \sum \tau_b(Q_b A) \qquad (A \in \mathscr{R}).$$

From (1),

$$\begin{aligned} \tau(C) &= \sum \tau_b(Q_b C) = \sum Q_b C = C, \\ \tau(CA) &= \sum \tau_b(Q_b CA) = \sum C\tau_b(Q_b A) = C\tau(A), \\ \tau(AA^*) &= \sum \tau_b((Q_b A)(Q_b A)^*) \\ &\leq (1 + \varepsilon) \sum \tau_b((Q_b A)^*(Q_b A)) = (1 + \varepsilon)\tau(A^*A) \end{aligned}$$

for each A in $\mathscr{R}$ and C in $\mathscr{C}$. It remains to prove that τ is ultraweakly continuous.

With ω a normal state of $\mathscr{C}$, and ω_b defined by $\omega_b(A) = \omega(\tau_b(Q_b A))$, ω_b is a positive normal functional on $\mathscr{R}$ and

$$\sum \|\omega_b\| = \sum \omega_b(I) = \sum \omega(\tau_b(Q_b)) = \sum \omega(Q_b) = 1.$$

Since the predual $\mathscr{R}_\sharp$ is a Banach space, $\sum \omega_b$ converges (in norm) to a normal state ρ of $\mathscr{R}$. Now

$$\rho(A) = \sum \omega(\tau_b(Q_b A)) = \omega(\tau(A)),$$

whence $\omega \circ \tau$ $(= \rho)$ is normal. Thus τ is ultraweakly continuous (see Remark 7.4.4). ■

We are now in a position to prove the existence and uniqueness of the center-valued trace in a finite von Neumann algebra.

8.2.8. THEOREM. *If $\mathscr{R}$ is a finite von Neumann algebra with center $\mathscr{C}$, there is a unique positive linear mapping τ from $\mathscr{R}$ into $\mathscr{C}$ such that*

(i) $\tau(AB) = \tau(BA)$ *for each A and B in $\mathscr{R}$,*
(ii) $\tau(C) = C$ *for each C in $\mathscr{C}$.*

Moreover, if $A \in \mathscr{R}$ and $C \in \mathscr{C}$,

(iii) $\tau(A) > 0$ *if* $A > 0$,
(iv) $\tau(CA) = C\tau(A)$,
(v) $\|\tau(A)\| \le \|A\|$,
(vi) τ *is ultraweakly continuous.*

Proof. Let $\{a_n\}$ be a strictly decreasing sequence of real numbers that converges to 1. By Lemma 8.2.7 we can choose ultraweakly continuous center states $\tau_1, \tau_2, \ldots$ on $\mathscr{R}$ such that

$$\tau_n(AA^*) \le a_n \tau_n(A^*A) \qquad (A \in \mathscr{R}). \tag{2}$$

We assert that, if $1 \le m < n$, then $a_m^2 \tau_m - \tau_n$ is a *positive* linear mapping from $\mathscr{R}$ into $\mathscr{C}$. For this, it is enough to prove that

$$a_m^2 \tau_m(E) \ge \tau_n(E) \tag{3}$$

for each projection E in $\mathscr{R}$. Moreover, since E is the sum of an orthogonal family of monic projections, while τ_m and τ_n are ultraweakly continuous, it suffices to prove (3) for monic projections. In this case, we can choose projections $E_1, \ldots, E_k$ in $\mathscr{R}$ and Q in $\mathscr{C}$ such that

$$E \sim E_1 \sim E_2 \sim \cdots \sim E_k, \qquad E_1 + E_2 + \cdots + E_k = Q.$$

From (2), with A a suitable partial isometry, we have

$$\tau_n(E) \le a_n \tau_n(E_j), \qquad \tau_m(E_j) \le a_m \tau_m(E) \qquad (j = 1, \ldots, k).$$

Thus

$$k\tau_n(E) \le a_n \sum_{j=1}^{k} \tau_n(E_j) = a_n \tau_n(Q) = a_n Q$$

$$= a_n \tau_m(Q) = a_n \sum_{j=1}^{k} \tau_m(E_j) \le k a_n a_m \tau_m(E) \le k a_m^2 \tau_m(E).$$

This proves our assertion that $a_m^2\tau_m - \tau_n \ge 0$.

When $1 \le m < n$, both τ_m and $a_m^2\tau_m - \tau_n$ are positive linear mappings from $\mathscr{R}$ into $\mathscr{C}$. By Lemma 8.2.2,

$$\begin{aligned}\|\tau_m - \tau_n\| &\le \|\tau_m - a_m^2\tau_m\| + \|a_m^2\tau_m - \tau_n\| \\ &= (a_m^2 - 1)\|\tau_m(I)\| + \|a_m^2\tau_m(I) - \tau_n(I)\| \\ &= 2(a_m^2 - 1)\end{aligned}$$

since $\tau_m(I) = \tau_n(I) = I$. Now $a_m \to 1$ as $m \to \infty$, so $\|\tau_m - \tau_n\| \to 0$ as $m, n \to \infty$. It follows that there is a bounded linear mapping $\tau\colon \mathscr{R} \to \mathscr{C}$ such that $\|\tau_n - \tau\| \to 0$.

Since each τ_n is a center state, so is τ. From (2), $\tau(AA^*) \le \tau(A^*A)$ (and so, $\tau(AA^*) = \tau(A^*A)$) for each A in $\mathscr{R}$. By Proposition 8.1.1, $\tau(AB) = \tau(BA)$ for each A and B in $\mathscr{R}$. If $\omega \in \mathscr{C}_\sharp$, $\omega \circ \tau$ is the norm limit of the linear functionals $\omega \circ \tau_n$. Since τ_n is ultraweakly continuous, $\omega \circ \tau_n \in \mathscr{R}_\sharp$, and so $\omega \circ \tau \in \mathscr{R}_\sharp$. Thus τ is ultraweakly continuous.

So far, we have proved the existence of a positive linear mapping τ, from $\mathscr{R}$ into $\mathscr{C}$, that satisfies conditions (i), (ii), (iv), (v), and (vi) in the theorem. In order to prove (iii), and the uniqueness of τ, we adopt a more arithmetic approach.

Suppose that F is a monic projection in $\mathscr{R}$, and choose projections $F_1, \ldots, F_k$ in $\mathscr{R}$ and Q in $\mathscr{C}$ so that

$$F \sim F_1 \sim F_2 \sim \cdots \sim F_k, \qquad F_1 + F_2 + \cdots + F_k = Q.$$

From (i), (ii), and Proposition 8.1.1, $\tau(F_j) = \tau(F)$, so

$$k\tau(F) = \sum_{j=1}^{k} \tau(F_j) = \tau(Q) = Q,$$

$$\tau(F) = \frac{1}{k} Q > 0. \tag{4}$$

If $A \in \mathscr{R}$ and $A > 0$, we can choose a positive real number a and a spectral projection E of A such that $A \ge aE > 0$. With F a monic subprojection of E, $A - aF \ge 0$, so $\tau(A - aF) \ge 0$. From (4), $\tau(A) \ge a\tau(F) > 0$, whence τ satisfies (iii).

If $\tau': \mathscr{R} \to \mathscr{C}$ is a positive linear mapping satisfying (i) and (ii), the argument used above to prove (4) applies also to τ'. Thus $\tau'(F) = \tau(F)$ for each monic projection F in $\mathscr{R}$. Every projection E in $\mathscr{R}$ is the sum of a family $\{E_b : b \in \mathbb{B}\}$ of monic projections. When $\mathbb{A}$ is a finite subset of $\mathbb{B}$,

$$\sum_{b \in \mathbb{A}} \tau(E_b) = \sum_{b \in \mathbb{A}} \tau'(E_b) = \tau'\left(\sum_{b \in \mathbb{A}} E_b\right) \leq \tau'(E).$$

This and the ultraweak continuity of τ now give

$$\tau(E) = \sum_{b \in \mathbb{B}} \tau(E_b) \leq \tau'(E).$$

By applying this conclusion to $I - E$, we have

$$I - \tau(E) = \tau(I - E) \leq \tau'(I - E) = I - \tau'(E), \qquad \tau'(E) \leq \tau(E).$$

Hence $\tau'(E) = \tau(E)$ for every projection E in $\mathscr{R}$. Since τ and τ' are bounded (Lemma 8.2.2), while $\mathscr{R}$ is the norm-closed linear span of its projections, it follows that $\tau'(A) = \tau(A)$ for each A in $\mathscr{R}$. ■

By using the properties of the center-valued trace, we obtain a new proof of the following result, which is a special case of (but essentially equivalent to) Proposition 6.3.10.

8.2.9. COROLLARY. *A finite von Neumann algebra $\mathscr{R}$, with countably decomposable center $\mathscr{C}$, is itself countably decomposable.*

Proof. By Corollary 5.5.17, $\mathscr{C}$ has a separating vector x, and $\langle Cx, x\rangle = \|C^{1/2}x\|^2 > 0$, whenever $C \in \mathscr{C}^+$ and $C \neq 0$. Suppose that τ is the center-valued trace on $\mathscr{R}$, and $\{E_b : b \in \mathbb{B}\}$ is an orthogonal family of non-zero projections in $\mathscr{R}$. For each finite subset $\mathbb{A}$ of $\mathbb{B}$,

$$\begin{aligned}\sum_{b \in \mathbb{A}} \langle \tau(E_b)x, x\rangle &= \left\langle \tau\left(\sum_{b \in \mathbb{A}} E_b\right)x, x\right\rangle \\ &\leq \langle \tau(I)x, x\rangle = \|x\|^2;\end{aligned}$$

so

$$\sum_{b \in \mathbb{B}} \langle \tau(E_b)x, x\rangle \leq \|x\|^2. \tag{5}$$

Since $E_b > 0$, we have $\tau(E_b) > 0$ and thus $\langle \tau(E_b)x, x\rangle > 0$ for each b in $\mathbb{B}$. This, with (5), implies that $\mathbb{B}$ is countable. ■

Bibliography: [17, 41, 44, 57, 82].

8.3. The Dixmier approximation theorem

Throughout this section, $\mathscr{R}$ is a von Neumann algebra with center $\mathscr{C}$ and unitary group $\mathscr{U}$. Given A in $\mathscr{R}$, we denote by $\mathrm{co}_{\mathscr{R}}(A)$ the convex hull of the set $\{UAU^* : U \in \mathscr{U}\}$, and by $\mathrm{co}_{\mathscr{R}}(A)^=$ its norm-closed convex hull. Our main purpose is to prove that, for every A in $\mathscr{R}$, $\mathrm{co}_{\mathscr{R}}(A)^=$ meets $\mathscr{C}$.

We denote by $\mathscr{D}$ the set of all mappings $\alpha : \mathscr{R} \to \mathscr{R}$ that can be defined by an equation of the form

$$\alpha(A) = \sum_{j=1}^{n} a_j U_j A U_j^* \qquad (A \in \mathscr{R}),$$

where $U_j \in \mathscr{U}$, $a_j > 0$ $(j = 1, \ldots, n)$, and $\sum_{j=1}^{n} a_j = 1$. It is clear that

(i) $\mathrm{co}_{\mathscr{R}}(A) = \{\alpha(A) : \alpha \in \mathscr{D}\}$ for each A in $\mathscr{R}$;
(ii) $\mathscr{C} = \{A \in \mathscr{R} : \alpha(A) = A$ for each α in $\mathscr{D}\}$.

Moreover, for each α, β in $\mathscr{D}$, A in $\mathscr{R}$, and C in $\mathscr{C}$,

(iii) α is a norm-decreasing linear operator on $\mathscr{R}$, and is continuous also in the strong-operator, weak-operator, and ultraweak topologies;
(iv) $\alpha\beta \in \mathscr{D}$, where $\alpha\beta$ is the composition of the mappings α and β;
(v) $\alpha(A^*) = \alpha(A)^*$, $\alpha(CA) = C\alpha(A)$, and $\alpha(A) \in \mathscr{R}^+$ if $A \in \mathscr{R}^+$;
(vi) $\mathrm{co}_{\mathscr{R}}(\alpha(A)) \subseteq \mathrm{co}_{\mathscr{R}}(A)$, $\mathrm{co}_{\mathscr{R}}(\alpha(A))^= \subseteq \mathrm{co}_{\mathscr{R}}(A)^= \subseteq \|A\|(\mathscr{R})_1$;
(vii) $\alpha(\mathrm{co}_{\mathscr{R}}(A)) \subseteq \mathrm{co}_{\mathscr{R}}(A)$, $\alpha(\mathrm{co}_{\mathscr{R}}(A)^=) \subseteq \mathrm{co}_{\mathscr{R}}(A)^=$.

8.3.1. LEMMA. *If $A = A^* \in \mathscr{R}$, there exist U in $\mathscr{U}$ and a self-adjoint element C in $\mathscr{C}$ such that*

$$\|\tfrac{1}{2}(A + UAU^*) - C\| \le \tfrac{3}{4}\|A\|.$$

Proof. We may assume that $\|A\| = 1$. With E the range projection of A^+, and $F = I - E$, we have

$$-F \le A \le E. \tag{1}$$

By the comparison theorem, there exist projections P and Q in $\mathscr{C}$, with sum I, such that $PE \precsim PF$, $QE \succsim QF$. Let E_1, E_2, F_1, F_2 be projections in $\mathscr{R}$ such that

$$PE \sim E_1 \le E_1 + E_2 = PF, \qquad QF \sim F_1 \le F_1 + F_2 = QE;$$

and let V, W be partial isometries in $\mathscr{R}$ that implement the two equivalences just mentioned. Since PE, E_1, E_2, QF, F_1, F_2 are mutually orthogonal projections with sum I, the operator

$$U = V + V^* + W + W^* + E_2 + F_2$$

is unitary, $U = U^*$, and

(2) $$UPEU^* = E_1, \qquad UE_1U^* = PE, \qquad UE_2U^* = E_2,$$

(3) $$UQFU^* = F_1, \qquad UF_1U^* = QF, \qquad UF_2U^* = F_2.$$

From (1), $-PF \le PA \le PE$; that is,

$$-E_1 - E_2 \le PA \le PE.$$

From this and (2),

$$-PE - E_2 \le UPAU^* \le E_1,$$

and thus

$$-\tfrac{1}{2}(PE + E_1) - E_2 \le \tfrac{1}{2}(PA + UPAU^*) \le \tfrac{1}{2}(PE + E_1).$$

Since $PE + E_1 + E_2 = P(E + F) = P$, these last inequalities imply that

$$-P \le \tfrac{1}{2}(PA + UPAU^*) \le \tfrac{1}{2}P,$$

(4) $$-\tfrac{3}{4}P \le \tfrac{1}{2}(PA + UPAU^*) + \tfrac{1}{4}P \le \tfrac{3}{4}P.$$

Again by (1), $-QF \le QA \le QE = F_1 + F_2$. By reasoning as above, but using (3) in place of (2), we obtain

(5) $$-\tfrac{3}{4}Q \le \tfrac{1}{2}(QA + UQAU^*) - \tfrac{1}{4}Q \le \tfrac{3}{4}Q.$$

From (4) and (5),

$$-\tfrac{3}{4}I \le \tfrac{1}{2}(A + UAU^*) - \tfrac{1}{4}(Q - P) \le \tfrac{3}{4}I,$$

whence $\|\tfrac{1}{2}(A + UAU^*) - \tfrac{1}{4}(Q - P)\| \le \tfrac{3}{4}$. This proves the required result, where $C = \tfrac{1}{4}(Q - P)$. ■

8.3.2. LEMMA. *If A is a self-adjoint element of $\mathscr{R}$ and $\varepsilon > 0$, there exist α in $\mathscr{D}$ and a self-adjoint element of C of $\mathscr{C}$ such that $\|\alpha(A) - C\| < \varepsilon$.*

Proof. It suffices to show that, for each $n = 1, 2, \ldots$, there exist α_n in $\mathscr{D}$ and a self-adjoint C_n in $\mathscr{C}$ such that

(6) $$\|\alpha_n(A) - C_n\| \le (\tfrac{3}{4})^n \|A\|.$$

The proof is by induction on n. The existence of suitable α_1 and C_1 follows from Lemma 8.3.1. When α_p and C_p have been constructed, the same lemma (applied to $\alpha_p(A) - C_p$) asserts the existence of α in $\mathscr{D}$ and a self-adjoint C in $\mathscr{C}$ such that

$$\|\alpha(\alpha_p(A) - C_p) - C\| \le \tfrac{3}{4}\|\alpha_p(A) - C_p\| \le (\tfrac{3}{4})^{p+1}\|A\|.$$

Since $\alpha(C_p) = C_p$, (6) is satisfied when $n = p + 1$ if we define $\alpha_{p+1} = \alpha\alpha_p$, $C_{p+1} = C_p + C$. ■

8.3.3. LEMMA. *If $A_1, \ldots, A_n \in \mathscr{R}$ and $\varepsilon > 0$, there exists α in $\mathscr{D}$ and $C_1, \ldots, C_n$ in $\mathscr{C}$ such that $\|\alpha(A_j) - C_j\| < \varepsilon$ $(j = 1, \ldots, n)$.*

Proof. Upon replacing each A_j by its self-adjoint and skew-adjoint parts, and ε by $\frac{1}{2}\varepsilon$, we reduce to the case in which $A_1, \ldots, A_n$ are self-adjoint. We prove the lemma, in this case, by induction on n. When $n = 1$, the result has already been proved in Lemma 8.3.2.

Suppose the result is known for the case of $n - 1$ self-adjoint operators. Then, we can choose β in $\mathscr{D}$ and $C_1, \ldots, C_{n-1}$ in $\mathscr{C}$ so that $\|\beta(A_j) - C_j\| < \varepsilon$ $(1 \leq j < n)$. From Lemma 8.3.2, applied to $\beta(A_n)$, there exist γ in $\mathscr{D}$ and C_n in $\mathscr{C}$ such that

$$\|\gamma\beta(A_n) - C_n\| < \varepsilon.$$

Since $\|\gamma\| = 1$ and $\gamma(C_j) = C_j$, we have

$$\|\gamma\beta(A_j) - C_j\| = \|\gamma(\beta(A_j) - C_j)\| \leq \|\beta(A_j) - C_j\| < \varepsilon \qquad (1 \leq j < n);$$

and the required conditions are satisfied, with $\alpha = \gamma\beta$. ■

8.3.4. PROPOSITION. *If $A_1, \ldots, A_n \in \mathscr{R}$, there exist $C_1, \ldots, C_n$ in $\mathscr{C}$ and $\alpha_1, \alpha_2, \alpha_3, \ldots$ in $\mathscr{D}$ such that*

$$\lim_{m \to \infty} \|\alpha_m(A_j) - C_j\| = 0 \qquad (j = 1, \ldots, n).$$

Proof. By induction on m, we construct β_m in $\mathscr{D}$ and $C_1^{(m)}, \ldots, C_n^{(m)}$ in $\mathscr{C}$ such that

$$(7) \quad \|\beta_m\beta_{m-1} \cdots \beta_1(A_j) - C_j^{(m)}\| < 2^{-m} \qquad (j = 1, \ldots, n; \quad m = 1, 2, \ldots).$$

Lemma 8.3.3 gives the starting case, $m = 1$, when applied to $A_1, \ldots, A_n$ with $\varepsilon = \frac{1}{2}$. When β_m and $C_j^{(m)}$ $(1 \leq j \leq n)$ have been constructed, the same lemma (applied to the operators $\beta_m\beta_{m-1} \cdots \beta_1(A_j)$, $1 \leq j \leq n$, with $\varepsilon = 2^{-m-1}$) proves the existence of suitable β_{m+1} and $C_j^{(m+1)}$ $(1 \leq j \leq n)$.

By (7),

$$\begin{aligned}
&\|C_j^{(m+1)} - C_j^{(m)}\| \\
&\quad \leq \|C_j^{(m+1)} - \beta_{m+1}\beta_m \cdots \beta_1(A_j)\| + \|\beta_{m+1}\beta_m \cdots \beta_1(A_j) - C_j^{(m)}\| \\
&\quad \leq 2^{-m-1} + \|\beta_{m+1}(\beta_m\beta_{m-1} \cdots \beta_1(A_j) - C_j^{(m)})\| \\
&\quad \leq 2^{-m-1} + \|\beta_m\beta_{m-1} \cdots \beta_1(A_j) - C_j^{(m)}\| \\
&\quad \leq 2^{-m-1} + 2^{-m} < 2^{1-m}
\end{aligned}$$

It follows that, for each $j = 1, \ldots, n$, $(C_j^{(m)})_{m \geq 1}$ is a Cauchy sequence and so converges to some C_j in $\mathscr{C}$. Moreover,

$$\|\beta_m \beta_{m-1} \ldots \beta_1(A_j) - C_j\| \leq \|\beta_m \beta_{m-1} \ldots \beta_1(A_j) - C_j^{(m)}\| + \|C_j^{(m)} - C_j\|$$
$$\leq 2^{-m} + \|C_j^{(m)} - C_j\| \to 0$$

as $m \to \infty$. This proves the proposition, with $\alpha_m = \beta_m \beta_{m-1} \ldots \beta_1$. ■

8.3.5. THEOREM (The Dixmier approximation theorem). *If $\mathscr{R}$ is a von Neumann algebra with center $\mathscr{C}$, and $A \in \mathscr{R}$, then $\mathrm{co}_{\mathscr{R}}(A)^=$ meets $\mathscr{C}$.*

Proof. By the preceding result, there exist $\alpha_1, \alpha_2, \ldots$ in $\mathscr{D}$ and C in $\mathscr{C}$ such that $\|\alpha_m(A) - C\| \to 0$. Hence $C \in \mathrm{co}_{\mathscr{R}}(A)^= \cap \mathscr{C}$. ■

8.3.6. THEOREM. *If $\mathscr{R}$ is a finite von Neumann algebra with center $\mathscr{C}$, and $A \in \mathscr{R}$, then $\mathrm{co}_{\mathscr{R}}(A)^= \cap \mathscr{C}$ consists of the single point $\tau(A)$, where τ is the center-valued trace on $\mathscr{R}$.*

Proof. Since $\tau(A) = \tau(U^*UA) = \tau(UAU^*)$ for each unitary U in $\mathscr{R}$, it follows by linearity and norm continuity of τ that $\tau(A) = \tau(B)$ whenever $B \in \mathrm{co}_{\mathscr{R}}(A)^=$. Hence $\tau(A) = \tau(C) = C$ for each C in $\mathrm{co}_{\mathscr{R}}(A)^= \cap \mathscr{C}$. This last set is non-empty, by the Dixmier approximation theorem, and so consists of the single element $\tau(A)$. ■

8.3.7. REMARK. In proving that $\mathrm{co}_{\mathscr{R}}(A)^= \cap \mathscr{C}$ reduces to a single point, when $\mathscr{R}$ is a finite von Neumann algebra, we have made use of the center-valued trace τ on $\mathscr{R}$. Alternatively, it is possible to show directly (without reference to τ) that $\mathrm{co}_{\mathscr{R}}(A)^=$ meets $\mathscr{C}$ at just one point, and to deduce from this the existence and properties of the center-valued trace. At the end of this section, we give an outline of a program that achieves this. ■

8.3.8. REMARK. If $\mathscr{R}$ is an infinite von Neumann algebra, then $\mathscr{R}$ has elements A such that $\mathrm{co}_{\mathscr{R}}(A)^= \cap \mathscr{C}$ contains more than one element (see Exercise 8.7.7). ■

8.3.9. PROPOSITION. *If $\mathscr{R}$ is a countably decomposable type III von Neumann algebra, $A \in \mathscr{R}$, and $A \neq 0$, then* $\mathrm{co}_{\mathscr{R}}(A)^=$ contains a non-zero *element C of $\mathscr{C}$.*

Proof. It suffices to show that there is a β in $\mathscr{D}$ such that $0 \notin \mathrm{co}_{\mathscr{R}}(\beta(A))^=$; for then, by the Dixmier approximation theorem, there is a (necessarily non-zero) element C of $\mathscr{C}$ for which $C \in \mathrm{co}_{\mathscr{R}}(\beta(A))^= \subseteq \mathrm{co}_{\mathscr{R}}(A)^=$.

We consider first the case in which A is self-adjoint. Upon replacing A by $-A$ if necessary, we may assume that $A^+ \neq 0$, and choose a positive real number s and a spectral projection E ($\neq 0$) of A such that $AE \geq sE$. Then

$$A = AE + A(I - E) \geq sE - t(I - E),$$

where $t = \|A\|$. With Q the central carrier of E, and $F = Q(I - E)$,

$$AQ \geq sE - tF, \qquad E + F = Q. \tag{8}$$

By Corollary 6.3.5 $E \sim Q \geq F$. Moreover, with n a positive integer for which $2(s + t) \leq ns$, repeated application of the halving lemma yields projections $E_1, \ldots, E_n$ in $\mathscr{R}$ such that

$$E_1 \sim E_2 \sim \cdots \sim E_n \sim E = E_1 + E_2 + \cdots + E_n,$$

since E is properly infinite. Since $F \precsim E \sim E_j$, we have $F \sim F_j \leq E_j$ for some projection F_j in $\mathscr{R}$. If V_j is a partial isometry in $\mathscr{R}$ from F to F_j, and

$$U_j = V_j + V_j^* + I - F - F_j$$

then U_j is unitary, and $U_j F U_j^* = F_j$, $U_j F_j U_j^* = F$, while $U_j G U_j^* = G$ for every subprojection G of $I - F - F_j$. Since

$$AQ \geq s(E - F_j) + sF_j - tF,$$

we have

$$U_j AQU_j^* \geq s(E - F_j) + sF - tF_j = sQ - (s + t)F_j.$$

With β in $\mathscr{D}$ defined by $\beta(S) = n^{-1}\sum_{j=1}^n U_j S U_j^*$, and $B = \beta(A)$, we have

$$\begin{aligned} BQ = \beta(A)Q = \beta(AQ) &= \frac{1}{n}\sum_{j=1}^n U_j AQU_j^* \\ &\geq sQ - \frac{s+t}{n}\sum_{j=1}^n F_j \geq \left(s - \frac{s+t}{n}\right)Q \geq \tfrac{1}{2}sQ. \end{aligned}$$

For every α in $\mathscr{D}$, $\alpha(B)Q = \alpha(BQ) \geq \frac{1}{2}s\alpha(Q) = \frac{1}{2}sQ$. Thus $TQ \geq \frac{1}{2}sQ$ for each T in $\mathrm{co}_{\mathscr{R}}(B)^=$, and therefore $0 \notin \mathrm{co}_{\mathscr{R}}(B)^= = \mathrm{co}_{\mathscr{R}}(\beta(A))^=$. This proves the required result in the case where A is self-adjoint.

For the general case, suppose that $A = H + iK$ with H and K self-adjoint in $\mathscr{R}$. Upon replacing A by iA if necessary, we may assume that $H \neq 0$. The preceding argument shows that $\mathrm{co}_{\mathscr{R}}(H)^=$ contains a non-zero element C_0 of $\mathscr{C}$. Choose β in $\mathscr{D}$ so that $\|\beta(H) - C_0\| < \|C_0\|$, and let $B = \beta(A)$. For each α in $\mathscr{D}$,

$$\begin{aligned} \|\tfrac{1}{2}[\alpha(B) + \alpha(B)^*] - C_0\| &= \|\tfrac{1}{2}[\alpha\beta(A) + \alpha\beta(A)^*] - C_0\| \\ &= \|\alpha\beta(\tfrac{1}{2}[A + A^*]) - C_0\| \\ &= \|\alpha\beta(H) - \alpha(C_0)\| \leq \|\beta(H) - C_0\|. \end{aligned}$$

Thus

$$\|\tfrac{1}{2}(T + T^*) - C_0\| \leq \|\beta(H) - C_0\| < \|C_0\|$$

for each T in $\mathrm{co}_{\mathscr{R}}(B)^=$, and therefore $0 \notin \mathrm{co}_{\mathscr{R}}(B)^= = \mathrm{co}_{\mathscr{R}}(\beta(A))^=$. ∎

We now establish the relation between the center-valued trace and central linear functionals (see the first paragraphs of Section 8.1) on a finite von Neumann algebra.

8.3.10. Proposition. *If $\mathscr{R}$ is a finite von Neumann algebra with center $\mathscr{C}$, then each ρ in $\mathscr{C}^\sharp$ extends uniquely to a central linear functional $\bar{\rho}$ in $\mathscr{R}^\sharp$. Moreover, $\bar{\rho} = \rho \circ \tau$, where τ is the center-valued trace on $\mathscr{R}$. Every central linear functional in $\mathscr{R}^\sharp$ arises in this way. Furthermore, $\|\bar{\rho}\| = \|\rho\|$, $\bar{\rho}$ is positive if ρ is positive, and $\bar{\rho} \in \mathscr{R}_\sharp$ if $\rho \in \mathscr{C}_\sharp$.*

Proof. Suppose that $\rho \in \mathscr{C}^\sharp$. It follows at once, from the properties of τ set out in Theorem 8.2.8, that $\rho \circ \tau$ is a norm-continuous central linear functional on $\mathscr{R}$, that $\rho \circ \tau | \mathscr{C} = \rho$, that $\rho \circ \tau$ is positive if ρ is positive, and that $\rho \circ \tau \in \mathscr{R}_\sharp$ if $\rho \in \mathscr{C}_\sharp$. Since τ is norm decreasing,

$$\|\rho \circ \tau\| \leq \|\rho\| = \|\rho \circ \tau | \mathscr{C}\| \leq \|\rho \circ \tau\|,$$

so $\|\rho\| = \|\rho \circ \tau\|$.

Suppose next that λ is a norm-continuous central linear functional on $\mathscr{R}$, and let $\rho = \lambda | \mathscr{C}$. Given A in $\mathscr{R}$, $\lambda(A) = \lambda(U^*UA) = \lambda(UAU^*)$ for each unitary U in $\mathscr{R}$. This, together with the linearity and norm continuity of λ, entails $\lambda(A) = \lambda(B)$ for each B in $\mathrm{co}_{\mathscr{R}}(A)^=$. By Theorem 8.3.6 we can take $B = \tau(A)$, so $\lambda(A) = \lambda(\tau(A)) = \rho(\tau(A))$. Hence $\lambda = \rho \circ \tau$. This proves that every norm-continuous central linear functional on $\mathscr{R}$ has the stated form, and also shows that the extension $\bar{\rho}$ is unique. ∎

Let $\mathscr{R}$ be a von Neumann algebra, with unitary group $\mathscr{U}$, acting on a Hilbert space $\mathscr{H}$. The Dixmier approximation theorem is equivalent to the assertion that, for each A in $\mathscr{R}$, the set $\mathrm{co}_{\mathscr{R}}(A)^=$ contains an operator T for which $UTU^* = T$ for all U in $\mathscr{U}$. We now consider the set $\mathrm{co}_{\mathscr{R}}(A)$, again defined as the convex hull of $\{UAU^* : U \in \mathscr{U}\}$, for each A in $\mathscr{B}(\mathscr{H})$. By analogy with the Dixmier approximation theorem (as interpreted above) one might hope that, for each A in $\mathscr{B}(\mathscr{H})$, the norm closure $\mathrm{co}_{\mathscr{R}}(A)^=$ (or, failing that, the weak-operator closure $\mathrm{co}_{\mathscr{R}}(A)^-$) meets the commutant $\mathscr{R}'$. It turns out that even the weaker statement, about $\mathrm{co}_{\mathscr{R}}(A)^-$, is false in general (see Exercise 8.7.29). In the positive direction we have the following result.

8.3.11. PROPOSITION. *Suppose that $\mathscr{R}$ is a von Neumann algebra acting on the Hilbert space $\mathscr{H}$, and there is a family $(\mathscr{R}_a)_{a\in\mathbb{A}}$ of finite-dimensional *subalgebras of $\mathscr{R}$ such that*

(i) *if $a, b \in \mathbb{A}$, there is an element c of $\mathbb{A}$ for which $\mathscr{R}_a \cup \mathscr{R}_b \subseteq \mathscr{R}_c$;*
(ii) $\mathscr{R} = (\bigcup_{a\in\mathbb{A}} \mathscr{R}_a)^-$.

Then, for each A in $\mathscr{B}(\mathscr{H})$, $\mathrm{co}_{\mathscr{R}}(A)^-$ meets $\mathscr{R}'$.

Proof. We may assume that $I \in \mathscr{R}_a$, whence the unitary group of $\mathscr{R}$ contains that of $\mathscr{R}_a$, for each a in $\mathbb{A}$.

We begin by proving that the unitary group of a finite-dimensional von Neumann algebra $\mathscr{S}$ has a *finite* subgroup whose linear span is $\mathscr{S}$. For this, suppose first that $\mathscr{S}$ is a type I_n factor, and let $\{E_{jk} : j, k = 1, \ldots, n\}$ be a self-adjoint system of matrix units for $\mathscr{S}$. With $S(n)$ the symmetric group of all permutations of the set $\{1, 2, \ldots, n\}$ and F the class of all mappings from $\{1, 2, \ldots, n\}$ into $\{1, -1\}$, define

$$V(f, \pi) = \sum_{j=1}^{n} f(j) E_{\pi(j)j} \qquad (f \in F, \quad \pi \in S(n)).$$

It is easily verified that $\{V(f, \pi) : f \in F, \pi \in S(n)\}$ is a finite subgroup $\mathscr{V}$ of the unitary group of $\mathscr{S}$. (In terms of matrices relative to $\{E_{jk}\}$, $\mathscr{V}$ is generated by the group of permutation matrices and the group of diagonal matrices with ± 1 at each diagonal entry.) The linear span of $\mathscr{V}$ contains each E_{jk}, and is therefore the whole of $\mathscr{S}$. Note that that $\mathscr{V} = -\mathscr{V}$.

A general finite-dimensional von Neumann algebra $\mathscr{S}$ is (* isomoprhic to) a finite direct sum $\sum_1^m \oplus \mathscr{S}_j$ of finite-dimensional factors $\mathscr{S}_1, \ldots, \mathscr{S}_m$. By the preceding paragraph, the unitary group of $\mathscr{S}_j$ has a finite subgroup $\mathscr{V}_j$ $(= -\mathscr{V}_j)$ whose linear span is $\mathscr{S}_j$. Thus

$$\mathscr{V} = \left\{ \sum_{j=1}^{m} \oplus V_j : V_j \in \mathscr{V}_j \right\}$$

is a finite subgroup of the unitary group of $\mathscr{S}$, and has linear span $\mathscr{S}$.

We show next that, for each a in $\mathbb{A}$, $\mathrm{co}_{\mathscr{R}}(A)$ meets $\mathscr{R}_a'$. For this, let $\mathscr{V}$ be a finite subgroup of the unitary group of $\mathscr{R}_a$, whose linear span is $\mathscr{R}_a$; and define $T = n^{-1} \sum_{V\in\mathscr{V}} VAV^*$, where n is the order of $\mathscr{V}$. Then $T \in \mathrm{co}_{\mathscr{R}}(A)$ and, since left translation by an element W of $\mathscr{V}$ permutes $\mathscr{V}$, we have

$$WTW^* = n^{-1} \sum_{V\in\mathscr{V}} (WV)T(WV)^* = T \qquad (W \in \mathscr{V}).$$

Thus $WT = TW$ for every W in $\mathscr{V}$; by linearity, T commutes with every element of $\mathscr{R}_a$, so $T \in \mathscr{R}_a' \cap \mathrm{co}_{\mathscr{R}}(A)$.

For each a in $\mathbb{A}$, the convex set $\mathscr{S}_a = \mathscr{R}'_a \cap \mathrm{co}_{\mathscr{R}}(A)^-$ is non-empty by the preceding paragraph, and is weak-operator compact since it is closed and bounded. When $a, b \in \mathbb{A}$, we can choose c in $\mathbb{A}$ so that $\mathscr{R}_a \cup \mathscr{R}_b \subseteq \mathscr{R}_c$, and then $\mathscr{S}_a \cap \mathscr{S}_b \supseteq \mathscr{S}_c$. Accordingly, the family $(\mathscr{S}_a)_{a \in \mathbb{A}}$ has the finite intersection property, and the compactness of $\mathscr{S}_a$ entails

$$\varnothing \neq \bigcap_{a \in \mathbb{A}} \mathscr{S}_a = \bigcap_{a \in \mathbb{A}} \mathscr{R}'_a \cap \mathrm{co}_{\mathscr{R}}(A)^- = \mathscr{R}' \cap \mathrm{co}_{\mathscr{R}}(A)^-. \quad \blacksquare$$

8.3.12. COROLLARY. *If $\mathscr{R}$ is an abelian von Neumann algebra acting on a Hilbert space $\mathscr{H}$, and $A \in \mathscr{B}(\mathscr{H})$, then $\mathrm{co}_{\mathscr{R}}(A)^-$ meets $\mathscr{R}'$.*

Proof. Let $(\mathscr{R}_a)_{a \in \mathbb{A}}$ be the family of all finite-dimensional * subalgebras of $\mathscr{R}$ containing I. If $a, b \in \mathbb{A}$, we can choose self-adjoint operators $S_1, \ldots, S_m, T_1, \ldots, T_n$ in $\mathscr{R}$, so that $\mathscr{R}_a$ is the linear span of $\{S_1, \ldots, S_m\}$ and $\mathscr{R}_b$ is the linear span of $\{T_1, \ldots, T_n\}$. Since $\mathscr{R}$ is abelian, it is easy to check that the linear span of $\{S_j T_k : 1 \leq j \leq m,\ 1 \leq k \leq n\}$ is a (finite-dimensional) * subalgebra of $\mathscr{R}$ that contains I, and so coincides with $\mathscr{R}_c$ for some c in $\mathbb{A}$. Clearly $\mathscr{R}_c \supseteq \mathscr{R}_a \cup \mathscr{R}_b$.

By the spectral theorem, each self-adjoint A in $\mathscr{R}$ is the norm limit of operators of the form $\sum_{j=1}^{n} a_j E_j$, with $a_1, \ldots, a_n$ scalars and $\{E_j\}$ an orthogonal family of projections in $\mathscr{R}$, with sum I. Since the linear span of such a family $\{E_j\}$ is a finite-dimensional * subalgebra of $\mathscr{R}$ containing I (and hence an $\mathscr{R}_b$), it follows that

$$A \in \left(\bigcup_{a \in \mathbb{A}} \mathscr{R}_a \right)^{=} \subseteq \left(\bigcup_{a \in \mathbb{A}} \mathscr{R}_a \right)^{-}.$$

Hence $\mathscr{R} = (\bigcup_{a \in \mathbb{A}} \mathscr{R}_a)^-$, and the required result follows from Proposition 8.3.11. $\blacksquare$

We conclude this section with a brief description of an alternative method for proving the existence and properties of the center-valued trace on a finite von Neumann algebra $\mathscr{R}$. This account is not essential to the logical structure of later parts of the book, and may be omitted if the reader wishes. By expressing $\mathscr{R}$ as a direct sum of algebras of types I_n $(n = 1, 2, \ldots)$ and II_1, it is not difficult to reduce the problem to the case in which $\mathscr{R}$ is either type I_n or type II_1. For algebras of type I_n, the existence and properties of the trace follow at once from the discussion of Example 8.1.3.

We assume henceforth that $\mathscr{R}$ is type II_1. Our starting point, in this case, is the Dixmier approximation theorem; and our first major objective is to show that, for each A in $\mathscr{R}$, $\mathrm{co}_{\mathscr{R}}(A)^{=} \cap \mathscr{C}$ consists of exactly one point. Of course, we now require a proof that (unlike the one given for Theorem 8.3.6) makes no reference to the center-valued trace on $\mathscr{R}$. It suffices to consider the case in which A is self-adjoint, since it is not difficult to verify that

$$\mathrm{co}_{\mathscr{R}}(H + iK)^{=} \cap \mathscr{C} \subseteq \{S + iT : S \in \mathrm{co}_{\mathscr{R}}(H)^{=} \cap \mathscr{C}, T \in \mathrm{co}_{\mathscr{R}}(K)^{=} \cap \mathscr{C}\}$$

when H and K are self-adjoint elements of $\mathscr{R}$.

Suppose, for the moment, that the following result is known.

LEMMA A. *Given any non-zero projection P in $\mathscr{C}$, there is a numerical trace ω on $\mathscr{R}$ for which $\omega(P) > 0$.*

From this we deduce that, if C is a non-zero self-adjoint element of $\mathscr{C}$, there is a numerical trace ω on $\mathscr{R}$ for which $\omega(C) \neq 0$. Indeed, we may suppose (upon replacing C by $-C$ if necessary) that $C^+ \neq 0$. We can then choose a spectral projection P of C and a real number a such that $CP \geq aP > 0$. From Lemma A, there is a numerical trace ω_0 on $\mathscr{R}$ such that $\omega_0(P) > 0$. With ω defined by $\omega(A) = \omega_0(AP)$, ω is a numerical trace on $\mathscr{R}$ and $\omega(C) = \omega_0(CP) \geq a\omega_0(P) > 0$.

Suppose now that A is a self-adjoint element of $\mathscr{R}$. A simple argument, similar to the proof of Theorem 8.3.6, shows that $\omega(A) = \omega(B)$ whenever ω is a numerical trace on $\mathscr{R}$ and $B \in \mathrm{co}_{\mathscr{R}}(A)^{=}$. It follows that, if $C_1, C_2 \in \mathrm{co}_{\mathscr{R}}(A)^{=} \cap \mathscr{C}$, then $\omega(C_1 - C_2) = 0$ for each numerical trace ω on $\mathscr{R}$. This, with the preceding paragraph, implies that $C_1 = C_2$. Thus $\mathrm{co}_{\mathscr{R}}(A)^{=} \cap \mathscr{C}$ reduces to a single point for each self-adjoint A in $\mathscr{R}$, and hence for every A in $\mathscr{R}$.

With A in $\mathscr{R}$, let $\tau(A)$ be the unique element of $\mathscr{C}$ in $\mathrm{co}_{\mathscr{R}}(A)^{=}$. Since

$$\mathrm{co}_{\mathscr{R}}(UAU^*) = \mathrm{co}_{\mathscr{R}}(A), \qquad \mathrm{co}_{\mathscr{R}}(C) = \{C\},$$

and

$$\mathrm{co}_{\mathscr{R}}(CA) = \{CB : B \in \mathrm{co}_{\mathscr{R}}(A)\},$$

whenever $A \in \mathscr{R}$, $C \in \mathscr{C}$, and U is unitary in $\mathscr{R}$, we have

$$\tau(UAU^*) = \tau(A), \qquad \tau(C) = C, \qquad \tau(CA) = C\tau(A).$$

By applying Proposition 8.3.4 to two elements A_1 and A_2 of $\mathscr{R}$, it follows easily that $\tau(A_1 + A_2) = \tau(A_1) + \tau(A_2)$. From this, it is readily verified that τ is a norm-decreasing positive linear mapping from $\mathscr{R}$ into $\mathscr{C}$. If $A \in \mathscr{R}$ and $A > 0$, the argument used in proving Theorem 8.2.8(iii) is again available to show that $\tau(A) > 0$. This proves the existence of the center-valued trace, and derives its main properties, with the exception of ultraweak continuity.

The uniqueness of τ is straightforward. Indeed, suppose that $\tau' : \mathscr{R} \to \mathscr{C}$ is a norm-continuous mapping such that

$$\tau'(UAU^*) = \tau'(A), \qquad \tau'(C) = C$$

whenever $A \in \mathscr{R}$, $C \in \mathscr{C}$, and U is a unitary operator in $\mathscr{R}$. A simple argument, similar to the proof of Theorem 8.3.6, shows that $\tau'(A) = \tau'(B)$ whenever $B \in \text{co}_{\mathscr{R}}(A)^=$. By taking $B = \tau(A)$, we obtain $\tau'(A) = \tau(A)$.

The ultraweak continuity of τ can be proved, within the present program, by an appeal to Theorem 7.1.12, after showing directly (from its other properties established above) that τ is completely additive on the set $\mathscr{P}$ of projections in $\mathscr{R}$. The proof of this complete additivity is based on simple facts about the mapping $\Delta = \tau | \mathscr{P}$: $\mathscr{P} \to \mathscr{C}$, the *center-valued dimension function* on $\mathscr{R}$, which is studied in Section 8.4. The necessary argument is set out, in detail, in Remarks 8.4.5 and 8.4.6.

It remains to prove Lemma A. This can be deduced, by a simple weak * compactness argument, from the following result.

LEMMA B. *Given a non-zero projection P in $\mathscr{C}$, and a positive integer n, there is a positive normal functional ω on $\mathscr{R}$ such that*

$$\omega(P) = \|\omega\| = 1, \qquad \omega(AA^*) \leq \left(1 + \frac{1}{n}\right)\omega(A^*A) \quad (A \in \mathscr{R}).$$

For this, it suffices to produce a positive normal functional ω_0 on $\mathscr{R}$ such that

$$\omega_0(P) > 0, \qquad \omega_0(AA^*) \leq \left(1 + \frac{1}{n}\right)\omega_0(A^*A) \quad (A \in \mathscr{R});$$

for then ω, defined by $\omega(A) = [\omega_0(P)]^{-1}\omega_0(PA)$, has the required properties. The existence of a suitable ω_0 can be deduced, by an argument analogous to the proof of Lemma 8.2.6, from the following result.

LEMMA C. *Given a non-zero projection P in $\mathscr{C}$, a positive normal functional ω on $\mathscr{R}$ such that $\omega(P) > 0$ and a positive integer n, there is a projection G in $\mathscr{R}$ such that $G \leq P$, $\omega(G) > 0$, and*

$$\omega(AA^*) \leq \left(1 + \frac{1}{n}\right)\omega(A^*A) \qquad (A \in G\mathscr{R}G).$$

The proof of Lemma C is closely analogous to that of Lemma 8.2.5, but a little simpler.

The program just described was based on the fact that $\mathrm{co}_{\mathscr{R}}(A)^=$ intersects $\mathscr{C}$ in a single point, for each A in $\mathscr{R}$, when $\mathscr{R}$ is finite. This was deduced above from Lemma A, but a minor modification of the argument permits the use of Lemma B instead (thus eliminating the need for Lemma A).

Bibliography: [17, 83, 104].

8.4. The dimension function

Suppose that $\mathscr{R}$ is a von Neumann algebra with center $\mathscr{C}$, and let $\mathscr{P}$ be the set of all projections in $\mathscr{R}$. By a *center-valued dimension function* on $\mathscr{R}$ we mean a mapping $\Delta\colon \mathscr{P} \to \mathscr{C}$ such that, if $E, F \in \mathscr{P}$ and Q is a projection in $\mathscr{C}$,

(i) $\Delta(E) > 0$ if $E \neq 0$;
(ii) $\Delta(E + F) = \Delta(E) + \Delta(F)$ if $EF = 0$;
(iii) $\Delta(E) = \Delta(F)$ if $E \sim F$;
(iv) $\Delta(Q) = Q$;
(v) $\Delta(QE) = Q\Delta(E)$.

If such a mapping Δ exists, then $\mathscr{R}$ is finite. Indeed, if E and F are projections in $\mathscr{R}$ such that $E \sim F \leq E$, properties (iii), (ii), and (i) yield

$$\Delta(F) = \Delta(E) = \Delta(F) + \Delta(E - F), \qquad \Delta(E - F) = 0, \qquad E - F = 0.$$

Conversely, suppose that $\mathscr{R}$ is a finite von Neumann algebra and τ is its center-valued trace. It follows at once, from the properties of τ set out in Theorem 8.2.8, that the restriction $\Delta = \tau\,|\,\mathscr{P}$ satsifies the five conditions above. Hence a von Neumann algebra $\mathscr{R}$ has a center-valued dimension function if and only if $\mathscr{R}$ is finite.

Our main purpose in this section, achieved in Theorems 8.4.3 and 8.4.4, is to prove the uniqueness of the dimension function in a finite von Neumann algebra, and to investigate its further properties. Before starting this program, we describe two examples that, in some measure, illustrate the theory that follows.

8.4.1. Example. Suppose that $\mathscr{H}$ is a Hilbert space with finite dimension n, and let $\mathscr{R} = \mathscr{B}(\mathscr{H})$. When E is a projection in $\mathscr{R}$, let $d(E)$ denote the dimension (in the elementary sense) of the subspace $E(\mathscr{H})$ of $\mathscr{H}$. We assert that $\mathscr{R}$ has a unique center-valued dimension function Δ, given by

$$\Delta(E) = \frac{d(E)}{n} I. \tag{1}$$

For this, note first that Δ, as defined by (1), is a center-valued dimension function on $\mathscr{R}$. Indeed, conditions (i), (ii), and (iii) above are easily verified, while (iv) and (v) result from the fact that $\mathscr{R}$ has only two central projections, $Q = 0, I$. If Δ' is another center-valued dimension function on $\mathscr{R}$, then Δ' takes the same value at all minimal projections in $\mathscr{R}$, since any two such projections are equivalent. Moreover, this common value of Δ' is a positive element of the center of $\mathscr{R}$, and so has the form cI, with $c > 0$. Each projection E in $\mathscr{R}$ is the sum of an orthogonal family of $d(E)$ minimal projections, and by properties (ii) and (iv) of Δ',

$$\Delta'(E) = d(E)cI, \qquad I = \Delta'(I) = d(I)cI = ncI.$$

Thus $c = n^{-1}$, and $\Delta'(E) = n^{-1}d(E)I = \Delta(E)$. ∎

8.4.2. Example. Suppose that $\mathscr{H}_1, \ldots, \mathscr{H}_k$ are Hilbert spaces with finite dimensions $n_1, \ldots, n_k$, respectively, and $\mathscr{R}$ is the von Neumann algebra $\sum_{j=1}^{k} \oplus \mathscr{B}(\mathscr{H}_j)$. Each projection E in $\mathscr{R}$ has the form $\{E_1, \ldots, E_k\}$, with E_j a projection in $\mathscr{B}(\mathscr{H}_j)$. We assert that $\mathscr{R}$ has a unique center-valued dimension function Δ, defined by

$$\Delta(\{E_1, \ldots, E_k\}) = \left\{\frac{d(E_1)}{n_1} I_1, \ldots, \frac{d(E_k)}{n_k} I_k\right\}, \tag{2}$$

where I_j is the identity operator on $\mathscr{H}_j$ and $d(E_j)$ is the dimension of the subspace $E_j(\mathscr{H}_j)$ of $\mathscr{H}_j$.

For this, observe that the center $\mathscr{C}$ of $\mathscr{R}$ consists of all elements of the form $\{c_1 I_1, \ldots, c_k I_k\}$ with $c_1, \ldots, c_k$ scalars. Moreover, two projections $\{E_1, \ldots, E_k\}$ and $\{F_1, \ldots, F_k\}$ in $\mathscr{R}$ are equivalent if and only if $d(E_j) = d(F_j)$ for $j = 1, \ldots, k$. From these remarks, it follows easily that Δ, as defined by (2), has the properties (i)–(v) required of a center-valued dimension function.

Suppose that Δ' is another center-valued dimension function on $\mathscr{R}$. With Q_j the central projection $\{0, \ldots, 0, I_j, 0, \ldots, 0\}$ and E a projection in $\mathscr{R}Q_j$,

$$\Delta'(E) = \Delta'(EQ_j) = \Delta'(E)Q_j;$$

so Δ' maps projections in $\mathscr{R}Q_j$ into $\mathscr{C}Q_j$. From this, it is easily verified that $\Delta' | \mathscr{R}Q_j$ is a center-valued dimension function on $\mathscr{R}Q_j$ ($\cong \mathscr{B}(\mathscr{H}_j)$). It now follows from the preceding example that

$$\Delta'(\{0, \ldots, 0, E_j, 0, \ldots, 0\}) = \left\{0, \ldots, 0, \frac{d(E_j)}{n_j} I_j, 0, \ldots, 0\right\}, \tag{3}$$

for each projection E_j in $\mathscr{B}(\mathscr{H}_j)$. With $E = \{E_1, \ldots, E_k\}$ a projection in $\mathscr{R}$, summation of (3) over $j = 1, \ldots, k$ yields

$$\Delta'(E) = \left\{\frac{d(E_1)}{n_1}I_1, \ldots, \frac{d(E_k)}{n_k}I_k\right\} = \Delta(E). \quad \blacksquare$$

We now investigate the properties of the dimension function in a finite von Neumann algebra.

8.4.3. THEOREM. *If $\mathscr{R}$ is a finite von Neumann algebra with center $\mathscr{C}$, and $\mathscr{P}$ is the set of projections in $\mathscr{R}$, there is a unique mapping $\Delta: \mathscr{P} \to \mathscr{C}$ such that*

(i) $\Delta(E) > 0$ *if* $E \in \mathscr{P}$ *and* $E \neq 0$;
(ii) $\Delta(E + F) = \Delta(E) + \Delta(F)$ *if* $E, F \in \mathscr{P}$ *and* $EF = 0$;
(iii) $\Delta(E) = \Delta(F)$ *if* $E, F \in \mathscr{P}$ *and* $E \sim F$;
(iv) $\Delta(Q) = Q$ *if* $Q \in \mathscr{P} \cap \mathscr{C}$.

Moreover, if $E, F \in \mathscr{P}$ *and* $Q \in \mathscr{P} \cap \mathscr{C}$,

(v) $\Delta(QE) = Q\Delta(E)$;
(vi) $E \sim F$ *if and only if* $\Delta(E) = \Delta(F)$;
(vii) $E \precsim F$ *if and only if* $\Delta(E) \leq \Delta(F)$;
(viii) *if* $(E_a)_{a \in \mathbb{A}}$ *is an orthogonal family of projections in* $\mathscr{R}$, *with sum* E, *then* $\Delta(E) = \sum_{a \in \mathbb{A}} \Delta(E_a)$.

Proof. With τ the center-valued trace on $\mathscr{R}$, and $\Delta = \tau | \mathscr{P}$, it follows from the properties of τ listed in Theorem 8.2.8 that Δ satisfies conditions (i)–(v) and (viii) above.

We prove next that Δ satisfies (vi) and (vii). For this, suppose that $E, F \in \mathscr{P}$. We already know, from (iii), that $\Delta(E) = \Delta(F)$ if $E \sim F$. If $E \prec F$, there is an E_1 in $\mathscr{P}$ for which $E \sim E_1 < F$; and, from (iii), (i), and (ii),

$$\Delta(E) = \Delta(E_1) < \Delta(E_1) + \Delta(F - E_1) = \Delta(F).$$

By combining these results, it follows that $\Delta(E) \leq \Delta(F)$ if $E \precsim F$. If $E \not\precsim F$, there is a central projection Q in $\mathscr{R}$ such that $QF \prec QE$, and the preceding argument shows that $\Delta(QF) < \Delta(QE)$. From (v), $Q\Delta(F) < Q\Delta(E)$, so $\Delta(E) \nleq \Delta(F)$. This proves (vii), and (vi) is an immediate consequence of (vii).

We have now established the existence of a mapping $\Delta: \mathscr{P} \to \mathscr{C}$ that has properties (i)–(viii). It remains to show that Δ is already uniquely determined by the first four conditions. Suppose that $\Delta': \mathscr{P} \to \mathscr{C}$ is another mapping that satisfies (i)–(iv). From (i) and (ii), $\Delta'(E) \geq \Delta'(F)$ whenever $E, F \in \mathscr{P}$ and $E \geq F$. With F a monic projection in $\mathscr{R}$, we can choose projections $F_1, \ldots, F_k$ in $\mathscr{R}$ and Q in $\mathscr{C}$ so that

$$F \sim F_1 \sim F_2 \sim \cdots \sim F_k, \qquad F_1 + F_2 + \cdots + F_k = Q.$$

From (iii), (ii), and (iv),

$$k\Delta'(F) = \Delta'(F_1) + \cdots + \Delta'(F_k) = \Delta'(Q) = Q,$$

so $\Delta'(F) = k^{-1}Q$. The same argument applies to Δ, so $\Delta'(F) = \Delta(F)$ for every monic projection F in $\mathscr{R}$. Each projection E in $\mathscr{R}$ is the sum of an orthogonal family $(E_b)_{b\in\mathbb{B}}$ of monic projections in $\mathscr{R}$. When $\mathbb{A}$ is a finite subset of $\mathbb{B}$, we have

$$\sum_{b\in\mathbb{A}} \Delta(E_b) = \sum_{b\in\mathbb{A}} \Delta'(E_b) = \Delta'\left(\sum_{b\in\mathbb{A}} E_b\right) \leq \Delta'(E).$$

This and the complete additivity of Δ now give

$$\Delta(E) = \sum_{b\in\mathbb{B}} \Delta(E_b) \leq \Delta'(E).$$

By applying this argument to $I - E$, we have

$$I - \Delta(E) = \Delta(I - E) \leq \Delta'(I - E) = I - \Delta'(E), \qquad \Delta'(E) \leq \Delta(E).$$

Hence $\Delta'(E) = \Delta(E)$ for each E in $\mathscr{P}$. ■

Our next objective is to give a description of the range of the center-valued dimension function Δ on a finite von Neumann algebra $\mathscr{R}$. For $n = 1, 2, 3, \ldots$, let Q_n be the largest projection, in the center $\mathscr{C}$ of $\mathscr{R}$, for which $\mathscr{R}Q_n$ is type I_n (or (0)); and let Q_0 $(= I - \sum_{n=1}^{\infty} Q_n)$ be the corresponding projection for type II_1. With $\mathscr{P}_n$ the set of projections in $\mathscr{R}Q_n$, it is an easy consequence of the uniqueness clause in Theorem 8.4.3 that $\Delta|\mathscr{P}_n$ is the dimension function on $\mathscr{R}Q_n$. For each projection E in $\mathscr{R}$,

$$\Delta(E) = \sum_{n=0}^{\infty} \Delta(E)Q_n, \qquad \Delta(E)Q_n = \Delta(EQ_n), \qquad EQ_n \in \mathscr{P}_n.$$

It follows that the range of Δ consists of all operators C in $\mathscr{C}$ such that CQ_n lies in the range of $\Delta|\mathscr{P}_n$ $(n = 0, 1, 2, \ldots)$. The following theorem permits a complete description of the range of $\Delta|\mathscr{P}_n$, and so also of Δ.

8.4.4. THEOREM. *Suppose that $\mathscr{R}$ is a finite von Neumann algebra with center $\mathscr{C}$, $\mathscr{P}$ is the set of projections in $\mathscr{R}$ and $\Delta: \mathscr{P} \to \mathscr{C}$ is the center-valued dimension function.*

(i) *If $\mathscr{R}$ is type I_n, the range of Δ consists of all operators of the form*

$$\sum_{j=1}^{n} \frac{j}{n} Q_j,$$

where $Q_1, \ldots, Q_n$ are pairwise orthogonal projections in $\mathscr{C}$.

(ii) *If $\mathscr{R}$ is type II_1, the range of Δ consists of all positive operators in the unit ball of $\mathscr{C}$.*

Proof. (i) If F is an abelian projection in a type I_n von Neumann algebra $\mathscr{R}$, with central carrier Q, then Q is the sum of n projections each equivalent to F, whence $\Delta(F) = n^{-1}\Delta(Q) = n^{-1}Q$. We can choose abelian projections $F_1, \ldots, F_n$ in $\mathscr{R}$, with sum I, each having central carrier I. With $Q_1, \ldots, Q_n$ pairwise orthogonal projections in $\mathscr{C}$, and E the projection $\sum_{j=1}^n Q_j(F_1 + \cdots + F_j)$ in $\mathscr{R}$, we have

$$\Delta(E) = \sum_{j=1}^{n} \frac{j}{n} Q_j.$$

Hence the range of Δ contains all operators of the form specified in (i).

Conversely, given any projection E in $\mathscr{R}$, we can choose an orthogonal family $\{Q_0, \ldots, Q_n\}$ of projections in $\mathscr{C}$, with sum I, such that Q_jE is the sum of j abelian projections each having central carrier Q_j (where $Q_0 = I - C_E$), from Corollary 6.5.5. Thus

$$\Delta(E) = \sum_{j=1}^{n} \Delta(Q_j E) = \sum_{j=1}^{n} \frac{j}{n} Q_j.$$

(ii) For each projection E in $\mathscr{R}$, we have $0 \leq \Delta(E) \leq \Delta(I) = I$, so the range of Δ is contained in the positive part $(\mathscr{C}^+)_1$ of the unit ball of $\mathscr{C}$.

Suppose next that $C \in (\mathscr{C}^+)_1$, and let $\{E_\lambda\}$ be the resolution of the identity for C. With C_n defined as

$$\sum_{r=1}^{2^n} 2^{-n}(r-1)(E_{r/2^n} - E_{(r-1)/2^n}),$$

$\{C_n\}$ is an increasing sequence in $(\mathscr{C}^+)_1$, with norm limit C, by Theorem 5.2.2.(v). From this, C can be expressed as

$$C = \sum_{j=1}^{\infty} 2^{-n(j)} P_j, \tag{4}$$

where $n(j)$ is a non-negative integer and P_j is a projection in $\mathscr{C}$ ($j = 1, 2, \ldots$). Indeed,

$$C = C_1 + \sum_{n=1}^{\infty} (C_{n+1} - C_n),$$

and each summand on the right-hand side is a finite sum of terms of the type appearing in the series in (4).

We assert that there is an orthogonal sequence $\{G_1, G_2, \ldots\}$ of projections in $\mathscr{R}$, such that $\Delta(G_j) = 2^{-n(j)}P_j$. For this, note first that (by Lemma 6.5.6, and since $\mathscr{R}$ is type II_1) P_j can be expressed as the sum of $2^{n(j)}$ equivalent projections in $\mathscr{R}$, any one, F_j, of which satisfies $\Delta(F_j) = 2^{-n(j)}P_j$.

We construct the G_j inductively, starting the process by taking G_1 to be F_1. When suitable $G_1, \ldots, G_{k-1}$ have been chosen,

$$\begin{aligned}\Delta(I - G_1 - \cdots - G_{k-1}) &= I - \sum_{j=1}^{k-1} \Delta(G_j)\\ &\geq C - \sum_{j=1}^{k-1} 2^{-n(j)}P_j\\ &= \sum_{j=k}^{\infty} 2^{-n(j)}P_j \geq 2^{-n(k)}P_k = \Delta(F_k).\end{aligned}$$

Hence $F_k \precsim I - G_1 - \cdots - G_{k-1}$, and there is a projection G_k in $\mathscr{R}$ for which

$$F_k \sim G_k \leq I - G_1 - \cdots - G_{k-1}, \qquad \Delta(G_k) = \Delta(F_k) = 2^{-n(k)}P_k.$$

This completes the inductive construction of the sequence $\{G_j\}$. With G the projection $\sum_{j=1}^{\infty} G_j$,

$$\Delta(G) = \sum_{j=1}^{\infty} \Delta(G_j) = \sum_{j=1}^{\infty} 2^{-n(j)}P_j = C.$$

It follows that the range of Δ is the whole of $(\mathscr{C}^+)_1$. ∎

The remainder of this section is concerned with a different approach to the complete additivity of the dimension function, and the ultraweak continuity of the trace, for a finite von Neumann algebra. In particular, it completes the alternative program, for constructing the trace and developing its properties, that was described at the end of Section 8.3. This material is not essential to the logical structure of later parts of the book, and may be omitted if the reader wishes.

8.4.5. Remark. In the proof of Theorem 8.4.3 (viii), the complete additivity of the dimension function Δ was obtained as a consequence of the ultraweak continuity of the trace τ. We now show that complete additivity can be deduced directly from some of the other properties of Δ, as set out in parts (i)–(v) of Theorem 8.4.3, without appeal to information about τ. This fact is used, in Remark 8.4.6, to give an alternative proof of the ultraweak continuity of τ. In the interests of simplicity we shall consider only the case (sufficient for our purpose in Remark 8.4.6) in which $\mathscr{R}$ is type II_1.

With $\mathscr{P}$ the set of projections in $\mathscr{R}$, and $\Delta : \mathscr{P} \to \mathscr{C}$ a mapping that satisfies conditions (i)–(v) in Theorem 8.4.3, one can show (as in the proof of that theorem) that Δ has also properties (vi) and (vii). Let $(E_b)_{b \in \mathbb{B}}$ be an orthogonal family of projections in $\mathscr{R}$, with sum E. For each finite subset $\mathbb{A}$ of $\mathbb{B}$, we have

$$\sum_{b \in \mathbb{A}} \Delta(E_b) \leq \sum_{b \in \mathbb{A}} \Delta(E_b) + \Delta\left(E - \sum_{b \in \mathbb{A}} E_b\right) = \Delta(E).$$

Since each $\Delta(E_b) \geq 0$, it follows that the sum $\sum_{b \in \mathbb{B}} \Delta(E_b)$ converges in the strong-operator topology, and

$$\sum_{b \in \mathbb{B}} \Delta(E_b) \leq \Delta(E). \tag{5}$$

In order to show that equality occurs in (5), and so establish the complete additivity of Δ, we have to prove that $\sum \Delta(E_b)x = \Delta(E)x$ for each vector x. For this, it suffices to show that $\sum \Delta(E_b)Q = \Delta(E)Q$, where Q is the cyclic projection in $\mathscr{C}$ with range $[\mathscr{C}'x]$. Moreover, the required equation can be rewritten in the form $\sum \Delta(E_b Q) = \Delta(EQ)$. The restriction $\Delta | \mathscr{P} \cap \mathscr{R}Q$ has range in $\mathscr{C}Q$, and inherits from Δ properties (i)–(vii) in Theorem 8.4.3. Upon replacing $\mathscr{R}$, $\mathscr{C}$, E_b, E by $\mathscr{R}Q$, $\mathscr{C}Q$, E_bQ, EQ, respectively, it suffices to prove that equality occurs in (5) under the additional assumption that x is cyclic for $\mathscr{C}'$, and hence separating for $\mathscr{C}$.

From (5), $\sum \langle \Delta(E_b)x, x\rangle \leq \langle \Delta(E)x, x\rangle$, so the set

$$\{b \in \mathbb{B} : \langle \Delta(E_b)x, x\rangle > 0\} \qquad (= \mathbb{A})$$

is countable. Since $\Delta(E_b) \in \mathscr{C}$, $\Delta(E_b) > 0$ if $E_b \neq 0$, and x is separating for $\mathscr{C}$, it follows that $\mathbb{A} = \{b \in \mathbb{B} : E_b \neq 0\}$. Upon suppressing zero terms from the family $\{E_b\}$, it now suffices to consider the case in which this family is replaced by an orthogonal *sequence* $\{E_1, E_2, \ldots\}$ of projections in $\mathscr{R}$, with sum E. In this case, (5) is replaced by

$$\sum_{n=1}^{\infty} \Delta(E_n) \leq \Delta(E), \tag{6}$$

and we want to show that equality occurs in (6).

Suppose the contrary. With Q a suitable spectral projection of $\Delta(E) - \sum \Delta(E_n)$, and k a suitable positive integer, we have

$$Q \in \mathscr{C}, \qquad Q \neq 0, \qquad \frac{1}{k}Q + \sum_{n=1}^{\infty} \Delta(E_n) \leq \Delta(E). \tag{7}$$

Since $\mathscr{R}$ is type II_1, Q can be expressed as the sum of k equivalent projections in $\mathscr{R}$, and any one, G, of these satisfies $\Delta(G) = k^{-1}Q$. We now assert the existence of an orthogonal sequence $\{F_0, F_1, F_2, \ldots\}$ of subprojections of E in $\mathscr{R}$ such that

$$F_0 \sim G, \qquad F_n \sim E_n \quad (n = 1, 2, \ldots). \tag{8}$$

Since $\Delta(G) = k^{-1}Q \leq \Delta(E)$, we have $G \precsim E$, and there is a projection F_0 in $\mathscr{R}$ such that $G \sim F_0 \leq E$. When suitable $F_0, \ldots, F_{n-1}$ have been chosen

$$\begin{aligned}\Delta(E - F_0 - F_1 - \cdots - F_{n-1}) &= \Delta(E) - \sum_{j=0}^{n-1} \Delta(F_j) \\ &= \Delta(E) - \frac{1}{k}Q - \sum_{j=1}^{n-1} \Delta(E_j) \\ &\geq \sum_{j=n}^{\infty} \Delta(E_j) \geq \Delta(E_n).\end{aligned}$$

Thus $E_n \precsim E - F_0 - F_1 - \cdots - F_{n-1}$, and there is a projection F_n in $\mathscr{R}$ for which $E_n \sim F_n \leq E - F_0 - F_1 - \cdots - F_{n-1}$. This permits an inductive construction of a sequence $\{F_n\}$ with the stated properties. Finally,

$$E = \sum_{n=1}^{\infty} E_n \sim \sum_{n=1}^{\infty} F_n < \sum_{n=0}^{\infty} F_n \leq E,$$

a contradiction, since $\mathscr{R}$ is finite. Hence equality occurs in (6), and Δ is completely additive. ■

8.4.6. Remark. At the end of Section 8.3, we described in outline an alternative method, based on the Dixmier approximation theorem and some related results, for constructing the center-valued trace τ on a finite von Neumann algebra (the problem was reduced at once to the case in which $\mathscr{R}$ is type II_1). The program was incomplete, in that it did not establish the ultraweak continuity of τ. The remaining properties of τ, as set out in Theorem 8.2.8, were proved. From these it follows at once that Δ, the restriction of τ to the set $\mathscr{P}$ of projections in $\mathscr{R}$, satisfies conditions (i)–(v) in Theorem 8.4.3. By Remark 8.4.5, Δ is completely additive. Hence τ is completely additive on projections. If ω is a normal state of $\mathscr{C}$, the state $\omega \circ \tau$ of $\mathscr{R}$ is completely additive, and is therefore normal. Accordingly, τ is ultraweakly continuous, since the normal states of $\mathscr{R}$ determine its ultraweak topology (Remark 7.4.4). ■

Bibliography: [56].

8.5. Tracial weights on factors

If $\mathscr{R}$ is a finite factor, our results (Theorem 8.2.8) concerning the existence, uniqueness, and properties of the center-valued trace τ reduce to assertions about a tracial state ρ_0. Since the center-valued dimension function is the restriction of τ to the set $\mathscr{P}$ of projections in $\mathscr{R}$, Theorem 8.4.4 describes the set of values taken by ρ_0 on $\mathscr{P}$. Specifically, $\mathscr{R}$ has a unique tracial state ρ_0; moreover, ρ_0 is faithful and normal. The range of $\rho_0 | \mathscr{P}$ is the set $\{0, 1/n, 2/n, \ldots, 1\}$ if $\mathscr{R}$ is type I_n, and is the whole interval $[0, 1]$ when $\mathscr{R}$ is type II_1. From the uniqueness of ρ_0, each numerical trace on $\mathscr{R}$ is a non-negative multiple of ρ_0.

In seeking analogous results for infinite factors, it is necessary to generalize the concept of "tracial state," since we have already observed in Section 8.1 that an infinite von Neumann algebra cannot have a faithful numerical trace. Our main purpose in this section is to introduce the notion of "tracial weight," and to show how the type of a factor can be determined from information about its tracial weights.

Suppose that $\mathscr{R}$ is a von Neumann algebra with unitary group $\mathscr{U}$. We recall from Definition 7.5.1 that a weight on $\mathscr{R}$ is a mapping ρ, from $\mathscr{R}^+$ into the interval $[0, \infty]$, such that

(i) $\rho(H + K) = \rho(H) + \rho(K) \qquad (H, K \in \mathscr{R}^+)$,
(ii) $\rho(aH) = a\rho(H) \qquad (H \in \mathscr{R}^+, a \geq 0)$.

We describe ρ as a *tracial weight* if, in addition,

(iii) $\rho(AA^*) = \rho(A^*A) \qquad (A \in \mathscr{R})$.

When ρ is a tracial weight, it follows from (iii), with $UH^{1/2}$ in place of A, that

(iv) $\rho(UHU^*) = \rho(H) \qquad (H \in \mathscr{R}^+, U \in \mathscr{U})$.

Moreover, if E and F are projections in $\mathscr{R}$, (iii) implies that

(v) $\rho(E) = \rho(F)$ if $E \sim F$;

and from this, it follows that

(vi) $\rho(E) \leq \rho(F)$ if $E \precsim F$.

Given any von Neumann algebra $\mathscr{R}$, there are at least two tracial weights, the first identically zero on $\mathscr{R}^+$, the second taking the value ∞ at every non-zero element of $\mathscr{R}^+$. Of course, neither of these is of interest; we shall be concerned mainly with tracial weights that are faithful, semi-finite, and normal, in the sense explained in Definition 7.5.1.

We recall from Lemma 7.5.2 that, if ρ is a weight on $\mathscr{R}$, the sets

$$\mathscr{N}_\rho = \{A \in \mathscr{R} : \rho(A^*A) < \infty\},$$
$$N_\rho = \{A \in \mathscr{R} : \rho(A^*A) = 0\}$$

are left ideals in $\mathscr{R}$. Moreover, the linear span $\mathscr{M}_\rho$ of the set $F_\rho = \{H \in \mathscr{R}^+ : \rho(H) < \infty\}$ satisfies $\mathscr{M}_\rho \cap \mathscr{R}^+ = F_\rho$, and coincides with the linear span $\mathscr{N}_\rho{}^*\mathscr{N}_\rho$ $(\subseteq \mathscr{N}_\rho \cap \mathscr{N}_\rho{}^*)$ of the set $\{A^*B : A, B \in \mathscr{N}_\rho\}$. The restriction $\rho \mid F_\rho$ extends, uniquely, to a positive hermitian linear functional (again denoted by ρ) on $\mathscr{M}_\rho$.

We note that, if ρ is a normal weight on $\mathscr{R}$, and $\{H_a\}$ is a bounded increasing net of operators in $\mathscr{R}^+$, with (strong-operator) limit H, then $\rho(H_a) \to_a \rho(H)$ (with the obvious conventions concerning ∞); in particular, ρ is completely additive on projections. Indeed, since $\rho(H_a) \leq \rho(H)$, it suffices to prove that, given a real number c such that $c < \rho(H)$, we have $\rho(H_a) > c$ for all sufficiently large a. Since ρ is normal, there is a family $(\rho_b)_{b \in \mathbb{B}}$ of positive normal functionals on $\mathscr{R}$, such that $\rho(A) = \sum_{b \in \mathbb{B}} \rho_b(A)$ for each A in $\mathscr{R}^+$. There is a finite subset $\mathbb{F}$ of the index set $\mathbb{B}$ such that $\sum_{b \in \mathbb{F}} \rho_b(H) > c$. Since $\rho_b(H_a) \to_a \rho_b(H)$, for each b in $\mathbb{F}$, we have

$$\rho(H_a) \geq \sum_{b \in \mathbb{F}} \rho_b(H_a) > c$$

for all sufficiently large a.

For tracial weights, we have the following result.

8.5.1. PROPOSITION. *Suppose that ρ is a tracial weight on a von Neumann algebra $\mathscr{R}$.*

(i) *$\mathscr{N}_\rho$, N_ρ, and $\mathscr{M}_\rho$ are (two-sided) ideals in $\mathscr{R}$.*
(ii) *$\rho(AB) = \rho(BA)$ when $A, B \in \mathscr{N}_\rho$.*
(iii) *$\rho(AB) = \rho(BA)$ when $A \in \mathscr{R}$ and $B \in \mathscr{M}_\rho$.*

Proof. (i) It is apparent, from the defining condition (iii) for tracial weights, that the left ideals $\mathscr{N}_\rho$ and N_ρ are self-adjoint; so they are two-sided ideals, and the same is true of $\mathscr{M}_\rho$ $(= \mathscr{N}_\rho{}^*\mathscr{N}_\rho)$.

(ii) Note first that $\rho(AB)$ (and, similarly, $\rho(BA)$) is defined when $A, B \in \mathscr{N}_\rho$; for $A \in \mathscr{N}_\rho{}^*$ and $AB \in \mathscr{N}_\rho{}^*\mathscr{N}_\rho = \mathscr{M}_\rho$. Since $\rho(AA^*) = \rho(A^*A) < \infty$, for all A in $\mathscr{N}_\rho$, it follows by polarization (see the proof of Lemma 7.5.2) that $\rho(A^*B) = \rho(BA^*)$ for all A and B in $\mathscr{N}_\rho$.

(iii) If $A \in \mathscr{R}$ and $B \in \mathscr{M}_\rho$, then $AB, BA \in \mathscr{M}_\rho$, whence $\rho(AB)$ and $\rho(BA)$ are defined (and finite). In order to prove that $\rho(AB) = \rho(BA)$, it suffices to consider the case in which $B \in F_\rho$ $(= \mathscr{M}_\rho \cap \mathscr{R}^+)$. In this case, $B^{1/2} \in \mathscr{N}_\rho$, so $AB^{1/2}, B^{1/2}A \in \mathscr{N}_\rho$; and from (ii),

$$\begin{aligned}\rho(AB) = \rho(AB^{1/2}B^{1/2}) &= \rho(B^{1/2}AB^{1/2}) \\ &= \rho(B^{1/2}B^{1/2}A) = \rho(BA). \quad \blacksquare\end{aligned}$$

8.5.2. PROPOSITION. (i) *If ρ is a semi-finite tracial weight on a von Neumann algebra $\mathscr{R}$, and A is a non-zero element of $\mathscr{R}^+$, there is a non-zero projection G in F_ρ, and a positive real number a, such that $A \geq aG$.*

(ii) *If ρ is a normal semi-finite tracial weight on a factor $\mathscr{R}$, and $\rho \neq 0$, then ρ is faithful. Moreover, a projection E in $\mathscr{R}$ is finite or infinite according as $\rho(E) < \infty$ or $\rho(E) = \infty$.*

Proof. (i) Since ρ is semi-finite, $\mathscr{N}_\rho$ $(= \mathscr{N}_\rho^* \supseteq \mathscr{M}_\rho)$ is weak-operator dense in $\mathscr{R}$. Since $A > 0$, we can choose B in $\mathscr{N}_\rho$ so that $A^{1/2}B \neq 0$; and, since $\mathscr{N}_\rho\mathscr{N}_\rho^* = \mathscr{N}_\rho^*\mathscr{N}_\rho = \mathscr{M}_\rho$, $0 \neq A^{1/2}BB^*A^{1/2} \in \mathscr{M}_\rho \cap \mathscr{R}^+ = F_\rho$. For a suitable spectral projection G of $A^{1/2}BB^*A^{1/2}$, and some positive real number c,

$$0 < cG \leq A^{1/2}BB^*A^{1/2} \leq \|B\|^2 A.$$

Hence $A \geq aG$, where $a = c\|B\|^{-2}$; and $0 \neq G \in F_\rho$.

(ii) If ρ is not faithful, there is a non-zero element G of $\mathscr{R}^+$ such that $\rho(G) = 0$. From (i), we may assume that G is a projection. Let $\{G_b\}$ be an orthogonal family of projections in $\mathscr{R}$, maximal subject to the condition that $G_b \sim G$ for each b. By maximality, $G \not\precsim I - \sum G_b$; and since $\mathscr{R}$ is a factor, $I - \sum G_b \prec G$. Hence

$$0 \leq \rho(I - \textstyle\sum G_b) \leq \rho(G_b) = \rho(G) = 0,$$

and the complete additivity of ρ entails

$$\rho(I) = \rho(I - \textstyle\sum G_b) + \sum \rho(G_b) = 0.$$

Thus $0 \leq \rho(A) \leq \|A\|\rho(I) = 0$ for each A in $\mathscr{R}^+$; and $\rho = 0$, a contradiction. Hence ρ is faithful.

An infinite projection E in $\mathscr{R}$ can be halved as $E = E_1 + E_2$, where $E \sim E_1 \sim E_2$. Since

$$0 < \rho(E) = \rho(E_1) + \rho(E_2) = 2\rho(E),$$

it follows that $\rho(E) = \infty$.

Suppose next that E is a (non-zero) finite projection in $\mathscr{R}$. From (i), E has a non-zero subprojection G $(\in \mathscr{R})$ such that $\rho(G) < \infty$. With $\{G_1, \ldots, G_n\}$ a (necessarily finite) maximal orthogonal family of subprojections of E, each equivalent to G, we have $E - \sum_{j=1}^n G_j \prec G$ (by maximality, and since $\mathscr{R}$ is a factor). Hence

$$\rho(E) = \rho\left(E - \sum_{j=1}^n G_j\right) + \sum_{j=1}^n \rho(G_j) \leq (n+1)\rho(G) < \infty. \quad \blacksquare$$

We now begin our investigation of tracial weights on the various types of factors.

8.5.3. PROPOSITION. *If $\mathscr{R}$ is a finite factor, there is a unique tracial state ρ_0 on $\mathscr{R}$; and ρ_0 is faithful and normal. If $\mathscr{P}$ is the set of all projections in $\mathscr{R}$, the range of the restriction $\rho_0|\mathscr{P}$ is the set $\{0, 1/n, 2/n, \ldots, 1\}$ when $\mathscr{R}$ is type I_n, and is the whole interval $[0, 1]$ when $\mathscr{R}$ is type II_1. Every semi-finite tracial weight on $\mathscr{R}$ is a non-negative scalar multiple of the restriction $\rho_0|\mathscr{R}^+$.*

Proof. All the statements in the proposition, except the final one, have already been noted in the first paragraph of this section, where it was observed also that every *numerical trace* on $\mathscr{R}$ is a non-negative multiple of ρ_0.

If ω is a semi-finite tracial weight on $\mathscr{R}$, the two-sided ideal $\mathscr{M}_\omega$ is weak-operator dense in $\mathscr{R}$. By Corollary 6.8.4, $\mathscr{M}_\omega = \mathscr{R}$. From Lemma 7.5.2, ω extends to a positive linear functional on $\mathscr{R}$; it is then a numerical trace on $\mathscr{R}$, by Proposition 8.5.1(iii), and is therefore a non-negative multiple of ρ_0. ■

8.5.4. PROPOSITION. *If $\mathscr{R}$ is a type III factor, there is no non-zero normal semi-finite tracial weight on $\mathscr{R}$.*

Proof. If such a weight ρ exists, it follows from Proposition 8.5.2(ii) that $\rho(E) = \infty$ for each non-zero projection E in $\mathscr{R}$, and this contradicts part (i) of that proposition. ■

8.5.5. PROPOSITION. *If $\mathscr{R}$ is a factor of type I_∞ or II_∞, there is a faithful normal semi-finite tracial weight ρ on $\mathscr{R}$. With $\mathscr{P}$ the set of all projections in $\mathscr{R}$, the range of the restriction $\rho|\mathscr{P}$ has the form $\{0, c, 2c, 3c, \ldots, \infty\}$ for some positive real number c when $\mathscr{R}$ is type I_∞, and is the whole interval $[0, \infty]$ when $\mathscr{R}$ is type II_∞. Each normal semi-finite tracial weight on $\mathscr{R}$ is a non-negative scalar mutiple of ρ.*

Proof. Let E be a non-zero finite projection in $\mathscr{R}$. By Proposition 6.3.12, there is an (infinite) orthogonal family $(E_b)_{b \in \mathbb{B}}$ of projections in $\mathscr{R}$, with sum I, each equivalent to E. When $b \in \mathbb{B}$, let V_b be a partial isometry in $\mathscr{R}$, from E_b to E. The von Neumann algebra $E\mathscr{R}E$ is a finite factor, and so has a (unique faithful normal) tracial state ω, by Proposition 8.5.3. When $A \in \mathscr{R}^+$ and $b \in \mathbb{B}$, we have $V_b A V_b^* \in (E\mathscr{R}E)^+$, and thus $0 \leq \omega(V_b A V_b^*) < \infty$. We now define a mapping $\rho: \mathscr{R}^+ \to [0, \infty]$ by

$$\rho(A) = \sum_{b \in \mathbb{B}} \omega(V_b A V_b^*) \qquad (A \in \mathscr{R}^+). \tag{1}$$

It is apparent that ρ is a weight on $\mathscr{R}$; and ρ is normal, since the mapping $A \to \omega(V_b A V_b^*)$ is a normal state ρ_b of $\mathscr{R}$, and $\rho(A) = \sum \rho_b(A)$ for all A in $\mathscr{R}^+$. Moreover, $\rho(E_b) = 1$ for each b in $\mathbb{B}$, and $\rho(E_{\mathbb{F}}) < \infty$ when $\mathbb{F}$ is a finite subset of $\mathbb{B}$, where $E_{\mathbb{F}} = \sum_{b \in \mathbb{F}} E_b$. Thus $A \in F_\rho$ whenever $0 \leq A \leq E_{\mathbb{F}}$, and the linear span $\mathcal{M}_\rho$ of F_ρ contains $E_{\mathbb{F}}\mathscr{R}E_{\mathbb{F}}$, for every finite subset $\mathbb{F}$ of $\mathbb{B}$. Since $\sum_{b\in\mathbb{B}} E_b = I$, $\mathcal{M}_\rho$ is weak-operator dense in $\mathscr{R}$; and ρ is semi-finite.

We prove next that ρ is a tracial weight; it then follows from Proposition 8.5.2(ii) that ρ is faithful (a fact that can be verified also by a simple direct argument). For each A in $\mathscr{R}$ and b in $\mathbb{B}$,

$$V_bAA^*V_b^* = \sum_{a\in\mathbb{B}} V_bAE_aA^*V_b^* = \sum_{a\in\mathbb{B}} (V_bAV_a^*)(V_bAV_a^*)^*.$$

Since $V_bAV_a^* \in E\mathscr{R}E$, while ω is a normal tracial state of $E\mathscr{R}E$,

$$\begin{aligned}\omega(V_bAA^*V_b^*) &= \sum_{a\in\mathbb{B}} \omega((V_bAV_a^*)(V_bAV_a^*)^*)\\ &= \sum_{a\in\mathbb{B}} \omega((V_bAV_a^*)^*(V_bAV_a^*))\\ &= \sum_{a\in\mathbb{B}} \omega(V_aA^*E_bAV_a^*).\end{aligned}$$

By summing over all b in $\mathbb{B}$, and inverting the order of the double summation on the right-hand side (which is permissible since the terms are non-negative), we obtain

$$\begin{aligned}\rho(AA^*) &= \sum_{b\in\mathbb{B}} \omega(V_bAA^*V_b^*) = \sum_{a\in\mathbb{B}}\sum_{b\in\mathbb{B}} \omega(V_aA^*E_bAV_a^*)\\ &= \sum_{a\in\mathbb{B}} \omega(V_aA^*AV_a^*) = \rho(A^*A).\end{aligned}$$

Hence ρ is a (faithful normal semi-finite) tracial weight on $\mathscr{R}$; and $\rho(I) = \sum_{b\in\mathbb{B}} \rho(E_b) = \sum_{b\in\mathbb{B}} 1 = \infty$.

Suppose that ρ' is another normal semi-finite tracial weight on $\mathscr{R}$. Since the projection E in $\mathscr{R}$ is finite, $\rho'(E) < \infty$ by Proposition 8.5.2(ii). Hence ρ' takes finite values throughout $(E\mathscr{R}E)^+$, and the restriction $\rho'|(E\mathscr{R}E)^+$ is a semi-finite tracial weight on the finite factor $E\mathscr{R}E$. By Proposition 8.5.3, there is a non-negative scalar a such that $\rho'|(E\mathscr{R}E)^+ = a\omega|(E\mathscr{R}E)^+$. For each A in $\mathscr{R}^+$,

$$A = \sum_{b\in\mathbb{B}} A^{1/2}E_bA^{1/2} = \sum_{b\in\mathbb{B}} (A^{1/2}V_b^*)(A^{1/2}V_b^*)^*.$$

Since ρ' is normal, and $V_bAV_b^* \in (E\mathscr{R}E)^+$, we have

$$\begin{aligned}\rho'(A) &= \sum_{b\in\mathbb{B}} \rho'((A^{1/2}V_b^*)(A^{1/2}V_b^*)^*)\\ &= \sum_{b\in\mathbb{B}} \rho'((A^{1/2}V_b^*)^*(A^{1/2}V_b^*))\\ &= \sum_{b\in\mathbb{B}} \rho'(V_bAV_b^*)\\ &= a\sum_{b\in\mathbb{B}} \omega(V_bAV_b^*) = a\rho(A);\end{aligned}$$

and thus, $\rho' = a\rho$.

If $\mathscr{R}$ is type I_∞, then ρ takes the same value c at each minimal projection in $\mathscr{R}$, since any two such projections are equivalent; and $0 < c < \infty$, by Proposition 8.5.2(ii). Each projection G in $\mathscr{R}$ is the sum of an orthogonal family of minimal projections, so $\rho(G)$ lies in the set $\{0, c, 2c, 3c, \ldots, \infty\}$. Since I is the sum of an infinite family of minimal projections, it follows (by considering also the finite subsums) that ρ takes each of the values $0, c, 2c, 3c, \ldots, \infty$ at suitable projections in $\mathscr{R}$.

Finally, suppose that $\mathscr{R}$ is type II_∞. Each non-negative real number has the form $n + c$, where n is a non-negative integer and $0 \le c < 1$. Since ω is the unique tracial state of the type II_1 factor $E\mathscr{R}E$, it follows from Proposition 8.5.3 that $\omega(F) = c$ for some projection F in $E\mathscr{R}E$. If $G_0, G_1, \ldots, G_n$ are distinct projections selected from the family $\{E_b\}$, $F \le E \sim G_0$ and thus $F \sim F_0 \le G_0$ for some projection F_0 in $\mathscr{R}$. It follows easily from (1) that $\rho(F_0) = \omega(F)$; so

$$\begin{aligned}\rho(F_0 + G_1 + G_2 + \cdots + G_n) &= \rho(F_0) + \sum_{j=1}^{n} \rho(G_n)\\ &= \omega(F) + n = c + n.\end{aligned}$$

Since, also, $\rho(I) = \infty$, we have shown that ρ takes each value in $[0, \infty]$ at a suitable projection in $\mathscr{R}$. ■

8.5.6. Remark. If $\mathscr{H}$ is an infinite-dimensional Hilbert space, it follows from Proposition 8.5.5 that there is a unique normal semi-finite tracial weight ρ_0 on the type I_∞ factor $\mathscr{B}(\mathscr{H})$ that takes the value 1 at each minimal projection in $\mathscr{B}(\mathscr{H})$. We assert that, if $(x_b)_{b \in \mathbb{B}}$ is any orthonormal basis of $\mathscr{H}$, then

$$\rho_0(A) = \sum_{b \in \mathbb{B}} \langle Ax_b, x_b \rangle \qquad (A \in \mathscr{B}(\mathscr{H})^+). \tag{2}$$

This is exactly what one would expect, by analogy with the finite-dimensional case discussed in Example 8.1.2.

To prove (2), let x be any unit vector in $\mathscr{H}$, let E and E_b be the one-dimensional projections whose ranges contain x and x_b, respectively, and let V_b be the partial isometry from E_b to E that carries x_b to x. This choice of E, E_b, V_b is consistent with the requirements of the proof of Proposition 8.5.5. As noted in the proof of that proposition, the tracial weight ρ defined by (1) takes the value 1 at each E_b (and hence at every minimal projection in $\mathscr{B}(\mathscr{H})$); so $\rho = \rho_0$. Now $V_b A V_b^* = \langle Ax_b, x_b \rangle E$, and the tracial state on $E\mathscr{B}(\mathscr{H})E$ satisfies $\omega(E) = 1$; so (2) is an immediate consequence of (1).

The ideal $\mathscr{M}_{\rho_0}$ is called the *trace class* in $\mathscr{B}(\mathscr{H})$; and its elements are described as *trace class operators* on $\mathscr{H}$. When $A \in F_{\rho_0}$, $\rho_0(A)$ is finite and the sum on the right-hand side of (2) converges. Since $\mathscr{M}_{\rho_0}$ is the linear span of F_{ρ_0}, (2) remains valid for every trace class operator A (with ρ_0 extended to a linear functional on $\mathscr{M}_{\rho_0}$). From Remark 7.1.10, the normal states on a von Neumann algebra $\mathscr{R}$ ($\subseteq \mathscr{B}(\mathscr{H})$) are precisely the mappings $A \to \rho_0(HA)$, where H is a positive trace class operator and $\rho_0(H) = 1$. ■

From the three preceding propositions, we have the following information concerning normal semi-finite tracial weights on factors.

8.5.7. THEOREM. *Suppose that $\mathscr{R}$ is a factor and $\mathscr{P}$ is the set of all projections in $\mathscr{R}$.*

If $\mathscr{R}$ is type III, *there is no non-zero normal semi-finite tracial weight on $\mathscr{R}$.*

If $\mathscr{R}$ is not type III, *there is a faithful normal semi-finite tracial weight ρ on $\mathscr{R}$, and every such weight is a positive scalar multiple of ρ. Moreover*

(i) *$\mathscr{R}$ is type I_n (with n a positive integer) if and only if the restriction $\rho \mid \mathscr{P}$ has range of the form $\{0, c, 2c, \ldots, nc\}$ for some positive c;*

(ii) *$\mathscr{R}$ is type I_∞ if and only if $\rho \mid \mathscr{P}$ has range of the form $\{0, c, 2c, 3c, \ldots, \infty\}$ for some positive c;*

(iii) *$\mathscr{R}$ is type II_1 if and only if $\rho \mid \mathscr{P}$ has range the closed interval $[0, c]$ for some positive c;*

(iv) *$\mathscr{R}$ is type II_∞ if and only if $\rho \mid \mathscr{P}$ has range $[0, \infty]$.*

8.5.8. REMARK. In factors of types I_n, I_∞, and II_1, the tracial weight ρ occurring in Theorem 8.5.7 can be uniquely determined by choice of the positive constant c. The usual values used are $c = 1/n$ when $\mathscr{R}$ is type I_n, $c = 1$ when $\mathscr{R}$ is type I_∞ or type II_1. For factors of types I_n or II_1, this normalizes ρ in such a way that $\rho(I) = 1$, and we obtain the tracial *state* of $\mathscr{R}$. However, there is no method by which the algebraic structure of a type II_∞ factor can be used to single out any particular tracial weight. Indeed, one can construct a type II_∞ factor $\mathscr{M}$ and a * automorphism θ of $\mathscr{M}$, such that $\rho \circ \theta \neq \rho$ when ρ is a faithful normal semi-finite tracial weight on $\mathscr{M}$ (see Proposition 13.1.10). ■

8.5.9. REMARK. We gather together a few simple additional facts, concerning a tracial weight ρ on a von Neumann algebra $\mathscr{R}$, which will be needed in later chapters. As noted in Section 7.5, the restriction $\rho \mid F_\rho$ extends to a positive hermitian linear functional (again denoted by ρ) on $\mathscr{M}_\rho$; and the equation $\langle A, B \rangle = \rho(B^*A)$ then defines a (positive) inner product on $\mathscr{N}_\rho$. Accordingly, if $\|B\|_2 = [\rho(B^*B)]^{1/2}$ when $B \in \mathscr{N}_\rho$, then $\| \ \|_2$

is a semi-norm on $\mathcal{N}_\rho$ (a norm, if ρ is faithful), and $\|B\|_2 = [\rho(BB^*)]^{1/2} = \|B^*\|_2$. If $A, C \in \mathcal{R}$ and $B \in \mathcal{N}_\rho$, then $ABC \in \mathcal{N}_\rho$ from Proposition 8.5.1(i), and

$$\|ABC\|_2 = [\rho(C^*B^*A^*ABC)]^{1/2} \leq [\|A\|^2\rho(C^*B^*BC)]^{1/2}$$
$$= [\|A\|^2\rho(BCC^*B^*)]^{1/2} \leq [\|A\|^2\rho(BB^*)\|C\|^2]^{1/2} = \|A\|\,\|B\|_2\,\|C\|;$$

so that $\|ABC\|_2 \leq \|A\|\,\|B\|_2\,\|C\|$.

If ρ is a numerical trace, then $\mathcal{N}_\rho = \mathcal{R}$, and $\|\ \|_2$ is defined throughout $\mathcal{R}$. When $\rho = \omega_x|\mathcal{R}$, where x is a trace vector for $\mathcal{R}$, $\|A\|_2 = [\omega_x(A^*A)]^{1/2} = \|Ax\|$.

Next, observe that a von Neumann algebra $\mathcal{R}$, with a faithful semi-finite tracial weight ρ, is semi-finite. Indeed, each non-zero projection E in $\mathcal{R}$ has a non-zero subprojection G in $\mathcal{R}$, such that $\rho(G) < \infty$; and G is finite in $\mathcal{R}$, since $G\mathcal{R}G$ has a faithful numerical trace $\rho|GRG$. Since each non-zero projection in $\mathcal{R}$ has a non-zero finite subprojection, $\mathcal{R}$ has no central portion of type III.

Conversely, we can show that a semi-finite von Neumann algebra $\mathcal{R}$ has a faithful normal semi-finite tracial weight. We sketch a proof. Note first that there is a family $\{Q_a\}$ of central projections in $\mathcal{R}$, with sum I, such that each Q_a is the sum of a family of equivalent finite projections (see Exercise 6.9.12). If each $\mathcal{R}Q_a$ has a faithful normal semi-finite tracial weight ρ_a, then $\mathcal{R}$ has such a weight ρ, defined on $\mathcal{R}^+$ by $\rho(A) = \sum \rho_a(AQ_a)$. By considering $\mathcal{R}Q_a$ in place of $\mathcal{R}$, we may now assume that there is an orthogonal family $\{E_b\}$ of equivalent finite projections in $\mathcal{R}$, with sum I. The required tracial weight can now be constructed as in the proof of Proposition 8.5.5. In the countably decomposable case, this construction is required (and given), as part of the proof of Lemma 9.2.19.

Finally, suppose that ρ is a normal tracial weight on a von Neumann algebra $\mathcal{R}$, and (as usual) denote also by ρ the extension of $\rho|F_\rho$ to a positive linear functional on the ideal $\mathcal{M}_\rho$. We assert that, when $A \in \mathcal{M}_\rho$, the linear functional $B \to \rho(AB)$ on $\mathcal{R}$ is ultraweakly continuous. For this, we may suppose that $A \in F_\rho$; so that $A^{1/2} \in \mathcal{N}_\rho$, and $\rho(AB) = \rho(A^{1/2}A^{1/2}B) = \rho(A^{1/2}BA^{1/2})$. If $\{B_a\}$ is an increasing net in $\mathcal{R}^+$ with strong-operator limit B_0, then $\{A^{1/2}B_aA^{1/2}\}$ increases to its limit $A^{1/2}B_0A^{1/2}$. Since ρ is a normal weight, from the discussion preceding Proposition 8.5.1, $\rho(A^{1/2}B_aA^{1/2}) \to_a \rho(A^{1/2}B_0A^{1/2})$; so the positive linear functional $B \to \rho(A^{1/2}BA^{1/2})$ $(= \rho(AB))$ on $\mathcal{R}$ is normal (and, therefore, ultraweakly continuous). ■

Bibliography: [23, 57].

8.6. Further examples of factors

In Section 6.7, we exhibited examples of factors of types II_1 and II_∞, having already noted that $\mathscr{B}(\mathscr{H})$ is a factor of type I_n (where n may be finite or infinite) when dim $\mathscr{H} = n$; but no factor of type III was produced. Our main purpose in the present section is to fill this gap. We describe a construction that can be used to obtain examples of factors from all types I_n, II_1, II_∞, and III. In proving that some of the factors produced below are of type III, we appeal to results in Section 8.5 concerning tracial weights; it is for this reason that the construction of type III factors has been deferred until the present stage. A more detailed examination of certain type III factors will be undertaken in Section 12.3.

An operator-theoretic construction. Throughout this subsection we suppose that $\mathscr{H}$ is a Hilbert space, $\mathscr{A}$ is a maximal abelian * subalgebra of $\mathscr{B}(\mathscr{H})$, G is a discrete group with unit e, and $U: g \to U(g)$ is a unitary representation of G on $\mathscr{H}$ (that is, U is a homomorphism from G into the group of all unitary operators on $\mathscr{H}$). *In addition, we assume, throughout, that the following two conditions are satisfied*:

(a) $U(g)\mathscr{A}U(g)^* = \mathscr{A}$ for each g in G;
(b) $\mathscr{A} \cap U(g)\mathscr{A} = \{0\}$ for all g ($\neq e$) in G.

From the first assumption (a), *G acts on* $\mathscr{A}$ in the following sense: for each g in G, the mapping $\alpha_g: A \to U(g)AU(g)^*$ is a * automorphism of $\mathscr{A}$ (that is, a * isomorphism from $\mathscr{A}$ onto $\mathscr{A}$), and $\alpha: g \to \alpha_g$ is a homomorphism from G into the group of * automorphisms of $\mathscr{A}$. The second assumption (b) is sometimes expressed as the assertion that *the action of G on* $\mathscr{A}$ *is free*, or that *G acts freely on* $\mathscr{A}$.

We shall say that *G acts ergodically on* $\mathscr{A}$ if the following condition is satisfied: if $A \in \mathscr{A}$ and $U(g)AU(g)^* = A$ for each g in G, then A is a scalar multiple of I.

Starting from the general situation described above, we construct a certain von Neumann algebra $\mathscr{R}$, and determine its commutant. We show that $\mathscr{R}$ is a factor if and only if G acts ergodically on $\mathscr{A}$, and in this case we give criteria for determining the type of $\mathscr{R}$. The next subsection is concerned with examples of particular $\mathscr{H}$, $\mathscr{A}$, G, and U satisfying the above conditions.

Let $\mathscr{K}$ be the Hilbert space $\sum_{g \in G} \oplus \mathscr{H}_g$, where each $\mathscr{H}_g$ is $\mathscr{H}$, so that $\mathscr{K}$ consists of all mappings $x: G \to \mathscr{H}$ for which $\sum_{g \in G} \|x(g)\|^2 < \infty$. With each T in $\mathscr{B}(\mathscr{K})$ we associate in the usual way a matrix $[T_{p,q}]_{p,q \in G}$ with entries $T_{p,q}$ in $\mathscr{B}(\mathscr{H})$, thus identifying $\mathscr{B}(\mathscr{K})$ with $n \otimes \mathscr{B}(\mathscr{H})$, where n is the cardinality of G. We define a * isomorphism Φ, from $\mathscr{B}(\mathscr{H})$ into $\mathscr{B}(\mathscr{K})$, by

$$\Phi(S) = \sum_{g \in G} \oplus S \qquad (S \in \mathscr{B}(\mathscr{H}));$$

so that $\Phi(S)$ has matrix $[\delta_{p,q}S]$, where $\delta_{p,q}$ is 0 or 1 according as $p \neq q$ or $p = q$. When $g \in G$, we denote by $V(g)$ the unitary operator on $\mathscr{K}$ that has matrix $[\delta_{p,gq}U(g)]$ (that is, $V(g)$ has matrix whose (p, q) entry is the indicated element $\delta_{p,gq}U(g)$). Simple matrix calculations show that

(1) $$V(g)V(h) = V(gh), \qquad V(g)\Phi(S) = \Phi(U(g)SU(g)^*)V(g),$$

when $S \in \mathscr{B}(\mathscr{H})$ and $g, h \in G$. The latter equation, together with condition (a) above, implies that

(2) $$V(g)\Phi(\mathscr{A})V(g)^* = \Phi(\mathscr{A}) \qquad (g \in G).$$

It is easily verified that $(V(g)x)(g') = U(g)x(g^{-1}g')$, where $x \in \mathscr{K}$ $(= l_2(G, \mathscr{H}))$. Indeed this equality may be used as the definition of $V(g)$ and (1), (2) proved in terms of it.

8.6.1. PROPOSITION. *The von Neumann algebra $\mathscr{R}$ generated by the operators $\Phi(A)$ (A in $\mathscr{A}$) and $V(g)$ (g in G) consists of all elements of $\mathscr{B}(\mathscr{K})$ that have matrices of the form $[U(pq^{-1})A(pq^{-1})]$, where $g \to A(g)$ is a mapping from G into $\mathscr{A}$. Its commutant $\mathscr{R}'$ consists of all elements of $\mathscr{B}(\mathscr{K})$ that have matrices of the form $[U(p)A'(q^{-1}p)U(p)^*]$, where $g \to A'(g)$ is a mapping from G into $\mathscr{A}$. Moreover, $\Phi(\mathscr{A})$ is a maximal abelian * subalgebra of $\mathscr{R}$; and $\mathscr{R}$ is a factor if and only if G acts ergodically on $\mathscr{A}$.*

Proof. Suppose that $T \in \mathscr{B}(\mathscr{K})$, and T has matrix $[T(p, q)]$. Simple matrix calculations show that T commutes with each $V(g)$ if and only if

(3) $$T(p, gq)U(g) = U(g)T(g^{-1}p, q) \qquad (g, p, q \in G).$$

Since $\mathscr{A}$ is maximal abelian, $\Phi(\mathscr{A})' = (\mathscr{A} \otimes I_n)' = n \otimes \mathscr{A}' = n \otimes \mathscr{A}$ (see Lemma 6.6.2); so T commutes with $\Phi(\mathscr{A})$ if and only if

(4) $$T(p, q) \in \mathscr{A} \qquad (p, q \in G).$$

If $T \in \mathscr{R}'$, then both (3) and (4) are satisfied. With $A'(g)$ defined as $T(e, g^{-1})$ when $g \in G$, we have $A'(g) \in \mathscr{A}$ by (4); and (3) gives

$$T(p, q) = U(p)T(e, p^{-1}q)U(p)^* = U(p)A'(q^{-1}p)U(p)^*.$$

Conversely, if there is a mapping $g \to A'(g): G \to \mathscr{A}$ such that $T(p, q) = U(p)A'(q^{-1}p)U(p)^*$, then a simple calculation yields (3), while (4) follows from our assumption that $U(g)\mathscr{A}U(g)^* = \mathscr{A}$. Hence T commutes with each $V(g)$ and each $\Phi(A)$, whence $T \in \mathscr{R}'$. This shows that $T \in \mathscr{R}'$ if and only if the matrix $[T(p, q)]$ has the form stated in the proposition.

The linear span $\mathscr{R}_0$ of the set $\{V(h)\Phi(A) : h \in G, A \in \mathscr{A}\}$ contains each $V(h)$ and each $\Phi(A)$, and is a * subalgebra of $\mathscr{B}(\mathscr{K})$, from (1) and (2); so $\mathscr{R} = \mathscr{R}_0^-$. Let $\mathscr{S}$ $(\subseteq \mathscr{B}(\mathscr{K}))$ be the linear subspace consisting of all operators with matrix of the form $[U(pq^{-1})A(pq^{-1})]$, where $g \to A(g)$ is a mapping from G into $\mathscr{A}$. When $h \in G$ and $A \in \mathscr{A}$, $V(h)\Phi(A)$ has a matrix of this form (with $A(g) = \delta_{g,h}A$), and thus lies in $\mathscr{S}$; so $\mathscr{R}_0 \subseteq \mathscr{S}$.

We prove next that $\mathscr{S}$ is weak-operator closed. For this, it suffices to note that $\mathscr{S}$ consists of all those T in $\mathscr{B}(\mathscr{K})$ whose matrices $[T(p,q)]$ satisfy the conditions

$$T(p,q) = T(pg,qg), \qquad U(qp^{-1})T(p,q) \in \mathscr{A} \qquad (g,p,q \in G)$$

(and then to use the fact that, for all p and q in G, the mapping $T \to T(p,q)\colon \mathscr{B}(\mathscr{K}) \to \mathscr{B}(\mathscr{H})$ is weak-operator continuous). Now direct calculation shows that the matrix $[U(pq^{-1})A(pq^{-1})]$ of an element of $\mathscr{S}$ satisfies the stated conditions. Conversely, if $[T(p,q)]$ satisfies those conditions, then $U(g^{-1})T(g,e)$ is an element $A(g)$ of $\mathscr{A}$ for each g in G; and $T \in \mathscr{S}$ since $T(p,q) = T(pq^{-1},e) = U(pq^{-1})A(pq^{-1})$. Hence the stated conditions characterize the matrices of elements of $\mathscr{S}$, and $\mathscr{S}$ is weak-operator closed.

From the two preceding paragraphs, $\mathscr{R} = \mathscr{R}_0^- \subseteq \mathscr{S}$. To prove that $\mathscr{R} = \mathscr{S}$, it now suffices to show that $ST' = T'S$ whenever $S \in \mathscr{S}$ and $T' \in \mathscr{R}'$, since this entails $\mathscr{S} \subseteq \mathscr{R}'' = \mathscr{R}$. We may choose mappings $g \to A(g)$, $g \to A'(g)\colon G \to \mathscr{A}$ so that S and T' have matrices $[U(pq^{-1})A(pq^{-1})]$ and $[U(p)A'(q^{-1}p)U(p)^*]$, respectively. The (p,q) matrix element for ST' is then given by

$$(ST')_{p,q} = \sum_{g \in G} U(pg^{-1})A(pg^{-1})U(g)A'(q^{-1}g)U(g)^*;$$

and the corresponding entry for $T'S$ is given by

$$(T'S)_{p,q} = \sum_{h \in G} U(p)A'(h^{-1}p)U(p)^*U(hq^{-1})A(hq^{-1}).$$

In the first equality, after using the fact that $A(pg^{-1})$ commutes with $U(g)A'(q^{-1}g)U(g)^*$ (because both lie in $\mathscr{A}$), we may substitute $qh^{-1}p$ for g, and sum over h instead of g. In this way, we obtain $(ST')_{p,q} = (T'S)_{p,q}$. Hence $ST' = T'S$, and $\mathscr{R} = \mathscr{S}$; and $\mathscr{R}$ consists of all elements of $\mathscr{B}(\mathscr{K})$ with matrices of the form stated in the proposition.

Suppose that $T \in \mathscr{R}$ and T commutes with the abelian * subalgebra $\Phi(\mathscr{A})$ of $\mathscr{R}$. From (4), the matrix $[T(p,q)]$ of T has all its entries in $\mathscr{A}$; moreover, since $T \in \mathscr{R}$, there is a mapping $g \to A(g)\colon G \to \mathscr{A}$ such that $T(p,q) = U(pq^{-1})A(pq^{-1})$. When $p \neq q$, since G acts freely on $\mathscr{A}$ (that is, satisfies condition (b) above), we have $T(p,q) = 0$, since $T(p,q) \in \mathscr{A} \cap U(pq^{-1})\mathscr{A} = \{0\}$. Since, also, $T(p,p) = A(e)$ for each p in G, it now follows that $T = \Phi(A(e)) \in \Phi(\mathscr{A})$. This shows that $\Phi(\mathscr{A})$ is a maximal abelian * subalgebra of $\mathscr{R}$.

An element T of $\mathscr{R}$ lies in the center $\mathscr{C}$ of $\mathscr{R}$ if and only if it commutes with $\Phi(\mathscr{A})$ and with each $V(g)$, since these generate $\mathscr{R}$. From the preceding paragraph, together with (1) and the fact that Φ is one-to-one,

$$\begin{aligned}\mathscr{C} &= \{T \in \Phi(\mathscr{A}) : T = V(g)TV(g)^* \text{ for each } g \text{ in } G\}\\ &= \{\Phi(A) : A \in \mathscr{A},\ \Phi(A) = V(g)\Phi(A)V(g)^* \text{ for each } g \text{ in } G\}\\ &= \{\Phi(A) : A \in \mathscr{A},\ A = U(g)AU(g)^* \text{ for each } g \text{ in } G\}.\end{aligned}$$

Accordingly, $\mathscr{R}$ is a factor if and only if the set

$$\{A \in \mathscr{A} : A = U(g)AU(g)^* \text{ for each } g \text{ in } G\}$$

consists of scalar multiples of I; that is, if and only if G acts ergodically on $\mathscr{A}$. ■

Before investigating the type of $\mathscr{R}$, when $\mathscr{R}$ is a factor, we need two auxiliary results. We retain the notation introduced above.

8.6.2. Lemma. *If T is a non-zero element of $\mathscr{R}^+$, and $\mathscr{W}$ is the unitary group of $\Phi(\mathscr{A})$, then the weak-operator closed convex hull $\mathscr{D}$ of the set $\{WTW^* : W \in \mathscr{W}\}$ contains a non-zero element of $\Phi(\mathscr{A})^+$.*

Proof. Since $\mathscr{D} \subseteq \mathscr{R}^+$, while $\mathscr{D}$ meets $\Phi(\mathscr{A})'$ by Corollary 8.3.12 and $\mathscr{R} \cap \Phi(\mathscr{A})' = \Phi(\mathscr{A})$ because $\Phi(\mathscr{A})$ is maximal abelian in $\mathscr{R}$, it follows that $\mathscr{D}$ meets $\Phi(\mathscr{A})^+$. It now suffices to show that $0 \notin \mathscr{D}$.

Since $T \in \mathscr{R}$, T has a matrix $[U(pq^{-1})A(pq^{-1})]$, where $A(g) \in \mathscr{A}$ for all g in G. Each W in $\mathscr{W}$ has a matrix of the form $[\delta_{p,q}W_0]$, where W_0 is a unitary operator in $\mathscr{A}$; and the matrix of WTW^* has $W_0A(e)W_0^*$ $(= A(e))$ at each diagonal position. From this, it follows that every operator in $\mathscr{D}$ has $A(e)$ in all the diagonal entries of its matrix. It now suffices to show that $A(e) \neq 0$.

The space $\mathscr{K}$ $(= \sum_{g \in G} \oplus \mathscr{H}_g)$, on which T acts, has an everywhere-dense subspace, the linear span of those vectors x in $\mathscr{K}$ that have a non-zero component (say x_0) in just one summand $\mathscr{H}_g$. Since $T > 0$, it follows that $T^{1/2}x \neq 0$ for some such vector x; and $A(e) \neq 0$, because

$$0 \neq \|T^{1/2}x\|^2 = \langle Tx, x\rangle = \langle A(e)x_0, x_0\rangle. \quad ■$$

8.6.3. Lemma. *If ρ is a normal semi-finite tracial weight on $\mathscr{R}$, the equation*

$$\rho_0(A) = \rho(\Phi(A)) \qquad (A \in \mathscr{A}^+) \tag{5}$$

defines a normal semi-finite weight ρ_0 on $\mathscr{A}$, and

$$\rho_0(U(g)AU(g)^*) = \rho_0(A) \qquad (A \in \mathscr{A}^+,\quad g \in G). \tag{6}$$

Conversely, if ρ_0 *is a normal semi-finite weight on* $\mathscr{A}$, *and satisfies* (6), *then* ρ_0 *can be obtained as in* (5) *from a normal semi-finite tracial weight* ρ *on* $\mathscr{R}$.

Proof. Suppose first that ρ is a normal semi-finite tracial weight on $\mathscr{R}$. Then ρ_0, as defined by (5), is a weight on $\mathscr{A}$. When $A \in \mathscr{A}^+$ and $g \in G$, it follows from (1) that

$$\begin{aligned}\rho_0(U(g)AU(g)^*) &= \rho(\Phi(U(g)AU(g)^*)) \\ &= \rho(V(g)\Phi(A)V(g)^*) = \rho(\Phi(A)) = \rho_0(A);\end{aligned}$$

so ρ_0 satisfies (6).

Since ρ is normal, there is a family $\{\rho_b\}$ of positive normal functionals on $\mathscr{R}$, such that $\rho(T) = \sum \rho_b(T)$ when $T \in \mathscr{R}^+$. For each index b, $\rho_b \circ \Phi$ is a positive normal functional on $\mathscr{A}$. Moreover,

$$\rho_0(A) = \rho(\Phi(A)) = \sum(\rho_b \circ \Phi)(A)$$

for all A in $\mathscr{A}^+$; so ρ_0 is normal.

We show next that ρ_0 is semi-finite (this is the most intricate part of the proof). To this end, note first that ρ_0 is a *tracial* weight (since $\mathscr{A}$ is abelian), so that $\mathscr{M}_{\rho_0}$ is an ideal in $\mathscr{A}$; and its weak-operator closure $\mathscr{M}_{\rho_0}^-$ has the form $\mathscr{A}(I - P)$ for some projection P in $\mathscr{A}$, by Theorem 6.8.8. We have to show that $P = 0$. Suppose the contrary; so that, from Proposition 8.5.2(i), there is an element T_0 of $\mathscr{R}$ for which

$$0 < T_0 \leq \Phi(P), \qquad \rho(T_0) < \infty.$$

Let

$$\mathscr{S} = \{S \in \mathscr{R} : 0 \leq S \leq \Phi(P),\ \rho(S) \leq \rho(T_0)\}.$$

Then $\mathscr{S}$ is convex; we assert also that $\mathscr{S}$ is weak-operator closed. For this last, recall that each of the positive normal functionals ρ_b (occurring in the preceding paragraph) can be expressed as a sum of functionals of the form $\omega_x | \mathscr{R}$. Hence there is a family $(x_d)_{d \in \mathbb{D}}$ of vectors, such that $\rho(T) = \sum_{d \in \mathbb{D}} \langle Tx_d, x_d \rangle$ when $T \in \mathscr{R}^+$. For each finite subset $\mathbb{F}$ of $\mathbb{D}$, the set

$$\mathscr{S}_{\mathbb{F}} = \left\{S \in \mathscr{R} : 0 \leq S \leq \Phi(P),\ \sum_{d \in \mathbb{F}} \langle Sx_d, x_d \rangle \leq \rho(T_0)\right\}$$

is weak-operator closed; and hence the same is true of the intersection $\mathscr{S}$ of all the sets $\mathscr{S}_{\mathbb{F}}$.

For each W in the unitary group $\mathscr{W}$ of $\Phi(\mathscr{A})$,

$$0 < WT_0W^* \leq W\Phi(P)W^* = \Phi(P), \qquad \rho(WT_0W^*) = \rho(T_0).$$

Accordingly, the convex weak-operator closed set $\mathscr{S}$ contains WT_0W^* for each W in $\mathscr{W}$, and so contains the weak-operator closed convex hull $\mathscr{D}$ of

$\{WT_0W^* : W \in \mathscr{W}\}$. By Lemma 8.6.2, there is a non-zero element A_0 of $\mathscr{A}^+$, such that $\Phi(A_0) \in \mathscr{S}$. Since $\Phi(A_0) \leq \Phi(P)$ and

$$\rho_0(A_0) = \rho(\Phi(A_0)) \leq \rho(T_0) < \infty,$$

we have $0 \neq A_0 \leq P$ and $A_0 \in F_{\rho_0} \subseteq \mathscr{M}_{\rho_0}$. We have now reached a contradiction, since

$$0 \neq A_0 = A_0P \in \mathscr{M}_{\rho_0} \subseteq \mathscr{M}_{\rho_0}^- = \mathscr{A}(I - P).$$

Thus $P = 0$, and ρ_0 is semi-finite.

So far, we have proved all the assertions contained in the first sentence of the lemma. Now suppose, conversely, that ρ_0 is a normal semi-finite weight on $\mathscr{A}$, and satisfies (6). Each T in $\mathscr{R}^+$ has a matrix $[T(p, q)]$ such that $T(p, q) = U(pq^{-1})A(pq^{-1})$, for some mapping $g \to A(g): G \to \mathscr{A}$. Since $T \geq 0$, the diagonal entry $T(e, e)$ $(= A(e))$ lies in $\mathscr{A}^+$; so we can define a mapping $\rho: \mathscr{R}^+ \to [0, \infty]$ by

$$\rho(T) = \rho_0(T(e, e)). \tag{7}$$

It is apparent that ρ is a weight on $\mathscr{R}$, and satisfies (5). Since ρ_0 is normal, there is a family $\{\omega_c\}$ of positive normal functionals on $\mathscr{A}$, such that $\rho_0(A) = \sum \omega_c(A)$ when $A \in \mathscr{A}^+$. Since $\rho(T) = \rho_0(T(e, e)) = \sum \omega_c(T(e, e))$ when $T \in \mathscr{R}^+$, and each of the mappings $T \to \omega_c(T(e, e))$ is a positive normal functional on $\mathscr{R}$, it follows that ρ is normal.

We prove next that ρ is a tracial weight. For this, suppose $T \in \mathscr{R}$, so that the matrix $[T(p, q)]$ of T has the form $[U(pq^{-1})A(pq^{-1})]$, with $A(g)$ in $\mathscr{A}$ for all g in G. The (e, e) matrix entries for TT^* and T^*T are given by

$$(TT^*)_{e,e} = \sum_{g \in G} T(e, g)T(e, g)^* = \sum_{g \in G} U(g^{-1})A(g^{-1})A(g^{-1})^*U(g^{-1})^*,$$

$$(T^*T)_{e,e} = \sum_{h \in G} T(h, e)^*T(h, e) = \sum_{h \in G} A(h)^*A(h) = \sum_{h \in G} A(h)A(h)^*.$$

From the normality of ρ_0, together with (7) and (6), it now follows that

$$\begin{aligned}\rho(TT^*) &= \sum_{g \in G} \rho_0(U(g^{-1})A(g^{-1})A(g^{-1})^*U(g^{-1})^*)\\ &= \sum_{h \in G} \rho_0(U(h)A(h)A(h)^*U(h)^*)\\ &= \sum_{h \in G} \rho_0(A(h)A(h)^*) = \rho(T^*T).\end{aligned}$$

Thus ρ is a tracial weight.

It remains to prove that ρ is semi-finite. For this, note first that $\Phi(A) \in F_\rho$ when $A \in F_{\rho_0}$, since $\rho(\Phi(A)) = \rho_0(A) < \infty$. Hence $\Phi(F_{\rho_0}) \subseteq F_\rho$; and by linearity, $\Phi(\mathscr{M}_{\rho_0}) \subseteq \mathscr{M}_\rho$. Since ρ_0 is a semi-finite (tracial) weight on

$\mathscr{A}$, $\mathscr{M}_{\rho_0}$ is an ideal (hence, a * subalgebra) in $\mathscr{A}$, and is weak-operator dense in $\mathscr{A}$. Accordingly there is a bounded net $\{A_a\}$ of elements of $\mathscr{M}_{\rho 0}$ that is weak-operator convergent to the identity I in $\mathscr{A}$. From this, the ideal $\mathscr{M}_\rho$ in $\mathscr{R}$ contains a net $\{\Phi(A_a)\}$ that is weak-operator convergent to I (in $\mathscr{R}$). Hence $\mathscr{M}_\rho$ is weak-operator dense in $\mathscr{R}$, and ρ is semi-finite. ■

In the following proposition, we consider the case in which G acts ergodically on $\mathscr{A}$ (so that $\mathscr{R}$ is a factor); and we give criteria for determining the type of $\mathscr{R}$. Once again, we retain the notation established above.

8.6.4. PROPOSITION. *Suppose that G acts ergodically on $\mathscr{A}$. Then $\mathscr{R}$ is a factor, and*

(i) *$\mathscr{R}$ is type* I *if and only if $\mathscr{A}$ has a minimal projection, and is then type* I_n, *where n is the cardinality of the set of all minimal projections in $\mathscr{A}$;*

(ii) *$\mathscr{R}$ is type* II *if and only if $\mathscr{A}$ has no minimal projection, and there is a normal semi-finite weight ρ_0 on $\mathscr{A}$ such that $\rho_0 \neq 0$ and*

$$\rho_0(U(g)AU(g)^*) = \rho_0(A) \qquad (A \in \mathscr{A}^+, \quad g \in G); \tag{8}$$

then $\mathscr{R}$ is type II_1 *or* II_∞ *according as $\rho_0(I) < \infty$* or *$\rho_0(I) = \infty$;*

(iii) *$\mathscr{R}$ is type* III *if and only if there is no non-zero normal semi-finite weight ρ_0 on $\mathscr{A}$ that satisfies* (8).

Proof. By Proposition 8.6.1, $\mathscr{R}$ is a factor. In determining its type, we shall use the criteria (in terms of tracial weights and minimal projections) that are given in Theorem 8.5.7 and Corollary 6.5.3.

Note that, in the circumstances considered in Lemma 8.6.3, $\rho \neq 0$ if and only if $\rho_0 \neq 0$, since $\rho(I) = \rho_0(I)$. Accordingly, $\mathscr{R}$ has a non-zero normal semi-finite tracial weight ρ (necessarily faithful, by Proposition 8.5.2(ii)) if and only if there is a non-zero normal semi-finite weight ρ_0 on $\mathscr{A}$ that satisfies (8); then, ρ can be chosen so that (5) holds. This already suffices to prove (iii), in view of Theorem 8.5.7.

We prove next that, if E is a minimal projection in $\mathscr{A}$, then $\Phi(E)$ $(= F)$ is a minimal projection in $\mathscr{R}$. Of course, F is a minimal projection in $\Phi(\mathscr{A})$. If G is a projection in $\mathscr{R}$, and $0 \le G \le F$, then G commutes with $\Phi(\mathscr{A})F$ $(= \{aF : a \in \mathbb{C}\})$ and with $\Phi(\mathscr{A})(I - F)$ $(= (I - F)\Phi(\mathscr{A}))$. Since $\Phi(\mathscr{A})$ is maximal abelian in $\mathscr{R}$, and G commutes with $\Phi(\mathscr{A})$, we now have $G \in \Phi(\mathscr{A})$ and $0 \le G \le F$; and $G = 0$ or F, from the minimality of F in $\Phi(\mathscr{A})$. Hence F is minimal in $\mathscr{R}$.

Suppose that $\mathscr{A}$ has a minimal projection. From the preceding paragraph, $\mathscr{R}$ has a minimal projection. By Corollary 6.5.3, $\mathscr{R}$ is type I_n, where n is the cardinality of any family of minimal projections in $\mathscr{R}$ with sum I. Since $\mathscr{A}$ is abelian, its minimal projections form an orthogonal family $\{E_a\}$. For each g in G, the * automorphism $A \to U(g)AU(g)^*$ of $\mathscr{A}$ permutes the members of the family $\{E_a\}$, and so leaves invariant the non-zero projection $\sum E_a$. Since G acts ergodically on $\mathscr{A}$, $\sum E_a = I$. From the preceding paragraph, it now follows that $\{\Phi(E_a)\}$ is an orthogonal family of minimal projections in $\mathscr{R}$, with sum I. Since $\mathscr{R}$ is type I_n, we can now identify n as the cardinality of the family $\{E_a\}$ of minimal projections in $\mathscr{A}$.

Conversely, suppose that $\mathscr{R}$ is type I. We shall show that $\mathscr{A}$ has a minimal projection (and this, together with the preceding paragraph, will complete the proof of (i)). By Theorem 8.5.7, $\mathscr{R}$ has a faithful normal semi-finite tracial weight ρ; and (5) defines a normal semi-finite weight ρ_0 on $\mathscr{A}$. From Proposition 8.5.2(i), there is a non-zero projection G in $\mathscr{A}$ such that $\rho_0(G) < \infty$. Since $\rho(\Phi(G)) = \rho_0(G)\ (< \infty)$, $\Phi(G)$ is a finite projection in $\mathscr{R}$, by Proposition 8.5.2(ii). Thus $\Phi(\mathscr{A})$ contains non-zero projections that are finite in $\mathscr{R}$. Each such projection F is the sum of a finite number, $n(F)$, of minimal projections in $\mathscr{R}$. By choosing F so that the positive integer $n(F)$ is minimized, we obtain a minimal projection in $\Phi(\mathscr{A})$ (and hence, a minimal projection in $\mathscr{A}$).

Since (i) and (iii) are now proved, and a factor is type II if and only if it is neither type I nor type III, it follows that $\mathscr{R}$ is type II if and only if $\mathscr{A}$ has no minimal projection, but has a non-zero normal semi-finite weight ρ_0 that satisfies (8). Then, ρ_0 gives rise to a (faithful) normal semi-finite tracial weight ρ on $\mathscr{R}$, such that $\rho(I) = \rho_0(I)$; and it follows from Theorem 8.5.7 that $\mathscr{R}$ is type II_1, if $\rho_0(I) < \infty$, and type II_∞ if $\rho_0(I) = \infty$. ∎

Measure-theoretic examples. In this subsection, m is a positive measure defined on a σ-algebra $\mathscr{S}$ of subsets of a set $S(\neq \varnothing)$, and G is a countable group (with unit e) of one-to-one mappings from S onto S, with composition of mappings as the group operation. *Throughout, the following three assumptions are in force.*

(A) *The measure space $(S, \mathscr{S}, m)$ is countably separated.* This means that there is a sequence $\{E_1, E_2, \ldots\}$ of sets, such that $(\varnothing \neq) E_j \in \mathscr{S}$ and $m(E_j) < \infty$ for all j, that separates the points of S in the following sense: if $s, t \in S$ and $s \neq t$, there is an integer j such that $s \in E_j$, $t \notin E_j$.

(B) *Each g in G preserves measurability and null sets.* In other words, if $g \in G$ and $X \subseteq S$, then $X \in \mathscr{S}$ if and only if $g(X) \in \mathscr{S}$; moreover, when $X \in \mathscr{S}$, $m(X) = 0$ if and only if $m(g(X)) = 0$.

(C) *G acts freely on S.* This means that, if $g \in G$ and $g \neq e$, the set $\{s \in S : g(s) = s\}$ is a null set.

From (A), it follows that $S = \bigcup_{j=1}^{\infty} E_j$, and that $\{s\} = \bigcap \{E_j : s \in E_j\}$ for each s in S. Moreover,

$$\{s \in S : g(s) \neq s\} = \bigcup_{j=1}^{\infty} \{E_j \backslash g^{-1}(E_j)\}.$$

Hence the measure space $(S, \mathscr{S}, m)$ is σ-finite, its points are measurable, and conditions (A) and (B) already imply the measurability of the set considered in (C).

We shall say that G *acts ergodically on* S if the following condition is satisfied: if $X \in \mathscr{S}$ and $m(g(X)\backslash X) = 0$ for each g in G, then either $m(X) = 0$ or $m(S\backslash X) = 0$.

Starting from the above measure-theoretic situation, we shall construct a Hilbert space $\mathscr{H}$, a maximal abelian * subalgebra $\mathscr{A}$ of $\mathscr{B}(\mathscr{H})$, and a unitary representation $U: g \to U_g$ of G on $\mathscr{H}$, in such a way that conditions (a) and (b) in the preceding section are satisfied, and G acts ergodically on $\mathscr{A}$ if and only if it acts ergodically on S. Before embarking on this program, we require some auxiliary results.

8.6.5. LEMMA. *Suppose that $X \in \mathscr{S}$ and $m(X) > 0$.*

(i) *There is a measurable subset Y of X for which $0 < m(Y) < \infty$.*

(ii) *If $g \in G\backslash\{e\}$, the set Y in part* (i) *can be chosen so that $g(Y) \cap Y = \varnothing$.*

(iii) *If $m(\{s\}) = 0$, for each s in S, the set Y in part* (i) *can be chosen so that $0 < m(Y) < m(X)$.*

Proof. (i) With $\{E_j\}$ the sequence occurring in condition (A) above,

$$X = \bigcup_{j=1}^{\infty} X \cap E_j, \qquad m(X \cap E_j) \leq m(E_j) < \infty.$$

Since $m(X) > 0$, we can choose j so that $m(X \cap E_j) > 0$, and then $X \cap E_j$ has the properties required of Y.

(ii) Since G acts freely and $g \in G\backslash\{e\}$, the set $X_0 = \{s \in X : g(s) \neq s\}$ has the same positive measure as X. Moreover, from conditions (A) and (B),

$$X_0 = \bigcup_{j=1}^{\infty} X \cap \{E_j \backslash g^{-1}(E_j)\},$$

and the sets on the right-hand side are measurable. For at least one value of j, the set $Y = X \cap \{E_j \backslash g^{-1}(E_j)\}$ has positive measure, and in addition $Y \cap g(Y) = \varnothing$ and $m(Y) \leq m(E_j) < \infty$.

(iii) We now assume that $m(\{s\}) = 0$ for each s in S, and prove by a contradiction argument that there is a measurable subset Y of X such that $0 < m(Y) < m(X)$.

If no such Y exists, it follows from (i) that $m(X) < \infty$; moreover, $m(Z)$ is 0 or $m(X)$ for each measurable subset Z of X. By augmenting the sequence $\{E_j\}$ occurring in condition (A) above, we may assume that each set $X\backslash E_j$

$(j = 1, 2, \ldots)$ appears as an E_k. For each j, $m(X \cap E_j)$ is 0 or $m(X)$, so there is a sequence $\{j(1), j(2), \ldots\}$ of integers such that

$$m(X \cap E_j) = \begin{cases} m(X) & (j = j(1), j(2), \ldots) \\ 0 & (j \notin \{j(n)\}). \end{cases}$$

Since

$$m\left(X \cap \bigcap_{n=1}^{\infty} E_{j(n)}\right) = m(X) - m\left(\bigcup_{n=1}^{\infty} (X \backslash E_{j(n)})\right)$$
$$= m(X) > 0,$$

while each point of S has measure zero, it follows that $X \cap \bigcap E_{j(n)}$ has two distinct elements s, t. From condition (A), we can choose j so that $s \in E_j$, $t \notin E_j$; and there is a k for which $E_k = X \backslash E_j$, whence $s \notin E_k$, $t \in E_k$. Since

$$0 < m(X) = m(X \cap E_j) + m(X \cap E_k),$$

it follows that one of the terms on the right-hand side is positive. Hence one of j, k appears in the sequence $\{j(n)\}$, and so one of s, t is not in $\bigcap E_{j(n)}$, a contradiction. This proves the existence of a set Y with the required properties. ■

8.6.6. LEMMA. *G acts ergodically on S if and only if the following condition is satisfied: if u is a bounded measurable complex-valued function on S, and $u(g(s)) = u(s)$ almost everywhere on S, for each g in G, then there is a complex number c such that $u(s) = c$ almost everywhere on S. If the word "bounded" is deleted, the assertion remains correct.*

Proof. Suppose first that the stated condition is satisfied (at least for bounded measurable functions). If X is a measurable set, and $m(g(X) \backslash X) = 0$ for each g in G, let u be the characteristic function of X; and note that $g^{-1}(X)$ has characteristic function $u \circ g$. For each g in G, $m(X \backslash g^{-1}(X)) = 0$, since g^{-1} preserves null sets, and also $m(g^{-1}(X) \backslash X) = 0$. Hence $u(g(s)) = u(s)$ almost everywhere on S, for each g in G. Accordingly, $u(s) = c$ almost everywhere on S, for some constant c; and c is 0 or 1. It follows that either $m(X) = 0$ or $m(S \backslash X) = 0$; and G acts ergodically on S.

Conversely, suppose that G acts ergodically on S. In order to verify the condition stated in the lemma (for either bounded or unbounded functions u), it suffices (by expressing u in terms of its real and imaginary parts) to consider only the case in which u is a real-valued function. Upon replacing $u(s)$ by arctan $u(s)$, we may suppose also that u is bounded. In this case, let l and L be the essential lower and upper bounds of u; that is

$$l = \max\{a : u(s) \geq a \text{ almost everywhere on } S\},$$

$$L = \min\{a : u(s) \leq a \text{ almost everywhere on } S\}.$$

Given that $u(g(s)) = u(s)$ almost everywhere on S, for each g in G, we have to show that $l = L$.

Suppose the contrary, and choose a real number b such that $l < b < L$. With X the measurable set $\{s \in S : u(s) \leq b\}$, both X and $S \setminus X$ have positive measure. Since G acts ergodically, there is an element g of G for which $m(g(X) \setminus X) > 0$. Moreover,

$$u(s) > b \geq u(g^{-1}(s)) \qquad (s \in g(X) \setminus X),$$

contradicting our assumption that $u(s) = u(g^{-1}(s))$ almost everywhere on S. Hence $l = L$, and $g(s) = l$ almost everywhere on S. ∎

8.6.7. Lemma. *For each g in G, there is a non-negative real-valued measurable function φ_g on S such that*

$$\int_S x(g(s))\, dm(s) = \int_S x(s)\varphi_g(s)\, dm(s)$$

(with the possibility that both sides are $+\infty$), for every non-negative measurable function x on S. Moreover, for each g and h in G,

$$\varphi_g(s) > 0, \qquad \varphi_{gh}(s) = \varphi_g(s)\varphi_h(g^{-1}(s)), \qquad \varphi_e(s) = 1$$

almost everywhere on S.

Proof. With g in G, it follows from condition (B) above that the equation $m_g(X) = m(g^{-1}(X))$ $(X \in \mathscr{S})$ defines a positive measure m_g on $\mathscr{S}$ that has the same null sets as m. With $\{E_j\}$ the sequence occurring in condition (A),

$$S = g(S) = \bigcup_{j=1}^{\infty} g(E_j), \qquad m_g(g(E_j)) = m(E_j) < \infty,$$

so m_g (as well as m) is σ-finite. By the Radon–Nikodým theorem there is a non-negative real-valued measurable function φ_g on S, such that

$$m(g^{-1}(X)) = m_g(X) = \int_X \varphi_g(s)\, dm(s) \qquad (X \in \mathscr{S}) \tag{9}$$

(with ∞ a possible value throughout). With X_0 the set $\{s \in S : \varphi_g(s) = 0\}$, it follows from (9) that $m(g^{-1}(X_0)) = 0$; so $m(X_0) = 0$, and $\varphi_g(s) > 0$ almost everywhere. Again from (9),

$$\int_S x(g(s))\, dm(s) = \int_S \varphi_g(s) x(s)\, dm(s) \tag{10}$$

when x is the characteristic function of a measurable set X, and by linearity this remains true when x is a non-negative simple function. The same equation is satisfied for every non-negative real-valued measurable function x, by the monotone convergence theorem, since x is the pointwise limit of an increasing sequence of simple functions.

With x the characteristic function of a measurable set X and h in G, $g^{-1}(X)$ has characteristic function $x \circ g$, so (9) and (10) yield

$$m(X) = \int_X \varphi_e(s)\, dm(s),$$

$$\begin{aligned} m_{gh}(X) &= m(h^{-1}g^{-1}(X)) = m_h(g^{-1}(X)) \\ &= \int_{g^{-1}(X)} \varphi_h(s)\, dm(s) \\ &= \int_S x(g(s))\, \varphi_h(s)\, dm(s) \\ &= \int_S x(s)\, \varphi_h(g^{-1}(s))\, \varphi_g(s)\, dm(s) \\ &= \int_X \varphi_h(g^{-1}(s))\, \varphi_g(s)\, dm(s). \end{aligned}$$

From the (essential) uniqueness of Radon–Nikodým derivatives, it follows that

$$\varphi_e(s) = 1, \qquad \varphi_{gh}(s) = \varphi_g(s)\, \varphi_h(g^{-1}(s))$$

almost everywhere on S. ■

Let $\mathscr{H}$ be the Hilbert space $L_2(S, \mathscr{S}, m)$. With each function u in L_∞ $(= L_\infty(S, \mathscr{S}, m))$, we associate the operator M_u (in $\mathscr{B}(\mathscr{H})$) of multiplication by u. Let

$$\mathscr{A} = \{M_u : u \in L_\infty\};$$

as noted in Example 5.1.6, $\mathscr{A}$ is a maximal abelian von Neumann subalgebra of $\mathscr{B}(\mathscr{H})$. The notation just introduced is used in the results that follow.

8.6.8. LEMMA. (i) *If $m(\{s\}) = 0$ for each s in S, then $\mathscr{A}$ has no minimal projection.*

(ii) *If $m(\{s_0\}) > 0$ for some s_0 in S, then $\mathscr{A}$ has a minimal projection.*

(iii) *If G acts ergodically on S, and $m(\{s_0\}) > 0$ for some s_0 in S, then S has a countable subset S_0 such that $m(S \setminus S_0) = 0$ and $m(\{s\}) > 0$ for each s in S_0. Moreover, the minimal projections in $\mathscr{A}$ can be put in (natural) one-to-one correspondence with the points of S_0, and also with the elements of G.*

Proof. For each measurable subset X of S, the operator of multiplication by the characteristic function of X is a projection $E(X)$ in $\mathscr{A}$, and $E(X) > 0$ if and only if $m(X) > 0$. Each projection in $\mathscr{A}$ arises in this way; and each subprojection of $E(X)$ in $\mathscr{A}$ has the form $E(Y)$, with Y a measurable subset of X. Moreover (given that $Y \subseteq X$), $0 < E(Y) < E(X)$ if and only if $0 < m(Y) < m(X)$. (For all this, see Example 2.5.12.)

If $m(\{s\}) = 0$ for each s in S, and $E(X)$ is a non-zero projection in $\mathscr{A}$, there is a measurable subset Y of X such that $0 < m(Y) < m(X)$, by Lemma 8.6.5(iii). Thus $0 < E(Y) < E(X)$, and $E(X)$ is not a minimal projection; so $\mathscr{A}$ has no minimal projections in this case. On the other hand, if $s_0 \in S$ and $m(\{s_0\}) > 0$, then $\{s_0\}$ has no proper subset of positive measure, and $E(\{s_0\})$ is a minimal projection in $\mathscr{A}$. This proves (i) and (ii).

Suppose now that G acts ergodically on S, while $s_0 \in S$ and $m(\{s_0\}) > 0$. Since G acts freely on S, $g(s_0) \neq s_0$ when $g \in G$ and $g \neq e$; so $g(s_0) \neq h(s_0)$ when g, h are different elements of G. Since G is countable, so is the set $\{g(s_0) : g \in G\}$ $(= S_0)$; moreover, $m(S_0) > 0$ and $g(S_0) = S_0$ for each g in G. From ergodicity, $m(S \backslash S_0) = 0$; but each point $g(s_0)$ of S_0 has positive measure, since g preserves null sets. Accordingly $\{E(\{s\}) : s \in S_0\}$ $(= \{E(\{g(s_0)\}) : g \in G\})$ is an orthogonal family of minimal projections in $\mathscr{A}$. Moreover, there are no other minimal projections in $\mathscr{A}$; for if $E(X)$ is a non-zero projection in $\mathscr{A}$, then $m(X) > 0$, so that X meets S_0 and $E(X) \geq E(\{s\})$ for some s in S_0. ∎

With x in $\mathscr{H}$ $(= L_2)$ and g in G, we can define a measurable function $U_g x$ on S by

$$(U_g x)(s) = [\varphi_g(s)]^{1/2} x(g^{-1}(s)),$$

where φ_g is the function introduced in Lemma 8.6.7. Since

$$\int_S |(U_g x)(s)|^2 \, dm(s) = \int_S \varphi_g(s) |x(g^{-1}(s))|^2 \, dm(s)$$

$$= \int_S |x(s)|^2 \, dm(s),$$

$U_g x \in \mathscr{H}$ and $U_g : \mathscr{H} \to \mathscr{H}$ is an isometric linear operator. Moreover, $U_e = I$ and $U_{gh} = U_g U_h$ for all g and h in G, since, from Lemma 8.6.7,

$$\begin{aligned}(U_g U_h x)(s) &= [\varphi_g(s)]^{1/2} (U_h x)(g^{-1}(s)) \\ &= [\varphi_g(s) \varphi_h(g^{-1}(s))]^{1/2} x(h^{-1} g^{-1}(s)) \\ &= [\varphi_{gh}(s)]^{1/2} x(h^{-1} g^{-1}(s)) = (U_{gh} x)(s).\end{aligned}$$

Hence $U : g \to U_g$ is a unitary representation of G on $\mathscr{H}$.

We now prove that, with $\mathscr{A}$ and U constructed as above, the basic conditions (a) and (b), set out in the preceding subsection, are satisfied. A straightforward calculation, using the fact that $\varphi_g(s)\varphi_{g^{-1}}(g^{-1}(s)) = \varphi_e(s) = 1$ almost everywhere (from Lemma 8.6.7), shows that

$$(11) \qquad U_g M_u U_g^* = M_{u \circ g^{-1}} \qquad (u \in L_\infty, \quad g \in G);$$

and from this, $U_g \mathscr{A} U_g^* = \mathscr{A}$. If $g \in G$, $g \neq e$, and $A \in \mathscr{A} \cap U_g \mathscr{A}$, we can choose v and w in L_∞ so that $A = M_v = U_g M_w$. If $A \neq 0$, the set $\{s \in S : v(s) \neq 0\}$ has positive measure, and by Lemma 8.6.5(ii) it has a measurable subset X such that $0 < m(X) < \infty$ and $X \cap g(X) = \varnothing$. If x is the characteristic function of X, then $x \in \mathscr{H}$; and for each s in X,

$$(M_v x)(s) = v(s)x(s) \neq 0,$$

$$(U_g M_w x)(s) = [\varphi_g(s)]^{1/2} w(g^{-1}(s)) x(g^{-1}(s)) = 0,$$

which is impossible, since $M_v = U_g M_w$. Thus $A = 0$, and $\mathscr{A} \cap U_g \mathscr{A} = \{0\}$.

We have now proved that

$$(12) \quad U_g \mathscr{A} U_g^* = \mathscr{A} \quad (g \in G), \qquad \mathscr{A} \cap U_g \mathscr{A} = \{0\} \quad (g \in G,\ g \neq e).$$

From the first of these conditions, each g in G gives rise to a * automorphism, $A \to U_g A U_g^*$, of $\mathscr{A}$. Concerning the ergodicity of this action of G on $\mathscr{A}$, we have the following result.

8.6.9. Proposition. *G acts ergodically on $\mathscr{A}$ if and only if G acts ergodically on S.*

Proof. G acts ergodically on $\mathscr{A}$ if and only if the set $\{A \in \mathscr{A} : U_g A U_g^* = A \text{ for each } g \text{ in } G\}$ consists of scalar multiples of I. From (11)

$$\begin{aligned}\{A \in \mathscr{A} : U_g A U_g^* &= A \text{ for each } g \text{ in } G\} \\ &= \{M_u : u \in L_\infty,\ M_u = M_{u \circ g^{-1}} \text{ for each } g \text{ in } G\} \\ &= \{M_u : u \in L_\infty,\ u(s) = u(g^{-1}(s)) \\ &\qquad \text{almost everywhere on } S \text{ for each } g \text{ in } G\}.\end{aligned}$$

Accordingly, G acts ergodically on $\mathscr{A}$ if and only if the following condition is satisfied: if $u \in L_\infty$ and $u(s) = u(g(s))$ almost everywhere on S, for each g in G, then there is a complex number c such that $u(s) = c$ almost everywhere on S. From Lemma 8.6.6, this last condition is satisfied if and only if G acts ergodically on S. ■

We have verified that the maximal abelian von Neumann subalgebra $\mathscr{A}$ of $\mathscr{B}(\mathscr{H})$ and the unitary representation $U : g \to U_g$ of G on $\mathscr{H}$ satisfy the conditions (a) and (b) set out at the beginning of the preceding subsection.

Let $\mathscr{R}$ be the von Neumann algebra considered in Proposition 8.6.1, so that $\mathscr{R}$ is a factor if and only if G acts ergodically on S, by Propositions 8.6.1 and 8.6.9. In the following result we interpret, in measure-theoretic terms, the criteria set out in Proposition 8.6.4 for determining the type of $\mathscr{R}$, when $\mathscr{R}$ is a factor. We retain the notation used above.

8.6.10. Proposition. *If G acts ergodically on S, then $\mathscr{R}$ is a factor, and its type is determined as follows.*

(i) *$\mathscr{R}$ is type* I *if and only if $m(\{s_0\}) > 0$ for some s_0 in S; it is then type* I_n, *where n is the (finite or countably infinite) cardinality of G.*

(ii) *$\mathscr{R}$ is type* II *if and only if $m(\{s\}) = 0$ for each s in S, and there is a non-zero σ-finite measure m_0, defined on the σ-algebra $\mathscr{S}$, invariant under G in the sense that $m_0(g(X)) = m_0(X)$ when $X \in \mathscr{S}$ and $g \in G$, and absolutely continuous with respect to m in the sense that $m_0(X) = 0$ when $X \in \mathscr{S}$ and $m(X) = 0$. When these conditions are satisfied, $\mathscr{R}$ is type II_1 or II_∞ according as $m_0(S) < \infty$ or $m_0(S) = \infty$.*

(iii) *$\mathscr{R}$ is type* III *if and only if there is no measure m_0 satisfying the conditions set out in* (ii).

Proof. We have already noted, in the discussion preceding the proposition, that $\mathscr{R}$ is a factor when G acts ergodically on S. In view of the information in Lemma 8.6.8, concerning minimal projections in $\mathscr{A}$, (i) is an immediate consequence of Proposition 8.6.4. Moreover, in order to prove (ii) and (iii), it suffices to show that there is a measure m_0 of the type described in (ii) if and only if $\mathscr{A}$ has a non-zero normal semi-finite weight ρ_0 such that $\rho_0(U_g A U_g^*) = \rho_0(A)$ whenever $A \in \mathscr{A}^+$ and $g \in G$ (and $\rho_0(I) < \infty$ if and only if $m_0(S) < \infty$).

Suppose first that such a weight ρ_0 exists. We can define a mapping $m_0 \colon \mathscr{S} \to [0, \infty]$ by $m_0(X) = \rho_0(M_x)$, where x is the characteristic function of the measurable set X. From the stated properties of ρ_0, it follows easily that m_0 is a non-zero measure on the σ-algebra $\mathscr{S}$, and $m_0(X) = 0$ when $m(X) = 0$. If $g \in G$, the set $g(X)$ has characteristic function $x \circ g^{-1}$; and from (11),

$$m_0(g(X)) = \rho_0(M_{x \circ g^{-1}}) = \rho_0(U_g M_x U_g^*) = \rho_0(M_x) = m_0(X).$$

It remains to prove that m_0 is σ-finite. Since ρ_0 is semi-finite, there is a non-zero projection F in $\mathscr{A}$ for which $\rho_0(F) < \infty$, by Proposition 8.5.2(i). Corresponding to F, there is a measurable subset Y of S, such that $m(Y) > 0$ and $m_0(Y) < \infty$; and for each g in G, we have $m_0(g(Y)) = m_0(Y)$ $(< \infty)$. The set $\bigcup_{g \in G} g(Y)$ $(= X)$ has positive (m-)measure, and $g(X) = X$ for each g in G. Since G acts ergodically on S, it follows that $m(S \setminus X) = 0$, and thus

$m_0(S \setminus X) = 0$. Accordingly, S is the union of countably many sets, $S \setminus X$ and $g(Y)$ with g in G, to each of which m_0 assigns finite measure; so m_0 is σ-finite.

Conversely, suppose that there is a measure m_0 with the properties set out in part (ii) of the proposition. By the Radon–Nikodým theorem there is a non-negative real-valued function φ on S such that

$$m_0(X) = \int_X \varphi(s)\, dm(s) \qquad (X \in \mathscr{S})$$

(with the possibility that both sides are infinite). Since m_0 is σ-finite, S is the disjoint union of a sequence $\{X_j\}$ of sets in $\mathscr{S}$, with $m_0(X_j) < \infty$. Note that if x_j is the characteristic function of X_j, and $y_j(s) = x_j(s)[\varphi(s)]^{1/2}$, then $y_j \in \mathscr{H}$ $(= L_2(S, \mathscr{S}, m))$, since

$$\int_S |y_j(s)|^2\, dm(s) = \int_S |x_j(s)|^2 \varphi(s)\, dm(s) = \int_{X_j} \varphi(s)\, dm(s) = m_0(X_j) < \infty.$$

Since m_0 is defined on the σ-algebra $\mathscr{S}$, and vanishes on the null sets of m, the equation

$$\rho_0(M_u) = \int_S u(s)\, dm_0(s) \qquad (u \in L_\infty(S, \mathscr{S}, m), \quad u \geq 0)$$

defines a mapping $\rho_0 \colon \mathscr{A}^+ \to [0, \infty]$. Moreover, ρ_0 is a non-zero (tracial) weight on $\mathscr{A}$, and is normal since

$$\begin{aligned}\rho_0(M_u) &= \int_S u(s)\varphi(s)\, dm(s)\\ &= \sum_{j=1}^{\infty} \int_{X_j} u(s)\varphi(s)\, dm(s) = \sum_{j=1}^{\infty} \langle M_u y_j, y_j \rangle.\end{aligned}$$

We assert also that ρ_0 is semi-finite. For this, let P_j be the projection in $\mathscr{A}$ corresponding to the measurable set X_j, so that $\rho_0(P_j) = m_0(X_j) < \infty$. The ideal $\mathscr{M}_{\rho_0}$ in $\mathscr{A}$ contains each P_j, and is therefore weak-operator dense in $\mathscr{A}$, since $\sum P_j = I$; and ρ_0 is semi-finite. Finally, the invariance of m_0 under G implies that

$$\begin{aligned}\rho_0(U_g M_u U_g^*) &= \rho_0(M_{u \circ g^{-1}})\\ &= \int_S u(g^{-1}(s))\, dm_0(s)\\ &= \int_S u(s)\, dm_0(s) = \rho_0(M_u),\end{aligned}$$

for each g in G and non-negative u in $L_\infty(S, \mathscr{S}, m)$. ■

8.6.11. EXAMPLE (A type I factor). Let S be a finite or countably infinite group, $\mathscr{S}$ the σ-algebra of all subsets of S, m the measure on $\mathscr{S}$ that assigns unit mass to each point of S, and G ($\cong S$) the group of all left translations $s \to ts: S \to S$ (with t in S). Conditions (A), (B), and (C) above are immediately verified, and G acts ergodically on S. From Proposition 8.6.10, the corresponding factor $\mathscr{R}$ is type I_n, where n is the cardinal of S. ■

8.6.12. EXAMPLE (A type II_1 factor). Let m be Lebesgue measure on the σ-algebra $\mathscr{S}$ of Borel subsets of the unit interval S ($= \{s \in \mathbb{R} : 0 \leq s < 1\}$), and let G be the group of all rational translations, modulo 1, of S. Thus the general member of G is a mapping $s \to \{s + r\}$, where r is rational and $\{a\}$ denotes the fractional part of the real number a.

Conditions (B) and (C) above are immediately verified, and (A) is satisfied with $\{E_j\}$ an enumeration of all half-open intervals $[a, b)$ in S with rational endpoints. The fact that G acts ergodically on S can be proved by identifying S with the circle group $\mathbb{T}_1$ by means of the exponential map, and appealing to the uniqueness of Haar measure on $\mathbb{T}_1$; for completeness we include below an elementary variant of this argument. Since

$$m(S) < \infty, \qquad m(\{s\}) = 0, \qquad m(g(X)) = m(X)$$

whenever $s \in S$, $X \in \mathscr{S}$, $g \in G$, it then follows from Proposition 8.6.10(ii) (with $m_0 = m$) that the corresponding factor $\mathscr{R}$ is type II_1.

To prove the ergodicity of G, suppose that $X_0 \in \mathscr{S}$, $m(X_0) > 0$, and $m(g(X_0) \setminus X_0) = 0$ for each g in G. The equation

$$m_1(X) = m(X \cap X_0) \qquad (X \in \mathscr{S})$$

defines a measure m_1 on $\mathscr{S}$, and $0 < m(X_0) = m_1(S) < \infty$. Since

$$\begin{aligned} m_1(g(X)) &= m(g(X) \cap X_0) = m(X \cap g^{-1}(X_0)) \\ &\leq m(X \cap X_0) + m(X \cap [g^{-1}(X_0) \setminus X_0]) \\ &= m(X \cap X_0) = m_1(X), \end{aligned}$$

it follows that $m_1(g(X)) = m_1(X)$ for each X in $\mathscr{S}$ and g in G. If n is a positive integer and X, Y are two rational intervals of the form $[a, b)$ in S, with $m(X) = m(Y) = 1/n$, there is a g in G for which $X = g(Y)$, and so $m_1(X) = m_1(Y)$. Thus m_1 takes the same value on all such intervals X, and by expressing S as the disjoint union of n such intervals, it follows that this constant value of m_1 is c/n ($= cm(X)$), where $c = m_1(S) = m(X_0) > 0$.

Since the σ-algebra $\mathscr{S}$ is generated by such intervals, of lengths $1, \frac{1}{2}, \frac{1}{3}, \ldots, m_1(X) = cm(X)$ for each X in $\mathscr{S}$. In particular,

$$m(S \setminus X_0) = c^{-1} m_1(S \setminus X_0) = c^{-1} m([S \setminus X_0] \cap X_0) = 0,$$

which proves the ergodicity of G. ■

8.6.13. Example (A type II_∞ factor). Let m be Lebesgue measure on the σ-algebra $\mathscr{S}$ of Borel subsets of the real line S, and let G be the group of all rational translations of S. Conditions (A), (B), and (C) and the ergodicity of G acting on S can be verified by arguments very similar to those used in the preceding example. Since

$$m(S) = \infty, \qquad m(\{s\}) = 0, \qquad m(g(X)) = m(X),$$

whenever $s \in S$, $X \in \mathscr{S}$, $g \in G$, it follows from Proposition 8.6.10(ii) (with $m_0 = m$) that the corresponding factor $\mathscr{R}$ is type II_∞. ■

8.6.14. Example (A type III factor). With S, $\mathscr{S}$, and m as in the preceding example, let G be the group of all mappings of the form $s \to as + b \colon S \to S$, where a ($\neq 0$) and b are rational. Conditions (A), (B), and (C) are verified as in the earlier examples. Since G contains the group G_0 of all rational translations of S, and the ergodicity of G_0 has been noted in Example 8.6.13, it follows that G itself acts ergodically.

We now show that the corresponding factor $\mathscr{R}$ is type III. By Proposition 8.6.10, we have to prove that there is no non-zero σ-finite measure m_0 on $\mathscr{S}$ such that, for g in G and X in $\mathscr{S}$,

$$(13) \qquad m_0(g(X)) = m_0(X), \qquad m_0(X) = 0 \quad \text{if} \quad m(X) = 0.$$

If such a measure m_0 exists, the Radon–Nikodým theorem implies that there is a non-negative real-valued measurable function φ on S such that

$$m_0(X) = \int_X \varphi(s)\, dm(s) \qquad (X \in \mathscr{S}).$$

Since both m and m_0 are invariant under translation by elements of G_0, it follows from the essential uniqueness of Radon–Nikodým derivatives that $\varphi(g(s)) = \varphi(s)$ almost everywhere on S, for each g in G_0. Since G_0 acts ergodically, it now results from Proposition 8.6.6 that there is a constant c such that $\varphi(s) = c$ almost everywhere. Thus $m_0 = cm$, $c \neq 0$ since $m_0 \neq 0$; and from (13), $m(g(X)) = m(X)$ for all g in G. This gives the contradiction, required to prove that no such m_0 exists, since m is not invariant under the mapping $s \to as + b$ when $|a| \neq 1$. ■

8.6.15. Remark. The procedure by which a type III factor was obtained in Example 8.6.14 illustrates a general method for the construction of such factors. Suppose that G_0 is a group of one-to-one mappings from a measure space $(S, \mathscr{S}, m)$ onto itself, acting ergodically on S, satisfying conditions (A), (B), and (C) of this subsection, and leaving m invariant. (Of course, the factor produced from the action of G_0 on S will not be type III, by Proposition 8.6.10(iii), with m_0 equal to m.) Suppose further that G_0 is a subgroup of a larger group G of one-to-one mappings from S onto S, and that G satisfies conditions (A), (B), and (C), but does not leave m invariant. Since G_0 acts ergodically on S, the same is true of G. We assert that the factor $\mathscr{R}$, arising from the action of G on S, is of type III.

In order to prove that $\mathscr{R}$ is type III, it suffices (by Proposition 8.6.10(iii)) to show that there is no non-zero σ-finite measure m_0, defined on $\mathscr{S}$, invariant under G, and absolutely continuous with respect to m. If such a measure m_0 did exist, it would not be a positive scalar multiple of m, since m is not invariant under G. Accordingly, we would have two essentially different (non-zero) measures, m_0 and m, both σ-finite and defined on $\mathscr{S}$, both invariant under G_0, and both absolutely continuous with respect to m. This leads to a contradiction, by use of a uniqueness theorem proved in the next paragraph; so no such measure m_0 exists, and $\mathscr{R}$ is of type III.

We now provide the necessary uniqueness theorem. We prove that, if G acts ergodically on $(S, \mathscr{S}, m)$, conditions (A), (B), and (C) are satisfied, and m_0 and m_0' are non-zero σ-finite measures on $\mathscr{S}$ that are invariant under G and absolutely continuous with respect to m, then m_0 is a positive scalar multiple of m_0'. While this can be proved, without difficulty, in purely measure-theoretic terms, it is also an easy consequence of our information concerning tracial weights on factors. From the proof of Proposition 8.6.10, m_0 and m_0' give rise to non-zero normal semi-finite weights ρ_0 and ρ_0' on $\mathscr{A}$ (each satisfying (6)), and

$$\rho_0(M_u) = \int_S u(s)\, dm_0(s), \qquad \rho_0'(M_u) = \int_S u(s)\, dm_0'(s)$$

for all u in $L_\infty^+(S, \mathscr{S}, m)$. There are corresponding (non-zero, hence faithful) normal semi-finite tracial weights ρ and ρ' on the *factor* arising from the action of G on S, these weights being constructed from ρ_0 and ρ_0' as in Lemma 8.6.3. From Theorem 8.5.7, ρ is a positive scalar multiple $c\rho'$ of ρ'; so $\rho_0 = c\rho_0'$, and $m_0 = cm_0'$. (The contradiction, in the preceding paragraph, is obtained by applying the present result with G_0 in place of G and m in place of m_0').

The foregoing discussion indicates a general method for constructing factors of type III. Suppose that G_0 acts ergodically on $(S, \mathscr{S}, m)$, satisfies conditions (A), (B), and (C), leaves m invariant, and gives rise to a factor of type II; if we can find a larger group G, acting on S and still satisfying conditions (A), (B), and (C) but not leaving m invariant, then the action of G on S leads to a factor of type III. (Note that it is never possible to find such a group G when the action of G_0 on S gives rise to a type I factor; for then, $m(\{s_0\}) > 0$ for some s_0 in S, and any factor produced (as in this section) from a group acting on S will be of type I.)

In using this procedure in Example 8.6.14, $(S, \mathscr{S}, m)$ and G_0 were the measure space and group previously used to construct a factor of type II_∞. In fact, m was Lebesgue measure on $\mathbb{R}$ $(= S)$, G_0 (leaving m invariant) was the group of rational translations, and G (not leaving m invariant) was the group of all mappings $s \to as + b: \mathbb{R} \to \mathbb{R}$, with a $(\neq 0)$ and b rational.

Bibliography: [56, 66].

8.7. Exercises

8.7.1. Let τ be the center-valued trace on a finite von Neumann algebra $\mathscr{R}$, and define a bounded linear operator $v: \mathscr{R} \to \mathscr{R}$ by $v(A) = A - \tau(A)$.

(i) Prove that v is a projection, and identify its range and null space.

(ii) By appeal to Lemma 6.5.6, or otherwise, show that $\|v\| = 2$ when $\mathscr{R}$ is of type II_1.

(iii) By use of Theorem 7.3.1, or otherwise, show that $\|v\| = 2n^{-1}(n-1)$ when $\mathscr{R}$ is a factor of type I_n (with $n < \infty$).

8.7.2. Suppose that $\mathfrak{A}_1$ and $\mathfrak{A}_2$ are self-adjoint subalgebras of a finite von Neumann algebra $\mathscr{R}$, τ is the center-valued trace on $\mathscr{R}$, and φ is a * isomorphism from $\mathfrak{A}_1$ onto $\mathfrak{A}_2$ such that $\tau(\varphi(A)) = \tau(A)$ for each A in $\mathfrak{A}_1$. For $j = 1, 2$, let $\mathscr{B}_j$ be the set of all operators of the form $C_0 + A_1C_1 + A_2C_2 + \cdots + A_nC_n$, with $\{C_0, \ldots, C_n\}$ a finite subset of the center $\mathscr{C}$ of $\mathscr{R}$ and $A_1, \ldots A_n$ in $\mathfrak{A}_j$.

(i) Show that $\mathscr{B}_j$ is a self-adjoint subalgebra of $\mathscr{R}$ that contains $\mathscr{C} \cup \mathfrak{A}_j$, and that φ extends to a * isomorphism $\bar{\varphi}$ from $\mathscr{B}_1$ onto $\mathscr{B}_2$ such that $\bar{\varphi}(C) = C$ and $\tau(\bar{\varphi}(B)) = \tau(B)$ whenever $C \in \mathscr{C}$ and $B \in \mathscr{B}_1$.

(ii) Let $\{Q_k\}$ be an orthogonal family of cyclic projections in $\mathscr{C}$ with sum I, and for each k let x_k be a unit vector such that Q_k has range $[\mathscr{C}'x_k]$. Show that the equation $\rho_k(R) = \langle \tau(R)x_k, x_k \rangle$ defines a faithful normal state ρ_k on the von Neumann algebra $\mathscr{R}Q_k$, and prove that the restriction $\bar{\varphi} \mid \mathscr{B}_1Q_k$ is a * isomorphism from $\mathscr{B}_1Q_k$ onto $\mathscr{B}_2Q_k$.

(iii) By using the result of Exercise 7.6.7, show that $\bar{\varphi} \mid \mathscr{B}_1Q_k$ extends to a * isomorphism from $\mathscr{B}_1^- Q_k$ onto $\mathscr{B}_2^- Q_k$. Deduce that $\bar{\varphi}$ extends to a * isomorphism from $\mathscr{B}_1^-$ onto $\mathscr{B}_2^-$.

8.7.3. Let τ be the center-valued trace on a finite von Neumann algebra $\mathscr{R}$ acting on a Hilbert space $\mathscr{H}$.

(i) Prove that, for each x in $\mathscr{H}$, there is a sequence $\{y_1, y_2, \ldots\}$ of elements of $\mathscr{H}$ such that $\sum \|y_n\|^2 < \infty$ and

$$\langle \tau(A)x, x\rangle = \sum \langle Ay_n, y_n\rangle \qquad (A \in \mathscr{R}).$$

(ii) Let $\mathscr{S}$ be the set of all vectors y in $\mathscr{H}$ with the following property: there is a vector x in $\mathscr{H}$ such that

$$\langle Ay, y\rangle \le \langle \tau(A)x, x\rangle \qquad (A \in \mathscr{R}^+).$$

Show that $\mathscr{S}$ is a separating set for $\mathscr{R}$.

(iii) Suppose that $\{R_a\}$ is a bounded net of elements of $\mathscr{R}$ and that the net $\{\tau(R_a^* R_a)\}$ is weak-operator convergent to 0. Show that $\|R_a A'y\| \to 0$ for each A' in $\mathscr{R}'$ and y in $\mathscr{S}$, and deduce that $\{R_a\}$ is strong-operator convergent to 0.

8.7.4. Suppose that $\mathscr{R}$ is a von Neumann algebra with center $\mathscr{C}$, φ is a * homomorphism from $\mathscr{R}$ onto a C^*-algebra $\mathfrak{A}$, and $\mathscr{U}(\mathfrak{A})$ is the unitary group of $\mathfrak{A}$. Prove that

(i) for each A in $\mathfrak{A}$, the norm-closed convex hull of the set $\{UAU^* : U \in \mathscr{U}(\mathfrak{A})\}$ meets $\varphi(\mathscr{C})$;
(ii) $\varphi(\mathscr{C})$ is the center of $\mathfrak{A}$.

8.7.5. Suppose that $\mathscr{R}$ is a von Neumann algebra with center $\mathscr{C}$, and $A_1, A_2 \in \mathscr{R}$. Show that (with the notation of Section 8.3) the set $\mathrm{co}_{\mathscr{R}}(A_1 + A_2)^= \cap \mathscr{C}$ is contained in the norm closure of the set

$$\{C_1 + C_2 : C_j \in \mathrm{co}_{\mathscr{R}}(A_j)^= \cap \mathscr{C}\ \ (j = 1, 2)\}.$$

8.7.6. Let $\mathscr{R}$ be a von Neumann algebra with center $\mathscr{C}$. With the notation of Section 8.3, show that if P is a projection in $\mathscr{C}$ and $\alpha \in \mathscr{D}$, there is an element α^P of $\mathscr{D}$ for which

$$\alpha^P(A) = P\alpha(A) + (I - P)A \qquad (A \in \mathscr{R}).$$

Hence show that if $P_1, \ldots, P_n$ are projections, with sum I, in $\mathscr{C}$ and $\beta_1, \ldots, \beta_n \in \mathscr{D}$, there is an element β of $\mathscr{D}$ for which

$$\beta(A) = P_1\beta_1(A) + \cdots + P_n\beta_n(A) \qquad (A \in \mathscr{R}).$$

Deduce that for all A in $\mathscr{R}$

$$\mathrm{co}_{\mathscr{R}}(A)^= \cap \mathscr{C} = \{C_1 + \cdots + C_n : C_j \in \mathrm{co}_{\mathscr{R}}(P_j A)^= \cap \mathscr{C}\quad (j = 1, \ldots, n)\},$$

and

$$\mathrm{co}_{\mathscr{R}}(P_j A)^= \cap \mathscr{C} = \{P_j C : C \in \mathrm{co}_{\mathscr{R}}(A)^= \cap \mathscr{C}\}.$$

8.7.7. Suppose that $\mathscr{R}$ is a von Neumann algebra with center $\mathscr{C}$, P is a projection in $\mathscr{C}$ that is properly infinite relative to $\mathscr{R}$, and E_0 is a subprojection of P in $\mathscr{R}$ such that $E_0 \sim P - E_0 \sim P$ (see Lemma 6.3.3).

(i) By expressing $P - E_0$ as the sum of n projections in $\mathscr{R}$, each of which is equivalent to E_0, and by arguing as in the proof of Proposition 8.3.9, show that

$$\frac{1}{n+1}P, \frac{n}{n+1}P \in \mathrm{co}_{\mathscr{R}}(E_0) \cap \mathscr{C} \qquad (n = 1, 2, \ldots).$$

Deduce that $cP \in \mathrm{co}_{\mathscr{R}}(E_0)^= \cap \mathscr{C}$ for all c in $[0, 1]$.

(ii) Prove that $\mathrm{co}_{\mathscr{R}}(E_0)^= \cap \mathscr{C}$ contains each operator C of the form $\sum_{j=1}^{k} c_j P_j$, where $c_1, \ldots, c_k$ are scalars in $[0, 1]$ and $P_1, \ldots, P_k$ are mutually orthogonal projections in $\mathscr{C}$ with sum P.

(iii) Deduce that

$$\mathrm{co}_{\mathscr{R}}(E_0)^= \cap \mathscr{C} = \{C \in \mathscr{C} : 0 \le C \le P\}.$$

8.7.8. Let $\mathscr{R}$ be a von Neumann algebra with center $\mathscr{C}$. Show that the following two conditions are equivalent.

(i) $\mathscr{R}$ is finite.

(ii) If $A, B \in \mathscr{R}$ and $X \in \mathrm{co}_{\mathscr{R}}(A)^= \cap \mathscr{C}$, $Y \in \mathrm{co}_{\mathscr{R}}(B)^= \cap \mathscr{C}$, then $X + Y \in \mathrm{co}_{\mathscr{R}}(A + B)^= \cap \mathscr{C}$.

8.7.9. Suppose that $\mathscr{R}$ is a properly infinite von Neumann algebra with center $\mathscr{C}$, and E is a projection in $\mathscr{R}$.

(i) Suppose that $F_1, \ldots, F_k$ are projections in $\mathscr{R}$, P is a projection in $\mathscr{C}$, $F_j \precsim E$ for $j = 1, \ldots, k$, and $F_1 \vee F_2 \vee \cdots \vee F_k = P$. Prove that $PE \sim P$.

(ii) Suppose that C is a non-zero element of $\mathrm{co}_{\mathscr{R}}(E)^= \cap \mathscr{C}$, and unitary operators $U_1, \ldots, U_k$ in $\mathscr{R}$ and positive scalars $a_1, \ldots, a_k$ with sum 1 are chosen so that

$$\|C - \sum_{j=1}^{k} a_j U_j E U_j^*\| < \|C\|.$$

Prove that $\bigvee_{j=1}^{k} U_j Q E U_j^* = Q$ for some non-zero projection Q in $\mathscr{C}$.

(iii) Prove that, if $\mathrm{co}_{\mathscr{R}}(E)^= \cap \mathscr{C} \neq \{0\}$, then $QE \sim Q$ for some non-zero projection Q in $\mathscr{C}$.

(iv) Prove that, if $\mathrm{co}_{\mathscr{R}}(E)^= \cap \mathscr{C} \neq \{C_E\}$, then $Q(C_E - E) \sim Q$ for some non-zero projection Q in $\mathscr{C}$.

(v) Show that there are central projections Q_1 and Q_2 maximal with respect to the properties that $Q_1 E \sim Q_1$ and $Q_2(I - E) \sim Q_2$. Prove that

$$\mathrm{co}_{\mathscr{R}}(E)^{=} \cap \mathscr{C} = \{I - Q_2 + C : C \in \mathscr{C},\ 0 \leq C \leq Q_1 Q_2\}.$$

[*Hint*. By use of (i) and the comparison theorem, show that $(I - Q_1)(I - Q_2) = 0$, so that

$$I = (I - Q_1) + (I - Q_2) + Q_1 Q_2.$$

Apply (iii) to $(I - Q_1)E$, (iv) to $(I - Q_2)E$, Exercise 8.7.7 to $Q_1 Q_2 E$, and Exercise 8.7.6 to E.]

Interpret the result of (v) in the case of an infinite factor.

8.7.10. Suppose that $\mathscr{R}$ is an infinite factor and $\{E_\lambda\}$ is the spectral resolution of a self-adjoint element A of $\mathscr{R}$. Let $\mathscr{C}$ denote the center $\{zI : z \in \mathbb{C}\}$ of $\mathscr{R}$, and define real numbers a, b by

$$a = \sup\{\lambda \in \mathbb{R} : E_\lambda \prec I\}, \qquad b = \inf\{\lambda \in \mathbb{R} : I - E_\lambda \prec I\}.$$

Show that

$$-\|A\| \leq a \leq b \leq \|A\|$$

and

$$\mathrm{co}_{\mathscr{R}}(A)^{=} \cap \mathscr{C} = \{cI : c \in \mathbb{R},\ a \leq c \leq b\}.$$

[*Hint*. Use the results of Exercise 8.7.9.]

8.7.11. Show that if A is a self-adjoint element of a countably decomposable type III factor $\mathscr{R}$, and $\mathscr{C}$ is $\mathscr{R} \cap \mathscr{R}'$ $(= \{zI : z \in \mathbb{C}\})$, then

$$\mathrm{co}_{\mathscr{R}}(A)^{=} \cap \mathscr{C} = \{cI : c \in \mathbb{R},\ m \leq c \leq M\}$$

where $[m, M]$ is the smallest interval containing $\mathrm{sp}(A)$.

8.7.12. Suppose that $\mathscr{R}$ is an infinite factor, $\{E_\lambda\}$ is the spectral resolution of a self-adjoint element A of $\mathscr{R}$, $\mathscr{C}$ is the center $\{zI : z \in \mathbb{C}\}$ of $\mathscr{R}$, $\mathscr{K}$ is the largest proper (norm-closed) two-sided ideal in $\mathscr{R}$ (see Exercise 6.9.51), and $\varphi : \mathscr{R} \to \mathscr{R}/\mathscr{K}$ is the quotient mapping. By using the result of Exercise 8.7.10, show that

$$\mathrm{co}_{\mathscr{R}}(A)^{=} \cap \mathscr{C} = \{cI : c \in \mathbb{R},\ m \leq c \leq M\},$$

where $[m, M]$ is the smallest interval containing $\mathrm{sp}(\varphi(A))$.

8.7.13. Interpret the results of Exercises 8.7.12 and 8.7.10 in the case of a countably decomposable factor $\mathscr{R}$ of type I_∞ or II_∞ (in particular, for the factor $\mathscr{B}(\mathscr{H})$, with $\mathscr{H}$ a separable Hilbert space), taking into account the information given in Corollary 6.3.5, Theorem 6.8.7, and Exercise 2.8.25.

8.7.14. Show that if $\mathscr{I}$ is a norm-closed two-sided ideal in a von Neumann algebra $\mathscr{R}$, and $R \in \mathscr{I}$, then $\mathrm{co}_{\mathscr{R}}(ARB)^{=} \subseteq \mathscr{I}$ for all A and B in $\mathscr{R}$.

8.7.15. Suppose that $\mathscr{R}$ is a von Neumann algebra with center $\mathscr{C}$, and $\mathscr{K}$ is a norm-closed ideal in $\mathscr{C}$. Let $j(\mathscr{K})$ be the norm-closed linear span $[\mathscr{R}\mathscr{K}]$ of the set $\mathscr{R}\mathscr{K}$ $(= \{RK : R \in \mathscr{R}, K \in \mathscr{K}\})$, and let

$$J(\mathscr{K}) = \{R \in \mathscr{R} : \mathrm{co}_{\mathscr{R}}(ARB)^{=} \cap \mathscr{C} \subseteq \mathscr{K} \quad (A, B \in \mathscr{R})\}.$$

Show that $j(\mathscr{K})$ and $J(\mathscr{K})$ are norm-closed two-sided ideals in $\mathscr{R}$, and

$$j(\mathscr{K}) \cap \mathscr{C} = J(\mathscr{K}) \cap \mathscr{C} = \mathscr{K}.$$

Prove also that, among all norm-closed two-sided ideals $\mathscr{I}$ in $\mathscr{R}$ such that $\mathscr{I} \cap \mathscr{C} = \mathscr{K}$, $j(\mathscr{K})$ is the smallest and $J(\mathscr{K})$ is the largest.

8.7.16. With the notation of Exercise 8.7.15, let $\mathscr{K}_1, \mathscr{K}_2, \mathscr{K}_a$ $(a \in \mathbb{A})$ be norm-closed ideals in $\mathscr{C}$, and let $\mathscr{I}$ be a two-sided ideal in $\mathscr{R}$. Prove that

(i) $J(\mathscr{K}_1) \subseteq J(\mathscr{K}_2)$ if and only if $\mathscr{K}_1 \subseteq \mathscr{K}_2$;
(ii) $J(\bigcap_{a \in \mathbb{A}} \mathscr{K}_a) = \bigcap_{a \in \mathbb{A}} J(\mathscr{K}_a)$;
(iii) $\mathscr{I}$ is a maximal two-sided ideal in $\mathscr{R}$ if and only if $\mathscr{I} = J(\mathscr{K})$ for some maximal ideal $\mathscr{K}$ in $\mathscr{C}$;
(iv) $\mathscr{I}$ is the intersection of a family of maximal two-sided ideals in $\mathscr{R}$ if and only if $\mathscr{I} = J(\mathscr{K})$ for some proper norm-closed ideal $\mathscr{K}$ in $\mathscr{C}$.

8.7.17. With the notation of Exercise 8.7.15, suppose that the von Neumann algebra $\mathscr{R}$ is finite. Prove that

$$J(\mathscr{K}) = \{R \in \mathscr{R} : \tau(R^*R) \in \mathscr{K}\},$$

where τ is the center-valued trace on $\mathscr{R}$.

8.7.18. Suppose that for each positive integer n, $\mathscr{R}_n$ is a finite von Neumann algebra with center $\mathscr{C}_n$, and τ_n is the center-valued trace on $\mathscr{R}_n$. Let $\mathscr{R}$ be the finite von Neumann algebra $\sum \oplus \mathscr{R}_n$, let $\mathscr{C}$ be the center $\sum \oplus \mathscr{C}_n$ of $\mathscr{R}$, and let $\mathscr{K}$ be the norm-closed ideal

$$\left\{\sum \oplus C_n \in \mathscr{C} : \lim_{n \to \infty} \|C_n\| = 0\right\}$$

in $\mathscr{C}$. With the notation of Exercise 8.7.15, show that

$$j(\mathscr{K}) = \left\{\sum \oplus R_n \in \mathscr{R} : \lim_{n \to \infty} \|R_n\| = 0\right\},$$

$$J(\mathscr{K}) = \left\{\sum \oplus R_n \in \mathscr{R} : \lim_{n \to \infty} \|\tau_n(R_n^* R_n)\| = 0\right\}.$$

By considering suitable projections in $\mathscr{R}$, show that $j(\mathscr{K}) \neq J(\mathscr{K})$ if *either* each $\mathscr{R}_n$ is of type II_1 *or* $\mathscr{R}_1$ is of type I_1, $\mathscr{R}_2$ is of type I_2, $\mathscr{R}_3$ is of type I_3, and so on.

8.7.19. With the notation of Exercise 8.7.15, suppose that the von Neumann algebra $\mathscr{R}$ is properly infinite, and that $\mathscr{K} = \{0\}$. Let E_0 be a projection in $\mathscr{R}$.

(i) By using the *results* of Exercises 8.7.15 and 6.9.51, show that $E_0 \in J(\{0\})$ if and only if $PE_0 \prec P$ for each non-zero projection P in $\mathscr{C}$.

(ii) Give a second proof of the assertion in (i), by using the *definition of* $J(\{0\})$ *given in Exercise* 8.7.15 and extending the reasoning indicated in Exercise 8.7.9.

8.7.20. With the notation of Exercise 8.7.15, suppose that the von Neumann algebra $\mathscr{R}$ is properly infinite and countably decomposable, $\mathscr{K}$ is $\{0\}$, and E_0 is a projection in $\mathscr{R}$. By using the result of Exercise 8.7.19, show that $E_0 \in J(\{0\})$ if and only if E_0 is finite. Deduce that $J(\{0\})$ is $\{0\}$ when $\mathscr{R}$ is a countably decomposable type III von Neumann algebra.

8.7.21. With the notation of Exercise 8.7.15 suppose that the von Neumann algebra $\mathscr{R}$ is properly infinite. Let E be a projection in $\mathscr{R}$, and let Q be the largest projection in $\mathscr{C}$ such that $QE \sim Q$. Show that $(I - Q)E \in J(\{0\})$, and that $E \in J(\mathscr{K})$ if and only if $Q \in \mathscr{K}$. Deduce that

$$J(\mathscr{K}) = j(\mathscr{K}) + J(\{0\}), \tag{1}$$

and that $J(\mathscr{K}) = j(\mathscr{K})$ when $\mathscr{R}$ is a countably decomposable type III von Neumann algebra.

Is (1) always satisfied when $\mathscr{K}$ is a norm-closed ideal in the center $\mathscr{C}$ of a *finite* von Neumann algebra?

8.7.22. Suppose that $\mathscr{R}$ is a properly infinite von Neumann algebra with center $\mathscr{C}$, $\mathscr{I}$ is a norm-closed two-sided ideal in $\mathscr{R}$, and (with the notation of Exercise 8.7.15) $J(\{0\}) \subseteq \mathscr{I}$. Show that $\mathscr{I} = J(\mathscr{I} \cap \mathscr{C})$.

8.7.23. Suppose that $\mathscr{R}$ and $\mathscr{S}$ are von Neumann algebras acting on a Hilbert space $\mathscr{H}$, and $\mathscr{R} \subseteq \mathscr{S}$. Let Φ be a positive linear mapping from $\mathscr{S}$ into $\mathscr{R}$, such that

$$\Phi(I) = I, \qquad \Phi(R_1 S R_2) = R_1 \Phi(S) R_2$$

whenever $R_1, R_2 \in \mathscr{R}$ and $S \in \mathscr{S}$. For all S in $\mathscr{S}$,

(i) prove that $\Phi(S^*) = \Phi(S)^*$;

(ii) by considering the element $[S - \Phi(S)]^*[S - \Phi(S)]$ of $\mathscr{S}$, show that $\Phi(S)^*\Phi(S) \leq \Phi(S^*S)$;

(iii) deduce that $\|\Phi(S)\| \leq \|S\|$.

[A mapping Φ with the properties set out in this exercise is described as a *conditional expectation* from $\mathscr{S}$ onto $\mathscr{R}$.]

8.7.24. Let $\mathscr{U}$ be the unitary group in a von Neumann algebra $\mathscr{R}$ acting on a Hilbert space $\mathscr{H}$. Suppose that $\text{co}_{\mathscr{R}}(T)^-$ meets the commutant $\mathscr{R}'$, for each T in $\mathscr{B}(\mathscr{H})$, where $\text{co}_{\mathscr{R}}(T)^-$ denotes the weak-operator closure of the convex hull $\text{co}_{\mathscr{R}}(T)$ of the set $\{UTU^* : U \in \mathscr{U}\}$.

Let $\mathscr{M}$ be the set of all positive linear mappings $\varphi\colon \mathscr{B}(\mathscr{H}) \to \mathscr{B}(\mathscr{H})$ such that

$$\varphi(I) = I, \qquad \varphi(R_1'TR_2') = R_1'\varphi(T)R_2', \qquad \varphi(T) \in \text{co}_{\mathscr{R}}(T)^-,$$

whenever $T \in \mathscr{B}(\mathscr{H})$ and $R_1', R_2' \in \mathscr{R}'$. Let $\mathscr{D}$ ($\subseteq \mathscr{M}$) be the set of all mappings $\alpha\colon \mathscr{B}(\mathscr{H}) \to \mathscr{B}(\mathscr{H})$ that can be defined by an equation of the form

$$\alpha(T) = \sum_{j=1}^{k} a_j U_j T U_j^*,$$

where $U_1, \ldots, U_k \in \mathscr{U}$ and $a_1, \ldots, a_k$ are positive scalars with sum 1.

(i) Prove that $\varphi(R') = R'$ whenever $\varphi \in \mathscr{M}$ and $R' \in \mathscr{R}'$, and that $\varphi_1 \circ \varphi_2 \in \mathscr{M}$ whenever $\varphi_1, \varphi_2 \in \mathscr{M}$.

(ii) Show that $\mathscr{M}$ can be viewed as a closed subset of the product topological space $\prod_{T \in \mathscr{B}(\mathscr{H})} X_T$, where X_T is $\text{co}_{\mathscr{R}}(T)^-$ with the weak-operator topology.

(iii) Suppose that $T_0 \in \mathscr{B}(\mathscr{H})$ and $A_0' \in \text{co}_{\mathscr{R}}(T_0)^- \cap \mathscr{R}'$. Show by a compactness argument that $A_0' = \psi(T_0)$ for some ψ in $\mathscr{M}$.

(iv) Show that if $T_1, \ldots, T_n \in \mathscr{B}(\mathscr{H})$, there is an element φ of $\mathscr{M}$ such that $\varphi(T_1), \ldots, \varphi(T_n) \in \mathscr{R}'$. [*Hint*. Use (i) and (iii) to give a proof by induction on n.]

(v) For each finite subset $\mathbb{F}$ of $\mathscr{B}(\mathscr{H})$, let

$$\mathscr{M}_{\mathbb{F}} = \{\varphi \in \mathscr{M} : \varphi(T) \in \mathscr{R}' \text{ whenever } T \in \mathbb{F}\}.$$

Show that the family of all such sets $\mathscr{M}_{\mathbb{F}}$ has non-empty intersection.

(vi) Deduce that there is a conditional expectation Φ from $\mathscr{B}(\mathscr{H})$ onto $\mathscr{R}'$, with the property that $\Phi(T) \in \text{co}_{\mathscr{R}}(T)^- \cap \mathscr{R}'$ for each T in $\mathscr{B}(\mathscr{H})$. Prove also that, if $T_0 \in \mathscr{B}(\mathscr{H})$ and $A_0' \in \text{co}_{\mathscr{R}}(T_0)^- \cap \mathscr{R}'$, then Φ can be chosen so that $\Phi(T_0) = A_0'$.

8.7.25. Let $\mathscr{R}$ be a von Neumann algebra of type I_n acting on a Hilbert space $\mathscr{H}$.

(i) Show that $\mathscr{R}$ satisfies the hypothesis of Proposition 8.3.11. [*Hint*. Use Theorem 6.6.5 and the continuity properties of * isomorphisms (Remark 7.4.4) to show that it suffices to consider the case in which $\mathscr{R}$ is $n \otimes \mathscr{C}$, where $\mathscr{C}$ is an abelian von Neumann algebra. Let $\mathbb{K}$ be a set with cardinality n, so that each element of $n \otimes \mathscr{C}$ is represented by a matrix $[C_{j,k}]_{j,k \in \mathbb{K}}$ with entries in $\mathscr{C}$. Let $\mathbb{A}$ be the set of all pairs $(\mathbb{F}, \mathscr{A})$, in which $\mathbb{F}$ is a finite subset of $\mathbb{K}$ and $\mathscr{A}$ is a finite-dimensional * subalgebra of $\mathscr{C}$. When $a = (\mathbb{F}, \mathscr{A}) \in \mathbb{A}$, let $\mathscr{R}_a$ be the set of all elements of $n \otimes \mathscr{C}$ with matrices $[C_{j,k}]$ such that $C_{j,k} \in \mathscr{A}$ for all j and k in $\mathbb{K}$ and $C_{j,k} = 0$ unless $j, k \in \mathbb{F}$.]

(ii) Use (i) and Exercise 8.7.24 to show that there is a conditional expectation Φ from $\mathscr{B}(\mathscr{H})$ onto $\mathscr{R}'$, with the property that $\Phi(T) \in \mathrm{co}_{\mathscr{R}}(T)^{-} \cap \mathscr{R}'$ for each T in $\mathscr{B}(\mathscr{H})$.

8.7.26. Show that the results of Exercise 8.7.25 remain valid for every type I von Neumann algebra $\mathscr{R}$ acting on a Hilbert space $\mathscr{H}$.

8.7.27. Suppose that $\mathscr{R}$ is a von Neumann algebra acting on a Hilbert space $\mathscr{H}$, and $\mathscr{A}$ is an abelian von Neumann subalgebra of $\mathscr{R}$.

(i) By adapting part of the proof of Lemma 8.2.3, show that the von Neumann algebra $\mathscr{A}'$ is of type I.

(ii) By using the result of Exercise 8.7.26, show that there exist conditional expectations, Φ from $\mathscr{B}(\mathscr{H})$ onto $\mathscr{A}$, and Ψ from $\mathscr{B}(\mathscr{H})$ onto $\mathscr{A}'$.

(iii) Prove also that there exist conditional expectations, Φ_0 from $\mathscr{R}$ onto $\mathscr{A}$, and Ψ_0 from $\mathscr{R}$ onto $\mathscr{R} \cap \mathscr{A}'$.

8.7.28. Suppose that $\mathscr{R}$ and $\mathscr{S}$ are von Neumann algebras acting on a Hilbert space $\mathscr{H}$, such that $\mathscr{R} \subseteq \mathscr{S}$ and $\mathscr{S}$ has a faithful normal tracial state τ.

(i) Show that if $H \in \mathscr{S}^+$, there is an element H_0 of $\mathscr{R}^+$ such that $\tau(HR) = \tau(H_0R)$ for each R in $\mathscr{R}$. [*Hint*. Apply Theorem 7.3.13 to suitable positive normal linear functionals on $\mathscr{R}$.]

(ii) Show that, for each element S of $\mathscr{S}$, there is a unique element $\varphi(S)$ of $\mathscr{R}$ such that $\tau(SR) = \tau(\varphi(S)R)$ for each R in $\mathscr{R}$.

(iii) Prove that the mapping $\varphi: \mathscr{S} \to \mathscr{R}$ defined by (ii) is an ultraweakly continuous conditional expectation from $\mathscr{S}$ onto $\mathscr{R}$, and is faithful in the sense that $\varphi(S) \neq 0$ when $0 \neq S \in \mathscr{S}^+$.

8.7.29. Suppose that G is a discrete group with unit element e. Adopt the notation of Section 6.7, and recall that the von Neumann algebra $\mathscr{L}_G$

has a faithful tracial state τ defined by $\tau(A) = \langle Ax_e, x_e\rangle$. When $z \in l_\infty(G)$, let M_z (in $\mathscr{B}(l_2(G))$ be the operator of multiplication by z.

(i) Suppose that Φ is a conditional expectation from $\mathscr{B}(l_2(G))$ onto $\mathscr{L}_G$, and define a linear functional ρ on $l_\infty(G)$ by

$$\rho(z) = \tau(\Phi(M_z)) \qquad (z \in l_\infty(G)).$$

Show that ρ is an invariant mean on G (in the sense explained in Exercise 3.5.7).

(ii) Suppose that ρ is an invariant mean on G. Show that ρ is a state of the C^*-algebra $l_\infty(G)$.

Suppose that $T \in \mathscr{B}(l_2(G))$. Given x and y in $l_2(G)$, define $z_{x,y}$ in $l_\infty(G)$ by

$$z_{x,y}(g) = \langle R_{x_g}^* T R_{x_g} x, y\rangle \qquad (g \in G).$$

Show that there is an element A_T of $\mathscr{B}(l_2(G))$ such that

$$\langle A_T x, y\rangle = \rho(z_{x,y}) \qquad (x, y \in l_2(G)),$$

and prove that

$$A_T \in \text{co}_{\mathscr{R}G}(T)^- \cap \mathscr{R}'_G.$$

(iii) From the results of (i), (ii), and Exercise 8.7.24, show that the following three conditions are equivalent:

(a) there is a conditional expectation from $\mathscr{B}(l_2(G))$ onto $\mathscr{L}_G$ $(= \mathscr{R}'_G)$;
(b) there is an invariant mean on G;
(c) for each T in $\mathscr{B}(l_2(G))$, $\text{co}_{\mathscr{R}G}(T)^-$ meets $\mathscr{R}'_G$.

8.7.30. Suppose that G and H are discrete groups, and adopt the notation of Section 6.7.

(i) Show that the von Neumann algebras $\mathscr{L}_G$ and $\mathscr{L}_H$ are unitarily equivalent if they are * isomorphic.

(ii) Show that $\mathscr{L}_G$ and $\mathscr{L}_H$ are not * isomorphic if G has an invariant mean and H has no invariant mean. [*Hint*. Use (i) and Exercise 8.7.29(iii).]

(iii) Re-prove Theorem 6.7.8 by showing that Π has an invariant mean and $\mathscr{F}_2$ has no invariant mean. [*Hint*. Let a and b be the two generators of $\mathscr{F}_2$, and let S be the set of reduced words in $\mathscr{F}_2$ that begin with a non-zero power of b. Use the fact that $\mathscr{F}_2 = S \cup bS$ and that S, aS, a^2S are disjoint.]

(iv) Show that no group containing $\mathscr{F}_2$ has an invariant mean. [*Hint*. Consider cosets of $\mathscr{F}_2$ and use the set S described in (iii).]

(v) Let G be $\Pi \oplus \mathscr{F}_2$. Show that no two of $\mathscr{L}_\Pi, \mathscr{L}_{\mathscr{F}_2}$, and $\mathscr{L}_G$, are * isomorphic. [*Hint*. Use Exercise 6.9.45.]

8.7.31. Let Δ be the center-valued dimension function on a finite von Neumann algebra $\mathscr{R}$. Prove that

$$\Delta(E \vee F) + \Delta(E \wedge F) = \Delta(E) + \Delta(F)$$

for all projections E and F in $\mathscr{R}$.

Deduce that $E \wedge F \neq 0$ if $\Delta(E) + \Delta(F) > I$, and that $E \vee F \neq I$ if $\Delta(E) + \Delta(F) < I$.

8.7.32. Suppose that Δ is the center-valued dimension function on a finite von Neumann algebra $\mathscr{R}$, E, F, and G are projections in $\mathscr{R}$, and $F \geq G$. Show that

$$\Delta(E \wedge F - E \wedge G) \leq \Delta(F - G).$$

8.7.33. Let Δ be the center-valued dimension function on a finite von Neumann algebra $\mathscr{R}$.

(i) Prove that Δ is weak-operator continuous on the set $\mathscr{P}$ of all projections in $\mathscr{R}$.

(ii) Suppose that $\{E_a\}$ is a net of projections in $\mathscr{R}$, and $\{\Delta(E_a)\}$ is weak-operator convergent to 0. Show that $\{E_a\}$ is strong-operator convergent to 0. [*Hint*. Use Exercise 8.7.3(iii).]

8.7.34. Suppose that E and F are projections in a von Neumann algebra $\mathscr{R}$, and E is the (strong-operator) limit of an increasing net $\{E_a\}$ of projections in $\mathscr{R}$.

(i) Prove that the net $\{E_a \vee F\}$ converges to $E \vee F$.

(ii) Prove that the net $\{E_a \wedge F\}$ converges, and that $\lim(E_a \wedge F) \leq E \wedge F$. Give an example in which $\lim(E_a \wedge F) < E \wedge F$, and show that $\lim(E_a \wedge F) = E \wedge F$ if $\mathscr{R}$ is finite.

(iii) State and prove the corresponding results for the case in which E is the limit of a *decreasing* net $\{E_a\}$ of projections in $\mathscr{R}$.

8.7.35. Suppose that E, F, and G are projections in a finite von Neumann algebra $\mathscr{R}$, E and F are the limits of increasing nets $\{E_a\}$ and $\{F_a\}$, respectively, of projections in $\mathscr{R}$, and G is the limit of a decreasing net $\{G_a\}$ of projections in $\mathscr{R}$ (the index set being the same for the three nets). From the relation

(1) $$E_a \wedge G_a - E \wedge G = (E_a \wedge G_a - E_a \wedge G) - (E \wedge G - E_a \wedge G)$$

and by using the results of the preceding three exercises, show that

(i) the net $\{E_a \wedge G_a\}$ is strong-operator convergent to $E \wedge G$;
(ii) $\{E_a \vee G_a\}$ is strong-operator convergent to $E \vee G$;
(iii) $\{E_a \wedge F_a\}$ is strong-operator convergent to $E \wedge F$.

8.7.36. Show that a normal tracial weight ρ on a factor $\mathscr{R}$ is either faithful or identically zero.

8.7.37. Show that a weight ρ on a von Neumann algebra $\mathscr{R}$ is tracial if and only if $\rho(UHU^*) = \rho(H)$ whenever $H \in \mathscr{R}^+$ and U is a unitary operator in $\mathscr{R}$.

8.7.38. Let ρ be the unique (faithful, normal) tracial state on a finite factor $\mathscr{R}$. Show that there is a finite orthogonal family $\{E_1, \ldots, E_k\}$ of equivalent cyclic projections in $\mathscr{R}$ with sum I. Prove that, for $j = 1, \ldots, k$, there is a vector x_j in the range of E_j such that

$$\rho(E_j A E_j) = \langle Ax_j, x_j \rangle \qquad (A \in \mathscr{R}).$$

Deduce that

$$\rho(A) = \sum_{j=1}^{k} \langle Ax_j, x_j \rangle \qquad (A \in \mathscr{R}).$$

8.7.39. Suppose that ρ is the unique (faithful, normal) tracial state on a finite factor $\mathscr{R}$, and define a norm $\| \ \|_2$ on $\mathscr{R}$ (as in Remark 8.5.9) by $\|A\|_2 = [\rho(A^*A)]^{1/2}$. By using the result of Exercise 8.7.38, show that the norm topology associated with $\| \ \|_2$ is coarser (weaker) than the strong-operator topology, but that these two topologies coincide on bounded subsets of $\mathscr{R}$.

8.7.40. With the notation of Exercise 8.7.39, let $\mathfrak{A}$ be a self-adjoint subalgebra of $\mathscr{R}$ (not necessarily containing I).

(i) Suppose that $A, A_1, A_2, \ldots$ are self-adjoint elements of $\mathscr{R}$ such that $\|A\| \leq 1$, $A_n \in \mathfrak{A}$ for $n = 1, 2, \ldots$, and $\|A - A_n\|_2 \to 0$ as $n \to \infty$. Define a continuous function $f : \mathbb{R} \to [-1, 1]$ by $f(t) = 2t(1 + t^2)^{-1}$, and note that the restriction $f \,|\, [-1, 1]$ has a continuous inverse mapping $g : [-1, 1] \to [-1, 1]$. Show that $\|f(A) - f(A_n)\|_2 \to 0$ as $n \to \infty$, and deduce that A $(= g(f(A)))$ lies in the strong-operator closure $\mathfrak{A}^-$ of $\mathfrak{A}$ in $\mathscr{R}$.

(ii) Deduce that the $\| \ \|_2$-closure of $\mathfrak{A}$ in $\mathscr{R}$ coincides with $\mathfrak{A}^-$.

8.7.41. Suppose that G is a countably infinite discrete group in which each element, other than the unit e, has an infinite conjugacy class. As in Section 6.7, define an orthonormal basis $\{x_g : g \in G\}$ of the Hilbert space $l_2(G)$ and a factor $\mathscr{L}_G$ of type II_1, acting on $l_2(G)$, for which each x_g is a trace vector.

(i) Show that G cannot be expressed as the union of a finite number of sets each of which is a right coset of some subgroup of G that has infinite index. [*Hint*. Suppose the contrary, and let $\{H_1, \ldots, H_n\}$ be a *minimal* set of subgroups of G, each with infinite index, such that G can be expressed as the union of a finite number of right cosets of $H_1, \ldots, H_n$. Choose some such expression for G, and by considering a right coset of H_1 that does *not* appear in that expression, show that H_1 (and hence, also, G) can be expressed as the union of a finite number of right cosets of $H_2, \ldots, H_n$, contradicting the minimality assumption.]

(ii) Suppose that $h, k \in G \backslash \{e\}$. Let

$$N_h = \{g \in G : g^{-1}hg = h\}, \qquad N_{h,k} = \{g \in G : g^{-1}hg = k\},$$

and note that N_h is a subgroup of G with infinite index and $N_{h,k}$ is either empty or a right coset of N_h. Deduce that if S is a finite subset of $G \backslash \{e\}$, there is an element g of G such that $g^{-1}Sg \cap S = \emptyset$.

(iii) Let $\{S_n\}$ be an increasing sequence of finite subsets of G, with union $G \backslash \{e\}$. For each positive integer n, choose an element g_n of G such that $g_n^{-1}S_n g_n \cap S_n = \emptyset$, and let U_n (in $\mathscr{L}_G$) be the unitary operator defined by

$$(U_n x)(g) = x(g_n^{-1}g) \qquad (g \in G, \quad x \in l_2(G)).$$

Show that

$$\lim_{n \to \infty} \langle U_n A U_n^* x_g, x_h \rangle = 0$$

whenever $A \in \mathscr{L}_G$, and g, h are distinct elements of G. Deduce that the sequence $\{U_n A U_n^*\}$ is weak-operator convergent to $\tau(A)I$, where τ is the (unique) tracial state of $\mathscr{L}_G$.

8.7.42. Let $\mathscr{H}$ be a separable infinite-dimensional Hilbert space, and let ρ_0 be the usual tracial weight on $\mathscr{B}(\mathscr{H})$ (see Remark 8.5.6). The purpose of this and the following three exercises is to show that $\mathscr{B}(\mathscr{H})$ has a semi-finite tracial weight τ (neither faithful nor normal) that is not a multiple of ρ_0, even when restricted to the subset of $\mathscr{B}(\mathscr{H})^+$ on which τ takes finite values.

Suppose that $\{y_1, y_2, y_3, \ldots\}$ is an orthonormal sequence in $\mathscr{H}$, $\{\lambda_1, \lambda_2, \lambda_3, \ldots\}$ is a decreasing sequence of non-negative real numbers, and A (in $\mathscr{B}(\mathscr{H})^+$) is defined by

$$Ax = \sum_{j=1}^{\infty} \lambda_j \langle x, y_j \rangle y_j \qquad (x \in \mathscr{H}).$$

Let $\{z_1, \ldots, z_m\}$ be a finite orthonormal system in $\mathscr{H}$. Prove that

(i) $\sum_{k=1}^{m} \langle Az_k, z_k \rangle \le \sum_{j=1}^{m} \lambda_j = \sum_{j=1}^{m} \langle Ay_j, y_j \rangle$;
(ii) $\sum_{k=1}^{m} \langle Az_k, z_k \rangle \ge \sum_{j=1}^{n} \lambda_j$

if the linear span $[z_1, \ldots, z_m]$ contains $y_1, \ldots, y_n$.

8.7.43. Suppose that $\mathscr{H}$ is a separable infinite-dimensional Hilbert space. For each positive compact linear operator A acting on $\mathscr{H}$, let $\{\lambda_j(A)\}$ be the sequence of non-zero eigenvalues of A, arranged in decreasing order and counted according to their multiplicities (as in Exercise 2.8.29(v)), and followed by a sequence of zeros if A has finite-dimensional range. Let

$$s_m(A) = \lambda_1(A) + \lambda_2(A) + \cdots + \lambda_m(A) \qquad (m = 1, 2, \ldots).$$

By using the results of Exercises 2.8.29 and 8.7.42, show that for all positive compact linear operators A and B,

(i) $s_m(A) \leq s_m(B)$ if $A \leq B$,
(ii) $s_m(A + B) \leq s_m(A) + s_m(B) \leq s_{2m}(A + B)$.

8.7.44. Let ρ be a pure state of the C^*-algebra l_∞ that vanishes on the ideal c_0 (see Exercise 4.6.56). Define positive linear mappings C, D, S, T from l_∞ into l_∞, each mapping the identity of l_∞ onto itself, by

$$C\{x_1, x_2, x_3, \ldots\} = \{x_1, \tfrac{1}{2}(x_1 + x_2), \tfrac{1}{3}(x_1 + x_2 + x_3), \ldots\},$$

$$D\{x_1, x_2, x_3, \ldots\} = \{x_1, x_1, x_2, x_2, x_3, x_3, \ldots\},$$

$$S\{x_1, x_2, x_3, \ldots\} = \{x_2, x_3, x_4, \ldots\},$$

$$T\{x_1, x_2, x_3, \ldots\} = \{x_1, \tfrac{1}{2}(x_1 + x_2), \tfrac{1}{4}(x_1 + \cdots + x_4), \tfrac{1}{8}(x_1 + \cdots + x_8), \ldots\}$$

and let μ be the state $\rho \circ C \circ T$ of l_∞. Prove that $T = STD$, that each of the operators C, D, S, T maps c_0 into c_0, and that $C - CS$ and $T - TS$ map l_∞ into c_0. Deduce that μ vanishes on c_0 and $\mu = \mu \circ S = \mu \circ D$.

8.7.45. With the notation of Exercises 8.7.43 and 8.7.44, let $\mathscr{F}$ be the set of all positive compact linear operators A acting on $\mathscr{H}$ for which the sequence $\{s_m(A)/\log(m + 1)\}_{m=1,2,\ldots}$ is bounded (and is therefore a positive element $G(A)$ of l_∞). Define a mapping $\tau: \mathscr{B}(\mathscr{H})^+ \to [0, \infty]$ by

$$\tau(A) = \begin{cases} \mu(G(A)) & (A \in \mathscr{F}), \\ \infty & (A \in \mathscr{B}(\mathscr{H})^+ \backslash \mathscr{F}). \end{cases}$$

Prove that

(i) $\tau(aA) = a\tau(A)$ when $A \in \mathscr{B}(\mathscr{H})^+$ and $a \geq 0$ (with the convention that $0 \cdot \infty = 0$);

(ii) $U\mathscr{F}U^* = \mathscr{F}$, and $G(UAU^*) = G(A)$, when U is a unitary operator acting on $\mathscr{H}$ and $A \in \mathscr{F}$;

(iii) $\tau(UAU^*) = \tau(A)$ when $A \in \mathscr{B}(\mathscr{H})^+$ and U is a unitary operator acting on $\mathscr{H}$;

(iv) if $A \in \mathscr{B}(\mathscr{H})^+$, $B \in \mathscr{F}$, and $A \leq B$, then $A \in \mathscr{F}$ [*Hint*. Deduce from the result of Exercise 4.6.41(iv) that A is compact.];

(v) if $A, B \in \mathscr{F}$, then $A + B \in \mathscr{F}$ and

$$D(G(A) + G(B)) - X_0 \leq G(A + B) \leq G(A) + G(B)$$

for some X_0 in the ideal c_0 in l_∞;

(vi) $\tau(A + B) = \tau(A) + \tau(B)$ when $A, B \in \mathscr{B}(\mathscr{H})^+$;

(vii) τ is a semi-finite tracial weight on $\mathscr{B}(\mathscr{H})$ [*Hint*. Use the result of Exercise 8.7.37];

(viii) τ is not normal, and is not a multiple of the usual tracial weight on $\mathscr{B}(\mathscr{H})$ (see Remark 8.5.6) even when restricted to the subset of $\mathscr{B}(\mathscr{H})^+$ on which τ takes finite values.

8.7.46. Suppose that $\mathscr{R}$ is a finite factor acting on a Hilbert space $\mathscr{H}$, and that x and u are generating unit trace vectors for $\mathscr{R}$. Let φ and ψ be the * anti-isomorphisms from $\mathscr{R}$ onto $\mathscr{R}'$ determined (see Theorem 7.2.15) by the conditions

$$\varphi(A)x = Ax, \qquad \psi(A)u = Au \qquad (A \in \mathscr{R}).$$

Let $\mathscr{A}_0$ be a maximal abelian von Neumann subalgebra of $\mathscr{R}$, and let $\mathscr{A}_\varphi$ and $\mathscr{A}_\psi$ be the abelian von Neumann algebras generated by $\mathscr{A}_0 \cup \varphi(\mathscr{A}_0)$ and $\mathscr{A}_0 \cup \psi(\mathscr{A}_0)$, respectively. Prove that

(i) there is a unitary operator U' in $\mathscr{R}'$ such that $U'x = u$;

(ii) $\psi(A) = U'\varphi(A)U'^*$ for each A in $\mathscr{R}$;

(iii) $\mathscr{A}_\psi = U'\mathscr{A}_\varphi U'^*$;

(iv) $\mathscr{A}_\psi$ is a maximal abelian subalgebra of $\mathscr{B}(\mathscr{H})$ if and only if the same is true of $\mathscr{A}_\varphi$. [In Exercises 8.7.48 and 8.7.49, we give examples to show that $\mathscr{A}_\psi$ can be, but need not be, a maximal abelian subalgebra of $\mathscr{B}(\mathscr{H})$, in the circumstances set out in the present exercise.]

8.7.47. With the notation of Section 8.6, *An operator-theoretic construction*, assume that conditions (a) and (b) are satisfied and that G acts ergodically on $\mathscr{A}$. Suppose that u_0 is a unit vector in $\mathscr{H}$, and $U(g)u_0 = u_0$ for each g in G.

(i) Prove that u_0 is a separating and generating vector for $\mathscr{A}$.

(ii) Let u be the vector $\sum_{g \in G} \oplus x_g$ in $\mathscr{K}$ $(= \sum_{g \in G} \oplus \mathscr{H}_g)$, where $x_e = u_0$ and $x_g = 0$ when $g \in G \backslash \{e\}$. Show that u is a generating trace vector for $\mathscr{R}$.

(iii) Let ψ be the * anti-isomorphism from $\mathscr{R}$ onto $\mathscr{R}'$ determined (see Theorem 7.2.15) by the condition

$$\psi(T)u = Tu \qquad (T \in \mathscr{R}).$$

Given that T (in $\mathscr{R}$) has matrix $[U(pq^{-1})A(pq^{-1})]$, where $A(g) \in \mathscr{A}$ for each g in G, find the matrix of $\psi(T)$.

(iv) Show that $\Phi(A)$ has matrix $[\delta_{p,q}A]$, and $\psi(\Phi(A))$ has matrix $[\delta_{p,q}U(p)AU(p)^*]$ for each A in $\mathscr{A}$.

(v) Suppose that $S \in \mathscr{B}(\mathscr{K})$ and S has matrix $[S_{p,q}]$. Show that $S \in \Phi(\mathscr{A})'$ if and only if $S_{p,q} \in \mathscr{A}$ for all p and q in G, and that $S \in \psi(\Phi(\mathscr{A}))'$ if and only if $U(p)^*S_{p,q}U(q) \in \mathscr{A}$ for all p and q in G.

(vi) Deduce from (v) that the von Neumann algebra $\mathscr{A}_\psi$ generated by $\Phi(\mathscr{A})$ (a maximal abelian subalgebra $\mathscr{A}_0$ of $\mathscr{R}$) and $\psi(\Phi(\mathscr{A}))$ (the maximal abelian subalgebra $\psi(\mathscr{A}_0)$ of $\mathscr{R}'$) is the maximal abelian subalgebra $\sum_{g\in G} \oplus \mathscr{A}$ of $\mathscr{B}(\mathscr{K})$.

8.7.48. By using Exercise 8.7.47 and Example 8.6.12, give an example of a factor $\mathscr{R}$ of type II_1 acting on a Hilbert space $\mathscr{H}$, a generating unit trace vector u for $\mathscr{R}$, and a maximal abelian von Neumann subalgebra $\mathscr{A}_0$ of $\mathscr{R}$, such that the von Neumann algebra $\mathscr{A}_\psi$ occurring in Exercise 8.7.46 is a maximal abelian subalgebra of $\mathscr{B}(\mathscr{H})$.

8.7.49. With the notation of Section 6.7, let G be the free group on two generators, a and b. Recall that $\mathscr{L}_G$ $(=\{L_y : y \in l_2(G),\ L_y \in \mathscr{B}(l_2(G))\})$ is a factor of type II_1 with commutant $\mathscr{R}_G$ $(=\{R_z : z \in l_2(G),\ R_z \in \mathscr{B}(l_2(G))\})$, and x_e is a generating unit trace vector for $\mathscr{L}_G$. Let ψ be the * anti-isomorphism from $\mathscr{L}_G$ onto $\mathscr{R}_G$ determined (see Theorem 7.2.15) by the condition $\psi(T)x_e = Tx_e$. Let $\mathscr{A}_0$ be the von Neumann algebra generated by L_{x_a}, and recall (Exercise 6.9.42) that $\mathscr{A}_0$ is a maximal abelian subalgebra of $\mathscr{L}_G$.

(i) Show that $\psi(L_y) = R_y$ when $y \in l_2(G)$ and $L_y \in \mathscr{B}(l_2(G))$.

(ii) Show that $\psi(\mathscr{A}_0)$ is the von Neumann subalgebra of $\mathscr{R}_G$ generated by R_{x_a}.

(iii) Suppose that $S \in \mathscr{B}(l_2(G))$, and let $[S(g,h)]$ be the matrix of S, defined by

$$S(g,h) = \langle Sx_h, x_g\rangle \qquad (g,h \in G).$$

Show that S commutes with L_{x_a} if and only if $S(g,h) = S(ag,ah)$ for all g and h in G, and that S commutes with R_{x_a} if and only if $S(g,h) = S(ga,ha)$ for all g and h in G.

(iv) Show that there is a partial isometry S_0 acting on $l_2(G)$, with matrix determined by the condition that $S_0(g,h)$ is 1 if there exist integers m and n such that $g = a^mba^n$ and $h = a^mbaba^n$, and $S(g,h)$ is 0 otherwise. Prove that S_0 commutes with L_{x_a} and R_{x_a}. Deduce that the von Neumann algebra $\mathscr{A}_\psi$ generated by $\mathscr{A}_0 \cup \psi(\mathscr{A}_0)$ is not maximal abelian in $\mathscr{B}(l_2(G))$.

8.7.50. With the notation of Section 8.6, *An operator-theoretic construction*, assume that conditions (a) and (b) of that section are satisfied, and let $\mathscr{W}$ be the unitary group of the maximal abelian subalgebra $\Phi(\mathscr{A})$ of $\mathscr{R}$. Prove that

(i) for each T in $\mathscr{R}$, the weak-operator closed convex hull $\text{co}_{\Phi(\mathscr{A})}(T)^-$ of the set $\{WTW^* : W \in \mathscr{W}\}$ meets $\Phi(\mathscr{A})$ in a single point [*Hint*. See the proof of Lemma 8.6.2.];

(ii) there is a unique ultraweakly continuous conditional expectation Ψ from $\mathscr{R}$ onto $\Phi(\mathscr{A})$, and Ψ is faithful.

8.7.51. Suppose that $\mathscr{R}$ is a von Neumann algebra with center $\mathscr{C}$, $H \in \mathscr{R}$, and $d = \inf\{\|H - C\| : C \in \mathscr{C}\}$. Show that the equation

$$\delta(A) = HA - AH \qquad (A \in \mathscr{R})$$

defines a derivation δ of $\mathscr{R}$. (The term "derivation" was defined in Exercise 4.6.65. When δ is obtained in the above manner from an element H of $\mathscr{R}$, it is described as an *inner* derivation of $\mathscr{R}$. We shall see, in Exercise 8.7.55, that every derivation of a von Neumann algebra is inner.)

Prove also that $d \le \|\delta\| \le 2d$. [*Hint*. Consider $\delta(U)U^*$, where U is a unitary element of $\mathscr{R}$.]

8.7.52. Let δ be a derivation of a von Neumann algebra $\mathscr{R}$, and recall from Exercises 4.6.65 and 7.6.15 that δ is bounded, and is weak-operator continuous on bounded subsets of $\mathscr{R}$.

(i) Suppose that P is a projection in the center $\mathscr{C}$ of $\mathscr{R}$. By using the relation $\delta(P) = \delta(P^2)$, show that $\delta(P) = 0$. Deduce that the restriction $\delta | \mathscr{R}P$ is a derivation δ_P of the von Neumann algebra $\mathscr{R}P$. Prove also that if the derivations δ_P of $\mathscr{R}P$ and δ_{I-P} of $\mathscr{R}(I - P)$ are both inner, then δ is inner.

(ii) Let E be a projection in $\mathscr{R}$. Show that the equation

$$\delta_E(EAE) = E\delta(EAE)E \qquad (A \in \mathscr{R})$$

defines a derivation δ_E of the von Neumann algebra $E\mathscr{R}E$, and $\|\delta_E\| \le \|\delta\|$.

(iii) Suppose that $\{E_a : a \in \mathbb{A}\}$ is an increasing net of projections in $\mathscr{R}$, and $\bigvee_{a \in \mathbb{A}} E_a = I$. Suppose also that, for each a in $\mathbb{A}$, there is an element H_a of $E_a\mathscr{R}E_a$ such that $\|H_a\| \le \|\delta\|$ and

$$\delta_{E_a}(A) = H_a A - AH_a \qquad (A \in E_a\mathscr{R}E_a).$$

Show that the net $\{H_a\}$ has a subnet that is weak-operator convergent to an element H of $\mathscr{R}$. Prove also that

$$\delta(A) = HA - AH \qquad (A \in \mathscr{R}).$$

[*Hint*. Consider first the case in which $A \in E_b\mathscr{R}E_b$ for some b in $\mathbb{A}$.]

8.7.53. Suppose that δ is a derivation of a countably decomposable finite von Neumann algebra $\mathcal{R}$. For each U in the unitary group $\mathcal{U}$ of $\mathcal{R}$, define an affine mapping $a_U : \mathcal{R} \to \mathcal{R}$ by

$$a_U(R) = \delta(U)U^* + URU^* \qquad (R \in \mathcal{R}).$$

(i) Prove that $a_{UV} = a_U \circ a_V$ for all U and V in $\mathcal{U}$.

(ii) Show that δ is inner if and only if there is an element H of $\mathcal{R}$ such that $a_U(H) = H$ for all U in $\mathcal{U}$.

(iii) Let $\mathcal{K}$ be the weak-operator closed convex hull of the set $\{\delta(U)U^* : U \in \mathcal{U}\}$. Show that $\mathcal{K}$ is weak-operator compact and $a_U(\mathcal{K}) \subseteq \mathcal{K}$ for all U in $\mathcal{U}$.

(iv) Let $\mathcal{F}$ be the family of all non-empty weak-operator compact convex subsets of $\mathcal{K}$ that are invariant under each of the mappings a_U $(U \in \mathcal{U})$. Show that $\mathcal{F}$, partially ordered by inclusion, has a minimal element $\mathcal{K}_0$.

(v) Show that $\mathcal{R}$ has a faithful normal tracial state ρ.

(vi) Let $M = \sup\{\|K\|_2 : K \in \mathcal{K}_0\}$, where $\| \ \|_2$ is the norm on $\mathcal{R}$ defined by $\|R\|_2 = [\rho(R^*R)]^{1/2}$. Given K_1 and K_2 in $\mathcal{K}_0$, show that

$$\|a_U(K_1) - a_U(K_2)\|_2 = \|K_1 - K_2\|_2 \qquad (U \in \mathcal{U}),$$

and that

$$K \in \mathcal{K}_0, \qquad \|K\|_2 \le [M^2 - \tfrac{1}{4}\|K_1 - K_2\|_2^2]^{1/2}$$

for each K in the weak-operator closed convex hull $\mathcal{K}_1$ of the set $\{a_U(\tfrac{1}{2}(K_1 + K_2)) : U \in \mathcal{U}\}$.

(vii) From the results of (vi) and the minimality property of $\mathcal{K}_0$, deduce that $\mathcal{K}_0$ consists of a single element H of $\mathcal{K}$. Prove that $\|H\| \le \|\delta\|$ and

$$\delta(R) = HR - RH \qquad (R \in \mathcal{R}).$$

8.7.54. Suppose that δ is a derivation of a countably decomposable type III von Neumann algebra $\mathcal{R}$. Let $\mathcal{C}$ and $\mathcal{U}$ denote the center and unitary group, respectively, of $\mathcal{R}$, and note that the results of Exercise 8.7.53(i)–(iv) remain valid in the present case.

(i) Show that the difference set

$$\mathcal{K}_0 - \mathcal{K}_0 = \{K_1 - K_2 : K_1, K_2 \in \mathcal{K}_0\}$$

is convex, weak-operator compact, and invariant under each of the mappings $R \to URU^* : \mathcal{R} \to \mathcal{R}$, where $U \in \mathcal{U}$.

(ii) Suppose that $\mathcal{K}_0$ does not consist of a single point. By use of (i) and Proposition 8.3.9, prove that there exist K_1, K_2 in $\mathcal{K}_0$ and C in $\mathcal{C}$ such that $K_1 - K_2 = C \neq 0$. Deduce that

$$a_U(K_1) - a_U(K_2) = C \qquad (U \in \mathcal{U}).$$

Let ω be a weak-operator continuous linear functional on $\mathscr{R}$ such that $\operatorname{Re}\omega(C) = 1$, and let $M = \sup\{\operatorname{Re}\omega(K) : K \in \mathscr{K}_0\}$. Prove that

$$K \in \mathscr{K}_0, \qquad \operatorname{Re}\omega(K) \leq M - 1$$

for each K in the weak-operator closed convex hull $\mathscr{K}_1$ of the set $\{a_U(K_2) : U \in \mathscr{U}\}$.

(iii) From the results of (ii) and the minimality property of $\mathscr{K}_0$, deduce that $\mathscr{K}_0$ consists of a single element H of $\mathscr{K}$. Prove that $\|H\| \leq \|\delta\|$ and

$$\delta(R) = HR - RH \qquad (R \in \mathscr{R}).$$

8.7.55. (i) Suppose that δ is a derivation of a von Neumann algebra $\mathscr{R}$. By using the results of the three preceding exercises, show that δ is inner.

(ii) Suppose that $\mathfrak{A}$ is a C^*-algebra acting on a Hilbert space $\mathscr{H}$, and δ is a derivation of $\mathfrak{A}$. By using (i) and Exercise 7.6.15, show that there is an element H of $\mathfrak{A}^-$ such that

$$\delta(A) = HA - AH \qquad (A \in \mathfrak{A}).$$

8.7.56. Let $\mathscr{R}$ be a von Neumann algebra of type I_n with n finite, $\mathscr{C}$ be the center of $\mathscr{R}$, and η be a center state of $\mathscr{R}$. Let $\{E_{jk}\}$ be a self-adjoint system of $n \times n$ matrix units for $\mathscr{R}$, and A_{kj} be $\eta(E_{jk})$. Let A be the element $\sum A_{jk}E_{jk}$ of $\mathscr{R}$, so that A corresponds to the matrix $[A_{jk}]$ under the * isomorphism of Theorem 6.6.5 between $\mathscr{R}$ and $n \otimes \mathscr{C}$.

(i) Prove that $\eta(B) = n\tau(AB)$ for each B in $\mathscr{R}$, where τ is the center-valued trace on $\mathscr{R}$, and deduce that η is ultraweakly continuous.

(ii) Show that A is self-adjoint and $\sum_{j=1}^{n} A_{jj} = I$.

(iii) By expressing A as $A^+ - A^-$ and considering $\eta(A^-)$, prove that $A \geq 0$.

(iv) Prove that there are n equivalent projections $G_1, \ldots, G_n$ abelian in $\mathscr{R}$ with sum I, and n positive operators $C_1, \ldots, C_n$ in $\mathscr{C}$ with sum I, such that

$$\eta(B) = \sum_{j=1}^{n} C_j\eta_j(B)$$

for each B in $\mathscr{R}$, where $\eta_j(B)$ is the (unique) element of $\mathscr{C}$ such that $G_jBG_j = \eta_j(B)G_j$. [*Hint*. Use Exercises 6.9.23 and 6.9.35, together with the * isomorphism between $\mathscr{R}$ and $n \otimes \mathscr{C}$.]

8.7.57. Let $\mathscr{R}$ be a finite von Neumann algebra acting on a Hilbert space $\mathscr{H}$ and τ be its center-valued trace. Suppose G is a projection in $\mathscr{R}$. Show that

(i) $\tau(G) \leq C_G$;

(ii) $C_{\tau(G)} = C_G$;

(iii) there is a self-adjoint element C^G affiliated with the center of $\mathscr{R}$ such that $C^G \hat{\cdot} \tau(G) = C_G$;

(iv) for some choices of $\mathscr{R}$ and G, C^G is not bounded.

8.7.58. Let $\mathscr{R}$ be a finite von Neumann algebra acting on a Hilbert space $\mathscr{H}$, τ be the center-valued trace on $\mathscr{R}$, and G be a non-zero projection in $\mathscr{R}$. Show that τ_0 is the center-valued trace on $G\mathscr{R}G$, where $\tau_0(S) = (\tau(S) \hat{\cdot} C^G)G$ for S in $G\mathscr{R}G$ and C^G is as in Exercise 8.7.57(iii).

8.7.59. Let $\mathscr{R}$ be a von Neumann algebra of type II_1 acting on a Hilbert space $\mathscr{H}$, and let $\mathscr{A}$ be a maximal abelian (self-adjoint) subalgebra of $\mathscr{R}$. Suppose H is a positive element in the unit ball of the center $\mathscr{C}$ of $\mathscr{R}$. Show that

(i) there is a projection E in $\mathscr{A}$ such that $\Delta(E) = H$, where Δ is the center-valued dimension function on $\mathscr{R}$ [*Hint*. Use Exercise 6.9.27.];

(ii) E (in (i)) can be chosen as a subprojection of a given projection G in $\mathscr{A}$ provided $H \le \Delta(G)$.

8.7.60. Let VH and WK be the polar decompositions of S and T, respectively, and suppose, $S, T \,\eta\, \mathscr{R}$, where $\mathscr{R}$ is a finite von Neumann algebra acting on a Hilbert space $\mathscr{H}$. Let E_n and F_n be the spectral projections for H and K, respectively, corresponding to the interval $[-n, n]$ for each positive integer n. Show that

(i) TF_n is a bounded everywhere-defined operator T_n in $\mathscr{R}$ and $\{G_n\}$ is an increasing sequence of projections in $\mathscr{R}$ with strong-operator limit I, where G_n has range $F_n(\mathscr{H}) \cap T_n^{-1}(E_n(\mathscr{H}))$ [*Hint*. Use Exercises 6.9.55 and 8.7.35(iii).];

(ii) ST has dense domain;

(iii) ST is preclosed and has a unique closed extension $S \hat{\cdot} T$ affiliated with $\mathscr{R}$ [*Hint*. Use Theorem 2.7.8 and Exercise 6.9.54.];

(iv) $S + T$ has a dense domain;

(v) $S + T$ is preclosed and has a unique closed extension $S \hat{+} T$ affiliated with $\mathscr{R}$;

(vi) if x_0 is a unit generating vector for $\mathscr{R}$ and z_0 is any other vector in $\mathscr{H}$, there is an operator T affiliated with $\mathscr{R}$ such that $Tx_0 = z_0$. (Compare Theorem 7.2.1′, Exercise 7.6.50(iv), and Theorem 5.6.15.)

CHAPTER 9

ALGEBRA AND COMMUTANT

Most of the material in this chapter is related, directly or indirectly, to one or both of two basic questions. To what extent does the algebraic structure of a von Neumann algebra influence the algebraic structure of its commutant? In what circumstances does the fact that two von Neumann algebras are * isomorphic imply that they are unitarily equivalent?

In Section 9.1, the first of these questions is considered, in relation to the algebraic classification of von Neumann algebras into types. It is shown that an algebra of type I (or II, or III) has a commutant of the same type. It is not the case, however, that algebra and commutant always have the same type when the finer classification into types I_n, II_1, II_∞, and III, is used; indeed, it is shown by examples that all combinations consistent with the preceding sentence do in fact occur as the types of a von Neumann algebra and its commutant.

The unitary implementation theorem implies that the algebraic structure of a von Neumann algebra with a separating and generating vector completely determines the algebraic structure of the commutant; for * isomorphism between two such algebras implies their unitary equivalence, and this entails unitary equivalence of the commutants. In Section 9.2, the relation between algebra and commutant in the presence of a separating and generating vector is studied in greater detail. It is shown that there is a conjugate-linear isometry on the underlying Hilbert space that implements a * anti-isomorphism between the algebra and its commutant. The analysis gives rise also to a one-parameter unitary group that implements a one-parameter group of * automorphisms of the algebra, the "modular automorphism group" associated with the separating and generating vector. Such automorphism groups arise also in the context of a von Neumann algebra with a specified faithful normal state (or faithful normal semi-finite weight). It will be seen, in Chapter XIII, that these groups provide a powerful tool in the detailed study of the structure of certain von Neumann algebras.

Section 9.3 contains a study of type I von Neumann algebras. It includes a necessary and sufficient condition for a * isomorphism between two such algebras to be unitarily implemented. The structure of type I algebras is determined in terms of matrix constructs with maximal abelian von Neumann algebras. In the case of separable Hilbert spaces, this leads to a complete classification of type I von Neumann algebras up to unitary equivalence, since Section 9.4 contains such a classification of maximal abelian algebra acting on separable Hilbert spaces.

In Section 9.5, the structure of type I von Neumann algebras is used as a tool in studying the classification, up to unitary equivalence, of the representations (on separable Hilbert spaces) of an abelian C^*-algebra. The problem is analyzed in terms of a measure-theoretic invariant (presented in two forms, as a null ideal sequence, and as a multiplicity function) associated with a given representation. As a special case, the theory provides unitary invariants for bounded normal operators acting on separable Hilbert spaces.

9.1. The type of the commutant

The main result of this section (Theorem 9.1.3) asserts that, if $\mathscr{R}$ is a von Neumann algebra acting on a Hilbert space $\mathscr{H}$, then the commutant $\mathscr{R}'$ is of type I (or II, or III) when $\mathscr{R}$ has the same property. Within the framework of this result, all other possibilities occur; for example, $\mathscr{R}$ can be type II_∞ with $\mathscr{R}'$ of type II_1. The section concludes with examples illustrating this point.

The relation between $\mathscr{R}$ and $\mathscr{R}'$ is one facet (perhaps, the cornerstone) of the *spatial* theory of von Neumann algebras, since it involves the action of $\mathscr{R}$ on the underlying Hilbert space. Spatial theory is here contrasted with the *algebraic* theory, which is concerned with the properties of $\mathscr{R}$ as an algebra with involution. The basic tools, used to link the algebraic and spatial aspects, are the theory of normal states, together with its implications concerning cyclic projections. This is already exemplified in the proof of the unitary implementation theorem (7.2.9). It is seen also in the present section, so we recall a number of results that will be needed. With x and y vectors, let E, F (in $\mathscr{R}$) and E', F' (in $\mathscr{R}'$) be the projections with ranges $[\mathscr{R}'x]$, $[\mathscr{R}'y]$, $[\mathscr{R}x]$, $[\mathscr{R}y]$, respectively. Then the central carrier of E in $\mathscr{R}$ coincides with that of E' in $\mathscr{R}'$, and has range $[\mathscr{R}\mathscr{R}'x]$; and $E \precsim F$ if and only if $E' \precsim F'$ (Propositions 5.5.2 and 5.5.13, Theorem 7.2.12). If $\mathscr{R}$ has a generating trace vector, then $\mathscr{R}'$ is finite (Theorem 7.2.15). If $\mathscr{R}$ is abelian and has a generating vector, then $\mathscr{R}' = \mathscr{R}$ (Corollary 7.2.16).

9.1.1. LEMMA. *If $\mathscr{R}$ is a finite von Neumann algebra acting on a Hilbert space $\mathscr{H}$, and x is a vector such that $[\mathscr{R}x] = [\mathscr{R}'x] = \mathscr{H}$, then $\mathscr{R}'$ is finite.*

Proof. We give two proofs of this result. The first makes use of the existence and properties of the center-valued trace τ in $\mathscr{R}$, as set out in Theorem 8.2.8. The second avoids any dependence on the theory developed in Chapter 8, but is in consequence somewhat more elaborate than the first.

For the first proof, we suppose that $\|x\| = 1$, and define a normal state ω by $\omega(A) = \langle \tau(A)x, x \rangle$, for A in $\mathscr{R}$. Since x is a generating vector for $\mathscr{R}'$, and hence separating for $\mathscr{R}$, it follows that $\omega_x(A) > 0$ and $\omega(A) > 0$ whenever $A \in \mathscr{R}^+$ and $A \neq 0$; so ω_x and ω both have support I. By Lemma 7.2.8, there is a unit vector y such that $\omega = \omega_y | \mathscr{R}$ and $[\mathscr{R}y] = \mathscr{H}$. With A, B in $\mathscr{R}$

$$\begin{aligned} \langle ABy, y \rangle = \omega(AB) &= \langle \tau(AB)x, x \rangle \\ &= \langle \tau(BA)x, x \rangle = \langle BAy, y \rangle. \end{aligned}$$

Hence y is a generating trace vector for $\mathscr{R}$, and $\mathscr{R}'$ is finite by Theorem 7.2.15.

For the second proof, we assume that $\mathscr{R}'$ is infinite, and show, in due course, that this assumption leads to a contradiction. Observe first that, if Q is a non-zero central projection in $\mathscr{R}$, then $\mathscr{R}Q$ satisfies the hypotheses of the lemma; for $\mathscr{R}Q$ is finite, and Qx is a generating vector for both $\mathscr{R}Q$ and $\mathscr{R}'Q$. Since $\mathscr{R}'$ is infinite, we can choose Q so that $\mathscr{R}'Q$ is properly infinite; and, upon replacing $\mathscr{R}$ by $\mathscr{R}Q$, we reduce to the case in which $\mathscr{R}'$ is properly infinite. This last implies, of course, that $\mathscr{R}'$ is not abelian, and it follows from Corollary 7.2.16 that $\mathscr{R}$ is not abelian.

Let E be a projection in $\mathscr{R}$ that is not in the center of $\mathscr{R}$, so that $0 < E < C_E$. With F_1 and F_2 the projections defined by $F_1 = C_E - E$, $F_2 = C_{F_1}E$, we have $F_1 \leq C_E$, hence $C_{F_1} \leq C_E$, and therefore

$$F_1 + F_2 = F_1 + C_{F_1}E = C_{F_1}(F_1 + E) = C_{F_1}C_E = C_{F_1},$$

while $C_{F_2} = C_{F_1}C_E = C_{F_1}$. Upon replacing $\mathscr{R}$ by $\mathscr{R}C_{F_1}$, we reduce to the case in which

$$F_1 + F_2 = I = C_{F_1} = C_{F_2}.$$

Observe that F_j has range $F_j(\mathscr{H}) = [F_j\mathscr{R}'x] = [\mathscr{R}'F_jx]$ $(j = 1, 2)$. If F_1', F_2' (in $\mathscr{R}'$) are the projections with ranges $[\mathscr{R}F_1x]$, $[\mathscr{R}F_2x]$, respectively, then

$$C_{F_1'} = C_{F_1} = I = C_{F_2} = C_{F_2'}.$$

We assert that F_1' (and, similarly, F_2') is finite in $\mathscr{R}'$. For this, suppose the contrary, and let P be a central projection in $\mathscr{R}'$ for which $F_1'P$ is properly infinite. Then both the projections P and $F_1'P$ in $\mathscr{R}'$ have central carrier P and are properly infinite; they are cyclic (hence countably decomposable), their ranges being $[\mathscr{R}Px]$ and $[\mathscr{R}PF_1x]$, respectively. By Corollary 6.3.5, $P \sim F_1'P$ (in $\mathscr{R}'$). Since P and F_1P have ranges $[\mathscr{R}'Px]$ and $[\mathscr{R}'PF_1x]$, it now follows from Theorem 7.2.12 that $P \sim F_1 P$ (in $\mathscr{R}$). However, $F_2 P \neq 0$, since $C_{F_2} = I$; and we have

$$P \sim F_1 P < F_1 P + F_2 P = P,$$

contradicting the finiteness of $\mathscr{R}$. This shows that F_1' and F_2' are finite in $\mathscr{R}'$.

Since $\mathscr{R}'$ is properly infinite, we can find projections G_1' and G_2' in $\mathscr{R}'$ such that

$$G_1' + G_2' = I \sim G_1' \sim G_2'.$$

Thus $F_j' \leq I \sim G_j'$, and there is a partial isometry V_j' in $\mathscr{R}'$, from F_j' to a subprojection F_j'' of G_j' $(j = 1, 2)$. With y_j defined as $V_j' F_j x$, we assert that

$$(1) \qquad F_j(\mathscr{H}) = [\mathscr{R}' y_j], \qquad F_j''(\mathscr{H}) = [\mathscr{R} y_j],$$

$$(2) \qquad \mathscr{H} = [\mathscr{R}'(y_1 + y_2)], \qquad (F_1'' + F_2'')(\mathscr{H}) = [\mathscr{R}(y_1 + y_2)].$$

Indeed,

$$F_j''(\mathscr{H}) = V_j' F_j'(\mathscr{H}) = [V_j' \mathscr{R} F_j x] = [\mathscr{R} V_j' F_j x] = [\mathscr{R} y_j],$$

and

$$\begin{aligned} F_j(\mathscr{H}) &= [\mathscr{R}' F_j x] \supseteq [\mathscr{R}' V_j' F_j x] = [\mathscr{R}' y_j] \supseteq [\mathscr{R}' V_j'^* y_j] \\ &= [\mathscr{R}' V_j'^* V_j' F_j x] = [\mathscr{R}' F_j' F_j x] = [\mathscr{R}' F_j x] = F_j(\mathscr{H}), \end{aligned}$$

which proves (1). For (2), observe that $F_1 F_2 = 0 = F_1'' F_2''$, while $F_j y_j = F_j'' y_j = y_j$ $(j = 1, 2)$. From this,

$$\begin{aligned} \mathscr{R}(y_1 + y_2) &= \{A(y_1 + y_2) : A \in \mathscr{R}\} \\ &\subseteq \{A_1 y_1 + A_2 y_2 : A_1, A_2 \in \mathscr{R}\} \\ &= \mathscr{R} y_1 + \mathscr{R} y_2 \\ &= \{(A_1 F_1 + A_2 F_2)(y_1 + y_2) : A_1, A_2 \in \mathscr{R}\} \subseteq \mathscr{R}(y_1 + y_2), \end{aligned}$$

so $\mathscr{R}(y_1 + y_2) = \mathscr{R} y_1 + \mathscr{R} y_2$. Since the subspaces $[\mathscr{R} y_1]$, $[\mathscr{R} y_2]$ are mutually orthogonal, by (1),

$$[\mathscr{R}(y_1 + y_2)] = [\mathscr{R} y_1] + [\mathscr{R} y_2] = F_1''(\mathscr{H}) + F_2''(\mathscr{H}) = (F_1'' + F_2'')(\mathscr{H}).$$

A similar argument shows that

$$[\mathscr{R}'(y_1 + y_2)] = [\mathscr{R}' y_1] + [\mathscr{R}' y_2] = F_1(\mathscr{H}) + F_2(\mathscr{H}) = \mathscr{H},$$

and so completes the proof of (2).

Since $[\mathscr{R}' x] = \mathscr{H} = [\mathscr{R}'(y_1 + y_2)]$, while $[\mathscr{R} x] = \mathscr{H}$ and $[\mathscr{R}(y_1 + y_2)]$ is the range of $F_1'' + F_2''$, it now follows from Theorem 7.2.12 that $F_1'' + F_2'' \sim I$ in $\mathscr{R}'$. This last is impossible, since I is infinite while $F_1'' + F_2''$ is finite. ∎

9.1.2. PROPOSITION. *If $\mathscr{R}$ is a von Neumann algebra acting on a Hilbert space $\mathscr{H}$, $x \in \mathscr{H}$, and E (in $\mathscr{R}$), E' (in $\mathscr{R}'$) are the projections with ranges $[\mathscr{R}'x]$ and $[\mathscr{R}x]$, respectively, then E' is abelian, or finite, or infinite, or properly infinite, relative to $\mathscr{R}'$, if and only if E has the corresponding property relative to $\mathscr{R}$.*

Proof. Since E and E' have the same central carrier Q, we may assume that $C_E = C_{E'} = I$, upon replacing $\mathscr{R}$ and $\mathscr{R}'$ by $\mathscr{R}Q$ and $\mathscr{R}'Q$. The mapping $A' \to A'E$, from $\mathscr{R}'$ onto $\mathscr{R}'E$, is then a * isomorphism, and carries E' to EE'. Thus E' is abelian (or finite, or infinite, or properly infinite) relative to $\mathscr{R}'$ if and only if the same is true of EE' relative to $\mathscr{R}'E$. Of course, E has any one of these four properties, relative to $\mathscr{R}$, if and only if it has the same property relative to $E\mathscr{R}E$. Moreover, $x \in E(\mathscr{H})$, and

$$[\mathscr{R}'Ex] = [\mathscr{R}'x] = E(\mathscr{H}), \qquad [E\mathscr{R}Ex] = [E\mathscr{R}x] = EE'(\mathscr{H}).$$

Upon replacing $\mathscr{H}$, $\mathscr{R}$, $\mathscr{R}'$, and E' by $E(\mathscr{H})$, $E\mathscr{R}E$, $\mathscr{R}'E$, and EE', we may assume that $E = I$. The same argument can then be applied, with the roles of $\mathscr{R}$ and $\mathscr{R}'$ reversed: upon replacing $\mathscr{R}$, $\mathscr{R}'$ by $\mathscr{R}E'$ and $E'\mathscr{R}'E'$, we may suppose that $E = E' = I$.

Following the preceding reductions, we assume henceforth that $[\mathscr{R}x] = [\mathscr{R}'x] = \mathscr{H}$. We have to show that $\mathscr{R}$ is abelian (or finite, or infinite, or properly infinite) if and only if $\mathscr{R}'$ has the same property.

If either $\mathscr{R}$ or $\mathscr{R}'$ is abelian, it results from Corollary 7.2.16 that $\mathscr{R} = \mathscr{R}'$, whence both are abelian. If one of $\mathscr{R}$, $\mathscr{R}'$ is finite, the other is finite by Lemma 9.1.1. In consequence, if either $\mathscr{R}$ or $\mathscr{R}'$ is infinite, then so is the other. From this last assertion, if one of $\mathscr{R}$, $\mathscr{R}'$ is properly infinite, then both are; for if Q is any non-zero central projection in $\mathscr{R}$, then $\mathscr{R}Q$, $\mathscr{R}'Q$ have a common generating vector Qx, and one of them is infinite, whence so is the other. ■

9.1.3. THEOREM. *If a von Neumann algebra $\mathscr{R}$ is type* I (*or type* II, *or type* III), *the same is true of its commutant $\mathscr{R}'$.*

Proof. Suppose first that $\mathscr{R}$ is type I, and let E be an abelian projection in $\mathscr{R}$ with central carrier I. *If $\mathscr{R}'$ is not type* I, there is a non-zero projection Q in $\mathscr{R} \cap \mathscr{R}'$ such that $\mathscr{R}'Q$ is either type II or type III; and $QE \neq 0$, since $C_E = I$. With x a unit vector in the range of QE, let F (in $\mathscr{R}Q$) and F' (in $\mathscr{R}'Q$) be the projections with ranges $[\mathscr{R}'x]$ and $[\mathscr{R}x]$, respectively. Since $F \leq QE$, F is abelian in $\mathscr{R}$ (hence, also, in $\mathscr{R}Q$). From Proposition 9.1.2, the non-zero projection F' is abelian (hence finite) in $\mathscr{R}'Q$, contrary to our assumption that $\mathscr{R}'Q$ is type II or type III. Hence $\mathscr{R}'$ is type I, when $\mathscr{R}$ is type I.

Suppose next that $\mathscr{R}$ is type II, and let E be a finite projection in $\mathscr{R}$ with central carrier I. *If* $\mathscr{R}'$ *is not type* II, there is a non-zero projection Q in $\mathscr{R} \cap \mathscr{R}'$ such that $\mathscr{R}'Q$ is type I or type III; and $QE \neq 0$, since $C_E = I$. Now $\mathscr{R}'Q$ has commutant $\mathscr{R}Q$ of type II, so the preceding paragraph shows that $\mathscr{R}'Q$ is not type I; and therefore $\mathscr{R}'Q$ is type III. With x a unit vector in the range of QE, and F, F' constructed as before, $F \leq QE$ and thus F is finite (in $\mathscr{R}$, hence in $\mathscr{R}Q$). From Proposition 9.1.2, the non-zero projection F' in $\mathscr{R}'Q$ is finite, contrary to our assertion that $\mathscr{R}'Q$ is type III. Hence $\mathscr{R}'$ is type II, when $\mathscr{R}$ is type II.

Suppose finally that $\mathscr{R}$ is type III. For each non-zero projection Q in $\mathscr{R} \cap \mathscr{R}'$, $\mathscr{R}'Q$ has commutant $\mathscr{R}Q$ of type III; and the preceding two paragraphs show that $\mathscr{R}'Q$ is neither type I nor type II. Since $\mathscr{R}'$ has no central restriction of either type I or type II, it follows that $\mathscr{R}'$ is type III. ■

9.1.4. COROLLARY. *A von Neumann algebra* $\mathscr{R}$ *is semi-finite if and only if its commutant is semi-finite.*

Proof. The von Neumann algebra $\mathscr{R}$ is semi-finite if and only if, for every projection P in $\mathscr{R} \cap \mathscr{R}'$, $\mathscr{R}P$ is not type III; and a similar assertion applies to $\mathscr{R}'$. By applying Theorem 9.1.3 to both $\mathscr{R}P$ and $\mathscr{R}'P$, it follows that $\mathscr{R}P$ is type III if and only if the same is true of $\mathscr{R}'P$. ■

9.1.5. EXAMPLES. Theorem 9.1.3 gives information about the commutant of a von Neumann algebra, in terms of the broad classification into types I, II, and III. It allows various possible combinations of differing types, for an algebra and its commutant, when the more detailed classification into types I_n ($n \leq \dim \mathscr{H}$), II_1, II_∞, and III is considered. We exhibit factor examples to show that all combinations consistent with Theorem 9.1.3 do in fact occur.

With m and n given cardinals, let $\mathscr{H}$ and $\mathscr{K}$ be Hilbert spaces of dimensions m and n, respectively, and let $\tilde{\mathscr{H}}$ be the direct sum $\sum \oplus \mathscr{H}_b$ of n copies of $\mathscr{H}$. As in Section 2.6, *Matrix representations*, we associate with each T in $\mathscr{B}(\tilde{\mathscr{H}})$ an $n \times n$ matrix $[T_{a,b}]$ whose entries $T_{a,b}$ lie in $\mathscr{B}(\mathscr{H})$. Let $\mathscr{R}$ be the von Neumann algebra denoted in Section 6.6 by $\mathscr{B}(\mathscr{H}) \otimes I_n$, which consists of all elements of $\mathscr{B}(\tilde{\mathscr{H}})$ having a diagonal matrix with the same element of $\mathscr{B}(\mathscr{H})$ at each diagonal position. Since $\mathscr{R}$ is * isomorphic to $\mathscr{B}(\mathscr{H})$, it is a factor of type I_m. By Lemma 6.6.2, $\mathscr{R}'$ is the algebra $n \otimes \{aI : a \in \mathbb{C}\}$, consisting of those elements of $\mathscr{B}(\tilde{\mathscr{H}})$ whose matrices have scalar multiples of the identity operator in each position. From this, together with Section 2.6, *Matrix representations*, there is an isomorphism U, from $\sum \oplus \mathscr{H}_b$ onto $\mathscr{H} \otimes \mathscr{K}$, such that

$$U\mathscr{R}'U^{-1} = \{I \otimes S : S \in \mathscr{B}(\mathscr{K})\}.$$

Accordingly, the mapping $S \to U^{-1}(I \otimes S)U$ is a * isomorphism from $\mathscr{B}(\mathscr{K})$ onto $\mathscr{R}'$, and $\mathscr{R}'$ is therefore a factor of type I_n.

The preceding paragraph exhibits a factor of type I_m with commutant of type I_n, where m and n are arbitrary preassigned cardinals. Factors of type II_1 with commutants also of type II_1 were constructed in Section 6.7 (see, in particular, Theorems 6.7.2(ii) and 6.7.5). Factors of type III (with commutant necessarily of type III) appear in Section 8.6. To exhibit the remaining possible type combinations, suppose that $\mathscr{R}$ is a factor which, together with its commutant, is of type II_1. By Theorem 6.7.10 and Lemma 6.6.2, $\aleph_0 \otimes \mathscr{R}$ is a factor $\mathscr{S}$ of type II_∞; while $\mathscr{S}'$ is $\mathscr{R}' \otimes I_{\aleph_0}$, which is * isomorphic to $\mathscr{R}'$ and is therefore of type II_1. Similarly $\mathscr{S} \otimes I_{\aleph_0}$ is a factor of type II_∞, with commutant $\aleph_0 \otimes \mathscr{S}'$ of type II_∞. ∎

The following result is needed in order to complete the discussion in Remark 7.2.10.

9.1.6. Proposition. *If $\mathscr{R}$ and $\mathscr{R}'$ are countably decomposable, properly infinite von Neumann algebras acting on the Hilbert space $\mathscr{H}$, then $\mathscr{R}$ has a joint generating and separating vector.*

Proof. From Proposition 5.5.18, there is a central projection P in $\mathscr{R}$ such that P is cyclic under $\mathscr{R}'$ and $I - P$ is cyclic under $\mathscr{R}$. Each of $\mathscr{R}P$, $\mathscr{R}'P$, $\mathscr{R}(I - P)$, and $\mathscr{R}'(I - P)$ is either 0 or a countably decomposable, properly infinite von Neumann algebra (acting on $P(\mathscr{H})$ or $(I - P)(\mathscr{H})$). Both $\mathscr{R}'P$ and $\mathscr{R}(I - P)$ have generating vectors. If we can establish that $\mathscr{R}'P$ has a separating vector, then this vector is generating for $\mathscr{R}P$ and the sum of a generating vector for $\mathscr{R}P$ and one for $\mathscr{R}(I - P)$ is a generating vector for $\mathscr{R}$. Hence, if we establish our proposition under the additional assumption that $\mathscr{R}$ has a generating vector, we shall have proved, at one time, that $\mathscr{R}'P$ has a separating vector, hence, that $\mathscr{R}$ has a generating vector (without assuming this), and, from what we shall just have proved, that $\mathscr{R}$ has a generating and separating vector.

We may assume, without loss of generality, that $\mathscr{R}$ has a generating vector x_0. If F is the projection in $\mathscr{R}$ with range $[\mathscr{R}'x_0]$, then F has central carrier I, since x_0 is generating for $\mathscr{R}$ (see Proposition 5.5.13). Since $[\mathscr{R}x_0]$ $(= \mathscr{H})$ is the range of the properly infinite projection I in $\mathscr{R}'$, F is properly infinite in $\mathscr{R}$, by Proposition 9.1.2. As I is countably decomposable in $\mathscr{R}$, $F \sim I$ in $\mathscr{R}$, from Corollary 6.3.5. If V is a partial isometry in $\mathscr{R}$ with initial projection F and final projection I, Vx_0 is a joint generating vector for $\mathscr{R}$ and $\mathscr{R}'$ (since $\mathscr{H} = [\mathscr{R}x_0] = [\mathscr{R}V^*Vx_0] \subseteq [\mathscr{R}Vx_0]$ and $[\mathscr{R}'Vx_0] = [V\mathscr{R}'x_0] = \mathscr{H}$). ∎

Bibliography: [56].

9.2. Modular theory

This section contains a study of the relation between a von Neumann algebra and its commutant when the algebra has a separating and generating vector. The first main result (Tomita's theorem, 9.2.9) includes the assertion that algebra and commutant are * anti-isomorphic in these circumstances.

One special situation of this kind has already been examined in Theorem 7.2.15, where it is shown that algebra and commutant are * anti-isomorphic when there is a generating (and necessarily separating) *trace* vector. This last result, together with its corollary (7.2.16) that an abelian von Neumann algebra with a generating vector is maximal abelian, plays a key role in the investigation in Section 9.1 of the type of the commutant. Nevertheless, it has limited applicability, since the assumption that there is a generating trace vector implies that both algebra and commutant are finite. The theory developed below, assuming only that there is a separating and generating vector (but dropping the trace condition), is much more subtle. It provides a far-reaching, and more readily applicable, generalization of the special case just mentioned; for example, it can be used, in place of the earlier anti-isomorphism theorem, to obtain the results of Section 9.1 without reference to traces or trace vectors. Some of its many further applications are described in Chapter 13.

We review some of the main features of the theory. Suppose that $\mathscr{R}$ is a von Neumann algebra acting on a Hilbert space $\mathscr{H}$, and u is a separating and generating vector for $\mathscr{R}$. We shall prove the existence of a conjugate-linear isometric mapping J from $\mathscr{H}$ onto $\mathscr{H}$, and a positive self-adjoint (in general, unbounded, but densely defined and invertible) operator Δ acting on $\mathscr{H}$, satisfying the conditions $J^2 = I$,

$$J\Delta^{1/2}Au = A^*u, \qquad J\Delta^{-1/2}A'u = A'^*u \qquad (A \in \mathscr{R}, \quad A' \in \mathscr{R}').$$

With considerably greater difficulty, we shall show also that

$$J\mathscr{R}J = \mathscr{R}', \qquad \Delta^{it}\mathscr{R}\Delta^{-it} = \mathscr{R} \qquad (t \in \mathbb{R}).$$

The mapping $\Phi: A \to JA^*J$ is a * anti-isomorphism from $\mathscr{R}$ onto $\mathscr{R}'$; for each real t, the unitary operator Δ^{it} implements a * automorphism $\sigma_t: A \to \Delta^{it}A\Delta^{-it}$ of $\mathscr{R}$. We refer to Δ as the *modular operator*, and to the one-parameter group $\{\sigma_t\}$ as the *modular automorphism group*, of $\mathscr{R}$ associated with the separating and generating vector u.

Consider, next, the more general situation of a von Neumann algebra $\mathscr{R}$ with a faithful normal state ω. The GNS construction applied to ω yields a faithful representation φ of $\mathscr{R}$, and $\varphi(\mathscr{R})$ is a von Neumann algebra with a separating and generating vector u such that $\omega = \omega_u \circ \varphi$. Let Δ be the modular operator of $\varphi(\mathscr{R})$, associated with u. Since $\Delta^{it}\varphi(\mathscr{R})\Delta^{-it} = \varphi(\mathscr{R})$, the equation

$$\sigma_t(A) = \varphi^{-1}(\Delta^{it}\varphi(A)\Delta^{-it}) \qquad (A \in \mathscr{R})$$

defines a one-parameter group $\{\sigma_t\}$ of * automorphisms of $\mathscr{R}$, the modular automorphism group of $\mathscr{R}$ corresponding to the faithful normal state ω. It turns out that there is an important analytic relation between the state ω and the automorphism group $\{\sigma_t\}$, expressed briefly by the statement that $\{\sigma_t\}$ satisfies (and is uniquely determined by) the *modular condition* relative to ω. This condition is that, for each A and B in $\mathscr{R}$, there is a complex-valued function f, bounded and continuous on the strip $\{z \in \mathbb{C} : 0 \le \operatorname{Im} z \le 1\}$ in the complex plane, and analytic on the interior of that strip, such that

$$f(t) = \omega(\sigma_t(A)B), \qquad f(t+i) = \omega(B\sigma_t(A)) \qquad (t \in \mathbb{R}).$$

The modular condition is frequently referred to as the "KMS boundary condition," owing to its occurrence as one mathematical formulation of a property introduced in statistical mechanics by Kubo, Martin, and Schwinger.

We assert that ω is a tracial state if and only if $\sigma_t = \iota$ (the identity automorphism of $\mathscr{R}$) for all real t. Indeed, if ω is tracial, we may suppose, upon replacing $\mathscr{R}$ by $\varphi(\mathscr{R})$, that $\omega = \omega_u | \mathscr{R}$, where u is a generating trace vector for $\mathscr{R}$. In this case, the whole theory reduces to that set out in Theorem 7.2.15; J is the continuous extension of the conjugate-linear isometry $Au \to A^*u : \mathscr{R}u \to \mathscr{R}u$, Δ is I, and each σ_t is ι. Since $\omega(AB) = \omega(BA)$, when $A, B \in \mathscr{R}$, the modular condition can be satisfied by taking for f a constant function. Conversely, suppose that ω is a faithful normal state of $\mathscr{R}$, and $\sigma_t = \iota$ for all real t. If $A, B \in \mathscr{R}$, and f is chosen to satisfy the requirements of the modular condition, then

$$f(t) = \omega(AB), \qquad f(t+i) = \omega(BA) \qquad (t \in \mathbb{R}).$$

It now follows easily (see the proof of Proposition 9.2.14) that f is constant on the strip $\{z \in \mathbb{C} : 0 \le z \le 1\}$; and from this, $\omega(AB) = \omega(BA)$, and ω is a tracial state.

In view of the preceding paragraph, the modular condition can be regarded as a generalization of the defining property, $\omega(AB) = \omega(BA)$, of a tracial state; and the modular operator and modular automorphism group can be viewed as providing a qualitative measure of the extent to which ω fails to be a tracial state.

The remainder of this section is divided into three parts. The first contains a detailed account of the theory just described, and concludes with a characterization (Theorem 9.2.21) of semi-finiteness, in von Neumann algebras, in terms of properties of modular automorphism groups. The second subsection is devoted to an alternative approach to Tomita's theorem. The final subsection is concerned with a more general version of modular theory, in which the faithful normal state ω is replaced by a faithful normal semi-finite weight.

A first approach to modular theory. Before beginning the detailed development of the results described above, we review some of the operator-theoretic tools required for this purpose. Throughout the remainder of this section, we make extensive use of the properties of conjugate-linear operators acting on a Hilbert space $\mathscr{H}$. By means of the conjugate Hilbert space $\bar{\mathscr{H}}$, introduced in the discussion preceding Definition 2.6.3, much of the theory of (bounded and unbounded) linear operators can be adapted so as to apply also in the conjugate-linear case. We recall that $\bar{\mathscr{H}}$ is the same *set* as $\mathscr{H}$, but has a Hilbert space structure different from (although derived from) that of $\mathscr{H}$. The norm topology on $\mathscr{H}$ coincides with the norm topology on $\bar{\mathscr{H}}$, and the same is true of the weak topologies. A mapping T, from $\mathscr{H}$ into another Hilbert space $\mathscr{K}$, has a variety of roles, in relation to the two Hilbert space structures on $\mathscr{H}$ and $\mathscr{K}$. It can be viewed as acting from $\mathscr{H}$ into $\mathscr{K}$, or from $\bar{\mathscr{H}}$ into $\bar{\mathscr{K}}$; it can be regarded also as acting from $\mathscr{H}$ into $\bar{\mathscr{K}}$, or from $\bar{\mathscr{H}}$ into $\mathscr{K}$. It may be necessary to consider a mapping in more than one of these roles, within a single discussion or argument. Linearity of $T: \mathscr{H} \to \mathscr{K}$ is equivalent to linearity of $T: \bar{\mathscr{H}} \to \bar{\mathscr{K}}$, and is also equivalent to conjugate-linearity of $T: \mathscr{H} \to \bar{\mathscr{K}}$ (or of $T: \bar{\mathscr{H}} \to \mathscr{K}$); and conjugate-linearity, in either of the first two roles, is equivalent to linearity in either of the last two. Norm continuity of T, in any one of the four roles, implies norm continuity in all of them.

Similar remarks apply to a mapping T from $\mathscr{D}$ into $\mathscr{K}$, where $\mathscr{D}$ (the domain of T) is a linear subspace of $\mathscr{H}$. Each such mapping again has four roles; for $\mathscr{D}$ can be regarded as a subspace of either $\mathscr{H}$ or $\bar{\mathscr{H}}$, while the range of T can be viewed as lying in either $\mathscr{K}$ or $\bar{\mathscr{K}}$. In accordance with the conventions already established for linear operators, we say that T is *densely defined* if $\mathscr{D}$ is dense in $\mathscr{H}$, that T is *closed* if its graph $\{(x, Tx) : x \in \mathscr{D}\}$ is a closed subset of $\mathscr{H} \oplus \mathscr{K}$, and that T is *preclosed* if it extends to a closed mapping. It is apparent that, if T is densely defined (or preclosed, or closed) in any one of its four roles just mentioned, then it retains that property in each of the four roles.

If $T: \mathscr{D} (\subseteq \mathscr{H}) \to \mathscr{K}$ is a densely defined linear operator, its adjoint T^* is a linear mapping $T^*: \mathscr{D}^* (\subseteq \mathscr{K}) \to \mathscr{H}$, where $\mathscr{D}^*$ is the subset

$$\{y \in \mathscr{K} : \text{there exists } z \text{ in } \mathscr{H} \text{ such that } \langle Tx, y\rangle = \langle x, z\rangle \ (x \in \mathscr{D})\},$$

and $T^*y = z$ when $y \in \mathscr{D}^*$. Since the inner products in $\bar{\mathscr{H}}$ and $\bar{\mathscr{K}}$ are obtained by complex conjugation from those in $\mathscr{H}$ and $\mathscr{K}$ (see the discussion preceding Definition 2.6.3), it follows that the same mapping T^*, now regarded as acting from $\mathscr{D}^*$ ($\subseteq \bar{\mathscr{K}}$) into $\bar{\mathscr{H}}$, is the adjoint of the linear operator $T: \mathscr{D}(\subseteq \bar{\mathscr{H}}) \to \bar{\mathscr{K}}$.

Suppose next that $T: \mathscr{D}(\subseteq \mathscr{H}) \to \mathscr{K}$ is densely defined and conjugate-linear. We denote by T^* the adjoint of the densely defined linear operator $T: \mathscr{D}(\subseteq \mathscr{H}) \to \bar{\mathscr{K}}$. Thus T^* is a linear mapping from its domain $\mathscr{D}^*$ ($\subseteq \bar{\mathscr{K}}$) into $\mathscr{H}$. Moreover, if we write $\langle\ ,\ \rangle^-$ for the inner product in $\bar{\mathscr{K}}$, then an element y of $\bar{\mathscr{K}}$ lies in $\mathscr{D}^*$ if and only if there is a vector z in $\mathscr{H}$, such that $\langle Tx, y\rangle^- = \langle x, z\rangle$ for all x in $\mathscr{D}$; and then, $T^*y = z$. Hence

$$\mathscr{D}^* = \{y \in \mathscr{K} : \text{there exists } z \text{ in } \mathscr{H} \text{ such that } \langle Tx, y\rangle = \langle z, x\rangle \ (x \in \mathscr{D})\};$$

and $T^*: \mathscr{D}^*(\subseteq \mathscr{K}) \to \mathscr{H}$ is a conjugate-linear mapping satisfying

$$\langle Tx, y\rangle = \langle T^*y, x\rangle \qquad (x \in \mathscr{D}, \quad y \in \mathscr{D}^*).$$

We refer to this conjugate-linear mapping as the *adjoint* of the conjugate-linear operator $T: \mathscr{D}(\subseteq \mathscr{H}) \to \mathscr{K}$. It follows from Theorem 2.7.8 that T is preclosed if and only if T^* is densely defined; and when this is so, the closure $\bar{T}$ of T coincides with T^{**} and has the same adjoint as T.

When $T: \mathscr{D}(\subseteq \mathscr{H}) \to \mathscr{K}$ is densely defined, and either linear or conjugate-linear, then in each of its four possible roles T has an adjoint operator. It is not difficult to verify that the adjoint is the same mapping in all four cases. From this, it follows that a linear self-adjoint operator $A: \mathscr{D}(\subseteq \mathscr{H}) \to \mathscr{H}$ is self-adjoint also when viewed as a linear operator on $\bar{\mathscr{H}}$. However, care is needed when applying function calculus in this situation; for if f is a complex-valued (continuous, or Borel) function, and we form the operator "$f(A)$" first with A viewed as an operator on $\mathscr{H}$, second with A acting on $\bar{\mathscr{H}}$, then the two mappings so obtained are not in general the same (see Exercise 9.6.8).

We consider next a closed, densely defined, conjugate-linear operator $T: \mathscr{D}(\subseteq \mathscr{H}) \to \mathscr{K}$. When viewed as a linear operator, from $\mathscr{D}(\subseteq \mathscr{H})$ into $\bar{\mathscr{K}}$, T has a polar decomposition VH, where V is a partial isometry from $\mathscr{H}$ into $\bar{\mathscr{K}}$, while $H(=(T^*T)^{1/2})$ is a (generally, unbounded) positive self-adjoint operator acting on $\mathscr{H}$. Thus $V: \mathscr{H} \to \mathscr{K}$ is *conjugate-linear* while $H: \mathscr{H} \to \mathscr{H}$ is *linear*.

Now suppose that $A: \mathscr{D}(\subseteq \mathscr{H}) \to \mathscr{H}$ is a self-adjoint operator with spectral resolution $\{E_\lambda\}$, while J is a conjugate-linear isometric mapping from $\mathscr{H}$ onto $\mathscr{H}$. Then J can be viewed as a unitary transformation from $\mathscr{H}$ onto $\bar{\mathscr{H}}$, and so has inverse J^*. From this, $JAJ^*: J(\mathscr{D})\ (\subseteq \bar{\mathscr{H}}) \to \bar{\mathscr{H}}$ is a (linear) self-adjoint operator, and has spectral resolution $\{JE_\lambda J^*\}$. Hence, $JAJ^*: J(\mathscr{D})\ (\subseteq \mathscr{H}) \to \mathscr{H}$ is a (linear) self-adjoint operator; and since the properties that characterize the spectral resolution (see Theorem 5.6.12(ii), (iii)) are unaffected by the transition from $\bar{\mathscr{H}}$ to $\mathscr{H}$, the spectral resolution is still $\{JE_\lambda J^*\}$ when JAJ^* is viewed as a self-adjoint operator acting in $\mathscr{H}$.

Throughout this section, we deal with the operator H^z, where H is a (generally unbounded) positive "invertible" operator on $\mathscr{H}$ and z is a complex number. We pause, here, to say precisely what is meant by this in terms of the discussion of Section 5.6, and to study some of the properties of the (unbounded-operator-valued) function $z \to H^z$. Recall (Theorems 5.6.18 and 5.6.19) that H and I generate an abelian von Neumann algebra $\mathscr{A}$ acting on $\mathscr{H}$, that $H\,\eta\,\mathscr{A}$, $H \in \mathscr{N}(\mathscr{A})$, where $\mathscr{N}(\mathscr{A})$ is the *algebra* of closed, densely defined operators affiliated with $\mathscr{A}$, that $\mathscr{A} \cong C(X)$ and $(C(X) \subseteq)\,\mathscr{N}(X) \cong \mathscr{N}(\mathscr{A})$, where X is an extremely disconnected compact Hausdorff space and $\mathscr{N}(X)$ is the algebra of *normal* functions on X. We defined H to be positive $(0 \le H)$ when $0 \le \langle Hx, x\rangle$ for each x in $\mathscr{D}(H)$ and proved that this is the case if and only if the (normal) function $\varphi(H)$ in $\mathscr{N}(X)$ representing H has range in $[0, \infty)$. (See Proposition 5.6.21 and the discussion preceding it.) We proved that, for each A in $\mathscr{N}(A)$, the range of the function $\varphi(A)$ representing A in $\mathscr{N}(X)$ is the spectrum, sp(A), of A, where $z_0 \in \text{sp}(A)$ when $A - z_0 I$ fails to have an inverse (in $\mathscr{N}(\mathscr{A})$) that lies in $\mathscr{A}$. (See Proposition 5.6.20 and the discussion preceding it.) We noted, too, that A has an inverse in $\mathscr{N}(\mathscr{A})$ provided $Ax \neq 0$ for x in $\mathscr{D}(\mathrm{A})$; and, in function terms, A has an inverse in $\mathscr{N}(\mathscr{A})$ exactly when $\varphi(A)$ vanishes on no non-null clopen subset of X. (See the comments following Proposition 5.6.20.) We say that A is *invertible* when it has an inverse in $\mathscr{N}(\mathscr{A})$ (that is, when $Ax \neq 0$ for x in $\mathscr{D}(A)$). More generally, we say of any densely defined operator that it is *invertible* if it is one-to-one and has dense range.

Now suppose that H is a positive invertible operator. We recall from Theorem 5.6.12(v) that, if $\{E_\lambda\}$ is the spectral resolution of H, then the function $\varphi(E_\lambda)$ (in $C(X)$) that represents E_λ is the characteristic function of the largest clopen subset X_λ of X on which $\varphi(H)$ takes values not exceeding λ. We have already noted that $\varphi(H)$ takes only non-negative values, and vanishes on no non-null clopen subset of X; so $E_\lambda = 0$ when $\lambda \le 0$ (in particular, $E_0 = 0$). If $E(S)$ denotes the spectral projection for H corresponding to a Borel subset S of $\mathbb{C}$ (see Theorem 5.6.26 and the remark following its proof), and $\{S_n\}$ is an increasing sequence of Borel sets with union $(0, \infty)$, then

$$I = I - E_0 = E((0, \infty)) = \lim_{n \to \infty} E(S_n).$$

In particular, $E((n^{-1}, n)) \uparrow I$ as $n \to \infty$.

We consider next the Borel-function calculus (see Remark 5.6.25 and Theorem 5.2.26) for a positive invertible operator H. We note that, if f and g are complex-valued Borel functions on $\mathbb{C}$ that coincide on $\text{sp}(H)\backslash\{0\}$ (in particular, if f and g coincide on the interval $(0, \infty)$), then $f(H) = g(H)$.

Indeed, $f \cdot e$ and $g \cdot e$ coincide on sp(H), where e is the characteristic function of $(0, \infty)$; moreover, $e(H) = E((0, \infty)) = I$, and thus

$$f(H) = f(H) \hat{\cdot} e(H) = (f \cdot e)(H) = (g \cdot e)(H) = g(H) \hat{\cdot} e(H) = g(H).$$

With z in $\mathbb{C}$, let f_z be the complex-valued Borel function on $\mathbb{C}$ defined as $\exp(z \operatorname{Log} u)$, for u in $\{w : w \in \mathbb{C},\ w \neq -|w|\}$ $(= \mathbb{C}_s)$, where "Log" is the principal value of the logarithm on $\mathbb{C}_s$, and as 0 when $u = -|u|$. Then $f_z \cdot f_{z'} = f_{z+z'}$ for all z and z' in $\mathbb{C}$. If H is a positive invertible operator (in $\mathscr{N}(\mathscr{A})$), we write H^z for $f_z(H)$ defined by the Borel-function calculus described in Remark 5.6.25. From Theorem 5.6.26,

$$H^{z_1} \hat{\cdot} H^{z_2} = f_{z_1}(H) \hat{\cdot} f_{z_2}(H) = (f_{z_1} \cdot f_{z_2})(H) = f_{z_1+z_2}(H) = H^{z_1+z_2}.$$

In particular, $H^{1/2}$ $(= f_{1/2}(H))$ is *the* square root of H, from Remark 5.6.32 (see also Theorem 5.6.15(iv)); and H^{-1} is the inverse (in $\mathscr{N}(\mathscr{A})$) to H. (Note for this that $H^0 = I$ when H is positive and invertible, since f_0 takes the value 1 throughout $\operatorname{sp}(H) \backslash \{0\}$.)

In Theorem 5.6.36, we noted that $t \to \exp itA$ is a (strong-operator continuous) one-parameter unitary group for each self-adjoint operator A on $\mathscr{H}$. If we replace A by $\log H$, where "log" is the Borel function on $\mathbb{C}$ defined as $\operatorname{Log} u$ when $u \in \mathbb{C}_s$ and as 0 when $u = -|u|$, then $\exp itA$ becomes H^{it}. For this, we apply the composite function rule, Corollary 5.6.29, of our Borel-function calculus, and note that $\exp(it \log)$ and f_{it} agree on $\operatorname{sp}(H) \backslash \{0\}$ $(\subseteq \mathbb{C}_s)$; so that

$$H^{it} = f_{it}(H) = (\exp \circ\ it \log)(H) = \exp(it \log H).$$

It follows that $t \to H^{it}$ is a (strong-operator continuous) one-parameter unitary group.

Since H is closed and one-to-one, its inverse H^{inv} (as a mapping) is closed. Moreover, H^{inv} is densely defined, for $\mathscr{D}(H^{\text{inv}})$ is the range of H and $H \hat{\cdot} H^{-1} = I$ (where $H^{-1} = f_{-1}(H)$). As $H \,\eta\, \mathscr{A}$, we have $H^{\text{inv}} \,\eta\, \mathscr{A}$. Thus H^{-1} and H^{inv} are both inverses to H in the algebra $\mathscr{N}(\mathscr{A})$; whence

$$H^{-1} = (H^{\text{inv}} \hat{\cdot} H) \hat{\cdot} H^{-1} = H^{\text{inv}} \hat{\cdot} (H \hat{\cdot} H^{-1}) = H^{\text{inv}}.$$

At the same time, the composite function rule assures us that $(H^{-1})^z = H^{-z}$, since H^{-1} is a positive invertible operator and $f_z \circ f_{-1}$ agrees with f_{-z} on $\mathbb{C}$. While $f_{-1} \circ f_z$ need not agree with f_{-z} on $\mathbb{C}$ (or even on $\mathbb{C}_s$ with 2 in place of z); $f_{-1} \circ f_z$ and f_{-z} do agree on the set of non-negative real numbers when z is a real number t. Thus

$$(H^{-1})^t = H^{-t} = (H^t)^{-1};$$

so that $(H^{-1})^t$, $(H^t)^{-1}$, and H^{-t} coincide with the inverse of H^t (in the sense of mappings).

We note some further elementary consequences of the theory developed in Section 5.6, and recall some of the results we shall need from that section. From Lemma 5.6.13, if $A\,\eta\,\mathscr{R}$ (so that $U'A = AU'$ for each unitary operator U' in $\mathscr{R}'$), then $T'A \subseteq AT'$ for each T' in $\mathscr{R}'$. In particular $T'\mathscr{D}(A) \subseteq \mathscr{D}(A)$. As applied in what follows, we shall study a positive invertible operator Δ. In this case, an operator such as $(\Delta^{-1} + I)^{-1}$ is an everywhere-defined bounded operator (for Δ^{-1} has positive spectrum, so that $(\Delta^{-1} - (-I))^{\text{inv}}$ is bounded and everywhere defined, and coincides with the Borel function $(\Delta^{-1} + I)^{-1}$ of Δ). If $\mathscr{A}$ is the abelian von Neumann algebra generated by Δ and I, $\Delta^z\,\eta\,\mathscr{A}$ and $(\Delta^{-1} + I)^{-1} \in \mathscr{A}$ so that $(\Delta^{-1} + I)^{-1}\mathscr{D}(\Delta^z) \subseteq \mathscr{D}(\Delta^z)$, for each complex z. Recall that, with A closed and B bounded and everywhere defined, AB is closed; so that $f(\Delta)g(\Delta)$ is densely defined and closed (since $g(\Delta) \in \mathscr{A}$ and $f(\Delta)\,\eta\,\mathscr{A}$), where f and g are Borel functions whose domains contain $\operatorname{sp}(\Delta)$ and g is bounded. Thus $(f \cdot g)(\Delta) = f(\Delta)\hat{\cdot}\, g(\Delta) = f(\Delta)g(\Delta)$. This last implies, for example, that

$$\Delta^z\Delta^{it} = \Delta^z \hat{\cdot}\, \Delta^{it} = \Delta^{z+it}$$

for all real t and complex z (whence, also,

$$\Delta^{it}\Delta^z = \Delta^z\Delta^{it} = \Delta^{z+it},$$

since $\Delta^z\,\eta\,\mathscr{A}$ and Δ^{it} is a unitary operator in $\mathscr{A}$ $(\subseteq \mathscr{A}')$); and also that the operator $\Delta^{-t}(\Delta^{-1} + I)^{-1}$ is everywhere defined and bounded, when $0 \le t \le 1$ (for it coincides with $\Delta^{-t}\hat{\cdot}(\Delta^{-1} + I)^{-1}$, and is therefore $g(\Delta)$, where g is the bounded Borel function $\lambda \to f_{-t}(\lambda)(f_{-1}(\lambda) + 1)^{-1} = \lambda^{1-t}(1 + \lambda)^{-1}$ on $\operatorname{sp}(\Delta)$).

Throughout the remainder of this section, $\mathscr{R}$ denotes a von Neumann algebra acting on a Hilbert space $\mathscr{H}$, and u is a unit vector that is separating and generating for $\mathscr{R}$. We define conjugate-linear mappings S_0 and F_0, with dense domains $\mathscr{R}u$ and $\mathscr{R}'u$, respectively, by

$$S_0Au = A^*u, \qquad F_0A'u = A'^*u \qquad (A \in \mathscr{R}, \quad A' \in \mathscr{R}').$$

9.2.1. Lemma. *The operator S_0 is preclosed, and its adjoint F is an extension of F_0. If S is the closure $\bar{S}_0$ of S_0, then $S^* = F$; moreover, $\mathscr{D}(S^2) = \mathscr{D}(S)$, $\mathscr{D}(F^2) = \mathscr{D}(F)$ and*

$$S^2y = y \quad (y \in \mathscr{D}(S)), \qquad F^2z = z \qquad (z \in \mathscr{D}(F)).$$

Proof. If $A \in \mathscr{R}$ and $A' \in \mathscr{R}'$,

$$\langle S_0Au, A'u\rangle = \langle A'^*A^*u, u\rangle = \langle A'^*u, Au\rangle.$$

Since $\mathscr{D}(S_0) = \mathscr{R}u$, it follows that $A'u \in \mathscr{D}(S_0^*)$ and $S_0^*A'u = A'^*u$; so S_0^* is an extension of F_0. Thus S_0^* is densely defined, and S_0 is preclosed. If $S = \bar{S}_0$, then $S^* = S_0^*$ by Theorem 2.7.8(i).

If $z \in \mathscr{D}(S^*)$,

$$\begin{aligned}\langle S_0 Au, S^*z\rangle &= \langle A^*u, S_0^*z\rangle \\ &= \langle z, S_0A^*u\rangle = \langle z, Au\rangle \qquad (A \in \mathscr{R}).\end{aligned}$$

Since $S^* = S_0^*$ and $\mathscr{D}(S_0) = \mathscr{R}u$, it now follows that $S^*z \in \mathscr{D}(S^*)$ and $(S^*)^2z = z$.

Finally, suppose that $y \in \mathscr{D}(S)$. From the preceding paragraph, whenever $z \in \mathscr{D}(S^*)$ we have $S^*z \in \mathscr{D}(S^*)$ and

$$\langle S^*z, Sy\rangle = \langle y, (S^*)^2z\rangle = \langle y, z\rangle.$$

Thus $Sy \in \mathscr{D}(S^{**}) = \mathscr{D}(S)$, and $S^2y = S^{**}Sy = y$. ■

9.2.2. Remark. While S is the closure of S_0. it is not clear from Lemma 9.2.1 whether or not F is the closure of F_0. In fact, it is (see Corollary 9.2.30); but we shall not need to use this aspect of the symmetry between S and F for the present. ■

Since S is a closed, densely defined, conjugate-linear operator, and $F = S^*$, it follows that FS and SF are positive self-adjoint (in general unbounded, but densely defined) linear operators. From Lemma 9.2.1, each of the operators S and F has range the same as its domain, is invertible, and coincides with its inverse. Accordingly, the positive self-adjoint operator Δ ($= FS = S^*S$) has an inverse Δ^{-1} ($= SF = SS^*$). Since S, Δ (and hence, also, $\Delta^{1/2}$) each has range dense in $\mathscr{H}$ (for $\Delta = \Delta^{1/2}\Delta^{1/2}$), the conjugate-linear partial isometry J, occurring in the polar decompositions of S (see Theorem 6.1.11), is in fact an isometry from $\mathscr{H}$ onto $\mathscr{H}$; and

$$S = J(S^*S)^{1/2} = J\Delta^{1/2} = (SS^*)^{1/2}J = \Delta^{-1/2}J.$$

By regarding J as a unitary transformation from $\mathscr{H}$ onto $\bar{\mathscr{H}}$, we deduce that $J^{-1} = J^*$. When $x \in \mathscr{D}(S)$ we have

$$x = S^2x = J\Delta^{1/2}\Delta^{-1/2}Jx = J^2x;$$

so $J^2 = I$ (and $J = J^{-1} = J^*$), since $\mathscr{D}(S)$ is dense in $\mathscr{H}$.

We noted that $S = J\Delta^{1/2} = \Delta^{-1/2}J$; from this, together with Lemma 6.1.10, $F = S^* = (J\Delta^{1/2})^* = \Delta^{1/2}J$, and $F = F^{-1} = (\Delta^{1/2}J)^{-1} = J\Delta^{-1/2}$.

Moreover, $\Delta^{-1/2} = J(J\Delta^{-1/2}) = J\Delta^{1/2}J$, and

$$\Delta^{-1} = SF = J\Delta^{1/2}\Delta^{1/2}J = J\Delta J.$$

If Δ has spectral resolution $\{E_\lambda\}$, then Δ^{-1} $(= J\Delta J = J\Delta J^*)$ has spectral resolution $\{JE_\lambda J\}$. Given any bounded Borel function f on $\mathbb{C}$, it results from Theorem 5.6.26 that, for all x in $\mathscr{H}$,

$$\begin{aligned}
\langle f(\Delta^{-1})x, x\rangle &= \int_{-\infty}^{\infty} f(\lambda)\, d\langle JE_\lambda Jx, x\rangle \\
&= \int_{-\infty}^{\infty} f(\lambda)\, d\langle Jx, E_\lambda Jx\rangle \\
&= \int_{-\infty}^{\infty} f(\lambda)\, d\langle E_\lambda Jx, Jx\rangle \\
&= \langle f(\Delta)Jx, Jx\rangle \\
&= \langle Jx, \bar{f}(\Delta)Jx\rangle \\
&= \langle J\bar{f}(\Delta)Jx, x\rangle,
\end{aligned}$$

where $\bar{f}(\lambda) = \overline{f(\lambda)}$; so $f(\Delta^{-1}) = J\bar{f}(\Delta)J$. With t a real number, we obtain $\Delta^{it} = J\Delta^{it}J$ and thus $\Delta^{it}J = J\Delta^{it}$, by taking f_{-it} for f.

Since $Su = Fu = u$, it follows that $\Delta u = FSu = u$. Thus, from Remark 5.6.32, $\Delta^{1/2}u = u$, and $Ju = J\Delta^{1/2}u = Su = u$.

The conclusions of the above argument are summarized in the following result.

9.2.3. Proposition. *There is a conjugate-linear isometric mapping J, from $\mathscr{H}$ onto $\mathscr{H}$, and an (in general, unbounded) invertible positive self-adjoint operator Δ acting on $\mathscr{H}$, such that $\Delta = FS$, $\Delta^{-1} = SF$, $J^2 = I$,*

$$S = J\Delta^{1/2} = \Delta^{-1/2}J, \qquad F = J\Delta^{-1/2} = \Delta^{1/2}J$$

and $J\Delta^{it} = \Delta^{it}J$ for all real t. Moreover, $Ju = \Delta u = u$.

As a first step toward the proof that $J\mathscr{R}J = \mathscr{R}'$ and $\Delta^{it}\mathscr{R}\Delta^{-it} = \mathscr{R}$ for all real t, we establish in the following two lemmas an initial (and rather less direct) link, in terms of J and Δ, between $\mathscr{R}$ and $\mathscr{R}'$.

9.2.4. Lemma. *If $A \in \mathscr{R}$, $r > 0$, and $y = (\Delta^{-1} + rI)^{-1}Au$, then $y = A'u$ for some A' in $\mathscr{R}'$.*

Proof. Since $(\Delta^{-1} + rI)^{-1}$ has range $\mathscr{D}(\Delta^{-1} + rI)$ $(= \mathscr{D}(\Delta^{-1}))$, we have $y \in \mathscr{D}(\Delta^{-1}) = \mathscr{D}(SF) \subseteq \mathscr{D}(F)$. Let z be Fy, and define linear operators Y_0

and Z_0, from $\mathscr{R}u$ into $\mathscr{H}$, by $Y_0Ru = Ry$, $Z_0Ru = Rz$ $(R \in \mathscr{R})$. Since u is separating for $\mathscr{R}$, Y_0 and Z_0 are well-defined; and since

$$\begin{aligned}\langle Y_0R_1u, R_2u\rangle &= \langle R_1y, R_2u\rangle = \langle y, R_1^*R_2u\rangle \\ &= \langle y, SR_2^*R_1u\rangle = \langle R_2^*R_1u, Fy\rangle \\ &= \langle R_1u, R_2z\rangle = \langle R_1u, Z_0R_2u\rangle\end{aligned}$$

for all R_1 and R_2 in $\mathscr{R}$, it follows that Y_0^* extends Z_0. Thus Y_0^* is densely defined, whence Y_0 is preclosed and has a closure Y. For each unitary U in $\mathscr{R}$, $U\mathscr{D}(Y_0) = \mathscr{D}(Y_0) = \mathscr{R}u$, and

$$Y_0URu = URy = UY_0Ru \qquad (R \in \mathscr{R});$$

so $Y_0U = UY_0$, $Y_0 = U^{-1}Y_0U$. From this, $Y = U^{-1}YU$, and Y is affiliated to $\mathscr{R}'$. Note also that $Yu = Y_0u = y$. It now suffices to show that Y is bounded; for then, $Y \in \mathscr{R}'$, and we can take $A' = Y$.

Suppose that Y is unbounded, and let its polar decomposition be VK. Then K is unbounded, so we can choose real numbers a, b such that $b > a > \|A\|/2r^{1/2}$ and $KE \neq 0$, where E is the spectral projection for K, corresponding to the interval $[a, b]$. Moreover, $V, E, KE \in \mathscr{R}'$ and, from Lemma 6.1.10,

$$Rz = Z_0Ru = Y_0^*Ru = Y^*Ru = KV^*Ru \qquad (R \in \mathscr{R}).$$

Thus

$$\begin{aligned}\|A\|^2\|Ez\|^2 \geq \|AEz\|^2 &= \|AEKV^*u\|^2 = \|A(KE)V^*u\|^2 = \|KEV^*Au\|^2 \\ &= \|KEV^*(\Delta^{-1} + rI)y\|2 \\ &\geq \|KEV^*(\Delta^{-1} + rI)y\|^2 - \|KEV^*(\Delta^{-1} - rI)y\|^2 \\ &= 4r\,\mathrm{Re}\langle KEV^*\Delta^{-1}y, KEV^*y\rangle = 4r\,\mathrm{Re}\langle \Delta^{-1}y, VK^2EV^*Yu\rangle \\ &= 4r\,\mathrm{Re}\langle \Delta^{-1}y, VK^3Eu\rangle = 4r\,\mathrm{Re}\langle SFy, VK^3Eu\rangle \\ &= 4r\,\mathrm{Re}\langle FV(K^3E)u, Fy\rangle = 4r\,\mathrm{Re}\langle K^3EV^*u, z\rangle \\ &= 4r\,\mathrm{Re}\langle K^2EKV^*u, z\rangle \\ &= 4r\,\mathrm{Re}\langle K^2Ez, z\rangle \\ &\geq 4ra^2\|Ez\|^2,\end{aligned}$$

because $K^2E \geq a^2E$. Since $4ra^2 > \|A\|^2$, it now follows that $Ez = 0$. For each R in $\mathscr{R}$,

$$0 = REz = ERz = EKV^*Ru = KEV^*Ru,$$

whence $KEV^* = 0$, since $\mathscr{R}u$ is everywhere dense in $\mathscr{H}$. From this, $KE = V^*VKE = V^*(KEV^*)^* = 0$, a contradiction. Hence, Y is bounded. ∎

In the work that follows, Lemma 9.2.4 could be replaced by a similar result, in which the roles of $\mathscr{R}$ and $\mathscr{R}'$ are exchanged, and Δ appears in place of Δ^{-1}. This latter result can be proved either by the method used above, or by means of a variant of the linear Radon–Nikodým theorem (see Lemma 9.2.31 and Exercise 9.6.12).

9.2.5. LEMMA. *If $A \in \mathscr{R}$, $r > 0$, and A' is chosen in $\mathscr{R}'$ so that $A'u = (\Delta^{-1} + rI)^{-1}Au$, then*

$$\langle Ax_1, x_2\rangle = \langle JA'^*J\Delta^{1/2}x_1, \Delta^{-1/2}x_2\rangle + r\langle JA'^*J\Delta^{-1/2}x_1, \Delta^{1/2}x_2\rangle$$

whenever $x_1, x_2 \in \mathscr{D}(\Delta^{1/2}) \cap \mathscr{D}(\Delta^{-1/2})$.

Proof. We may apply, to the (positive, invertible) operator Δ, the results set out in the discussion preceding Lemma 9.2.1. We shall make use of the properties of Δ^t $(=f_t(\Delta))$, for real values of t. We recall also that the operator $(\Delta^{-1} + I)^{-1}$ is everywhere defined and bounded (indeed, so is $\Delta^{-t}(\Delta^{-1} + I)^{-1}$, when $0 \le t \le 1$); and $(\Delta^{-1} + I)^{-1}\mathscr{D}(\Delta^t) \subseteq \mathscr{D}(\Delta^t)$, for all real t. By using these results, together with Lemma 9.2.4, we show first that, when $x \in \mathscr{D}(\Delta^{1/2}) \cap \mathscr{D}(\Delta^{-1/2})$, there is a sequence $\{B_n'\}$ in $\mathscr{R}'$ such that $B_n'u \in \mathscr{D}(\Delta^{1/2})$ (of course, $B_n'u \in \mathscr{D}(F) = \mathscr{D}(\Delta^{-1/2})$), and

$$(1) \qquad B_n'u \to x, \qquad \Delta^{1/2}B_n'u \to \Delta^{1/2}x, \qquad \Delta^{-1/2}B_n'u \to \Delta^{-1/2}x.$$

Indeed, since u is a generating vector for $\mathscr{R}$, there is a sequence $\{B_n\}$ in $\mathscr{R}$ such that $B_n^*u \to J\Delta^{-1/2}x + J\Delta^{1/2}x$; and since

$$\Delta^{1/2}B_nu = J^2\Delta^{1/2}B_nu = JSB_nu = JB_n^*u,$$

it follows that

$$\Delta^{1/2}B_nu \to \Delta^{-1/2}x + \Delta^{1/2}x.$$

Since $\Delta^{1/2}$ has inverse $\Delta^{-1/2}$ (as a mapping) and $\Delta^{-1} = (\Delta^{-1})^{1/2}(\Delta^{-1})^{1/2} = \Delta^{-1/2}\Delta^{-1/2}$, while $x \in \mathscr{D}(\Delta^{-1/2}) \cap \mathscr{D}(\Delta^{1/2})$, we have $\Delta^{1/2}x \in \mathscr{D}(\Delta^{-1}) \subseteq \mathscr{D}(\Delta^{-1/2})$, and $\Delta^{-t}\Delta^{1/2}x = \Delta^{1/2-t}x$ when t is $0, \frac{1}{2}$, or 1. Accordingly, the last displayed equation can be rewritten in the form

$$\Delta^{1/2}B_nu \to (\Delta^{-1} + I)\Delta^{1/2}x;$$

and since the operator $\Delta^{-t}(\Delta^{-1} + I)^{-1}$ is everywhere defined and bounded,

$$(2) \qquad \Delta^{-t}(\Delta^{-1} + I)^{-1}\Delta^{1/2}B_nu \to \Delta^{1/2-t}x \qquad (t = 0, \tfrac{1}{2}, 1).$$

By Lemma 9.2.4, we can choose B_n' in $\mathscr{R}'$ so that

$$B_n'u = (\Delta^{-1} + I)^{-1}B_nu;$$

and $B_n'u \in (\Delta^{-1} + I)^{-1}\mathscr{D}(S) = (\Delta^{-1} + I)^{-1}\mathscr{D}(\Delta^{1/2}) \subseteq \mathscr{D}(\Delta^{1/2})$. If $\mathscr{A}$ denotes the von Neumann algebra generated by Δ, then $\Delta^{1/2}\ \eta\ \mathscr{A}$, $(\Delta^{-1} + I)^{-1} \in \mathscr{A}$ $(\subseteq \mathscr{A}')$, and thus $(\Delta^{-1} + I)^{-1}\Delta^{1/2} \subseteq \Delta^{1/2}(\Delta^{-1} + I)^{-1}$; so

$$\begin{aligned}\Delta^{-t}(\Delta^{-1} + I)^{-1}\Delta^{1/2}B_n u &= \Delta^{-t}\Delta^{1/2}(\Delta^{-1} + I)^{-1}B_n u\\ &= \Delta^{1/2-t}B_n'u \qquad (t = 0, \tfrac{1}{2}, 1)\end{aligned}$$

(since $B_n'u \in \mathscr{D}(\Delta^{-1/2}) \cap \mathscr{D}(\Delta^{1/2})$). This, together with (2), proves (1).

From the preceding argument, it now suffices to prove the lemma in the case where $x_j = B_j'u \in \mathscr{D}(\Delta^{1/2})$, for $j = 1, 2$. In this case,

$$\begin{aligned}\langle AB_1'u, B_2'u\rangle &= \langle Au, B_1'^*B_2'u\rangle\\ &= \langle(\Delta^{-1} + rI)A'u, B_1'^*B_2'u\rangle\\ &= \langle SFA'u + rA'u, B_1'^*B_2'u\rangle\\ &= \langle SA'^*u, B_1'^*B_2'u\rangle + r\langle B_1'A'u, B_2'u\rangle\\ &= \langle FB_1'^*B_2'u, A'^*u\rangle + r\langle FA'^*B_1'^*u, B_2'u\rangle\\ &= \langle B_1'u, B_2'A'^*u\rangle + r\langle FA'^*FB_1'u, B_2'u\rangle\\ &= \langle B_1'u, FA'FB_2'u\rangle + r\langle FA'^*FB_1'u, B_2'u\rangle\\ &= \langle B_1'u, \Delta^{1/2}JA'J\Delta^{-1/2}B_2'u\rangle + r\langle\Delta^{1/2}JA'^*J\Delta^{-1/2}B_1'u, B_2'u\rangle\\ &= \langle JA'^*J\Delta^{1/2}B_1'u, \Delta^{-1/2}B_2'u\rangle + r\langle JA'^*J\Delta^{-1/2}B_1'u, \Delta^{1/2}B_2'u\rangle,\end{aligned}$$

since $B_j'u \in \mathscr{D}(\Delta^{1/2})$ and $(JA'J)^* = JA'^*J$. ■

9.2.6. Remark. We now indicate (without detailed proofs) some considerations that motivate the next few stages of the argument, although not contributing to its formal structure. Since Δ is an invertible positive self-adjoint operator, $\{\Delta^{it} : t \in \mathbb{R}\}$ is a strong-operator continuous one-parameter group of unitary operators; and when $B \in \mathscr{B}(\mathscr{H})$, the function $\Delta^{it}B\Delta^{-it}$ of t is strong-operator continuous and bounded on $\mathbb{R}$. When $f \in L_1(\mathbb{R})$, we can define a bounded linear operator $\Lambda_f: \mathscr{B}(\mathscr{H}) \to \mathscr{B}(\mathscr{H})$ by

$$\Lambda_f(B) = \int_{\mathbb{R}} f(t)\Delta^{it}B\Delta^{-it}\,dt.$$

(Compare this with the construction described in the second paragraph of Theorem 4.5.9.) It is not difficult to check that the mapping $f \to \Lambda_f$ is a norm-reducing homomorphism, from the Banach algebra $L_1(\mathbb{R})$ (with convolution multiplication) into $\mathscr{B}(\mathscr{B}(\mathscr{H}))$.

If it is known that $J\mathscr{R}J = \mathscr{R}'$ (equivalently, $J\mathscr{R}'J = \mathscr{R}$) and $\Delta^{it}\mathscr{R}\Delta^{-it} = \mathscr{R}$ $(t \in \mathbb{R})$, it follows that $\Lambda_f(A) \in J\mathscr{R}'J$ whenever $A \in \mathscr{R}$ and

$f \in L_1(\mathbb{R})$. Conversely, if we can show that $\Lambda_f(\mathscr{R}) \subseteq J\mathscr{R}'J$, for a large class of L_1 functions f, it is not unreasonable to hope that this will lead to a proof that $\Delta^{it}\mathscr{R}\Delta^{-it} \subseteq J\mathscr{R}'J$ $(t \in \mathbb{R})$; and it is then not difficult to verify that $\mathscr{R} = J\mathscr{R}'J$ and $\Delta^{it}\mathscr{R}\Delta^{-it} = \mathscr{R}$.

We prove below (Lemma 9.2.8) that, if $r > 0$, $B, C \in \mathscr{B}(\mathscr{H})$, and

$$(3) \qquad \langle Bx_1, x_2\rangle = \langle C\Delta^{1/2}x_1, \Delta^{-1/2}x_2\rangle + r\langle C\Delta^{-1/2}x_1, \Delta^{1/2}x_2\rangle$$

whenever $x_1, x_2 \in \mathscr{D}(\Delta^{1/2}) \cap \mathscr{D}(\Delta^{-1/2})$, then

$$(4) \qquad C = \int_{\mathbb{R}} f_r(t)\Delta^{it}B\Delta^{-it}\,dt = \Lambda_{f_r}(B),$$

where f_r is a specified L_1 function. When $B \in \mathscr{R}$, it results from Lemma 9.2.5 that (3) has a solution C in $J\mathscr{R}'J$. Since $C = \Lambda_{f_r}(B)$, it follows that $\Lambda_{f_r}(\mathscr{R}) \subseteq J\mathscr{R}'J$; so $\{f_r : r > 0\}$ is a class of L_1 functions with the property suggested in the preceding paragraph. ■

9.2.7. LEMMA. *When $r, s > 0$,*

$$\int_{\mathbb{R}} \frac{r^{it-1/2}s^{it}\,dt}{e^{\pi t} + e^{-\pi t}} = \frac{1}{s^{-1/2} + rs^{1/2}}.$$

Proof. The function $f(z) = s^{iz}/(e^{\pi z} - e^{-\pi z})$ is analytic except for poles at integral multiples of i; and it has residue $1/2\pi$ at 0. With $z = a + ib$, $f(z) \to 0$ as $|a| \to \infty$, uniformly for any bounded set of values of b. By integrating f around the rectangle with vertices at $\pm R \pm \frac{1}{2}i$, and then letting $R \to \infty$, we obtain

$$\begin{aligned} 2\pi i\left(\frac{1}{2\pi}\right) &= \int_{\mathbb{R}} [f(t - \tfrac{1}{2}i) - f(t + \tfrac{1}{2}i)]\,dt \\ &= \int_{\mathbb{R}} \left[\frac{s^{it+1/2}}{e^{\pi t - i\pi/2} - e^{-\pi t + i\pi/2}} - \frac{s^{it-1/2}}{e^{\pi t + i\pi/2} - e^{-\pi t - i\pi/2}}\right] dt \\ &= i(s^{1/2} + s^{-1/2}) \int_{\mathbb{R}} \frac{s^{it}\,dt}{e^{\pi t} + e^{-\pi t}}. \end{aligned}$$

Thus

$$\int_{\mathbb{R}} \frac{s^{it}\,dt}{e^{\pi t} + e^{-\pi t}} = \frac{1}{s^{1/2} + s^{-1/2}}.$$

Upon replacing s by rs, and then multiplying throughout by $r^{-1/2}$, we obtain the required result. ■

We now show how to "solve for C" an operator equation of type (3).

9.2.8. LEMMA. *Suppose that D is an invertible (not necessarily bounded) positive self-adjoint operator acting on a Hilbert space $\mathscr{H}$, $r > 0$, $B, C \in \mathscr{B}(\mathscr{H})$, and*

$$(5)\qquad \langle Bx_1, x_2\rangle = \langle CD^{1/2}x_1, D^{-1/2}x_2\rangle + r\langle CD^{-1/2}x_1, D^{1/2}x_2\rangle$$

whenever $x_1, x_2 \in \mathscr{D}(D^{1/2}) \cap \mathscr{D}(D^{-1/2})$. Then

$$(6)\qquad \langle Cy_1, y_2\rangle = \int_{\mathbb{R}} \frac{r^{it-1/2}}{e^{\pi t} + e^{-\pi t}} \langle D^{it}BD^{-it}y_1, y_2\rangle\, dt$$

for all y_1 and y_2 in $\mathscr{H}$.

Proof. Since $\{D^{it} : t \in \mathbb{R}\}$ is a strong-operator continuous one-parameter unitary group, while multiplication is strong-operator continuous on bounded subsets of $\mathscr{B}(\mathscr{H})$, it follows that the bounded function $\langle D^{it}BD^{-it}y_1, y_2\rangle$ of t is continuous on $\mathbb{R}$, whenever $y_1, y_2 \in \mathscr{H}$. Hence the right-hand side of (6) exists as a Lebesgue integral.

The remainder of the proof is divided into three stages. First, we consider the case in which D has the form $\sum_{j=1}^{m} a_j F_j$, where $a_1, \ldots, a_m$ are positive real numbers and $\{F_1, \ldots, F_m\}$ is an orthogonal family of projections with sum I. In these circumstances, (5) can be written in the form

$$\begin{aligned} B &= D^{-1/2}CD^{1/2} + rD^{1/2}CD^{-1/2} \\ &= \sum_{j,k=1}^{m} (a_j^{-1/2}a_k^{1/2} + ra_j^{1/2}a_k^{-1/2})F_jCF_k, \end{aligned}$$

since $D^{1/2} = \sum a_j^{1/2}F_j$ and $D^{-1/2} = \sum a_k^{-1/2}F_k$. From this,

$$F_jBF_k = (a_j^{-1/2}a_k^{1/2} + ra_j^{1/2}a_k^{-1/2})F_jCF_k \qquad (j, k = 1, \ldots, m),$$

and

$$C = \sum_{j,k=1}^{m} F_jCF_k = \sum_{j,k=1}^{m} \left[\left(\frac{a_j}{a_k}\right)^{-1/2} + r\left(\frac{a_j}{a_k}\right)^{1/2}\right]^{-1} F_jBF_k.$$

By Lemma 9.2.7

$$\begin{aligned} \langle Cy_1, y_2\rangle &= \sum_{j,k=1}^{m} \langle F_jBF_ky_1, y_2\rangle \int_{\mathbb{R}} \left(\frac{a_j}{a_k}\right)^{it} \frac{r^{it-1/2}}{e^{\pi t} + e^{-\pi t}}\, dt \\ &= \int_{\mathbb{R}} \frac{r^{it-1/2}}{e^{\pi t} + e^{-\pi t}} \left[\sum_{j,k=1}^{m} \left(\frac{a_j}{a_k}\right)^{it} \langle F_jBF_ky_1, y_2\rangle\right] dt \\ &= \int_{\mathbb{R}} \frac{r^{it-1/2}}{e^{\pi t} + e^{-\pi t}} \langle D^{it}BD^{-it}y_1, y_2\rangle\, dt \end{aligned}$$

for all y_1, y_2 in $\mathscr{H}$. This proves (6) in the first (very special) case.

Suppose next that D is bounded and has a bounded inverse. We may choose positive real numbers a, b so that $aI \leq D \leq bI$. From the spectral theorem, D is the norm limit of a sequence $\{D_n\}$ of operators of the type considered in the preceding paragraph, each satisfying $aI \leq D_n \leq bI$ and lying in the (abelian) von Neumann algebra generated by D. Given any continuous complex-valued function f on $[a, b]$, by passing to the function representation of this algebra and using the uniform continuity of f, it follows that $f(D)$ is the norm limit of the sequence $\{f(D_n)\}$. This shows, in particular, that D^z is the norm limit of $\{D_n^z\}$ for all complex z. With B_n defined by

$$B_n = D_n^{-1/2}CD_n^{1/2} + rD_n^{1/2}CD_n^{-1/2},$$

the sequence $\{B_n\}$ is norm convergent; and from (5)

$$\lim B_n = D^{-1/2}CD^{1/2} + rD^{1/2}CD^{-1/2} = B.$$

From the preceding paragraph, with B_n and D_n in place of B and D,

$$\langle Cy_1, y_2\rangle = \int_{\mathbb{R}} \frac{r^{it-1/2}}{e^{\pi t} + e^{-\pi t}} \langle D_n^{it}B_nD_n^{-it}y_1, y_2\rangle\, dt,$$

for all y_1, y_2 in $\mathscr{H}$. When $n \to \infty$, $D_n^{it}B_nD_n^{-it} \to D^{it}BD^{-it}$, and (6) follows, by the dominated convergence theorem.

Finally, we consider the general case, in which D is unbounded. For each positive integer n, let E_n be the spectral projection for D, corresponding to the interval $[n^{-1}, n]$. The increasing sequence $\{E_n\}$ is strong-operator convergent to I, since D is positive and invertible. For a given choice of n, let D_0, B_0, and C_0 be the restrictions to $E_n(\mathscr{H})$ of the operators D, E_nBE_n, and E_nCE_n, respectively. For all x_1, x_2 in $\mathscr{H}$, $E_nx_1, E_nx_2 \in \mathscr{D}(D^{1/2}) \cap \mathscr{D}(D^{-1/2})$, and it results from (5) that

$$\begin{aligned}\langle B_0E_nx_1, E_nx_2\rangle = {} & \langle C_0D_0^{1/2}E_nx_1, D_0^{-1/2}E_nx_2\rangle \\ & + r\langle C_0D_0^{-1/2}E_nx_1, D_0^{1/2}E_nx_2\rangle.\end{aligned}$$

Since D_0 is bounded and has bounded inverse, we may deduce from the preceding paragraph that

$$\langle C_0y_1, y_2\rangle = \int_{\mathbb{R}} \frac{r^{it-1/2}}{e^{\pi t} + e^{-\pi t}} \langle D_0^{it}B_0D_0^{-it}y_1, y_2\rangle\, dt$$

for all y_1, y_2 in $E_n(\mathscr{H})$; that is

$$\langle CE_ny_1, E_ny_2\rangle = \int_{\mathbb{R}} \frac{r^{it-1/2}}{e^{\pi t} + e^{-\pi t}} \langle D^{it}BD^{-it}E_ny_1, E_ny_2\rangle\, dt$$

whenever $y_1, y_2 \in \mathscr{H}$. When $n \to \infty$, we obtain (6), by the dominated convergence theorem. ■

We are now in a position to prove the basic result of this section. We show first that

$$\Delta^{it}\mathscr{R}\Delta^{-it} \subseteq J\mathscr{R}'J \qquad (t \in \mathbb{R}). \tag{7}$$

For this, since the subspace $J\mathscr{R}'J$ of $\mathscr{B}(\mathscr{H})$ is weak-operator closed, it suffices to show that $\omega(\Delta^{it}A\Delta^{-it}) = 0$ ($t \in \mathbb{R}$) whenever $A \in \mathscr{R}$ and ω is a weak-operator continuous linear functional on $\mathscr{B}(\mathscr{H})$ that vanishes on $J\mathscr{R}'J$.

Suppose, then, that $A \in \mathscr{R}$. With $r > 0$, it results from Lemmas 9.2.4 and 9.2.5 that there is an element A_r' of $\mathscr{R}'$ for which $A_r'u = (\Delta^{-1} + rI)^{-1}Au$, and

$$\langle Ax_1, x_2\rangle = \langle JA_r'^*J\Delta^{1/2}x_1, \Delta^{-1/2}x_2\rangle + r\langle JA_r'^*J\Delta^{-1/2}x_1, \Delta^{1/2}x_2\rangle$$

for all x_1, x_2 in $\mathscr{D}(\Delta^{1/2}) \cap \mathscr{D}(\Delta^{-1/2})$. By Lemma 9.2.8,

$$\langle JA_r'^*Jy_1, y_2\rangle = \int_{\mathbb{R}} \frac{r^{it-1/2}}{e^{\pi t} + e^{-\pi t}}\langle \Delta^{it}A\Delta^{-it}y_1, y_2\rangle\, dt$$

whenever $y_1, y_2 \in \mathscr{H}$. By taking finite sums of equations of this last type, corresponding to different pairs (y_1, y_2), it now follows that

$$\omega(JA_r'^*J) = \int_{\mathbb{R}} \frac{r^{it-1/2}}{e^{\pi t} + e^{-\pi t}}\omega(\Delta^{it}A\Delta^{-it})\, dt$$

for every weak-operator continuous linear functional ω on $\mathscr{B}(\mathscr{H})$. If ω vanishes on $J\mathscr{R}'J$, the left-hand side of the last equation is zero, so

$$\int_{\mathbb{R}} \frac{r^{it-1/2}}{e^{\pi t} + e^{-\pi t}}\omega(\Delta^{it}A\Delta^{-it})\, dt = 0 \qquad (r > 0).$$

Upon multiplying throughout by $r^{1/2}$, and then taking $r = e^u$, it follows that the continuous L_1 function

$$f(t) = \frac{\omega(\Delta^{it}A\Delta^{-it})}{e^{\pi t} + e^{-\pi t}} \qquad (t \in \mathbb{R})$$

has zero Fourier transform, and so itself vanishes identically; that is, $\omega(\Delta^{it}A\Delta^{-it}) = 0$ ($t \in \mathbb{R}$). This completes the proof of (7).

With $t = 0$ in (7), we obtain $\mathscr{R} \subseteq J\mathscr{R}'J$, whence $J\mathscr{R}J \subseteq \mathscr{R}'$. We now establish the reverse inclusions. For this, suppose that $A \in \mathscr{R}$ and $A', B' \in \mathscr{R}'$. Since $Ju = u$, $JAJ \in \mathscr{R}'$ and $(JAJ)^* = JA^*J$, it follows that

$$JAJA'u = FA'^*JA^*Ju = FA'^*JA^*u.$$

Since, also, $JF = J\Delta^{1/2}J = SJ$, we obtain

$$\begin{aligned}\langle B'JA'u, A^*u\rangle &= \langle AJA'u, B'^*u\rangle \\ &= \langle J(JAJ)A'u, FB'u\rangle = \langle JFA'^*JA^*u, FB'u\rangle \\ &= \langle SJA'^*JA^*u, FB'u\rangle = \langle B'u, S^2JA'^*JA^*u\rangle \\ &= \langle B'u, JA'^*JA^*u\rangle = \langle JA'JB'u, A^*u\rangle.\end{aligned}$$

Since $\mathscr{R}u$ is everywhere dense in $\mathscr{H}$, it now follows that

$$B'JA'u = JA'JB'u \qquad (A', B' \in \mathscr{R}').$$

When $A', B', C' \in \mathscr{R}'$, we may apply the last equation, with either $A'C'$ or C' in place of A'. This gives

$$\begin{aligned}(B'JA'J)JC'u = B'JA'C'u &= JA'C'JB'u \\ &= JA'JJC'JB'u = (JA'JB')JC'u.\end{aligned}$$

Since $J\mathscr{R}'u$ is everywhere dense in $\mathscr{H}$, we deduce that $B'JA'J = JA'JB'$, whence $JA'J \in \mathscr{R}'' = \mathscr{R}$. Thus $J\mathscr{R}'J \subseteq \mathscr{R}$; the reverse inclusion has already been proved, so $J\mathscr{R}'J = \mathscr{R}$ (and $\mathscr{R}' = J\mathscr{R}J$).

It now follows from (7) that $\Delta^{it}\mathscr{R}\Delta^{-it} \subseteq \mathscr{R}$ for all real t; and upon replacing t by $-t$, we deduce that $\Delta^{it}\mathscr{R}\Delta^{-it} = \mathscr{R}$.

In the following theorem, we summarize the main results obtained so far in this section.

9.2.9. THEOREM (Tomita's theorem). *Suppose that a von Neumann algebra $\mathscr{R}$, acting on a Hilbert space $\mathscr{H}$, has a separating and generating vector u. Then the conjugate-linear operator $S_0: \mathscr{R}u\ (\subseteq \mathscr{H}) \to \mathscr{H}$, defined by $S_0Au = A^*u$ ($A \in \mathscr{R}$) is preclosed, and its adjoint F satisfies $FA'u = A'^*u$ ($A' \in \mathscr{R}'$). The closure S ($= F^*$) of S_0 has polar decomposition $S = J\Delta^{1/2}$, in which J is a conjugate-linear isometry from $\mathscr{H}$ onto $\mathscr{H}$, Δ ($= FS$) is an (in general, unbounded) positive self-adjoint operator with an (in general, unbounded) inverse Δ^{-1} ($= SF$), and $J^2 = I$, $J\Delta^{it} = \Delta^{it}J$ ($t \in \mathbb{R}$). Moreover, $Ju = \Delta u = u$, and*

$$J\mathscr{R}J = \mathscr{R}', \qquad \Delta^{it}\mathscr{R}\Delta^{-it} = \mathscr{R} \qquad (t \in \mathbb{R}).$$

In the circumstances described in Theorem 9.2.9, we refer to Δ as the *modular operator* for $\mathscr{R}$, corresponding to the separating and generating vector u. For each real t, the equation $\sigma_t(A) = \Delta^{it}A\Delta^{-it}$ defines a * automorphism σ_t of $\mathscr{R}$, and we refer to the one-parameter group $\{\sigma_t : t \in \mathbb{R}\}$ as the *modular automorphism group*. Since Δ^{it} is strong-operator continuous, as a function of t, the same is true of $\sigma_t(A)$ for each A in $\mathscr{R}$.

9.2.10. Definition. A one-parameter group $\{\alpha_t : t \in \mathbb{R}\}$ of * automorphisms of a von Neumann algebra $\mathscr{R}$ satisfies the *modular condition* relative to a state ω of $\mathscr{R}$ if, given any elements A and B of $\mathscr{R}$, there is a complex-valued function f, bounded and continuous on the strip $\{z \in \mathbb{C} : 0 \leq \operatorname{Im} z \leq 1\}$, and analytic on the interior of that strip, such that

$$f(t) = \omega(\alpha_t(A)B), \qquad f(t+i) = \omega(B\alpha_t(A)) \qquad (t \in \mathbb{R}). \quad \blacksquare$$

In connection with the modular condition, we shall frequently make use of the following simple result on analytic functions.

9.2.11. Lemma. *Suppose that $a < b < c$, and f is a complex-valued function that is bounded and continuous on the closed strip $\{z \in \mathbb{C} : a \leq \operatorname{Im} z \leq c\}$ and is analytic on each of the open strips*

$$\Omega_1 = \{z \in \mathbb{C} : a < \operatorname{Im} z < b\}, \qquad \Omega_2 = \{z \in \mathbb{C} : b < \operatorname{Im} z < c\}.$$

Then f is analytic on the open strip

$$\Omega_3 = \{z \in \mathbb{C} : a < \operatorname{Im} z < c\}.$$

Proof. Suppose that $r > 0$. For $j = 1, 2, 3$, let Γ_j be the contour formed by traversing the counter-clockwise boundary of the open rectangle $R_j = \{z \in \Omega_j : -r < \operatorname{Re} z < r\}$. The function F, defined by

$$F(z) = \frac{1}{2\pi i} \oint_{\Gamma_3} \frac{f(w)\, dw}{w - z} \qquad (z \in R_3),$$

is analytic on R_3. Moreover, when $z \in R_1$,

$$\begin{aligned} F(z) &= \frac{1}{2\pi i} \oint_{\Gamma_1} \frac{f(w)\, dw}{w - z} + \frac{1}{2\pi i} \oint_{\Gamma_2} \frac{f(w)\, dw}{w - z} \\ &= \qquad f(z) \qquad + \qquad 0 \qquad = f(z), \end{aligned}$$

by Cauchy's integral formula and Cauchy's theorem; and a similar argument applies when $z \in R_2$. Accordingly, the functions f and F coincide on R_3, except possibly on the line $\operatorname{Im} z = b$. By continuity, they coincide throughout the region R_3; and f is therefore analytic on $R_3 = \{z \in \Omega_3 : -r < \operatorname{Re} z < r\}$. Since this has been proved for every positive r, f is analytic on Ω_3. $\blacksquare$

9.2.12. Lemma. *If $x, y \in \mathscr{D}(\Delta^{1/2})$, the function $f(z) = \langle \Delta^z x, y \rangle$ is defined, bounded, and continuous on the strip $\Omega = \{z \in \mathbb{C} : 0 \leq \operatorname{Re} z \leq \frac{1}{2}\}$, and is analytic on the interior of Ω.*

Proof. By polarization, it suffices to prove the result when $x = y$, so we confine attention to this case. With $\{E_\lambda\}$ the spectral resolution for Δ, from Theorem 5.6.26,

$$\int_{[0,\infty)} d\langle E_\lambda x, x\rangle = \|x\|^2 < \infty, \qquad \int_{[0,\infty)} \lambda\, d\langle E_\lambda x, x\rangle = \|\Delta^{1/2}x\|^2 < \infty.$$

When $z = a + ib \in \Omega$, and $\lambda > 0$, we have

$$|\lambda^z|^2 = \lambda^{2a} \le \max(1, \lambda) < 1 + \lambda,$$

since $0 \le a \le \frac{1}{2}$. Hence

$$\int_{[0,\infty)} |\lambda^z|^2\, d\langle E_\lambda x, x\rangle \le \int_{[0,\infty)} (1 + \lambda)\, d\langle E_\lambda x, x\rangle$$
$$\le \|x\|^2 + \|\Delta^{1/2}x\|^2 < \infty.$$

This shows that $x \in \mathscr{D}(\Delta^z)$, and $\|\Delta^z x\|$ is bounded, for z in Ω. Accordingly, the function $f(z) = \langle \Delta^z x, x\rangle$ is defined and bounded throughout Ω, and

$$f(z) = \int_{[0,\infty)} \lambda^z\, d\langle E_\lambda x, x\rangle.$$

For $n = 1, 2, 3, \ldots$, we can define a complex-valued function f_n on $\mathbb{C}$ by

$$f_n(z) = \int_{[n^{-1}, n]} \lambda^z\, d\langle E_\lambda x, x\rangle.$$

Upon expanding λ^z ($= \exp z \log \lambda$) as a power series in z, which converges uniformly for λ in $[n^{-1}, n]$, it follows that f_n has a power-series expansion, and is therefore analytic, throughout $\mathbb{C}$. Moreover, when $z = a + ib \in \Omega$, $|\lambda^z| = \lambda^a \le 1 + \lambda$, and

$$|f(z) - f_n(z)| = \left|\int_{[0,n^{-1}) \cup (n,\infty)} \lambda^z\, d\langle E_\lambda x, x\rangle\right|$$
$$\le \int_{[0,n^{-1}) \cup (n,\infty)} (1 + \lambda)\, d\langle E_\lambda x, x\rangle.$$

The right-hand side is independent of z, and converges to zero when $n \to \infty$; so $f_n(z) \to f(z)$, uniformly on Ω. Since each f_n is analytic on $\mathbb{C}$, it now follows that f is continuous on Ω and analytic on its interior. ■

9.2.13. THEOREM. *If $\{\sigma_t : t \in \mathbb{R}\}$ is the modular automorphism group of a von Neumann algebra $\mathscr{R}$, corresponding to a separating and generating vector u, then $\{\sigma_t\}$ satisfies the modular condition relative to the vector state $\omega_u | \mathscr{R}$.*

Proof. When $A, B \in \mathcal{R}$, since $\Delta^{-it}u = u$ (because $\Delta u = u$),

$$\omega_u(\sigma_t(A)B) = \langle \Delta^{it} A \Delta^{-it} Bu, u \rangle = \langle \Delta^{-it} Bu, A^* \Delta^{-it} u \rangle = \langle \Delta^{-it} Bu, A^* u \rangle,$$

$$\omega_u(B\sigma_t(A)) = \langle B\Delta^{it} A \Delta^{-it} u, u \rangle = \langle \Delta^{it} Au, B^* u \rangle.$$

Since $\mathcal{R}u \subseteq \mathcal{D}(\Delta^{1/2})$, it follows from Lemma 9.2.12 that each of the functions

$$g(z) = \langle \Delta^{-iz} Bu, A^* u \rangle \qquad (0 \leq \operatorname{Im} z \leq \tfrac{1}{2}),$$

$$h(z) = \langle \Delta^{1+iz} Au, B^* u \rangle \qquad (\tfrac{1}{2} \leq \operatorname{Im} z \leq 1)$$

is bounded and continuous on the strip where it is defined, and is analytic on the interior of that strip. Since, also,

$$\begin{aligned} g(t + \tfrac{1}{2}i) &= \langle \Delta^{-it} \Delta^{1/2} Bu, A^* u \rangle = \langle \Delta^{-it} J \cdot J\Delta^{1/2} Bu, SAu \rangle \\ &= \langle J\Delta^{-it} SBu, J\Delta^{1/2} Au \rangle = \langle \Delta^{1/2} Au, \Delta^{-it} B^* u \rangle \\ &= \langle \Delta^{1/2+it} Au, B^* u \rangle = h(t + \tfrac{1}{2}i) \qquad (t \in \mathbb{R}), \end{aligned}$$

it follows that g and h can be combined to form a single function f, bounded and continuous on the strip $\{z \in \mathbb{C} : 0 \leq \operatorname{Im} z \leq 1\}$. By Lemma 9.2.11, f is analytic on the interior of this strip.

Finally, for all real t,

$$f(t) = g(t) = \langle \Delta^{-it} Bu, A^* u \rangle = \omega_u(\sigma_t(A)B),$$

$$f(t + i) = h(t + i) = \langle \Delta^{it} Au, B^* u \rangle = \omega_u(B\sigma_t(A)). \quad \blacksquare$$

We now investigate some of the consequences of the modular condition.

9.2.14. PROPOSITION. *Suppose that ω is a state of a von Neumann algebra $\mathcal{R}$, $\{\alpha_t : t \in \mathbb{R}\}$ is a one-parameter group of * automorphisms of $\mathcal{R}$ that satisfies the modular condition relative to ω, and $H \in \mathcal{R}$.*

(i) *ω is invariant under $\{\alpha_t\}$, in the sense that $\omega(\alpha_t(A)) = \omega(A)$ for each A in $\mathcal{R}$ and t in $\mathbb{R}$.*

(ii) *If $\alpha_t(H) = H$ for all real t, then $\omega(AH) = \omega(HA)$ for each A in $\mathcal{R}$.*

(iii) *If ω is faithful, and $\omega(AH) = \omega(HA)$ for each A in $\mathcal{R}$, then $\alpha_t(H) = H$ for all real t.*

(iv) *If ω is faithful, then $\alpha_t(C) = C$ for all real t and for each C in the center of $\mathcal{R}$.*

Proof. Suppose that $H \in \mathcal{R}$ and $\omega(AH) = \omega(HA)$ for each A in $\mathcal{R}$. Given A in $\mathcal{R}$, it results from the modular condition that there is a complex-valued function f that is bounded and continuous on the closed strip $\{z \in \mathbb{C} : 0 \leq \operatorname{Im} z \leq 1\}$, analytic on the interior of that strip, and satisfying

$$f(t) = \omega(\alpha_t(A)H), \qquad f(t + i) = \omega(H\alpha_t(A)).$$

By our assumption concerning H, $f(t) = f(t+i)$ for all real t. Thus f extends to a function, defined, bounded and continuous on $\mathbb{C}$, with period i, that is analytic on each of the open strips $\{z \in \mathbb{C} : n < \text{Im}\, z < n+1\}$ ($n = 0, \pm 1, \pm 2, \ldots$). From Lemma 9.2.11, this extended function is analytic on $\mathbb{C}$, and is therefore constant by Liouville's theorem. Accordingly, f is constant, and

$$\omega(\alpha_t(A)H) = f(t) = f(0) = \omega(AH) \qquad (t \in \mathbb{R}).$$

Since the above argument applies, in particular, with I in place of H, it follows that $\omega(\alpha_t(A)) = \omega(A)$; so (i) is proved.

Suppose next that ω is faithful and $\omega(AH) = \omega(HA)$ $(A \in \mathscr{R})$. From (i), together with the first paragraph of this proof,

$$\omega(\alpha_t(A)\alpha_t(H)) = \omega(\alpha_t(AH)) = \omega(AH) = \omega(\alpha_t(A)H).$$

Since $\alpha_t(\mathscr{R}) = \mathscr{R}$, it now follows that $\omega(B[\alpha_t(H) - H]) = 0$ for each B in $\mathscr{R}$. In particular,

$$\omega([\alpha_t(H) - H]^*[\alpha_t(H) - H]) = 0;$$

so $\alpha_t(H) = H$, since ω is faithful. This proves (iii); and (iv) follows at once, upon replacing H by a general element C of the center of $\mathscr{R}$.

It remains to prove (ii). Suppose that $A \in \mathscr{R}$, and $\alpha_t(H) = H$ for all real t. From the modular condition, there is a complex-valued function g, satisfying the usual boundedness, continuity, and analyticity conditions, for which

$$g(t) = \omega(\alpha_t(H)A) = \omega(HA), \qquad g(t+i) = \omega(A\alpha_t(H)) = \omega(AH) \qquad (t \in \mathbb{R}).$$

A suitable non-zero multiple of g has constant *real* value on $\mathbb{R}$, and so has an analytic continuation across $\mathbb{R}$, by the Schwarz reflection principle. The extended function is analytic on a domain D that contains both $\mathbb{R}$ and the open strip $\{z \in \mathbb{C} : 0 < \text{Im}\, z < 1\}$. It is constant on $\mathbb{R}$, and is therefore constant on D. Accordingly, g is constant on the open strip, and by continuity is constant on its closure. Hence

$$\omega(HA) = g(t) = g(t+i) = \omega(AH). \quad \blacksquare$$

9.2.15. Corollary. *Suppose that ω is a state of a von Neumann algebra $\mathscr{R}$, and $\{\alpha_t : t \in \mathscr{R}\}$ is a one-parameter group of * automorphisms of $\mathscr{R}$ that satisfies the modular condition relative to ω. If each α_t is the identity automorphism ι of $\mathscr{R}$, then ω is a tracial state. Conversely, if ω is a faithful tracial state, then $\alpha_t = \iota$ for all real t.*

Proof. This follows at once from parts (ii) and (iii) of Proposition 9.2.14. $\blacksquare$

If ω is a faithful normal state of the von Neumann algebra $\mathscr{R}$, the GNS construction applied to ω yields a faithful representation φ with generating unit vector u such that $\omega_u \circ \varphi = \omega$. From Corollary 7.1.7, $\varphi(\mathscr{R})$ is a von Neumann algebra acting on the representation space $\mathscr{H}$. Since ω is faithful, u is separating for $\mathscr{R}$. If Δ is the modular operator for $\varphi(\mathscr{R})$ corresponding to u and $\sigma_t(A) = \varphi^{-1}(\Delta^{it}\varphi(A)\Delta^{-it})$, then, from Theorem 9.2.13, $\{\sigma_t\}$ is a one-parameter group of * automorphisms of $\mathscr{R}$ satisfying the modular condition relative to ω. If $\{\alpha_t\}$ is a one-parameter group of * automorphisms of $\mathscr{R}$ satisfying the modular condition relative to ω, we show, in the theorem that follows this discussion, that $\alpha_t = \sigma_t$ for each real t. Of course $\varphi \circ \alpha_t \circ \varphi^{-1}$ is a * automorphism β_t of $\varphi(\mathscr{R})$ and $\{\beta_t\}$ is a one-parameter group of * automorphisms of $\varphi(\mathscr{R})$ satisfying the modular condition relative to $\omega_u | \varphi(\mathscr{R})$. Thus we may assume that $\mathscr{R}$ acts on $\mathscr{H}$ and $\alpha_t = \beta_t$. By Proposition 9.2.14(i), $\omega_u | \mathscr{R}$ is invariant under $\{\alpha_t\}$, so

$$\|\alpha_t(A)u\| = [\omega_u(\alpha_t(A^*A))]^{1/2} = [\omega_u(A^*A)]^{1/2} = \|Au\|$$

for each A in $\mathscr{R}$. From this, and since $\mathscr{R}u$ is dense in $\mathscr{H}$, there is a unitary operator U_t on $\mathscr{H}$, for which $U_t Au = \alpha_t(A)u$ $(A \in \mathscr{R})$. Since $\alpha_{s+t} = \alpha_s\alpha_t$, it follows that $U_{s+t} = U_s U_t$ for all real s and t. When $A, B \in \mathscr{R}$,

$$U_t ABu = \alpha_t(AB)u = \alpha_t(A)\alpha_t(B)u = \alpha_t(A)U_t Bu;$$

so $U_t A = \alpha_t(A)U_t$ and $\alpha_t(A) = U_t A U_t^*$. Moreover,

$$\langle U_t Au, Bu\rangle = \langle B^*\alpha_t(A)u, u\rangle = \omega_u(B^*\alpha_t(A)) \qquad (t \in \mathbb{R});$$

and the right-hand side is continuous as a function of t, since $\{\alpha_t\}$ satisfies the modular condition relative to $\omega_u | \mathscr{R}$. Thus, with $|t|$ small,

$$\langle [2I - (U_t + U_{-t})]Au, Au\rangle \qquad (= \|(U_t - I)Au\|^2)$$

is small. Accordingly, the one-parameter unitary group $\{U_t\}$ is strong-operator continuous, and the function $\langle \alpha_t(A)x, y\rangle$ $(= \langle AU_{-t}x, U_{-t}y\rangle)$ of t is continuous on $\mathbb{R}$ for all A in $\mathscr{R}$ and x, y in $\mathscr{H}$. By Stone's theorem (5.6.36), there is a (generally, unbounded) self-adjoint operator $H\colon \mathscr{D}(H)\ (\subseteq \mathscr{H}) \to \mathscr{H}$ such that

$$U_t = \exp itH \quad (t \in \mathbb{R}), \qquad iHx = \lim_{t\to 0} \frac{U_t x - x}{t} \quad (x \in \mathscr{D}(H)),$$

and $\mathscr{D}(H)$ consists precisely of those x in $\mathscr{H}$ for which the indicated limit exists in the norm topology on $\mathscr{H}$.

It now suffices to prove that $\exp H = \Delta$, for then $\Delta^{it} = \exp itH = U_t$ since $f_{it}(\exp \lambda) = \exp it\lambda$ for λ in $\mathrm{sp}(H)$; and

$$\alpha_t(A) = U_t A U_t^* = \Delta^{it} A \Delta^{-it} = \sigma_t(A) \qquad (A \in \mathscr{R}).$$

To effect this proof, we shall recast the modular condition, by means of Fourier transforms, in terms that link Δ and $\{\alpha_t\}$. For this purpose, we introduce the class $\mathscr{F}$ of all continuous functions $g\colon \mathbb{R} \to \mathbb{R}$ that vanish outside a compact interval (depending on g) and satisfy the condition

$$\int_{\mathbb{R}} |\check{g}(t + is)|\, dt < \infty \qquad (s \in \mathbb{R}),$$

where $\check{g}$ is the entire function of a complex variable z, defined by

$$\check{g}(z) = \frac{1}{2\pi} \int_{\mathbb{R}} e^{-i\lambda z} g(\lambda)\, d\lambda.$$

When $g \in \mathscr{F}$, $\check{g}$ is an extension to the complex domain of the "inverse Fourier transform" of g, and the ordinary Fourier transform $\hat{g}$ is given by $\hat{g}(t) = 2\pi\check{g}(-t)$. Since

$$|\check{g}(t + is)| \le \frac{1}{2\pi} \int_{\mathbb{R}} e^{\lambda s} |g(\lambda)|\, d\lambda,$$

$\check{g}$ is bounded on each region of the complex plane in which $\operatorname{Im} z$ is bounded. Also, for each fixed s, $\check{g}(t + is) \to 0$ as $|t| \to \infty$, by applying the Riemann–Lebesgue lemma (Corollary 3.2.28(iii)) to the (real) Fourier transform of the function $g(\lambda) \exp s\lambda$ of λ. Moreover,

$$\check{g}(z + i) = \frac{1}{2\pi} \int_{\mathbb{R}} e^{-i\lambda z} e^{\lambda} g(\lambda)\, d\lambda = \check{h}(z), \tag{8}$$

where h (in $\mathscr{F}$) is defined by $h(\lambda) = g(\lambda) \exp \lambda$. Since

$$\int_{\mathbb{R}} |\hat{g}(t)|\, dt = 2\pi \int_{\mathbb{R}} |\check{g}(-t)|\, dt = 2\pi \int_{\mathbb{R}} |\check{g}(t)|\, dt < \infty,$$

it follows by Fourier inversion (Theorem 3.2.30), taking into account the continuity of g, that

$$g(\lambda) = \frac{1}{2\pi} \int_{\mathbb{R}} e^{-it\lambda} \hat{g}(t)\, dt = \int_{\mathbb{R}} e^{it\lambda} \check{g}(t)\, dt \qquad (\lambda \in \mathbb{R}).$$

In other words, if k lies in the set $\check{\mathscr{F}}$ $(= \{\check{g} : g \in \mathscr{F}\})$, then the unique element g of $\mathscr{F}$ for which $\check{g} = k$ is given by $g = \hat{k}$ (the Fourier transform of the L_1 function $k \,|\, \mathbb{R}$).

When $k \in \check{\mathscr{F}}$ and $A \in \mathscr{R}$, we define A_k to be the operator associated with the bounded conjugate-bilinear functional $(x, y) \to \int_{\mathbb{R}} k(t) \langle \alpha_t(A)x, y \rangle\, dt$ on $\mathscr{H} \times \mathscr{H}$. Since a weak-operator continuous linear functional ρ on $\mathscr{B}(\mathscr{H})$ is a finite sum of mappings of the form $T \to \langle Tx, y \rangle$, it follows that $\rho(A_k) = \int_{\mathbb{R}} k(t)\rho(\alpha_t(A))\,dt$. Hence $\rho(A_k) = 0$ if ρ vanishes on $\mathscr{R}$, and thus $A_k \in \mathscr{R}$ (by the Hahn–Banach theorem). Note also that

$$\langle A_k u, y\rangle = \int_{\mathbb{R}} k(t)\langle \alpha_t(A)u, y\rangle\,dt = \int_{\mathbb{R}} k(t)\langle U_t A u, y\rangle\,dt$$

$$= \int_{\mathbb{R}} k(t)\langle e^{itH} A u, y\rangle\,dt = \langle \hat{k}(H) A u, y\rangle,$$

by Stone's theorem (5.6.36); so that

$$\text{(9)} \qquad A_k u = \hat{k}(H) A u \qquad (A \in \mathscr{R}, \quad k \in \check{\mathscr{F}}).$$

Before continuing the formal proof that $\Delta = \exp H$, we give a heuristic argument that suggests how the modular condition should be reinterpreted. Given A and B in $\mathscr{R}$, there is a function f, bounded and continuous on the strip $\{z \in \mathbb{C} : 0 \le z \le 1\}$ and analytic on the interior of that strip, such that

$$\langle \alpha_t(A)Bu, u\rangle = f(t), \qquad \langle B\alpha_t(A)u, u\rangle = f(t+i) \qquad (t \in \mathbb{R}).$$

Suppose (for the purposes of our heuristic argument) that $f = \check{g}$, where $g \in \mathscr{F}$. Then $f(t) = \check{g}(t)$, and from (8), $f(t+i) = \check{g}(t+i) = \check{h}(t)$, where $h \in \mathscr{F}$ and $h(\lambda) = g(\lambda)\exp\lambda$. Thus $\check{g}(t) = \langle \alpha_t(A)Bu, u\rangle$, $\check{h}(t) = \langle B\alpha_t(A)u, u\rangle$, and

$$g(\lambda) = \int_{\mathbb{R}} e^{it\lambda}\check{g}(t)\,dt = \int_{\mathbb{R}} e^{it\lambda}\langle \alpha_t(A)Bu, u\rangle\,dt,$$

$$h(\lambda) = \int_{\mathbb{R}} e^{it\lambda}\check{h}(t)\,dt = \int_{\mathbb{R}} e^{it\lambda}\langle B\alpha_t(A)u, u\rangle\,dt.$$

Since $h(\lambda) = g(\lambda)\exp\lambda$,

$$\text{(10)} \qquad e^{\lambda}\int_{\mathbb{R}} e^{it\lambda}\langle \alpha_t(A)Bu, u\rangle\,dt = \int_{\mathbb{R}} e^{it\lambda}\langle B\alpha_t(A)u, u\rangle\,dt.$$

Now suppose that $l \in \check{\mathscr{F}}$, and that k (in $\check{\mathscr{F}}$) is given by $k(z) = l(z+i)$; so that (compare (8)) $\hat{k}(\lambda) = \hat{l}(\lambda)\exp\lambda$. With $-\lambda$ in place of λ, the last equation gives $\check{k}(\lambda)\exp\lambda = \check{l}(\lambda)$. Upon multiplying both sides of (10) by $\check{k}(\lambda)$, and integrating over $\mathbb{R}$ with respect to λ, we obtain (heuristically)

$$\int_{\mathbb{R}}\int_{\mathbb{R}} \check{l}(\lambda)e^{it\lambda}\langle \alpha_t(A)Bu, u\rangle\,dt\,d\lambda = \int_{\mathbb{R}}\int_{\mathbb{R}} \check{k}(\lambda)e^{it\lambda}\langle B\alpha_t(A)u, u\rangle\,dt\,d\lambda;$$

that is

$$\int_{\mathbb{R}}\left[\int_{\mathbb{R}} e^{it\lambda}\check{l}(\lambda)\,d\lambda\right]\langle \alpha_t(A)Bu, u\rangle\,dt = \int_{\mathbb{R}}\left[\int_{\mathbb{R}} e^{it\lambda}\check{k}(\lambda)\,d\lambda\right]\langle \alpha_t(A)u, B^*u\rangle\,dt,$$

whence

$$\int_{\mathbb{R}} l(t)\langle\alpha_t(A)Bu, u\rangle\, dt = \int_{\mathbb{R}} k(t)\langle\alpha_t(A)u, B^*u\rangle\, dt.$$

Thus $\langle A_l Bu, u\rangle = \langle A_k u, B^*u\rangle$; equivalently

$$\langle Bu, A_l^*u\rangle = \langle A_k u, B^*u\rangle \qquad (B \in \mathscr{R}). \tag{11}$$

The heuristic arguments set out in the preceding paragraph will shortly be replaced by a detailed proof of (11). In the meantime, we assume that (11) is satisfied, and on this basis show how the proof that $\Delta = \exp H$ can be completed. Since F is the adjoint of the mapping $Bu \to B^*u\colon \mathscr{R}u \to \mathscr{R}u$, it follows from (11) that $A_k u \in \mathscr{D}(F)$ and $FA_k u = A_l^* u = SA_l u$, whence $A_k u = FSA_l u$ (from Lemma 9.2.1). Hence, from Proposition 9.2.3,

$$A_l u \in \mathscr{D}(\Delta), \qquad A_k u = \Delta A_l u, \tag{12}$$

whenever $A \in \mathscr{R}$, $l \in \check{\mathscr{F}}$, and k (in $\check{\mathscr{F}}$) is defined by $k(z) = l(z + i)$. This is our reinterpretation of the modular condition.

From (9) and (12), and since $\hat{k}(\lambda) = \hat{l}(\lambda)\exp\lambda$,

$$\hat{k}(H)Au = A_k u = \Delta A_l u = \Delta\hat{l}(H)Au = \Delta e^{-H}\hat{k}(H)Au,$$

whenever $k \in \check{\mathscr{F}}$ and $A \in \mathscr{R}$; equivalently (with $k = \check{g}$),

$$\Delta^{-1}g(H)Au = e^{-H}g(H)Au,$$

for each A in $\mathscr{R}$ and g in $\mathscr{F}$. Now $(\exp - H)g(H) \in \mathscr{B}(\mathscr{H})$, and $\Delta^{-1}g(H)$ is closed (since $g(H) \in \mathscr{B}(\mathscr{H})$). Thus $\Delta^{-1}g(H) \in \mathscr{B}(\mathscr{H})$ and $\Delta^{-1}g(H) = (\exp -H)g(H)$.

For each positive integer n and λ in $\mathbb{R}$, define $g_n(\lambda)$ to be $1 - (|\lambda|/n)$ when $|\lambda| \le n$ and 0 when $n < |\lambda|$. A computation (noted at the beginning of the proof of Theorem 3.2.30) yields

$$\check{g}_n(z) = \frac{1 - \cos nz}{n\pi z^2} \qquad (z \ne 0);$$

so that $g_n \in \mathscr{F}$. Since $0 \le g_n(\lambda) \le 1$ and $g_n(\lambda) \uparrow 1$ for each real λ, $g_n(H) \uparrow I$ (by σ-normality of the mapping $f \to f(H)$). If $x \in \mathscr{D}(\exp -H)$, $g_n(H)x \to x$ and

$$\begin{aligned}\Delta^{-1}g_n(H)x &= (\exp - H)g_n(H)x\\ &= g_n(H)(\exp -H)x \to (\exp -H)x.\end{aligned}$$

Since Δ^{-1} is closed, $x \in \mathscr{D}(\Delta^{-1})$ and $\Delta^{-1}x = (\exp -H)x$. Thus $\exp -H \subseteq \Delta^{-1}$. As $\exp -H$ and Δ^{-1} are self-adjoint, $\Delta^{-1} = \exp -H$ and $\Delta = \exp H$.

It now remains to prove (11), when $A, B \in \mathscr{R}$, $l \in \check{\mathscr{F}}$, and k (in $\check{\mathscr{F}}$) is defined by $k(z) = l(z + i)$. Choose f satisfying the requirements of the modular condition for A and B. If Γ is the contour formed by the (counter-clockwise) rectangle with vertices $\pm R$ and $\pm R + i$, then

$$0 = \int_{\Gamma} l(z)f(z)\,dz = \int_{-R}^{R} l(t)f(t)\,dt - \int_{-R}^{R} l(t+i)f(t+i)\,dt + i\varepsilon(R),$$

where

$$\varepsilon(R) = \int_0^1 l(R+is)f(R+is)\,ds - \int_0^1 l(-R+is)f(-R+is)\,ds.$$

Now f is bounded on the strip $\{z \in \mathbb{C} : 0 \leq \operatorname{Im} z \leq 1\}$; the same is true of l, and in addition $l(\pm R + is) \to 0$ as $R \to \infty$, as noted in our discussion (preceding (8)) of functions in $\check{\mathscr{F}}$. The dominated convergence theorem applies and $\varepsilon(R_n) \to 0$ for each sequence $\{R_n\}$ tending to ∞. It follows that

$$\int_{\mathbb{R}} l(t)f(t)\,dt = \int_{\mathbb{R}} l(t+i)f(t+i)\,dt.$$

Since $f(t) = \langle \alpha_t(A)Bu, u\rangle$, $f(t+i) = \langle \alpha_t(A)u, B^*u\rangle$, and $l(t+i) = k(t)$, the preceding equation can be rewritten in the form

$$\int_{\mathbb{R}} l(t)\langle \alpha_t(A)Bu, u\rangle\,dt = \int_{\mathbb{R}} k(t)\langle \alpha_t(A)u, B^*u\rangle\,dt.$$

Thus $\langle A_l Bu, u\rangle = \langle A_k u, B^*u\rangle$, and (11) is proved.

We now summarize, as a theorem, the results of the preceding discussion.

9.2.16. THEOREM. *If ω is a faithful normal state of a von Neumann algebra $\mathscr{R}$, there is a unique one-parameter group $\{\alpha_t\}$ of * automorphisms of $\mathscr{R}$ that satisfies the modular condition relative to ω.*

In the circumstances set out in Theorem 9.2.16, we refer to $\{\alpha_t\}$ as the *modular automorphism group of $\mathscr{R}$, corresponding to the faithful normal state ω*. Note that $\alpha_t(A)$ is strong-operator continuous as a function of t, for each A in $\mathscr{R}$. Indeed, *weak*-operator continuity follows from the fact that

$$\alpha_t(A) = \varphi^{-1}(\Delta^{it}\varphi(A)\Delta^{-it}),$$

where φ is the faithful representation of $\mathscr{R}$ engendered by the state ω and Δ is the modular operator corresponding to a separating and generating vector of the von Neumann algebra $\varphi(\mathscr{R})$; for φ^{-1} is weak-operator continuous on bounded subsets of $\varphi(\mathscr{R})$, by Remark 7.4.4. To establish *strong*-operator continuity, we may replace A by a unitary operator U in $\mathscr{R}$, since $\mathscr{R}$ is the linear span of its unitary group. For each x in the underlying Hilbert space $\mathscr{H}$, $\|\alpha_t(U)x\| = \|x\|$ for all real t; and the weakly continuous mapping $t \to \alpha_t(U)x$: $\mathbb{R} \to \mathscr{H}$ is norm continuous, by Proposition 2.3.5.

The *centralizer* of a state ω of a C^*-algebra $\mathfrak{A}$ is defined to be the set

$$\{H \in \mathfrak{A} : \omega(AH) = \omega(HA)\ (A \in \mathfrak{A})\}.$$

From the fact that ω is norm continuous and hermitian, it is easy to verify that its centralizer is a C^*-subalgebra of $\mathfrak{A}$. When ω is a faithful normal state of a von Neumann algebra $\mathscr{R}$, it follows from Proposition 9.2.14(ii) and (iii) that the centralizer of ω consists exactly of those elements of $\mathscr{R}$ that are fixed under the action of the modular automorphism group of $\mathscr{R}$ relative to ω.

We now show that a slightly weakened form of the modular condition suffices to determine the modular automorphism group.

9.2.17. LEMMA. *Suppose that ω is a faithful normal state of a von Neumann algebra $\mathscr{R}$, $\{\alpha_t\}$ is a one-parameter group of * automorphisms of $\mathscr{R}$, and $\mathscr{R} = \mathfrak{A}^-$, where $\mathfrak{A}$ is a self-adjoint subalgebra of $\mathscr{R}$ with the following property: given any A and B in $\mathfrak{A}$, there is a complex-valued function f, bounded and continuous on the strip $\Omega = \{z \in \mathbb{C} : 0 \le \operatorname{Im} z \le 1\}$, and analytic on the interior of Ω, such that*

$$f(t) = \omega(\alpha_t(A)B), \qquad f(t+i) = \omega(B\alpha_t(A)) \qquad (t \in \mathbb{R}).$$

Then $\{\alpha_t\}$ is the modular automorphism group of $\mathscr{R}$, corresponding to ω.

Proof. We use approximation arguments to show that $\{\alpha_t\}$ satisfies the modular condition relative to ω.

Suppose first that $A \in \mathfrak{A}$ and $B = B^* \in \mathscr{R}$. Let $\{B_a : a \in \mathbb{A}\}$ be a bounded net of self-adjoint elements of $\mathfrak{A}$, that is strong-operator convergent to B. For each a in $\mathbb{A}$, we may choose a function f_a corresponding, as in the statement of the lemma, to the elements A, B_a of $\mathfrak{A}$. We shall show that the net $\{f_a\}$ of functions converges uniformly on Ω. To this end, note that

$$\begin{aligned} |f_a(t) - f_b(t)| &= |\omega(\alpha_t(A)(B_a - B_b))| \\ &\le [\omega(\alpha_t(AA^*))]^{1/2}[\omega((B_a - B_b)^2)]^{1/2} \\ &\le \|A\|\,[\omega((B_a - B_b)^2)]^{1/2}, \end{aligned}$$

and similarly

$$|f_a(t+i) - f_b(t+i)| \le \|A\|\,[\omega((B_a - B_b)^2)]^{1/2} \qquad (t \in \mathbb{R}).$$

Since $f_a - f_b$ is bounded and continuous on Ω, and analytic on its interior, it now follows from (a variant of) the maximum modulus principle that

$$|f_a(z) - f_b(z)| \leq \|A\| [\omega((B_a - B_b)^2)]^{1/2} \qquad (z \in \Omega).$$

From the strong-operator continuity of ω, and of operator multiplication (on bounded sets), the right-hand side of the last inequality is small, when a and b are sufficiently large in $\mathbb{A}$. Thus $\{f_a\}$ converges, uniformly on Ω.

The limit function f is bounded and continuous on Ω, and analytic on its interior; and we obtain

$$f(t) = \omega(\alpha_t(A)B), \qquad f(t+i) = \omega(B\alpha_t(A)) \qquad (t \in \mathbb{R})$$

by taking limits (over a) in the corresponding relations for f_a. By linearity, such a function f exists for all A in $\mathfrak{A}$ and B (not necessarily self-adjoint) in $\mathscr{R}$.

When $A \in \mathfrak{A}$ and $B = I$, the corresponding function f satisfies $f(t) = f(t+i) = \omega(\alpha_t(A))$ $(t \in \mathbb{R})$. It follows, just as in the proof of Proposition 9.2.14, that f is constant on Ω. Since $f(t) = f(0)$, $\omega(\alpha_t(A)) = \omega(A)$; from the ultraweak continuity of ω and α_t (see Remark 7.4.4), the last equation is satisfied for all A in $\mathscr{R}$ $(= \mathfrak{A}^-)$, so ω is invariant under $\{\alpha_t\}$.

Suppose next that $A, B \in \mathscr{R}$ and $A = A^*$. Let $\{A_a : a \in \mathbb{A}\}$ be a bounded net of self-adjoint elements of $\mathfrak{A}$ that is strong-operator convergent to A. From the preceding argument, we may choose complex-valued functions f_a $(a \in \mathbb{A})$, bounded and continuous on Ω, analytic on the interior of Ω, with boundary values $f_a(t) = \omega(\alpha_t(A_a)B)$, $f_a(t+i) = \omega(B\alpha_t(A_a))$. Just as before, we can show that the net $\{f_a\}$ converges uniformly on Ω, that the limit function f is bounded and continuous on Ω and analytic on its interior, and that $f(t) = \omega(\alpha_t(A)B)$, $f(t+i) = \omega(B\alpha_t(A))$ for all real t. Moreover, by linearity, such a function f exists for all A and B in $\mathscr{R}$ (with A no longer assumed self-adjoint); and $\{\alpha_t\}$ satisfies the modular condition, relative to ω. At one point only, a slight addition is needed to the previous approximation argument; in estimating the boundary values of $f_a - f_b$, we use the invariance of ω under $\{\alpha_t\}$ to conclude that $\omega(\alpha_t((A_a - A_b)^2)) = \omega((A_a - A_b)^2)$. ∎

Our next objective is to give a criterion for semi-finiteness of a von Neumann algebra $\mathscr{R}$, in terms of its modular automorphism group relative to a faithful normal state. This is achieved in Theorem 9.2.21 after some preparatory lemmas. By assuming the existence of a faithful normal state, attention is confined to countably decomposable algebras; but this restriction is removed when the theory is extended so as to apply to weights.

9.2.18. LEMMA. *Suppose that* ω *is a faithful normal state and* ρ *is a faithful normal tracial state on a von Neumann algebra* $\mathscr{R}$. *Then there is a positive element* K *in the unit ball of* $\mathscr{R}$ *such that*

$$\rho((I - K)A) = \omega(KA) = \omega(AK) \qquad (A \in \mathscr{R}).$$

Proof. From Theorem 7.3.13 there is a positive element K in the unit ball of $\mathscr{R}$, such that

$$2\rho(A) = (\omega + \rho)(KA + AK) \qquad (A \in \mathscr{R}).$$

Since $\rho(AK) = \rho(KA)$, we can rewrite the last equation in the form

$$2\rho((I - K)A) = \omega(KA + AK) \qquad (A \in \mathscr{R}). \tag{13}$$

Upon replacing A, first by KA and then by AK, we obtain

$$2\rho(K(I - K)A) = \omega(K^2A + KAK),$$

$$2\rho(K(I - K)A) = 2\rho((I - K)AK) = \omega(KAK + AK^2).$$

Accordingly, $\omega(K^2A) = \omega(AK^2)$ for each A in $\mathscr{R}$. From this, $\omega(KA) = \omega(AK)$ $(A \in \mathscr{R})$; for the centralizer of ω is a C^*-subalgebra of $\mathscr{R}$ that contains K^2 and therefore contains its positive square root K. It now follows from (13) that

$$\rho((I - K)A) = \omega(KA) = \omega(AK) \qquad (A \in \mathscr{R}). \quad \blacksquare$$

We recall, from Section 8.5, some notation and concepts that will be used in the next lemma. With $\mathscr{R}$ a von Neumann algebra and ρ a tracial weight on $\mathscr{R}^+$, the linear span of the set $F_\rho = \{A \in \mathscr{R}^+ : \rho(A) < \infty\}$ is a two-sided ideal $\mathscr{M}_\rho$ in $\mathscr{R}$. The restriction $\rho | F_\rho$ extends uniquely to a positive linear functional (also denoted by ρ) on $\mathscr{M}_\rho$, and $\rho(AB) = \rho(BA)$ when $A \in \mathscr{M}_\rho$ and $B \in \mathscr{R}$. If ρ is normal and $A \in \mathscr{M}_\rho$, the mapping $B \to \rho(AB)$ is an ultraweakly continuous linear functional on $\mathscr{R}$ (see Remark 8.5.9).

9.2.19. LEMMA. *If* ω *is a faithful normal state of a semi-finite von Neumann algebra* $\mathscr{R}$, *there is a normal semi-finite faithful tracial weight* ρ *on* $\mathscr{R}$, *and a positive element* K *in the unit ball of* $\mathscr{R}$, *such that* $I - K \in F_\rho$ *and*

$$\rho((I - K)A) = \omega(KA) = \omega(AK) \qquad (A \in \mathscr{R}).$$

Moreover, both K *and* $I - K$ *are one-to-one mappings.*

Proof. It suffices to consider separately the two cases in which $\mathscr{R}$ (with a faithful normal state ω) is *either* finite *or* properly infinite but semi-finite; for the general semi-finite $\mathscr{R}$ is a direct sum of two algebras, one of each of these kinds.

If $\mathscr{R}$ is finite, it has a center-valued trace τ, and $\omega \circ \tau$ is a faithful normal tracial state ρ on $\mathscr{R}$. From Lemma 9.2.18 there is a positive element K in the unit ball of $\mathscr{R}$, such that

$$\rho((I - K)A) = \omega(KA) = \omega(AK) \qquad (A \in \mathscr{R});$$

and $I - K \in \mathscr{R}^+ = F_\rho$. With P and Q the null projections of K and $I - K$, respectively, $KP = (I - K)Q = 0$, whence

$$\rho(P) = \rho((I - K)P) = \omega(KP) = 0,$$

$$\omega(Q) = \omega(KQ) = \rho((I - K)Q) = 0.$$

Since ρ and ω are faithful, $P = Q = 0$; so K and $I - K$ are both one-to-one.

If $\mathscr{R}$ is properly infinite (but semi-finite), let E be a finite projection in $\mathscr{R}$ that has central carrier I. Since $\mathscr{R}$ has a faithful normal state, it is countably decomposable. From Proposition 6.3.12 there is an orthogonal family $\{E_b\}$ of projections in $\mathscr{R}$, each equivalent to E, with sum I. The finite von Neumann algebra $E\mathscr{R}E$ has a faithful normal state $t\omega | E\mathscr{R}E$, and so has a faithful normal tracial state ρ_0, by the preceding paragraph. For each b, let V_b be a partial isometry in $\mathscr{R}$, from E_b to E. With $\rho\colon \mathscr{R}^+ \to [0, \infty]$ defined by

$$\rho(A) = \sum \rho_0(V_b A V_b^*) \qquad (A \in \mathscr{R}^+),$$

it is a matter of routine manipulation (as in the proof of Proposition 8.5.5) to verify that ρ is a semi-finite normal tracial weight. We assert that ρ is faithful. For this, suppose that $A \in \mathscr{R}^+$ and $\rho(A) = 0$. Then $\rho_0(V_b A V_b^*) = 0$ for each b, and therefore $V_b A V_b^* = 0$ since ρ_0 is faithful. Accordingly,

$$(A^{1/2}E_b)^*(A^{1/2}E_b) = E_b A E_b = V_b^* V_b A V_b^* V_b = 0,$$

and $A^{1/2}E_b = 0$ for all b. Hence $A = \sum A^{1/2}A^{1/2}E_b = 0$; and ρ is faithful.

The finite subsums of $\sum E_b$ can be viewed as an increasing net $\{F_a : a \in \mathbb{A}\}$ of projections in $\mathscr{R}$, strong-operator convergent to I, such that $\rho(F_a) < \infty$ for each a in $\mathbb{A}$. The restriction $r\rho | F_a \mathscr{R} F_a$ is a faithful normal tracial state, and $s\omega | F_a \mathscr{R} F_a$ is a faithful normal state, of the von Neumann algebra $F_a \mathscr{R} F_a$. From Lemma 9.2.18 there is a positive element K_a in the unit ball of $F_a \mathscr{R} F_a$, such that

$$(14) \qquad \rho((F_a - K_a)F_a A F_a) = \omega(K_a F_a A F_a) = \omega(F_a A F_a K_a) \qquad (A \in \mathscr{R}).$$

Since the positive part $(\mathscr{R}^+)_1$ of the unit ball of $\mathscr{R}$ is weak-operator compact, we may assume (upon replacing $\{K_a : a \in \mathbb{A}\}$ by a cofinal subnet, and $\{F_a : a \in \mathbb{A}\}$ by the corresponding subnet) that $\{K_a\}$ is weak-operator convergent to an element K of $(\mathscr{R}^+)_1$.

Suppose that $a, b \in \mathbb{A}$ and $a \leq b$. Since $F_a \leq F_b \in \mathcal{M}_\rho$ and $F_b K_b = K_b = K_b F_b$, it follows from (14) that

$$\begin{aligned}\rho((I - K_b)AF_a) &= \rho((I - K_b)AF_aF_b) = \rho(F_b(I - K_b)AF_aF_b)\\ &= \rho((F_b - K_b)F_bAF_aF_b) = \omega(K_bF_bAF_aF_b)\\ &= \omega(K_bAF_a)\end{aligned}$$

for all A in $\mathcal{R}$. Since both ω and the linear functional $R \to \rho(RF_a)$ on $\mathcal{R}$ are ultraweakly continuous, we can take limits over b to obtain

$$(15) \qquad \rho((I - K)AF_a) = \omega(KAF_a) \qquad (A \in \mathcal{R}).$$

Since ρ is a tracial weight and $F_a \in \mathcal{M}_\rho$, it follows from (15), with $A = I$, that

$$\rho((I - K)^{1/2}F_a(I - K)^{1/2}) = \rho((I - K)F_a) = \omega(KF_a).$$

The net $\{(I - K)^{1/2}F_a(I - K)^{1/2}\}$ increases to $I - K$, so the normality of ρ and ω now entails (from a discussion preceding Proposition 8.5.1)

$$\rho(I - K) = \omega(K) < \infty.$$

Hence $I - K \in F_\rho \subseteq \mathcal{M}_\rho$, and the linear functional $R \to \rho((I - K)R)$ on $\mathcal{R}$ is ultraweakly continuous. By taking limits over a, in (15), we now obtain

$$\rho((I - K)A) = \omega(KA) \qquad (A \in \mathcal{R}).$$

From this,

$$\rho(A^*(I - K)) = \rho((I - K)A^*) = \omega(KA^*),$$

whence (by taking complex conjugates of both sides)

$$\rho((I - K)A) = \omega(AK) \qquad (A \in \mathcal{R}).$$

The argument, used above (in the case where $\mathcal{R}$ is finite) to show that K and $I - K$ are one-to-one, applies also in the present situation. ■

9.2.20. Lemma. *Suppose that ω is a faithful normal state of a semi-finite von Neumann algebra $\mathcal{R}$, $\{\sigma_t\}$ is the corresponding modular automorphism group, ρ and K satisfy the conclusions of Lemma 9.2.19, and H is the (in general, unbounded) invertible positive operator $K^{-1}(I - K)$. Then $H \,\eta\, \mathcal{R}$, and*

$$\sigma_t(A) = H^{it}AH^{-it} \qquad (A \in \mathcal{R}, \quad t \in \mathbb{R}).$$

Proof. Since $0 \le K \le I$, while K is one-to-one and $I - K$ is bounded and one-to-one, it follows that K^{-1} is positive and invertible, $K^{-1}(I - K) = K^{-1}\hat{\cdot}(I - K)\,\eta\,\mathscr{R}$ and $K^{-1}(I - K)$ is positive and invertible. Thus H^{it} is a unitary operator in $\mathscr{R}$, for all real t, and the equation

$$\alpha_t(A) = H^{it}AH^{-it} \qquad (A \in \mathscr{R})$$

defines a one-parameter group $\{\alpha_t\}$ of * automorphisms of $\mathscr{R}$. We have to show that $\alpha_t = \sigma_t$; and for this, it suffices to verify that $\{\alpha_t\}$ satisfies the weakened form of the modular condition required in Lemma 9.2.17.

For each positive integer n (≥ 3), let E_n be the spectral projection for K, corresponding to the interval $[n^{-1}, 1 - n^{-1}]$. Then $\{E_n\}$ is an increasing sequence of projections in $\mathscr{R}$; and $\lim E_n = I$, since $0 \le K \le I$ and K, $I - K$ are one-to-one (whence the spectral projections for K corresponding to the one-point sets $\{0\}, \{1\}$ are 0). Thus $\bigcup E_n \mathscr{R} E_n$ is a strong-operator dense * subalgebra $\mathfrak{A}$ of $\mathscr{R}$.

If $\mathscr{H}$ denotes the Hilbert space on which $\mathscr{R}$ acts, the restriction $H_n = H \mid E_n(\mathscr{H})$ is a bounded positive operator with a bounded inverse. Accordingly, $(H_n)^{iz}$ is defined, and has a norm-convergent power series expansion

$$(H_n)^{iz} = \exp(iz \log H_n) = \sum_{m=0}^{\infty} \frac{i^m z^m (\log H_n)^m}{m!}$$

for all complex z. Moreover, $\|(H_n)^{iz}\|$ is bounded on any subset of the complex plane on which $\operatorname{Im} z$ is bounded; and $(H_n)^{iz} = H^{iz}E_n \mid E_n(\mathscr{H})$ (Corollary 5.6.31).

Given A and B in $\mathfrak{A}$, we can choose n (≥ 3) so that $A \in E_n \mathscr{R} E_n$. From the preceding paragraph, the equation

$$f(z) = \omega(H^{iz}E_n AH^{-iz}E_n B)$$

defines an entire function f that is bounded on each strip of finite width parallel to the real axis. Since $K \in \mathscr{R}$, and by virtue of our choice of E_n, we can find elements C_n and D_n of $E_n \mathscr{R} E_n$ such that

$$E_n = (I - K)C_n = D_n K = KD_n.$$

Note that, since the restrictions $K \mid E_n(\mathscr{H})$ and $(I - K) \mid E_n(\mathscr{H})$ are bounded operators with bounded inverses, while $H = K^{-1}(I - K)$, it follows that

$H^{-1}(I-K)C_n = KC_n$ and $KHE_n = (I-K)E_n$. For all real t,

$$\begin{aligned} f(t) &= \omega(H^{it}E_nAH^{-it}E_nB) \\ &= \omega(H^{it}E_nAE_nH^{-it}B) \\ &= \omega(H^{it}AH^{-it}B) = \omega(\alpha_t(A)B) \end{aligned}$$

and

$$\begin{aligned} f(t+i) &= \omega(H^{it}H^{-1}E_nAH^{-it}HE_nB) \\ &= \omega(H^{it}H^{-1}(I-K)C_nAE_nHE_nH^{-it}B) \\ &= \omega(H^{it}KC_nAD_nKHE_nH^{-it}B) \\ &= \omega(KH^{it}C_nAD_n(I-K)E_nH^{-it}B) \\ &= \rho((I-K)H^{it}C_nAD_nE_nH^{-it}(I-K)B) \\ &= \rho((I-K)B(I-K)H^{it}C_nAD_nH^{-it}) \\ &= \omega(B(I-K)H^{it}C_nAD_nH^{-it}K) \\ &= \omega(BH^{it}(I-K)C_nAD_nKH^{-it}) \\ &= \omega(BH^{it}E_nAE_nH^{-it}) \\ &= \omega(BH^{it}AH^{-it}) = \omega(B\alpha_t(A)). \end{aligned}$$

From Lemma 9.2.17, $\{\alpha_t\}$ coincides with the modular automorphism group $\{\sigma_t\}$ of $\mathscr{R}$, corresponding to ω. ∎

9.2.21. THEOREM. *Suppose that ω is a faithful normal state of a von Neumann algebra $\mathscr{R}$, and $\{\sigma_t\}$ is the corresponding modular automorphism group. Then $\mathscr{R}$ is semi-finite if and only if there is a positive (in general, unbounded) invertible self-adjoint operator H, affiliated with $\mathscr{R}$, such that*

$$\sigma_t(A) = H^{it}AH^{-it} \qquad (A \in \mathscr{R}, \quad t \in \mathbb{R}).$$

Proof. Suppose first that there is an operator H with the stated properties. For each positive integer n, let E_n be the spectral projection for H, corresponding to the interval $[n^{-1}, n]$. Then $E_n \in \mathscr{R}$, and $\sigma_t(E_n) = E_n$ for all real t, since E_n commutes with H^{it}. Moreover, E_n increases with n; and $\lim E_n = I$, since H is positive and invertible. The restriction H_n of H to the range of E_n is a bounded positive operator and has a bounded inverse; so $(H_n)^z$ has a power series expansion $\sum z^m(\log H_n)^m/m!$, valid for all complex z.

Suppose that $A, B \in E_n\mathscr{R}E_n$. Since $\{\sigma_t\}$ satisfies the modular condition relative to ω, there is a complex-valued function f, bounded and continuous on the strip $\Omega = \{z \in \mathbb{C} : 0 \le \operatorname{Im} z \le 1\}$, analytic on the interior of Ω, and satisfying

$$f(t) = \omega(\sigma_t(A)B), \qquad f(t+i) = \omega(B\sigma_t(A)) \qquad (t \in \mathbb{R}).$$

With ω_n the restriction $\omega \mid E_n \mathscr{R} E_n$, it follows from the preceding paragraph that the equation

$$F(z) = \omega_n(H_n^{iz} A H_n^{-iz} B) = \omega(H^{iz} A H^{-iz} B)$$

defines an entire function F. Moreover, $F - f$ is continuous on Ω, analytic on the interior of Ω, and vanishes on $\mathbb{R}$. By the Schwarz reflection principle, $F - f$ has an analytic continuation across $\mathbb{R}$; and since the extended function vanishes on $\mathbb{R}$, it is zero throughout its domain of definition. In particular

$$\begin{aligned} \omega_n(H_n^{-1} A H_n B) &= F(i) \\ &= f(i) = \omega(BA) = \omega_n(BA). \end{aligned}$$

Upon writing AH_n^{-1} in place of A, we obtain

$$\omega_n(H_n^{-1} AB) = \omega_n(BAH_n^{-1}) \qquad (A, B \in E_n \mathscr{R} E_n). \tag{16}$$

With B replaced by E_n, this last equation yields $\omega_n(H_n^{-1}A) = \omega_n(AH_n^{-1})$ for each A in $E_n \mathscr{R} E_n$. Accordingly, the centralizer of ω_n is a C^*-subalgebra of $E_n \mathscr{R} E_n$ that contains H_n^{-1}, and therefore contains $H_n^{-1/2}$. From this, (16) can be rewritten in the form

$$\omega_n(H_n^{-1/2} ABH_n^{-1/2}) = \omega_n(H_n^{-1/2} BAH_n^{-1/2}) \qquad (A, B \in E_n \mathscr{R} E_n).$$

Accordingly, $E_n \mathscr{R} E_n$ has a faithful tracial state, $A \to \omega_n(H_n^{-1/2} AH_n^{-1/2})$, and is therefore a finite von Neumann algebra; so E_n is a finite projection in $\mathscr{R}$.

With Q a non-zero central projection in $\mathscr{R}$, QE_n is a finite projection in $\mathscr{R}Q$, and is non-zero for large n since $Q = \lim QE_n$. This shows that the von Neumann algebra $\mathscr{R}Q$ is not type III. Since $\mathscr{R}$ has no central portion of type III, it is semi-finite.

The preceding argument shows that the existence of an operator H, with the properties set out in the statement of the theorem, implies that $\mathscr{R}$ is semi-finite. The converse implication follows at once from Lemma 9.2.20. ■

A * automorphism α of a C^*-algebra $\mathfrak{A}$ is said to be *inner* if there is a unitary element U of $\mathfrak{A}$ such that $\alpha(A) = UAU^*$ for each A in $\mathfrak{A}$; it is described as *outer* if there is no such U.

Suppose that ω is a faithful normal state of a von Neumann algebra $\mathscr{R}$, and $\{\sigma_t\}$ is the corresponding modular automorphism group. If $\mathscr{R}$ is semi-finite, it results from Theorem 9.2.21 that each σ_t is inner (being implemented by the unitary operator H^{it} in $\mathscr{R}$). Conversely, it can be shown [95] that, if each σ_t is inner, and if the predual of $\mathscr{R}$ is a separable Banach space, then $\mathscr{R}$ is semi-finite. This converse result, however, is not by any means an immediate consequence of Theorem 9.2.21. Since σ_t is inner, there is a unitary operator U_t in $\mathscr{R}$ that implements it. It is a problem of topological group cohomology to show that it is possible to choose the operators U_t in such a way that they form a strong-operator continuous unitary group in $\mathscr{R}$ (whence $U_t = H^{it}$ and $H \,\eta\, \mathscr{R}$). Compare Exercise 14.4.16(i).

Tomita's theorem—a second approach. We present another method of proving Tomita's theorem (9.2.9) based on the following observation (which we describe in loose, heuristic terms). With the notation introduced to this point and A, B, C in $\mathscr{R}$,

$$SASBCu = BCA^*u = BSASCu.$$

In some sense, then, SAS commutes with $\mathscr{R}$ and $S\mathscr{R}S$ $(= S\mathscr{R}S^{-1})$ is in $\mathscr{R}'$. Roughly, S implements a "similarity" of the self-adjoint algebra $\mathscr{R}$ onto the self-adjoint algebra $\mathscr{R}'$. Replacing the operator algebras $\mathscr{R}$ and $\mathscr{R}'$ by (bounded) self-adjoint operators A and B, we can draw special conclusions. Suppose that $TAT^{-1} = B$, where T is a bounded operator with bounded inverse. Let UH be the polar decomposition of T. Then

$$HAH^{-1} = U^{-1}BU = (U^{-1}BU)^* = H^{-1}AH;$$

so that $H^2A = AH^2$. Since H is a (norm) limit of polynomials in H^2, $HA = AH$ and $UAU^{-1} = B$.

Replacing A and B by C^*-algebras $\mathfrak{A}$ and $\mathscr{B}$, the same argument yields that $H^2\mathfrak{A}H^{-2} = \mathfrak{A}$. A more sophisticated argument [27] tells us that, then, $H\mathfrak{A}H^{-1} = \mathfrak{A}$ so that $U\mathfrak{A}U^{-1} = \mathscr{B}$ (as well as $H^{it}\mathfrak{A}H^{-it} = \mathfrak{A}$). The resemblance of these conclusions to those of Tomita's theorem (9.2.9), viz., $J\mathscr{R}J^{-1} = \mathscr{R}'$ and $\Delta^{it}\mathscr{R}\Delta^{-it} = \mathscr{R}$, under the circumstances that (loosely!) $S\mathscr{R}S^{-1} = \mathscr{R}'$ and $S = J\Delta^{1/2}$, needs no further comment. The difficulties with this parallel lie in the looseness of the identity, $S\mathscr{R}S^{-1} = \mathscr{R}'$, the fact that (in general) S is unbounded, and, to a lesser extent, in the fact that S is conjugate-linear, rather than linear. Surmounting these difficulties may be viewed as the task required to prove Tomita's theorem. It is this task that we undertake in the discussion that follows. We begin with that part of the "unbounded similarity theory" needed for the proof of Tomita's theorem.

In dealing with operators of the form TAT^{-1} appearing in our unbounded similarities, the following lemma will be of great technical use. In most of its applications H will be a positive, self-adjoint, invertible operator. With $E([m^{-1}, m])$ the spectral projection for H corresponding to the interval $[m^{-1}, m]$, $\{E([m^{-1}, m])\}$ is a monotone increasing sequence with least upper bound I (for $Hx = 0$ only if $x = 0$). Thus $\{E([m^{-1}, m])\}$ is a bounding sequence and $\bigcup_{m=1}^{\infty} E([m^{-1}, m])(\mathscr{H})$ is a core (we denote by $\mathscr{D}_0(H)$) for H and, indeed, for each $f(H)$, where f is a Borel function whose domain contains $\mathrm{sp}(H)$ and which is bounded on finite closed subintervals of the positive real axis. In particular $\mathscr{D}_0(H)$ is a core for H^z for all complex z. (See Lemma 5.6.14 and the discussion preceding Lemma 9.2.1, in connection with these observations.) Under the conditions described in the lemma that follows, we can pass from the core given there to the core $\mathscr{D}_0(H)$.

9.2.22. LEMMA. *If H and K are closed, densely defined, linear operators on the complex Hilbert space $\mathscr{H}$, $\mathscr{D}_0$ is a core for H, A is a bounded, everywhere-defined operator, and KAH is defined and bounded on $\mathscr{D}_0$, then KAH has domain $\mathscr{D}(H)$ and KAH is a bounded extension of $KAH \mid \mathscr{D}_0$. In addition $(KAH)^*$ is in $\mathscr{B}(\mathscr{H})$ and $(KAH)^* \mid \mathscr{D}(K^*) = H^*A^*K^*$.*

Proof. Suppose h_0 is a unit vector in $\mathscr{D}(H)$. Since $\mathscr{D}_0$ is a core for H, there is a sequence $\{h_n\}$ of unit vectors in $\mathscr{D}_0$ such that $h_n \to h_0$ and $Hh_n \to Hh_0$. As $A \in \mathscr{B}(\mathscr{H})$, $AHh_n \to AHh_0$. Now $KAH \mid \mathscr{D}_0$ is bounded, by hypothesis, so that $KAHh_n$ converges to some vector in $\mathscr{H}$. Since K is closed, $AHh_0 \in \mathscr{D}(K)$ and $KAHh_n \to KAHh_0$. Thus $\mathscr{D}(KAH) = \mathscr{D}(H)$ and $\|KAHh_0\| \le \|KAH \mid \mathscr{D}_0\|$. Hence KAH is a bounded extension of $KAH \mid \mathscr{D}_0$ to $\mathscr{D}(H)$ and $\|KAH\| = \|KAH \mid \mathscr{D}_0\|$ $(= b)$.

With x in $\mathscr{D}(H)$ and y in $\mathscr{H}$, $|\langle KAHx, y\rangle| \le b\|x\| \cdot \|y\|$; so that $y \in \mathscr{D}((KAH)^*)$, and $\langle x, (KAH)^*y\rangle = \langle KAHx, y\rangle$. Thus $\mathscr{D}((KAH)^*) = \mathscr{H}$ and $\|(KAH)^*y\| \le b\|y\|$; so that $(KAH)^*$ is bounded. If we restrict y to $\mathscr{D}(K^*)$, then $\langle KAHx, y\rangle = \langle Hx, A^*K^*y\rangle$. Hence $A^*K^*y \in \mathscr{D}(H^*)$ and $\langle KAHx, y\rangle = \langle x, H^*A^*K^*y\rangle$; so that $(KAH)^*y = H^*A^*K^*y$. ■

9.2.23. LEMMA. *If H is a positive, self-adjoint, invertible operator on $\mathscr{H}$, $\mathscr{D}_0$ is a core for H^{-1}, and A is a bounded, everywhere-defined operator on $\mathscr{H}$ such that $\mathscr{D}_0 \subseteq \mathscr{D}(H^jAH^{-j}) \cap \mathscr{D}(H^jA^*H^{-j})$ and $H^jAH^{-j} \mid \mathscr{D}_0$, $H^jA^*H^{-j} \mid \mathscr{D}_0$ are bounded for each integer j in $[0, n]$, where n is a positive integer, then $\mathscr{D}(H^zAH^{-z}) = \mathscr{D}(H^{-z})$ and H^zAH^{-z} is bounded with (unique) bounded extension $\varphi_z(A)$ to $\mathscr{H}$ for each complex z in the closure of the strip $\{z : -n < \operatorname{Re} z < n\}$ $(= \mathscr{S})$. If x_0 and y_0 are unit vectors in $\mathscr{H}$, $z \to \langle \varphi_z(A)x_0, y_0\rangle$ is analytic on $\mathscr{S}$, continuous on its closure $\mathscr{S}^-$, and bounded on $\mathscr{S}^-$ by $\max\{\|H^nAH^{-n}\|, \|H^{-n}AH^n\|\}$ $(= b)$.*

Proof. Suppose we have proved that H^jAH^{-j} is bounded and $\mathscr{D}(H^jAH^{-j}) = \mathscr{D}(H^{-j})$ for some integer j in $[0, n)$. If x is a unit vector in $\mathscr{D}(H^{-j-1})$ $(\subseteq \mathscr{D}(H^{-1}))$, we can choose $\{x_n\}$, a sequence of unit vectors in $\mathscr{D}_0$, such that $x_n \to x$ and $H^{-1}x_n \to H^{-1}x$. By assumption, $H^{-1}x \in \mathscr{D}(H^jAH^{-j})$ $(= \mathscr{D}(H^{-j}))$, H^jAH^{-j} is bounded and $\mathscr{D}_0 \subseteq \mathscr{D}(H^{j+1}AH^{-j-1})$. Thus $H^jAH^{-j}(H^{-1}x_n) \to H^jAH^{-j}(H^{-1}x)$. Now $H^{j+1}AH^{-j-1} \mid \mathscr{D}_0$ is bounded, by hypothesis, so that $\{H^{j+1}AH^{-j-1}x_n\}$ is a Cauchy sequence and converges to some vector in $\mathscr{H}$. Since H is closed, $H^{j+1}AH^{-j-1}x_n = H(H^jAH^{-j}(H^{-1}x_n)) \to H(H^jAH^{-j}(H^{-1}x)) = H^{j+1}AH^{-j-1}x$. It follows that

$\|H^{j+1}AH^{-j-1}x\| \le \|H^{j+1}AH^{-j-1}|\mathscr{D}_0\|$. Since A $(= H^0AH^0)$ is bounded and $\mathscr{D}(A) = \mathscr{D}(H^0) = \mathscr{D}(I) = \mathscr{H}$, H^jAH^{-j} (and $H^jA^*H^{-j}$) are bounded with domain $\mathscr{D}(H^{-j})$ for all integers j in $[0, n]$. From Lemma 9.2.22 (with $\mathscr{D}(H^{-j})$ in place of $\mathscr{D}_0$), $(H^jA^*H^{-j})^*$ is bounded and everywhere defined and $(H^jA^*H^{-j})^*|\mathscr{D}(H^j) = H^{-j}AH^j$ for $j = 0, 1, \ldots, n$. In particular $H^jAH^{-j}|\mathscr{D}_0(H)$ is bounded for each integer j in $[-n, n]$ (and $\mathscr{D}_0(H) \subseteq \mathscr{D}(H^jAH^{-j})$), since $\mathscr{D}(H^jAH^{-j}) = \mathscr{D}(H^{-j})$.

Let E_m be the spectral projection for H corresponding to $[m^{-1}, m]$, where m is a positive integer. If $\mathscr{H}_m = E_m(\mathscr{H})$, then $\mathscr{D}_0(H) = \bigcup_{m=1}^{\infty} \mathscr{H}_m$. The operator $H|\mathscr{H}_m$ is a bounded positive operator H_m with bounded inverse. Thus H_m^z is bounded with bounded inverse H_m^{-z}, for each complex z. In particular, H_m^z maps $\mathscr{H}_m$ onto itself. Hence, with x in $\mathscr{H}_m$, there is a y in $\mathscr{H}_m$ such that $H^{-z}x = H^{-n}y$. Since $\mathscr{D}_0(H) \subseteq \mathscr{D}(H^nAH^{-n})$, $AH^{-z}x = AH^{-n}y \in \mathscr{D}(H^n)$. If $0 \le t \le n$, where $z = t + is$, then $\mathscr{D}(H^n) \subseteq \mathscr{D}(H^t) = \mathscr{D}(H^z)$ so that $\mathscr{D}_0(H) \subseteq \mathscr{D}(H^zAH^{-z})$. (See the discussion preceding Lemma 9.2.1.) If $-n \le t \le 0$, then

$$\mathscr{D}(H^{-n}) = \mathscr{D}((H^{-1})^n) \subseteq \mathscr{D}((H^{-1})^{-t}) = \mathscr{D}(H^t) = \mathscr{D}(H^z).$$

Now $AH^{-z}x = AH^ny' \in \mathscr{D}(H^{-n})$ for some y' in $\mathscr{H}_m$, since $\mathscr{D}_0(H) \subseteq \mathscr{D}(H^{-n}AH^n)$. Thus $\mathscr{D}_0(H) \subseteq \mathscr{D}(H^zAH^{-z})$ for all z in $\mathscr{S}^-$.

If $z \in \mathscr{S}^-$ and x, y are unit vectors in $\mathscr{H}_m$, then

$$\langle H^zAH^{-z}x, y\rangle = \langle E_mH^zAH^{-z}E_mx, y\rangle = \langle H_m^zE_mAH_m^{-z}x, y\rangle;$$

and $z \to \langle H_m^zE_mAH_m^{-z}x, y\rangle$ is entire. (Compare the first paragraph of the proof of Theorem 9.2.21.) In addition, when $z = t + is \in \mathscr{S}^-$,

$$|\langle H^zAH^{-z}x, y\rangle| \le \|H_m^zE_mAE_mH_m^{-z}\| \le m^{2|t|}\|A\| \le m^{2n}\|A\|.$$

For all z in $\mathscr{S}^-$ and each pair of unit vectors x, y in $\mathscr{H}_m$, $|\langle H^zAH^{-z}x, y\rangle| \le b$ from (a variant of) the maximum modulus principle, since $|\langle H^{n+is}AH^{-n-is}x, y\rangle| \le \|H^nAH^{-n}\|$ and $|\langle H^{-n+is}AH^{n-is}x, y\rangle| \le \|H^{-n}AH^n\|$. Now $\mathscr{H}_m \subseteq \mathscr{H}_{m+1}$ and $\mathscr{D}_0(H)$ is dense in $\mathscr{H}$; so that $\|H^zAH^{-z}x\| \le b$ for each unit vector x in $\mathscr{D}_0(H)$. As $\mathscr{D}_0(H)$ is a core for H^{-z}, we may apply Lemma 9.2.22 to conclude that $\mathscr{D}(H^zAH^{-z}) = \mathscr{D}(H^{-z})$ and that H^zAH^{-z} has a (unique) bounded extension $\varphi_z(A)$ to $\mathscr{H}$ satisfying $\|\varphi_z(A)\| \le b$.

Let $\{x_n\}, \{y_n\}$ be sequences of unit vectors in $\mathscr{D}_0(H)$ with limits x_0 and y_0, respectively. Then

$$\begin{aligned}
&|\langle \varphi_z(A)x_0, y_0\rangle - \langle H^zAH^{-z}x_n, y_n\rangle| \\
&\quad \le |\langle \varphi_z(A)x_0, y_0\rangle - \langle \varphi_z(A)x_n, y_0\rangle| + |\langle \varphi_z(A)x_n, y_0\rangle - \langle H^zAH^{-z}x_n, y_n\rangle| \\
&\quad \le b(\|x_0 - x_n\| + \|y_0 - y_n\|) \to 0, \qquad n \to \infty,
\end{aligned}$$

for all z in $\mathscr{S}^-$. Thus $z \to \langle \varphi_z(A)x_0, y_0\rangle$ is analytic on $\mathscr{S}$, continuous on $\mathscr{S}^-$, and bounded on $\mathscr{S}^-$ by b. ■

9.2.24. COROLLARY. *If H is a positive, self-adjoint, invertible operator on $\mathscr{H}$, $\mathscr{D}_0$ is a core for H^{-1}, and A is a bounded, everywhere-defined operator on $\mathscr{H}$ such that $\mathscr{D}_0 \subseteq \mathscr{D}(H^jAH^{-j}) \cap \mathscr{D}(H^jA^*H^{-j})$ and $H^jAH^{-j}|\mathscr{D}_0$, $H^jA^*H^{-j}|\mathscr{D}_0$ are bounded for each positive integer j, then $\mathscr{D}(H^zAH^{-z}) = \mathscr{D}(H^{-z})$ and H^zAH^{-z} is bounded with (unique) bounded extension $\varphi_z(A)$ to $\mathscr{H}$ for each complex z. If x_0 and y_0 are unit vectors in $\mathscr{H}$, $z \to \langle\varphi_z(A)x_0, y_0\rangle$ is an entire function bounded on the strip $\{z: -n \le \operatorname{Re} z \le n\}$ by $\max\{\|H^nAH^{-n}\|, \|H^{-n}AH^n\|\}$.*

9.2.25. LEMMA. *If H is a positive, self-adjoint, invertible operator on $\mathscr{H}$, $\mathscr{D}_0$ is a core for H^{-1}, $\mathfrak{A}_0$ is a self-adjoint algebra of operators on $\mathscr{H}$ such that for each A in $\mathfrak{A}_0$, $\mathscr{D}_0 \subseteq \mathscr{D}(HAH^{-1})$ and $HAH^{-1}|\mathscr{D}_0$ has a (unique) bounded extension $\varphi(A)$ to $\mathscr{H}$ in $\mathfrak{A}_0$ satisfying $\|\varphi^n(A)\| \le k_A^n$ for each positive integer n and some constant k_A (depending on A), then $\mathscr{D}(H^zAH^{-z}) = \mathscr{D}(H^{-z})$, H^zAH^{-z} is bounded for each complex z and each A in $\mathfrak{A}_0$, and its (unique) bounded extension $\varphi_z(A)$ to $\mathscr{H}$ lies in $\mathfrak{A}_0''$.*

Proof. From Lemma 9.2.22, $\mathscr{D}(HAH^{-1}) = \mathscr{D}(H^{-1})$ and HAH^{-1} has a bounded extension $\varphi(A)$ to $\mathscr{H}$ for each A in $\mathfrak{A}_0$. Suppose we have proved that $H^jAH^{-j} \subseteq \varphi^j(A)$ and $\mathscr{D}(H^jAH^{-j}) = \mathscr{D}(H^{-j})$ for some positive integer j. Then $H^{j+1}AH^{-j-1} \subseteq H\varphi^j(A)H^{-1}$. By hypothesis, $\mathscr{D}_0 \subseteq \mathscr{D}(H\varphi^j(A)H^{-1})$ and $H\varphi^j(A)H^{-1}|\mathscr{D}_0$ has a bounded extension $\varphi^{j+1}(A)$ $(= \varphi(\varphi^j(A)))$ to $\mathscr{H}$. Again, from Lemma 9.2.22, $\mathscr{D}(H\varphi^j(H)H^{-1}) = \mathscr{D}(H^{-1})$ and $H\varphi^j(A)H^{-1}$ is bounded. With x in $\mathscr{D}(H^{-j-1})$, $x \in \mathscr{D}(H^{-1}) = \mathscr{D}(H\varphi^j(A)H^{-1})$, $H^{-1}x \in \mathscr{D}(H^{-j}) = \mathscr{D}(H^jAH^{-j})$ and $H^jAH^{-j}(H^{-1}x) = \varphi^j(A)(H^{-1}x) \in \mathscr{D}(H)$. Thus $\varphi^{j+1}(A)x = H\varphi^j(A)H^{-1}x = H^{j+1}AH^{-j-1}x$, $\mathscr{D}(H^{j+1}AH^{-j-1}) = \mathscr{D}(H^{-j-1})$, and $H^{j+1}AH^{-j-1} \subseteq \varphi^{j+1}(A)$. By induction, $\mathscr{D}(H^nAH^{-n}) = \mathscr{D}(H^{-n})$ and $H^nAH^{-n} \subseteq \varphi^n(A)$ for each positive integer n. If we replace $\mathscr{D}_0$ by $\mathscr{D}_0(H)$ in the statement of Corollary 9.2.24, we conclude that $\mathscr{D}(H^zAH^{-z}) = \mathscr{D}(H^{-z})$ and that H^zAH^{-z} is bounded for all complex z. Moreover, for each pair of unit vectors x, y in $\mathscr{H}$, $z \to \langle\varphi_z(A)x, y\rangle$ is entire and

$$|\langle\varphi_z(A)x, y\rangle| \le \max\{\|H^nAH^{-n}\|, \|H^{-n}AH^n\|\}$$

when $|\operatorname{Re} z| \le n$. Now $\|H^nAH^{-n}\| = \|\varphi^n(A)\| \le k_A^n$ and $\|H^{-n}AH^n\| = \|(H^{-n}AH^n)^*\| = \|H^nA^*H^{-n}\| \le k_{A^*}^n$ for each positive integer n, from Lemma 9.2.22; so that $|\langle\varphi_z(A)x, y\rangle| \le C_A^n$ when $|\operatorname{Re} z| \le n$, where $C_A = \max\{k_A, k_{A^*}, 1\}$.

If $\mathfrak{A}_0'$ contains no projections other than 0 and I, then $\varphi_z(A) \in \mathscr{B}(\mathscr{H}) = \mathfrak{A}_0''$. Suppose E' is a projection in $\mathfrak{A}_0'$ distinct from 0 and I. If we show that $(I - E')\varphi_z(A)E' = 0$, then $(I - E')\varphi_z(A)E' = 0 = E'\varphi_z(A)(I - E')$, and $E'\varphi_z(A) = \varphi_z(A)E'$. From this, $\varphi_z(A) \in \mathfrak{A}_0''$. It will suffice to show that $f(z) = 0$ for z in $\mathbb{C}_r$ $(= \{z : 0 < \operatorname{Re} z\})$, where $f(z) = C_A^{-(z+1)}\langle \varphi_z(A)x_0, y_0\rangle$, $x_0 \in E'(\mathscr{H})$, $y_0 \in (I - E')(\mathscr{H})$, $\|x_0\| = \|y_0\| = 1$. We prove this by showing that f has a zero of infinite order at 1. If $z = t + is \in \mathbb{C}_r$, there is an integer n such that $0 < t < n \le t + 1$. In this case,

$$|f(z)| = C_A^{-(t+1)}|\langle \varphi_z(A)x_0, y_0\rangle| \le C_A^{n-t-1} \le 1.$$

Moreover, $\langle \varphi_n(A)x_0, y_0\rangle = \langle \varphi^n(A)E'x_0, (I - E')y_0\rangle = 0$ for each positive integer n, since $\varphi^n(A) \in \mathfrak{A}_0$. Thus $f(n) = 0$ for each positive integer n. Hence $f(z) = (z - 1)g(z)$, where g is bounded and analytic on $\mathbb{C}_r$. Suppose we have shown that $f(z) = (z - 1)^k h(z)$ for some positive integer k, where h is bounded and analytic on $\mathbb{C}_r$. We wish to show that $h(1) = 0$. We know that h vanishes at $2, 3, \ldots$. On replacing h by an appropriate scalar multiple, we may assume that $|h(z)| \le 1$ for z in $\mathbb{C}_r$. Let $F_n(z)$ be $(2 - z)(3 - z)\cdots(n - z)/n!$. With ε positive, $1 - \varepsilon \le |F_n(z)|$ for all z sufficiently near the imaginary axis, so that $|h(z)/F_n(z)| \le (1 - \varepsilon)^{-1}$ for z on the perimeter of a half-disk of large radius with center on the positive real axis near 0 and diameter parallel to the imaginary axis. From the maximum modulus principle (and the fact that ε is arbitrary), $|h(z)/F_n(z)| \le 1$ for all z in $\mathbb{C}_r$. In particular, $|h(1)| \le |F_n(1)| = 1/n$. It follows that $h(1) = 0$ and that 1 is a zero of infinite order for f. Hence $\langle \varphi_z(A)x_0, y_0\rangle = 0$ for each z in $\mathbb{C}_r$. Since $z \to \langle \varphi_z(A)x_0, y_0\rangle$ is entire, it vanishes on $\mathbb{C}$. ∎

9.2.26. THEOREM. *Let $\mathfrak{A}_0$ and $\mathscr{B}_0$ be self-adjoint algebras of operators on complex Hilbert spaces $\mathscr{H}$ and $\mathscr{K}$, respectively; and let T be a closed, invertible, linear or conjugate-linear transformation from $\mathscr{H}$ into $\mathscr{K}$. If T^{-1} has a core $\mathscr{D}_1$ such that*

(i) *$\mathscr{D}_1 \subseteq \mathscr{D}(TAT^{-1})$ and $TAT^{-1} | \mathscr{D}_1$ has a (unique) bounded extension in $\mathscr{B}_0$ for each A in $\mathfrak{A}_0$,*

(ii) *each B in $\mathscr{B}_0$ is such an extension,*

(iii) *$\|H^nAH^{-n}\| \le k_A^n$ for each positive integer n and some constant k_A (depending on A in $\mathfrak{A}_0$), where UH is the polar decomposition of T,*

then $U\mathfrak{A}_0''U^{-1} = \mathscr{B}_0''$, $\mathscr{D}(H^zAH^{-z}) = \mathscr{D}(H^{-z})$, and H^zAH^{-z} has a (unique) bounded extension $\varphi_z(A)$ in $\mathfrak{A}_0''$ for each A in $\mathfrak{A}_0$. In particular, $t \to H^{it}$ is a strong-operator continuous, one-parameter unitary group that gives rise to a one-parameter group of automorphisms of $\mathfrak{A}_0''$.

Proof. From our hypothesis, $U^{-1}(\mathscr{D}_1)$ $(= \mathscr{D}_0)$ is a core for H^{-1} such that $\mathscr{D}_0 \subseteq \mathscr{D}(HAH^{-1})$ and $HAH^{-1} | \mathscr{D}_0$ has a unique bounded extension in the self-adjoint algebra $U^{-1}\mathscr{B}_0U$ for each A in $\mathfrak{A}_0$. From Lemma 9.2.22, $\mathscr{D}(HAH^{-1}) = \mathscr{D}(H^{-1})$, $(HAH^{-1})^*$ is a bounded, everywhere-defined operator, $\mathscr{D}(H^{-1}A^*H) = \mathscr{D}(H)$, and $(HAH^{-1})^* | \mathscr{D}(H) = H^{-1}A^*H$. Moreover,

$(HAH^{-1})^* \in U^{-1}\mathscr{B}_0 U$; so that $U(HAH^{-1})^*U^{-1}$ (in $\mathscr{B}_0$) is the extension of $UHA_0H^{-1}U^{-1}|\mathscr{D}_1$ to $\mathscr{K}$ for some A_0 in $\mathfrak{A}_0$. Thus $(HAH^{-1})^*$ is the unique bounded extension of $HA_0H^{-1}|\mathscr{D}_0$. Again from Lemma 9.2.22, $\mathscr{D}(HA_0H^{-1}) = \mathscr{D}(H^{-1})$ and $(HAH^{-1})^*$ is a bounded extension of both $H^{-1}A^*H$ and HA_0H^{-1}. As $\mathscr{D}_0(H) \subseteq \mathscr{D}(H) \cap \mathscr{D}(H^{-1})$, $H^{-1}A^*H|\mathscr{D}_0(H) = HA_0H^{-1}|\mathscr{D}_0(H)$. With x in $\mathscr{D}_0(H)$, $Hx \in \mathscr{D}_0(H)$ so that $H^{-1}A^*H^2x = HA_0x$, and $H^{-2}A^*H^2x = A_0x$. Thus $\mathscr{D}_0(H) \subseteq \mathscr{D}(H^{-2}A^*H^2)$ and $H^{-2}A^*H^2|\mathscr{D}_0(H) = A_0|\mathscr{D}_0(H)$. From Lemma 9.2.22, $(H^{-2}A^*H^2)^*$ is bounded and everywhere defined, $\mathscr{D}(H^2AH^{-2}) = \mathscr{D}(H^{-2})$ $(\supseteq \mathscr{D}_0(H))$ and $(H^{-2}A^*H^2)^*|\mathscr{D}(H^{-2}) = H^2AH^{-2}$. Thus $A_0^* = (A_0|\mathscr{D}_0(H))^* = (H^{-2}A^*H^2|\mathscr{D}_0(H))^* = (H^{-2}A^*H^2)^*$; and $H^2AH^{-2}|\mathscr{D}_0(H) = A_0^*|\mathscr{D}_0(H)$. From the first part of the argument of Lemma 9.2.25, $\mathscr{D}(H^{2n}AH^{-2n}) = \mathscr{D}(H^{-2n})$ and $H^{2n}AH^{-2n}$ is bounded for each positive integer n. By hypothesis $\|H^{2n}AH^{-2n}\| \leq k_A^{2n}$ for each such n. Lemma 9.2.25 applies, now, with H^2 in place of H, $\mathscr{D}_0(H)$ in place of $\mathscr{D}_0$, and k_A^2 in place of k_A. We conclude that $\mathscr{D}(H^{2z}AH^{-2z}) = \mathscr{D}(H^{-2z})$ and $H^{2z}AH^{-2z}$ has a (unique) bounded extension $\varphi_{2z}(A)$ in $\mathfrak{A}_0''$ for each complex z (so that the same is true with z in place of $2z$). In particular, $H^{it}AH^{-it} \in \mathfrak{A}_0''$ for each A in $\mathfrak{A}_0$—hence for each A in $\mathfrak{A}_0''$. At the same time, the (unique) bounded extension $\varphi(A)$ of $HAH^{-1}|\mathscr{D}_0$ is in $\mathfrak{A}_0''$. Since $U\varphi(A)U^{-1}|\mathscr{D}_1 = TAT^{-1}|\mathscr{D}_1$ and, by assumption, $TAT^{-1}|\mathscr{D}_1$ has a (unique) bounded extension to $\mathscr{K}$ in $\mathscr{B}_0$; $U\varphi(A)U^{-1} \in \mathscr{B}_0$. On the other hand, given B in $\mathscr{B}_0$, by hypothesis, there is an A in $\mathfrak{A}_0$ such that B is the unique extension of $TAT^{-1}|\mathscr{D}_1$ $(= U\varphi(A)U^{-1}|\mathscr{D}_1)$. Hence $B = U\varphi(A)U^{-1}$; and $U^{-1}BU = \varphi(A) \in \mathfrak{A}_0''$. Thus $U^{-1}\mathscr{B}_0''U \subseteq \mathfrak{A}_0''$.

We note next that the hypotheses apply with the roles of T and $\mathfrak{A}_0$ interchanged with those of T^{-1} and $\mathscr{B}_0$, from which we can conclude, as above, that $U\mathfrak{A}_0''U^{-1} \subseteq \mathscr{B}_0'' \subseteq U\mathfrak{A}_0''U^{-1}$, and, hence, that $U\mathfrak{A}_0''U^{-1} = \mathscr{B}_0''$. To see this, note that, with B and $\varphi(A)$ as above,

$$T^{-1}BT|\mathscr{D}_0(H) = H^{-1}U^{-1}BUH|\mathscr{D}_0(H) = H^{-1}\varphi(A)H|\mathscr{D}_0(H) = A|\mathscr{D}_0(H).$$

That is, $T^{-1}BT|\mathscr{D}_0(H)$ has a bounded extension A in $\mathfrak{A}_0$ and each A in $\mathfrak{A}_0$ is such an extension. For the growth condition on the bound, note that $T = UH = KU$, where $K = (TT^*)^{1/2}$, from Theorem 6.1.11; so that $K = UHU^{-1}$. Thus

$$K^nBK^{-n} = UH^nU^{-1}(U\varphi(A)U^{-1})UH^{-n}U^{-1} = UH^n\varphi(A)H^{-n}U^{-1};$$

so that $K^nBK^{-n}|\mathscr{D}_0(K)$ is bounded and

$$\|K^nBK^{-n}|\mathscr{D}_0(K)\| = \|H^{n+1}AH^{-n-1}|\mathscr{D}_0(H)\| \leq k_A^{n+1}$$

for all positive integers n, which establishes the desired symmetry between the roles of T and $\mathfrak{A}_0$ and those of T^{-1} and $\mathscr{B}_0$. ∎

In order to apply Theorem 9.2.26 to prove Tomita's theorem (9.2.9) in the manner outlined at the beginning of this subsection, we must find strong-operator-dense, self-adjoint subalgebras $\mathfrak{A}_0$ and $\mathscr{B}_0$ of $\mathscr{R}$ and $\mathscr{R}'$, respectively, that S transforms onto one another. Note that if $Au = A'u$ with A in $\mathscr{R}$ and A' in $\mathscr{R}'$, then, for each B in $\mathscr{R}$, $SA^*SBu = BAu = A'Bu$; so that SA^*S ($= SA^*S^{-1}$) has A' as its (unique) bounded extension to $\mathscr{H}$. In the case where $Au = A'u$, we say that A and A' are *reflections* of one another (about u). Conversely, if SA^*S has a bounded extension A' to $\mathscr{H}$, then A' lies in $\mathscr{R}'$ and $Au = A'u$. Thus, we must locate a large family of operators A in $\mathscr{R}$ and A' in $\mathscr{R}'$ that are reflections of one another. Since we want $\mathfrak{A}_0$ and $\mathscr{B}_0$ to be self-adjoint algebras and have S transform $\mathfrak{A}_0$ onto $\mathscr{B}_0$, we must, in fact, find sequences of operators $(\ldots, A'_{-3}, A_{-2}, A'_{-1}, A_0, A'_1, A_2, \ldots)$ such that $A_{2n} \in \mathscr{R}$, $A'_{2n+1} \in \mathscr{R}'$, $A'_{2n-1}u = A^*_{2n}u$, and $A_{2n}u = A'^*_{2n+1}u$ for each integer n. We call such a sequence of operators a *weak reflection sequence*. For the purposes of the growth condition of Theorem 9.2.26, we shall also want $\|A_n\| \le k^{|n|}$ and $\|A'_m\| \le k^{|m|}$ for some constant k. We call such a sequence a *reflection sequence* (for $\mathscr{R}$, $\mathscr{R}'$, and u).

If A and B in $\mathscr{R}$ are in reflection sequences $(\ldots, A'_{-1}, A_0, A'_1, \ldots)$ and $(\ldots, B'_{-1}, B_0, B'_1, \ldots)$, respectively, renumbering, we may assume that $A = A_0$ and $B = B_0$. (Of course the constants k and k', corresponding to these reflection sequences, must be modified appropriately.) Then $aA + B$ belongs to the reflection sequence

$$(\ldots, \bar{a}A'_{-1} + B'_{-1}, aA_0 + B_0, \bar{a}A'_1 + B'_1, aA_2 + B_2, \ldots),$$

while AB belongs to the reflection sequence

$$(\ldots, A_{-2}B_{-2}, A'_{-1}B'_{-1}, A_0B_0, A'_1B'_1, \ldots)$$

and A^* belongs to the "adjoint" reflection sequence

$$(\ldots, A^*_2, A'^*_1, A^*_0, A'^*_{-1}, A^*_{-2}, \ldots).$$

We summarize this discussion as follows.

9.2.27. Lemma. *The subset $\mathfrak{A}_0$ consisting of those elements of $\mathscr{R}$ that belong to a reflection sequence is a self-adjoint subalgebra of $\mathscr{R}$.*

With A_0 in the reflection sequence $(\ldots, A'_{-1}, A_0, A'_1, \ldots)$, $A^*_0u = A'_{-1}u$ so that $SA_0S^{-1} | \mathscr{R}u = A'_{-1} | \mathscr{R}u$. Since $\mathscr{R}u$ is a core for S (and $S = S^{-1}$, from Lemma 9.2.1), from Lemma 9.2.22, $\mathscr{D}(SA_0S^{-1}) = \mathscr{D}(S^{-1}) = \mathscr{D}(\Delta^{1/2})$ and $SA_0S^{-1} = A'_{-1} | \mathscr{D}(\Delta^{1/2})$. If we knew that $F = \bar{F}_0$ (recall that F is defined as S^*, and $F_0 \subseteq F$) the same argument (using the facts that $A'^*_{-1}u = A_{-2}u$ and $\mathscr{D}(F^{-1}) = \mathscr{D}(S^{*-1}) = \mathscr{D}(\Delta^{-1/2})$) would show that $\mathscr{D}(FA'_{-1}F^{-1}) = \mathscr{D}(F^{-1}) = \mathscr{D}(\Delta^{-1/2})$ and that $FA'_{-1}F^{-1} = A_{-2} | \mathscr{D}(\Delta^{-1/2})$.

It would follow that, with x in $\mathscr{D}(\Delta^{-1})$ $(\subseteq \mathscr{D}(\Delta^{-1/2}))$, $F^{-1}x \in \mathscr{D}(S^{-1}) = \mathscr{D}(SA_0S^{-1})$, that $S^{-1}F^{-1}x = \Delta^{-1}x$, and that

$$A_{-2}x = FA'_{-1}F^{-1}x = FSA_0S^{-1}F^{-1}x = \Delta A_0\Delta^{-1}x.$$

We would have, from this, that $\mathscr{D}(\Delta A_0\Delta^{-1}) = \mathscr{D}(\Delta^{-1})$, that A_{-2} is the unique bounded extension of $\Delta A_0\Delta^{-1}$, and that $\Delta A_0\Delta^{-1}|\mathscr{D}_0(\Delta) = A_{-2}|\mathscr{D}_0(\Delta)$. By the same token, $\Delta A_{-2}\Delta^{-1}|\mathscr{D}_0(\Delta) = A_{-4}|\mathscr{D}_0(\Delta)$; and $\Delta A_{-2}\Delta^{-1}|\mathscr{D}_0(\Delta) = \Delta^2A_0\Delta^{-2}|\mathscr{D}_0(\Delta)$. Thus $\Delta^nA_0\Delta^{-n}|\mathscr{D}_0(\Delta) = A_{-2n}|\mathscr{D}_0(\Delta)$ so that $\|\Delta^nA_0\Delta^{-n}\| \le k^{2n}$ for each positive integer n. Since $SA_{2n}S^{-1} = A'_{2n-1}|\mathscr{D}(S^{-1})$; and each element of $\mathscr{B}_0$ is the extension of some SA_0S^{-1} with A_0 in $\mathfrak{A}_0$, Theorem 9.2.26 applies with S in place of T, $\mathscr{R}u$ for $\mathscr{D}_1$, $\Delta^{1/2}$ for H, and J for U, and we conclude that $\Delta^zA\Delta^{-z}$ has a (unique) bounded extension in $\mathfrak{A}_0''$ for each A in $\mathfrak{A}_0$ and $J\mathfrak{A}_0''J^{-1} = \mathscr{B}_0''$. Once we show that $\mathfrak{A}_0$ and $\mathscr{B}_0$ are strong-operator dense in $\mathscr{R}$ and $\mathscr{R}'$, respectively (and prove the symmetric roles of S and F in the process), Theorem 9.2.9 will follow.

Given vectors x and y in H, L_x and R_y defined on $\mathscr{R}'u$ and $\mathscr{R}u$, respectively, by

$$L_xA'u = A'x, \qquad R_yAu = Ay$$

satisfy

$$L_xB'A'u = B'L_xA'u, \qquad R_yBAu = BR_yAu.$$

Loosely, then, L_x commutes with $\mathscr{R}'$ and R_y commutes with $\mathscr{R}$. In general, L_x and R_y are not preclosed.

9.2.28. LEMMA. *If $x \in \mathscr{D}(F_0^*)$, there is a preclosed operator L_x with closure (denoted, again, by L_x) affiliated with $\mathscr{R}$, having $\mathscr{R}'u$ as a core, and satisfying $L_xA'u = A'x$. Moreover, $\mathscr{R}'u \subseteq \mathscr{D}(L_x) \cap \mathscr{D}(L_x^*)$ and $L_x^*B'u = B'F_0^*x$ for each B' in $\mathscr{R}'$. If $y \in \mathscr{D}(S_0^*)$, there is a preclosed operator R_y with closure (denoted, again, by R_y) affiliated with $\mathscr{R}'$, having $\mathscr{R}u$ as a core, and satisfying $R_yAu = Ay$. Moreover, $\mathscr{R}u \subseteq \mathscr{D}(R_y) \cap \mathscr{D}(R_y^*)$ and $R_y^*Bu = BS_0^*y$.*

Proof. With A' and B' in $\mathscr{R}'$,

$$\langle L_xA'u, B'u\rangle = \langle x, F_0B'^*A'u\rangle = \langle A'u, B'F_0^*x\rangle.$$

Hence $B'u \in \mathscr{D}(L_x^*)$; and, from Theorem 2.7.8(ii), L_x has a closure (denoted, again, by L_x). Moreover, $L_x^*B'u = B'F_0^*x$. Since $U'^*L_xU'|\mathscr{R}'u = L_x|\mathscr{R}'u$ for each unitary operator U' in $\mathscr{R}'$ and $\mathscr{R}'u$ is a core for L_x, $L_x\eta\mathscr{R}$, from Remark 5.6.3. A similar argument applies to R_y. ∎

9.2.29. LEMMA. *If $T\,\eta\,\mathscr{R}$ and $u \in \mathscr{D}(T) \cap \mathscr{D}(T^*)$, then $Tu \in \mathscr{D}(S)$ and $STu = T^*u$. If $T'\,\eta\,\mathscr{R}'$ and $u \in \mathscr{D}(T') \cap \mathscr{D}(T'^*)$, then $T'u \in \mathscr{D}(\bar{F}_0)$ and $\bar{F}_0T'u = T'^*u$.*

Proof. With VH the polar decomposition of T, $V \in \mathscr{R}$ and $H \,\eta\, \mathscr{R}$. Let E_n be the spectral projection for H corresponding to $[-n, n]$ (so that E_n tends to I in the strong-operator topology and $E_n \in \mathscr{R}$). Let H_n be HE_n. Then H_n is bounded, defined everywhere, in $\mathscr{R}$, and $E_n H \subseteq HE_n$. Now $VH_n u \in \mathscr{D}(S)$, and $SVH_n u = H_n V^*u$. Since $u \in \mathscr{D}(T^*)$ and $T^* = HV^*$, $V^*u \in \mathscr{D}(H)$ and $H_n V^*u = E_n HV^*u \to HV^*u = T^*u$. As $u \in \mathscr{D}(T) = \mathscr{D}(H)$, we have $H_n u = E_n Hu \to Hu$; so that $VH_n u \to VHu = Tu$. Hence (Tu, T^*u) is in the graph of S, since S is closed. It follows that $Tu \in \mathscr{D}(S)$ and $STu = T^*u$. Similarly $T'u \in \mathscr{D}(\bar{F}_0)$ and $\bar{F}_0 T'u = T'^*u$. ■

9.2.30. Corollary. *The closure $\bar{F}_0$ of F_0 is F. Moreover, $S = F^*$ and $F = S^*$.*

Proof. From Lemma 9.2.1, $\bar{F}_0 \subseteq F = S_0^* = S^*$ so that $S = S^{**} = F^* \subseteq \bar{F}_0^* = F_0^*$ (compare Theorem 2.7.8(i)). If $x \in \mathscr{D}(F_0^*)$, then $L_x \,\eta\, \mathscr{R}$ and $\mathscr{R}'u \subseteq \mathscr{D}(L_x) \cap \mathscr{D}(L_x^*)$ (from Lemma 9.2.28). In particular $u \in \mathscr{D}(L_x) \cap \mathscr{D}(L_x^*)$; so that $x = L_x u \in \mathscr{D}(S) = \mathscr{D}(F^*)$. Hence $S = F^* = \bar{F}_0^*$ and $\bar{F}_0 = \bar{F}_0^{**} = F = F^{**} = S^*$. ■

We prove, next, the lemma that provides the "bridge" from $\mathscr{R}$ to $\mathscr{R}'$ in terms of S and its polar decomposition $J\Delta^{1/2}$. Interchanging the roles of $\mathscr{R}$ and $\mathscr{R}'$ and defining $\tilde{S}A'u$ to be A'^*u and $\tilde{F}Au$ to be A^*u, we have $\tilde{\Delta} = \tilde{F}\tilde{S} = S^{-1}F^{-1} = (FS)^{-1} = \Delta^{-1}$. Thus, the second assertion of the "bridging lemma", which follows, is a consequence of the first when applied to $\mathscr{R}$ and $\mathscr{R}'$ with their roles reversed. (It extends Lemma 9.2.4.)

9.2.31. Lemma. *If $x = (\Delta - aI)^{-1}A_0'u$, where $a \neq |a|$ and $A_0' \in \mathscr{R}'$, then $L_x \in \mathscr{R}$ and $\|L_x\| \leq a_0\|A_0'\|$, where $a_0 = (2|a| - 2\,\mathrm{Re}\, a)^{-1/2}$. If $y = (\Delta^{-1} - aI)^{-1}A_0 u$, where $A_0 \in \mathscr{R}$, then $R_y \in \mathscr{R}'$ and $\|R_y\| \leq a_0\|A_0\|$.*

Proof. Since Δ is positive, $\Delta(\Delta - aI)^{-1}$ is bounded and everywhere defined. Thus $x \in \mathscr{D}(\Delta) \subseteq \mathscr{D}(\Delta^{1/2}) = \mathscr{D}(S) = \mathscr{D}(F_0^*)$. From Lemma 9.2.28, $L_x \,\eta\, \mathscr{R}$. Let UH and KU be the polar decompositions of L_x. Then U, viewed as a unitary transformation from the closure of the range of H onto the closure of the range of K effects a unitary equivalence ($H = U^*KU$) between H and K viewed as operators on these closures. Thus, if M and N are spectral projections for H and K corresponding to the same (arbitrary, but fixed) closed, finite subinterval of $(a_0\|A_0'\|, \infty)$, then U, M, and N are in $\mathscr{R}$, and $UHM = KNU$. Thus

$$SNx = SNL_x u = SNKUu = U^*KNu = MHU^*u = ML_x^*u = MSx.$$

(This last equality results from Lemma 9.2.28 and Corollary 9.2.30.) If $N \neq 0$, then $Nu \neq 0$. By choice of N,

$$\begin{aligned}\|A_0'\|^2 \|Nu\|^2 & \\ &< a_0^{-2}\|KNu\|^2 = a_0^{-2}\|U^*KNu\|^2 = a_0^{-2}\|MSx\|^2 \\ &= a_0^{-2}\langle MSx, Sx\rangle = a_0^{-2}\langle SNx, Sx\rangle = a_0^{-2}\langle \Delta x, Nx\rangle \\ &= 2|a|\langle Nx, N\Delta x\rangle - 2\,\mathrm{Re}\langle aNx, N\Delta x\rangle \\ &\leq 2|a|\,\|Nx\|\cdot\|N\Delta x\| - 2\,\mathrm{Re}\langle aNx, N\Delta x\rangle \\ &\leq \|N\Delta x\|^2 + |a|^2\|Nx\|^2 - 2\,\mathrm{Re}\langle aNx, N\Delta x\rangle \\ &= \|N(\Delta - aI)x\|^2 = \|NA_0'u\|^2 \leq \|A_0'\|^2\|Nu\|^2.\end{aligned}$$

Thus $N = 0$, L_x is bounded, and $\|L_x\| \leq a_0\|A_0'\|$. ■

We are looking for operators in (weak) reflection sequences, and, in a preliminary way, for operators A_0 and A_0' that are reflections of one another (about u). Equivalently we want to find vectors in $\mathscr{R}u \cap \mathscr{R}'u$. We say that $(\ldots, y_{-2}, y_{-1}, y_0, y_1, y_2, \ldots)$ is a (weak) reflection sequence of vectors when $y_{2n} = A_{2n}u, y_{2n+1} = A_{2n+1}'u$ for each integer n, and $(\ldots, A_{-1}', A_0, A_1', \ldots)$ is a (weak) reflection sequence (of operators).

9.2.32. LEMMA. *The vector y_0 lies in a weak reflection sequence of vectors if and only if $y_0 \in \mathscr{D}(\Delta^n)$ and $\Delta^n y_0 \in \mathscr{R}u \cap \mathscr{R}'u$ for each integer n.*

Proof. Suppose, first, that $y_0 = A_0u$ and that $(\ldots, A_{-1}', A_0, A_1', \ldots)$ is a weak reflection sequence. From the discussion following Lemma 9.2.27, $\mathscr{D}(\Delta^n A_0 \Delta^{-n}) = \mathscr{D}(\Delta^{-n})$ and $\Delta^n A_0 \Delta^{-n} = A_{-2n}|\mathscr{D}(\Delta^{-n})$. (That discussion is applicable without reservation now that we have proved $F = \bar{F}_0$ in Cor. 9.2.30.) From Proposition 9.2.3, $\Delta u = u$, so that $u \in \mathscr{D}(\Delta^{-n})$ and $\Delta^{-n}u = u$. Thus $u \in \mathscr{D}(\Delta^n A_0 \Delta^{-n})$ and $A'^*_{-2n+1}u = A_{-2n}u = \Delta^n A_0 \Delta^{-n}u = \Delta^n A_0 u = \Delta^n y_0$. It follows that $y_0 \in \mathscr{D}(\Delta^n)$ and $\Delta^n y_0 \in \mathscr{R}u \cap \mathscr{R}'u$ for each integer n.

Suppose, now, that $y_0 \in \mathscr{D}(\Delta^n)$ and $\Delta^n y_0 \in \mathscr{R}u \cap \mathscr{R}'u$ for each integer n. Then $\Delta^{-n}y_0 = Au = A'^*u$ with A in $\mathscr{R}$ and A' in $\mathscr{R}'$. From the preceding, we see that we must define A_{2n} to be A and A_{2n+1}' to be A'. Of course, $y_0 = A_0u = A_1'^*u$ and $A_{2n}u = A_{2n+1}'^*u$, from this definition. Since

$$A_{2n}u = \Delta^{-n}y_0 = \Delta^{-1}\Delta^{-n+1}y_0 = SFA_{2n-1}'^*u = SA_{2n-1}'u,$$

we have

$$A_{2n-1}'u = SA_{2n}u = A_{2n}^*u.$$

Hence $(\ldots, A_{-1}', A_0, A_1', \ldots)$ is a weak reflection sequence. ■

If $E(k^{-1}, k)$ is the spectral projection for Δ corresponding to the interval (k^{-1}, k), then, since $e(\lambda^{-1}) = e(\lambda)$ with λ in $\mathrm{sp}(\Delta)\backslash\{0\}$ and e the characteristic function of this interval, $E(k^{-1}, k)$ is the spectral projection for Δ^{-1} corresponding to (k^{-1}, k). In the lemmas that follow, we show that vectors in $\mathscr{R}u \cap E(k^{-1}, k)(\mathscr{H})$ lie in a reflection sequence and that the set of these vectors, with k taking values in $(1, \infty)$, is a core for $\Delta^{1/2}$.

We assume the following lemma temporarily. It will be proved with the aid of the bridging lemma (9.2.31) and some analysis of the special functions involved.

9.2.33. LEMMA. *If $f_a(t) = \exp(-|t - a|)$ for real a and $A \in \mathscr{R}$, then $f_a(\log \Delta)Au = Bu$ where $B \in \mathscr{R}$ and $\|B\| \leq \|A\|$.*

9.2.34. LEMMA. *If $A_0 u \in E(k^{-1}, k)(\mathscr{H})$ for some k greater than 1 and $A_0 \in \mathscr{R}$, then $\Delta^n A_0 u = A_n u$, where $A_n \in \mathscr{R}$ and $\|A_n\| \leq k^{|n|}\|A_0\|$. In addition $A_0 u = A'u$, where $A' \in \mathscr{R}'$ and $\|A'\| \leq k^{1/2}\|A_0\|$. The statements obtained by interchanging $\mathscr{R}$ and $\mathscr{R}'$ in the preceding assertions are also valid.*

Proof. With "log" the restriction of Log to $[0, \infty)$ (see the discussion preceding Lemma 9.2.1, where Log is defined as a Borel function on $\mathbb{C}$) for all t in $[k^{-1}, k]$,

$$k \exp(-|\log t - \log k|) = t.$$

Thus, from Corollaries 5.6.29 and 5.6.31,

$$\begin{aligned} kf_{\log k}(\log \Delta)E(k^{-1}, k) \,|\, E(k^{-1}, k)(\mathscr{H}) &= (kf_{\log k} \circ \log)(\Delta E(k^{-1}, k)) \,|\, E(k^{-1}, k)(\mathscr{H}) \\ &= \Delta E(k^{-1}, k) \,|\, E(k^{-1}, k)(\mathscr{H}); \end{aligned}$$

so that

$$\Delta A_0 u = kf_{\log k}(\log \Delta)A_0 u = A_1 u,$$

where $A_1 \in \mathscr{R}$ and $\|A_1\| \leq k\|A_0\|$. (The last equality uses Lemma 9.2.33.) In the same way, with t in $[k^{-1}, k]$,

$$k \exp(-|\log(t^{-1}) - \log k|) = k \exp(-|\log t + \log k|) = t^{-1},$$

so that

$$\Delta^{-1} A_0 u = kf_{-\log k}(\log \Delta)A_0 u = A_{-1} u,$$

with A_{-1} in $\mathscr{R}$ and $\|A_{-1}\| \leq k\|A_0\|$. Since $A_1 u \in \mathscr{R}u \cap E(k^{-1}, k)(\mathscr{H})$, it follows, from what we have proved, that $\Delta A_1 u = A_2 u$, where $A_2 \in \mathscr{R}$ and $\|A_2\| \leq k^2\|A_0\|$. In addition $A_2 u \in \mathscr{R}u \cap E(k^{-1}, k)(\mathscr{H})$. Continuing in this way, we construct A_n with the desired properties.

We noted, prior to proving Lemma 9.2.31, that reversing the roles of $\mathscr{R}$ and $\mathscr{R}'$ corresponds to replacing Δ by Δ^{-1}. This establishes the first assertion of our lemma with $\mathscr{R}'$ in place of $\mathscr{R}$. From the bridging lemma (9.2.31), $(kI + \Delta^{-1})^{-1}A_0u = A_0'u$, where $A_0' \in \mathscr{R}'$ and $\|A_0'\| \leq (4k)^{-1/2}\|A_0\|$. Thus $A_0u = (kI + \Delta^{-1})A_0'u = kA_0'u + A_1'u$, where $A_1' \in \mathscr{R}'$ and $\|A_1'\| \leq k\|A_0'\|$. (Note, for this, that $A_0'u = (kI + \Delta^{-1})^{-1}A_0u \in E(k^{-1}, k)(\mathscr{H})$ and apply what we have proved thus far.) If we let A' be $kA_0' + A_1'$, the second assertion of the lemma follows. ■

It follows from Lemmas 9.2.32 and 9.2.34 that each vector in $\mathscr{R}u \cap E(k^{-1}, k)(\mathscr{H})$ or in $\mathscr{R}'u \cap E(k^{-1}, k)(\mathscr{H})$ lies in a reflection sequence.

9.2.35. LEMMA. *The linear manifold $\bigcup_{n=2}^{\infty} \mathscr{R}u \cap E(n^{-1}, n)(\mathscr{H})$ $(= \mathscr{D})$ is a core for $\Delta^{1/2}$.*

Proof. Suppose $A \in \mathscr{R}$ and

$$g_n(t) = e^{-|t|} - (e^n + e^{-n})^{-1}(e^{-|t-n|} + e^{-|t+n|})$$

with n an integer greater than 1. Then $\{g_n\}$ is an increasing sequence of positive functions vanishing outside (but nowhere zero on) the interval $(-n, n)$ and converging at each t to $\exp(-|t|)$. (Note, for this, that $g_n(t) = g_n(-t)$, so that we may assume $0 \leq t$; and write

$$g_n(t) = e^{-t}\left[1 - \frac{e^{2t} + 1}{e^{2n} + 1}\right]$$

when $0 \leq t \leq n$.) From Lemma 9.2.33, $g_n(\log \Delta)Au = Bu$, where $B \in \mathscr{R}$. With e_n the characteristic function of $(\exp - n, \exp n)$, the Borel functions $g_n \circ \log$ and $(g_n \circ \log)e_n$ coincide on $\operatorname{sp}\Delta\backslash\{0\}$ (since g_n vanishes outside of $(-n, n)$). Hence $g_n(\log \Delta) = g_n(\log \Delta)E_n$, where $E_n = e_n(\Delta)$; and

$$g_n(\log \Delta)Au = g_n(\log \Delta)E_n Au = E_n g_n(\log \Delta)Au \in \mathscr{D},$$

when $n = 2, 3, \ldots$ and $A \in \mathscr{R}$. If $h(t) = g_n(\log t)^{-1}$ for t in $(\exp -n, \exp n)$ and $h(t) = 0$ for other (real) t, then $(g_n \circ \log) \cdot e_n \cdot h = e_n$; so that

$$g_n(\log \Delta)E_n \hat{\cdot} h(\Delta) = E_n.$$

It follows that $g_n(\log \Delta)$ maps $E_n(\mathscr{H})$ onto a dense subset of itself. As $g_n(\log \Delta)$ is bounded (g_n is bounded) and $\{E_n Au : A \in \mathscr{R}\}$ is dense in $E_n(\mathscr{H})$, $g_n(\log \Delta)$ maps this set onto a dense subset of $E_n(\mathscr{H})$ and, as noted, into $\mathscr{D}$. Thus $\mathscr{D} \cap E_n(\mathscr{H})$ $(= \mathscr{D}_n)$ is dense in $E_n(\mathscr{H})$. With y in $E_n(\mathscr{H})$, let y_m in $\mathscr{D}_n$ tend to y. Then $\Delta^{1/2}y_m \to \Delta^{1/2}y$ and $(y, \Delta^{1/2}y) \in \mathscr{G}(\Delta^{1/2}|\mathscr{D})^-$. Since $\bigcup_{n=2}^{\infty} E_n(\mathscr{H})$ is a core for $\Delta^{1/2}$, $\mathscr{D}$ is a core for $\Delta^{1/2}$. ■

Proof of Lemma 9.2.33. If

$$h_a(t) = [\cosh(t - a)]^{-1}(= 2[e^{t-a} + e^{a-t}]^{-1}),$$

then, when $t \in \text{sp}\,\Delta\backslash\{0\}$,

$$h_a(\log t) = 2(e^{-a}t + e^{a}t^{-1})^{-1} = 2i(t + ie^{a})^{-1}(t^{-1} + ie^{-a})^{-1}.$$

Thus

$$h_a(\log \Delta) = 2(e^{-a}\Delta + e^{a}\Delta^{-1})^{-1} = 2i(\Delta + ie^{a}I)^{-1}(\Delta^{-1} + ie^{-a}I)^{-1}.$$

From the bridging lemma, with A in $\mathscr{R}$, we have $h_a(\log \Delta)Au = B_0 u$, where $B_0 \in \mathscr{R}$ and $\|B_0\| \le \|A\|$. As we shall note,

$$e^{-|t|} = \sum_{n=1}^{\infty} a_n[\cosh t]^{-(2n-1)}, \tag{17}$$

for all real t; and convergence is uniform on $\mathbb{R}$, where $0 < a_n$ and $\sum_{n=1}^{\infty} a_n = 1$. Assuming this, for the moment, we have, with t in $\text{sp}\,\Delta\backslash\{0\}$,

$$f_a(\log t) = e^{-|\log t - a|} = \sum_{n=1}^{\infty} a_n[h_a(\log t)]^{2n-1};$$

so that

$$f_a(\log \Delta) = \sum_{n=1}^{\infty} a_n[h_a(\log \Delta)]^{2n-1},$$

where convergence is in the operator-norm topology. Thus, for each A in $\mathscr{R}$,

$$f_a(\log \Delta)Au = \sum_{n=1}^{\infty} a_n[h_a(\log \Delta)]^{2n-1}Au = \sum_{n=1}^{\infty} a_n B_n u,$$

where $B_n \in \mathscr{R}$ and $\|B_n\| \le \|A\|$. Since $0 < a_n$ and $\sum a_n = 1$; $\sum_{n=1}^{\infty} a_n B_n$ converges in norm to an operator B in $\mathscr{R}$ and $\|B\| \le \|A\|$.

We conclude this argument by verifying (17). Note that

$$(\cosh t)^{-1} = \frac{2}{e^t + e^{-t}} = \frac{2e^{-|t|}}{(e^{-|t|})^2 + 1} = \frac{2s}{s^2 + 1} \qquad (= f(s)),$$

where $s = \exp(-|t|)$. We want to express s as a power series in $2s(s^2 + 1)^{-1}$, in particular, as a function, the inverse to f, of $2s(s^2 + 1)^{-1}$. Now f is an order-preserving homeomorphism of $[-1, 1]$ onto $[-1, 1]$ and its inverse g assigns $r^{-1}[1 - (1 - r^2)^{1/2}]$ to r in $[-1, 1]\backslash\{0\}$ and 0 to 0. Employing the binomial expansion for $(1 - r^2)^{1/2}$, we can represent $g(r)$ as $\frac{1}{2}r + \sum_{n=2}^{\infty} a_n r^{2n-1}$ on $(-1, 1)$, where

$$a_n = \frac{1 \cdot 3 \cdot 5 \cdots (2n - 3)}{2 \cdot 4 \cdots 2n}.$$

In particular, $0 < a_n$ for $n = 2, 3, \ldots$. Let a_1 be $\frac{1}{2}$ and let $g_n(r)$ be $\sum_{k=1}^{n} a_k r^{2k-1}$ for r in $[0, 1]$. Then, for r in $[0, 1)$, $g_n(r) \le g(r)$; so that $g_n(1) \le g(1)$, by continuity of g_n and g. Thus, by monotonicity of g_n for all n and r in $[0, 1]$, $g_n(r) \le g_n(1) = \sum_{k=1}^{n} a_k \le g(1) = 1$. For r in $[0, 1)$, then, $g(r) \le \sum_{n=1}^{\infty} a_n \le 1$. Since $g(r)$ tends to 1 as r tends to 1^-, $\sum_{n=1}^{\infty} a_n = 1$. As $0 < a_n$ and $\sum a_n$ converges, $\sum a_n r^{2n-1}$ converges uniformly and absolutely on $[-1, 1]$. It follows that

$$e^{-|t|} = s = g\left(\frac{2s}{s^2 + 1}\right) = g([\cosh t]^{-1}) = \sum_{n=1}^{\infty} a_n [\cosh t]^{-(2n-1)}$$

and that convergence is uniform on $\mathbb{R}$. ∎

To conclude the proof of Tomita's theorem (9.2.9), we must show that $\mathfrak{A}_0$ and, hence, $\mathscr{B}_0$ are strong-operator dense in $\mathscr{R}$ and $\mathscr{R}'$, respectively. From Lemma 9.2.35, the transform of u by some of the operators in $\mathfrak{A}_0$ is a core for $\Delta^{1/2}$ so that $\mathfrak{A}_0 u$ is a core for $\Delta^{1/2}$. The following density theorem completes the argument.

9.2.36. THEOREM. *If $\mathscr{R}$ is a von Neumann algebra acting on a Hilbert space $\mathscr{H}$, $\mathscr{R}_0$ is a self-adjoint subalgebra of $\mathscr{R}$, and u is a unit separating and generating vector for $\mathscr{R}$, then the following statements are equivalent*:

(i) *$\mathscr{R}_0$ is strong-operator dense in $\mathscr{R}$;*
(ii) *$\mathscr{R}_0^{\mathrm{h}} u$ is dense in $\mathscr{R}^{\mathrm{h}} u$, where $\mathscr{R}_0^{\mathrm{h}}$ and $\mathscr{R}^{\mathrm{h}}$ are the sets of self-adjoint operators in $\mathscr{R}_0$ and $\mathscr{R}$;*
(iii) *$\mathscr{R}_0 u$ is a core for $\Delta^{1/2}$.*

Proof. (i) ⇒ (ii) Since $\mathscr{R}_0$ is weak-operator dense in $\mathscr{R}$ and the adjoint operation is weak-operator continuous, $\mathscr{R}_0^{\mathrm{h}}$ is weak-operator dense in $\mathscr{R}^{\mathrm{h}}$. As $\mathscr{R}_0^{\mathrm{h}}$ and $\mathscr{R}^{\mathrm{h}}$ are convex, $\mathscr{R}_0^{\mathrm{h}}$ is strong-operator dense in $\mathscr{R}^{\mathrm{h}}$.

(ii) ⇒ (iii) Since $\mathscr{R}u$ is a core for $\Delta^{1/2}$, given A in $\mathscr{R}$, it will suffice to find operators A_n in $\mathscr{R}_0$ such that $A_n u \to Au$ and $\Delta^{1/2} A_n u$ $(= JSA_n u = JA_n^* u) \to \Delta^{1/2} Au$ $(= JA^* u)$, or, equivalently (since J is an isometry), such that $A_n^* u \to A^* u$. Now $A = H_1 + iH_2$, with H_1 and H_2 self-adjoint operators in $\mathscr{R}$. By assumption, there are self-adjoint operators K_{1n} and K_{2n} in $\mathscr{R}_0$ such that $K_{1n} u \to H_1 u$ and $K_{2n} u \to H_2 u$. If $A_n = K_{1n} + iK_{2n}$, then $A_n \in \mathscr{R}_0$, $A_n u \to Au$, and $A_n^* u \to A^* u$.

(iii) ⇒ (i) Since $\mathscr{R}_0 u$ is a core for $\Delta^{1/2}$, it is, in particular, dense in $\mathscr{H}$. Thus the maximal projection (see Proposition 5.1.8) in $\mathscr{R}_0^-$ is I and

$$\mathscr{R}_0'' \subseteq (\mathscr{R}_0^-)'' = \mathscr{R}_0^- \subseteq \mathscr{R}.$$

If we show that $\mathscr{R}_0' \subseteq \mathscr{R}'$, then $\mathscr{R} = \mathscr{R}'' \subseteq \mathscr{R}_0''$ and $\mathscr{R}_0^- = \mathscr{R}$. We prove that $\mathscr{R}_0' \subseteq \mathscr{R}'$ by showing that each self-adjoint H' in $\mathscr{R}_0'$ lies in $\mathscr{R}'$. If $x \in \mathscr{D}(\Delta^{1/2})$, by assumption, there is a sequence $\{A_n\}$ in $\mathscr{R}_0$ such that $A_n u \to x$ and $\Delta^{1/2} A_n u$ $(= JA_n^* u) \to \Delta^{1/2} x$. Thus

$$\langle SA_n u, H'u \rangle = \langle J\Delta^{1/2} A_n u, H'u \rangle \to \langle J\Delta^{1/2} x, H'u \rangle = \langle Sx, H'u \rangle$$

and

$$\langle SA_n u, H'u \rangle = \langle A_n^* u, H'u \rangle = \langle H'u, A_n u \rangle \to \langle H'u, x \rangle.$$

Hence $\langle Sx, H'u \rangle = \langle H'u, x \rangle$ for all x in $\mathscr{D}(\Delta^{1/2})$ $(= \mathscr{D}(S))$; so that $H'u \in \mathscr{D}(S^*)$ $(= \mathscr{D}(F))$ and $FH'u = H'u$. It follows, from Lemma 9.2.28, that $R_{H'u} \,\eta\, \mathscr{R}'$. If $A \in \mathscr{R}_0$, then

$$R_{H'u} Au = AH'u = H'Au,$$

since $H' \in \mathscr{R}_0'$. Thus H' $(= R_{H'u}) \in \mathscr{R}'$. ■

A further extension of modular theory. So far, we have developed modular theory in the context of a von Neumann algebra $\mathscr{R}$ with a faithful normal state ρ. By the GNS construction, ρ gives rise to a faithful representation π of $\mathscr{R}$; moreover, $\pi(\mathscr{R})$ is a von Neumann algebra acting on a Hilbert space $\mathscr{H}_\rho$, with a separating and generating unit vector u such that $\rho = \omega_u \circ \pi$. The modular operator Δ and the conjugate-linear isometry J, constructed for the von Neumann algebra $\pi(\mathscr{R})$ and its separating and generating vector u, act on the Hilbert space $\mathscr{H}_\rho$ and satisfy

$$(18) \qquad J\pi(\mathscr{R})J = \pi(\mathscr{R})', \qquad \Delta^{it}\pi(\mathscr{R})\Delta^{-it} = \pi(\mathscr{R}) \qquad (t \in \mathbb{R}).$$

The modular automorphism group $\{\sigma_t\}$, defined by

$$(19) \qquad \pi(\sigma_t(A)) = \Delta^{it}\pi(A)\Delta^{-it} \qquad (A \in \mathscr{R}, \quad t \in \mathbb{R}),$$

is the unique one-parameter group of * automorphisms of $\mathscr{R}$ that satisfies the modular condition relative to the state ρ. The relations between modular automorphism groups derived from different faithful normal states will be studied in Theorem 13.1.9. By the unitary implementation theorem (7.2.9) the representation π is uniquely determined (up to unitary equivalence), and is, in that sense, independent of the choice of the state ρ (see Exercise 9.6.26, where a more general result is obtained).

A von Neumann algebra $\mathscr{R}$ has a faithful normal state if and only if $\mathscr{R}$ is countably decomposable (Exercise 7.6.46); so the theory just described is applicable only in the countably decomposable case. In the present subsection we indicate (without giving all proofs in detail) how the theory can be extended by using weights in place of states. When generalized in this way, modular theory applies to all von Neumann algebras, and becomes a more powerful tool even in the countably decomposable case.

We begin by describing some of the main features of the extended theory. Every von Neumann algebra $\mathscr{R}$ has a faithful normal semi-finite weight ρ (Exercise 7.6.46). By the construction defined in Theorem 7.5.3, ρ gives rise to a faithful representation π of $\mathscr{R}$, and $\pi(\mathscr{R})$ is a von Neumann algebra acting on a Hilbert space $\mathscr{H}_\rho$. In fact, $\mathscr{H}_\rho$ is the completion of the pre-Hilbert space consisting of the left ideal $\mathscr{N}_\rho$ $(= \{A \in \mathscr{R} : \rho(A^*A) < \infty\})$ in $\mathscr{R}$ with the inner product $\langle\ ,\ \rangle_\rho$ defined by $\langle A, B\rangle_\rho = \rho(B^*A)$. (Note that the left ideal N_ρ $(= \{A \in \mathscr{R} : \rho(A^*A) = 0\})$ is $\{0\}$ since ρ is faithful.) In this context, it is possible to prove the following version of Tomita's theorem.

9.2.37. THEOREM. *Suppose that ρ is a faithful normal semi-finite weight on a von Neumann algebra $\mathscr{R}$, and $\pi: \mathscr{R} \to \mathscr{B}(\mathscr{H}_\rho)$ is the faithful normal representation of $\mathscr{R}$ constructed in Theorem* 7.5.3. *Then the conjugate-linear mapping $A \to A^*$, with domain $\mathscr{N}_\rho \cap \mathscr{N}_\rho^*$, is a preclosed densely defined operator acting on $\mathscr{H}_\rho$ and its closure S has polar decomposition $J\Delta^{1/2}$ in which Δ is an invertible positive operator acting on $\mathscr{H}_\rho$, J is a conjugate-linear isometry acting on $\mathscr{H}_\rho$, $J^2 = I$, and* (18) *is satisfied.*

The representation π is uniquely determined, up to unitary equivalence (Exercise 9.6.26) and is, in that sense, independent of the choice of the weight ρ. The *modular automorphism group* $\{\sigma_t\}$, *of* $\mathscr{R}$ *relative to* ρ, is defined by (19). Corresponding to Theorem 9.2.16, it is possible, in the present context, to prove the following uniqueness result.

9.2.38. THEOREM. *Under the conditions of Theorem* 9.2.37, *the modular automorphism group of $\mathscr{R}$ relative to ρ is the unique one-parameter group $\{\alpha_t\}$ of * automorphisms of $\mathscr{R}$ such that*

(a) *$\rho(\alpha_t(A)) = \rho(A)$ for all A in $\mathscr{R}^+$ and t in $\mathbb{R}$;*

(b) *given any elements A, B of $\mathscr{N}_\rho \cap \mathscr{N}_\rho^*$, there is a complex-valued function f, bounded and continuous on the strip $\{z \in \mathbb{C} : 0 \le \operatorname{Im} z \le 1\}$ and analytic on the interior of that strip, such that*

$$f(t) = \rho(\alpha_t(A)B), \qquad f(t+i) = \rho(B\alpha_t(A)) \qquad (t \in \mathbb{R}).$$

Taken together, conditions (a) and (b) in Theorem 9.2.38 constitute the *modular condition* relative to ρ, in a form appropriate to the present context. In the case of a *state* ρ, (a) can be omitted since it is a consequence of (b), by Proposition 9.2.14(i). In formulating the boundary condition (b) when ρ is a *weight*, however, it is necessary to ensure that $\alpha_t(A)B$ and $B\alpha_t(A)$ lie in the linear space $\mathcal{M}_\rho$ on which ρ takes finite values (see Lemma 7.5.2); and condition (a) implies that $\alpha_t(A)B$, $B\alpha_t(A)$ both lie in $\mathcal{N}_\rho^*\mathcal{N}_\rho$ ($\subseteq \mathcal{M}_\rho$), when $A, B \in \mathcal{N}_\rho \cap \mathcal{N}_\rho^*$.

We now outline the arguments needed to establish the theory just described. Throughout, ρ is a faithful normal semi-finite weight on a von Neumann algebra $\mathcal{R}$, and the subsets F_ρ, $\mathcal{M}_\rho$, $\mathcal{N}_\rho$, and N_ρ ($= \{0\}$) of $\mathcal{R}$ are defined as in Section 7.5. The Hilbert space $\mathcal{H}_\rho$ and the (faithful, normal) representation π of $\mathcal{R}$ (with $\pi(\mathcal{R})$ a von Neumann algebra acting on $\mathcal{H}_\rho$) are constructed as in Theorem 7.5.3, and the norm and inner product in $\mathcal{H}_\rho$ are denoted by $\| \ \|_\rho$ and $\langle \ , \ \rangle_\rho$, respectively. Thus $\mathcal{N}_\rho$ is a dense subspace of $\mathcal{H}_\rho$, and

$$(20) \qquad \langle B, C\rangle_\rho = \rho(C^*B), \qquad \pi(A)B = AB \qquad (B, C \in \mathcal{N}_\rho, \quad A \in \mathcal{R}).$$

We denote by $\mathfrak{A}$ the self-adjoint subalgebra $\mathcal{N}_\rho \cap \mathcal{N}_\rho^*$ of $\mathcal{R}$.

9.2.39. Proposition. $\mathfrak{A}^- = \mathcal{R}$, $\pi(\mathfrak{A})^- = \pi(\mathcal{R})$.

Proof. From Lemma 7.5.2, $\mathcal{N}_\rho$ is a left ideal in $\mathcal{R}$, $\mathcal{N}_\rho^*$ is a right ideal, and

$$\mathfrak{A} = \mathcal{N}_\rho \cap \mathcal{N}_\rho^* \supseteq \mathcal{N}_\rho^*\mathcal{N}_\rho = \mathcal{M}_\rho.$$

Since ρ is semi-finite, $\mathcal{M}_\rho$ (and hence, also, $\mathfrak{A}$) is weak-operator dense in $\mathcal{R}$; so $\mathfrak{A}^- = \mathcal{R}$. By the Kaplansky density theorem, and since π is weak-operator continuous on the unit ball $(\mathcal{R})_1$,

$$\pi((\mathfrak{A})_1)^- \supseteq \pi((\mathfrak{A})_1^-) = \pi((\mathcal{R})_1),$$

and thus $\pi(\mathfrak{A})^- = \pi(\mathcal{R})$. ■

We now consider some relations between the algebra $\mathfrak{A}$ ($\subseteq \mathcal{N}_\rho \subseteq \mathcal{H}_\rho$) and the Hilbert space $\mathcal{H}_\rho$.

9.2.40. Proposition. (i) *For each A in $\mathfrak{A}$, the mapping $B \to AB: \mathfrak{A} \to \mathfrak{A}$ is bounded relative to $\| \ \|_\rho$.*

(ii) *When $A, B, C \in \mathfrak{A}$, $\langle AB, C\rangle_\rho = \langle B, A^*C\rangle_\rho$.*

(iii) *The linear span $\mathfrak{A}^2$ of the set $\{AB : A, B \in \mathfrak{A}\}$ is dense in $\mathcal{H}_\rho$.*

(iv) *The mapping $A \to A^*$, considered as a conjugate-linear operator with domain $\mathfrak{A}$ dense in $\mathcal{H}_\rho$, is preclosed.*

Proof. Both (i) and (ii) are immediate consequences of (20).

(iii) In fact, our argument will prove the stronger result that the set $\{AB : A, B \in \mathfrak{A}\}$ is dense in $\mathcal{H}_\rho$. Since π is isometric and $\pi(\mathfrak{A})^- = \pi(\mathcal{R})$, it follows from the Kaplansky density theorem that there is a net $\{H_j\}$, consisting of self-adjoint operators in the unit ball of $\mathfrak{A}$, such that $\{\pi(H_j)\}$ is strong-operator convergent to I. For each S in $\mathcal{N}_\rho$ ($\subseteq \mathscr{H}_\rho$) and each index j, we have $H_j \in \mathfrak{A}$ and

$$H_j S \in \mathfrak{A}\mathcal{N}_\rho \subseteq \mathcal{N}_\rho^* \mathcal{N}_\rho \subseteq \mathcal{N}_\rho^* \cap \mathcal{N}_\rho = \mathfrak{A};$$

so $H_j^2 S \in \mathfrak{A}^2$. Moreover,

$$\|H_j^2 S - S\|_\rho = \|\pi(H_j)^2 S - S\|_\rho \to 0,$$

since the net $\{\pi(H_j)^2\}$ is strong-operator convergent to I. Hence the closure of $\mathfrak{A}^2$ in $\mathscr{H}_\rho$ contains each S in the dense subspace $\mathcal{N}_\rho$, and so coincides with $\mathscr{H}_\rho$.

(iv) Suppose that $\{A_n\}$ is a sequence in $\mathfrak{A}$, $x \in \mathscr{H}_\rho$, and

$$\|A_n\|_\rho \to 0, \qquad \|A_n^* - x\|_\rho \to 0 \tag{21}$$

as $n \to \infty$. We have to show that $x = 0$.

Given any positive real number ε, we can choose a positive integer $N(\varepsilon)$ such that $\|A_m^* - A_n^*\|_\rho < \varepsilon$ whenever $m, n \geq N(\varepsilon)$. Thus

$$\lim_{n\to\infty} \rho(A_n^* A_n) = 0, \qquad \rho((A_m - A_n)(A_m - A_n)^*) < \varepsilon^2 \qquad (m, n \geq N(\varepsilon)). \tag{22}$$

Since ρ is a normal weight on $\mathscr{R}$, it is the sum of a family of positive normal functionals, each of which can in turn be expressed as a sum of vector functionals $\omega_y | \mathscr{R}$. Accordingly, there is a family $\{y_a : a \in \mathbb{A}\}$ of vectors in the Hilbert space $\mathscr{H}$ on which $\mathscr{R}$ acts, such that

$$\rho(A) = \sum_{a\in\mathbb{A}} \langle Ay_a, y_a \rangle \qquad (A \in \mathscr{R}^+).$$

Thus

$$\rho(A^*A) = \sum_{a\in\mathbb{A}} \|Ay_a\|^2 \qquad (A \in \mathscr{R}).$$

Since ρ is faithful, the family $\{y_a\}$ is separating for $\mathscr{R}$, and is therefore generating for $\mathscr{R}'$.

From (22),

$$\lim_{n\to\infty} \sum_{a\in\mathbb{A}} \|A_n y_a\|^2 = 0,$$

$$\sum_{a\in\mathbb{A}} \|A_m^* y_a - A_n^* y_a\|^2 < \varepsilon^2 \qquad (m, n \geq N(\varepsilon)). \tag{23}$$

It follows that, for each a in $\mathbb{A}$, $\{A_n y_a\}$ converges to 0, and $\{A_n^* y_a\}$ is a Cauchy sequence and so converges to an element z_a of $\mathscr{H}$. Moreover, if $a, b \in \mathbb{A}$ and $A' \in \mathscr{R}'$,

$$\begin{aligned}\langle z_b, A'y_a \rangle &= \lim_{n\to\infty} \langle A_n^* y_b, A'y_a \rangle \\ &= \lim_{n\to\infty} \langle y_b, A'A_n y_a \rangle = 0.\end{aligned}$$

From this, and since the family $\{y_a\}$ is generating for $\mathscr{R}'$, $z_b = 0$. Thus, for each b in $\mathbb{A}$, the sequence $\{A_n^* y_b\}$ converges to 0.

It follows from (23) that, for each finite subset $\mathbb{F}$ of $\mathbb{A}$,

$$\sum_{a \in \mathbb{F}} \|A_m^* y_a - A_n^* y_a\|^2 < \varepsilon^2 \qquad (m, n \geq N(\varepsilon)).$$

When $m \to \infty$, we obtain

$$\sum_{a \in \mathbb{F}} \|A_n^* y_a\|^2 \leq \varepsilon^2 \qquad (n \geq N(\varepsilon)).$$

Since this last inequality is satisfied for all finite subsets $\mathbb{F}$ of $\mathbb{A}$,

$$\sum_{a \in \mathbb{A}} \|A_n^* y_a\|^2 \leq \varepsilon^2 \qquad (n \geq N(\varepsilon)).$$

Thus

$$\begin{aligned} \|A_n^*\|_\rho &= [\rho(A_n A_n^*)]^{1/2} \\ &= \left[\sum_{a \in \mathbb{A}} \|A_n^* y_a\|^2\right]^{1/2} \to 0 \end{aligned}$$

as $n \to \infty$, and $x = 0$ from (21). ∎

The following definition is motivated by the result of Proposition 9.2.40.

9.2.41. DEFINITION. Suppose that $\mathfrak{A}_0$ is an associative linear algebra over the complex field, with an involution * and a positive definite inner product $\langle\ ,\ \rangle_0$. Let $\mathscr{H}_0$ be the Hilbert space obtained as the completion of $\mathfrak{A}_0$ relative to the norm associated with $\langle\ ,\ \rangle_0$. Then $\mathfrak{A}_0$ is a *left Hilbert algebra* if the following four conditions are satisfied:

(i) for each A in $\mathfrak{A}_0$, the mapping $B \to AB\colon \mathfrak{A}_0 \to \mathfrak{A}_0$ is bounded;
(ii) when $A, B, C \in \mathfrak{A}_0$, $\langle AB, C\rangle_0 = \langle B, A^*C\rangle_0$;
(iii) the linear span $\mathfrak{A}_0^2$ of the set $\{AB\colon A, B \in \mathfrak{A}_0\}$ is dense in $\mathscr{H}_0$;
(iv) as a densely defined conjugate-linear operator with domain $\mathfrak{A}_0$ in the Hilbert space $\mathscr{H}_0$, the mapping $A \to A^*\colon \mathfrak{A}_0 \to \mathfrak{A}_0$ is preclosed.

The *left von Neumann algebra* of a left Hilbert algebra $\mathfrak{A}_0$ is the von Neumann algebra $\mathscr{L}(\mathfrak{A}_0)$ generated by $\{\pi_0(A) : A \in \mathfrak{A}_0\}$, where (for each A in $\mathfrak{A}_0$) $\pi_0(A)$ denotes the unique bounded linear operator acting on $\mathscr{H}_0$ that extends the mapping $B \to AB\colon \mathfrak{A}_0 \to \mathfrak{A}_0$. ∎

From Proposition 9.2.40, the algebra $\mathfrak{A}$ constructed from $\mathscr{R}$ and ρ is a left Hilbert algebra, the corresponding Hilbert space being $\mathscr{H}_\rho$. It follows from (20) that, when $A \in \mathfrak{A}$, $\pi(A)$ is the unique bounded linear operator acting on $\mathscr{H}_\rho$ that extends the mapping $B \to AB: \mathfrak{A} \to \mathfrak{A}$. From this, and since $\pi(\mathfrak{A})^- = \pi(\mathscr{R})$, the left von Neumann algebra $\mathscr{L}(\mathfrak{A})$ is $\pi(\mathscr{R})$. In [10, 11, 15], modular theory is developed in the context of a left Hilbert algebra and its left von Neumann algebra. The treatment sketched below is based on the methods of [15], as applied to the left Hilbert algebra $\mathfrak{A}$ constructed from $\mathscr{R}$ and ρ.

By Proposition 9.2.40(iii), (iv), the mapping $S_0: A \to A^*: \mathfrak{A}^2 \to \mathfrak{A}^2$ can be viewed as a preclosed conjugate-linear operator with domain $\mathscr{D}(S_0)$ ($= \mathfrak{A}^2$) dense in $\mathscr{H}_\rho$. Moreover, S_0 is an involution, in the sense that

$$\mathscr{D}(S_0^2) = \mathscr{D}(S_0), \qquad S_0^2 x = x \qquad (x \in \mathscr{D}(S_0)).$$

Let

$$F = S_0^*, \qquad S = F^* = S_0^{**},$$

so that S is the closure $\bar{S}_0$ of S_0. It is easy to verify that both F and S are involutions. Indeed, suppose that $y \in \mathscr{D}(F)$ ($= \mathscr{D}(S_0^*)$). For each x in $\mathscr{D}(S_0)$, we have $S_0 x \in \mathscr{D}(S_0)$, and

$$\langle S_0 x, Fy \rangle = \langle S_0 x, S_0^* y \rangle = \langle y, S_0^2 x \rangle = \langle y, x \rangle;$$

so $Fy \in \mathscr{D}(S_0^*)$ ($= \mathscr{D}(F)$) and $F^2 y = S_0^* F y = y$. This proves that F is an involution. We obtain the same result for S ($= F^*$) by repeating the argument of the three preceding sentences, with F, S in place of S_0, F, respectively.

The choice of $\mathfrak{A}^2$ (rather than $\mathfrak{A}$) as the domain of S_0 is a matter of technical convenience; we shall prove later that $\mathfrak{A} \subseteq \mathscr{D}(S)$ and $SA = A^*$ for all A in $\mathfrak{A}$ (see the discussion following (48)).

We assert that there is a conjugate-linear isometric mapping J from $\mathscr{H}_\rho$ onto $\mathscr{H}_\rho$, and an invertible positive operator Δ acting on $\mathscr{H}_\rho$, such that

$$J^2 = I, \qquad \Delta = FS, \qquad \Delta^{-1} = SF, \qquad J\Delta^{it} = \Delta^{it} J$$

for all real t, and

$$S = J\Delta^{1/2} = \Delta^{-1/2} J, \qquad F = J\Delta^{-1/2} = \Delta^{1/2} J.$$

Indeed, the arguments used to prove Proposition 9.2.3 are applicable in the present context.

We now digress, to consider how the present discussion is related to our first approach to modular theory, in the special case in which the weight ρ is a vector state $\omega_u | \mathscr{R}$ associated with a separating and generating vector u for $\mathscr{R}$. In this case, $\mathscr{N}_\rho$ is $\mathscr{R}$ and the inner product $\langle \ , \ \rangle_\rho$ on $\mathscr{N}_\rho$ is given by

$$\langle A, B \rangle_\rho = \rho(B^*A) = \langle B^*Au, u \rangle = \langle Au, Bu \rangle \qquad (A, B \in \mathscr{R}).$$

The equation $U_0 A = Au$ ($A \in \mathscr{R}$) defines an isometric linear mapping U_0 from $\mathscr{R}$ (a dense subspace of $\mathscr{H}_\rho$) onto $\mathscr{R}u$ (a dense subspace of the Hilbert space $\mathscr{H}$ on which $\mathscr{R}$ acts), and U_0 extends by continuity to a unitary transformation from $\mathscr{H}_\rho$ onto $\mathscr{H}$. Since $\mathfrak{A} = \mathscr{N}_\rho \cap \mathscr{N}_\rho^* = \mathscr{R}$, we have $\mathfrak{A}^2 = \mathscr{R}$, and S_0 (as defined in the present discussion) is the mapping $A \to A^*: \mathscr{R} \to \mathscr{R}$. It is readily verified that

$$(24) \qquad U\pi(A)U^* = A, \qquad (US_0U^*)(Au) = A^*u,$$

for each A in $\mathscr{R}$. By means of the unitary transformation U, $\mathscr{H}_\rho$ can be identified with $\mathscr{H}$; π then corresponds to the identity representation of $\mathscr{R}$ acting on $\mathscr{H}$, $\mathfrak{A}$ $(\subseteq \mathscr{H}_\rho)$ corresponds to the subspace $\mathscr{R}u$ of $\mathscr{H}$, and S_0 corresponds to the operator $Au \to A^*u: \mathscr{R}u \to \mathscr{R}u$ that was considered in our first approach to modular theory (and denoted, in that approach, by S_0). In the first approach, the subspace $\mathscr{R}'u$ of $\mathscr{H}$ had a significant role. This suggests that we should find a characterization of the corresponding subspace $U^*R'u$ of $\mathscr{H}_\rho$, in a form that can then be used in developing modular theory in the more general setting.

If $y \in \mathscr{H}$, the equation

$$(25) \qquad R_y Au = Ay \qquad (A \in \mathscr{R})$$

defines a linear operator R_y with domain $\mathscr{R}u$ dense in $\mathscr{H}$, and R_y commutes with $\mathscr{R}$, in the sense that

$$(R_y A)(Bu) = (AR_y)(Bu) \qquad (A, B \in \mathscr{R})$$

(see the discussion preceding Lemma 9.2.28). If R_y is bounded, it extends by continuity to an element A' of $\mathscr{R}'$, and $y = R_y u = A'u \in \mathscr{R}'u$. Conversely, if $y = A'u$ for some A' in $\mathscr{R}'$, then

$$R_y Au = Ay = AA'u = A'Au \qquad (A \in \mathscr{R}),$$

and R_y is the bounded operator $A'|\mathscr{R}u$. Hence

$$(26) \qquad \mathscr{R}'u = \{y \in \mathscr{H} : R_y \text{ is bounded}\}.$$

From (24), (25), and (26), it follows that U^*R_yU is the mapping $A \to \pi(A)U^*y$, with domain $\mathscr{R}$ $(= \mathfrak{A} \subseteq \mathscr{H}_\rho)$, and that $U^*\mathscr{R}'u$ is the subspace

$$(27) \qquad \{z \in \mathscr{H}_\rho : \text{the mapping } A \to \pi(A)z : \mathfrak{A} \to \mathscr{H}_\rho \text{ is bounded}\}$$

of $\mathscr{H}_\rho$. When we revert to modular theory in the more general setting, it turns out that this subspace of $\mathscr{H}_\rho$ is too large for our purposes. We therefore seek an appropriate modification of the descriptions, (26) and (27), of $\mathscr{R}'u$ and $U^*\mathscr{R}'u$, in the special case now under consideration. We have already noted that US_0U^* is the operator that, in Lemma 9.2.1, was denoted by S_0. From that lemma, the domain of $(US_0U^*)^*$ $(= UFU^*)$

contains $\mathscr{R}'u$. Accordingly, (26) remains valid when $\mathscr{H}$ is replaced by $\mathscr{D}(UFU^*)$, and $U^*\mathscr{R}'u$ is the subspace

$$\{z \in \mathscr{D}(F) : \text{the mapping } A \to \pi(A)z : \mathfrak{A} \to \mathscr{H}_\rho \text{ is bounded}\}$$

of $\mathscr{H}_\rho$.

We now resume the development of the general theory, with $\mathscr{H}_\rho$, π, and $\mathfrak{A}$ constructed from a faithful normal semi-finite weight ρ on a von Neumann algebra $\mathscr{R}$. Motivated by the preceding discussion, we introduce the subspace

$$\mathfrak{A}' = \{z \in \mathscr{D}(F) : \text{the mapping } A \to \pi(A)z : \mathfrak{A} \to \mathscr{H}_\rho \text{ is bounded}\} \tag{28}$$

of $\mathscr{H}_\rho$. When $z \in \mathfrak{A}'$, the mapping $A \to \pi(A)z$ extends uniquely to a bounded linear operator $\pi'(z)$ acting on $\mathscr{H}_\rho$, and

$$\pi'(z)A = \pi(A)z \qquad (A \in \mathfrak{A}, \quad z \in \mathfrak{A}'). \tag{29}$$

Since $\mathfrak{A}$ is dense in $\mathscr{H}_\rho$, and

$$\begin{aligned}\pi'(z)\pi(A)B &= \pi'(z)(AB) = \pi(AB)z \\ &= \pi(A)\pi(B)z = \pi(A)\pi'(z)B\end{aligned}$$

whenever $A, B \in \mathfrak{A}$ and $z \in \mathfrak{A}'$, we have

$$\pi'(z) \in \pi(\mathfrak{A})' = \pi(\mathscr{R})' \qquad (z \in \mathfrak{A}').$$

It follows that $\pi'(\mathfrak{A}') \subseteq \pi(\mathscr{R})'$. In Proposition 9.2.42, we prove that $\pi'(\mathfrak{A}')$ is a self-adjoint subalgebra of $\pi(\mathscr{R})'$, and we shall show in due course that $\pi'(\mathfrak{A}')^- = \pi(\mathscr{R})'$ (see the discussion following (46)). At that stage, we shall verify that (18) is satisfied in the present context, and prove the closely related results that $J\mathfrak{A} = \mathfrak{A}'$ and $\Delta^{it}\mathfrak{A} = \mathfrak{A}$ for all real t.

An intuitive understanding of the formulae and algebraic manipulations occurring in the following proposition and its proof is best obtained by reference to the special case (just considered) in which ρ is the vector state arising from a separating and generating vector u for $\mathscr{R}$. In that case, $\mathscr{H}$ and $\mathscr{H}_\rho$ can be identified by means of the unitary operator U described above; then $\mathfrak{A}' = \mathscr{R}'u$, and $\pi'(A'u) = A'$ for each A' in $\mathscr{R}'$.

9.2.42. Proposition. (i) *If $z, z_1, z_2 \in \mathfrak{A}'$, then $Fz, \pi'(z_1)z_2 \in \mathfrak{A}'$.*

(ii) *With its linear structure as a subspace of $\mathscr{H}_\rho$, with multiplication defined by $z_1 z_2 = \pi'(z_1)z_2$, and with the involution $z \to Fz$, $\mathfrak{A}'$ becomes an associative linear algebra with involution and π' is a * homomorphism from $\mathfrak{A}'$ into $\pi(\mathscr{R})'$.*

(iii) *If $T \in \pi(\mathscr{R})'$ and $z_1, z_2 \in \mathfrak{A}'$, then $\pi'(z_1)Tz_2 \in \mathfrak{A}'$ and*

$$\pi'(\pi'(z_1)Tz_2) = \pi'(z_1)T\pi'(z_2), \qquad F\pi'(z_1)Tz_2 = \pi'(z_2)^*T^*Fz_1.$$

Proof. (i) Suppose that $z \in \mathfrak{A}'$. When $A, B \in \mathfrak{A}$, we have

$$\begin{aligned}\langle \pi(A)Fz, B\rangle_\rho &= \langle Fz, \pi(A^*)B\rangle_\rho = \langle S_0^* z, A^*B\rangle_\rho \\ &= \langle S_0(A^*B), z\rangle_\rho = \langle B^*A, z\rangle_\rho = \langle \pi(B^*)A, z\rangle_\rho \\ &= \langle A, \pi(B)z\rangle_\rho = \langle A, \pi'(z)B\rangle_\rho = \langle \pi'(z)^*A, B\rangle_\rho .\end{aligned}$$

Since $\mathfrak{A}$ is dense in $\mathscr{H}_\rho$, it follows that $\pi(A)Fz = \pi'(z)^*A$. Hence the mapping $A \to \pi(A)Fz : \mathfrak{A} \to \mathscr{H}_\rho$ extends to the bounded linear operator $\pi'(z)^*$ acting on $\mathscr{H}_\rho$; since, also, $Fz \in \mathscr{D}(F)$, we conclude (see (28) and (29)) that

$$(30) \qquad Fz \in \mathfrak{A}', \qquad \pi'(Fz) = \pi'(z)^* \qquad (z \in \mathfrak{A}').$$

This includes the first assertion of (i); the remainder of (i) follows from (iii), with $T = I$.

(iii) Suppose that $z_1, z_2 \in \mathfrak{A}'$ and $T \in \pi(\mathscr{R})'$. For each A in $\mathfrak{A}$, $\pi(A)$ commutes with T and $\pi'(z_1)$ $(\in \pi(\mathscr{R})')$, so

$$(31) \qquad \pi(A)\pi'(z_1)Tz_2 = \pi'(z_1)T\pi(A)z_2 = \pi'(z_1)T\pi'(z_2)A.$$

When $A, B \in \mathfrak{A}$, $S_0(AB) = B^*A^* = \pi(B)^*A^*$, and by use of (30) and (31) we obtain

$$\begin{aligned}\langle S_0(AB), \pi'(z_1)Tz_2\rangle_\rho &= \langle A^*, \pi(B)\pi'(z_1)Tz_2\rangle_\rho \\ &= \langle A^*, \pi'(z_1)T\pi'(z_2)B\rangle_\rho = \langle \pi'(Fz_2)T^*\pi'(Fz_1)A^*, B\rangle_\rho \\ &= \langle \pi(A^*)\pi'(Fz_2)T^*Fz_1, B\rangle_\rho = \langle \pi'(z_2)^* T^*Fz_1, AB\rangle_\rho .\end{aligned}$$

Since $F = S_0^*$ and $\mathscr{D}(S_0) = \mathfrak{A}^2$, it now follows that

$$\pi'(z_1)Tz_2 \in \mathscr{D}(F), \qquad F\pi'(z_1)Tz_2 = \pi'(z_2)^* T^*Fz_1 .$$

Moreover, by (31), the mapping $A \to \pi(A)\pi'(z_1)Tz_2 : \mathfrak{A} \to \mathscr{H}_\rho$ extends to the bounded linear operator $\pi'(z_1)T\pi'(z_2)$ acting on $\mathscr{H}_\rho$. Thus

$$\pi'(z_1)Tz_2 \in \mathfrak{A}', \qquad \pi'(\pi'(z_1)Tz_2) = \pi'(z_1)T\pi'(z_2).$$

This proves (iii). From (iii), with $T = I$, and (30) we have

$$(32) \quad \pi'(z_1)z_2 \in \mathfrak{A}', \quad F\pi'(z_1)z_2 = \pi'(Fz_2)Fz_1, \quad \pi'(\pi'(z_1)z_2) = \pi'(z_1)\pi'(z_2).$$

(ii) From (32), the equation $z_1z_2 = \pi'(z_1)z_2$ defines a bilinear mapping $(z_1, z_2) \to z_1z_2 : \mathfrak{A}' \times \mathfrak{A}' \to \mathfrak{A}'$. With this multiplication $\mathfrak{A}'$ becomes an associative linear algebra since (again by (32))

$$\begin{aligned}(z_1z_2)z_3 &= \pi'(z_1z_2)z_3 = \pi'(\pi'(z_1)z_2)z_3 \\ &= \pi'(z_1)\pi'(z_2)z_3 = \pi'(z_1)(z_2z_3) = z_1(z_2z_3).\end{aligned}$$

The conjugate-linear mapping $F|\mathfrak{A}'$ is an involution on $\mathfrak{A}'$, since $F(\mathfrak{A}') \subseteq \mathfrak{A}'$, $F^2z = z$, and (by (32)) $F(z_1z_2) = (Fz_2)(Fz_1)$, when $z, z_1, z_2 \in \mathfrak{A}'$. Moreover, the linear mapping $\pi' : \mathfrak{A}' \to \pi(\mathscr{R})'$ is a * homomorphism since (by (32) and (30))

$$\pi'(z_1z_2) = \pi'(z_1)\pi'(z_2), \qquad \pi'(Fz) = \pi'(z)^*. \quad \blacksquare$$

Proposition 9.2.42(ii) motivates many of the algebraic manipulations that follow; for an intuitive understanding of these arguments, it is very desirable to view the mapping $\pi' : \mathfrak{A}' \to \mathscr{B}(\mathscr{H}_\rho)$ as a * homomorphism between algebras with involution.

With each z in $\mathfrak{A}'$ we have associated an element $\pi'(z)$ of $\pi(\mathscr{R})'$, determined by (29). More generally, we prove next that each z in $\mathscr{D}(F)$ gives rise to a closed densely defined operator $\pi'(z)$ affiliated with $\pi(\mathscr{R})'$, the notation being consistent with our previous use of $\pi'(z)$ when $z \in \mathfrak{A}'$. From the first few lines of the proof of Proposition 9.2.42,

$$\langle \pi(A)Fz, B\rangle_\rho = \langle A, \pi(B)z\rangle_\rho \qquad (A, B \in \mathfrak{A}, \quad z \in \mathscr{D}(F)).$$

From this, it follows that the linear operator $B \to \pi(B)z$, with domain $\mathfrak{A}$ dense in $\mathscr{H}_\rho$, is preclosed; for its adjoint extends the mapping $A \to \pi(A)Fz : \mathfrak{A} \to \mathscr{H}_\rho$, and is therefore densely defined.

When $z \in \mathscr{D}(F)$, we denote by $\pi'(z)$ the closure of the linear operator $B \to \pi(B)z : \mathfrak{A} \to \mathscr{H}_\rho$. Thus

$$\text{(33)} \qquad \pi'(z)A = \pi(A)z \qquad (A \in \mathfrak{A}, \quad z \in \mathscr{D}(F)),$$

and (from the preceding paragraph)

$$\text{(34)} \qquad \pi'(Fz) \subseteq \pi'(z)^* \qquad (z \in \mathscr{D}(F)).$$

We assert that

$$\text{(35)} \qquad \pi'(z)\ \eta\ \pi(\mathscr{R})' \qquad (z \in \mathscr{D}(F));$$

equivalently, that $T\pi'(z) \subseteq \pi'(z)T$ for all T in $\pi(\mathscr{R})$. From Lemma 5.6.13, and since $\pi(\mathfrak{A})^- = \pi(\mathscr{R})$, it suffices to prove that

$$\pi(A)\pi'(z) \subseteq \pi'(z)\pi(A)$$

for each A in $\mathfrak{A}$. To this end, suppose that $x \in \mathscr{D}(\pi'(z))$. Since $\mathfrak{A}$ is a core for $\pi'(z)$, there is a sequence $\{B_n\}$ in $\mathfrak{A}$ such that $\|x - B_n\|_\rho \to 0$ and $\|\pi'(z)x - \pi'(z)B_n\|_\rho \to 0$. Since $\pi'(z)$ is closed, $\pi(A)B_n \to \pi(A)x$, and

$$\begin{aligned}\pi'(z)\pi(A)B_n &= \pi'(z)(AB_n) = \pi(AB_n)z \\ &= \pi(A)\pi(B_n)z = \pi(A)\pi'(z)B_n \to \pi(A)\pi'(z)x,\end{aligned}$$

it follows that $\pi(A)x \in \mathscr{D}(\pi'(z))$ and $\pi'(z)\pi(A)x = \pi(A)\pi'(z)x$. Thus $\pi(A)\pi'(z) \subseteq \pi'(z)\pi(A)$, and (35) is proved.

Note that (28) can now be rewritten in the form

$$\mathfrak{A}' = \{z \in \mathscr{D}(F) : \pi'(z) \text{ is bounded}\}. \tag{36}$$

The statements and proofs of the two following results are analogous to those of Lemmas 9.2.4, 9.2.5.

9.2.43. LEMMA. *If $A \in \mathfrak{A}$, $r > 0$, and $y = (\Delta^{-1} + rI)^{-1}A$, then $y \in \mathfrak{A}'$.*

Proof. It is clear that $y \in \mathscr{D}(\Delta^{-1}) = \mathscr{D}(SF) \subseteq \mathscr{D}(F)$; by (36), it suffices to show that $\pi'(y)$ is bounded. Let VK be the polar decomposition of $\pi'(y)$. From Lemma 6.1.10 and (34), $KV^* = \pi'(y)^* \supseteq \pi'(z)$, where $z = Fy$.

If $\pi'(y)$ is unbounded, so is K, and we can choose real numbers a, b such that $b > a > \|\pi(A)\|/2r^{1/2}$ and $KE \neq 0$, where E is the spectral projection for K corresponding to the interval $[a, b]$. Since $\pi'(y)\,\eta\,\pi(\mathscr{R})'$, we have $V, E, KE \in \pi(\mathscr{R})'$. Thus

$$\begin{aligned}\pi(A)Ez &= E\pi(A)z = E\pi'(z)A \\ &= EKV^*A = KEV^*(\Delta^{-1} + rI)y,\end{aligned}$$

and

$$\begin{aligned}\|\pi(A)\|^2 \|Ez\|_\rho^2 \geq \|\pi(A)Ez\|_\rho^2 &= \|KEV^*(\Delta^{-1} + rI)y\|_\rho^2 \\ &\geq \|KEV^*(\Delta^{-1} + rI)y\|_\rho^2 - \|KEV^*(\Delta^{-1} - rI)y\|_\rho^2 \\ &= 4r \operatorname{Re}\langle KEV^*\Delta^{-1}y, KEV^*y\rangle_\rho \\ &= 4r \operatorname{Re}\langle SFy, VK^2EV^*y\rangle_\rho .\end{aligned}$$

We assert that

$$VK^2EV^*y \in \mathscr{D}(F), \qquad FVK^2EV^*y = K^2Ez. \tag{37}$$

Once this has been established, it follows from the preceding inequalities that

$$\begin{aligned}\|\pi(A)\|^2 \|Ez\|_\rho^2 &\geq 4r \operatorname{Re}\langle SFy, VK^2EV^*y\rangle_\rho \\ &= 4r \operatorname{Re}\langle FVK^2EV^*y, Fy\rangle_\rho = 4r \operatorname{Re}\langle K^2Ez, z\rangle_\rho \geq 4ra^2 \|Ez\|_\rho^2 .\end{aligned}$$

Since $\|\pi(A)\|^2 < 4ra^2$, it follows that $Ez = 0$. Now $\mathfrak{A}$ is dense in $\mathscr{H}_\rho$, and

$$KEV^*B = E\pi'(z)B = E\pi(B)z = \pi(B)Ez = 0 \qquad (B \in \mathfrak{A}),$$

so $KEV^* = 0$. Hence

$$KE = V^*V(KE) = V^*(KEV^*)^* = 0,$$

a contradiction. Thus $\pi'(y)$ is bounded, and $y \in \mathfrak{A}'$.

It remains to prove (37). Given any $B, C \in \mathfrak{A}$, $S_0(BC) = \pi(C)^*B^*$, and thus

$$\begin{aligned}\langle S_0(BC), VK^2EV^*y\rangle_\rho &= \langle B^*, \pi(C)VK^2EV^*y\rangle_\rho \\ &= \langle B^*, VK^2EV^*\pi(C)y\rangle_\rho = \langle B^*, VK^2EV^*\pi'(y)C\rangle_\rho \\ &= \langle B^*, \pi'(y)KEV^*VKC\rangle_\rho = \langle \pi'(z)B^*, KEKC\rangle_\rho \\ &= \langle \pi(B)^*z, K^2EC\rangle_\rho = \langle K^2Ez, \pi(B)C\rangle_\rho \\ &= \langle K^2Ez, BC\rangle_\rho.\end{aligned}$$

Since $\mathscr{D}(S_0) = \mathfrak{A}^2$ and $F = S_0^*$, this proves (37). ■

9.2.44. LEMMA. *If $A \in \mathfrak{A}$, $r > 0$, and y is the element $(\Delta^{-1} + rI)^{-1}A$ of $\mathfrak{A}'$, then*

$$\begin{aligned}\langle \pi(A)x_1, x_2\rangle_\rho &= \langle J\pi'(y)^*J\Delta^{1/2}x_1, \Delta^{-1/2}x_2\rangle_\rho \\ &\quad + r\langle J\pi'(y)^*J\Delta^{-1/2}x_1, \Delta^{1/2}x_2\rangle_\rho\end{aligned}$$

for all x_1 and x_2 in $\mathscr{D}(\Delta^{1/2}) \cap \mathscr{D}(\Delta^{-1/2})$.

Proof. The argument is in two parts. Since each is closely analogous to the corresponding stage in the proof of Lemma 9.2.5, we give only a brief outline.

The first step is an approximation argument. Given any x in $\mathscr{D}(\Delta^{1/2}) \cap \mathscr{D}(\Delta^{-1/2})$, there is a sequence $\{B_n\}$ in the dense subspace $\mathfrak{A}^2$ of $\mathscr{H}_\rho$ such that

$$B_n^* \to J\Delta^{-1/2}x + J\Delta^{1/2}x$$

in the norm topology of $\mathscr{H}_\rho$. With z_n defined as $(\Delta^{-1} + I)^{-1}B_n$, $z_n \in \mathfrak{A}'$ by Lemma 9.2.43, and it is not difficult to show that

$$z_n \in \mathscr{D}(\Delta^{1/2}) \cap \mathscr{D}(\Delta^{-1/2}), \qquad \Delta^{1/2-t}z_n \to \Delta^{1/2-t}x \qquad (t = 0, \tfrac{1}{2}, 1).$$

Since both x_1 and x_2 can be approximated in the manner indicated in the preceding paragraph, it now suffices to prove the lemma under the additional assumption that $x_1, x_2 \in \mathfrak{A}'$. In this case, the result follows from a simple algebraic manipulation, similar to the one occurring in the final paragraph of the proof of Lemma 9.2.5. Indeed, the earlier argument can be "translated" in the manner suggested by Proposition 9.2.42(ii), replacing $B_j'u$ by x_j, $B_j'^*B_k'u$ by $\pi'(Fx_j)x_k$, and so on. ■

In our first approach to modular theory (in the case of a von Neumann algebra $\mathscr{R}$ with a separating and generating vector u) we were able to show (see (7)) that $\Delta^{it}\mathscr{R}\Delta^{-it} \subseteq J\mathscr{R}'J$ by a fairly straightforward application of

Lemmas 9.2.4, 9.2.5, and 9.2.8. From this, $J\Delta^{it}A\Delta^{-it}J \in \mathscr{R}'$ when $A \in \mathscr{R}$, and (since $Ju = \Delta u = u$)

$$J\Delta^{it}A\Delta^{-it}Ju = J\Delta^{it}Au. \tag{38}$$

A little further algebraic manipulation was then sufficient to show that $\Delta^{it}\mathscr{R}\Delta^{-it} = \mathscr{R}$ and $J\mathscr{R}J = \mathscr{R}'$. In the present context, a similar argument (based on Lemmas 9.2.43, 9.2.44, and 9.2.8) can be used to show that $\Delta^{it}\pi(\mathscr{R})\Delta^{-it} \subseteq J\pi(\mathscr{R})'J$; but this argument needs slight adaptation so as to include an analogue of (38),

$$J\Delta^{it}A \in \mathfrak{A}', \qquad \pi'(J\Delta^{it}A) = J\Delta^{it}\pi(A)\Delta^{-it}J \qquad (A \in \mathfrak{A}), \tag{39}$$

before we can go on to show that $\Delta^{it}\pi(\mathscr{R})\Delta^{it} = \pi(\mathscr{R})$ and $J\pi(\mathscr{R})J = \pi(\mathscr{R})'$. Our next objective, then, is to prove (39).

Suppose that $A \in \mathfrak{A}$. For each positive real number r, we can define y_r in $\mathfrak{A}'$ by $y_r = (\Delta^{-1} + rI)^{-1}A$. From Lemmas 9.2.44 and 9.2.8, we have

$$\langle J\pi'(y_r)^*Jx_1, x_2\rangle_\rho = \int_{\mathbb{R}} \frac{r^{it-1/2}}{e^{\pi t} + e^{-\pi t}}\langle \Delta^{it}\pi(A)\Delta^{-it}x_1, x_2\rangle_\rho\, dt$$

for all x_1, x_2 in $\mathscr{H}_\rho$. By taking $x_1 = JB$ and $x_2 = Jx$, where $x \in \mathscr{H}_\rho$ and $B \in \mathfrak{A}$, we obtain

$$\begin{aligned}\int_{\mathbb{R}} &\frac{r^{it-1/2}}{e^{\pi t} + e^{-\pi t}}\langle x, J\Delta^{it}\pi(A)\Delta^{-it}JB\rangle_\rho\, dt \\ &= \langle x, \pi'(y_r)^*B\rangle_\rho = \langle x, \pi'(Fy_r)B\rangle_\rho = \langle x, \pi(B)Fy_r\rangle_\rho \\ &= \langle x, \pi(B)J\Delta^{-1/2}y_r\rangle_\rho = \langle x, \pi(B)J\Delta^{-1/2}(\Delta^{-1} + rI)^{-1}A\rangle_\rho.\end{aligned}$$

Hence, for all x in $\mathscr{H}_\rho$ and B in $\mathfrak{A}$,

$$\begin{aligned}\int_{\mathbb{R}} &\frac{r^{it-1/2}}{e^{\pi t} + e^{-\pi t}}\langle x, J\Delta^{it}\pi(A)\Delta^{-it}JB\rangle_\rho\, dt \\ &= \langle \Delta^{-1/2}(\Delta^{-1} + rI)^{-1}A, J\pi(B)^*x\rangle_\rho.\end{aligned} \tag{40}$$

We now assert that

$$\langle \Delta^{-1/2}(\Delta^{-1} + rI)^{-1}x_1, x_2\rangle_\rho = \int_{\mathbb{R}} \frac{r^{it-1/2}}{e^{\pi t} + e^{-\pi t}}\langle \Delta^{it}x_1, x_2\rangle_\rho\, dt \tag{41}$$

for all x_1, x_2 in $\mathscr{H}_\rho$. This can be deduced from Lemma 9.2.7, by an approximation argument similar to (but simpler than) the proof of Lemma 9.2.8. An alternative proof runs as follows. By polarization, it suffices to consider the case in which $x_1 = x_2$ $(= x)$. If $\{E_\lambda\}$ is the spectral resolution of Δ, then $E_\lambda = 0$ when $\lambda \le 0$, and it follows from Theorem 5.6.26 and Lemma 9.2.7 that

$$\begin{aligned}
&\langle \Delta^{-1/2}(\Delta^{-1} + rI)^{-1}x, x\rangle_\rho \\
&= \int_0^\infty \frac{\lambda^{-1/2}}{\lambda^{-1} + r}\, d\langle E_\lambda x, x\rangle_\rho \\
&= \int_0^\infty \frac{1}{\lambda^{-1/2} + r\lambda^{1/2}}\, d\langle E_\lambda x, x\rangle_\rho = \int_0^\infty \left[\int_{\mathbb{R}} \frac{r^{it-1/2}\lambda^{it}}{e^{\pi t} + e^{-\pi t}}\, dt\right] d\langle E_\lambda x, x\rangle_\rho \\
&= \int_{\mathbb{R}} \frac{r^{it-1/2}}{e^{\pi t} + e^{-\pi t}}\left[\int_0^\infty \lambda^{it}\, d\langle E_\lambda x, x\rangle_\rho\right] dt = \int_{\mathbb{R}} \frac{r^{it-1/2}}{e^{\pi t} + e^{-\pi t}} \langle \Delta^{it}x, x\rangle_\rho\, dt.
\end{aligned}$$

From (41),

$$\begin{aligned}
\langle \Delta^{-1/2}(\Delta^{-1} + rI)^{-1}A, J\pi(B)^*x\rangle_\rho &= \int_{\mathbb{R}} \frac{r^{it-1/2}}{e^{\pi t} + e^{-\pi t}} \langle \Delta^{it}A, J\pi(B)^*x\rangle_\rho\, dt \\
&= \int_{\mathbb{R}} \frac{r^{it-1/2}}{e^{\pi t} + e^{-\pi t}} \langle x, \pi(B)J\Delta^{it}A\rangle_\rho\, dt.
\end{aligned}$$

This, with (40), shows that

$$\int_{\mathbb{R}} \frac{r^{it-1/2}}{e^{\pi t} + e^{-\pi t}} \langle x, J\Delta^{it}\pi(A)\Delta^{-it}JB - \pi(B)J\Delta^{it}A\rangle_\rho\, dt = 0$$

for all x in $\mathscr{H}_\rho$, B in $\mathfrak{A}$, and r (> 0). By writing $r = e^u$, it follows that (for all x in $\mathscr{H}_\rho$ and B in $\mathfrak{A}$) the continuous L_1 function

$$\frac{1}{e^{\pi t} + e^{-\pi t}} \langle x, J\Delta^{it}\pi(A)\Delta^{-it}JB - \pi(B)J\Delta^{it}A\rangle_\rho$$

of t has zero for its Fourier transform, and so vanishes for all real t. Thus

(42) $$J\Delta^{it}\pi(A)\Delta^{-it}JB = \pi(B)J\Delta^{it}A \qquad (A, B \in \mathfrak{A}, \quad t \in \mathbb{R}).$$

From (42), the mapping $B \to \pi(B)J\Delta^{it}A\colon \mathfrak{A} \to \mathscr{H}_\rho$ extends to the bounded linear operator $J\Delta^{it}\pi(A)\Delta^{-it}J$ acting on $\mathscr{H}_\rho$. In order to complete the proof of (39), it now suffices to show that $J\Delta^{it}A \in \mathscr{D}(F)$. We assert that

(43) $$J\Delta^{it}A \in \mathscr{D}(F), \qquad FJ\Delta^{it}A = J\Delta^{it}A^* \qquad (A \in \mathfrak{A}).$$

Indeed, since $F = S_0^*$ and $\mathscr{D}(S_0) = \mathfrak{A}^2$, (43) is equivalent to the condition

$$\langle C^*B^*, J\Delta^{it}A\rangle_\rho = \langle J\Delta^{it}A^*, BC\rangle_\rho \qquad (A, B, C \in \mathfrak{A}),$$

which can be rewritten in the form

$$\langle B^*, \pi(C)J\Delta^{it}A\rangle_\rho = \langle \pi(B)^*J\Delta^{it}A^*, C\rangle_\rho \qquad (A, B, C \in \mathfrak{A}).$$

This last equation is a simple consequence of (42), so (43) (and hence, also, (39)) is proved.

From (39),

$$J\Delta^{it}\mathfrak{A} \subseteq \mathfrak{A}', \qquad J\Delta^{it}\pi(\mathfrak{A})\Delta^{-it}J \subseteq \pi'(\mathfrak{A}'). \tag{44}$$

Since $\pi(\mathfrak{A})^{-} = \pi(\mathscr{R})$, it now follows that

$$J\Delta^{it}\pi(\mathscr{R})\Delta^{-it}J \subseteq \pi'(\mathfrak{A}')^{-} \subseteq \pi(\mathscr{R})' \qquad (t \in \mathbb{R}). \tag{45}$$

With $t = 0$, we obtain $J\mathfrak{A} \subseteq \mathfrak{A}'$ (so $\mathfrak{A}'$ is dense in $\mathscr{H}_\rho$) and

$$J\pi(\mathscr{R})J \subseteq \pi'(\mathfrak{A}')^{-} \subseteq \pi(\mathscr{R})'. \tag{46}$$

We shall prove that equality holds throughout (46). For this, note first that $I = JIJ \in J\pi(\mathscr{R})J \subseteq \pi'(\mathfrak{A}')^{-}$; so there is a net $\{z_j\}$ in $\mathfrak{A}'$ such that $\{\pi'(z_j)\}$ is bounded and strong-operator convergent to I. If $T \in \pi(\mathscr{R})'$, then $\pi'(z_j)T\pi'(z_j) \in \pi'(\mathfrak{A}')$ by Proposition 9.2.42(iii), and

$$T = \lim \pi'(z_j)T\pi'(z_j) \in \pi'(\mathfrak{A}')^{-}.$$

Thus $\pi(\mathscr{R})' \subseteq \pi'(\mathfrak{A}')^{-}$; from (46), $\pi(\mathscr{R})' = \pi'(\mathfrak{A}')^{-}$.

We can now "translate" the argument set out in the third paragraph preceding Theorem 9.2.9; the main tools for this purpose are Proposition 9.2.42(ii), together with (39) and (43) in the case $t = 0$. In this way, we prove that

$$\pi'(y)Jx = J\pi'(x)Jy, \qquad (J\pi'(x)J)\pi'(y) = \pi'(y)(J\pi'(x)J), \tag{47}$$

whenever $x, y \in \mathfrak{A}'$. From the second of these relations, and since $\pi'(\mathfrak{A}')^{-} = \pi(\mathscr{R})'$, it follows that

$$J\pi(\mathscr{R})'J \subseteq \pi(\mathscr{R})'' = \pi(\mathscr{R}), \qquad \pi(\mathscr{R})' \subseteq J\pi(\mathscr{R})J.$$

Hence equality holds throughout (46).

From $J\pi(\mathscr{R})J = \pi(\mathscr{R})'$ and (45), it follows easily that $\Delta^{it}\pi(\mathscr{R})\Delta^{-it} = \pi(\mathscr{R})$ for all real t. Note also that, since $J\mathfrak{A} \subseteq \mathfrak{A}' \subseteq \mathscr{D}(F)$, we have $\mathfrak{A} \subseteq \mathscr{D}(FJ) = \mathscr{D}(\Delta^{1/2})$, and thus

$$\mathfrak{A} \subseteq \mathscr{D}(\Delta^{1/2}) = \mathscr{D}(S). \tag{48}$$

The operator $S_1 \colon A \to A^*$ (with domain $\mathfrak{A}$ dense in $\mathscr{H}_\rho$) is preclosed (Proposition 9.2.40(iv)) and extends S_0, so $S = \bar{S}_0 \subseteq \bar{S}_1$. From this, together with (48), $S \supseteq \bar{S}_1 \,|\, \mathfrak{A} = S_1$; so $S = \bar{S} \supseteq \bar{S}_1$, and therefore $S = \bar{S}_1$. This establishes Tomita's theorem (9.2.37), in the context of a von Neumann algebra $\mathscr{R}$ and the representation π associated with a faithful normal semi-finite weight ρ on $\mathscr{R}$.

Our next objective is to prove that the modular automorphism group $\{\sigma_t\}$ of $\mathscr{R}$ relative to ρ, defined by (19), satisfies the modular condition in the form set out in Theorem 9.2.38(a), (b). To this end, we shall need to show that

(49) $$J\mathfrak{A} = \mathfrak{A}', \qquad \Delta^{it}\mathfrak{A} = \mathfrak{A} \qquad (t \in \mathbb{R}).$$

We require two preparatory results.

9.2.45. Lemma. *If $y \in \mathscr{H}_\rho$ and $\langle \pi(A)y, y\rangle_\rho \le \rho(A)$ for each A in $\mathscr{R}^+$, there is a vector x in $\mathscr{H}_\rho$ such that*

$$x \in \mathfrak{A}', \qquad x = Fx, \qquad \langle \pi(A)x, x\rangle_\rho = \langle \pi(A)y, y\rangle_\rho \qquad (A \in \mathscr{R}).$$

Proof. When $A \in \mathfrak{A}$,

$$\|\pi(A)y\|_\rho^2 = \langle \pi(A^*A)y, y\rangle_\rho \le \rho(A^*A) = \|A\|_\rho^2;$$

so the mapping $A \to \pi(A)y\colon \mathfrak{A} \to \mathscr{H}_\rho$ is bounded, and extends by continuity to a bounded linear operator T acting on $\mathscr{H}_\rho$. Since $\mathfrak{A}$ is dense in $\mathscr{H}_\rho$, $\pi(\mathfrak{A})^- = \pi(\mathscr{R})$, and $\pi(A)TB = \pi(AB)y = T(AB) = T\pi(A)B$ for all A and B in $\mathfrak{A}$, it follows that $T \in \pi(\mathscr{R})'$. If T has polar decomposition VH, then $V, H \in \pi(\mathscr{R})'$, and the range space

$$[T(\mathscr{H}_\rho)] = [T(\mathfrak{A})] = [\pi(\mathfrak{A})y] = [\pi(\mathscr{R})y]$$

of VV^* contains y. Thus

$$\langle \pi(A)y, y\rangle_\rho = \langle \pi(A)y, VV^*y\rangle_\rho = \langle \pi(A)x, x\rangle_\rho \qquad (A \in \mathscr{R}),$$

where $x = V^*y$.

Since

$$\pi(A)x = \pi(A)V^*y = V^*\pi(A)y = V^*TA = HA \qquad (A \in \mathfrak{A}),$$

the mapping $A \to \pi(A)x\colon \mathfrak{A} \to \mathscr{H}_\rho$ extends to the bounded operator H. Moreover, when $A, B \in \mathfrak{A}$,

$$\begin{aligned}\langle S_0(AB), x\rangle_\rho &= \langle \pi(B)^*A^*, x\rangle_\rho = \langle A^*, \pi(B)x\rangle_\rho \\ &= \langle A^*, HB\rangle_\rho = \langle HA^*, B\rangle_\rho = \langle \pi(A^*)x, B\rangle_\rho = \langle x, AB\rangle_\rho.\end{aligned}$$

Since $F = S_0^*$ and $\mathscr{D}(S_0) = \mathfrak{A}^2$, it now follows that $x \in \mathscr{D}(F)$ and $Fx = x$. Finally, $x \in \mathfrak{A}'$ since $\pi'(x)$ is the bounded operator H. ∎

9.2.46. Lemma. *There is a family $\{x_b\}$ of elements of $\mathfrak{A}'$ such that $Fx_b = x_b$ for each index b, and*

$$\rho(R) = \sum_b \langle \pi(R)x_b, x_b\rangle_\rho \qquad (R \in \mathscr{R}^+).$$

Proof. Since π is a * isomorphism from $\mathscr{R}$ onto $\pi(\mathscr{R})$, $\rho \circ \pi^{-1}$ is a normal weight on $\pi(\mathscr{R})$. Hence there is a family $\{y_b\}$ of vectors in $\mathscr{H}_\rho$, such that

$$\rho(\pi^{-1}(T)) = \sum_b \langle Ty_b, y_b\rangle_\rho \qquad (T \in \pi(\mathscr{R})^+);$$

equivalently,

$$\rho(R) = \sum_b \langle \pi(R)y_b, y_b\rangle_\rho \qquad (R \in \mathscr{R}^+).$$

It follows from Lemma 9.2.45 that, for each index b, there is a vector x_b in $\mathfrak{A}'$ such that $x_b = Fx_b$ and

$$\langle \pi(R)x_b, x_b\rangle_\rho = \langle \pi(R)y_b, y_b\rangle_\rho \qquad (R \in \mathscr{R}). \quad \blacksquare$$

From (39), $J\Delta^{it}\mathfrak{A} \subseteq \mathfrak{A}'$ for all real t; in particular, $J\mathfrak{A} \subseteq \mathfrak{A}'$. If we prove that $\mathfrak{A}' \subseteq J\mathfrak{A}$, then it follows that $\mathfrak{A}' = J\mathfrak{A}$, that $J\Delta^{it}\mathfrak{A} \subseteq J\mathfrak{A}$ and $\Delta^{it}\mathfrak{A} \subseteq \mathfrak{A}$ for all real t, and hence that $\Delta^{it}\mathfrak{A} = \mathfrak{A}$ for all real t. Accordingly, in order to prove (49), it remains only to show that $\mathfrak{A}' \subseteq J\mathfrak{A}$; and for this, it suffices to show that $J\mathfrak{A}$ contains each vector x satisfying $x = Fx \in \mathfrak{A}'$. Given such an x, $\pi'(x)$ is a self-adjoint element of $\pi(\mathscr{R})'$, so $J\pi'(x)J$ is a self-adjoint element $\pi(R)$ of $\pi(\mathscr{R})$. From (47), we have

$$\pi(R)y = \pi'(y)Jx \qquad (y \in \mathfrak{A}'). \tag{50}$$

We shall show that $R \in \mathfrak{A}$ and $x = JR$.

Let $\{x_b : b \in \mathbb{B}\}$ be a family of vectors with the properties set out in Lemma 9.2.46. For each finite subset $\mathbb{F}$ of $\mathbb{B}$, and each A in $\mathfrak{A}$, we have

$$\begin{aligned}\sum_{b \in \mathbb{F}} \|\pi'(x_b)A\|_\rho^2 &= \sum_{b \in \mathbb{F}} \|\pi(A)x_b\|_\rho^2 \\ &= \sum_{b \in \mathbb{F}} \langle \pi(A^*A)x_b, x_b\rangle_\rho \\ &\le \rho(A^*A) = \|A\|_\rho^2.\end{aligned}$$

Since $\mathfrak{A}$ is dense in $\mathscr{H}_\rho$, it follows that

$$\sum_{b \in \mathbb{F}} \|\pi'(x_b)u\|_\rho^2 \le \|u\|_\rho^2 \qquad (u \in \mathscr{H}_\rho).$$

This last inequality is satisfied for every finite subset $\mathbb{F}$ of $\mathbb{B}$, so

$$\sum_{b \in \mathbb{B}} \|\pi'(x_b)u\|_\rho^2 \le \|u\|_\rho^2 \qquad (u \in \mathscr{H}_\rho). \tag{51}$$

From (50) and (51),

$$\rho(R^2) = \sum_{b \in \mathbb{B}} \langle \pi(R^2)x_b, x_b \rangle_\rho = \sum_{b \in \mathbb{B}} \|\pi(R)x_b\|_\rho^2$$
$$= \sum_{b \in \mathbb{B}} \|\pi'(x_b)Jx\|_\rho^2 \leq \|Jx\|_\rho^2 = \|x\|_\rho^2 < \infty.$$

Since R is self-adjoint and $\rho(R^2) < \infty$, it follows that $R \in \mathcal{N}_\rho \cap \mathcal{N}_\rho^* = \mathfrak{A}$. Moreover, (50) can now be rewritten in the form

$$\pi'(y)R = \pi'(y)Jx \qquad (y \in \mathfrak{A}');$$

this implies that $TR = TJx$ for each T in $\pi'(\mathfrak{A}')^-$ $(= \pi(\mathscr{R})')$, and hence that $R = Jx$ (since we can take T to be I). Thus $x = JR \in J\mathfrak{A}$, $\mathfrak{A}' \subseteq J\mathfrak{A}$, and (49) is proved.

We now consider the modular automorphism group $\{\sigma_t\}$ of $\mathscr{R}$ relative to ρ, and show that it satisfies the modular condition. If $A, B \in \mathfrak{A}$ and $t \in \mathbb{R}$, then $\Delta^{it}A \in \mathfrak{A}$ and $J\Delta^{it}A, JB \in \mathfrak{A}'$, by (49). From (19), (39), and (47) with $J\Delta^{it}A$ for x and JB for y, we have

$$\pi(\sigma_t(A))JB = \Delta^{it}\pi(A)\Delta^{-it}JB = J\pi'(J\Delta^{it}A)B$$
$$= \pi'(JB)\Delta^{it}A = \pi(\Delta^{it}A)JB.$$

Since π is faithful and $\mathfrak{A}$ is dense in $\mathscr{H}_\rho$, it follows that

(52) $$\sigma_t(A) = \Delta^{it}A \qquad (A \in \mathfrak{A}).$$

In order to prove that

$$\rho(\sigma_t(R)) = \rho(R) \qquad (R \in \mathscr{R}^+, \quad t \in \mathbb{R}),$$

it suffices to consider the case in which $\rho(R) < \infty$. In that case, $R^{1/2}$ is a self-adjoint element A of $\mathfrak{A}$, and $\sigma_t(A) = \Delta^{it}A \in \mathfrak{A}$ $(\subseteq \mathscr{H}_\rho)$ for all real t. Since Δ^{it} is a unitary operator on $\mathscr{H}_\rho$, $\|\sigma_t(A)\|_\rho^2 = \|A\|_\rho^2$; that is, $\rho(\sigma_t(R)) = \rho(R)$.

The preceding paragraph shows that $\{\sigma_t\}$ satisfies part (a) of the modular condition relative to ρ (see Theorem 9.2.38). Part (b) can be verified by the argument already used in proving Theorem 9.2.13, since $\mathcal{N}_\rho \cap \mathcal{N}_\rho^* = \mathfrak{A} \subseteq \mathscr{D}(\Delta^{1/2})$ (see (48)) and

$$\rho(\sigma_t(A)B) = \rho(A\sigma_{-t}(B)) = \langle \sigma_{-t}(B), A^* \rangle_\rho = \langle \Delta^{-it}B, A^* \rangle_\rho,$$
$$\rho(B\sigma_t(A)) = \langle \sigma_t(A), B^* \rangle_\rho = \langle \Delta^{it}A, B^* \rangle_\rho,$$

when $A, B \in \mathfrak{A}$.

The arguments required to establish the uniqueness assertion in Theorem 9.2.38 are closely analogous, except at one point, to those already used in proving Theorem 9.2.16; so we give only a brief outline. Suppose that a one-parameter group $\{\alpha_t\}$ of * automorphisms of $\mathscr{R}$ satisfies conditions (a) and (b) of Theorem 9.2.38. Since $\rho \circ \alpha_t = \rho$, each α_t preserves the subsets F_ρ, $\mathscr{M}_\rho$, $\mathscr{N}_\rho$, and $\mathscr{N}_\rho \cap \mathscr{N}_\rho^*$ $(= \mathfrak{A})$ of $\mathscr{R}$, associated with the weight ρ. Moreover, $\|\alpha_t(A)\|_\rho = \|A\|_\rho$ when $A \in \mathfrak{A}$ $(\subseteq \mathscr{H}_\rho)$; so there is a unitary operator U_t acting on $\mathscr{H}_\rho$, such that

$$U_t A = \alpha_t(A) \qquad (A \in \mathfrak{A} \subseteq \mathscr{H}_\rho, \quad t \in \mathbb{R}).$$

Straightforward calculation shows that $\{U_t\}$ is a one-parameter unitary group, and $\pi(\alpha_t(A)) = U_t \pi(A) U_{-t}$ when $A \in \mathfrak{A}$ (since $\pi(\alpha_t(A))B = \alpha_t(A)B = U_t(A\alpha_{-t}(B)) = U_t(AU_{-t}(B)) = U_t(\pi(A)U_{-t}(B)) = U_t\pi(A)U_{-t}(B)$ for each B in $\mathfrak{A}$). From ultraweak continuity of π and α_t, and since $\mathscr{R} = \mathfrak{A}^-$,

$$\pi(\alpha_t(R)) = U_t \pi(R) U_{-t} \qquad (R \in \mathscr{R}).$$

Since $\{\alpha_t\}$ satisfies condition (b) in Theorem 9.2.38, the mapping

$$t \to \langle U_t A, B\rangle_\rho = \rho(B^*\alpha_t(A))$$

is continuous on $\mathbb{R}$, when $A, B \in \mathfrak{A}$. From this, since $\mathfrak{A}$ is dense in $\mathscr{H}_\rho$, and since

$$\|(U_t - I)A\|_\rho^2 = \langle [2I - U_t - U_{-t}]A, A\rangle_\rho,$$

it follows that the one-parameter group $\{U_t\}$ is strong-operator continuous. Since the * isomorphism $\pi^{-1}\colon \pi(\mathscr{R}) \to \mathscr{R}$ is ultraweakly continuous, the same is true of the mapping

$$t \to \alpha_t(R) = \pi^{-1}(U_t \pi(R) U_{-t})\colon \mathbb{R} \to \mathscr{R}$$

for each fixed R in $\mathscr{R}$. By Stone's theorem, $U_t = \exp itH$, where the infinitesimal generator H is a self-adjoint operator with domain dense in $\mathscr{H}_\rho$. We have to show that $\exp H = \Delta$ (whence $U_t = \Delta^{it}$ and $\alpha_t = \sigma_t$).

For each A in $\mathscr{R}$ and k in $L_1(\mathbb{R})$, the mapping

$$(x, y) \to \int_{\mathbb{R}} k(t)\langle \alpha_t(A)x, y\rangle\, dt$$

is a bounded conjugate-bilinear functional on $\mathscr{H}$ (the Hilbert space on which $\mathscr{R}$ acts), and the corresponding operator is an element A_k of $\mathscr{R}$. We require the following result.

9.2.47. Lemma. *If $k \in L_1(\mathbb{R})$ and $A, B \in \mathfrak{A}$, then $A_k \in \mathfrak{A}$ and*

$$\rho(BA_k) = \int_{\mathbb{R}} k(t)\rho(B\alpha_t(A))\, dt, \qquad \rho(A_k B) = \int_{\mathbb{R}} k(t)\rho(\alpha_t(A)B)\, dt.$$

Suppose, for the moment, that this lemma has been proved. When $A, B \in \mathfrak{A}$ and $k \in L_1(\mathbb{R})$,

$$\langle A_k, B\rangle_\rho = \rho(B^*A_k) = \int_{\mathbb{R}} k(t)\rho(B^*\alpha_t(A))\,dt$$

and $\rho(B^*\alpha_t(A)) = \langle \alpha_t(A), B\rangle_\rho = \langle U_t A, B\rangle_\rho = \langle e^{itH}A, B\rangle_\rho$. By Stone's theorem (5.6.36), $\langle A_k, B\rangle_\rho = \langle \hat{k}(H)A, B\rangle_\rho$; so

$$A_k = \hat{k}(H)A \qquad (A \in \mathfrak{A} \subseteq \mathscr{H}_\rho, \quad k \in L_1(\mathbb{R})).$$

Now suppose that l lies in the function class $\check{\mathscr{F}}$ used in proving Theorem 9.2.16, and define k in $\check{\mathscr{F}}$ by $k(z) = l(z + i)$, so that $\hat{k}(\lambda) = \hat{l}(\lambda)\exp\lambda$. Given A and B in $\mathfrak{A}$, let f be the function occurring in condition (b) of Theorem 9.2.38. By integrating $l(z)f(z)$ round the (counter-clockwise) rectangle with vertices $\pm R$, $\pm R + i$, and then taking limits as $R \to \infty$ (as in the proof of Theorem 9.2.16) and using Lemma 9.2.47, we obtain $\rho(A_l B) = \rho(BA_k)$; that is

$$\langle B, A_l^*\rangle_\rho = \langle A_k, B^*\rangle_\rho.$$

Since $S = \bar{S}_1$ and $F = S^* = S_1^*$ (where S_1 is defined in the discussion following (48)), it now follows that $A_k \in \mathscr{D}(F)$, $FA_k = A_l^* = SA_l$, $A_l \in \mathscr{D}(\Delta)$ and $\Delta A_l = A_k$. By "translating" the argument used (following (12)) in proving Theorem 9.2.16, we now conclude that $\Delta = \exp H$.

It remains to prove Lemma 9.2.47. By linearity, it suffices to consider the case in which A and B are self-adjoint and $k(t) \geq 0$ for all real t. In this case, A_k is self-adjoint. There is a family $\{y_a : a \in \mathbb{A}\}$ of vectors in $\mathscr{H}$, such that

$$\rho(R) = \sum_{a \in \mathbb{A}} \langle Ry_a, y_a\rangle \qquad (R \in \mathscr{R}^+),$$

and this formula remains valid (with the sum on the right-hand side converging) when $R \in \mathscr{M}_\rho$. When $a \in \mathbb{A}$, the bounded function $\|\alpha_t(A)y_a\|$ of t is continuous on $\mathbb{R}$, since (by the argument set out in the paragraph following Theorem 9.2.16) the mapping $t \to \alpha_t(A)$ is strong-operator continuous. For each unit vector u in $\mathscr{H}$,

$$|\langle A_k y_a, u\rangle| = \left|\int_{\mathbb{R}} k(t)\langle \alpha_t(A)y_a, u\rangle\,dt\right| \leq \int_{\mathbb{R}} k(t)\,\|\alpha_t(A)y_a\|\,dt;$$

so $\|A_k y_a\| \leq \int_{\mathbb{R}} k(t)\,\|\alpha_t(A)y_a\|\,dt$, and

$$\text{(53)} \qquad \langle A_k^2 y_a, y_a\rangle = \|A_k y_a\|^2 \leq \int_{\mathbb{R}}\int_{\mathbb{R}} k(s)k(t)\,\|\alpha_s(A)y_a\|\,\|\alpha_t(A)y_a\|\,ds\,dt.$$

If $\mathbb{F}$ is a finite subset of $\mathbb{A}$,

$$\begin{aligned}\sum_{a\in\mathbb{F}} \|\alpha_s(A)y_a\|\,\|\alpha_t(A)y_a\| &\le \left[\sum_{a\in\mathbb{F}} \|\alpha_s(A)y_a\|^2\right]^{1/2}\left[\sum_{a\in\mathbb{F}} \|\alpha_t(A)y_a\|^2\right]^{1/2}\\ &\le \left[\sum_{a\in\mathbb{A}} \langle\alpha_s(A^2)y_a, y_a\rangle\right]^{1/2}\left[\sum_{a\in\mathbb{A}} \langle\alpha_t(A^2)y_a, y_a\rangle\right]^{1/2}\\ &= [\rho(\alpha_s(A^2))\rho(\alpha_t(A^2))]^{1/2} = \rho(A^2).\end{aligned}$$

This, together with (53), yields

$$\sum_{a\in\mathbb{F}} \langle A_k^2 y_a, y_a\rangle \le \int_{\mathbb{R}}\int_{\mathbb{R}} k(s)k(t)\rho(A^2)\,ds\,dt = \rho(A^2)\left[\int_{\mathbb{R}} k(t)\,dt\right]^2$$

for every finite subset $\mathbb{F}$ of $\mathbb{A}$. Thus

$$\rho(A_k^2) = \sum_{a\in\mathbb{A}} \langle A_k^2 y_a, y_a\rangle \le \rho(A^2)\left[\int_{\mathbb{R}} k(t)\,dt\right]^2 < \infty.$$

Since A_k is self-adjoint, it now follows that $A_k \in \mathcal{N}_\rho \cap \mathcal{N}_\rho^* (= \mathfrak{A})$.

Since A, B, and A_k are self-adjoint elements of $\mathfrak{A}$, we have $A_k B, \alpha_t(A)B \in \mathcal{M}_\rho$, and

$$\rho(A_k B) = \sum_{a\in\mathbb{A}} \langle A_k B y_a, y_a\rangle = \sum_{a\in\mathbb{A}} \int_{\mathbb{R}} k(t)\langle\alpha_t(A)By_a, y_a\rangle\,dt,$$

$$\rho(\alpha_t(A)B) = \sum_{a\in\mathbb{A}} \langle\alpha_t(A)By_a, y_a\rangle.$$

In order to show that $\rho(A_k B) = \int_{\mathbb{R}} k(t)\rho(\alpha_t(A)B)\,dt$, we have to prove that

$$\sum_{a\in\mathbb{A}} \int_{\mathbb{R}} k(t)\langle\alpha_t(A)By_a, y_a\rangle\,dt = \int_{\mathbb{R}} k(t)\left[\sum_{a\in\mathbb{A}} \langle\alpha_t(A)By_a, y_a\rangle\right]dt. \tag{54}$$

Now $\sum_{a\in\mathbb{A}} \|By_a\|^2 = \sum_{a\in\mathbb{A}} \langle B^2 y_a, y_a\rangle = \rho(B^2) < \infty$, and

$$\sum_{a\in\mathbb{A}} \|\alpha_t(A)y_a\|^2 = \rho(\alpha_t(A^2)) = \rho(A^2) < \infty. \tag{55}$$

Since, also

$$\begin{aligned}&\sum_{a\in\mathbb{A}} |\langle\alpha_t(A)By_a, y_a\rangle|\\ &\quad= \sum_{a\in\mathbb{A}} |\langle By_a, \alpha_t(A)y_a\rangle| \le \sum_{a\in\mathbb{A}} \|By_a\|\,\|\alpha_t(A)y_a\|\\ &\quad\le \left[\sum_{a\in\mathbb{A}} \|By_a\|^2\right]^{1/2}\left[\sum_{a\in\mathbb{A}} \|\alpha_t(A)y_a\|^2\right]^{1/2} = [\rho(A^2)\rho(B^2)]^{1/2},\end{aligned}$$

it follows that the finite subsums of $\sum_{a \in \mathbb{A}} k(t)\langle \alpha_t(A)By_a, y_a\rangle$ are all dominated in absolute value by $[\rho(A^2)\rho(B^2)]^{1/2}k(t)$. Since $k \in L_1(\mathbb{R})$, (54) will now follow from the dominated convergence theorem if we show that both sides of the equation are unchanged when $\mathbb{A}$ is replaced by a suitable *countable* subset $\mathbb{B}$.

When $t \in \mathbb{R}$, let $\mathbb{B}_t$ be the subset $\{a \in \mathbb{A} : \alpha_t(A)y_a \neq 0\}$ of $\mathbb{A}$; also, let $\mathbb{B} = \bigcup \{\mathbb{B}_t : t \text{ rational}\}$. In view of (55), each $\mathbb{B}_t$ is countable, and hence so is $\mathbb{B}$. When $a \in \mathbb{A}\backslash\mathbb{B}$, $\alpha_t(A)y_a = 0$ for all rational t; by continuity this remains true for all real t, and

$$0 = \langle By_a, \alpha_t(A)y_a\rangle = \langle \alpha_t(A)By_a, y_a\rangle \qquad (t \in \mathbb{R}).$$

Hence both sides of (54) are unaltered, when $\mathbb{A}$ is replaced by its countable subset $\mathbb{B}$. As noted above, this completes the proof of (54), and so shows that $\rho(A_k B) = \int_{\mathbb{R}} k(t)\rho(\alpha_t(A)B)\,dt$. By taking complex conjugates of both sides of this last equation, we obtain the corresponding expression for $\rho(BA_k)$.

Bibliography: [10, 11, 14, 15, 33, 45, 75, 95, 99, 103, 105].

9.3. Unitary equivalence of type I algebras

This section is concerned with the spatial structure of type I von Neumann algebras, including the problems of deciding whether two given algebras of type I are unitarily equivalent, and whether a given * isomorphism between two such algebras is unitarily implemented. After straightforward reductions discussed in the next few paragraphs, it suffices to answer these questions in the case where both von Neumann algebras have the same type I_m, and their commutants have the same type I_n. In this case we show (Theorems 9.3.3 and 9.3.4) that the algebras are unitarily equivalent if and only if their centers are * isomorphic, and that every * isomorphism is unitarily implemented. The structure of such algebras is determined (Theorem 9.3.2), in terms of maximal abelian von Neumann algebras, together with the matrix constructs introduced in Section 6.6. For the case of separable Hilbert spaces, a complete description of the maximal abelian von Neumann algebras is given in Section 9.4.

Suppose that $\mathscr{R}$ is a type I von Neumann algebra acting on a Hilbert space $\mathscr{H}$, and $\mathscr{C}$ is the common center of $\mathscr{R}$ and $\mathscr{R}'$. There are projections P_m $(1 \leq m \leq \dim \mathscr{H})$ in $\mathscr{C}$, with sum I, such that $\mathscr{R}P_m$ is type I_m unless $P_m = 0$. By Theorem 9.1.3, $\mathscr{R}'$ is also of type I, so we can choose the corresponding projections P'_n $(1 \leq n \leq \dim \mathscr{H})$ for $\mathscr{R}'$. With P_{mn} defined to be $P_m P'_n$, P_{mn} is a projection in $\mathscr{C}$ $(1 \leq m, n \leq \dim \mathscr{H})$ and $\sum P_{mn} = I$ (so $\mathscr{R}$ is unitarily equivalent to the direct sum of the von Neumann algebras $\mathscr{R}P_{mn}$); moreover,

$\mathscr{R}P_{mn}$ is of type I_m, with commutant $\mathscr{R}'P_{mn}$ of type I_n, unless $P_{mn} = 0$. It is convenient to define P_{mn} to be 0 when either m or n exceeds dim $\mathscr{H}$; and it is apparent that the properties just stated (including $\sum P_{mn} = I$) determine the P_{mn} uniquely.

Now suppose that $\mathscr{S}$ is a type I von Neumann algebra acting on a Hilbert space $\mathscr{K}$. The process just used, to construct central projections P_{mn} in $\mathscr{R}$, yields the corresponding (unique) projections Q_{mn} in the center of $\mathscr{S}$. We assert that $\mathscr{R}$ and $\mathscr{S}$ are unitarily equivalent if and only if $\mathscr{R}P_{mn}$ and $\mathscr{S}Q_{mn}$ are unitarily equivalent for all m and n. Indeed, if U is an isomorphism from $\mathscr{H}$ onto $\mathscr{K}$, and $\mathscr{S} = U\mathscr{R}U^{-1}$, then the * isomorphism $\psi: A \to UAU^{-1}$ from $\mathscr{B}(\mathscr{H})$ onto $\mathscr{B}(\mathscr{K})$ carries $\mathscr{R}$ onto $\mathscr{S}$, $\mathscr{R}'$ onto $\mathscr{S}'$, $\mathscr{R}P_{mn}$ onto $\mathscr{S}\psi(P_{mn})$, and $\mathscr{R}'P_{mn}$ onto $\mathscr{S}'\psi(P_{mn})$. From this, $\mathscr{S}\psi(P_{mn})$ and $\mathscr{S}'\psi(P_{mn})$ are of types I_m and I_n, respectively; and, of course, the $\psi(P_{mn})$ are central projections in $\mathscr{S}$, with sum I. Accordingly, $Q_{mn} = \psi(P_{mn}) = UP_{mn}U^{-1}$, from uniqueness of the Q_{mn}; and U restricts to a unitary transformation U_{mn} from $P_{mn}(\mathscr{H})$ onto $Q_{mn}(\mathscr{K})$. Since $U\mathscr{R}P_{mn}U^{-1} = \psi(\mathscr{R}P_{mn}) = \mathscr{S}Q_{mn}$, U_{mn} implements a unitary equivalence between $\mathscr{R}P_{mn}$ and $\mathscr{S}Q_{mn}$. Conversely, if (for each m and n) there is a unitary operator U_{mn}, from $P_{mn}(\mathscr{H})$ onto $Q_{mn}(\mathscr{K})$, that implements a unitary equivalence between $\mathscr{R}P_{mn}$ and $\mathscr{S}Q_{mn}$, then $\mathscr{S} = U\mathscr{R}U^{-1}$, where U is the isomorphism from $\mathscr{H}$ onto $\mathscr{K}$ whose restriction to $P_{mn}(\mathscr{H})$ is U_{mn}. In this way, the unitary equivalence problem for $\mathscr{R}$ and $\mathscr{S}$ is reduced to the corresponding question about $\mathscr{R}P_{mn}$ and $\mathscr{S}Q_{mn}$. Theorem 9.3.3 tells us that the latter two algebras are unitarily equivalent if and only if their centers are * isomorphic.

Suppose next that φ is a * isomorphism from $\mathscr{R}$ onto $\mathscr{S}$. If φ is implemented by a unitary operator U from $\mathscr{H}$ onto $\mathscr{K}$, the argument set out in the preceding paragraph shows that $Q_{mn} = UP_{mn}U^{-1} = \varphi(P_{mn})$ for each m and n. Conversely, if $Q_{mn} = \varphi(P_{mn})$ for each m and n, the restriction $\varphi \mid \mathscr{R}P_{mn}$ is a * isomorphism from $\mathscr{R}P_{mn}$ onto $\mathscr{S}Q_{mn}$. From Theorem 9.3.4, $\varphi \mid \mathscr{R}P_{mn}$ is implemented by a unitary transformation U_{mn} from $P_{mn}(\mathscr{H})$ onto $Q_{mn}(\mathscr{K})$; so φ is implemented by the isomorphism, from $\mathscr{H}$ onto $\mathscr{K}$, whose restriction to $P_{mn}(\mathscr{H})$ is U_{mn}. Accordingly, once Theorem 9.3.4 is established, it follows that φ is unitarily implemented if and only if $\varphi(P_{mn}) = Q_{mn}$ for each m and n.

The remainder of this section is concerned with algebras of type I_m with commutants of type I_n. We shall make use of the following result, which has already been noted (Remark 7.2.10) in the countably decomposable case.

9.3.1. THEOREM. *If $\mathscr{A}_j$ is a maximal abelian von Neumann algebra acting on a Hilbert space $\mathscr{H}_j$ ($j = 1, 2$) and φ is a * isomorphism from $\mathscr{A}_1$ onto $\mathscr{A}_2$, there is a unitary operator U from $\mathscr{H}_1$ onto $\mathscr{H}_2$ such that $\varphi(A) = UAU^{-1}$ for each A in $\mathscr{A}_1$.*

Proof. Let $\{P_a\}$ be an orthogonal family of cyclic (hence countably decomposable) projections in $\mathscr{A}_1$, with sum I. With Q_a defined as $\varphi(P_a)$, Q_a is a countably decomposable projection in $\mathscr{A}_2$, and $\sum Q_a = I$. The countably decomposable von Neumann algebras $\mathscr{A}_1 P_a$ and $\mathscr{A}_2 Q_a$ are clearly maximal abelian (as algebras acting on the ranges of P_a and Q_a, respectively), and the restriction $\varphi \mid \mathscr{A}_1 P_a$ is a * isomorphism from $\mathscr{A}_1 P_a$ onto $\mathscr{A}_2 Q_a$. By Remark 7.2.10, there is a unitary transformation U_a, from $P_a(\mathscr{H}_1)$ onto $Q_a(\mathscr{H}_2)$, such that $U_a A P_a U_a^{-1} = \varphi(AP_a) = \varphi(A)Q_a \quad (A \in \mathscr{A}_1)$. Accordingly, $\varphi(A) = UAU^{-1}$ $(A \in \mathscr{A}_1)$, where U is the isomorphism from $\mathscr{H}_1$ onto $\mathscr{H}_2$ whose restriction to $P_a(\mathscr{H}_1)$ is U_a, for each a. ∎

In the following theorem, we make us of the matrix algebra constructs $n \otimes \mathscr{R}$ and $\mathscr{R} \otimes I_n$, described in Section 6.6. Note that, from Lemma 6.6.2, $n \otimes \mathscr{R}$ has center $(n \otimes \mathscr{R}) \cap (\mathscr{R}' \otimes I_n)$; and this is $\mathscr{C} \otimes I_n$, where $\mathscr{C}$ is the center of $\mathscr{R}$. Thus the centers of $\mathscr{R}$ and $n \otimes \mathscr{R}$ are * isomorphic.

9.3.2. THEOREM. *If $\mathscr{R}$ is a von Neumann algebra of type I_m, with commutant of type I_n, then $\mathscr{R}$ is unitarily equivalent to $m \otimes (\mathscr{A} \otimes I_n)$, where $\mathscr{A}$ is a maximal abelian von Neumann algebra acting on a Hilbert space $\mathscr{K}$. Moreover, $\mathscr{A}$ is * isomorphic to the center of $\mathscr{R}$, and is uniquely determined up to unitary equivalence.*

Proof. Let $(E_a)_{a \in \mathbb{A}}$ be a family of m equivalent abelian projections in $\mathscr{R}$, with sum I, and let E be any projection in $\mathscr{R}$ equivalent to the E_a (for example, E could be an E_a). Let $\mathscr{L}$ be the Hilbert space $E(\mathscr{H})$, and $\mathscr{S}$ the abelian von Neumann algebra $E\mathscr{R}E$ acting on $\mathscr{L}$. Since E has central carrier I, the mapping $A' \to A'E$ is a * isomorphism from $\mathscr{R}'$ onto $\mathscr{R}'E$ $(= \mathscr{S}')$; so $\mathscr{S}'$ is of type I_n.

For each a, let V_a be a partial isometry in $\mathscr{R}$, with initial and final projections E_a and E, respectively. Then V_a restricts to a unitary transformation from $E_a(\mathscr{H})$ onto $\mathscr{L}$, and the equation $Ux = \sum_{a \in \mathbb{A}} \oplus V_a x$ $(x \in \mathscr{H})$ defines a unitary operator U from $\mathscr{H}$ onto $\sum_{a \in \mathbb{A}} \oplus \mathscr{L}$. For each A' in $\mathscr{R}'$ and x in $\mathscr{H}$,

$$UA'x = \sum \oplus V_a A'x = \sum \oplus A'EV_a x = (\sum \oplus A'E)Ux.$$

Thus $UA'U^{-1}$ is $\sum \oplus A'E$, the operator on $\sum \oplus \mathscr{L}$ whose matrix has $A'E$ in each diagonal position and zeros elsewhere; so $U\mathscr{R}'U^{-1} = \mathscr{R}'E \otimes I_m = \mathscr{S}' \otimes I_m$. From Lemma 6.6.2,

$$U\mathscr{R}U^{-1} = (U\mathscr{R}'U^{-1})' = (\mathscr{S}' \otimes I_m)' = m \otimes \mathscr{S}.$$

Since $\mathscr{S}'$ is of type I_n, there is a family $(E_b')_{b \in \mathbb{B}}$ of n equivalent abelian projections in $\mathscr{S}'$, with sum I. Let E' be any projection in $\mathscr{S}'$ equivalent to the E_b. Let $\mathscr{K}$ be the Hilbert space $E'(\mathscr{L})$ and $\mathscr{A}$ the von Neumann algebra $E'\mathscr{S}'E'$ acting on $\mathscr{K}$. Since $\mathscr{A}$ is abelian, $\mathscr{A} \subseteq \mathscr{A}'$; moreover, since $\mathscr{A}'$ is the abelian von Neumann algebra $\mathscr{S}E'$, $\mathscr{A}' \subseteq \mathscr{A}'' = \mathscr{A}$. Thus $\mathscr{A} = \mathscr{A}'$ $(= E'\mathscr{S}'E' = \mathscr{S}E')$, and $\mathscr{A}$ is maximal abelian.

For each b in $\mathbb{B}$, let V_b' be a partial isometry in $\mathscr{S}'$, with initial and final projections E_b' and E', respectively. Reasoning similar to that used above shows that the equation $Vy = \sum_{b \in \mathbb{B}} \oplus V_b'y$ $(y \in \mathscr{L})$ defines a unitary operator V from $\mathscr{L}$ onto $\sum_{b \in \mathbb{B}} \oplus \mathscr{K}$, and

$$V\mathscr{S}V^{-1} = \mathscr{S}E' \otimes I_n = \mathscr{A} \otimes I_n.$$

With W the isomorphism $\sum_{a \in \mathbb{A}} \oplus V$, from $\sum_{a \in \mathbb{A}} \oplus \mathscr{L}$ onto $\sum_{a \in \mathbb{A}} \oplus \{\sum_{b \in \mathbb{B}} \oplus \mathscr{K}\}$,

$$WU\mathscr{R}U^{-1}W^{-1} = W(m \otimes \mathscr{S})W^{-1} = m \otimes V\mathscr{S}V^{-1} = m \otimes (\mathscr{A} \otimes I_n).$$

So far, we have proved the existence of a maximal abelian von Neumann algebra $\mathscr{A}$, such that $\mathscr{R}$ is unitarily equivalent to $m \otimes (\mathscr{A} \otimes I_n)$. Given any such algebra $\mathscr{A}$, the center of $\mathscr{R}$ is * isomorphic to that of $m \otimes (\mathscr{A} \otimes I_n)$, hence to the center of $\mathscr{A} \otimes I_n$ (which is $\mathscr{A} \otimes I_n$), and thus to $\mathscr{A}$. Accordingly, any two such algebras $\mathscr{A}$ are * isomorphic, and by Theorem 9.3.1 they are unitarily equivalent. ■

9.3.3. THEOREM. *If $\mathscr{R}_j$ is a von Neumann algebra of type I_m acting on a Hilbert space $\mathscr{H}_j$, with commutant $\mathscr{R}_j'$ of type I_n $(j = 1, 2)$, and the centers of $\mathscr{R}_1$ and $\mathscr{R}_2$ are * isomorphic, there is a unitary operator U from $\mathscr{H}_1$ onto $\mathscr{H}_2$ such that $\mathscr{R}_2 = U\mathscr{R}_1U^{-1}$. If $\mathscr{R}_j$ has center $\mathscr{C}_j$ $(j = 1, 2)$, and φ is a * isomorphism from $\mathscr{C}_1$ onto $\mathscr{C}_2$, then U may be chosen so that $\varphi(C) = UCU^{-1}$ for each C in $\mathscr{C}_1$.*

Proof. Let $\mathbb{A}$ and $\mathbb{B}$ be sets with cardinality m and n, respectively. For $j = 1, 2$, by Theorem 9.3.2, there is a maximal abelian von Neumann algebra $\mathscr{A}_j$, acting on a Hilbert space $\mathscr{K}_j$, such that $\mathscr{R}_j$ is unitarily equivalent to the algebra $m \otimes (\mathscr{A}_j \otimes I_n)$ acting on the space $\mathscr{L}_j = \sum_{a \in \mathbb{A}} \oplus \{\sum_{b \in \mathbb{B}} \oplus \mathscr{K}_j\}$. Accordingly, it suffices to prove the theorem under the additional assumption that $\mathscr{R}_j = m \otimes (\mathscr{A}_j \otimes \mathrm{I}_n)$. In this case, the center $\mathscr{C}_j$ of $\mathscr{R}_j$ is $(\mathscr{A}_j \otimes I_n) \otimes I_m$, and consists of all operators of the form $\sum_{a \in \mathbb{A}} \oplus \{\sum_{b \in \mathbb{B}} \oplus A\}$, with A in $\mathscr{A}_j$. With φ a * isomorphism from $\mathscr{C}_1$ onto $\mathscr{C}_2$, there is a * isomorphism ψ from $\mathscr{A}_1$ onto $\mathscr{A}_2$ such that

$$\varphi\left(\sum_{a \in \mathbb{A}} \oplus \left\{\sum_{b \in \mathbb{B}} \oplus A\right\}\right) = \sum_{a \in \mathbb{A}} \oplus \left\{\sum_{b \in \mathbb{B}} \oplus \psi(A)\right\} \tag{1}$$

for each A in $\mathscr{A}_1$. From Theorem 9.3.1, there is a unitary operator V, from $\mathscr{K}_1$ onto $\mathscr{K}_2$, such that $\psi(A) = VAV^{-1}$ $(A \in \mathscr{A}_1)$. With U the unitary operator $\sum_{a\in\mathbb{A}} \oplus \{\sum_{b\in\mathbb{B}} \oplus V\}$, from $\mathscr{L}_1$ onto $\mathscr{L}_2$,

$$\begin{aligned} U\mathscr{R}_1U^{-1} = U\{m \otimes (\mathscr{A}_1 \otimes I_n)\}U^{-1} &= m \otimes (V\mathscr{A}_1V^{-1} \otimes I_n) \\ &= m \otimes (\mathscr{A}_2 \otimes I_n) = \mathscr{R}_2. \end{aligned}$$

Moreover, since V implements ψ, it follows from (1) that U implements the * isomorphism φ from $\mathscr{C}_1$ onto $\mathscr{C}_2$. ■

9.3.4. THEOREM. *If $\mathscr{R}$ and $\mathscr{S}$ are von Neumann algebras of type I_m, acting on Hilbert spaces $\mathscr{H}$ and $\mathscr{K}$, respectively, with commutants $\mathscr{R}'$ and $\mathscr{S}'$ of type I_n, and φ is a * isomorphism from $\mathscr{R}$ onto $\mathscr{S}$, there is a unitary operator U from $\mathscr{H}$ onto $\mathscr{K}$ such that $\varphi(A) = UAU^{-1}$ $(A \in \mathscr{R})$.*

Proof. Let $\{E_a\}$ be a set of m equivalent abelian projections with sum I in $\mathscr{R}$, let E be a member of this family, and let V_a be a partial isometry in $\mathscr{R}$, with E_a and E as initial and final projections. With F_a, F, and W_a defined to be $\varphi(E_a)$, $\varphi(E)$, and $\varphi(V_a)$, respectively, $\{F_a\}$ is a family of equivalent abelian projections with sum I in $\mathscr{S}$, F is a member of the family, and W_a is a partial isometry in $\mathscr{S}$, from F_a to F.

Suppose that the * isomorphism $\varphi \,|\, E\mathscr{R}E$, from $E\mathscr{R}E$ onto $F\mathscr{S}F$, is implemented by a unitary operator U_0 from $E(\mathscr{H})$ onto $F(\mathscr{K})$. For each a, the restriction to $E_a(\mathscr{H})$ of $W_a^*U_0V_a$ is a unitary transformation from $E_a(\mathscr{H})$ onto $F_a(\mathscr{K})$. Let U be the unitary operator, from $\mathscr{H}$ onto $\mathscr{K}$, whose restriction to $E_a(\mathscr{H})$ coincides with $W_a^*U_0V_a$. For any indices a and b, and each A in $\mathscr{R}$,

$$\begin{aligned} F_a\varphi(A)F_b &= \varphi(E_aAE_b) = \varphi(V_a^*EV_aAV_b^*EV_b) \\ &= \varphi(V_a^*)\varphi(EV_aAV_b^*E)\varphi(V_b) = W_a^*U_0EV_aAV_b^*EU_0^*W_b \\ &= F_aW_a^*U_0V_aAV_b^*U_0^*W_bF_b = F_aUAU^*F_b. \end{aligned}$$

Upon summing over a and b, we have $\varphi(A) = UAU^*$ $(A \in \mathscr{R})$.

It remains to prove the existence of a unitary operator U_0 with the stated properties. Now the von Neumann algebras $E\mathscr{R}E$ and $F\mathscr{S}F$ are abelian (type I_1), their commutants $\mathscr{R}'E$ and $\mathscr{S}'F$ are of type I_n, and we have to show that a given * isomorphism from $E\mathscr{R}E$ onto $F\mathscr{R}F$ is unitarily implemented. This is an immediate consequence of the final assertion in Theorem 9.3.3 (in the case $m = 1$). ■

It is apparent that an *inner* * automorphism of a von Neumann algebra $\mathscr{R}$ maps each element of the center of $\mathscr{R}$ onto itself. Certain outer * automorphisms of von Neumann algebras also leave the center elementwise fixed (see Exercise 6.9.43(ii)); but we show, as a simple application of Theorem 9.3.4, that this does not occur in the type I case.

9.3.5. COROLLARY. *If φ is a * automorphism of a type* I *von Neumann algebra $\mathscr{R}$, and $\varphi(C) = C$ for each C in the center of $\mathscr{R}$, then φ is inner.*

Proof. By Theorem 9.1.3, $\mathscr{R}'$ is type I, so there is an abelian projection E' in $\mathscr{R}'$ that has central carrier I. The mapping $A \to AE'$ is a * isomorphism from $\mathscr{R}$ onto $\mathscr{R}E'$, and the commutant $E'\mathscr{R}'E'$ of $\mathscr{R}E'$ is abelian. Moreover, the property of $\mathscr{R}$ to be proved is preserved by * isomorphisms between von Neumann algebras. Upon replacing $\mathscr{R}$ by $\mathscr{R}E'$, we may suppose that $\mathscr{R}'$ is abelian; whence $\mathscr{R}' \subseteq (\mathscr{R}')' = \mathscr{R}$, and $\mathscr{R}'$ is the center of $\mathscr{R}$.

There is a family $\{P_n : 1 \le n \le \dim \mathscr{H}\}$ of central projections in $\mathscr{R}$, with sum I, such that $\mathscr{R}P_n$ is of type I_n unless $P_n = 0$; and $\mathscr{R}'P_n$ is of type I_1 (abelian). Since $\varphi(P_n) = P_n$, the restriction $\varphi \mid \mathscr{R}P_n$ is a * automorphism of $\mathscr{R}P_n$. From Theorem 9.3.4, there is a unitary operator U_n acting on $P_n(\mathscr{H})$ that implements $\varphi \mid \mathscr{R}P_n$. For A in $\mathscr{R}$,

$$\varphi(A) = \sum \varphi(A)P_n = \sum \varphi(AP_n) = \sum U_n AP_n U_n^{-1} = UAU^{-1},$$

where U is the unitary operator on $\mathscr{H}$ whose restriction to $P_n(\mathscr{H})$ is U_n for each n. With A in $\mathscr{R}'$, A lies in the center of $\mathscr{R}$, so $UA = UAU^{-1}U = \varphi(A)U = AU$. Thus $U \in \mathscr{R}'' = \mathscr{R}$, and φ is inner. ■

Bibliography: [48, 56, 58].

9.4. Abelian von Neumann algebras

In Section 9.3 we showed that every type I von Neumann algebra $\mathscr{R}$ can be decomposed as a direct sum of algebras of type I_m with commutants of type I_n; and we described (Theorem 9.3.2) the structure of the direct summands in terms of matrix constructs and maximal abelian von Neumann algebras. When $\mathscr{R}$ is abelian (type I_1), only terms in which $m = 1$ occur in the above decomposition. Accordingly, an abelian von Neumann algebra $\mathscr{A}$, acting on a Hilbert space $\mathscr{H}$, is unitarily equivalent to a direct sum $\sum_{n \le \dim \mathscr{H}} \oplus \mathscr{A}_n \otimes I_n$, where $\mathscr{A}_n$ is (zero or) a maximal abelian von Neumann algebra acting on a Hilbert space $\mathscr{H}_n$ (and $\mathscr{A}_n \otimes I_n$ has commutant of type I_n). From this, $\mathscr{A}$ is * isomorphic to the maximal abelian von Neumann algebra $\sum \oplus \mathscr{A}_n$; any other maximal abelian algebra, * isomorphic to $\mathscr{A}$, is unitarily equivalent to $\sum \oplus \mathscr{A}_n$, by Theorem 9.3.1.

In view of the preceding paragraph, questions concerning the structure of abelian von Neumann algebras reduce at once to the maximal abelian case. If the original algebra acts on a separable Hilbert space, this is true also of the various maximal abelian algebras that arise. Our purpose in this section, achieved in Theorem 9.4.1, is to describe, up to unitary equivalence, all maximal abelian von Neumann algebras acting on separable Hilbert spaces. The simple examples that follow form the basis of the description.

Let $\mathscr{H}_c$ denote the separable Hilbert space $L_2(S, \mathscr{S}, m)$ ($= L_2$), where m is Lebesgue measure on the σ-algebra $\mathscr{S}$ of Borel subsets of the unit interval S. With f in L_∞, let M_f denote the operator on L_2 of multiplication by f. We have seen (Example 5.1.6 and Lemma 8.6.8(i)) that the set $\{M_f : f \in L_\infty\}$ is a maximal abelian von Neumann algebra $\mathscr{A}_c$ acting on $\mathscr{H}_c$, that has no minimal projections. The absence of minimal projections can intuitively be regarded as a continuity property of $\mathscr{A}_c$, and the suffix c is intended to suggest continuity—but not, of course, in the technical sense introduced in Section 6.5 (preceding Corollary 6.5.3).

If n is a positive integer or $\aleph_0$, let S_n be a set with n elements, and let $\mathscr{H}_n$ be the n-dimensional Hilbert space $l_2(S_n)$. For each bounded function f on S_n, multiplication by f is a bounded linear operator M_f on $\mathscr{H}_n$; the set $\{M_f : f \in l_\infty(S_n)\}$ is a maximal abelian von Neumann algebra $\mathscr{A}_n$ acting on $\mathscr{H}_n$. This can be proved by checking that $\mathscr{A}_n = \mathscr{A}_n'$, an easy consequence of the fact that $\mathscr{A}_n$ consists of all bounded operators whose matrices (relative to the usual orthonormal basis in $l_2(S_n)$) are diagonal. It can also be deduced from Example 5.1.6, since $l_2(S_n)$ is $L_2(S_n, \mathscr{S}_n, m_n)$, where $\mathscr{S}_n$ is the σ-algebra of all subsets of S_n, and m_n is the measure that assigns unit mass to each point. The projections in $\mathscr{A}_n$ are multiplications by characteristic functions of subsets of S_n, minimal projections corresponding to single points. With f in $l_\infty(S_n)$, M_f is $\sum f(s)E_s$, where E_s is the minimal projection corresponding to s. Accordingly, $\mathscr{A}_n$ has precisely n minimal projections, and these generate $\mathscr{A}_n$.

From the preceding paragraphs, it follows that each of $\mathscr{A}_c$, $\mathscr{A}_j$ $(1 \le j \le \aleph_0)$, and $\mathscr{A}_c \oplus \mathscr{A}_k$ $(1 \le k \le \aleph_0)$ is a maximal abelian von Neumann algebra acting on a separable Hilbert space. Moreover, no two of them are * isomorphic, since $\mathscr{A}_c$ has no minimal projections, $\mathscr{A}_j$ has precisely j minimal projections and is generated by them, and $\mathscr{A}_c \oplus \mathscr{A}_k$ has precisely k minimal projections and is not generated by them.

With the notation just introduced, we can now state the main result of this section.

9.4.1. THEOREM. *Each maximal abelian von Neumann algebra, acting on a separable Hilbert space, is unitarily equivalent to exactly one of the algebras* $\mathscr{A}_c$, $\mathscr{A}_j$ $(1 \le j \le \aleph_0)$, $\mathscr{A}_c \oplus \mathscr{A}_k$ $(1 \le k \le \aleph_0)$.

Proof. Let $\mathscr{R}$ be a maximal abelian von Neumann algebra, acting on a separable Hilbert space $\mathscr{H}$. We consider, first, the case in which $\mathscr{R}$ has no minimal projections, and show in this case that $\mathscr{R}$ is unitarily equivalent to $\mathscr{A}_c$. The following observations, concerning $\mathscr{A}_c$ and $\mathscr{H}_c$, motivate the proof. In $\mathscr{H}_c$ (the L_2 space associated with Lebesgue measure on the unit interval $[0, 1]$) there is a separating and generating unit vector u for $\mathscr{A}_c$, defined by $u(s) = 1$ $(0 \le s \le 1)$. With t in $[0, 1]$, the operator of multiplication by the characteristic function g_t of $[0, t]$ is a projection G_t in $\mathscr{A}_c$, and $G_s \leq G_t$ if $s \leq t$. Since $G_t u = g_t$, the linear span of $\{G_t u : 0 \leq t \leq 1\}$ consists of all step functions on $[0, 1]$, and is therefore dense in $\mathscr{H}_c$. Moreover,

$$\langle G_s u, G_t u \rangle = \langle g_s, g_t \rangle = \min(s, t) \qquad (s, t \in [0, 1]).$$

We shall construct a vector in $\mathscr{H}$, and projections in $\mathscr{R}$, that have properties analogous to those, just described, of u and G_s $(0 \leq s \leq 1)$.

Since $\mathscr{H}$ is separable, $\mathscr{R}$ is countably decomposable, and so (Corollary 5.5.17) has a separating and generating unit vector x (this will correspond to u).

We assert that $\mathscr{R}$ is generated, as a von Neumann algebra, by a sequence of projections. Indeed, if $\mathscr{P}$ is the set of all projections in $\mathscr{R}$, the subset $\{Px : P \in \mathscr{P}\}$ of the separable Hilbert space $\mathscr{H}$ has a countable dense subset $\{P_n x : n = 1, 2, 3, \ldots\}$, where $P_1, P_2, P_3, \ldots \in \mathscr{P}$. With P in $\mathscr{P}$, there is a sequence $\{n(1), n(2), n(3), \ldots\}$ of positive integers such that $Px = \lim P_{n(r)} x$, whence

$$PAx = APx = \lim AP_{n(r)}x = \lim P_{n(r)}Ax$$

for each A in $\mathscr{R}$. Since $\mathscr{R}x$ is dense in $\mathscr{H}$, and $P, P_{n(r)}$ lie in the unit ball of $\mathscr{R}$, it now follows that $P_{n(r)} \to P$ in the strong-operator topology. From this, the von Neumann algebra generated by $\{P_1, P_2, P_3, \ldots\}$ contains every projection in $\mathscr{R}$, and so coincides with $\mathscr{R}$.

The set $\mathscr{P}$ is partially ordered, by the usual order relation on projections. We can construct finite totally ordered subsets $\mathscr{F}_1, \mathscr{F}_2, \ldots$ of $\mathscr{P}$, such that $\mathscr{F}_1 \subseteq \mathscr{F}_2 \subseteq \cdots$, and the linear span of $\mathscr{F}_n$ contains P_n. Indeed, we start the process by taking $\mathscr{F}_1 = \{0, P_1, I\}$. When a suitable set $\mathscr{F}_n = \{E_0, E_1, \ldots, E_m\}$ has been constructed, with $0 = E_0 < E_1 < \cdots < E_m = I$, we have

$$E_r \leq E_r + (E_{r+1} - E_r)P_{n+1} \leq E_{r+1}, \qquad P_{n+1} = \sum_{r=0}^{m-1} (E_{r+1} - E_r)P_{n+1},$$

and it suffices to define $\mathscr{F}_{n+1}$ to be the set

$$\mathscr{F}_n \cup \{E_r + (E_{r+1} - E_r)P_{n+1} : r = 0, \ldots, m - 1\}.$$

With $\mathscr{F}_1, \mathscr{F}_2, \ldots$ obtained in this way, $\bigcup \mathscr{F}_n$ is a totally ordered subset $\mathscr{F}_\infty$ of $\mathscr{P}$, whose linear span contains each P_n. The class $\mathscr{C}$, of all totally ordered subsets of $\mathscr{P}$ that contain $\mathscr{F}_\infty$, is itself partially ordered by inclusion. If a subclass $\mathscr{C}_0$ of $\mathscr{C}$ is totally ordered by inclusion, its union is the least upper bound of $\mathscr{C}_0$ in $\mathscr{C}$. By Zorn's lemma, $\mathscr{C}$ has a maximal element $\mathscr{F}$. Since EF is E or F, when $E, F \in \mathscr{F}$, and $\mathscr{F}_\infty \subseteq \mathscr{F}$, the linear span of $\mathscr{F}$ is a self-adjoint algebra that contains each P_n. Accordingly, $\mathscr{F}$ is a maximal totally ordered family of projections in $\mathscr{R}$, whose strong-operator-closed linear span is $\mathscr{R}$.

Suppose that $\mathscr{F}_0 \subseteq \mathscr{F}$. With E in $\mathscr{F}$, *either* $E \geq F$ for all F in $\mathscr{F}_0$ (and $E \geq \bigvee \mathscr{F}_0$) *or* $E \leq F$ for some F in $\mathscr{F}_0$ (and $E \leq \bigvee \mathscr{F}_0$). Accordingly, the family $\mathscr{F} \cup \{\bigvee \mathscr{F}_0\}$ is totally ordered. By maximality of $\mathscr{F}$, it follows that $\bigvee \mathscr{F}_0 \in \mathscr{F}$ (and, similarly, $\bigwedge \mathscr{F}_0 \in \mathscr{F}$). Since $\mathscr{F}_0$ is totally ordered, $\bigvee \mathscr{F}_0$ and $\bigwedge \mathscr{F}_0$ lie in the strong-operator closure of $\mathscr{F}_0$.

The equation $\varphi(F) = \langle Fx, x \rangle$ $(F \in \mathscr{F})$ defines a mapping φ from $\mathscr{F}$ into $[0, 1]$. We assert that φ is one-to-one, order preserving, and has range the whole of $[0, 1]$. Indeed, if E, F are distinct elements of $\mathscr{F}$, we may suppose that $E < F$, since x is a separating vector for $\mathscr{R}$,

$$\varphi(F) - \varphi(E) = \langle (F - E)x, x \rangle = \|(F - E)x\|^2 > 0.$$

This shows that φ is one-to-one and order preserving. The range of φ contains 0 and 1, since $0, I \in \mathscr{F}$. If $0 < t < 1$, $\mathscr{F}$ is the disjoint union of the sets

$$\mathscr{F}_0 = \{F \in \mathscr{F} : \varphi(F) < t\}, \qquad \mathscr{F}_1 = \{F \in \mathscr{F} : \varphi(F) \geq t\}.$$

Define F_0 and F_1 in $\mathscr{F}$ by $F_0 = \bigvee \mathscr{F}_0$, $F_1 = \bigwedge \mathscr{F}_1$. Since the mapping $A \to \langle Ax, x \rangle$ is strong-operator continuous, and F_j lies in the strong-operator closure of $\mathscr{F}_j$ $(j = 0, 1)$, it follows that

$$\varphi(F_0) \leq t \leq \varphi(F_1) \tag{1}$$

(and $F_0 \leq F_1$, since φ preserves order).

If $F_0 < F_1$, there is a projection E in $\mathscr{R}$ such that $0 < E < F_1 - F_0$, since $\mathscr{R}$ has no minimal projections. Since $\mathscr{F} = \mathscr{F}_0 \cup \mathscr{F}_1$, and $E_0 < F_0 + E < E_1$ whenever $E_0 \in \mathscr{F}_0$ and $E_1 \in \mathscr{F}_1$, it now follows that $\mathscr{F} \cup \{F_0 + E\}$ is totally ordered and properly contains $\mathscr{F}$, contrary to the maximality of $\mathscr{F}$. Hence $F_0 = F_1$, whence $\varphi(F_0) = t$, from (1); so φ has range $[0, 1]$

The inverse of φ is a strictly increasing mapping, $s \to E_s$, from $[0, 1]$ onto $\mathscr{F}$, and $\langle E_s x, x \rangle = s$. When $s, t \in [0, 1]$,

$$\begin{aligned}\langle E_s x, E_t x \rangle &= \langle E_t E_s x, x \rangle \\ &= \langle E_{\min(s,t)} x, x \rangle = \min(s, t) = \langle G_s u, G_t u \rangle.\end{aligned}$$

It follows that

$$\|\sum_{r=1}^{q} a_r E_{s(r)} x\| = \|\sum_{r=1}^{q} a_r G_{s(r)} u\|, \tag{2}$$

whenever $s(1), \ldots, s(q) \in [0, 1]$ and $a_1, \ldots, a_q$ are scalars. Since the linear span of $\mathscr{F}$ $(= \{E_s : 0 \leq s \leq 1\})$ is strong-operator dense in $\mathscr{R}$, linear combinations of vectors $E_s x$ $(0 \leq s \leq 1)$ are dense in $\mathscr{R}x$, and hence in $\mathscr{H}$. From this, and the corresponding property of $G_s u$ (already noted), together with

(2), there is a unitary operator U from $\mathscr{H}$ onto $\mathscr{H}_c$, such that $UE_s x = G_s u$ $(0 \le s \le 1)$. For s and t in $[0, 1]$,

$$\begin{aligned} UE_s E_t x &= UE_{\min(s,t)} x = G_{\min(s,t)} u \\ &= G_s G_t u = G_s UE_t x. \end{aligned}$$

Since the vectors $E_t x$ $(0 \le t \le 1)$ generate $\mathscr{H}$, $UE_s = G_s U$, and $UE_s U^{-1} = G_s \in \mathscr{A}_c$. Hence $U\mathscr{F}U^{-1} \subseteq \mathscr{A}_c$, $U\mathscr{R}U^{-1} \subseteq \mathscr{A}_c$ (because $\mathscr{F}$ generates $\mathscr{R}$), and finally $U\mathscr{R}U^{-1} = \mathscr{A}_c$ (because $U\mathscr{R}U^{-1}$ is *maximal* abelian). This completes the proof that $\mathscr{R}$ is unitarily equivalent to $\mathscr{A}_c$, when $\mathscr{R}$ has no minimal projections.

Suppose next that $\mathscr{R}$ contains some minimal projections, and denote the set of all such projections by $\mathscr{E}$. With E in $\mathscr{E}$ and A in $\mathscr{R}$, $AE = EA = EAE = aE$ for some scalar a, by Proposition 6.4.3. If x is a non-zero vector in $E(\mathscr{H})$, E dominates (and so coincides with) the cyclic projection with range $[\mathscr{R}'x]$. Since

$$\mathscr{R}'x = \mathscr{R}x = \mathscr{R}Ex = \{ax : a \in \mathbb{C}\},$$

$E(\mathscr{H})$ is one-dimensional (this can be deduced, also, from Proposition 6.4.4). With E, F distinct elements of $\mathscr{E}$, the projection EF is a scalar multiple of both E and F, and is therefore 0. From the preceding remarks, and since $\mathscr{H}$ is separable, the elements of $\mathscr{E}$ form an orthogonal family of n one-dimensional projections, where n is a cardinal such that $1 \le n \le \aleph_0$. We can therefore index $\mathscr{E}$ as $\{E_s : s \in S_n\}$, where S_n is the set used above in constructing $\mathscr{A}_n$.

Let Q be the projection $\sum E_s$ in $\mathscr{R}$. If A is in $\mathscr{R}$, $AQ = \sum AE_s = \sum f(s)E_s$, where $AE_s = E_s AE_s = f(s)E_s$ and $|f(s)| \le \|A\|$; whence $\mathscr{R}Q$ consists of all operators of the form $\sum f(s)E_s$, where $f \in l_\infty(S_n)$. With e_s a unit vector in the range of E_s, for each s in S_n, the family $\{e_s\}$ is an orthonormal basis of $Q(\mathscr{H})$, and the equation

$$Ux = \sum_{s \in S_n} x(s)e_s \qquad (x \in l_2(S_n))$$

defines a unitary transformation U from $l_2(S_n)$ onto $Q(\mathscr{H})$. Since

$$\begin{aligned} UM_f x &= \sum_{s \in S_n} (M_f x)(s)e_s = \sum_{s \in S_n} f(s)x(s)e_s \\ &= \left(\sum_{s \in S_n} f(s)E_s\right)\left(\sum_{s \in S_n} x(s)e_s\right) = \left(\sum_{s \in S_n} f(s)E_s\right)Ux, \end{aligned}$$

when $f \in l_\infty(S_n)$ and $x \in l_2(S_n)$, it follows that $UM_f U^{-1} = \sum f(s)E_s$. Accordingly, $U\mathscr{A}_n U^{-1} = \mathscr{R}Q$; so $\mathscr{R}Q$ is unitarily equivalent to $\mathscr{A}_n$.

If $Q = I$, $\mathscr{R}$ is unitarily equivalent to $\mathscr{A}_n$. If $Q \neq I$, $\mathscr{R}(I - Q)$ is a maximal abelian algebra with no minimal projections, acting on a separable Hilbert space, and is therefore unitarily equivalent to $\mathscr{A}_c$. In this case, $\mathscr{R}$ is unitarily equivalent to $\mathscr{R}(I - Q) \oplus \mathscr{R}Q$, and hence to $\mathscr{A}_c \oplus \mathscr{A}_n$.

We have now shown that $\mathscr{R}$ is unitarily equivalent to one of the maximal abelian von Neumann algebras listed in the theorem—to precisely one since, as already noted, no two of them are * isomorphic. ■

9.4.2. REMARK. With Theorem 9.4.1 available to us, we are in a position to describe a particularly effective way of dealing with (unbounded) self-adjoint operators. The discussion that ensues may be considered an extension of the comments leading to Theorem 5.6.4. (This material can be reviewed by the reader with profit at this time.) If A is an unbounded self-adjoint operator acting on a Hilbert space $\mathscr{H}$, we produced a spectral resolution for A by studying the (bounded) inverses of $A \pm iI$. These inverses, which we denoted by T_+ and T_-, are adjoints of one another and commute. The abelian von Neumann algebra they generate lies in a maximal abelian von Neumann algebra $\mathscr{A}$ (on $\mathscr{H}$). If we assume that $\mathscr{H}$ is separable, Theorem 9.4.1 applies and $(\mathscr{H}, \mathscr{A})$ is unitarily equivalent to $L_2(S, \mathscr{S}, m)$ and its multiplication algebra—where $(S, \mathscr{S}, m)$ is one of the special measure spaces discussed in this section. Since $A \,\eta\, \mathscr{A}$, $A = M_g$ for some real-valued, measurable function g, finite almost everywhere, from Theorem 5.6.4. There is no difficulty, now, in forming the spectral resolution of A and Borel functions of A by resolving g and composing g with Borel functions. ■

Bibliography: [63].

9.5. Spectral multiplicity

This section is concerned with the problem of determining whether two given representations of an abelian C^*-algebra are equivalent, and (as a particular case) with the classical question of deciding whether two given normal operators are unitarily equivalent. It turns out that the problem can be divided into two separate parts. The first concerns the possibility of extending a given * isomorphism, between two abelian C^*-algebras of operators, to a * isomorphism between their weak-operator closures. This extension problem is measure-theoretic in nature, and its solution requires the introduction of a measure-theoretic invariant, associated with a representation of an abelian C^*-algebra. The second part of the problem is that of deciding whether a given * isomorphism, between two abelian von

Neumann algebras, is unitarily implemented. The solution to this last question has already been completely worked out in Section 9.3, in the more general context of type I von Neumann algebras. In the present setting, it can be expressed more conveniently, in a form related to the measure-theoretic invariant just mentioned.

Although the program just described can be carried out in full generality, the non-separable case involves rather elaborate measure-theoretic technicalities. Accordingly, in the interests of simplicity, we shall confine attention to representations that act on separable Hilbert spaces. We begin by describing, in rather more detail, the way in which the problem can be divided into two parts, and the nature of the invariants used.

Suppose that $\mathscr{A}$ is an abelian C^*-algebra and, for $j = 1, 2$, φ_j is a representation of $\mathscr{A}$ on a Hilbert space $\mathscr{H}_j$. If φ_1 and φ_2 are equivalent, there is a unitary transformation U from $\mathscr{H}_1$ onto $\mathscr{H}_2$, such that

$$\varphi_2(A) = U\varphi_1(A)U^{-1} \qquad (A \in \mathscr{A}).$$

Since $U\varphi_1(\mathscr{A})U^{-1} = \varphi_2(\mathscr{A})$, we have $U\varphi_1(\mathscr{A})^- U^{-1} = \varphi_2(\mathscr{A})^-$; and U implements a * isomorphism α, from $\varphi_1(\mathscr{A})^-$ onto $\varphi_2(\mathscr{A})^-$, such that

$$\alpha(\varphi_1(A)) = \varphi_2(A) \qquad (A \in \mathscr{A}). \tag{1}$$

Conversely, if there is such a * isomorphism $\alpha\colon \varphi_1(\mathscr{A})^- \to \varphi_2(\mathscr{A})^-$, and if α is unitarily implemented, then φ_1 and φ_2 are equivalent. Accordingly, our original problem, concerning equivalence of φ_1 and φ_2, reduces to two questions; first, the *extension problem* (Is there a * isomorphism α of the type just described?); and second, the *implementation problem* (Is α unitarily implemented?). The extension problem is aptly named. In order that it should have an affirmative answer, it is necessary that φ_1 and φ_2 have the same kernel. However, if the kernels of φ_1 and φ_2 coincide, the equation $\alpha_0(\varphi_1(A)) = \varphi_2(A)$ $(A \in \mathscr{A})$ defines (unambiguously) a * isomorphism α_0 from $\varphi_1(\mathscr{A})$ onto $\varphi_2(\mathscr{A})$; and the extension problem concerns the possibility of extending α_0 to a * isomorphism α between the weak-operator closures, $\varphi_1(\mathscr{A})^-$ and $\varphi_2(\mathscr{A})^-$.

In describing the invariants introduced below, it is convenient to use the function representation of $\mathscr{A}$, so reducing to the case in which $\mathscr{A} = C(S)$ for some compact Hausdorff space S. With φ a representation of $\mathscr{A}$, acting on a separable Hilbert space $\mathscr{H}$, the abelian von Neumann algebra $\varphi(\mathscr{A})^-$ contains a (uniquely determined) orthogonal family $\{P_n : 1 \le n \le \aleph_0\}$ of projections, with sum I, such that $\varphi(\mathscr{A})'P_n$ is type I_n unless $P_n = 0$. For each vector x in $\mathscr{H}$, $\omega_x \circ \varphi$ is a positive linear functional on $\mathscr{A}$ $(= C(S))$, and so corresponds to a (unique) regular Borel measure μ_x on S. Since $\mathscr{H}$ is separable, $\varphi(\mathscr{A})^-$ has separating vectors; and it turns out that all measures of the form μ_u, where u is a separating vector for $\varphi(\mathscr{A})^-$, have the same family of null sets. This family $\mathscr{N}_\varphi$, the *null ideal* of φ, proves to be a complete invariant for the extension problem, for representations acting on separable Hilbert spaces (in other words, the extension problem has an affirmative answer if and only if φ_1 and φ_2 have the same null ideal). For $j = 1, 2, 3, \ldots$, we may consider the representation ψ_j: $A \to \varphi(A)Q_j$ of $\mathscr{A}$ on $Q_j(\mathscr{H})$, where

$$Q_j = \sum_{j \le n \le \aleph_0} P_n \qquad (j = 1, 2, 3, \ldots).$$

The null ideals of these representations form an increasing sequence, the *null ideal sequence* of φ, the first member of this sequence being $\mathscr{N}_\varphi$. If two representations φ_1 and φ_2 of $\mathscr{A}$ (acting on separable Hilbert spaces) have the same null ideal sequence, they have the same null ideal, whence the extension problem has an affirmative answer. Moreover, it is possible to show that the corresponding * isomorphism α: $\varphi_1(\mathscr{A})^- \to \varphi_2(\mathscr{A})^-$ satisfies the criteria (set out in Section 9.3) for unitary implementation; so φ_1 and φ_2 are equivalent. Accordingly, the null ideal sequence is a complete invariant, for equivalence of representations of $\mathscr{A}$ acting on separable Hilbert spaces. There is an alternative formulation of this result, in which the null ideal sequence is replaced by another (essentially equivalent) invariant, the *multiplicity function* of φ.

We now begin the detailed development of the theory. We suppose throughout that S is a compact Hausdorff space, and denote by $\mathscr{S}$ the σ-algebra of all Borel subsets of S. By a *σ-ideal* in $\mathscr{S}$, we mean a subset $\mathscr{N}$ of $\mathscr{S}$ that is closed under countable unions and has the following property: if $N \in \mathscr{N}$ and $Y \in \mathscr{S}$, then $N \cap Y \in \mathscr{N}$. With each regular Borel measure μ on S we associate a σ-ideal, the family of all μ-null sets; σ-ideals arising in this way are described as *null ideals* in $\mathscr{S}$. If the Hilbert space $L_2(S, \mathscr{S}, \mu)$ is separable, the corresponding null ideal $\mathscr{N}$ is said to be of *separable type*. In this case, $L_2(S, \mathscr{S}, v)$ is separable for every measure v whose null sets include the members of $\mathscr{N}$; for the Radon–Nikodým derivative $dv/d\mu$ is a non-negative-valued element h of $L_1(S, \mathscr{S}, \mu)$, and the mapping $g \to gh^{1/2}$ is an isometric linear mapping from $L_2(S, \mathscr{S}, v)$ into $L_2(S, \mathscr{S}, \mu)$.

A *null ideal sequence* (of *separable type*) in $\mathscr{S}$ is an increasing sequence $\{\mathscr{N}_j\}$ of σ-ideals in $\mathscr{S}$, the first of which, $\mathscr{N}_1$, is a null ideal (of separable type). An essentially equivalent concept is that of a *multiplicity function* on $\mathscr{S}$, by which we mean a mapping m, from $\mathscr{S}$ into the set $\{0, 1, 2, \ldots, \aleph_0\}$, such that

(i) $m(Y_1) \le m(Y_2)$ if $Y_1, Y_2 \in \mathscr{S}$ and $Y_1 \subseteq Y_2$;
(ii) $m(\bigcup_{j=1}^{\infty} Y_j) = \sup_j m(Y_j)$ if $Y_1, Y_2, \ldots \in \mathscr{S}$;
(iii) the σ-ideal $\{Y \in \mathscr{S} : m(Y) = 0\}$ is a null ideal.

With every multiplicity function m on $\mathscr{S}$, we can associate a null ideal sequence $\{\mathscr{N}_j\}$ in $\mathscr{S}$, defined by

$$\mathscr{N}_j = \{Y \in \mathscr{S} : m(Y) < j\} \qquad (j = 1, 2, \ldots).$$

Conversely, every null ideal sequence $\{\mathscr{N}_j\}$ arises in the above manner from a unique multiplicity function m, defined by

$$m(Y) = \begin{cases} \aleph_0 & \text{if} \quad Y \in \mathscr{S}, \quad Y \notin \bigcup_{j=1}^{\infty} \mathscr{N}_j \\ \\ k - 1 & \text{if} \quad Y \in \bigcup_{j=1}^{\infty} \mathscr{N}_j \quad \text{and} \quad k = \min\{j: Y \in \mathscr{N}_j\}. \end{cases}$$

In this way, we establish a one-to-one correspondence between null ideal sequences and multiplicity functions.

If m is a multiplicity function on $\mathscr{S}$, $Y \in \mathscr{S}$ and $m(Y) = j$, we say that Y has *uniform multiplicity* j if, for every Borel subset Z of Y, $m(Z)$ is either j or 0. Note that a Borel set Y has uniform multiplicity 0 if $m(Y) = 0$.

The following result provides further information about the structure of null ideal sequences and multiplicity functions.

9.5.1. PROPOSITION. *Suppose that $\mathscr{S}$ is the σ-algebra of all Borel subsets of a compact Hausdorff space S.*

(i) *If μ is a regular Borel measure on S, $\{Y_1, Y_2, Y_3, \ldots\}$ is a decreasing sequence in $\mathscr{S}$ with $Y_1 = S$, and*

(2) $$\mathscr{N}_j = \{Y \in \mathscr{S} : \mu(Y \cap Y_j) = 0\} \qquad (j = 1, 2, 3, \ldots),$$

then $\{\mathscr{N}_j\}$ is a null ideal sequence in $\mathscr{S}$. The corresponding multiplicity function m on $\mathscr{S}$ is given by

(3) $$m(Y) = \begin{cases} 0 & \text{if} \quad \mu(Y) = 0 \\ \sup\{j : \mu(Y \cap Y_j) \neq 0\} & \text{if} \quad \mu(Y) > 0. \end{cases}$$

If $Y \in \mathscr{S}$ and j is a positive integer, then Y has uniform multiplicity j if and only if

$$\mu(Y) = \mu(Y \cap (Y_j \setminus Y_{j+1})) > 0;$$

and Y has uniform multiplicity $\aleph_0$ if and only if

$$\mu(Y) = \mu\left(Y \cap \bigcap_{j=1}^{\infty} Y_j\right) > 0.$$

(ii) *Every null ideal sequence arises from the construction described in* (i). *More precisely, if $\{\mathscr{N}_j\}$ is a null ideal sequence in $\mathscr{S}$, there is a decreasing sequence $\{Y_1, Y_2, Y_3, \ldots\}$ of Borel sets, with $Y_1 = S$, that has the following* *property*: *if* μ *is any regular Borel measure on* S, *whose family of null sets is* $\mathcal{N}_1$, *then* (2) *holds. Moreover, if* $\{Y_j'\}$ *is another sequence of Borel sets that has the properties just ascribed to* $\{Y_j\}$, *then*

$$Y_j \setminus Y_j', Y_j' \setminus Y_j \in \mathcal{N}_1 \qquad (j = 1, 2, 3, \ldots);$$

and for every measure μ *of the type just described*, Y_j *and* Y_j' *differ only by a* μ-*null set*.

Proof. (i) Under the conditions set out in (i), it is apparent that (2) defines a null ideal sequence $\{\mathcal{N}_j\}$ in $\mathcal{S}$, and that the corresponding multiplicity function m on $\mathcal{S}$ is given by (3).

Suppose that $Y \in \mathcal{S}$ and j is a positive integer. Observe that Y is the disjoint union of the three sets

$$Y \setminus Y_j, \qquad Y \cap (Y_j \setminus Y_{j+1}), \qquad Y \cap Y_{j+1}.$$

If Y has uniform multiplicity j, $Y \setminus Y_j$ has multiplicity j or 0; by (3), $m(Y \setminus Y_j) < j$, so $m(Y \setminus Y_j) = 0$ and $\mu(Y \setminus Y_j) = 0$. Moreover, $\mu(Y \cap Y_{j+1}) = 0$, since otherwise $m(Y) > j$, by (3); and $\mu(Y) > 0$ since $m(Y) \neq 0$. Thus $\mu(Y) = \mu(Y \cap (Y_j \setminus Y_{j+1})) > 0$, if Y has uniform multiplicity j; and the converse implication is apparent from (3).

If Y has uniform multiplicity $\aleph_0$, then $\mu(Y) > 0$ and, for $j = 1, 2, 3, \ldots$, the subset $Y \setminus Y_j$ of Y has multiplicity 0 or $\aleph_0$, but has multiplicity at most $j - 1$, by (3). Hence $m(Y \setminus Y_j) = 0$, $\mu(Y \setminus Y_j) = 0$ and

$$\mu\left(Y \setminus \bigcap_{j=1}^{\infty} Y_j\right) = \mu\left(\bigcup_{j=1}^{\infty} (Y \setminus Y_j)\right) = 0.$$

Accordingly, $\mu(Y) = \mu(Y \cap \bigcap_{j=1}^{\infty} Y_j) > 0$, when Y has uniform multiplicity $\aleph_0$; and the reverse implication is apparent from (3).

(ii) Suppose that $\{\mathcal{N}_j\}$ is a null ideal sequence in $\mathcal{S}$. Then $\mathcal{N}_1$ is a null ideal, so we may choose a regular Borel measure μ on S whose null sets are precisely the members of $\mathcal{N}_1$. We shall give an inductive construction for Borel sets $Y_1, Y_2, Y_3, \ldots$ satisfying the required conditions. With $Y_1 = S$, the σ-ideal $\mathcal{N}_1$ is already expressed in the form (2).

Suppose next that $k > 1$, and that $Y_1, \ldots, Y_{k-1}$ are Borel sets, such that $Y_1 \supseteq Y_2 \supseteq \cdots \supseteq Y_{k-1}$ and (2) is satisfied when $1 \le j < k$. Let $\{X_a\}$ be a disjoint family of Borel subsets of Y_{k-1}, maximal subject to the conditions that $X_a \in \mathcal{N}_k$ and $\mu(X_a) > 0$. (If there are no such sets X_a, interpret $\bigcup X_a$ as $\varnothing$ in the argument that follows.) Since $\mu(S) < \infty$, the family $\{X_a\}$ is at most countably infinite, so $\bigcup X_a$ is a Borel subset X of Y_{k-1}; and $X \in \mathcal{N}_k$, since $\mathcal{N}_k$ is a σ-ideal. Let Y_k be $Y_{k-1} \setminus X$, whence Y_{k-1} is the disjoint union of Borel sets Y_k and X.

With N in $\mathscr{S}$, N is the disjoint union of the three sets

$$N \backslash Y_{k-1}, \qquad N \cap X, \qquad N \cap Y_k.$$

From (2), with $j = k - 1$, we have $N \backslash Y_{k-1} \in \mathscr{N}_{k-1} \subseteq \mathscr{N}_k$; and $N \cap X \in \mathscr{N}_k$ since $X \in \mathscr{N}_k$. Accordingly, $N \in \mathscr{N}_k$ if and only if $N \cap Y_k \in \mathscr{N}_k$. If $N \cap Y_k \in \mathscr{N}_k$, then $\mu(N \cap Y_k) = 0$, since otherwise $N \cap Y_k$ could be added to the family $\{X_a\}$, contrary to the maximality assumption. Conversely, if $\mu(N \cap Y_k) = 0$, we have $N \cap Y_k \in \mathscr{N}_1 \subseteq \mathscr{N}_k$. From the preceding argument, $N \in \mathscr{N}_k$ if and only if $\mu(N \cap Y_k) = 0$. This shows that (2) is satisfied when $j = k$, and so completes the inductive construction of the sequence $\{Y_j\}$. It is apparent that (2) can be rewritten in the form

$$\text{(4)} \qquad \mathscr{N}_j = \{Y \in \mathscr{S} : Y \cap Y_j \in \mathscr{N}_1\} \qquad (j = 1, 2, 3, \ldots);$$

whence (2) remains true when μ is replaced by any other measure having the same family, $\mathscr{N}_1$, of null sets.

Suppose that $\{Y_j'\}$ is another sequence of Borel subsets of S, and that (2) (and hence, also, (4)) is satisfied with Y_j' in place of Y_j. By applying (4), first as it stands and then with Y_j' for Y_j, we deduce first that $Y_j' \backslash Y_j \in \mathscr{N}_j$, then that $Y_j' \backslash Y_j \in \mathscr{N}_1$; and a similar argument shows that $Y_j \backslash Y_j' \in \mathscr{N}_1$. Thus Y_j and Y_j' differ only by a μ-null set, when μ is any measure that has null ideal $\mathscr{N}_1$. ■

Our next objective is to construct the null ideal, null ideal sequence, and multiplicity function associated with a representation φ of $C(S)$, acting on a separable Hilbert space $\mathscr{H}$. The range of φ is an abelian C^*-subalgebra

$$\mathscr{C} = \{\varphi(f) : f \in C(S)\},$$

of $\mathscr{B}(\mathscr{H})$; so the weak-operator closure $\mathscr{C}^-$ is an abelian von Neumann algebra. Since $\mathscr{H}$ is separable, $\mathscr{C}^-$ is countably decomposable, and so has separating vectors (Corollary 5.5.17). For each x in $\mathscr{H}$, the mapping

$$\omega_x \circ \varphi : f \to \omega_x(\varphi(f)) = \langle \varphi(f)x, x \rangle$$

is a positive linear functional on $C(S)$; so there is a (unique) regular Borel measure μ_x on S, such that

$$\text{(5)} \qquad \langle \varphi(f)x, x \rangle = \int_S f \, d\mu_x \qquad (x \in \mathscr{H}, \quad f \in C(S)).$$

Since

$$\|\varphi(f)x\|^2 = \langle \varphi(f)x, \varphi(f)x \rangle = \langle \varphi(|f|^2)x, x \rangle = \int_S |f|^2 \, d\mu_x,$$

while (equivalence classes of) continuous functions on S form an everywhere-dense subspace of the Hilbert space $L_2(S, \mathscr{S}, \mu_x)$, there is an isometric linear mapping U, from $L_2(S, \mathscr{S}, \mu_x)$ into $\mathscr{H}$, such that $Uf = \varphi(f)x$ when $f \in C(S)$. Since $\mathscr{H}$ is separable, the same is true of $L_2(S, \mathscr{S}, \mu_x)$.

The measures μ_x have already been encountered in Theorem 5.2.6(v), and $\mu_x(Y) = \langle E(Y)x, x\rangle$ for each Borel subset Y of S, where $Y \to E(Y)\colon \mathscr{S} \to \mathscr{C}^-$ is the projection-valued measure described in that theorem.

The notation just introduced ($\varphi, \mathscr{H}, \mathscr{C}, \mu_x, E(Y)$) will be used throughout the remainder of this section.

9.5.2. PROPOSITION. *Suppose that $x, u \in \mathscr{H}$, and u is a separating vector for $\mathscr{C}^-$.*

(i) *The kernel of φ consists of all μ_u-null functions in $C(S)$.*
(ii) *A Borel subset Y of S is μ_x-null if and only if $E(Y)x = 0$.*
(iii) *Each μ_u-null subset of S is also μ_x-null.*
(iv) *If x is a separating vector for $\mathscr{C}^-$, then μ_x and μ_u have the same null sets.*

Proof. (i) Since $\varphi(f) \in \mathscr{C} \subseteq \mathscr{C}^-$ when $f \in C(S)$, while u is a separating vector for $\mathscr{C}^-$, the kernel of φ consists of those f for which $\varphi(f)u = 0$. Now

$$\|\varphi(f)u\|^2 = \int_S |f|^2 \, d\mu_u,$$

and the right-hand side vanishes if and only if f is μ_u-null.

(ii) This follows from the fact that

$$\mu_x(Y) = \langle E(Y)x, x\rangle = \|E(Y)x\|^2.$$

(iii) If $Y \in \mathscr{S}$ and $\mu_u(Y) = 0$, then $E(Y) \in \mathscr{C}^-$ and, from (ii), $E(Y)u = 0$. Since u is a separating vector for $\mathscr{C}^-$, $E(Y) = 0$; so $E(Y)x = 0$ and, again by (ii), $\mu_x(Y) = 0$.

(iv) This is an immediate consequence of (iii), since the roles of x and u in (iii) can be reversed when x, as well as u, is a separating vector for $\mathscr{C}^-$. ∎

By the *null ideal* of the representation φ, we mean the family $\mathscr{N}_\varphi$ of all null sets of the measure μ_u, where u is a separating vector for $\mathscr{C}^-$. By Proposition 9.5.2(iv), this does not depend on the particular choice of separating vector. It follows from the remarks following (5) that $\mathscr{N}_\varphi$ is a null ideal of separable type.

We denote by $L_\infty(\varphi)$ the set of all Borel measurable complex-valued functions on S that are bounded on the complement of some set in $\mathscr{N}_\varphi$, and by $N(\varphi)$ its subset, consisting of Borel functions that vanish on the complement of some set in $\mathcal{N}_\varphi$. Thus $L_\infty(\varphi) = L_\infty(S, \mathcal{S}, \mu_u)$, and $N(\varphi)$ consists of all μ_u-null functions, whenever u is a separating vector for $\mathcal{C}^-$.

In the present context, it is convenient to drop our usual convention, and to distinguish between the set $L_\infty(\varphi)$ of *functions* and the set $L_\infty(\varphi)/N(\varphi)$ of *equivalence classes* of functions (modulo null functions). Note that $L_\infty(\varphi)$ is an abelian algebra over the complex field, has an involution (complex conjugation), contains $C(S)$ as a self-adjoint subalgebra, and contains $N(\varphi)$ as a self-adjoint ideal.

Our approach to the extension problem hinges on the following result.

9.5.3. PROPOSITION. *There is a unique mapping $\bar{\varphi}: L_\infty(\varphi) \to \mathcal{B}(\mathcal{H})$ such that*

$$\langle \bar{\varphi}(f)x, x\rangle = \int_S f\, d\mu_x \qquad (x \in \mathcal{H}, \quad f \in L_\infty(\varphi)). \tag{6}$$

*Moreover, $\bar{\varphi}$ is a * homomorphism, with kernel $N(\varphi)$ and range $\mathcal{C}^-$; and $\bar{\varphi}(f) = \varphi(f)$ when $f \in C(S)$. Each projection in $\mathcal{C}^-$ has the form $E(Y)$, where Y is a Borel set in S; and $E(Y) = \bar{\varphi}(q)$, where q is the characteristic function of Y.*

Proof. Let u be a separating vector for $\mathcal{C}^-$, so that $L_\infty(\varphi)$ is $L_\infty(S, \mathcal{S}, \mu_u)$ and $N(\varphi)$ is the set of all μ_u-null functions. The quotient space $L_\infty(\varphi)/N(\varphi)$, with the ($\mu_u$-) essential supremum norm, will be identified in the usual way with the Banach dual space of $L_1(S, \mathcal{S}, \mu_u)/N(\varphi)$, and will be considered also as a C^*-algebra (with multiplication defined pointwise, and with complex conjugation for involution). We shall consider, as well, the C^*-algebra $\mathcal{B}$ of all *bounded* complex-valued Borel functions on S (with pointwise algebraic structure, complex conjugation as involution, and the supremum norm). When $f \in \mathcal{B}$, we denote by $[f]$ the element $f + N(\varphi)$ of $L_\infty(\varphi)/N(\varphi)$. The mapping $f \to [f]$ is a * homomorphism from $\mathcal{B}$ onto $L_\infty(\varphi)/N(\varphi)$, with kernel $\mathcal{B} \cap N(\varphi)$. The projections in $L_\infty(\varphi)/N(\varphi)$ are precisely the elements $[q]$, in which q is the characteristic function of a Borel set Y in S.

There is a norm-dense * subalgebra $\mathcal{B}_0$ of $\mathcal{B}$, consisting of the Borel simple functions (that is, those elements of $\mathcal{B}$ that take only finitely many distinct values). Each f in $\mathcal{B}_0$ can be expressed (in many ways) as a linear combination $\sum_{j=1}^n a_j q_j$, in which each q_j is the characteristic function of a Borel set Y_j. Note that, for x in $\mathcal{H}$,

$$\int_S f\, d\mu_x = \sum_{j=1}^n a_j \mu_x(Y_j) = \sum_{j=1}^n a_j \langle E(Y_j)x, x\rangle = \left\langle \left(\sum_{j=1}^n a_j E(Y_j) \right) x, x \right\rangle.$$

If $f = 0$, then $\langle (\sum_{j=1}^n a_j E(Y_j))x, x\rangle = 0$ for each x in $\mathcal{H}$, whence $\sum_{j=1}^n a_j E(Y_j) = 0$.

From the previous paragraph, it follows that we can define (unambiguously) an adjoint-preserving linear mapping $\varphi_0: \mathscr{B}_0 \to \mathscr{C}^-$, by setting $\varphi_0(f) = \sum_{j=1}^n a_j E(Y_j)$ when $f = \sum_{j=1}^n a_j q_j$ and $q_1, \ldots, q_n$ are the characteristic functions of Borel sets $Y_1, \ldots, Y_n$ in S. (Note, in particular, that $\varphi_0(q) = E(Y)$ when q is the characteristic function of a Borel set Y in S.) Moreover,

$$\langle \varphi_0(f)x, x \rangle = \int_S f \, d\mu_x \qquad (x \in \mathscr{H}) \tag{7}$$

for each f in $\mathscr{B}_0$. If $f, g \in \mathscr{B}_0$, we can express S as the disjoint union of Borel sets $Y_1, \ldots, Y_n$, on each of which both f and g are constant; and the projections $E(Y_1), \ldots, E(Y_n)$ are mutually orthogonal (some may be 0) and have sum I. With q_j the characteristic function of Y_j, $f = \sum a_j q_j$ and $g = \sum b_j q_j$, for suitable complex numbers a_j, b_j. Then,

$$\|\varphi_0(f)\| = \|\sum_{j=1}^n a_j E(Y_j)\| \leq \max_j |a_j| = \sup_{s \in S} |f(s)| = \|f\|;$$

and since $fg = \sum a_j b_j q_j$,

$$\varphi_0(fg) = \sum_{j=1}^n a_j b_j E(Y_j) = \left(\sum_{j=1}^n a_j E(Y_j) \right) \left(\sum_{j=1}^n b_j E(Y_j) \right) = \varphi_0(f)\varphi_0(g).$$

Accordingly, φ_0 is a bounded * homomorphism from $\mathscr{B}_0$ into $\mathscr{C}^-$.

Since $\mathscr{B}_0$ is everywhere dense in $\mathscr{B}$, φ_0 extends by continuity to a * homomorphism (again denoted by φ_0) from $\mathscr{B}$ into $\mathscr{C}^-$. Moreover, since convergence in $\mathscr{B}$ is uniform convergence on S, (7) remains valid for each f in $\mathscr{B}$. From this,

$$\begin{aligned} \|\varphi_0(f)u\|^2 &= \langle \varphi_0(f)^* \varphi_0(f)u, u \rangle \\ &= \langle \varphi_0(|f|^2)u, u \rangle = \int_S |f|^2 \, d\mu_u; \end{aligned}$$

and since u is a separating vector for $\mathscr{C}^-$, $\varphi_0(f) = 0$ if and only if f is μ_u-null. Thus $\varphi_0: \mathscr{B} \to \mathscr{C}^-$ has kernel $\mathscr{B} \cap N(\varphi)$.

Each element of $L_\infty(\varphi)$ can be expressed in the form $g + h$, with g in $\mathscr{B}$ and h in $N(\varphi)$. If $g_1 + h_1 = g_2 + h_2$, where $g_j \in \mathscr{B}$ and $h_j \in N(\varphi)$, for $j = 1, 2$, then $g_1 - g_2 = h_2 - h_1 \in \mathscr{B} \cap N(\varphi)$; and from the preceding paragraph, $\varphi_0(g_1 - g_2) = 0$, whence $\varphi_0(g_1) = \varphi_0(g_2)$. It now follows that the equation

$$\bar{\varphi}(g + h) = \varphi_0(g) \qquad (g \in \mathscr{B}, \quad h \in N(\varphi)) \tag{8}$$

defines (unambiguously) a * homomorphism $\bar{\varphi}$ from $L_\infty(\varphi)$ into $\mathscr{C}^-$. Since $\varphi_0(g) = 0$ only when $g \in \mathscr{B} \cap N(\varphi)$, $\bar{\varphi}$ has kernel $N(\varphi)$. Since elements h of $N(\varphi)$ are μ_u-null, and are therefore μ_x-null for each x in $\mathscr{H}$ by Proposition 9.5.2(iii),

$$\langle \bar{\varphi}(g + h)x, x \rangle = \langle \varphi_0(g)x, x \rangle = \int_S g \, d\mu_x = \int_S (g + h) \, d\mu_x$$

when $g \in \mathscr{B}$ and $h \in N(\varphi)$. This proves (6). From (5) and (6), $\langle \bar{\varphi}(f)x, x \rangle = \langle \varphi(f)x, x \rangle$ for each x in $\mathscr{H}$, when $f \in C(S)$; so $\bar{\varphi}(f) = \varphi(f)$, in this case.

So far, we have shown that φ extends to a * homomorphism $\bar{\varphi}$ from $L_\infty(\varphi)$ into $\mathscr{C}^-$, and that $\bar{\varphi}$ satisfies (6) and has kernel $N(\varphi)$. Note that, for each f in $L_\infty(\varphi)$, the condition (6) determines the operator $\bar{\varphi}(f)$ uniquely, since it specifies the value of $\langle \bar{\varphi}(f)x, x \rangle$ for each x in $\mathscr{H}$. Accordingly, $\bar{\varphi}$ is uniquely determined by (6), among mappings from $L_\infty(\varphi)$ into $\mathscr{B}(\mathscr{H})$.

Our next objective is to show that the range of $\bar{\varphi}$ is the whole of $\mathscr{C}^-$. Since

$$\mathscr{C}^- \supseteq \varphi_0(\mathscr{B}) = \bar{\varphi}(L_\infty(\varphi)) \supseteq \bar{\varphi}(C(S)) = \varphi(C(S)) = \mathscr{C},$$

it suffices (by the Kaplansky density theorem) to show that the unit ball of $\varphi_0(\mathscr{B})$ is weak-operator compact. From the first paragraph of this proof, the * homomorphism $f \to [f]$ from $\mathscr{B}$ onto $L_\infty(\varphi)/N(\varphi)$ has the same kernel, $\mathscr{B} \cap N(\varphi)$, as the * homomorphism $\varphi_0 \colon \mathscr{B} \to \mathscr{C}^-$. Hence the equation

$$\psi([f]) = \varphi_0(f) \qquad (f \in \mathscr{B}) \tag{9}$$

defines a * isomorphism ψ from the C^*-algebra $L_\infty(\varphi)/N(\varphi)$ into $\mathscr{C}^-$, and the range of ψ is $\varphi_0(\mathscr{B})$ $(= \bar{\varphi}(L_\infty(\varphi)))$. By Theorem 4.1.8(iii), ψ is an isometry, and so carries the unit ball B of $L_\infty(\varphi)/N(\varphi)$ onto that of $\varphi_0(\mathscr{B})$. By the Alaoglu–Bourbaki theorem (1.6.5(i)), B is compact as a subset of $L_\infty(\varphi)/N(\varphi)$ in its weak * topology as the Banach dual space of $L_1(S, \mathscr{S}, \mu_u)/N(\varphi)$. In order to show that the unit ball of $\varphi_0(\mathscr{B})$ is weak-operator compact (and hence, that $\bar{\varphi}$ has range $\mathscr{C}^-$), it now suffices to show that ψ is continuous, from $L_\infty(\varphi)/N(\varphi)$ with the weak * topology, into $\mathscr{C}^-$ with the weak-operator topology. When $x \in \mathscr{H}$, every μ_u-null set is μ_x-null; the Radon–Nikodým derivative $d\mu_x/d\mu_u$ is an element h of (the predual) $L_1(S, \mathscr{S}, \mu_u)$, and

$$\omega_x(\psi([f])) = \langle \varphi_0(f)x, x \rangle = \int_S f \, d\mu_x = \int_S fh \, d\mu_u \qquad (f \in \mathscr{B}).$$

This shows that the linear functional $\omega_x \circ \psi$ on $L_\infty(\varphi)/N(\varphi)$ is weak * continuous. Hence ψ has the required continuity property, and $\bar{\varphi}$ (thus, also, ψ) has range $\mathscr{C}^-$.

Since ψ is a * isomorphism from $L_\infty(\varphi)/N(\varphi)$ onto $\mathscr{C}^-$, it follows from the final assertion in the first paragraph of this proof that every projection in $\mathscr{C}^-$ has the form $\psi([q])$, where q is the characteristic function of a Borel set Y in S. Since $q \in \mathscr{B}$, it follows from (9) and (8) that

$$\psi([q]) = \varphi_0(q) = \bar{\varphi}(q);$$

and we have already noted that $\varphi_0(q) = E(Y)$. ■

Our solution of the extension problem is set out in the following result.

9.5.4. THEOREM. *Suppose that S is a compact Hausdorff space and, for $k = 1, 2$, φ_k is a representation of $C(S)$ on a separable Hilbert space $\mathscr{H}_k$ that has range $\mathscr{C}_k$. In order that there exist a * isomorphism α, from $\mathscr{C}_1^-$ onto $\mathscr{C}_2^-$, such that $\alpha(\varphi_1(f)) = \varphi_2(f)$ whenever $f \in C(S)$, it is necessary and sufficient that φ_1 and φ_2 have the same null ideal.*

Proof. Suppose first that φ_1 and φ_2 have the same null ideal. Then $N(\varphi_1) = N(\varphi_2)$ and $L_\infty(\varphi_1) = L_\infty(\varphi_2)$, so we can denote these spaces simply by N and L_∞. From Proposition 9.5.3, φ_k extends to a * homomorphism $\bar{\varphi}_k \colon L_\infty \to \mathscr{C}_k^-$, that has range $\mathscr{C}_k^-$ and kernel N. Since $\bar{\varphi}_1$ and $\bar{\varphi}_2$ have the same kernel, the equation

$$\alpha(\bar{\varphi}_1(f)) = \bar{\varphi}_2(f) \qquad (f \in L_\infty)$$

defines (unambiguously) a * isomorphism α from $\mathscr{C}_1^-$ onto $\mathscr{C}_2^-$, and $\alpha(\varphi_1(f)) = \varphi_2(f)$ when $f \in C(S)$.

Conversely, suppose there is a * isomorphism α with the properties set out in the theorem. For x in $\mathscr{H}_1$ and y in $\mathscr{H}_2$, let μ_x and ν_y be the regular Borel measures on S such that

$$\langle \varphi_1(f)x, x\rangle = \int_S f\, d\mu_x, \qquad \langle \varphi_2(f)y, y\rangle = \int_S f\, d\nu_y \qquad (f \in C(S)).$$

With v a separating vector for $\mathscr{C}_2^-$, the positive linear functional $\omega_v \circ \alpha$ on $\mathscr{C}_1^-$ is ultraweakly continuous, and so has the form $\omega_u | \mathscr{C}_1^-$, for some u in $\mathscr{H}_1$ (Remark 7.4.4, Theorem 7.2.3). If $A \in \mathscr{C}_1^-$ and $Au = 0$, then

$$0 = \omega_u(A^*A) = \omega_v(\alpha(A^*A)) = \|\alpha(A)v\|^2;$$

whence $\alpha(A) = 0$ and $A = 0$, since v is a separating vector for $\mathscr{C}_2^-$. It follows that u is a separating vector for $\mathscr{C}_1^-$. For each f in $C(S)$,

$$\begin{aligned}\int_S f\, d\mu_u &= \omega_u(\varphi_1(f)) = \omega_v(\alpha(\varphi_1(f))) \\ &= \omega_v(\varphi_2(f)) = \int_S f\, d\nu_v.\end{aligned}$$

Thus $\mu_u = \nu_v$, so μ_u and ν_v have the same null sets; that is, φ_1 and φ_2 have the same null ideal. ∎

Before defining the null-ideal sequence of a representation φ of $C(S)$, we prove two preparatory results. The first, although not absolutely essential, is included for convenience and completeness. We denote by $\bar{\varphi}: L_\infty(\varphi) \to \mathscr{C}^-$ the * homomorphism occurring in Proposition 9.5.3.

9.5.5. **Proposition.** *Suppose that $\mathscr{N}_\varphi$ is the null ideal of the representation φ of $C(S)$, acting on the separable Hilbert space $\mathscr{H}$, and μ is a regular Borel measure on S.*

(i) *μ has the form μ_x, for some x in $\mathscr{H}$, if and only if each element of $\mathscr{N}_\varphi$ is a μ-null set.*

(ii) *μ has the form μ_x, with x a separating vector for $\mathscr{C}^-$, if and only if*

$$\mathscr{N}_\varphi = \{N \in \mathscr{S} : \mu(N) = 0\}.$$

Proof. Let u be a separating vector for $\mathscr{C}^-$, so that $\mathscr{N}_\varphi$ is the family of all μ_u-null sets. In both parts of the present proposition, the implication in one direction has already been proved (Proposition 9.5.2(iii), (iv)), so it suffices to establish the reverse implication.

Suppose that each element of $\mathscr{N}_\varphi$ is a μ-null set; that is, μ_u-null sets are μ-null. The Radon–Nikodým derivative $d\mu/d\mu_u$ is an element h of $L_1(S, \mathscr{S}, \mu_u)$, and $h(s) \geq 0$ $(s \in S)$. With $f_1, f_2, f_3, \ldots$ in $L_\infty(\varphi)$ $(= L_\infty(S, \mathscr{S}, \mu_u))$ defined by

$$f_n(s) = \begin{cases} (h(s))^{1/2} & \text{if } \ n - 1 \leq h(s) < n \\ 0 & \text{otherwise,} \end{cases}$$

we have $h(s) = \sum (f_n(s))^2$ and $f_m f_n = 0$ when $m \neq n$. With x_n defined as $\bar{\varphi}(f_n)u$, and g in $C(S)$,

$$\begin{aligned} \langle \varphi(g)x_m, x_n \rangle &= \langle \bar{\varphi}(g)\bar{\varphi}(f_m)u, \bar{\varphi}(f_n)u \rangle \\ &= \langle \bar{\varphi}(g f_m f_n)u, u \rangle \\ &= \begin{cases} 0 & \text{if } \ m \neq n \\ \displaystyle\int_S g f_n^2 \, d\mu_u & \text{if } \ m = n. \end{cases} \end{aligned}$$

By taking $g(s) = 1$ $(s \in S)$, it follows that $\{x_n\}$ is an orthogonal sequence in $\mathscr{H}$, and that

$$\sum \|x_n\|^2 = \sum \int_S f_n^2 \, d\mu_u = \int_S h \, d\mu_u = \mu(S) < \infty.$$

Accordingly, the series $\sum x_n$ converges to an element x of $\mathscr{H}$. For each g in $C(S)$,

$$\langle \varphi(g)x, x\rangle = \lim_{n\to\infty} \sum_{j,k=1}^{n} \langle \varphi(g)x_j, x_k\rangle = \lim_{n\to\infty} \sum_{j=1}^{n} \int_S g f_j^2 \, d\mu_u .$$

By the dominated convergence theorem,

$$\langle \varphi(g)x, x\rangle = \int_S gh \, d\mu_u = \int_S g \, d\mu;$$

so $\mu = \mu_x$.

Now suppose, further, that $\mathscr{N}_\varphi$ is the family of all μ-null sets; that is, μ_u and μ_x have the same null sets. If $A \in \mathscr{C}^-$ and $Ax = 0$, it results from Proposition 9.5.3 that $A = \bar{\varphi}(f)$ for some f in $L_\infty(\varphi)$. Since

$$\int_S |f|^2 \, d\mu_x = \|\bar{\varphi}(f)x\|^2 = \|Ax\|^2 = 0,$$

it follows that f is μ_x-null, equivalently μ_u-null. Thus $f \in N(\varphi)$, and $A = \bar{\varphi}(f) = 0$; so x is a separating vector for $\mathscr{C}^-$. ∎

If Y is a Borel subset of S, the projection $E(Y)$ lies in $\mathscr{C}^-$ ($\subseteq \mathscr{C}'$), and we can consider the representation, $f \to \varphi(f)E(Y)$, of $C(S)$ on the range of $E(Y)$. In the following lemma, we describe the null ideal of this representation.

9.5.6. Lemma. *If Y is a Borel subset of S, the representation $\psi: f \to \varphi(f)E(Y)$, of $C(S)$ on the range of $E(Y)$, has null ideal $\mathscr{N}_\psi$ given by*

$$\mathscr{N}_\psi = \{N \in \mathscr{S} : N \cap Y \in \mathscr{N}_\varphi\}.$$

Proof. Let u be a separating vector for $\mathscr{C}^-$, so that $L_\infty(\varphi) = L_\infty(S, \mathscr{S}, \mu_u)$ and $N(\varphi)$ consists of all μ_u-null functions. Write Q in place of $E(Y)$, so that $Q = \bar{\varphi}(q)$, where q is the characteristic function of Y. The weak-operator closure $\mathscr{C}^-Q$ of the range $\mathscr{C}Q$ of the representation ψ has Qu as a separating vector. Accordingly, $\mathscr{N}_\psi$ is the ideal of all null sets of the measure ν defined by

$$\int_S f \, d\nu = \langle \psi(f)Qu, Qu\rangle \qquad (f \in C(S)).$$

Now

$$\int_S f \, d\nu = \langle \varphi(f)\bar{\varphi}(q)u, \bar{\varphi}(q)u\rangle = \langle \bar{\varphi}(fq)u, u\rangle$$

$$= \int_S fq \, d\mu_u = \int_Y f \, d\mu_u \qquad (f \in C(S)).$$

Thus $\nu(N) = \mu_u(N \cap Y)$, for each Borel set N, and

$$\mathcal{N}_\psi = \{N \in \mathcal{S} : \nu(N) = 0\} = \{N \in \mathcal{S} : \mu_u(N \cap Y) = 0\}$$
$$= \{N \in \mathcal{S} : N \cap Y \in \mathcal{N}_\varphi\}. \quad \blacksquare$$

Since $\mathscr{C}'$ is a type I von Neumann algebra acting on a separable Hilbert space $\mathscr{H}$, its center $\mathscr{C}^-$ contains a (unique) orthogonal family $\{P_n : 1 \leq n \leq \aleph_0\}$ of projections, with sum I, such that $\mathscr{C}'P_n$ is type I_n unless $P_n = 0$. For $j = 1, 2, \ldots$, let

$$Q_j = \sum_{j \leq n \leq \aleph_0} P_j,$$

and let $\mathcal{N}_j$ be the null ideal of the representation $\psi_j \colon f \to \varphi(f)Q_j$ of $C(S)$ on $Q_j(\mathscr{H})$. By use of Proposition 9.5.3 we shall show, in the next paragraph, that $\{\mathcal{N}_j\}$ is a null ideal sequence in $\mathcal{S}$ (with an associated multiplicity function m on $\mathcal{S}$). We refer to $\{\mathcal{N}_j\}$ and m as the *null ideal sequence of the representation* φ and the *multiplicity function* of φ.

To verify that $\{\mathcal{N}_j\}$ is a null ideal sequence, note first that $Q_j = E(Y_j)$ for some Borel subset Y_j of S. Since $Q_1 = I$, we may take Y_1 to be S. Moreover,

$$Q_j = Q_1 Q_2 \cdots Q_j = E(Y_1)E(Y_2) \cdots E(Y_j) = E(Y_1 \cap Y_2 \cap \cdots \cap Y_j);$$

and upon replacing Y_j by $Y_1 \cap Y_2 \cap \cdots \cap Y_j$, we may suppose that

$$S = Y_1 \supseteq Y_2 \supseteq Y_3 \supseteq \cdots.$$

Since ψ_j is the representation $f \to \varphi(f)E(Y_j)$, it follows from Lemma 9.5.6 that

$$\mathcal{N}_j = \{N \in \mathcal{S} : N \cap Y_j \in \mathcal{N}_\varphi\} = \{N \in \mathcal{S} : \mu_u(N \cap Y_j) = 0\},$$

where u is a separating vector for $\mathscr{C}^-$. It now results from Proposition 9.5.1 that $\{\mathcal{N}_j\}$ is a null ideal sequence in $\mathcal{S}$.

9.5.7. THEOREM. *If S is a compact Hausdorff space and, for $k = 1, 2$, φ_k is a representation of $C(S)$ on a separable Hilbert space $\mathscr{H}_k$, then φ_1 and φ_2 are equivalent if and only if they have the same null ideal sequence.*

Proof. All the constructs used, in defining the null ideal sequence of a representation of $C(S)$, are preserved by (unitary) equivalence of representations. Thus, equivalent representations have the same null ideal sequence.

Conversely, suppose that φ_1 and φ_2 have the same null ideal sequence, $\{\mathcal{N}_j\}$. For $k = 1, 2$, let $\mathscr{C}_k$ denote the range of φ_k, let $\bar{\varphi}_k \colon L_\infty(\varphi_k) \to \mathscr{C}_k^-$ be the * homomorphism constructed as in Proposition 9.5.3, and let $Y \to E_k(Y)$ be the projection-valued measure corresponding to φ_k. Let $\{P_n^{(k)} : 1 \leq n \leq \aleph_0\}$

be the orthogonal family of projections in $\mathscr{C}_k^-$, with sum I, such that $\mathscr{C}_k' P_n^{(k)}$ is type I_n unless $P_n^{(k)} = 0$; and let

$$Q_j^{(k)} = \sum_{j \le n \le \aleph_0} P_n^{(k)} \qquad (j = 1, 2, 3, \ldots).$$

Just as in the discussion preceding the statement of the theorem, there is a decreasing sequence $(Y_j^{(k)})_{j \ge 1}$ of Borel subsets of S, with $Y_1^{(k)} = S$, such that

$$Q_j^{(k)} = E_k(Y_j^{(k)}) = \bar{\varphi}_k(q_j^{(k)}) \qquad (j = 1, 2, 3, \ldots),$$

where $q_j^{(k)}$ is the characteristic function of $Y_j^{(k)}$. Moreover, since $\{\mathscr{N}_j\}$ is the null ideal sequence of φ_k, we have

(10) $$\mathscr{N}_j = \{N \in \mathscr{S} : N \cap Y_j^{(k)} \in \mathscr{N}_1\} \qquad (j = 1, 2, 3, \ldots).$$

Since $\mathscr{N}_1$ is a null ideal, there is a regular Borel measure μ on S, whose null sets are precisely the members of $\mathscr{N}_1$; and we can rewrite (10) in the form

(11) $$\mathscr{N}_j = \{N \in \mathscr{S} : \mu(N \cap Y_j^{(k)}) = 0\} \qquad (j = 1, 2, 3, \ldots).$$

Since φ_1 and φ_2 have the same null ideal, $\mathscr{N}_1$, it follows as in the proof of Theorem 9.5.4 that $L_\infty(\varphi_1) = L_\infty(\varphi_2)$ $(=L_\infty)$, and that there is a * isomorphism α, from $\mathscr{C}_1^-$ onto $\mathscr{C}_2^-$, such that

$$\alpha(\bar{\varphi}_1(f)) = \bar{\varphi}_2(f) \qquad (f \in L_\infty).$$

We shall show that $\alpha(P_j^{(1)}) = P_j^{(2)}$ $(1 \le j \le \aleph_0)$. In view of the theory developed in Section 9.3, this implies that α is unitarily implemented, whence (since $\alpha(\varphi_1(f)) = \varphi_2(f)$ when $f \in C(S)$) φ_1 and φ_2 are equivalent.

Since (11) holds for $k = 1, 2$, it results from Proposition 9.5.1(ii) that, for $j = 1, 2, 3, \ldots$,

$$Y_j^{(1)} \setminus Y_j^{(2)} \in \mathscr{N}_1, \qquad Y_j^{(2)} \setminus Y_j^{(1)} \in \mathscr{N}_1.$$

From this, and since φ_1 has null ideal $\mathscr{N}_1$, it follows that $q_j^{(1)} - q_j^{(2)}$ lies in $N(\varphi_1)$, the kernel of $\bar{\varphi}_1$. Thus

$$\alpha(Q_j^{(1)}) = \alpha(\bar{\varphi}_1(q_j^{(1)})) = \alpha(\bar{\varphi}_1(q_j^{(2)})) = \bar{\varphi}_2(q_j^{(2)}) = Q_j^{(2)},$$

$$P_j^{(2)} = Q_j^{(2)} - Q_{j+1}^{(2)} = \alpha(Q_j^{(1)} - Q_{j+1}^{(1)}) = \alpha(P_j^{(1)}) \qquad (j = 1, 2, 3, \ldots),$$

$$P_{\aleph_0}^{(2)} = I - \sum_{j=1}^{\infty} P_j^{(2)} = \alpha\left(I - \sum_{j=1}^{\infty} P_j^{(1)}\right) = \alpha(P_{\aleph_0}^{(1)}).$$

As noted above, this implies that φ_1 and φ_2 are equivalent. ∎

9.5.8. Corollary. *If S is a compact Hausdorff space and, for $k = 1, 2$, φ_k is a representation of $C(S)$ acting on a separable Hilbert space $\mathscr{H}_k$, then φ_1 and φ_2 are equivalent if and only if they have the same multiplicity function.*

Proof. This follows at once from Theorem 9.5.7, in view of the one-to-one correspondence between null ideal sequences and multiplicity functions. ∎

9.5.9. PROPOSITION. *Suppose that m is the multiplicity function of a representation φ of $C(S)$ on a separable Hilbert space $\mathscr{H}$, $\mathscr{C}$ is the range of φ, Y is a Borel subset of S, and $1 \le n \le \aleph_0$. Then Y has uniform multiplicity n, relative to m, if and only if the von Neumann algebra $\mathscr{C}'E(Y)$ is of type I_n.*

Proof. We shall use the notation introduced in the discussion preceding the statement of Theorem 9.5.7; in addition, the characteristic functions of the Borel sets Y, Y_j will be denoted by q, q_j, so that $E(Y) = \bar{\varphi}(q)$ and $Q_j = E(Y_j) = \bar{\varphi}(q_j)$ for $j = 1, 2, \ldots$. Since

$$P_{\aleph_0} = \bigwedge_{j=1}^{\infty} Q_j, \qquad P_n = Q_n - Q_{n+1} \qquad (n = 1, 2, \ldots),$$

it follows that $P_n = \bar{\varphi}(p_n)$ $(1 \le n \le \aleph_0)$, where p_n is the characteristic function $q_n - q_{n+1}$ of $Y_n \setminus Y_{n+1}$ when $n < \aleph_0$, while $p_{\aleph_0}$ is the characteristic function $\lim_{j\to\infty} q_j$ of $\bigcap Y_j$. Now $\mathscr{C}'E(Y)$ is type I_n if and only if $0 < E(Y) = E(Y)P_n$; that is, $0 < \bar{\varphi}(q) = \bar{\varphi}(qp_n)$. This occurs if and only if q is not μ_u-null, and, μ_u-almost everywhere, $q = qp_n$. Hence $\mathscr{C}'E(Y)$ is type I_n if and only if

$$\mu_u(Y) = \begin{cases} \mu_u(Y \cap (Y_n \setminus Y_{n+1})) > 0 & \text{if} \quad n < \aleph_0 \\ \mu_u(Y \cap \bigcap Y_j) > 0 & \text{if} \quad n = \aleph_0. \end{cases}$$

However, this is precisely the condition, set out in Proposition 9.5.1(i), under which Y has uniform multiplicity n. ∎

We say that a representation φ of $C(S)$ has *uniform multiplicity n*, where $1 \le n \le \aleph_0$, if S has uniform multiplicity n, relative to the multiplicity function of φ. By Proposition 9.5.9, this occurs if and only if the commutant $\mathscr{C}'$ of the range $\mathscr{C}$ of φ is of type I_n.

Further information concerning representations with uniform multiplicity is set out in Theorem 9.5.12 below.

9.5.10. EXAMPLE. We show that every null ideal sequence $\{\mathscr{N}_j\}$ of separable type in $\mathscr{S}$ is the null ideal sequence of a representation φ of $C(S)$, acting on a separable Hilbert space; and we give an explicit construction for φ.

From Proposition 9.5.1, there is a regular Borel measure μ on S, and a decreasing sequence $\{Y_j\}$ of Borel subsets of S, such that $Y_1 = S$ and

$$\mathscr{N}_j = \{N \in \mathscr{S} : \mu(N \cap Y_j) = 0\}.$$

Since the family $\mathcal{N}_1$ of all μ-null sets is a null ideal of separable type, the Hilbert space $L_2(S, \mathcal{S}, \mu)$ $(= L_2)$ is separable. Accordingly, the same is true of its closed subspace

$$\mathcal{H}_j = \{x \in L_2 : x(s) = 0 \ \mu\text{-almost everywhere on } S \backslash Y_j\},$$

and of the Hilbert space direct sum $\mathcal{H} = \sum_{j=1}^{\infty} \oplus \mathcal{H}_j$. When $f \in C(S)$, we can define a bounded linear operator $\varphi(f)$, acting on $\mathcal{H}$, by

$$\varphi(f)\{x_1, x_2, x_3, \ldots\} = \{fx_1, fx_2, fx_3, \ldots\}.$$

It is apparent that φ is a representation of $C(S)$, acting on $\mathcal{H}$, and we show in due course that the null ideal sequence of φ is $\{\mathcal{N}_j\}$.

For each f in $L_\infty(S, \mathcal{S}, \mu)$ $(= L_\infty)$, let $\Phi(f) : L_2 \to L_2$ be the bounded linear operator of multiplication by f. Thus Φ is a * homomorphism from L_∞ onto a maximal abelian von Neumann subalgebra $\mathcal{A}$ of $\mathcal{B}(L_2)$ (Example 5.1.6), and the restriction $\Phi \mid C(S)$ is a representation Φ_0 of $C(S)$ on L_2. The range $\mathcal{A}_0$ of Φ_0 is a * subalgebra of $\mathcal{A}$, so $\mathcal{A}_0^- \subseteq \mathcal{A}$. Conversely, each element of $\mathcal{A}$ has the form $\Phi(f)$, with f in L_∞, and there is a bounded sequence $\{f_n\}$ in $C(S)$ such that $f_n(s) \to f(s)$ μ-almost everywhere. As noted in the final paragraph of Example 2.5.12, it follows from the dominated convergence theorem that the sequence $\{\Phi(f_n)\}$ in $\mathcal{A}_0$ is strong-operator convergent to $\Phi(f)$; so $\Phi(f) \in \mathcal{A}_0^-$, and $\mathcal{A} \subseteq \mathcal{A}_0^-$. Hence $A_0^- = \mathcal{A}$, and so $\mathcal{A}_0^-$ is maximal abelian.

Let $e_1, e_2, e_3, \ldots, e_\infty$ be the characteristic functions of the sets $Y_1, Y_2, Y_3, \ldots, \bigcap Y_j$, respectively. Then $e_j \in L_\infty$, $\Phi(e_j)$ is a projection E_j in $\mathcal{A}$ $(1 \le j \le \infty)$, $E_1 = I$, and the decreasing sequence $\{E_1, E_2, E_3, \ldots\}$ is strong-operator convergent to E_∞. Moreover, $\mathcal{H}_j = E_j(L_2)$ $(j = 1, 2, \ldots)$ and

$$\varphi(f) = \sum_{1 \le j < \infty} \oplus \Phi(f)E_j \qquad (f \in C(S)).$$

The range $\mathcal{C}$ of φ consists of all operators of the form $\sum \oplus AE_j$, with A in $\mathcal{A}_0$; so

$$\mathcal{C}^- = \left\{ \sum_{1 \le j < \infty} \oplus AE_j : A \in \mathcal{A} \right\}.$$

Let $F_j = E_j - E_{j+1}$ $(j = 1, 2, \ldots)$, $F_\infty = E_\infty$, and note that $\{F_n : 1 \le n \le \infty\}$ is an orthogonal family of projections, with sum I, in $\mathcal{A}$. Accordingly, if

$$P_n = \sum_{1 \le j < \infty} \oplus F_n E_j \qquad (1 \le n \le \infty),$$

then $\{P_n : 1 \le n \le \infty\}$ is an orthogonal family of projections, with sum I, in $\mathscr{C}^-$. Since

$$F_n E_j = \begin{cases} F_n & (1 \le j \le n \le \infty) \\ 0 & (1 \le n < j \le \infty), \end{cases}$$

$\mathscr{C}^- P_n$ is the von Neumann algebra

$$\left\{ \sum_{1 \le j \le n} \oplus AF_n : A \in \mathscr{A} \right\};$$

that is, in the notation of Section 6.6, $\mathscr{C}^- P_n = \mathscr{A}F_n \otimes I_n$. Since $\mathscr{A}F_n$ is maximal abelian, $\mathscr{C}'P_n = n \otimes \mathscr{A}F_n$, a von Neumann algebra of type I_n (unless F_n, and hence also P_n, is 0).

From the preceding discussion, it follows that the jth member of the null ideal sequence of φ is the null ideal of the representation $\psi_j : f \to \varphi(f)Q_j$ of $C(S)$, where

$$Q_j = \sum_{j \le n \le \infty} P_n \qquad (j = 1, 2, 3, \ldots).$$

Since e_1 ($\in L_\infty \subseteq L_2$) is a separating vector for $\mathscr{A}$ ($= \mathscr{A}E_1$), the vector $u = \{e_1, 0, 0, 0, \ldots\}$ in $\mathscr{H}$ is separating for $\mathscr{C}^-$, and $Q_j u$ is a separating vector for the weak-operator closure $\mathscr{C}^- Q_j$ of the range of ψ_j. Moreover,

$$Q_j u = \sum_{j \le n \le \infty} P_n u = \left\{ \sum_{j \le n \le \infty} F_n e_1, 0, 0, 0, \ldots \right\}$$
$$= \{E_j e_1, 0, 0, 0, \ldots\} = \{e_j, 0, 0, 0, \ldots\}.$$

For f in $C(S)$,

$$\langle \psi_j(f) Q_j u, Q_j u \rangle = \langle \varphi(f) Q_j u, Q_j u \rangle$$
$$= \langle f e_j, e_j \rangle = \int_S f e_j \, d\mu = \int_S f \, d\nu_j,$$

where ν_j is the regular Borel measure defined by $\nu_j(Y) = \mu(Y \cap Y_j)$ $(Y \in \mathscr{S})$. Accordingly, the null ideal of ψ_j is

$$\{N \in \mathscr{S} : \nu_j(N) = 0\} = \{N \in \mathscr{S} : \mu(N \cap Y_j) = 0\} = \mathscr{N}_j;$$

and the null ideal sequence of φ is $\{\mathscr{N}_j\}$. ∎

9.5.11. REMARK. Every representation φ of $C(S)$, acting on a separable Hilbert space, is equivalent to one of the type described in Example 9.5.10. Indeed, since the null ideal $\mathscr{N}_\varphi$ is of separable type, the same is true of the null ideal sequence $\{\mathscr{N}_j\}$ of φ; and by Theorem 9.5.7, φ is equivalent to the representation constructed from $\{\mathscr{N}_j\}$ as in Example 9.5.10. ∎

9.5.12. THEOREM. *Suppose that μ is a regular measure on the σ-algebra $\mathscr{S}$ of all Borel subsets of a compact Hausdorff space S, and L_2 (= $L_2(S, \mathscr{S}, \mu)$) and $\mathscr{K}$ are separable Hilbert spaces. When $f \in C(S)$, let $M_f\colon L_2 \to L_2$ be the operator of multiplication by f. Then the representation*

$$\psi\colon f \to M_f \otimes I : C(S) \to \mathscr{B}(L_2 \otimes \mathscr{K})$$

has uniform multiplicity n, where n = dim $\mathscr{K}$. *Moreover, every representation φ of C(S), acting on a separable Hilbert space and having uniform multiplicity, is equivalent to one of the type just described.*

Proof. Let $\{\mathscr{N}_j\}$ be the null ideal sequence in which $\mathscr{N}_j$ consists of all μ-null sets when $j \le n$ and (if $n < \aleph_0$) $\mathscr{N}_j = \mathscr{S}$ when $j > n$. The corresponding multiplicity function m is given by

$$m(Y) = \begin{cases} n & \text{if} \quad Y \in \mathscr{S}, \quad \mu(Y) > 0 \\ 0 & \text{if} \quad Y \in \mathscr{S}, \quad \mu(Y) = 0. \end{cases} \tag{12}$$

Thus S has uniform multiplicity n, relative to m; so the representation φ of $C(S)$, constructed from $\{\mathscr{N}_j\}$ as in Example 9.5.10, has uniform multiplicity n. In that construction, we can take $Y_j = S$ when $j \le n$ and (if $n < \aleph_0$) $Y_j = \varnothing$ when $j > n$. Accordingly, when $f \in C(S)$, $\varphi(f)$ is the direct sum of n copies of M_f, and acts on the Hilbert direct sum $\mathscr{H}$ of n copies of L_2. From Section 2.6, *Tensor products and the Hilbert–Schmidt class* (in particular, 2.6(17)), it now follows that there is an isomorphism U from $\mathscr{H}$ onto $L_2 \otimes \mathscr{K}$ such that

$$U\varphi(f)U^{-1} = M_f \otimes I = \psi(f) \qquad (f \in C(S)).$$

Hence ψ is equivalent to φ, and so has uniform multiplicity n.

Conversely, suppose that φ is a representation of $C(S)$, acting on a separable Hilbert space, and has uniform multiplicity n. The multiplicity function m of φ can be described, as in Proposition 9.5.1, in terms of a measure μ and a decreasing sequence $\{Y_j\}$ of Borel subsets of S, with $Y_1 = S$. From that proposition, and since S has uniform multiplicity n relative to m, it follows that (after adjusting by μ-null sets) we can take $Y_j = S$ when $j \le n$ and (if $n < \aleph_0$) $Y_j = \varnothing$ when $j > n$. Accordingly, φ has the same multiplicity function, defined by (12) above, as does ψ; and φ is equivalent to ψ, by Corollary 9.5.8. ■

9.5.13. REMARK. The results of this section can be applied to the problem of unitary equivalence for normal operators. Suppose that A is a normal operator acting on a separable Hilbert space $\mathscr{H}_A$. The spectrum sp A is a compact subset S of the complex plane, and the function calculus for A is a representation

$$\varphi_A\colon f \to f(A) : C(S) \to \mathscr{B}(\mathscr{H}_A)$$

of $C(S)$. We define the null ideal sequence and multiplicity function of A to be those of the representation φ_A.

Suppose next that A and B are normal operators acting on separable Hilbert spaces $\mathscr{H}_A$ and $\mathscr{H}_B$, respectively. If $B = UAU^{-1}$, where U is an isomorphism from $\mathscr{H}_A$ onto $\mathscr{H}_B$, then A and B have the same spectrum S. When f is a polynomial, $f(s) = \sum a_{jk} s^j \bar{s}^k$ $(s \in S)$, we have

$$f(B) = \sum a_{jk} B^j B^{*k} = U(\sum a_{jk} A^j A^{*k})U^{-1} = Uf(A)U^{-1};$$

so $\varphi_B(f) = U\varphi_A(f)U^{-1}$. This remains true for all f in $C(S)$, since φ_A, φ_B are norm continuous, and polynomials in $s, \bar{s}$ are everywhere dense in $C(S)$; so φ_A and φ_B are equivalent. Conversely, if $\text{sp}\, A = \text{sp}\, B = S$, and φ_A, φ_B are equivalent representations of $C(S)$, there is an isomorphism U, from $\mathscr{H}_A$ onto $\mathscr{H}_B$, such that $\varphi_B(f) = U\varphi_A(f)U^{-1}$ $(f \in C(S))$. With f the identity mapping on S, we obtain $B = UAU^{-1}$.

From the preceding paragraph, two normal operators A and B, acting on separable Hilbert spaces, are unitarily equivalent if and only if they have the same spectrum S and the representations φ_A and φ_B of $C(S)$ are equivalent. It now follows from Theorem 9.5.7 and Corollary 9.5.8 that A and B are unitarily equivalent if and only if they have the same spectrum and the same null ideal sequence— equivalently, if and only if they have the same spectrum and multiplicity function. ■

Bibliography: [35, 36, 38, 43, 59, 60, 71, 88, 102].

9.6. Exercises

9.6.1. Prove that a von Neumann algebra $\mathscr{R}$ is of type I if and only if $\mathscr{R}$ is * isomorphic to a von Neumann algebra with abelian commutant.

9.6.2. Suppose that $\mathscr{R}$ is a finite von Neumann algebra acting on a Hilbert space $\mathscr{H}$, with a separating and generating vector x. Suppose $y \in \mathscr{H}$ and $T \in \mathscr{R}$. Prove that

(i) y is a generating vector for $\mathscr{R}$ if and only if y is a separating vector for $\mathscr{R}$;

(ii) Tx is a separating and generating vector for $\mathscr{R}$ if and only if the range projection of T is I.

9.6.3. Let $\mathscr{R}$ be a finite von Neumann algebra acting on a Hilbert space $\mathscr{H}$, with a separating and generating vector. Observe, from Lemma 9.1.1 and its proof, that $\mathscr{R}'$ is finite and $\mathscr{R}$ has a generating trace vector x_0.

Denote by φ the * anti-isomorphism from $\mathscr{R}$ onto $\mathscr{R}'$ described in Theorem 7.2.15, so that $\varphi(A)x_0 = Ax_0$ for each A in $\mathscr{R}$. Let τ and τ' be the center-valued traces on $\mathscr{R}$ and $\mathscr{R}'$, respectively.

(i) Show that $\varphi(C) = C$ when $C \in \mathscr{R} \cap \mathscr{R}'$.

(ii) Prove that $\tau(A) = \tau'(\varphi(A))$ for each A in $\mathscr{R}$.

(iii) Let E be a projection in $\mathscr{R}$. Show that E has range $[\mathscr{R}'Ex_0]$ and $\varphi(E)$ has range $[\mathscr{R}Ex_0]$.

(iv) Let E and E' be projections in $\mathscr{R}$ and $\mathscr{R}'$, respectively. Show that $\tau(E) = \tau'(E')$ if and only if there is a vector y such that E has range $[\mathscr{R}'y]$ and E' has range $[\mathscr{R}y]$. [*Hint*. Use (ii), (iii), and Theorems 7.2.12 and 8.4.3(vi).]

9.6.4. Let $\mathscr{R}$ be a finite factor with a generating vector x_0, so that $\mathscr{R}'$ is finite by Proposition 9.1.2. Let E be the projection in $\mathscr{R}$ with range $[\mathscr{R}'x_0]$, and denote by $\tau, \tau', \tau_1, \tau_1'$ the unique tracial states of the factors $\mathscr{R}, \mathscr{R}'$, $E\mathscr{R}E$, $\mathscr{R}'E$, respectively. Let c_0 be the positive real number $\tau(E)$.

(i) Prove that

$$c_0\tau_1(ERE) = \tau(ERE), \qquad \tau'(R') = \tau_1'(R'E)$$

for all R in $\mathscr{R}$ and R' in $\mathscr{R}'$.

(ii) Suppose that F and F' are projections in $\mathscr{R}$ and $\mathscr{R}'$, respectively. Prove that $\tau(F) = c_0\tau'(F')$ if and only if there is a vector y such that F has range $[\mathscr{R}'y]$ and F' has range $[\mathscr{R}y]$. [*Hint*. Show that it is sufficient to consider the case in which $F \le E$, and in that case apply the result of Exercise 9.6.3 to the projections F (in $E\mathscr{R}E$) and $F'E$ (in $\mathscr{R}'E$).]

9.6.5. Suppose that $\mathscr{R}$ is a finite factor with finite commutant $\mathscr{R}'$, and denote by τ and τ' the unique tracial states of $\mathscr{R}$ and $\mathscr{R}'$, respectively. Prove that there is a positive real number c with the following property: if F ($\in \mathscr{R}$) and F' ($\in \mathscr{R}'$) are projections, then $\tau(F) = c\tau'(F')$ if and only if there is a vector y such that F has range $[\mathscr{R}'y]$ and F' has range $[\mathscr{R}y]$. [We call c the *coupling* (or *linking*) *constant* of $\mathscr{R}$.] [*Hint*. Use Corollary 8.2.9, Proposition 5.5.18, and Exercise 9.6.4.]

9.6.6. Let $\mathscr{R}$ be a von Neumann algebra acting on a Hilbert space $\mathscr{H}$, with center $\mathscr{C}$ and with a generating vector x_0, and suppose that both $\mathscr{R}$ and $\mathscr{R}'$ are finite. Let E be the projection in $\mathscr{R}$ with range $[\mathscr{R}'x_0]$; denote by $\tau, \tau', \tau_1, \tau_1'$ the center-valued traces on the finite von Neumann algebras $\mathscr{R}, \mathscr{R}'$, $E\mathscr{R}E$, $\mathscr{R}'E$, respectively; and let C_0 be $\tau(E)$. Note, from Propositions

5.5.13 and 5.5.5, that $C_E = I$ and the mapping $\varphi: R' \to R'E$ is a * isomorphism from $\mathscr{R}'$ onto $\mathscr{R}'E$.

(i) Prove that C_0 has range dense in $\mathscr{H}$.
(ii) Prove that

$$C_0\tau_1(ERE) = \tau(ERE)E, \qquad \varphi(\tau'(R')) = \tau_1'(\varphi(R'))$$

for all R in $\mathscr{R}$ and R' in $\mathscr{R}'$. [*Hint*. To prove the first relation, let Q_n be the spectral projection for C_0 corresponding to the interval $[n^{-1}, \infty)$ for $n = 1, 2, \ldots$. Show that it suffices to prove the equation for R in $\mathscr{R}Q_n$, and so reduce to the case in which C_0 has a bounded inverse.]

(iii) Suppose that F and F' are projections in $\mathscr{R}$ and $\mathscr{R}'$, respectively. Prove that $\tau(F) = C_0\tau'(F')$ if and only if there is a vector y such that F has range $[\mathscr{R}'y]$ and F' has range $[\mathscr{R}y]$. [*Hint*. Upgrade the argument needed to solve Exercise 9.6.4(ii).]

9.6.7. Let $\mathscr{R}$ be a von Neumann algebra acting on a Hilbert space $\mathscr{H}$, with countably decomposable center $\mathscr{C}$, and suppose that both $\mathscr{R}$ and $\mathscr{R}'$ are finite. Denote by τ, τ' the center-valued traces on $\mathscr{R}, \mathscr{R}'$, respectively.

(i) Deduce from Corollary 8.2.9 and Proposition 5.5.18 that there is a projection Q in $\mathscr{C}$ such that the von Neumann algebras $\mathscr{R}Q$ and $\mathscr{R}'(I - Q)$ have generating vectors x_1 and x_2, respectively.

(ii) Let E (in $\mathscr{R}Q$) and E' (in $\mathscr{R}'(I - Q)$) be the projections with ranges $[\mathscr{R}'x_1]$ and $[\mathscr{R}x_2]$, respectively. Define C_1 in $\mathscr{C}Q$ and C_2 in $\mathscr{C}(I - Q)$ by $C_1 = \tau(E)$, $C_2 = \tau'(E')$. Suppose that F and F' are projections in $\mathscr{R}$ and $\mathscr{R}'$, respectively. Deduce from Exercise 9.6.6 that F and F' have a common generating vector if and only if

$$\tau(F)Q = C_1\tau'(F')Q, \qquad C_2\tau(F)(I - Q) = \tau'(F')(I - Q).$$

(iii) Prove that there is a unique invertible element C_0, in the algebra $\mathscr{N}(\mathscr{C})$ of operators affiliated to $\mathscr{C}$, with the following property: if F and F' are projections in $\mathscr{R}$ and $\mathscr{R}'$, respectively, then $\tau(F) = C_0 \hat{\cdot}\, \tau'(F')$ if and only if F and F' have a common generating vector. (We call C_0 the *coupling* (or *linking*) *operator* of $\mathscr{R}$.)

(iv) Show that a projection F in $\mathscr{R}$ is cyclic if and only if $\tau(F) \le C_0$.

9.6.8. Let $\mathscr{H}$ be a Hilbert space, and denote by $\bar{\mathscr{H}}$ the same set $\mathscr{H}$ with the conjugate Hilbert space structure (see the discussion preceding Definition 2.6.3). Suppose that $A \in \mathscr{B}(\mathscr{H})$, and recall that the same mapping A can be viewed as an element $\bar{A}$ of $\mathscr{B}(\bar{\mathscr{H}})$.

(i) Show that the mapping $A \to \bar{A}$ is a multiplicative adjoint-preserving conjugate-linear isometry from $\mathscr{B}(\mathscr{H})$ onto $\mathscr{B}(\bar{\mathscr{H}})$.

(ii) Show that $\operatorname{sp}(\bar{A}) = \{\bar{\lambda} : \lambda \in \operatorname{sp}(A)\}$.

(iii) Suppose that A is self-adjoint. Prove that $\bar{A}$ is self-adjoint, and $\mathrm{sp}(\bar{A}) = \mathrm{sp}(A)$. Show also that, for each f in $C(\mathrm{sp}(A))$, $f(A)$ is the same mapping as $\bar{f}(\bar{A})$, where $\bar{f}$ is defined by $\bar{f}(t) = \overline{f(t)}$.
(iv) Give an example of an element f of $C(\mathrm{sp}(A))$ such that $f(\bar{A})$ is not the same mapping as $f(A)$.

9.6.9. With the notation of Exercise 9.6.8, let $\mathscr{A}$ ($\subseteq \mathscr{B}(\mathscr{H})$) be an abelian von Neumann algebra.

(i) Show that the mapping $A \to \bar{A} : \mathscr{B}(\mathscr{H}) \to \mathscr{B}(\bar{\mathscr{H}})$ is a homeomorphism with respect both to the weak-operator topologies and to the strong-operator topologies. Deduce that the set $\{\bar{A} : A \in \mathscr{A}\}$ is an abelian von Neumann algebra $\bar{\mathscr{A}}$ acting on $\bar{\mathscr{H}}$.
(ii) Suppose that T is a closed densely defined linear operator on $\mathscr{H}$. Show that the same mapping T can be viewed as a closed densely defined linear operator $\bar{T}$ on $\bar{\mathscr{H}}$. Prove that $\bar{T}$ is self-adjoint, or positive, if and only if T has the same property, and that $\bar{T}\,\eta\,\bar{\mathscr{A}}$ if and only if $T\,\eta\,\mathscr{A}$. (Note that $\bar{A}$ does not indicate the closure of A in the notation of this exercise.)
(iii) Show that the restriction to $\mathscr{N}(\mathscr{A})$ of the mapping $T \to \bar{T}$ is a one-to-one mapping from $\mathscr{N}(\mathscr{A})$ onto $\mathscr{N}(\bar{\mathscr{A}})$; is conjugate-linear, multiplicative, and adjoint-preserving relative to the * algebra structure on $\mathscr{N}(\mathscr{A})$ (see Theorem 5.6.15); and is an isomorphism for the order structure on $\mathscr{N}(\mathscr{A})$ (see the discussion preceding Lemma 5.6.22).
(iv) Suppose that $T\,\eta\,\mathscr{A}$ and T is self-adjoint. Show that $\mathrm{sp}(T) = \mathrm{sp}(\bar{T})$ and that $f(T)$ is the same mapping as $\bar{f}(\bar{T})$ when f is a Borel function on $\mathbb{C}$. [*Hint*. Apply Theorem 5.6.27 to the mapping

$$\psi: f \to \overline{\bar{f}(\bar{T})} : \mathscr{B}_u \to \mathscr{N}(\bar{\mathscr{A}}).]$$

9.6.10. Use the equation $\Delta^{-1} = J\Delta J$ and the result of Exercise 9.6.9 to prove that $f(\Delta^{-1}) = J\bar{f}(\Delta)J$ for every Borel function f on $\mathbb{C}$ (see the discussion preceding Proposition 9.2.3).

9.6.11. Suppose that $\mathscr{R}$ is a von Neumann algebra acting on a Hilbert space $\mathscr{H}$, with a generating trace vector x_0, and refer to Lemma 7.2.14, Theorem 7.2.15, and Exercise 7.6.49 for information including the following: there is an involution J of $\mathscr{H}$ and a * anti-isomorphism φ from $\mathscr{R}$ onto $\mathscr{R}'$ such that

$$JAx_0 = A^*x_0, \qquad \varphi(A)x_0 = Ax_0, \qquad \varphi(A) = JA^*J$$

for each A in $\mathscr{R}$. Suppose that HV is the polar decomposition of an element T of $\mathscr{R}$ that has a bounded inverse, and $u = Tx_0$.

(i) Prove that u is a separating and generating vector for $\mathscr{R}$.

(ii) Let Δ_u, J_u, and $\{\sigma_t\}$ be the modular operator, involution, and modular automorphism group associated with the separating and generating vector u of $\mathscr{R}$. Prove that

$$J_u \Delta_u^{1/2} = JV^*\varphi(V^*)\varphi(V^*H^{-1}V)H,$$

and deduce that

$$J_u = JV^*\varphi(V^*) = \varphi(V)VJ, \qquad \Delta_u = \varphi(V^*H^{-2}V)H^2,$$
$$\sigma_t(A) = H^{2it}AH^{-2it} \qquad (A \in \mathscr{R}, \quad t \in \mathbb{R}).$$

9.6.12. Suppose that u is a separating and generating vector for a von Neumann algebra $\mathscr{R}$ acting on a Hilbert space $\mathscr{H}$, $A' \in \mathscr{R}'$, and $r > 0$. Define linear functions ω, ω_C $(C \in \mathscr{R})$ on $\mathscr{R}$ by

$$\omega(R) = \langle RA'u, u\rangle, \qquad \omega_C(R) = \langle CRu, u\rangle + r\langle RCu, u\rangle \qquad (R \in \mathscr{R}),$$

and let

$$\mathscr{K} = \{\omega_C : C \in \mathscr{R}, \|C\| \le k\},$$

where $k = \frac{1}{2}r^{-1/2}\|A'\|$.

(i) Show that $\mathscr{K}$ is a convex subset of the Banach dual space $\mathscr{R}^\#$ of $\mathscr{R}$, and is compact in the weak * topology on $\mathscr{R}^\#$.

(ii) Show that $\omega \in \mathscr{K}$. [*Hint*. Suppose the contrary, and deduce that there is an element B of $\mathscr{R}$ such that

$$\operatorname{Re}\omega(B) > 1, \qquad \operatorname{Re}\omega_C(B) \le 1 \qquad (C \in \mathscr{R}, \quad \|C\| \le k).$$

Let B have polar decomposition VK, and obtain a contradiction by taking $C = kV^*$.]

(iii) Deduce from (ii) that there is an element A of $\mathscr{R}$ such that $\|A\| \le k$ and

$$\langle RA'u, u\rangle = \langle ARu, u\rangle + r\langle RAu, u\rangle \qquad (R \in \mathscr{R}).$$

(iv) Prove that $(\Delta + rI)^{-1}A'u = Au$, where Δ is the modular operator associated with the separating and generating vector u for $\mathscr{R}$.

9.6.13. Suppose that D is an invertible (not necessarily bounded) positive self-adjoint operator on a Hilbert space $\mathscr{H}$, $r > 0$, and $B \in \mathscr{B}(\mathscr{H})$.

(i) Show that there is an element C of $\mathscr{B}(\mathscr{H})$ such that

$$\langle Cy_1, y_2\rangle = \int_{\mathbb{R}} \frac{r^{it-1/2}}{e^{\pi t} + e^{-\pi t}} \langle D^{it}BD^{-it}y_1, y_2\rangle \, dt$$

for all y_1 and y_2 in $\mathscr{H}$. Prove also that $\|C\| \le \frac{1}{2}r^{-1/2}\|B\|$.

(ii) By an argument analogous to the proof of Lemma 9.2.8, show that

$$\langle Bx_1, x_2\rangle = \langle CD^{1/2}x_1, D^{-1/2}x_2\rangle + r\langle CD^{-1/2}x_1, D^{1/2}x_2\rangle$$

for all x_1 and x_2 in $\mathscr{D}(D^{1/2}) \cap \mathscr{D}(D^{-1/2})$.

9.6.14. Suppose that $\{\sigma_t\}$ is a one-parameter group of * automorphisms of a von Neumann algebra $\mathscr{R}$; $\rho, \rho_1, \rho_2, \rho_3, \ldots$ are positive linear functionals on $\mathscr{R}$; $\|\rho - \rho_n\| \to 0$ as $n \to \infty$; and $\{\sigma_t\}$ satisfies the modular condition relative to ρ_n for each $n = 1, 2, 3, \ldots$. Prove that $\{\sigma_t\}$ satisfies the modular condition relative to ρ.

9.6.15. Suppose that $\mathscr{R}$ is a von Neumann algebra with center $\mathscr{C}$ and $\{\sigma_t\}$ is a one-parameter group of * automorphisms of $\mathscr{R}$.

(i) Suppose that $\{\sigma_t\}$ satisfies the modular condition relative to a positive linear functional ρ on $\mathscr{R}$, $C \in \mathscr{C}^+$, and ρ_0 is the positive linear functional on $\mathscr{R}$ defined by

$$\rho_0(A) = \rho(CA) \qquad (A \in \mathscr{R}).$$

Show that $\{\sigma_t\}$ satisfies the modular condition relative to ρ_0.

(ii) Suppose that $\{\sigma_t\}$ satisfies the modular condition relative to $\omega_u | \mathscr{R}$, where the vector u lies in the domain $\mathscr{D}(C)$ of an operator C affiliated to $\mathscr{C}$. Show that $\{\sigma_t\}$ satisfies the modular condition relative to $\omega_{Cu} | \mathscr{R}$. [*Hint*. Use (i) and Exercise 9.6.14.]

9.6.16. Let $\{\sigma_t\}$ be the modular automorphism group corresponding to a faithful normal state ω of a von Neumann algebra $\mathscr{R}$. Suppose also that $\{\sigma_t\}$ satisfies the modular condition relative to a positive normal linear functional ω_0 on $\mathscr{R}$, and $\omega_0 \leq \omega$.

(i) Prove that, if $H, K \in \mathscr{R}^+$ and $\omega(HAH) = \omega(KAK)$ for each A in $\mathscr{R}$, then $H = K$. [*Hint*. By using the first part of the discussion following Corollary 9.2.15, reduce to the case in which $\omega = \omega_u | \mathscr{R}$, where u is a separating (and generating) vector for $\mathscr{R}$. In this case, use Exercise 7.6.23 to show that $Ku = V'Hu$ and $Hu = V'^*Ku$ for some partial isometry V' in $\mathscr{R}'$. Prove that

$$\|K^{1/2}u\|^2 = \langle V'H^{1/2}u, H^{1/2}u \rangle = \|H^{1/2}u\|^2.$$

Deduce that $V'H^{1/2}u = H^{1/2}u$ and that $Ku = Hu$.]

(ii) By the Sakai–Radon–Nikodým theorem (7.3.6) there is a positive operator H in the unit ball of $\mathscr{R}$ such that

$$\omega_0(A) = \omega(HAH) \qquad (A \in \mathscr{R}).$$

Show that $\omega_0(A) = \omega(\sigma_t(H)A\sigma_t(H))$, and deduce that

$$\sigma_t(H) = H, \qquad \omega(AH) = \omega(HA) \qquad (A \in \mathscr{R}, \quad t \in \mathbb{R}).$$

[*Hint*. Apply Proposition 9.2.14 to ω and ω_0.]

(iii) Suppose that $A, B \in \mathscr{R}$. By using the modular condition, first for A, B, ω_0 and then for A, BH^2, ω, show that $\omega((H^2B - BH^2)A) = 0$.

(iv) Deduce that H^2 is a positive element C in the unit ball of the center of $\mathscr{R}$ and $\omega_0(A) = \omega(CA)$ for each A in $\mathscr{R}$. (Compare this with Exercise 9.6.15(i).)

9.6.17. Suppose that $\mathscr{R}$ is a von Neumann algebra with center $\mathscr{C}$, acting on a Hilbert space $\mathscr{H}$, $\{\sigma_t\}$ is the modular automorphism group corresponding to a separating and generating vector u for $\mathscr{R}$, ω is a positive normal linear functional on $\mathscr{R}$, and $\{\sigma_t\}$ satisfies the modular condition with respect to ω (as well as $\omega_u | \mathscr{R}$).

(i) Prove that $\omega + \omega_u | \mathscr{R} = \omega_v | \mathscr{R}$ for some separating and generating vector v for $\mathscr{R}$.

(ii) Use Exercise 9.6.16 to prove the existence of elements S, T of $\mathscr{C}^+$ such that $\omega_u | \mathscr{R} = \omega_{Sv} | \mathscr{R}$, $\omega = \omega_{Tv} | \mathscr{R}$, and S has null space $\{0\}$ and range dense in $\mathscr{H}$.

(iii) Deduce from Exercise 7.6.23 that $u = V'Sv$ for some unitary operator V' in $\mathscr{R}'$.

(iv) Let $S^{-1}: S(\mathscr{H}) \to \mathscr{H}$ be the inverse of the mapping $S: \mathscr{H} \to S(\mathscr{H})$. Show that S^{-1}, T, and $T \hat{\cdot} S^{-1} (= C)$ are positive elements of the algebra $\mathscr{N}(\mathscr{C})$ (see Theorem 5.6.15). Prove also that $u \in \mathscr{D}(C)$ and $\omega = \omega_{Cu} | \mathscr{R}$. (Compare this with Exercise 9.6.15(ii).)

9.6.18. With the notation used in the development of modular theory in the context of a von Neumann algebra $\mathscr{R}$ and a faithful normal semi-finite weight ρ on $\mathscr{R}$ (starting in the paragraph preceding Proposition 9.2.39), suppose that U is a unitary operator in the center $\mathscr{C}$ of $\mathscr{R}$. Prove that

$$\pi(U)S_0\pi(U) = S_0,$$

where S_0 is the conjugate-linear mapping $A \to A^* : \mathfrak{A}^2 \to \mathfrak{A}^2$. Hence show that $\pi(U)J\pi(U) = J$, and deduce that

$$J\pi(C)J = \pi(C^*) \qquad (C \in \mathscr{C}).$$

9.6.19. Suppose that $\mathscr{R}$ is a von Neumann algebra with center $\mathscr{C}$, acting on a Hilbert space $\mathscr{H}$, and $J: \mathscr{H} \to \mathscr{H}$ is a conjugate-linear isometry such that $J^2 = I$, $J\mathscr{R}J = \mathscr{R}'$, and $JCJ = C^*$ for each C in $\mathscr{C}$. Let E be a non-zero projection in $\mathscr{R}$, and let E' be the projection JEJ in $\mathscr{R}'$. Prove that

(i) if $y \in \mathscr{H}$ and E has range $[\mathscr{R}'y]$, then E' has range $[\mathscr{R}Jy]$;
(ii) $C_E = C_{E'}$;
(iii) EE' is a non-zero projection and its range is invariant under J;
(iv) $x = Jx = Ex$ for some unit vector x in $\mathscr{H}$.

9.6.20. Suppose that $\mathscr{R}$ is a von Neumann algebra with center $\mathscr{C}$, acting on a Hilbert space $\mathscr{H}$, and $J\colon \mathscr{H} \to \mathscr{H}$ is a conjugate-linear isometry such that $J^2 = I$, $J\mathscr{R}J = \mathscr{R}'$, and $JCJ = C^*$ for each C in $\mathscr{C}$. Let E ($\neq 0$) be a countably decomposable projection in $\mathscr{R}$. Prove that there is a vector x in $\mathscr{H}$ such that $Jx = x$, E has range $[\mathscr{R}'x]$, and JEJ has range $[\mathscr{R}x]$. [*Hint*. By using Exercise 9.6.19, show that there is a (finite or infinite) sequence $\{x_1, x_2, \ldots\}$ of unit vectors in $\mathscr{H}$ such that $x_n = Jx_n = Ex_n$ for each n, the subspaces $[\mathscr{R}'x_n]$ $(n = 1, 2, \ldots)$ are mutually orthogonal, and $\bigvee[\mathscr{R}'x_n]$ is the range of E. Prove that the subspaces $[\mathscr{R}x_n]$ $(n = 1, 2, \ldots)$ are mutually orthogonal, and $\bigvee[\mathscr{R}x_n]$ is the range of JEJ. Follow the reasoning used in the first paragraph of the proof of Proposition 5.5.18.]

9.6.21. Suppose that $\mathscr{R}$ is a von Neumann algebra with center $\mathscr{C}$. Show that there is an orthogonal family $\{Q_a\}$ of projections in $\mathscr{C}$ such that $\sum Q_a = I$ and each Q_a is the sum of an orthogonal family of equivalent cyclic projections in $\mathscr{R}$. Prove also that, if both $\mathscr{R}$ and $\mathscr{R}'$ are properly infinite, the family $\{Q_a\}$ can be chosen in such a way that each Q_a is the sum of an orthogonal family of infinitely many equivalent properly infinite cyclic projections in $\mathscr{R}$. (Compare Exercises 6.9.12 and 6.9.13.)

9.6.22. Suppose that $\mathscr{R}$ and $\mathscr{S}$ are von Neumann algebras acting on Hilbert spaces $\mathscr{H}$ and $\mathscr{K}$, respectively, and φ is a * isomorphism from $\mathscr{R}$ onto $\mathscr{S}$. Suppose also that $\{Q_a : a \in \mathbb{A}\}$ is an orthogonal family of central projections in $\mathscr{R}$ with sum I, and, for each a in $\mathbb{A}$, the * isomorphism $\varphi \mid \mathscr{R}Q_a$ from $\mathscr{R}Q_a$ onto $\mathscr{S}\varphi(Q_a)$ is implemented by a unitary transformation from $Q_a(\mathscr{H})$ onto $\varphi(Q_a)(\mathscr{K})$. Prove that φ is implemented by a unitary transformation from $\mathscr{H}$ onto $\mathscr{K}$.

9.6.23. Suppose that $\mathscr{H}$ and $\mathscr{K}$ are Hilbert spaces, $\mathscr{R}$ ($\subseteq \mathscr{B}(\mathscr{H})$) and $\mathscr{S}$ ($\subseteq \mathscr{B}(\mathscr{K})$) are von Neumann algebras, $\mathscr{R}$ has a generating vector x and E (in $\mathscr{R}$) is the projection with range $[\mathscr{R}'x]$, $\mathscr{S}$ has a generating vector y and F (in $\mathscr{S}$) is the projection with range $[\mathscr{S}'y]$, φ is a * isomorphism from $\mathscr{R}$ onto $\mathscr{S}$, and $\varphi(E) = F$. Show that the * isomorphism $\varphi \mid E\mathscr{R}E$ from $E\mathscr{R}E$ onto $F\mathscr{S}F$ is implemented by a unitary transformation U_0 from $E(\mathscr{H})$ onto $F(\mathscr{K})$. Deduce that φ is implemented by a unitary transformation U from $\mathscr{H}$ onto $\mathscr{K}$. [*Hint*. Let $\{E_a : a \in \mathbb{A}\}$ be an orthogonal family of projections in $\mathscr{R}$, maximal subject to the condition that $0 < E_a \precsim E$ for each a in $\mathbb{A}$. Let V_a be a partial isometry in $\mathscr{R}$, with initial projection E_a and final projection a subprojection of E. Show that $\sum E_a = I$, and define U by requiring $U \mid E_a(\mathscr{H}) = \varphi(V_a^*)U_0 V_a$ for each a in $\mathbb{A}$.]

9.6.24. Suppose that $\mathscr{H}$ and $\mathscr{K}$ are Hilbert spaces, $\mathscr{R}$ ($\subseteq \mathscr{B}(\mathscr{H})$) and $\mathscr{S}$ ($\subseteq \mathscr{B}(\mathscr{K})$) are von Neumann algebras, $x \in \mathscr{H}$, and $y \in \mathscr{K}$. Let E (in $\mathscr{R}$), E' (in

$\mathscr{R}'$), F (in $\mathscr{S}$), and F' (in $\mathscr{S}'$) be the projections with ranges $[\mathscr{R}'x]$, $[\mathscr{R}x]$, $[\mathscr{S}'y]$, and $[\mathscr{S}y]$, respectively. Suppose that there exist orthogonal families of projections, $\{E'_a : a \in \mathbb{A}\}$ and $\{F'_a : a \in \mathbb{A}\}$, indexed by the same set $\mathbb{A}$ and both having sum I, such that

$$E'_a \in \mathscr{R}', \qquad E'_a \sim E', \qquad F'_a \in \mathscr{S}', \qquad F'_a \sim F' \qquad (a \in \mathbb{A}).$$

Show that, if φ is a * isomorphism from $\mathscr{R}$ onto $\mathscr{S}$, and $\varphi(E) = F$, then φ is implemented by a unitary transformation U from $\mathscr{H}$ onto $\mathscr{K}$. [*Hint*. By using Exercise 9.6.23, show that there is a unitary transformation U_0 from $E'(\mathscr{H})$ onto $F'(\mathscr{K})$ such that $\varphi(A)F' = U_0AE'U_0^*$ for each A in $\mathscr{R}$. Choose partial isometries V'_a in $\mathscr{R}'$ and W'_a in $\mathscr{S}'$, that implement the equivalences $E' \sim E'_a$ and $F' \sim F'_a$, respectively, and define U by the condition $UE'_a = W'_aU_0V'^*_a$.]

9.6.25. Suppose that, for $j = 1, 2$, $\mathscr{R}_j$ is a von Neumann algebra with center $\mathscr{C}_j$, acting on a Hilbert space $\mathscr{H}_j$, and $J_j\colon \mathscr{H}_j \to \mathscr{H}_j$ is a conjugate-linear isometry such that $J_j^2 = I$, $J_j\mathscr{R}_jJ_j = \mathscr{R}'_j$, and $J_jCJ_j = C^*$ for each C in $\mathscr{C}_j$. Let φ be a * isomorphism from $\mathscr{R}_1$ onto $\mathscr{R}_2$. Prove that φ is implemented by a unitary transformation from $\mathscr{H}_1$ onto $\mathscr{H}_2$. [*Hint*. By using Exercises 9.6.21 and 9.6.22, reduce to the case in which $\mathscr{R}_1$ contains an orthogonal family $\{E_a\}$ of equivalent cyclic projections with sum I. Define $F_a = \varphi(E_a)$, $E'_a = J_1E_aJ_1$, $F'_a = J_2F_aJ_2$, and use Exercises 9.6.20 and 9.6.24.]

9.6.26. Suppose that ρ_1 and ρ_2 are faithful normal semi-finite weights on a von Neumann algebra $\mathscr{R}$, and π_1 and π_2 are the corresponding representations of $\mathscr{R}$, constructed as in Theorem 7.5.3. By using Theorem 9.2.37 and Exercises 9.6.18 and 9.6.25, show that π_1 and π_2 are equivalent.

9.6.27. Suppose that $\mathscr{R}$ is a finite von Neumann algebra with center $\mathscr{C}$, acting on a Hilbert space $\mathscr{H}$, φ is a * automorphism of $\mathscr{R}$, and $\varphi(C) = C$ for each C in $\mathscr{C}$. Prove that there is a unitary operator U acting on $\mathscr{H}$ such that $\varphi(A) = UAU^*$ for each A in $\mathscr{R}$. [*Hint*. By using Exercises 9.6.21 (applied to $\mathscr{R}'$) and 9.6.22, reduce to the case in which $\mathscr{R}'$ contains an orthogonal family of equivalent cyclic projections with sum I. Show that $\tau \circ \varphi = \tau$, where τ is the center-valued trace on $\mathscr{R}$, and deduce that $\varphi(E) \sim E$ for each projection E in $\mathscr{R}$. Use Exercise 9.6.24.]

9.6.28. Suppose that $\mathscr{R}$ is a von Neumann algebra with center $\mathscr{C}$, acting on a Hilbert space $\mathscr{H}$, and both $\mathscr{R}$ and $\mathscr{R}'$ are properly infinite. Let φ be a * automorphism of $\mathscr{R}$ such that $\varphi(C) = C$ for all C in $\mathscr{C}$. Prove that there is a unitary operator U acting on $\mathscr{H}$ such that $\varphi(A) = UAU^*$ for each

A in $\mathscr{R}$. [*Hint*. By use of Exercises 9.6.21 (applied to $\mathscr{R}'$) and 9.6.22, reduce to the case in which $\mathscr{R}'$ contains an orthogonal family of infinitely many equivalent properly infinite cyclic projections with sum I. Use Exercise 9.6.24 together with Corollary 6.3.5.]

9.6.29. Suppose that $\mathscr{R}$ is a von Neumann algebra with center $\mathscr{C}$, acting on a Hilbert space $\mathscr{H}$, and there is no projection P in $\mathscr{C}$ such that the von Neumann algebras $\mathscr{R}P$ and $\mathscr{R}'P$ are of types II_∞ and II_1, respectively. Let φ be a * automorphism of $\mathscr{R}$ such that $\varphi(C) = C$ for each C in $\mathscr{C}$. By using Exercises 9.6.27 and 9.6.28 and Corollary 9.3.5, show that there is a unitary operator U acting on $\mathscr{H}$ such that $\varphi(A) = UAU^*$ for each A in $\mathscr{R}$. (The exclusion of algebras that have a direct summand of type II_∞ with commutant of type II_1 is necessary, in the formulation of this exercise. In Exercise 13.4.3, we exhibit a factor $\mathscr{R}$ of type II_∞ with commutant of type II_1, and a * automorphism of $\mathscr{R}$ that is not unitarily implemented. Exercise 9.6.33 provides additional information about * automorphisms of an algebra of type II_∞ with commutant of type II_1.)

9.6.30. Suppose that, for $j = 1, 2$, $\mathscr{R}_j$ is a von Neumann algebra with countably decomposable center $\mathscr{C}_j$, acting on a Hilbert space $\mathscr{H}_j$, both $\mathscr{R}_j$ and $\mathscr{R}_j'$ are finite, C_j (in $\mathscr{N}(\mathscr{C}_j)$) is the coupling operator of $\mathscr{R}_j$ (see Exercise 9.6.7), and τ_j is the center-valued trace on $\mathscr{R}_j$. Let φ be a * isomorphism from $\mathscr{R}_1$ onto $\mathscr{R}_2$.

(i) Prove that $\varphi \circ \tau_1 = \tau_2 \circ \varphi$.

(ii) By using Theorems 3.4.3 and 5.6.19, prove that the * isomorphism $\varphi | \mathscr{C}_1$, from $\mathscr{C}_1$ onto $\mathscr{C}_2$, extends to a * isomorphism ψ from $\mathscr{N}(\mathscr{C}_1)$ onto $\mathscr{N}(\mathscr{C}_2)$.

(iii) Suppose that a projection Q in $\mathscr{C}_1$ is the sum of an orthogonal family $\{E_1', \ldots, E_k'\}$ of equivalent cyclic projections in $\mathscr{R}_1'$. Choose x_1 in $\mathscr{H}_1$ so that E_1' has range $[\mathscr{R}_1 x_1]$, let E (in $\mathscr{R}_1$) be the projection with range $[\mathscr{R}_1' x_1]$, and let F be $\varphi(E)$. Show that, if $\psi(C_1) = C_2$, there is an orthogonal family $\{F_1', \ldots, F_k'\}$ of equivalent cyclic projections in $\mathscr{R}_2'$ with sum $\varphi(Q)$, and a vector x_2 in $\mathscr{H}_2$ such that F has range $[\mathscr{R}_2' x_2]$ and F_1' has range $[\mathscr{R}_2 x_2]$.

(iv) By using (iii), together with Exercises 9.6.21, 9.6.22, and 9.6.24, show that φ is implemented by a unitary transformation from $\mathscr{H}_1$ onto $\mathscr{H}_2$ if and only if $\psi(C_1) = C_2$.

9.6.31. Suppose that, for $j = 1, 2$, $\mathscr{R}_j$ is a finite von Neumann algebra acting on a separable Hilbert space $\mathscr{H}_j$, and $\mathscr{R}_j'$ is properly infinite. Prove that

(i) $\mathscr{R}_j$ has a separating vector x_j [*Hint*. Proposition 5.5.18.];

(ii) $\mathscr{R}_j'$ contains an (infinite) orthogonal sequence of projections, with sum I, each equivalent to the cyclic projection with range $[\mathscr{R}_j x_j]$ [*Hint*. Use Proposition 6.3.12.];

(iii) if φ is a * isomorphism from $\mathscr{R}_1$ onto $\mathscr{R}_2$, then φ is implemented by a unitary transformation from $\mathscr{H}_1$ onto $\mathscr{H}_2$.

9.6.32. Suppose that, for $j = 1, 2$, $\mathscr{R}_j$ is a von Neumann algebra acting on a separable Hilbert space $\mathscr{H}_j$, and $\mathscr{R}_j'$ is properly infinite. Let φ be a * isomorphism from $\mathscr{R}_1$ onto $\mathscr{R}_2$. By using Exercise 9.6.31 and Remark 7.2.10, show that φ is implemented by a unitary transformation from $\mathscr{H}_1$ onto $\mathscr{H}_2$.

9.6.33. Suppose that, for $j = 1, 2$, $\mathscr{R}_j$ is a properly infinite von Neumann algebra acting on a separable Hilbert space $\mathscr{H}_j$, and $\mathscr{R}_j'$ is finite. Note that $\mathscr{R}_j$ has a generating vector x_j (Exercise 9.6.31(i)), and let E_j be the projection in $\mathscr{R}_j$ with range $[\mathscr{R}_j' x_j]$. Show that a * isomorphism φ from $\mathscr{R}_1$ onto $\mathscr{R}_2$ is implemented by a unitary transformation from $\mathscr{H}_1$ onto $\mathscr{H}_2$ if and only if $\varphi(E_1) \sim E_2$. (For an example in which the condition $\varphi(E_1) \sim E_2$ is *not* satisfied, we refer to Exercise 13.4.3.)

9.6.34. Suppose that m and n are cardinals, $\mathscr{A}$ is a maximal abelian von Neumann algebra acting on a Hilbert space $\mathscr{K}$, and $\mathscr{R}$ is the von Neumann algebra $m \otimes (\mathscr{A} \otimes I_n)$. Prove that $\mathscr{R}$ is of type I_m and $\mathscr{R}'$ is of type I_n (in other words, prove the *converse* of Theorem 9.3.2).

9.6.35. Suppose that $\mathscr{R}$ is a von Neumann algebra of type I_m, with commutant $\mathscr{R}'$ of type I_n. Show that the center $\mathscr{C}$ of $\mathscr{R}$ has commutant $\mathscr{C}'$ of type I_{mn}.

9.6.36. Let $\mathscr{R}$ and $\mathscr{S}$ be finite type I von Neumann algebras both acting on the same Hilbert space $\mathscr{H}$ and having the same center $\mathscr{C}$, and let φ be a * isomorphism from $\mathscr{R}$ onto $\mathscr{S}$ such that $\varphi(C) = C$ for each C in $\mathscr{C}$. Prove that there is a unitary operator U in $\mathscr{C}'$ such that $\varphi(A) = UAU^*$ for each A in $\mathscr{R}$. [*Hint*. Use Exercise 9.6.35 and the discussion preceding Theorem 9.3.1.]

9.6.37. Let $\mathscr{R}$ be a finite type I von Neumann algebra with center $\mathscr{C}$ and let $\mathscr{B}$ be a von Neumann algebra such that $\mathscr{C} \subseteq \mathscr{B} \subseteq \mathscr{R}$. Prove that there is an orthogonal family $\{F_j\}$ of projections in $\mathscr{B}$, with sum I, such that $F_j \mathscr{B} F_j = \mathscr{C} F_j$ for each index j. [*Hint*. Use Corollary 6.5.5.]

9.6.38. Suppose that $\mathscr{R}$ is a finite type I von Neumann algebra with center $\mathscr{C}$ and, for $j = 1, 2$, $\mathscr{A}_j$ is an abelian von Neumann algebra such that $\mathscr{C} \subseteq \mathscr{A}_j \subseteq \mathscr{R}$. Let φ be a * isomorphism from $\mathscr{A}_1$ onto $\mathscr{A}_2$ such that $\varphi(C) = C$ for each C in $\mathscr{C}$, and suppose also that $\tau(\varphi(A)) = \tau(A)$ for each A in $\mathscr{A}_1$, where τ is the center-valued trace on $\mathscr{R}$. Show that there is a unitary operator V in $\mathscr{R}$ such that $\varphi(A) = VAV^*$ for each A in $\mathscr{A}_1$. (Note that this exercise prepares for, and is subsumed in, the one that follows.) [*Hint.* Apply Exercise 9.6.37, with $\mathscr{A}_1$ in place of $\mathscr{B}$, so obtaining an orthogonal family $\{F_j\}$ of projections in $\mathscr{A}_1$. Show that $F_j \sim \varphi(F_j)$ (in $\mathscr{R}$) for each index j. Let V be $\sum V_j$, where V_j is a partial isometry in $\mathscr{R}$ that implements the equivalence of F_j and $\varphi(F_j)$.]

9.6.39. Suppose that $\mathfrak{A}_1$ and $\mathfrak{A}_2$ are self-adjoint subalgebras of a finite type I von Neumann algebra $\mathscr{R}$, and φ is a * isomorphism from $\mathfrak{A}_1$ onto $\mathfrak{A}_2$ such that $\tau(\varphi(A)) = \tau(A)$ for each A in $\mathfrak{A}_1$, where τ is the center-valued trace on $\mathscr{R}$. Show that there is a unitary operator U in $\mathscr{R}$ such that $\varphi(A) = UAU^*$ for each A in $\mathfrak{A}_1$. [*Hint.* By use of Exercise 9.6.1, reduce to the case in which $\mathscr{R}'$ is abelian. Then use Exercise 8.7.2 to reduce to the case in which $\mathfrak{A}_1$ and $\mathfrak{A}_2$ are von Neumann algebras that contain the center $\mathscr{C}$ $(= \mathscr{R}')$ of $\mathscr{R}$, and $\varphi(C) = C$ for each C in $\mathscr{C}$. By applying Exercise 9.6.38, with $\mathscr{A}_j$ the center of $\mathfrak{A}_j$, reduce further to the case in which $\mathfrak{A}_1$ and $\mathfrak{A}_2$ have the same center. Complete the argument by using Exercise 9.6.36.]

9.6.40. Suppose that $\{P_1, P_2, P_3, \ldots\}$ is a sequence of projections in an abelian von Neumann algebra $\mathscr{A}$. Let $\mathfrak{A}_0$ be the C^*-subalgebra of $\mathscr{A}$ generated by the positive operator A_0, where $A_0 = \sum_{n=1}^{\infty} 3^{-n} P_n$. Show that $P_n \in \mathfrak{A}_0$ $(n = 1, 2, 3, \ldots)$. [*Hint.* Let $\mathscr{A}_0$ be the C^*-algebra generated by $\{A_0, P_1, P_2, \ldots\}$, and let X be a compact Hausdorff space such that $\mathscr{A}_0 \cong C(X)$. Use Theorem 4.1.8(ii) and Example 4.4.9 to find a continuous function f such that $P_1 = f(A_0)$.]

9.6.41. Suppose that $\mathscr{A}$ is an abelian von Neumann algebra acting on a separable Hilbert space. Show that $\mathscr{A}$ contains a positive operator that generates $\mathscr{A}$ as a von Neumann algebra. [*Hint.* Use part of the proof of Theorem 9.4.1, together with Exercise 9.6.40.]

9.6.42. By noting that every abelian von Neumann algebra acting on a separable Hilbert space is * isomorphic to one of the algebras described in Theorem 9.4.1, give another proof of the result of Exercise 9.6.41.

9.6.43. Let $\mathscr{A}$ be the abelian von Neumann algebra generated by a bounded normal operator A acting on a separable Hilbert space $\mathscr{H}$, and let $\mathscr{B}(\mathrm{sp}(A))$ be the algebra of all bounded Borel functions on $\mathrm{sp}(A)$. Recall from

Remark 9.5.13 that the (continuous) function calculus for A is a representation $\varphi_A: f \to f(A)$ of $C(\mathrm{sp}(A))$ on $\mathscr{H}$. Let $\bar{\varphi}_A: L_\infty(\varphi_A) \to \mathscr{B}(\mathscr{H})$ be the extension of φ_A described in Proposition 9.5.3. Show that $\bar{\varphi}_A(g) = g(A)$ for each g in $\mathscr{B}(\mathrm{sp}(A))$, and deduce that

$$\mathscr{A} = \{g(A) : g \in \mathscr{B}(\mathrm{sp}(A))\}.$$

9.6.44. Suppose that $\{A_a : a \in \mathbb{A}\}$ is a commuting family of bounded self-adjoint operators acting on a separable Hilbert space $\mathscr{H}$. Show that there is a bounded positive operator H acting on $\mathscr{H}$, and a family $\{g_a : a \in \mathbb{A}\}$ of bounded Borel functions on $\mathrm{sp}(H)$, such that

$$A_a = g_a(H) \qquad (a \in \mathbb{A}).$$

9.6.45. Let $\mathfrak{A}$ be the abelian C^*-algebra $C(X)$, where X is a compact Hausdorff space. Show that $\mathfrak{A}$ is generated (as a C^*-algebra) by a single self-adjoint element if and only if X is homeomorphic to a compact subset of $\mathbb{R}$. Deduce that the statement of Exercise 9.6.41 does not remain true if the term "von Neumann algebra" is replaced by "C^*-algebra" at both occurrences.

9.6.46. Suppose that S is a set, and for each bounded complex-valued function f on S, denote by M_f the bounded linear operator, acting on $l_2(S)$, of multiplication by f. Let $\mathscr{A}$ be the abelian von Neumann algebra $\{M_f : f \in l_\infty(S)\}$. Prove that

(i) if the cardinality of S exceeds that of $\mathbb{R}$, there is no self-adjoint element A_0 of $\mathscr{A}$ that generates $\mathscr{A}$ as a von Neumann algebra (compare this with Exercise 9.6.41);

(ii) if S is the interval $[0, 1]$, ι is the identity mapping on S, and A_0 is the self-adjoint element M_ι of $\mathscr{A}$, then A_0 generates $\mathscr{A}$ as a von Neumann algebra, but

$$\mathscr{A} \neq \{g(A_0) : g \in \mathscr{B}(\mathrm{sp}(A_0))\}$$

(compare this with Exercise 9.6.43).

9.6.47. Let S be the compact subset $\{0, 1, 1/2, 1/3, \ldots, 1/n, \ldots\}$ of $\mathbb{R}$, and let S_0 be $S \backslash \{0\}$. Let φ and φ_0 be the representations of $C(S)$ in which the elements of $C(S)$ act by pointwise multiplication on the separable Hilbert spaces $l_2(S)$ and $l_2(S_0)$, respectively.

(i) Show that φ and φ_0 are faithful.

(ii) Prove that the weak-operator closures of the C^*-algebras $\varphi(C(S))$ $(= \mathfrak{A})$ and $\varphi_0(C(S))$ $(= \mathfrak{A}_0)$ are maximal abelian von Neumann algebras, and there is a unitary transformation V from $l_2(S)$ onto $l_2(S_0)$ such that $V\mathfrak{A}^- V^* = \mathfrak{A}_0^-$.

(iii) Determine the null ideals of φ and φ_0, and deduce that φ and φ_0 are not equivalent.

(iv) With the notation of Proposition 9.5.3, determine the mappings $\bar{\varphi}$ and $\bar{\varphi}_0$, and observe that $\bar{\varphi}$ is one-to-one but $\bar{\varphi}_0$ is not.

(v) Let α be an automorphism of $C(S)$, and let η be the homeomorphism of S such that $\alpha(f) = f \circ \eta$ for each f in $C(S)$ (see Theorem 3.4.3). Show that $\eta(0) = 0$, and deduce that there exist unitary operators U and U_0 acting on $l_2(S)$ and $l_2(S_0)$, respectively, such that

$$\varphi(\alpha(f)) = U\varphi(f)U^*, \qquad \varphi_0(\alpha(f)) = U_0\varphi_0(f)U_0^* \qquad (f \in C(S)).$$

(vi) Show that there is no unitary transformation W from $l_2(S)$ onto $l_2(S_0)$ such that $W\mathfrak{A}W^* = \mathfrak{A}_0$.

(vii) Deduce that the abelian C^*-algebras $\mathfrak{A}$ (acting on $l_2(S)$) and $\mathfrak{A}_0$ (acting on $l_2(S_0)$) are * isomorphic, their weak-operator closures are unitarily equivalent, there is a * isomorphism from $\mathfrak{A}$ onto $\mathfrak{A}_0$ that extends to a * homomorphism from $\mathfrak{A}^-$ onto $\mathfrak{A}_0^-$, but there is no * isomorphism from $\mathfrak{A}$ onto $\mathfrak{A}_0$ that extends to a * isomorphism from $\mathfrak{A}^-$ onto $\mathfrak{A}_0^-$.

9.6.48. The purpose of this exercise is to examine the conclusions of Proposition 9.5.3 as they apply to the representation considered in Example 9.5.10. Suppose that S is a compact Hausdorff space, and φ is the representation of $C(S)$ constructed (as in Example 9.5.10) from a null ideal sequence $\{\mathcal{N}_j\}$ of separable type. Let $\mathscr{C}$ be the C^*-algebra $\varphi(C(S))$, and let $\bar{\varphi}: L_\infty(\varphi) \to \mathscr{C}^-$ be the * homomorphism associated with φ as in Proposition 9.5.3. With the notation used in Example 9.5.10, prove that

(i) $L_\infty(\varphi) = L_\infty(S, \mathscr{S}, \mu)$;

(ii) when $g \in L_\infty(\varphi)$ and $\{x_1, x_2, \ldots\}$ is an element of the Hilbert space $\sum_{j=1}^{\infty} \oplus \mathscr{H}_j \,(= \mathscr{H})$ on which $\varphi(C(S))$ acts,

$$\bar{\varphi}(g)\{x_1, x_2, \ldots\} = \{gx_1, gx_2, \ldots\};$$

(iii) If Y is a Borel subset of S and $E(Y)$ is the projection occurring in Proposition 9.5.3, then

$$E(Y)\{x_1, x_2, \ldots\} = \{qx_1, qx_2, \ldots\}$$

whenever $\{x_1, x_2, \ldots\} \in \mathscr{H}$, where q is the characteristic function of Y;

(iv) $Q_j = E(Y_j)$.

9.6.49. (i) Suppose that S is a compact subset of the complex plane $\mathbb{C}$, μ is a (regular) measure defined on the σ-algebra $\mathscr{S}$ of Borel subsets of S, and $\{Y_j\}$ is a decreasing sequence of Borel subsets of S, with $Y_1 = S$. Let $\mathscr{H}$ be $\sum_{j=1}^{\infty} \oplus \mathscr{H}_j$, where $\mathscr{H}_j$ is the closed subspace of $L_2 \; (= L_2(S, \mathscr{S}, \mu))$

consisting of the functions in L_2 that vanish almost everywhere on the complement of Y_j. Let ι be the identity mapping on S, and let A_0 be the element of $\mathscr{B}(\mathscr{H})$ defined as follows:

$$A_0\{x_1, x_2, \ldots\} = \{\iota x_1, \iota x_2, \ldots\}$$

whenever $\{x_1, x_2, \ldots\} \in \sum_{j=1}^{\infty} \oplus \mathscr{H}_j$. Show that A_0 is normal.

(ii) Suppose that A is a bounded normal operator acting on a separable Hilbert space $\mathscr{H}_A$. By use of Remark 9.5.11 and the first paragraph of Remark 9.5.13, show that it is possible to choose S $(= \mathrm{sp}(A))$, μ, and $\{Y_j\}$, satisfying the conditions set out in (i), in such a way that A is unitarily equivalent to the operator A_0 described in (i).

9.6.50. Suppose that A is a bounded normal operator acting on a separable Hilbert space $\mathscr{H}$. We say that A has *uniform multiplicity* n, where $1 \le n \le \aleph_0$, if the (continuous) function calculus for A has uniform multiplicity n when viewed (as in Remark 9.5.13) as a representation φ_A of $C(\mathrm{sp}(A))$ on $\mathscr{H}$.

(i) Show that A has uniform multiplicity n if and only if the abelian C^*-algebra $\mathfrak{A}$ generated by A, A^*, and I has commutant $\mathfrak{A}'$ of type I_n.

(ii) Suppose that S is a compact subset of the complex plane $\mathbb{C}$, μ is a (regular) measure on the σ-algebra $\mathscr{S}$ of Borel subsets of S, ι is the identity mapping on S, M_ι is the bounded operator acting on L_2 $(= L_2(S, \mathscr{S}, \mu))$ of multiplication by ι, and $\mathscr{K}$ is a Hilbert space of dimension n. Show that the bounded operator $M_\iota \otimes I$, acting on $L_2 \otimes \mathscr{K}$, is normal and has uniform multiplicity n.

(iii) Show that, if A has uniform multiplicity n and $S = \mathrm{sp}(A)$, then μ can be chosen in such a way that the conditions set out in (ii) are satisfied and A is unitarily equivalent to the operator $M_\iota \otimes I$ described in (ii).

9.6.51. Let $\mathscr{R}$ be a von Neumann algebra with center $\mathscr{C}$, acting on a Hilbert space $\mathscr{H}$, and let J be a conjugate-linear isometry of $\mathscr{H}$ onto $\mathscr{H}$ such that $J^2 = I$, $J\mathscr{R}J = \mathscr{R}'$, and $JCJ = C^*$ for each C in $\mathscr{C}$.

(i) Show that $A \to JA^*J$ is a * anti-isomorphism of $\mathscr{R}$ onto $\mathscr{R}'$.

(ii) Let ψ be a * anti-isomorphism of $\mathscr{R}$ onto $\mathscr{R}'$. Show that there is a unitary operator U on $\mathscr{H}$ such that (the conjugate-linear isometry) JU implements ψ:

$$JUA^*(JU)^* = \psi(A) \qquad (A \in \mathscr{R}).$$

[*Hint*. Use Exercise 9.6.25.]

9.6.52. Let $\mathscr{R}$ be a von Neumann algebra acting on a Hilbert space $\mathscr{H}$ with generating and separating vector u. Let S_0, S, F_0, F, J, and Δ be the operators defined in Section 9.2. Suppose J' is a conjugate-linear isometry of $\mathscr{H}$ into $\mathscr{H}$ such that

$$J'u = u, \qquad J'^2 = I, \qquad J'\mathscr{R}J' = \mathscr{R}', \qquad \langle AJ'AJ'u, u\rangle \geq 0 \qquad (A \in \mathscr{R}).$$

Let $H_0 Au$ be $J'A^*u$ $(A \in \mathscr{R})$ and U be $J'J$. Show that

(i) $0 \leq \langle AJAJu, u\rangle \quad (A \in \mathscr{R})$;

(ii) $H_0 = J'S_0$, and H_0 has closure $J'S$ $(= H)$;

(iii) $\langle Hx, x\rangle \geq 0$ for each x in $\mathscr{D}(H)$ $(= \mathscr{D}(\Delta^{1/2}))$ and H is symmetric [*Hint*. Use Exercise 7.6.52(i).];

(iv) H is self-adjoint [*Hint*. Note that $H^* = S_0^* J'$ and that $J'\mathscr{R}'u$ $(= \mathscr{R}u)$ is a core for H^* as well as for H.];

(v) H is positive and $H = U\Delta^{1/2}$;

(vi) $J = J'$, and J is the unique operator with the properties assumed for J'.

9.6.53. Let $\mathscr{R}$ be a von Neumann algebra acting on a Hilbert space $\mathscr{H}$, u be a separating and generating vector for $\mathscr{R}$, and S, F, J and Δ have the meanings attributed to them in Section 9.2. With x a vector in $\mathscr{H}$, let $\varphi_x(A)$ be $\langle Au, x\rangle$ for each A in $\mathscr{R}$, and $\varphi_x'(A')$ be $\langle A'u, x\rangle$ for each A' in $\mathscr{R}'$. Show that

(i) $x \in \mathscr{D}(S)$ $(= \mathscr{D}(F_0^*) = \mathscr{D}(\Delta^{1/2}))$ and that $Sx = x$ for a vector x in $\mathscr{H}$ if and only if the (normal) linear functional φ_x' on $\mathscr{R}'$ is hermitian; symmetrically, $y \in \mathscr{D}(F)$ and $Fy = y$ if and only if φ_y is hermitian;

(ii) $\varphi_x' \geq 0$ if and only if $x = Hu$ for some positive H affiliated with $\mathscr{R}$ [*Hint*. Use L_x in combination with Exercise 7.6.55, and Lemma 9.2.28.];

(iii) the set of vectors x in $\mathscr{H}$ such that $\varphi_x' \geq 0$ is a (norm-)closed cone $\mathscr{V}_u^0$ in $\mathscr{H}$ (see p. 212) and conclude (by symmetry) that the same is true of the set $\mathscr{V}_u^{1/2}$ of vectors x in $\mathscr{H}$ such that $\varphi_x \geq 0$;

(iv) $\mathscr{V}_u^0$ and $\mathscr{V}_u^{1/2}$ (of (iii)) are dual cones, that is, $w \in \mathscr{V}_u^0$ if and only if $\langle w, v\rangle \geq 0$ for each v in $\mathscr{V}_u^{1/2}$, and $v \in \mathscr{V}_u^{1/2}$ if and only if $\langle w, v\rangle \geq 0$ for each w in $\mathscr{V}_u^0$;

(v) $\mathscr{V}_u^0$ is the norm closure of $\mathscr{R}^+u$ and $\mathscr{V}_u^{1/2}$ is the norm closure of $\mathscr{R}'^+u$;

(vi) $\Delta^{1/2}\mathscr{R}^+u = \mathscr{R}'^+u$, $\Delta^{-1/2}\mathscr{R}'^+u = \mathscr{R}^+u$, and deduce that $\mathscr{V}_u^{1/2}, \mathscr{V}_u^0$ are the norm closures of $\Delta^{1/2}\mathscr{R}^+u$, $\Delta^{-1/2}\mathscr{R}'^+u$, respectively.

9.6.54. With the notation and assumptions of Exercise 9.6.53 and with ω a normal state of $\mathscr{R}$, show that

(i) there is a vector v in $\mathscr{V}_u^0$ such that $\omega_v | \mathscr{R} = \omega$ [*Hint*. Use Theorem 7.2.3 to find a vector z such that $\omega = \omega_z | \mathscr{R}$ and then use Theorem 7.3.2 to study φ'_z.];

(ii) $$\|v - u\| = \inf\{\|z - u\| : \omega_z | \mathscr{R} = \omega\},$$

where v is as in (i) [*Hint*. Use Exercise 9.6.53(ii) to express v as Hu, with H a positive operator affiliated with $\mathscr{R}$; use Exercise 7.6.23(ii) to prove that if $\omega_z | \mathscr{R} = \omega$ for some vector z, then $\operatorname{Re}\langle z, u\rangle \le \langle Hu, u\rangle$.];

(iii) the vector v in (i) is unique. [*Hint*. Prove that $\operatorname{Re}\langle z, u\rangle = \langle Hu, u\rangle$, with the notation of the preceding hint, only when $z = Hu$ and then use the formula of (ii).]

9.6.55. Let $\mathscr{R}$ be a von Neumann algebra acting on a Hilbert space $\mathscr{H}$, u be a separating and generating vector for $\mathscr{R}$, and S, F, J and Δ have the meanings attributed to them in Section 9.2. With a in $[0, \frac{1}{2}]$, let $\mathscr{V}_u^a$ be the norm closure of $\{\Delta^a Au : A \in \mathscr{R}^+\}$. (The notation $\mathscr{V}_u^0$ and $\mathscr{V}_u^{1/2}$ of Exercise 9.6.53 is in agreement with the definition of $\mathscr{V}_u^a$ by virtue of Exercise 9.6.53(v), (vi).) Let a' be $\frac{1}{2} - a$. Show that

(i) $\mathscr{V}_u^a$ is a (closed) cone and

$$J\Delta^a Au = \Delta^{a'} Au \quad (A \in \mathscr{R}^+), \qquad J\mathscr{V}_u^a = \mathscr{V}_u^{a'};$$

(ii) $\mathscr{W}_u^0$, the real-linear span of $\mathscr{V}_u^0$, is contained in the domain of $\Delta^{1/2}$ and

$$\Delta^{1/2} y = Jy, \qquad \|\Delta^{1/2} y\| = \|y\| \qquad (y \in \mathscr{W}_u^0)$$

[*Hint*. Establish the first equation for y in $\mathscr{R}^+ u$; use the facts that J is an isometry and $\Delta^{1/2}$ is closed.];

(iii) $\|Hx\| \le \|Kx\|$ when $x \in \mathscr{D}(H) \cap \mathscr{D}(K)$ and H and K are self-adjoint operators affiliated with an abelian von Neumann algebra such that $H^2 \le K^2$ [*Hint*. Use a common bounding sequencc of projections for H, K, H^2, K^2.];

(iv) $$\|\Delta^a y\| \le 2^{1/2}\|y\| \qquad (y \in \mathscr{W}_u^0)$$

[*Hint*. Express Δ^a as $\Delta^a(I - E) \hat{+} \Delta^a E$, where E is the spectral projection for Δ corresponding to $[0, 1]$. Note that $\|\Delta^a E\| \le 1$ and that $\Delta^a(I - E) \le \Delta^{1/2}(I - E)$. Use (iii).];

(v) $\Delta^a \mathscr{V}_u^0$ is dense in $\mathscr{V}_u^a$ [*Hint*. Use (iv).];

(vi) $\mathscr{V}_u^a$ and $\mathscr{V}_u^{a'}$ are dual cones; in particular $\mathscr{V}_u^{1/4}$ is *self-dual*. [*Hint*. If $\langle y, x\rangle \ge 0$ for each x in $\mathscr{V}_u^a$, use Theorems 3.2.30 and 5.6.36 to show that $h_n(\ln \Delta)y$ is in the dual cone to $\mathscr{V}_u^a$ (with h_n as defined in the proof of Theorem 3.2.30). Prove that $(y_n =) h_n(\ln \Delta) y \to y$, $y_n \in \mathscr{D}(\Delta^{-a'})$, and $\Delta^{-a'} y_n \in \mathscr{V}_u^0$. Use (v).]

9.6.56. We adopt the notation of Exercise 9.6.55, but write $\mathscr{V}_u$ in place of $\mathscr{V}_u^{1/4}$. Let $\mathfrak{A}_0$ and $\mathscr{B}_0$ be the (strong-operator-dense) * subalgebras of $\mathscr{R}$ and $\mathscr{R}'$, respectively, consisting of elements in reflection sequences. (See Subsection 9.2, *Tomita's theorem—a second approach*.) Show that

(i) $A_0JA_0Ju \in \mathscr{V}_u$ $(A_0 \in \mathfrak{A}_0)$ [*Hint*. Recall that $JA_0u = \Delta^{1/2}A_0^*u$, and use the fact $\Delta^{-1/4}A_0\Delta^{1/4}$ has a (unique) extension in $\mathscr{R}$, by Theorem 9.2.26 and the discussion following it.];

(ii) $AJAJu \in \mathscr{V}_u$ $(A \in \mathscr{R})$ [*Hint*. Use (i) and the Kaplansky density theorem.];

(iii) $\{\Delta^{1/4}A_0^2u : A_0 \in (\mathfrak{A}_0)_h\}$ is dense in $\mathscr{V}_u$ [*Hint*. Use Corollary 5.3.6 (especially its proof) and Exercise 9.6.55(iv).];

(iv) $\{AJAJu : A \in \mathscr{R}\} = \{A'JA'Ju : A' \in \mathscr{R}'\}$, and this set is dense in $\mathscr{V}_u$ [*Hint*. Use (ii) and (iii), and prove that $BJBJu = \Delta^{1/4}A_0^2u$, where B (in $\mathscr{R}$) is the extension of $\Delta^{1/4}A_0\Delta^{-1/4}$ and $A_0 \in (\mathfrak{A}_0)_h$.];

(v) $AJAJ\mathscr{V}_u \subseteq \mathscr{V}_u$ $(A \in R)$.

9.6.57. We adopt the notation of Exercise 9.6.56. Suppose $x \in \mathscr{V}_u$. Show that

(i) $Jx = x$[*Hint*. Use Exercise 9.6.55(i).];

(ii) $JE = E'J$ and $JEE' = EE'J$, where E and E' are the projections with ranges $[\mathscr{R}'x]$ and $[\mathscr{R}x]$, respectively;

(iii) x is separating for $\mathscr{R}$ if and only if x is generating for $\mathscr{R}$;

(iv) if x is separating for $\mathscr{R}$ and J' is the modular conjugation corresponding to x, then $J' = J$. [*Hint*. Use Exercise 9.6.52(vi).]

9.6.58. We adopt the notation of Exercise 9.6.56. Let x, y, and v be vectors in $\mathscr{V}_u$ and suppose that $0 = \langle x, y\rangle = \langle v, u\rangle$. Let E and E' be the projections with ranges $[\mathscr{R}'x]$ and $[\mathscr{R}x]$, respectively. Show that

(i) $v = 0$ [*Hint*. Choose A_n in $\mathscr{R}^+$ such that $\{\Delta^{1/4}A_n^2u\}$ tends to v and show that $\{\|A_nu\|\}$ tends to 0. Choose B' in $\mathscr{B}_0$ and use the fact that $\Delta^{1/4}B'\Delta^{-1/4}$ has a (unique) extension in $\mathscr{R}'$ to prove that $\langle v, B'u\rangle = 0$.];

(ii) JEE' $(= J')$ is the modular conjugation for $E\mathscr{R}EE'$ $(= \mathscr{S})$ acting on $EE'(\mathscr{H})$ with generating and separating vector x [*Hint*. Use Exercises 9.6.57(i), (ii) and 9.6.52.];

(iii) $\mathscr{V}_x' \subseteq \mathscr{V}_u$, where $\mathscr{V}_x'$ is the self-dual cone for $\{\mathscr{S}, x\}$ (corresponding to $\mathscr{V}_u$ for $\{\mathscr{R}, u\}$) [*Hint*. Use Exercise 9.6.56(iv), (v).];

(iv) $EE'y = 0$. [*Hint*. Show that $EE'y \in \mathscr{V}_x'$ by using (iii) and the fact that $\mathscr{V}_x'$ and $\mathscr{V}_u$ are self-dual. Then use (i).]

9.6.59. With the notation of Exercise 9.6.58, let F and F' be the projections with ranges $[\mathscr{R}'y]$ and $[\mathscr{R}y]$, respectively. Let z be Ey, let z' be

$E'y$, and let M, M', N, and N' be the projections with ranges $[\mathscr{R}'z]$, $[\mathscr{R}z]$, $[\mathscr{R}'z']$, and $[\mathscr{R}z']$, respectively. Show that

(i) $JM = N'J$, $JN = M'J$, $C_M = C_{M'} = C_N = C_{N'}$, $M \le E$, and $N' \le E'$ [*Hint*. Note that $Jz = z'$, and that $JPJ = P$ for each central projection P in $\mathscr{R}$ by Exercise 9.6.18.];

(ii) if $z \ne 0$, there is a non-zero partial isometry U in $\mathscr{R}$ such that $U^*U \le M$, $UU^* \le N$, and $U^*Uz \ne 0$;

(iii) if $z \ne 0$ and G' is the projection with range $[\mathscr{R}U^*Uz]$, in the notation of (ii), there is a non-zero partial isometry V' in $\mathscr{R}'$ such that $V'^*V' \le G' \le M'$, $V'V'^* \le N'$, and $V'^*V'U^*Uz \ne 0$;

(iv) if $z \ne 0$, then $UV'z$ is a non-zero vector in $NN'(\mathscr{H})$, and there is an A in $N\mathscr{R}N$ such that

$$0 < \langle UV'z, Az'\rangle = -\tfrac{1}{2}\langle BJBJy, y\rangle,$$

where $B = A^*U - JV'J \in \mathscr{R}$, in the notation of (iii) [*Hint*. Note that $AN'z' = Az'$ and z' is a generating vector for $N\mathscr{R}NN'$ acting on $NN'(\mathscr{H})$. By using $E'V' = V'$, $UE = U$, and $JE' = EJ$, prove that

$$\langle UV'y, Ay\rangle = \langle UV'z, Az'\rangle, \qquad \langle y, V'JV'y\rangle = 0, \qquad \langle UJA^*Uy, Ay\rangle = 0.];$$

(v) $Ey = E'y = 0$ and $EF = E'F' = 0$. [*Hint*. Note that $0 \le \langle BJBJy, y\rangle$ and use the conclusions of (ii), (iii), and (iv).]

9.6.60. With the notation of Exercise 9.6.56, let $\mathscr{H}_r$ be $\{x : Jx = x\}$. Show that

(i) $\mathscr{H}_r$ is a real Hilbert space relative to the structure imposed by $\mathscr{H}$;

(ii) each element of $\mathscr{H}$ has a decomposition $x_r + ix_i$ with x_r and x_i in $\mathscr{H}_r$ and this decomposition is unique;

(iii) each element of $\mathscr{H}_r$ has a unique decomposition $x_+ - x_-$, where x_+ and x_- are orthogonal vectors in $\mathscr{V}_u$ [*Hint*. Use Proposition 2.2.1 to find x_+, and recall that $\mathscr{V}_u$ is self-dual.];

(iv) if $x, y \in \mathscr{V}_u$, then

$$\|x - y\|^2 \le \|\omega_x|\mathscr{R} - \omega_y|\mathscr{R}\| \le \|x - y\|\,\|x + y\|,$$

and conclude that $\omega_x|\mathscr{R} = \omega_y|\mathscr{R}$ if and only if $x = y$ [*Hint*. Use (iii) to express $x - y$ as $v - w$, where v and w are in $\mathscr{V}_u$ and $\langle v, w\rangle = 0$. Let E and F be the projections with ranges $[\mathscr{R}'v]$ and $[\mathscr{R}'w]$, respectively. Consider $(\omega_x - \omega_y)(E - F)$, and use the facts that $EF = 0$ (Exercise 9.6.59(v)) and $0 \le \langle w, x\rangle$, $0 \le \langle v, y\rangle$.];

(v) $\{\omega_x|\mathscr{R} : x \in \mathscr{V}_u\}$ is norm closed in $\mathscr{R}^\sharp$.

9.6.61. Let H be a positive invertible (possibly unbounded) operator on a Hilbert space $\mathscr{H}$. Show that

(i) $H^{1/4}(I + H^{1/2})^{-1}$ is a bounded, everywhere-defined operator on $\mathscr{H}$ and is equal to $(H^{1/4} \hat{+} H^{-1/4})^{-1}$;

(ii) with x and y in $\mathscr{H}$,

$$\langle (H^{1/4} \hat{+} H^{-1/4})^{-1}x, y\rangle = \int_{\mathbb{R}} (e^{\pi t} + e^{-\pi t})^{-1} \langle H^{it/2}x, y\rangle \, dt$$

[*Hint*. Use Lemma 9.2.7 and argue as in the proof of Lemma 9.2.8.];

(iii) $\Delta^{1/4}(I + \Delta^{1/2})^{-1}\mathscr{V}_u^a \subseteq \mathscr{V}_u^a$ for each a in $[0, \frac{1}{2}]$, with the notation of Exercise 9.6.55. [*Hint*. Use (ii), Exercise 9.6.55(vi), and note that the $\Delta^{it}\mathscr{V}_u^a \subseteq \mathscr{V}_u^a$ for each real t.]

9.6.62. With the notation of Exercise 9.6.56, let ω be a normal linear functional on $\mathscr{R}$ such that $0 \le \omega \le \omega_u | \mathscr{R}$. From Proposition 7.3.5, there is an operator H' in $(\mathscr{R}'^+)_1$ such that $\omega = \omega_{u,H'u} | \mathscr{R}$.

(i) Suppose $x \in \mathscr{D}(\Delta^{-1/2}) \cap \mathscr{V}_u$ and

$$(*) \qquad \omega = \tfrac{1}{2}(\omega_{u,x} + \omega_{x,u}) | \mathscr{R}.$$

Show that $x = 2(I + \Delta^{1/2})^{-1}H'u$. [*Hint*. Note that $x \in \mathscr{D}(F)$ and $Fx = \Delta^{1/2}Jx = \Delta^{1/2}x$.]

(ii) With x as in (i), show that

$$(**) \qquad x = 2\Delta^{1/4}(I + \Delta^{1/2})^{-1}\Delta^{-1/4}H'u.$$

[*Hint*. Use Exercise 9.6.61(i).]

(iii) With x defined by (**), show that $\Delta^{-1/4}H'u \in \mathscr{V}_u$ and that $x \in \mathscr{V}_u$. [*Hint*. Recall that $J\Delta^{-1/4}J = \Delta^{1/4}$, $JH'J \in \mathscr{R}$, and $J\mathscr{V}_u = \mathscr{V}_u$. Use Exercise 9.6.61(iii).]

(iv) Define x by (**). Show that $x \in \mathscr{D}(\Delta^{-1/2}) \cap \mathscr{V}_u$ and satisfies (*). [*Hint*. Note that $x = 2(I + \Delta^{1/2})^{-1}H'u$ and that $(I + \Delta^{1/2})^{-1}\Delta^{-1/2} \subseteq \Delta^{-1/2}(I + \Delta^{1/2})^{-1}$.]

(v) With x as in (iv), show that

$$u - x = 2\Delta^{1/4}(I + \Delta^{1/2})^{-1}\Delta^{-1/4}(I - H')u \in \mathscr{V}_u.$$

[*Hint*. Note that $(I + \Delta^{1/2})u = 2u$ and proceed as in (iii) (with $I - H'$ in place of H').]

9.6.63. With the notation of Exercise 9.6.62, show that

(i) there is a y in $\mathscr{V}_u$ such that $u - y \in \mathscr{V}_u$ and

$$\omega_u \mid \mathscr{R} - \omega = \tfrac{1}{2}(\omega_{u,y} + \omega_{y,u}) \mid \mathscr{R}$$

[*Hint*. Use Exercise 9.6.62(iv) and (v).];

(ii) $$u - \tfrac{1}{2}y\,(= z) \in \mathscr{V}_u, \qquad \omega_z \mid \mathscr{R} - \omega = \omega_{(1/2)y} \mid \mathscr{R}$$

and

$$\|\omega_u \mid \mathscr{R} - \omega\| = \langle u, y \rangle$$

where y is as in (i) [*Hint*. Recall that $\mathscr{V}_u$ is a cone and is self-dual.];

(iii) with y and z as in (ii),

$$\|\omega_z \mid \mathscr{R} - \omega\| \le \tfrac{1}{4}\|\omega_u \mid \mathscr{R} - \omega\|$$

[*Hint*. Note that $0 \le \langle y, y \rangle \le \langle u, y \rangle$.];

(iv) with x' in $\mathscr{V}_u$ such that $\omega \le \omega_{x'} \mid \mathscr{R}$, E and E' the projections whose ranges are $[\mathscr{R}'x']$ and $[\mathscr{R}x']$, respectively, and $\mathscr{R}_0$ the von Neumann algebra $E\mathscr{R}EE'$ acting on $EE'(\mathscr{H})\,(= \mathscr{H}_0)$, we have that x' is generating and separating for $\mathscr{R}_0$, $\mathscr{V}_{x'} \subseteq \mathscr{V}_u$, the equation

$$\omega_0(EAEE') = \omega(A) \qquad (A \in \mathscr{R})$$

defines a positive normal linear functional ω_0 on $\mathscr{R}_0$, and there is a vector z' in $\mathscr{V}_{x'}$ such that

$$\|\omega_{z'} \mid \mathscr{R} - \omega\| \le \tfrac{1}{4}\|\omega_{x'} \mid \mathscr{R} - \omega\|,$$

and $\omega \le \omega_{z'} \mid \mathscr{R}$ [*Hint*. Use Exercise 9.6.58(ii) and (iii) and apply the result of (iii) to ω_0 and $(\mathscr{R}_0, x')$.];

(v) there is a sequence $\{u(n)\}$ in $\mathscr{V}_u$ with u as $u(0)$ such that, $\omega \le \omega_{u(n)} \mid \mathscr{R}$,

$$\|\omega_{u(n)} \mid \mathscr{R} - \omega\| \le \tfrac{1}{4}\|\omega_{u(n-1)} \mid \mathscr{R} - \omega\|,$$

and $\{u(n)\}$ converges to some v in $\mathscr{V}_u$ such that $\omega_v \mid \mathscr{R} = \omega$ [*Hint*. Use z of (ii) as $u(1)$ and apply the conclusion of (iv), inductively, to construct $u(n)$. Use Exercise 9.6.60(iv) to show that $\{u(n)\}$ is Cauchy convergent.];

(vi) the set of (normal) linear functionals ω' on $\mathscr{R}$ such that $0 \le a\omega' \le \omega_u \mid \mathscr{R}$ for some positive a is a norm-dense subset of the set of all vector functionals on $\mathscr{R}$, and each positive normal linear functional on $\mathscr{R}$ has a representation as $\omega_{v'} \mid \mathscr{R}$ for a unique v' in $\mathscr{V}_u$. [*Hint*. Note that $\omega_{A'u} \mid \mathscr{R}$ is such an ω' for each A' in $\mathscr{R}'$. Use Exercise 9.6.60(iv) and (v).]

9.6.64. Let $\mathscr{R}$ and $\mathscr{S}$ be von Neumann algebras acting on Hilbert spaces $\mathscr{H}$ and $\mathscr{K}$ with separating and generating unit vectors u and v, respectively, and let φ be a * isomorphism of $\mathscr{R}$ onto $\mathscr{S}$. Let $\mathscr{V}_u$ and $\mathscr{V}_v$ be the respective self-dual cones for $\mathscr{R}$ and $\mathscr{S}$ corresponding to u and v. (See Exercise 9.6.55(vi).)

(i) With x in $\mathscr{V}_u$ a separating or generating vector for $\mathscr{R}$, show that $\mathscr{V}_x = \mathscr{V}_u$. [*Hint*. Use Exercises 9.6.58(iii), 9.6.57(iv), 9.6.60(iii).]

(ii) Show that there is a unique unitary transformation U of $\mathscr{H}$ onto $\mathscr{K}$ such that $UAU^{-1} = \varphi(A)$ and $Uu' = v$ for some vector u' in $\mathscr{V}_u$. [*Hint*. Use Exercise 9.6.63(vi) with the functional $(\omega_v | \mathscr{S}) \circ \varphi$ and apply Exercise 7.6.23.]

(iii) With U as in (ii), show that $U\mathscr{V}_u = \mathscr{V}_v$. [*Hint*. Use (i).]

(iv) With ω and ω' normal states of $\mathscr{R}$ and $\mathscr{S}$, respectively, denote by u_ω and $v_{\omega'}$ the (unique) vectors in $\mathscr{V}_u$ and $\mathscr{V}_v$ whose corresponding vector states are ω and ω', respectively. Show that $Uu_{\omega' \circ \varphi} = v_{\omega'}$ for each normal state ω' of $\mathscr{S}$, where U is as in (ii). [*Hint*. Note that $U^{-1}v_{\omega'} \in \mathscr{V}_u$ from (iii).]

(v) Suppose $\mathscr{R} = \mathscr{S}$ and $\mathscr{H} = \mathscr{K}$. Show that there is a unique unitary operator U' in $\mathscr{R}'$ such that $U'u_\omega = v_\omega$ for each normal state ω of $\mathscr{R}$.

(vi) With the assumption of (v), J the modular conjugation for $(\mathscr{R}, u)$ and J' the modular conjugation for $(\mathscr{R}, v)$, show that there is a unitary operator V in $\mathscr{R}$ such that $VAV^* = JJ'AJ'J$ for all A in $\mathscr{R}$. [*Hint*. Use Exercise 9.6.51(i). Find u' in $\mathscr{V}_u$ such that $\omega_{u'} | \mathscr{R} = \omega_v | \mathscr{R}$. Note that $U'u' = v$ and $U'JU'^* = J'$, where U' is as in (v). Use Exercises 9.6.63(vi) and 9.6.57(iv) for this.]

9.6.65. Let $\mathscr{R}$ be a von Neumann algebra acting on a Hilbert space $\mathscr{H}$ with a generating and separating vector u. Let ψ be a * automorphism of $\mathscr{R}$ and U_ψ be the (unique) unitary operator on $\mathscr{H}$ (described in Exercise 9.6.64(ii) and (iii)) such that $U_\psi \mathscr{V}_u = \mathscr{V}_u$ and

$$\psi(A) = U_\psi A U_\psi^* \qquad (A \in \mathscr{R}).$$

With ψ' another * automorphism of $\mathscr{R}$, show that $U_{\psi\psi'} = U_\psi U_{\psi'}$ and conclude that the mapping $\psi \to U_\psi$ is a unitary representation of the group of * automorphisms of $\mathscr{R}$.

CHAPTER 10

SPECIAL REPRESENTATIONS OF C^*-ALGEBRAS

This chapter is devoted to a study of representations of C^*-algebras, with the emphasis on the properties of certain particular types of representation, but with applications to a number of more general questions. Some information on this subject has already been given in Section 4.5, in connection with the proof of thc Gelfand–Neumark theorem. There, the GNS construction provided a close link between states and cyclic representations of a C^*-algebra; and a suitable direct sum of cyclic representations, the *universal representation*, was shown to be faithful. Special properties of this universal representation are established in Section 10.1, and used to obtain further information about ideals and quotients of C^*-algebras. By exploiting the close relation between an arbitrary representation of a C^*-algebra and its universal representation, several criteria are given for ultraweak continuity of a linear functional on such an algebra. *Irreducible representations* are introduced in Section 10.2, and characterized as the representations produced by the GNS construction from pure states. The transitivity properties of irreducible algebras, described in Section 5.4, are used in determining when two pure states give rise to equivalent irreducible representations, and in investigating the connections between ideals and pure states. Section 10.3 is concerned with the concepts of *disjointness* and *quasi-equivalence*, for representations of C^*-algebras. A criterion is given for the disjointness of a family of representations, in terms of the von Neumann algebra generated by the direct sum representation, and is used to establish the main property of the *reduced atomic representation*. Finally, Section 10.4 provides illustrations of the preceding theory, through a study of the representations of certain special classes of C^*-algebras.

10.1. The universal representation

With each state ρ of a C^*-algebra $\mathfrak{A}$, we can associate, by means of the GNS construction, a representation π_ρ of $\mathfrak{A}$ on a Hilbert space $\mathscr{H}_\rho$, and a unit cyclic vector x_ρ for π_ρ, so that

$$\rho(A) = \langle \pi_\rho(A)x_\rho, x_\rho \rangle \qquad (A \in \mathfrak{A}).$$

With $\mathscr{S}$ the state space of $\mathfrak{A}$, the representation

$$\Phi = \sum_{\rho \in \mathscr{S}} \oplus \pi_\rho,$$

on the Hilbert space $\mathscr{H}_\Phi = \sum_{\rho \in \mathscr{S}} \oplus \mathscr{H}_\rho$, is called the *universal representation* of $\mathfrak{A}$. We noted, in Remark 4.5.8, that Φ is a faithful representation, and that each state of the C^*-algebra $\Phi(\mathfrak{A})$ is a vector state (equivalently, each state of $\mathfrak{A}$ has the form $\omega_x \circ \Phi$, with x a unit vector in $\mathscr{H}_\Phi$). In the present section, we derive some further properties of the universal representation, and apply these to the study of other representations of $\mathfrak{A}$.

10.1.1. PROPOSITION. *If Φ is the universal representation of a C^*-algebra $\mathfrak{A}$, each bounded linear functional ρ on $\Phi(\mathfrak{A})$ is weak-operator continuous and extends, uniquely and without change of norm, to a weak-operator continuous linear functional $\bar{\rho}$ on $\Phi(\mathfrak{A})^-$. If ρ is hermitian, or positive, the same is true of $\bar{\rho}$. The mapping $\rho \to \bar{\rho}$ is an isometric isomorphism from the dual space $\Phi(\mathfrak{A})^\#$ onto the predual of $\Phi(\mathfrak{A})^-$.*

Proof. Each state of $\Phi(\mathfrak{A})$ is a vector state, $\omega_x | \Phi(\mathfrak{A})$, and so extends to a vector state, $\omega_x | \Phi(\mathfrak{A})^-$, of $\Phi(\mathfrak{A})^-$. Each bounded linear functional ρ on $\Phi(\mathfrak{A})$ is a finite linear combination of states of $\Phi(\mathfrak{A})$, with real coefficients if ρ is hermitian, positive coefficients if ρ is positive. Thus ρ extends to the corresponding finite linear combination of vector states of $\Phi(\mathfrak{A})^-$. This proves the existence of an extension of ρ to a weak-operator continuous linear functional $\bar{\rho}$ on $\Phi(\mathfrak{A})^-$, with $\bar{\rho}$ hermitian, or positive, when ρ has this property; and $\bar{\rho}$ is unique, by continuity. The weak-operator closed set $\{A \in \Phi(\mathfrak{A})^- : |\bar{\rho}(A)| \le \|\rho\|\}$ contains the unit ball of $\Phi(\mathfrak{A})$, and so contains the unit ball of $\Phi(\mathfrak{A})^-$, by the Kaplansky density theorem. From this, $\|\bar{\rho}\| \le \|\rho\|$, and therefore $\|\bar{\rho}\| = \|\rho\|$.

It is apparent that the mapping $\rho \to \bar{\rho}$ is an isometric isomorphism from $\Phi(\mathfrak{A})^\#$ into the predual $[\Phi(\mathfrak{A})^-]_\#$. With ω in this predual, and ρ the element $\omega | \Phi(\mathfrak{A})$ of $\Phi(\mathfrak{A})^\#$, $\bar{\rho}$ and ω coincide on $\Phi(\mathfrak{A})$ and are ultraweakly continuous on $\Phi(\mathfrak{A})^-$, so $\bar{\rho} = \omega$. ■

10.1.2. COROLLARY. *If Φ is the universal representation of a C^*-algebra $\mathfrak{A}$, the weak-operator topology on $\Phi(\mathfrak{A})^-$ coincides with the ultraweak topology. Moreover, the weak-operator and ultraweak topologies on $\Phi(\mathfrak{A})$ both coincide with the weak topology of $\Phi(\mathfrak{A})$ as a Banach space.*

Proof. Each of the three topologies on $\Phi(\mathfrak{A})$ mentioned in the corollary is the weak topology induced by a certain set of linear functionals on $\Phi(\mathfrak{A})$. In order to prove that the topologies coincide, it suffices to show that the corresponding sets of linear functionals are the same. A similar comment applies to the two topologies on $\Phi(\mathfrak{A})^-$.

A weak-operator continuous linear functional on $\Phi(\mathfrak{A})^-$ is ultraweakly continuous. Conversely, from Proposition 10.1.1, each ultraweakly continuous linear functional on $\Phi(\mathfrak{A})^-$ has the form $\bar{\rho}$, with ρ in $\Phi(\mathfrak{A})^\#$, and is therefore weak-operator continuous. Accordingly, the weak-operator and ultraweak topologies coincide on $\Phi(\mathfrak{A})^-$ (hence, also, on $\Phi(\mathfrak{A})$).

From Proposition 10.1.1, the weak-operator continuous linear functionals on $\Phi(\mathfrak{A})$ are precisely the bounded linear functionals; so the weak-operator and weak topologies on $\Phi(\mathfrak{A})$ coincide. ■

10.1.3. REMARK. For completeness, we note that some of the conclusions of Proposition 10.1.1 and Corollary 10.1.2 can be strengthened. With ω an ultraweakly continuous state of $\Phi(\mathfrak{A})^-$, $\omega = \bar{\rho}$ for some state ρ of $\Phi(\mathfrak{A})$. Since $\rho = \omega_x | \Phi(\mathfrak{A})$ for some unit vector x in $\mathscr{H}_\Phi$, it is apparent that $\omega = \omega_x | \Phi(\mathfrak{A})^-$. Hence each ultraweakly continuous state of $\Phi(\mathfrak{A})^-$ is a vector state; by Corollary 7.3.3 (and its proof), each ultraweakly continuous linear functional on $\Phi(\mathfrak{A})^-$ has the form $\omega_{y,z} | \Phi(\mathfrak{A})^-$, with y and z in $\mathscr{H}_\Phi$. Each bounded linear functional on $\Phi(\mathfrak{A})$ extends to an ultraweakly continuous linear functional on $\Phi(\mathfrak{A})^-$, by Proposition 10.1.1, and so has the form $\omega_{y,z} | \Phi(\mathfrak{A})$. ■

10.1.4. PROPOSITION. *Suppose that $\mathfrak{A}$ is a C*-algebra, $\Phi\colon \mathfrak{A} \to \mathscr{B}(\mathscr{H}_\Phi)$ is its universal representation, $\mathscr{K}$ is a convex subset of $\Phi(\mathfrak{A})$ and $\mathscr{K}^-$ is the ultraweak closure of $\mathscr{K}$ in $\mathscr{B}(\mathscr{H}_\Phi)$. Then the strong-operator and weak-operator closures of $\mathscr{K}$ in $\mathscr{B}(\mathscr{H}_\Phi)$ both coincide with $\mathscr{K}^-$, while the norm closure of $\mathscr{K}$ is $\Phi(\mathfrak{A}) \cap \mathscr{K}^-$.*

Proof. For each of the topologies under consideration, $\Phi(\mathfrak{A})^-$ is closed, so the closure of $\mathscr{K}$ in $\mathscr{B}(\mathscr{H})$ is the same as its closure in $\Phi(\mathfrak{A})^-$. From this, together with Corollary 10.1.2, $\mathscr{K}^-$ is the weak-operator closure of $\mathscr{K}$, and is also the strong-operator closure of $\mathscr{K}$ by Theorem 5.1.2. The norm closure $\mathscr{K}^=$ of $\mathscr{K}$ is a subset of $\Phi(\mathfrak{A})$, and coincides with the closure of $\mathscr{K}$ in $\Phi(\mathfrak{A})$ with its weak topology as a Banach space, by Theorem 1.3.4. It now follows from Corollary 10.1.2 that $\mathscr{K}^=$ is the ultraweak closure of $\mathscr{K}$ relative to $\Phi(\mathfrak{A})$; that is, $\mathscr{K}^= = \Phi(\mathfrak{A}) \cap \mathscr{K}^-$. ■

The final clause of Proposition 10.1.4, relating the ultraweak and norm closures of $\mathscr{K}$, is used in proving the following assertion about ideals; it has several other applications (see, for example, Exercise 10.5.2).

10.1.5. PROPOSITION. *If Φ is the universal representation of a C*-algebra $\mathfrak{A}$, and $\mathscr{K}$ is a norm-closed left ideal in $\Phi(\mathfrak{A})$, there is a projection E in $\Phi(\mathfrak{A})^-$ such that*

$$\mathscr{K} = \Phi(\mathfrak{A}) \cap \Phi(\mathfrak{A})^- E = \Phi(\mathfrak{A}) \cap \Phi(\mathfrak{A})E, \qquad \mathscr{K}^- = \Phi(\mathfrak{A})^- E.$$

If $\mathscr{K}$ is a norm-closed two-sided ideal in $\Phi(\mathfrak{A})$, E lies in the center of $\Phi(\mathfrak{A})^-$.

Proof. From the (separate) continuity of multiplication in the weak-operator topology, the weak-operator closure $\mathscr{K}^-$ of $\mathscr{K}$ is a left ideal in $\Phi(\mathfrak{A})^-$, and is a two-sided ideal in $\Phi(\mathfrak{A})^-$ if $\mathscr{K}$ is a two-sided ideal in $\Phi(\mathfrak{A})$. By Theorem 6.8.8, $\mathscr{K}^- = \Phi(\mathfrak{A})^- E$ for some projection E in $\Phi(\mathfrak{A})^-$; and E lies in the center of $\Phi(\mathfrak{A})^-$ if $\mathscr{K}$ is a two-sided ideal. Since $\mathscr{K}$ is a norm-closed subset of $\Phi(\mathfrak{A})$, it results from Proposition 10.1.4 that

$$\mathscr{K} = \Phi(\mathfrak{A}) \cap \mathscr{K}^- = \Phi(\mathfrak{A}) \cap \Phi(\mathfrak{A})^- E.$$

If $A \in \Phi(\mathfrak{A}) \cap \Phi(\mathfrak{A})^- E$, then $A = AE \in \Phi(\mathfrak{A})E$ and it follows that $A \in \Phi(\mathfrak{A}) \cap \Phi(\mathfrak{A})E$. This shows that

$$\Phi(\mathfrak{A}) \cap \Phi(\mathfrak{A})^- E \subseteq \Phi(\mathfrak{A}) \cap \Phi(\mathfrak{A})E,$$

and the reverse inclusion is apparent. ■

Of course, there is a result, parallel to Proposition 10.1.5, for right ideals. Note also that this proposition implies that a closed two-sided ideal in a C^*-algebra is self-adjoint, a fact proved previously in Corollary 4.2.10 (compare Theorem 4.1.9 and the comment following it).

Before continuing our study of the universal representation of a C^*-algebra $\mathfrak{A}$, we digress to obtain further information about ideals, quotients, and general representations of $\mathfrak{A}$. Our first objective is to show that, if $\mathscr{K}$ is a closed two-sided ideal in $\mathfrak{A}$, the quotient $\mathfrak{A}/\mathscr{K}$ has a natural C^*-algebra structure. This is proved below (Theorem 10.1.7) by use of Proposition 10.1.5, together with the following auxiliary result (another approach was indicated in Exercise 4.6.60).

10.1.6. LEMMA. *Suppose that $\mathfrak{A}$ is a C^*-algebra acting on a Hilbert space $\mathscr{H}$, E is a projection in the center of $\mathfrak{A}^-$, and $\mathscr{K}$ is the two-sided ideal $\mathfrak{A} \cap \mathfrak{A}^- E$ in $\mathfrak{A}$. For each A in $\mathfrak{A}$, the distance $d(A, \mathscr{K})$ from A to $\mathscr{K}$ is given by*

$$d(A, \mathscr{K}) = \|A(I - E)\|.$$

Moreover, there is an element K_0 of $\mathscr{K}$ such that

$$d(A, \mathscr{K}) = \|A - K_0\|.$$

Proof. For each K in $\mathscr{K}$, $K(I - E) = 0$, and

$$\|A - K\| \geq \|(A - K)(I - E)\| = \|A(I - E)\|;$$

so

$$d(A, \mathscr{K}) \geq \|A(I - E)\|. \tag{1}$$

To prove the reverse inequality, we identify operators on $\mathscr{H} \oplus \mathscr{H}$ with 2×2 matrices having entries in $\mathscr{B}(\mathscr{H})$, and denote by $\mathscr{B}$ the C^*-algebra consisting of those operators on $\mathscr{H} \oplus \mathscr{H}$ whose matrices have all entries in $\mathfrak{A}$. With

$$B = \begin{bmatrix} 0 & A \\ A^* & 0 \end{bmatrix}, \qquad F = \begin{bmatrix} E & 0 \\ 0 & E \end{bmatrix},$$

B is a self-adjoint element of $\mathscr{B}$, F is a projection in the center of $\mathscr{B}^-$, and $\|A(I - E)\| = \|B(I - F)\|$. These matrices are introduced so that we can study A by means of the function calculus for B. If p is a polynomial,

$$p(t) = a_1 t + a_2 t^2 + \cdots + a_n t^n,$$

with real coefficients and zero constant term, we have

$$\begin{aligned} p(B)(I - F) &= (a_1 B + a_2 B^2 + \cdots + a_n B^n)(I - F) \\ &= a_1 B(I - F) + a_2[B(I - F)]^2 + \cdots + a_n[B(I - F)]^n \\ &= p(B(I - F)). \end{aligned}$$

Let $f\colon \mathbb{R} \to \mathbb{R}$ be a continuous function such that

$$f(t) = t \qquad (|t| \leq \|B(I - F)\|)$$

and $|f(t)| \leq \|B(I - F)\|$ for all t. Since f can be approximated uniformly on the compact interval $[-\|B\|, \|B\|]$ by polynomials of the type just considered, we have

$$f(B)(I - F) = f(B(I - F)) = B(I - F),$$

$$B - f(B) = (B - f(B))F. \tag{2}$$

Since $B - f(B) \in \mathscr{B}$, the upper right-hand entry in its matrix is an element K_0 of $\mathfrak{A}$. By writing (2) in terms of matrices, it follows that $K_0 = K_0 E$, so

$$K_0 \in \mathfrak{A} \cap \mathfrak{A}^- E = \mathscr{K}.$$

Also, $A - K_0$ is the upper right-hand entry in the matrix of $f(B)$, so

$$\begin{aligned} \|A - K_0\| \leq \|f(B)\| &= \sup\{|f(t)| : t \in \operatorname{sp}(B)\} \\ &\leq \|B(I - F)\| = \|A(I - E)\|. \end{aligned}$$

This, together with (1), shows that

$$\|A - K_0\| = d(A, \mathscr{K}) = \|A(I - E)\|. \qquad \blacksquare$$

10.1.7. THEOREM. *Suppose that $\mathscr{K}$ is a proper closed two-sided ideal in a C^*-algebra $\mathfrak{A}$, and φ is the canonical mapping from $\mathfrak{A}$ onto the quotient* *algebra* $\mathfrak{A}/\mathscr{K}$. *Then* $\mathfrak{A}/\mathscr{K}$ *has an involution defined by* $\varphi(A)^* = \varphi(A^*)$ $(A \in \mathfrak{A})$. *With this involution, and the usual quotient norm,* $\mathfrak{A}/\mathscr{K}$ *is a* C^*-*algebra; and* φ *carries the closed unit ball of* $\mathfrak{A}$ *onto that of* $\mathfrak{A}/\mathscr{K}$.

Proof. Since the universal representation Φ of $\mathfrak{A}$ is an (isometric) * isomorphism, it suffices to prove the corresponding assertions concerning $\Phi(\mathfrak{A})$ and its closed two-sided ideal $\Phi(\mathscr{K})$. In view of this, together with Proposition 10.1.5, we reduce to the case in which $\mathfrak{A}$ acts on a Hilbert space and $\mathfrak{A}^-$ contains a central projection E such that $\mathscr{K} = \mathfrak{A} \cap \mathfrak{A}^- E$.

If $A, B \in \mathfrak{A}$ and $\varphi(A) = \varphi(B)$, then $A - B \in \mathscr{K}$; since $\mathscr{K}$ is self-adjoint, $A^* - B^* \in \mathscr{K}$, and $\varphi(A^*) = \varphi(B^*)$. Accordingly, the mapping $\varphi(A) \to \varphi(A^*)$ is well defined, and it is clearly an involution on $\mathfrak{A}/\mathscr{K}$.

With its usual quotient norm, $\mathfrak{A}/\mathscr{K}$ is a Banach algebra; and by Lemma 10.1.6,

$$\|\varphi(A)\| = \|A(I - E)\| \qquad (A \in \mathfrak{A}).$$

Thus

$$\begin{aligned}\|\varphi(A^*)\varphi(A)\| &= \|\varphi(A^*A)\| = \|A^*A(I - E)\| \\ &= \|[A(I - E)]^*A(I - E)\| = \|A(I - E)\|^2 = \|\varphi(A)\|^2\end{aligned}$$

for each A in $\mathfrak{A}$. This shows that $\mathfrak{A}/\mathscr{K}$, with the involution just described, is a C^*-algebra. With B in the closed unit ball of $\mathfrak{A}/\mathscr{K}$, $B = \varphi(A)$ for some A in $\mathfrak{A}$. By Lemma 10.1.6, we can choose K_0 in $\mathscr{K}$ so that

$$\|A - K_0\| = \|\varphi(A)\| = \|B\| \le 1;$$

and $\varphi(A - K_0) = \varphi(A) = B$. ■

Note that, in the circumstances described in Theorem 10.1.7, the quotient mapping φ is a * homomorphism from $\mathfrak{A}$ onto $\mathfrak{A}/\mathscr{K}$.

The first assertion, in the following result, has already been proved in Theorem 4.1.9; it is now deduced (along with additional information) from Theorem 10.1.7.

10.1.8. COROLLARY. *If* $\mathfrak{A}$ *and* $\mathscr{B}$ *are* C^*-*algebras and* ψ *is a * homomorphism from* $\mathfrak{A}$ *into* $\mathscr{B}$, *then* $\psi(\mathfrak{A})$ *is a* C^*-*subalgebra of* $\mathscr{B}$, *and* ψ *carries the closed unit ball of* $\mathfrak{A}$ *onto that of* $\psi(\mathfrak{A})$. *If* $\pi: \mathfrak{A} \to \mathscr{B}(\mathscr{H}_\pi)$ *is a representation of* $\mathfrak{A}$, *then* $\pi(\mathfrak{A})$ *is a* C^*-*algebra of operators acting on* $\mathscr{H}_\pi$.

Proof. By Theorem 4.1.8(i), ψ is norm continuous, so the kernel $\mathscr{K}$ of ψ is a closed two-sided ideal in $\mathfrak{A}$. Moreover, $\psi = \theta\varphi$, where $\varphi: \mathfrak{A} \to \mathfrak{A}/\mathscr{K}$ is the quotient mapping and $\theta: \mathfrak{A}/\mathscr{K} \to \mathscr{B}$ is a one-to-one continuous linear mapping. Since $\mathfrak{A}/\mathscr{K}$ is a C^*-algebra, by Theorem 10.1.7, and ψ, φ are * homomorphisms, it follows that θ is a * isomorphism. By Theorem 4.1.8(iii), $\theta(\mathfrak{A}/\mathscr{K})$ $(=\psi(\mathfrak{A}))$ is a C^*-subalgebra of $\mathscr{B}$; and since θ is isometric, the assertion about closed unit balls is an immediate consequence of the final statement in Theorem 10.1.7. With ψ replaced by π, the preceding argument shows that $\pi(\mathfrak{A})$ is a C^*-subalgebra of $\mathscr{B}(\mathscr{H}_\pi)$. ■

10.1.9. COROLLARY. *Suppose that $\mathfrak{A}$ is a C^*-algebra, $\mathscr{B}$ is a C^*-subalgebra of $\mathfrak{A}$, and $\mathscr{K}$ is a closed two-sided ideal in $\mathfrak{A}$. Then the set*

$$\mathscr{B} + \mathscr{K} = \{B + K : B \in \mathscr{B}, K \in \mathscr{K}\}$$

*is a C^*-subalgebra of $\mathfrak{A}$, and there is a natural * isomorphism between the C^*-algebras $(\mathscr{B} + \mathscr{K})/\mathscr{K}$ and $\mathscr{B}/\mathscr{B} \cap \mathscr{K}$.*

Proof. With $\varphi: \mathfrak{A} \to \mathfrak{A}/\mathscr{K}$ the quotient mapping, the restriction $\varphi \mid \mathscr{B}: \mathscr{B} \to \mathfrak{A}/\mathscr{K}$ is a * homomorphism, so $\varphi(\mathscr{B})$ is closed in $\mathfrak{A}/\mathscr{K}$ by Corollary 10.1.8. Since φ is continuous, while

$$\mathscr{B} + \mathscr{K} = \varphi^{-1}(\varphi(\mathscr{B})),$$

it now follows that $\mathscr{B} + \mathscr{K}$ is closed in $\mathfrak{A}$, and is thus a C^*-subalgebra of $\mathfrak{A}$.
Again, $\varphi \mid \mathscr{B}$ is a * homomorphism of $\mathscr{B}$ onto $(\mathscr{B} + \mathscr{K})/\mathscr{K}$ with kernel $\mathscr{B} \cap \mathscr{K}$; and it induces a * isomorphism of $\mathscr{B}/\mathscr{B} \cap \mathscr{K}$ onto $(\mathscr{B} + \mathscr{K})/\mathscr{K}$. ■

We now revert to the study of the universal representation of a C^*-algebra, and for this purpose require the following auxiliary result.

10.1.10. LEMMA. *Suppose that $\mathscr{H}$ and $\mathscr{K}$ are Hilbert spaces, $\mathfrak{A}$ is a C^*-subalgebra of $\mathscr{B}(\mathscr{H})$, $\eta: \mathfrak{A} \to \mathscr{B}(\mathscr{K})$ is a linear mapping, and the restriction $\eta \mid (\mathfrak{A})_1$, of η to the closed unit ball $(\mathfrak{A})_1$ of $\mathfrak{A}$, is ultraweakly continuous. Then η is ultraweakly continuous on $\mathfrak{A}$, is bounded, and extends (uniquely, and without change of norm) to an ultraweakly continuous linear mapping $\bar{\eta}: \mathfrak{A}^- \to \mathscr{B}(\mathscr{K})$. If η is a * homomorphism, then $\bar{\eta}$ is a * homomorphism and $\bar{\eta}(\mathfrak{A}^-) = \eta(\mathfrak{A})^-$.*

Proof. For all x and y in $\mathscr{K}$, the linear functional $\omega_{x,y} \circ \eta$ on $\mathfrak{A}$ is ultraweakly continuous (and so, also, norm continuous) on $(\mathfrak{A})_1$, and is therefore bounded. Accordingly, the supremum

$$\sup\{|\langle\eta(A)x, y\rangle| : A \in (\mathfrak{A})_1\}$$

is finite. Since each bounded linear functional on $\mathscr{K}$ corresponds (as in Riesz's representation theorem) to an element y of $\mathscr{K}$, it now follows from Theorem 1.8.12 (a variant of the principle of uniform boundedness) that the subset $\{\eta(A) : A \in (\mathfrak{A})_1\}$ of $\mathscr{B}(\mathscr{K})$ is bounded. Hence η is a bounded linear mapping.

In proving that there is an ultraweakly continuous linear extension $\bar{\eta} : \mathfrak{A}^- \to \mathscr{B}(\mathscr{K})$, we may assume that $\|\eta\| = 1$. For each positive real number r, η maps the ball $(\mathfrak{A})_r$ $(=r(\mathfrak{A})_1)$ into $(\mathscr{B}(\mathscr{K}))_r$, and $\eta \mid (\mathfrak{A})_r$ is (uniformly) ultraweakly continuous. Since $(\mathfrak{A})_r$ is ultraweakly dense in $(\mathfrak{A}^-)_r$, and $(\mathscr{B}(\mathscr{K}))_r$ is ultraweakly compact, it follows that $\eta \mid (\mathfrak{A})_r$ extends (uniquely) to an ultraweakly continuous mapping $\eta_r : (\mathfrak{A}^-)_r \to (\mathscr{B}(\mathscr{K}))_r$. Thus, when $r \le s$, $\eta_r(A) = \eta_s(A)$ for all A in $(\mathfrak{A}^-)_r$. Accordingly, there is a mapping $\bar{\eta} : \mathfrak{A}^- \to \mathscr{B}(\mathscr{K})$, such that $\bar{\eta} \mid (\mathfrak{A}^-)_r = \eta_r$ $(r > 0)$.

It is apparent that $\bar{\eta}$ extends η and is norm decreasing, and that the restriction $\bar{\eta} \mid (\mathfrak{A}^-)_r$ is ultraweakly continuous for each positive r. Given A, B in $\mathfrak{A}^-$, and scalars s, t, let

$$r = \max\{\|A\|, \|B\|, |s|\,\|A\| + |t|\,\|B\|\}.$$

There are nets $\{A_a\}$, $\{B_b\}$ in $\mathfrak{A}$, converging ultraweakly to A, B, respectively, such that $\|A_a\| \le \|A\|$, $\|B_b\| \le \|B\|$. Then A_a, B_b, $sA_a + tB_b$ all lie in $(\mathfrak{A}^-)_r$, and

$$\bar{\eta}(sA_a + tB_b) = s\bar{\eta}(A_a) + t\bar{\eta}(B_b),$$

by the linearity of η $(=\bar{\eta} \mid \mathfrak{A})$. When first $A_a \to A$ and then $B_b \to B$, it results from the ultraweak continuity of $\bar{\eta} \mid (\mathfrak{A}^-)_r$ that

$$\bar{\eta}(sA + tB) = s\bar{\eta}(A) + t\bar{\eta}(B).$$

This proves the linearity of $\bar{\eta}$.

Since $\bar{\eta}$ is norm decreasing and extends η, $\|\eta\| \le \|\bar{\eta}\| \le 1 = \|\eta\|$, so $\|\bar{\eta}\| = \|\eta\|$. If ω is an ultraweakly continuous linear functional on $\mathscr{B}(\mathscr{K})$, the composite mapping $\omega \circ \bar{\eta}$ is ultraweakly continuous on $(\mathfrak{A}^-)_r$ for each positive r. By Proposition 7.4.5 (and Remark 7.4.4), $\omega \circ \bar{\eta}$ is ultraweakly continuous on $\mathfrak{A}^-$. Hence $\bar{\eta}$ is ultraweakly continuous on $\mathfrak{A}^-$, and η $(=\bar{\eta} \mid \mathfrak{A})$ is ultraweakly continuous on $\mathfrak{A}$. We have now proved the existence of the required extension $\bar{\eta}$, and its uniqueness is apparent.

Suppose now that η is a * homomorphism. With A, B in $\mathfrak{A}^-$, we can choose nets $\{A_a\}$, $\{B_b\}$ as above, and

$$\bar{\eta}(A_a^*) = \bar{\eta}(A_a)^*, \qquad \bar{\eta}(A_a B_b) = \bar{\eta}(A_a)\bar{\eta}(B_b).$$

When first $A_a \to A$ and then $B_b \to B$, it results from the ultraweak continuity of $\bar{\eta}$ and the adjoint operation, and the separate ultraweak continuity of operator multiplication, that

$$\bar{\eta}(A^*) = \bar{\eta}(A)^*, \qquad \bar{\eta}(AB) = \bar{\eta}(A)\bar{\eta}(B).$$

Hence $\bar{\eta}$ is a * homomorphism. By Corollary 10.1.8, $\eta(\mathfrak{A})$ is a C^*-subalgebra $\mathscr{B}$ of $\mathscr{B}(\mathscr{K})$, and $\eta((\mathfrak{A})_1) = (\mathscr{B})_1$. Since $\bar{\eta}$ is ultraweakly continuous and $(\mathfrak{A}^-)_1$ is ultraweakly compact, $\bar{\eta}((\mathfrak{A}^-)_1)$ is an ultraweakly compact set that contains $(\mathscr{B})_1$ and so contains its closure $(\mathscr{B}^-)_1$. Thus $\bar{\eta}(\mathfrak{A}^-) \supseteq \mathscr{B}^-$; since the reverse inclusion is apparent, from ultraweak continuity of $\bar{\eta}$, we have $\bar{\eta}(\mathfrak{A}^-) = \mathscr{B}^- = \eta(\mathfrak{A})^-$. ■

10.1.11. COROLLARY. *Suppose that $\mathfrak{A}$ is a C^*-algebra acting on a Hilbert space $\mathscr{H}$.*

(i) *If ω is a linear functional on $\mathfrak{A}$, and the restriction $\omega|(\mathfrak{A})_1$ is ultraweakly continuous, then ω is ultraweakly continuous on $\mathfrak{A}$ and extends (uniquely and without change of norm) to an ultraweakly continuous linear functional $\bar{\omega}$ on $\mathfrak{A}^-$.*

(ii) *The set of all ultraweakly continuous linear functionals on $\mathfrak{A}$ is a norm-closed subspace of the Banach dual space $\mathfrak{A}^\#$.*

Proof. (i) This follows by applying Lemma 10.1.10 to the mapping $\eta: A \to \omega(A)I : \mathfrak{A} \to \mathscr{B}(\mathscr{K})$, with $\mathscr{K}$ a one-dimensional Hilbert space.

(ii) With $(\mathfrak{A}^-)_\#$ the predual of $\mathfrak{A}^-$, and $\mathscr{L}$ the set of all ultraweakly continuous linear functionals on $\mathfrak{A}$, it follows from (i) that the mapping $\rho \to \rho|\mathfrak{A} : (\mathfrak{A}^-)_\# \to \mathfrak{A}^\#$ is linear, isometric, and has range $\mathscr{L}$. Since $(\mathfrak{A}^-)_\#$ is a Banach space, the same is true of $\mathscr{L}$; so $\mathscr{L}$ is a norm-closed subspace of $\mathfrak{A}^\#$. ■

Corollary 10.1.11 extends certain properties of ultraweakly continuous linear functionals, previously established for von Neumann algebras, to the context of represented C^*-algebras. It should be noted, however, that (in contrast with the von Neumann algebra case) * isomorphisms between represented C^*-algebras are not, in general, ultraweakly continuous (Exercise 10.5.30).

The following theorem describes the relationship between a general representation of a C^*-algebra and its universal representation. It is this connection that justifies the description "universal."

10.1.12. THEOREM. *If π is a representation of a C^*-algebra $\mathfrak{A}$, and Φ is the universal representation of $\mathfrak{A}$, there is a projection P in the center of $\Phi(\mathfrak{A})^-$, and a * isomorphism α from the von Neumann algebra $\Phi(\mathfrak{A})^-P$ onto $\pi(\mathfrak{A})^-$, such that $\pi(A) = \alpha(\Phi(A)P)$ for each A in $\mathfrak{A}$.*

Proof. Since Φ is a faithful representation, $\pi \circ \Phi^{-1}$ is a * homomorphism β from $\Phi(\mathfrak{A})$ onto $\pi(\mathfrak{A})$, and is therefore norm decreasing. With ω an ultraweakly continuous linear functional on $\pi(\mathfrak{A})$, the linear functional $\omega \circ \beta$ on $\Phi(\mathfrak{A})$ is bounded, and is therefore ultraweakly continuous by

Proposition 10.1.1. From this, β is ultraweakly continuous, and so extends to an ultraweakly continuous * homomorphism $\bar{\beta}$ from $\Phi(\mathfrak{A})^-$ onto $\pi(\mathfrak{A})^-$, by Lemma 10.1.10. The kernel of $\bar{\beta}$ is an ultraweakly closed two-sided ideal in $\Phi(\mathfrak{A})^-$, and by Theorem 6.8.8 it has the form $\Phi(\mathfrak{A})^-Q$, for some projection Q in the center of $\Phi(\mathfrak{A})^-$. With P defined as $I - Q$,

$$\bar{\beta}(A) = \bar{\beta}(AP + AQ) = \bar{\beta}(AP) \qquad (A \in \Phi(\mathfrak{A})^-),$$

so the restriction, $\alpha = \bar{\beta} \mid \Phi(\mathfrak{A})^-P$, has the same range, $\pi(\mathfrak{A})^-$, as $\bar{\beta}$. Moreover, α is one-to-one, and is therefore a * isomorphism from $\Phi(\mathfrak{A})^-P$ onto $\pi(\mathfrak{A})^-$, since the kernel $\Phi(\mathfrak{A})^-Q$ of $\bar{\beta}$ meets $\Phi(\mathfrak{A})^-P$ only at 0. Finally, for each A in $\mathfrak{A}$,

$$\begin{aligned}\alpha(\Phi(A)P) &= \bar{\beta}(\Phi(A)P) = \bar{\beta}(\Phi(A)P + \Phi(A)Q) \\ &= \bar{\beta}(\Phi(A)) = \beta(\Phi(A)) = \pi \circ \Phi^{-1}(\Phi(A)) = \pi(A). \quad \blacksquare\end{aligned}$$

The conclusions of the preceding theorem can be expressed conveniently by means of a diagram. With Φ the universal representation of a C^*-algebra $\mathfrak{A}$, and π another representation, we can choose a central projection P in $\Phi(\mathfrak{A})^-$ and a * isomorphism α from $\Phi(\mathfrak{A})^-P$ onto $\pi(\mathfrak{A})^-$, and set up the following commutative diagram, in which ι denotes an inclusion mapping, α_0 is the * isomorphism $\alpha \mid \Phi(\mathfrak{A})P$ from $\Phi(\mathfrak{A})P$ onto $\pi(\mathfrak{A})$, and ψ is the * homomorphism $A \to AP$ from $\Phi(\mathfrak{A})$ onto $\Phi(\mathfrak{A})P$:

$$\begin{array}{ccccc}
\Phi(\mathfrak{A}) & \xrightarrow{\psi} & \Phi(\mathfrak{A})P & \xrightarrow{\iota} & \Phi(\mathfrak{A})^-P \\
\downarrow{\scriptstyle \Phi^{-1}} & & \downarrow{\scriptstyle \alpha_0} & & \downarrow{\scriptstyle \alpha} \\
\mathfrak{A} & \xrightarrow{\pi} & \pi(\mathfrak{A}) & \xrightarrow{\iota} & \pi(\mathfrak{A})^-
\end{array}$$

From Remark 7.4.4, α is ultraweakly bicontinuous, and therefore the same is true of α_0. Moreover, ψ is ultraweakly continuous, and is a * isomorphism (and hence isometric) if π is a faithful representation.

The conclusions of Theorem 10.1.12 will be discussed further (Remark 10.3.2) in connection with quasi-equivalence and disjointness of representations. In the meantime, we use the theorem to obtain the following characterization of ultraweak continuity of linear mappings between operator algebras.

10.1.13. THEOREM. *Suppose that $\mathscr{H}$ and $\mathscr{K}$ are Hilbert spaces, $\mathfrak{A}$ is a C^*-algebra acting on $\mathscr{H}$, Φ is the universal representation of $\mathfrak{A}$, and P is the central projection in $\Phi(\mathfrak{A})^-$ that occurs in Theorem 10.1.12 when $\pi\colon \mathfrak{A} \to \mathscr{B}(\mathscr{H})$ is the inclusion mapping. If $\eta\colon \mathfrak{A} \to \mathscr{B}(\mathscr{K})$ is a bounded linear* *mapping, there is a unique ultraweakly continuous linear mapping* $\bar{\eta}_P\colon \Phi(\mathfrak{A})^- \to \mathscr{B}(\mathscr{K})$ *such that*

$$\eta(A) = \bar{\eta}_P(\Phi(A)) \qquad (A \in \mathfrak{A}); \tag{3}$$

and η *is ultraweakly continuous if and only if, also,*

$$\eta(A) = \bar{\eta}_P(\Phi(A)P) \qquad (A \in \mathfrak{A}). \tag{4}$$

Proof. Since Φ is isometric, $\eta \circ \Phi^{-1}$ is a bounded linear mapping $\eta_P\colon \Phi(\mathfrak{A}) \to \mathscr{B}(\mathscr{K})$. With ω an ultraweakly continuous linear functional on $\mathscr{B}(\mathscr{K})$, $\omega \circ \eta_P$ is a bounded linear functional on $\Phi(\mathfrak{A})$, and is therefore ultraweakly continuous; so η_P is ultraweakly continuous. For a linear mapping $\bar{\eta}_P\colon \Phi(\mathfrak{A})^- \to \mathscr{B}(\mathscr{K})$, (3) is satisfied if and only if $\bar{\eta}_P | \Phi(\mathfrak{A}) = \eta_P$; and by Lemma 10.1.10, there is a unique ultraweakly continuous mapping $\bar{\eta}_P$ with this property.

If η is ultraweakly continuous, it extends to an ultraweakly continuous linear mapping $\bar{\eta}\colon \mathfrak{A}^- \to \mathscr{B}(\mathscr{K})$. Since $\pi\colon \mathfrak{A} \to \mathscr{B}(\mathscr{H})$ is the inclusion mapping, the * isomorphism α in Theorem 10.1.12 satisfies

$$A = \pi(A) = \alpha(\Phi(A)P) \qquad (A \in \mathfrak{A}).$$

From this, together with (3),

$$\bar{\eta}_P(\Phi(A)) = \eta(A) = \bar{\eta}(A) = \bar{\eta}(\alpha(\Phi(A)P)) \qquad (A \in \mathfrak{A}).$$

Accordingly,

$$\bar{\eta}_P(B) = \bar{\eta}(\alpha(BP))$$

for each B in $\Phi(\mathfrak{A})$; and the ultraweak continuity of $\bar{\eta}_P$, $\bar{\eta}$, and α ensures that this remains valid for all B in $\Phi(\mathfrak{A})^-$. Upon replacing B by BP, the right-hand side is unchanged, so $\bar{\eta}_P(BP) = \bar{\eta}_P(B)$ when $B \in \Phi(\mathfrak{A})^-$. In particular,

$$\eta(A) = \bar{\eta}_P(\Phi(A)) = \bar{\eta}_P(\Phi(A)P) \qquad (A \in \mathfrak{A});$$

so ultraweak continuity of η entails (4).

Conversely, suppose that (4) is satisfied. Since $\alpha(\Phi(A)P) = A$ $(A \in \mathfrak{A})$, it follows from (4) that η is the restriction $\bar{\eta}_P \circ \alpha^{-1} | \mathfrak{A}$, and the ultraweak continuity of $\bar{\eta}_P$ and α^{-1} implies the same continuity for η. ■

We now introduce some notation required for a more detailed study of the Banach dual space $\mathfrak{A}^\#$ of a C^*-algebra $\mathfrak{A}$. With Φ the universal representation of $\mathfrak{A}$, and ρ in $\mathfrak{A}^\#$, it results from Proposition 10.1.1 that $\rho \circ \Phi^{-1}$ extends uniquely to an ultraweakly continuous linear functional $\tilde{\rho}$ on $\Phi(\mathfrak{A})^-$. The mapping $\rho \to \tilde{\rho}$ is linear and isometric, $\tilde{\rho}$ is positive or hermitian when ρ has the same property, and $\rho(A) = \tilde{\rho}(\Phi(A))$ for each A in

$\mathfrak{A}$. Given S in $\Phi(\mathfrak{A})^-$, we can define bounded linear functionals $S\rho$ and ρS on $\mathfrak{A}$ by

$$(S\rho)(A) = \tilde{\rho}(\Phi(A)S), \qquad (\rho S)(A) = \tilde{\rho}(S\Phi(A)) \qquad (A \in \mathfrak{A}).$$

Note that

$$(\widetilde{S\rho})(\Phi(A)) = (S\rho)(A) = \tilde{\rho}(\Phi(A)S), \qquad (\widetilde{\rho S})(\Phi(A)) = \tilde{\rho}(S\Phi(A))$$

for each A in $\mathfrak{A}$. Thus

$$(\widetilde{S\rho})(A) = \tilde{\rho}(AS), \qquad (\widetilde{\rho S})(A) = \tilde{\rho}(SA)$$

for all A in $\Phi(\mathfrak{A})$ and hence, by ultraweak continuity, for all A in $\Phi(\mathfrak{A})^-$. Routine arguments now show that the mappings

$$(\rho, S) \to S\rho, \qquad (\rho, S) \to \rho S : \ \mathfrak{A}^\# \times \Phi(\mathfrak{A})^- \to \mathfrak{A}^\#$$

are bounded and bilinear, that $I\rho = \rho I = \rho$, and that

$$(ST)\rho = S(T\rho), \qquad (S\rho)T = S(\rho T), \qquad (\rho S)T = \rho(ST).$$

With S in the center of $\Phi(\mathfrak{A})^-$, $S\rho = \rho S$ and $S\rho$ is hermitian if ρ is hermitian and S is self-adjoint. For positive S in the center of $\Phi(\mathfrak{A})^-$, and positive ρ in $\mathfrak{A}^\#$, $S\rho$ is positive.

The above notation is used in the following characterization (essentially a reformulation of Theorem 10.1.13) of the ultraweakly continuous linear functionals on a C^*-algebra acting on a Hilbert space.

10.1.14. PROPOSITION. *Suppose that $\mathfrak{A}$ is a C^*-algebra acting on a Hilbert space $\mathscr{H}$, Φ is its universal representation, and P is the central projection in $\Phi(\mathfrak{A})^-$ that occurs in Theorem* 10.1.12 *when $\pi\colon \mathfrak{A} \to \mathscr{B}(\mathscr{H})$ is the inclusion mapping. Then a bounded linear functional ρ on $\mathfrak{A}$ is ultraweakly continuous if and only if $\rho = P\rho$.*

Proof. From the discussion preceding this proposition, $\rho(A) = \tilde{\rho}(\Phi(A))$ for each A in $\mathfrak{A}$, and $\tilde{\rho}$ is an ultraweakly continuous linear functional on $\Phi(\mathfrak{A})^-$. With $\mathscr{K}$ a one-dimensional Hilbert space, we can identify $\mathscr{B}(\mathscr{K})$ with the scalar field, and consider ρ as a bounded linear mapping from $\mathfrak{A}$ into $\mathscr{B}(\mathscr{K})$. The mapping $\bar{\rho}_P$ that then occurs in Theorem 10.1.13, is $\tilde{\rho}$; and, therefore, ρ is ultraweakly continuous if and only if $\rho(A) = \tilde{\rho}(\Phi(A)P) = (P\rho)(A)$ for each A in $\mathfrak{A}$. ■

In the circumstances considered in Proposition 10.1.14, the mapping $\rho \to P\rho$ is a norm-decreasing projection from $\mathfrak{A}^\#$ onto a closed subspace $\mathfrak{A}^{\mathrm{u}}$ consisting of all the ultraweakly continuous linear functionals on $\mathfrak{A}$. Non-zero elements of the complementary closed subspace

$$\mathfrak{A}^{\mathrm{s}} = \{\rho \in \mathfrak{A}^\# : P\rho = 0\}$$

are described as *singular* functionals on $\mathfrak{A}$. A more intuitive interpretation of the term "singular" is provided by Proposition 10.1.17 below. In the meantime, the following theorem sets out some of the immediate consequences of the definition just given. We emphasize that the terms "singular" and "ultraweakly continuous," as applied to bounded linear functionals on a C^*-algebra $\mathfrak{A}$, are meaningful only when $\mathfrak{A}$ is presented as an algebra of operators acting on a specific Hilbert space.

10.1.15. THEOREM. *Suppose that $\mathfrak{A}$ is a C^*-algebra acting on a Hilbert space $\mathscr{H}$.*

(i) *The ultraweakly continuous linear functionals on $\mathfrak{A}$ form a norm-closed subspace $\mathfrak{A}^{u}$ of the Banach dual space $\mathfrak{A}^{\#}$.*

(ii) *The singular elements of $\mathfrak{A}^{\#}$, together with 0, form a norm-closed subspace $\mathfrak{A}^{s}$ of $\mathfrak{A}^{\#}$.*

(iii) *Each ρ in $\mathfrak{A}^{\#}$ can be expressed, uniquely, in the form $\rho = \rho_u + \rho_s$, with ρ_u in $\mathfrak{A}^{u}$ and ρ_s in $\mathfrak{A}^{s}$. Moreover, $\|\rho\| = \|\rho_u\| + \|\rho_s\|$; and if ρ is positive, or hermitian, the same is true of ρ_u and ρ_s.*

(iv) *If Φ is the universal representation of $\mathfrak{A}$, and $S \in \Phi(\mathfrak{A})^-$, the mappings $\rho \to S\rho$, $\rho \to \rho S : \mathfrak{A}^{\#} \to \mathfrak{A}^{\#}$ leave $\mathfrak{A}^{u}$ and $\mathfrak{A}^{s}$ invariant.*

Proof. We continue to use the notation established in Proposition 10.1.14 and the discussion that precedes it. We have already noted that $\mathfrak{A}^{u}$ and $\mathfrak{A}^{s}$ are complementary closed subspaces of $\mathfrak{A}^{\#}$, and that the corresponding projection from $\mathfrak{A}^{\#}$ onto $\mathfrak{A}^{u}$ is the mapping $\rho \to P\rho$. This proves (i), (ii), and the first assertion in (iii), with

$$\rho_u = P\rho, \qquad \rho_s = (I - P)\rho.$$

Since P and $I - P$ are positive elements of the center of $\Phi(\mathfrak{A})^-$, ρ_u and ρ_s are positive, or hermitian, when ρ has this property. Given A_1 and A_2 in the unit ball $(\mathfrak{A})_1$, $\Phi(A_1)P + \Phi(A_2)(I - P)$ lies in $(\Phi(\mathfrak{A})^-)_1$, so

$$\begin{aligned}\|\rho\| = \|\tilde{\rho}\| &\geq |\tilde{\rho}(\Phi(A_1)P + \Phi(A_2)(I - P))| \\ &= |(P\rho)(A_1) + (\rho - P\rho)(A_2)| = |\rho_u(A_1) + \rho_s(A_2)|.\end{aligned}$$

By taking the upper bound of the right-hand side, as A_1 and A_2 vary in $(\mathfrak{A})_1$, we obtain

$$\|\rho\| \geq \|\rho_u\| + \|\rho_s\|.$$

Since the reverse inequality is apparent, this completes the proof of (iii).

With ρ in $\mathfrak{A}^{\#}$ and S in $\Phi(\mathfrak{A})^-$,

$$P(S\rho) = (PS)\rho = (SP)\rho = S(P\rho), \qquad P(\rho S) = (P\rho)S.$$

Accordingly, the mappings $\rho \to S\rho$, $\rho \to \rho S : \mathfrak{A}^\# \to \mathfrak{A}^\#$ both commute with the projection $\rho \to P\rho$ from $\mathfrak{A}^\#$ onto $\mathfrak{A}^u$, and hence leave invariant both its range space $\mathfrak{A}^u$ and its null space $\mathfrak{A}^s$. ■

10.1.16. COROLLARY. *If $\mathfrak{A}$ is a C^*-algebra acting on a Hilbert space $\mathscr{H}$, ρ is a singular element of $\mathfrak{A}^\#$, $B, C \in \mathfrak{A}$, and ρ_0 in $\mathfrak{A}^\#$ is defined by*

$$\rho_0(A) = \rho(BAC) \qquad (A \in \mathfrak{A}),$$

then ρ_0 is singular or zero.

Proof. With Φ the universal representation of $\mathfrak{A}$, it is not difficult to check that $\rho_0 = C_0 \rho B_0$, where $B_0 = \Phi(B)$ and $C_0 = \Phi(C)$. Since $\rho \in \mathfrak{A}^s$, it now follows from Theorem 10.1.15(iv) that $\rho_0 \in \mathfrak{A}^s$. ■

10.1.17. PROPOSITION. *Suppose that $\mathfrak{A}$ is a C^*-algebra acting on a Hilbert space $\mathscr{H}$.*

(i) *A state ρ of $\mathfrak{A}$ is singular if and only if there is no non-zero ultraweakly continuous linear functional ω on $\mathfrak{A}$ such that $0 \le \omega \le \rho$.*

(ii) *A non-zero bounded linear functional on $\mathfrak{A}$ is singular if and only if it is a finite linear combination of singular states of $\mathfrak{A}$.*

Proof. (i) Each state ρ of $\mathfrak{A}$ can be expressed in the form $\rho_u + \rho_s$, as in Theorem 10.1.15(iii), and $\rho_u, \rho_s \ge 0$. If ρ is not singular, ρ_u is a non-zero ultraweakly continuous linear functional ω on $\mathfrak{A}$, satisfying $0 \le \omega \le \rho$.

Conversely, suppose that there is such a functional ω. Then $\omega = P\omega$ (where P has its usual meaning), and since $\rho - \omega \ge 0$, we have

$$P\rho - \omega = P(\rho - \omega) \ge 0.$$

Thus $P\rho \ge \omega > 0$, whence $P\rho \ne 0$ and ρ is not singular.

(ii) Each bounded linear functional ρ on $\mathfrak{A}$ is a finite linear combination $\sum a_j \rho_j$ of states ρ_j of $\mathfrak{A}$. If ρ is singular,

$$\rho = (I - P)\rho = \sum a_j(I - P)\rho_j.$$

Since $\rho_j \ge 0$, $(I - P)\rho_j$ is (zero or) a multiple of a singular state. Accordingly, ρ is a finite linear combination of singular states.

Conversely, since $\mathfrak{A}^s$ is a subspace of $\mathfrak{A}^\#$, a non-zero finite linear combination of singular states of $\mathfrak{A}$ is itself a singular functional on $\mathfrak{A}$. ■

Each state ρ of a C^*-algebra $\mathfrak{A}$ has the form $\omega_x \circ \Phi$, where $\Phi: \mathfrak{A} \to \mathscr{B}(\mathscr{H}_\Phi)$ is the universal representation of $\mathfrak{A}$ and x is a unit vector in $\mathscr{H}_\Phi$. We now give conditions, on x, for ρ to be ultraweakly continuous, or singular, when $\mathfrak{A}$ acts on a Hilbert space.

10.1.18. PROPOSITION. *Suppose that* $\mathfrak{A}$ *is a* C*-*algebra acting on a Hilbert space* $\mathscr{H}$, $\Phi\colon \mathfrak{A} \to \mathscr{B}(\mathscr{H}_\Phi)$ *is its universal representation,* P *is the central projection in* $\Phi(\mathfrak{A})^-$ *that occurs in Theorem* 10.1.12 *when* $\pi\colon \mathfrak{A} \to \mathscr{B}(\mathscr{H})$ *is the inclusion mapping,* x *is a unit vector in* $\mathscr{H}_\Phi$, *and* $\rho = \omega_x \circ \Phi$. *Then* ρ *is ultraweakly continuous if and only if* $Px = x$, *and is singular if and only if* $Px = 0$.

Proof. We can decompose ρ in the form $\rho_u + \rho_s$, as in Theorem 10.1.15(iii), and ρ_u, ρ_s are positive. Since $\rho \circ \Phi^{-1}$ is $\omega_x | \Phi(\mathfrak{A})$, the ultraweakly continuous extension $\tilde{\rho}$ of $\rho \circ \Phi^{-1}$ to $\Phi(\mathfrak{A})^-$ is $\omega_x | \Phi(\mathfrak{A})^-$. Thus

$$\|\rho_u\| = \rho_u(I) = (P\rho)(I) = \tilde{\rho}(P) = \omega_x(P) = \|Px\|^2,$$

$$\|\rho_s\| = \rho_s(I) = (\rho - P\rho)(I) = \tilde{\rho}(I - P) = \|x - Px\|^2.$$

Finally, ρ is ultraweakly continuous if and only if $\rho_s = 0$, and singular if and only if $\rho_u = 0$. ■

It is apparent that an ultraweakly continuous linear functional on a von Neumann algebra $\mathscr{R}$ is completely additive on projections. As an application of the theory just developed, we now prove, conversely, that complete additivity on projections implies ultraweak continuity for a bounded linear functional on $\mathscr{R}$. For *positive* linear functionals, this has already been proved (Theorem 7.1.12 and Remark 7.4.4).

10.1.19. THEOREM. *If* ρ *is a bounded linear functional on a von Neumann algebra* $\mathscr{R}$, *and* $\rho(\sum E_a) = \sum \rho(E_a)$ *whenever* $\{E_a\}$ *is an orthogonal family of projections in* $\mathscr{R}$, *then* ρ *is ultraweakly continuous.*

Proof. We can express ρ in the form $\rho_1 + i\rho_2$, with ρ_1 and ρ_2 hermitian functionals, and it is apparent that ρ_1 and ρ_2 are also completely additive on projections. Accordingly, we may suppose at the outset that ρ is hermitian. We can then decompose ρ in the form $\rho_u + \rho_s$, as in Theorem 10.1.15(iii), and ρ_u, ρ_s are both hermitian. Since ρ and ρ_u are completely additive on projections (ρ_u, because it is ultraweakly continuous), the same is true of ρ_s. We have to show that $\rho_s = 0$.

We suppose that $\rho_s \neq 0$, so that ρ_s is hermitian, singular, and completely additive; and in due course, we obtain a contradiction. Since $\rho_s \neq 0$, there is a projection E in $\mathscr{R}$ such that $\rho_s(E) \neq 0$. Upon replacing ρ by $-\rho$ (and hence ρ_s by $-\rho_s$) if necessary, we may suppose that $\rho_s(E) > 0$. Let $\{F_a\}$ be a maximal orthogonal family of non-zero subprojections of E in $\mathscr{R}$, for which $\rho_s(F_a) \leq 0$, and let $F = \sum F_a$. Accordingly, $F \in \mathscr{R}$, F is a subprojection of E, and the complete additivity of ρ_s entails

$$\rho_s(F) = \sum \rho_s(F_a) \leq 0 < \rho_s(E).$$

Thus $E - F$ is a projection G in $\mathscr{R}$, and $\rho_s(G) > 0$. From the maximality of the family $\{F_a\}$, $\rho_s(G_1) > 0$ for each non-zero projection G_1 in $G\mathscr{R}G$, so the restriction $\rho_s | G\mathscr{R}G$ is a positive linear functional. However, $\rho_s | G\mathscr{R}G$ inherits complete additivity from ρ_s, and is therefore ultraweakly continuous by Theorem 7.1.12 and Remark 7.4.4.

The equation $\tau(A) = \rho_s(GAG)$ defines a bounded linear functional τ on $\mathscr{R}$, and $\tau \neq 0$ since $\tau(G) = \rho_s(G) > 0$. From Corollary 10.1.16, and from ultraweak continuity of $\rho_s | G\mathscr{R}G$, τ is both singular and ultraweakly continuous, a contradiction. ∎

10.1.20. COROLLARY. *If ρ is a bounded linear functional on a von Neumann algebra $\mathscr{R}$, and the restriction $\rho | \mathscr{A}$ is ultraweakly continuous whenever $\mathscr{A}$ is a maximal abelian * subalgebra of $\mathscr{R}$, then ρ is ultraweakly continuous on $\mathscr{R}$.*

Proof. An orthogonal family $\{E_a\}$ of projections in $\mathscr{R}$ generates an abelian von Neumann subalgebra $\mathscr{A}_0$ of $\mathscr{R}$, and $\mathscr{A}_0$ contains $\sum E_a$ as well as each E_a. There is at least one maximal abelian * subalgebra $\mathscr{A}$ of $\mathscr{R}$ that contains $\mathscr{A}_0$, and the ultraweak continuity of $\rho | \mathscr{A}$ entails $\rho(\sum E_a) = \sum \rho(E_a)$. Thus ρ is completely additive on projections, and is ultraweakly continuous by Theorem 10.1.19. ∎

With $\mathfrak{A}$ a C^*-algebra and Φ its universal representation, the second adjoint mapping $\Phi^{\#\#}$ is an isometric isomorphism from the bidual space $\mathfrak{A}^{\#\#}$ onto $\Phi(\mathfrak{A})^{\#\#}$. We conclude this section by showing that $\Phi(\mathfrak{A})^{\#\#}$ (and hence $\mathfrak{A}^{\#\#}$) can be identified with the von Neumann algebra $\Phi(\mathfrak{A})^-$.

10.1.21. PROPOSITION. *Let Φ be the universal representation of a C^*-algebra $\mathfrak{A}$, and for each bounded linear functional ρ on $\Phi(\mathfrak{A})$, let $\bar{\rho}$ denote its (unique) extension to an ultraweakly continuous linear functional on $\Phi(\mathfrak{A})^-$. Then for each A in $\Phi(\mathfrak{A})^-$, the equation*

$$\hat{A}(\rho) = \bar{\rho}(A) \qquad (\rho \in \Phi(\mathfrak{A})^{\#})$$

defines a bounded linear functional $\hat{A}$ on $\Phi(\mathfrak{A})^{\#}$. The mapping $A \to \hat{A}$ is an isometric isomorphism from $\Phi(\mathfrak{A})^-$ onto the bidual space $\Phi(\mathfrak{A})^{\#\#}$, and its restriction to $\Phi(\mathfrak{A})$ is the canonical embedding of $\Phi(\mathfrak{A})$ in $\Phi(\mathfrak{A})^{\#\#}$.

Proof. From Theorem 7.4.2, the equation

$$\tilde{A}(\omega) = \omega(A) \qquad (A \in \Phi(\mathfrak{A})^-, \quad \omega \in (\Phi(\mathfrak{A})^-)_{\#})$$

defines an isometric isomorphism $\beta\colon A \to \tilde{A}$ from $\Phi(\mathfrak{A})^-$ onto the Banach dual space, $(\Phi(\mathfrak{A})^-)_{\#}^{\#}$, of $(\Phi(\mathfrak{A})^-)_{\#}$. The adjoint α of the isometric isomorphism, $\rho \to \bar{\rho}$, of $\Phi(\mathfrak{A})^{\#}$ onto $(\Phi(\mathfrak{A})^-)_{\#}$ described in Proposition 10.1.1 is an isometric isomorphism of $(\Phi(\mathfrak{A})^-)^\#_\#$ onto $\Phi(\mathfrak{A})^{\#\#}$. The composition $\alpha \circ \beta$ is an isometric isomorphism $A \to \hat{A}$ from $\Phi(\mathfrak{A})^-$ onto $\Phi(\mathfrak{A})^{\#\#}$; moreover, with A in $\Phi(\mathfrak{A})^-$ and ρ in $\Phi(\mathfrak{A})^\#$,

$$\hat{A}(\rho) = [(\alpha \circ \beta)(A)](\rho) = \beta(A)(\bar{\rho}) = \bar{\rho}(A).$$

When $A \in \Phi(\mathfrak{A})$, $\hat{A}(\rho) = \rho(A)$ for each ρ in $\Phi(\mathfrak{A})^\#$, so the mapping $A \to \hat{A} : \Phi(\mathfrak{A}) \to \Phi(\mathfrak{A})^{\#\#}$ is the canonical embedding. ■

Bibliography: [28, 85, 86, 92, 93].

10.2. Irreducible representations

In Section 5.4 we studied irreducible C^*-algebras of operators acting on Hilbert spaces. For such algebras, we proved the equivalence of topological irreducibility, algebraic irreducibility, and certain concepts of transitivity.

Suppose now that $\pi\colon \mathfrak{A} \to \mathscr{B}(\mathscr{H}_\pi)$ is a representation of a C^*-algebra $\mathfrak{A}$. By Corollary 10.1.8, $\pi(\mathfrak{A})$ is a C^*-subalgebra of $\mathscr{B}(\mathscr{H}_\pi)$. We say that π is *irreducible* if $\{0\}$ and $\mathscr{H}_\pi$ are the only closed subspaces of $\mathscr{H}_\pi$ invariant under each operator in $\pi(\mathfrak{A})$. By Theorem 5.4.1, π is irreducible if and only if the commutant $\pi(\mathfrak{A})'$ consists of scalars or, equivalently, $\pi(\mathfrak{A})'' = \mathscr{B}(\mathscr{H}_\pi)$. By Corollary 5.4.4, π is irreducible if and only if $\{0\}$ and $\mathscr{H}_\pi$ are the only linear manifolds (not assumed closed) in $\mathscr{H}_\pi$, invariant under each operator in $\pi(\mathfrak{A})$. From the transitivity properties of irreducible algebras, we obtain the following result.

10.2.1. THEOREM. *If $\pi\colon \mathfrak{A} \to \mathscr{B}(\mathscr{H}_\pi)$ is an irreducible representation of a C^*-algebra $\mathfrak{A}$, $\{x_1, \ldots, x_n\}$ is a linearly independent set in $\mathscr{H}_\pi$, and $y_1, \ldots, y_n \in \mathscr{H}_\pi$, then there is an element A of $\mathfrak{A}$ such that $\pi(A)x_j = y_j$ $(j = 1, \ldots, n)$. If $y_j = Bx_j$ $(j = 1, \ldots, n)$ for some self-adjoint B in $\mathscr{B}(\mathscr{H}_\pi)$, A can be chosen self-adjoint. If $y_j = Vx_j$ $(j = 1, \ldots, n)$ for some unitary operator V on $\mathscr{H}_\pi$, A can be chosen to be a unitary element of the form* exp iH, *with H self-adjoint in $\mathfrak{A}$.*

Proof. By Theorem 5.4.3, there is an element A_0 in $\pi(\mathfrak{A})$ such that $A_0x_j = y_j$ $(j = 1, \ldots, n)$; and $A_0 = \pi(A)$ for some A in $\mathfrak{A}$. Moreover, if $y_j = Bx_j$ $(j = 1, \ldots, n)$ for some self-adjoint B in $\mathscr{B}(\mathscr{H}_\pi)$, A_0 can be chosen self-adjoint. Then,

$$\tilde{A}_0 = \tfrac{1}{2}(A_0 + A_0^*) = \pi(\tfrac{1}{2}(A + A^*)),$$

and upon replacing A by $\frac{1}{2}(A + A^*)$, we can take A self-adjoint.

If $y_j = Vx_j$ $(j = 1, \ldots, n)$ for some unitary operator V on $\mathscr{H}_\pi$, it follows from Theorem 5.4.5 that there is a unitary U_0 in $\pi(\mathfrak{A})$, of the form $\exp iH_0$

for some self-adjoint H_0 in $\pi(\mathfrak{A})$, such that $U_0 x_j = y_j$ $(j = 1, \ldots, n)$. There is a self-adjoint element H of $\mathfrak{A}$ for which $\pi(H) = H_0$, $\exp iH$ is a unitary element U of $\mathfrak{A}$, and

$$\pi(U) = \pi(\exp iH) = \exp i\pi(H) = \exp iH_0 = U_0. \quad \blacksquare$$

10.2.2. REMARK. If $\mathscr{R}$ is a von Neumann algebra, each unitary element of $\mathscr{R}$ has the form $\exp iH$, with H self-adjoint in $\mathscr{R}$ (Theorem 5.2.5); but for general C^*-algebras, this is not necessarily the case. In proving the last part of the above theorem, we used the fact that, if π is a representation of a C^*-algebra $\mathfrak{A}$, then an "exponential" unitary U_0 in $\pi(\mathfrak{A})$ can be expressed as $\pi(U)$, with U an exponential unitary in $\mathfrak{A}$. For a general unitary element U_0 of $\pi(\mathfrak{A})$, there may be no unitary U in $\mathfrak{A}$ such that $\pi(U) = U_0$ (see Exercise 4.6.3). It is thus essential, in proving the above theorem, to know that the unitary U occurring in Theorem 5.4.5 is an exponential. $\blacksquare$

If $\pi\colon \mathfrak{A} \to \mathscr{B}(\mathscr{H}_\pi)$ is a representation of a C^*-algebra $\mathfrak{A}$ and x is a non-zero vector in $\mathscr{H}_\pi$, the non-zero closed subspace $[\pi(\mathfrak{A})x]$ is invariant under each operator in $\pi(\mathfrak{A})$. If π is irreducible, $[\pi(\mathfrak{A})x] = \mathscr{H}_\pi$; in fact, since π is algebraically irreducible, we have the stronger result that $\pi(\mathfrak{A})x = \mathscr{H}_\pi$. Conversely, it is apparent that π is irreducible if $[\pi(\mathfrak{A})x] = \mathscr{H}_\pi$ for each non-zero vector x in $\mathscr{H}_\pi$. Thus, π is irreducible if and only if every non-zero vector in $\mathscr{H}_\pi$ is cyclic for π.

If π is irreducible, so is every representation equivalent to π; moreover, if x is a unit vector in $\mathscr{H}_\pi$ (necessarily cyclic for π), and ρ is the state $\omega_x \circ \pi$ of $\mathfrak{A}$, it results from Proposition 4.5.3 that π is equivalent to the representation obtained from ρ by means of the GNS construction. Since each irreducible representation of $\mathfrak{A}$ is equivalent to one that arises in this way from a state, it is natural to ask which states of $\mathfrak{A}$ give rise to irreducible representations, and under what conditions two such states give rise to equivalent irreducible representations. The next two theorems deal with these questions.

10.2.3. THEOREM. *If ρ is a state of a C^*-algebra $\mathfrak{A}$, the representation π_ρ obtained from ρ by means of the* GNS *construction is irreducible if and only if ρ is pure.*

Proof. We shall write x for the unit vector, usually denoted by x_ρ, occurring in the GNS construction.

Suppose first that π_ρ is irreducible. If $\rho = a\mu + b\nu$, where μ, ν are states of $\mathfrak{A}$ and a, b are positive real numbers with sum 1, then

$$\begin{aligned}|\mu(A)|^2 &\le \mu(A^*A) \le a^{-1}\rho(A^*A)\\ &= a^{-1}\langle \pi_\rho(A^*A)x, x\rangle = a^{-1}\|\pi_\rho(A)x\|^2 \le a^{-1}\|\pi_\rho(A)\|^2\end{aligned}$$

for each A in $\mathfrak{A}$. Thus $\mu(A) = 0$ if $\pi_\rho(A) = 0$; and the equation

$$\omega(\pi_\rho(A)) = \mu(A) \qquad (A \in \mathfrak{A})$$

defines a linear functional ω on $\pi_\rho(\mathfrak{A})$. Since $\omega(I) = \mu(I) = 1$, while the preceding inequalities give

$$0 \leq \mu(A^*A) = \omega(\pi_\rho(A^*A)) \leq a^{-1}\langle \pi_\rho(A^*A)x, x\rangle \qquad (A \in \mathfrak{A}),$$

it follows that ω is a state of $\pi_\rho(\mathfrak{A})$, and that $\omega \leq a^{-1}\omega_x | \pi_\rho(\mathfrak{A})$. From Proposition 7.3.5, $\omega = \omega_{Sx} | \pi_\rho(\mathfrak{A})$ for some element S of $\pi_\rho(\mathfrak{A})'$. However, since π_ρ is irreducible, $\pi_\rho(\mathfrak{A})' = \{cI : c \in \mathbb{C}\}$; so S is a scalar, and ω is a positive multiple of $\omega_x | \pi_\rho(\mathfrak{A})$. Since $\omega(I) = \omega_x(I) = 1$, it follows that $\omega = \omega_x | \pi_\rho(\mathfrak{A})$; so

$$\mu = \omega \circ \pi_\rho = \omega_x \circ \pi_\rho = \rho,$$

and ρ is a pure state.

Conversely, suppose that ρ is a pure state of $\mathfrak{A}$. If $\omega_x | \pi_\rho(\mathfrak{A})$ is a convex combination of distinct states μ and ν of $\pi_\rho(\mathfrak{A})$, then ρ $(=\omega_x \circ \pi_\rho)$ is a convex combination of the distinct states $\mu \circ \pi_\rho$ and $\nu \circ \pi_\rho$ of $\mathfrak{A}$, contrary to the assumption that ρ is pure. Thus $\omega_x | \pi_\rho(\mathfrak{A})$ is a pure state of $\pi_\rho(\mathfrak{A})$.

Suppose that S' is a positive element of the unit ball of $\pi_\rho(\mathfrak{A})'$, and let $y = (S')^{1/2}x$. Then, with A a positive element of $\pi_\rho(\mathfrak{A})$,

$$0 \leq \omega_y(A) = \langle AS'x, x\rangle \leq \langle Ax, x\rangle = \omega_x(A);$$

so that $\omega_y | \pi_\rho(\mathfrak{A})$ and $(\omega_x - \omega_y) | \pi_\rho(\mathfrak{A})$ are non-negative multiples $a\sigma$ and $b\tau$, respectively, of states σ and τ of $\pi_\rho(\mathfrak{A})$. Moreover,

$$\omega_x | \pi_\rho(\mathfrak{A}) = a\sigma + b\tau;$$

and by evaluating both sides of this equation at I, we obtain $a + b = 1$. Since $\omega_x | \pi_\rho(\mathfrak{A})$ is a pure state, it now follows that

$$\omega_y | \pi_\rho(\mathfrak{A}) = a\sigma = a\omega_x | \pi_\rho(\mathfrak{A}).$$

For A and B in $\pi_\rho(\mathfrak{A})$,

$$\begin{aligned}\langle S'Ax, Bx\rangle &= \langle B^*A(S')^{1/2}x, (S')^{1/2}x\rangle = \langle B^*Ay, y\rangle \\ &= \omega_y(B^*A) = a\omega_x(B^*A) = a\langle Ax, Bx\rangle.\end{aligned}$$

Since $\pi_\rho(\mathfrak{A})x$ is dense in $\mathscr{H}_\rho$, it now follows that $S' = aI$. Accordingly, the positive part of the unit ball of $\pi_\rho(\mathfrak{A})'$, and hence the whole of $\pi_\rho(\mathfrak{A})'$, consists of scalars; and π_ρ is irreducible. ∎

10.2.4. Corollary. *If A is a non-zero element of a C^*-algebra $\mathfrak{A}$, there is an irreducible representation π of $\mathfrak{A}$ such that $\pi(A) \neq 0$.*

Proof. From Proposition 4.5.5 there is a pure state ρ of $\mathfrak{A}$ such that $\pi_\rho(A) \neq 0$. ■

10.2.5. COROLLARY. *If $\pi: \mathfrak{A} \to \mathscr{B}(\mathscr{H}_\pi)$ is an irreducible representation of a C^*-algebra $\mathfrak{A}$ and x is a unit vector in $\mathscr{H}_\pi$, then the state $\omega_x \circ \pi$ of $\mathfrak{A}$ is pure.*

Proof. Since π is irreducible, x is a cyclic vector for π. By Proposition 4.5.3 (the essential uniqueness of the GNS construction), π is equivalent to the representation π_ρ associated with the state $\rho = \omega_x \circ \pi$. Irreducibility of π implies that π_ρ is irreducible, and ρ is pure by Theorem 10.2.3. ■

10.2.6. THEOREM. *If ρ, τ are pure states of a C^*-algebra $\mathfrak{A}$, and π_ρ, π_τ are the corresponding irreducible representations obtained by means of the GNS construction, then π_ρ and π_τ are equivalent if and only if there is a unitary U in $\mathfrak{A}$ such that*

$$\rho(A) = \tau(U^*AU) \qquad (A \in \mathfrak{A}).$$

When this condition is satisfied, U can be chosen in the form $\exp iH$, *with H self-adjoint in $\mathfrak{A}$.*

Proof. If there is a unitary U in $\mathfrak{A}$ that satisfies the stated equation,

$$\rho(A) = \tau(U^*AU) = \langle \pi_\tau(U^*AU)x_\tau, x_\tau \rangle = \langle \pi_\tau(A)x, x \rangle,$$

where x is the unit vector $\pi_\tau(U)x_\tau$ in $\mathscr{H}_\tau$. Hence $\rho = \omega_x \circ \pi_\tau$, and x is a cyclic vector for π_τ since π_τ is irreducible. From the (essential) uniqueness of the GNS construction (Proposition 4.5.3), π_τ is equivalent to π_ρ.

Conversely, suppose that π_ρ and π_τ are equivalent, and let W be an isomorphism from $\mathscr{H}_\rho$ onto $\mathscr{H}_\tau$, such that

$$\pi_\rho(A) = W^*\pi_\tau(A)W \qquad (A \in \mathfrak{A}).$$

Since x_τ and Wx_ρ are unit vectors in $\mathscr{H}_\tau$, there is a unitary operator V on $\mathscr{H}_\tau$ such that $Vx_\tau = Wx_\rho$. Since π_τ is irreducible, it follows from Theorem 10.2.1 that there is a unitary U in $\mathfrak{A}$ (with $U = \exp iH$ for some self-adjoint H in $\mathfrak{A}$), such that $\pi_\tau(U)x_\tau = Wx_\rho$. For each A in $\mathfrak{A}$,

$$\begin{aligned} \rho(A) &= \langle \pi_\rho(A)x_\rho, x_\rho \rangle = \langle W^*\pi_\tau(A)Wx_\rho, x_\rho \rangle \\ &= \langle \pi_\tau(A)Wx_\rho, Wx_\rho \rangle = \langle \pi_\tau(A)\pi_\tau(U)x_\tau, \pi_\tau(U)x_\tau \rangle \\ &= \langle \pi_\tau(U^*AU)x_\tau, x_\tau \rangle = \tau(U^*AU). \quad ■ \end{aligned}$$

The remainder of this section is concerned with the relation between states and one-sided ideals of operator algebras. We first recall from Section

4.5 some of the notation and results associated with the GNS construction. Suppose that ρ is a state of a C^*-algebra $\mathfrak{A}$, $\mathcal{N}_\rho$ is its null space $\rho^{-1}(0)$, and $\mathcal{L}_\rho$ is its left kernel,

$$\mathcal{L}_\rho = \{A \in \mathfrak{A} : \rho(A^*A) = 0\}.$$

Then $\mathcal{N}_\rho$ is a closed self-adjoint subspace of $\mathfrak{A}$, $\mathcal{L}_\rho$ is a closed left ideal in $\mathfrak{A}$; and $\mathcal{L}_\rho \subseteq \mathcal{N}_\rho$, since it follows from the Cauchy–Schwarz inequality that ρ vanishes on $\mathcal{L}_\rho$. The equation

$$(1) \qquad \langle A + \mathcal{L}_\rho, B + \mathcal{L}_\rho \rangle = \rho(B^*A) \qquad (A, B \in \mathfrak{A})$$

defines a definite inner product on the quotient space $\mathfrak{A}/\mathcal{L}_\rho$. The completion of $\mathfrak{A}/\mathcal{L}_\rho$, relative to this inner product, is a Hilbert space $\mathcal{H}_\rho$; and the GNS construction produces a representation $\pi_\rho : \mathfrak{A} \to \mathcal{B}(\mathcal{H}_\rho)$ such that

$$\pi_\rho(A)(B + \mathcal{L}_\rho) = AB + \mathcal{L}_\rho \qquad (A, B \in \mathfrak{A}).$$

Suppose that $\mathcal{K}$ is a left ideal in $\mathfrak{A}$. If ρ vanishes on $\mathcal{K}$, then $A^*A \in \mathcal{K}$, hence $\rho(A^*A) = 0$ and $A \in \mathcal{L}_\rho$, for each A in $\mathcal{K}$; so $\mathcal{K} \subseteq \mathcal{L}_\rho$. Conversely, if $\mathcal{K} \subseteq \mathcal{L}_\rho$, then $\mathcal{K} \subseteq \mathcal{N}_\rho = \rho^{-1}(0)$. Hence ρ vanishes on a left ideal $\mathcal{K}$ in $\mathfrak{A}$ if and only if $\mathcal{K} \subseteq \mathcal{L}_\rho$.

The notation just described will be used throughout the remainder of this section.

10.2.7. THEOREM. *If ρ is a pure state of a C^*-algebra $\mathfrak{A}$, and $\mathcal{L}_\rho$ is the left kernel of ρ, the quotient space $\mathfrak{A}/\mathcal{L}_\rho$ is complete relative to the inner product defined by* (1).

Proof. Since π_ρ is (topologically, hence algebraically) irreducible, while $\mathfrak{A}/\mathcal{L}_\rho$ is a non-zero linear manifold in $\mathcal{H}_\rho$ and is invariant under each operator in $\pi_\rho(\mathfrak{A})$, it follows that $\mathfrak{A}/\mathcal{L}_\rho = \mathcal{H}_\rho$. ∎

For each state ρ of $\mathfrak{A}$, $\mathcal{N}_\rho$ is self-adjoint and contains $\mathcal{L}_\rho$, so $\mathcal{N}_\rho \supseteq \mathcal{L}_\rho + \mathcal{L}_\rho^*$.

10.2.8. THEOREM. *If ρ is a state of a C^*-algebra $\mathfrak{A}$, $\mathcal{N}_\rho$ its null space, and $\mathcal{L}_\rho$ its left kernel, then the following three conditions are equivalent:*

(i) *ρ is pure.*
(ii) *$\mathcal{N}_\rho = \mathcal{L}_\rho + \mathcal{L}_\rho^*$.*
(iii) *ρ is the only state of $\mathfrak{A}$ that vanishes on $\mathcal{L}_\rho$.*

Proof. Suppose first that ρ is pure. With B in $\mathcal{N}_\rho$,

$$\langle I + \mathcal{L}_\rho, B^* + \mathcal{L}_\rho \rangle = \rho(B) = 0;$$

so the projection E, from $\mathscr{H}_\rho$ onto the one-dimensional subspace containing $B^* + \mathscr{L}_\rho$, satisfies

$$E(I + \mathscr{L}_\rho) = 0, \qquad E(B^* + \mathscr{L}_\rho) = B^* + \mathscr{L}_\rho.$$

Since π_ρ is irreducible, it follows from Theorem 10.2.1 that there is a self-adjoint A in $\mathfrak{A}$ such that

$$\pi_\rho(A)(I + \mathscr{L}_\rho) = 0, \qquad \pi_\rho(A)(B^* + \mathscr{L}_\rho) = B^* + \mathscr{L}_\rho;$$

equivalently

$$A \in \mathscr{L}_\rho, \qquad B^* - AB^* \in \mathscr{L}_\rho.$$

Thus $BA \in \mathscr{L}_\rho$, $B - BA \in \mathscr{L}_\rho^*$,

$$B = BA + (B - BA) \in \mathscr{L}_\rho + \mathscr{L}_\rho^*;$$

and therefore $\mathscr{N}_\rho \subseteq \mathscr{L}_\rho + \mathscr{L}_\rho^*$. Since the reverse inclusion has already been noted, $\mathscr{N}_\rho = \mathscr{L}_\rho + \mathscr{L}_\rho^*$. Thus (i) implies (ii).

Suppose next that $\mathscr{N}_\rho = \mathscr{L}_\rho + \mathscr{L}_\rho^*$. If τ is a state of $\mathfrak{A}$, and vanishes on $\mathscr{L}_\rho$, then τ is hermitian and so vanishes also on $\mathscr{L}_\rho^*$. From this, τ vanishes on the null space $\mathscr{L}_\rho + \mathscr{L}_\rho^*$ of ρ, and is therefore a scalar multiple of ρ. Since $\tau(I) = \rho(I) = 1$, it now follows that $\tau = \rho$. Thus (ii) implies (iii).

Finally, suppose that ρ is the only state of $\mathfrak{A}$ that vanishes on $\mathscr{L}_\rho$. If $\rho = a\mu + b\nu$, where μ, ν are states of $\mathfrak{A}$ and a, b are positive real numbers with sum 1,

$$|\mu(A)|^2 \le \mu(A^*A) \le a^{-1}\rho(A^*A) = 0 \qquad (A \in \mathscr{L}_\rho).$$

Hence μ (and, similarly, ν) vanishes on $\mathscr{L}_\rho$, and therefore $\mu = \nu = \rho$. Thus ρ is a pure state, so (iii) implies (i). ■

10.2.9. LEMMA. *If $\mathscr{I}$ and $\mathscr{J}$ are closed left ideals in a C^*-algebra $\mathfrak{A}$, and $\mathscr{I} \subsetneqq \mathscr{J}$, there is a pure state of $\mathfrak{A}$ that vanishes on $\mathscr{I}$ but does not vanish on $\mathscr{J}$.*

Proof. We may suppose that $\mathfrak{A}$ is given, acting on a Hilbert space $\mathscr{H}$, in its universal representation. By Proposition 10.1.5, there are projections E and F in $\mathfrak{A}^-$ such that

$$\mathscr{I}^- = \mathfrak{A}^-E, \qquad \mathscr{I} = \mathfrak{A} \cap \mathfrak{A}^-E,$$

$$\mathscr{J}^- = \mathfrak{A}^-F, \qquad \mathscr{J} = \mathfrak{A} \cap \mathfrak{A}^-F.$$

Since $\mathscr{I} \subsetneqq \mathscr{J}$ and $\mathscr{I}^- \subseteq \mathscr{J}^-$, it is apparent that $\mathfrak{A}^-E \subsetneqq \mathfrak{A}^-F$, so $E < F$. With x a unit vector in the range of the projection $F - E$, ω_x vanishes on $\mathscr{I}$, does not vanish on $\mathscr{J}^-$ since $\omega_x(F) = 1$, and by continuity does not vanish on $\mathscr{J}$.

With $\mathscr{S}$ the state space of $\mathfrak{A}$, the set

$$\mathscr{S}_0 = \{\rho \in \mathscr{S} : \rho | \mathscr{I} = 0\}$$

is convex and weak * compact, and has an element $\omega_x | \mathfrak{A}$ that does not vanish on $\mathscr{J}$. By the Krein–Milman theorem, $\mathscr{S}_0$ is the weak * closed convex hull of its extreme points, and so has at least one extreme point ρ for which $\rho | \mathscr{J} \neq 0$. Moreover, $\rho | \mathscr{I} = 0$ since $\rho \in \mathscr{S}_0$, so it remains only to prove that ρ is a pure state of $\mathfrak{A}$.

Suppose that $\rho = a_1\rho_1 + a_2\rho_2$, where $\rho_1, \rho_2 \in \mathscr{S}$ and a_1, a_2 are positive real numbers with sum 1. For A in $\mathscr{I}$, we have $A^*A \in \mathscr{I}$, and therefore

$$|\rho_j(A)|^2 \leq \rho_j(A^*A) \leq a_j^{-1}\rho(A^*A) = 0.$$

Hence $\rho_j | \mathscr{I} = 0$, equivalently $\rho_j \in \mathscr{S}_0$ $(j = 1, 2)$. Since ρ is an extreme point of $\mathscr{S}_0$, it follows that $\rho_1 = \rho_2 = \rho$. Accordingly, ρ is an extreme point of $\mathscr{S}$; that is, ρ is a pure state. ∎

10.2.10. THEOREM. *Suppose that $\mathfrak{A}$ is a C*-algebra.*

(i) *If ρ is a state of $\mathfrak{A}$, its left kernel $\mathscr{L}_\rho$ is a maximal left ideal of $\mathfrak{A}$ if and only if ρ is pure.*

(ii) *If $\mathscr{K}$ is a maximal left ideal in $\mathfrak{A}$, there is a unique pure state ρ of $\mathfrak{A}$ such that $\mathscr{K} = \mathscr{L}_\rho$.*

(iii) *Each proper closed left ideal in $\mathfrak{A}$ is the intersection of the maximal left ideals containing it.*

Proof. (i) Suppose first that ρ is a state of $\mathfrak{A}$ and $\mathscr{L}_\rho$ is a maximal left ideal in $\mathfrak{A}$. From Lemma 10.2.9, with $\mathscr{I} = \mathscr{L}_\rho$ and $\mathscr{J} = \mathfrak{A}$, there is a pure state τ of $\mathfrak{A}$ that vanishes on $\mathscr{L}_\rho$. Accordingly, $\mathscr{L}_\rho \subseteq \mathscr{L}_\tau$, and the maximality of $\mathscr{L}_\rho$ entails $\mathscr{L}_\rho = \mathscr{L}_\tau$. Thus ρ vanishes on $\mathscr{L}_\tau$; by Theorem 10.2.8 (applied to the pure state τ), it now follows that $\rho = \tau$, so ρ is pure.

Conversely, suppose that ρ is a pure state. If $\mathscr{K}$ is a closed left ideal in $\mathfrak{A}$, and $\mathscr{L}_\rho \subseteq \mathscr{K} \subsetneqq \mathfrak{A}$, it follows from Lemma 10.2.9 (with $\mathscr{I} = \mathscr{K}$ and $\mathscr{J} = \mathfrak{A}$) that there is a pure state τ of $\mathfrak{A}$ that vanishes on $\mathscr{K}$. In particular, τ vanishes on $\mathscr{L}_\rho$, whence $\tau = \rho$ by Theorem 10.2.8. Thus ρ vanishes on $\mathscr{K}$; hence $\mathscr{K} \subseteq \mathscr{L}_\rho$, and therefore $\mathscr{K} = \mathscr{L}_\rho$. This shows that $\mathscr{L}_\rho$ is a maximal left ideal in $\mathfrak{A}$.

(ii), (iii) With $\mathscr{K}$ a proper closed left ideal in $\mathfrak{A}$, let $\mathscr{S}_0$ be the set of pure states ρ of $\mathfrak{A}$ such that $\mathscr{K} \subseteq \mathscr{L}_\rho$, and let

$$\mathscr{J} = \bigcap \{\mathscr{L}_\rho : \rho \in \mathscr{S}_0\}$$

(with the interpretation that $\mathscr{J} = \mathfrak{A}$ if $\mathscr{S}_0$ is empty). It is clear that $\mathscr{J}$ is a closed left ideal in $\mathfrak{A}$, and that $\mathscr{K} \subseteq \mathscr{J}$. If ρ is a pure state of $\mathfrak{A}$ that vanishes on $\mathscr{K}$, then $\mathscr{K} \subseteq \mathscr{L}_\rho$ and hence $\rho \in \mathscr{S}_0$; so $\mathscr{J} \subseteq \mathscr{L}_\rho$, and ρ vanishes also on $\mathscr{J}$. It now follows from Lemma 10.2.9 that

$$\text{(2)} \qquad \mathscr{K} = \mathscr{J} = \bigcap \{\mathscr{L}_\rho : \rho \in \mathscr{S}_0\};$$

and since $\mathscr{K}$ is a proper ideal, $\mathscr{S}_0$ is not empty.

From the preceding paragraph, together with part (i) of the theorem, each proper closed left ideal $\mathscr{K}$ in $\mathfrak{A}$ is the intersection of a family of maximal left ideals; so $\mathscr{K}$ is the intersection of the set of all maximal left ideals that contain it.

With $\mathscr{K}$ a maximal left ideal, it follows from (2) that $\mathscr{K} \subseteq \mathscr{L}_\rho$ for some pure state ρ of $\mathfrak{A}$; and $\mathscr{K} = \mathscr{L}_p$, by maximality. If τ is any pure state of $\mathfrak{A}$ for which $\mathscr{K} = \mathscr{L}_\tau$, then τ vanishes on $\mathscr{K}$ $(=\mathscr{L}_\rho)$, and so $\tau = \rho$, by Theorem 10.2.8. ■

Bibliography: [30, 42, 84].

10.3. Disjoint representations

Suppose that $\mathfrak{A}$ is a C^*-algebra and $\pi: \mathfrak{A} \to \mathscr{B}(\mathscr{H})$ is a representation of $\mathfrak{A}$. If a closed non-zero subspace $\mathscr{K}$ of $\mathscr{H}$ is invariant under each operator in $\pi(\mathfrak{A})$, the mapping

$$A \to \pi(A) | \mathscr{K} : \quad \mathfrak{A} \to \mathscr{B}(\mathscr{K})$$

is a representation of $\mathfrak{A}$ on $\mathscr{K}$. A representation obtained in this way, by composing π with restriction to a subspace $\mathscr{K}$ invariant under $\pi(\mathfrak{A})$, is described as a *subrepresentation* of π. Of course, the orthogonal complement $\mathscr{K}^\perp$ is invariant under $\pi(\mathfrak{A})$, and gives rise to another subrepresentation if $\mathscr{K} \neq \mathscr{H}$.

The concept of subrepresentation is closely related to the notion of direct sum of representations. Suppose that $\mathscr{H}$ is the direct sum of a family $\{\mathscr{H}_a\}$ of Hilbert spaces, and π is the direct sum $\sum_a \oplus \pi_a$ of representations $\pi_a: \mathfrak{A} \to \mathscr{B}(\mathscr{H}_a)$. If $\mathscr{H}_a$ is identified in the usual way with a subspace of $\mathscr{H}$, it is then invariant under each operator in $\pi(\mathfrak{A})$, and π_a is identified with the subrepresentation of π corresponding to $\mathscr{H}_a$. More precisely, in terms of the notation introduced in Section 2.6 (*Direct sums*), there is an isomorphism U_a from $\mathscr{H}_a$ onto a subspace $\mathscr{H}_a'$ of $\mathscr{H}$, with $\mathscr{H}_a'$ invariant under each operator in $\pi(\mathfrak{A})$, and

$$\pi(A) | \mathscr{H}_a' = U_a \pi_a(A) U_a^* \qquad (A \in \mathfrak{A});$$

so that U_a implements an equivalence between π_a and the subrepresentation of π corresponding to $\mathscr{H}_a'$. In a similar way, if $\{\mathscr{K}_a\}$ is a pairwise orthogonal family of closed subspaces of $\mathscr{H}$, such that each $\mathscr{K}_a$ is invariant under $\pi(\mathfrak{A})$

and $\bigvee \mathscr{K}_a = \mathscr{H}$, then the usual isomorphism from $\mathscr{H}$ onto $\sum_a \oplus \mathscr{K}_a$ implements an equivalence between π and the direct sum of its subrepresentations on the subspaces $\mathscr{K}_a$. Thus π can be identified with the direct sum of these subrepresentations.

With $\pi: \mathfrak{A} \to \mathscr{B}(\mathscr{H})$ a representation of the C^*-algebra $\mathfrak{A}$, the closed subspaces of $\mathscr{H}$ invariant under $\pi(\mathfrak{A})$ correspond to projections in the commutant $\pi(\mathfrak{A})'$. With E' such a projection, and $\mathscr{K}$ its range,

$$\pi(A)E' \mid \mathscr{K} = \pi(A) \mid \mathscr{K}, \qquad \pi(A)E' \mid \mathscr{K}^{\perp} = 0,$$

for each A in $\mathfrak{A}$. In view of this it is often convenient to describe the subrepresention of π on the subspace $\mathscr{K}$ as "the representation $A \to \pi(A)E'$." It is understood, in this context, that $\pi(A)E'$ is to be viewed as an operator acting on the range $\mathscr{K}$ of E'.

10.3.1. Definition. Two representations $\varphi: \mathfrak{A} \to \mathscr{B}(\mathscr{H}_\varphi)$, $\psi: \mathfrak{A} \to \mathscr{B}(\mathscr{H}_\psi)$ of a C^*-algebra $\mathfrak{A}$ are said to be *disjoint* if no subrepresentation of φ is equivalent to a subrepresentation of ψ. They are described as *quasi-equivalent* if there is a * isomorphism α from $\varphi(\mathfrak{A})^-$ onto $\psi(\mathfrak{A})^-$, such that $\alpha(\varphi(A)) = \psi(A)$ for each A in $\mathfrak{A}$. ∎

The relation between the two concepts just introduced (as being opposites, in a certain sense) is clarified in Corollary 10.3.4 below, and another characterization of quasi-equivalence is given in Proposition 10.3.13. It is evident that "quasi-equivalence" is an equivalence relation on the set of representations of $\mathfrak{A}$. Equivalent representations are quasi-equivalent. Indeed, if U is an isomorphism from $\mathscr{H}_\varphi$ onto $\mathscr{H}_\psi$, and

$$\psi(A) = U\varphi(A)U^* \qquad (A \in \mathfrak{A}),$$

then $U\varphi(\mathfrak{A})U^* = \psi(\mathfrak{A})$; hence $U\varphi(\mathfrak{A})^- U^* = \psi(\mathfrak{A})^-$. So U implements a * isomorphism α, from $\varphi(\mathfrak{A})^-$ onto $\psi(\mathfrak{A})^-$, and $\alpha(\varphi(A)) = \psi(A)$ when $A \in \mathfrak{A}$.

The concept of quasi-equivalence has already been encountered (though not named) in Section 9.5, since the extension problem considered there is precisely that of determining when two given representations of an abelian C^*-algebra are quasi-equivalent.

10.3.2. Remark. Suppose that $\mathfrak{A}$ is a C^*-algebra and Φ is its universal representation. The universal property of Φ, as set out in Theorem 10.1.12, amounts to the assertion that each representation of $\mathfrak{A}$ is quasi-equivalent to a subrepresentation $A \to \Phi(A)P$ of Φ, P being a projection in the center of $\Phi(\mathfrak{A})^-$, and hence in (the center of) $\Phi(\mathfrak{A})'$. Note also that each cyclic representation is equivalent to a subrepresentation of Φ. For this, recall that Φ is the direct sum of the representations π_ρ that arise through the GNS construction from states ρ of $\mathfrak{A}$. If a representation φ of $\mathfrak{A}$ has a unit cyclic vector x, and ρ is the state $\omega_x \circ \varphi$, it results from the (essential) uniqueness of the GNS construction that φ is equivalent to π_ρ, and hence to a subrepresentation of Φ. ■

10.3.3. THEOREM. *Suppose that $\mathfrak{A}$ is a C^*-algebra, $\pi: \mathfrak{A} \to \mathscr{B}(\mathscr{H}_\pi)$ is a representation of $\mathfrak{A}$, E' and F' are non-zero projections in $\pi(\mathfrak{A})'$ (with central carriers $C_{E'}$, $C_{F'}$) and*

$$\varphi: A \to \pi(A)E', \qquad \psi: A \to \pi(A)F'$$

are the corresponding subrepresentations of π.

(i) *φ and ψ are equivalent if and only if E' and F' are equivalent projections in $\pi(\mathfrak{A})'$.*

(ii) *φ and ψ are quasi-equivalent if and only if $C_{E'} = C_{F'}$.*

(iii) *φ and ψ are disjoint if and only if $C_{E'}C_{F'} = 0$.*

Proof. (i) If φ and ψ are equivalent, there is an isomorphism W, from the Hilbert space $E'(\mathscr{H}_\pi)$ onto $F'(\mathscr{H}_\pi)$, such that

$$\psi(A)W = W\varphi(A) \qquad (A \in \mathfrak{A}).$$

The operator WE' can be regarded as a partial isometry V' acting on $\mathscr{H}_\pi$, with initial and final projections E' and F', respectively. For x in $\mathscr{H}_\pi$ and A in $\mathfrak{A}$,

$$\begin{aligned}\pi(A)V'x &= \pi(A)F'WE'x = \psi(A)WE'x \\ &= W\varphi(A)E'x = W\pi(A)E'x = WE'\pi(A)x = V'\pi(A)x.\end{aligned}$$

Thus $V' \in \pi(\mathfrak{A})'$, and therefore the projections E' and F' are equivalent in $\pi(\mathfrak{A})'$.

Conversely, if $E' \sim F'$, there is a partial isometry V', from E' to F', in $\pi(\mathfrak{A})'$. For each A in $\mathfrak{A}$,

$$\pi(A)F'V' = \pi(A)V'E' = V'\pi(A)E'.$$

By restriction, V' gives rise to an isomorphism W from $E'(\mathscr{H}_\pi)$ onto $F'(\mathscr{H}_\pi)$, and the above equations entail

$$\psi(A)W = W\varphi(A) \qquad (A \in \mathfrak{A});$$

so φ and ψ are equivalent.

(ii) If $C_{E'} = C_{F'}$ ($= P$, say), the mappings

$$\beta: \mathscr{R}P \to \mathscr{R}E' : \pi(\mathfrak{A})^- P \to \pi(\mathfrak{A})^- E' = \varphi(\mathfrak{A})^-,$$

$$\gamma: \mathscr{R}P \to \mathscr{R}F' : \pi(\mathfrak{A})^- P \to \pi(\mathfrak{A})^- F' = \psi(\mathfrak{A})^-$$

are * isomorphisms, by Proposition 5.5.5. Thus $\gamma \circ \beta^{-1}$ is a * isomorphism α from $\varphi(\mathfrak{A})^-$ onto $\psi(\mathfrak{A})^-$. Since

$$\begin{aligned}\alpha(\varphi(A)) &= \gamma \circ \beta^{-1}(\pi(A)E') \\ &= \gamma(\pi(A)P) = \pi(A)F' = \psi(A) \qquad (A \in \mathfrak{A}),\end{aligned}$$

φ and ψ are quasi-equivalent.

Conversely, suppose that φ and ψ are quasi-equivalent, and let α be a * isomorphism, from $\varphi(\mathfrak{A})^-$ $(=\pi(\mathfrak{A})^-E')$ onto $\psi(\mathfrak{A})^-$ $(=\pi(\mathfrak{A})^-F')$ such that $\alpha(\varphi(\mathrm{A})) = \psi(\mathrm{A})$ when $A \in \mathfrak{A}$. Then

$$\alpha(\pi(A)E') = \pi(A)F' \qquad (A \in \mathfrak{A}),$$

and since α is ultraweakly continuous, by Remark 7.4.4,

$$\alpha(PE') = PF' \qquad (P \in \pi(\mathfrak{A})^-).$$

With P a projection in the center of $\pi(\mathfrak{A})'$, the preceding equation applies since $P \in \pi(\mathfrak{A})^-$, and shows that $PE' = 0$ if and only if $PF' = 0$. Thus $C_{E'} = C_{F'}$.

(iii) The representations φ and ψ fail to be disjoint if and only if they have equivalent subrepresentations. Since

$$\varphi(\mathfrak{A})' = (\pi(\mathfrak{A})E')' = E'\pi(\mathfrak{A})'E',$$

subrepresentations of φ have the form

$$A \to \varphi(A)E_0' = \pi(A)E_0',$$

where E_0' is a non-zero projection in $E'\pi(\mathfrak{A})'E'$, that is, a non-zero subprojection of E' in $\pi(\mathfrak{A})'$. From this and the corresponding statement for ψ, together with part (i) of the present theorem, φ and ψ have equivalent subrepresentations precisely when E' and F' have equivalent non-zero subprojections in $\pi(\mathfrak{A})'$. By Proposition 6.1.8, this occurs if and only if $C_{E'}C_{F'} \neq 0$. ■

10.3.4. COROLLARY. *Suppose that φ and ψ are representations of a C^*-algebra $\mathfrak{A}$.*

(i) *φ and ψ are disjoint if and only if they have no quasi-equivalent subrepresentations*

(ii) *φ and ψ are quasi-equivalent if and only if φ has no subrepresentation disjoint from ψ, and ψ has no subrepresentation disjoint from φ.*

Proof. With π the representation $\varphi \oplus \psi$, φ and ψ are (equivalent to) subrepresentations of π. We may therefore assume throughout that

$$\varphi(A) = \pi(A)E', \qquad \psi(A) = \pi(A)F',$$

where π is a representation of $\mathfrak{A}$ and E', F' are non-zero projections in $\pi(\mathfrak{A})'$. Subrepresentations of φ and ψ then have the form

$$\varphi_0 : A \to \pi(A)E'_0, \qquad \psi_0 : A \to \pi(A)F'_0,$$

respectively, where E'_0 ($\leq E'$) and F'_0 ($\leq F'$) are also non-zero projections in $\pi(\mathfrak{A})'$.

(i) It is clear that $C_{E'}C_{F'} = 0$ if and only if it is impossible to choose non-zero projections E'_0 ($\leq E'$) and F'_0 ($\leq F'$) in $\pi(\mathfrak{A})'$ that have the same central carrier. In view of Theorem 10.3.3, this proves (i).

(ii) It is clear that $C_{E'} = C_{F'}$ if and only if

$$C_{E'_0}C_{F'} \neq 0 \neq C_{E'}C_{F'_0}$$

whenever E'_0 ($\leq E'$) and F'_0 ($\leq F'$) are non-zero projections in $\pi(\mathfrak{A})'$. In view of Theorem 10.3.3, this proves (ii). ∎

We now give a criterion for pairwise disjointness of a family $\{\pi_a\}$ of representations of a C^*-algebra $\mathfrak{A}$, in terms of the direct sum $\pi = \sum_a \oplus \pi_a$. Since

$$\pi(A) = \sum_a \oplus \pi_a(A) \in \sum_a \oplus \pi_a(\mathfrak{A})^- \qquad (A \in \mathfrak{A}),$$

if follows that $\pi(\mathfrak{A}) \subseteq \sum_a \oplus \pi_a(\mathfrak{A})^-$, and hence that

$$\pi(\mathfrak{A})^- \subseteq \sum_a \oplus \pi_a(\mathfrak{A})^-$$

In general, the inclusion is strict. From the Kaplansky density theorem, and since π_a maps the unit ball of $\mathfrak{A}$ onto that of $\pi_a(\mathfrak{A})$, it follows that $\sum_a \oplus \pi_a(\mathfrak{A})^-$ is the ultraweak closure of the C^*-algebra consisting of all operators of the form $\sum_a \oplus \pi_a(A_a)$, with $\{A_a\}$ a bounded family of elements of $\mathfrak{A}$. On the other hand, $\pi(\mathfrak{A})^-$ is the ultraweak closure of the C^*-algebra $\pi(\mathfrak{A})$, consisting of all operators $\sum_a \oplus \pi_a(A)$, where the *same* element A of $\mathfrak{A}$ is used in each component.

10.3.5. THEOREM. *A family $\{\pi_a\}$ of representations of a C^*-algebra $\mathfrak{A}$ is pairwise disjoint if and only if*

$$\pi(\mathfrak{A})^- = \sum_a \oplus \pi_a(\mathfrak{A})^-, \tag{1}$$

where $\pi = \sum_a \oplus \pi_a$.

Proof. If $\pi_a(\mathfrak{A})$ acts on the Hilbert space $\mathscr{H}_a$, then $\pi(\mathfrak{A})$ acts on $\mathscr{H} = \sum_a \oplus \mathscr{H}_a$, $\pi(\mathfrak{A})'$ contains the projection E'_a from $\mathscr{H}$ onto the subspace (corresponding to) $\mathscr{H}_a$, and π_a is (equivalent to) the subrepresentation $A \to \pi(A)E'_a$ of π.

With C_a the central carrier of E'_a, it follows from Theorem 10.3.3(iii) that $\{\pi_a\}$ consists of pairwise-disjoint representations if and only if different C_a are mutually orthogonal. Since $\{E'_a\}$ is an orthogonal family of projections with sum I, and $E'_a \le C_a$, the family $\{C_a\}$ is pairwise orthogonal if and only if $E'_a = C_a$ for each a. Now $E'_a = C_a$ if and only if E'_a lies in the center $\mathscr{C}$ of $\pi(\mathfrak{A})'$. Since $E'_a \in \pi(\mathfrak{A})'$ and $\mathscr{C} = \pi(\mathfrak{A})^- \cap \pi(\mathfrak{A})'$, it follows that $E'_a = C_a$ if and only if $E'_a \in \pi(\mathfrak{A})^-$.

The preceding argument shows that $\{\pi_a\}$ consists of pairwise-disjoint representations if and only if each E'_a lies in $\pi(\mathfrak{A})^-$. It remains to show that this last condition is fulfilled if and only if (1) is satisfied.

It is apparent that $E'_a \in \sum_a \oplus \pi_a(\mathfrak{A})^-$, so each E'_a lies in $\pi(\mathfrak{A})^-$ when (1) is satisfied. Conversely, suppose that $\pi(\mathfrak{A})^-$ contains each E'_a. With $\{A_a\}$ a bounded family of elements of $\mathfrak{A}$, and T the operator $\sum_a \oplus \pi_a(A_a)$, $TE'_a = \pi(A_a)E'_a \in \pi(\mathfrak{A})^-$, and thus

$$T = \sum_a TE'_a \in \pi(\mathfrak{A})^-.$$

Since $\sum_a \oplus \pi_a(\mathfrak{A})^-$ is the ultraweak closure of the set of all such operators T, it now follows that

$$\sum_a \oplus \pi_a(\mathfrak{A})^- \subseteq \pi(\mathfrak{A})^-;$$

and this proves (1), since the reverse inclusion has already been established. ∎

10.3.6. Corollary. *If ρ and τ are states of a C^*-algebra $\mathfrak{A}$, and the representations π_ρ and π_τ obtained by the* GNS *construction are disjoint, then* $\|\rho - \tau\| = 2$.

Proof. With $\pi = \pi_\rho \oplus \pi_\tau$, it follows from Theorem 10.3.5 that

$$\pi(\mathfrak{A})^- = \pi_\rho(\mathfrak{A})^- \oplus \pi_\tau(\mathfrak{A})^-.$$

From this, together with the Kaplansky density theorem, there is a net $\{T_a\}$ in the unit ball of $\pi(\mathfrak{A})$ that converges ultraweakly to the element $I \oplus (-I)$ of $\pi_\rho(\mathfrak{A})^- \oplus \pi_\tau(\mathfrak{A})^-$. We can set $T_a = \pi(A_a)$, with A_a in the unit ball of $\mathfrak{A}$, and then

$$\pi_\rho(A_a) \underset{a}{\rightarrow} I, \qquad \pi_\tau(A_a) \underset{a}{\rightarrow} -I,$$

ultraweakly. Thus

$$\begin{aligned} 2 = \|\rho\| + \|\tau\| &\ge \|\rho - \tau\| \ge |\rho(A_a) - \tau(A_a)| \\ &= |\langle \pi_\rho(A_a)x_\rho, x_\rho\rangle - \langle \pi_\tau(A_a)x_\tau, x_\tau\rangle| \\ &\underset{a}{\rightarrow} \|x_\rho\|^2 + \|x_\tau\|^2 = 2, \end{aligned}$$

and therefore $\|\rho - \tau\| = 2$. ∎

We now apply the preceding theory to families of irreducible representations.

10.3.7. PROPOSITION. *Suppose that φ and ψ are irreducible representations of a C^*-algebra $\mathfrak{A}$.*

(i) *φ and ψ are equivalent if and only if they are quasi-equivalent.*
(ii) *φ and ψ are inequivalent if and only if they are disjoint.*

Proof. Since an irreducible representation has no subrepresentation other than itself, it is apparent from Definition 10.3.1 that, for such representations, the concepts of inequivalence and disjointness coincide. This proves (ii).

If φ and ψ are quasi-equivalent, they are not disjoint (Corollary 10.3.4(i)), and are therefore equivalent by part (ii) of the present theorem. This proves (i), since we have already noted that equivalence of representations implies quasi-equivalence. ∎

10.3.8. COROLLARY. *If ρ and τ are pure states of a C^*-algebra $\mathfrak{A}$, and the irreducible representations π_ρ and π_τ obtained from the* GNS *construction are inequivalent, then $\|\rho - \tau\| = 2$.*

Proof. Since π_ρ and π_τ are inequivalent, they are disjoint, by Proposition 10.3.7(ii). Accordingly, the present result is a special case of Corollary 10.3.6. ∎

10.3.9. COROLLARY. *If $\{\pi_a\}$ is a family of irreducible representations of a C^*-algebra $\mathfrak{A}$, no two of which are equivalent, and $\pi = \sum_a \oplus \pi_a$, then*

$$\pi(\mathfrak{A})^- = \sum_a \oplus \mathscr{B}(\mathscr{H}_a),$$

where $\mathscr{H}_a$ is the Hilbert space on which $\pi_a(\mathfrak{A})$ acts.

Proof. Since the π_a are irreducible and pairwise inequivalent, they are pairwise disjoint by Proposition 10.3.7(ii), and $\pi_a(\mathfrak{A})^- = \mathscr{B}(\mathscr{H}_a)$. The present result is therefore an immediate consequence of Theorem 10.3.5. ∎

The universal representation of a C^*-algebra has been used extensively in earlier sections of this chapter. We now introduce another important representation, the *reduced atomic representation*.

By partitioning the irreducible representations of a C^*-algebra $\mathfrak{A}$ into equivalence classes, and selecting one member from each class, we obtain a maximal family $(\pi_a)_{a\in\mathbb{A}}$ of pairwise-inequivalent irreducible representations. Every irreducible representation of $\mathfrak{A}$ is equivalent to exactly one π_a; and we may assume, if we wish, that the index set $\mathbb{A}$ is the set of equivalence classes of irreducible representations. Any other maximal family of pairwise-inequivalent irreducible representations can be indexed by the same set, as $(\varphi_a)_{a\in\mathbb{A}}$, in such a way that φ_a is equivalent to π_a, for each a in $\mathbb{A}$. If the equivalence between φ_a and π_a is implemented by an isomorphism V_a between the underlying Hilbert spaces, then $\sum_a \oplus V_a$ implements an equivalence between the representations $\sum_a \oplus \varphi_a$ and $\sum_a \oplus \pi_a$.

By the *reduced atomic representation* of $\mathfrak{A}$ we mean a representation Ψ of the form $\sum_a \oplus \pi_a$, where $\{\pi_a\}$ is a maximal family of pairwise-inequivalent irreducible representations of $\mathfrak{A}$. Because of its dependence on the choice of the family $\{\pi_a\}$, Ψ is not unique (and should, strictly speaking, be described as "a reduced atomic representation"). However, this degree of ambiguity is unimportant, since it results from the preceding paragraph that any two such representations are equivalent. The main properties of the reduced atomic representation are summarized in the following result.

10.3.10. PROPOSITION. *Suppose that $\mathfrak{A}$ is a C^*-algebra and $\Psi: \mathfrak{A} \to \mathscr{B}(\mathscr{H})$ is its reduced atomic representation. Then Ψ is faithful, and $\mathscr{H}$ can be expressed as a direct sum $\sum_a \oplus \mathscr{H}_a$, in such a way that*

(i) $\Psi(\mathfrak{A})^- = \sum_a \oplus \mathscr{B}(\mathscr{H}_a)$,

(ii) *each irreducible representation of $\mathfrak{A}$ is equivalent to the subrepresentation of Ψ on just one of the subspaces $\mathscr{H}_a$.*

Proof. We have $\Psi = \sum_a \oplus \pi_a$, where $\{\pi_a\}$ is a maximal family of pairwise-inequivalent irreducible representations of $\mathfrak{A}$. Thus $\mathscr{H} = \sum_a \oplus \mathscr{H}_a$, where $\mathscr{H}_a$ is the Hilbert space on which $\pi_a(\mathfrak{A})$ acts. Each irreducible representation of $\mathfrak{A}$ is equivalent to just one π_a, and so to the subrepresentation obtained by composing Ψ with restriction to the corresponding subspace $\mathscr{H}_a$. This proves (ii), and (i) is an immediate consequence of Corollary 10.3.9.

If $A \in \mathfrak{A}$ and $\Psi(A) = 0$, then $\pi_a(A) = 0$ for each a. With π an irreducible representation of $\mathfrak{A}$, π is equivalent to some π_a, hence π has the same kernel as π_a, and so $\pi(A) = 0$. From Corollary 10.2.4, $A = 0$; so Ψ is faithful. ∎

10.3.11. DEFINITION. A *factor* (or *primary*) representation of a C^*-algebra $\mathfrak{A}$ is a representation $\pi: \mathfrak{A} \to \mathscr{B}(\mathscr{H}_\pi)$ such that the von Neumann algebra $\pi(\mathfrak{A})^-$ is a factor. A state ρ of $\mathfrak{A}$ is a *factor* (or *primary*) state if the representation obtained from ρ by the GNS construction is primary. ∎

10.3.12. PROPOSITION. (i) *A representation π of a C^*-algebra $\mathfrak{A}$ is primary if and only if every subrepresentation of π is quasi-equivalent to π.*

(ii) *Two primary representations, φ and ψ, of a C^*-algebra $\mathfrak{A}$ are either disjoint or quasi-equivalent.*

Proof. (i) The von Neumann algebra $\pi(\mathfrak{A})^-$ is a factor (that is, π is primary) if and only if its commutant $\pi(\mathfrak{A})'$ is a factor; and this occurs if and only if every non-zero projection in $\pi(\mathfrak{A})'$ has central carrier I. It now follows from Theorem 10.3.3(ii) that π is primary if and only if every subrepresentation of π is quasi-equivalent to π.

(ii) By Corollary 10.3.4(i), if the primary representations φ and ψ are not disjoint, there are subrepresentations φ_0 (of φ) and ψ_0 (of ψ), with φ_0 quasi-equivalent to ψ_0. It then follows, from part (i) of the present proposition, that all four representations φ, φ_0, ψ_0, ψ are quasi-equivalent. ∎

We conclude this section with another characterization of quasi-equivalence; loosely speaking, it amounts to the assertion that two representations of a C^*-algebra are quasi-equivalent if and only if they have the same normal states.

With each representation φ of a C^*-algebra $\mathfrak{A}$, we can associate a subset

$$\mathcal{N}(\varphi) = \{\omega \circ \varphi : \omega \text{ a normal state of } \varphi(\mathfrak{A})^-\}$$

of the state space of $\mathfrak{A}$. Since the normal states of $\varphi(\mathfrak{A})^-$ separate the points of $\varphi(\mathfrak{A})^-$, the set

$$\{A \in \mathfrak{A} : \rho(A) = 0 \text{ for each } \rho \text{ in } \mathcal{N}(\varphi)\}$$

is precisely the kernel of φ.

10.3.13. Proposition. *Two representations, φ and ψ, of a C^*-algebra $\mathfrak{A}$ are quasi-equivalent if and only if $\mathcal{N}(\varphi) = \mathcal{N}(\psi)$.*

Proof. If φ and ψ are quasi-equivalent, let $\alpha\colon \varphi(\mathfrak{A})^- \to \psi(\mathfrak{A})^-$ be a * isomorphism with the properties set out in Definition 10.3.1. Since both α and its inverse are ultraweakly continuous, the mapping $\omega \to \omega \circ \alpha$ carries the normal states of $\psi(\mathfrak{A})^-$ onto those of $\varphi(\mathfrak{A})^-$. Thus

$$\begin{aligned}\mathcal{N}(\varphi) &= \{\omega_1 \circ \varphi : \omega_1 \text{ a normal state of } \varphi(\mathfrak{A})^-\}\\ &= \{\omega \circ \alpha \circ \varphi : \omega \text{ a normal state of } \psi(\mathfrak{A})^-\}\\ &= \{\omega \circ \psi : \omega \text{ a normal state of } \psi(\mathfrak{A})^-\}\\ &= \mathcal{N}(\psi).\end{aligned}$$

Conversely, if $\mathcal{N}(\varphi) = \mathcal{N}(\psi)$, it follows from the remark preceding Proposition 10.3.13 that φ and ψ have the same kernel. Accordingly, the equation $\alpha_0(\varphi(A)) = \psi(A)$ defines (unambiguously) a * isomorphism α_0 from $\varphi(\mathfrak{A})$ onto $\psi(\mathfrak{A})$. Given a normal state ω of $\psi(\mathfrak{A})^-$, $\omega \circ \psi \in \mathcal{N}(\psi)$ $(=\mathcal{N}(\varphi))$, so $\omega \circ \psi = \omega_1 \circ \varphi$ for some normal state ω_1 of $\varphi(\mathfrak{A})^-$. Moreover,

$$(\omega \circ \alpha_0)(\varphi(A)) = \omega(\psi(A)) = \omega_1(\varphi(A)) \qquad (A \in \mathfrak{A}),$$

and $\omega \circ \alpha_0$ $(=\omega_1)$ is ultraweakly continuous. Thus α_0 (and, similarly, α_0^{-1}) is ultraweakly continuous. From Lemma 10.1.10, α_0 extends to an ultraweakly continuous * homomorphism α from $\varphi(\mathfrak{A})^-$ onto $\psi(\mathfrak{A})^-$, and α_0^{-1} extends to an ultraweakly continuous * homomorphism β from $\psi(\mathfrak{A})^-$ onto $\varphi(\mathfrak{A})^-$. Now $\beta(\alpha(B)) = B$ when $B \in \varphi(\mathfrak{A})$, and by continuity this remains true for all B in $\varphi(\mathfrak{A})^-$. Hence α is one to one, and is a * isomorphism from $\varphi(\mathfrak{A})^-$ onto $\psi(\mathfrak{A})^-$. Since

$$\alpha(\varphi(A)) = \alpha_0(\varphi(A)) = \psi(A),$$

for all A in $\mathfrak{A}$, φ and ψ are quasi-equivalent. ∎

10.3.14. Proposition. *If φ is a primary representation of a C*-algebra $\mathfrak{A}$ and $\tau \in \mathcal{N}(\varphi)$, then φ is quasi-equivalent to the representation π_τ obtained from τ by means of the* GNS *construction.*

Proof. Upon replacing φ by a quasi-equivalent representation, we may suppose that

$$\varphi(A) = \Phi(A)P \qquad (A \in \mathfrak{A}),$$

where $\Phi\colon \mathfrak{A} \to \mathcal{B}(\mathcal{H}_\Phi)$ is the universal representation of $\mathfrak{A}$ and P is a projection in the center of $\Phi(\mathfrak{A})^-$ (see Remark 10.3.2). Since $\tau \in \mathcal{N}(\varphi)$, there is a normal state ω of $\Phi(\mathfrak{A})^-P$ $(=\varphi(\mathfrak{A})^-)$ such that $\tau = \omega \circ \varphi$. At the same time, there is a unit vector x in $\mathcal{H}_\Phi$ such that $\tau = \omega_x \circ \Phi$; and π_τ is (unitarily equivalent to) the representation $A \to \Phi(A)E'$, where E' is the projection (in $\Phi(\mathfrak{A})'$) with range $[\Phi(\mathfrak{A})x]$. Since

$$\omega_x(\Phi(A)) = \tau(A) = \omega(\varphi(A)) = \omega(\Phi(A)P) \qquad (A \in \mathfrak{A}),$$

it follows from ultraweak continuity of ω_x and ω that $\omega_x(B) = \omega(BP)$ for each B in $\Phi(\mathfrak{A})^-$. By taking P for B, we obtain

$$\|Px\|^2 = \omega_x(P) = \omega(P) = \tau(I) = 1 = \|x\|^2.$$

Thus x lies in the range of P, $E' \leq P$, and π_τ is (unitarily equivalent to) a subrepresentation of φ. From Proposition 10.3.12(i), φ and π_τ are quasi-equivalent. ∎

Bibliography: [30, 53]

10.4. Examples

This section is concerned with representations of certain types of C^*-algebras. Through these examples, it illustrates the general theory developed in earlier parts of the present chapter. It has four subsections, each devoted to a class of C^*-algebras: abelian algebras; algebras, acting on a Hilbert space $\mathscr{H}$, and containing the minimal closed two-sided ideal $\mathscr{K}$ in $\mathscr{B}(\mathscr{H})$, the compact operators; $\mathscr{B}(\mathscr{H})$ itself, and its quotient $\mathscr{B}(\mathscr{H})/\mathscr{K}$, the Calkin algebra; and finally, certain inductive limits of finite-dimensional C^*-algebras, the uniformly matricial algebras.

Abelian C^*-algebras. In this subsection we describe the irreducible representations, reduced atomic representation, and universal representation of an abelian C^*-algebra.

We recall from Section 4.4 that the pure states of an abelian C^*-algebra $\mathfrak{A}$ are precisely the non-zero multiplicative linear functionals on $\mathfrak{A}$, and that the set $\mathscr{P}$ of all pure states is a compact Hausdorff space relative to the weak * topology. The function representation

$$A \to \hat{A} : \mathfrak{A} \to C(\mathscr{P}), \tag{1}$$

defined by $\hat{A}(\rho) = \rho(A)$ $(A \in \mathfrak{A}, \rho \in \mathscr{P})$, is a * isomorphism from $\mathfrak{A}$ onto $C(\mathscr{P})$.

If $\pi\colon \mathfrak{A} \to \mathscr{B}(\mathscr{H})$ is an irreducible representation of $\mathfrak{A}$, $\mathscr{B}(\mathscr{H}) = \pi(\mathfrak{A})^-$ and thus $\mathscr{B}(\mathscr{H})$ is abelian, so $\mathscr{H}$ is one-dimensional. From this, $\mathscr{B}(\mathscr{H}) = \{aI : a \in \mathbb{C}\}$; and, for each A in $\mathfrak{A}$, $\pi(A)$ has the form $\rho(A)I$, with $\rho(A)$ a scalar. Since π is a representation, ρ is a non-zero multiplicative functional on $\mathfrak{A}$. Accordingly, each irreducible representation of $\mathfrak{A}$ has the form

$$\pi\colon A \to \rho(A)I : \mathfrak{A} \to \mathscr{B}(\mathscr{H}), \tag{2}$$

with $\mathscr{H}$ a one-dimensional Hilbert space and ρ in $\mathscr{P}$. Conversely, given such ρ and $\mathscr{H}$, it is apparent that the mapping π, defined by (2), is an irreducible representation of $\mathfrak{A}$.

With $\mathscr{H}_1$, $\mathscr{H}_2$ one-dimensional Hilbert spaces, and ρ_1, ρ_2 in $\mathscr{P}$, we assert that the two corresponding irreducible representations

$$\pi_j\colon A \to \rho_j(A)I : \mathfrak{A} \to \mathscr{B}(\mathscr{H}_j)$$

$(j = 1, 2)$ are equivalent if and only if $\rho_1 = \rho_2$. Indeed, if $\rho_1 = \rho_2$, each isomorphism U from $\mathscr{H}_1$ onto $\mathscr{H}_2$ implements an equivalence between π_1 and π_2. Conversely, if π_1 and π_2 are equivalent, there is an isomorphism U from $\mathscr{H}_1$ onto $\mathscr{H}_2$, such that

$$\begin{aligned}\rho_2(A)I = \pi_2(A) &= U\pi_1(A)U^* \\ &= U(\rho_1(A)I)U^* = \rho_1(A)I \qquad (A \in \mathfrak{A}),\end{aligned}$$

and thus $\rho_1 = \rho_2$.

The preceding paragraphs show that the irreducible representations of $\mathfrak{A}$ act on one-dimensional Hilbert spaces, and that there is a natural one-to-one correspondence between the equivalence classes of irreducible representations and the pure states of $\mathfrak{A}$. These results can be regarded as illustrations of the general theory developed in Section 10.2. We noted, there, that each pure state ρ of $\mathfrak{A}$ gives rise to an irreducible representation π_ρ, through the GNS construction, and that every irreducible representation of $\mathfrak{A}$ is equivalent to one obtained in this way. With ρ and τ two pure states of $\mathfrak{A}$, π_ρ and π_τ are equivalent if and only if there is a unitary element U of $\mathfrak{A}$ such that

$$\rho(A) = \tau(U^*AU) \qquad (A \in \mathfrak{A}),$$

by Theorem 10.2.6. Since $\mathfrak{A}$ is abelian, this last condition reduces to the assertion that $\rho = \tau$; so π_ρ and π_τ are inequivalent when $\rho \neq \tau$. Since ρ is multiplicative,

$$\rho(A^*A) = \rho(A^*)\rho(A) = |\rho(A)|^2 \qquad (A \in \mathfrak{A}),$$

so the left kernel

$$\mathscr{L}_\rho = \{A \in \mathfrak{A} : \rho(A^*A) = 0\}$$

coincides with the null space $\mathscr{N}_\rho$ of ρ, and thus has codimension one in $\mathfrak{A}$. The Hilbert space $\mathscr{H}_\rho$, on which $\pi_\rho(\mathfrak{A})$ acts, is $\mathfrak{A}/\mathscr{L}_\rho$ (with a suitable inner product), and is therefore one dimensional. With A, B in $\mathfrak{A}$

$$\rho(AB - \rho(A)B) = \rho(AB) - \rho(A)\rho(B) = 0,$$

so $AB - \rho(A)B \in \mathscr{L}_\rho$. Hence

$$\pi_\rho(A)(B + \mathscr{L}_\rho) = AB + \mathscr{L}_\rho = \rho(A)B + \mathscr{L}_\rho = \rho(A)(B + \mathscr{L}_\rho),$$

and therefore $\pi_\rho(A) = \rho(A)I$.

We show next that the reduced atomic representation of $\mathfrak{A}$ is closely related to the function representation (1). Each bounded complex-valued function f on $\mathscr{P}$ gives rise to a multiplication operator $M(f)$, which acts on the Hilbert space $l_2(\mathscr{P})$, and is defined by

$$(M(f)x)(\rho) = f(\rho)x(\rho) \qquad (x \in l_2(\mathscr{P}), \quad \rho \in \mathscr{P}).$$

With A in $\mathfrak{A}$, $\hat{A}$ is a continuous (hence bounded) function on $\mathscr{P}$; and the mapping

$$A \to M(\hat{A})$$

is a representation of $\mathfrak{A}$ on $l_2(\mathscr{P})$. We assert that it is the reduced atomic representation. For this, we associate with each ρ in $\mathscr{P}$ the irreducible representation

$$\psi_\rho : A \to \rho(A)I \; : \; \mathfrak{A} \to \mathscr{B}(\mathbb{C})$$

of $\mathfrak{A}$ on the Hilbert space $\mathbb{C}$. From the preceding discussion, $(\psi_\rho)_{\rho \in \mathscr{P}}$ is a maximal family of pairwise-inequivalent irreducible representations of $\mathfrak{A}$, and thus

$$\Psi = \sum_{\rho \in \mathscr{P}} \oplus \, \psi_\rho$$

is the reduced atomic representation. The space, $\sum_{\rho \in \mathscr{P}} \oplus \, \mathbb{C}$, on which $\Psi(\mathfrak{A})$ acts, is $l_2(\mathscr{P})$. When $x \in l_2(\mathscr{P})$ and $A \in \mathfrak{A}$,

$$\begin{aligned}(\Psi(A)x)(\rho) &= \psi_\rho(A)x(\rho) = \rho(A)x(\rho) \\ &= \hat{A}(\rho)x(\rho) = (M(\hat{A})x)(\rho) \qquad (\rho \in \mathscr{P});\end{aligned}$$

so $\Psi(A) = M(\hat{A})$.

We conclude this subsection by describing the universal representation Φ of $\mathfrak{A}$, in terms of multiplication operators on certain L_2 spaces. For this, let $\mathscr{S}$ be the state space of $\mathfrak{A}$, so that

$$\Phi = \sum_{\rho \in \mathscr{S}} \oplus \, \pi_\rho,$$

with π_ρ the representation obtained from ρ by means of the GNS construction. Let $\mathscr{M}$ be the set of all positive regular Borel measures μ on $\mathscr{P}$ for which $\mu(\mathscr{P}) = 1$. Since the function representation (1) is a * isomorphism, while states of $C(\mathscr{P})$ are in one-to-one correspondence with elements of $\mathscr{M}$ (see Remark 1.7.6), there is a one-to-one mapping $\rho \to \mu_\rho$, from $\mathscr{S}$ onto $\mathscr{M}$, such that

$$\rho(A) = \int_{\mathscr{P}} \hat{A}(\tau) \, d\mu_\rho(\tau) \qquad (A \in \mathfrak{A}).$$

With ρ in $\mathscr{S}$, let μ be μ_ρ, $\mathscr{K}_\mu$ the Hilbert space $L_2(\mathscr{P}, \mu)$, and u_μ the unit vector in $\mathscr{K}_\mu$ defined by

$$u_\mu(\tau) = 1 \qquad (\tau \in \mathscr{P}).$$

When $A \in \mathfrak{A}$, $\hat{A}$ is a continuous (hence bounded and μ-measurable) function on $\mathscr{P}$, and so gives rise to a multiplication operator $M_\mu(A)$, acting on $\mathscr{K}_\mu$, and defined by

$$(M_\mu(A)x)(\tau) = \hat{A}(\tau)x(\tau) \qquad (x \in \mathscr{K}_\mu, \quad \tau \in \mathscr{P}).$$

The mapping $M_\mu: A \to M_\mu(A)$ is a representation of $\mathfrak{A}$ on $\mathscr{K}_\mu$, and

$$\begin{aligned}\langle M_\mu(A)u_\mu, u_\mu\rangle &= \int_{\mathscr{P}} (M_\mu(A)u_\mu)(\tau)\overline{u_\mu(\tau)}\, d\mu(\tau)\\ &= \int_{\mathscr{P}} \hat{A}(\tau)\, d\mu(\tau) = \rho(A) \qquad (A \in \mathfrak{A}).\end{aligned}$$

Moreover, u_μ is a cyclic vector for the representation M_μ, since

$$\begin{aligned}M_\mu(\mathfrak{A})u_\mu &= \{M_\mu(A)u_\mu : A \in \mathfrak{A}\}\\ &= \{\hat{A} : A \in \mathfrak{A}\} = C(\mathscr{P}),\end{aligned}$$

and $C(\mathscr{P})$ is everywhere dense in $L_2(\mathscr{P}, \mu)$. It now follows, from the (essential) uniqueness of the GNS construction (Proposition 4.5.3) that π_ρ is equivalent to M_μ. Accordingly, the universal representation of $\mathfrak{A}$, $\sum_{\rho \in \mathscr{S}} \oplus \pi_\rho$, is equivalent to $\sum_{\mu \in \mathscr{M}} \oplus M_\mu$.

Compact operators. In this subsection, $\mathscr{H}$ is an infinite-dimensional Hilbert space, $\mathscr{I}$ is the set of all bounded linear operators acting on $\mathscr{H}$ that have finite-dimensional range spaces, and $\mathscr{K}$ is the norm closure of $\mathscr{I}$ in $\mathscr{B}(\mathscr{H})$. Elements of $\mathscr{K}$ are described as *compact linear operators* (this terminology being motivated by results set out in Exercises 2.8.20(v) and 2.8.25(i)). Since $\mathscr{I}$ is a proper two-sided ideal in $\mathscr{B}(\mathscr{H})$, the same is true of $\mathscr{K}$. Our purpose, in this subsection, is to study representations of some C^*-algebras related to $\mathscr{K}$.

Since the finite projections in the von Neumann algebra $\mathscr{B}(\mathscr{H})$ are precisely those with finite-dimensional range spaces, it follows from Theorem 6.8.3 that every non-zero two-sided ideal in $\mathscr{B}(\mathscr{H})$ contains $\mathscr{I}$. Accordingly, every non-zero closed two-sided ideal in $\mathscr{B}(\mathscr{H})$ contains $\mathscr{K}$. From this, together with Corollary 4.2.10, $\mathscr{K}$ itself has no closed ideals other than $\{0\}$ or $\mathscr{K}$, since such an ideal in $\mathscr{K}$ would be an ideal also in $\mathscr{B}(\mathscr{H})$. Both $\mathscr{I}$ and $\mathscr{K}$ are self-adjoint, from Proposition 6.8.9, and have weak-operator closure $\mathscr{B}(\mathscr{H})$ by Remark 6.8.10. The projections in $\mathscr{K}$ have finite-dimensional range spaces, and so lie in $\mathscr{I}$. Indeed, if E is a projection in $\mathscr{K}$, we can choose T in $\mathscr{I}$ so that $\|E - T\| < 1$. Since

$$\|E - ETE\| = \|E(E - T)E\| < 1,$$

the restricted operator $ETE \,|\, E(\mathscr{H})$ is invertible and has finite-dimensional range space, and so E is finite dimensional. If $\mathscr{H}$ is separable, $\mathscr{K}$ is the only closed two-sided ideal in $\mathscr{B}(\mathscr{H})$ other than $\{0\}$ and $\mathscr{B}(\mathscr{H})$, by Theorem 6.8.7.

The ideal theory, developed in Section 6.8 for a general von Neumann algebra $\mathscr{R}$, was based on the comparison theory of projections in $\mathscr{R}$. When $\mathscr{R}$ is $\mathscr{B}(\mathscr{H})$, comparison theory is straightforward, and the arguments set out in Section 6.8 provide very simple proofs of the results just stated concerning the ideals $\mathscr{I}$ and $\mathscr{K}$.

10.4.1. PROPOSITION. *If ρ is a bounded linear functional on a subspace $\mathscr{M}$ of $\mathscr{B}(\mathscr{H})$, and $\mathscr{K} \subseteq \mathscr{M}$, then ρ is ultraweakly continuous if and only if $\|\rho\| = \|\rho \mid \mathscr{K}\|$.*

Proof. Suppose first that ρ is ultraweakly continuous. Let $\{E_a\}$ be an increasing net of projections in $\mathscr{K}$, for which $\bigvee E_a = I$; for example, $\{E_a\}$ could be the net of all projections from $\mathscr{H}$ onto finite-dimensional subspaces, ordered by inclusion of their range spaces. For each T in $\mathscr{M}$,

$$TE_a \in \mathscr{K}, \qquad \|TE_a\| \leq \|T\|,$$

$\{TE_a\}$ converges ultraweakly to T, and therefore

$$\rho(T) = \lim_a \rho(TE_a), \qquad |\rho(T)| \leq \sup_a |\rho(TE_a)| \leq \|\rho \mid \mathscr{K}\| \, \|T\|.$$

It follows that $\|\rho\| \leq \|\rho \mid \mathscr{K}\|$, whence $\|\rho\| = \|\rho \mid \mathscr{K}\|$.

In proving the converse implication, we may suppose that $\|\rho\| = \|\rho \mid \mathscr{K}\| = 1$. By the Hahn–Banach theorem, ρ extends without change of norm to a bounded linear functional τ on $\mathscr{B}(\mathscr{H})$; and

$$\|\tau \mid \mathscr{K}\| = \|\rho \mid \mathscr{K}\| = \|\rho\| = \|\tau\| = 1.$$

It now suffices to show that τ is a norm limit of weak-operator continuous linear functionals on $\mathscr{B}(\mathscr{H})$, since this implies ultraweak continuity of τ, and hence of ρ $(=\tau \mid \mathscr{M})$, by Theorem 10.1.15(i).

Since τ is norm continuous and $\mathscr{K}$ is the norm closure of $\mathscr{I}$, $\|\tau \mid \mathscr{I}\| = \|\tau \mid \mathscr{K}\| = 1$. Given any positive ε, we can choose S in the unit ball of $\mathscr{I}$, such that $\tau(S)$ is real and

$$\tau(S) > 1 - \tfrac{1}{2}\varepsilon^2.$$

The range projection E of S^* is finite dimensional, since $S^* \in \mathscr{I}$, and $ES^* = S^*$, $SE = S$. With A in the unit ball of $\mathscr{B}(\mathscr{H})$, let a be a scalar such that

$$|a| = 1, \qquad |\tau(A(I - E))| = a\tau(A(I - E)).$$

Then

$$\begin{aligned}1 - \tfrac{1}{2}\varepsilon^2 + \varepsilon|\tau(A(I-E))| &= 1 - \tfrac{1}{2}\varepsilon^2 + \tau(\varepsilon a A(I-E))\\ &< \tau(S + \varepsilon a A(I-E))\\ &= \tau(SE + \varepsilon a A(I-E))\\ &\le \|SE + \varepsilon a A(I-E)\|\\ &= \|[SE + \varepsilon a A(I-E)][ES^* + \varepsilon\bar{a}(I-E)A^*]\|^{1/2}\\ &= \|SES^* + \varepsilon^2 A(I-E)A^*\|^{1/2}\\ &\le (1+\varepsilon^2)^{1/2} < 1 + \tfrac{1}{2}\varepsilon^2.\end{aligned}$$

Hence

$$|\tau(A - AE)| < \varepsilon \qquad (A \in \mathscr{B}(\mathscr{H}),\quad \|A\| \le 1).$$

The last inequality implies that

$$\|\tau - \tau_\varepsilon\| < \varepsilon, \tag{3}$$

where τ_ε is the linear functional defined on $\mathscr{B}(\mathscr{H})$ by $\tau_\varepsilon(A) = \tau(AE)$. It now suffices to prove that τ_ε is weak-operator continuous.

With $x_1, \ldots, x_n$ an orthonormal basis of the finite-dimensional subspace $E(\mathscr{H})$, and x in $\mathscr{H}$,

$$Ex = \sum_{j=1}^{n} \langle x, x_j\rangle x_j,$$

$$\|AEx\| = \|\sum_{j=1}^{n} \langle x, x_j\rangle Ax_j\| \le \|x\| \sum_{j=1}^{n} \|Ax_j\|.$$

Thus $\|AE\| \le \sum \|Ax_j\|$, and therefore

$$|\tau_\varepsilon(A)| = |\tau(AE)| \le \|AE\| \le \sum_{j=1}^{n} \|Ax_j\|.$$

It now follows that τ_ε is strong-operator continuous, and therefore weak-operator continuous by Lemma 7.1.3. ∎

10.4.2. COROLLARY. *If $\mathscr{L}$ is a subspace of $\mathscr{K}$, each bounded linear functional on $\mathscr{L}$ is ultraweakly continuous.*

Proof. Each bounded linear functional ρ on $\mathscr{L}$ extends to a bounded linear functional $\bar{\rho}$ on $\mathscr{K}$. By Proposition 10.4.1, with $\mathscr{M} = \mathscr{K}$, $\bar{\rho}$ is ultraweakly continuous, whence so is ρ. ∎

10.4.3. PROPOSITION. *If $\mathfrak{A}$ is a C^*-subalgebra of $\mathscr{B}(\mathscr{H})$, and $\mathscr{K} \subseteq \mathfrak{A}$, each bounded linear functional ρ on $\mathfrak{A}$ can be expressed uniquely in the form $\rho_1 + \rho_2$, where ρ_1 is an ultraweakly continuous linear functional on $\mathfrak{A}$, and ρ_2 is a bounded linear functional on $\mathfrak{A}$ that vanishes on $\mathscr{K}$. If ρ is positive, then so are ρ_1 and ρ_2.*

Proof. By the Hahn–Banach theorem, the linear functional $\rho | \mathscr{K}$ on $\mathscr{K}$ extends, without change of norm, to a bounded linear functional ρ_1 on $\mathfrak{A}$. Since $\rho_1 | \mathscr{K} = \rho | \mathscr{K}$, $\rho - \rho_1$ is a bounded linear functional ρ_2 on $\mathfrak{A}$, $\rho_2 | \mathscr{K} = 0$, and $\rho = \rho_1 + \rho_2$. Since

$$\|\rho_1\| = \|\rho | \mathscr{K}\| = \|\rho_1 | \mathscr{K}\|,$$

ρ_1 is ultraweakly continuous by Proposition 10.4.1.

If ρ has another decomposition $\tau_1 + \tau_2$, where τ_1 is ultraweakly continuous and $\tau_2 | \mathscr{K} = 0$, then $\rho_1 - \tau_1$ is ultraweakly continuous, and vanishes on $\mathscr{K}$ since $\rho_1 - \tau_1 = \tau_2 - \rho_2$. From Proposition 10.4.1

$$\|\rho_1 - \tau_1\| = \|(\rho_1 - \tau_1) | \mathscr{K}\| = 0;$$

so $\rho_1 = \tau_1$, and $\rho_2 = \tau_2$.

Finally, suppose that $\rho \geq 0$, and let $\{E_a\}$ be an increasing net of projections in $\mathscr{K}$, for which $\bigvee E_a = I$. For each A in $\mathfrak{A}$,

$$A^*E_aA \in \mathscr{K} \cap \mathfrak{A}^+, \qquad A^*A - A^*E_aA = A^*(I - E_a)A \in \mathfrak{A}^+,$$

and $\{A^*E_aA\}$ converges ultraweakly to A^*A. Since ρ_1 is ultraweakly continuous and $\rho_1 | \mathscr{K} = \rho | \mathscr{K}$,

$$\rho_1(A^*A) = \lim_a \rho_1(A^*E_aA) = \lim_a \rho(A^*E_aA) \geq 0,$$

$$\rho_2(A^*A) = \rho(A^*A) - \rho_1(A^*A) = \lim_a \rho(A^*A - A^*\mathrm{E}_aA) \geq 0.$$

It follows that ρ_1 and ρ_2 are positive when ρ is positive. ■

Under the conditions considered in Proposition 10.4.3, ρ_1 and ρ_2 are the ultraweakly continuous and singular parts of ρ, which occur in the decomposition discussed in Theorem 10.1.15(iii) (see Exercise 10.5.51).

10.4.4. COROLLARY. *If $\mathfrak{A}$ is a C^*-subalgebra of $\mathscr{B}(\mathscr{H})$, $\mathscr{K} \subseteq \mathfrak{A}$, and ρ is a pure state of $\mathfrak{A}$, then either $\rho | \mathscr{K} = 0$, or $\rho = \omega_x | \mathfrak{A}$ for some unit vector x in $\mathscr{H}$.*

Proof. Since ρ extends to a pure state of $\mathscr{B}(\mathscr{H})$, we may assume that $\mathfrak{A} = \mathscr{B}(\mathscr{H})$. From Proposition 10.4.3, ρ is a convex combination $s\rho_1 + t\rho_2$ of states ρ_1 and ρ_2 of $\mathscr{B}(\mathscr{H})$, such that $s\rho_1$ is ultraweakly continuous and $t\rho_2 | \mathscr{K} = 0$.

If $\rho | \mathscr{K} \neq 0$, then $s \neq 0$. Since ρ is an extreme point of the state space of $\mathscr{B}(\mathscr{H})$, it follows that $\rho = \rho_1$, whence ρ is ultraweakly continuous (hence, normal). From Theorem 7.1.12, there is a sequence $\{x_n\}$ of unit vectors in $\mathscr{H}$, and a sequence $\{c_n\}$ of non-negative real numbers with sum 1, such that

$$\rho = \sum_{n=1}^{\infty} c_n \omega_{x_n}.$$

We may assume that $0 < c_1 \leq 1$, and the last equation can then be written in the form

$$\rho = c_1 \omega_{x_1} + (1 - c_1)\tau,$$

with τ a state of $\mathscr{B}(\mathscr{H})$. Since ρ is a pure state, it now follows that $\rho = \omega_{x_1}$. ■

10.4.5. REMARK. We assert that, if $\mathfrak{A}$ is a C^*-subalgebra of $\mathscr{B}(\mathscr{H})$, and $\mathscr{K} \subseteq \mathfrak{A}$, then both possibilities for pure states of $\mathfrak{A}$, as set out in Corollary 10.4.4, do in fact occur. Each vector state of $\mathfrak{A}$ is a pure state, and there are also pure states that vanish on $\mathscr{K}$.

For this note first that, since $\mathfrak{A}^- \supseteq \mathscr{K}^- = \mathscr{B}(\mathscr{H})$, the given representation of $\mathfrak{A}$, acting on $\mathscr{H}$, is irreducible—a fact that can also be proved by an elementary direct argument. With x a unit vector in $\mathscr{H}$, the state $\omega_x | \mathfrak{A}$ is pure by Corollary 10.2.5. Observe next that $\mathscr{K}$ is a closed two-sided ideal in $\mathfrak{A}$, whence $\mathfrak{A}/\mathscr{K}$ is a C^*-algebra. With ψ an irreducible representation of $\mathfrak{A}/\mathscr{K}$ and φ the canonical mapping from $\mathfrak{A}$ onto $\mathfrak{A}/\mathscr{K}$, $\psi \circ \varphi$ is an irreducible representation π of $\mathfrak{A}$, and $\pi(\mathscr{K}) = \{0\}$. For some pure state ρ of $\mathfrak{A}$, π_ρ is equivalent to π, hence $\pi_\rho(\mathscr{K}) = (0)$, and

$$\rho(A) = \langle \pi_\rho(A)x_\rho, x_\rho \rangle = 0 \qquad (A \in \mathscr{K}). \quad ■$$

If $\mathfrak{A}$ is a C^*-subalgebra of $\mathscr{B}(\mathscr{H})$, and $\mathscr{K} \subseteq \mathfrak{A}$, the inclusion mapping $\iota: \mathfrak{A} \to \mathscr{B}(\mathscr{H})$ is an irreducible representation of $\mathfrak{A}$. Our next result shows that, up to equivalence, it is the only faithful irreducible representation.

10.4.6. THEOREM. *If $\mathfrak{A}$ is a C^*-subalgebra of $\mathscr{B}(\mathscr{H})$, $\mathscr{K} \subseteq \mathfrak{A}$, and $\pi: \mathfrak{A} \to \mathscr{B}(\mathscr{H}_\pi)$ is an irreducible representation of $\mathfrak{A}$, then either $\pi(\mathscr{K}) = \{0\}$, or π is equivalent to the given representation of $\mathfrak{A}$ on $\mathscr{H}$.*

Proof. For some pure state ρ of $\mathfrak{A}$, the corresponding representation π_ρ is equivalent to π; so we may assume that $\pi = \pi_\rho$. From Corollary 10.4.4, either $\rho | \mathscr{K} = 0$, or $\rho = \omega_x | \mathfrak{A}$ for some unit vector x in $\mathscr{H}$.

Consider first the case in which $\rho | \mathscr{K} = 0$. With A, B in $\mathfrak{A}$ and K in $\mathscr{K}$, $B^*KA \in \mathscr{K}$ and therefore

$$\langle \pi_\rho(K)\pi_\rho(A)x_\rho, \pi_\rho(B)x_\rho \rangle = \langle \pi_\rho(B^*KA)x_\rho, x_\rho \rangle = \rho(B^*KA) = 0.$$

Since $\pi_\rho(\mathfrak{A})x_\rho$ is everywhere dense in $\mathscr{H}_\rho$, it follows that $\pi_\rho(K) = 0$; so π_ρ vanishes on $\mathscr{K}$.

If $\rho = \omega_x | \mathfrak{A}$ for some unit vector x in $\mathscr{H}$, then $[\mathfrak{A}x] = \mathscr{H}$; and it follows from the (essential) uniqueness of the GNS construction that π_ρ is equivalent to the given representation of $\mathfrak{A}$ on $\mathscr{H}$. ■

10.4.7. THEOREM. *Suppose that $\mathfrak{A}$ is a C^*-subalgebra of $\mathscr{B}(\mathscr{H})$ and $\mathscr{K} \subseteq \mathfrak{A}$. Then each representation π of $\mathfrak{A}$ is equivalent to a direct sum $\pi_1 \oplus \pi_2$, where π_1 is (zero or) a representation of the form*

$$A \to A \otimes I : \mathfrak{A} \to \mathscr{B}(\mathscr{H} \otimes \mathscr{H}_1)$$

for some Hilbert space $\mathscr{H}_1$, and π_2 is (zero or) a representation that vanishes on $\mathscr{K}$.

Proof. We show first that, if $\varphi: \mathfrak{A} \to \mathscr{B}(\mathscr{H}_\varphi)$ is a representation of $\mathfrak{A}$, and $\varphi(\mathscr{K}) \neq \{0\}$, then φ has a subrepresentation equivalent to the given representation

$$\pi_0: A \to A : \mathfrak{A} \to \mathscr{B}(\mathscr{H}).$$

For this, note first that the closed two-sided ideal

$$\{K \in \mathscr{K} : \varphi(K) = 0\}$$

in $\mathscr{K}$ is not the whole of $\mathscr{K}$ (since $\varphi(\mathscr{K}) \neq \{0\}$), and therefore consists of 0 alone. Let E be the projection from $\mathscr{H}$ onto a one-dimensional subspace, so that $E \in \mathscr{K}$ $(\subseteq \mathfrak{A})$, $E \neq 0$, and therefore $\varphi(E) \neq 0$. Let x and y be unit vectors in the range spaces of the projections E and $\varphi(E)$, respectively, and let $\mathscr{L}$ be the closed subspace $[\varphi(\mathfrak{A})y]$ of $\mathscr{H}_\varphi$. Then x is cyclic for π_0, and y is cyclic for the subrepresentation

$$\varphi_0: A \to \varphi(A) | \mathscr{L}$$

of φ. Since E is a one-dimensional projection,

$$EAE = \langle Ax, x \rangle E \qquad (A \in \mathfrak{A}),$$

and

$$\begin{aligned} \omega_y(\varphi_0(A)) &= \langle \varphi(A)y, y \rangle = \langle \varphi(A)\varphi(E)y, \varphi(E)y \rangle \\ &= \langle \varphi(EAE)y, y \rangle = \langle Ax, x \rangle \langle \varphi(E)y, y \rangle \\ &= \langle Ax, x \rangle = \omega_x(A) = \omega_x(\pi_0(A)). \end{aligned}$$

It now follows, from the (essential) uniqueness of the GNS construction, that the subrepresentation φ_0 of φ is equivalent to π_0.

If $\pi(\mathscr{K}) = \{0\}$, the conclusion of the theorem is apparent, with $\pi_1 = 0$ and $\pi_2 = \pi$. We therefore suppose that $\pi(\mathscr{K}) \neq \{0\}$, and deduce from the preceding paragraph that π has at least one subrepresentation equivalent to π_0. Let $(E'_a)_{a \in \mathbb{A}}$ be a maximal orthogonal family of projections in $\pi(\mathfrak{A})'$, such that each of the subrepresentations

$$\varphi_a \colon A \to \pi(A)E'_a$$

of π is equivalent to π_0. With F' the projection $I - \sum E'_a$ in $\pi(\mathfrak{A})'$, π is equivalent to the direct sum

$$(\sum \oplus \varphi_a) \oplus \pi_2,$$

where π_2 is the subrepresentation $A \to \pi(A)F'$ of π. From maximality of the family $\{E'_a\}$, π_2 has no subrepresentation equivalent to π_0, and it follows from the preceding paragraph that $\pi_2(\mathscr{K}) = \{0\}$. Since each φ_a is equivalent to π_0, π is equivalent to the representation

$$A \to \left(\sum_{a \in \mathbb{A}} \oplus A \right) \oplus \pi_2(A).$$

If $\mathscr{H}_1$ is a Hilbert space whose dimension is the cardinality of the index set $\mathbb{A}$, we can identify $\sum_{a \in \mathbb{A}} \oplus \mathscr{H}$ with $\mathscr{H} \otimes \mathscr{H}_1$, as in Remark 2.6.8, in such a way that $\sum_{a \in \mathbb{A}} \oplus A$ corresponds to $A \otimes I$ (see 2.6(17)). Thus π is equivalent to $\pi_1 \oplus \pi_2$, where π_1 is the representation

$$A \to A \otimes I : \mathfrak{A} \to \mathscr{B}(\mathscr{H} \otimes \mathscr{H}_1). \quad \blacksquare$$

10.4.8. Example. We prove that the C^*-algebra $\mathfrak{A}$, consisting of all operators of the form $cI + K$, with K in $\mathscr{K}$ and c in $\mathbb{C}$, has just two equivalence classes of irreducible representations; and we give the form of an arbitrary representation of $\mathfrak{A}$.

To verify that $\mathfrak{A}$, as just defined, is indeed a C^*-subalgebra of $\mathscr{B}(\mathscr{H})$, observe that $\mathfrak{A}$ is a subalgebra of $\mathscr{B}(\mathscr{H})$ since $\mathscr{K}$ is an ideal in $\mathscr{B}(\mathscr{H})$, is self-adjoint since $\mathscr{K}$ is self-adjoint, and is closed, by Corollary 1.5.4, since $\mathscr{K}$ is a closed subspace of $\mathscr{B}(\mathscr{H})$.

Since $\mathfrak{A}$ acts irreducibly on $\mathscr{H}$, the inclusion mapping

$$A \to A : \mathfrak{A} \to \mathscr{B}(\mathscr{H})$$

is a faithful irreducible representation π_1 of $\mathfrak{A}$. With $I_{\mathbb{C}}$ the identity operator on the scalar field $\mathbb{C}$ (considered as a one-dimensional Hilbert space), the equation

$$\pi_2(cI + K) = cI_{\mathbb{C}} \qquad (K \in \mathscr{K}, \quad c \in \mathbb{C})$$

defines an irreducible representation $\pi_2: \mathfrak{A} \to \mathscr{B}(\mathbb{C})$. Since $\pi_2(\mathscr{K}) = \{0\}$, π_2 is not faithful; from this (alternatively, because π_1 and π_2 act on Hilbert spaces of different dimensions) it is apparent that π_1 and π_2 are inequivalent.

We assert that every irreducible representation of $\mathfrak{A}$ is equivalent either to π_1 or to π_2, whence the reduced atomic representation of $\mathfrak{A}$ is

$$\pi_1 \oplus \pi_2 : cI + K \to (cI + K) \oplus cI_{\mathbb{C}} : \mathfrak{A} \to \mathscr{B}(\mathscr{H} \oplus \mathbb{C}).$$

For this, suppose that $\pi: \mathfrak{A} \to \mathscr{B}(\mathscr{H}_\pi)$ is an irreducible representation of $\mathfrak{A}$ that is not equivalent to π_1. By Theorem 10.4.6, $\pi(\mathscr{K}) = \{0\}$, so

$$\pi(cI + K) = \pi(cI) = cI \qquad (K \in \mathscr{K}, \quad c \in \mathbb{C}).$$

Since $\pi(\mathfrak{A})$ consists of scalars and acts irreducibly, $\mathscr{H}_\pi$ is one dimensional. It is apparent that every isomorphism from $\mathscr{H}_\pi$ onto the one-dimensional Hilbert space $\mathbb{C}$, implements an equivalence between π and π_2.

Finally, we assert that every representation of $\mathfrak{A}$ has the form

$$cI + K \to (cI + K) \otimes I \oplus cI : \mathfrak{A} \to \mathscr{B}(\mathscr{H} \otimes \mathscr{H}_1 \oplus \mathscr{H}_2)$$

for suitable Hilbert spaces $\mathscr{H}_1$ and $\mathscr{H}_2$. Indeed, this is an immediate consequence of Theorem 10.4.7, since it is apparent that every representation of $\mathfrak{A}$ that vanishes on $\mathscr{K}$ has the form

$$cI + K \to cI : \mathfrak{A} \to \mathscr{B}(\mathscr{H}_2),$$

for a suitable Hilbert space $\mathscr{H}_2$. ∎

The next two results describe situations in which C^*-algebras larger than $\mathscr{K}$ occur.

10.4.9. PROPOSITION. *If $\mathfrak{A}$ is a C^*-subalgebra of $\mathscr{B}(\mathscr{H})$, the same is true of the set*

$$\mathfrak{A} + \mathscr{K} = \{A + K : A \in \mathfrak{A}, K \in \mathscr{K}\}.$$

Proof. This is a special case of the first conclusion in Corollary 10.1.9. ∎

10.4.10. PROPOSITION. *If $\mathfrak{A}$ is a C^*-algebra that acts irreducibly on $\mathscr{H}$, and $\mathfrak{A} \cap \mathscr{K} \neq \{0\}$, then $\mathscr{K} \subseteq \mathfrak{A}$.*

Proof. Since $\mathscr{K}$ is the norm closure of the ideal $\mathscr{I}$ of all bounded linear operators having finite-dimensional range spaces, it suffices to prove that $\mathscr{I} \subseteq \mathfrak{A}$.

We assert first that $\mathfrak{A}$ contains a non-zero element H of $\mathcal{I} \cap \mathcal{B}(\mathcal{H})^+$. To this end, let S be R^*R, where R is a non-zero element of $\mathfrak{A} \cap \mathcal{K}$, and choose a real number a such that $0 < a < \|S\|$. We can find continuous functions f and g on the interval $[0, \|S\|]$, taking values in $[0,1]$, such that

$$f(s) = 0 \quad (0 \le s \le a), \qquad f(\|S\|) = 1,$$

$$g(0) = 0, \qquad g(s)f(s) = f(s) \quad (0 \le s \le \|S\|).$$

Since f and g are uniform limits of polynomials with zero constant terms, while $S \in \mathcal{K} \cap \mathfrak{A}$, it follows that $f(S)$, $g(S) \in \mathcal{K} \cap \mathfrak{A}^+$. Since $f(S) = g(S)f(S)$, the range projection E of $f(S)$ satisfies $E = g(S)E$, so $E \in \mathcal{K}$. Thus E is finite dimensional, and our assertion is proved, with $H = f(S)$. Note also that $H = EHE$, since H is self-adjoint and has range projection E.

We claim next that $\mathfrak{A}$ contains a one-dimensional projection. For this, observe that H can be regarded as the direct sum of EHE (acting on the finite-dimensional space $E(\mathcal{H})$) and 0 (acting on $E(\mathcal{H})^\perp$), and so has finite spectrum. Let p be a polynomial such that $p(0) = 0$ and p takes the value 1 on the remainder of $sp(H)$. Since $p(H) = Ep(H)E$, $p(H)$ is a subprojection of E, and is thus a finite-dimensional projection E_0 ($\neq 0$) in $\mathfrak{A}$ (as it happens, $E_0 = E$, but we shall not need this fact). Let $x_1, \ldots, x_n$ be an orthonormal basis of $E_0(\mathcal{H})$, and let F be the one-dimensional projection whose range contains x_1. Since $\mathfrak{A}$ acts irreducibly on $\mathcal{H}$, it follows from Theorem 5.4.3 that $\mathfrak{A}$ has an element A such that $Ax_j = Fx_j$ ($j = 1, \ldots, n$); and

$$F = FE_0 = AE_0 \in \mathfrak{A}.$$

With C in $\mathcal{I}$, let $y_1, \ldots, y_m$ be an orthonormal basis of the range space of C. By Theorem 5.4.5, there are unitary operators $U_1, \ldots, U_m$ in $\mathfrak{A}$, such that $U_jx_1 = y_j$ ($j = 1, \ldots, m$). Accordingly, $\mathfrak{A}$ contains the one-dimensional projection $U_jFU_j^*$ whose range contains y_j, and so contains

$$G = \sum_{j=1}^{m} U_jFU_j^*,$$

the range projection of C. By Theorem 5.4.3, $\mathfrak{A}$ has an element A such that $Ay_j = C^*y_j$ ($j = 1, \ldots, m$); and

$$C^*G = AG \in \mathfrak{A}, \qquad C = GC = (C^*G)^* \in \mathfrak{A}.$$

Hence $\mathcal{I} \subseteq \mathfrak{A}$, and therefore $\mathcal{K} \subseteq \mathfrak{A}$. ∎

$\mathcal{B}(\mathcal{H})$ and the Calkin algebra. Suppose that $\mathcal{H}$ is a separable infinite-dimensional Hilbert space, and $\mathcal{K}$ is the closed two-sided ideal in $\mathcal{B}(\mathcal{H})$ consisting of the compact linear operators acting on $\mathcal{H}$. This subsection is concerned with the C^*-algebra $\mathcal{B}(\mathcal{H})$ and its quotient algebra $\mathfrak{A} = \mathcal{B}(\mathcal{H})/\mathcal{K}$,

the *Calkin algebra*. Since $\mathscr{B}(\mathscr{H})$ has no closed two-sided ideals other than $\{0\}$, $\mathscr{K}$, and $\mathscr{B}(\mathscr{H})$ (see Theorem 6.8.7), $\mathfrak{A}$ is a simple C^*-algebra.

From Corollary 10.4.4, the pure states of $\mathscr{B}(\mathscr{H})$ fall into two broad classes, vector states and pure states that vanish on $\mathscr{K}$. Even those that vanish on $\mathscr{K}$ are limits of weak * convergent nets of vector states, by Corollary 4.3.10. From Theorem 10.4.6, all faithful irreducible representations of $\mathscr{B}(\mathscr{H})$ are equivalent to its given representation acting on $\mathscr{H}$, and the remaining irreducible representations vanish on $\mathscr{K}$.

If $\pi: \mathscr{B}(\mathscr{H}) \to \mathscr{B}(\mathscr{H}_\pi)$ is a representation of $\mathscr{B}(\mathscr{H})$, and $\pi(\mathscr{K}) = \{0\}$, the mapping

$$\pi': A + \mathscr{K} \to \pi(A) : \mathfrak{A} \to \mathscr{B}(\mathscr{H}_\pi)$$

is well defined, and is a representation of $\mathfrak{A}$. Accordingly, if φ is the canonical mapping from $\mathscr{B}(\mathscr{H})$ onto the Calkin algebra $\mathfrak{A}$ $(=\mathscr{B}(\mathscr{H})/\mathscr{K})$, the equation $\pi = \pi' \circ \varphi$ gives a one-to-one correspondence between representations π' of $\mathfrak{A}$ and representations π of $\mathscr{B}(\mathscr{H})$ that vanish on $\mathscr{K}$. Two such representations π of $\mathscr{B}(\mathscr{H})$ are equivalent if and only if the two corresponding representations π' of $\mathfrak{A}$ are equivalent; moreover, since $\pi(\mathscr{B}(\mathscr{H})) = \pi'(\mathfrak{A})$, π is irreducible if and only if π' is irreducible.

In view of the two preceding paragraphs, the problem of finding (equivalence classes of) irreducible representations of $\mathscr{B}(\mathscr{H})$ reduces at once to the corresponding question for the Calkin algebra $\mathfrak{A}$. In the results that follow, we obtain some information about representations of these algebras, including the fact that each has exactly 2^c inequivalent irreducible representations, where c is the cardinal of the continuum. A complete analysis of the representations of these algebras is beyond the scope of present knowledge.

10.4.11. PROPOSITION. *If π is a representation of $\mathscr{B}(\mathscr{H})$ on a Hilbert space $\mathscr{H}_\pi$, and $\pi(\mathscr{K}) = \{0\}$, then $\mathscr{H}_\pi$ is not separable.*

Proof. It suffices to construct a family $(E_u)_{u\in\mathbb{R}}$ of commuting projections in $\mathscr{B}(\mathscr{H})$, such that $E_u \notin \mathscr{K}$ but $E_uE_v \in \mathscr{K}$ whenever $u \neq v$. Indeed, since $\mathscr{B}(\mathscr{H})$ has no proper ideal larger than $\mathscr{K}$, the kernel of π is $\mathscr{K}$, and the above conditions entail

$$\pi(E_u) \neq 0, \qquad \pi(E_u)\pi(E_v) = 0 \qquad (u \neq v).$$

Thus $(\pi(E_u))_{u\in\mathbb{R}}$ is an uncountable family of non-zero pairwise orthogonal projections acting on $\mathscr{H}_\pi$, and so $\mathscr{H}_\pi$ is not separable.

With $(x_a)_{a\in\mathbb{A}}$ an orthonormal basis of $\mathscr{H}$, the index set $\mathbb{A}$ is countably infinite; we may assume that $\mathbb{A}$ consists of all points (m, n) in $\mathbb{R}^2$ with integer coordinates. We can associate with each subset $\mathbb{B}$ of $\mathbb{A}$ the projection $E(\mathbb{B})$ onto the closed subspace of $\mathscr{H}$ generated by $\{x_b : b \in \mathbb{B}\}$, and it is clear that $E(\mathbb{B})E(\mathbb{B}') = E(\mathbb{B} \cap \mathbb{B}')$. For each real number u, let S_u be the strip

$$\{(s, t) \in \mathbb{R}^2 : |s + ut| < 1\}$$

in the plane, and let $\mathbb{B}_u = \mathbb{A} \cap S_u$. Then each $\mathbb{B}_u$ is an infinite subset of $\mathbb{A}$ since, given any integer t, there is an integer s such that $|s + tu| < 1$ (and $(s, t) \in \mathbb{B}_u$). However, $\mathbb{B}_u \cap \mathbb{B}_v$ is finite when $u \neq v$, since $S_u \cap S_v$ is bounded. From this, it follows easily that the projections

$$E_u = E(\mathbb{B}_u) \qquad (u \in \mathbb{R})$$

satisfy the conditions required above. ∎

10.4.12. Corollary. *If π is a representation of the Calkin algebra on a Hilbert space $\mathscr{H}_\pi$, then $\mathscr{H}_\pi$ is not separable.*

Proof. This follows by applying Proposition 10.4.11 to the representation $\pi \circ \varphi$ of $\mathscr{B}(\mathscr{H})$, where $\varphi\colon \mathscr{B}(\mathscr{H}) \to \mathscr{B}(\mathscr{H})/\mathscr{K}$ is the quotient mapping. ∎

The following result is a particular case of Theorem 10.4.7.

10.4.13. Proposition. *Each representation π of $\mathscr{B}(\mathscr{H})$ is equivalent to a direct sum $\pi_1 \oplus \pi_2$, where π_1 is (zero or) a representation of the form*

$$A \to A \otimes I : \mathscr{B}(\mathscr{H}) \to \mathscr{B}(\mathscr{H} \otimes \mathscr{H}_1)$$

for some Hilbert space $\mathscr{H}_1$, and π_2 is (zero or) a representation that vanishes on $\mathscr{K}$.

10.4.14. Corollary. *Each representation of $\mathscr{B}(\mathscr{H})$ on a separable Hilbert space is equivalent to one of the form*

$$A \to A \otimes I : \mathscr{B}(\mathscr{H}) \to \mathscr{B}(\mathscr{H} \otimes \mathscr{H}_1)$$

for some separable Hilbert space $\mathscr{H}_1$.

Proof. This follows at once from Propositions 10.4.13 and 10.4.11. ∎

10.4.15. Proposition. *There are exactly 2^c distinct equivalence classes of irreducible representations of $\mathscr{B}(\mathscr{H})$, where c is the cardinal of the continuum.*

Proof. With $\{x_1, x_2, \ldots\}$ an orthonormal basis of $\mathscr{H}$, an element of $\mathscr{B}(\mathscr{H})$ is uniquely determined by a countable family of real numbers, the real and imaginary parts of each entry in its matrix relative to the basis. Hence

$$\text{card}\,\mathscr{B}(\mathscr{H}) \leq c^{\aleph_0} = (2^{\aleph_0})^{\aleph_0} = 2^{\aleph_0\aleph_0} = 2^{\aleph_0} = c.$$

By considering multiples of the identity operator, it follows now that card $\mathscr{B}(\mathscr{H}) = c$. The same argument shows that $\mathscr{B}(\mathscr{H})$ has precisely c self-adjoint, or unitary, elements.

We prove next that the set $\mathscr{P}$ of all pure states of $\mathscr{B}(\mathscr{H})$ has cardinality 2^c. Since a pure state is determined by its (real) values at each of the c self-adjoint elements of $\mathscr{B}(\mathscr{H})$, it is apparent that

$$\text{card}\,\mathscr{P} \leq c^c = (2^{\aleph_0})^c = 2^{\aleph_0 c} = 2^c.$$

To prove the reverse inequality, consider the product space

$$S = \prod_{0 \leq t \leq 1} I_t,$$

where each I_t is a copy of the compact interval $[0, 1]$. With the product topology, S is a compact Hausdorff space; and

$$\text{card}\,S = c^c = 2^c.$$

Moreover, S consists of all functions s: $[0, 1] \to [0, 1]$, and has a countable dense subset $\{s_1, s_2, s_3, \ldots\}$ obtained by enumerating all polynomials with rational coefficients that map $[0, 1]$ into $[0, 1]$. With f in $C(S)$, let T_f be the element of $\mathscr{B}(\mathscr{H})$ whose matrix relative to the orthonormal basis $\{x_n\}$ is diagonal and has $f(s_n)$ in the (n, n) position. The mapping

$$f \to T_f : C(S) \to \mathscr{B}(\mathscr{H})$$

is a * isomorphism, and its range is an abelian C^*-subalgebra $\mathscr{A}$ of $\mathscr{B}(\mathscr{H})$. Since $C(S)$ has 2^c distinct pure states (corresponding to the points of S), the same is true of $\mathscr{A}$. Each pure state of $\mathscr{A}$ extends to a pure state of $\mathscr{B}(\mathscr{H})$, so card $\mathscr{P} \geq 2^c$, whence card $\mathscr{P} = 2^c$.

Let m be the cardinal number of the set of all equivalence classes of irreducible representations of $\mathscr{B}(\mathscr{H})$. We introduce an equivalence relation on $\mathscr{P}$, in which two pure states ρ and τ are equivalent if and only if there is a unitary operator U on $\mathscr{H}$ such that $\rho(A) = \tau(U^*AU)$ for each A in $\mathscr{B}(\mathscr{H})$. By Theorem 10.2.6, there is a one-to-one correspondence between equivalence classes of pure states and equivalence classes of irreducible representations; so $\mathscr{P}$ is partitioned into exactly m equivalence classes. Since there are c unitary operators acting on $\mathscr{H}$, each equivalence class of pure states has at least one, and at most c, members. Thus

$$m \leq \text{card}\,\mathscr{P} = 2^c \leq mc = \max(m, c);$$

and since $c < 2^c$, it follows that $m = 2^c$. ∎

10.4.16. COROLLARY. *There are exactly 2^c distinct equivalence classes of irreducible representations of the Calkin algebra, where c is the cardinal of the continuum.*

Proof. There is a one-to-one correspondence between equivalence classes of irreducible representations of the Calkin algebra, and all but one (consisting of faithful representations) of the equivalence classes of irreducible representations of $\mathscr{B}(\mathscr{H})$. Accordingly, the result follows from Proposition 10.4.15. ∎

Uniformly matricial algebras. A C^*-algebra $\mathfrak{A}$ (with unit I) is said to be *uniformly matricial* if there is a sequence $\{\mathfrak{A}_j\}$ of C^*-subalgebras of $\mathfrak{A}$, and a sequence $\{n_j\}$ of positive integers, such that $\mathfrak{A}_j$ is * isomorphic to the algebra of all $n_j \times n_j$ complex matrices,

$$I \in \mathfrak{A}_1 \subsetneqq \mathfrak{A}_2 \subsetneqq \mathfrak{A}_3 \subsetneqq \cdots,$$

and $\mathfrak{A}$ is the norm closure of $\bigcup \mathfrak{A}_j$. We then describe $\mathfrak{A}$, in more detail, as uniformly matricial *of type* $\{n_j\}$, and refer to the sequence $\{\mathfrak{A}_j\}$ as a *generating nest of type* $\{n_j\}$ *for* $\mathfrak{A}$.

We prove in this subsection that a uniformly matricial C^*-algebra $\mathfrak{A}$ of type $\{n_j\}$ exists if and only if the sequence $\{n_j\}$ is strictly increasing and n_j divides n_{j+1} $(j = 1, 2, \ldots)$. Moreover, when these conditions are satisfied, $\mathfrak{A}$ is unique (up to * isomorphism) and is a simple C^*-algebra. Finally, we give examples of inequivalent irreducible representations of the uniformly matricial algebra of type $\{2^j\}$. This last algebra is frequently known as the *CAR algebra* (or *fermion algebra*), in view of its occurrence, in statistical mechanics, in connection with representations of the canonical anticommutation relations for systems of fermions.

Uniformly matricial algebras (and their representations) are the subject of an extensive literature—in which they are usually termed UHF (uniformly hyperfinite) algebras. By applying algebraic techniques in the finite-dimensional subalgebras that constitute a generating nest, and then carrying out a (perhaps delicate) limiting process, it is possible to solve for uniformly matricial algebras (and for some similar, but more general, algebras) certain problems that are much less tractable in the context of arbitrary C^*-algebras. The present subsection provides only the briefest introduction to the subject, and for further information we refer to Chapter 12.

We require some simple auxiliary results concerning finite-dimensional algebras. With k a positive integer, we shall describe as a *factor of type* I_k any C^*-algebra $\mathfrak{A}$ that is * isomorphic to the algebra $M_k(\mathbb{C})$ of all $k \times k$ complex matrices. This differs (slightly) from our previous usage, in that $\mathfrak{A}$ is here an abstract C^*-algebra, not attached to any particular Hilbert space (but * isomorphic to the algebra of all linear operators on a k-dimensional Hilbert space). A factor $\mathfrak{A}$ of type I_k contains a self-adjoint system $\{E_{rs}: r, s = 1, \ldots, k\}$ of matrix units, and is the linear span of this system; indeed, it suffices to take $E_{rs} = \varphi(e_{rs})$, where $\varphi: M_k(\mathbb{C}) \to \mathfrak{A}$ is a * isomorphism, and e_{rs} is the $k \times k$ matrix with 1 in the (r, s) position and zeros elsewhere. Conversely, if a C^*-algebra $\mathfrak{A}$ contains a self-adjoint system $\{E_{rs}: r, s = 1, \ldots, k\}$ of matrix units, and $\sum E_{rr} = I$, then the linear span of this system is a C^*-subalgebra $\mathscr{B}$ of $\mathfrak{A}$ and is a factor of type I_k, the mapping

$$\varphi: [a_{rs}] \to \sum_{r,s=1}^{k} a_{rs} E_{rs}$$

being a * isomorphism from $M_k(\mathbb{C})$ onto $\mathscr{B}$.

If j, k are positive integers, j divides k, and $k/j = q$, then each $k \times k$ complex matrix has a block decomposition as a $j \times j$ matrix with entries in $M_q(\mathbb{C})$. With $A = [a_{rs}]$ in $M_j(\mathbb{C})$, let $\varphi(A)$ be the element of $M_k(\mathbb{C})$ that has block decomposition $[a_{rs} I_q]$, where I_q is the $q \times q$ identity matrix. Then φ is a * isomorphism from $M_j(\mathbb{C})$ into $M_k(\mathbb{C})$; moreover, if $\{e_{rs}^{(1)}: r, s = 1, \ldots, j\}$ and $\{e_{rs}^{(2)}: r, s = 1, \ldots, k\}$ denote the usual systems of matrix units in $M_j(\mathbb{C})$ and $M_k(\mathbb{C})$, respectively, we have

$$\varphi(e_{rs}^{(1)}) = \sum_{t=1}^{q} e_{(r-1)q+t,(s-1)q+t}^{(2)}.$$

These considerations motivate the following result. We remind the reader of our convention that a * homomorphism φ of one C^*-algebra into another is assumed to map the unit of one onto that of the other (see the discussion preceding Proposition 4.1.1).

10.4.17. PROPOSITION. *Suppose that j, k are positive integers and $\mathfrak{A}_1$, $\mathfrak{A}_2$ are factors of types* I_j, I_k, *respectively.*

(i) *There is a * isomorphism from $\mathfrak{A}_1$ into $\mathfrak{A}_2$ if and only if j divides k.*

(ii) *If $\mathfrak{A}_1$ is a C^*-subalgebra of $\mathfrak{A}_2$, then j divides k; moreover, if $q = k/j$ and $\{E_{rs}^{(1)}: r, s = 1, \ldots, j\}$ is a self-adjoint system of matrix units in $\mathfrak{A}_1$, there is a self-adjoint system $\{E_{rs}^{(2)}: r, s = 1, \ldots, k\}$ of matrix units in $\mathfrak{A}_2$ such that*

$$E_{rs}^{(1)} = \sum_{t=1}^{q} E_{(r-1)q+t,(s-1)q+t}^{(2)} \qquad (r, s = 1, \ldots, j).$$

(iii) *If also $\mathscr{B}_1$ and $\mathscr{B}_2$ are factors of types* I_j, I_k, *respectively, while $\mathscr{B}_1$ is a C^*-subalgebra of $\mathscr{B}_2$ and $\mathfrak{A}_1$ is a C^*-subalgebra of $\mathfrak{A}_2$, then each * isomorphism from $\mathfrak{A}_1$ onto $\mathscr{B}_1$ extends to a * isomorphism from $\mathfrak{A}_2$ onto $\mathscr{B}_2$.*

Proof. (i) If j divides k, there is a * isomorphism from $M_j(\mathbb{C})$ into $M_k(\mathbb{C})$, as noted in the discussion preceding the proposition. Since $\mathfrak{A}_1$ is * isomorphic to $M_j(\mathbb{C})$, and $\mathfrak{A}_2$ to $M_k(\mathbb{C})$, there is a * isomorphism from $\mathfrak{A}_1$ into $\mathfrak{A}_2$.

Conversely, suppose that $\mathfrak{A}_2$ has a C^*-subalgebra $\mathfrak{A}_0$ that is * isomorphic to $\mathfrak{A}_1$, and let $\{E_{rs}^{(1)}: r, s = 1, \ldots, j\}$ be a self-adjoint system of matrix units in $\mathfrak{A}_0$. Since $\mathfrak{A}_2$ is a factor of type I_k, the projection $E_{11}^{(1)}$ in $\mathfrak{A}_2$ is the sum of a finite orthogonal family $\{G_1, \ldots, G_q\}$ of minimal projections in $\mathfrak{A}_2$. We can now find further orthogonal families of q minimal projections in $\mathfrak{A}_2$, $\{G_{q+1}, \ldots, G_{2q}\}$ with sum $E_{22}^{(1)}$, $\{G_{2q+1}, \ldots, G_{3q}\}$ with sum $E_{33}^{(1)}, \ldots,$ $\{G_{jq-q+1}, \ldots, G_{jq}\}$ with sum $E_{jj}^{(1)}$. Indeed, $E_{r1}^{(1)}$ is a partial isometry from $E_{11}^{(1)}$ to $E_{rr}^{(1)}$, so $E_{r1}^{(1)}G_t$ is a partial isometry from G_t to a minimal projection G_{rq-q+t} in $\mathfrak{A}_2$ $(t = 1, \ldots, q)$, and $E_{rr}^{(1)}$ is the sum of the orthogonal family $\{G_{rq-q+1}, \ldots, G_{rq}\}$. Since

$$I = \sum_{r=1}^{j} E_{rr}^{(1)} = \sum_{s=1}^{jq} G_s,$$

while $\mathfrak{A}_2$ is type I_k, it follows that $k = jq$.

(ii) If $\mathfrak{A}_1$ is a C^*-subalgebra of $\mathfrak{A}_2$, the above reasoning applies with $\mathfrak{A}_1$ in place of $\mathfrak{A}_0$. Thus, we have already proved that j divides k; and, given a system $\{E_{rs}^{(1)}: r, s = 1, \ldots, j\}$ of matrix units in $\mathfrak{A}_1$, we continue using the preceding notation, to construct the required matrix units in $\mathfrak{A}_2$.

Let $V_1 = G_1$, and when $2 \le t \le q$ choose a partial isometry V_t, from G_1 to G_t, in $\mathfrak{A}_2$; there is such a V_t, since G_1 and G_t are minimal projections in $\mathfrak{A}_2$. When $2 \le r \le j$ and $1 \le t \le q$, $E_{r1}^{(1)}V_t$ is a partial isometry, V_{rq-q+t}, from G_1 to G_{rq-q+t}, in $\mathfrak{A}_2$. In this way, we obtain a family $\{V_1, \ldots, V_k\}$ of partial isometries in $\mathfrak{A}_2$, such that V_r has initial and final projections G_1 and G_r, respectively, and $V_1 = G_1$. With $E_{rs}^{(2)}$ defined as $V_rV_s^*$, it follows from Lemma 6.6.4 that $\{E_{rs}^{(2)}: r, s = 1, \ldots, k\}$ is a self-adjoint system of matrix units in $\mathfrak{A}_2$, since $\sum G_r = I$. Moreover,

$$\begin{aligned}\sum_{t=1}^{q} E_{rq-q+t,sq-q+t}^{(2)} &= \sum_{t=1}^{q} V_{rq-q+t}V_{sq-q+t}^* \\ &= \sum_{t=1}^{q} E_{r1}^{(1)}V_tV_t^*E_{1s}^{(1)} = \sum_{t=1}^{q} E_{r1}^{(1)}G_tE_{1s}^{(1)} \\ &= E_{r1}^{(1)}E_{11}^{(1)}E_{1s}^{(1)} = E_{rs}^{(1)} \qquad (r, s = 1, \ldots, j).\end{aligned}$$

(iii) Since $\mathfrak{A}_1$ is a C^*-subalgebra of $\mathfrak{A}_2$, we can choose self-adjoint systems of matrix units denoted by $\{E_{rs}^{(1)}: r, s = 1, \ldots, j\}$ for $\mathfrak{A}_1$ and $\{E_{rs}^{(2)}: r, s = 1, \ldots, k\}$ for $\mathfrak{A}_2$, that are related as in (ii). With φ a * isomorphism from $\mathfrak{A}_1$ onto $\mathscr{B}_1$, and $F_{rs}^{(1)} = \varphi(E_{rs}^{(1)})$, $\{F_{rs}^{(1)}: r, s = 1, \ldots, j\}$ is a self-adjoint system of matrix units for the C^*-subalgebra $\mathscr{B}_1$ of $\mathscr{B}_2$. Let $\{F_{rs}^{(2)} : r, s = 1, \ldots, k\}$ be a self-adjoint system of matrix units for $\mathscr{B}_2$, related to the units $F_{rs}^{(1)}$ as in (ii). There is a * isomorphism ψ from $\mathfrak{A}_2$ onto $\mathscr{B}_2$, such that $\psi(E_{rs}^{(2)}) = F_{rs}^{(2)}$ $(r, s = 1, \ldots, k)$. Since

$$\psi(E_{rs}^{(1)}) = \sum_{t=1}^{q} \psi(E_{rq-q+t,sq-q+t}^{(2)})$$

$$= \sum_{t=1}^{q} F_{rq-q+t,sq-q+t}^{(2)} = F_{rs}^{(1)} = \varphi(E_{rs}^{(1)})$$

$(r, s = 1, \ldots, j)$, it follows that the restriction $\psi \mid \mathfrak{A}_1$ is φ. ∎

10.4.18. PROPOSITION. *There is a uniformly matricial C^*-algebra of type $\{n_j\}$ if and only if the sequence $\{n_j\}$ of positive integers is strictly increasing, and n_j divides n_{j+1} $(j = 1, 2, \ldots)$. When these conditions are satisfied, all uniformly matricial algebras of type $\{n_j\}$ are * isomorphic, and are simple C^*-algebras.*

Proof. If $\mathfrak{A}$ is a uniformly matricial algebra of type $\{n_j\}$, it has a generating nest $\{\mathfrak{A}_j\}$, with $\mathfrak{A}_j$ a factor of type I_{n_j}. Since $\mathfrak{A}_j \subsetneqq \mathfrak{A}_{j+1}$, it follows that $n_j < n_{j+1}$, and n_j divides n_{j+1} by Proposition 10.4.17(ii).

Conversely, suppose that $n_j < n_{j+1}$ and n_j divides n_{j+1} $(j = 1, 2, \ldots)$, and let $\mathscr{M}_j$ be the algebra of all $n_j \times n_j$ complex matrices. The set $\mathscr{M}$ of all bounded sequences $\{A_r\}$ with A_r in $\mathscr{M}_r$ $(r = 1, 2, \ldots)$ is a C^*-algebra when provided with pointwise-algebraic operations and involution, and the norm defined by

$$\|\{A_r\}\| = \sup\{\|A_r\| : r = 1, 2, \ldots\}.$$

The subset $\mathscr{K}$ of $\mathscr{M}$, consisting of those sequences $\{A_r\}$ such that $\|A_r\| \to 0$ as $r \to \infty$, is a closed two-sided ideal in $\mathscr{M}$. For each $r = 1, 2, \ldots$, there is a * isomorphism φ_r from $\mathscr{M}_r$ into $\mathscr{M}_{r+1}$, by Proposition 10.4.17(i). The mapping θ_j from $\mathscr{M}_j$ into $\mathscr{M}$, defined by

$$\theta_j(A) = \{0, 0, \ldots, 0, A, \varphi_j(A), \varphi_{j+1}\varphi_j(A), \varphi_{j+2}\varphi_{j+1}\varphi_j(A), \ldots\}$$

(with A in the jth position) is linear, multiplicative, and adjoint preserving; and $\theta_j(I) - I \in \mathscr{K}$. With π the canonical mapping from $\mathscr{M}$ onto $\mathscr{M}/\mathscr{K}$, $\pi\theta_j$ is a * isomorphism from $\mathscr{M}_j$ onto a C^*-subalgebra $\mathfrak{A}_j$ of $\mathscr{M}/\mathscr{K}$. For A in $\mathscr{M}_j$,

$$\theta_j(A) - \theta_{j+1}\varphi_j(A) = \{0, 0, \ldots, 0, A, 0, 0, \ldots\} \in \mathscr{K};$$

so $\pi\theta_j(A) = \pi\theta_{j+1}(\varphi_j(A)) \in \pi\theta_{j+1}(\mathscr{M}_{j+1}) = \mathfrak{A}_{j+1}$, and thus $\mathfrak{A}_j \subseteq \mathfrak{A}_{j+1}$. Since, also, $\mathfrak{A}_j$ is a factor of type I_{n_j}, it follows that the norm closure of $\bigcup \mathfrak{A}_j$ in $\mathscr{M}/\mathscr{K}$ is a uniformly matricial algebra of type $\{n_j\}$.

If $\mathfrak{A}$ and $\mathscr{B}$ are two uniformly matricial algebras of type $\{n_j\}$, we can choose generating nests of type $\{n_j\}$, $\{\mathfrak{A}_j\}$ for $\mathfrak{A}$ and $\{\mathscr{B}_j\}$ for $\mathscr{B}$. Repeated application of Proposition 10.4.17(iii) permits the construction of * isomorphisms $\psi_j\colon \mathfrak{A}_j \to \mathscr{B}_j$ $(j = 1, 2, \ldots)$ such that $\psi_{j+1} | \mathfrak{A}_j = \psi_j$. There is then a unique mapping ψ, from $\bigcup \mathfrak{A}_j$ onto $\bigcup \mathscr{B}_j$, such that $\psi | \mathfrak{A}_j = \psi_j$ $(j = 1, 2, \ldots)$. Now $\psi(I) = I$, and ψ is linear, multiplicative, adjoint preserving, and isometric, since each ψ_j has these properties. Accordingly, ψ extends by continuity to a * isomorphism from $\mathfrak{A}$ onto $\mathscr{B}$.

Suppose that $\mathscr{I}$ $(\neq \mathfrak{A})$ is a closed two-sided ideal in $\mathfrak{A}$, and $\pi\colon \mathfrak{A} \to \mathfrak{A}/\mathscr{I}$ is the canonical * homomorphism. Since $\mathfrak{A}_j$ is a simple C^*-algebra, its closed two-sided ideal $\mathscr{I} \cap \mathfrak{A}_j$ is $\{0\}$ or $\mathfrak{A}_j$. However, $I \notin \mathscr{I}$, so $\mathscr{I} \cap \mathfrak{A}_j = \{0\}$. From this, the restriction $\pi | \mathfrak{A}_j$ is a * isomorphism, and is therefore isometric. Hence $\pi | \bigcup \mathfrak{A}_j$ is isometric, and by continuity, π is isometric throughout $\mathfrak{A}$. Thus the kernel $\mathscr{I}$ of π is $\{0\}$, and $\mathfrak{A}$ is a simple C^*-algebra. ∎

10.4.19. EXAMPLE. We produce examples of irreducible representations of the CAR algebra $\mathfrak{A}$ (the uniformly matricial algebra of type $\{2^j\}$). We show that our construction gives c inequivalent irreducible representations, where c is the cardinal of the continuum.

If $\{\mathfrak{A}_j\}$ is a generating nest of type $\{2^j\}$ for $\mathfrak{A}$, it follows by repeated application of Proposition 10.4.17(ii) that we can find a self-adjoint system $\{E_{ab}^{(j)} : a, b = 1, \ldots, 2^j\}$ of matrix units in $\mathfrak{A}_j$ $(j = 1, 2, \ldots)$, so that

(4) $$E_{ab}^{(j)} = E_{2a-1,2b-1}^{(j+1)} + E_{2a,2b}^{(j+1)}.$$

In order to construct a representation of $\mathfrak{A}$, we shall produce operators $F_{ab}^{(j)}$, acting on an L_2 space, with algebraic properties resembling those of the matrix units $E_{ab}^{(j)}$. For this, let S be the half-open interval $[0, 1)$, and let D be the subset consisting of dyadic rational numbers (those with denominator 2^j for some non-negative integer j) in S. With addition (modulo 1) as binary operation, S is a group and D a subgroup. For each d in D, let $g_d\colon S \to S$ be translation by d, so that

$$g_d(s) = \begin{cases} s + d & (0 \le s < 1 - d) \\ s + d - 1 & (1 - d \le s < 1). \end{cases}$$

Then $\{g_d : d \in D\}$ is a group G of transformations of S, and preserves the σ-algebra $\mathscr{S}$ of Borel subsets of S. Let m be a σ-finite positive measure on $\mathscr{S}$ that is invariant under G. For $j = 1, 2, \ldots$, we can define bounded linear operators $F_{ab}^{(j)}$ $(a, b = 1, \ldots, 2^j)$, acting on the Hilbert space L_2 $(= L_2(S, \mathscr{S}, m))$, by

$$(F_{ab}^{(j)}x)(s) = \begin{cases} x(s + (b - a)/2^j) & ((a - 1)/2^j \le s < a/2^j) \\ 0 & \text{elsewhere on } S. \end{cases}$$

Indeed, $F_{aa}^{(j)}$ is the projection corresponding to multiplication by the characteristic function of the interval $[(a-1)2^{-j}, a2^{-j})$; and since m is invariant under translation (modulo 1) by $(b-a)2^{-j}$, $F_{ab}^{(j)}$ is a partial isometry from $F_{bb}^{(j)}$ to $F_{aa}^{(j)}$. Routine calculation shows that $\{F_{ab}^{(j)} : a, b = 1, \ldots, 2^j\}$ is a self-adjoint system of matrix units in $\mathscr{B}(L_2)$, that $\sum F_{aa}^{(j)} = I$, and that

(5) $$F_{ab}^{(j)} = F_{2a-1,2b-1}^{(j+1)} + F_{2a,2b}^{(j+1)}.$$

For $j = 1, 2, \ldots,$ there is a * isomorphism $\varphi_j\colon \mathfrak{A}_j \to \mathscr{B}(L_2)$ such that $\varphi_j(E_{ab}^{(j)}) = F_{ab}^{(j)}$ $(a, b = 1, \ldots, 2^j)$, and it follows from (4), (5) that $\varphi_{j+1} \mid \mathfrak{A}_j = \varphi_j$. There is a unique mapping $\varphi\colon \bigcup \mathfrak{A}_j \to \mathscr{B}(L_2)$, such that $\varphi \mid \mathfrak{A}_j = \varphi_j$ $(j = 1, 2, \ldots)$. Moreover, $\varphi(I) = I$, and φ is linear, multiplicative, adjoint preserving, and isometric, since the same is true of each φ_j. Hence φ extends by continuity to a * isomorphism π from $\mathfrak{A}$ into $\mathscr{B}(L_2)$.

In this way, we obtain a representation π of $\mathfrak{A}$ on L_2, such that

$$\pi(E_{ab}^{(j)}) = F_{ab}^{(j)} \qquad (j = 1, 2, \ldots; \quad a, b = 1, \ldots, 2^j).$$

We now identify the commutant $\pi(\mathfrak{A})'$, and determine the circumstances under which π is irreducible. With f in L_∞ $(=L_\infty(S, \mathscr{S}, m))$, we write M_f for the bounded linear operator (acting on L_2) of multiplication by f. When f is the characteristic function of a Borel set Y, M_f is a projection, which we denote also by $P(Y)$. As noted in Example 5.1.6, the set $\mathscr{A} = \{M_f : f \in L_\infty\}$ is a maximal abelian von Neumann algebra acting on L_2. Moreover, from Example 2.5.12, $\{P(Y) : Y \in \mathscr{S}\}$ is the set of all projections in $\mathscr{A}$, and

(6) $$P(Y_1 \setminus Y_2) = P(Y_1 \setminus Y_1 \cap Y_2) = P(Y_1) - P(Y_1)P(Y_2),$$

(7) $$P\left(\bigcup_1^\infty Y_j\right) = \bigvee_1^\infty P(Y_j),$$

whenever $Y_1, Y_2, \ldots \in \mathscr{S}$. With $\mathscr{S}_0$ the family $\{Y \in \mathscr{S} : P(Y) \in \pi(\mathfrak{A})^-\}$ of subsets of S, it follows from (6) and (7) that $Y_1 \setminus Y_2 \in \mathscr{S}_0$ and $\bigcup_1^\infty Y_n \in \mathscr{S}_0$ whenever $Y_1, Y_2, \ldots \in \mathscr{S}_0$; that is, $\mathscr{S}_0$ is a σ-algebra. Moreover, since

$$P([(a-1)2^{-j}, a2^{-j})) = F_{aa}^{(j)} = \pi(E_{aa}^{(j)}) \in \pi(\mathfrak{A})^-,$$

$\mathscr{S}_0$ contains all the intervals $[(a-1)2^{-j}, a2^{-j})$ $(j = 1, 2, \ldots; a = 1, \ldots, 2^j)$, and therefore coincides with the σ-algebra $\mathscr{S}$ of *all* Borel subsets of S. Accordingly, $P(Y) \in \pi(\mathfrak{A})^-$ for all Y in $\mathscr{S}$, $\pi(\mathfrak{A})^-$ contains all the projections in $\mathscr{A}$ and so contains $\mathscr{A}$, whence $\pi(\mathfrak{A})' \subseteq \mathscr{A}' = \mathscr{A}$. From this, and since $\pi(\mathfrak{A})$ is generated by the operators $F_{ab}^{(j)}$ $(=\pi(E_{ab}^{(j)}))$, it follows that

(8) $$\pi(\mathfrak{A})' = \{M_f : f \in L_\infty, M_f \text{ commutes with each } F_{ab}^{(j)}\}.$$

A routine calculation shows that M_f commutes with a particular $F_{ab}^{(j)}$ if and only if

$$f(s) = f\left(s + \frac{b-a}{2^j}\right)$$

for almost all s in $[(a-1)2^{-j}, a2^{-j})$. Thus M_f commutes with all the operators $F_{ab}^{(j)}$ if and only if $f(s) = f(g_d(s))$ almost everywhere on S, whenever $d \in D$. From this, together with (8),

$$\pi(\mathfrak{A})' = \{M_f : f \in L_\infty, f = f \circ g \text{ almost everywhere on } S\ (g \in G)\}.$$

If G acts ergodically on $(S, \mathscr{S}, m)$, $\pi(\mathfrak{A})'$ consists of scalars and hence π is irreducible, for, by Lemma 8.6.6, an L_∞ function f, such that $f = f \circ g$ almost everywhere $(g \in G)$, is a constant (modulo null functions). If G does not act ergodically on $(S, \mathscr{S}, m)$, there is a Borel subset Y of S for which

$$m(Y) \neq 0 \neq m(S \setminus Y), \qquad m(g(Y) \setminus Y) = 0 \quad (g \in G).$$

With f the characteristic function of Y, $f \in L_\infty$ and $f = f \circ g$ almost everywhere $(g \in G)$; and $P(Y)$ $(= M_f)$ is a projection, not 0 or I, in $\pi(\mathfrak{A})'$. Thus π is irreducible if and only if G acts ergodically on $(S, \mathscr{S}, m)$.

By means of the above construction, a G-invariant measure m on $\mathscr{S}$ gives rise to a representation π of $\mathfrak{A}$. We show next that the set

$$A(m) = \{s \in S : m(\{s\}) \neq 0\}$$

of all "atoms" of m is determined by the equivalence class of the representation π. For this, suppose that $s \in S$. For each positive integer j, we can choose $a(j)$ in $\{1, 2, 3, \ldots, 2^j\}$ so that

$$(a(j) - 1)2^{-j} \leq s < a(j)2^{-j}.$$

With $P_s^{(j)}$ the projection $E_{a(j)a(j)}^{(j)}$ in $\mathfrak{A}$, and $Y_s^{(j)}$ the interval $[(a(j)-1)2^{-j}, a(j)2^{-j})$,

$$\pi(P_s^{(j)}) = F_{a(j)a(j)}^{(j)} = P(Y_s^{(j)}),$$

$$\bigwedge_{j=1}^{\infty} \pi(P_s^{(j)}) = P\left(\bigcap_{j=1}^{\infty} Y_s^{(j)}\right) = P(\{s\}).$$

Now s is an atom for m if and only if $P(\{s\}) \neq 0$; so

$$A(m) = \left\{s \in S : \bigwedge_{j=1}^{\infty} \pi(P_s^{(j)}) \neq 0\right\}. \tag{9}$$

If two G-invariant measures m_1 and m_2 on $\mathscr{S}$ give rise to equivalent representations π_1 and π_2 of $\mathfrak{A}$, and $U\colon L_2(S, \mathscr{S}, m_1) \to L_2(S, \mathscr{S}, m_2)$ is an isomorphism that implements the equivalence, then

$$\bigwedge_{j=1}^{\infty} \pi_2(P_s^{(j)}) = \bigwedge_{j=1}^{\infty} U\pi_1(P_s^{(j)})U^* = U\left(\bigwedge_{j=1}^{\infty} \pi_1(P_s^{(j)})\right)U^*;$$

and from (9), $A(m_1) = A(m_2)$.

We conclude with some examples of measures on $\mathscr{S}$, suitable for use in the above construction. With s in S, let m_s be the measure that has a unit mass at each point of the "orbit" $G(s) = \{g(s) : g \in G\}$, and assigns zero measure to $S \backslash G(s)$. It is apparent that m_s is G-invariant and ergodic, so the corresponding representation π_s of $\mathfrak{A}$ is irreducible. Since $A(m_s) = G(s)$, two points s and t of S give rise to inequivalent representations π_s and π_t, when the orbits $G(s)$ and $G(t)$ are different (of course, $m_s = m_t$ and therefore $\pi_s = \pi_t$, when $G(s) = G(t)$). Since the orbits are just the cosets of the countable subgroup D of S, it is readily verified that there are c distinct orbits; so we have constructed c inequivalent irreducible representations of $\mathfrak{A}$.

When m is Lebesgue measure on $\mathscr{S}$, m is G-invariant, and a slight modification of the argument used in the final paragraph of Example 8.6.12 shows that G acts ergodically on $(S, \mathscr{S}, m)$. The corresponding representation π of $\mathfrak{A}$ is therefore irreducible, and, since $A(m)$ is empty, π is not equivalent to any of the representations π_s $(s \in S)$ constructed above. ■

Bibliography: [2, 3, 7, 29, 49, 73, 81].

10.5. Exercises

10.5.1. Suppose that β is a * automorphism of a C^*-algebra $\mathfrak{A}$, $\mathscr{S}$ is the state space of $\mathfrak{A}$, $\Phi\colon \mathfrak{A} \to \mathscr{B}(\mathscr{H}_\Phi)$ is the universal representation of $\mathfrak{A}$, and $\pi_\rho\colon \mathfrak{A} \to \mathscr{B}(\mathscr{H}_\rho)$ is the representation engendered by ρ, when $\rho \in \mathscr{S}$. Show that

(i) the mapping $\rho \to \rho \circ \beta : \mathscr{S} \to \mathscr{S}$ is one-to-one and has range $\mathscr{S}$;

(ii) for each ρ in $\mathscr{S}$, the representations $\pi_\rho \circ \beta$ and $\pi_{\rho\circ\beta}$ of $\mathfrak{A}$ are equivalent;

(iii) there is a unitary operator U acting on $\mathscr{H}_\Phi$ such that $U\Phi(A)U^* = \Phi(\beta(A))$ for each A in $\mathfrak{A}$;

(iv) the * automorphism $\Phi\beta\Phi^{-1}$ of $\Phi(\mathfrak{A})$ extends uniquely to a * automorphism $\bar{\beta}$ of the von Neumann algebra $\Phi(\mathfrak{A})^-$.

10.5.2. Suppose that $\mathfrak{A}$ is a C^*-algebra, $\mathscr{U}$ is the unitary group of $\mathfrak{A}$, and $\mathscr{U}_e$ $(\subseteq \mathscr{U})$ is the set of all exponential unitary elements of $\mathfrak{A}$; that is,

$$\mathscr{U}_e = \{\exp iH : H = H^* \in \mathfrak{A}\}.$$

Suppose $A \in \mathfrak{A}$, and let $\mathscr{K}$ and $\mathscr{K}_e$ $(\subseteq \mathscr{K})$ be the norm-closed convex hulls of the sets $\{UAU^*: U \in \mathscr{U}\}$ and $\{UAU^*: U \in \mathscr{U}_e\}$, respectively.

(i) By using Theorems 5.2.5 and 5.3.5 and Proposition 5.3.2, prove that $\varphi(\mathscr{U}_e)$ is strong-operator dense in $\varphi(\mathscr{U})$ for each faithful representation φ of $\mathfrak{A}$.

(ii) By using (i) and Proposition 10.1.4, prove that $\mathscr{K}_e = \mathscr{K}$.

10.5.3. When $\mathscr{R}$ is a von Neumann algebra, let $[\text{co}\,\mathscr{U}(\mathscr{R})]^=$ denote the norm closure of the convex hull $\text{co}\,\mathscr{U}(\mathscr{R})$ of the unitary group $\mathscr{U}(\mathscr{R})$ of $\mathscr{R}$.

(i) Show that $A^*, AB \in [\text{co}\,\mathscr{U}(\mathscr{R})]^=$ whenever $A, B \in [\text{co}\,\mathscr{U}(\mathscr{R})]^=$, that $[\text{co}\,\mathscr{U}(\mathscr{R})]^= \subseteq (\mathscr{R})_1$, and that $\text{co}\,\mathscr{U}(\mathscr{R})$ contains each self-adjoint element of $(\mathscr{R})_1$. [*Hint*. For the last result, see the proof of Theorem 4.1.7.]

(ii) Suppose that P is a central projection in $\mathscr{R}$. Show that

$$A + (I - P),\ P + B,\ A + B \in [\text{co}\,\mathscr{U}(\mathscr{R})]^=$$

whenever $A \in [\text{co}\,\mathscr{U}(\mathscr{R}P)]^=$ and $B \in [\text{co}\,\mathscr{U}(\mathscr{R}(I - P))]^=$.

(iii) Show that $\text{co}\,\mathscr{U}(\mathscr{R})$ contains each partial isometry V in $\mathscr{R}$ such that $I - V^*V \sim I - VV^*$.

(iv) Deduce from (i) and (iii) that $\text{co}\,\mathscr{U}(\mathscr{R}) = (\mathscr{R})_1$ when $\mathscr{R}$ is finite.

(v) Suppose that $\mathscr{R}$ is properly infinite, $V \in \mathscr{R}$, and V is an isometry (that is, $V^*V = I$). By choosing projections $E_1, \ldots, E_n$ in $\mathscr{R}$ such that

$$E_1 \sim E_2 \sim \cdots \sim E_n \sim I = E_1 + E_2 + \cdots + E_n,$$

and considering the partial isometries $V_1, \ldots, V_n$, where $V_j = V(I - E_j)$, deduce from (iii) that $V \in [\text{co}\,\mathscr{U}(\mathscr{R})]^=$.

(vi) Suppose that $\mathscr{R}$ is properly infinite, V is a partial isometry in $\mathscr{R}$, and $I - V^*V \precsim I - VV^*$. Show that there is a partial isometry W in $\mathscr{R}$ such that $V + W$ and $V - W$ are isometries. Deduce from (v) that $V \in [\text{co}\,\mathscr{U}(\mathscr{R})]^=$.

(vii) Suppose that $\mathscr{R}$ is properly infinite and V is a partial isometry in $\mathscr{R}$. By applying the comparison theorem to $I - V^*V$ and $I - VV^*$, and using (vi), (i), and (ii), show that $V \in [\text{co}\,\mathscr{U}(\mathscr{R})]^=$. Deduce that $[\text{co}\,\mathscr{U}(\mathscr{R})]^= = (\mathscr{R})_1$.

(viii) Show that $[\text{co}\,\mathscr{U}(\mathscr{R})]^= = (\mathscr{R})_1$ for every von Neumann algebra $\mathscr{R}$.

10.5.4. Suppose that $\mathfrak{A}$ is a C^*-algebra and $\mathscr{U}_e$ is the set

$$\{\exp iH : H = H^* \in \mathfrak{A}\}$$

of all exponential unitary elements of $\mathfrak{A}$. Show that the norm-closed convex hull $[\mathrm{co}\,\mathscr{U}_e]^=$ of $\mathscr{U}_e$ is the unit ball $(\mathfrak{A})_1$. (This result is known as the Russo–Dye theorem. Further information on this subject can be found in Exercises 10.5.91–10.5.100.) [*Hint.* It suffices to prove the corresponding result for $\Phi(\mathfrak{A})$, where Φ is the universal representation of $\mathfrak{A}$. To this end, use Propositions 10.1.4 and 5.3.2 and Theorems 5.2.5 and 5.3.5.]

10.5.5. Suppose that $\mathfrak{A}$ is a C^*-algebra, $A \in \mathfrak{A}$, and $\|A\| < 1$. Let D be the disk $\{\lambda \in \mathbb{C} : |\lambda|\,\|A\| < 1\}$, and define a holomorphic function $f\colon D \to \mathfrak{A}$ by

$$f(\lambda) = (I - AA^*)^{-1/2}(\lambda I + A)(I + \lambda A^*)^{-1}(I - A^*A)^{1/2}.$$

Show that

(i) $Ag(A^*A) = g(AA^*)A$ for every continuous complex-valued function g on $[0, 1]$. [*Hint.* Consider first the case in which g is a polynomial.];

(ii) if $|\lambda| = 1$,

$$(I + \lambda A^*)(\lambda I + A)^{-1} = A^* + (I - A^*A)(\lambda I + A)^{-1},$$

$$(\lambda I + A)^{-1}(I + \lambda A^*) = A^* + (\lambda I + A)^{-1}(I - AA^*);$$

(iii) $f(0) = A$, and $f(\lambda)$ is unitary when $|\lambda| = 1$;

(iv) $$A = \frac{1}{2\pi}\int_0^{2\pi} f(e^{i\theta})\,d\theta.$$

Deduce that the norm-closed convex hull of the unitary group $\mathscr{U}(\mathfrak{A})$ of $\mathfrak{A}$ is the unit ball $(\mathfrak{A})_1$. (Compare this with Exercises 10.5.4 and 10.5.92(v).)

10.5.6. Suppose that $\mathscr{K}$ is a norm-closed two-sided ideal in a C^*-algebra $\mathfrak{A}$ and $\{V_\lambda : \lambda \in \Lambda\}$ is an increasing two-sided approximately identiy for $\mathscr{K}$ (see Proposition 4.2.12 and Exercises 4.6.35 and 4.6.36). Let $\Phi\colon \mathfrak{A} \to \mathscr{B}(\mathscr{H}_\Phi)$ be the universal representation of $\mathfrak{A}$, so that $\Phi(\mathscr{K})$ is a norm-closed two-sided ideal in $\Phi(\mathfrak{A})$, and (see Proposition 10.1.5) let E be a projection in the center of $\Phi(\mathfrak{A})^-$ such that

$$\Phi(\mathscr{K}) = \Phi(\mathfrak{A}) \cap \Phi(\mathfrak{A})^- E = \Phi(\mathfrak{A}) \cap \Phi(\mathfrak{A})E, \qquad \Phi(\mathscr{K})^- = \Phi(\mathfrak{A})^- E.$$

(i) Show that the increasing net $\{\Phi(V_\lambda)\}$ is strong-operator convergent to E.

(ii) Suppose that $\mu \in \Lambda$ and $A_1, \ldots, A_n \in \mathfrak{A}$. Let $\mathscr{J}_\mu$ be the set $\{V_\lambda : \lambda \in \Lambda, \lambda \geq \mu\}$, and let $\mathscr{C}_\mu$ be the set of all finite convex combinations of elements of $\mathscr{J}_\mu$. Show that 0 lies in the strong-operator closure of the set

$$\left\{\sum_{j=1}^{n} \oplus\, \Phi(VA_j - A_jV) : V \in \mathscr{J}_\mu\right\}$$

of operators acting on the Hilbert space $\sum_{j=1}^{n} \oplus \mathscr{H}_{\Phi}$. By reasoning as in the proof of Proposition 10.1.4, deduce that 0 lies in the norm closure of the set

$$\left\{\sum_{j=1}^{n} \oplus \Phi(HA_j - A_jH) : H \in \mathscr{C}_{\mu}\right\}.$$

Hence show that there exists H in $\mathscr{C}_{\mu}$ such that

$$\|HA_j - A_jH\| \leq n^{-1} \qquad (j = 1, \ldots, n).$$

(iii) Let Γ be the set of all triples $(\mathbb{F}, \mu, H)$ in which $\mathbb{F}$ is a finite subset of $\mathfrak{A}$, $\mu \in \Lambda$, $H \in \mathscr{C}_{\mu}$, and $\|HA - AH\| \leq n^{-1}$ for all A in $\mathbb{F}$, where n is the number of elements in $\mathbb{F}$. Show that Γ is directed by the binary relation $\leq$, in which $(\mathbb{F}_1, \mu_1, H_1) \leq (\mathbb{F}_2, \mu_2, H_2)$ if and only if $\mathbb{F}_1 \subseteq \mathbb{F}_2$, $\mu_1 \leq \mu_2$, and $H_1 \leq H_2$.

(iv) Define a mapping $\gamma \to H_{\gamma} : \Gamma \to \mathscr{K}$ as follows: H_{γ} is H if γ is $(\mathbb{F}, \mu, H)$. Show that $\{H_{\gamma}\}$ is an increasing two-sided approximate identity for $\mathscr{K}$ with the additional property that

$$\lim_{\gamma} \|H_{\gamma}A - AH_{\gamma}\| = 0$$

for each A in $\mathfrak{A}$. (An approximate identity with this additional property is described as *quasi-central*.)

10.5.7. Suppose that $\mathscr{H}$ is a Hilbert space, $A_1, \ldots, A_n \in \mathscr{B}(\mathscr{H})^+$, and $A_1 + \cdots + A_n \leq I$.

(i) Show that $\|\sum_{j=1}^{n} A_jx_j\|^2 \leq \sum_{j=1}^{n}\langle A_jx_j, x_j\rangle$ whenever $x_1, \ldots, x_n \in \mathscr{H}$. [*Hint*. Define an inner product on $\mathscr{H} \oplus \mathscr{H} \oplus \cdots \oplus \mathscr{H}$ (n copies) by

$$\langle\{y_1, \ldots, y_n\}, \{z_1, \ldots, z_n\}\rangle = \sum_{j=1}^{n} \langle A_jy_j, z_j\rangle,$$

and apply the Schwarz inequality to the vectors $\{z, z, \ldots, z\}$ and $\{x_1, x_2, \ldots, x_n\}$, where $z = \sum_{j=1}^{n} A_jx_j$.]

(ii) Deduce that

$$\left(\sum_{j=1}^{n} a_jA_j\right)^*\left(\sum_{j=1}^{n} a_jA_j\right) \leq \sum_{j=1}^{n} |a_j|^2 A_j$$

for all complex numbers $a_1, \ldots, a_n$.

10.5.8. Suppose that $\mathscr{R}$ and $\mathscr{S}$ are von Neumann algebras, η is a positive linear mapping from $\mathscr{R}$ into $\mathscr{S}$, and $\|\eta(I)\| \le 1$.

(i) Let $\{E_1, \ldots, E_n\}$ be an orthogonal family of projections in $\mathscr{R}$, with sum I. By using the result of Exercise 10.5.7(ii), with $\eta(E_j)$ in place of A_j, show that

$$\eta\left(\sum_{j=1}^{n} a_j E_j\right)^* \eta\left(\sum_{j=1}^{n} a_j E_j\right) \le \eta\left(\left(\sum_{j=1}^{n} a_j E_j\right)^*\left(\sum_{j=1}^{n} a_j E_j\right)\right)$$

for all complex numbers $a_1, \ldots, a_n$.

(ii) Deduce that $\eta(A)^*\eta(A) \le \eta(A^*A)$ for each normal element A of $\mathscr{R}$.

10.5.9. Suppose that $\mathfrak{A}$ and $\mathscr{B}$ are C^*-algebras, η is a positive linear mapping from $\mathfrak{A}$ into $\mathscr{B}$, and $\|\eta(I)\| \le 1$. Prove that $\eta(A)^*\eta(A) \le \eta(A^*A)$ for each normal element A of $\mathfrak{A}$. (This result is known as the generalized Schwarz inequality.) [*Hint*. Let Φ be the universal representation of $\mathfrak{A}$, and let φ be a faithful representation of $\mathscr{B}$. Show that the mapping

$$\eta_0 = \varphi \circ \eta \circ \Phi^{-1} : \Phi(\mathfrak{A}) \to \varphi(\mathscr{B})$$

extends to a positive linear mapping $\bar{\eta}_0\colon \Phi(\mathfrak{A})^- \to \varphi(\mathscr{B})^-$, and $\|\bar{\eta}_0(I)\| \le 1$. Apply the result of Exercise 10.5.8(ii) to $\bar{\eta}_0$.]

10.5.10. Suppose that $\mathfrak{A}$ and $\mathscr{B}$ are C^*-algebras, and η is a positive linear mapping from $\mathfrak{A}$ into $\mathscr{B}$. Show that η is bounded and $\|\eta\| = \|\eta(I)\|$. [*Hint*. Use Exercises 10.5.4 and 10.5.9.]

10.5.11. (i) Suppose that $\mathfrak{A}$ is a C^*-algebra of operators acting on a Hilbert space $\mathscr{H}$ and VH is the polar decomposition of an element A of $\mathfrak{A}$. Show that $A(H + n^{-1}I)^{-1/2} \in \mathfrak{A}$ for each positive integer n, and deduce that $VH^{1/2} \in \mathfrak{A}$.

(ii) Suppose that $\mathscr{K}$ is a norm-closed left ideal in a C^*-algebra $\mathfrak{A}$. By taking $\mathfrak{A}$ in its universal representation, deduce from (i) and Proposition 10.1.5 that each element of $\mathscr{K}$ can be expressed in the form BC, with B in $\mathscr{K}$ and C in $\mathscr{K}^+$.

10.5.12. Suppose that $\mathfrak{A}$ is a C^*-algebra acting on a Hilbert space $\mathscr{H}$, $\mathfrak{X}^\#$ is the Banach dual space of a Banach space $\mathfrak{X}$, $\eta\colon \mathfrak{A} \to \mathfrak{X}^\#$ is a norm-continuous linear mapping, and η is continuous also relative to the ultraweak topology on $\mathfrak{A}$ and the weak * topology on $\mathfrak{X}^\#$. Show that

(i) for each x in $\mathfrak{X}$, the equation

$$\rho_x(A) = (\eta(A))(x) \qquad (A \in \mathfrak{A})$$

defines an ultraweakly continuous linear functional ρ_x on $\mathfrak{A}$, and $\|\rho_x\| \leq \|\eta\| \, \|x\|$;

(ii) ρ_x extends to an ultraweakly continuous linear functional $\bar{\rho}_x$ on $\mathfrak{A}^-$, and $\|\bar{\rho}_x\| = \|\rho_x\|$;

(iii) for each A in $\mathfrak{A}^-$, the equation

$$(\bar{\eta}(A))(x) = \bar{\rho}_x(A) \qquad (x \in \mathfrak{X})$$

defines an element $\bar{\eta}(A)$ of $\mathfrak{X}^\#$;

(iv) $\bar{\eta}$ is a bounded linear mapping from $\mathfrak{A}^-$ into $\mathfrak{X}^\#$, $\bar{\eta}(A) = \eta(A)$ when $A \in \mathfrak{A}$, $\|\bar{\eta}\| = \|\eta\|$, and $\bar{\eta}$ is continuous relative to the ultraweak topology on $\mathfrak{A}^-$ and the weak * topology on $\mathfrak{X}^\#$.

10.5.13. Suppose that $\mathfrak{A}$ is a C^*-algebra acting on a Hilbert space $\mathscr{H}$, $\mathfrak{X}$ is a Banach space, and the Banach dual space $\mathfrak{X}^\#$ is a Banach $\mathfrak{A}$-module in the sense of Exercise 4.6.66. We describe $\mathfrak{X}^\#$ as a *dual* $\mathfrak{A}$-*module* if, for each A_0 in $\mathfrak{A}$, the linear mappings

$$\rho \to A_0\rho, \qquad \rho \to \rho A_0 \; : \; \mathfrak{X}^\# \to \mathfrak{X}^\#$$

are weak * continuous. If, further, for each ρ_0 in $\mathfrak{X}^\#$, the linear mappings

$$A \to A\rho_0, \qquad A \to \rho_0 A \; : \; \mathfrak{A} \to \mathfrak{X}^\#$$

are continuous relative to the ultraweak topology on $\mathfrak{A}$ and the weak * topology on $\mathfrak{X}^\#$, we say that $\mathfrak{X}^\#$ is a *dual normal* $\mathfrak{A}$-*module*.

Now suppose that $\mathfrak{X}^\#$ is a dual normal $\mathfrak{A}^-$-module (hence, by restriction, a dual normal $\mathfrak{A}$-module). Let $\delta: \mathfrak{A} \to \mathfrak{X}^\#$ be a derivation, and recall from Exercise 4.6.66 that δ is necessarily norm continuous. Let Φ be the universal representation of $\mathfrak{A}$, and choose P and α as in Theorem 10.1.12, with $\pi: \mathfrak{A} \to \mathscr{B}(\mathscr{H})$ the inclusion mapping. Show that

(i) $\mathfrak{X}^\#$ becomes a dual normal $\Phi(\mathfrak{A})^-$-module when the action of $\Phi(\mathfrak{A})^-$ on $\mathfrak{X}^\#$ is defined by

$$S \cdot \rho = \alpha(SP)\rho, \qquad \rho \cdot S = \rho\alpha(SP) \qquad (S \in \Phi(\mathfrak{A})^-, \quad \rho \in \mathfrak{X}^\#),$$

and that $P \cdot \rho = \rho \cdot P = \rho$ for each ρ in $\mathfrak{X}^\#$;

(ii) the equation $\delta_P(S) = \delta(\alpha(SP))$ defines a derivation $\delta_P: \Phi(\mathfrak{A}) \to \mathfrak{X}^\#$;

(iii) δ_P extends to a derivation $\bar{\delta}_P: \Phi(\mathfrak{A})^- \to \mathfrak{X}^\#$ that is continuous relative to the ultraweak topology on $\Phi(\mathfrak{A})^-$ and the weak * topology on $\mathfrak{X}^\#$;

(iv) $\bar{\delta}_P(P) = 0$, and $\delta(A) = \bar{\delta}_P(\alpha^{-1}(A))$ for each A in $\mathfrak{A}$;

(v) δ is continuous relative to the ultraweak topology on $\mathfrak{A}$ and the weak * topology on $\mathfrak{X}^\#$, and extends to a derivation $\bar{\delta}_P \circ \alpha^{-1}: \mathfrak{A}^- \to \mathfrak{X}^\#$.

10.5.14. Suppose that $\mathfrak{A}$ is a C^*-algebra, β is a * automorphism of $\mathfrak{A}$, and $\|\beta - \iota\| < 2$, where ι is the identity mapping on $\mathfrak{A}$. Let $\Phi: \mathfrak{A} \to \mathscr{B}(\mathscr{H}_\Phi)$ be the universal representation of $\mathfrak{A}$, and let $\bar{\beta}$ be the * automorphism of $\Phi(\mathfrak{A})^-$ occurring in Exercise 10.5.1. Let $\pi: \mathfrak{A} \to \mathscr{B}(\mathscr{H}_\pi)$ be a faithful representation of $\mathfrak{A}$, and choose P, α as in Theorem 10.1.12. Show that

(i) $\|\bar{\beta} - \bar{\iota}\| = \|\beta - \iota\|$;
(ii) $\bar{\beta}(P) = P$;
(iii) the restriction $\bar{\beta} | \Phi(\mathfrak{A})^- P$ is a * automorphism $\tilde{\beta}$ of the von Neumann algebra $\Phi(\mathfrak{A})^- P$;
(iv) the * automorphism $\pi\beta\pi^{-1}$ of $\pi(\mathfrak{A})$ is a homeomorphism of $\pi(\mathfrak{A})$ with its ultraweak topology, and extends to a * automorphism $\alpha\tilde{\beta}\alpha^{-1}$ of $\pi(\mathfrak{A})^-$.

10.5.15. Suppose that ρ is a singular positive linear functional on a von Neumann algebra $\mathscr{R}$ and E is a non-zero projection in $\mathscr{R}$. Show that there is a non-zero subprojection F of E in $\mathscr{R}$ such that $\rho(F) = 0$. [*Hint*. Choose a vector x such that $x = Ex$ and $\|x\|^2 > \rho(E)$. Let G be $\sum G_b$, where $\{G_b\}$ is an orthogonal family of projections in $\mathscr{R}$, maximal subject to the conditions that $G_b \le E$ and $\omega_x(G_b) \le \rho(G_b)$ for each index b. Show that $\omega_x(G) \le \rho(G)$, and deduce that $E - G$ is a non-zero projection F in $\mathscr{R}$ such that $\rho | F\mathscr{R}F \le \omega_x | F\mathscr{R}F$. Use Proposition 7.3.5 and Corollary 10.1.16 to show that $\rho(F) = 0$.]

10.5.16. Suppose that $\mathscr{R}$ is a von Neumann algebra, ρ is a faithful state of $\mathscr{R}$, and $\rho_u + \rho_s$ is the decomposition of ρ into its ultraweakly continuous and singular parts (see Theorem 10.1.15(iii)). Show that ρ_u is a faithful normal positive linear functional on $\mathscr{R}$. [*Hint*. Use Exercise 10.5.15.]

10.5.17. Suppose that $\mathfrak{A}$ is an infinite- dimensional C^*-algebra, and let Φ be the universal representation of $\mathfrak{A}$.

(i) Prove that $\Phi(\mathfrak{A})^-$ contains an infinite orthogonal sequence $\{E_1, E_2, \ldots\}$ of non-zero projections. [*Hint*. Use Exercise 4.6.13.]
(ii) Prove that $\Phi(\mathfrak{A})^-$ has a norm-closed subspace that is isometrically isomorphic to the Banach space l_∞. Deduce that, as a Banach space, $\Phi(\mathfrak{A})^-$ is not reflexive. [*Hint*. Use (i) and Exercises 1.9.11 and 1.9.24.]
(iii) Prove that, as a Banach space, $\mathfrak{A}$ is not reflexive. [*Hint*. Use (ii), Proposition 10.1.21, and Exercise 1.9.12.]

10.5.18. Suppose that $\mathfrak{A}^\#$ is the Banach dual space of a C^*-algebra $\mathfrak{A}$ and Φ is the universal representation of $\mathfrak{A}$. When $\rho \in \mathfrak{A}^\#$, let $\tilde{\rho}$ be the unique ultraweakly continuous extension to $\Phi(\mathfrak{A})^-$ of the linear functional

$\rho \circ \Phi^{-1}$ on $\Phi(\mathfrak{A})$ (see the discussion preceding Proposition 10.1.14). When $S \in \Phi(\mathfrak{A})^-$, define a linear functional $\tilde{S}$ on $\mathfrak{A}^\#$ by the equation

$$\tilde{S}(\rho) = \tilde{\rho}(S) \qquad (\rho \in \mathfrak{A}^\#).$$

Show that the mapping $S \to \tilde{S}$ is an isometric isomorphism from $\Phi(\mathfrak{A})^-$ onto $\mathfrak{A}^{\#\#}$. [*Hint*. Note that Φ^{-1} is an isometric isomorphism from $\Phi(\mathfrak{A})$ onto $\mathfrak{A}$, and use Proposition 10.1.21.]

10.5.19. With the notation of the discussion preceding Proposition 10.1.14, prove that

$$(\widetilde{S\rho})(T) = \tilde{\rho}(TS), \qquad (\widetilde{\rho S})(T) = \tilde{\rho}(ST)$$

whenever $S, T \in \Phi(\mathfrak{A})^-$ and $\rho \in \mathfrak{A}^\#$.

10.5.20. Suppose that Φ is the universal representation of a C^*-algebra $\mathfrak{A}$, and define mapppings

$$(\rho, S) \to S\rho, \qquad (\rho, S) \to \rho S : \mathfrak{A}^\# \times \Phi(\mathfrak{A})^- \to \mathfrak{A}^\#$$

as in the discussion preceding Proposition 10.1.14. Show that

(i) $\mathfrak{A}^\#$ is a Banach $\Phi(\mathfrak{A})^-$-module, and a dual $\Phi(\mathfrak{A})$-module in the sense of Exercise 10.5.13;

(ii) if $\mathfrak{X}_0$ is a norm-closed subspace of $\mathfrak{A}^\#$ with the property that $B\rho$, $\rho B \in \mathfrak{X}_0$ whenever $B \in \Phi(\mathfrak{A})$ and $\rho \in \mathfrak{X}_0$, then $S\rho$, $\rho S \in \mathfrak{X}_0$ whenever $S \in \Phi(\mathfrak{A})^-$ and $\rho \in \mathfrak{X}_0$. [*Hint*. By the Hahn–Banach theorem it suffices to show that $\Omega(S\rho) = 0 = \Omega(\rho S)$ whenever $\rho \in \mathfrak{X}_0$, $S \in \Phi(\mathfrak{A})^-$, $\Omega \in \mathfrak{A}^{\#\#}$, and $\Omega | \mathfrak{X}_0 = 0$. Use Exercises 10.5.18 and 10.5.19.]

10.5.21. Suppose that $\mathfrak{A}$ and $\mathscr{B}$ are C^*-algebras and $\eta: \mathfrak{A} \to \mathscr{B}$ is a linear mapping such that

$$\eta(A^*) = \eta(A)^*, \qquad \eta(AB + BA) = \eta(A)\eta(B) + \eta(B)\eta(A)$$

for all A and B in $\mathfrak{A}$. (We refer to such a mapping η as a *Jordan * homomorphism* from $\mathfrak{A}$ into $\mathscr{B}$; compare Exercise 7.6.18.) When $A, B \in \mathfrak{A}$, denote by $[A, B]$ the commutator $AB - BA$. Establish the identities

$$ABA + BAB = (A + B)^3 - A^3 - B^3 - (A^2B + BA^2) - (B^2A + AB^2),$$

$$ABC + CBA = (A + C)B(A + C) - ABA - CBC$$

$$[[A, B], C] = ABC + CBA - (BAC + CAB),$$

$$[A, B]^2 = A(BAB) + (BAB)A - AB^2A - BA^2B$$

for all A, B, C in $\mathfrak{A}$, and show that

(i) $\eta(A^n) = \eta(A)^n$ $(n = 1, 2, \ldots)$;
(ii) $\eta(ABA) = \eta(A)\eta(B)\eta(A)$;
(iii) $\eta(ABC + CBA) = \eta(A)\eta(B)\eta(C) + \eta(C)\eta(B)\eta(A)$;
(iv) $\eta([[A, B], C]) = [[\eta(A), \eta(B)], \eta(C)]$;
(v) $\eta([A, B]^2) = [\eta(A), \eta(B)]^2$.

10.5.22. With the notation of Exercise 10.5.21, suppose that $\mathscr{B}_0$ is the smallest norm-closed subalgebra of $\mathscr{B}$ that contains $\eta(\mathfrak{A})$ (we do not assume that $I \in \mathscr{B}_0$), and let $\mathscr{C}$ and $\mathscr{C}_0$ be the centers of $\mathfrak{A}$ and $\mathscr{B}_0$, respectively. Show that

(i) $\mathscr{B}_0$ is a self-adjoint subalgebra of $\mathscr{B}$;
(ii) if $A, B \in \mathfrak{A}$ and $[A, B] = 0$, then $[\eta(A), \eta(B)] \in \mathscr{C}_0$;
(iii) if $A, B \in \mathfrak{A}$ and $[A, B] = 0$, then $[\eta(A), \eta(B)] = 0$, and $\eta(AB) = \eta(A)\eta(B)$;
(iv) $\eta(\mathscr{C}) \subseteq \mathscr{C}_0$;
(v) $\eta(I)$ is a projection in $\mathscr{B}$ and is the unit element of $\mathscr{B}_0$;
(vi) if E is a projection in $\mathfrak{A}$, then $\eta(E)$ is a projection in $\mathscr{B}_0$;
(vii) if E, F are mutually orthogonal projections in $\mathfrak{A}$, then $\eta(E)$, $\eta(F)$ are mutually orthogonal projections in $\mathscr{B}_0$;
(viii) if P is a central projection in $\mathfrak{A}$, then $\eta(P)$ is a central projection in $\mathscr{B}_0$, and $\eta(AP) = \eta(A)\eta(P)$ for each A in $\mathfrak{A}$.

10.5.23. With the notation of Exercise 10.5.22, suppose that $n \geq 2$, $\{E_{jk} : j, k = 1, \ldots, n\}$ is a self-adjoint system of matrix units in $\mathfrak{A}$, and $\sum_{j=1}^n E_{jj} = I$. Let I_0 be the unit element $\eta(I)$ of $\mathscr{B}_0$, and define

$$F_{jk} = \eta(E_{jk}) \qquad (j, k = 1, \ldots, n),$$

$$G_{jk} = F_{jj}F_{jk}F_{kk}, \qquad H_{jk} = F_{jj}F_{kj}F_{kk} \qquad (j, k = 1, \ldots, n; \quad j \neq k).$$

(i) Show that $F_{kj} = F_{jk}^*$ $(j, k = 1, \ldots, n)$, and that $G_{kj} = G_{jk}^*$, $H_{kj} = H_{jk}^*$ when $j \neq k$.

(ii) By using the relation $E_{jk} = E_{jj}E_{jk}E_{kk} + E_{kk}E_{jk}E_{jj}$, show that

$$F_{jk} = G_{jk} + H_{kj} \qquad (j, k = 1, \ldots, n; \quad j \neq k).$$

(iii) Show that $F_{11}, \ldots, F_{nn}$ are mutually orthogonal projections in $\mathscr{B}_0$ with sum I_0, and that

$$F_{jj}F_{jk} = F_{jk}F_{kk} = G_{jk}, \qquad F_{kk}F_{jk} = F_{jk}F_{jj} = H_{kj} \qquad (j \neq k).$$

(iv) Suppose that j, k, l are distinct elements of $\{1, 2, \ldots, n\}$. Show that $F_{jj}F_{kl} = 0$, and

$$G_{jk}G_{kl} = F_{jj}F_{jk}F_{kl} = F_{jj}(F_{jk}F_{kl} + F_{kl}F_{jk}) = G_{jl}.$$

Prove also that $H_{jk}H_{kl} = H_{jl}$.

(v) Show that $G_{jk}G_{kj} = G_{jl}G_{lj}$ and $H_{jk}H_{kj} = H_{jl}H_{lj}$ if j, k, l are distinct elements of $\{1, 2, \ldots, n\}$.

(vi) For $j = 1, \ldots, n$, define G_{jj} and H_{jj} to be $G_{jk}G_{kj}$ and $H_{jk}H_{kj}$, respectively, where $k \neq j$ (and note that, from (v), the definitions are independent of the choice of k). Show that

$$G_{jj} = F_{jk}F_{kj}F_{jj}, \qquad H_{jj} = F_{kj}F_{jk}F_{jj}$$

and deduce that

$$F_{jj} = G_{jj} + H_{jj}.$$

(vii) Show that $G_{jk}H_{kj} = 0 = H_{jk}G_{kj}$, when $j \neq k$. Deduce that

$$G_{jk}H_{lm} = 0 = H_{jk}G_{lm} \qquad (j, k, l, m = 1, \ldots, n).$$

(viii) Show that $\{G_{jk} : j, k = 1, \ldots, n\}$ and $\{H_{jk} : j, k = 1, \ldots, n\}$ are self-adjoint systems of matrix units in $\mathscr{B}_0$, and

$$G_{11} + \cdots + G_{nn} + H_{11} + \cdots + H_{nn} = I_0.$$

10.5.24. With the notation of Exercise 10.5.23, let

$$G = G_{11} + \cdots + G_{nn}, \qquad H = H_{11} + \cdots + H_{nn},$$

and let $\mathscr{D}$ be the set of all elements of $\mathfrak{A}$ that commute with all the matrix units E_{jk} $(j, k = 1, \ldots, n)$. Show that

(i) each element A of $\mathfrak{A}$ can be expressed uniquely in the form

$$A = \sum_{j,k=1}^{n} D_{jk}E_{jk}$$

with all the coefficients D_{jk} *in* $\mathscr{D}$, and (compare Lemma 6.6.3) $D_{jk} = \sum_{l=1}^{n} E_{lj}AE_{kl}$;

(ii) for each D in $\mathscr{D}$, $\eta(D)$ commutes with all the elements F_{jk}, G_{jk}, H_{jk};

(iii) if $A \in \mathfrak{A}$, and A is expressed as in (i), then

$$\eta(A) = \sum_{j,k=1}^{n} \eta(D_{jk})F_{jk};$$

(iv) $G, H \in \mathscr{C}_0$, and $G + H = I_0$;

(v) if $A, B \in \mathscr{D}$ and $j \neq k$, then

$$ABE_{jj} + BAE_{kk} = (AE_{jk} + BE_{kj})^2,$$

$$\eta(AB)F_{jj} + \eta(BA)F_{kk} = \eta(A)\eta(B)F_{jk}F_{kj} + \eta(B)\eta(A)F_{kj}F_{jk},$$

$$\eta(AB)G_{jk} = \eta(A)\eta(B)G_{jk}, \qquad \eta(AB)H_{jk} = \eta(B)\eta(A)H_{jk};$$

(vi) for all A and B in $\mathfrak{A}$,

$$\eta(AB)G = \eta(A)G\eta(B)G, \qquad \eta(AB)H = \eta(B)H\eta(A)H;$$

(vii) the equations

$$\eta_1(A) = \eta(A)G, \qquad \eta_2(A) = \eta(A)H$$

define a * homomorphism $\eta_1: \mathfrak{A} \to \mathscr{B}_0 G\ (\subseteq \mathscr{B})$ and a * anti-homomorphism $\eta_2: \mathfrak{A} \to \mathscr{B}_0 H\ (\subseteq \mathscr{B})$, and $\eta = \eta_1 + \eta_2$.

10.5.25. Suppose that $\mathscr{R}$ and $\mathscr{S}$ are von Neumann algebras and $\eta: \mathscr{R} \to \mathscr{S}$ is an ultraweakly continuous Jordan * homomorphism. Let $\mathscr{B}_0$ be the smallest norm-closed subalgebra of $\mathscr{S}$ that contains $\eta(\mathscr{R})$, let $\mathscr{S}_0$ be $\mathscr{B}_0^-$, and let I_0 be $\eta(I)$. Denote by $\mathscr{C}$, $\mathscr{C}_0$, and $\mathscr{Z}_0$, the centers of $\mathscr{R}$, $\mathscr{B}_0$, and $\mathscr{S}_0$, respectively.

(i) Show that $\mathscr{B}_0$ and $\mathscr{S}_0$ are self-adjoint subalgebras of $\mathscr{S}$, and $\mathscr{C}_0^- \subseteq \mathscr{Z}_0$.

(ii) Show that I_0 is a projection in $\mathscr{S}$, and is the unit element of $\mathscr{B}_0$ and $\mathscr{S}_0$.

(iii) For each positive integer n, let P_n be the largest projection in $\mathscr{C}$ such that $\mathscr{R}P_n$ is of type I_n unless $P_n = 0$, and let P_0 be $I - \sum_{n=1}^{\infty} P_n$. Let Q_n be $\eta(P_n)$ for $n \geq 0$. Show that $\{Q_n\}$ is an orthogonal family of projections in $\mathscr{Z}_0$, and $\sum_{n=0}^{\infty} Q_n = I_0$. Note that, for $n \geq 2$, the von Neumann algebra $\mathscr{R}P_n$ contains a self-adjoint system of $n \times n$ matrix units, in which the n diagonal elements have sum P_n. Prove that $\mathscr{R}P_0$ contains a self-adjoint system of 2×2 matrix units, in which the two diagonal elements have sum P_0. Deduce from Exercise 10.5.24 that, for all $n \geq 0$, there are projections G_n and H_n in $\mathscr{Z}_0$, with sum Q_n, such that the mapping $A \to \eta(A)G_n : \mathscr{R}P_n \to \mathscr{S}_0 G_n$ is a * homomorphism and the mapping $A \to \eta(A)H_n : \mathscr{R}P_n \to \mathscr{S}_0 H_n$ is a * anti-homomorphism.

(iv) Prove that there are projections G and H in $\mathscr{Z}_0$ such that $G + H = I_0$ and the equations

$$\eta_1(A) = \eta(A)G, \qquad \eta_2(A) = \eta(A)H \qquad (A \in \mathscr{R})$$

define a * homomorphism $\eta_1: \mathscr{R} \to \mathscr{S}_0 G\ (\subseteq \mathscr{S})$ and a * anti-homomorphism $\eta_2: \mathscr{R} \to \mathscr{S}_0 H\ (\subseteq \mathscr{S})$ for which $\eta_1 + \eta_2 = \eta$.

(v) Conversely, suppose that $\xi: \mathscr{R} \to \mathscr{S}$ is a hermitian linear mapping, and there exist projections I_0, G, H in $\mathscr{S}$ such that $G + H = I_0$,

$$I_0\xi(A) = \xi(A)I_0 = \xi(A), \qquad G\xi(A) = \xi(A)G, \qquad H\xi(A) = \xi(A)H$$

for each A in $\mathscr{R}$, the mapping $A \to \xi(A)G : \mathscr{R} \to \mathscr{S}$ is a * homomorphism, and the mapping $A \to \xi(A)H : \mathscr{R} \to \mathscr{S}$ is a * anti-homomorphism. Show that ξ is a Jordan * homomorphism.

10.5.26. Suppose that $\mathscr{R}$ and $\mathscr{S}$ are von Neumann algebras and η is a Jordan * isomorphism from $\mathscr{R}$ onto $\mathscr{S}$. Show that there exist central projections P_1, P_2 in $\mathscr{R}$ and Q_1, Q_2 in $\mathscr{S}$ such that $P_1 + P_2 = I$, $Q_1 + Q_2 = I$, $\eta(P_1) = Q_1$, $\eta(P_2) = Q_2$, $\eta \mid \mathscr{R}P_1$ is a * isomorphism from $\mathscr{R}P_1$ onto $\mathscr{S}Q_1$, and $\eta \mid \mathscr{R}P_2$ is a * anti-isomorphism from $\mathscr{R}P_2$ onto $\mathscr{S}Q_2$. (Since $\mathscr{R}$ and $\mathscr{S}$ are unitarily equivalent to $\mathscr{R}P_1 \oplus \mathscr{R}P_2$ and $\mathscr{S}Q_1 \oplus \mathscr{S}Q_2$, respectively, this result can be stated in the form that a Jordan * isomorphism from one von Neumann algebra onto another is a direct sum of a * isomorphism and a * anti-isomorphism.)

10.5.27. Suppose that $\mathscr{R}$ is a factor, $\mathscr{S}$ is a von Neumann algebra, and η is a Jordan * isomorphism from $\mathscr{R}$ onto $\mathscr{S}$. Show that $\mathscr{S}$ is a factor and η is either a * isomorphism or a * anti-isomorphism.

10.5.28. Suppose that $\mathfrak{A}$ is a C^*-algebra, $\mathscr{B}$ is a C^*-algebra of operators acting on a Hilbert space $\mathscr{H}$, and η is a hermitian bounded linear mapping from $\mathfrak{A}$ onto $\mathscr{B}$. Let Φ be the universal representation of $\mathfrak{A}$, and let $\bar{\eta}: \Phi(\mathfrak{A})^- \to \mathscr{B}(\mathscr{H})$ be the unique ultraweakly continuous linear mapping that extends $\eta \circ \Phi^{-1}: \Phi(\mathfrak{A}) \to \mathscr{B} \subseteq \mathscr{B}(\mathscr{H})$ (see Theorem 10.1.13). Show that $\bar{\eta}$ has range $\mathscr{B}^-$, and that $\bar{\eta}$ is a Jordan * homomorphism if η is a Jordan * homomorphism. By applying the result of Exercise 10.5.25 to $\bar{\eta}$, deduce that η is a Jordan * homomorphism if and only if there is a projection P in the center of $\mathscr{B}^-$ such that

$$\eta(AB)P = \eta(A)\eta(B)P, \qquad \eta(AB)(I - P) = \eta(B)\eta(A)(I - P)$$

for all A and B in $\mathfrak{A}$. (Note that, in the case of a Jordan * isomorphism from $\mathfrak{A}$ onto $\mathscr{B}$, the present exercise augments the information obtained in Exercises 7.6.16–7.6.18 about isometries between operator algebras.)

10.5.29. Let η be a bounded linear isomorphism of one C^*-algebra $\mathfrak{A}$ onto another C^*-algebra $\mathscr{B}$ and let Φ and Ψ be the universal representations of $\mathfrak{A}$ and $\mathscr{B}$, respectively.

(i) Show that $\Psi \circ \eta \circ \Phi^{-1}$ extends uniquely to a linear *isomorphism* $\bar{\eta}$ of $\Phi(\mathfrak{A})^-$ onto $\Psi(\mathscr{B})^-$ that is an ultraweak homeomorphism.

(ii) Deduce that $\bar{\eta}$ is a Jordan * isomorphism when η is a Jordan * isomorphism.

10.5.30. Use the pattern of the construction of the Cantor set to construct a closed nowhere-dense subset S' of $[0, 1]$ having Lebesgue measure $\frac{1}{2}$ (remove the centered open interval of length $\frac{1}{4}$ from $[0, 1]$ and so on). Let S be $[0, 1]\backslash S'$, $\mathscr{H}$ be $L_2(0, 1)$, and $\mathscr{K}$ be $L_2(S)$ (relative to Lebesgue measure), and let $\mathfrak{A}$ and $\mathscr{B}$ be the (abelian) C^*-algebras consisting of all multiplication operators corresponding to continuous functions (on $[0, 1]$) on $\mathscr{H}$ and $\mathscr{K}$, respectively. Let φ be the mapping on $\mathfrak{A}$ arising from restricting a continuous function on $[0, 1]$ to S. Show that

(i) φ is a * isomorphism of $\mathfrak{A}$ onto $\mathscr{B}$;

(ii) φ extends to an ultraweakly continuous * homomorphism $\bar{\varphi}$ of $\mathfrak{A}^-$ onto $\mathscr{B}^-$;

(iii) $\bar{\varphi}$ is not an isomorphism (compare this conclusion with those of Exercise 10.5.29);

(iv) the * isomorphism φ^{-1} from $\mathscr{B}$ onto $\mathfrak{A}$ is not ultraweakly continuous (see the discussion following Corollary 10.1.11).

10.5.31. Let $\mathfrak{A}$ and $\mathscr{B}$ be C^*-algebras, and let η be a Jordan * isomorphism of $\mathfrak{A}$ onto $\mathscr{B}$. Show that

(i) η is a linear order isomorphism of $\mathfrak{A}$ onto $\mathscr{B}$ (that is, η and η^{-1} are positive linear mappings);

(ii) η is bounded;

(iii) η is an isometry.

10.5.32. (i) Show that a linear order isomorphism of one C^*-algebra onto another that maps I onto I is a Jordan * isomorphism. [*Hint*. Use Exercise 10.5.9.]

(ii) Find an example of a positive linear mapping of one C^*-algebra onto another that is a linear isomorphism but whose inverse is not a positive linear mapping. [*Hint*. Consider $\mathbb{C}^2$ for each of the C^*-algebras.]

10.5.33. If η is a linear order isomorphism of one C^*algebra $\mathfrak{A}$ *into* another C^*-algebra $\mathscr{B}$ and $\eta(I) = I$, must η be a Jordan * homomorphism? Proof? Counterexample?

10.5.34. Let $\mathscr{H}$ and $\mathscr{K}$ be Hilbert spaces, $\mathfrak{A}$ and $\mathscr{B}$ be self-adjoint subalgebras of $\mathscr{B}(\mathscr{H})$ and $\mathscr{B}(\mathscr{K})$, respectively, and $\mathscr{S}_w(\mathfrak{A})$, $\mathscr{S}_w(\mathscr{B})$ be the (convex) families of weak-operator continuous states on $\mathfrak{A}$ and $\mathscr{B}$, respectively.

(i) Show that an affine mapping φ' of $\mathscr{S}_w(\mathfrak{A})$ into $\mathscr{S}_w(\mathscr{B})$($\varphi'(a\omega_1 + (1 - a)\omega_2) = a\varphi'(\omega_1) + (1 - a)\varphi'(\omega_2)$ for each a in $[0, 1]$ and ω_1, ω_2 in $\mathscr{S}_w(\mathfrak{A})$) has a unique linear extension φ mapping $\mathfrak{A}_w^{\#}$, the linear space of all weak-operator continuous linear functionals on $\mathfrak{A}$, into $\mathscr{B}^\#_{\mathrm{w}}$.

(ii) With φ as constructed in (i), show that there is a weak-operator continuous, positive linear mapping α of $\mathscr{B}^-$ into $\mathfrak{A}^-$ such that $\omega'(\alpha(B)) = \varphi(\omega)'(B)$ for each B in $\mathscr{B}^-$ and each ω in $\mathfrak{A}^\#_{\mathrm{w}}$, where ω' is the (unique) weak-operator continuous extension of ω from $\mathfrak{A}$ to $\mathfrak{A}^-$; and such that $\alpha(I) = I$.

(iii) When φ' is an affine isomorphism of $\mathscr{S}_{\mathrm{w}}(\mathfrak{A})$ onto $\mathscr{S}_{\mathrm{w}}(\mathscr{B})$, show that α (in (ii)) is a Jordan * isomorphism of $\mathscr{B}^-$ onto $\mathfrak{A}^-$.

10.5.35. Let $\mathfrak{A}$ acting on a Hilbert space $\mathscr{H}$ be the universal representation of the C^*-algebra $\mathfrak{A}$. With ω in $\mathfrak{A}^\#$, let ω' denote the (unique) weak-operator continuous extension of ω from $\mathfrak{A}$ to $\mathfrak{A}^-$. For B in $\mathfrak{A}^-$, let $\hat{B}$ denote the function on $\mathfrak{A}^\#$ that assigns $\omega'(B)$ to ω. For a given B in $\mathfrak{A}^-$, assume that $\hat{B}\,|\,\mathscr{S}(\mathfrak{A})$ is weak * continuous. Show that

(i) $\hat{B}\,|\,\mathscr{C}$ is weak * continuous, where

$$\mathscr{C} = \{a\rho : a \in [0, 1],\ \rho \in \mathscr{S}(\mathfrak{A})\};$$

(ii) $\hat{B}\,|\,(\mathfrak{A}^\#)_1$ is weak * continuous [*Hint*. Recall that the hermitian elements of $(\mathfrak{A}^\#)_1$ are contained in $\mathscr{C} - \mathscr{C}$.];

(iii) $B \in \mathfrak{A}$. [*Hint*. Use Exercise 1.9.15.]

10.5.36. Let φ be a weak * continuous affine isomorphism of the state space $\mathscr{S}(\mathfrak{A})$ of one C^*-algebra $\mathfrak{A}$ onto the state space $\mathscr{S}(\mathscr{B})$ of another C^*-algebra $\mathscr{B}$. Show that there is a Jordan * isomorphism α of $\mathscr{B}$ onto $\mathfrak{A}$ such that $\rho(\alpha(B)) = \varphi(\rho)(B)$ for each ρ in $\mathscr{S}(\mathfrak{A})$ and each B in $\mathscr{B}$. [*Hint*. Use the universal representations of $\mathfrak{A}$ and $\mathscr{B}$ and the results of Exercises 10.5.34 and 10.5.35].

10.5.37. Let $\mathscr{M}$ be a factor acting on a separable Hilbert space $\mathscr{H}$.

(i) Suppose ρ is a normal pure state of $\mathscr{M}$. Show that $\mathscr{M}$ is of type I.

(ii) If $\mathscr{M}$ is of type I, is each pure state of $\mathscr{M}$ normal? Proof? Counterexample?

10.5.38. Let Φ be the universal representation of a C^*-algebra $\mathfrak{A}$ on $\mathscr{H}_\Phi$. Suppose $\{\rho_n\}$ is a sequence of states of $\mathfrak{A}$ tending in norm to ρ_0 and P_n is the support of $\bar{\rho}_n|\mathscr{C}$, where $\bar{\rho}_n$ is the unique ultraweakly continuous linear extension of $\rho_n \circ \Phi^{-1}$ from $\Phi(\mathfrak{A})$ to $\Phi(\mathfrak{A})^-$ and $\mathscr{C}$ is the center of $\Phi(\mathfrak{A})^-$. Show that $\{P_n P_0\}$ is strong-operator convergent to P_0. [*Hint*. Use Exercises 7.6.14, 7.6.28, and 7.6.21.]

10.5.39. Let Φ be the universal representation of a C^*-algebra $\mathfrak{A}$ on $\mathscr{H}_\Phi$. With ρ a state of $\mathfrak{A}$, we say that ρ is of a certain type (I_n, II_1, II_∞, or III) when $\pi_\rho(\mathfrak{A})^-$ is a von Neumann algebra of that type. Let P_n, P_{c_1}, $P_{c\infty}$, P_∞, be the maximal central projections in $\Phi(\mathfrak{A})^-$ such that $\Phi(\mathfrak{A})^- P_n$, $\Phi(\mathfrak{A})^- P_{c_1}$, $\Phi(\mathfrak{A})^- P_{c\infty}$, and $\Phi(\mathfrak{A})^- P_\infty$, are (0) or of types I_n, II_1, II_∞, and III, respectively. Show that

(i) a state ρ of $\mathfrak{A}$ is of type I_n, II_1, II_∞, or III if and only if $\bar{\rho}(P_n)$, $\bar{\rho}(P_{c_1})$, $\bar{\rho}(P_{c\infty})$, or $\bar{\rho}(P_\infty)$ is 1, respectively, where $\bar{\rho}$ is the ultraweakly continuous extension of $\rho \circ \Phi^{-1}$ to $\Phi(\mathfrak{A})^-$;
(ii) the set of states of $\mathfrak{A}$ of a given type is a norm-closed convex subset of $\mathfrak{A}^\#$.

10.5.40. Find an example of a countably generated (norm-separable) C^*-algebra whose set of states of type I is not weak * closed in $\mathfrak{A}^\#$.

10.5.41. Let π_1, π_2, and π_3 be representations of a C^*-algebra $\mathfrak{A}$ such that π_1 is quasi-equivalent to a subrepresentation of π_2 (we say that π_1 is *quasi-subequivalent* to π_2 in this case, and write $\pi_1 \precsim_q \pi_2$) and π_2 is quasi-equivalent to π_3. Show that $\pi_1 \precsim_q \pi_3$.

10.5.42. Let π_1 and π_2 be representations of a C^*-algebra $\mathfrak{A}$. Let Φ be the universal representation of $\mathfrak{A}$ on $\mathscr{H}_\Phi$, and let P_1 and P_2 be the central projections in $\Phi(\mathfrak{A})^-$ corresponding to π_1 and π_2, respectively, as described in Theorem 10.1.12. Show that

(i) $\pi_1 \precsim_q \pi_2$ if and only if $P_1 \le P_2$;
(ii) the set of quasi-equivalence classes of representations of $\mathfrak{A}$ is partially ordered by $\precsim_q$;
(iii) the quasi-equivalence class of π is minimal relative to $\precsim_q$ if and only if π is primary.

10.5.43. Let π_1 and π_2 be representations of a C^*-algebra $\mathfrak{A}$ quasi-equivalent to the representations π_1' and π_2', respectively. Show that π_1 and π_2 are disjoint if and only if π_1' and π_2' are disjoint.

10.5.44. Let π_1 and π_2 be representations of a C^*-algebra $\mathfrak{A}$, Φ be the universal representation of $\mathfrak{A}$, and P_1 and P_2 be the central projections in $\Phi(\mathfrak{A})^-$ corresponding to π_1 and π_2 (as in Theorem 10.1.12). Show that

(i) π_1 and π_2 are disjoint if and only if $P_1 P_2 = 0$;
(ii) $\mathscr{Q}(\mathfrak{A})$, the set of quasi-equivalence classes of representations of $\mathfrak{A}$ partially ordered by $\precsim_q$ (we adopt the terminology and notation of Exercise 10.5.41 and include the 0 mapping of $\mathfrak{A}$ as an element of $\mathscr{Q}(\mathfrak{A})$ although this

mapping has been excluded as a representation of $\mathfrak{A}$), is a lattice isomorphic to the lattice of projections in the center of $\Phi(\mathfrak{A})^-$ (through the mapping that assigns P_1 to the quasi-equivalence class of π_1);

(iii) the quasi-equivalence classes of π_1 and π_2 have the 0 mapping as their greatest lower bound in $\mathcal{Q}(\mathfrak{A})$ if and only if π_1 and π_2 are disjoint.

10.5.45. Let ρ and η be states of the C^*-algebra $\mathfrak{A}$. Write $\rho \sim_q \eta$ and $\rho \precsim_q \eta$ when $\pi_\rho \sim_q \pi_\eta$ (π_ρ and π_η are quasi-equivalent) respectively $\pi_\rho \precsim_q \pi_\eta$ (π_ρ is quasi-subequivalent to π_η in the terminology of Exercise 10.5.41). Let $\{\rho_n\}$ be a sequence of states of a C^*-algebra $\mathfrak{A}$ and suppose that $\rho_{n+1} \precsim_q \rho_n$ for n in $\{1, 2, \ldots\}$ and that $\{\rho_n\}$ tends in norm to ρ_0. Show that

(i) $\rho_0 \precsim_q \rho_n$ for each n in $\{1, 2, \ldots\}$;

(ii) ρ_0 is a factor state quasi-equivalent to each ρ_n when each ρ_n is a factor state;

(iii) ρ_0 need not be quasi-equivalent to ρ_n when $\rho_1 \sim_q \rho_2 \sim_q \rho_3 \sim_q \cdots$.

10.5.46. Let ρ_0 be the norm limit of a sequence $\{\rho_n\}$ of factor states of a C^*-algebra $\mathfrak{A}$. Show that ρ_0 is a factor state quasi-equivalent to all but a finite number of $\{\rho_n\}$. Deduce that the set of factor states of $\mathfrak{A}$ is norm closed.

10.5.47. Use Exercise 7.6.34 to prove that a norm limit of a sequence of pure states of a C*-algebra is a pure state equivalent to all but a finite number of the pure states in the sequence. (By "equivalence" of states, we mean equivalence of the corresponding representations.)

10.5.48. Suppose $\mathfrak{A}$ is a C^*-algebra acting irreducibly on the Hilbert space $\mathscr{H}$, x and y are unit vectors in $\mathscr{H}$, and $\mathscr{H}_0$ is the two-dimensional subspace of $\mathscr{H}$ generated by x and y. Let E and F be the one-dimensional projections in $\mathscr{B}(\mathscr{H}_0)$ with ranges $[x]$ and $[y]$, respectively.

(i) Show that $\|\omega_x | \mathfrak{A} - \omega_y | \mathfrak{A}\| = \|\omega_x - \omega_y\| = \|(\omega_x - \omega_y) | \mathscr{B}(\mathscr{H}_0)\| = \operatorname{tr}(|E - F|)$, where "tr" denotes the (non-normalized) trace on $\mathscr{B}(\mathscr{H}_0)$ (viewed as the algebra of complex 2×2 matrices).

(ii) Show that

$$\|\omega_x | \mathfrak{A} - \omega_y | \mathfrak{A}\| = 2[1 - |\langle x, y\rangle|^2]^{1/2}$$

[*Hint*. Note that x and y are eigenvectors for $(E - F)^2$ and deal with the cases $\langle x, y\rangle = 0$ and $\langle x, y\rangle \neq 0$ separately.]

10.5.49. Use Exercise 10.5.48 to prove again that a norm limit of pure states of a C^*-algebra is a pure state (and hence that the family of pure states is norm closed).

10.5.50. Use Theorem 10.2.3 and the result of Exercise 10.5.48 to show once again (see Exercise 4.6.26(ii)) that a C^*-algebra $\mathfrak{A}$ is abelian if there is a positive real number δ such that $\|\rho_1 - \rho_2\| \geq \delta$ whenever ρ_1 and ρ_2 are distinct pure states of $\mathfrak{A}$.

10.5.51. Suppose that $\mathscr{H}$ is a Hilbert space, $\mathscr{K}$ $(\subseteq \mathscr{B}(\mathscr{H}))$ is the ideal of all compact linear operators, $\mathfrak{A}$ is a C^*-algebra such that $\mathscr{K} \subseteq \mathfrak{A} \subseteq \mathscr{B}(\mathscr{H})$, and ρ $(\neq 0)$ is a bounded linear functional on $\mathfrak{A}$ that vanishes on $\mathscr{K}$. Show that ρ is singular, in the sense of the discussion preceding Theorem 10.1.15. [*Hint*. By considering the linear functional induced by ρ on $\mathfrak{A}/\mathscr{K}$, show that ρ is a linear combination of states of $\mathfrak{A}$ that vanish on $\mathscr{K}$. Then use Proposition 10.1.17.]

10.5.52. Let $\mathfrak{A}$ be a C^*-algebra acting on a Hilbert space $\mathscr{H}$, and let $\mathscr{K}$ be the ideal of compact operators. Suppose ρ is a positive linear functional on $\mathfrak{A}$ such that $\|\rho\| = \|\rho \mid \mathfrak{A} \cap \mathscr{K}\|$. Show that

(i) there is an increasing sequence $\{E_n\}$ of finite-dimensional projections E_n in $\mathfrak{A}$ such that $\{\rho(E_n)\}$ tends to $\|\rho\|$;
(ii) $\|\rho - \rho_n\| \to 0$, where $\rho_n(A) = \rho(E_n A E_n)$.

10.5.53. Let $\mathfrak{A}$ be a C^*-algebra, $\mathscr{I}$ be a norm-closed, two-sided ideal in $\mathfrak{A}$, and ρ be a positive linear functional on $\mathfrak{A}$. Show that

(i) $\rho = \rho_1 + \rho_2$ with ρ_1 and ρ_2 positive linear functionals on $\mathfrak{A}$ such that $\|\rho_1\| = \|\rho_1 \mid \mathscr{I}\|$ and $\rho_2 \mid \mathscr{I} = 0$;
(ii) the decomposition of (i) is unique. [*Hint*. Use Proposition 10.1.5.]

10.5.54. Let $\mathscr{H}$ be an infinite-dimensional Hilbert space and $\mathscr{K}$ be the ideal of compact operators on $\mathscr{H}$. Show that

(i) the vector state space (that is, the weak * closure of the set of vector states) of $\mathscr{B}(\mathscr{H})$ coincides with the pure state space (that is, the weak * closure of the set of pure states) of $\mathscr{B}(\mathscr{H})$;
(ii) the set of states of $\mathscr{B}(\mathscr{H})$ that annihilate $\mathscr{K}$ is a non-null weak * compact convex subset $\mathscr{K}_+^\perp$ of $\mathscr{B}(\mathscr{H})^\#$ whose extreme points are pure states of $\mathscr{B}(\mathscr{H})$.

10.5.55. Let $\mathscr{H}$ be a Hilbert space and $\mathscr{K}$ be the ideal of compact operators on $\mathscr{H}$. Let ρ be in $\mathscr{K}_+^\perp$, the set of states of $\mathscr{B}(\mathscr{H})$ that annihilate $\mathscr{K}$. Show that

(i) if ρ is a pure state of $\mathscr{B}(\mathscr{H})$ and E is a projection such that $(I - E)(\mathscr{H})$ is finite dimensional, for each finite set of operators $A_1, \ldots, A_m$ in $\mathscr{B}(\mathscr{H})$ and each positive ε, there is a unit vector x in $E(\mathscr{H})$ such that

$$|\omega_x(A_j) - \rho(A_j)| < \varepsilon \qquad (j \in \{1, \ldots, m\})$$

[*Hint.* Use Corollary 4.3.10.];

(ii) ρ is a weak * limit of vector states of $\mathscr{B}(\mathscr{H})$ and ρ is in the pure state space of $\mathscr{B}(\mathscr{H})$;

(iii) $a\omega_x + (1 - a)\rho$ is a weak * limit of vector states of $\mathscr{B}(\mathscr{H})$ and is in the pure state space of $\mathscr{B}(\mathscr{H})$. [*Hint.* Use Exercise 4.6.69(ii).]

10.5.56. Let $\mathfrak{A}$ be a C^*-algebra acting on a Hilbert space $\mathscr{H}$, and let $\mathscr{K}$ be the ideal of compact operators on $\mathscr{H}$. Show that

(i) $a\omega_x | \mathfrak{A} + (1 - a)\rho$ is a weak * limit of vector states of $\mathfrak{A}$ for each unit vector x in $\mathscr{H}$ and each a in $[0, 1]$, where ρ is a state of $\mathfrak{A}$ that is 0 on $\mathfrak{A} \cap \mathscr{K}$ [*Hint.* Use the quotient mapping of $\mathscr{B}(\mathscr{H})$ onto $\mathscr{B}(\mathscr{H})/\mathscr{K}$ and use Exercise 10.5.55.];

(ii) each weak * limit of vector states of $\mathfrak{A}$ has the form $a\omega_x | \mathfrak{A} + (1 - a)\rho$, where x is a unit vector in $\mathscr{H}$, ρ is a state of $\mathfrak{A}$ that is 0 on $\mathfrak{A} \cap \mathscr{K}$, and $a \in [0, 1]$. [*Hint.* Use Exercises 10.5.52 and 7.6.33.]

10.5.57. Let $\mathfrak{A}$ be a C^*-algebra acting on a Hilbert space $\mathscr{H}$.

(i) Suppose $\mathscr{H}$ is two dimensional and $\mathfrak{A}$ is a maximal abelian subalgebra of $\mathscr{B}(\mathscr{H})$. Show that $\mathfrak{A}$ does not separate the set of vector states of $\mathscr{B}(\mathscr{H})$.

(ii) Suppose $\mathfrak{A}$ acts irreducibly on $\mathscr{H}$. Show that $\mathfrak{A}$ does separate the set of vector states of $\mathscr{B}(\mathscr{H})$.

(iii) Suppose $\mathfrak{A}$ separates the set of vector states of $\mathscr{B}(\mathscr{H})$. Show that $\mathfrak{A}$ acts irreducibly on $\mathscr{H}$.

(iv) Suppose $\mathscr{B}$ is a C^*-algebra containing $\mathfrak{A}$ and acting irreducibly on $\mathscr{H}$. Show that $\mathfrak{A}$ acts irreducibly on $\mathscr{H}$ if $\mathfrak{A}$ separates the set of vector states of $\mathscr{B}$.

10.5.58. Let $\mathfrak{A}$ be a C^*-algebra acting irreducibly on a Hilbert space $\mathscr{H}$, $\mathscr{K}$ be the ideal of compact operators on $\mathscr{H}$, and $\mathscr{B}$ be a C^*-subalgebra of $\mathfrak{A}$ that separates the pure state space of $\mathfrak{A}$. Show that

(i) $\mathscr{B}$ acts irreducibly on $\mathscr{H}$;

(ii) $\varphi(\mathscr{B}) = \varphi(\mathfrak{A})$, where φ is the quotient mapping of $\mathfrak{A}$ onto $\mathfrak{A}/\mathfrak{A} \cap \mathscr{K}$ [*Hint.* Use Exercises 4.6.70 and 10.5.56.];

(iii) $\mathfrak{A} = \mathscr{B}$ if $\mathfrak{A} \cap \mathscr{K} = (0)$;

(iv) $\mathfrak{A} = \mathscr{B} + \mathscr{K}$ if $\mathscr{K} \subseteq \mathfrak{A}$;

(v) $\mathfrak{A} = \mathscr{B}$. [*Hint*. Note that if $\mathscr{B} \cap \mathscr{K} = (0)$ and $\mathscr{K} \subseteq \mathfrak{A}$, then $\omega_x|\mathscr{B}$ extends to a state of $\mathfrak{A}$ that annihilates $\mathscr{K}$.]

10.5.59. Let $\mathfrak{A}$ be a simple C^*-algebra and $\mathscr{B}$ be a C^*-algebra contained in $\mathfrak{A}$ that separates the pure state space of $\mathfrak{A}$. Show that $\mathfrak{A} = \mathscr{B}$.

10.5.60. Suppose that β is a * automorphism of a C^*-algebra $\mathfrak{A}$, and $\Psi: \mathfrak{A} \to \mathscr{B}(\mathscr{H}_\Psi)$ is the reduced atomic representation of $\mathfrak{A}$. Given any representation $\varphi: \mathfrak{A} \to \mathscr{B}(\mathscr{H}_\varphi)$ of $\mathfrak{A}$, we can consider also the representation $\varphi \circ \beta: \mathfrak{A} \to \mathscr{B}(\mathscr{H}_\varphi)$. Show that

(i) if φ and ψ are equivalent representations of $\mathfrak{A}$, then $\varphi \circ \beta$ and $\psi \circ \beta$ are equivalent representations;

(ii) if φ is an irreducible representation of $\mathfrak{A}$, then $\varphi \circ \beta$ is an irreducible representation;

(iii) there is a unitary operator U acting on $\mathscr{H}_\Psi$ such that $\Psi(\beta(A)) = U\Psi(A)U^*$ for each A in $\mathfrak{A}$;

(iv) the * automorphism $\Psi\beta\Psi^{-1}$ of $\Psi(\mathfrak{A})$ extends uniquely to a * automorphism $\bar{\beta}$ of the von Neumann algebra $\Psi(\mathfrak{A})^-$.

10.5.61. Suppose that $\mathfrak{A}$ is a C^*-algebra, δ is a derivation of $\mathfrak{A}$, and $\delta(A^*) = \delta(A)^*$ for each A in $\mathfrak{A}$ (in these circumstances, we refer to δ as a * derivation of $\mathfrak{A}$). Recall, from Exercise 4.6.65, that δ is bounded. Let α be $\exp \delta$, in the sense of the holomorphic function calculus for δ in the Banach algebra $\mathscr{B}(\mathfrak{A})$ of all bounded linear operators acting on $\mathfrak{A}$, and note that $\alpha = \sum_{n=0}^{\infty} (n!)^{-1}\delta^n$, by Theorem 3.3.5.

(i) Establish the "Leibniz formula,"

$$\delta^n(AB) = \sum_{r=0}^{n} \binom{n}{r}\delta^r(A)\delta^{n-r}(B) \qquad (A, B \in \mathfrak{A};\quad n = 1, 2, \ldots),$$

where $\binom{n}{r}$ is the binomial coefficient $n!/r!(n-r)!$.

(ii) Prove that α is a * automorphism of $\mathfrak{A}$.

(iii) Suppose that $\mathfrak{A}$ acts on a Hilbert space $\mathscr{H}$, H is a self-adjoint element of $\mathscr{B}(\mathscr{H})$, and

$$\delta(A) = i(HA - AH) \qquad (A \in \mathfrak{A}).$$

Prove that

$$\delta^n(A) = i^n \sum_{r=0}^{n} (-1)^r \binom{n}{r} H^{n-r}AH^r \qquad (A \in \mathfrak{A},\quad n = 1, 2, \ldots),$$

and that

$$\alpha(A) = UAU^* \qquad (A \in \mathscr{R}),$$

where U is the unitary operator $\exp iH$.

[In Exercise 10.5.72, we prove that a * automorphism α of a C^*-algebra $\mathfrak{A}$ has the form $\exp\delta$, where δ is a * derivation of $\mathfrak{A}$, if $\|\alpha - \iota\| < 2$, where ι is the identity mapping on $\mathfrak{A}$; Exercises 10.5.65–10.5.71 lead up to this result. From Exercises 10.5.63 and 6.9.43(ii), it follows that a * automorphism of a von Neumann algebra cannot necessarily be expressed as the exponential of a * derivation.]

10.5.62. Suppose that δ is a * derivation of a C^*-algebra $\mathfrak{A}$ and α is the * automorphism $\exp\delta$ of $\mathfrak{A}$. Show that α is "universally weakly inner" in the following sense: if φ is a faithful representation of $\mathfrak{A}$, the * automorphism $\varphi\alpha\varphi^{-1}$ of $\varphi(\mathfrak{A})$ is implemented by a unitary operator in $\varphi(\mathfrak{A})^-$. [*Hint*. Use Exercise 8.7.55(ii) and 10.5.61.]

10.5.63. Show that a * automorphism α of a von Neumann algebra $\mathscr{R}$ is inner if and only if α has the form $\exp\delta$, where δ is a * derivation of $\mathscr{R}$.

10.5.64. Suppose that $\mathfrak{A}$ is a C^*-algebra, δ is a bounded linear operator from $\mathfrak{A}$ into $\mathfrak{A}$, and $\exp t\delta$ is a * automorphism α_t of $\mathfrak{A}$ for each real number t. Show that δ is a * derivation of $\mathfrak{A}$.

10.5.65. Suppose that $\mathscr{B}$ is a C^*-algebra, ι is the identity mapping on $\mathscr{B}$, β is a bounded linear operator from $\mathscr{B}$ into $\mathscr{B}$, and $\mathbb{C}\backslash\mathrm{sp}(\beta)$ is connected, where $\mathrm{sp}(\beta)$ denotes the spectrum of β in the Banach algebra of all bounded linear operators acting on $\mathscr{B}$. Let $\mathfrak{A}$ be a norm-closed subspace of $\mathscr{B}$, and suppose that $\beta(\mathfrak{A}) \subseteq \mathfrak{A}$.

(i) Show that $(z\iota - \beta)^{-1}(\mathfrak{A}) \subseteq \mathfrak{A}$ when $|z| > \|\beta\|$.

(ii) Suppose that $A \in \mathfrak{A}$ and ρ is a bounded linear functional on $\mathscr{B}$ that vanishes on $\mathfrak{A}$. Prove that

$$\rho((z\iota - \beta)^{-1}(A)) = 0 \qquad (z \in \mathbb{C}\backslash\mathrm{sp}(\beta)),$$

and deduce that $(z\iota - \beta)^{-1}(\mathfrak{A}) \subseteq \mathfrak{A}$ for all z in $\mathbb{C}\backslash\mathrm{sp}(\beta)$.

(iii) Suppose that f is a complex-valued function holomorphic on an open set containing $\mathrm{sp}(\beta)$, and let $f(\beta)$ be the bounded linear operator on $\mathscr{B}$ that corresponds to f in the holomorphic function calculus for β. Show that $(f(\beta))(\mathfrak{A}) \subseteq \mathfrak{A}$.

10.5.66. Suppose that $\mathscr{B}$ is a C^*-algebra, H is a self-adjoint element of $\mathscr{B}$, and $\|H\| < \frac{1}{2}\pi$. Define bounded linear operators L_H, R_H, δ, and β, acting on $\mathscr{B}$, by

$$L_H B = HB, \qquad R_H B = BH, \qquad \delta(B) = i(HB - BH) \qquad (B \in \mathscr{B}),$$

and $\beta = \exp\delta$. (For these operators, "functions" are to be interpreted in terms of holomorphic function calculus within the Banach algebra of all bounded linear operators acting on $\mathscr{B}$, and "sp" will denote spectrum relative to that algebra.)

(i) Prove that $\mathrm{sp}(L_H) \subseteq \mathrm{sp}_{\mathscr{B}}(H)$, $\mathrm{sp}(R_H) \subseteq \mathrm{sp}_{\mathscr{B}}(H)$.
(ii) Deduce from (i) that

$$\begin{aligned}\mathrm{sp}(\delta) &\subseteq \{i(s-t) : s, t \in \mathrm{sp}_{\mathscr{B}}(H)\}\\ &\subseteq \{iu : u \in \mathbb{R}, |u| \le 2\|H\|\},\\ \mathrm{sp}(\beta) &\subseteq \{\exp iu : u \in \mathbb{R}, |u| \le 2\|H\|\}\\ &\subseteq \{z \in \mathbb{C} : |z| = 1, z \ne -1\}.\end{aligned}$$

[*Hint*. Use Exercise 3.5.24 and the spectral mapping theorem (3.3.6).]
(iii) Let "log" denote the principal value of the logarithm in

$$\mathbb{C}_s = \{z \in \mathbb{C} : z \ne -|z|\},$$

the plane slit along the negative real axis; that is,

$$\log re^{iu} = \log r + iu \qquad (r > 0, \quad -\pi < u < \pi).$$

Prove that $\delta = \log\beta$. [*Hint*. Use Theorem 3.3.8.]
(iv) Suppose that $\mathfrak{A}$ is a norm-closed subspace of $\mathscr{B}$, and $\beta(\mathfrak{A}) \subseteq \mathfrak{A}$. By using Exercise 10.5.65(iii), show that $\delta(\mathfrak{A}) \subseteq \mathfrak{A}$.

10.5.67. Suppose that V is a unitary operator acting on a Hilbert space $\mathscr{H}$, α is the * automorphism of $\mathscr{B}(\mathscr{H})$ defined by

$$\alpha(A) = VAV^* \qquad (A \in \mathscr{B}(\mathscr{H})),$$

and x is a unit vector in $\mathscr{H}$.

(i) Show that $\|(VE - EV)x\| = (1 - |\langle Vx, x\rangle|^2)^{1/2}$, where E is the projection from $\mathscr{H}$ onto the one-dimensional subspace containing x.
(ii) By considering $[\alpha(2E - I) - 2E + I]Vx$, prove that

$$\|\alpha - \iota\| \ge 2(1 - |\langle Vx, x\rangle|^2)^{1/2},$$

where ι is the identity mapping on $\mathscr{B}(\mathscr{H})$.

10.5.68. Suppose that V is a unitary operator acting on a Hilbert space $\mathscr{H}$, α is the * automorphism of $\mathscr{B}(\mathscr{H})$ defined by

$$\alpha(A) = VAV^* \qquad (A \in \mathscr{B}(\mathscr{H})),$$

$\lambda_1, \ldots, \lambda_n$ are distinct elements of $\mathrm{sp}(V)$, and $c = a_1\lambda_1 + \cdots + a_n\lambda_n$, where $a_1, \ldots, a_n$ are positive real numbers with sum 1.

(i) Suppose that $\varepsilon > 0$; $\mathcal{O}_1, \ldots, \mathcal{O}_n$ are disjoint open sets in $\mathbb{C}$, each with diameter less than ε, such that $\lambda_j \in \mathcal{O}_j$; and E_j is the spectral projection for V corresponding to $\mathcal{O}_j$. Show that

$$E_j \neq 0, \qquad \|VE_j - \lambda_j E_j\| < \varepsilon \quad (j = 1, \ldots, n).$$

Deduce that if $x = \sum_{j=1}^{n} a_j^{1/2} y_j$, where y_j is a unit vector in the range of E_j, then

$$\|x\| = 1, \qquad |\langle Vx, x\rangle - c| < \varepsilon.$$

(ii) By using Exercise 10.5.67 and (i), show that

$$\|\alpha - \iota\| \geq 2(1 - |c|^2)^{1/2},$$

and deduce that

$$|c| \geq \tfrac{1}{2}(4 - \|\alpha - \iota\|^2)^{1/2}.$$

10.5.69. Suppose that $\mathscr{H}$ is a Hilbert space, α is a * automorphism of $\mathscr{B}(\mathscr{H})$, and $\|\alpha - \iota\| < 2$, where ι is the identity mapping on $\mathscr{B}(\mathscr{H})$.

(i) By using Corollary 9.3.5, show that there is a unitary operator V acting on $\mathscr{H}$ such that

$$\alpha(A) = VAV^* \qquad (A \in \mathscr{B}(\mathscr{H})).$$

(ii) Deduce from Exercise 10.5.68 that

$$|c| \geq \tfrac{1}{2}(4 - \|\alpha - \iota\|^2)^{1/2} \quad (>0)$$

for all c in the (closed) convex hull of sp(V).

(iii) Let c_0 be the point closest to 0 in the (closed) convex hull of sp(V), and let U be the unitary operator $c_0^{-1}|c_0|V$. Show that

$$\alpha(A) = UAU^* \qquad (A \in \mathscr{B}(\mathscr{H}))$$

and

$$\text{sp}(U) \subseteq \{z \in \mathbb{C} : |z| = 1, \text{Re}\, z \geq \tfrac{1}{2}(4 - \|\alpha - \iota\|^2)^{1/2}\}.$$

[This condition on sp(U) can be interpreted geometrically as saying that sp(U) lies in the arc of the unit circle symmetric about 1 with endpoints midway (on the circle) between 1 and the points on the circle at (straight-line) distance $\|\alpha - \iota\|$ from 1. In view of the assumption that $\|\alpha - \iota\| < 2$, it follows that sp(U) is contained in the "open right half-plane" $\{z \in \mathbb{C} : \text{Re}\, z > 0\}$.]

10.5.70. Suppose that $\{\mathscr{H}_a : a \in \mathbb{A}\}$ is a family of Hilbert spaces, $\mathscr{H}$ is $\sum \oplus \mathscr{H}_a$, $\mathfrak{A}$ is a C^*-algebra of operators acting on $\mathscr{H}$, and $\mathfrak{A}^-$ is

$\sum \oplus \mathscr{B}(\mathscr{H}_a)$. Let α be a * automorphism of $\mathfrak{A}$ such that $\|\alpha - \iota\| < 2$, where ι is the identity mapping on $\mathfrak{A}$.

(i) By using the result of Exercise 10.5.14, with $\pi: \mathfrak{A} \to \mathscr{B}(\mathscr{H})$ the inclusion mapping, show that there exist automorphisms $\tilde{\alpha}$ of $\mathfrak{A}^-$ and α_a of $\mathscr{B}(\mathscr{H}_a)$, for each a in $\mathbb{A}$, such that $\tilde{\alpha}$ extends α,

$$\|\alpha_a - \iota_a\| \leq \|\tilde{\alpha} - \tilde{\iota}\| = \|\alpha - \iota\| < 2$$

(where $\tilde{\iota}$, ι_a are the identity mappings on $\mathfrak{A}^-$, $\mathscr{B}(\mathscr{H}_a)$, respectively), and

$$\tilde{\alpha}(\sum \oplus A_a) = \sum \oplus \alpha_a(A_a)$$

for each element $\sum \oplus A_a$ of $\mathfrak{A}^-$ $(=\sum \oplus \mathscr{B}(\mathscr{H}_a))$.

(ii) Show that there is a unitary element U of $\mathfrak{A}^-$ such that

$$\alpha(A) = UAU^* \qquad (A \in \mathfrak{A})$$

and

$$\mathrm{sp}(U) \subseteq \{z \in \mathbb{C} : |z| = 1, \operatorname{Re} z \geq \tfrac{1}{2}(4 - \|\alpha - \iota\|^2)^{1/2}\}.$$

10.5.71. Suppose that $\mathfrak{A}$ is a C^*-algebra acting on a Hilbert space $\mathscr{H}$, α is a * automorphism of $\mathfrak{A}$, and U is a unitary operator acting on $\mathscr{H}$ such that

$$\alpha(A) = UAU^* \qquad (A \in \mathfrak{A})$$

and

$$\mathrm{sp}(U) \subseteq \{z \in \mathbb{C} : |z| = 1, \operatorname{Re} z > 0\}.$$

Prove that

(i) there is a self-adjoint element H of $\mathscr{B}(\mathscr{H})$ such that $\|H\| < \frac{1}{2}\pi$ and $\exp iH = U$;

(ii) the equations

$$\bar{\alpha}(B) = UBU^*, \qquad \bar{\delta}(B) = i(HB - BH) \qquad (B \in \mathscr{B}(\mathscr{H}))$$

define a * automorphism $\bar{\alpha}$ and a * derivation $\bar{\delta}$ of $\mathscr{B}(\mathscr{H})$, such that $\bar{\alpha} = \exp \bar{\delta}$ and $\bar{\alpha} | \mathfrak{A} = \alpha$;

(iii) $\bar{\delta}(\mathfrak{A}) \subseteq \mathfrak{A}$, and $\bar{\delta} | \mathfrak{A}$ is a * derivation δ of $\mathfrak{A}$ such that $\exp \delta = \alpha$. [*Hint*. Use Exercise 10.5.66(iv).]

10.5.72. Suppose that $\mathfrak{A}$ is a C^*-algebra, α is a * automorphism of $\mathfrak{A}$, and $\|\alpha - \iota\| < 2$, where ι is the identity mapping on $\mathfrak{A}$. Show that there is a * derivation δ of $\mathfrak{A}$ such that $\alpha = \exp \delta$. [*Hint*. It is sufficient to consider the case in which $\mathfrak{A}$ is given, acting on a Hilbert space $\mathscr{H}$, in its reduced atomic representation. In this case, use Exercises 10.5.70 and 10.5.71.]

10.5.73. Suppose that α is a * automorphism of a von Neumann algebra $\mathscr{R}$, and $\|\alpha - \iota\| < 2$, where ι is the identity mapping on $\mathscr{R}$. Show that α is an inner automorphism of $\mathscr{R}$.

10.5.74. Suppose that $\mathfrak{A}$ is a C^*-algebra and aut($\mathfrak{A}$) is the set of all * automorphisms of $\mathfrak{A}$.

(i) Show that aut($\mathfrak{A}$) is a subgroup of the (multiplicative) group of invertible elements in the Banach algebra $\mathscr{B}(\mathfrak{A})$ of all bounded linear operators from $\mathfrak{A}$ into $\mathfrak{A}$. Deduce that aut($\mathfrak{A}$), with its (relative) norm topology as a subset of $\mathscr{B}(\mathfrak{A})$, is a topological group. (A set G that is both a group and also a Hausdorff topological space is described as a *topological group* if the mappings

$$(g, h) \to gh : G \times G \to G \qquad \text{and} \qquad g \to g^{-1} : G \to G$$

are continuous.)

(ii) Show that $\|\alpha - \beta\| \leq 2$ for all α and β in aut($\mathfrak{A}$).

(iii) Suppose that $\alpha \in$ aut($\mathfrak{A}$) and $\|\alpha - \iota\| < 2$, where ι is the unit element of aut($\mathfrak{A}$). Deduce from Exercises 10.5.72 and 10.5.62 that α lies on a (norm-continuous) one-parameter subgroup of aut($\mathfrak{A}$), and is universally weakly inner. (By a *one-parameter subgroup* of a topological group G, we mean a continuous homomorphism $t \to g_t$ from the additive group $\mathbb{R}$ into G; we refer to "the one-parameter subgroup $\{g_t\}$ of G.")

(iv) Let aut$_\iota$($\mathfrak{A}$) be the subgroup of aut($\mathfrak{A}$) generated (algebraically) by the set $\{\alpha \in$ aut($\mathfrak{A}$)$: \|\alpha - \iota\| < 2\}$. Show that aut$_\iota$($\mathfrak{A}$) is a connected open subgroup of aut($\mathfrak{A}$), and deduce that aut$_\iota$($\mathfrak{A}$) is the connected component of ι in aut($\mathfrak{A}$). [*Hint*. By considering cosets of aut$_\iota$($\mathfrak{A}$), show that aut$_\iota$($\mathfrak{A}$) is closed as well as open.]

(v) Show that aut$_\iota$($\mathfrak{A}$) is the subgroup of aut($\mathfrak{A}$) generated (algebraically) by the one-parameter subgroups of aut($\mathfrak{A}$), and each element of aut$_\iota$($\mathfrak{A}$) is universally weakly inner.

10.5.75. Suppose that $\mathfrak{A}$ is a C^*-algebra and aut($\mathfrak{A}$) is the topological group considered in Exercise 10.5.74. Show that

(i) aut($\mathfrak{A}$) is discrete if and only if $\mathfrak{A}$ is abelian;

(ii) aut($\mathfrak{A}$) is connected if $\mathfrak{A}$ is a type I factor;

(iii) aut($\mathfrak{A}$) is neither discrete nor connected if $\mathfrak{A}$ is the type II_1 factor $\mathscr{L}_{\mathscr{F}_2}$ considered in Exercise 6.9.43(ii).

10.5.76. Suppose that $\mathscr{H}$ is a Hilbert space and δ is a * derivation of $\mathscr{B}(\mathscr{H})$. Let K be an element of $\mathscr{B}(\mathscr{H})$ such that $\delta(A) = i(KA - AK)$ for each A in $\mathscr{B}(\mathscr{H})$ (see Exercise 8.7.55(i)). Show that, if c is a suitably chosen real number and $H = \frac{1}{2}(K + K^*) - cI$, then H is self-adjoint,

$$\delta(A) = i(HA - AH) \qquad (A \in \mathscr{B}(\mathscr{H})),$$

sp(H) contains both $\|H\|$ and $-\|H\|$, and $\|H\| = \frac{1}{2}\|\delta\|$. [*Hint*. For the last assertion, suppose that $\varepsilon > 0$, and let E_ε, F_ε be the spectral projections for H, corresponding to the intervals $[\|H\| - \varepsilon, \|H\|]$, $[-\|H\|, -\|H\| + \varepsilon]$, respectively. Consider $\delta(V_\varepsilon)$, where V_ε is a non-zero partial isometry such that $V_\varepsilon^* V_\varepsilon \leq E_\varepsilon$, $V_\varepsilon V_\varepsilon^* \leq F_\varepsilon$.]

10.5.77. Suppose that $\mathfrak{A}$ is a simple C^*-algebra (containing I) acting on a Hilbert space $\mathscr{H}$, δ is a * derivation of $\mathfrak{A}$, H is a self-adjoint element of $\mathscr{B}(\mathscr{H})$, and

$$\|H\| = \tfrac{1}{2}\|\delta\|, \qquad \delta(A) = i(HA - AH) \qquad (A \in \mathfrak{A}).$$

(i) Let F' be a non-zero projection in $\mathfrak{A}'$. Show that the mapping $A \rightarrow AF'$ is a faithful representation of $\mathfrak{A}$ on the Hilbert space $F'(\mathscr{H})$. Prove that

$$\delta(A)F' = i(F'HF'A - AF'HF') \qquad (A \in \mathfrak{A}),$$

and deduce that $\|F'HF'\| = \|H\|$.

(ii) Suppose that E' is a projection in $\mathfrak{A}'$, and let S be $(I - E')HE'$. Show that $S \in \mathfrak{A}'$. Deduce from (i) that $S = 0$. [*Hint*. If $S \neq 0$, let F' be the spectral projection for S^*S, corresponding to the interval $[\varepsilon, \infty)$, where $0 < \varepsilon < \|S^*S\|$. Deduce from (i) that

$$\begin{aligned}\|F'HE'HF'\| \geq \|H\|^2 &\geq \|F'H^2F'\| \\ &= \|F'HE'HF' + F'S^*SF'\|,\end{aligned}$$

and hence obtain a contradiction.]

(iii) Show that $H \in \mathfrak{A}^-$.

(iv) By considering the universal representation of the C^*-algebra $\mathscr{B}$ generated by $\mathfrak{A}$ and H, and using Proposition 10.1.4, show that $H \in \mathfrak{A}$.

10.5.78. Suppose that $\mathfrak{A}$ is a simple C^*-algebra (with unit), and δ is a derivation of $\mathfrak{A}$. Prove that δ is inner. [*Hint*. Show that it suffices to consider only * derivations. By taking an irreducible representation of $\mathfrak{A}$, reduce to the case in which $\mathfrak{A} \subseteq \mathscr{B}(\mathscr{H})$ and $\mathfrak{A}^- = \mathscr{B}(\mathscr{H})$, for some Hilbert space $\mathscr{H}$. In this case, use Exercises 7.6.15, 10.5.76, and 10.5.77.]

10.5.79. Find an example of a derivation of a C^*-algebra such that the derivation is not inner.

10.5.80. Let $\mathfrak{A}_0$ be an abelian C^*-algebra acting on a Hilbert space $\mathscr{H}$ such that $\mathfrak{A}_0 \cap \mathscr{K} = (0)$, where $\mathscr{K}$ is the ideal of compact operators on $\mathscr{H}$, and let $\mathfrak{A}$ be the C^*-algebra $\mathfrak{A}_0 + \mathscr{K}$. Show that

(i) the center $\mathscr{C}$ of $\mathfrak{A}$ consists of scalar multiples of I;

(ii) $\mathrm{co}_{\mathfrak{A}}(A)^= \cap \mathscr{C} = \varnothing$ for each A in $\mathfrak{A}$ not of the form $aI + B$ with B in $\mathscr{K}$, where $\mathrm{co}_{\mathfrak{A}}(A)^=$ is the norm-closed convex hull of $\{UAU^* : U \in \mathscr{U}\}$ and $\mathscr{U}$ is the group of unitary operators in $\mathfrak{A}$ (compare this conclusion with Theorem 8.3.5) [*Hint*. Consider the quotient mapping of $\mathfrak{A}$ onto $\mathfrak{A}/\mathscr{K}$.];

(iii) there is an $\mathfrak{A}_0$ and an A in $\mathfrak{A}$ not of the form $aI + B$ with B in $\mathscr{K}$.

10.5.81. Suppose that $\mathfrak{A}$ is a C^*-algebra. A subset $\mathscr{P}$ of $\mathfrak{A}$ is described as a *primitive ideal* if $\mathscr{P}$ is the kernel $\pi^{-1}(0)$ of some irreducible representation π of $\mathfrak{A}$. (This implies that $\mathscr{P}$ is a closed two-sided ideal, and $\mathscr{P} \neq \mathfrak{A}$.) The set of all primitive ideals in $\mathfrak{A}$ is denoted by $\mathrm{prim}(\mathfrak{A})$.

(i) Suppose that $\mathscr{I}$ is a closed two-sided ideal in $\mathfrak{A}$, and $\mathscr{I} \neq \mathfrak{A}$. Show that

$$\mathscr{I} = \bigcap \{\mathscr{P} : \mathscr{P} \in \mathrm{prim}(\mathfrak{A}), \mathscr{I} \subseteq \mathscr{P}\}.$$

[*Hint*. Apply Corollary 10.2.4 to the C^*-algebra $\mathfrak{A}/\mathscr{I}$.]

(ii) For each $\mathscr{P}$ in $\mathrm{prim}(\mathfrak{A})$, let $\varphi_{\mathscr{P}}\colon \mathfrak{A} \to \mathfrak{A}/\mathscr{P}$ be the quotient mapping. Show that

$$\|A\| = \sup\{\|\varphi_{\mathscr{P}}(A)\| : \mathscr{P} \in \mathrm{prim}(\mathfrak{A})\} \qquad (A \in \mathfrak{A}).$$

[*Hint*. Consider the mapping $A \to \sum_{\mathscr{P} \in \mathrm{prim}(\mathfrak{A})} \oplus\, \varphi_{\mathscr{P}}(A)$ from $\mathfrak{A}$ into the algebra $\sum_{\mathscr{P} \in \mathrm{prim}(\mathfrak{A})} \oplus\, \mathfrak{A}/\mathscr{P}$ (see Exercise 3.5.3).]

(iii) Show that a maximal (proper) two-sided ideal in $\mathfrak{A}$ is primitive.

(iv) Suppose that $\mathscr{I}_1$ and $\mathscr{I}_2$ are closed two-sided ideals in $\mathfrak{A}$, $\mathscr{P} \in \mathrm{prim}(\mathfrak{A})$, and $A_1A_2 \in \mathscr{P}$ whenever $A_1 \in \mathscr{I}_1$ and $A_2 \in \mathscr{I}_2$. Show that $\mathscr{P}$ contains at least one of $\mathscr{I}_1$ and $\mathscr{I}_2$. [*Hint*. Let $\pi\colon \mathfrak{A} \to \mathscr{B}(\mathscr{H})$ be an irreducible representation that has kernel $\mathscr{P}$, and consider the closed subspace $\mathscr{K}$ of $\mathscr{H}$ generated by the set $\{\pi(A_2)x : A_2 \in \mathscr{I}_2, x \in \mathscr{H}\}$.]

10.5.82. Suppose that $\mathfrak{A}$ is a C^*-algebra and $\mathrm{prim}(\mathfrak{A})$ is the set of all primitive ideals in $\mathfrak{A}$. Given any subset $\mathscr{S}$ of $\mathfrak{A}$, the *hull* $h(\mathscr{S})$ of $\mathscr{S}$ is defined by

$$h(\mathscr{S}) = \{\mathscr{P} \in \mathrm{prim}(\mathfrak{A}) : \mathscr{S} \subseteq \mathscr{P}\}.$$

Show that

(i) $\{h(\mathscr{S}) : \mathscr{S} \subseteq \mathfrak{A}\}$ is the family of all closed sets in a topology on prim($\mathfrak{A}$);

(ii) if $\mathscr{F} \subseteq \text{prim}(\mathfrak{A})$ and $\mathscr{I}_0 = \bigcap \{\mathscr{P} : \mathscr{P} \in \mathscr{F}\}$, then the closure $\mathscr{F}^-$ of $\mathscr{F}$ is given by $\mathscr{F}^- = h(\mathscr{I}_0)$;

(iii) if $\mathscr{P} \in \text{prim}(\mathfrak{A})$, then the one-point set $\{\mathscr{P}\}$ is closed in prim($\mathfrak{A}$) if and only if $\mathscr{P}$ is a maximal (proper) two-sided ideal in $\mathfrak{A}$;

(iv) if $\mathscr{P}_1, \mathscr{P}_2 \in \text{prim}(\mathfrak{A})$ and $\mathscr{P}_1 \neq \mathscr{P}_2$, there is an open set in prim($\mathfrak{A}$) that contains just one of $\mathscr{P}_1, \mathscr{P}_2$;

(v) prim($\mathfrak{A}$) is compact in the following sense: if $\bigcap_{a \in \mathbb{A}} \Gamma_a = \varnothing$, where each Γ_a is a closed subset of prim($\mathfrak{A}$), then $\bigcap_{a \in \mathbb{F}} \Gamma_a = \varnothing$ for some finite subset $\mathbb{F}$ of $\mathbb{A}$;

(vi) prim($\mathfrak{A}$) has just two elements, and the topology described in (i) is not Hausdorff, if $\mathfrak{A} = \mathscr{B}(\mathscr{H})$, where $\mathscr{H}$ is a separable infinite-dimensional Hilbert space.

[The topology in (i) is called the *Jacobson topology* on prim($\mathfrak{A}$); with this topology, prim($\mathfrak{A}$) is called the *primitive ideal space*, or *primitive spectrum*, of $\mathfrak{A}$. The result of (iv) can be expressed as the assertion that prim($\mathfrak{A}$) is a T_0-space.]

10.5.83. Suppose that $\mathfrak{A}$ is a C^*-algebra, and f: prim($\mathfrak{A}$) $\to [0, 1]$ is a continuous function on the primitive ideal space prim($\mathfrak{A}$). For each $\mathscr{P}$ in prim($\mathfrak{A}$), let $\varphi_{\mathscr{P}}: \mathfrak{A} \to \mathfrak{A}/\mathscr{P}$ be the quotient mapping. The following results are subsumed in Exercise 10.5.84, but are required for the solution of that exercise.

(i) Define open subsets $\mathscr{O}_0, \ldots, \mathscr{O}_n$ of prim($\mathfrak{A}$) by

$$\mathscr{O}_j = \left\{\mathscr{P} \in \text{prim}(\mathfrak{A}) : \frac{j-1}{n} < f(\mathscr{P}) < \frac{j+1}{n}\right\},$$

and let

$$\mathscr{I}_j = \bigcap \{\mathscr{P} : \mathscr{P} \in \text{prim}(\mathfrak{A}) \backslash \mathscr{O}_j\}$$

Show that

$$\mathscr{O}_j = \{\mathscr{P} \in \text{prim}(\mathfrak{A}) : \mathscr{I}_j \nsubseteq \mathscr{P}\}, \qquad \mathscr{I}_0 + \cdots + \mathscr{I}_n = \mathfrak{A}.$$

(ii) Suppose that $A \in \mathfrak{A}^+$. By using (i) and Exercise 4.6.64, show that there exist $A_0, \ldots, A_n$ such that

$$A = A_0 + \cdots + A_n, \qquad A_j \in \mathscr{I}_j^+ \qquad (j = 0, \ldots, n).$$

Prove that

$$\|\varphi_{\mathscr{P}}(B) - f(\mathscr{P})\varphi_{\mathscr{P}}(A)\| < \frac{2}{n}\|A\| \qquad (\mathscr{P} \in \mathrm{prim}(\mathfrak{A})),$$

where B ($\in \mathfrak{A}$) is defined by

$$B = \sum_{j=0}^{n} \frac{j}{n} A_j.$$

10.5.84. Suppose that $\mathfrak{A}$ is a C^*-algebra, $A \in \mathfrak{A}$, and f is a (bounded) continuous complex-valued function on the primitive ideal space prim($\mathfrak{A}$). For each $\mathscr{P}$ in prim($\mathfrak{A}$), let $\varphi_{\mathscr{P}}: \mathfrak{A} \to \mathfrak{A}/\mathscr{P}$ be the quotient mapping.

(i) Suppose that $\varepsilon > 0$. By using the result of Exercise 10.5.83(ii), show that there is an element B_ε of $\mathfrak{A}$ such that

$$\|\varphi_{\mathscr{P}}(B_\varepsilon) - f(\mathscr{P})\varphi_{\mathscr{P}}(A)\| < \varepsilon \qquad (\mathscr{P} \in \mathrm{prim}(\mathfrak{A})).$$

(ii) Deduce that there is an element B of $\mathfrak{A}$ such that

$$\varphi_{\mathscr{P}}(B) = f(\mathscr{P})\varphi_{\mathscr{P}}(A) \qquad (\mathscr{P} \in \mathrm{prim}(\mathfrak{A})).$$

[The result in (ii) is known as the Dauns–Hoffman theorem.]

10.5.85. Let $\mathscr{B}$ be a C^*-subalgebra of the C^*-algebra $\mathfrak{A}$, and let φ_0 be an idempotent ($\varphi_0 \circ \varphi_0 = \varphi_0$) linear mapping of $\mathfrak{A}$ onto $\mathscr{B}$ such that $\|\varphi_0\| = 1$. Suppose $\mathfrak{A}$ acting on $\mathscr{H}$ is the universal representation of $\mathfrak{A}$ and, in this representation, E is a projection in $\mathscr{B}^-$. Show that

(i) φ_0 is a positive linear mapping of $\mathfrak{A}$ onto $\mathscr{B}$ such that $\varphi_0(I) = I$ [*Hint*. Compose φ_0 with states of $\mathscr{B}$ and use Theorem 4.3.2.];

(ii) φ_0 extends uniquely to an ultraweakly continuous idempotent linear mapping φ of $\mathfrak{A}^-$ onto $\mathscr{B}^-$ such that $\|\varphi\| = 1$, and φ is a positive linear mapping;

(iii) $\omega_x \circ \varphi$ is a state of $\mathfrak{A}^-$ definite on E when x is a unit vector in either $E(\mathscr{H})$ or $(I - E)(\mathscr{H})$ (see Exercise 4.6.16);

(iv) $E\varphi(EA)E = E\varphi(AE)E = E\varphi(A)E$, $E\varphi(EAE)E = E\varphi(A)E$, and $(I - E)\varphi(EA)(I - E) = (I - E)\varphi(AE)(I - E) = 0$ for each A in $\mathfrak{A}^-$;

(v) $\varphi(EAE) = E\varphi(A)E$ for each A in $\mathfrak{A}^-$ [*Hint*. Recall that, with A self-adjoint, $-\|A\|E \leq EAE \leq \|A\|E$.];

(vi) $\varphi(EA(I - E)) = (I - E)\varphi(EA(I - E))E + E\varphi(EA(I - E))(I - E)$ for each A in $\mathfrak{A}^-$.

10.5.86. With the notation and assumptions of Exercise 10.5.85, show that

(i) $\|ET(I-E)+(I-E)SE\| = \max\{\|ET(I-E)\|, \|(I-E)SE\|\}$ for all T and S in $\mathscr{B}(\mathscr{H})$;

(ii) $(I-E)\varphi(EA(I-E))E = 0$ [*Hint*. Suppose the contrary and show that, for each A in $\mathfrak{A}^-$ and all large integers n,

$$n\|(I-E)\varphi(EA(I-E))E\| = \|\varphi(EA(I-E)) + (n-1)(I-E)\varphi(EA(I-E))E\|$$

(use (i) and Exercise 10.5.85(vi)), and deduce a contradiction from this equality.];

(iii) $\varphi(EA) = E\varphi(A)$ and $\varphi(AE) = \varphi(A)E$ for each A in $\mathfrak{A}^-$;

(iv) $\varphi(BA) = B\varphi(A)$ and $\varphi(AB) = \varphi(A)B$ for each A in $\mathfrak{A}^-$ and each B in $\mathscr{B}^-$;

(v) φ is a conditional expectation from $\mathfrak{A}^-$ onto $\mathscr{B}^-$ in the sense of Exercise 8.7.23;

(vi) φ_0 is a conditional expectation from $\mathfrak{A}$ onto $\mathscr{B}$ (that is, the conditions on Φ in Exercise 8.7.23 are fulfilled for φ_0 with φ_0, $\mathfrak{A}$, and $\mathscr{B}$, in place of Φ, $\mathscr{S}$, and $\mathscr{R}$).

10.5.87. Suppose the C^*-algebra $\mathfrak{A}$ is (linearly isomorphic and isometric to) the norm dual of a Banach space $\mathfrak{A}_\#$, and let η be the natural injection of $\mathfrak{A}_\#$ into $\mathfrak{A}^\#$.

(i) Let v be an element of $\mathfrak{A}^{\#\#}$. Show that $v \circ \eta = A$ for a unique A in $\mathfrak{A}$ (viewed as linear functionals on $\mathfrak{A}_\#$).

(ii) Let $\mathfrak{A}$ acting on $\mathscr{H}$ be the universal representation of $\mathfrak{A}$, and let $A \to \hat{A}$ be the (isometric linear) isomorphism (of Proposition 10.1.21) between $\mathfrak{A}^-$ and $\mathfrak{A}^{\#\#}$. Let $\varphi(A)$ be the element in $\mathfrak{A}$ (obtained in (i)) such that $\hat{A} \circ \eta = \varphi(A)$, where $A \in \mathfrak{A}^-$. Show that φ is an idempotent mapping of $\mathfrak{A}^-$ onto $\mathfrak{A}$ and $\|\varphi\| = 1$. Deduce that φ is a conditional expectation from $\mathfrak{A}^-$ onto $\mathfrak{A}$.

(iii) Let $\mathscr{K}$ be $\varphi^{-1}(0)$. Show that $\mathscr{K}$ is a weak-operator-closed two-sided ideal in $\mathfrak{A}^-$.

(iv) Let P be the central projection in $\mathfrak{A}^-$ (of Theorem 6.8.8) such that $\mathscr{K} = \mathfrak{A}^-P$. Show that $\mathfrak{A}^-(I-P) = \mathfrak{A}(I-P)$. [*Hint*. Note that $A - \varphi(A) \in \mathfrak{A}^-P$ for each A in $\mathfrak{A}^-$.]

(v) Show that $\mathfrak{A}(I-P)$ is * isomorphic to $\mathfrak{A}$ and conclude that $\mathfrak{A}$ is a W^*-algebra. (Compare this exercise with Exercises 7.6.41–7.6.45.)

10.5.88. Let $\mathfrak{A}$ be the uniformly matricial (the CAR) algebra of type $\{2^j\}$, and let $\{\mathfrak{A}_j\}$ be a generating nest of type $\{2^j\}$. (The existence of such an algebra $\mathfrak{A}$ is proved in Proposition 10.4.18). With $\mathscr{S}$ a subset of $\mathfrak{A}$, let $\mathscr{S}^c$ be the set of those T in $\mathfrak{A}$ that commute with every S in $\mathscr{S}$, and let $\mathscr{B}_j$ be $\mathfrak{A}_{j-1}^c \cap \mathfrak{A}_j$ for j in $\{1, 2, \ldots\}$, where $\mathfrak{A}_0 = \{cI : c \in \mathbb{C}\}$.

(i) Show that each $\mathscr{B}_j$ is a C^*-subalgebra of $\mathfrak{A}$ (containing I) * isomorphic to the algebra of 2×2 complex matrices.

(ii) Show that $B_jB_k = B_kB_j$ when $B_j \in \mathscr{B}_j$, $B_k \in \mathscr{B}_k$, and $j \neq k$; $\mathfrak{A}_j$ is the linear span of products $B_1, \ldots, B_j$ with B_k in $\mathscr{B}_k$; $\mathfrak{A}$ is the norm closure of the linear span of products $B_1, \ldots, B_n$ with B_k in $\mathscr{B}_k$ and n in $\{1, 2, \ldots\}$.

(iii) In each $\mathscr{B}_j$, choose a 2×2 self-adjoint system of matrix units and let $\sigma_x^{(j)}, \sigma_y^{(j)}, \sigma_z^{(j)}$ be the elements of $\mathscr{B}_j$ whose matrix representations relative to the chosen matrix units are

$$\begin{pmatrix} 1 & 0 \\ 0 & -1 \end{pmatrix}, \quad \begin{pmatrix} 0 & i \\ -i & 0 \end{pmatrix}, \quad \begin{pmatrix} 0 & 1 \\ 1 & 0 \end{pmatrix},$$

respectively. (These are the "Pauli spin matrices.") Show that $\sigma_x^{(j)}, \sigma_y^{(j)}, \sigma_z^{(j)}$ generate $\mathscr{B}_j$ as an algebra and that $\{\sigma_x^{(j)}, \sigma_y^{(j)}, \sigma_z^{(j)} : j = 1, 2, \ldots\}$ generate $\mathfrak{A}$ as a C^*-algebra.

(iv) Let C_j be $\sigma_z^{(1)} \cdots \sigma_z^{(j-1)}(\sigma_x^{(j)} - i\sigma_y^{(j)})/2$. Show that

$$(*) \qquad \begin{aligned} C_jC_k + C_kC_j &= 0 \qquad (j, k = 1, 2, \ldots), \\ C_jC_k^* + C_k^*C_j &= 0 \qquad (j \neq k), \\ C_jC_j^* + C_j^*C_j &= I \qquad (j = 1, 2, \ldots). \end{aligned}$$

[The equations $(*)$ are referred to as the "canonical anticommutation relations"—abbreviated "CAR"—and the set of elements $\{C_j\}$ is said to "satisfy the CAR."]

(v) Show that

$$\sigma_z^{(j)} = 2C_j^*C_j - I, \ \sigma_x^{(j)} = \sigma_z^{(1)} \cdots \sigma_z^{(j-1)}(C_j + C_j^*), \ \sigma_y^{(j)} = i\sigma_z^{(1)} \cdots \sigma_z^{(j-1)}(C_j - C_j^*).$$

Conclude that the set $\{C_1, \ldots, C_j\}$ generates $\mathfrak{A}_j$ as a (finite-dimensional) C^*-algebra and that $\{C_1, C_2, \ldots\}$ generates $\mathfrak{A}$ as a C^*-algebra.

(iv) Let $\mathscr{W}$ be the set of products $C_{m(1)}^* \cdots C_{m(h)}^* C_{n(1)} \cdots C_{n(k)}$, where $0 < m(1) < m(2) < \cdots < m(h) \leq j$, $0 < n(1) < \cdots < n(k) \leq j$, and $h, k \in \{0, \ldots, j\}$. (Such a product is said to be "Wick ordered." If $h = 0$, the product is $C_{n(1)} \cdots C_{n(k)}$. If $k = 0$, the product is $C_{m(1)}^* \cdots C_{m(h)}^*$. If h and k are 0, the product is I.) Show that $\mathscr{W}$ is a (linear-space) basis for $\mathfrak{A}_j$.

10.5.89. With the notation of Exercise 10.5.88, let π_0 be a representation of $\mathfrak{A}$ on the Hilbert space $\mathscr{H}$. Then $\{\pi_0(C_j)\}$ is a family of operators on $\mathscr{H}$ that satisfy the CAR. We say that a family of operators acting on a Hilbert space and satisfying the CAR is a "representation of the CAR." Thus each representation of the CAR algebra $\mathfrak{A}$ gives rise to a representation of the CAR. Show that each representation of the CAR arises, in this way, from a representation of the CAR algebra.

10.5.90. With the terminology of Exercise 10.5.89, we call a representation $\{\pi(C_j)\}$ of the CAR on a Hilbert space $\mathscr{H}$ "irreducible" when no closed subspace of $\mathscr{H}$ other than $\{0\}$ and $\mathscr{H}$ is invariant under all the operators in $\{\pi(C_j), \pi(C_j)^*\}$. Show that there are an uncountable infinity of (unitarily) inequivalent irreducible representations of the CAR.

10.5.91. With the notation of Exercise 10.5.3, show that co $\mathscr{U}(\mathscr{R}) \neq (\mathscr{R})_1$ if $\mathscr{R}$ is not finite. Conclude that co $\mathscr{U}(\mathscr{R}) = (\mathscr{R})_1$ if and only if $\mathscr{R}$ is finite. [*Hint*. Use Theorem 7.3.1.]

10.5.92. Let $\mathfrak{A}$ be a C^*-algebra, $\mathscr{U}(\mathfrak{A})$ its unitary group, co $\mathscr{U}(\mathfrak{A})$ the convex hull of $\mathscr{U}(\mathfrak{A})$, and S an element of $\mathfrak{A}$ such that $\|S\| < 1$. Show that

(i) each invertible element of $(\mathfrak{A})_1$ is the midpoint of two elements of $\mathscr{U}(\mathfrak{A})$ [*Hint*. Use the proof of Theorem 4.1.7 and "polar decomposition."];

(ii) for each U in $\mathscr{U}(\mathfrak{A})$, $(S + U)/2$ is the midpoint of two elements of $\mathscr{U}(\mathfrak{A})$;

(iii) for each U in $\mathscr{U}(\mathfrak{A})$, $U + (n - 1)S = \sum_{k=1}^{n} U_k$ for some $U_1, \ldots, U_n$ in $\mathscr{U}(\mathfrak{A})$;

(iv) $S = n^{-1}\sum_{k=1}^{n} U_k$ for some $U_1, \ldots, U_n$ in $\mathscr{U}(\mathfrak{A})$ when $\|S\| < 1 - 2n^{-1}$ (we say that S is the *mean* of the n unitary elements $U_1, \ldots, U_n$) [*Hint*. Note that $(n - 1)^{-1}(nS - I)$ has norm less than 1 and use it in place of S in (iii)—use I in place of U.];

(v) the open unit ball of $\mathfrak{A}$ is contained in co $\mathscr{U}(\mathfrak{A})$ and conclude again (see Exercise 10.5.4) that $[\text{co}\, \mathscr{U}(\mathfrak{A})]^= = (\mathfrak{A})_1$;

(vi) each T in $\mathfrak{A}$ is some positive multiple of a sum of three unitary elements of $\mathfrak{A}$.

10.5.93. With the notation of Exercise 10.5.92, show that

(i) $\frac{1}{2}[\mathscr{U}(\mathfrak{A}) + \mathscr{U}(\mathfrak{A})] = [a\mathscr{U}(\mathfrak{A}) + (1 - a)\mathscr{U}(\mathfrak{A}) : 0 \leq a \leq 1]$;

(ii) when $\mathfrak{A}$ is a von Neumann algebra, the sets in (i) coincide with

$$\{UH : U \in \mathscr{U}(\mathfrak{A}), 0 \leq H \leq I, H \in \mathfrak{A}\}.$$

[*Hint*. Note that, with U_1, U_2 in $\mathscr{U}(\mathfrak{A})$, $U_1 + U_2 = U_1(I + U_1^*U_2)$ and $I + U_1^*U_2$ is a normal operator.]

10.5.94. Let $\mathfrak{A}$ be a C^*-algebra. Show that each convex combination of 2^n or fewer elements of $\mathscr{U}(\mathfrak{A})$ is a mean of 2^n elements of $\mathscr{U}(\mathfrak{A})$. [*Hint*. Use Exercise 10.5.93(i) and argue by induction.]

10.5.95. Let $\mathscr{H}$ be a Hilbert space and V be an isometry of $\mathscr{H}$ $(V^*V = I)$ such that $VV^* = E < I$. Show that

(i) sp V is the closed unit disk in $\mathbb{C}$ [*Hint*. Let e_0 be a unit vector in $(I - E)(\mathscr{H})$, e_k be $V^k e_0$ for k in $\{1, 2, \ldots\}$, and argue, as in Example 3.2.18, that λ is an eigenvalue for V^* when $|\lambda| < 1$.];

(ii) if $U_1, \ldots, U_n$ are unitary operators acting on $\mathscr{H}$, then

$$\|V - n^{-1}\sum_{k=1}^{n} U_k\| \geq 2n^{-1}$$

[*Hint*. Assume the contrary, study $\|U_1^*V - n^{-1}I\|$, and use (i).];

(iii) S_n is a mean of n, but not fewer than n, unitary elements of $\mathscr{R}$ when V lies in the von Neumann algebra $\mathscr{R}$ and $S_n = a_n V$, where

$$1 - 2(n-1)^{-1} < a_n < 1 - 2n^{-1}$$

[*Hint*. Use Exercise 10.5.92(iv) and (ii).];

(iv) there is no number n such that each element of $(\mathscr{R})_1^{\circ}$ can be expressed as a convex combination of n or fewer elements of $\mathscr{U}(\mathscr{R})$, when $\mathscr{R}$ is an infinite von Neumann algebra and $(\mathscr{R})_1^{\circ} = \{A \in \mathscr{R}: \|A\| < 1\}$.

10.5.96. Let $\mathfrak{A}$ be a C^*-algebra, $\mathscr{U}(\mathfrak{A})$ be its unitary group, and A be a self-adjoint element of $\mathfrak{A}$. Let $\mathscr{S}_a$ be $[-1, -(1-2a)] \cup [(1-2a), 1]$, where $0 \leq a \leq \frac{1}{2}$. Show that

(i) sp $A \subseteq \mathscr{S}_a$ if $A = aU_1 + (1-a)U_2$ for some U_1 and U_2 in $\mathscr{U}(\mathfrak{A})$ [*Hint*. With λ in sp A, use Exercises 4.6.16 and 4.6.31 to find a state ρ of $\mathfrak{A}$, definite on A, such that $\rho(A) = \lambda$.];

(ii) ξ is a continuous mapping of $\mathscr{S}_a$ into $\mathbb{C}_1^+$ such that $|1 + a(1-a)^{-1}\xi(t)| = |t(1-a)^{-1}|$, where $\mathbb{C}_1^+$ is the set of complex numbers of modulus 1 with non-negative imaginary part, $\xi(t) = \theta(g(|t(1-a)^{-1}|))$ for t in $\mathscr{S}_a$, $\theta(r)$ is the (unique) element of $\mathbb{C}_1^+$ with real part r, and g is the inverse mapping to $t \to |1 + a(1-a)^{-1}\theta(t)|$ of $[-1, 1]$ onto the closed interval $[(1-2a)(1-a)^{-1}, (1-a)^{-1}]$;

(iii) f_1 and f_2 are continuous mappings of $\mathscr{S}_a$ into the complex numbers of modulus 1 and $af_1(t) + (1-a)f_2(t) = t$ for t in $\mathscr{S}_a$, where $f_1(t) = t + i(1-t^2)^{1/2}$ and $f_2(t) = t - i(1-t^2)^{1/2}$ when $a = \frac{1}{2}$, $f_2(t) = t(1-a)^{-1}(1 + a(1-a)^{-1}\xi(t))^{-1}$ and $f_1(t) = f_2(t)\xi(t)$ when $a \neq \frac{1}{2}$;

(iv) $A = aU_1 + (1-a)U_2$ for some U_1 and U_2 in $\mathscr{U}(\mathfrak{A})$ when sp $A \subseteq \mathscr{S}_a$;

(v) $a\mathscr{U}(\mathfrak{A}) + (1-a)\mathscr{U}(\mathfrak{A})$

$$= \{UH : U \in \mathscr{U}(\mathfrak{A}),\ H = H^* \in \mathfrak{A},\ \operatorname{sp} H \subseteq \mathscr{S}_a\}$$

when $0 \le a < \frac{1}{2}$;

(vi) $a\mathscr{U}(\mathfrak{A}) + (1-a)\mathscr{U}(\mathfrak{A}) \subseteq b\mathscr{U}(\mathfrak{A}) + (1-b)\mathscr{U}(\mathfrak{A})$
when $0 \le a \le b \le \frac{1}{2}$; in particular, each convex combination of two unitary elements of $\mathfrak{A}$ is a mean of two unitary elements of $\mathfrak{A}$. (See Exercise 10.5.93(i).)

10.5.97. Suppose T is an element of the unit ball of a C^*-algebra $\mathfrak{A}$ with unitary group $\mathscr{U}(\mathfrak{A})$ and $\inf\{\|T - U\| : U \in \mathscr{U}(\mathfrak{A})\} \le 2a$, where $0 \le a < \frac{1}{2}$. Show that $T \in a\mathscr{U}(\mathfrak{A}) + (1-a)\mathscr{U}(\mathfrak{A})$. [*Hint*. Use Exercises 4.6.16 and 4.6.31 to establish that $\operatorname{sp} H \subseteq [1-2b, 1]$ when $a < b < \frac{1}{2}$, where $H = (T^*T)^{1/2}$.]

10.5.98. Let $\mathfrak{A}$ be a C^*-algebra and S be an element of $\mathfrak{A}$ such that $\|S\| \le 1 - \varepsilon$, where $0 < \varepsilon < (n+1)^{-1}$. Let $b_1V_1 + \cdots + b_nV_n$ be a convex combination of unitary elements $V_1, \ldots, V_n$ of $\mathfrak{A}$ such that

$$\|S - (b_1V_1 + \cdots + b_nV_n)\| < \varepsilon^2(1-\varepsilon)^{-1}$$

and let T be $b^{-1}[S - (1-\varepsilon)(b_1V_1 + \cdots + b_{n-1}V_{n-1})]$, where $b = \varepsilon + (1-\varepsilon)b_n$. Show that

(i) $\|T\| \le 1$;

(ii) $\|T - V_n\| \le 2\varepsilon b^{-1} < 1$, where $n^{-1} \le b_n$;

(iii) $T = (1 - \varepsilon b^{-1})U_n + \varepsilon b^{-1}U_{n+1}$ for some U_n and U_{n+1} in $\mathscr{U}(\mathfrak{A})$ [*Hint*. Use Exercise 10.5.97.];

(iv) S is a convex combination $a_1U_1 + \cdots + a_nU_n + \varepsilon U_{n+1}$ of unitary elements $U_1, \ldots, U_{n+1}$ of $\mathfrak{A}$;

(v) $(\mathfrak{A})_1^{\circ} \cap (\mathrm{co}_n\,\mathscr{U}(\mathfrak{A}))^= = (\mathfrak{A})_1^{\circ} \cap \mathrm{co}_{n+}\,\mathscr{U}(\mathfrak{A})$, where n is a positive integer, $\mathrm{co}_n\,\mathscr{U}(\mathfrak{A})$ is the set of convex combinations of n (or fewer) elements of $\mathscr{U}(\mathfrak{A})$, $\mathrm{co}_{n+}\,\mathscr{U}(\mathfrak{A})$ is the set of convex combinations of $n+1$ elements of $\mathscr{U}(\mathfrak{A})$ in which the last coefficient may be chosen less than a preassigned positive ε, and $(\mathfrak{A})_1^{\circ} = \{A \in \mathfrak{A} : \|A\| < 1\}$.

10.5.99. Let $\mathfrak{A}$ be a C^*-algebra and $\mathfrak{A}_{\mathrm{inv}}$ be its group of invertible elements. Show that the following statements are equivalent:

(i) $\mathfrak{A}_{\mathrm{inv}}$ is norm dense in $\mathfrak{A}$;

(ii) $\frac{1}{2}(\mathscr{U}(\mathfrak{A}) + \mathscr{U}(\mathfrak{A}))$ is norm dense in $(\mathfrak{A})_1$;

(iii) $(\mathfrak{A})_1^{\circ} \subseteq \mathrm{co}_{2+}\,\mathscr{U}(\mathfrak{A})$.

10.5.100. Let T be an element that is not a convex combination of fewer than n unitary elements of $\mathfrak{A}$, where $n \geq 3$. Suppose that we have $T = a_1U_1 + \cdots + a_nU_n$, where $a_1, \ldots, a_n$ are non-negative real numbers with sum 1, and $U_1, \ldots, U_n$ are unitary elements in $\mathfrak{A}$. Show that

(i) $a_i \leq a_j + a_k$ if $j \neq k$;
(ii) $(n-1)^{-1} \leq a_j + a_k$ if $j \neq k$;
(iii) $a_j \leq 2(n+1)^{-1}$ for all j.

CHAPTER 11

TENSOR PRODUCTS

This chapter is concerned with various concepts of tensor product for families of operator algebras. The simplest construct of this type, considered in Section 11.1, is the "represented C^*-algebra tensor product," a natural development from the study in Section 2.6 of tensor products of Hilbert spaces and of bounded linear operators, and applicable to finite families of C^*-algebras acting on specified Hilbert spaces. A similar process, for forming the "von Neumann algebra tensor product" of a finite family of von Neumann algebras, is considered in Section 11.2.

For finite families of abstract C^*-algebras, there are various possible tensor product constructions, which in general yield different tensor product algebras. They are studied in Section 11.3, along with the special class of "nuclear" C^*-algebras for which all the tensor products coincide. The most important of the various constructions, the "spatial tensor product," is closely related to the "represented C^*-algebra tensor product." It is developed further, so as to apply to infinite families of C^*-algebras, in Section 11.4.

In all the situations described above, it is shown that states (or * isomorphisms) of the component algebras can be combined to produce states (or * isomorphisms) of the tensor product algebra. For certain infinite tensor products, a characterization of the primary states is given. For finite families of von Neumann algebras, our analysis includes a description of the commutant and center of the tensor product algebra, and of its type decomposition.

11.1. Tensor products of represented C^*-algebras

In this section, we consider only *represented* C^*-algebras; that is, C^*-algebras consisting of operators acting on specified Hilbert spaces. Given a finite number $\mathfrak{A}_1, \ldots, \mathfrak{A}_n$ of such algebras, we define their tensor product $\mathfrak{A}_1 \otimes \cdots \otimes \mathfrak{A}_n$ (again, a represented C^*-algebra), and show how states (or * homomorphisms) of the component algebras $\mathfrak{A}_1, \ldots, \mathfrak{A}_n$ can be combined to form states (or * homomorphisms) of $\mathfrak{A}_1 \otimes \cdots \otimes \mathfrak{A}_n$. We include a number of examples, in which it is possible to give an explicit description of the tensor product algebra.

We shall see later (Section 11.3) that the results just described can be used to introduce a concept (but not the only possible one) of tensor product, for finite families of *abstract* C^*-algebras. The consistency result required for this purpose is contained in Theorem 11.1.3.

Suppose, then, that $\mathfrak{A}_1, \ldots, \mathfrak{A}_n$ are C^*-algebras, acting on Hilbert spaces $\mathscr{H}_1, \ldots, \mathscr{H}_n$, respectively. Let $\mathscr{H}$ be the Hilbert space $\mathscr{H}_1 \otimes \cdots \otimes \mathscr{H}_n$, and denote by $\mathfrak{A}_0$ the subset of $\mathscr{B}(\mathscr{H})$ that consists of all finite sums of operators of the form $A_1 \otimes \cdots \otimes A_n$, with A_j in $\mathfrak{A}_j$ $(1 \leq j \leq n)$. From the elementary properties of tensor products of bounded linear operators, as set out in the discussion following Proposition 2.6.12, it is apparent that $\mathfrak{A}_0$ is a * subalgebra (containing the identity operator) of $\mathscr{B}(\mathscr{H})$. Its norm closure is a C^*-algebra acting on $\mathscr{H}$, the (*represented C^*-algebra*) *tensor product* $\mathfrak{A}_1 \otimes \cdots \otimes \mathfrak{A}_n$ of the represented C^*-algebras $\mathfrak{A}_1, \ldots, \mathfrak{A}_n$. Observe that, if $\mathscr{B}_j$ is a C^*-subalgebra of $\mathfrak{A}_j$ $(1 \leq j \leq n)$, then $\mathscr{B}_1 \otimes \cdots \otimes \mathscr{B}_n$ is a C^*-subalgebra of $\mathfrak{A}_1 \otimes \cdots \otimes \mathfrak{A}_n$.

Suppose that, for $j = 1, \ldots, n$, $\mathfrak{A}_j$ is the C^*-algebra generated by a subset $\mathscr{S}_j$ (containing I) of $\mathscr{B}(\mathscr{H}_j)$. We assert that, in these circumstances, $\mathfrak{A}_1 \otimes \cdots \otimes \mathfrak{A}_n$ is the C^*-algebra generated by the set

$$\mathscr{S} = \{S_1 \otimes \cdots \otimes S_n : S_1 \in \mathscr{S}_1, \ldots, S_n \in \mathscr{S}_n\}.$$

Indeed, let $\mathfrak{A}$ be the C^*-algebra generated by $\mathscr{S}$. For each j, the mapping

$$A_j \to I \otimes \cdots \otimes I \otimes A_j \otimes I \otimes \cdots \otimes I : \mathscr{B}(\mathscr{H}_j) \to \mathscr{B}(\mathscr{H}_1 \otimes \cdots \otimes \mathscr{H}_n)$$

is an (isometric) * isomorphism. The inverse image of $\mathfrak{A}$, under this mapping, is a C^*-algebra that contains $\mathscr{S}_j$ and so contains $\mathfrak{A}_j$. Thus

$$I \otimes \cdots \otimes I \otimes A_j \otimes I \otimes \cdots \otimes I \in \mathfrak{A} \qquad (A_j \in \mathfrak{A}_j; \quad j = 1, \ldots, n),$$

and therefore $A_1 \otimes \cdots \otimes A_n \in \mathfrak{A}$ whenever $A_1 \in \mathfrak{A}_1, \ldots, A_n \in \mathfrak{A}_n$. This implies that $\mathfrak{A}_1 \otimes \cdots \otimes \mathfrak{A}_n \subseteq \mathfrak{A}$, and the reverse inclusion is clear.

We have already noted (in Proposition 2.6.5 and the discussion following Proposition 2.6.12) that, if $1 \leq r \leq n$, there is a unitary operator U from $\mathscr{H}_1 \otimes \cdots \otimes \mathscr{H}_n$ onto $(\mathscr{H}_1 \otimes \cdots \otimes \mathscr{H}_r) \otimes (\mathscr{H}_{r+1} \otimes \cdots \otimes \mathscr{H}_n)$, determined by the condition

$$U(x_1 \otimes \cdots \otimes x_n) = (x_1 \otimes \cdots \otimes x_r) \otimes (x_{r+1} \otimes \cdots \otimes x_n);$$

moreover,

$$U(A_1 \otimes \cdots \otimes A_n)U^* = (A_1 \otimes \cdots \otimes A_r) \otimes (A_{r+1} \otimes \cdots \otimes A_n).$$

From this, together with the preceding paragraph,

$$U(\mathfrak{A}_1 \otimes \cdots \otimes \mathfrak{A}_n)U^* = (\mathfrak{A}_1 \otimes \cdots \otimes \mathfrak{A}_r) \otimes (\mathfrak{A}_{r+1} \otimes \cdots \otimes \mathfrak{A}_n).$$

This establishes the "associativity" of the tensor product of represented C^*-algebras, and we shall frequently identify

$$(\mathfrak{A}_1 \otimes \cdots \otimes \mathfrak{A}_r) \otimes (\mathfrak{A}_{r+1} \otimes \cdots \otimes \mathfrak{A}_n)$$

with $\mathfrak{A}_1 \otimes \cdots \otimes \mathfrak{A}_n$ by means of the unitary equivalence just described. By use of this associativity, questions concerning tensor products of n C^*-algebras can often be reduced immediately to the case in which $n = 2$.

11.1.1. PROPOSITION. *Suppose that $\mathfrak{A}_j$ is a C^*-algebra acting on a Hilbert space $\mathscr{H}_j$, and ρ_j is a state of $\mathfrak{A}_j$ $(j = 1, \ldots, n)$. Then there is a unique state ρ of $\mathfrak{A}_1 \otimes \cdots \otimes \mathfrak{A}_n$ such that*

$$\rho(A_1 \otimes A_2 \otimes \cdots \otimes A_n) = \rho_1(A_1)\rho_2(A_2)\cdots\rho_n(A_n)$$

whenever $A_1 \in \mathfrak{A}_1, \ldots, A_n \in \mathfrak{A}_n$.

Proof. Since states are norm continuous, while $\mathfrak{A}_1 \otimes \cdots \otimes \mathfrak{A}_n$ is generated, as a norm-closed linear space, by operators of the form $A_1 \otimes \cdots \otimes A_n$, it is apparent that there is at most one state ρ with the required property. The existence of such a state is proved in three stages.

First, suppose that ρ_j is the vector state of $\mathfrak{A}_j$ arising from a unit vector x_j in $\mathscr{H}_j$ $(j = 1, \ldots, n)$. In this case, it suffices to take $\rho = \omega_x | \mathfrak{A}_1 \otimes \cdots \otimes \mathfrak{A}_n$, where $x = x_1 \otimes \cdots \otimes x_n$, since

$$\begin{aligned}\langle (A_1 \otimes \cdots \otimes A_n)x, x\rangle &= \langle A_1x_1 \otimes \cdots \otimes A_nx_n, x_1 \otimes \cdots \otimes x_n\rangle \\ &= \langle A_1x_1, x_1\rangle\langle A_2x_2, x_2\rangle \cdots \langle A_nx_n, x_n\rangle \\ &= \rho_1(A_1)\rho_2(A_2)\cdots\rho_n(A_n).\end{aligned}$$

Second, suppose that, when $j = 1, \ldots, n$, ρ_j is a convex combination,

$$\rho_j = \sum_{k=1}^{m(j)} c_{jk}\omega_{jk}$$

of vector states ω_{jk} of $\mathfrak{A}_j$. When $k(1), \ldots, k(n)$ are integers satisfying

$$1 \le k(j) \le m(j) \quad (j = 1, \ldots, n),$$

it follows from the preceding paragraph that there is a vector state $\rho_{k(1)\cdots k(n)}$ of $\mathfrak{A}_1 \otimes \cdots \otimes \mathfrak{A}_n$ such that

$$\rho_{k(1)\cdots k(n)}(A_1 \otimes \cdots \otimes A_n) = \omega_{1k(1)}(A_1)\omega_{2k(2)}(A_2)\cdots\omega_{nk(n)}(A_n).$$

With ρ the state $\sum c_{1k(1)}c_{2k(2)}\cdots c_{nk(n)}\rho_{k(1)\cdots k(n)}$ of $\mathfrak{A}_1 \otimes \cdots \otimes \mathfrak{A}_n$,

$$\begin{aligned}
\rho(A_1 \otimes \cdots \otimes A_n) &\\
&= \sum_{k(1)=1}^{m(1)} \sum_{k(2)=2}^{m(2)} \cdots \sum_{k(n)=1}^{m(n)} c_{1k(1)}c_{2k(2)} \cdots c_{nk(n)}\omega_{1k(1)}(A_1)\omega_{2k(2)}(A_2)\cdots\omega_{nk(n)}(A_n)\\
&= \left(\sum_{k(1)=1}^{m(1)} c_{1k(1)}\omega_{1k(1)}(A_1)\right)\left(\sum_{k(2)=1}^{m(2)} c_{2k(2)}\omega_{2k(2)}(A_2)\right)\cdots\left(\sum_{k(n)=1}^{m(n)} c_{nk(n)}\omega_{nk(n)}(A_n)\right)\\
&= \rho_1(A_1)\rho_2(A_2)\cdots\rho_n(A_n) \qquad (A_1 \in \mathfrak{A}_1, \ldots, A_n \in \mathfrak{A}_n).
\end{aligned}$$

Finally, we consider the general case. Let $\mathscr{S}_j$ be the state space of $\mathfrak{A}_j$, and let $\mathscr{V}_j$ be the subset of $\mathscr{S}_j$ consisting of convex combinations of vector states. From Corollary 4.3.10, $\mathscr{V}_j$ is everywhere dense in $\mathscr{S}_j$ (in the weak* topology), so $\mathscr{V}_1 \times \cdots \times \mathscr{V}_n$ is everywhere dense in $\mathscr{S}_1 \times \cdots \times \mathscr{S}_n$ (provided with the product weak* topology). With ρ_j a state of $\mathfrak{A}_j$ ($j = 1, \ldots, n$), there is a net $\{(\rho_{1a}, \ldots, \rho_{na}) : a \in \mathbb{A}\}$ in $\mathscr{V}_1 \times \cdots \times \mathscr{V}_n$ that converges to $(\rho_1, \ldots, \rho_n)$. From the preceding paragraph, there is a state μ_a of $\mathfrak{A}_1 \otimes \cdots \otimes \mathfrak{A}_n$ such that

$$\mu_a(A_1 \otimes \cdots \otimes A_n) = \rho_{1a}(A_1)\rho_{2a}(A_2)\cdots\rho_{na}(A_n). \tag{1}$$

By compactness, the net $\{\mu_a : a \in \mathbb{A}\}$ has a cofinal subnet, weak* convergent to a state ρ of $\mathfrak{A}_1 \otimes \cdots \otimes \mathfrak{A}_n$. By taking limits in (1), following this subnet, we obtain

$$\rho(A_1 \otimes \cdots \otimes A_n) = \rho_1(A_1)\rho_2(A_2)\cdots\rho_n(A_n). \quad \blacksquare$$

The state ρ occurring in Proposition 11.1.1 is denoted by $\rho_1 \otimes \cdots \otimes \rho_n$; and states of this type are described as *product states* of $\mathfrak{A}_1 \otimes \cdots \otimes \mathfrak{A}_n$. As noted in the second paragraph of the proof, when ρ_j is the vector state of $\mathfrak{A}_j$ arising from a unit vector x_j ($j = 1, \ldots, n$), ρ is the vector state arising from $x_1 \otimes \cdots \otimes x_n$.

Given a product state ρ of $\mathfrak{A}_1 \otimes \cdots \otimes \mathfrak{A}_n$, the "factors" $\rho_1, \ldots, \rho_n$ are uniquely determined, since

$$\rho_j(A_j) = \rho(I \otimes \cdots \otimes I \otimes A_j \otimes I \otimes \cdots \otimes I) \qquad (A_j \in \mathfrak{A}_j).$$

11.1.2. PROPOSITION. *Let $\mathfrak{A}_1, \ldots, \mathfrak{A}_n$ be C^*-algebras acting on Hilbert spaces $\mathscr{H}_1, \ldots, \mathscr{H}_n$, respectively, and let $\mathfrak{A}_0$ be the * subalgebra of $\mathfrak{A}_1 \otimes \cdots \otimes \mathfrak{A}_n$ that consists of all finite sums of operators of the form $A_1 \otimes \cdots \otimes A_n$, with A_j in $\mathfrak{A}_j$ ($j = 1, \ldots, n$). Then, for each A in $\mathfrak{A}_1 \otimes \cdots \otimes \mathfrak{A}_n$,*

$$\|A\|^2 = \sup \frac{\rho(S^*A^*AS)}{\rho(S^*S)},$$

where the supremum is taken over all pairs (ρ, S) *in which* ρ *is a product state of* $\mathfrak{A}_1 \otimes \cdots \otimes \mathfrak{A}_n$, $S \in \mathfrak{A}_0$, *and* $\rho(S^*S) \neq 0$.

Proof. With ρ an arbitrary state of $\mathfrak{A}_1 \otimes \cdots \otimes \mathfrak{A}_n$, and S any element of this algebra for which $\rho(S^*S) \neq 0$, the equation $\mu(A) = \rho(S^*AS)/\rho(S^*S)$ defines a state μ of $\mathfrak{A}_1 \otimes \cdots \otimes \mathfrak{A}_n$; so

$$\frac{\rho(S^*A^*AS)}{\rho(S^*S)} = \mu(A^*A) \leq \|A^*A\| = \|A\|^2.$$

In particular,

$$\sup \frac{\rho(S^*A^*AS)}{\rho(S^*S)} \leq \|A\|^2, \tag{2}$$

where the supremum is interpreted as in the statement of the proposition.

For each $j = 1, \ldots, n$, let $\{E'_{ja} : a \in \mathbb{A}_j\}$ be an orthogonal family of cyclic projections, with sum I, in the von Neumann algebra $\mathfrak{A}'_j$; and let E'_{ja} have range $[\mathfrak{A}_j x_{ja}]$, where x_{ja} is a unit vector in $\mathscr{H}_j$. For each $a = (a(1), \ldots, a(n))$ in the product $\mathbb{A} = \mathbb{A}_1 \times \cdots \times \mathbb{A}_n$, define E'_a as $E'_{1a(1)} \otimes \cdots \otimes E'_{na(n)}$, x_a as $x_{1a(1)} \otimes \cdots \otimes x_{na(n)}$, and let ρ_a be the vector state of $\mathfrak{A}_1 \otimes \cdots \otimes \mathfrak{A}_n$ associated with x_a (so that ρ_a is a product state).

It is apparent that $\{E'_a : a \in \mathbb{A}\}$ is an orthogonal family of projections acting on $\mathscr{H}_1 \otimes \cdots \otimes \mathscr{H}_n$. We assert also that $\sum E'_a = I$, and that E'_a is the cyclic projection in $(\mathfrak{A}_1 \otimes \cdots \otimes \mathfrak{A}_n)'$, corresponding to the vector x_a. For this, suppose that $y_1 \in \mathscr{H}_1, \ldots, y_n \in \mathscr{H}_n$. For each j, y_j is the limit in norm of finite sums of vectors of the form $E'_{ja(j)} y_j$, with $a(j)$ in $\mathbb{A}_j$; so $y_1 \otimes \cdots \otimes y_n$ is the norm limit of finite sums of vectors

$$E'_{1a(1)} y_1 \otimes \cdots \otimes E'_{na(n)} y_n = E'_a (y_1 \otimes \cdots \otimes y_n),$$

where $a = (a(1), \ldots, a(n)) \in \mathbb{A}$. This shows that the range of $\sum E'_a$ contains every simple tensor $y_1 \otimes \cdots \otimes y_n$, whence $\sum E'_a = I$. Suppose next that $a = (a(1), \ldots, a(n)) \in \mathbb{A}$. Clearly E'_a commutes with $A_1 \otimes \cdots \otimes A_n$ whenever $A_1 \in \mathfrak{A}_1, \ldots, A_n \in \mathfrak{A}_n$, and so lies in $(\mathfrak{A}_1 \otimes \cdots \otimes \mathfrak{A}_n)'$. Since, also, $E'_a x_a = x_a$, it follows that E'_a contains the cyclic projection corresponding to x_a. Each vector u in the range of E'_a can be approximated in norm by a finite sum $\sum_k y_{1k} \otimes \cdots \otimes y_{nk}$ of simple tensors. Now

$$\begin{aligned} \|u - \sum_k E'_{1a(1)} y_{1k} \otimes \cdots \otimes E'_{na(n)} y_{nk}\| &= \|E'_a(u - \sum_k y_{1k} \otimes \cdots \otimes y_{nk})\| \\ &\leq \|u - \sum_k y_{1k} \otimes \cdots \otimes y_{nk}\|, \end{aligned}$$

and each $E'_{ja(j)} y_{jk}$ can be approximated in norm by a vector of the form $A_{jk} x_{ja(j)}$, with A_{jk} in $\mathfrak{A}_j$. Accordingly, u can be approximated in norm by a vector

$$\sum_k A_{1k}x_{1a(1)} \otimes \cdots \otimes A_{nk}x_{na(n)} = Sx_a,$$

where $S = \sum_k A_{1k} \otimes \cdots \otimes A_{nk} \in \mathfrak{A}_1 \otimes \cdots \otimes \mathfrak{A}_n$. Thus E'_a coincides with the cyclic projection in $(\mathfrak{A}_1 \otimes \cdots \otimes \mathfrak{A}_n)'$, corresponding to x_a.

Suppose that $A \in \mathfrak{A}_1 \otimes \cdots \otimes \mathfrak{A}_n$. From the properties of the family $\{E'_a\}$ set out in the preceding paragraph, and since $\mathfrak{A}_0$ is everywhere dense in $\mathfrak{A}_1 \otimes \cdots \otimes \mathfrak{A}_n$,

$$\begin{aligned}\|A\| &= \sup\{\|AE'_a\| : a \in \mathbb{A}\}\\ &= \sup\left\{\frac{\|ASx_a\|}{\|Sx_a\|} : a \in \mathbb{A}, \quad S \in \mathfrak{A}_1 \otimes \cdots \otimes \mathfrak{A}_n, \quad Sx_a \neq 0\right\}\\ &= \sup\left\{\frac{\|ASx_a\|}{\|Sx_a\|} : a \in \mathbb{A}, \quad S \in \mathfrak{A}_0, \quad Sx_a \neq 0\right\}.\end{aligned}$$

Since ρ_a is the vector state arising from x_a, we can rewrite the above equation in the form

$$\|A\|^2 = \sup\left\{\frac{\rho_a(S^*A^*AS)}{\rho_a(S^*S)} : a \in \mathbb{A}, \quad S \in \mathfrak{A}_0, \quad \rho_a(S^*S) \neq 0\right\}.$$

Since each ρ_a is a product state, this shows that equality occurs in (2) ■

11.1.3. THEOREM. *Suppose that, for* $j = 1, \ldots, n$, $\mathfrak{A}_j$ *and* $\mathscr{B}_j$ *are* C^*-*algebras, acting on Hilbert spaces* $\mathscr{H}_j$ *and* $\mathscr{K}_j$, *respectively, and* φ_j *is a* * *homomorphism from* $\mathfrak{A}_j$ *into* $\mathscr{B}_j$. *Then there is a unique* * *homomorphism* φ, *from* $\mathfrak{A}_1 \otimes \cdots \otimes \mathfrak{A}_n$ *into* $\mathscr{B}_1 \otimes \cdots \otimes \mathscr{B}_n$, *such that*

$$\varphi(A_1 \otimes \cdots \otimes A_n) = \varphi_1(A_1) \otimes \cdots \otimes \varphi_n(A) \qquad (A_1 \in \mathfrak{A}_1, \ldots, A_n \in \mathfrak{A}_n).$$

If each φ_j *is a* * *isomorphism, then so is* φ. *If* $\varphi_j(\mathfrak{A}_j) = \mathscr{B}_j$ $(j = 1, \ldots, n)$, *then* $\varphi(\mathfrak{A}_1 \otimes \cdots \otimes \mathfrak{A}_n) = \mathscr{B}_1 \otimes \cdots \otimes \mathscr{B}_n$.

Proof. Since $\varphi_j(\mathfrak{A}_j)$ is a C^*-subalgebra of $\mathscr{B}_j$ from Theorem 4.1.9 and Corollary 10.1.8, we have that $\varphi_1(\mathfrak{A}_1) \otimes \cdots \otimes \varphi_n(\mathfrak{A}_n)$ is a C^*-subalgebra of $\mathscr{B}_1 \otimes \cdots \otimes \mathscr{B}_n$. Accordingly, we may replace $\mathscr{B}_j$ by φ_j $(\mathfrak{A}_j)$, and assume henceforth that $\varphi_j(\mathfrak{A}_j) = \mathscr{B}_j$.

For each state ρ_j of $\mathscr{B}_j$, $\rho_j \circ \varphi_j$ is a state of $\mathfrak{A}_j$; and every state of $\mathfrak{A}_j$ arises in this way, if φ_j is a * isomorphism. Accordingly, with each product state $(\rho =)\rho_1 \otimes \cdots \otimes \rho_n$ of $\mathscr{B}_1 \otimes \cdots \otimes \mathscr{B}_n$, we can associate a product state $(\varphi^*\rho =)(\rho_1 \circ \varphi_1) \otimes \cdots \otimes (\rho_n \circ \varphi_n)$ of $\mathfrak{A}_1 \otimes \cdots \otimes \mathfrak{A}_n$; and every product state of $\mathfrak{A}_1 \otimes \cdots \otimes \mathfrak{A}_n$ arises in this way, if each φ_j is a * isomorphism.

Let $\mathfrak{A}_0$ be the everywhere-dense * subalgebra of $\mathfrak{A}_1 \otimes \cdots \otimes \mathfrak{A}_n$, consisting of finite sums of operators of the form $A_1 \otimes \cdots \otimes A_n$; and let $\mathscr{B}_0$ be the corresponding * subalgebra of $\mathscr{B}_1 \otimes \cdots \otimes \mathscr{B}_n$. Suppose that

$$A = \sum_{j=1}^{p} A_1 \otimes \cdots \otimes A_n, \qquad S = \sum_{k=1}^{q} S_{1k} \otimes \cdots \otimes S_{nk}$$

(both elements of $\mathfrak{A}_0$); and define B, T in $\mathscr{B}_0$ by

$$B = \sum_{j=1}^{p} \varphi_1(A_{1j}) \otimes \cdots \otimes \varphi_n(A_{nj}), \qquad T = \sum_{k=1}^{q} \varphi_1(S_{1k}) \otimes \cdots \otimes \varphi_n(S_{nk}).$$

Then

$$S^*A^*AS = \sum_{j,l=1}^{p} \sum_{k,m=1}^{q} S_{1m}^* A_{1l}^* A_{1j} S_{1k} \otimes \cdots \otimes S_{nm}^* A_{nl}^* A_{nj} S_{nk},$$

and

$$T^*B^*BT = \sum_{j,l=1}^{p} \sum_{k,m=1}^{q} \varphi_1(S_{1m}^* A_{1l}^* A_{1j} S_{1k}) \otimes \cdots \otimes \varphi_n(S_{nm}^* A_{nl}^* A_{nj} S_{nk}).$$

From this it follows that, if ρ is a product state $\rho_1 \otimes \cdots \otimes \rho_n$ of $\mathscr{B}_1 \otimes \cdots \otimes \mathscr{B}_n$, then

$$\rho(T^*B^*BT) = (\varphi^*\rho)(S^*A^*AS);$$

and similarly, $\rho(T^*T) = (\varphi^*\rho)(S^*S)$.

When q and the operators S_{jk} are allowed to vary, S takes all values in $\mathfrak{A}_0$ and T takes all values in $\mathscr{B}_0$. When ρ runs through all product states of $\mathscr{B}_1 \otimes \cdots \otimes \mathscr{B}_n$, $\varphi^*\rho$ runs through certain product states of $\mathfrak{A}_1 \otimes \cdots \otimes \mathfrak{A}_n$ (all of them, when each φ_j is a * isomorphism). It now follows from Proposition 11.1.2 that $\|B\| \leq \|A\|$, with equality when each φ_j is a * isomorphism.

The preceding argument shows that, if $A_{jk} \in \mathfrak{A}_j$ $(k = 1, \ldots, p; j = 1, \ldots, n)$, then

$$\text{(3)} \qquad \left\| \sum_{k=1}^{p} \varphi_1(A_{1k}) \otimes \cdots \otimes \varphi_n(A_{nk}) \right\| \leq \left\| \sum_{k=1}^{p} A_{1k} \otimes \cdots \otimes A_{nk} \right\|;$$

and equality occurs in (3) when each φ_j is a * isomorphism. From this, it follows easily that the equation

$$\varphi_0\left(\sum_{k=1}^{p} A_{1k} \otimes \cdots \otimes A_{nk} \right) = \sum_{k=1}^{p} \varphi_1(A_{1k}) \otimes \cdots \otimes \varphi_n(A_{nk})$$

defines (unambiguously) a non-decreasing, linear, multiplicative, and adjoint-preserving mapping φ_0, from $\mathfrak{A}_0$ onto $\mathscr{B}_0$. By continuity, φ_0 extends to a * homomorphism φ from $\mathfrak{A}_1 \otimes \cdots \otimes \mathfrak{A}_n$ into $\mathscr{B}_1 \otimes \cdots \otimes \mathscr{B}_n$. The range of φ is the whole of $\mathscr{B}_1 \otimes \cdots \otimes \mathscr{B}_n$ (recall our present assumption that $\varphi_j(\mathfrak{A}_j) = \mathscr{B}_j$), since it is norm closed and contains the everywhere-dense subset $\mathscr{B}_0$. If each φ_j is a * isomorphism, φ_0 is isometric; so in this case φ is an isometry, and is therefore a * isomorphism.

We have now established the existence of a * homomorphism φ satisfying the required conditions. Since these conditions specify the values of φ on the everywhere-dense subset $\mathfrak{A}_0$ of $\mathfrak{A}_1 \otimes \cdots \otimes \mathfrak{A}_n$, while * homomorphisms between C^*-algebras are continuous, there is only one such φ. ■

The * homomorphism φ occurring in Theorem 11.1.3 is denoted by $\varphi_1 \otimes \cdots \otimes \varphi_n$.

11.1.4. Example. Suppose that $\mathscr{H}$ and $\mathscr{K}$ are Hilbert spaces, $\mathfrak{A}$ ($\subseteq \mathscr{B}(\mathscr{H})$) is a C^*-algebra, and $\mathbb{C}_{\mathscr{K}}$ is the one-dimensional C^*-algebra consisting of scalar multiples of the identity operator $I_{\mathscr{K}}$ on $\mathscr{K}$. We assert that

$$\mathfrak{A} \otimes \mathbb{C}_{\mathscr{K}} = \{A \otimes I_{\mathscr{K}} : A \in \mathfrak{A}\}.$$

Indeed, the set on the right-hand side contains $A \otimes C$ for each A in $\mathfrak{A}$ and C in $\mathbb{C}_{\mathscr{K}}$, and is a C^*-algebra since it is the range of the * isomorphism

$$A \to A \otimes I_{\mathscr{K}} : \mathfrak{A} \to \mathscr{B}(\mathscr{H} \otimes \mathscr{K}).$$

If $\{y_b : b \in \mathbb{B}\}$ is an orthonormal basis of $\mathscr{K}$, the equation $U(\sum \oplus x_b) = \sum x_b \otimes y_b$ defines a unitary operator U from $\sum_{b \in \mathbb{B}} \oplus \mathscr{H}$ onto $\mathscr{H} \otimes \mathscr{K}$. Moreover, when $A \in \mathscr{B}(\mathscr{H})$, $U^{-1}(A \otimes I_{\mathscr{K}})U$ is the operator on $\sum_{b \in \mathbb{B}} \oplus \mathscr{H}$, whose matrix (in the sense of Section 2.6, *Matrix representations*) is diagonal, with A in each diagonal position. Accordingly, $U^{-1}(\mathfrak{A} \otimes \mathbb{C}_{\mathscr{K}})U$ is the C^*-algebra consisting of all operators acting on $\sum_{b \in \mathbb{B}} \oplus \mathscr{H}$, whose matrices have the form $[\delta_{ab} A]$, with A in $\mathfrak{A}$.

If $\mathfrak{A}$ is replaced by a von Neumann algebra $\mathscr{R}$, the preceding paragraph identifies $U^{-1}(\mathscr{R} \otimes \mathbb{C}_{\mathscr{K}})U$ as the von Neumann algebra $\mathscr{R} \otimes I_n$, where the latter notation is taken from Section 6.6 and n is the (not necessarily finite) dimension of $\mathscr{K}$. Accordingly, $\mathscr{R} \otimes \mathbb{C}_{\mathscr{K}}$ is itself a von Neumann algebra. This result can also be obtained as an immediate consequence of the Kaplansky density theorem, together with the weak-operator continuity (on bounded sets) of the mapping $A \to A \otimes I$.

Finally, we consider briefly the case in which $\mathfrak{A} = \mathscr{R} = \mathscr{B}(\mathscr{H})$. Since $\mathscr{B}(\mathscr{H}) \otimes \mathbb{C}_{\mathscr{K}} = U(\mathscr{B}(\mathscr{H}) \otimes I_n)U^{-1}$, and $\mathscr{B}(\mathscr{H}) \otimes I_n$ is a type I factor (Example 9.1.5), it follows that $\mathscr{B}(\mathscr{H}) \otimes \mathbb{C}_{\mathscr{K}}$ is a type I factor with commutant $U(\mathscr{B}(\mathscr{H}) \otimes I_n)'U^{-1}$. Moreover (Lemma 6.6.2), $\mathscr{B}(\mathscr{H}) \otimes I_n$ has commutant $n \otimes \mathscr{B}(\mathscr{H})'$ ($= n \otimes \mathbb{C}_{\mathscr{H}}$) consisting of all bounded operators, acting on $\sum_{b \in \mathbb{B}} \oplus \mathscr{H}$, whose matrices have scalar multiples of the identity in each entry.

It now results from Section 2.6, *Matrix representations* (see the discussion immediately preceding Proposition 2.6.13), that

$$U(\mathscr{B}(\mathscr{H}) \otimes I_n)'U^{-1} = \{I_{\mathscr{H}} \otimes B : B \in \mathscr{B}(\mathscr{K})\} = \mathbb{C}_{\mathscr{H}} \otimes \mathscr{B}(\mathscr{K});$$

so $\mathscr{B}(\mathscr{H}) \otimes \mathbb{C}_{\mathscr{K}}$ has commutant $\mathbb{C}_{\mathscr{H}} \otimes \mathscr{B}(\mathscr{K})$. ∎

11.1.5. EXAMPLE. Suppose that $\mathfrak{A}$ is a C^*-algebra acting on a Hilbert space $\mathscr{H}$, and $\mathscr{K}$ is a finite-dimensional Hilbert space. With $n = \dim \mathscr{K}$, let $\{y_1, \ldots, y_n\}$ be an orthonormal basis of $\mathscr{K}$, and let $\{E_{jk} : j, k = 1, \ldots, n\}$ be the system of matrix units for $\mathscr{B}(\mathscr{K})$, defined by $E_{jk}x = \langle x, y_k \rangle y_j$. We assert that each element of $\mathfrak{A} \otimes \mathscr{B}(\mathscr{K})$ can be expressed uniquely in the form $\sum A_{jk} \otimes E_{jk}$, with A_{jk} in $\mathfrak{A}$ ($j, k = 1, \ldots, n$). Moreover, $\mathfrak{A} \otimes \mathscr{B}(\mathscr{K})$ is unitarily equivalent to the C^*-algebra, acting on $\mathscr{H} \oplus \cdots \oplus \mathscr{H}$ (n copies), consisting of those operators whose $n \times n$ matrices have all entries in $\mathfrak{A}$. In particular $\mathfrak{A} \otimes \mathscr{B}(\mathscr{K})$ is a von Neumann algebra, if $\mathfrak{A}$ is a von Neumann algebra (and $\dim \mathscr{K} = n < \infty$), since it is unitarily equivalent to $n \otimes \mathfrak{A}$.

To establish these results, let $\mathscr{C}$ be the * algebra consisting of all operators of the form $\sum_1^m A_j \otimes B_j$, with $A_1, \ldots, A_m$ in $\mathfrak{A}$ and $B_1, \ldots, B_m$ in $\mathscr{B}(\mathscr{K})$. Since $\mathscr{B}(\mathscr{K})$ is the linear span of the matrix units E_{jk}, every element S of $\mathscr{C}$ can be expressed as

$$S = \sum_{j,k=1}^{n} A_{jk} \otimes E_{jk}, \tag{4}$$

with each A_{jk} in $\mathfrak{A}$. Given S_1 and S_2 in $\mathscr{C}$, say

$$S_1 = \sum_{j,k=1}^{n} A_{jk}^{(1)} \otimes E_{jk}, \qquad S_2 = \sum_{j,k=1}^{n} A_{jk}^{(2)} \otimes E_{jk},$$

we have (see 2.6(16))

$$\begin{aligned} \|A_{jk}^{(1)} - A_{jk}^{(2)}\| &= \|(A_{jk}^{(1)} - A_{jk}^{(2)}) \otimes E_{jk}\| \\ &= \|(I \otimes E_{jj})(S_1 - S_2)(I \otimes E_{kk})\| \le \|S_1 - S_2\|. \end{aligned}$$

With $S_1 = S_2 = S$, it follows that the operators A_{jk} in (4) are uniquely determined. Moreover, if $\{S_m\}$ is a Cauchy sequence in $\mathscr{C}$, say

$$S_m = \sum_{j,k=1}^{n} A_{jk}^{(m)} \otimes E_{jk} \qquad (m = 1, 2, \ldots),$$

then, for each $j, k = 1, \ldots, n$, $(A_{jk}^{(m)})_{m \ge 1}$ is a Cauchy sequence, and so converges to an element $A_{jk}^{(0)}$ of $\mathfrak{A}$. Accordingly $\{S_m\}$ has a limit, $\sum A_{jk}^{(0)} \otimes E_{jk}$, in $\mathscr{C}$. This shows that $\mathscr{C}$ is complete; so $\mathfrak{A} \otimes \mathscr{B}(\mathscr{K}) = \mathscr{C}$, and each element of $\mathfrak{A} \otimes \mathscr{B}(\mathscr{K})$ can be expressed (uniquely) in the form (4).

Let U be the unitary operator, from $\mathscr{H} \oplus \cdots \oplus \mathscr{H}$ (n copies) onto $\mathscr{H} \otimes \mathscr{K}$, defined by $U(x_1, \ldots, x_n) = x_1 \otimes y_1 + \cdots + x_n \otimes y_n$. If an element S of $\mathfrak{A} \otimes \mathscr{B}(\mathscr{K})$ is expressed in the form (4),

$$U^{-1}SU(x_1, \ldots, x_n) = U^{-1}\left(\sum_{j,k=1}^{n} A_{jk} \otimes E_{jk}\right)(x_1 \otimes y_1 + \cdots + x_n \otimes y_n)$$

$$= U^{-1}\left(\sum_{j=1}^{n}\left(\sum_{k=1}^{n} A_{jk}x_k\right) \otimes y_j\right) = (z_1, \ldots, z_n),$$

where $z_j = \sum A_{jk}x_k$. Thus $U^{-1}SU$ is the operator, acting on $\mathscr{H} \oplus \cdots \oplus \mathscr{H}$, whose matrix is $[A_{jk}]$; and $U^{-1}(\mathfrak{A} \otimes \mathscr{B}(\mathscr{K}))U$ consists of all operators whose matrices have each entry in $\mathfrak{A}$. ∎

11.1.6. EXAMPLE. We assert that, if $\mathscr{H}$ and $\mathscr{K}$ are Hilbert spaces, at least one of which is finite dimensional, then $\mathscr{B}(\mathscr{H}) \otimes \mathscr{B}(\mathscr{K}) = \mathscr{B}(\mathscr{H} \otimes \mathscr{K})$.

For this, we may suppose that $\dim \mathscr{K} = n < \infty$. From the preceding example, with $\mathfrak{A} = \mathscr{B}(\mathscr{H})$, $\mathscr{B}(\mathscr{H}) \otimes \mathscr{B}(\mathscr{K})$ is unitarily equivalent to $n \otimes \mathscr{B}(\mathscr{H})$. Since *every* bounded linear operator acting on $\mathscr{H} \oplus \cdots \oplus \mathscr{H}$ (n copies) can be represented by an $n \times n$ matrix with entries in $\mathscr{B}(\mathscr{H})$, $n \otimes \mathscr{B}(\mathscr{H})$ is $\mathscr{B}(\mathscr{H} \oplus \cdots \oplus \mathscr{H})$; and thus $\mathscr{B}(\mathscr{H}) \otimes \mathscr{B}(\mathscr{H}) = \mathscr{B}(\mathscr{H} \otimes \mathscr{K})$.

In Example 11.2.2, we shall obtain a similar result, without restriction on the dimensions of $\mathscr{H}$ or $\mathscr{K}$, in which $\otimes$ is replaced by a "von Neumann algebra tensor product." ∎

11.1.7. EXAMPLE. Suppose that $\mathfrak{A}$ is an (abstract) C^*-algebra, π is a faithful representation of $\mathfrak{A}$ on a Hilbert space $\mathscr{H}$, and S is a compact Hausdorff space. Let π_{m} be the faithful representation of $C(S)$ on $l_2(S)$ in which continuous functions operate by multiplication; that is,

$$(\pi_{\mathrm{m}}(f)x)(s) = f(s)x(s) \qquad (s \in S),$$

when $f \in C(S)$ and $x \in l_2(S)$. We shall give a description, up to unitary equivalence, of the tensor product of the represented C^*-algebras $\pi(\mathfrak{A})$ and $\pi_{\mathrm{m}}(C(S))$.

The set $C(S, \mathfrak{A})$, of all norm-continuous mappings F from S into $\mathfrak{A}$, has a pointwise * algebra structure, and is a C^*-algebra relative to the norm defined by $\|F\| = \sup\{\|F(s)\| : s \in S\}$. It has a faithful representation π_0, on the Hilbert space $\mathscr{H}_0 = \sum_{s \in S} \oplus \mathscr{H}$, in which

$$\pi_0(F) = \sum_{s \in S} \oplus\, \pi(F(s)).$$

We shall show that, for a suitable unitary operator U from $\mathscr{H}_0$ onto $\mathscr{H} \otimes l_2(S)$,

$$\pi(\mathfrak{A}) \otimes \pi_{\mathrm{m}}(C(S)) = U\pi_0(C(S, \mathfrak{A}))U^{-1}; \tag{5}$$

and

$$(6)\qquad \pi(A) \otimes \pi_{\mathrm{m}}(f) = U\pi_0(f(\ )A)U^{-1} \qquad (A \in \mathfrak{A},\ \ f \in C(S)),$$

where $f(\ )A$ denotes the element F of $C(S, \mathfrak{A})$ defined by $F(s) = f(s)A$.

To this end, let $\{y_s : s \in S\}$ be the orthonormal basis of $l_2(S)$, in which y_s is the function taking the value 1 at s and 0 elsewhere on S; and note that $\pi_{\mathrm{m}}(f)y_s = f(s)y_s$, for all f in $C(S)$. The equation

$$Uw = \sum_{s \in S} w(s) \otimes y_s \qquad \left(w = \sum_{s \in S} \oplus\, w(s) \in \mathscr{H}_0\right)$$

defines a unitary operator U, from $\mathscr{H}_0$ onto $\mathscr{H} \otimes \mathscr{K}$. When $A \in \mathfrak{A}, f \in C(S)$, and $w \in \mathscr{H}_0$,

$$\begin{aligned}
&(\pi(A) \otimes \pi_{\mathrm{m}}(f))Uw \\
&\quad = (\pi(A) \otimes \pi_{\mathrm{m}}(f))\left(\sum_{s \in S} w(s) \otimes y_s\right) \\
&\quad = \sum_{s \in S} \pi(A)w(s) \otimes \pi_{\mathrm{m}}(f)y_s = \sum_{s \in S} f(s)\pi(A)w(s) \otimes y_s \\
&\quad = U\left(\sum_{s \in S} \oplus\, \pi(f(s)A)w(s)\right) = U\pi_0(f(\ )A)w.
\end{aligned}$$

Hence $(\pi(A) \otimes \pi_{\mathrm{m}}(f))U = U\pi_0(f(\ )A)$, and (6) is proved.

From (6), the C^*-algebra $U\pi_0(C(S, \mathfrak{A}))U^{-1}$ contains $\pi(A) \otimes \pi_{\mathrm{m}}(f)$, whenever $A \in \mathfrak{A}$ and $f \in C(S)$; so

$$U\pi_0(C(S, \mathfrak{A}))U^{-1} \supseteq \pi(\mathfrak{A}) \otimes \pi_{\mathrm{m}}(C(S)).$$

It remains to show that $U\pi_0(F)U^{-1} \in \pi(\mathfrak{A}) \otimes \pi_{\mathrm{m}}(C(S))$, whenever $F \in C(S, \mathfrak{A})$. Given any positive real number ε, it results from the norm continuity of F and the compactness of S that there is a finite open covering $\{G_1, \ldots, G_n\}$ of S, such that

$$\|F(s) - F(t)\| < \varepsilon \qquad (s, t \in G_j;\ \ j = 1, \ldots, n).$$

We can choose continuous real-valued functions $f_1, \ldots, f_n$ on S, such that $0 \le f_j(s) \le 1$, $\sum f_j(s) = 1$ $(s \in S)$ and f_j vanishes on $S \setminus G_j$. For $j = 1, \ldots, n$, choose s_j in G_j and let $A_j = F(s_j)$. Then

$$\begin{aligned}
\|F(s) - \sum_{j=1}^{n} f_j(s)A_j\| &= \|\sum_{j=1}^{n} f_j(s)[F(s) - A_j]\| \\
&= \|\sum_{j=1}^{n} f_j(s)[F(s) - F(s_j)]\| \le \sum_{j=1}^{n} f_j(s)\|F(s) - F(s_j)\|
\end{aligned}$$

for each s in S. Now $f_j(s) = 0$ when $s \notin G_j$ and $\|F(s) - F(s_j)\| < \varepsilon$ when $s \in G_j$, so

$$\|F(s) - \sum_{j=1}^{n} f_j(s)A_j\| \leq \sum_{j=1}^{n} \varepsilon f_j(s) = \varepsilon \qquad (s \in S).$$

Thus $\|F - \sum f_j(\ )A_j\| \leq \varepsilon$; and, by (6),

$$\|U\pi_0(F)U^{-1} - \sum_{j=1}^{n} \pi(A_j) \otimes \pi_{\mathrm{m}}(f_j)\| = \|U\pi_0\left(F - \sum_{j=1}^{n} f_j(\ )A_j\right)U^{-1}\|$$
$$\leq \varepsilon.$$

This shows that $U\pi_0(F)U^{-1}$ is a norm limit of operators having the form $\sum_1^n \pi(A_j) \otimes \pi_{\mathrm{m}}(f_j)$, and so lies in $\pi(\mathfrak{A}) \otimes \pi_{\mathrm{m}}(C(S))$. ∎

In discussing the tensor product $\mathscr{H} \otimes \mathscr{K}$ of two Hilbert spaces, we obtained a necessary and sufficient condition that a sum $\sum_1^n x_j \otimes y_j$ of simple tensors should be zero (Proposition 2.6.6); and we deduced from this that the set of all finite sums of simple tensors can be identified with the *algebraic* tensor product of $\mathscr{H}$ and $\mathscr{K}$ (Remark 2.6.7). The following result, concerning tensor products of represented C^*-algebras, is similar in both form and purpose (see also the discussion at the end of Section 11.3, *The spatial tensor product*).

11.1.8. PROPOSITION. *Suppose that $\mathfrak{A}$ and $\mathscr{B}$ are C^*-algebras, acting on Hilbert spaces $\mathscr{H}$ and $\mathscr{K}$, respectively, and $\mathscr{C}$ is the * subalgebra of $\mathfrak{A} \otimes \mathscr{B}$ that consists of all finite sums of operators of the form $A \otimes B$, with A in $\mathfrak{A}$ and B in $\mathscr{B}$.*

(i) *If $A_1, \ldots, A_n \in \mathfrak{A}$ and $B_1, \ldots, B_n \in \mathscr{B}$, then $\sum_1^n A_j \otimes B_j = 0$ if and only if there is an $n \times n$ complex matrix $[c_{jk}]$ such that*

$$\sum_{j=1}^{n} c_{jk}A_j = 0 \qquad (k = 1, \ldots, n),$$

$$\sum_{k=1}^{n} c_{jk}B_k = B_j \qquad (j = 1, \ldots, n).$$

(ii) *If L is a bilinear mapping from $\mathfrak{A} \times \mathscr{B}$ into a complex vector space $\mathscr{L}$, there is a unique linear mapping T, from $\mathscr{C}$ into $\mathscr{L}$, such that $L(A, B) = T(A \otimes B)$ for all A in $\mathfrak{A}$ and B in $\mathscr{B}$.*

Proof. (i) Let $\mathscr{R}$ be the factor $\mathscr{B}(\mathscr{H}) \otimes \mathbb{C}_{\mathscr{K}}$, so that $\mathscr{R}' = \mathbb{C}_{\mathscr{H}} \otimes \mathscr{B}(\mathscr{K})$ (Example 11.1.4). If we define

$$R_j = A_j \otimes I,\ R'_j = I \otimes B_j \qquad (j = 1, \ldots, n),$$

then $R_j \in \mathscr{R}$, $R'_j \in \mathscr{R}'$, and $\sum A_j \otimes B_j = \sum R_j R'_j$. The required result is now an immediate consequence of Theorem 5.5.4.

(ii) The reasoning used to prove Proposition 2.6.6(ii) applies also in the present situation. ∎

11.2. Tensor products of von Neumann algebras

Suppose that $\mathscr{R}_1, \ldots, \mathscr{R}_n$ are von Neumann algebras, acting on Hilbert spaces $\mathscr{H}_1, \ldots, \mathscr{H}_n$, respectively; and denote by $\mathscr{R}_0$ the * algebra, acting on the Hilbert space $\mathscr{H} = \mathscr{H}_1 \otimes \cdots \otimes \mathscr{H}_n$, that consists of all finite sums of operators of the form $A_1 \otimes \cdots \otimes A_n$, where $A_j \in \mathscr{R}_j$ for $j = 1, \ldots, n$. One concept of tensor product, for $\mathscr{R}_1, \ldots, \mathscr{R}_n$, is already available from Section 11.1, the (represented C^*-algebra) tensor product $\mathscr{R}_1 \otimes \cdots \otimes \mathscr{R}_n$. The present section is concerned with the (*von Neumann algebra*) *tensor product*, $\mathscr{R}_1 \bar{\otimes} \cdots \bar{\otimes} \mathscr{R}_n$, which is defined to be the von Neumann algebra $\mathscr{R}_0^-$ generated by $\mathscr{R}_0$; that is,

$$\mathscr{R}_1 \bar{\otimes} \cdots \bar{\otimes} \mathscr{R}_n = (\mathscr{R}_1 \otimes \cdots \otimes \mathscr{R}_n)^-.$$

In the first subsection, we provide a number of examples that illustrate the way in which some of our earlier results on von Neumann algebras can be expressed in terms of tensor products. We show how normal states (or * isomorphisms) of the component algebras $\mathscr{R}_1, \ldots, \mathscr{R}_n$ can be combined to form normal states (or * isomorphisms) of $\mathscr{R}_1 \bar{\otimes} \cdots \bar{\otimes} \mathscr{R}_n$; it turns out that the * isomorphism class of $\mathscr{R}_1 \bar{\otimes} \cdots \bar{\otimes} \mathscr{R}_n$ is determined by the * isomorphism classes of the von Neumann algebras $\mathscr{R}_1, \ldots, \mathscr{R}_n$. In the second subsection, the commutant of $\mathscr{R}_1 \bar{\otimes} \cdots \bar{\otimes} \mathscr{R}_n$ is identified as $\mathscr{R}'_1 \bar{\otimes} \cdots \bar{\otimes} \mathscr{R}'_n$; and, from this "commutation theorem," we deduce the corresponding result for the center of $\mathscr{R}_1 \bar{\otimes} \cdots \bar{\otimes} \mathscr{R}_n$. In the third subsection, we show how the type of the tensor product von Neumann algebra can be determined from the types of the component algebras. The fourth (and final) subsection provides a brief account of tensor products of unbounded operators. The theory is applied to determine how the modular structure of $\mathscr{R}_1 \bar{\otimes} \cdots \bar{\otimes} \mathscr{R}_n$ is related to the modular structures of the von Neumann algebras $\mathscr{R}_1, \ldots, \mathscr{R}_n$, when each of these has a separating generating vector. These results are used in giving a second proof of the commutation theorem.

Elementary properties. Suppose that, for $j = 1, \ldots, n$, $\mathscr{R}_j$ is the von Neumann algebra generated by a subset $\mathscr{S}_j$ (containing I) of $\mathscr{B}(\mathscr{H}_j)$. We assert that, in these circumstances, $\mathscr{R}_1 \bar{\otimes} \cdots \bar{\otimes} \mathscr{R}_n$ is the von Neumann algebra generated by the set

$$\mathscr{S} = \{S_1 \otimes \cdots \otimes S_n : S_1 \in \mathscr{S}_1, \ldots, S_n \in \mathscr{S}_n\}.$$

Indeed, let $\mathscr{R}$ be the von Neumann algebra generated by $\mathscr{S}$. For each j, the mapping

$$A_j \to I \otimes \cdots \otimes I \otimes A_j \otimes I \otimes \cdots \otimes I : \mathscr{B}(\mathscr{H}_j) \to \mathscr{B}(\mathscr{H}_1 \otimes \cdots \otimes \mathscr{H}_n)$$

is an (isometric) * isomorphism and is strong-operator continuous on bounded sets. The inverse image of $\mathscr{R}$, under this mapping, is a * algebra with strong-operator-closed unit ball (hence, a von Neumann algebra), that contains $\mathscr{S}_j$ and so contains $\mathscr{R}_j$. Thus

$$I \otimes \cdots \otimes I \otimes A_j \otimes I \otimes \cdots \otimes I \in \mathscr{R} \qquad (A_j \in \mathscr{R}_j;\ \ j = 1, \ldots, n),$$

and therefore $A_1 \otimes \cdots \otimes A_n \in \mathscr{R}$ whenever $A_1 \in \mathscr{R}_1, \ \ldots, \ A_n \in \mathscr{R}_n$. This implies that $\mathscr{R}_1 \mathbin{\bar{\otimes}} \cdots \mathbin{\bar{\otimes}} \mathscr{R}_n \subseteq \mathscr{R}$, and the reverse inclusion is clear.

We have already seen, in Section 11.1, that the natural unitary operator U, from $\mathscr{H}_1 \otimes \cdots \otimes \mathscr{H}_n$ onto $(\mathscr{H}_1 \otimes \cdots \otimes \mathscr{H}_r) \otimes (\mathscr{H}_{r+1} \otimes \cdots \otimes \mathscr{H}_n)$, satisfies

$$U(A_1 \otimes \cdots \otimes A_n)U^* = (A_1 \otimes \cdots \otimes A_r) \otimes (A_{r+1} \otimes \cdots \otimes A_n)$$

and

$$U(\mathscr{R}_1 \otimes \cdots \otimes \mathscr{R}_n)U^* = (\mathscr{R}_1 \otimes \cdots \otimes \mathscr{R}_r) \otimes (\mathscr{R}_{r+1} \otimes \cdots \otimes \mathscr{R}_n).$$

Upon forming the weak-operator closure of each side of the last equation, and using the result of the preceding paragraph, we obtain

$$\begin{aligned} U(\mathscr{R}_1 \mathbin{\bar{\otimes}} \cdots \mathbin{\bar{\otimes}} \mathscr{R}_n)U^* &= (\mathscr{R}_1 \otimes \cdots \otimes \mathscr{R}_r)^- \mathbin{\bar{\otimes}} (\mathscr{R}_{r+1} \otimes \cdots \otimes \mathscr{R}_n)^- \\ &= (\mathscr{R}_1 \mathbin{\bar{\otimes}} \cdots \mathbin{\bar{\otimes}} \mathscr{R}_r) \mathbin{\bar{\otimes}} (\mathscr{R}_{r+1} \mathbin{\bar{\otimes}} \cdots \mathbin{\bar{\otimes}} \mathscr{R}_n). \end{aligned}$$

The associativity of the von Neumann algebra tensor product, $\bar{\otimes}$, follows.

11.2.1. Example. Suppose that $\mathscr{H}$ and $\mathscr{K}$ are Hilbert spaces, $\mathscr{R}(\subseteq \mathscr{B}(\mathscr{H}))$ is a von Neumann algebra, and $\mathbb{C}_{\mathscr{K}}$ is the one-dimensional von Neumann algebra consisting of scalar multiples of the identity operator $I_{\mathscr{K}}$ on $\mathscr{K}$. From the discussion of $\mathscr{R} \otimes \mathbb{C}_{\mathscr{K}}$ in Example 11.1.4, it follows at once that

$$\mathscr{R} \mathbin{\bar{\otimes}} \mathbb{C}_{\mathscr{K}} = \mathscr{R} \otimes \mathbb{C}_{\mathscr{K}} = \{A \otimes I_{\mathscr{K}} : A \in \mathscr{R}\},$$

and that $\mathscr{R} \mathbin{\bar{\otimes}} \mathbb{C}_{\mathscr{K}}$ is unitarily equivalent to the von Neumann algebra $\mathscr{R} \otimes I_n$, where $n = \dim \mathscr{K}$. ■

11.2.2. Example. Suppose that $\mathscr{R}$ is a von Neumann algebra acting on a Hilbert space $\mathscr{H}$, $\mathscr{K}$ is another Hilbert space, and $n = \dim \mathscr{K}$. We assert that $\mathscr{R} \mathbin{\bar{\otimes}} \mathscr{B}(\mathscr{K})$ is unitarily equivalent to the von Neumann algebra $n \otimes \mathscr{R}$

defined in Section 6.6, and that $(\mathscr{R} \mathbin{\overline{\otimes}} \mathscr{B}(\mathscr{K}))' = \mathscr{R}' \mathbin{\overline{\otimes}} \mathbb{C}_{\mathscr{K}}$ (this last relation is a special case of the commutation theorem, 11.2.16, proved later).

To this end, let $\{y_b : b \in \mathbb{B}\}$ be an orthonormal basis of $\mathscr{K}$ (so that $\mathbb{B}$ has cardinality n), and let $\mathscr{H}_0$ be $\sum_{b \in \mathbb{B}} \oplus \mathscr{H}$. With each T in $\mathscr{B}(\mathscr{H}_0)$, we associate in the usual way (see Section 2.6, *Matrix representations*) a matrix $[T_{ab}]$, indexed by $\mathbb{B}$ and with entries in $\mathscr{B}(\mathscr{H})$. The equation

$$Ux = \sum_{b \in \mathbb{B}} x_b \otimes y_b \qquad \left(x = \sum_{b \in \mathbb{B}} \oplus x_b \in \mathscr{H}_0\right)$$

defines a unitary transformation U, from $\mathscr{H}_0$ onto $\mathscr{H} \otimes \mathscr{K}$. Moreover, when $A \in \mathscr{B}(\mathscr{H})$ and $B \in \mathscr{B}(\mathscr{K})$, $U^{-1}(A \otimes I)U$ has matrix $[\delta_{ab} A]$, and $U^{-1}(I \otimes B)U$ has matrix $[s_{ab} I]$, where $[s_{ab}]$ is the (numerical) matrix of B relative to the orthonormal basis $\{y_b\}$ of $\mathscr{K}$; so U^{-1} $(A \otimes B)U$ has matrix $[s_{ab} A]$.

The von Neumann algebra $n \otimes \mathscr{R}$ consists of those bounded operators on $\mathscr{H}_0$ whose matrices have each entry in $\mathscr{R}$. From the preceding paragraph, $U^{-1}(A \otimes B)U \in n \otimes \mathscr{R}$ whenever $A \in \mathscr{R}$, $B \in \mathscr{B}(\mathscr{K})$; so

$$U^{-1}(\mathscr{R} \mathbin{\overline{\otimes}} \mathscr{B}(\mathscr{K}))U \subseteq n \otimes \mathscr{R}.$$

To prove the reverse inclusion, suppose that $T \in n \otimes \mathscr{R}$, so that T has matrix $[A_{ab}]$ with all coefficients in $\mathscr{R}$. Let $\{E_{ab} : a, b \in \mathbb{B}\}$ be the system of matrix units for $\mathscr{B}(\mathscr{K})$, defined by $E_{ab}x = \langle x, y_b \rangle y_a$, so that the (numerical) matrix of E_{ab} has 1 in the (a, b) entry and zeros elsewhere. With $\mathbb{F}$ a finite subset of $\mathbb{B}$, the matrix of the operator

$$U^{-1}\left(\sum_{a, b \in \mathbb{F}} A_{ab} \otimes E_{ab}\right)U$$

has A_{ab} in the (a, b) position when $a, b \in \mathbb{F}$, and zeros elsewhere. Hence

$$U^{-1}\left(\sum_{a, b \in \mathbb{F}} A_{ab} \otimes E_{ab}\right)U = E(\mathbb{F})TE(\mathbb{F}),$$

where $E(\mathbb{F})$ is the projection in $n \otimes \mathscr{R}$ whose matrix has I in the (b, b) position when $b \in \mathbb{F}$, and zeros elsewhere. Accordingly,

$$E(\mathbb{F})TE(\mathbb{F}) \in U^{-1}(\mathscr{R} \mathbin{\overline{\otimes}} \mathscr{B}(\mathscr{K}))U$$

for every finite subset $\mathbb{F}$ of $\mathbb{B}$. When these subsets are directed by inclusion, the net $\{E(\mathbb{F})TE(\mathbb{F})\}$ is strong-operator convergent to T. Thus

$$T \in U^{-1}(\mathscr{R} \mathbin{\overline{\otimes}} \mathscr{B}(\mathscr{K}))U,$$

and $n \otimes \mathscr{R} = U^{-1}(\mathscr{R} \mathbin{\overline{\otimes}} \mathscr{B}(\mathscr{K}))U$.

From Lemma 6.6.2, $(n \otimes \mathscr{R})'$ is the von Neumann algebra $\mathscr{R}' \otimes I_n$ consisting of those operators acting on $\mathscr{H}_0$ having matrices of the form $[\delta_{ab} A']$, with A' in $\mathscr{R}'$. Accordingly,

$$\begin{aligned}(n \otimes \mathscr{R})' &= \mathscr{R}' \otimes I_n \\ &= \{U^{-1}(A' \otimes I)U : A' \in \mathscr{R}'\} \\ &= U^{-1}(\mathscr{R}' \mathbin{\overline{\otimes}} \mathbb{C}_{\mathscr{K}})U.\end{aligned}$$

Since $\mathscr{R} \mathbin{\overline{\otimes}} \mathscr{B}(\mathscr{K}) = U(n \otimes \mathscr{R})U^{-1}$, it now follows that $(\mathscr{R} \mathbin{\overline{\otimes}} \mathscr{B}(\mathscr{K}))' = \mathscr{R}' \mathbin{\overline{\otimes}} \mathbb{C}_{\mathscr{K}}$.

With $\mathscr{R} = \mathscr{B}(\mathscr{H})$, the preceding discussion shows that $\mathscr{B}(\mathscr{H}) \mathbin{\overline{\otimes}} \mathscr{B}(\mathscr{K})$ is unitarily equivalent to $n \otimes \mathscr{B}(\mathscr{H})$; and this last algebra is $\mathscr{B}(\mathscr{H}_0)$, since *every* bounded operator acting on $\mathscr{H}_0$ is represented by some $n \times n$ matrix of elements of $\mathscr{B}(\mathscr{H})$. Thus,

$$\mathscr{B}(\mathscr{H}) \mathbin{\overline{\otimes}} \mathscr{B}(\mathscr{K}) = \mathscr{B}(\mathscr{H} \otimes \mathscr{K}). \quad \blacksquare$$

11.2.3. Remark. In view of the unitary equivalences established in the two preceding examples, we shall sometimes identify $\mathscr{R} \mathbin{\overline{\otimes}} \mathbb{C}_{\mathscr{K}}$ with $\mathscr{R} \otimes I_n$, and $\mathscr{R} \mathbin{\overline{\otimes}} \mathscr{B}(\mathscr{K})$ with $n \otimes \mathscr{R}$. ■

11.2.4. Example. Suppose that $\mathscr{R}$ is a type I_m von Neumann algebra, with commutant $\mathscr{R}'$ of type I_n. We assert that $\mathscr{R}$ is unitarily equivalent to an algebra $\mathscr{B}(\mathscr{H}) \mathbin{\overline{\otimes}} \mathscr{A} \mathbin{\overline{\otimes}} \mathbb{C}_{\mathscr{K}}$, where $\mathscr{H}$ and $\mathscr{K}$ are Hilbert spaces having dimensions m and n, respectively, and $\mathscr{A}$ is a maximal abelian von Neumann algebra * isomorphic to the center of $\mathscr{R}$; moreover, $\mathscr{A}$ is uniquely determined, up to unitary equivalence. Indeed, this is simply a rephrasing of Theorem 9.3.2, taking into account the unitary equivalences discussed in Examples 11.2.1 and 11.2.2.

It follows that every type I von Neumann algebra is unitarily equivalent to one of the form

$$\sum_j \oplus \mathscr{B}(\mathscr{H}_j) \mathbin{\overline{\otimes}} \mathscr{A}_j \mathbin{\overline{\otimes}} \mathbb{C}_{\mathscr{K}_j},$$

where each $\mathscr{A}_j$ is a maximal abelian von Neumann algebra. ■

11.2.5. Example. We assert that every type I factor $\mathscr{R}$ is unitarily equivalent to one of the form $\mathscr{B}(\mathscr{H}) \mathbin{\overline{\otimes}} \mathbb{C}_{\mathscr{K}}$. Indeed, $\mathscr{R}$ is type I_m and $\mathscr{R}'$ is type I_n, for suitable cardinals m, n. From the preceding example, $\mathscr{R}$ is unitarily equivalent to $\mathscr{B}(\mathscr{H}) \mathbin{\overline{\otimes}} \mathscr{A} \mathbin{\overline{\otimes}} \mathbb{C}_{\mathscr{K}}$, where $\dim \mathscr{H} = m$, $\dim \mathscr{K} = n$, and the von Neumann algebra $\mathscr{A}$ is both maximal abelian and one-dimensional. Accordingly, $\mathscr{A} = \mathscr{B}(\mathscr{L})$, with $\mathscr{L}$ a one-dimensional Hilbert space; and $\mathscr{R}$ is unitarily equivalent to $\mathscr{B}(\mathscr{H}) \mathbin{\overline{\otimes}} \mathbb{C}_{\mathscr{K}}$. If U is a unitary operator that implements this equivalence then, since $(\mathscr{B}(\mathscr{H}) \mathbin{\overline{\otimes}} \mathbb{C}_{\mathscr{K}})'$ is $\mathbb{C}_{\mathscr{H}} \mathbin{\overline{\otimes}} \mathscr{B}(\mathscr{K})$,

$$U\mathscr{R}U^{-1} = \mathscr{B}(\mathscr{H}) \mathbin{\overline{\otimes}} \mathbb{C}_{\mathscr{K}}, \qquad U\mathscr{R}'U^{-1} = \mathbb{C}_{\mathscr{H}} \mathbin{\overline{\otimes}} \mathscr{B}(\mathscr{K}).$$

If $\mathscr{M}(\subseteq \mathscr{B}(\mathscr{H}))$ and $\mathscr{N}(\subseteq \mathscr{B}(\mathscr{K}))$ are von Neumann algebras, then $\mathscr{M} \mathbin{\overline{\otimes}} \mathbb{C}_{\mathscr{K}}$ and $\mathbb{C}_{\mathscr{H}} \mathbin{\overline{\otimes}} \mathscr{N}$ are commuting von Neumann algebras acting on $\mathscr{H} \otimes \mathscr{K}$. Given a von Neumann algebra $\mathscr{R}$, one can ask whether it is possible to choose $\mathscr{M}$, $\mathscr{N}$, and a unitary operator U, such that

$$U\mathscr{R}U^{-1} = \mathscr{M} \mathbin{\overline{\otimes}} \mathbb{C}_{\mathscr{K}}, \qquad U\mathscr{R}'U^{-1} = \mathbb{C}_{\mathscr{H}} \mathbin{\overline{\otimes}} \mathscr{N}.$$

We have just seen that this *is* possible, with $\mathscr{M}$ as $\mathscr{B}(\mathscr{H})$ and $\mathscr{N}$ as $\mathscr{B}(\mathscr{K})$, when $\mathscr{R}$ is a type I factor. In fact, it is possible *only* in this case. Indeed, if $\mathscr{M}$, $\mathscr{N}$, U satisfy the stated conditions, then

$$\mathscr{M} \mathbin{\overline{\otimes}} \mathbb{C}_{\mathscr{K}} = (\mathbb{C}_{\mathscr{H}} \mathbin{\overline{\otimes}} \mathscr{N})' = \mathscr{B}(\mathscr{H}) \mathbin{\overline{\otimes}} \mathscr{N}' \supseteq \mathscr{B}(\mathscr{H}) \mathbin{\overline{\otimes}} \mathbb{C}_{\mathscr{K}};$$

whence $\mathscr{M} = \mathscr{B}(\mathscr{H})$, $U\mathscr{R}U^{-1} = \mathscr{B}(\mathscr{H}) \mathbin{\overline{\otimes}} \mathbb{C}_{\mathscr{K}}$, and $\mathscr{R}$ is a type I factor. ∎

11.2.6. Example. In Section 6.7, we associated, with each discrete group G, a finite von Neumann algebra $\mathscr{L}_G$; and we proved that $\mathscr{L}_G$ is a factor of type II_1, when G is an i.c.c. group. We show here that $\mathscr{L}_{G\times H}$ is unitarily equivalent to $\mathscr{L}_G \mathbin{\overline{\otimes}} \mathscr{L}_H$, when $G \times H$ is the direct product of discrete groups G and H. When G and H are i.c.c. groups, so is $G \times H$; so the (von Neumann algebra) tensor product of the type II_1 factors $\mathscr{L}_G$ and $\mathscr{L}_H$ is another such factor. This exemplifies general results proved later (Proposition 11.2.20, Corollary 11.2.17).

To establish these results, let $\{x_g : g \in G\}$ be the orthonormal basis of $l_2(G)$ in which the function x_g takes the value 1 at g and 0 elsewhere on G. For each g in G, let L_g be the unitary operator, acting on $l_2(G)$ (and denoted by L_{x_g} in Section 6.7), defined by $L_g x = x_g * x$ $(x \in l_2(G))$. We shall use similar notation in relation to the groups H and $G \times H$. From Example 2.6.10, there is a unitary operator U, from $l_2(G) \otimes l_2(H)$ onto $l_2(G \times H)$, such that

$$(U(x \otimes y))(g, h) = x(g)y(h) \qquad (g \in G, \quad h \in H)$$

whenever $x \in l_2(G)$ and $y \in l_2(H)$; in particular,

$$U(x_g \otimes x_h) = x_{(g,h)}.$$

When $(g, h), (f, k) \in G \times H$,

$$\begin{aligned} U(L_g \otimes L_h)U^{-1}x_{(f,k)} &= U(L_g \otimes L_h)(x_f \otimes x_k) \\ &= U(L_g x_f \otimes L_h x_k) = U(x_{gf} \otimes x_{hk}) = x_{(gf,hk)} \\ &= L_{(g,h)}x_{(f,k)}; \end{aligned}$$

so

$$U(L_g \otimes L_h)U^{-1} = L_{(g,h)}.$$

Since the sets $\{L_g : g \in G\}$, $\{L_h : h \in H\}$, and $\{L_{(g,h)} : (g, h) \in G \times H\}$ generate the von Neumann algebras $\mathscr{L}_G$, $\mathscr{L}_H$, and $\mathscr{L}_{G \times H}$, respectively (Theorem 6.7.2(iii)), it now follows that

$$U(\mathscr{L}_G \mathbin{\overline{\otimes}} \mathscr{L}_H)U^{-1} = \mathscr{L}_{G \times H}. \quad \blacksquare$$

We now turn to the general theory of tensor products of von Neumann algebras. Before considering * isomorphisms between such algebras, we require some preliminary results concerning normal states of the tensor product algebra.

11.2.7. Proposition. *If ω_j is a normal state of a von Neumann algebra $\mathscr{R}_j$ $(j = 1, \ldots, n)$, there is a unique normal state ω of $\mathscr{R}_1 \mathbin{\overline{\otimes}} \cdots \mathbin{\overline{\otimes}} \mathscr{R}_n$, such that*

$$\omega(A_1 \otimes A_2 \otimes \cdots \otimes A_n) = \omega_1(A_1)\omega_2(A_2) \cdots \omega_n(A_n)$$

whenever $A_1 \in \mathscr{R}_1, \ldots, A_n \in \mathscr{R}_n$.

Proof. There is a sequence $\{x_{j1}, x_{j2}, \ldots\}$ of vectors, in the Hilbert space on which $\mathscr{R}_j$ acts, such that

$$\sum_{k=1}^{\infty} \|x_{jk}\|^2 = 1, \qquad \omega_j(A) = \sum_{k=1}^{\infty} \langle Ax_{jk}, x_{jk} \rangle \quad (A \in \mathscr{R}_j).$$

Since

$$\begin{aligned}
&\sum_{k(1)} \sum_{k(2)} \cdots \sum_{k(n)} \|x_{1k(1)} \otimes x_{2k(2)} \otimes \cdots \otimes x_{nk(n)}\|^2 \\
&\quad = \sum_{k(1)} \sum_{k(2)} \cdots \sum_{k(n)} \|x_{1k(1)}\|^2 \, \|x_{2k(2)}\|^2 \cdots \|x_{nk(n)}\|^2 \\
&\quad = \left(\sum_{k(1)} \|x_{1k(1)}\|^2\right)\left(\sum_{k(2)} \|x_{2k(2)}\|^2\right) \cdots \left(\sum_{k(n)} \|x_{nk(n)}\|^2\right) = 1,
\end{aligned}$$

the equation

$$\omega(A) = \sum_{k(1)} \sum_{k(2)} \cdots \sum_{k(n)} \langle A(x_{1k(1)} \otimes \cdots \otimes x_{nk(n)}), \; x_{1k(1)} \otimes \cdots \otimes x_{nk(n)} \rangle$$

defines a normal state ω of $\mathscr{R}_1 \bar{\otimes} \cdots \bar{\otimes} \mathscr{R}_n$. When $A_1 \in \mathscr{R}_1, \ldots, A_n \in \mathscr{R}_n$,

$$\begin{aligned}\omega(A_1 \otimes \cdots \otimes A_n) &= \sum_{k(1)} \sum_{k(2)} \cdots \sum_{k(n)} \langle A_1 x_{1k(1)}, x_{1k(1)} \rangle \langle A_2 x_{2k(2)}, x_{2k(2)} \rangle \cdots \langle A_n x_{nk(n)}, x_{nk(n)} \rangle \\ &= \left(\sum_{k(1)} \langle A_1 x_{1k(1)}, x_{1k(1)} \rangle \right) \left(\sum_{k(2)} \langle A_2 x_{2k(2)}, x_{2k(2)} \rangle \right) \cdots \left(\sum_{k(n)} \langle A_n x_{nk(n)}, x_{nk(n)} \rangle \right) \\ &= \omega_1(A_1)\omega_2(A_2) \cdots \omega_n(A_n)\end{aligned}$$

(since the series occurring here are absolutely convergent). This proves the existence of a normal state ω with the required properties; and since these properties determine the values taken by ω on an ultraweakly dense subset of $\mathscr{R}_1 \bar{\otimes} \cdots \bar{\otimes} \mathscr{R}_n$, there is only one such normal state. ■

The state ω occurring in Proposition 11.2.7 is denoted by $\omega_1 \bar{\otimes} \cdots \bar{\otimes} \omega_n$; and states of this type are described as *normal product states* of $\mathscr{R}_1 \bar{\otimes} \cdots \bar{\otimes} \mathscr{R}_n$. It is apparent that $\omega_1 \bar{\otimes} \cdots \bar{\otimes} \omega_n$ is the unique ultraweakly continuous extension, to $\mathscr{R}_1 \bar{\otimes} \cdots \bar{\otimes} \mathscr{R}_n$, of the product state $\omega_1 \otimes \cdots \otimes \omega_n$ of the C*-algebra $\mathscr{R}_1 \otimes \cdots \otimes \mathscr{R}_n$.

11.2.8. Proposition. *If $\mathscr{R}_1, \ldots, \mathscr{R}_n$ are von Neumann algebras and $\mathscr{R} = \mathscr{R}_1 \bar{\otimes} \cdots \bar{\otimes} \mathscr{R}_n$, then the predual $\mathscr{R}_\sharp$ is the norm-closed subspace of the dual space $\mathscr{R}^\sharp$ generated by the set of all normal product states of $\mathscr{R}$.*

Proof. Let $\mathscr{L}(\subseteq \mathscr{R}^\sharp)$ be the set of all finite linear combinations of normal product states of $\mathscr{R}$. Denote by $\mathscr{H}_j$ the Hilbert space on which $\mathscr{R}_j$ acts, and let $\mathscr{H}$ be $\mathscr{H}_1 \otimes \cdots \otimes \mathscr{H}_n$. It is apparent that $\mathscr{R}_\sharp$ contains $\mathscr{L}^=$, the norm closure of $\mathscr{L}$; and we have to prove the reverse inclusion.

Each ω in $\mathscr{R}_\sharp$ can be expressed as the sum of a norm-convergent series $\sum_k \omega_{u(k)v(k)} \mid \mathscr{R}$, where $u(k), v(k) \in \mathscr{H}$; so it suffices to show that $\omega_{uv} \mid \mathscr{R} \in \mathscr{L}^=$ whenever $u, v \in \mathscr{H}$. Upon approximating u and v (in norm) by finite sums, x and y, of simple tensors, ω_{uv} is approximated in norm by ω_{xy}; moreover, ω_{xy} is a finite linear combination of linear functionals of the form

(1) $$\omega_{x(1)\otimes \cdots \otimes x(n),\, y(1)\otimes \cdots \otimes y(n)} \mid \mathscr{R},$$

with $x(j)$ and $y(j)$ in $\mathscr{H}_j$, for $j = 1, \ldots, n$. Accordingly, it is sufficient to show that $\mathscr{L}$ contains all linear functionals of the type occurring in (1). By polarization, $\omega_{x(j)y(j)} \mid \mathscr{R}_j$ can be expressed as a linear combination,

$$\omega_{x(j)y(j)} \mid \mathscr{R}_j = \sum_{k=1}^{4} a_{jk} \omega_{jk},$$

of vector states $\omega_{j1}, \ldots, \omega_{j4}$ of $\mathscr{R}_j$. When $A_1 \in \mathscr{R}_1, \ldots, A_n \in \mathscr{R}_n$,

$$\begin{aligned}
&\omega_{x(1)\otimes\cdots\otimes x(n),\, y(1)\otimes\cdots\otimes y(n)}(A_1 \otimes \cdots \otimes A_n)\\
&\quad = \langle A_1 x(1), y(1)\rangle\langle A_2 x(2), y(2)\rangle \cdots \langle A_n x(n), y(n)\rangle\\
&\quad = \sum_{k(1)\cdots k(n)=1}^{4} a_{1k(1)}a_{2k(2)}\cdots a_{nk(n)}\omega_{1k(1)}(A_1)\omega_{2k(2)}(A_2)\cdots\omega_{nk(n)}(A_n)\\
&\quad = \sum_{k(1)\cdots k(n)=1}^{4} a_{1k(1)}a_{2k(2)}\cdots a_{nk(n)}(\omega_{1k(1)} \mathbin{\bar{\otimes}} \cdots \mathbin{\bar{\otimes}} \omega_{nk(n)})(A_1 \otimes \cdots \otimes A_n).
\end{aligned}$$

Thus

$$\begin{aligned}
&\omega_{x(1)\otimes\cdots\otimes x(n),\, y(1)\otimes\cdots\otimes y(n)} \mid \mathscr{R}\\
&\qquad = \sum_{k(1)\cdots k(n)=1}^{4} a_{1k(1)}a_{2k(2)}\cdots a_{nk(n)}(\omega_{1k(1)} \mathbin{\bar{\otimes}} \cdots \mathbin{\bar{\otimes}} \omega_{nk(n)}) \qquad (\in \mathscr{L}),
\end{aligned}$$

since the linear functionals occurring in this last equation are ultraweakly continuous and coincide on an ultraweakly dense subset of $\mathscr{R}$. ■

But see Exercise 11.5.12 in connection with Proposition 11.2.8.

11.2.9. THEOREM. *Suppose that, for $j = 1, \ldots, n$, $\mathscr{R}_j$ and $\mathscr{S}_j$ are von Neumann algebras and φ_j is an ultraweakly continuous * homomorphism from $\mathscr{R}_j$ into $\mathscr{S}_j$. Then there is a unique ultraweakly continuous * homomorphism φ, from $\mathscr{R}_1 \mathbin{\bar{\otimes}} \cdots \mathbin{\bar{\otimes}} \mathscr{R}_n$ into $\mathscr{S}_1 \mathbin{\bar{\otimes}} \cdots \mathbin{\bar{\otimes}} \mathscr{S}_n$, such that*

$$\varphi(A_1 \otimes \cdots \otimes A_n) = \varphi_1(A_1) \otimes \cdots \otimes \varphi_n(A_n) \qquad (A_1 \in \mathscr{R}_1, \ldots, A_n \in \mathscr{R}_n).$$

If $\varphi_j(\mathscr{R}_j) = \mathscr{S}_j$, for each j, then $\varphi(\mathscr{R}_1 \mathbin{\bar{\otimes}} \cdots \mathbin{\bar{\otimes}} \mathscr{R}_n) = \mathscr{S}_1 \mathbin{\bar{\otimes}} \cdots \mathbin{\bar{\otimes}} \mathscr{S}_n$.

Proof. We recall first that, if $\mathscr{R}$ and $\mathscr{S}$ are von Neumann algebras (acting on Hilbert spaces $\mathscr{H}$ and $\mathscr{K}$, respectively), and $\eta\colon \mathscr{R} \to \mathscr{S}$ is an ultraweakly continuous * homomorphism, then $\eta(\mathscr{R})$ is a von Neumann subalgebra of $\mathscr{S}$. This is a special case of Lemma 10.1.10 (with $\mathfrak{A} = \mathfrak{A}^- = \mathscr{R}$, and $\bar{\eta} = \eta$); it is also a straightforward consequence of the Kaplansky density theorem, together with the fact (Corollary 10.1.8) that η maps the closed unit ball $(\mathscr{R})_1$ onto $(\eta(\mathscr{R}))_1$.

Upon replacing $\mathscr{S}_j$ by its von Neumann subalgebra $\varphi_j(\mathscr{R}_j)$, we may assume that $\mathscr{S}_j = \varphi_j(\mathscr{R}_j)$ $(j = 1, \ldots, n)$. From Theorem 11.1.3 there is a * homomorphism φ_0, from the represented C^*-algebra $\mathscr{R}_1 \otimes \cdots \otimes \mathscr{R}_n$ onto $\mathscr{S}_1 \otimes \cdots \otimes \mathscr{S}_n$, such that

$$\varphi_0(A_1 \otimes \cdots \otimes A_n) = \varphi_1(A_1) \otimes \cdots \otimes \varphi_n(A_n) \qquad (A_1 \in \mathscr{R}_1, \ldots, A_n \in \mathscr{R}_n).$$

It now suffices to prove that φ_0 extends to an ultraweakly continuous * homomorphism φ, from $\mathscr{R}_1 \mathbin{\bar{\otimes}} \cdots \mathbin{\bar{\otimes}} \mathscr{R}_n$ onto $\mathscr{S}_1 \mathbin{\bar{\otimes}} \cdots \mathbin{\bar{\otimes}} \mathscr{S}_n$. From Lemma 10.1.10, it is enough to show that φ_0 is ultraweakly continuous.

Suppose that ω_j is a normal state of $\mathscr{S}_j$ ($j = 1, \ldots, n$), and ω is the normal product state $\omega_1 \mathbin{\bar{\otimes}} \cdots \mathbin{\bar{\otimes}} \omega_n$ of $\mathscr{S}_1 \mathbin{\bar{\otimes}} \cdots \mathbin{\bar{\otimes}} \mathscr{S}_n$. Since φ_j is ultraweakly continuous, $\omega_j \circ \varphi_j$ is a normal state σ_j of $\mathscr{R}_j$, and $\sigma_1 \mathbin{\bar{\otimes}} \cdots \mathbin{\bar{\otimes}} \sigma_n$ is a normal product state σ of $\mathscr{R}_1 \mathbin{\bar{\otimes}} \cdots \mathbin{\bar{\otimes}} \mathscr{R}_n$. When $A_1 \in \mathscr{R}_1, \ldots, A_n \in \mathscr{R}_n$,

$$\begin{aligned}
\sigma(A_1 \otimes \cdots \otimes A_n) &= \sigma_1(A_1)\sigma_2(A_2) \cdots \sigma_n(A_n) \\
&= \omega_1(\varphi_1(A_1))\omega_2(\varphi_2(A_2)) \cdots \omega_n(\varphi_n(A_n)) \\
&= \omega(\varphi_1(A_1) \otimes \cdots \otimes \varphi_n(A_n)) \\
&= \omega(\varphi_0(A_1 \otimes \cdots \otimes A_n))
\end{aligned}$$

By linearity and norm continuity of φ_0, it follows that $\sigma(A) = \omega(\varphi_0(A))$ for each A in $\mathscr{R}_1 \otimes \cdots \otimes \mathscr{R}_n$.

The preceding paragraph shows that the linear functional $\omega \circ \varphi_0$ on $\mathscr{R}_1 \otimes \cdots \otimes \mathscr{R}_n$ is ultraweakly continuous (in fact, the restriction of a normal product state on $\mathscr{R}_1 \mathbin{\bar{\otimes}} \cdots \mathbin{\bar{\otimes}} \mathscr{R}_n$), whenever ω is a normal product state of $\mathscr{S}_1 \mathbin{\bar{\otimes}} \cdots \mathbin{\bar{\otimes}} \mathscr{S}_n$. By Proposition 11.2.8, each normal state ρ of $\mathscr{S}_1 \mathbin{\bar{\otimes}} \cdots \mathbin{\bar{\otimes}} \mathscr{S}_n$ is the limit of a norm convergent sequence $\{\rho_n\}$, each ρ_n being a finite linear combination of normal product states. The linear functionals $\rho_n \circ \varphi_0$ on $\mathscr{R}_1 \otimes \cdots \otimes \mathscr{R}_n$ are ultraweakly continuous, by the above argument, and converge in norm to $\rho \circ \varphi_0$. By Theorem 10.1.15(i), $\rho \circ \varphi_0$ is ultraweakly continuous; and hence, so is φ_0. ■

The * homomorphism φ occurring in Theorem 11.2.9 is denoted by $\varphi_1 \mathbin{\bar{\otimes}} \cdots \mathbin{\bar{\otimes}} \varphi_n$.

11.2.10. THEOREM. *If $\mathscr{R}_j$ and $\mathscr{S}_j$ are von Neumann algebras, and φ_j is a * isomorphism from $\mathscr{R}_j$ onto $\mathscr{S}_j$ ($j = 1, \ldots, n$), there is a unique * isomorphism φ, from $\mathscr{R}_1 \mathbin{\bar{\otimes}} \cdots \mathbin{\bar{\otimes}} \mathscr{R}_n$ onto $\mathscr{S}_1 \mathbin{\bar{\otimes}} \cdots \mathbin{\bar{\otimes}} \mathscr{S}_n$, such that*

$$\varphi(A_1 \otimes \cdots \otimes A_n) = \varphi_1(A_1) \otimes \cdots \otimes \varphi_n(A_n) \qquad (A_1 \in \mathscr{R}_1, \ldots, A_n \in \mathscr{R}_n).$$

Proof. Each φ_j is ultraweakly bicontinuous (Remark 7.4.4), so there are ultraweakly continuous * homomorphisms

$$\begin{aligned}
\varphi &= \varphi_1 \mathbin{\bar{\otimes}} \cdots \mathbin{\bar{\otimes}} \varphi_n : \mathscr{R}_1 \mathbin{\bar{\otimes}} \cdots \mathbin{\bar{\otimes}} \mathscr{R}_n \text{ onto } \mathscr{S}_1 \mathbin{\bar{\otimes}} \cdots \mathbin{\bar{\otimes}} \mathscr{S}_n, \\
\psi &= \varphi_1^{-1} \mathbin{\bar{\otimes}} \cdots \mathbin{\bar{\otimes}} \varphi_n^{-1} : \mathscr{S}_1 \mathbin{\bar{\otimes}} \cdots \mathbin{\bar{\otimes}} \mathscr{S}_n \text{ onto } \mathscr{R}_1 \mathbin{\bar{\otimes}} \cdots \mathbin{\bar{\otimes}} \mathscr{R}_n,
\end{aligned}$$

with the properties set out in Theorem 11.2.9. When $A_1 \in \mathscr{R}_1, \ldots, A_n \in \mathscr{R}_n$

$$\psi(\varphi(A_1 \otimes \cdots \otimes A_n)) = \psi(\varphi_1(A_1) \otimes \cdots \otimes \varphi_n(A_n)) = A_1 \otimes \cdots \otimes A_n.$$

From the linearity and ultraweak continuity of φ and ψ, $\psi(\varphi(A)) = A$ for all A in $\mathscr{R}_1 \mathbin{\overline{\otimes}} \cdots \mathbin{\overline{\otimes}} \mathscr{R}_n$; so φ is one-to-one. This proves the existence of a * isomorphism φ with the stated properties; and, since any such φ is ultraweakly continuous, it is apparent that φ is unique. ∎

By identifying $n \otimes \mathscr{R}$ with $\mathscr{R} \mathbin{\overline{\otimes}} \mathscr{B}(\mathscr{L})$, where the Hilbert space $\mathscr{L}$ has dimension n (Remark 11.2.3), the final assertion of Lemma 6.6.2 is seen to be a special case of Theorem 11.2.10.

The commutation theorem. If $\mathscr{R}$ and $\mathscr{S}$ are von Neumann algebras, acting on Hilbert spaces $\mathscr{H}$ and $\mathscr{K}$, respectively, it is apparent that the operators $A \otimes B$ and $A' \otimes B'$ commute, when $A \in \mathscr{R}$, $B \in \mathscr{S}$, $A' \in \mathscr{R}'$, and $B' \in \mathscr{S}'$; so $\mathscr{R}' \mathbin{\overline{\otimes}} \mathscr{S}' \subseteq (\mathscr{R} \mathbin{\overline{\otimes}} \mathscr{S})'$, equivalently, $\mathscr{R} \mathbin{\overline{\otimes}} \mathscr{S} \subseteq (\mathscr{R}' \mathbin{\overline{\otimes}} \mathscr{S}')'$. Our next objective, attained in Theorem 11.2.16, is to show that $(\mathscr{R} \mathbin{\overline{\otimes}} \mathscr{S})' = \mathscr{R}' \mathbin{\overline{\otimes}} \mathscr{S}'$. From this "commutation theorem," and by appeal to the associativity of $\overline{\otimes}$, it is easy to deduce the corresponding result,

$$(\mathscr{R}_1 \mathbin{\overline{\otimes}} \cdots \mathbin{\overline{\otimes}} \mathscr{R}_n)' = \mathscr{R}_1' \mathbin{\overline{\otimes}} \cdots \mathbin{\overline{\otimes}} \mathscr{R}_n',$$

for a tensor product of n von Neumann algebras.

In proving the commutation theorem, we consider first the case in which the von Neumann algebras have generating vectors. We require some auxiliary results and some additional notation. If $\mathscr{R}$ is a von Neumann algebra acting on a Hilbert space $\mathscr{H}$, the set of all self-adjoint elements of $\mathscr{R}$ will be denoted by $\mathscr{R}^{\mathrm{h}}$. We write $\mathscr{H}_{\mathrm{r}}$ for the *real* Hilbert space that consists of the set $\mathscr{H}$, with vector addition and multiplication of vectors by real numbers the same as in $\mathscr{H}$, and with the inner product $\langle\ ,\ \rangle_{\mathrm{r}}$ defined by

$$\langle x, y\rangle_{\mathrm{r}} = \mathrm{Re}\langle x, y\rangle (= \tfrac{1}{2}[\langle x, y\rangle + \langle y, x\rangle]) \qquad (x, y \in \mathscr{H}).$$

Since the norm of $\mathscr{H}_{\mathrm{r}}$ coincides with that of $\mathscr{H}$, a subset Y of $\mathscr{H}$ has the same closure $Y^=$ in $\mathscr{H}_{\mathrm{r}}$ as in $\mathscr{H}$. The symbol $Y^{\perp}$ will be used to denote the subspace orthogonal to Y in the *real* Hilbert space. Note that the linear mapping $x \to ix : \mathscr{H}_{\mathrm{r}} \to \mathscr{H}_{\mathrm{r}}$ preserves both norm and inner products in $\mathscr{H}_{\mathrm{r}}$, that $(iY)^{\perp} = i(Y^{\perp})$, and that $\mathscr{R}^{\mathrm{h}}u$ is a (real-)linear subspace of $\mathscr{H}_{\mathrm{r}}$, when $u \in \mathscr{H}$.

11.2.11. Proposition. *If $\mathscr{M}$ and $\mathscr{N}$ are von Neumann algebras acting on a Hilbert space $\mathscr{H}$, such that $\mathscr{M} \subseteq \mathscr{N}'$ and $\mathscr{M}$ has a generating vector u, then the following three conditions are equivalent:*

(i) $\mathscr{M} = \mathscr{N}'$;

(ii) *$\mathscr{M}^{\mathrm{h}}u + i\mathscr{N}^{\mathrm{h}}u$ is everywhere dense in $\mathscr{H}$;*

(iii) $(\mathscr{M}^{\mathrm{h}}u)^{\perp} = (i\mathscr{N}^{\mathrm{h}}u)^=$.

Proof. We assert first that

$$(2) \qquad (i(\mathscr{M}')^{\mathrm{h}}u)^{=} \subseteq (\mathscr{M}^{\mathrm{h}}u)^{\perp}.$$

Indeed, if $A \in \mathscr{M}^{\mathrm{h}}$ and $B \in (\mathscr{M}')^{\mathrm{h}}$, then AB $(= BA)$ is self-adjoint, $\langle Bu, Au\rangle$ $(= \langle ABu, u\rangle)$ is real, and

$$\langle iBu, Au\rangle_{\mathrm{r}} = \mathrm{Re}\, i\langle Bu, Au\rangle = 0.$$

This shows that $i(\mathscr{M}')^{\mathrm{h}}u \subseteq (\mathscr{M}^{\mathrm{h}}u)^{\perp}$, and so proves (2).

Since $\mathscr{N} \subseteq \mathscr{M}'$, it results from (2) that

$$i\mathscr{N}^{\mathrm{h}}u \subseteq i\,(\mathscr{M}')^{\mathrm{h}}u \subseteq (\mathscr{M}^{\mathrm{h}}u)^{\perp}.$$

Accordingly, the closed (real-linear) subspaces $(i\mathscr{N}^{\mathrm{h}}u)^{=}$ and $(\mathscr{M}^{\mathrm{h}}u)^{=}$ of $\mathscr{H}_{\mathrm{r}}$ are mutually orthogonal, and

$$\begin{aligned}(\mathscr{M}^{\mathrm{h}}u + i\mathscr{N}^{\mathrm{h}}u)^{=} &= (\mathscr{M}^{\mathrm{h}}u)^{=} + (i\,\mathscr{N}^{\mathrm{h}}\,u)^{=}\\ &\subseteq (\mathscr{M}^{\mathrm{h}}u)^{=} + (\mathscr{M}^{\mathrm{h}}u)^{\perp} = \mathscr{H}_{\mathrm{r}},\end{aligned}$$

with equality throughout only when $(i\mathscr{N}^{\mathrm{h}}u)^{=} = (\mathscr{M}^{\mathrm{h}}u)^{\perp}$. This proves the equivalence of (ii) and (iii).

Suppose now that (ii) and (iii) are satisfied. If $A' \in (\mathscr{M}')^{\mathrm{h}}$, it results from (2) and (iii) that

$$iA'u \in (\mathscr{M}^{\mathrm{h}}u)^{\perp} = (i\mathscr{N}^{\mathrm{h}}u)^{=},$$

whence $A'u \in (\mathscr{N}^{\mathrm{h}}u)^{=}$. Accordingly, there is a sequence $\{B_n\}$ in $\mathscr{N}^{\mathrm{h}}$, such that $A'u = \lim B_n u$. When $B' \in \mathscr{N}'$ and $A_1, A_2 \in \mathscr{M}(\subseteq \mathscr{N}')$,

$$\begin{aligned}\langle B'A'A_1u, A_2u\rangle &= \langle B'A_1A'u, A_2u\rangle = \lim_{n\to\infty} \langle B'A_1B_nu, A_2u\rangle\\ &= \lim_{n\to\infty} \langle B'A_1u, A_2B_nu\rangle = \langle B'A_1u, A_2A'u\rangle\\ &= \langle A'B'A_1u, A_2u\rangle.\end{aligned}$$

Since u is a generating vector for $\mathscr{M}$, it follows that $B'A' = A'B'$, whenever $B' \in \mathscr{N}'$; so $A' \in \mathscr{N}'' = \mathscr{N}$. This shows that $\mathscr{M}' \subseteq \mathscr{N}$, whence $\mathscr{M} \supseteq \mathscr{N}'$; and the reverse inclusion is given. Hence the (equivalent) conditions (ii) and (iii) imply (i).

It now suffices to show that (i) implies (ii); so we have to prove that the real-linear subspace $(\mathscr{N}')^{\mathrm{h}}u + i\mathscr{N}^{\mathrm{h}}u$ is everywhere dense in $\mathscr{H}$, when u is a generating vector for $\mathscr{N}'$. Suppose that $v \in \mathscr{H}$, and v is orthogonal (in the real Hilbert space $\mathscr{H}_{\mathrm{r}}$) to both $(\mathscr{N}')^{\mathrm{h}}u$ and $i\mathscr{N}^{\mathrm{h}}u$; we must show that $v = 0$. For this, consider the von Neumann algebra $\mathscr{R}$ $(= 2 \otimes \mathscr{N})$, which acts on $\mathscr{H} \oplus \mathscr{H}$ and consists of those operators whose 2×2 matrices have entries in $\mathscr{N}$; and recall from Section 6.6 that $\mathscr{R}'$ ($= \mathscr{N}' \otimes I_2$) is the set of all operators with 2×2 diagonal matrices having the same element of $\mathscr{N}'$ at each diagonal position. With w the vector (u, v) in $\mathscr{H} \oplus \mathscr{H}$, and E the cyclic projection in $\mathscr{R}$ with range $[\mathscr{R}'w]$, the matrix of E has the form

$$\begin{bmatrix} P & R \\ R^* & Q \end{bmatrix},$$

where $P, Q, R \in \mathscr{N}$, $0 \le P \le I$, and $0 \le Q \le I$. Since $Ew = w$,

$$Pu + Rv = u. \tag{3}$$

For each self-adjoint element A' of $\mathscr{N}'$, v is orthogonal (in $\mathscr{H}_r$) to $A'u$; so

$$0 = \mathrm{Re}\langle v, A'u\rangle = \langle v, A'u\rangle + \langle A'u, v\rangle = \langle v, A'u\rangle + \langle u, A'v\rangle.$$

By decomposing a general element of $\mathscr{N}'$ into its real and imaginary parts, it now follows that $\langle v, A'u\rangle + \langle u, A'v\rangle = 0$ for all A' in $\mathscr{N}'$; equivalently, $\langle w_0, R'w\rangle = 0$ for all R' in $\mathscr{R}'$, where w_0 is the vector (v, u) in $\mathscr{H} \oplus \mathscr{H}$. Thus w_0 is orthogonal to the range of E; whence $Ew_0 = 0$, and

$$Pv + Ru = 0. \tag{4}$$

For each self-adjoint element A of $\mathscr{N}$, v is orthogonal (in $\mathscr{H}_r$) to iAu; so

$$0 = \mathrm{Re}\langle v, iAu\rangle = \mathrm{Im}\langle v, Au\rangle,$$

$$0 = \langle v, Au\rangle - \langle Au, v\rangle = \langle v, Au\rangle - \langle u, Av\rangle.$$

Thus

$$\langle v, Au\rangle = \langle u, Av\rangle \qquad (A \in \mathscr{N}); \tag{5}$$

in particular, this applies with P or R in place of A. From (3), (5), and (4)

$$\begin{aligned} \|(I - P)^{1/2}u\|^2 &= \langle u, u - Pu\rangle = \langle u, Rv\rangle \\ &= \langle v, Ru\rangle = -\langle v, Pv\rangle = -\|P^{1/2}v\|^2, \end{aligned}$$

and $(I - P)u = Pv = 0$. Now u is generating for $\mathscr{N}'$, hence separating for $\mathscr{N}$; so $P = I$, and $v = 0$. ∎

We shall use Proposition 11.2.11, with $\mathscr{M} = \mathscr{R} \,\bar{\otimes}\, \mathscr{S}$ and $\mathscr{N} = \mathscr{R}' \,\bar{\otimes}\, \mathscr{S}'$, to show that $(\mathscr{R} \,\bar{\otimes}\, \mathscr{S})' = \mathscr{R}' \,\bar{\otimes}\, \mathscr{S}'$ when each of the von Neumann algebras $\mathscr{R}$ and $\mathscr{S}$ has a generating vector. For this, we first require one more preparatory result. If $\mathscr{H}$ and $\mathscr{K}$ are Hilbert spaces, and Y ($\subseteq \mathscr{H}$), Z ($\subseteq \mathscr{K}$) are real-linear subspaces, we denote by $Y \odot Z$ the real-linear subspace of $\mathscr{H} \otimes \mathscr{K}$ that consists of finite sums of simple tensors $y \otimes z$, with y in Y and z in Z.

11.2.12. LEMMA. *If $\mathscr{H}$, $\mathscr{K}$ are Hilbert spaces and Y $(\subseteq \mathscr{H})$, Z $(\subseteq \mathscr{K})$ are real-linear subspaces, such that $Y + iY$ is everywhere dense in $\mathscr{H}$ and $Z + iZ$ is everywhere dense in $\mathscr{K}$, then $Y \odot Z + i(Y^{\perp} \odot Z^{\perp})$ is everywhere dense in $\mathscr{H} \otimes \mathscr{K}$.*

Proof. Note first that $Y \odot Z + i(Y^{\perp} \odot Z^{\perp})$ is a (real-) linear subspace of the real Hilbert space $(\mathscr{H} \otimes \mathscr{K})_r$. Suppose that $w \in \mathscr{H} \otimes \mathscr{K}$, and w is orthogonal (in $(\mathscr{H} \otimes \mathscr{K})_r$) to both $Y \odot Z$ and $i(Y^{\perp} \odot Z^{\perp})$. We have to show that $w = 0$.

The equation $b(u, v) = \langle u \otimes v, w \rangle$ defines a bounded bilinear functional b on $\mathscr{H} \times \mathscr{K}$, and this can be viewed as a conjugate-bilinear functional on $\mathscr{H} \times \bar{\mathscr{K}}$. Accordingly, there is a bounded conjugate-linear mapping $T : \mathscr{H} \to \mathscr{K}$ (see Theorem 2.4.1 and the introductory discussion of *A first approach to modular theory*, Section 9.2) such that

$$\langle Tu, v \rangle = \langle u \otimes v, w \rangle \qquad (u \in \mathscr{H}, \quad v \in \mathscr{K}). \tag{6}$$

Upon taking real parts throughout this last equation, and recalling that the conjugate-linear operator T^* satisfies $\langle T^*v, u \rangle = \langle Tu, v \rangle$, we obtain

$$\langle Tu, v \rangle_r = \langle T^*v, u \rangle_r = \langle u \otimes v, w \rangle_r \qquad (u \in \mathscr{H}, \quad v \in \mathscr{K}).$$

From this, and since w is orthogonal (in $(\mathscr{H} \otimes \mathscr{K})_r$) to $Y \odot Z$ and $i(Y^{\perp} \odot Z^{\perp})$, it follows that $\langle Tu, v \rangle_r = 0 = \langle T^*v, u \rangle_r$ when *either* $u \in Y$, $v \in Z$ (and so, by continuity, when $u \in Y^{=}$ and $v \in Z^{=}$) *or* when $u \in iY^{\perp}$, $v \in Z^{\perp}$. Thus

$$\begin{aligned} &T(Y^{=}) \subseteq Z^{\perp}, \qquad T(iY^{\perp}) \subseteq (Z^{\perp})^{\perp} = Z^{=}, \\ &T^*(Z^{=}) \subseteq Y^{\perp}, \qquad T^*(Z^{\perp}) \subseteq (iY^{\perp})^{\perp} = iY^{=}. \end{aligned} \tag{7}$$

From these relations, together with the conjugate linearity of T and T^*,

$$(T^*T)(iY^{\perp}) \subseteq T^*(Z^{=}) \subseteq Y^{\perp},$$

$$(TT^*)(Z^{\perp}) \subseteq T(iY^{=}) = iT(Y^{=}) \subseteq iZ^{\perp}.$$

Hence the positive linear operators T^*T, TT^* satisfy

$$(T^*T)(Y^{\perp}) \subseteq iY^{\perp}, \qquad (TT^*)(Z^{\perp}) \subseteq iZ^{\perp}, \tag{8}$$

$$(T^*T)^2(Y^{\perp}) \subseteq Y^{\perp}, \qquad (TT^*)^2(Z^{\perp}) \subseteq Z^{\perp}. \tag{9}$$

Since T^*T is the norm limit of a sequence of polynomials (with real coefficients) in $(T^*T)^2$, it results from (9) that $(T^*T)(Y^{\perp}) \subseteq Y^{\perp}$. This, with (8), yields

$$\begin{aligned} (T^*T)(Y^{\perp}) \subseteq Y^{\perp} \cap iY^{\perp} &= Y^{\perp} \cap (iY)^{\perp} \\ &= (Y + iY)^{\perp} = \mathscr{H}^{\perp} = \{0\}; \end{aligned}$$

and a similar argument shows that $(TT^*)(Z^\perp) = \{0\}$. Hence $T(Y^\perp) = \{0\}$, $T^*(Z^\perp) = \{0\}$; and from (7),

$$(T^*T)(\mathscr{H}) = (T^*T)(Y^= + Y^\perp) = (T^*T)(Y^=)$$
$$\subseteq T^*(Z^\perp) = \{0\}.$$

This shows that $T = 0$; from (6), $\langle u \otimes v, w\rangle = 0$ whenever $u \in \mathscr{H}$ and $v \in \mathscr{K}$, so $w = 0$. ■

11.2.13. Proposition. *If $\mathscr{R}$ and $\mathscr{S}$ are von Neumann algebras, each having a generating vector, then $(\mathscr{R} \mathbin{\overline{\otimes}} \mathscr{S})' = \mathscr{R}' \mathbin{\overline{\otimes}} \mathscr{S}'$.*

Proof. Suppose that $\mathscr{R}$ acts on $\mathscr{H}$ and has a generating vector v, while $\mathscr{S}$ acts on $\mathscr{K}$ and has a generating vector w. We may define real-linear subspaces $Y\ (\subseteq \mathscr{H})$ and $Z\ (\subseteq \mathscr{K})$ by $Y = \mathscr{R}^{\text{h}}v$, $Z = \mathscr{S}^{\text{h}}w$. Since $Y + iY\ (= \mathscr{R}v)$ is everywhere dense in $\mathscr{H}$, and $Z + iZ\ (= \mathscr{S}w)$ is everywhere dense in $\mathscr{K}$, it results from Lemma 11.2.12 that $Y \odot Z + i(Y^\perp \odot Z^\perp)$ is everywhere dense in $\mathscr{H} \otimes \mathscr{K}$.

Let $\mathscr{M} = \mathscr{R} \mathbin{\overline{\otimes}} \mathscr{S}$, $\mathscr{N} = \mathscr{R}' \mathbin{\overline{\otimes}} \mathscr{S}'$, and $u = v \otimes w$. Then $\mathscr{M}u$ contains all simple tensors $x \otimes y$, with x in $\mathscr{R}v$ and y in $\mathscr{S}w$. Since $[\mathscr{R}v] = \mathscr{H}$ and $[\mathscr{S}w] = \mathscr{K}$, it follows that $[\mathscr{M}u] = \mathscr{H} \otimes \mathscr{K}$. Moreover, $\mathscr{M} \subseteq \mathscr{N}'$; so $\mathscr{M}, \mathscr{N}, u$ satisfy the hypothesis of Proposition 11.2.11. Accordingly, in order to deduce that $\mathscr{M} = \mathscr{N}'$ (whence $(\mathscr{R} \mathbin{\overline{\otimes}} \mathscr{S})' = \mathscr{M}' = \mathscr{N} = \mathscr{R}' \mathbin{\overline{\otimes}} \mathscr{S}'$), it now suffices to show that $\mathscr{M}^{\text{h}}u + i\mathscr{N}^{\text{h}}u$ is everywhere dense in $\mathscr{H} \otimes \mathscr{K}$.

When $A \in \mathscr{R}^{\text{h}}$ and $B \in \mathscr{S}^{\text{h}}$,

$$Av \otimes Bw = (A \otimes B)(v \otimes w) = (A \otimes B)u \in \mathscr{M}^{\text{h}}u.$$

Thus

$$\mathscr{M}^{\text{h}}u \supseteq \mathscr{R}^{\text{h}}v \odot \mathscr{S}^{\text{h}}w\ (= Y \odot Z);$$

and a similar argument shows that $\mathscr{N}^{\text{h}}u \supseteq (\mathscr{R}')^{\text{h}}v \odot (\mathscr{S}')^{\text{h}}w$. By considering norm closures, we obtain

$$(\mathscr{N}^{\text{h}}u)^= \supseteq ((\mathscr{R}')^{\text{h}}v)^= \odot ((\mathscr{S}')\ w)^=.$$

From Proposition 11.2.11 (with $\mathscr{R}, \mathscr{R}', v$ in place of $\mathscr{M}, \mathscr{N}, u$),

$$Y^\perp = (\mathscr{R}^{\text{h}}v)^\perp = (i(\mathscr{R}')^{\text{h}}v)^=;$$

and similarly, $Z^\perp = (i(\mathscr{S}')^{\text{h}}w)^=$. Hence

$$(\mathscr{N}^{\text{h}}u)^= \supseteq Y^\perp \odot Z^\perp;$$

so the norm closure of $\mathscr{M}^{\text{h}}u + i\mathscr{N}^{\text{h}}u$ contains that of $Y \odot Z + i(Y^\perp \odot Z^\perp)$, which is the whole of $\mathscr{H} \otimes \mathscr{K}$. ■

In deducing the full commutation theorem from the special case just established, we require two further lemmas.

11.2.14. LEMMA. *Suppose that* $\mathscr{R}$ *and* $\mathscr{S}$ *are von Neumann algebras acting on Hilbert spaces* $\mathscr{H}$ *and* $\mathscr{K}$, *respectively, and* E $(\in \mathscr{R})$ *and* F $(\in \mathscr{S})$ *are projections.*

(i) *If* C_E *and* C_F *denote the central carriers of* E *and* F *(relative to* $\mathscr{R}$ *and* $\mathscr{S}$, *respectively), then the projection* $E \otimes F$ *lies in both the von Neumann algebras* $\mathscr{R} \,\overline{\otimes}\, \mathscr{S}$, $(\mathscr{R}' \,\overline{\otimes}\, \mathscr{S}')'$, *and in each of them has central carrier* $C_E \otimes C_F$.

(ii) *There is a unitary operator* U, *from* $E(\mathscr{H}) \otimes F(\mathscr{K})$ *onto the space* $(E \otimes F)(\mathscr{H} \otimes \mathscr{K})$, *such that*

$$U(E\mathscr{R}E \,\overline{\otimes}\, F\mathscr{S}F)U^* = (E \otimes F)(\mathscr{R} \,\overline{\otimes}\, \mathscr{S})(E \otimes F)$$

$$U(\mathscr{R}'E \,\overline{\otimes}\, \mathscr{S}'F)U^* = (\mathscr{R}' \,\overline{\otimes}\, \mathscr{S}')(E \otimes F).$$

Proof. (i) It is apparent that $E \otimes F \in \mathscr{R} \,\overline{\otimes}\, \mathscr{S} \subseteq (\mathscr{R}' \,\overline{\otimes}\, \mathscr{S}')'$; we denote by P and Q the central carriers of $E \otimes F$ in $\mathscr{R} \,\overline{\otimes}\, \mathscr{S}$ and $(\mathscr{R}' \,\overline{\otimes}\, \mathscr{S}')'$, respectively. Now $C_E \in \mathscr{R} \cap \mathscr{R}'$, $C_F \in \mathscr{S} \cap \mathscr{S}'$, so $C_E \otimes C_F$ lies in the centers of both $\mathscr{R} \,\overline{\otimes}\, \mathscr{S}$ and $\mathscr{R}' \,\overline{\otimes}\, \mathscr{S}'$ (and thus, also, in the center of $(\mathscr{R}' \,\overline{\otimes}\, \mathscr{S}')'$). Since $E \otimes F = (C_E \otimes C_F)(E \otimes F)$, while

$$\begin{aligned} P(\mathscr{H} \otimes \mathscr{K}) &= [(\mathscr{R} \,\overline{\otimes}\, \mathscr{S})(E \otimes F)(\mathscr{H} \otimes \mathscr{K})] \\ &\subseteq [(\mathscr{R}' \,\overline{\otimes}\, \mathscr{S}')'(E \otimes F)(\mathscr{H} \otimes \mathscr{K})] = Q(\mathscr{H} \otimes \mathscr{K}), \end{aligned}$$

it follows that $P \le Q \le C_E \otimes C_F$.

To complete the proof of (i), it now suffices to show that $C_E \otimes C_F \le P$. With w in the range of $C_E \otimes C_F$, and $\varepsilon > 0$, we can choose $u_1, \ldots, u_m$ in $\mathscr{H}$ and $v_1, \ldots, v_m$ in $\mathscr{K}$ so that $\|w - \sum u_j \otimes v_j\| < \varepsilon$. Now

$$\|w - \sum C_E u_j \otimes C_F v_j\| = \|(C_E \otimes C_F)(w - \sum u_j \otimes v_j)\| < \varepsilon.$$

Moreover, each $C_E u_j$ can be approximated in norm by a finite sum of vectors of the form AEx, with A in $\mathscr{R}$ and x in $\mathscr{H}$; and a similar remark applies to the vectors $C_F v_j$. Accordingly, we can choose $x_1, \ldots, x_n$ in $\mathscr{H}$, $y_1, \ldots, y_n$ in $\mathscr{K}$, $A_1, \ldots, A_n$ in $\mathscr{R}$, and $B_1, \ldots, B_n$ in $\mathscr{S}$, so that $\|w - w_0\| < \varepsilon$, where

$$\begin{aligned} w_0 = \sum_{k=1}^{n} A_k E x_k \otimes B_k F y_k &= \sum_{k=1}^{n} (A_k \otimes B_k)(E \otimes F)(x_k \otimes y_k) \\ &\in [(\mathscr{R} \,\overline{\otimes}\, \mathscr{S})(E \otimes F)(\mathscr{H} \otimes \mathscr{K})] = P(\mathscr{H} \otimes \mathscr{K}) \end{aligned}$$

This shows that $w \in P(\mathscr{H} \otimes \mathscr{K})$; so $C_E \otimes C_F \le P$.

(ii) If $x_1, \ldots, x_m, u_1, \ldots, u_n \in E(\mathscr{H})$ and $y_1, \ldots, y_m, v_1, \ldots, v_n \in F(\mathscr{K})$, the inner product

$$\langle \sum_{j=1}^{m} x_j \otimes y_j, \sum_{k=1}^{n} u_k \otimes v_k \rangle \; (= \sum_j \sum_k \langle x_j, u_k \rangle \langle y_j, v_k \rangle)$$

is the same, whether the symbols $\sum x_j \otimes y_j, \sum u_k \otimes v_k$ denote two vectors in $E(\mathscr{H}) \otimes F(\mathscr{K})$ or two vectors in $\mathscr{H} \otimes \mathscr{K}$. When they are considered as elements of $\mathscr{H} \otimes \mathscr{K}$,

$$(E \otimes F)(\sum x_j \otimes y_j) = \sum Ex_j \otimes Fy_j = \sum x_j \otimes y_j,$$

so $\sum x_j \otimes y_j \in (E \otimes F)(\mathscr{H} \otimes \mathscr{K})$. Accordingly, there is an isometric linear mapping U, from $E(\mathscr{H}) \otimes F(\mathscr{K})$ into $(E \otimes F)(\mathscr{H} \otimes \mathscr{K})$, defined by the condition

$$U(x \otimes y) = x \otimes y \in (E \otimes F)(\mathscr{H} \otimes \mathscr{K}) \qquad (x \in E(\mathscr{H}), \quad y \in F(\mathscr{K})).$$

Given v in $(E \otimes F)(\mathscr{H} \otimes \mathscr{K})$, and $\varepsilon > 0$, we can choose $z_1, \ldots, z_p$ in $\mathscr{H}$ and $w_1, \ldots, w_p$ in $\mathscr{K}$, so that $\|v - \sum z_j \otimes w_j\| < \varepsilon$. If $v_0 = \sum Ez_j \otimes Fw_j$, then v_0 lies in the range of U, and

$$\|v - v_0\| = \|(E \otimes F)(v - \sum z_j \otimes w_j)\| < \varepsilon.$$

It follows that the range of U is both closed and everywhere dense in $(E \otimes F)(\mathscr{H} \otimes \mathscr{K})$; and U is a unitary operator, from $E(\mathscr{H}) \otimes F(\mathscr{K})$ onto $(E \otimes F)(\mathscr{H} \otimes \mathscr{K})$.

Suppose that $A \in \mathscr{R}$, $A' \in \mathscr{R}'$, $B \in \mathscr{S}$, $B' \in \mathscr{S}'$. Given x in $E(\mathscr{H})$ and y in $F(\mathscr{K})$, and considering $x \otimes y$ as an element of $E(\mathscr{H}) \otimes F(\mathscr{K})$, straightforward calculation yields

$$(E \otimes F)(A \otimes B)(E \otimes F)U(x \otimes y) = U(EAE \otimes FBF)(x \otimes y),$$

$$(A' \otimes B')(E \otimes F)U(x \otimes y) = U(A'E \otimes B'F)(x \otimes y).$$

Thus $(E \otimes F)(A \otimes B)(E \otimes F) = U(EAE \otimes FBF)U^*$, $(A' \otimes B')(E \otimes F) = U(A'E \otimes B'F)U^*$; and

$$U(E\mathscr{R}E \,\overline{\otimes}\, F\mathscr{S}F)U^* = (E \otimes F)(\mathscr{R} \,\overline{\otimes}\, \mathscr{S})(E \otimes F),$$

$$U(\mathscr{R}'E \,\overline{\otimes}\, \mathscr{S}'F)U^* = (\mathscr{R}' \,\overline{\otimes}\, \mathscr{S}')(E \otimes F). \quad \blacksquare$$

11.2.15. Lemma. *Suppose that $\mathscr{M}$ and $\mathscr{N}$ are von Neumann algebras acting on the same Hilbert space, $\mathscr{M} \subseteq \mathscr{N}$, $\{E_a\}$ is a family of projections in $\mathscr{M}$, and the central carrier P_a of E_a, relative to $\mathscr{M}$, is also its central carrier in $\mathscr{N}$. If $\bigvee P_a = I$, and $E_a \mathscr{M} E_a = E_a \mathscr{N} E_a$ for all a, then $\mathscr{M} = \mathscr{N}$.*

Proof. Note first that, while $\mathscr{M}' \supseteq \mathscr{N}'$,

$$\mathscr{M}'E_a = (E_a \mathscr{M} E_a)' = (E_a \mathscr{N} E_a)' = \mathscr{N}'E_a$$

for each a. Thus $\mathscr{M}'P_a \supseteq \mathscr{N}'P_a$, but the * isomorphism

$$\varphi_a : A'P_a \to A'E_a : \ \mathscr{M}'P_a \to \mathscr{M}'E_a$$

carries $\mathscr{N}'P_a$ onto $\mathscr{N}'E_a$ $(= \mathscr{M}'E_a = \varphi_a\ (\mathscr{M}'P_a))$. Hence $\mathscr{N}'P_a = \mathscr{M}'P_a$; and, by taking commutants, we obtain $\mathscr{N}P_a = \mathscr{M}P_a$.

If $S = S^* \in \mathscr{N}$, then $SP_a \in \mathscr{N}P_a = \mathscr{M}P_a \subseteq \mathscr{M}$. Accordingly, the set

$$\{Z \in \mathscr{M} \cap \mathscr{N}' : SZ \in \mathscr{M}\}$$

is a weak-operator closed * subalgebra of the center $\mathscr{M} \cap \mathscr{M}'$ of $\mathscr{M}$ that contains each P_a and, so, contains $\bigvee P_a$ $(= I)$. Hence $S \in \mathscr{M}$, whenever $S = S^* \in \mathscr{N}$; and thus $\mathscr{M} = \mathscr{N}$. ∎

11.2.16. THEOREM. *If* $\mathscr{R}$ *and* $\mathscr{S}$ *are von Neumann algebras, then* $(\mathscr{R} \mathbin{\overline{\otimes}} \mathscr{S})' = \mathscr{R}' \mathbin{\overline{\otimes}} \mathscr{S}'$.

Proof. Let $\{E_a'\}$ be an orthogonal family of cyclic projections in $\mathscr{R}'$, with sum I; and let $\{F_b'\}$ be such a family in $\mathscr{S}'$. For each pair (a, b) of suffixes, the von Neumann algebras $\mathscr{R}E_a'$ and $\mathscr{S}F_b'$ have generating vectors; so, by Proposition 11.2.13,

$$E_a'\mathscr{R}'E_a' \mathbin{\overline{\otimes}} F_b'\mathscr{S}'F_b' = (\mathscr{R}E_a')' \mathbin{\overline{\otimes}} (\mathscr{S}F_b')' = (\mathscr{R}E_a' \mathbin{\overline{\otimes}} \mathscr{S}F_b')'.$$

It now follows from Lemma 11.2.14(ii) (with the roles of algebra and commutant exchanged) that

$$\begin{aligned}(E_a' \otimes F_b')(\mathscr{R}' \mathbin{\overline{\otimes}} \mathscr{S}')(E_a' \otimes F_b') &= ((\mathscr{R} \mathbin{\overline{\otimes}} \mathscr{S})(E_a' \otimes F_b'))' \\ &= (E_a' \otimes F_b')(\mathscr{R} \mathbin{\overline{\otimes}} \mathscr{S})'(E_a' \otimes F_b').\end{aligned}$$

By Lemma 11.2.14(i), $E_a' \otimes F_b'$ has the same central carrier, P_{ab}, relative to $\mathscr{R}' \mathbin{\overline{\otimes}} \mathscr{S}'$ and $(\mathscr{R} \mathbin{\overline{\otimes}} \mathscr{S})'$. Moreover $\bigvee P_{ab} = I$, since

$$\begin{aligned}\bigvee_{a,b} P_{ab} \geq \bigvee_{a,b} E_a' \otimes F_b' &= \sum_{a,b} (E_a' \otimes I)(I \otimes F_b') \\ &= \left(\sum_a E_a' \otimes I\right)\left(\sum_b I \otimes F_b'\right),\end{aligned}$$

and the right-hand side is I (from strong-operator continuity, on bounded sets, of the mappings $A \to A \otimes I$, $B \to I \otimes B$). The required result now follows from Lemma 11.2.15, where $\mathscr{M} = \mathscr{R}' \mathbin{\overline{\otimes}} \mathscr{S}'$, $\mathscr{N} = (\mathscr{R} \mathbin{\overline{\otimes}} \mathscr{S})'$, and with $\{E_a' \otimes F_b'\}$ in place of $\{E_a\}$. ∎

11.2.17. Corollary. *If $\mathscr{R}$ and $\mathscr{S}$ are von Neumann algebras with centers $\mathscr{Z}_{\mathscr{R}}$ and $\mathscr{Z}_{\mathscr{S}}$, respectively, then $\mathscr{R} \bar{\otimes} \mathscr{S}$ has center $\mathscr{Z}_{\mathscr{R}} \bar{\otimes} \mathscr{Z}_{\mathscr{S}}$. In particular, $\mathscr{R} \bar{\otimes} \mathscr{S}$ is a factor if both $\mathscr{R}$ and $\mathscr{S}$ are factors.*

Proof. With $\mathscr{M}$ the von Neumann algebra $\mathscr{R} \bar{\otimes} \mathscr{S}$, and $\mathscr{Z}$ its center, $\mathscr{Z}'$ is generated (as a von Neumann algebra) by $\mathscr{M} \cup \mathscr{M}'$. Now $\mathscr{M}$ is generated by $\{A \otimes B : A \in \mathscr{R},\ B \in \mathscr{S}\}$, and $\mathscr{M}'$ $(= \mathscr{R}' \bar{\otimes} \mathscr{S}')$ is generated by $\{A' \otimes B' : A' \in \mathscr{R}',\ B' \in \mathscr{S}'\}$; so $\mathscr{Z}'$ is generated by

$$\{AA' \otimes BB' : A \in \mathscr{R},\ A' \in \mathscr{R}',\ B \in \mathscr{S},\ B' \in \mathscr{S}'\}.$$

Since $\{AA' : A \in \mathscr{R},\ A' \in \mathscr{R}'\}$ generates $\mathscr{Z}'_{\mathscr{R}}$, and $\{BB' : B \in \mathscr{S},\ B' \in \mathscr{S}'\}$ generates $\mathscr{Z}'_{\mathscr{S}}$, it now follows that $\mathscr{Z}' = \mathscr{Z}'_{\mathscr{R}} \bar{\otimes} \mathscr{Z}'_{\mathscr{S}}$. Upon taking commutants, we obtain $\mathscr{Z} = \mathscr{Z}_{\mathscr{R}} \bar{\otimes} \mathscr{Z}_{\mathscr{S}}$. ∎

11.2.18. Corollary. *If $\mathscr{A}_1$ and $\mathscr{A}_2$ are maximal abelian von Neumann algebras, acting on Hilbert spaces $\mathscr{H}_1$ and $\mathscr{H}_2$, respectively, then $\mathscr{A}_1 \bar{\otimes} \mathscr{A}_2$ is a maximal abelian von Neumann algebra acting on $\mathscr{H}_1 \otimes \mathscr{H}_2$.*

Proof. Since $\mathscr{A}_1 = \mathscr{A}'_1$ and $\mathscr{A}_2 = \mathscr{A}'_2$, it follows from the commutation theorem that

$$(\mathscr{A}_1 \bar{\otimes} \mathscr{A}_2)' = \mathscr{A}'_1 \bar{\otimes} \mathscr{A}'_2 = \mathscr{A}_1 \bar{\otimes} \mathscr{A}_2. \quad ∎$$

The type of tensor products. In this subsection, we determine the type decomposition of $\mathscr{R} \bar{\otimes} \mathscr{S}$ in terms of the type decompositions of the von Neumann algebras $\mathscr{R}$ and $\mathscr{S}$. Suppose that $\{P_a\}$ is an orthogonal family of central projections in $\mathscr{R}$, with sum I, such that each of the von Neumann algebras $\mathscr{R}P_a$ is of a specified type (either I_n, or II_1, or II_∞ or III); and let $\{Q_b\}$ be such a family for $\mathscr{S}$. Then $\{P_a \otimes Q_b\}$ is an orthogonal family of central projections in $\mathscr{R} \bar{\otimes} \mathscr{S}$, with sum I. If we show that each of the von Neumann algebras $(\mathscr{R} \bar{\otimes} \mathscr{S})(P_a \otimes Q_b)$ is of just one type (and specify which), the type decomposition of $\mathscr{R} \bar{\otimes} \mathscr{S}$ is determined. Now $(\mathscr{R} \bar{\otimes} \mathscr{S})(P_a \otimes Q_b)$ is unitarily equivalent to $\mathscr{R}P_a \bar{\otimes} \mathscr{S}Q_b$ (Lemma 11.2.14), and the types of $\mathscr{R}P_a$ and $\mathscr{S}Q_b$ are specified. Accordingly, it suffices to determine the type of a von Neumann algebra tensor product $\mathscr{R} \bar{\otimes} \mathscr{S}$ (as just one of I_n, II_1, II_∞, III), given that each of $\mathscr{R}$ and $\mathscr{S}$ is of just one, specified type.

The net effect of the results proved below is to establish the "multiplication table" (Table 11.1), in which the entry in a given row and column gives the type of $\mathscr{R} \bar{\otimes} \mathscr{S}$, when the types of $\mathscr{R}$ and $\mathscr{S}$ are specified by that row and column, respectively.

11.2.19. Proposition. *If $\mathscr{R}$ and $\mathscr{S}$ are von Neumann algebras of types I_m and I_n, respectively (where the cardinals m and n need not be finite), then $\mathscr{R} \bar{\otimes} \mathscr{S}$ is of type I_{mn}.*

TABLE 11.1

TABLE SHOWING THE TYPE OF $\mathscr{R} \bar{\otimes} \mathscr{S}$

	Type of $\mathscr{S}$				
Type of $\mathscr{R}$	I_n, n finite	I_n, n infinite	II_1	II_∞	III
I_m, m finite	I_{mn}	I_{mn}	II_1	II_∞	III
I_m, m infinite	I_{mn}	I_{mn}	II_∞	II_∞	III
II_1	II_1	II_∞	II_1	II_∞	III
II_∞	II_∞	II_∞	II_∞	II_∞	III
III	III	III	III	III	III

Proof. There is a family $\{E_a : a \in \mathbb{A}\}$ of abelian projections in $\mathscr{R}$, such that $\mathbb{A}$ has cardinality m, $\sum E_a = I$, and each E_a has central carrier I; and there is a similar family $\{F_b : b \in \mathbb{B}\}$, where $\mathbb{B}$ has cardinality n, for $\mathscr{S}$. By Lemma 11.2.14, each of the projections $E_a \otimes F_b$ in $\mathscr{R} \bar{\otimes} \mathscr{S}$ has central carrier I, and is abelian since $(E_a \otimes F_b)(\mathscr{R} \bar{\otimes} \mathscr{S})(E_a \otimes F_b)$ is unitarily equivalent to the tensor product $E_a\mathscr{R}E_a \bar{\otimes} F_b\mathscr{S}F_b$ of abelian von Neumann algebras. From this, and since the family $\{E_a \otimes F_b : a \in \mathbb{A},\ b \in \mathbb{B}\}$ has sum I and cardinality mn, it follows that $\mathscr{R} \bar{\otimes} \mathscr{S}$ is of type I_{mn}. ∎

11.2.20. PROPOSITION. *If $\mathscr{R}$ and $\mathscr{S}$ are finite von Neumann algebras, then $\mathscr{R} \bar{\otimes} \mathscr{S}$ is finite. If, further, at least one of $\mathscr{R}$ and $\mathscr{S}$ is of type* II_1, *then $\mathscr{R} \bar{\otimes} \mathscr{S}$ is of type* II_1.

Proof. Let $\{P_a\}$ be an orthogonal family of projections, each cyclic in the center of $\mathscr{R}$, with sum I; and let $\{Q_b\}$ be such a family for $\mathscr{S}$. The projections $P_a \otimes Q_b$ are central in $\mathscr{R} \bar{\otimes} \mathscr{S}$, and $\sum P_a \otimes Q_b = I$. In order to prove that $\mathscr{R} \bar{\otimes} \mathscr{S}$ is finite (or of type II_1), it suffices to show that each of the algebras $(\mathscr{R} \bar{\otimes} \mathscr{S})(P_a \otimes Q_b)$ is finite (or of type II_1). By Lemma 11.2.14(ii), this is equivalent to proving that $\mathscr{R}P_a \bar{\otimes} \mathscr{S}Q_b$ is finite (or of type II_1). The center of $\mathscr{R}P_a$ has a separating vector x, since P_a is cyclic in $\mathscr{R} \cap \mathscr{R}'$. With τ the center-valued trace on $\mathscr{R}P_a$, $\omega_x \circ \tau$ is a faithful normal tracial state on $\mathscr{R}P_a$. Let $\mathscr{R}_a$ be a von Neumann algebra, * isomorphic to $\mathscr{R}P_a$, in which every normal state is a vector state (for example, $\mathscr{R}_a$ could be the universal normal representation of $\mathscr{R}P_a$, or $\mathscr{R}P_a \bar{\otimes} \mathbb{C}_{\mathscr{L}}$, where $\mathscr{L}$ is an infinite-dimensional Hilbert space). Then $\mathscr{R}_a$ has a faithful normal tracial state, and this is necessarily a vector state associated with a separating trace vector for $\mathscr{R}_a$. Similarly, we can find a von Neumann algebra $\mathscr{S}_b$, * isomorphic to $\mathscr{S}Q_b$, that has a separating trace vector. Since $\mathscr{R}P_a \mathbin{\overline{\otimes}} \mathscr{S}Q_b$ is * isomorphic to $\mathscr{R}_a \mathbin{\overline{\otimes}} \mathscr{S}_b$ (Theorem 11.2.10), it now remains to show that that $\mathscr{R}_a \mathbin{\overline{\otimes}} \mathscr{S}_b$ is finite (or of type II_1).

In view of the preceding reductions, it suffices to consider the case in which $\mathscr{R}$ and $\mathscr{S}$ have separating trace vectors, u and v, respectively. Since u is generating for $\mathscr{R}'$, and v for $\mathscr{S}'$, $u \otimes v$ is a generating vector for $\mathscr{R}' \mathbin{\overline{\otimes}} \mathscr{S}'$, and is therefore separating for $\mathscr{R} \mathbin{\overline{\otimes}} \mathscr{S}$. Moreover, if $A_1, A_2 \in \mathscr{R}$ and $B_1, B_2 \in \mathscr{S}$

$$\begin{aligned}
&\langle (A_1 \otimes B_1)(A_2 \otimes B_2)(u \otimes v), u \otimes v \rangle \\
&\quad = \langle A_1 A_2 u, u \rangle \langle B_1 B_2 v, v \rangle \\
&\quad = \langle A_2 A_1 u, u \rangle \langle B_2 B_1 v, v \rangle = \langle (A_2 \otimes B_2)(A_1 \otimes B_1)(u \otimes v), u \otimes v \rangle.
\end{aligned}$$

Since $\mathscr{R} \mathbin{\overline{\otimes}} \mathscr{S}$ is the weak-operator closed linear span of operators of the form $A \otimes B$, with A in $\mathscr{R}$ and B in $\mathscr{S}$, it now follows that $u \otimes v$ is a separating trace vector for $\mathscr{R} \mathbin{\overline{\otimes}} \mathscr{S}$. Accordingly, $\mathscr{R} \mathbin{\overline{\otimes}} \mathscr{S}$ has a faithful tracial state $\omega_{u \otimes v}$, and is therefore finite.

Now suppose further that at least one of $\mathscr{R}$ and $\mathscr{S}$ (say $\mathscr{R}$) is of type II_1. If the finite von Neumann algebra $\mathscr{R} \mathbin{\overline{\otimes}} \mathscr{S}$ is not of type II_1, there is a positive integer n and a non-zero central projection Q in $\mathscr{R} \mathbin{\overline{\otimes}} \mathscr{S}$, such that $(\mathscr{R} \mathbin{\overline{\otimes}} \mathscr{S})Q$ is of type I_n; we show that this situation leads to a contradiction. Since $\mathscr{R}$ is type II_1, there is a family $\{E_0, E_1, \ldots, E_n\}$ of $n + 1$ equivalent projections in $\mathscr{R}$ with sum I. Since E_0 has central carrier I in $\mathscr{R}$, $E_0 \otimes I$ has central carrier I in $\mathscr{R} \mathbin{\overline{\otimes}} \mathscr{S}$; so $(E_0 \otimes I)Q$ is a non-zero projection in $(\mathscr{R} \mathbin{\overline{\otimes}} \mathscr{S})Q$, and so contains an abelian projection F_0 $(\neq 0)$. With V_j $(\in \mathscr{R})$ a partial isometry from E_0 to E_j, $(V_j \otimes I)F_0$ is a partial isometry, in $(\mathscr{R} \mathbin{\overline{\otimes}} \mathscr{S})Q$, from F_0 to a subprojection F_j of $(E_j \otimes I)Q$. Accordingly, $\{F_0, F_1, \ldots, F_n\}$ is an orthogonal family of $n + 1$ equivalent abelian projections in a type I_n von Neumann algebra; this, however, is impossible, and the contradiction results from our assumption that $\mathscr{R} \mathbin{\overline{\otimes}} \mathscr{S}$ is not of type II_1. ■

11.2.21. Proposition. *If $\mathscr{R}$ and $\mathscr{S}$ are semi-finite von Neumann algebras, then $\mathscr{R} \mathbin{\overline{\otimes}} \mathscr{S}$ is semi-finite. If, further, at least one of $\mathscr{R}$ and $\mathscr{S}$ is of type* II, *then $\mathscr{R} \mathbin{\overline{\otimes}} \mathscr{S}$ is of type* II.

Proof. Suppose that E is a finite projection, with central carrier I, in $\mathscr{R}$; and let F be such a projection in $\mathscr{S}$. Since the von Neumann algebras $E\mathscr{R}E$ and $F\mathscr{S}F$ are finite, the same is true of $E\mathscr{R}E \mathbin{\overline{\otimes}} F\mathscr{S}F$ (Proposition 11.2.20), and of the unitarily equivalent algebra $(E \otimes F)(\mathscr{R} \mathbin{\overline{\otimes}} \mathscr{S})(E \otimes F)$. Thus $E \otimes F$ is a finite projection, with central carrier I, in $\mathscr{R} \mathbin{\overline{\otimes}} \mathscr{S}$; and $\mathscr{R} \mathbin{\overline{\otimes}} \mathscr{S}$ is semi-finite.

If at least one of $\mathcal{R}$ and $\mathcal{S}$ is of type II, then at least one of $E\mathcal{R}E$ and $F\mathcal{S}F$ is of type II_1. By Proposition 11.2.20, $E\mathcal{R}E \mathbin{\bar{\otimes}} F\mathcal{S}F$ (and hence also $(E \otimes F)(\mathcal{R} \mathbin{\bar{\otimes}} \mathcal{S})(E \otimes F)$) is of type II_1. From this, and since $E \otimes F$ has central carrier I in $\mathcal{R} \mathbin{\bar{\otimes}} \mathcal{S}$, it follows that $\mathcal{R} \mathbin{\bar{\otimes}} \mathcal{S}$ is of type II. ∎

11.2.22. PROPOSITION. *If $\mathcal{R}$ and $\mathcal{S}$ are von Neumann algebras, at least one of which is properly infinite, then $\mathcal{R} \mathbin{\bar{\otimes}} \mathcal{S}$ is properly infinite.*

Proof. We may suppose that $\mathcal{R}$ is properly infinite. From the halving lemma (6.3.3), there are projections E_1, E_2 in $\mathcal{R}$ such that $E_1 \sim E_2 \sim I = E_1 + E_2$. If Q is a non-zero central projection in $\mathcal{R} \mathbin{\bar{\otimes}} \mathcal{S}$, then (relative to $\mathcal{R} \mathbin{\bar{\otimes}} \mathcal{S}$)

$$Q(E_1 \otimes I) \sim Q(E_2 \otimes I) \sim Q = Q(E_1 \otimes I) + Q(E_2 \otimes I).$$

Hence each non-zero central projection Q in $\mathcal{R} \mathbin{\bar{\otimes}} \mathcal{S}$ is infinite. ∎

The preceding four propositions establish all the entries in Table 11.1 except those in which at least one of $\mathcal{R}$ and $\mathcal{S}$ is of type III. The entries in which both $\mathcal{R}$ and $\mathcal{S}$ are of type I are justified by Proposition 11.2.19. When at least one of $\mathcal{R}$ and $\mathcal{S}$ is of type II, while the other is either of type I or of type II, then $\mathcal{R} \mathbin{\bar{\otimes}} \mathcal{S}$ is type II (Proposition 11.2.21), is finite if both $\mathcal{R}$ and $\mathcal{S}$ are finite (Proposition 11.2.20) but is properly infinite when at least one of $\mathcal{R}$ and $\mathcal{S}$ is properly infinite (Proposition 11.2.22); and this establishes the corresponding entries in the table. To justify the remaining entries, we have to show that $\mathcal{R} \mathbin{\bar{\otimes}} \mathcal{S}$ is of type III whenever at least one of $\mathcal{R}$ and $\mathcal{S}$ is of type III. This is proved in Proposition 11.2.26 below, after some preparatory results.

11.2.23. PROPOSITION. *A projection E in a von Neumann algebra $\mathcal{R}$ is finite if and only if the mapping $A \to EA^*$ is strong-operator continuous on the unit ball $(\mathcal{R})_1$.*

Proof. If E is infinite, it has a properly infinite subprojection F_0 ($\in \mathcal{R}$), by Proposition 6.3.7. By repeated application of the halving lemma (6.3.3) we can find sequences $\{F_1, F_2, \ldots\}$, $\{G_1, G_2, \ldots\}$ of projections in $\mathcal{R}$, such that

$$F_j \sim G_j \sim F_{j-1} = F_j + G_j \qquad (j = 1, 2, \ldots);$$

and $\{G_j\}$ is an orthogonal sequence of subprojections of F_0, each equivalent to F_0. For $j = 1, 2, \ldots$, let V_j be a partial isometry in $\mathcal{R}$, from G_j to F_0. For each x in $\mathcal{H}$, the Hilbert space on which $\mathcal{R}$ acts, $\|V_j x\| = \|G_j x\| \to 0$ as $j \to \infty$; so the sequence $\{V_j\}$ in $(\mathcal{R})_1$ is strong-operator convergent to 0. With x_0 a unit vector in the range of F_0,

$$\|EV_j^* x_0\| = \|V_j^* x_0\| = \|x_0\| = 1 \qquad (j = 1, 2, \ldots);$$

whence $\{EV_j^*\}$ is not strong-operator convergent to 0. Accordingly, the mapping $A \to EA^*$ is not strong-operator continuous on $(\mathscr{R})_1$.

Now suppose that E is finite. Since $EA^* = E(AC_E)^*$, we may replace $\mathscr{R}$ by $\mathscr{R}C_E$, and thus reduce to the case in which $C_E = I$. Let $\{E_a\}$ be an orthogonal family of projections in $\mathscr{R}$, maximal subject to the condition that $0 < E_a \precsim E$ for each a. This maximality assumption implies that the projection $(F =)I - \sum E_a$ has no non-zero subprojection equivalent to a subprojection of E. Since $C_E = I$, it follows (Proposition 6.1.8) that $C_F = 0$, whence $F = 0$ and $\sum E_a = I$.

Let $\{A_j\}$ be a net of elements of $(\mathscr{R})_1$, strong-operator convergent to 0. We have to prove that $\{EA_j^*\}$ is strong-operator convergent to 0. For this, it suffices to show that $\|EA_j^*x\| \to 0$ whenever $x \in E_a(\mathscr{H})$ for some a; for finite sums of such vectors x form an everywhere-dense subspace of $\mathscr{H}$, and so serve to determine the strong-operator topology on $(\mathscr{R})_1$. With x in $E_a(\mathscr{H})$, and V_a $(\in \mathscr{R})$ a partial isometry from E_a to a subprojection of E,

$$\begin{aligned}\|EA_j^*x\| &= \|EA_j^*E_ax\| \\ &= \|EA_j^*V_a^*EV_ax\| = \|(EV_aA_jE)^*EV_ax\|;\end{aligned}$$

and the right-hand side will tend to zero if the net $\{(EV_aA_jE)^*\}$ (indexed by j) is strong-operator convergent to 0. Since the net $\{EV_aA_jE\}$ is strong-operator convergent to 0, it now suffices to establish the strong-operator continuity of the adjoint operation on the unit ball of the finite von Neumann algebra $E\mathscr{R}E$.

Upon replacing $\mathscr{R}$ by $E\mathscr{R}E$, we may now suppose that $\mathscr{R}$ is finite and $E = I$. Let $\mathscr{K}$ be the set of vectors x in $\mathscr{H}$ with the following property: there is a normal tracial state ω of $\mathscr{R}$, such that $\omega_x | \mathscr{R} \le \omega$. We assert that $\mathscr{K}$ is separating for $\mathscr{R}$. For this, suppose that $A \in \mathscr{R}$ and $A \ne 0$. With τ the center-valued trace on $\mathscr{R}$, $\tau(A^*A) > 0$, and thus $\omega_y\ (\tau(A^*A)) > 0$ for some y in $\mathscr{H}$. Since $\omega_y \circ \tau$ is a normal tracial state, there is a sequence $\{x(n)\}$ of vectors (necessarily in $\mathscr{K}$) such that $\sum \|x(n)\|^2 < \infty$ and $\omega_y \circ \tau = \sum \omega_{x(n)} | \mathscr{R}$. Moreover,

$$0 < \omega_y(\tau(A^*A)) = \sum \omega_{x(n)}(A^*A) = \sum \|Ax(n)\|^2;$$

so $Ax(n) \ne 0$ for some value of n. This shows that $\mathscr{K}$ is separating for $\mathscr{R}$, hence generating for $\mathscr{R}'$.

Once again, suppose that $\{A_j\}$ is a net of elements of $(\mathscr{R})_1$, strong-operator convergent to 0. In order to prove that the bounded net $\{A_j^*\}$ is strong-operator convergent to 0, it suffices to show that $\|A_j^*A'x\| \to 0$ whenever $x \in \mathscr{K}$ and $A' \in \mathscr{R}'$; for finite sums of such vectors $A'x$ form an everywhere-dense subset of $\mathscr{H}$. Let ω be a normal tracial state of $\mathscr{R}$, such that $\omega_x | \mathscr{R} \le \omega$.

Since $\{A_j\}$ is strong-operator convergent to 0, $\{A_j^*A_j\}$ is weak-operator convergent to 0, and thus $\omega(A_j^*A_j) \to 0$. Hence

$$\|A_j^*A'x\| = \|A'A_j^*x\| \le \|A'\|\,\|A_j^*x\| = \|A'\|[\omega_x(A_jA_j^*)]^{1/2}$$
$$\le \|A'\|[\omega(A_jA_j^*)]^{1/2} = \|A'\|[\omega(A_j^*A_j)]^{1/2} \to 0. \quad \blacksquare$$

Suppose that $\mathscr{M}$ and $\mathscr{N}$ are von Neumann algebras acting on a Hilbert space $\mathscr{H}$, and $\mathscr{N} \subseteq \mathscr{M}$. By a *conditional expectation* from $\mathscr{M}$ onto $\mathscr{N}$, we mean a positive linear mapping $\Phi : \mathscr{M} \to \mathscr{N}$ such that

(i) $\Phi(I) = I$,

(ii) $\Phi(ASB) = A\Phi(S)B$,

(iii) $\Phi(S)^*\Phi(S) \le \Phi(S^*S)$

whenever $S \in \mathscr{M}$ and $A, B \in \mathscr{N}$.

Such a mapping Φ is necessarily a projection, with norm 1, from $\mathscr{M}$ onto $\mathscr{N}$. Indeed, by taking $S = B = I$ in (ii), we obtain $\Phi(A) = A\Phi(I)$, for each A in $\mathscr{N}$. The inequality $S^*S \le \|S\|^2I$ entails $\Phi(S^*S) \le \|S\|^2\Phi(I) = \|S\|^2I$: so by (iii),

$$\|\Phi(S)x\|^2 = \langle\Phi(S)^*\Phi(S)x, x\rangle \le \langle\Phi(S^*S)x, x\rangle$$
$$\le \|S\|^2\langle x, x\rangle = \|S\|^2\|x\|^2 \qquad (x \in \mathscr{H}),$$

and $\|\Phi(S)\| \le \|S\|$. Condition (iii) has been included in the above definition for convenience; it can be deduced from (i) and (ii) (see Exercise 8.7.23), but we shall not need to make use of this fact.

Note also that, for a conditional expectation Φ, weak-operator continuity (on $\mathscr{M}$, or on $(\mathscr{M})_1$) entails strong-operator continuity on the same set. Indeed, if a net $\{S_a\}$ (in $\mathscr{M}$, or in $(\mathscr{M})_1$) is strong-operator convergent to 0, then $\{S_a^*S_a\}$ is weak-operator convergent to 0, whence the same is true of $\{\Phi(S_a^*S_a)\}$ by the assumed weak-operator continuity of Φ; so, for each x in $\mathscr{H}$,

$$\|\Phi(S_a)x\|^2 = \langle\Phi(S_a)^*\Phi(S_a)x, x\rangle$$
$$\le \langle\Phi(S_a^*S_a)x, x\rangle \to 0.$$

It can be shown that every projection of norm 1, from a von Neumann algebra $\mathscr{M}$ onto a von Neumann subalgebra $\mathscr{N}$, is a conditional expectation [100]. (Compare Exercises 10.5.85 and 10.5.86.)

11.2.24. PROPOSITION. *If $\mathscr{R}$ and $\mathscr{S}$ are von Neumann algebras, acting on Hilbert spaces $\mathscr{H}$ and $\mathscr{K}$, respectively, there is a family $\{\Phi_z : z \in \mathscr{K}, \|z\| = 1\}$ of conditional expectations from $\mathscr{R} \mathbin{\bar{\otimes}} \mathscr{S}$ onto $\mathscr{R} \mathbin{\bar{\otimes}} \mathbb{C}_{\mathscr{K}}$, each weak-operator* *continuous on the unit ball* $(\mathscr{R} \bar{\otimes} \mathscr{S})_1$, *with the following property: if* $T \in (\mathscr{R} \bar{\otimes} \mathscr{S})^+$ *and* $T \neq 0$, *then* $\Phi_z(T) \neq 0$ *for some* z.

Proof. Given a unit vector z in $\mathscr{K}$, we can associate with each element T of $\mathscr{R} \bar{\otimes} \mathscr{S}$ a conjugate-bilinear functional b_{zT} on $\mathscr{H}$, defined by

$$b_{zT}(x, y) = \langle T(x \otimes z), y \otimes z \rangle \qquad (x, y, \in \mathscr{H}).$$

Since $|b_{zT}(x, y)| \leq \|T\| \|x \otimes z\| \|y \otimes z\| = \|T\| \|x\| \|y\|$, b_{zT} is bounded, with $\|b_{zT}\| \leq \|T\|$. Accordingly, b_{zT} corresponds to a bounded linear operator $\Psi_z(T)$ acting on $\mathscr{H}$, and we have

$$\|\Psi_z(T)\| \leq \|T\|, \qquad \langle \Psi_z(T)x, y \rangle = \langle T(x \otimes z), y \otimes z \rangle \qquad (x, y \in \mathscr{H}).$$

It is apparent from the last equation that Ψ_z is weak-operator continuous, and is a positive linear mapping from $\mathscr{R} \bar{\otimes} \mathscr{S}$ into $\mathscr{B}(\mathscr{H})$. Since

$$\langle \Psi_z(I)x, y \rangle = \langle x \otimes z, y \otimes z \rangle = \langle x, y \rangle \langle z, z \rangle = \langle x, y \rangle,$$

for all x and y in $\mathscr{H}$, it follows that $\Psi_z(I) = I$.

We now show that

$$\text{(10)} \qquad \Psi_z(T) \in \mathscr{R}, \qquad \Psi_z((A \otimes I)T(B \otimes I)) = A\Psi_z(T)B$$

whenever $T \in \mathscr{R} \bar{\otimes} \mathscr{S}$ and $A, B \in \mathscr{R}$. In view of the linearity and weak-operator continuity of Ψ_z, it suffices to establish these equations when T has the form $R \otimes S$, with R in $\mathscr{R}$ and S in $\mathscr{S}$. Now

$$\begin{aligned} \langle \Psi_z(R \otimes S)x, y) &= \langle (R \otimes S)(x \otimes z), y \otimes z \rangle \\ &= \langle Sz, z \rangle \langle Rx, y \rangle \qquad (x, y \in \mathscr{H}) \end{aligned}$$

and thus

$$\Psi_z(R \otimes S) = \langle Sz, z \rangle R \in \mathscr{R}.$$

Moreover

$$\begin{aligned} \Psi_z((A \otimes I)(R \otimes S)(B \otimes I)) &= \Psi_z(ARB \otimes S) \\ &= \langle Sz, z \rangle ARB = A\Psi_z(R \otimes S)B. \end{aligned}$$

This proves (10).

Observe next that $\Psi_z(T)^*\Psi_z(T) \leq \Psi_z(T^*T)$, for each T in $\mathscr{R} \bar{\otimes} \mathscr{S}$. Indeed, given x in $\mathscr{H}$, there is a unit vector y in $\mathscr{H}$ such that

$$\begin{aligned} \langle \Psi_z(T)^*\Psi_z(T)x, x \rangle &= \|\Psi_z(T)x\|^2 \\ &= \langle \Psi_z(T)x, y \rangle^2 = \langle T(x \otimes z), y \otimes z \rangle^2 \\ &\leq \|T(x \otimes z)\|^2 \|y \otimes z\|^2 \\ &= \langle T^*T(x \otimes z), x \otimes z \rangle = \langle \Psi_z(T^*T)x, x \rangle. \end{aligned}$$

From the properties of Ψ_z established above, together with the weak-operator continuity on $(\mathscr{R})_1$ of the mapping $A \to A \otimes I$, it follows that the equation $\Phi_z(T) = \Psi_z(T) \otimes I$ defines a conditional expectation Φ_z, from $\mathscr{R} \mathbin{\bar{\otimes}} \mathscr{S}$ onto $\mathscr{R} \mathbin{\bar{\otimes}} \mathbb{C}_{\mathscr{K}}$, that is weak-operator continuous on $(\mathscr{R} \mathbin{\bar{\otimes}} \mathscr{S})_1$. If $T \in (\mathscr{R} \mathbin{\bar{\otimes}} \mathscr{S})^+$ and $T \neq 0$, we can choose unit vectors, x in $\mathscr{H}$ and z in $\mathscr{K}$, so that $T^{1/2}(x \otimes z) \neq 0$. Then, $0 \neq \|T^{1/2}(x \otimes z)\|^2 = \langle T(x \otimes z), x \otimes z\rangle = \langle \Psi_z(T)x, x\rangle$; so $\Phi_z(T) = \Psi_z(T) \otimes I \neq 0$. ■

11.2.25. PROPOSITION. *Suppose that* $\mathscr{M}$ *and* $\mathscr{N}$ *are von Neumann algebras acting on a Hilbert space* $\mathscr{H}$, $\mathscr{N} \subseteq \mathscr{M}$, *and* $\mathscr{N}$ *is of type* III. *Suppose also that there is a family* $\{\Phi_a\}$ *of conditional expectations from* $\mathscr{M}$ *onto* $\mathscr{N}$, *such that*

(i) *each* Φ_a *is weak-operator continuous on* $(\mathscr{M})_1$;
(ii) *if* $S \in \mathscr{M}^+$ *and* $\Phi_a(S) = 0$ *for every* a, *then* $S = 0$.

Then $\mathscr{M}$ *is of type* III.

Proof. We suppose that $\mathscr{M}$ is not of type III, and in due course obtain a contradiction. With E a non-zero finite projection in $\mathscr{M}$, $\Phi_a(E) \neq 0$ for some a, and we can choose A in $\mathscr{N}^+$ so that $A\Phi_a(E)$ is a non-zero projection F in $\mathscr{N}$.

Suppose that $\{S_b\}$ is a net of elements of $(\mathscr{N})_1$, strong-operator convergent to 0. Since, also, $S_b \in (\mathscr{M})_1$, while E is a finite projection in $\mathscr{M}$, it results from Proposition 11.2.23 that $ES_b^* \to 0$ in the strong-operator topology. From this. $\Phi_a(ES_b^*) \to 0$ in the same topology, since the weak-operator continuity of Φ_a on $(\mathscr{M})_1$ entails also strong-operator continuity. As

$$FS_b^* = A\Phi_a(E)S_b^* = A\Phi_a(ES_b^*) \to 0,$$

it now follows that the mapping $S \to FS^*$ is strong-operator continuous on $(\mathscr{N})_1$. By Proposition 11.2.23, the non-zero projection F in $\mathscr{N}$ is finite, a contradiction since $\mathscr{N}$ is type III. ■

11.2.26. PROPOSITION. *If* $\mathscr{R}$ *and* $\mathscr{S}$ *are von Neumann algebras, at least one of which is of type* III, *then* $\mathscr{R} \mathbin{\bar{\otimes}} \mathscr{S}$ *is of type* III.

Proof. If $\mathscr{R}$ is of type III, we may apply Proposition 11.2.25, where $\mathscr{M} = \mathscr{R} \mathbin{\bar{\otimes}} \mathscr{S}$ and $\mathscr{N} = \mathscr{R} \mathbin{\bar{\otimes}} \mathbb{C}_{\mathscr{K}}$, to deduce that $\mathscr{R} \mathbin{\bar{\otimes}} \mathscr{S}$ is of type III. The existence of a suitable family of conditional expectations from $\mathscr{M}$ onto $\mathscr{N}$ is assured by Proposition 11.2.24.

A similar argument applies if $\mathscr{S}$ is of type III. ■

Tensor products of unbounded operators. In Proposition 2.6.12 and the comments following it, we introduce the concept of the tensor product

$A_1 \otimes \cdots \otimes A_n$ of bounded operators acting on Hilbert spaces $\mathscr{H}_1, \ldots, \mathscr{H}_n$. If $T_1, \ldots, T_n$ are (not necessarily bounded) linear operators on these spaces, we can define an operator $T_1 \odot \cdots \odot T_n$ on $\mathscr{D}(T_1) \odot \cdots \odot \mathscr{D}(T_n)$, the algebraic tensor product of the linear spaces $\mathscr{D}(T_1), \ldots, \mathscr{D}(T_n)$, by means of the formula

$$(T_1 \odot \cdots \odot T_n)\left(\sum_{k=1}^{m} x_{1k} \otimes \cdots \otimes x_{nk}\right) = \sum_{k=1}^{m} T_1 x_{1k} \otimes \cdots \otimes T_n x_{nk},$$

where $x_{jk} \in \mathscr{D}(T_j)$ for $k = 1, \ldots, m$. If $T_1 \subseteq S_1, \ldots, T_n \subseteq S_n$, then, clearly,

$$T_1 \odot \cdots \odot T_n \subseteq S_1 \odot \cdots \odot S_n. \tag{11}$$

In particular, if $T_1, \ldots, T_n$ are densely defined and preclosed (so that $T_1 \odot \cdots \odot T_n$ is densely defined), then $T_1 \odot \cdots \odot T_n \subseteq \overline{T}_1 \odot \cdots \odot \overline{T}_n$. However, $\overline{T}_1 \odot \cdots \odot \overline{T}_n$ may not be closed (choose I for each T_j). In the proposition that follows, we show that $T_1 \odot \cdots \odot T_n$ is preclosed. It is convenient to reserve the notation $T_1 \otimes \cdots \otimes T_n$ for the closure of $T_1 \odot \cdots \odot T_n$, in this case. We assume that preclosed operators are densely defined.

11.2.27. PROPOSITION. *If $T_1, \ldots, T_n$ are densely defined operators on Hilbert spaces $\mathscr{H}_1, \ldots, \mathscr{H}_n$, then $T_1 \odot \cdots \odot T_n$ is a densely defined operator on $\mathscr{H}_1 \odot \cdots \odot \mathscr{H}_n$ and*

$$T_1^* \odot \cdots \odot T_n^* \subseteq (T_1 \odot \cdots \odot T_n)^*.$$

If $T_1, \ldots, T_n$ are preclosed, then $T_1 \odot \cdots \odot T_n$ is preclosed.

Proof. If $y_1 \in \mathscr{D}(T_1^*), \ldots, y_n \in \mathscr{D}(T_n^*)$ and $z \in \mathscr{D}(T_1 \odot \cdots \odot T_n)$, then $z = \sum_{k=1}^{m} x_{1k} \otimes \cdots \otimes x_{nk}$, where $x_{jk} \in \mathscr{D}(T_j)$ $(j = 1, \ldots, n;\ k = 1, \ldots, m)$; and

$$\begin{aligned}
&\langle (T_1 \odot \cdots \odot T_n)z, y_1 \otimes \cdots \otimes y_n \rangle \\
&= \sum_{k=1}^{m} \langle T_1 x_{1k} \otimes \cdots \otimes T_n x_{nk}, y_1 \otimes \cdots \otimes y_n \rangle \\
&= \sum_{k=1}^{m} \langle T_1 x_{1k}, y_1 \rangle \cdots \langle T_n x_{nk}, y_n \rangle \\
&= \sum_{k=1}^{m} \langle x_{1k}, T_1^* y_1 \rangle \cdots \langle x_{nk}, T_n^* y_n \rangle \\
&= \sum_{k=1}^{m} \langle x_{1k} \otimes \cdots \otimes x_{nk}, T_1^* y_1 \otimes \cdots \otimes T_n^* y_n \rangle \\
&= \langle z, (T_1^* \odot \cdots \odot T_n^*)(y_1 \otimes \cdots \otimes y_n) \rangle.
\end{aligned}$$

Thus $y_1 \otimes \cdots \otimes y_n \in \mathscr{D}((T_1 \odot \cdots \odot T_n)^*)$ and

$$(T_1 \odot \cdots \odot T_n)^*(y_1 \otimes \cdots \otimes y_n) = (T_1^* \odot \cdots \odot T_n^*)(y_1 \otimes \cdots \otimes y_n).$$

It follows that

$$\mathscr{D}(T_1^* \odot \cdots \odot T_n^*) \subseteq \mathscr{D}((T_1 \odot \cdots \odot T_n)^*),$$

and our first assertion is established.

If $T_1, \ldots, T_n$ are preclosed, then $T_1^*, \ldots, T_n^*$, $T_1^* \odot \cdots \odot T_n^*$, and $(T_1 \odot \cdots \odot T_n)^*$ are densely defined; so that $T_1 \odot \cdots \odot T_n$ is preclosed, from Theorem 2.7.8(ii). ■

11.2.28. DEFINITION. If $T_1, \ldots, T_n$ are densely defined, preclosed operators, we denote the closure of $T_1 \odot \cdots \odot T_n$ by $T_1 \otimes \cdots \otimes T_n$ and refer to it as the *tensor product* of $T_1, \ldots, T_n$. ■

When $T_1 \subseteq S_1, \ldots, T_n \subseteq S_n$ with $S_1, \ldots, S_n, T_1, \ldots, T_n$ densely defined, preclosed operators, from (11) we have

$$T_1 \otimes \cdots \otimes T_n \subseteq S_1 \otimes \cdots \otimes S_n. \tag{12}$$

11.2.29. LEMMA. *If $T_1, \ldots, T_n$ are densely defined, closed operators on Hilbert spaces $\mathscr{H}_1, \ldots, \mathscr{H}_n$ and $\mathscr{D}_1, \ldots, \mathscr{D}_n$ are cores for $T_1, \ldots, T_n$, respectively, then $\mathscr{D}_1 \odot \cdots \odot \mathscr{D}_n$ is a core for $T_1 \otimes \cdots \otimes T_n$.*

Proof. If $z \in \mathscr{D}(T_1 \otimes \cdots \otimes T_n)$, there is a sequence of vectors z_j in $\mathscr{D}(T_1 \odot \cdots \odot T_n)$ $(= \mathscr{D}(T_1) \odot \cdots \odot \mathscr{D}(T_n))$ such that $\{z_j\}$ tends to z and $\{(T_1 \odot \cdots \odot T_n)z_j\}$ tends to $(T_1 \otimes \cdots \otimes T_n)z$. Now

$$z_j = \sum_{k=1}^{m_j} x_{1k}^{(j)} \otimes \cdots \otimes x_{nk}^{(j)} \qquad (j = 1, 2, \ldots),$$

where $x_{ik}^{(j)} \in \mathscr{D}(T_i)$, $i = 1, \ldots, n$; $k = 1, \ldots, m_j$; $j = 1, 2, \ldots$. Since $\mathscr{D}_i$ is a core for T_i, there are vectors $y_{ik}^{(j)}$ in $\mathscr{D}_i$ (near $x_{ik}^{(j)}$) such that

$$\|w_j - z_j\| < \frac{1}{j}; \qquad \|(T_1 \odot \cdots \odot T_n)(w_j - z_j)\| < \frac{1}{j} < \quad (j = 1, 2, \ldots),$$

where $w_j = \sum_{k=1}^{m_j} y_{1k}^{(j)} \otimes \cdots \otimes y_{nk}^{(j)}$, $j = 1, 2, \ldots$. Thus $\{w_j\}$ tends to z and $\{(T_1 \odot \cdots \odot T_n)w_j\}$ tends to $(T_1 \otimes \cdots \otimes T_n)z$ as j tends to ∞. Our lemma follows, when we note that $w_j \in \mathscr{D}_1 \odot \cdots \odot \mathscr{D}_n$. ■

11.2.30. COROLLARY. *If $T_1, \ldots, T_n$ are densely defined, preclosed operators then*

$$T_1 \otimes \cdots \otimes T_n = \bar{T}_1 \otimes \cdots \otimes \bar{T}_n. \tag{13}$$

Proof. The preceding lemma, applied to the cores $\mathscr{D}(T_1), \ldots, \mathscr{D}(T_n)$ for $\bar{T}_1, \ldots, \bar{T}_n$, yields the fact that $\mathscr{D}(T_1) \odot \cdots \odot \mathscr{D}(T_n)$ is a core for $\bar{T}_1 \otimes \cdots \otimes \bar{T}_n$. Now

$$T_1 \odot \cdots \odot T_n \subseteq \bar{T}_1 \odot \cdots \odot \bar{T}_n \subseteq \bar{T}_1 \otimes \cdots \otimes \bar{T}_n$$

and $\bar{T}_1 \otimes \cdots \otimes \bar{T}_n$ is closed. Thus

$$T_1 \otimes \cdots \otimes T_n \subseteq \bar{T}_1 \otimes \cdots \otimes \bar{T}_n. \tag{14}$$

Since $\mathscr{D}(T_1) \odot \cdots \odot \mathscr{D}(T_n)$ is a submanifold of $\mathscr{D}(T_1 \otimes \cdots \otimes T_n)$ and a core for $\bar{T}_i \otimes \cdots \otimes \bar{T}_n$, (13) follows from (14). ■

11.2.31 REMARK. If $A_1, \ldots, A_n$, $B_1, \ldots, B_n$ are operators on the Hilbert spaces $\mathscr{H}_1, \ldots, \mathscr{H}_n$, then

$$A_1B_1 \odot \cdots \odot A_nB_n \subseteq (A_1 \odot \cdots \odot A_n)(B_1 \odot \cdots \odot B_n). \tag{15}$$

To see this, it suffices to note that if $x_j \in \mathscr{D}(A_jB_j)$, $j = 1, \ldots, n$, then $x_1 \otimes \cdots \otimes x_n \in \mathscr{D}((A_1 \odot \cdots \odot A_n)(B_1 \odot \cdots \odot B_n))$, and that

$$\begin{aligned}(A_1B_1 \odot \cdots \odot A_nB_n)&(x_1 \otimes \cdots \otimes x_n)\\ &= A_1B_1x_1 \otimes \cdots \otimes A_nB_nx_n\\ &= (A_1 \odot \cdots \odot A_n)(B_1 \odot \cdots \odot B_n)(x_1 \otimes \cdots \otimes x_n);\end{aligned}$$

since $\{x_1 \otimes \cdots \otimes x_n\}$ is a set of linear generators for $\mathscr{D}(A_1B_1 \odot \cdots \odot A_nB_n)$. The fact that the inclusion in (15) may be strict is illustrated by noting that $(A \odot B)(0 \odot I)$ is defined and 0 on $\mathscr{H}_1 \odot \mathscr{H}_2$ while $(A \cdot 0 \odot B \cdot I)$ is defined and 0 only on $\mathscr{H}_1 \odot \mathscr{D}(B)$. In the case where $T_1, \ldots, T_n$ are everywhere-defined, bounded operators, we have equality,

$$(T_1 \odot \cdots \odot T_n)(B_1 \odot \cdots \odot B_n) = T_1B_1 \odot \cdots \odot T_nB_n; \tag{16}$$

for, with this assumption,

$$\begin{aligned}\mathscr{D}((T_1 \odot \cdots \odot T_n)(B_1 \odot \cdots \odot B_n)) &= \mathscr{D}(B_1) \odot \cdots \odot \mathscr{D}(B_n)\\ &= \mathscr{D}(T_1B_1) \odot \cdots \odot \mathscr{D}(T_nB_n)\\ &= \mathscr{D}(T_1B_1 \odot \cdots \odot T_nB_n),\end{aligned}$$

and (16) follows from (15). ■

11.2.32. LEMMA. *If $T_j \in \mathscr{B}(\mathscr{H}_j)$ and B_j is a preclosed operator on the Hilbert space $\mathscr{H}_j$, then*

$$(T_1 \otimes \cdots \otimes T_n)(B_1 \otimes \cdots \otimes B_n) \subseteq T_1B_1 \otimes \cdots \otimes T_nB_n \tag{17}$$

and

$$B_1T_1 \otimes \cdots \otimes B_nT_n \subseteq (B_1 \otimes \cdots \otimes B_n)(T_1 \otimes \cdots \otimes T_n). \tag{18}$$

Proof. If $z \in \mathscr{D}(B_1 \otimes \cdots \otimes B_n)$, there is a sequence of vectors z_n in $\mathscr{D}(B_1) \odot \cdots \odot \mathscr{D}(B_n)$ for which we have $z_n \to z$ and $(B_1 \odot \cdots \odot B_n)z_n \to (B_1 \otimes \cdots \otimes B_n)z$. Since $T_1 \odot \cdots \odot T_n$ is bounded with (unique) everywhere-defined, bounded extension $T_1 \otimes \cdots \otimes T_n$, we have

$$\begin{aligned}(T_1B_1 \odot \cdots \odot T_nB_n)z_n &= (T_1 \odot \cdots \odot T_n)(B_1 \odot \cdots \odot B_n)z_n \\ &\to (T_1 \otimes \cdots \otimes T_n)(B_1 \otimes \cdots \otimes B_n)z.\end{aligned}$$

Thus $z \in \mathscr{D}(T_1B_1 \otimes \cdots \otimes T_nB_n)$ and

$$(T_1B_1 \otimes \cdots \otimes T_nB_n)z = (T_1 \otimes \cdots \otimes T_n)(B_1 \otimes \cdots \otimes B_n)z,$$

from which we conclude (17).

Again, from Remark 11.2.31,

$$\begin{aligned}B_1T_1 \odot \cdots \odot B_nT_n &\subseteq (B_1 \odot \cdots \odot B_n)(T_1 \odot \cdots \odot T_n) \\ &\subseteq (B_1 \otimes \cdots \otimes B_n)(T_1 \otimes \cdots \otimes T_n).\end{aligned}$$

Now $(B_1 \otimes \cdots \otimes B_n)(T_1 \otimes \cdots \otimes T_n)$ is closed since $B_1 \otimes \cdots \otimes B_n$ is closed and $T_1 \otimes \cdots \otimes T_n \in \mathscr{B}(\mathscr{H}_1 \otimes \cdots \otimes \mathscr{H}_n)$; and (18) follows. ∎

11.2.33. Proposition. *If $H_1, \ldots, H_n$ are self-adjoint (positive) operators acting on Hilbert spaces $\mathscr{H}_1, \ldots, \mathscr{H}_n$, respectively, then $H_1 \otimes \cdots \otimes H_n$ $(= H)$ is a self-adjoint (positive) operator on $\mathscr{H}_1 \otimes \cdots \otimes \mathscr{H}_n (= \mathscr{H})$.*

Proof. From the discussion following Lemma 5.6.13 and Theorem 5.6.15(i), there is a bounding sequence $\{E_{kj}\}$ of projections on $\mathscr{H}_k$ for H_k, $k = 1, \ldots, n$. Let E_j be $E_{1j} \otimes \cdots \otimes E_{nj}$ and note, from the preceding lemma, that

$$\begin{aligned}E_jH &= (E_{1j} \otimes \cdots \otimes E_{nj})(H_1 \otimes \cdots \otimes H_n) \subseteq E_{1j}H_1 \otimes \cdots \otimes E_{nj}H_n \\ &\subseteq H_1E_{1j} \otimes \cdots \otimes H_nE_{nj} \subseteq (H_1 \otimes \cdots \otimes H_n)(E_{1j} \otimes \cdots \otimes E_{nj}) = HE_j.\end{aligned}$$

By choice of E_{kj}, H_kE_{kj} is a bounded, self-adjoint operator on H_k, so that $H_1E_{1j} \otimes \cdots \otimes H_nE_{nj} = HE_j$ and HE_j is a bounded, self-adjoint operator on $\mathscr{H}$. Since $\{E_{kj}\}$ is an increasing sequence with strong-operator limit I, for $k = 1, \ldots, n$, the same is true for $\{E_j\}$, and $\{E_j\}$ is a bounding sequence for H. From Lemma 5.6.14, $\bigcup_{j=1}^{\infty} E_j(\mathscr{H})$ is a core for H. From Lemma 5.6.1, H is self-adjoint.

If $H_1, \ldots, H_n$ are positive operators, they have positive square roots $H_1^{1/2}, \ldots, H_n^{1/2}$. From Remark 11.2.31,

$$\begin{aligned}H_1 \odot \cdots \odot H_n = H_1^{1/2} \cdot H_1^{1/2} \odot \cdots \odot H_n^{1/2} \cdot H_n^{1/2} &\subseteq (H_1^{1/2} \odot \cdots \odot H_n^{1/2})^2 \\ &\subseteq (H_1^{1/2} \otimes \cdots \otimes H_n^{1/2})^2.\end{aligned}$$

From what we have proved, $H_1 \otimes \cdots \otimes H_n$ and $H_1^{1/2} \otimes \cdots \otimes H_n^{1/2}$ are self-adjoint. Thus $(H_1^{1/2} \otimes \cdots \otimes H_n^{1/2})^2$ is self adjoint and, in particular, closed. Therefore

$$H_1 \otimes \cdots \otimes H_n \subseteq (H_1^{1/2} \otimes \cdots \otimes H_n^{1/2})^2.$$

By taking adjoints of both sides of this inclusion, we have that $H_1 \otimes \cdots \otimes H_n = (H_1^{1/2} \otimes \cdots \otimes H_n^{1/2})^2$ (self-adjoint operators are maximal symmetric), and $H_1 \otimes \cdots \otimes H_n$ is positive. ■

11.2.34. Lemma. *If $A_1, \ldots, A_n$ are preclosed operators with range projections $E_1, \ldots, E_n$ on Hilbert spaces $\mathscr{H}_1, \ldots, \mathscr{H}_n$, then $E_1 \otimes \cdots \otimes E_n$ is the range projection of $A_1 \otimes \cdots \otimes A_n$.*

Proof. From Lemma 11.2.32,

$$(E_1 \otimes \cdots \otimes E_n)(A_1 \otimes \cdots \otimes A_n) \subseteq E_1A_1 \otimes \cdots \otimes E_nA_n = A_1 \otimes \cdots \otimes A_n.$$

But $\mathscr{D}((E_1 \otimes \cdots \otimes E_n)(A_1 \otimes \cdots \otimes A_n)) = \mathscr{D}(A_1 \otimes \cdots \otimes A_n)$ so that $(E_1 \otimes \cdots \otimes E_n)(A_1 \otimes \cdots \otimes A_n) = A_1 \otimes \cdots \otimes A_n$ and $E_1 \otimes \cdots \otimes E_n$ contains the range projection of $A_1 \otimes \cdots \otimes A_n$. If $y_j \in E_j(\mathscr{H}_j)$, there is an x_j in $\mathscr{D}(A_j)$ with Ax_j near y_j; so that $A_1x_1 \otimes \cdots \otimes A_nx_n$ is near $y_1 \otimes \cdots \otimes y_n$. Now $A_1x_1 \otimes \cdots \otimes A_nx_n$ $(=(A_1 \otimes \cdots \otimes A_n)(x_1 \otimes \cdots \otimes x_n))$ is in the range of $A_1 \otimes \cdots \otimes A_n$; and vectors of the form $y_1 \otimes \cdots \otimes y_n$ generate $(E_1 \otimes \cdots \otimes E_n)(\mathscr{H}_1 \odot \cdots \odot \mathscr{H}_n)$, a dense submanifold of the range of $E_1 \otimes \cdots \otimes E_n$. Thus $E_1 \otimes \cdots \otimes E_n$ is contained in the range projection of $A_1 \otimes \cdots \otimes A_n$. Combining this with the reverse inclusion, which has been established, we conclude that $E_1 \otimes \cdots \otimes E_n$ is the range projection of $A_1 \otimes \cdots \otimes A_n$. ■

11.2.35. Proposition. *If $A_1, \ldots, A_n$ are densely defined closed linear (or conjugate-linear) operators on Hilbert spaces $\mathscr{H}_1, \ldots, \mathscr{H}_n$ and $V_1H_1, \ldots, V_nH_n$ are their respective polar decompositions, then $(V_1 \otimes \cdots \otimes V_n)(H_1 \otimes \cdots \otimes H_n)$ is the polar decomposition of $A_1 \otimes \cdots \otimes A_n$.*

Proof. If E_j and F_j are the range projections of A_j^* and A_j, respectively, then E_j is the range projection of H_j, $V_j^*V_j = E_j$, and $V_jV_j^* = F_j$. Thus

$$(V_1 \otimes \cdots \otimes V_n)^*(V_1 \otimes \cdots \otimes V_n) = V_1^*V_1 \otimes \cdots \otimes V_n^*V_n = E_1 \otimes \cdots \otimes E_n.$$

From the preceding lemma, $E_1 \otimes \cdots \otimes E_n$ is the range projection of $H_1 \otimes \cdots \otimes H_n$ and $F_1 \otimes \cdots \otimes F_n$ is the range projection of $A_1 \otimes \cdots \otimes A_n$. From Proposition 11.2.33, $H_1 \otimes \cdots \otimes H_n$ is a positive operator. Now $V_1 \otimes \cdots \otimes V_n$ is a partial isometry with initial projection the range projection of $H_1 \otimes \cdots \otimes H_n$ and final projection the range projection of

$A_1 \otimes \cdots \otimes A_n$. It follows that $(V_1 \otimes \cdots \otimes V_n)(H_1 \otimes \cdots \otimes H_n)$ is closed. From Remark 11.2.31 and Lemma 11.2.32,

$$\begin{aligned} A_1 \odot \cdots \odot A_n = V_1H_1 \odot \cdots \odot V_nH_n &\subseteq (V_1 \odot \cdots \odot V_n)(H_1 \odot \cdots \odot H_n) \\ &\subseteq (V_1 \otimes \cdots \otimes V_n)(H_1 \otimes \cdots \otimes H_n) \\ &\subseteq V_1H_1 \otimes \cdots \otimes V_nH_n = A_1 \otimes \cdots \otimes A_n. \end{aligned}$$

As $A_1 \otimes \cdots \otimes A_n$ is the closure of $A_1 \odot \cdots \odot A_n$ and, as we have noted, $(V_1 \otimes \cdots \otimes V_n)(H_1 \otimes \cdots \otimes H_n)$ is closed, we have

$$A_1 \otimes \cdots \otimes A_n = (V_1 \otimes \cdots \otimes V_n)(H_1 \otimes \cdots \otimes H_n). \quad ■$$

11.2.36. REMARK. If $\mathscr{R}_1, \ldots, \mathscr{R}_n$ are von Neumann algebras acting on Hilbert spaces $\mathscr{H}_1, \ldots, \mathscr{H}_n$ and $u_1, \ldots, u_n$ are separating and generating vectors for $\mathscr{R}_1, \ldots, \mathscr{R}_n$, respectively, there is a modular structure, $(S_k, F_k, J_k, \Delta_k)$, associated with each $\mathscr{R}_k$, u_k (see Section 9.2.) Let $\mathscr{H}$ be $\mathscr{H}_1 \otimes \cdots \otimes \mathscr{H}_n$, $\mathscr{R}$ be $\mathscr{R}_1 \bar{\otimes} \cdots \bar{\otimes} \mathscr{R}_n$, u be $u_1 \otimes \cdots \otimes u_n$, S be $S_1 \otimes \cdots \otimes S_n$, F be $F_1 \otimes \cdots \otimes F_n$, J be $J_1 \otimes \cdots \otimes J_n$, Δ be $\Delta_1 \otimes \cdots \otimes \Delta_n$. Then u is generating for the von Neumann algebra $\mathscr{R}$ acting on $\mathscr{H}$. We shall identify (S, F, J, Δ) as the modular structure for $\mathscr{R}$; and, using this identification, produce another proof of the commutation theorem (11.2.16). Of course ($\mathscr{R}_0' =$) $\mathscr{R}_1' \bar{\otimes} \cdots \bar{\otimes} \mathscr{R}_n' \subseteq \mathscr{R}'$; and u is generating for $\mathscr{R}_1' \bar{\otimes} \cdots \bar{\otimes} \mathscr{R}_n'$, since $u_1, \ldots, u_n$ are separating and generating for $\mathscr{R}_1', \ldots, \mathscr{R}_n'$. The modular structure for $\mathscr{R}_k'$, u_k interchanges S_k and F_k and replaces Δ_k by Δ_k^{-1}. (See the discussion preceding Lemma 9.2.31.)

Let S_0Au be A^*u for A in $\mathscr{R}$ (as in Section 9.2). We know that S_0 is preclosed. If $A_k \in \mathscr{R}_k$, $k = 1, \ldots, n$,

$$\begin{aligned} S_0(A_1 \otimes \cdots \otimes A_n)u &= (A_1 \otimes \cdots \otimes A_n)^*u = A_1^*u_1 \otimes \cdots \otimes A_n^*u_n \\ &= S_1A_1u_1 \otimes \cdots \otimes S_nA_nu_n = S(A_1 \otimes \cdots \otimes A_n)u. \end{aligned}$$

Let $\mathscr{S}$ be the algebra generated by $\{A_1 \otimes \cdots \otimes A_n\}$; so that $\mathscr{S}$ is a weak (and strong)-operator dense subalgebra of $\mathscr{R}$. Since $\mathscr{R}_ku_k$ is a core for S_k, by construction, $\mathscr{S}u\,(=\mathscr{R}_1u_1 \odot \cdots \odot \mathscr{R}_nu_n)$ is a core for $S\,(=S_1 \otimes \cdots \otimes S_n)$, from Lemma 11.2.29. We show that $\mathscr{S}u$ is a core for $\bar{S}_0$; so that $S = \bar{S}_0$, since S_0 and S agree on $\mathscr{S}u$. Note, for this, that $\mathscr{S}^{\mathrm{h}}$ (the set of self-adjoint elements in $\mathscr{S}$) is weak (and from Theorem 5.1.2, strong)-operator dense in $\mathscr{R}^{\mathrm{h}}$. Thus, given B in $\mathscr{R}$, there is a sequence $\{B_n\}$ in $\mathscr{S}$ such that $B_nu \to Bu$ and $S_0B_nu = B_n^*u \to B^*u = S_0Bu$. It follows that $\mathscr{S}u$ is a core for $\bar{S}_0$, since $\mathscr{R}u$ is; and $S = \bar{S}_0$.

With this identification of S for $\mathscr{R}$, u as $S_1 \otimes \cdots \otimes S_n$, we can apply Proposition 11.2.35 and conclude that (S, J, Δ) is the modular structure for $\mathscr{R}$, u. At the same time. (F, J, Δ^{-1}) is the modular structure for $\mathscr{R}'_0$ $(=\mathscr{R}'_1 \mathbin{\bar{\otimes}} \cdots \mathbin{\bar{\otimes}} \mathscr{R}'_n)$. Now $\mathscr{R} \subseteq \mathscr{R}_0$ and $\mathscr{R}u$ is a core for $\Delta^{1/2}$, since Δ is the modular operator for $\mathscr{R}$ as well as for $\mathscr{R}_0$. Thus, from Theorem 9.2.36(iii), $\mathscr{R}=\mathscr{R}_0$ and $\mathscr{R}'=\mathscr{R}'_0$, that is,

$$(\mathscr{R}_1 \mathbin{\bar{\otimes}} \cdots \mathbin{\bar{\otimes}} \mathscr{R}_n)' = \mathscr{R}'_1 \mathbin{\bar{\otimes}} \cdots \mathbin{\bar{\otimes}} \mathscr{R}'_n.$$

We recognize this as the commutation theorem, Theorem 11.2.16. We have proved it, once again, through this study of tensor products of unbounded operators and the identification of the modular structure for the tensor products of von Neumann algebras when we reduce the general case to the present case where each $\mathscr{R}_k$ has a separating and generating vector u_k. This reduction is accomplished by a more vigorous use of Lemmas 11.2.14 and 11.2.15. Note that in the present argument, none of the deeper parts of modular theory appears—the fact that S_0 is preclosed, its definition, and the density result, Theorem 9.2.36, suffice. In our earlier proof of the commutation theorem, Proposition 11.2.11 is the density result used. ∎

11.2.37. Proposition. *If $A_1, \ldots, A_n$ are preclosed operators on Hilbert spaces $\mathscr{H}_1, \ldots, \mathscr{H}_n$, then*

$$(A_1 \otimes \cdots \otimes A_n)^* = A_1^* \otimes \cdots \otimes A_n^*. \tag{19}$$

Proof. From Corollary 11.2.30,

$$(A_1 \otimes \cdots \otimes A_n)^* = (\bar{A}_1 \otimes \cdots \otimes \bar{A}_n)^*,$$

and from Theorem 2.7.8(i), $A_j^* = \bar{A}_j^*$. Hence, we may suppose, without loss of generality, that each A_j is closed. Let V_jH_j be the polar decomposition of A_j, so that $V_j^*V_j = E_j$ and $V_jV_j^* = F_j$, where E_j, F_j are the range projections of A_j^* (as well as H_j) and A_j, respectively. From Proposition 11.2.35,

$$A_1 \otimes \cdots \otimes A_n = (V_1 \otimes \cdots \otimes V_n)(H_1 \otimes \cdots \otimes H_n), \tag{20}$$

and (20) is the polar decomposition of $A_1 \otimes \cdots \otimes A_n$. From Proposition 11.2.33, $H_1 \otimes \cdots \otimes H_n$ is self-adjoint; and from Lemma 11.2.34, we have that $E_1 \otimes \cdots \otimes E_n$ is its range projection. Thus

$$(E_1 \otimes \cdots \otimes E_n)(H_1 \otimes \cdots \otimes H_n) = H_1 \otimes \cdots \otimes H_n. \tag{21}$$

By applying Lemma 6.1.10 to (20) and (21), we have

$$(A_1 \otimes \cdots \otimes A_n)^* = (H_1 \otimes \cdots \otimes H_n)(V_1 \otimes \cdots \otimes V_n)^* \tag{22}$$

and

$$(H_1 \otimes \cdots \otimes H_n)(E_1 \otimes \cdots \otimes E_n) = H_1 \otimes \cdots \otimes H_n. \tag{23}$$

We multiply both sides of (22) by $V_1 \otimes \cdots \otimes V_n$ to obtain

$$(24)\quad (A_1 \otimes \cdots \otimes A_n)^*(V_1 \otimes \cdots \otimes V_n) = (H_1 \otimes \cdots \otimes H_n)(E_1 \otimes \cdots \otimes E_n) = H_1 \otimes \cdots \otimes H_n,$$

from (23). Again from Lemma 6.1.10, $A_j^* = H_j V_j^*$ so that $A_j^* V_j = H_j E_j = H_j$. Thus, from Lemma 11.2.32 (18),

(25)

$$H_1 \otimes \cdots \otimes H_n = A_1^* V_1 \otimes \cdots \otimes A_n^* V_n \subseteq (A_1^* \otimes \cdots \otimes A_n^*)(V_1 \otimes \cdots \otimes V_n).$$

On combining (24) and (25), we have

$$(26)\quad (A_1 \otimes \cdots \otimes A_n)^*(V_1 \otimes \cdots \otimes V_n) \subseteq (A_1^* \otimes \cdots \otimes A_n^*)(V_1 \otimes \cdots \otimes V_n).$$

By multiplying both sides of (26) with $(V_1 \otimes \cdots \otimes V_n)^*$, we have

$$(27)\quad (A_1 \otimes \cdots \otimes A_n)^*(F_1 \otimes \cdots \otimes F_n) \subseteq (A_1^* \otimes \cdots \otimes A_n^*)(F_1 \otimes \cdots \otimes F_n).$$

Since $F_1 \otimes \cdots \otimes F_n$ is the range projection of $A_1 \otimes \cdots \otimes A_n$;

$$(28)\quad A_1 \otimes \cdots \otimes A_n = (F_1 \otimes \cdots \otimes F_n)(A_1 \otimes \cdots \otimes A_n).$$

From Lemma 6.1.10 applied to (28), we have,

$$(29)\quad (A_1 \otimes \cdots \otimes A_n)^* = (A_1 \otimes \cdots \otimes A_n)^*(F_1 \otimes \cdots \otimes F_n);$$

so that, combining (27) and (29),

$$(30)\quad (A_1 \otimes \cdots \otimes A_n)^* \subseteq (A_1^* \otimes \cdots \otimes A_n^*)(F_1 \otimes \cdots \otimes F_n).$$

By taking adjoints of both sides of (30), we have

(31)

$$(F_1 \otimes \cdots \otimes F_n)(A_1^* \otimes \cdots \otimes A_n^*)^* \subseteq [(A_1^* \otimes \cdots \otimes A_n^*)(F_1 \otimes \cdots \otimes F_n)]^* \subseteq A_1 \otimes \cdots \otimes A_n.$$

We apply the inclusion (31) with A_j in place of A_j^* and E_j in place of F_j, to obtain

$$(32)\quad (E_1 \otimes \cdots \otimes E_n)(A_1 \otimes \cdots \otimes A_n)^* \subseteq A_1^* \otimes \cdots \otimes A_n^*.$$

In particular,

$$(33)\quad \mathscr{D}((A_1 \otimes \cdots \otimes A_n)^*) \subseteq \mathscr{D}(A_1^* \otimes \cdots \otimes A_n^*).$$

From Theorem 2.7.8(i) and Proposition 11.2.27, we have

$$(34)\quad A_1^* \otimes \cdots \otimes A_n^* \subseteq (A_1 \otimes \cdots \otimes A_n)^*.$$

Combining (33) and (34) yields (19). ∎

The preceding proposition can be used to establish the form of the polar decomposition of $A_1 \otimes \cdots \otimes A_n$ companion to the one developed in Proposition 11.2.35.

11.2.38. COROLLARY. *If $A_1, \ldots, A_n$ are closed operators on Hilbert spaces $\mathscr{H}_1, \ldots, \mathscr{H}_n$ and $K_j V_j$ is a polar decomposition for A_j then*

$$(K_1 \otimes \cdots \otimes K_n)(V_1 \otimes \cdots \otimes V_n)$$

is a polar decomposition for $A_1 \otimes \cdots \otimes A_n$.

Proof. From our hypothesis, $V_j^* K_j$ is a polar decomposition for A_j^*, and, from Proposition 11.2.35,

$$A_1^* \otimes \cdots \otimes A_n^* = (V_1^* \otimes \cdots \otimes V_n^*)(K_1 \otimes \cdots \otimes K_n) \tag{35}$$

By taking adjoints of both sides of (35), using Lemma 6.1.10, and applying Proposition 11.2.37, we have

$$A_1 \otimes \cdots \otimes A_n = (K_1 \otimes \cdots \otimes K_n)(V_1 \otimes \cdots \otimes V_n). \tag{36}$$ ■

11.2.39. REMARK If H_j is a positive operator on $\mathscr{H}_j$, then H_j generates an abelian von Neumann algebra $\mathscr{A}_j$. (See Theorem 5.6.18 and the comment following it.) The set,

$$\{I_1 \otimes \cdots \otimes I_{j-1} \otimes A_j \otimes I_{j+1} \otimes \cdots \otimes I_n : A_j \in \mathscr{A}_j\} \ (= \tilde{\mathscr{A}}_j),$$

is an abelian von Neumann algebra on $\mathscr{H}_1 \otimes \cdots \otimes \mathscr{H}_n \ (= \mathscr{H})$ and

$$I_1 \otimes \cdots \otimes I_{j-1} \otimes H_j \otimes I_{j+1} \otimes \cdots \otimes I_n \ (= \tilde{H}_j)$$

is affiliated with $\tilde{\mathscr{A}}_j$. The algebras $\tilde{\mathscr{A}}_1, \ldots, \tilde{\mathscr{A}}_n$ commute and generate the abelian von Neumann algebra $\mathscr{A}_1 \bar{\otimes} \cdots \bar{\otimes} \mathscr{A}_n \ (= \mathscr{A})$. The operators $\tilde{H}_j$ are positive operators, as is $H_1 \otimes \cdots \otimes H_n \ (= H)$, from Proposition 11.2.33. Remark 5.6.34 tells us that $\tilde{H}_1 + \cdots + \tilde{H}_n$ is a positive self-adjoint operator affiliated with $\mathscr{A}$ (a fact that is useful in some mathematical formulations of quantum dynamics).

Assuming only that each H_j is self-adjoint (not necessarily positive) and applying (15) we have

$$H_1 \odot \cdots \odot H_n \subseteq \tilde{H}_1 \cdots \tilde{H}_n \subseteq \tilde{H}_1 \hat{\cdot} \cdots \hat{\cdot} \tilde{H}_n,$$

so that $H_1 \otimes \cdots \otimes H_n \subseteq \tilde{H}_1 \hat{\cdot} \cdots \hat{\cdot} \tilde{H}_n$. Since $H_1 \otimes \cdots \otimes H_n$ and $\tilde{H}_1 \hat{\cdot} \cdots \hat{\cdot} \tilde{H}_n$ are self-adjoint

$$H_1 \otimes \cdots \otimes H_n = \tilde{H}_1 \hat{\cdot} \cdots \hat{\cdot} \tilde{H}_n \, \eta \, \mathscr{A}.$$ ■

Bibliography: [21, 55, 56, 74, 78, 82, 95, 96, 99].

11.3. Tensor products of abstract C^*-algebras

In this section, we consider various ways in which, given two (abstract) C^*-algebras $\mathfrak{A}$ and $\mathscr{B}$, it is possible to form another such algebra that can reasonably be regarded as the tensor product of $\mathfrak{A}$ and $\mathscr{B}$. This is equivalent to the determination of certain norms (the "C^*-norms") on the algebraic tensor product $\mathfrak{A} \odot \mathscr{B}$. In general, there are many such norms, including a largest and a smallest one. Certain C^*-algebras $\mathfrak{A}$ have the property that, for every choice of $\mathscr{B}$, $\mathfrak{A} \odot \mathscr{B}$ has only one C^*-norm; and the C^*-algebra tensor product is unique in this case. Such algebras $\mathfrak{A}$, described as "nuclear" C^*-algebras, include all abelian and uniformly matricial C^*-algebras.

The spatial tensor product. In this subsection, we describe the simplest type of tensor product for abstract C^*-algebras, which is derived from the "represented C^*-algebra tensor product" introduced in Section 11.1. It is clear, from Theorem 11.1.3, that it is possible to define a tensor product of abstract C^*-algbras in this manner; for, by that theorem, the * isomorphism class of a tensor product of represented C^*-algebras is determined by the * isomorphism classes of the component algebras. This is formalized in the following result.

11.3.1. PROPOSITION. *Suppose that $\mathfrak{A}_1, \ldots, \mathfrak{A}_n$ are (abstract) C^*-algebras*

(i) *There is a C^*-algebra $\mathfrak{A}$, and a multilinear mapping p from $\mathfrak{A}_1 \times \cdots \times \mathfrak{A}_n$ into $\mathfrak{A}$, with the following property: if π_j is a faithful representation of $\mathfrak{A}_j$ ($j = 1, \ldots, n$), there is a * isomorphism φ, from $\mathfrak{A}$ onto (the represented C^*-algebra tensor product) $\pi_1(\mathfrak{A}_1) \otimes \cdots \otimes \pi_n(\mathfrak{A}_n)$, such that*

$$\varphi(p(A_1, \ldots, A_n)) = \pi_1(A_1) \otimes \cdots \otimes \pi_n(A_n) \qquad (A_1 \in \mathfrak{A}_1, \ldots, A_n \in \mathfrak{A}_n).$$

(ii) *If $\mathfrak{A}_0$ and p_0 also have the properties ascribed in* (i) *to $\mathfrak{A}$ and p, respectively, there is a * isomorphism ψ from $\mathfrak{A}$ onto $\mathfrak{A}_0$, such that*

$$\psi(p(A_1, \ldots, A_n)) = p_0(A_1, \ldots, A_n) \qquad (A_1 \in \mathfrak{A}_1, \ldots, A_n \in \mathfrak{A}_n).$$

Proof. (i) Choose and fix faithful representations $\Phi_1, \ldots, \Phi_n$ of $\mathfrak{A}_1, \ldots, \mathfrak{A}_n$, respectively; for example, they could be the universal representations. Let $\mathfrak{A}$ be $\Phi_1(\mathfrak{A}_1) \otimes \cdots \otimes \Phi_n(\mathfrak{A}_n)$, and define $p\colon \mathfrak{A}_1 \times \cdots \times \mathfrak{A}_n \to \mathfrak{A}$ by

$$p(A_1, \ldots, A_n) = \Phi_1(A_1) \otimes \cdots \otimes \Phi_n(A_n).$$

If π_j is a faithful representation of $\mathfrak{A}_j$ ($j = 1, \ldots, n$), then $\pi_j \circ \Phi_j^{-1}$ is a * isomorphism φ_j from $\Phi_j(\mathfrak{A}_j)$ onto $\pi_j(\mathfrak{A}_j)$. With φ the * isomorphism $\varphi_1 \otimes \cdots \otimes \varphi_n$, from $\mathfrak{A}(= \Phi_1(\mathfrak{A}_1) \otimes \cdots \otimes \Phi_n(\mathfrak{A}_n))$ onto $\pi_1(\mathfrak{A}_1) \otimes \cdots \otimes \pi_n(\mathfrak{A}_n)$,

$$\begin{aligned}\varphi(p(A_1, \ldots, A_n)) &= \varphi(\Phi_1(A_1) \otimes \cdots \otimes \Phi_n(A_n)) \\ &= \varphi_1(\Phi_1(A_1)) \otimes \cdots \otimes \varphi_n(\Phi_n(A_n)) \\ &= \pi_1(A_1) \otimes \cdots \otimes \pi_n(A_n).\end{aligned}$$

(ii) Under the conditions set out in part (ii) of the proposition, and with π_j a faithful representation of $\mathfrak{A}_j$ $(j = 1, \ldots, n)$, we can choose * isomorphisms φ and φ_0, from $\mathfrak{A}$ and $\mathfrak{A}_0$ (respectively) onto $\pi_1(\mathfrak{A}_1) \otimes \cdots \otimes \pi_n(\mathfrak{A}_n)$, such that

$$\varphi(p(A_1, \ldots, A_n)) = \pi_1(A_1) \otimes \cdots \otimes \pi_n(A_n) = \varphi_0(p_0(A_1, \ldots, A_n)),$$

whenever $A_1 \in \mathfrak{A}_1, \ldots, A_n \in \mathfrak{A}_n$. The * isomorphism $\psi = \varphi_0^{-1} \circ \varphi$, from $\mathfrak{A}$ onto $\mathfrak{A}_0$, has the stated properties. ∎

The C^*-algebra $\mathfrak{A}$ occurring in Proposition 11.3.1 is described as the *spatial tensor product* of the C^*-algebras $\mathfrak{A}_1, \ldots, \mathfrak{A}_n$, and is denoted by $\mathfrak{A}_1 \otimes \cdots \otimes \mathfrak{A}_n$. When $A_1 \in \mathfrak{A}_1, \ldots, A_n \in \mathfrak{A}_n$, we usually write $A_1 \otimes \cdots \otimes A_n$ for the element $p(A_1, \ldots, A_n)$ of $\mathfrak{A}$. The pair $(\mathfrak{A}, p)$ is unique, up to * isomorphism, in a sense made precise in part (ii) of the proposition.

If each $\mathfrak{A}_j$ is a represented C^*-algebra, acting on a Hilbert space $\mathscr{H}_j$, the identity mapping on $\mathfrak{A}_j$ is a faithful representation π_j of $\mathfrak{A}_j$. It results from Proposition 11.3.1 that the spatial tensor product $\mathfrak{A}$ of $\mathfrak{A}_1, \ldots, \mathfrak{A}_n$ is * isomorphic to their "represented C^*-algebra tensor product," acting on $\mathscr{H}_1 \otimes \cdots \otimes \mathscr{H}_n$. Accordingly, we may identify these two tensor product algebras, and there is no ambiguity in using the same symbol, $\mathfrak{A}_1 \otimes \cdots \otimes \mathfrak{A}_n$, for both.

The theory developed in Section 11.1 now extends, without difficulty, to the context of spatial tensor products of (abstract) C^*-algebras $\mathfrak{A}_1, \ldots, \mathfrak{A}_n$; indeed, it suffices to take a faithful representation π_j of $\mathfrak{A}_j$, for each j, and to apply the results of Section 11.1 to the represented C^*-algebras $\pi_1(\mathfrak{A}_1), \ldots, \pi_n(\mathfrak{A}_n)$. We review the main aspects of the theory so obtained. When $A_j \in \mathfrak{A}_j$ $(j = 1, \ldots, n)$,

$$\|A_1 \otimes A_2 \otimes \cdots \otimes A_n\| = \|A_1\| \, \|A_2\| \cdots \|A_n\|.$$

The spatial tensor product is *associative*, in the sense that there is a * isomorphism, from $\mathfrak{A}_1 \otimes \cdots \otimes \mathfrak{A}_n$ onto $(\mathfrak{A}_1 \otimes \cdots \otimes \mathfrak{A}_r) \otimes (\mathfrak{A}_{r+1} \otimes \cdots \otimes \mathfrak{A}_n)$ that carries $A_1 \otimes \cdots \otimes A_n$ onto $(A_1 \otimes \cdots \otimes A_r) \otimes (A_{r+1} \otimes \cdots \otimes A_n)$. If ρ_j is a state of $\mathfrak{A}_j$ $(j = 1, \ldots, n)$, there is a *product state* $\rho = \rho_1 \otimes \cdots \otimes \rho_n$ of $\mathfrak{A}_1 \otimes \cdots \otimes \mathfrak{A}_n$, uniquely determined by the condition

$$\rho(A_1 \otimes A_2 \otimes \cdots \otimes A_n) = \rho_1(A_1)\rho_2(A_2) \cdots \rho_n(A_n).$$

The * subalgebra $\mathfrak{A}_0$ of $\mathfrak{A}_1 \otimes \cdots \otimes \mathfrak{A}_n$ that consists of all finite sums of simple tensors $A_1 \otimes \cdots \otimes A_n$ is everywhere dense in $\mathfrak{A}_1 \otimes \cdots \otimes \mathfrak{A}_n$. Moreover, for each A in $\mathfrak{A}_1 \otimes \cdots \otimes \mathfrak{A}_n$,

$$\|A\|^2 = \sup \frac{\rho(S^*A^*AS)}{\rho(S^*S)}, \tag{1}$$

where the supremum is taken over all pairs (ρ, S) in which ρ is a product state of $\mathfrak{A}_1 \otimes \cdots \otimes \mathfrak{A}_n$, $S \in \mathfrak{A}_0$, and $\rho(S^*S) \neq 0$. If φ_j is a * homomorphism from $\mathfrak{A}_j$ into another C^*-algebra $\mathscr{B}_j$ $(j = 1, \ldots, n)$, there is a * homomorphism $\varphi = \varphi_1 \otimes \cdots \otimes \varphi_n$, from $\mathfrak{A}_1 \otimes \cdots \otimes \mathfrak{A}_n$ into $\mathscr{B}_1 \otimes \cdots \otimes \mathscr{B}_n$, uniquely determined by the condition

$$\varphi(A_1 \otimes \cdots \otimes A_n) = \varphi_1(A_1) \otimes \cdots \otimes \varphi_n(A_n);$$

and if φ_j is a * isomorphism (or has range $\mathscr{B}_j$) for each j, then φ is a * isomorphism (or has range $\mathscr{B}_1 \otimes \cdots \otimes \mathscr{B}_n$).

We describe next some results, concerning states and representations of $\mathfrak{A}_1 \otimes \cdots \otimes \mathfrak{A}_n$ that were not foreshadowed in Section 11.1. Suppose that, for $j = 1, \ldots, n$, $\varphi_j : \mathfrak{A}_j \to \mathscr{B}(\mathscr{H}_j)$ is a representation. If we regard φ_j as a * homomorphism from $\mathfrak{A}_j$ onto the represented C^*-algebra $\varphi_j(\mathfrak{A}_j)$, and identify the "spatial" and "represented" tensor products of $\varphi_1(\mathfrak{A}_1), \ldots, \varphi_n(\mathfrak{A}_n)$, the * homomorphism $\varphi_1 \otimes \cdots \otimes \varphi_n$ is a representation of $\mathfrak{A}_1 \otimes \cdots \otimes \mathfrak{A}_n$ on $\mathscr{H}_1 \otimes \cdots \otimes \mathscr{H}_n$, with range $\varphi_1(\mathfrak{A}_1) \otimes \cdots \otimes \varphi_n(\mathfrak{A}_n)$.

By a *tracial state* of a C^*-algebra $\mathfrak{A}$, we mean a state ρ of $\mathfrak{A}$ such that $\rho(AB) = \rho(BA)$ for all A and B in $\mathfrak{A}$; of course, this is in agreement with our previous use of the term in the case of von Neumann algebras.

11.3.2. Proposition. *Suppose that, for $j = 1, \ldots, n$, $\mathfrak{A}_j$ is a C^*-algebra, $\varphi_j : \mathfrak{A}_j \to \mathscr{B}(\mathscr{H}_j)$ is a representation, and ρ_j is a state of $\mathfrak{A}_j$. Let $\mathfrak{A}$ be $\mathfrak{A}_1 \otimes \cdots \otimes \mathfrak{A}_n$, φ be $\varphi_1 \otimes \cdots \otimes \varphi_n$, and ρ be $\rho_1 \otimes \cdots \otimes \rho_n$.*

(i) *φ is irreducible if and only if each φ_j is irreducible.*
(ii) *ρ is pure if and only if each ρ_j is pure.*
(iii) *ρ is tracial if and only if each ρ_j is tracial.*

Proof. (i) Since

$$\begin{aligned}\varphi(\mathfrak{A})^- &= [\varphi_1(\mathfrak{A}_1) \otimes \cdots \otimes \varphi_n(\mathfrak{A}_n)]^- \\ &= \varphi_1(\mathfrak{A}_1)^- \mathbin{\bar{\otimes}} \cdots \mathbin{\bar{\otimes}} \varphi_n(\mathfrak{A}_n)^-,\end{aligned}$$

it follows from the commutation theorem (11.2.16) that

$$\varphi(\mathfrak{A})' = \varphi_1(\mathfrak{A}_1)' \mathbin{\bar{\otimes}} \cdots \mathbin{\bar{\otimes}} \varphi_n(\mathfrak{A}_n)'.$$

Accordingly, $\varphi(\mathfrak{A})'$ consists of scalars if and only if, for each j, $\varphi_j(\mathfrak{A}_j)'$ consists of scalars.

(ii) Let $\pi_j : \mathfrak{A}_j \to \mathscr{B}(\mathscr{K}_j)$ be the representation arising from ρ_j by the GNS construction, and let x_j be a cyclic unit vector for π_j, such that $\rho_j(A_j) = \langle \pi_j(A_j)x_j, x_j \rangle$ for all A_j in $\mathfrak{A}_j$. With π the representation $\pi_1 \otimes \cdots \otimes \pi_n$ of $\mathfrak{A}$, and x the cyclic unit vector $x_1 \otimes \cdots \otimes x_n$ for $\pi(\mathfrak{A})$,

$$\rho(A_1 \otimes \cdots \otimes A_n) = \langle \pi(A_1 \otimes \cdots \otimes A_n)x, x \rangle$$

whenever $A_1 \in \mathfrak{A}_1, \ldots, A_n \in \mathfrak{A}_n$. Hence $\rho = \omega_x \circ \pi$, and π is (unitarily equivalent to) the representation engendered by ρ. Accordingly, ρ is pure if and only if π is irreducible. By part (i), however, π is irreducible if and only if each π_j is irreducible; equivalently, if and only if each ρ_j is pure.

(iii) If each ρ_j is tracial, it is apparent that $\rho(AB) = \rho(BA)$ whenever A and B are simple tensors in $\mathfrak{A}$ $(=\mathfrak{A}_1 \otimes \cdots \otimes \mathfrak{A}_n)$. From the linearity and norm continuity of ρ, this remains true for all A and B in $\mathfrak{A}$; so ρ is tracial.

Conversely, suppose that ρ is tracial. Given A_j and B_j in $\mathfrak{A}_j$,

$$\rho_j(A_jB_j) = \rho(AB) = \rho(BA) = \rho_j(B_jA_j),$$

where $A = I \otimes \cdots \otimes I \otimes A_j \otimes I \otimes \cdots \otimes I$, $B = I \otimes \cdots \otimes I \otimes B_j \otimes I \otimes \cdots \otimes I$. Thus ρ_j is tracial, for each $j = 1, \ldots, n$. ∎

We remark that part (i) of Proposition 11.3.2 can be proved without appeal to the commutation theorem, with only a little additional effort.

By using various examples discussed in Section 11.1, we now give an explicit description of the spatial tensor product $\mathfrak{A} \otimes \mathscr{B}$, in certain cases.

For each positive integer n, the C^*-algebra $M_n(\mathbb{C})$ of all $n \times n$ complex matrices has a faithful representation as the algebra of all (bounded) linear operators acting on an n-dimensional Hilbert space. It follows easily from the discussion of Example 11.1.5 that, for any C^*-algebra $\mathfrak{A}$, $\mathfrak{A} \otimes M_n(\mathbb{C})$ is * isomorphic to the algebra $M_n(\mathfrak{A})$ of all $n \times n$ matrices with entries in $\mathfrak{A}$; if $A \in \mathfrak{A}$, and B is an $n \times n$ complex matrix $[b_{jk}]$, the simple tensor $A \otimes B$ in $\mathfrak{A} \otimes M_n(\mathbb{C})$ corresponds to the matrix $[b_{jk}A]$ in $M_n(\mathfrak{A})$. In particular (Example 11.1.6), $M_p(\mathbb{C}) \otimes M_q(\mathbb{C})$ is * isomorphic to $M_{pq}(\mathbb{C})$.

If $\mathfrak{A}$ is a C^*-algebra and S is a compact Hausdorff space, it is easy to deduce from Example 11.1.7 that $\mathfrak{A} \otimes C(S)$ is * isomorphic to the C^*-algebra $C(S, \mathfrak{A})$ of all norm-continuous functions F from S into $\mathfrak{A}$, in such a way that a simple tensor $A \otimes f$ (with f in $C(S)$ and A in $\mathfrak{A}$) corresponds to the function $F: S \to \mathfrak{A}$ defined by $F(s) = f(s)A$. When X and Y are compact Hausdorff spaces, there is a * isomorphism from $C(X) \otimes C(Y)$ onto $C(X \times Y)$ that carries a simple tensor $f \otimes g$ to the function h defined by $h(x, y) = f(x)g(y)$. Indeed, $C(X) \otimes C(Y)$ is * isomorphic to $C(Y, C(X))$, from the first sentence in this paragraph; and the equation

$$(\varphi(F))(x, y) = (F(y))(x) \qquad (F \in C(Y, C(X)),\ x \in X,\ y \in Y)$$

defines a * isomorphism φ from $C(Y, C(X))$ onto $C(X \times Y)$.

When $\mathfrak{A}$ and $\mathscr{B}$ are C^*-algebras, their *algebraic tensor product* $\mathfrak{A} \odot \mathscr{B}$ is defined to be the * subalgebra $\mathscr{C}_0$ of the spatial tensor product $\mathfrak{A} \otimes \mathscr{B}$ that consists of all finite sums $\sum A_j \otimes B_j$ of simple tensors. The condition that such a sum should vanish, as set out in Proposition 11.1.8 for tensor products of represented C^*-algebras, extends at once to the present case. The same applies to the "universal" property, that every bilinear mapping from $\mathfrak{A} \times \mathscr{B}$ into a complex vector space can be factored (uniquely) through $\mathscr{C}_0$. Accordingly, the above definition of the algebraic tensor product is in agreement with the usual conventions in elementary linear algebra.

From Proposition 11.1.8 we note, in particular, that $A \otimes B$ is a non-zero element of $\mathfrak{A} \odot \mathscr{B}$, when $A(\in \mathfrak{A})$ and $B(\in \mathscr{B})$ are both non-zero.

The spatial tensor product and the algebraic tensor product are both well behaved in relation to the formation of C^*-subalgebras. To be more specific, suppose that $\mathfrak{A}$ and $\mathscr{B}$ are C^*-algebras and that $\mathfrak{A}_1(\subseteq \mathfrak{A})$ and $\mathscr{B}_1(\subseteq \mathscr{B})$ are C^*-subalgebras. Let $\mathscr{C}_1$ be the * subalgebra of $\mathfrak{A} \odot \mathscr{B}$ that consists of all finite sums of simple tensors $A_1 \otimes B_1$, with A_1 in $\mathfrak{A}_1$ and B_1 in $\mathscr{B}_1$; and let $\mathscr{C}$ be the closure of $\mathscr{C}_1$ in $\mathfrak{A} \otimes \mathscr{B}$. We assert that there is a * isomorphism from $\mathfrak{A}_1 \otimes \mathscr{B}_1$ into $\mathfrak{A} \otimes \mathscr{B}$ that maps $A_1 \otimes B_1$ in $\mathfrak{A}_1 \otimes \mathscr{B}_1$ onto $A_1 \otimes B_1$ in $\mathfrak{A} \otimes \mathscr{B}$, and hence carries $\mathfrak{A}_1 \odot \mathscr{B}_1$ onto $\mathscr{C}_1$ and $\mathfrak{A}_1 \otimes \mathscr{B}_1$ onto $\mathscr{C}$. In this way, $\mathfrak{A}_1 \odot \mathscr{B}_1$ can be identified with the subalgebra $\mathscr{C}_1$ of $\mathfrak{A} \odot \mathscr{B}$; and $\mathfrak{A}_1 \otimes \mathscr{B}_1$, with the subalgebra $\mathscr{C}$ of $\mathfrak{A} \otimes \mathscr{B}$. Indeed, the inclusion mappings, $\alpha: \mathfrak{A}_1 \to \mathfrak{A}$ and $\beta: \mathscr{B}_1 \to \mathscr{B}$, are * isomorphisms; and it is immediately verified that the * isomorphism $\alpha \otimes \beta: \mathfrak{A}_1 \otimes \mathscr{B}_1 \to \mathfrak{A} \otimes \mathscr{B}$ has the stated properties.

C^*-norms on $\mathfrak{A} \odot \mathscr{B}$. Suppose that $\mathfrak{A}$ and $\mathscr{B}$ are C^*-algebras, and $\mathfrak{A} \odot \mathscr{B}$ is their algebraic tensor product. A C^*-algebra $\mathscr{C}$ can reasonably be regarded as a "C^*-algebra tensor product" of $\mathfrak{A}$ and $\mathscr{B}$ if it contains $\mathfrak{A} \odot \mathscr{B}$ as an everywhere-dense * subalgebra. Given such a C^*-algebra $\mathscr{C}$, its norm restricts to a norm $|||\ |||$ on $\mathfrak{A} \odot \mathscr{B}$, such that

(i) $(\mathfrak{A} \odot \mathscr{B}, |||\ |||)$ is a normed algebra,
(ii) $|||S^*S||| = |||S|||^2$ for each S in $\mathfrak{A} \odot \mathscr{B}$.

Moreover, $\mathscr{C}$ can be identified with the completion of $(\mathfrak{A} \odot \mathscr{B}, |||\ |||)$. Conversely, if a norm $|||\ |||$ on $\mathfrak{A} \odot \mathscr{B}$ satisfies conditions (i) and (ii), the completion of $(\mathfrak{A} \odot \mathscr{B}, |||\ |||)$ is a Banach algebra $\mathscr{C}$; the involution on $\mathfrak{A} \odot \mathscr{B}$ is isometric, by (ii), and so extends (still satisfying the C^*-condition) to an involution of $\mathscr{C}$. Accordingly, $\mathscr{C}$ is a C^*-algebra, and contains $\mathfrak{A} \odot \mathscr{B}$ as an everywhere-dense * subalgebra.

By a *C^*-norm* on $\mathfrak{A} \odot \mathscr{B}$, we mean a norm that satisfies conditions (i) and (ii) above. It is sometimes convenient to use a letter, say α, for such a norm, and to write $\alpha(S)$ rather than $|||S|||$. When α is a C^*-norm on $\mathfrak{A} \odot \mathscr{B}$, the completion of $(\mathfrak{A} \odot \mathscr{B}, \alpha)$ will be denoted by $\mathfrak{A} \otimes_\alpha \mathscr{B}$; and the same symbol α will be used for the norm on $\mathfrak{A} \otimes_\alpha \mathscr{B}$.

The spatial tensor product $\mathfrak{A} \otimes \mathscr{B}$ contains $\mathfrak{A} \odot \mathscr{B}$ as an everywhere-dense * subalgebra. By restricting its norm, we obtain a C^*-norm σ on $\mathfrak{A} \odot \mathscr{B}$, the *spatial C^*-norm*. However, we shall continue to write $\mathfrak{A} \otimes \mathscr{B}$, rather than $\mathfrak{A} \otimes_\sigma \mathscr{B}$, for the spatial tensor product.

We shall see in due course (Corollary 11.3.10) that every C^*-norm α on $\mathfrak{A} \odot \mathscr{B}$ is a *cross-norm*; by which we mean that $\alpha(A \otimes B) = \|A\| \, \|B\|$ for all A in $\mathfrak{A}$ and B in $\mathscr{B}$. In the meantime, we require some partial results, involving a slight extension of the concept of C^*-norm. By a *C^*-semi-norm* on $\mathfrak{A} \odot \mathscr{B}$, we mean a semi-norm α on $\mathfrak{A} \odot \mathscr{B}$ such that $\alpha(I) = 1$, $\alpha(ST) \leq \alpha(S)\alpha(T)$, and $\alpha(S^*S) = [\alpha(S)]^2$ for all S and T in $\mathfrak{A} \odot \mathscr{B}$. Thus a C^*-semi-norm that vanishes only at the zero element of $\mathfrak{A} \odot \mathscr{B}$ is a C^*-norm in the sense defined above. If α and β are C^*-semi-norms on $\mathfrak{A} \odot \mathscr{B}$, then so is γ, where $\gamma(S) = \max\{\alpha(S), \beta(S)\}$ for each S in $\mathfrak{A} \odot \mathscr{B}$.

11.3.3. LEMMA. *Suppose that $\mathfrak{A}$ and $\mathscr{B}$ are C^*-algebras.*

(i) *If α is a C^*-norm on $\mathfrak{A} \odot \mathscr{B}$, then*

$$\alpha(A \otimes I) = \|A\|, \qquad \alpha(I \otimes B) = \|B\| \qquad (A \in \mathfrak{A}, \quad B \in \mathscr{B});$$

and the sets $\{A \otimes I : A \in \mathfrak{A}\}$, $\{I \otimes B : B \in \mathscr{B}\}$ are C^*-subalgebras of $\mathfrak{A} \otimes_\alpha \mathscr{B}$.

(ii) *If α is a C^*-semi-norm on $\mathfrak{A} \odot \mathscr{B}$, then $\alpha(A \otimes B) \leq \|A\| \, \|B\|$ for all A in $\mathfrak{A}$ and B in $\mathscr{B}$.*

(iii) *If $\{\alpha_a\}$ is a family of C^*-semi-norms on $\mathfrak{A} \odot \mathscr{B}$, the equation*

$$\beta(S) = \sup_a \alpha_a(S) \qquad (S \in \mathfrak{A} \odot \mathscr{B})$$

defines a C^*-semi-norm β on $\mathfrak{A} \odot \mathscr{B}$.

Proof. (i) When α is a C^*-norm on $\mathfrak{A} \odot \mathscr{B}$, the mappings

$$A \to A \otimes I : \mathfrak{A} \to \mathfrak{A} \otimes_\alpha \mathscr{B}$$

and $B \to I \otimes B : \mathscr{B} \to \mathfrak{A} \otimes_\alpha \mathscr{B}$ are * isomorphisms; so they are isometries, and their ranges are C^*-subalgebras of $\mathfrak{A} \otimes_\alpha \mathscr{B}$.

(ii) With α a C^*-semi-norm on $\mathfrak{A} \odot \mathscr{B}$, and σ the spatial C^*-norm, the equation

$$\gamma(S) = \max\{\alpha(S), \sigma(S)\} \qquad (S \in \mathfrak{A} \odot \mathscr{B})$$

defines a C^*-norm γ. By applying to γ the results of part (i), we obtain

$$\begin{aligned} \alpha(A \otimes B) \leq \gamma(A \otimes B) &= \gamma((A \otimes I)(I \otimes B)) \\ &\leq \gamma(A \otimes I)\gamma(I \otimes B) = \|A\| \, \|B\|. \end{aligned}$$

(iii) If $S = \sum_{j=1}^n A_j \otimes B_j$, it follows from (ii) that, for each index a,

$$\alpha_a(S) \leq \sum_{j=1}^n \alpha_a(A_j \otimes B_j) \leq \sum_{j=1}^n \|A_j\| \, \|B_j\|.$$

Thus $\beta(S) \le \sum_{j=1}^{n} \|A_j\| \, \|B_j\| < \infty$; and it now follows easily that β is a C^*-semi-norm on $\mathfrak{A} \odot \mathscr{B}$. ■

Suppose that $\mathfrak{A}$ and $\mathscr{B}$ are C^*-algebras, and we have representations, φ (of $\mathfrak{A}$) and ψ (of $\mathscr{B}$). We describe (φ, ψ) as a *commuting pair of representations of* $(\mathfrak{A}, \mathscr{B})$ if $\varphi(\mathfrak{A})$, $\psi(\mathscr{B})$ act on the same Hilbert space and $\varphi(A)$ commutes with $\psi(B)$ whenever $A \in \mathfrak{A}$ and $B \in \mathscr{B}$.

We now show that there is a largest C^*-semi-norm on $\mathfrak{A} \odot \mathscr{B}$.

11.3.4. THEOREM. *If $\mathfrak{A}$ and $\mathscr{B}$ are C^*-algebras, there is a C^*-norm μ on $\mathfrak{A} \odot \mathscr{B}$, such that $\mu(S) \ge \alpha(S)$ for all S in $\mathfrak{A} \odot \mathscr{B}$ and all C^*-semi-norms α on $\mathfrak{A} \odot \mathscr{B}$. Moreover, if $S = \sum_{j=1}^{n} A_j \otimes B_j$, then*

$$\mu(S) = \sup \left\| \sum_{j=1}^{n} \varphi(A_j)\psi(B_j) \right\|, \tag{2}$$

where the supremum is taken over all commuting pairs (φ, ψ) of representations of $(\mathfrak{A}, \mathscr{B})$. There exist commuting pairs (φ_0, ψ_0) of faithful representations of $(\mathfrak{A}, \mathscr{B})$ with the property that (for each $S = \sum_1^n A_j \otimes B_j$ in $\mathfrak{A} \odot \mathscr{B}$)

$$\mu(S) = \left\| \sum_{j=1}^{n} \varphi_0(A_j)\psi_0(B_j) \right\|.$$

Proof. From Lemma 11.3.3 (iii), there is a largest C^*-semi-norm μ on $\mathfrak{A} \odot \mathscr{B}$, defined by

$$\mu(S) = \sup_{\alpha} \alpha(S) \qquad (S \in \mathfrak{A} \odot \mathscr{B}),$$

the supremum being taken over the set of all C^*-semi-norms α on $\mathfrak{A} \odot \mathscr{B}$. Since one possible choice of α is the spatial C^*-norm σ, we have $\mu(S) \ge \sigma(S) > 0$ for each non-zero element S of $\mathfrak{A} \odot \mathscr{B}$. Thus μ is a C^*-norm.

Suppose that (φ, ψ) is a commuting pair of representations of $(\mathfrak{A}, \mathscr{B})$, each acting on a Hilbert space $\mathscr{H}$. From the universal property of the algebraic tensor product, the bilinear mapping

$$(A, B) \to \varphi(A)\psi(B) \colon \mathfrak{A} \times \mathscr{B} \to \mathscr{B}(\mathscr{H})$$

factors through $\mathfrak{A} \odot \mathscr{B}$; that is, there is a linear mapping $\pi \colon \mathfrak{A} \odot \mathscr{B} \to \mathscr{B}(\mathscr{H})$ such that

$$\pi(A \otimes B) = \varphi(A)\psi(B) \qquad (A \in \mathfrak{A}, \quad B \in \mathscr{B}).$$

Since $\varphi(\mathfrak{A})$ commutes with $\psi(\mathscr{B})$, it is easily verified that π is a * homomorphism; so the equation $\alpha(S) = \|\pi(S)\|$ defines a C^*-semi-norm α on

$\mathfrak{A} \odot \mathscr{B}$. Since μ is the greatest C^*-semi-norm, $\mu(S) \geq \alpha(S)$; that is

$$\mu\left(\sum_{j=1}^{n} A_j \otimes B_j\right) \geq \left\|\sum_{j=1}^{n} \varphi(A_j)\psi(B_j)\right\|. \tag{3}$$

Finally, let π_0 be a faithful representation of $\mathfrak{A} \otimes_\mu \mathscr{B}$, and define

$$\varphi_0(A) = \pi_0(A \otimes I), \qquad \psi_0(B) = \pi_0(I \otimes B).$$

Then (φ_0, ψ_0) is a commuting pair of (faithful) representations of $(\mathfrak{A}, \mathscr{B})$; and, since π_0 is isometric,

$$\begin{aligned}\mu\left(\sum_{j=1}^{n} A_j \otimes B_j\right) &= \left\|\pi_0\left(\sum_{j=1}^{n} A_j \otimes B_j\right)\right\| \\ &= \left\|\sum_{j=1}^{n} \pi_0(A_j \otimes I)\pi_0(I \otimes B_j)\right\| = \left\|\sum_{j=1}^{n} \varphi_0(A_j)\psi_0(B_j)\right\|.\end{aligned}$$

This, together with (3), establishes the required formulae for μ. ■

Given representations $\varphi: \mathfrak{A} \to \mathscr{B}(\mathscr{H})$ and $\psi: \mathscr{B} \to \mathscr{B}(\mathscr{K})$, the representation $\varphi \otimes \psi$ of the (spatial) tensor product $\mathfrak{A} \otimes \mathscr{B}$ does not increase norm; moreover, from the definition of the spatial C^*-norm σ, $\varphi \otimes \psi$ is isometric if φ and ψ are faithful. We can define a commuting pair (φ_1, ψ_1) of representations of $(\mathfrak{A}, \mathscr{B})$ on $\mathscr{H} \otimes \mathscr{K}$ by

$$\varphi_1(A) = \varphi(A) \otimes I, \qquad \psi_1(B) = I \otimes \psi(B). \tag{4}$$

When $A_1, \ldots, A_n \in \mathfrak{A}$ and $B_1, \ldots, B_n \in \mathscr{B}$, we have

$$\begin{aligned}\left\|\sum_{j=1}^{n} \varphi_1(A_j)\psi_1(B_j)\right\| &= \left\|\sum_{j=1}^{n} \varphi(A_j) \otimes \psi(B_j)\right\| \\ &= \left\|(\varphi \otimes \psi)\left(\sum_{j=1}^{n} A_j \otimes B_j\right)\right\| \leq \sigma\left(\sum_{j=1}^{n} A_j \otimes B_j\right),\end{aligned}$$

with equality if φ and ψ are faithful. While the largest C^*-norm μ on $\mathfrak{A} \odot \mathscr{B}$ is derived, as in (2), from the set of *all* commuting pairs of representations, the spatial C^*-norm σ relates to pairs (φ_1, ψ_1) of the special form (4). For certain pairs $(\mathfrak{A}, \mathscr{B})$ of C^*-algebras, these two C^*-norms are distinct (Example 11.3.14).

Our next objective is to show that σ is always the smallest C^*-norm on $\mathfrak{A} \odot \mathscr{B}$ (Theorem 11.3.9), and that σ and μ coincide if at least one of the C^*-algebras $\mathfrak{A}$, $\mathscr{B}$ is abelian (Theorem 11.3.7). We consider first the case in which both are abelian. The key to all these results is the existence of product states of $\mathfrak{A} \otimes_\alpha \mathscr{B}$ for every C^*-norm α. With λ a state of $\mathfrak{A}$ and ν a state of $\mathscr{B}$, we can form the product state $\lambda \otimes \nu$ of the spatial tensor product $\mathfrak{A} \otimes \mathscr{B}$. By restriction, this yields a linear functional $\lambda \odot \nu$ on $\mathfrak{A} \odot \mathscr{B}$, such that

$$(\lambda \odot \nu)(I) = 1, \qquad (\lambda \odot \nu)(A \otimes B) = \lambda(A)\nu(B), \qquad (\lambda \odot \nu)(S^*S) \geq 0,$$

whenever $A \in \mathfrak{A}$, $B \in \mathscr{B}$, and $S \in \mathfrak{A} \odot \mathscr{B}$. If it happens that

$$(5) \qquad |(\lambda \odot \nu)(S)| \leq \alpha(S) \qquad (S \in \mathfrak{A} \odot \mathscr{B}),$$

then $\lambda \odot \nu$ extends uniquely, by continuity, to a bounded linear functional $\lambda \otimes_\alpha \nu$ of norm 1 on $\mathfrak{A} \otimes_\alpha \mathscr{B}$; and $\lambda \otimes_\alpha \nu$ is a state, since $(\lambda \otimes_\alpha \nu)(I) = 1$. We shall see eventually that the condition (5) is automatically satisfied for every C^*-norm α.

11.3.5. LEMMA. *If $\mathfrak{A}$ and $\mathscr{B}$ are abelian C^*-algebras, there is only one C^*-norm on $\mathfrak{A} \odot \mathscr{B}$.*

Proof. If α is a C^*-norm on $\mathfrak{A} \odot \mathscr{B}$, the C^*-algebra $\mathfrak{A} \otimes_\alpha \mathscr{B}$ is abelian, since it is the closure of its abelian subalgebra $\mathfrak{A} \odot \mathscr{B}$. Accordingly, the pure state space $\mathscr{P}$ of $\mathfrak{A} \otimes_\alpha \mathscr{B}$ consists of all non-zero multiplicative linear functionals on $\mathfrak{A} \otimes_\alpha \mathscr{B}$; and

$$(6) \qquad \alpha(S) = \sup\{|\rho(S)|: \rho \in \mathscr{P}\} \qquad (S \in \mathfrak{A} \otimes_\alpha \mathscr{B}).$$

We now determine $\mathscr{P}$ in terms of the pure state spaces, $\mathscr{P}(\mathfrak{A})$ of $\mathfrak{A}$ and $\mathscr{P}(\mathscr{B})$ of $\mathscr{B}$. When $\lambda \in \mathscr{P}(\mathfrak{A})$ and $\nu \in \mathscr{P}(\mathscr{B})$, both λ and ν are multiplicative, and the same is true of the linear functional $\lambda \odot \nu$ on $\mathfrak{A} \odot \mathscr{B}$. If it happens that $|(\lambda \odot \nu)(S)| \leq \alpha(S)$ for each S in $\mathfrak{A} \odot \mathscr{B}$, then $\lambda \odot \nu$ extends (uniquely, by continuity) to a non-zero multiplicative linear functional $\lambda \otimes_\alpha \nu$ on $\mathfrak{A} \otimes_\alpha \mathscr{B}$. We assert that every non-zero multiplicative linear functional ρ on $\mathfrak{A} \otimes_\alpha \mathscr{B}$ arises in this way. Indeed, given such a ρ, we can define multiplicative linear functionals, λ on $\mathfrak{A}$ and ν on $\mathscr{B}$, by

$$\lambda(A) = \rho(A \otimes I), \qquad \nu(B) = \rho(I \otimes B).$$

Thus $\lambda \in \mathscr{P}(\mathfrak{A})$, $\nu \in \mathscr{P}(\mathscr{B})$; and, since

$$\begin{aligned} \rho(A \otimes B) = \rho((A \otimes I)(I \otimes B)) &= \rho(A \otimes I)\rho(I \otimes B) \\ &= \lambda(A)\nu(B) = (\lambda \odot \nu)(A \otimes B), \end{aligned}$$

it follows that $\rho(S) = (\lambda \odot \nu)(S)$ for each S in $\mathfrak{A} \odot \mathscr{B}$. Hence

$$|(\lambda \odot \nu)(S)| = |\rho(S)| \leq \alpha(S),$$

and ρ is the unique continuous extension, $\lambda \otimes_\alpha \nu$, of $\lambda \odot \nu$ to $\mathfrak{A} \otimes_\alpha \mathscr{B}$.

Let

$$\mathscr{E}_\alpha = \{(\lambda, \nu) \in \mathscr{P}(\mathfrak{A}) \times \mathscr{P}(\mathscr{B}): |(\lambda \odot \nu)(S)| \leq \alpha(S) \ (S \in \mathfrak{A} \odot \mathscr{B})\}.$$

Then $\mathscr{E}_\alpha$ is a closed subset of $\mathscr{P}(\mathfrak{A}) \times \mathscr{P}(\mathscr{B})$, in the product weak* topology. From the preceding paragraph,

$$\mathscr{P} = \{\lambda \otimes_\alpha \nu : (\lambda, \nu) \in \mathscr{E}_\alpha\}. \tag{7}$$

We assert that $\mathscr{E}_\alpha$ is the whole of $\mathscr{P}(\mathfrak{A}) \times \mathscr{P}(\mathscr{B})$. For this, suppose the contrary, so that $\mathscr{P}(\mathfrak{A}) \times \mathscr{P}(\mathscr{B}) \backslash \mathscr{E}_\alpha$ is a non-empty open subset of $\mathscr{P}(\mathfrak{A}) \times \mathscr{P}(\mathscr{B})$. Accordingly, there are non-empty open sets, $\mathscr{U}$ in $\mathscr{P}(\mathfrak{A})$ and $\mathscr{V}$ in $\mathscr{P}(\mathscr{B})$, such that

$$\mathscr{U} \times \mathscr{V} \subseteq \mathscr{P}(\mathfrak{A}) \times \mathscr{P}(\mathscr{B}) \backslash \mathscr{E}_\alpha.$$

From the representations of $\mathfrak{A}$ and $\mathscr{B}$ as continuous functions, on $\mathscr{P}(\mathfrak{A})$ and $\mathscr{P}(\mathscr{B})$, respectively, it follows that there are non-zero elements, A in $\mathfrak{A}$ and B in $\mathscr{B}$, such that

$$\lambda(A) = \nu(B) = 0 \qquad (\lambda \in \mathscr{P}(\mathfrak{A}) \backslash \mathscr{U}, \quad \nu \in \mathscr{P}(\mathscr{B}) \backslash \mathscr{V}).$$

Hence $(\lambda \odot \nu)(A \otimes B) = \lambda(A)\nu(B) = 0$, unless $(\lambda, \nu) \in \mathscr{U} \times \mathscr{V}$; so

$$(\lambda \odot \nu)(A \otimes B) = 0 \qquad ((\lambda, \nu) \in \mathscr{E}_\alpha).$$

It now results from (6) and (7) that

$$\begin{aligned}\alpha(A \otimes B) &= \sup\{|\rho(A \otimes B)| : \rho \in \mathscr{P}\}\\ &= \sup\{|(\lambda \otimes_\alpha \nu)(A \otimes B)| : (\lambda, \nu) \in \mathscr{E}_\alpha\}\\ &= \sup\{|(\lambda \odot \nu)(A \otimes B)| : (\lambda, \nu) \in \mathscr{E}_\alpha\} = 0;\end{aligned}$$

a contradiction, since α is a norm on $\mathfrak{A} \odot \mathscr{B}$ and $A \otimes B \neq 0$. This proves our assertion that $\mathscr{E}_a = \mathscr{P}(\mathfrak{A}) \times \mathscr{P}(\mathscr{B})$.

With S in $\mathfrak{A} \odot \mathscr{B}$, it now follows from (6) and (7) that

$$\begin{aligned}\alpha(S) &= \sup\{|(\lambda \otimes_\alpha \nu)(S)| : \lambda \in \mathscr{P}(\mathfrak{A}), \nu \in \mathscr{P}(\mathscr{B})\}\\ &= \sup\{|(\lambda \odot \nu)(S)| : \lambda \in \mathscr{P}(\mathfrak{A}), \nu \in \mathscr{P}(\mathscr{B})\},\end{aligned}$$

and the last expression is independent of α. ■

11.3.6. LEMMA. *Suppose that $\mathfrak{A}$ and $\mathscr{B}$ are C^*-algebras, α is a C^*-norm on $\mathfrak{A} \odot \mathscr{B}$, and λ is a state of $\mathfrak{A}$. Then there is a state ρ of $\mathfrak{A} \otimes_\alpha \mathscr{B}$ such that $\rho(A \otimes I) = \lambda(A) (A \in \mathfrak{A})$. If λ is a pure state of $\mathfrak{A}$, then given any such ρ, there is a state ν of $\mathscr{B}$ such that $\rho(A \otimes B) = \lambda(A)\nu(B)$ $(A \in \mathfrak{A}, B \in \mathscr{B})$.*

Proof. The equation $\lambda_0(A \otimes I) = \lambda(A)$ defines a state λ_0 of the C^*-subalgebra $\{A \otimes I : A \in \mathfrak{A}\}$ of $\mathfrak{A} \otimes_\alpha \mathscr{B}$; and it suffices to choose, for ρ, any extension of λ_0 to a state of $\mathfrak{A} \otimes_\alpha \mathscr{B}$.

Now suppose that λ is a pure state of $\mathfrak{A}$. If $B \in \mathscr{B}$ and $0 \leq B \leq I$,

$$\lambda(A) = \rho(A \otimes I) = \rho(A \otimes B) + \rho(A \otimes (I - B)) = \lambda_1(A) + \lambda_2(A) \qquad (A \in \mathfrak{A}),$$

where λ_1 and λ_2 are the positive linear functionals on $\mathfrak{A}$ defined by $\lambda_1(A) = \rho(A \otimes B)$, $\lambda_2(A) = \rho(A \otimes (I - B))$. Since λ is a pure state, λ_1 is a non-negative multiple $k\lambda$ of λ, and $k = k\lambda(I) = \lambda_1(I) = \rho(I \otimes B)$. Accordingly,

$$\rho(A \otimes B) = \lambda_1(A) = k\lambda(A) = \lambda(A)\rho(I \otimes B) \qquad (A \in \mathfrak{A}).$$

We have now shown that $\rho(A \otimes B) = \lambda(A)\rho(I \otimes B)$, whenever $A \in \mathfrak{A}$, $B \in \mathscr{B}$, $0 \le B \le I$; and by linearity, this remains true for all A in $\mathfrak{A}$ and B in $\mathscr{B}$. It now suffices to define a state ν of $\mathscr{B}$ by $\nu(B) = \rho(I \otimes B)$. ∎

11.3.7. THEOREM. *If $\mathfrak{A}$ and $\mathscr{B}$ are C^*-algebras, at least one of which is abelian, there is only one C^*-norm on $\mathfrak{A} \odot \mathscr{B}$.*

Proof. Suppose that $\mathfrak{A}$ is abelian, and α is a C^*-norm on $\mathfrak{A} \odot \mathscr{B}$. With λ in the pure state space $\mathscr{P}(\mathfrak{A})$ of $\mathfrak{A}$, define

$$\mathscr{S}_\lambda = \{\nu \in \mathscr{S}(\mathscr{B}) : |(\lambda \odot \nu)(S)| \le \alpha(S) \text{ for all } S \text{ in } \mathfrak{A} \odot \mathscr{B}\},$$

where $\mathscr{S}(\mathscr{B})$ is the state space of $\mathscr{B}$. Then $\mathscr{S}_\lambda$ is a weak* compact convex subset of $\mathscr{S}(\mathscr{B})$; and for each ν in $\mathscr{S}_\lambda$, $\lambda \odot \nu$ extends, uniquely, to a state $\lambda \otimes_\alpha \nu$ of $\mathfrak{A} \otimes_\alpha \mathscr{B}$.

We assert that $\mathscr{S}_\lambda$ is the whole of $\mathscr{S}(\mathscr{B})$. For this, it suffices (by Theorem 4.3.9) to show that

$$(8) \qquad \|C_0\| = \sup\{|\nu(C_0)| : \nu \in \mathscr{S}_\lambda\}$$

for each self-adjoint element C_0 of $\mathscr{B}$. Let $\mathscr{C}$ be the abelian C^*-subalgebra of $\mathscr{B}$ that is generated by I and C_0. There is a multiplicative linear functional ν_0 on $\mathscr{C}$, such that $|\nu_0(C_0)| = \|C_0\|$. Now $\mathfrak{A} \odot \mathscr{C}$ can be identified with the * subalgebra of $\mathfrak{A} \odot \mathscr{B}$, that consists of all finite sums of simple tensors $A \otimes C$, with A in $\mathfrak{A}$ and C in $\mathscr{C}$. By Lemma 11.3.5, the restriction of α to $\mathfrak{A} \odot \mathscr{C}$ coincides with the spatial C^*-norm σ on $\mathfrak{A} \odot \mathscr{C}$. Accordingly, the closure $\mathscr{A}$ of $\mathfrak{A} \odot \mathscr{C}$ in $\mathfrak{A} \otimes_\alpha \mathscr{B}$ can be identified with the completion of $(\mathfrak{A} \odot \mathscr{C}, \sigma)$; that is, with the *spatial* tensor product $\mathfrak{A} \otimes \mathscr{C}$. We can now form the product state $\lambda \otimes \nu_0$ of $\mathfrak{A} \otimes \mathscr{C}$ $(=\mathscr{A})$, and extend this to a state ρ of $\mathfrak{A} \otimes_\alpha \mathscr{B}$. For each A in $\mathfrak{A}$,

$$\rho(A \otimes I) = (\lambda \otimes \nu_0)(A \otimes I) = \lambda(A).$$

By Lemma 11.3.6, there is a state ν of $\mathscr{B}$ such that $\rho(A \otimes \mathscr{B}) = \lambda(A)\nu(B)$, whenever $A \in \mathfrak{A}$ and $B \in \mathscr{B}$. Since

$$|(\lambda \odot \nu)(S)| = |\rho(S)| \le \alpha(S) \qquad (S \in \mathfrak{A} \odot \mathscr{B}),$$

it follows that $\nu \in \mathscr{S}_\lambda$. However,

$$\begin{aligned} |\nu(C_0)| &= |\lambda(I)\nu(C_0)| = |\rho(I \otimes C_0)| \\ &= |(\lambda \otimes \nu_0)(I \otimes C_0)| = |\nu_0(C_0)| = \|C_0\|. \end{aligned}$$

This proves (8), and establishes our claim that $\mathscr{S}_\lambda = \mathscr{S}(\mathscr{B})$.

The preceding argument shows that, given any pure state λ of $\mathfrak{A}$, and any state ν of $\mathscr{B}$, the linear functional $\lambda \odot \nu$ on $\mathfrak{A} \odot \mathscr{B}$ extends to a state $\lambda \otimes_\alpha \nu$ of $\mathfrak{A} \otimes_\alpha \mathscr{B}$. We prove next that every pure state of $\mathfrak{A} \otimes_\alpha \mathscr{B}$ arises in this way. For this, note first that, since $\mathfrak{A}$ is abelian, each element $A \otimes I$ of $\mathfrak{A} \otimes_\alpha \mathscr{B}$ lies in the center. A pure state ρ of $\mathfrak{A} \otimes_\alpha \mathscr{B}$ is multiplicative on the center (Proposition 4.3.14), so the equation $\lambda(A) = \rho(A \otimes I)$ defines a multiplicative linear functional (that is, a pure state) λ of $\mathfrak{A}$. From Lemma 11.3.6 there is a state ν of $\mathscr{B}$ such that $\rho(A \otimes B) = \lambda(A)\nu(B)$, for all A in $\mathfrak{A}$ and B in $\mathscr{B}$; and $\rho = \lambda \otimes_\alpha \nu$.

We have now shown that

$$\{\lambda \otimes_\alpha \nu : \lambda \in \mathscr{P}(\mathfrak{A}),\ \nu \in \mathscr{S}(\mathscr{B})\}$$

is a set of states of $\mathfrak{A} \otimes_\alpha \mathscr{B}$ that contains all the pure states. From this,

$$\alpha(H) = \sup\{(\lambda \otimes_\alpha \nu)(H) : \lambda \in \mathscr{P}(\mathfrak{A}),\ \nu \in \mathscr{S}(\mathscr{B})\},$$

for every positive element H of $\mathfrak{A} \otimes_\alpha \mathscr{B}$. When $S \in \mathfrak{A} \odot \mathscr{B}$,

$$\begin{aligned}\alpha(S) &= [\alpha(S^*S)]^{1/2} \\ &= [\sup\{(\lambda \otimes_\alpha \nu)(S^*S) : \lambda \in \mathscr{P}(\mathfrak{A}),\ \nu \in \mathscr{S}(\mathscr{B})\}]^{1/2} \\ &= [\sup\{(\lambda \odot \nu)(S^*S) : \lambda \in \mathscr{P}(\mathfrak{A}),\ \nu \in \mathscr{S}(\mathscr{B})\}]^{1/2};\end{aligned}$$

and the last expression is independent of α. ■

11.3.8. PROPOSITION. *If $\mathfrak{A}$ and $\mathscr{B}$ are C^*-algebras, α is a C^*-norm on $\mathfrak{A} \odot \mathscr{B}$, λ is a state of $\mathfrak{A}$ and ν is a state of $\mathscr{B}$, there is a unique state $\lambda \otimes_\alpha \nu$ of $\mathfrak{A} \otimes_a \beta$ such that*

$$(\lambda \otimes_a \nu)(A \otimes B) = \lambda(A)\nu(B) \qquad (A \in \mathfrak{A}, \quad B \in \mathscr{B}).$$

Proof. Suppose first that λ is a *pure* state of $\mathfrak{A}$, and consider the weak * compact convex subset,

$$\mathscr{S}_\lambda = \{\nu \in \mathscr{S}(\mathscr{B}) : |(\lambda \odot \nu)(S)| \leq \alpha(S) \text{ for all } S \text{ in } \mathfrak{A} \odot \mathscr{B}\},$$

of the state space $\mathscr{S}(\mathscr{B})$. By following the second paragraph of the proof of Theorem 11.3.7, it follows that $\mathscr{S}_\lambda$ is the whole of $\mathscr{S}(\mathscr{B})$; the argument applies verbatim, except that we appeal to Theorem 11.3.7 (rather than Lemma 11.3.5) in concluding that α restricts to the spatial C^*-norm on $\mathfrak{A} \odot \mathscr{C}$.

Next, suppose that ν is a state of $\mathscr{B}$, and consider the weak * compact convex subset,

$$\mathscr{T}_\nu = \{\lambda \in \mathscr{S}(\mathfrak{A}) : |(\lambda \odot \nu)(S)| \leq \alpha(S) \text{ for all } S \text{ in } \mathfrak{A} \odot \mathscr{B}\},$$

of $\mathscr{S}(\mathfrak{A})$. By the preceding paragraph, $\mathscr{T}_\nu$ contains each pure state of $\mathfrak{A}$, and is therefore the whole of $\mathscr{S}(\mathfrak{A})$.

We have now shown that, for all states λ of $\mathfrak{A}$ and ν of $\mathscr{B}$,

$$|(\lambda \odot \nu)(S)| \le \alpha(S) \qquad (S \in \mathfrak{A} \odot \mathscr{B});$$

and from this, $\lambda \odot \nu$ extends, uniquely by continuity, to a state $\lambda \otimes_\alpha \nu$ of $\mathfrak{A} \otimes_\alpha \mathscr{B}$. ∎

11.3.9. THEOREM. *If $\mathfrak{A}$ and $\mathscr{B}$ are C^*-algebras, α is a C^*-norm on $\mathfrak{A} \odot \mathscr{B}$ and σ is the spatial C^*-norm, then $\alpha(T) \ge \sigma(T)$ for every T in $\mathfrak{A} \odot \mathscr{B}$.*

Proof. Given states λ of $\mathfrak{A}$ and ν of $\mathscr{B}$, we can form the state $\lambda \otimes_\alpha \nu$ of $\mathfrak{A} \otimes_\alpha \mathscr{B}$. When $S \in \mathfrak{A} \odot \mathscr{B}$ and $(\lambda \odot \nu)(S^*S) \ne 0$, the equation

$$\eta(R) = \frac{(\lambda \otimes_\alpha \nu)(S^*RS)}{(\lambda \otimes_\alpha \nu)(S^*S)} \qquad (R \in \mathfrak{A} \otimes_\alpha \mathscr{B})$$

defines a state η of $\mathfrak{A} \otimes_\alpha \mathscr{B}$. Accordingly, with T in $\mathfrak{A} \odot \mathscr{B}$,

$$\begin{aligned}[\alpha(T)]^2 &= \alpha(T^*T) \\ &\ge \frac{(\lambda \otimes_\alpha \nu)(S^*T^*TS)}{(\lambda \otimes_\alpha \nu)(S^*S)} = \frac{(\lambda \odot \nu)(S^*T^*TS)}{(\lambda \odot \nu)(S^*S)}.\end{aligned}$$

By varying λ in $\mathscr{S}(\mathfrak{A})$, ν in $\mathscr{S}(\mathscr{B})$ and S in $\mathfrak{A} \odot \mathscr{B}$ (subject to the condition $(\lambda \odot \nu)(S^*S) \ne 0$), it now follows from the formula (1) for the spatial C^*-norm that $\alpha(T) \ge \sigma(T)$. ∎

11.3.10. COROLLARY. *If $\mathfrak{A}$ and $\mathscr{B}$ are C^*-algebras, every C^*-norm on $\mathfrak{A} \odot \mathscr{B}$ is a cross-norm.*

Proof. Suppose that α is a C^*-norm on $\mathfrak{A} \odot \mathscr{B}$, and $A \in \mathfrak{A}$, $B \in \mathscr{B}$. From our discussion of the spatial tensor product (following the proof of Proposition 11.3.1), it follows that the spatial C^*-norm σ is a cross-norm. Accordingly, from Theorem 11.3.9 and Lemma 11.3.3(ii)),

$$\|A\| \, \|B\| = \sigma(A \otimes B) \le \alpha(A \otimes B) \le \|A\| \, \|B\|;$$

so $\alpha(A \otimes B) = \|A\| \, \|B\|$. ∎

Nuclear C^*-algebras. A C^*-algebra $\mathfrak{A}$ is said to be *nuclear* if, for every C^*-algebra $\mathscr{B}$, there is only one C^*-norm on $\mathfrak{A} \odot \mathscr{B}$ (equivalently, the largest C^*-norm on $\mathfrak{A} \odot \mathscr{B}$ coincides with the spatial C^*-norm).

11.3.11. LEMMA. *If $\mathfrak{A}$ is a finite-dimensional C^*-algebra, then $\mathfrak{A}$ is nuclear; and, for every C^*-algebra $\mathscr{B}$, $\mathfrak{A} \odot \mathscr{B}$ is complete relative to its unique C^*-norm.*

Proof. If $\pi\colon \mathfrak{A} \to \mathscr{B}(\mathscr{H})$ is a faithful representation, then $\pi(\mathfrak{A})$ is a finite-dimensional von Neumann algebra. The center of $\pi(\mathfrak{A})$ contains a finite orthogonal family $\{P_1, \ldots, P_k\}$ of minimal projections, with sum I. For each

j, $\pi(\mathfrak{A})P_j$ is a finite-dimensional factor, and is therefore of type I_n for some positive integer n. From this, it follows easily that $\mathfrak{A}$ contains a finite orthogonal family $\{E_1, \ldots, E_m\}$ of minimal projections, with sum I; and that, for $p, q = 1, \ldots, m$, $E_p\mathfrak{A}E_q$ is at most one dimensional. We may choose V_{pq} in $\mathfrak{A}$, so that either $V_{pq} = 0$ or $\|V_{pq}\| = 1$, while $E_p\mathfrak{A}E_q$ consists of all scalar multiples of V_{pq}.

Now suppose that $\mathscr{B}$ is a C^*-algebra and α is a C^*-norm on $\mathfrak{A} \odot \mathscr{B}$. Define projections $F_1, \ldots, F_m$ in $\mathfrak{A} \odot \mathscr{B}$, with sum I, by $F_p = E_p \otimes I$. If S is an element $\sum_1^n A_j \otimes B_j$ of $\mathfrak{A} \odot \mathscr{B}$,

$$F_p S F_q = \sum_{j=1}^{n} E_p A_j E_q \otimes B_j = V_{pq} \otimes B$$

for some B in $\mathscr{B}$, since each $E_p A_j E_q$ is a scalar multiple of V_{pq}. Accordingly

$$F_p(\mathfrak{A} \odot \mathscr{B})F_q = \{V_{pq} \otimes B : B \in \mathscr{B}\}.$$

We assert that, as a subset of $(\mathfrak{A} \odot \mathscr{B}, \alpha)$, $F_p(\mathfrak{A} \odot \mathscr{B})F_q$ is complete. Since this is apparent when $V_{pq} = 0$, we may assume that $\|V_{pq}\| = 1$. In this case, the mapping $B \to V_{pq} \otimes B$, from $\mathscr{B}$ into $(\mathfrak{A} \odot \mathscr{B}, \alpha)$ is an isometry, since α is a cross-norm; so its range, $F_p(\mathfrak{A} \odot \mathscr{B})F_q$, is complete.

Suppose that $\{S_r\}$ is a Cauchy sequence in $(\mathfrak{A} \odot \mathscr{B}, \alpha)$. For each $p, q = 1, \ldots, m$, $\{F_p S_r F_q\}$ is a Cauchy sequence in the complete subset $F_p(\mathfrak{A} \odot \mathscr{B})F_q$, and so converges to an element T_{pq} of $\mathfrak{A} \odot \mathscr{B}$. Since $S_r = \sum F_p S_r F_q$, $\{S_r\}$ converges (relative to α) to $\sum T_{pq}$. Thus $(\mathfrak{A} \odot \mathscr{B}, \alpha)$ is complete.

Given two C^*-norms, α and β, on $\mathfrak{A} \odot \mathscr{B}$, we can regard the identity mapping on $\mathfrak{A} \odot \mathscr{B}$ as a * isomorphism between the C^*-algebras $(\mathfrak{A} \odot \mathscr{B}, \alpha)$ and $(\mathfrak{A} \odot \mathscr{B}, \beta)$. As such, it is an isometry; so $\alpha = \beta$. ∎

11.3.12. PROPOSITION. *Suppose that a C^*-algebra $\mathfrak{A}$ has a family $\{\mathfrak{A}_a : a \in \mathbb{A}\}$ of C^*-subalgebras, each containing the identity of $\mathfrak{A}$, such that*

(i) *each $\mathfrak{A}_a$ is nuclear;*
(ii) *given any a and b in $\mathbb{A}$, there is a c in $\mathbb{A}$ such that*

$$\mathfrak{A}_a \cup \mathfrak{A}_b \subseteq \mathfrak{A}_c;$$

(iii) *$\bigcup \mathfrak{A}_a$ is everywhere dense in $\mathfrak{A}$.*

Then $\mathfrak{A}$ is nuclear.

Proof. Suppose that $\mathscr{B}$ is a C^*-algebra, and α, β are C^*-norms on $\mathfrak{A} \odot \mathscr{B}$. If S is an element $\sum_1^m A_j \otimes B_j$ of $\mathfrak{A} \odot \mathscr{B}$, then, for $j = 1, \ldots, m$, we can choose a sequence $\{A_{j1}, A_{j2}, A_{j3}, \ldots\}$ of elements of $\bigcup \mathfrak{A}_a$ that converges to A_j. Let

$$S_n = \sum_{j=1}^{m} A_{jn} \otimes B_j,$$

and note that

$$\begin{aligned}\alpha(S - S_n) &= \alpha\left(\sum_{j=1}^{m} (A_j - A_{jn}) \otimes B_j\right) \\ &\leq \sum_{j=1}^{m} \alpha((A_j - A_{jn}) \otimes B_j) \\ &= \sum_{j=1}^{m} \|A_j - A_{jn}\| \, \|B_j\| \to 0\end{aligned}$$

as $n \to \infty$. The same argument applies also to β, so

$$\alpha(S) = \lim \alpha(S_n), \qquad \beta(S) = \lim \beta(S_n).$$

For a in $\mathbb{A}$, we may identify $\mathfrak{A}_a \odot \mathscr{B}$, in the usual way, with a subalgebra of $\mathfrak{A} \odot \mathscr{B}$. Since α and β restrict to C^*-norms on $\mathfrak{A}_a \odot \mathscr{B}$, while $\mathfrak{A}_a$ is nuclear, it follows that α and β coincide on $\mathfrak{A}_a \odot \mathscr{B}$. With n in $\mathbb{N}$, $A_{1n}, A_{2n}, \ldots, A_{mn} \in \bigcup \mathfrak{A}_a$; so, from (ii), we can choose a in $\mathbb{A}$ so that $A_{1n}, A_{2n}, \ldots, A_{mn} \in \mathfrak{A}_a$. Then, $S_n \in \mathfrak{A}_a \odot \mathscr{B}$, and therefore $\alpha(S_n) = \beta(S_n)$. When $n \to \infty$, we obtain $\alpha(S) = \beta(S)$; so $\alpha = \beta$. ■

11.3.13. THEOREM. *Abelian C^*-algebras and uniformly matricial C^*-algebras are nuclear.*

Proof. The assertion concerning abelian C^*-algebras is a restatement of Theorem 11.3.7; for matricial C^*-algebras, it suffices to combine Lemma 11.3.11 and Proposition 11.3.12. ■

11.3.14. EXAMPLE. We exhibit C^*-algebras $\mathscr{L}_0$ and $\mathscr{R}_0$ such that $\mathscr{L}_0 \odot \mathscr{R}_0$ has more than one C^*-norm; so that $\mathscr{L}_0$ and $\mathscr{R}_0$ are not nuclear.

To this end, let $\mathscr{L}$ be a factor of type II_1, acting on a Hilbert space $\mathscr{H}$, with commutant $\mathscr{R}$ also of type II_1. Suppose that $\mathscr{L}_0$ is a C^*-subalgebra of $\mathscr{L}$, $\mathscr{R}_0$ is a C^*-subalgebra of $\mathscr{R}$, and $\mathfrak{A}$ is the C^*-subalgebra of $\mathscr{B}(\mathscr{H})$ that is generated by $\mathscr{L}_0 \cup \mathscr{R}_0$. We prove below that $\mathfrak{A}$ is * isomorphic to $\mathscr{L}_0 \otimes_\alpha \mathscr{R}_0$ for some C^*-norm α on $\mathscr{L}_0 \odot \mathscr{R}_0$; and we observe that the spatial tensor product $\mathscr{L}_0 \otimes \mathscr{R}_0$ has a faithful tracial state. We then show that, for suitably chosen $\mathscr{L}$, $\mathscr{R}$, $\mathscr{L}_0$, and $\mathscr{R}_0$ (satisfying the above conditions), $\mathfrak{A}$ has no faithful tracial state. Hence, $\mathscr{L}_0 \otimes \mathscr{R}_0$ and $\mathscr{L}_0 \otimes_\alpha \mathscr{R}_0$ are not * isomorphic, and α does not coincide with the spatial C^*-norm on $\mathscr{L}_0 \odot \mathscr{R}_0$.

Since $\mathscr{L}_0$ and $\mathscr{R}_0$ act on $\mathscr{H}$, $\mathscr{L}_0 \otimes \mathscr{R}_0$ can be identified with their "represented" tensor product acting on $\mathscr{H} \otimes \mathscr{H}$, which is a C^*-subalgebra of the von Neumann algebra tensor product $\mathscr{L} \mathbin{\overline{\otimes}} \mathscr{R}$. Since $\mathscr{L} \mathbin{\overline{\otimes}} \mathscr{R}$ is a factor of type II_1, it has a faithful tracial state, whence the same is true of $\mathscr{L}_0 \otimes \mathscr{R}_0$.

Let $\mathfrak{A}_0$ be the subset of $\mathscr{B}(\mathscr{H})$, consisting of all finite sums of operators of the form LR, with L in $\mathscr{L}_0$ and R in $\mathscr{R}_0$. Then $\mathfrak{A}_0$ is a * subalgebra of $\mathscr{B}(\mathscr{H})$ with norm closure $\mathfrak{A}$. The bilinear mapping $(L, R) \to LR: \mathscr{L}_0 \times \mathscr{R}_0 \to \mathfrak{A}_0$ factors through $\mathscr{L}_0 \odot \mathscr{R}_0$, by the universal property of the algebraic tensor product; so there is a linear mapping $\varphi: \mathscr{L}_0 \odot \mathscr{R}_0 \to \mathfrak{A}_0$ such that $\varphi(L \otimes R) = LR$. It is immediately verified that φ carries $\mathscr{L}_0 \odot \mathscr{R}_0$ *onto* $\mathfrak{A}_0$, and is a * homomorphism (since $\mathscr{L}_0$ and $\mathscr{R}_0$ commute). We assert that φ is one-to-one. Indeed, suppose that A is an element $\sum_1^n L_j \otimes R_j$ of $\mathscr{L}_0 \odot \mathscr{R}_0$, and $\varphi(A) = 0$. Since $\mathscr{L}$ is a factor, while $L_1, \ldots, L_n \in \mathscr{L}$, $R_1, \ldots, R_n \in \mathscr{R}$ $(=\mathscr{L}')$ and $\sum_1^n L_j R_j = \varphi(A) = 0$, it follows from Theorem 5.5.4 that there is an $n \times n$ complex matrix $[c_{jk}]$ such that

$$\sum_{j=1}^{n} c_{jk} L_j = 0 \qquad (k = 1, \ldots, n),$$

$$\sum_{k=1}^{n} c_{jk} R_k = R_j \qquad (j = 1, \ldots, n).$$

Accordingly, $A = \sum_1^n L_j \otimes R_j = 0$; and φ is one-to-one.

From the preceding paragraph, we can define a C^*-norm α on $\mathscr{L}_0 \odot \mathscr{R}_0$ by $\alpha(A) = \|\varphi(A)\|$. Then, φ is an isometric * isomorphism from $(\mathscr{L}_0 \odot \mathscr{R}_0, \alpha)$ onto $\mathfrak{A}_0$, and extends by continuity to a * isomorphism from $\mathscr{L}_0 \otimes_\alpha \mathscr{R}_0$ onto $\mathfrak{A}$.

We shall show that $\mathscr{L}, \mathscr{R}, \mathscr{L}_0$, and $\mathscr{R}_0$ can be chosen (satisfying the above conditions) in such a way that $\mathfrak{A}$ contains a projection P and an infinite sequence U_j of unitaries, such that the projections $\{U_j P U_j^*\}$ are mutually orthogonal. This suffices to prove that $\mathfrak{A}$ has no faithful tracial state. Indeed, if τ is a tracial state of $\mathfrak{A}$,

$$\tau(I) \geq \tau\left(\sum_{j=1}^{n} U_j P U_j^*\right) = n\tau(P) \qquad (n = 1, 2, \ldots);$$

so that $\tau(P) = 0$, and τ is not faithful. In the example constructed below, P has one-dimensional range and $\mathfrak{A}$ acts irreducibly on $\mathscr{H}$; so by Proposition 10.4.10, $\mathfrak{A}$ contains all the compact operators acting on $\mathscr{H}$.

To construct this example, let $\mathscr{F}$ be the free group on two generators, a_1 and a_2; and let $\{x_g: g \in \mathscr{F}\}$ be the orthonormal basis of $l_2(\mathscr{F})$, in which $x_g(h) = \delta_{gh}$ for all g and h in $\mathscr{F}$. When $g \in \mathscr{F}$, define unitary operators L_g and R_g, acting on $l_2(\mathscr{F})$, by $L_g x = x_g * x$, $R_g x = x * x_g$; equivalently,

$$(L_g x)(h) = x(g^{-1}h), \qquad (R_g x)(h) = x(hg^{-1}) \qquad (g, h \in \mathscr{F}).$$

From Section 6.7, the von Neumann algebras $\mathscr{L}_{\mathscr{F}}$ and $\mathscr{R}_{\mathscr{F}}$, generated by the sets $\{L_g: g \in \mathscr{F}\}$ and $\{R_g: g \in \mathscr{F}\}$, respectively, are factors of type II_1; and $(\mathscr{L}_{\mathscr{F}})' = \mathscr{R}_{\mathscr{F}}$. Let $\mathscr{L}_0$ and $\mathscr{R}_0$ be any C^*-subalgebras such that

$$L_g \in \mathscr{L}_0 \subseteq \mathscr{L}_{\mathscr{F}}, \qquad R_g \in \mathscr{R}_0 \subseteq \mathscr{R}_{\mathscr{F}} \qquad (g \in \mathscr{F});$$

and let $\mathfrak{A}$ be the C^*-algebra generated by $\mathscr{L}_0 \cup \mathscr{R}_0$.

We shall prove that $\mathfrak{A}$ contains the projection P_e whose range is the one-dimensional subspace $[x_e]$, where as usual e denotes the unit element of $\mathscr{F}$. For this, let

$$A = L_{a_1}R_{a_1}^* + R_{a_1}L_{a_1}^* + L_{a_2}R_{a_2}^* + R_{a_2}L_{a_2}^*.$$

Then

$$A = A^* \in \mathfrak{A}, \qquad Ax_e = 4x_e;$$

from this, and since A is the sum of four unitaries,

$$\|A\| = 4, \qquad P_eA = AP_e = 4P_e. \tag{9}$$

We assert next that, if $\varepsilon = \frac{1}{25}$, then

$$A(I - P_e) \leq (4 - \varepsilon^2)(I - P_e). \tag{10}$$

To this end it suffices (since A commutes with P_e) to show that

$$\langle Ax, x\rangle \leq 4 - \varepsilon^2$$

whenever x is a unit vector in the range of $I - P_e$. For such x,

$$x(e) = \langle x, x_e\rangle = \langle (I - P_e)x, P_ex_e\rangle = 0.$$

Since

$$\langle Ax, x\rangle = 2\,\mathrm{Re}[\langle R_{a_1}L_{a_1}^*x, x\rangle + \langle R_{a_2}L_{a_2}^*x, x\rangle],$$

while

$$\begin{aligned}\sum_{g\in\mathscr{F}} |x(g) - x(a_jga_j^{-1})|^2 &= \|x - R_{a_j}L_{a_j}^*x\|^2\\ &= \|x\|^2 + \|R_{a_j}L_{a_j}^*x\|^2 - 2\,\mathrm{Re}\langle R_{a_j}L_{a_j}^*x, x\rangle\\ &= 2[1 - \mathrm{Re}\langle R_{a_j}L_{a_j}^*x, x\rangle] \qquad (j = 1, 2),\end{aligned}$$

it follows that

$$\langle Ax, x\rangle = 4 - \sum_{j=1}^{2}\sum_{g\in\mathscr{F}} |x(g) - x(a_jga_j^{-1})|^2.$$

If we now define $\mu(S) = \sum_{g \in S} |x(g)|^2$ for each subset S of $\mathscr{F}$, and use the argument set out in the second paragraph of the proof of Theorem 6.7.8, we deduce that

$$\sum_{g \in \mathscr{F}} |x(g) - x(a_j g a_j^{-1})|^2 > \frac{1}{(25)^2} = \varepsilon^2,$$

for at least one of the values, $j = 1, 2$. Hence $\langle Ax, x \rangle \leq 4 - \varepsilon^2$, and (10) is proved.

With T in $\mathfrak{A}$ defined by

$$T = \tfrac{1}{8}(4I + A),$$

it results from (9) and (10) that

$$P_e T = T P_e = P_e, \qquad 0 \leq T(I - P_e) \leq \eta(I - P_e),$$

where $\eta = (8 - \varepsilon^2)/8 < 1$. Accordingly, for each positive integer n,

$$T^n P_e = P_e, \qquad 0 \leq T^n(I - P_e) \leq \eta^n(I - P_e).$$

Hence

$$\|T^n - P_e\| = \|T^n - T^n P_e\| \leq \eta^n \to 0$$

as $n \to \infty$; and $P_e = \lim T^n \in \mathfrak{A}$.

For each g in $\mathscr{F}$, $L_g P_e L_g^*$ is the projection P_g whose range is the one-dimensional subspace $[x_g]$; so $\{L_g P_e L_g^* : g \in \mathscr{F}\}$ is an orthogonal family. Thus the conditions required above are satisfied, with P_e for P, and with $\{U_1, U_2, \ldots\}$ an enumeration of the unitaries $\{L_g : g \in \mathscr{F}\}$.

Observe, finally, that $\mathfrak{A}$ acts irreducibly on $l_2(\mathscr{F})$. Indeed, if $A' \in \mathfrak{A}'$, then A' commutes with each L_g and R_g; so

$$\mathfrak{A}' \subseteq (\mathscr{L}_{\mathscr{F}})' \cap (\mathscr{R}_{\mathscr{F}})' = (\mathscr{L}_{\mathscr{F}})' \cap \mathscr{L}_{\mathscr{F}} = \{aI : a \in \mathbb{C}\}. \quad \blacksquare$$

Bibliography: [8, 9, 25, 34, 51, 52, 94].

11.4. Infinite tensor products of C^*-algebras

In passing from finite to infinite tensor products, we shall make use of the concept of inductive limit, as applied to a directed system of C^*-algebras.

Let $\{\mathfrak{A}_f : f \in \mathbb{F}\}$ be a family of C^*-algebras (with unit I_f in $\mathfrak{A}_f$), in which the index set $\mathbb{F}$ is directed by a binary relation $\leq$. Suppose that, whenever $f, g \in \mathbb{F}$ and $f \leq g$, there is specified a * isomorphism Φ_{gf}, from $\mathfrak{A}_f$ into $\mathfrak{A}_g$ (with $\Phi_{gf}(I_f) = I_g$); and finally, suppose that $\Phi_{hg} \circ \Phi_{gf} = \Phi_{hf}$ whenever $f, g, h \in \mathbb{F}$ and $f \leq g \leq h$. In these circumstances we say that the C^*-algebras

$\{\mathfrak{A}_f : f \in \mathbb{F}\}$, together with the * isomorphisms $\{\Phi_{gf} : f, g \in \mathbb{F}, f \leq g\}$, constitute a *directed system of C^*-algebras*. Note that Φ_{ff} is the identity mapping on $\mathfrak{A}_f$. Indeed, Φ_{ff} is one-to-one; and, when $A \in \mathfrak{A}_f$, we have

$$\begin{aligned}\Phi_{ff}(A - \Phi_{ff}(A)) &= \Phi_{ff}(A) - (\Phi_{ff} \circ \Phi_{ff})(A) \\ &= \Phi_{ff}(A) - \Phi_{ff}(A) = 0,\end{aligned}$$

whence $A = \Phi_{ff}(A)$.

The simplest examples of such directed systems are those in which $\{\mathfrak{A}_f : f \in \mathbb{F}\}$ is a *net* of C^*-subalgebras of a given C^*-algebra $\mathfrak{A}$. By this, we mean that each $\mathfrak{A}_f$ is a C^*-subalgebra (containing the unit I) of $\mathfrak{A}$, that $\mathbb{F}$ is directed by $\leq$, and that $\mathfrak{A}_f \subseteq \mathfrak{A}_g$ whenever $f \leq g$. With $\Phi_{gf} : \mathfrak{A}_f \to \mathfrak{A}_g$ the inclusion mapping (whenever $f \leq g$), we obtain a directed system of C^*-algebras. Given such a net, $\{\mathfrak{A}_f : f \in \mathbb{F}\}$, the norm closure of $\bigcup \mathfrak{A}_f$ is itself a C^*-subalgebra of $\mathfrak{A}$ (and could have been used, in place of $\mathfrak{A}$, at the outset).

The following proposition shows that every directed system of C^*-algebras can be embedded, essentially uniquely, as a net of C^*-subalgebras of a suitable (minimal) C^*-algebra.

11.4.1. PROPOSITION. *Suppose that the C^*-algebras $\{\mathfrak{A}_f : f \in \mathbb{F}\}$, and the * isomorphism $\Phi_{gf} : \mathfrak{A}_f \to \mathfrak{A}_g$ ($f, g \in \mathbb{F}$; $f \leq g$), together form a directed system.*

(i) *There is a C^*-algebra $\mathfrak{A}$ and, for each f in $\mathbb{F}$, a * isomorphism φ_f from $\mathfrak{A}_f$ into $\mathfrak{A}$ (carrying the unit of $\mathfrak{A}_f$ onto that of $\mathfrak{A}$), such that $\varphi_f = \varphi_g \circ \Phi_{gf}$ when $f \leq g$, and $\bigcup \{\varphi_f(\mathfrak{A}_f) : f \in \mathbb{F}\}$ is everywhere dense in $\mathfrak{A}$.*

(ii) *The C^*-algebra $\mathfrak{A}$ occurring in* (i) *is uniquely determined, up to * isomorphism; if $\mathscr{B}$ is a C^*-algebra, $\psi_f : \mathfrak{A}_f \to \mathscr{B}$ is a * isomorphism (for each f in $\mathbb{F}$), and conditions analogous to those in* (i) *are satisfied, then there is a * isomorphism θ from $\mathfrak{A}$ onto $\mathscr{B}$, such that $\psi_f = \theta \circ \varphi_f$ for each f in $\mathbb{F}$.*

Proof. (i) The set $\mathscr{S}$, consisting of all families $\{A_h : h \in \mathbb{F}\}$ in which $A_h \in \mathfrak{A}_h$ and $\sup\|A_h\| < \infty$, is a C^*-algebra relative to the pointwise * algebra structure and the supremum norm. It has a norm-closed two-sided ideal $\mathscr{K}$ that consists of all elements $\{A_h\}$ of $\mathscr{S}$ for which the net $\{\|A_h\| : h \in \mathbb{F}\}$ converges to 0.

When $f \in \mathbb{F}$, we define a mapping $\theta_f : \mathfrak{A}_f \to \mathscr{S}$ as follows: when $A \in \mathfrak{A}_f$, $\theta_f(A)$ is the family $\{A_h : h \in \mathbb{F}\}$ for which

$$A_h = \begin{cases} \Phi_{hf}(A) & (h \geq f) \\ 0 & (\text{otherwise}). \end{cases}$$

Observe that θ_f is linear, multiplicative, and adjoint-preserving, and has the following additional properties:

(a) $\theta_f(A) \notin \mathscr{K}$ when A is a non-zero element of $\mathfrak{A}_f$;
(b) $\theta_f(A) - \theta_g(\Phi_{gf}(A)) \in \mathscr{K}$ whenever $f \leq g$ and $A \in \mathfrak{A}_f$;
(c) $\theta_f(I_f) - I \in \mathscr{K}$, where I is the unit $\{I_h : h \in \mathbb{F}\}$ of $\mathscr{S}$.

For (a), it suffices to note that, if $\theta_f(A) = \{A_h : h \in \mathbb{F}\}$, then $\|A_h\| = \|\Phi_{hf}(A)\| = \|A\|$, when $h \geq f$. For (b), observe that $\theta_f(A) - \theta_g(\Phi_{gf}(A))$ has zero "h coordinate" when $h \geq g$ (and hence lies in $\mathscr{K}$), because $\Phi_{hf} = \Phi_{hg} \circ \Phi_{gf}$. Finally, $\theta_f(I_f) - I$ has zero "h coordinate" when $h \geq f$, since $\Phi_{hf}(I_f) = I_h$.

From the preceding paragraph, the mapping $A \rightarrow \theta_f(A) + \mathscr{K}$ is a * isomorphism φ_f (preserving unit elements) from $\mathfrak{A}_f$ into the quotient C^*-algebra $\mathscr{S}/\mathscr{K}$, and $\varphi_f = \varphi_g \circ \Phi_{gf}$ when $f \leq g$. This last condition implies that $\varphi_f(\mathfrak{A}_f) \subseteq \varphi_g(\mathfrak{A}_g)$ when $f \leq g$; so that $\{\varphi_f(\mathfrak{A}_f) : f \in \mathbb{F}\}$ is a net of C^*-subalgebras of $\mathscr{S}/\mathscr{K}$, and the norm closure of $\bigcup \varphi_f(\mathfrak{A}_f)$ is itself a C^*-subalgebra $\mathfrak{A}$ of $\mathscr{S}/\mathscr{K}$. With this choice of $\mathfrak{A}$, all the conditions required in (i) are satisfied.

(ii) Under the conditions set out in (ii), $\psi_f \circ \varphi_f^{-1}$ is a * isomorphism from $\varphi_f(\mathfrak{A}_f)$ onto $\psi_f(\mathfrak{A}_f)$, for each f in $\mathbb{F}$. When $f \leq g$, $\psi_g \circ \varphi_g^{-1}$ extends $\psi_f \circ \varphi_f^{-1}$, since for every A in $\mathfrak{A}_f$,

$$\begin{aligned}(\psi_g \circ \varphi_g^{-1})(\varphi_f(A)) &= \psi_g(\Phi_{gf}(A)) \\ &= \psi_f(A) = (\psi_f \circ \varphi_f^{-1})(\varphi_f(A)).\end{aligned}$$

Accordingly, there is a mapping θ_0, from $\bigcup \varphi_f(\mathfrak{A}_f)$ onto $\bigcup \psi_f(\mathfrak{A}_f)$, such that $\theta_0 | \varphi_f(\mathfrak{A}_f) = \psi_f \circ \varphi_f^{-1}$ for each f. Moreover, θ_0 is linear, multiplicative, adjoint-preserving, and isometric (since the same is true of each $\psi_f \circ \varphi_f^{-1}$); and the domain and range of θ_0 are everywhere dense in $\mathscr{U}$ and $\mathscr{B}$, respectively. Thus θ_0 extends by continuity to a * isomorphism θ from $\mathfrak{A}$ onto $\mathscr{B}$; and $\psi_f = \theta \circ \varphi_f$ since $\theta | \varphi_f(\mathfrak{A}_f) = \psi_f \circ \varphi_f^{-1}$. ∎

The C^*-algebra $\mathfrak{A}$ occurring in Proposition 11.4.1 is called the *inductive limit* of the directed system $\{\mathfrak{A}_f : f \in \mathbb{F}\}$. We have noted during the proof that $\{\varphi_f(\mathfrak{A}_f) : f \in \mathbb{F}\}$ is a net of C^*-subalgebra of $\mathfrak{A}$, whose union is everywhere dense in $\mathfrak{A}$. Upon identifying $\mathfrak{A}_f$ with $\varphi_f(\mathfrak{A}_f)$ by means of φ_f, the * isomorphisms $\Phi_{gf} (f \leq g)$ become inclusion mappings, and the original directed system can be replaced by this net.

11.4.2. PROPOSITION. *If $\mathfrak{A}$ is the inductive limit of a directed system of simple C^*-algebras, then $\mathfrak{A}$ is simple.*

Proof. There is a net $\{\mathfrak{A}_f : f \in \mathbb{F}\}$ of simple C^*-subalgebras of $\mathfrak{A}$, with union everywhere dense in $\mathfrak{A}$. If $\mathscr{I}$ is a closed two-sided ideal in $\mathfrak{A}$, and $\mathscr{I} \neq \mathfrak{A}$, let $\varphi : \mathfrak{A} \rightarrow \mathfrak{A}/\mathscr{I}$ be the quotient map. Since $\mathfrak{A}_f$ is simple and $I \notin \mathscr{I}$, it follows that $\mathfrak{A}_f \cap \mathscr{I} = \{0\}$. Accordingly, the restriction

$$\varphi | \mathfrak{A}_f : \mathfrak{A}_f \rightarrow \mathfrak{A}/\mathscr{I}$$

is a * isomorphism, and is therefore isometric. From this, φ is isometric on $\bigcup \mathfrak{A}_f$ and hence, by continuity, on the whole of $\mathfrak{A}$; so $\mathscr{I} = \varphi^{-1}(0) = \{0\}$. ∎

We now define the tensor product of an (in general, infinite) family $\{\mathfrak{A}_a : a \in \mathbb{A}\}$ of C^*-algebras, as being the inductive limit of the (spatial) tensor products of finite subfamilies. The set $\mathbb{F}$ of all finite subsets F of $\mathbb{A}$ is directed by the inclusion relation $\subseteq$. With each F in $\mathbb{F}$, we can associate the spatial tensor product $\mathfrak{A}_F$ of the finite family $\{\mathfrak{A}_a : a \in F\}$. If $F, G \in \mathbb{F}$ and $F \subseteq G$, the associativity of the tensor product gives rise to a natural * isomorphism θ_{GF} from $\mathfrak{A}_F \otimes \mathfrak{A}_{G \setminus F}$ onto $\mathfrak{A}_G$; and the equation $\Phi_{GF}(A) = \theta_{GF}(A \otimes I)$ defines a * isomorphism from $\mathfrak{A}_F$ into $\mathfrak{A}_G$. By a further appeal to associativity of the tensor product, we deduce that $\Phi_{HF} = \Phi_{HG} \circ \Phi_{GF}$, when $F, G, H \in \mathbb{F}$ and $F \subseteq G \subseteq H$. Accordingly, the family $\{\mathfrak{A}_F : F \in \mathbb{F}\}$, with the * isomorphisms Φ_{GF} just described, is a directed system of C^*-algebras. The inductive limit of this system is a C^*-algebra $\mathfrak{A}$, called the *tensor product of the family* $\{\mathfrak{A}_a : a \in \mathbb{A}\}$, and denoted by $\bigotimes_{a \in \mathbb{A}} \mathfrak{A}_a$.

From Proposition 11.4.1, $\mathfrak{A}$ is characterized (up to * isomorphism) by the existence of a * isomorphism φ_F from $\mathfrak{A}_F$ onto a C^*-subalgebra $\mathfrak{A}(F)$ of $\mathfrak{A}$, for each F in $\mathbb{F}$, such that

$$\varphi_F = \varphi_G \circ \Phi_{GF} \qquad (F, G \in \mathbb{F}, \quad F \subseteq G),$$

and $\cup\{\mathfrak{A}(F) : F \in \mathbb{F}\}$ is everywhere dense in $\mathfrak{A}$. When F consists of a single element b of $\mathbb{A}$, we write φ_b in place of φ_F, $\mathfrak{A}(b)$ in place of $\mathfrak{A}(F)$, and refer to $\mathfrak{A}(b)$ as the *canonical image of* $\mathfrak{A}_b$ *in* $\mathfrak{A}(= \bigotimes_{a \in \mathbb{A}} \mathfrak{A}_a)$.

Our next objective is to give a characterization of the tensor product algebra, in terms of the properties of these canonical images $\mathfrak{A}(a)$ $(a \in \mathbb{A})$ of the component algebras $\mathfrak{A}_a$. For this, suppose that F is a finite subset of $\mathbb{A}$, with (distinct) elements $a(1), \ldots, a(n)$. Then $\mathfrak{A}_F = \mathfrak{A}_{a(1)} \otimes \cdots \otimes \mathfrak{A}_{a(n)}$; and, for $j = 1, \ldots, n$, the natural * isomorphism from $\mathfrak{A}_{a(j)}$ into $\mathfrak{A}_F$ is (in the notation used above) $\Phi_{F\{a(j)\}}$, corresponding to the elements $\{a(j)\}$ and F (with $\{a(j)\} \subseteq F$) of $\mathbb{F}$. Moreover, $\varphi_{a(1)}^{-1} \otimes \cdots \otimes \varphi_{a(n)}^{-1}$ is a * isomorphism from $\mathfrak{A}(a(1)) \otimes \cdots \otimes \mathfrak{A}(a(n))$ onto $\mathfrak{A}_F$, and thus $\varphi_F \circ (\varphi_{a(1)}^{-1} \otimes \cdots \otimes \varphi_{a(n)}^{-1})$ is a * isomorphism ψ_F from $\mathfrak{A}(a(1)) \otimes \cdots \otimes \mathfrak{A}(a(n))$ onto $\mathfrak{A}(F)$. If $A_j \in \mathfrak{A}(a(j))$ $(j = 1, \ldots, n)$, then $A_j = \varphi_{a(j)}(B_j)$ for some B_j in $\mathfrak{A}_{a(j)}$; and we have

$$A_j = \varphi_{a(j)}(B_j) = \varphi_F(\Phi_{F\{a(j)\}}(B_j)) = \varphi_F(I \otimes \cdots \otimes I \otimes B_j \otimes I \otimes \cdots \otimes I),$$

$$\begin{aligned} A_1 A_2 \cdots A_n &= \varphi_F(B_1 \otimes \cdots \otimes B_n) \\ &= \varphi_F(\varphi_{a(1)}^{-1}(A_1) \otimes \cdots \otimes \varphi_{a(n)}^{-1}(A_n)) \\ &= \varphi_F((\varphi_{a(1)}^{-1} \otimes \cdots \otimes \varphi_{a(n)}^{-1})(A_1 \otimes \cdots \otimes A_n)) = \psi_F(A_1 \otimes \cdots \otimes A_n). \end{aligned}$$

In particular, since the operators $I \otimes \cdots \otimes I \otimes A_j \otimes I \otimes \cdots \otimes I$ (with A_j in the jth position), $j = 1, \ldots, n$, all commute, their images $A_1, \ldots, A_n$ under ψ_F are commuting elements of $\mathfrak{A}$; so the canonical images $\mathfrak{A}(a)(a \in \mathbb{A})$ are *mutually commuting* C^*-subalgebras of $\mathfrak{A}$. Since $\mathfrak{A}(a(1)) \otimes \cdots \otimes \mathfrak{A}(a(n))$ is the closed linear span of the set of all simple tensors, the range $\mathfrak{A}(F)$ of ψ_F is the closed linear span of operator products $A_1A_2 \cdots A_n$, with A_j in $\mathfrak{A}(a(j))$ for $j = 1, \ldots, n$; in particular, $\mathfrak{A}(F)$ is the C^*-subalgebra of $\mathfrak{A}$ generated by $\mathfrak{A}(a(1)) \cup \cdots \cup \mathfrak{A}(a(n))$. From this, and since $\cup\, \mathfrak{A}(F)$ is everywhere dense in $\mathfrak{A}$, the C^*-algebra generated by $\cup\{\mathfrak{A}(a): a \in \mathbb{A}\}$ is the whole of $\mathfrak{A}$.

We show now that the properties just described can be used to characterize the tensor product $\bigotimes_{a \in \mathbb{A}} \mathfrak{A}_a$ up to * isomorphism.

11.4.3. PROPOSITION. *Suppose that $\mathfrak{A}$ and $\mathfrak{A}_a$ $(a \in \mathbb{A})$ are C^*-algebras. Then $\mathfrak{A}$ is (* isomorphic to) the tensor product algebra $\bigotimes_{a \in \mathbb{A}} \mathfrak{A}_a$ if and only if there is a family $\{\mathfrak{A}(a): a \in \mathbb{A}\}$ of (mutually commuting) C^*-subalgebras of $\mathfrak{A}$, such that*

(i) *$\mathfrak{A}$(a) contains the unit of $\mathfrak{A}$, and is * isomorphic to $\mathfrak{A}_a$;*
(ii) *the C^*-algebra generated by $\cup\{\mathfrak{A}(a): a \in \mathbb{A}\}$ is the whole of $\mathfrak{A}$;*
(iii) *if F is a finite set, consisting of distinct elements $a(1), \ldots, a(n)$ of $\mathbb{A}$, there is a * isomorphism ψ_F from $\mathfrak{A}(a(1)) \otimes \cdots \otimes \mathfrak{A}(a(n))$ into $\mathfrak{A}$, such that*

$$\psi_F(A_1 \otimes A_2 \otimes \cdots \otimes A_n) = A_1A_2, \ldots, A_n$$

whenever $A_1 \in \mathfrak{A}(a(1)), \ldots, A_n \in \mathfrak{A}(a(n))$.

When these conditions are satisfied, $\mathfrak{A}(a)$ is the canonical image of $\mathfrak{A}_a$ in $\mathfrak{A}$, and the range of ψ_F is the C^*-subalgebra $\mathfrak{A}(F)$ of $\mathfrak{A}$ generated by $\bigcup\{\mathfrak{A}(a): a \in F\}$. Moreover, $\mathfrak{A}(F) \subseteq \mathfrak{A}(G)$ when F, G are finite subsets of $\mathbb{A}$ and $F \subseteq G$; and $\bigcup \mathfrak{A}(F)$ is everywhere dense in $\mathfrak{A}$.

Proof. When $\mathfrak{A}$ is (* isomorphic to) $\bigotimes_{a \in \mathbb{A}} \mathfrak{A}_a$, the existence of a family $\{\mathfrak{A}(a): a \in \mathbb{A}\}$, with the stated properties, has already been established in the discussion preceding the proposition. Conversely, suppose that $\mathfrak{A}$ has such a family of C^*-subalgebras, and let $\mathbb{F}$ be the set of all finite subsets of $\mathbb{A}$. For each a in $\mathbb{A}$, let θ_a be a * isomorphism from $\mathfrak{A}_a$ onto $\mathfrak{A}(a)$; and, when $F = \{a(1), \ldots, a(n)\} \in \mathbb{F}$, let φ_F be the * isomorphism $\psi_F \circ (\theta_{a(1)} \otimes \cdots \otimes \theta_{a(n)})$ from $\mathfrak{A}_{a(1)} \otimes \cdots \otimes \mathfrak{A}_{a(n)}$ $(=\mathfrak{A}_F)$ into $\mathfrak{A}$. Then $\varphi_F(\mathfrak{A}_F)$ coincides with the range $\mathfrak{A}(F)$ of ψ_F, and it is apparent from condition (iii) that this is the C^*-subalgebra of $\mathfrak{A}$ that is generated by $\mathfrak{A}(a(1)) \cup \cdots \cup \mathfrak{A}(a(n))$ (in particular, $\varphi_F(\mathfrak{A}_F)$ is $\mathfrak{A}(a)$ when $F = \{a\}$). Accordingly, $\mathfrak{A}(F) \subseteq \mathfrak{A}(G)$ when $F \subseteq G$; and from condition (ii), $\bigcup\{\varphi_F(\mathfrak{A}_F): F \in \mathbb{F}\}$ is everywhere dense in $\mathfrak{A}$.

Suppose now that $G = \{a(1), \ldots, a(n), \ldots, a(m)\}$, a finite subset of $\mathbb{A}$ that contains F $(=\{a(1), \ldots, a(n)\})$; and let Φ_{GF} be the natural * isomorphism from $\mathfrak{A}_F$ into $\mathfrak{A}_G$. When $A_j \in \mathfrak{A}_{a(j)}$ $(j = 1, \ldots, n)$,

$$\begin{aligned}\varphi_G(\Phi_{GF}(A_1 \otimes \cdots \otimes A_n)) &= \varphi_G(A_1 \otimes \cdots \otimes A_n \otimes I \otimes \cdots \otimes I)\\ &= \psi_G(\theta_{a(1)}(A_1) \otimes \cdots \otimes \theta_{a(n)}(A_n) \otimes \theta_{a(n+1)}(I) \otimes \cdots \otimes \theta_{a(m)}(I))\\ &= \theta_{a(1)}(A_1)\theta_{a(2)}(A_2) \cdots \theta_{a(n)}(A_n)\theta_{a(n+1)}(I) \cdots \theta_{a(m)}(I)\\ &= \theta_{a(1)}(A_1)\theta_{a(2)}(A_2) \cdots \theta_{a(n)}(A_n) = \varphi_F(A_1 \otimes \cdots \otimes A_n);\end{aligned}$$

so $\varphi_G \circ \Phi_{GF} = \varphi_F$.

We have now shown that the * isomorphisms $\varphi_F : \mathfrak{A}_F \to \mathfrak{A}$ satisfy the conditions that, by Proposition 11.4.1, characterize the tensor product algebra $\bigotimes_{a \in \mathbb{A}} \mathfrak{A}_a$ up to * isomorphism. ■

11.4.4. Remark. Suppose that $\mathfrak{A}$ is a C^*-algebra, and is generated by a family $\{\mathfrak{A}(a) : a \in \mathbb{A}\}$ of mutually commuting C^*-subalgebras, each containing the unit of $\mathfrak{A}$. Intuitively, Proposition 11.4.3 asserts that $\mathfrak{A}$ can be identified with $\bigotimes_{a \in \mathbb{A}} \mathfrak{A}(a)$ if and only if, for each *finite* subset F of $\mathbb{A}$, the C^*-subalgebra of $\mathfrak{A}$ that is generated by the algebras $\mathfrak{A}(a)$ $(a \in F)$ can be naturally identified with their spatial tensor product. We shall see later (Proposition 11.4.9, Corollary 11.4.10) that, in certain cases, this condition is automatically satisfied. ■

We now consider certain * isomorphisms, and states, of infinite-tensor-product C^*-algebras. It is convenient to express the results in the terminology used in Proposition 11.4.3.

11.4.5. Proposition. *Suppose that* $\{\mathfrak{A}_a : a \in \mathbb{A}\}$ *and* $\{\mathscr{B}_a : a \in \mathbb{A}\}$ *are families of* C^*-*algebras,* $\mathfrak{A} = \bigotimes_{a \in \mathbb{A}} \mathfrak{A}_a$ *and* $\mathscr{B} = \bigotimes_{a \in \mathbb{A}} \mathscr{B}_a$. *Let* $\mathfrak{A}(a)$ *denote the canonical image of* $\mathfrak{A}_a$ *in* $\mathfrak{A}$, $\mathscr{B}(a)$ *that of* $\mathscr{B}_a$ *in* $\mathscr{B}$.

(i) *If* $\mathfrak{A}_a$ *is * isomorphic to* $\mathscr{B}_a$ *for each a in* $\mathbb{A}$, *then* $\mathfrak{A}$ *is * isomorphic to* $\mathscr{B}$.

(ii) *If* θ_a *is a * isomorphism from* $\mathfrak{A}(a)$ *onto* $\mathscr{B}(a)$ *for each a in* $\mathbb{A}$, *there is a * isomorphism* θ *from* $\mathfrak{A}$ *onto* $\mathscr{B}$, *such that* $\theta | \mathfrak{A}(a) = \theta_a (a \in \mathbb{A})$.

Proof. Part (i) is an immediate consequence of Proposition 11.4.3 (it can also be deduced at once from part (ii) of the present proposition).

It is not difficult to prove (ii) by using the inductive limit structure and reasoning as in the proof of Proposition 11.4.1. However, an argument based on Proposition 11.4.3 seems more intuitive. For each finite subset $F = \{a(1), \ldots, a(n)\}$ of $\mathbb{A}$, there are * isomorphisms

$$\psi_F : \mathfrak{A}(a(1)) \otimes \cdots \otimes \mathfrak{A}(a(n)) \to \mathfrak{A},$$

$$\chi_F : \mathscr{B}(a(1)) \otimes \cdots \otimes \mathscr{B}(a(n)) \to \mathscr{B}$$

with the properties set out in Proposition 11.4.3. The range $\mathfrak{A}(F)$ of ψ_F is the C^*-subalgebra of $\mathfrak{A}$ that is generated by $\bigcup\{\mathfrak{A}(a): a \in F\}$; and a similar remark applies to the range $\mathscr{B}(F)$ of χ_F.

The mapping $\chi_F \circ (\theta_{a(1)} \otimes \cdots \otimes \theta_{a(n)}) \circ \psi_F^{-1}$ is a * isomorphism θ_F from $\mathfrak{A}(F)$ onto $\mathscr{B}(F)$, and

$$\theta_F(A_1 A_2 \cdots A_n) = \theta_{a(1)}(A_1)\theta_{a(2)}(A_2) \cdots \theta_{a(n)}(A_n)$$

whenever $A_1 \in \mathfrak{A}(a(1)), \ldots, A_n \in \mathfrak{A}(a(n))$. Upon replacing all but one of $A_1, \ldots, A_n$ by I, we deduce that $\theta_F \mid \mathfrak{A}(a) = \theta_a$ for each a in F. If $F \subseteq G$, then $\mathfrak{A}(F) \subseteq \mathfrak{A}(G)$ and $\theta_F \mid \mathfrak{A}(a) = \theta_G \mid \mathfrak{A}(a) = \theta_a (a \in F)$; since $\mathfrak{A}(F)$ is generated by $\bigcup\{\mathfrak{A}(a): a \in F\}$, it now follows that $\theta_F = \theta_G \mid \mathfrak{A}(F)$.

From the preceding paragraph, there is a mapping θ_0, from $\bigcup \mathfrak{A}(F)$ onto $\bigcup \mathscr{B}(F)$, such that $\theta_0 \mid \mathfrak{A}(F) = \theta_F$ for each finite subset F of $\mathbb{A}$; in particular, $\theta_0 \mid \mathfrak{A}(a) = \theta_a$, when $a \in \mathbb{A}$. Moreover, θ_0 is linear, multiplicative, adjoint-preserving, and isometric, since each θ_F has these properties. Accordingly, θ_0 extends by continuity to a * isomorphism θ from $\mathfrak{A}$ onto $\mathscr{B}$. ■

The isomorphism θ occurring in Proposition 11.4.5(ii) is denoted by $\bigotimes_{a \in \mathbb{A}} \theta_a$.

11.4.6. PROPOSITION. *Suppose that* $\{\mathfrak{A}_a : a \in \mathbb{A}\}$ *is a family of* C^*-*algebras,* $\mathfrak{A} = \bigotimes_{a \in \mathbb{A}} \mathfrak{A}_a$, *and, for each* a *in* $\mathbb{A}$, ρ_a *is a state of* $\mathfrak{A}(a)$, *the canonical image of* $\mathfrak{A}_a$ *in* $\mathfrak{A}$. *Then there is a unique state* ρ *of* $\mathfrak{A}$ *such that*

$$\rho(A_1 A_2 \cdots A_n) = \rho_{a(1)}(A_1)\rho_{a(2)}(A_2) \cdots \rho_{a(n)}(A_n)$$

whenever $a(1), \ldots, a(n)$ *are distinct elements of* $\mathbb{A}$ *and* $A_j \in \mathfrak{A}(a(j))$ $(j = 1, \ldots, n)$.

Proof. We shall use the notation occurring in the statement of Proposition 11.4.3. For each finite subset $F = \{a(1), \ldots, a(n)\}$ of $\mathbb{A}$, we can form the product state $\rho_{a(1)} \otimes \cdots \otimes \rho_{a(n)}$ of $\mathfrak{A}(a(1)) \otimes \cdots \otimes \mathfrak{A}(a(n))$. The equation

$$\rho_F = (\rho_{a(1)} \otimes \cdots \otimes \rho_{a(n)}) \circ \psi_F^{-1}$$

defines a state ρ_F of $\mathfrak{A}(F)$, and

$$\rho_F(A_1 A_2 \cdots A_n) = \rho_{a(1)}(A_1)\rho_{a(2)}(A_2) \cdots \rho_{a(n)}(A_n)$$

whenever $A_1 \in \mathfrak{A}(a(1)), \ldots, A_n \in \mathfrak{A}(a(n))$. Note that ρ_F is the only state of $\mathfrak{A}(F)$ with this property, since operator products $A_1 A_2 \cdots A_n$ of the type just considered generate $\mathfrak{A}(F)$ as a norm-closed linear subspace of $\mathfrak{A}$. Accordingly, the condition required of the state ρ of $\mathfrak{A}$ is that $\rho \mid \mathfrak{A}(F) = \rho_F$ for each F; and since $\bigcup \mathfrak{A}(F)$ is everywhere dense in $\mathfrak{A}$, there is at most one such ρ.

By taking $A_{n+1} = \cdots = A_m = I$, it follows that

$$\rho_G(A_1 A_2 \cdots A_n) = \rho_F(A_1 A_2 \cdots A_n),$$

whenever G is a subset $\{a(1), \ldots, a(n), \ldots, a(m)\}$ of $\mathbb{A}$ that contains $F(=\{a(1), \ldots, a(n)\})$, and $A_j \in \mathfrak{A}(a(j))$ $(j = 1, \ldots, n)$; so $\rho_G \mid \mathfrak{A}(F) = \rho_F$. From this, there is a complex-valued function ρ_0, defined throughout $\bigcup \mathfrak{A}(F)$, such that $\rho_0 \mid \mathfrak{A}(F) = \rho_F$ for each F. Moreover, ρ_0 is a bounded linear functional, and $\|\rho_0\| = \rho_0(I) = 1$, since each ρ_F has these properties. Thus ρ_0 extends, by continuity, to a state ρ of $\mathfrak{A}$; and $\rho \mid \mathfrak{A}(F) = \rho_F$ for each F. ∎

The state ρ occurring in Proposition 11.4.6 is denoted by $\bigotimes_{a \in \mathbb{A}} \rho_a$; and such states are described as *product states* of $\bigotimes_{a \in \mathbb{A}} \mathfrak{A}_a$. Given a product state ρ, the component states ρ_a are uniquely determined, since $\rho_a = \rho \mid \mathfrak{A}(a)$.

11.4.7. PROPOSITION. *Under the conditions set out in Proposition* 11.4.6, *the product state* $\rho = \bigotimes_{a \in \mathbb{A}} \rho_a$ *of* $(\mathfrak{A} =)\ \bigotimes_{a \in \mathbb{A}} \mathfrak{A}_a$ *is pure if and only if each* ρ_a *is pure, and is tracial if and only if each* ρ_a *is tracial.*

Proof. If ρ_b is not pure, for some b in $\mathbb{A}$, it is a convex combination of two different states, $\rho_b^{(1)}$ and $\rho_b^{(2)}$, of $\mathfrak{A}(b)$. Thus $\bigotimes_{a \in \mathbb{A}} \rho_a$ is not pure, since it is a convex combination of the distinct product states obtained upon replacing ρ_b by $\rho_b^{(j)} (j = 1, 2)$.

Suppose next that each ρ_a is pure. In proving Proposition 11.4.6, it has been noted that $\rho \mid \mathfrak{A}(F) = \rho_F$, for each finite subset $F = \{a(1), \ldots, a(n)\}$ of $\mathbb{A}$, where ρ_F is the state $(\rho_{a(1)} \otimes \cdots \otimes \rho_{a(n)}) \circ \psi_F^{-1}$ of $\mathfrak{A}(F)$. From Proposition 11.3.2(ii), each ρ_F is pure. If $\rho = s\sigma_1 + (1 - s)\sigma_2$, where $0 < s < 1$ and σ_1, σ_2 are states of $\mathfrak{A}$, restriction to $\mathfrak{A}(F)$ yields

$$\rho_F = s(\sigma_1 \mid \mathfrak{A}(F)) + (1 - s)(\sigma_2 \mid \mathfrak{A}(F)).$$

Since ρ_F is pure, $\sigma_1 \mid \mathfrak{A}(F) = \sigma_2 \mid \mathfrak{A}(F) = \rho_F = \rho \mid \mathfrak{A}(F)$. Accordingly, σ_1, σ_2, and ρ coincide on the everywhere-dense subset $\bigcup \mathfrak{A}(F)$ of $\mathfrak{A}$; whence $\sigma_1 = \sigma_2 = \rho$, and ρ is pure.

If ρ is tracial, restriction to $\mathfrak{A}(a)$ shows that ρ_a is tracial for each a in $\mathbb{A}$. Conversely, if each ρ_a is tracial, then each ρ_F is tracial (Proposition 11.3.2(iii)), and thus $\rho(AB) = \rho(BA)$ for all A and B in $\bigcup \mathfrak{A}(F)$; and by continuity, this remains true for all A and B in $\mathfrak{A}$. ∎

In establishing that certain C^*-algebras can be viewed as infinite tensor products, we shall need the following auxiliary result.

11.4.8. LEMMA. *If* $\mathcal{M}$ *is a type* I *factor acting on a Hilbert space* $\mathcal{H}$, $\mathcal{B}$ *is a* C^*-*subalgebra of* $\mathcal{M}'$, *and* $\mathfrak{A}$ *is the* C^*-*algebra generated by* $\mathcal{M} \cup \mathcal{B}$, *there is a* * *isomorphism* φ *from* $\mathcal{M} \otimes \mathcal{B}$ *onto* $\mathfrak{A}$, *such that* $\varphi(A \otimes B) = AB$ $(A \in \mathcal{M}, B \in \mathcal{B})$.

Proof. From Examples 11.2.5 and 11.2.1, we can choose Hilbert spaces $\mathscr{K}$, $\mathscr{L}$, and a unitary operator U from $\mathscr{H}$ onto $\mathscr{K} \otimes \mathscr{L}$, so that $U\mathscr{M}U^* = \mathscr{B}(\mathscr{K}) \mathbin{\overline{\otimes}} \mathbb{C}_{\mathscr{L}} = \mathscr{B}(\mathscr{K}) \otimes \mathbb{C}_{\mathscr{L}}$. Since $U\mathscr{B}U^* \subseteq (U\mathscr{M}U^*)' = \mathbb{C}_{\mathscr{K}} \otimes \mathscr{B}(\mathscr{L})$, there is a C^*-algebra $\mathscr{B}_0$ acting on $\mathscr{L}$, such that $U\mathscr{B}U^* = \mathbb{C}_{\mathscr{K}} \otimes \mathscr{B}_0$. Thus $U\mathfrak{A}U^*$, the C^*-algebra generated by $U\mathscr{M}U^* \cup U\mathscr{B}U^*$, is $\mathscr{B}(\mathscr{K}) \otimes \mathscr{B}_0$. We therefore have * isomorphisms

$$\psi: A \to UAU^*: \mathfrak{A} \to \mathscr{B}(\mathscr{K}) \otimes \mathscr{B}_0,$$

$$\theta_1: S \to U^*(S \otimes I)U: \mathscr{B}(\mathscr{K}) \to \mathscr{M},$$

$$\theta_2: T \to U^*(I \otimes T)U: \mathscr{B}_0 \to \mathscr{B}.$$

When $A \in \mathscr{M}$ and $B \in \mathscr{B}$,

$$UAU^* = U(\theta_1 \circ \theta_1^{-1})(A)U^* = \theta_1^{-1}(A) \otimes I, \qquad UBU^* = I \otimes \theta_2^{-1}(B);$$

so $\psi(AB) = UABU^* = \theta_1^{-1}(A) \otimes \theta_2^{-1}(B)$, and

$$(\theta_1 \otimes \theta_2)(\psi(AB)) = A \otimes B.$$

It now suffices to take, for φ, the inverse of the * isomorphism

$$(\theta_1 \otimes \theta_2) \circ \psi: \mathfrak{A} \to \mathscr{M} \otimes \mathscr{B}. \quad \blacksquare$$

11.4.9. PROPOSITION. *Suppose that $\mathscr{H}$ is a Hilbert space, $\{\mathscr{M}_a: a \in \mathbb{A}\}$ is a family of mutually commuting type* I *subfactors of $\mathscr{B}(\mathscr{H})$ (each containing the unit of $\mathscr{B}(\mathscr{H})$) and $\mathfrak{A}$ is the C^*-algebra generated by $\cup \mathscr{M}_a$. Then there is a * isomorphism from $\mathfrak{A}$ onto $\bigotimes_{a \in \mathbb{A}} \mathscr{M}_a$ that carries each $\mathscr{M}_a$ onto its canonical image in $\bigotimes_{a \in \mathbb{A}} \mathscr{M}_a$.*

Proof. We shall use Proposition 11.4.3, where $\mathfrak{A}_a = \mathfrak{A}(a) = \mathscr{M}_a$. It is apparent that conditions (i) and (ii) in that proposition are satisfied; so it suffices to verify (iii).

Suppose that $a(1), \ldots, a(n)$ are distinct elements of $\mathbb{A}$. We have to establish the existence of a * isomorphism ψ, from $\mathscr{M}_{a(1)} \otimes \cdots \otimes \mathscr{M}_{a(n)}$ into $\mathfrak{A}$, such that

$$\psi(A_1 \otimes A_2 \otimes \cdots \otimes A_n) = A_1 A_2 \cdots A_n$$

when $A_j \in \mathscr{M}_{a(j)}$ $(j = 1, \ldots, n)$. We prove this by induction on n, noting first that the result is evident when $n = 1$. If it has been proved for sets of $n - 1$ elements of $\mathbb{A}$, there is a * isomorphism ψ_0, from $\mathscr{M}_{a(2)} \otimes \cdots \otimes \mathscr{M}_{a(n)}$ into $\mathfrak{A}$, such that

$$\psi_0(A_2 \otimes A_3 \otimes \cdots \otimes A_n) = A_2 A_3 \cdots A_n$$

when $A_j \in \mathscr{M}_{a(j)}$ $(j = 2, \ldots, n)$. The range of ψ_0 is the C^*-algebra $\mathscr{B}$ generated by $\mathscr{M}_{a(2)} \cup \cdots \cup \mathscr{M}_{a(n)}$, and is contained in $\mathscr{M}'_{a(1)}$. Since $\mathscr{M}_{a(1)} \cup \mathscr{B} \subseteq \mathfrak{A}$, it now follows from Lemma 11.4.8 that there is a * isomorphism φ, from $\mathscr{M}_{a(1)} \otimes \mathscr{B}$ into $\mathfrak{A}$, such that $\varphi(A \otimes B) = AB$ when $A \in \mathscr{M}_{a(1)}$ and $B \in \mathscr{B}$. If we identify $\mathscr{M}_{a(1)} \otimes \cdots \otimes \mathscr{M}_{a(n)}$ in the usual way with

$$\mathscr{M}_{a(1)} \otimes (\mathscr{M}_{a(2)} \otimes \cdots \otimes \mathscr{M}_{a(n)}),$$

and write ι for the identity mapping on $\mathscr{M}_{a(1)}$, the * isomorphism

$$\psi = \varphi \circ (\iota \otimes \psi_0): \mathscr{M}_{a(1)} \otimes \cdots \otimes \mathscr{M}_{a(n)} \to \mathfrak{A}$$

has the required property. ■

In Section 10.4, *Uniformly matricial algebras*, we used the term "factor of type I_k," where k is a positive integer, to describe any C^*-algebra * isomorphic to the algebra $M_k(\mathbb{C})$ of all $k \times k$ complex matrices. Without specifying k, we shall refer to such algebras as "finite type I factors." Of course, they are weak-operator closed (and so become finite type I factors in the von Neumann algebra sense) in every representation.

11.4.10. Corollary. *If a C^*-algebra $\mathfrak{A}$ is generated by a family $\{\mathscr{M}_a : a \in \mathbb{A}\}$ of mutually commuting C^*-subalgebras, and each $\mathscr{M}_a$ is a finite type* I *factor (that contains the unit of $\mathfrak{A}$), there is a * isomorphism from $\mathfrak{A}$ onto $\bigotimes_{a \in \mathbb{A}} \mathscr{M}_a$ that carries each $\mathscr{M}_a$ onto its canonical image in $\bigotimes_{a \in \mathbb{A}} \mathscr{M}_a$.*

Proof. By taking a faithful representation of $\mathfrak{A}$, we reduce to a situation already covered by Proposition 11.4.9. ■

We conclude this section by considering states ρ, and the corresponding representations π_ρ, of a C^*-algebra $\mathfrak{A}$ of the form $\bigotimes_{a \in \mathbb{A}} \mathscr{M}_a$, where each $\mathscr{M}_a$ is a finite type I factor. The main result (Theorem 11.4.15) is a characterization of the primary states of such an algebra. For this, we require some auxiliary results.

11.4.11. Lemma. *Suppose that $\mathscr{M}$ is a type* I *factor acting on a Hilbert space $\mathscr{H}$.*

(i) *If $\mathscr{R}$ is a von Neumann algebra, $\mathscr{M} \subseteq \mathscr{R} \subseteq \mathscr{B}(\mathscr{H})$, and $\mathscr{N} = \mathscr{M}' \cap \mathscr{R}$, then $\mathscr{R}$ is generated (as a von Neumann algebra) by $\mathscr{M} \cup \mathscr{N}$. If $\omega_{\mathscr{M}}$ and $\omega_{\mathscr{N}}$ are normal states of $\mathscr{M}$ and $\mathscr{N}$, respectively, there is a unique normal state ω of $\mathscr{R}$ such that $\omega(MN) = \omega_{\mathscr{M}}(M)\omega_{\mathscr{N}}(N)$ for all M in $\mathscr{M}$ and N in $\mathscr{N}$.*

(ii) *Suppose that $\mathscr{S}$, $\mathscr{S}_a (a \in \mathbb{A})$ are von Neumann subalgebras of $\mathscr{M}'$, and $\mathscr{S} = \bigcap \mathscr{S}_a$. If $\mathscr{R}$, $\mathscr{R}_a (a \in \mathbb{A})$ are the von Neumann algebras generated by $\mathscr{M} \cup \mathscr{S}$ and $\mathscr{M} \cup \mathscr{S}_a$, respectively, then $\mathscr{R} = \bigcap \mathscr{R}_a$.*

Proof. In view of Example 11.2.5, we may assume that $\mathscr{H} = \mathscr{K} \otimes \mathscr{L}$ and $\mathscr{M} = \mathscr{B}(\mathscr{K}) \mathbin{\bar{\otimes}} \mathbb{C}_{\mathscr{L}}$.

(i) Since $\mathscr{R}' \subseteq \mathscr{M}' = \mathbb{C}_{\mathscr{K}} \mathbin{\bar{\otimes}} \mathscr{B}(\mathscr{L})$, it follows that $\mathscr{R}' = \mathbb{C}_{\mathscr{K}} \mathbin{\bar{\otimes}} \mathscr{T}$, for some von Neumann algebra $\mathscr{T}$ acting on $\mathscr{L}$; so $\mathscr{R} = \mathscr{R}'' = \mathscr{B}(\mathscr{K}) \mathbin{\bar{\otimes}} \mathscr{T}'$. Since $\mathscr{N} = \mathscr{M}' \cap \mathscr{R}$, $\mathscr{N}'$ is the von Neumann algebra generated by

$$\mathscr{M} \cup \mathscr{R}' = (\mathscr{B}(\mathscr{K}) \mathbin{\bar{\otimes}} \mathbb{C}_{\mathscr{L}}) \cup (\mathbb{C}_{\mathscr{K}} \mathbin{\bar{\otimes}} \mathscr{T});$$

so $\mathscr{N}' = \mathscr{B}(\mathscr{K}) \mathbin{\bar{\otimes}} \mathscr{T}$, and $\mathscr{N} = \mathscr{N}'' = \mathbb{C}_{\mathscr{K}} \mathbin{\bar{\otimes}} \mathscr{T}'$. Thus $\mathscr{R}$ $(=\mathscr{B}(\mathscr{K}) \mathbin{\bar{\otimes}} \mathscr{T}')$ is generated as a von Neumann algebra by $\mathscr{M}$ $(=\mathscr{B}(\mathscr{K}) \mathbin{\bar{\otimes}} \mathbb{C}_{\mathscr{L}})$ and $\mathscr{N}$ $(= \mathbb{C}_{\mathscr{K}} \mathbin{\bar{\otimes}} \mathscr{T}')$.

Given normal states $\omega_{\mathscr{M}}$ of $\mathscr{M}$ and $\omega_{\mathscr{N}}$ of $\mathscr{N}$, we can define normal states ω_1 of $\mathscr{B}(\mathscr{K})$ and ω_2 of $\mathscr{T}'$ by $\omega_1(B) = \omega_{\mathscr{M}}(B \otimes I)$, $\omega_2(T') = \omega_{\mathscr{N}}(I \otimes T')$. We require a normal state ω of $\mathscr{R}$ $(=\mathscr{B}(\mathscr{K}) \mathbin{\bar{\otimes}} \mathscr{T}')$, such that $\omega(B \otimes T') = \omega_1(B)\omega_2(T')$ for all B in $\mathscr{B}(\mathscr{K})$ and T' in $\mathscr{T}'$. There is just one such state ω (Proposition 11.2.7), the normal product state $\omega_1 \mathbin{\bar{\otimes}} \omega_2$.

(ii) Since $\mathscr{S}$, $\mathscr{S}_a \subseteq \mathscr{M}' = \mathbb{C}_{\mathscr{K}} \mathbin{\bar{\otimes}} \mathscr{B}(\mathscr{L})$, there are von Neumann algebras $\mathscr{T}$, $\mathscr{T}_a (a \in \mathbb{A})$, acting on $\mathscr{L}$, such that $\mathscr{S} = \mathbb{C}_{\mathscr{K}} \mathbin{\bar{\otimes}} \mathscr{T}$, $\mathscr{S}_a = \mathbb{C}_{\mathscr{K}} \mathbin{\bar{\otimes}} \mathscr{T}_a$. Moreover, $\mathscr{T} = \bigcap \mathscr{T}_a$, $\mathscr{R} = \mathscr{B}(\mathscr{K}) \mathbin{\bar{\otimes}} \mathscr{T}$, and $\mathscr{R}_a = \mathscr{B}(\mathscr{K}) \mathbin{\bar{\otimes}} \mathscr{T}_a$. The commutant of $\bigcap \mathscr{R}_a$ is the von Neumann algebra generated by

$$\bigcup \mathscr{R}'_a = \bigcup \mathbb{C}_{\mathscr{K}} \mathbin{\bar{\otimes}} \mathscr{T}'_a = \{I \otimes T' : T \in \bigcup \mathscr{T}'_a\};$$

and the von Neumann algebra generated by $\bigcup \mathscr{T}'_a$ is $\mathscr{T}'$, since $\mathscr{T} = \bigcap \mathscr{T}_a$. Thus $(\bigcap \mathscr{R}_a)' = \mathbb{C}_{\mathscr{K}} \mathbin{\bar{\otimes}} \mathscr{T}'$, and so $\bigcap \mathscr{R}_a = \mathscr{B}(\mathscr{K}) \mathbin{\bar{\otimes}} \mathscr{T} = \mathscr{R}$. ∎

11.4.12. LEMMA. *Suppose that $\{\mathscr{R}_a : a \in \mathbb{A}\}$ is a family of von Neumann algebras acting on a Hilbert space $\mathscr{H}$, in which the index set $\mathbb{A}$ is directed by a partial ordering $\leq$, and $\mathscr{R}_b \subseteq \mathscr{R}_a$ when $a \leq b$. If $\mathscr{R}$ is the von Neumann algebra generated by $\bigcup \mathscr{R}_a$ and ω is an ultraweakly continuous linear functional on $\mathscr{R}$ that vanishes on $\bigcap \mathscr{R}_a$, then* $\lim_a \|\omega \mid \mathscr{R}_a\| = 0$. *In particular, if ω_1 and ω_2 are normal states of $\mathscr{R}$ and $\bigcap \mathscr{R}_a$ consists of scalars, then* $\lim_a \|(\omega_1 - \omega_2) | \mathscr{R}_a\| = 0$.

Proof. When $a, b \in \mathbb{A}$ and $a \leq b$, we have $0 \leq \|\omega \mid \mathscr{R}_b\| \leq \|\omega \mid \mathscr{R}_a\|$, since $\mathscr{R}_b \subseteq \mathscr{R}_a$; so the net $\{\|\omega | \mathscr{R}_a\| : a \in \mathbb{A}\}$ converges, with its greatest lower bound l as its limit.

If $l > 0$, the set $X_a = \{A \in \mathscr{R}_a : \|A\| \leq 1, |\omega(A)| \geq \frac{1}{2}l\}$ is non-empty and ultraweakly compact. The family $\{X_a : a \in \mathbb{A}\}$ has the finite intersection property, since $X_b \subseteq X_a$ when $a \leq b$, and so has non-void intersection. With A in $\bigcap X_a$, we have $A \in \cap \mathscr{R}_a$ and $|\omega(A)| \geq \frac{1}{2}l > 0$, contradicting our assumption that $\omega \mid \bigcap \mathscr{R}_a = 0$. Hence $l = 0$. ∎

Suppose that $\mathscr{M}$ is a type I factor acting on a Hilbert space $\mathscr{H}$, $\mathscr{R}$ is a von Neumann algebra such that $\mathscr{M} \subseteq \mathscr{R} \subseteq \mathscr{B}(\mathscr{H})$, ω is a normal state of $\mathscr{R}$, and

$\mathcal{N} = \mathcal{M}' \cap \mathcal{R}$. From Lemma 11.4.11(i), with $\omega_{\mathcal{M}} = \omega|\mathcal{M}$ and $\omega_{\mathcal{N}} = \omega|\mathcal{N}$, there is a unique normal state $\omega^\times$ of $\mathcal{R}$, such that $\omega^\times(MN) = \omega(M)\omega(N)$ whenever $M \in \mathcal{M}$ and $N \in \mathcal{N}$. We call $\omega^\times$ the *normal factorization of* ω *relative to* $\mathcal{M}$.

11.4.13. PROPOSITION. *Let $\mathfrak{A}$ be a C*-algebra acting on a Hilbert space $\mathscr{H}$ with a unit cyclic vector x. Suppose that there is a net $\{\mathcal{M}_a : a \in \mathbb{A}\}$ of type I subfactors of $\mathscr{B}(\mathscr{H})$, such that $\bigcup \mathcal{M}_a$ has norm closure $\mathfrak{A}$. When a, $b \in \mathbb{A}$ and $b \le a$ (so $\mathcal{M}_b \subseteq \mathcal{M}_a$), let $\omega_b^\times$ denote the normal factorization of $\omega_x|\mathfrak{A}^-$ relative to $\mathcal{M}_b$, and let $\mathcal{R}_{ab}$ be the von Neumann algebra generated by $\mathcal{M}_b \cup (\mathcal{M}_a' \cap \mathfrak{A}^-)$. Then $\mathfrak{A}^-$ is a factor if and only if*

$$\lim_a \|(\omega_x - \omega_b^\times)\,|\mathcal{R}_{ab}\| = 0$$

for each b in $\mathbb{A}$.

Proof. Given b in $\mathbb{A}$, $\mathfrak{A}^-$ is the von Neumann algebra generated by $\cup\{\mathcal{M}_a : a \ge b\}$, so $\mathfrak{A}' = \cap\{\mathcal{M}_a' : a \ge b\}$. Accordingly, the center $\mathscr{C}$ of $\mathfrak{A}^-$ is given by

$$\mathscr{C} = \mathfrak{A}' \cap \mathfrak{A}^- = \cap\{(\mathcal{M}_a' \cap \mathfrak{A}^-) : a \ge b\}.$$

Since $\mathcal{M}_a' \cap \mathfrak{A}^- \subseteq \mathcal{M}_b'$ when $a \ge b$, it now follows from Lemma 11.4.11(ii) that $\bigcap_a\{\mathcal{R}_{ab} : a \ge b\}$ is the von Neumann algebra generated by $\mathcal{M}_b \cup \mathscr{C}$.

If $\mathfrak{A}^-$ is a factor, $\mathscr{C}$ consists of scalars, and the preceding paragraph shows that $\bigcap_a\{\mathcal{R}_{ab} : a \ge b\} = \mathcal{M}_b$. Since ω_x and $\omega_b^\times$ coincide on $\mathcal{M}_b$,

$$\lim_a \|(\omega_x - \omega_b^\times)\,|\mathcal{R}_{ab}\| = 0$$

by Lemma 11.4.12.

Conversely, suppose that $\lim_a \|(\omega_x - \omega_b^\times)|\mathcal{R}_{ab}\| = 0$ for each b in $\mathbb{A}$. If $C \in \mathscr{C}$, then $C \in \mathcal{R}_{ab}$ whenever a, $b \in \mathbb{A}$ and $a \ge b$. When $A \in \mathcal{M}_b$ $(\subseteq \mathcal{R}_{ab})$, as $C \in \mathcal{M}_b' \cap \mathfrak{A}^-$, we have

$$\begin{aligned}|\omega_x(AC) - \omega_x(A)\omega_x(C)| &= |(\omega_x - \omega_b^\times)(AC)| \\ &\le \|(\omega_x - \omega_b^\times)|\mathcal{R}_{ab}\|\,\|AC\| \underset{a}{\to} 0;\end{aligned}$$

so $\omega_x(AC) = \omega_x(A)\omega_x(C)$ $(A \in \mathcal{M}_b; b \in \mathbb{A})$. Since $\bigcup \mathcal{M}_b$ is ultraweakly dense in $\mathfrak{A}^-$, the same equation is satisfied for all A in $\mathfrak{A}^-$. With $A = [C - \omega_x(C)I]^*$, we obtain

$$\begin{aligned}\|[C - \omega_x(C)I]x\|^2 &= \omega_x(A[C - \omega_x(C)I]) \\ &= \omega_x(AC) - \omega_x(A)\omega_x(C) = 0.\end{aligned}$$

Since x is a generating vector for $\mathfrak{A}^-$, it is separating for $\mathscr{C}$ $(\subseteq \mathfrak{A}')$; so $C = \omega_x(C)I$, and $\mathfrak{A}^-$ is a factor. ■

Proposition 11.4.13 can be viewed as providing a criterion for deciding whether the state $\rho = \omega_x \mid \mathfrak{A}$ of $\mathfrak{A}$ is primary; for the given representation of $\mathfrak{A}$ acting on $\mathscr{H}$ is unitarily equivalent to the representation engendered by ρ. However, the criterion is expressed in terms of $\bar{\rho}\ (= \omega_x | \mathfrak{A}^-)$, the extension of ρ to a normal state of $\mathfrak{A}^-$. When the $\mathscr{M}_a$ are *finite* type I factors, the conditions can be reformulated in terms of ρ itself. For this, we need the following preliminary result.

11.4.14. LEMMA. *If $\mathscr{M}$ is a finite type* I *factor acting on a Hilbert space $\mathscr{H}$, $\mathscr{B}$ is a C^*-subalgebra of $\mathscr{M}'$, and $\mathfrak{A}$ is the C^*-algebra generated by $\mathscr{M} \cup \mathscr{B}$, then*

$$\mathscr{B} = \mathscr{M}' \cap \mathfrak{A}, \qquad \mathscr{B}^- = \mathscr{M}' \cap \mathfrak{A}^-.$$

Proof. From Examples 11.2.5 and 11.2.1, we may assume that $\mathscr{H} = \mathscr{K} \otimes \mathscr{L}$ and $\mathscr{M} = \mathscr{B}(\mathscr{K})\, \bar{\otimes}\, \mathbb{C}_{\mathscr{L}} = \mathscr{B}(\mathscr{K}) \otimes \mathbb{C}_{\mathscr{L}}$, where $\mathscr{K}$ and $\mathscr{L}$ are Hilbert spaces and $\dim \mathscr{K} < \infty$. Since $\mathscr{B} \subseteq \mathscr{M}' = \mathbb{C}_{\mathscr{K}} \otimes \mathscr{B}(\mathscr{L})$, there is a C^*-algebra $\mathscr{B}_0$ acting on $\mathscr{L}$, such that $\mathscr{B} = \mathbb{C}_{\mathscr{K}} \otimes \mathscr{B}_0$; and $\mathscr{B}^- = \mathbb{C}_{\mathscr{K}} \otimes \mathscr{B}_0^-$. Since $\mathfrak{A}$ is the C^*-algebra generated by $\mathscr{M}\ (=\mathscr{B}(\mathscr{K}) \otimes \mathbb{C}_{\mathscr{L}})$ and $\mathscr{B}\ (=\mathbb{C}_{\mathscr{K}} \otimes \mathscr{B}_0)$, it follows that $\mathfrak{A} = \mathscr{B}(\mathscr{K}) \otimes \mathscr{B}_0$. Moreover, $\mathfrak{A}^- = \mathscr{B}(\mathscr{K}) \otimes \mathscr{B}_0^-$, since this (represented C^*-algebra) tensor product is already weak-operator closed (Example 11.1.5).

We assert that, for an arbitrary C^*-algebra $\mathscr{S}$ acting on $\mathscr{L}$,

$$\mathbb{C}_{\mathscr{K}} \otimes \mathscr{S} = (\mathbb{C}_{\mathscr{K}} \otimes \mathscr{B}(\mathscr{L})) \cap (\mathscr{B}(\mathscr{K}) \otimes \mathscr{S})$$

(and by using this relation, with $\mathscr{B}_0, \mathscr{B}_0^-$ in place of $\mathscr{S}$, we obtain the required conclusion). For this, we make use of the representations of these tensor products as matrix algebras (Examples 11.1.4 and 11.1.5). With $k = \dim \mathscr{K}$, $\mathbb{C}_{\mathscr{K}} \otimes \mathscr{B}(\mathscr{L})$ is the algebra of all $k \times k$ diagonal matrices with the same element of $\mathscr{B}(\mathscr{L})$ in each diagonal position, while $\mathscr{B}(\mathscr{K}) \otimes \mathscr{S}$ is the algebra of all $k \times k$ matrices with entries in $\mathscr{S}$. Their intersection is therefore the algebra of $k \times k$ diagonal matrices with the same element of $\mathscr{S}$ in each diagonal place; and this is $\mathbb{C}_{\mathscr{K}} \otimes \mathscr{S}$. ■

Suppose that ρ is a state of a C^*-algebra $\mathfrak{A}$, $\mathscr{M}$ and $\mathscr{B}$ are mutually commuting C^*-subalgebras (that contain the unit of $\mathfrak{A}$), $\mathscr{M} \cup \mathscr{B}$ generates $\mathfrak{A}$ as a C^*-algebra, and $\mathscr{M}$ is a finite type I factor. Upon taking a faithful representation of $\mathfrak{A}$, it follows from Lemma 11.4.8 that there is a * isomorphism φ, from $\mathscr{M} \otimes \mathscr{B}$ onto $\mathfrak{A}$, such that $\varphi(M \otimes B) = MB\ (M \in \mathscr{M}, B \in \mathscr{B})$. Under this isomorphism, the product state $(\rho \mid \mathscr{M}) \otimes (\rho \mid \mathscr{B})$ of $\mathscr{M} \otimes \mathscr{B}$ corresponds to a state $\rho^\times$ of $\mathfrak{A}$, uniquely determined by the condition $\rho^\times(MB) = \rho(M)\rho(B)\ (M \in \mathscr{M},\ B \in \mathscr{B})$. We call $\rho^\times$ the *factorization of* ρ *relative to* $(\mathscr{M}, \mathscr{B})$.

11.4.15. THEOREM. *Suppose* $\mathfrak{A}$ *is a* C^**-algebra with unit* I, $\{\mathcal{M}_a : a \in \mathbb{A}\}$ *is a family of mutually commuting finite type* I *factors such that* $I \in \mathcal{M}_a \subseteq \mathfrak{A}$, $\cup \mathcal{M}_a$ *generates* $\mathfrak{A}$ *as a* C^**-algebra, and* ρ *is a state of* $\mathfrak{A}$. *When* F, G *are finite subsets of* $\mathbb{A}$, *such that* $G \subseteq F$, *let* $\mathcal{M}_F$ *and* $\mathcal{M}_F^c$ *be the* C^**-subalgebras of* $\mathfrak{A}$ *that are generated by* $\cup \{\mathcal{M}_a : a \in F\}$ *and* $\cup \{\mathcal{M}_a : a \in \mathbb{A} \backslash F\}$ *(respectively), and let* $\mathfrak{A}_{FG}$ *be the* C^**-subalgebra generated by* $\mathcal{M}_G \cup \mathcal{M}_F^c$.

(i) $\mathcal{M}_F$ *is a finite type* I *factor,* $\mathcal{M}_F \cup \mathcal{M}_F^c$ *generates* $\mathfrak{A}$ *as a* C^**-algebra, and* $\mathcal{M}_F^c$ *is the relative commutant* $\{A \in \mathfrak{A} : AM = MA (M \in \mathcal{M}_F)\}$ of $\mathcal{M}_F$ in $\mathfrak{A}$.

(ii) $\mathfrak{A}$ *is a simple* C^**-algebra.*

(iii) *If* $\rho_G^\times$ *denotes the factorization of* ρ *relative to* $(\mathcal{M}_G, \mathcal{M}_G^c)$, *then* ρ *is primary if and only if*

$$\lim_F \|(\rho - \rho_G^\times) \mid \mathfrak{A}_{FG}\| = 0$$

for each finite subset G *of* $\mathbb{A}$, *the limit being taken over the family (directed by inclusion) of finite subsets* F *for which* $G \subseteq F \subseteq \mathbb{A}$.

Proof. By Corollary 11.4.10, $\mathfrak{A}$ is * isomorphic to $\bigotimes_{a \in \mathbb{A}} \mathcal{M}_a$, in such a way that $\mathcal{M}_a$ corresponds to its canonical image in the tensor product algebra. Accordingly, for each finite subset F of $\mathbb{A}$, $\mathcal{M}_F$ is * isomorphic to the spatial tensor product of the family $\{\mathcal{M}_a : a \in F\}$. Since each $\mathcal{M}_a$ is a finite type I factor, the same is true of $\mathcal{M}_F$ (Example 11.1.6); in particular, therefore, $\mathcal{M}_F$ is simple. If $\mathbb{F}$ is the family of all finite subsets of $\mathbb{A}$ (directed by $\subseteq$), $\mathfrak{A}$ is * isomorphic to the inductive limit of the net $\{\mathcal{M}_F : F \in \mathbb{F}\}$, since it is the norm closure of $\cup \mathcal{M}_F$; so $\mathfrak{A}$ is simple by Proposition 11.4.2. It follows that the representation π_ρ, engendered by the state ρ, is faithful. Upon identifying $\mathfrak{A}$ with $\pi_\rho(\mathfrak{A})$, we may assume that $\mathfrak{A}$ acts on a Hilbert space $\mathcal{H}$, and has a unit cyclic vector x such that $\rho = \omega_x \mid \mathfrak{A}$. Moreover, ρ is primary if and only if $\mathfrak{A}^-$ is a factor.

The C^*-algebra generated by $\mathcal{M}_F \cup \mathcal{M}_F^c$ contains each $\mathcal{M}_a (a \in \mathbb{A})$, and is therfore the whole of $\mathfrak{A}$. Since $\mathcal{M}_F^c \subseteq \mathcal{M}_F'$, it follows from Lemma 11.4.14 that $\mathcal{M}_F^c = \mathcal{M}_F' \cap \mathfrak{A}$ (the relative commutant of $\mathcal{M}_F$ in $\mathfrak{A}$), and that $(\mathcal{M}_F^c)^- = \mathcal{M}_F' \cap \mathfrak{A}^-$. Hence $\mathfrak{A}_{FG}^-$ is the von Neumann algebra $\mathcal{R}_{FG}$ generated by $\mathcal{M}_G \cup (\mathcal{M}_F' \cap \mathfrak{A}^-)$.

The normal factorization $\omega_G^\times$ of $\omega_x \mid \mathfrak{A}^-$, relative to $\mathcal{M}_G$, satisfies $\omega_G^\times(MB) = \omega_x(M)\omega_x(B)$ whenever $M \in \mathcal{M}_G$ and $B \in \mathcal{M}_G' \cap \mathfrak{A}^-$. In particular, therefore, $\omega_G^\times(MB) = \rho(M)\rho(B)$ when $M \in \mathcal{M}_G$ and $B \in \mathcal{M}_G^c$; and thus, $\omega_G^\times \mid \mathfrak{A}$ coincides with the factorization $\rho_G^\times$ of ρ relative to $(\mathcal{M}_G, \mathcal{M}_G^c)$. Since $\omega_x \mid \mathfrak{A}^-$ and $\omega_G^\times$ are normal states of $\mathfrak{A}^-$, and $\mathfrak{A}_{FG}^- = \mathcal{R}_{FG}$, it follows from the Kaplansky density theorem that

$$\|(\rho - \rho_G^\times) \mid \mathfrak{A}_{FG}\| = \|(\omega_x - \omega_G^\times) \mid \mathfrak{A}_{FG}\| = \|(\omega_x - \omega_G^\times) \mid \mathcal{R}_{FG}\|.$$

From Proposition 11.4.13 (with $(\mathbb{A}, \leq)$ replaced by $(\mathbb{F}, \subseteq)$), $\mathfrak{A}^-$ is a factor (equivalently, ρ is primary) if and only if

$$\lim_F \|(\rho - \rho_G^{\times}) \mid \mathfrak{A}_{FG}\| = 0$$

for each G in $\mathbb{F}$. ■

11.4.16. REMARK. We observe that a product state $\rho = \bigotimes_{a \in \mathbb{A}} \rho_a$ of a tensor product C^*-algebra $\mathfrak{A} = \bigotimes_{a \in \mathbb{A}} \mathcal{M}_a$ is primary, if each $\mathcal{M}_a$ is a finite type I factor. For this, we may suppose that the $\mathcal{M}_a$ are mutually commuting C^*-subalgebras of $\mathfrak{A}$ (each containing its unit), and that $\mathfrak{A}$ is generated as a C^*-algebra by $\bigcup \mathcal{M}_a$. To verify that ρ is primary, it suffices to show that it satisfies the condition set out in Theorem 11.4.15(iii). In fact, we prove the stronger assertion that $\rho = \rho_F^{\times}$ for each finite subset F of $\mathbb{A}$. Indeed, it is apparent that $\rho(AB) = \rho(A)\rho(B)$ when $A = A_1 A_2 \cdots A_m$, $B = B_1 B_2 \cdots B_n$, the A_j being chosen from different $\mathcal{M}_a$ (with a's in F), and the B_k from different $\mathcal{M}_b$ (with the b's in $\mathbb{A} \backslash F$). Since $\mathcal{M}_F$ is the norm-closed linear subspace of $\mathfrak{A}$ generated by such products A, and $\mathcal{M}_F^c$ is generated similarly by such products B, it follows that $\rho(AB) = \rho(A)\rho(B)$ whenever $A \in \mathcal{M}_F$ and $B \in \mathcal{M}_F^c$; so $\rho = \rho_F^{\times}$. ■

Bibliography: [4, 6, 65].

11.5. Exercises

11.5.1. Suppose that $\mathfrak{A}$ is an abelian C^*-algebra, and $\mathscr{B}$ is a C^*-algebra with center $\mathscr{C}$. Identify $\mathfrak{A} \otimes \mathscr{C}$, in the usual way, as a C^*-subalgebra of $\mathfrak{A} \otimes \mathscr{B}$ (see the final paragraph of Subsection 11.3, *The spatial tensor product*). Show that $\mathfrak{A} \otimes \mathscr{C}$ is the center of $\mathfrak{A} \otimes \mathscr{B}$. [*Hint*. Identify $\mathfrak{A}$ with $C(S)$ for some compact Hausdorff space S, and $\mathfrak{A} \otimes \mathscr{B}$ with $C(S, \mathscr{B})$.]

11.5.2. Suppose that $\mathscr{R}$ is a von Neumann algebra with center $\mathscr{Z}$ and $\mathscr{B}$ is a C^*-algebra with center $\mathscr{C}$.

(i) Show that $\mathscr{Z} \otimes \mathscr{B}$ contains the center of $\mathscr{R} \otimes \mathscr{B}$. [*Hint*. Suppose that S lies in the center of $\mathscr{R} \otimes \mathscr{B}$. Given any positive ε, approximate S in norm to within ε by an element $\sum_{j=1}^{m} R_j \otimes B_j$ of $\mathscr{R} \odot \mathscr{B}$, and deduce from Proposition 8.3.4 that S lies at distance not more than ε from $\mathscr{Z} \odot \mathscr{B}$.]

(ii) By using (i) and Exercise 11.5.1, show that $\mathscr{R} \otimes \mathscr{B}$ has center $\mathscr{Z} \otimes \mathscr{C}$.

11.5.3. Suppose that, for $j = 1, 2$, $\mathscr{C}_j$ is a C^*-subalgebra of an abelian C^*-algebra $\mathscr{Z}_j$. Let $\mathscr{C}$ be the C^*-subalgebra $(\mathscr{C}_1 \otimes \mathscr{Z}_2) \cap (\mathscr{Z}_1 \otimes \mathscr{C}_2)$ of $\mathscr{Z}_1 \otimes \mathscr{Z}_2$.

(i) Show that each pure state of $\mathscr{C}$ has the form $\rho_1 \otimes \rho_2 | \mathscr{C}$, where ρ_1 and ρ_2 are pure states of $\mathscr{Z}_1$ and $\mathscr{Z}_2$, respectively.

(ii) Suppose that ρ_1 and τ_1 are pure states of $\mathscr{Z}_1$, ρ_2 and τ_2 are pure states of $\mathscr{Z}_2$, and $\rho_1 \otimes \rho_2 | \mathscr{C}_1 \otimes \mathscr{C}_2 = \tau_1 \otimes \tau_2 | \mathscr{C}_1 \otimes \mathscr{C}_2$. Prove that $\rho_1 \otimes \rho_2 | \mathscr{C} = \tau_1 \otimes \tau_2 | \mathscr{C}$.

(iii) From (i), (ii), and the Stone–Weierstrass theorem, deduce that $\mathscr{C} = \mathscr{C}_1 \otimes \mathscr{C}_2$.

11.5.4. Suppose that $\mathfrak{A}_1$ and $\mathfrak{A}_2$ are C^*-algebras with centers $\mathscr{C}_1$ and $\mathscr{C}_2$, respectively. Show that $\mathfrak{A}_1 \otimes \mathfrak{A}_2$ has center $\mathscr{C}_1 \otimes \mathscr{C}_2$. [*Hint.* We may assume that $\mathfrak{A}_1$ and $\mathfrak{A}_2$ act on Hilbert spaces $\mathscr{H}_1$ and $\mathscr{H}_2$, respectively. Let $\mathscr{Z}_j$ denote the center of $\mathfrak{A}_j^-$. Show that the center of $\mathfrak{A}_1 \otimes \mathfrak{A}_2$ is contained in the intersection of the centers of $\mathfrak{A}_1 \otimes \mathfrak{A}_2^-$ and $\mathfrak{A}_1^- \otimes \mathfrak{A}_2$, and use the results of Exercises 11.5.2(ii) and 11.5.3(ii).]

11.5.5. Suppose that $\mathfrak{A}$ and $\mathscr{B}$ are simple C^*-algebras. Let π be an irreducible representation of $\mathfrak{A} \otimes \mathscr{B}$ on a Hilbert space $\mathscr{H}$.

(i) Show that the mappings

$$\pi_1 : A \to \pi(A \otimes I) : \mathfrak{A} \to \mathscr{B}(\mathscr{H}),$$

$$\pi_2 : B \to \pi(I \otimes B) : \mathscr{B} \to \mathscr{B}(\mathscr{H}),$$

are faithful representations of $\mathfrak{A}$ and $\mathscr{B}$, respectively. Prove also that $\pi_1(\mathfrak{A})^-$ and $\pi_2(\mathscr{B})^-$ are factors, and $\pi_2(\mathscr{B})^- \subseteq \pi_1(\mathfrak{A})'$.

(ii) Suppose that $A_1, \ldots, A_n \in \mathfrak{A}$, $B_1, \ldots, B_n \in \mathscr{B}$, and

$$\pi\left(\sum_{j=1}^{n} A_j \otimes B_j\right) = 0.$$

Show that $\sum_{j=1}^{n} A_j \otimes B_j = 0$. [*Hint.* Apply Theorem 5.5.4 (with the factor $\pi_1(\mathfrak{A})^-$ in place of $\mathscr{R}$) and Proposition 11.1.8.]

(iii) Show that the mapping $\pi | \mathfrak{A} \odot \mathscr{B}$ is an isometry, and deduce that π is a faithful representation of $\mathfrak{A} \otimes \mathscr{B}$. [*Hint.* Use Theorem 11.3.9.]

(iv) Deduce that $\mathfrak{A} \otimes \mathscr{B}$ is a simple C^*-algebra.

11.5.6. Suppose that $\mathfrak{A}$ and $\mathscr{B}$ are C^*-algebras and α is a C^*-norm on $\mathfrak{A} \odot \mathscr{B}$. Show that the identity mapping on $\mathfrak{A} \odot \mathscr{B}$ extends to a * homomorphism from $\mathfrak{A} \otimes_\alpha \mathscr{B}$ onto the spatial tensor product $\mathfrak{A} \otimes \mathscr{B}$. Deduce that $\mathfrak{A} \otimes_\alpha \mathscr{B}$ is simple if and only if both $\mathfrak{A}$ and $\mathscr{B}$ are simple and α is the spatial C^*-norm on $\mathfrak{A} \odot \mathscr{B}$.

11.5.7. Show that the tensor product $\mathscr{B}(\mathscr{H}) \otimes \mathscr{B}(\mathscr{K})$ of the (represented) C^*-algebras $\mathscr{B}(\mathscr{H})$ and $\mathscr{B}(\mathscr{K})$ is not the whole of $\mathscr{B}(\mathscr{H} \otimes \mathscr{K})$ when both the Hilbert spaces $\mathscr{H}$ and $\mathscr{K}$ are infinite dimensional. [*Hint*. Let $\{y_b : b \in \mathbb{B}\}$ be an orthonormal basis of $\mathscr{K}$, and define a unitary transformation U from $\sum_{b \in \mathbb{B}} \oplus \mathscr{H}$ onto $\mathscr{H} \otimes \mathscr{K}$ by the equation $U(\sum \oplus x_b) = \sum x_b \otimes y_b$. Represent an element T_0 of $\mathscr{B}(\mathscr{H} \otimes \mathscr{K})$ by the matrix $[T_{ab}]$ of the bounded operator $U^{-1} T_0 U$ acting on $\sum_{b \in \mathbb{B}} \oplus \mathscr{H}$, as in the discussion preceding Proposition 2.6.13. Show that, when $T_0 \in \mathscr{B}(\mathscr{H}) \otimes \mathscr{B}(\mathscr{K})$, there is a (norm-) compact subset of $\mathscr{B}(\mathscr{H})$ that contains all the entries T_{ab} in the matrix representing T_0.]

11.5.8. Suppose that $\mathfrak{A}$ and $\mathscr{B}$ are C^*-algebras and α is a C^*-norm on $\mathfrak{A} \odot \mathscr{B}$. When $A \in \mathfrak{A}$ and $B \in \mathscr{B}$, write $A \otimes B$ and $A \otimes_\alpha B$ to denote the corresponding simple tensors in the C^*-algebras $\mathfrak{A} \otimes \mathscr{B}$ and $\mathfrak{A} \otimes_\alpha \mathscr{B}$, respectively. Let π be a representation of $\mathfrak{A} \otimes_\alpha \mathscr{B}$ on a Hilbert space $\mathscr{H}$, and define representations π_1 of $\mathfrak{A}$ and π_2 of $\mathscr{B}$ by

$$\pi_1(A) = \pi(A \otimes_\alpha I), \qquad \pi_2(B) = \pi(I \otimes_\alpha B).$$

Suppose that $\pi_1(\mathfrak{A})^-$ is a factor of type I. Show that

(i) there exist Hilbert spaces $\mathscr{H}_1$ and $\mathscr{H}_2$, and a unitary transformation U from $\mathscr{H}$ onto $\mathscr{H}_1 \otimes \mathscr{H}_2$, such that

$$U\pi_1(\mathfrak{A})^- U^* = \mathscr{B}(\mathscr{H}_1) \otimes \mathbb{C}_{\mathscr{H}_2},$$
$$U\pi_2(\mathscr{B})U^* \subseteq \mathbb{C}_{\mathscr{H}_1} \otimes \mathscr{B}(\mathscr{H}_2);$$

(ii) there exist representations φ_1 (of $\mathfrak{A}$ on $\mathscr{H}_1$) and φ_2 (of $\mathscr{B}$ on $\mathscr{H}_2$) such that

$$U\pi(A \otimes_\alpha B)U^* = \varphi_1(A) \otimes \varphi_2(B) \qquad (A \in \mathfrak{A}, \quad B \in \mathscr{B});$$

(iii)
$$\left\|\pi\left(\sum_{j=1}^{n} A_j \otimes_\alpha B_j\right)\right\| \leq \sigma\left(\sum_{j=1}^{n} A_j \otimes B_j\right)$$

whenever $A_1, \ldots, A_n \in \mathfrak{A}$ and $B_1, \ldots, B_n \in \mathscr{B}$, where σ denotes the spatial C^*-norm on $\mathfrak{A} \odot \mathscr{B}$.

11.5.9. Let $\mathfrak{A}$ be a C^*-algebra with the following property: for each representation φ of $\mathfrak{A}$, the von Neumann algebra $\varphi(\mathfrak{A})^-$ is of type I. (In these circumstances we describe $\mathfrak{A}$ as a *type* I C^**-algebra.*) Prove that $\mathfrak{A}$ is nuclear. [*Hint.* Suppose that $\mathscr{B}$ is a C^*-algebra and α is a C^*-norm on $\mathfrak{A} \odot \mathscr{B}$. Given any irreducible representation π of $\mathfrak{A} \otimes_\alpha \mathscr{B}$, define representations π_1 of $\mathfrak{A}$ and π_2 of $\mathscr{B}$ as in Exercise 11.5.8. Prove that $\pi_1(\mathfrak{A})^-$ is a factor (necessarily of type I). By using the result of Exercise 11.5.8(iii), show that α coincides with the spatial C^*-norm σ on $\mathfrak{A} \odot \mathscr{B}$.]

11.5.10. Suppose that $\{e_1, e_2\}$ and $\{f_1, f_2\}$ are orthonormal bases of two-dimensional Hilbert spaces $\mathscr{H}$ and $\mathscr{K}$, respectively. Let P be the projection from $\mathscr{H} \otimes \mathscr{K}$ onto the one-dimensional subspace that contains $e_1 \otimes f_1 + e_2 \otimes f_2$, so that

$$P \in \mathscr{B}(\mathscr{H} \otimes \mathscr{K})^+ = (\mathscr{B}(\mathscr{H}) \otimes \mathscr{B}(\mathscr{K}))^+ = (\mathscr{B}(\mathscr{H}) \mathbin{\overline{\otimes}} \mathscr{B}(\mathscr{K}))^+.$$

By considering

$$\langle P(e_1 \otimes f_2), e_1 \otimes f_2\rangle, \qquad \langle P(e_2 \otimes f_1), e_2 \otimes f_1\rangle, \qquad \langle P(e_1 \otimes f_1), e_2 \otimes f_2\rangle,$$

show that P does not lie in the norm closure (equivalently, the weak-operator closure) of the set of all operators of the form $\sum_{j=1}^k A_j \otimes B_j$, where $A_1, \ldots, A_k \in \mathscr{B}(\mathscr{H})^+$ and $B_1, \ldots, B_k \in \mathscr{B}(\mathscr{K})^+$.

11.5.11. Give an example of C^*-algebras $\mathfrak{A}$ and $\mathscr{B}$ such that the state space of $\mathfrak{A} \otimes \mathscr{B}$ is not the norm-closed convex hull of the set of all product states of $\mathfrak{A} \otimes \mathscr{B}$. [*Hint.* With the notation of Exercise 11.5.10, let $\mathfrak{A} = \mathscr{B}(\mathscr{H})$, $\mathscr{B} = \mathscr{B}(\mathscr{K})$, and let ρ be the state ω_x of $\mathscr{B}(\mathscr{H} \otimes \mathscr{K})$ $(= \mathscr{B}(\mathscr{H}) \otimes \mathscr{B}(\mathscr{K}))$, where x is the unit vector $2^{-1/2}(e_1 \otimes f_1 + e_2 \otimes f_2)$. Show that each convex combination ρ_0 of product states of $\mathscr{B}(\mathscr{H}) \otimes \mathscr{B}(\mathscr{K})$ can be expressed in the form

$$\rho_0 = \sum_{r=1}^{l} \omega_{x_r} \otimes \omega_{y_r},$$

where $x_1, \ldots, x_l \in \mathscr{H}$ and $y_1, \ldots, y_l \in \mathscr{K}$. Let

$$x_r = c_r^{(1)} e_1 + c_r^{(2)} e_2, \qquad y_r = d_r^{(1)} f_1 + d_r^{(2)} f_2,$$

and identify elements of $\mathscr{B}(\mathscr{H})$ and $\mathscr{B}(\mathscr{K})$ with their matrices relative to the orthonormal bases $\{e_1, e_2\}$ and $\{f_1, f_2\}$, respectively. By considering the values taken by ρ and ρ_0 at

$$\begin{bmatrix} 1 & 0 \\ 0 & 0 \end{bmatrix} \otimes \begin{bmatrix} 0 & 0 \\ 0 & 1 \end{bmatrix}, \quad \begin{bmatrix} 0 & 0 \\ 0 & 1 \end{bmatrix} \otimes \begin{bmatrix} 1 & 0 \\ 0 & 0 \end{bmatrix}, \quad \begin{bmatrix} 0 & 1 \\ 0 & 0 \end{bmatrix} \otimes \begin{bmatrix} 0 & 1 \\ 0 & 0 \end{bmatrix},$$

show that $\|\rho - \rho_0\| \geq \frac{1}{4}$.]

11.5.12. Give an example of von Neumann algebras $\mathscr{R}$ and $\mathscr{S}$ and a normal state ω of $\mathscr{R} \mathbin{\bar{\otimes}} \mathscr{S}$ that is not in the weak * closed convex hull of the set of all normal product states of $\mathscr{R} \mathbin{\bar{\otimes}} \mathscr{S}$. [*Hint*. Proceed as in Exercise 11.5.11.]

11.5.13. In Example 11.1.7, we identify (in effect) the C^*-algebra $C(S, \mathfrak{A})$ of norm-continuous mappings (provided with pointwise operations and supremum norm) of a compact Hausdorff space S into a C^*-algebra $\mathfrak{A}$ with $\mathfrak{A} \otimes C(S)$. Show that such a mapping represents a positive element of $\mathfrak{A} \otimes C(S)$ if and only if its value at each point of S is a positive element of $\mathfrak{A}$.

11.5.14. Suppose $\mathfrak{A}_1$, $\mathfrak{A}_2$, $\mathscr{B}_1$, $\mathscr{B}_2$ are C^*-algebras and η_1, η_2 are bounded linear mappings of $\mathfrak{A}_1$ into $\mathscr{B}_1$ and $\mathfrak{A}_2$ into $\mathscr{B}_2$, respectively. Must the (unique) linear mapping η_0 of $\mathfrak{A}_1 \odot \mathfrak{A}_2$ into $\mathscr{B}_1 \odot \mathscr{B}_2$ satisfying $\eta_0(A_1 \otimes A_2) = \eta_1(A_1) \otimes \eta_2(A_2)$ for all A_1 in $\mathfrak{A}_1$ and A_2 in $\mathfrak{A}_2$ be bounded when $\mathfrak{A}_1 \odot \mathfrak{A}_2$ and $\mathscr{B}_1 \odot \mathscr{B}_2$ are provided with C^*-norms? Proof? Counter-example? [*Hint*. Consider Example 11.3.14.]

11.5.15. Let η be the mapping of $M_n(\mathbb{C})$ into itself that assigns to each matrix $[a_{jk}]$ its transpose matrix (whose (j, k) entry is a_{kj}). Show that

(i) η is a * anti-automorphism of $M_n(\mathbb{C})$;

(ii) η is a positive linear mapping of $M_n(\mathbb{C})$ into itself (see the discussion preceding Lemma 8.2.2);

(iii) when $n \geq 2$, the (unique) linear mapping $\eta \otimes \iota$ of $M_n(\mathbb{C}) \otimes M_2(\mathbb{C})$ into itself that assigns $\eta(A) \otimes B$ to $A \otimes B$ is *not* a positive linear mapping. (*Hint*. Express an element T of $M_n(\mathbb{C}) \otimes M_2(\mathbb{C})$ as a 2×2 matrix with entries from $M_n(\mathbb{C})$ and note that $(\eta \otimes \iota)(T)$ has representing matrix obtained from that of T by transposing each block of the 2×2 matrix. Choose for T a positive matrix that has 0 at all entries not in the upper, principal, $(n + 1) \times (n + 1)$ block and a non-zero scalar at the $(n, n + 1)$ entry. Recall that if a positive matrix has 0 at some diagonal entry, then 0 is at each entry of the corresponding row and column—see Exercise 4.6.11.]

11.5.16. A positive linear mapping η of a C^*-algebra $\mathfrak{A}$ is said to be *completely positive* when, for each positive integer n, $\eta \otimes \iota_n$, the (unique) linear mapping whose value at $A \otimes B$ is $\eta(A) \otimes B$ for each A in $\mathfrak{A}$ and each B in $M_n(\mathbb{C})$, is positive. Show that

(i) η is completely positive when η is a * homomorphism;

(ii) η is completely positive when $\eta(A) = TAT^*$ for each A in $\mathfrak{A}$, where $\mathfrak{A}$ acts on the Hilbert space $\mathscr{H}$ and T is a given bounded linear transformation of $\mathscr{H}$ into another Hilbert space $\mathscr{K}$;

(iii) η is completely positive when η is a composition of completely positive mappings;

(iv) η is completely positive when $\eta(A) = T\varphi(A)T^*$, where φ is a * homomorphism of $\mathfrak{A}$ into $\mathscr{B}(\mathscr{H})$ and T is a bounded linear transformation of the Hilbert space $\mathscr{H}$ into the Hilbert space $\mathscr{K}$;

(v) not each positive linear mapping of a C^*-algebra is completely positive. [*Hint*. See Exercise 11.5.15.]

11.5.17. Let η be a completely positive mapping (see Exercise 11.5.16) of a C^*-algebra $\mathfrak{A}$ into $\mathscr{B}(\mathscr{H})$ for some Hilbert space $\mathscr{H}$ and let $\{e_a\}_{a\in\mathbb{A}}$ be an orthonormal basis for $\mathscr{H}$. Denote by $\tilde{\mathfrak{A}}$ the linear space of functions from $\mathbb{A}$ to $\mathfrak{A}$ that take the value 0 at all but a finite number of elements of $\mathbb{A}$, where $\tilde{\mathfrak{A}}$ is provided with pointwise addition and scalar multiplication (so that $\tilde{\mathfrak{A}}$ is the *restricted direct sum* of $\mathfrak{A}$ with itself over the index set $\mathbb{A}$). Show that

(i)
$$\langle \tilde{A}, \tilde{A}' \rangle = \sum_{a, a' \in \mathbb{A}} \langle \eta(A_{a'}'^{*} A_a) e_a, e_{a'} \rangle$$

defines an inner product (see p. 75) on $\tilde{\mathfrak{A}}$, where $\tilde{A} = \{A_a\}_{a\in\mathbb{A}}$ and $\tilde{A}' = \{A'_{a'}\}_{a'\in\mathbb{A}}$ [*Hint*. To show that $\langle \tilde{A}, \tilde{A} \rangle \geq 0$, use the fact that the $n \times n$ matrix whose (j, k) entry is $A_j^* A_k$ is a positive element of $\mathfrak{A} \otimes M_n(\mathbb{C})$, where $A_1, \ldots A_n$ are the non-zero coordinates of $\tilde{A}$.];

(ii) $0 = \langle \tilde{A}, \tilde{B} \rangle = \langle \tilde{B}, \tilde{A} \rangle$ for each $\tilde{B}$ in $\tilde{\mathfrak{A}}$, when $\langle \tilde{A}, \tilde{A} \rangle = 0$, $\tilde{\mathscr{L}}$ is a linear space, where

$$\tilde{\mathscr{L}} = \{\tilde{A} \in \tilde{\mathfrak{A}} : \langle \tilde{A}, \tilde{A} \rangle = 0\},$$

and

$$\langle \tilde{A} + \tilde{\mathscr{L}}, \tilde{B} + \tilde{\mathscr{L}} \rangle_0 = \langle \tilde{A}, \tilde{B} \rangle$$

defines a definite inner product on $\mathscr{K}_0$, the quotient space $\tilde{\mathfrak{A}}/\tilde{\mathscr{L}}$;

(iii) $0 \leq \langle \tilde{B}, \tilde{B} \rangle \leq \|A\|^2 \langle \tilde{A}, \tilde{A} \rangle$, where $\tilde{A} = \{A_a\}_{a\in\mathbb{A}}$ and $\tilde{B} = \{AA_a\}_{a\in\mathbb{A}}$, and conclude that $\varphi_0(A)$ is a bounded linear mapping of $\mathscr{K}_0$ into $\mathscr{K}_0$, where

$$\varphi_0(A)(\tilde{A} + \tilde{\mathscr{L}}) = \tilde{B} + \tilde{\mathscr{L}}$$

[*Hint*. Let T be the $n \times n$ matrix whose non-zero entries are in the first row, and this first row consists of the non-zero coordinates of $\tilde{A}$. Let R and S be the $n \times n$ matrices whose only non-zero entries are their (1, 1) entries, and these are A^*A and $\|A\|^2 I$, respectively. Use the fact that $T^*RT \leq T^*ST$ in conjunction with the complete positivity of η.];

(iv) φ is a representation of $\mathfrak{A}$ on $\mathscr{K}$, where $\varphi(A)$ is the (unique) bounded extension of $\varphi_0(A)$ from $\mathscr{K}_0$ to $\mathscr{K}$, and $\mathscr{K}$ is the completion of $\mathscr{K}_0$ relative to $\langle\ ,\ \rangle_0$;

(v) $\{\tilde{I}_a + \tilde{\mathscr{L}}\}_{a\in\mathbb{A}}$ is an orthonormal set in $\mathscr{K}$, when $\eta(I) = I$, where $\tilde{I}_a$ is the element of $\tilde{\mathfrak{A}}$ with I at the a coordinate and 0 at all others;

(vi) $V^*\varphi(A)V = \eta(A)$ $(A \in \mathfrak{A})$, when $\eta(I) = I$, where V is the (unique) isometry of $\mathscr{H}$ into $\mathscr{K}$ such that $Ve_a = \tilde{I}_a + \tilde{\mathscr{L}}$ for each a in $\mathbb{A}$.

11.5.18. Adopt the notation and assumptions of Exercise 11.5.17 (exclusive of the assumption that $\eta(I) = I$). Let $\mathscr{H}_0$ be the dense linear manifold in $\mathscr{H}$ consisting of finite linear combinations of $\{e_a\}_{a \in \mathbb{A}}$, and let $T_0(\sum_{a \in \mathbb{A}_0} r_a e_a)$ be $\sum_{a \in \mathbb{A}_0} r_a(\tilde{I}_a + \tilde{\mathscr{L}})$ for each finite subset $\mathbb{A}_0$ of $\mathbb{A}$. Show that

(i) T_0 is a bounded linear transformation [*Hint.* Note that $\langle \tilde{I}_a + \tilde{\mathscr{L}}, I_{a'} + \tilde{\mathscr{L}} \rangle = \langle \eta(I)e_a, e_{a'} \rangle$.];

(ii) $T^*\varphi(A)T = \eta(A)$ $(A \in \mathfrak{A})$, where T is the (unique) bounded extension of T_0 from $\mathscr{H}_0$ to $\mathscr{H}$;

(iii) when $\eta(I) = I$, there is a Hilbert space $\mathscr{H}'$ containing $\mathscr{H}$ and a representation φ' of $\mathfrak{A}$ on $\mathscr{H}'$ such that $E\varphi'(A)E = \eta(A)E$ for each A in $\mathfrak{A}$, where E is the projection of $\mathscr{H}'$ on $\mathscr{H}$. [*Hint.* Let $\mathscr{K}_1$ be the range of V in $\mathscr{K}$, $\mathscr{K}_2$ be the orthogonal complement of $\mathscr{K}_1$ in $\mathscr{K}$, $\mathscr{H}'$ be $\mathscr{H} \oplus \mathscr{K}_2$, and $U(x, y)$ be $Vx + y$ for x in $\mathscr{H}$ and y in $\mathscr{K}_2$. Identify $\mathscr{H}$ with $\{(x, 0) : x \in \mathscr{H}\}$ and $\eta(A)(x, 0)$ with $(\eta(A)x, 0)$. Define $\varphi'(A)$ to be $U^*\varphi(A)U$.]

11.5.19. Let η be a positive linear mapping of a C^*-algebra $\mathfrak{A}$ into $\mathscr{B}(\mathscr{H})$ for some Hilbert space $\mathscr{H}$; let φ and φ' be representations of $\mathfrak{A}$ on Hilbert spaces $\mathscr{K}$ and $\mathscr{K}'$, respectively; and let T and T' be bounded linear transformations of $\mathscr{H}$ into $\mathscr{K}$ and $\mathscr{K}'$, respectively, such that $T^*\varphi(A)T = \eta(A)$ and $T'^*\varphi'(A)T' = \eta(A)$ for each A in $\mathfrak{A}$ (as described in Exercises 11.5.17 and 11.5.18). Let $\mathscr{K}_0$ and $\mathscr{K}_0'$ be the closure of the ranges of T and T', respectively, and let E and E' be the projections of $\mathscr{K}$ and $\mathscr{K}'$ onto $\mathscr{K}_0$ and $\mathscr{K}'_0$, respectively. Show that there is a unitary transformation U of $\mathscr{K}_0$ onto $\mathscr{K}'_0$ such that, for each A in $\mathfrak{A}$,

$$T' = UT, \qquad E\varphi(A)E = U^*E'\varphi'(A)E'U.$$

11.5.20. Let $\mathfrak{A}$ be a C^*-algebra acting on a Hilbert space $\mathscr{H}$. Show that

(i) the matrix $[\langle x_k, x_j \rangle]$ whose j, k entry is $\langle x_k, x_j \rangle$ is positive, where $x_1, \ldots, x_n$ are vectors in $\mathscr{H}$ [*Hint.* Let $\{e_1, \ldots, e_n\}$ be an orthonormal basis for an n-dimensional subspace of $\mathscr{H}$ containing $x_1, \ldots, x_n$. Note that $[\langle x_k, x_j \rangle]$ is the matrix of T^*T relative to $\{e_1, \ldots, e_n\}$, where $Te_j = x_j$.];

(ii) the matrix $[A_j^*A_k]$ whose j, k entry is $A_j^*A_k$ is in $M_n(\mathfrak{A})^+$ for each set of n elements $\{A_1, \ldots, A_n\}$ in $\mathfrak{A}$ and conclude that the matrix all of whose entries are a given positive A in $\mathfrak{A}$ is in $M_n(\mathfrak{A})^+$ [*Hint.* Consider the matrix whose first row is $A_1, \ldots, A_n$ and all of whose other entries are 0.];

(iii) each positive element of $M_n(\mathfrak{A})$ is a sum of matrices of the form $[A_j^*A_k]$ [*Hint.* Express a positive element of $M_n(\mathfrak{A})$ as T^*T with T in $M_n(\mathfrak{A})$.];

(iv) $[A_{jk}] \to [\langle A_{jk}x_k, x_j\rangle]$: $M_n(\mathfrak{A}) \to M_n(\mathbb{C})$ is a positive linear mapping for each set of n vectors $\{x_1, \ldots, x_n\}$ in $\mathscr{H}$ and conclude that $[\langle A_{jk}x, x\rangle] \geq 0$ for each x in $\mathscr{H}$ when $[A_{jk}] \geq 0$ [*Hint*. From (iii), it suffices to show that $[\langle A_j^*A_kx_k, x_j\rangle] \geq 0$ for each set of n elements $A_1, \ldots, A_n$ in $\mathfrak{A}$. Use (i).];

(v) if $[A_{jk}]$ is in $M_n(\mathfrak{A})^+$ and $[B_{jk}]$ is in $M_n(\mathfrak{A}')^+$, then $[A_{jk}B_{jk}] \geq 0$. [*Hint*. Use (iii) and (ii).]

11.5.21. Define an *n-state* of a C^*-algebra $\mathfrak{A}$ to be a matrix $[\rho_{jk}]$ of linear functionals on $\mathfrak{A}$ such that $[\rho_{jk}(A_{jk})] \geq 0$ when $[A_{jk}] \in M_n(\mathfrak{A})^+$ and $\rho_{jj}(I) = 1$ for j in $\{1, \ldots, n\}$. Show that

(i) if $\mathfrak{A}$ acts on a Hilbert space $\mathscr{H}$ and $\{x_1, \ldots, x_n\}$ is a set of n unit vectors in $\mathscr{H}$, then $[\omega_{x_k, x_j} \mid \mathfrak{A}]$ is an n-state of $\mathfrak{A}$ [*Hint*. Use Exercise 11.5.20(iv).];

(ii) a linear mapping η of $\mathfrak{A}$ into a C^*-algebra $\mathscr{B}$, such that $\eta(I) = I$, is completely positive, if and only if $[\rho_{jk} \circ \eta]$ is an n-state of $\mathfrak{A}$ for each n-state $[\rho_{jk}]$ of $\mathscr{B}$ [*Hint*. Note that $\langle [B_{jk}]\tilde{x}, \tilde{x}\rangle = \langle [\langle B_{jk}x_k, x_j\rangle]\tilde{a}, \tilde{a}\rangle$, where $\tilde{x} = \{x_1, \ldots, x_n\}$ and $\tilde{a} = \{1, 1, \ldots, 1\}$.];

(iii) $[\rho_{jk}]$ is an n-state of $\mathfrak{A}$ when each ρ_{jk} is the same state ρ of $\mathfrak{A}$ [*Hint*. Use the GNS construction and (i).];

(iv) a positive linear mapping of $\mathfrak{A}$ into an abelian C^*-algebra $\mathscr{B}$ is completely positive. [*Hint*. Note that $[B_{jk}] \in M_n(\mathscr{B})^+$ if and only if $[\rho(B_{jk})] \geq 0$ for each pure state ρ of $\mathscr{B}$. Then use (iii).]

11.5.22. Let $\mathfrak{A}$ be an abelian C^*-algebra and $[\rho_{jk}]$ be an $n \times n$ matrix of linear functionals on $\mathfrak{A}$.

(i) Find a representation π of $\mathfrak{A}$ on a Hilbert space $\mathscr{H}$ with a cyclic vector u and a matrix of operators H_{jk} in $\pi(\mathfrak{A})^-$ such that $\rho_{jk}(A) = \langle \pi(A)H_{jk}u, u\rangle$ for each A in $\mathfrak{A}$. [*Hint*. Express each ρ_{jk} as $\eta_{jk} + i\tau_{jk}$ with η_{jk} and τ_{jk} hermitian, and let ρ be $\sum_{j,k=1}^n (\eta_{jk}^+ + \eta_{jk}^- + \tau_{jk}^+ + \tau_{jk}^-)$. Use the GNS representation corresponding to ρ as π together with Proposition 7.3.5 and Corollary 7.2.16.]

(ii) Suppose $[\rho_{jk}(A)] \geq 0$ for each A in $\mathfrak{A}^+$. Show that $[H_{jk}(p)] \geq 0$ for each p in X, where $\pi(\mathfrak{A})^- \cong C(X)$, and we denote by the same symbol an element of $\pi(\mathfrak{A})^-$ and the function representing it. Conclude that $[H_{jk}] \geq 0$. [*Hint*. Suppose the contrary and find a non-null clopen subset X_0 (with corresponding projection E) of X and a vector $\{a_1, \ldots, a_n\}$ $(=\tilde{a})$ in $\mathbb{C}^n$ such that $\langle [H_{jk}(p)]\tilde{a}, \tilde{a}\rangle < 0$ for each p in X_0. Deduce the contradiction $0 > \sum_{j,k=1}^n \bar{a}_j a_k EH_{jk}$ and

$$0 > \sum_{j,k=1}^n \bar{a}_j a_k \langle EH_{jk}u, u\rangle = \langle [\langle EH_{jk}u, u\rangle]\tilde{a}, \tilde{a}\rangle \geq 0$$

—the first inequality uses Exercise 11.5.21(iii); the last inequality follows from the choice of H_{jk} and the present hypothesis.]

(iii) Show that $[\rho_{jk}]$ is an n-state of $\mathfrak{A}$ (*abelian*) if and only if $[\rho_{jk}(A)] \geq 0$ for each A in $\mathfrak{A}^+$ and $\rho_{jj}(I) = 1$. [*Hint*. Recall Exercise 11.5.20(ii), (v), and (iv).]

(iv) Show that each positive linear mapping of $\mathfrak{A}$ (*abelian*) is completely positive.

11.5.23. Let η be a positive linear mapping of a C^*-algebra $\mathfrak{A}$ into a C^*-algebra $\mathscr{B}$ such that $\eta(I) = I$. Show that

(i) $ET^*ETE \leq ET^*TE$ when $E, T \in \mathscr{B}(\mathscr{K})$ for some Hilbert space $\mathscr{K}$ and E is a projection;

(ii) $\eta(A)^*\eta(A) \leq \eta(A^*A)$ for each normal operator A in $\mathfrak{A}$. [*Hint*. Use Exercises 11.5.18(iii) and 11.5.22(iv) with (i).] (This provides another approach to Exercise 10.5.9.)

11.5.24. Let $\{A_n\}$ be a sequence of positive operators on a Hilbert space $\mathscr{H}$ with $\sum_{n=1}^{\infty} A_n$ weak-operator convergent to I.

(i) Find a positive linear mapping η of $C(X)$ into $\mathscr{B}(\mathscr{H})$ such that $\eta(1) = I$ and $\eta(f_n) = A_n$, where X is the compact subset $\{0, 1/n : n = 1, 2, \ldots\}$ of $\mathbb{R}$ and f_n takes the value 1 at $1/n$ and 0 at other points.

(ii) Find a Hilbert space $\mathscr{K}$ containing $\mathscr{H}$ and a sequence $\{E_n\}$ of projections on $\mathscr{K}$ with sum I such that $EE_nE = A_nE$ for each n, where E is the projection in $\mathscr{B}(\mathscr{K})$ with range $\mathscr{H}$. [*Hint*. Use Exercises 11.5.18(iii) and 11.5.22(iv).]

11.5.25. Let $\mathscr{R}$ be a von Neumann algebra acting on a Hilbert space $\mathscr{H}$. Show that

(i) the mapping $x \otimes y \to y \otimes x$ $(x, y \in \mathscr{H})$ extends to a unitary operator U on $\mathscr{H} \otimes \mathscr{H}$;

(ii) U (in (i)) is self-adjoint;

(iii) the mapping $A \otimes B \to B \otimes A$ $(A, B \in \mathscr{R})$ extends to a * automorphism of $\mathscr{R} \mathbin{\overline{\otimes}} \mathscr{R}$.

11.5.26. Let $\{\mathscr{H}_f : f \in \mathbb{F}\}$ be a family of Hilbert spaces in which the index set $\mathbb{F}$ is directed by $\leq$. Suppose that, whenever $f, g \in \mathbb{F}$ and $f \leq g$, there is an isometric linear mapping U_{gf} from $\mathscr{H}_f$ into $\mathscr{H}_g$, and $U_{hg}U_{gf} = U_{hf}$ whenever $f, g, h \in \mathbb{F}$ and $f \leq g \leq h$.

(i) Prove that U_{ff} is the identity mapping on $\mathscr{H}_f$.

(ii) Construct a Hilbert space $\mathscr{H}$ and, for each f in $\mathbb{F}$, an isometric linear mapping U_f from $\mathscr{H}_f$ into $\mathscr{H}$, in such a way that $U_f = U_g U_{gf}$ whenever $f, g \in \mathbb{F}$ and $f \le g$, and $\bigcup\{U_f(\mathscr{H}_f): f \in \mathbb{F}\}$ is everywhere dense in $\mathscr{H}$. [*Hint*. Let $\mathfrak{X}$ be the Banach space consisting of all families $\{x_h : h \in \mathbb{F}\}$ in which $x_h \in \mathscr{H}_h$ and $\sup\{\|x_h\| : h \in \mathbb{F}\} < \infty$ (with pointwise-linear structure and the supremum norm). Let $\mathfrak{X}_0$ be the closed subspace of $\mathfrak{X}$ consisting of those families $\{x_h : h \in \mathbb{F}\}$ for which the net $\{\|x_h\| : h \in \mathbb{F}\}$ converges to 0, and let $Q : \mathfrak{X} \to \mathfrak{X}/\mathfrak{X}_0$ be the quotient mapping. Adapt the argument used in proving Proposition 11.4.1(i).]

(iii) Suppose that $\mathscr{K}$ is a Hilbert space, V_f is an isometric linear mapping from $\mathscr{H}_f$ into $\mathscr{K}$, for each f in $\mathbb{F}$, $V_f = V_g U_{gf}$ whenever $f, g \in \mathbb{F}$ and $f \le g$, and $\bigcup\{V_f(\mathscr{H}_f): f \in \mathbb{F}\}$ is everywhere dense in $\mathscr{K}$. Show that there is a unitary transformation W from $\mathscr{H}$ onto $\mathscr{K}$ such that $V_f = WU_f$ for each f in $\mathbb{F}$.

Note. In the circumstances set out above, we say that the Hilbert spaces $\mathscr{H}_f$ $(f \in \mathbb{F})$ and the isometries U_{gf} $(f, g \in \mathbb{F}, f \le g)$ together constitute a *directed system of Hilbert spaces*; the Hilbert space $\mathscr{H}$ occurring in (ii) (together with the isometries U_f $(f \in \mathbb{F})$) is the *inductive limit* of the directed system. The effect of (iii) is to show that the constructs in (ii) are unique up to unitary equivalence.

11.5.27. Suppose that $\mathbb{F}$ is a directed set, and the Hilbert spaces $\mathscr{H}_f (f \in \mathbb{F})$ and isometric linear mappings $U_{gf}: \mathscr{H}_f \to \mathscr{H}_g (f, g \in \mathbb{F}, f \le g)$ together constitute a directed system of Hilbert spaces. Construct the (inductive limit) Hilbert space $\mathscr{H}$ and the isometric linear mappings $U_f: \mathscr{H}_f \to \mathscr{H}$, as in Exercise 11.5.26(ii). Suppose also that $A_f \in \mathscr{B}(\mathscr{H}_f)$ for each f in $\mathbb{F}$, $\sup\{\|A_f\| : f \in \mathbb{F}\} < \infty$, and there is an element f_0 of $\mathbb{F}$ such that $A_g U_{gf} = U_{gf} A_f$ whenever $f, g \in \mathbb{F}$ and $f_0 \le f \le g$. Show that there is a unique element A of $\mathscr{B}(\mathscr{H})$ such that $AU_f = U_f A_f$ whenever $f_0 \le f \in \mathbb{F}$. (We refer to A as the *inductive limit* of the family $\{A_f : f \in \mathbb{F}\}$ of bounded operators.)

11.5.28. Suppose that $\mathbb{F}$ is a directed set, and the Hilbert spaces $\mathscr{H}_f$ $(f \in \mathbb{F})$ and isometric linear mappings $U_{gf} : \mathscr{H}_f \to \mathscr{H}_g$ $(f, g \in \mathbb{F}, f \le g)$ together constitute a directed system of Hilbert spaces. Suppose also that the C^*-algebras $\mathfrak{A}_f$ $(f \in \mathbb{F})$ and the * isomorphisms

$$\Phi_{gf} : \mathfrak{A}_f \to \mathfrak{A}_g \ (f, g \in \mathbb{F}, f \le g)$$

together constitute a directed system of C^*-algebras. Suppose, finally, that π_f is a representation of $\mathfrak{A}_f$ on $\mathscr{H}_f$, for each f in $\mathbb{F}$, and

$$\pi_g(\Phi_{gf}(A))U_{gf} = U_{gf}\pi_f(A) \qquad (A \in \mathfrak{A}_f)$$

whenever $f, g \in \mathbb{F}$ and $f \le g$. Construct the inductive limit Hilbert space $\mathscr{H}$ and the isometric linear mappings $U_f : \mathscr{H}_f \to \mathscr{H}$ as in Exercise 11.5.26(ii), and construct the inductive limit C^*-algebra $\mathfrak{A}$ and the * isomorphisms $\varphi_f : \mathfrak{A}_f \to \mathfrak{A}$ as in Proposition 11.4.1(i). Prove that there is a unique representation π of $\mathfrak{A}$ on $\mathscr{H}$ such that

$$\pi(\varphi_f(A))U_f = U_f\pi_f(A) \qquad (A \in \mathfrak{A}_f)$$

for each f in $\mathbb{F}$. (We refer to π as the *inductive limit* of the family $\{\varphi_f : f \in \mathbb{F}\}$ of representations.)

11.5.29. Suppose that $\{\mathscr{H}_a : a \in \mathbb{A}\}$ is a family of Hilbert spaces and u_a is a unit vector in $\mathscr{H}_a$ for each a in $\mathbb{A}$. Note that the family $\mathbb{F}$ of all finite subsets of $\mathbb{A}$ is directed by the inclusion relation $\subseteq$. When F is an element $\{a_1, \ldots, a_n\}$ of $\mathbb{F}$, define a Hilbert space $\mathscr{H}_F$ and a unit vector u_F in $\mathscr{H}_F$ by

$$\mathscr{H}_F = \mathscr{H}_{a_1} \otimes \cdots \otimes \mathscr{H}_{a_n}, \qquad u_F = u_{a_1} \otimes \cdots \otimes u_{a_n}.$$

(i) Suppose that $F, G \in \mathbb{F}$ and

$$F = \{a_1, \ldots, a_n\} \subseteq \{a_1, \ldots, a_n, a_{n+1}, \ldots, a_m\} = G.$$

Show that there is a unique isometric linear mapping $U_{GF} : \mathscr{H}_F \to \mathscr{H}_G$ such that

$$U_{GF}(x_{a_1} \otimes \cdots \otimes x_{a_n}) = x_{a_1} \otimes \cdots \otimes x_{a_n} \otimes u_{a_{n+1}} \otimes \cdots \otimes u_{a_m}$$

whenever $x_{a_j} \in \mathscr{H}_{a_j}$ for j in $\{1, \ldots, n\}$.

(ii) Show that the Hilbert spaces $\mathscr{H}_F$ $(F \in \mathbb{F})$ and the isometric linear mappings U_{GF} $(F, G \in \mathbb{F}, F \subseteq G)$ together constitute a directed system of Hilbert spaces.

(iii) Suppose that the Hilbert space $\mathscr{H}$, together with the isometric linear mappings $U_F : \mathscr{H}_F \to \mathscr{H}$ $(F \in \mathbb{F})$, is the inductive limit of the directed system occurring in (ii). Show that there is a unit vector u in $\mathscr{H}$ such that $U_F u_F = u$ for each F in $\mathbb{F}$. (We refer to the inductive limit Hilbert space $\mathscr{H}$ as the *tensor product* of the family $\{\mathscr{H}_a : a \in \mathbb{A}\}$, and denote it by $\bigotimes_{a \in \mathbb{A}} \mathscr{H}_a$. Strictly speaking, we should use a nomenclature and notation that recognizes the dependence of this construction on the choice of u_a.)

11.5.30. With the notation of Exercise 11.5.29, suppose that π_a is a representation of a C^*-algebra $\mathfrak{A}_a$ on the Hilbert space $\mathscr{H}_a$ for each a in $\mathbb{A}$. Construct the directed system of C^*-algebras consisting of C^*-algebras $\mathfrak{A}_F$ $(F \in \mathbb{F})$ and * isomorphisms $\Phi_{GF} : \mathfrak{A}_F \to \mathfrak{A}_G$ $(F, G \in \mathbb{F}, F \subseteq G)$, as in the discussion following Proposition 11.4.2, so that the inductive limit of the system is a C^*-algebra $\mathfrak{A}$ $(= \bigotimes_{a \in \mathbb{A}} \mathfrak{A}_a)$ together with * isomorphisms $\varphi_F : \mathfrak{A}_F \to \mathfrak{A}$ $(F \in \mathbb{F})$ satisfying conditions analogous to those set out in Proposition 11.4.1(i). When $F \in \mathbb{F}$, the tensor product of the family $\{\pi_a : a \in F\}$ is a representation π_F of $\mathfrak{A}_F$ on $\mathscr{H}_F$.

(i) Prove that

$$\pi_G(\Phi_{GF}(A))U_{GF} = U_{GF}\pi_F(A) \qquad (A \in \mathfrak{A}_F)$$

whenever $F, G \in \mathbb{F}$ and $F \subseteq G$.

(ii) Deduce that there is a unique representation π of $\mathfrak{A}$ $(= \bigotimes_{a \in \mathbb{A}} \mathfrak{A}_a)$ on $\mathscr{H}$ $(= \bigotimes_{a \in \mathbb{A}} \mathscr{H}_a)$ such that

$$\pi(\varphi_F(A))U_F = U_F\pi_F(A) \qquad (A \in \mathfrak{A}_F)$$

for each F in $\mathbb{F}$.

(iii) Prove that $\omega_u \circ \pi \circ \varphi_F = \omega_{u_F} \circ \pi_F$ $(F \in \mathbb{F})$.

(iv) When $a \in \mathbb{A}$ and $F = \{a\} \in \mathbb{F}$, we write φ_a in place of φ_F, so that φ_a is a * isomorphism from $\mathfrak{A}_a$ into $\mathfrak{A}$. Define states ρ of $\mathfrak{A}$ and ρ_a of $\varphi_a(\mathfrak{A}_a)$ $(\subseteq \mathfrak{A})$ by $\rho = \omega_u \circ \pi$, $\rho_a = \omega_{u_a} \circ \pi_a \circ \varphi_a^{-1}$. Show that ρ is the product state $\bigotimes_{a \in \mathbb{A}} \rho_a$.

(v) Show that u is a cyclic vector for π if, for each a in $\mathbb{A}$, u_a is a cyclic vector for π_a.

CHAPTER 12

APPROXIMATION BY MATRIX ALGEBRAS

The examples of special classes of C^*-algebras and their representations studied in Section 10.4 include *uniformly matricial algebras*. To recall, such an algebra has a norm-dense subalgebra that is the ascending union of distinct C^*-subalgebras each * isomorphic to some $M_n(\mathbb{C})$ and each containing the identity operator. In the present chapter, we study these algebras in more detail. We shall see (Theorem 12.1.1) that the isomorphism classes of such algebras are characterized by a function that assigns to each prime the maximum power to which it divides the order of one of the matrix algebras in a generating nest. In particular there are many isomorphism classes of uniformly matricial algebras.

In Section 12.1 we show that a finite von Neumann algebra that contains an ultraweakly dense uniformly matricial C^*-algebra is a factor; and, by contrast with the uniformly matricial C^*-algebras, in Section 12.2, we show that all such finite factors are * isomorphic (Theorem 12.2.1).

The information resulting from these investigations will allow us to describe a continuum of non-isomorphic (matricial) factors of type III. In Section 12.3, we study states of matricial C^*-algebras and the GNS representations corresponding to them. We establish criteria for primary states to be quasi-equivalent (Proposition 12.3.2) and for states to be transforms of one another under a * automorphism (Theorem 12.3.6) showing, in particular, that this is the case for each pair of pure states (Theorem 12.3.4). The last part of the third section deals with a special class of states of the matricial C^*-algebra generated by a commuting family of factors of type I_2 (the CAR algebra). These (*product*) states are parametrized by a number a in $[0, 1]$. The states corresponding to 0 and 1 are pure (and, so, give rise to irreducible GNS representations). The other parameters correspond to states that have GNS representations whose ultraweak closures can be (explicitly) realized in terms of the group-measure-theoretic construction of Section 8.6. Using the criteria developed in that section, we show that these ultraweak closures are factors of type III when $a \neq 0, \frac{1}{2}, 1$ (and of type II_1 when $a = \frac{1}{2}$). The section closes with a proof that these factors are not * isomorphic for distinct values of the parameter a in $(0, \frac{1}{2})$ (Theorem 12.3.14).

12.1. Isomorphism of uniformly matricial algebras

With the aid of the constructions described in Chapter 11, we can establish the existence and uniqueness (up to * isomorphism) of the uniformly matricial C^*-algebra $\mathfrak{A}$ of type $\{n_j\}$ in a manner different from that occurring in Proposition 10.4.18. Suppose $\{\mathfrak{A}_j\}$ is a generating nest for $\mathfrak{A}$. By Example 11.1.6, $\mathfrak{A}_j \cong M_{m_j}(\mathbb{C}) \otimes \mathfrak{A}_{j-1}$, where m_j is n_j/n_{j-1} when $j \in \{2, 3, \ldots\}$ ($\mathfrak{A}_0$ is $\mathbb{C}I$ and m_1 is n_1). From this, $\mathfrak{A}_j$ is generated (as a C^*-algebra) by $\mathfrak{A}_{j-1} \cup \mathscr{B}_j$, where $\mathscr{B}_j (\cong M_{m_j}(\mathbb{C}))$ is the commutant of $\mathfrak{A}_{j-1}$ in $\mathfrak{A}_j$. Hence $\mathfrak{A}$ is the norm closure of the algebra generated by a mutually commuting family $\{\mathscr{B}_j\}$ of C^*-subalgebras $\mathscr{B}_j$ such that $\mathscr{B}_j \cong M_{m_j}(\mathbb{C})$. This view of $\mathfrak{A}$ as the infinite C^*-tensor product of factors of type I_{m_j} (see Corollary 11.4.10) permits us to construct the uniformly matricial C^*-algebra of type $\{n_j\}$. The uniqueness of this algebra follows from Proposition 11.4.5.

With the tensor product formulation in view, if $m_j = m_j' m_j''$, we can decompose $\mathscr{B}_j$ as a tensor product of factors of types $\mathrm{I}_{m_j'}$ and $\mathrm{I}_{m_j''}$. Moreover, the subalgebras $\mathscr{B}_j$ can be "regrouped" to form a different generating nest. Thus, while assuming that $\{n_j\}$ and $\{n_j'\}$ are identical is sufficient to guarantee that the corresponding uniformly matricial C^*-algebras are * isomorphic, this condition is far from necessary. The tensor-product, number-theoretic discussion we have been following indicates that the appropriate isomorphism invariant for uniformly matricial C^*-algebras is the sequence $\{a_n\}$, where p_n is the nth prime and $p_n^{a_n}$ is the largest power of p_n that divides any n_j (a_n can be 0 or ∞, with the obvious interpretation). A uniformly matricial C^*-algebra $\mathfrak{A}$ with invariant $\{a_n\}$ contains a_n C^*-subalgebras isomorphic to $M_{p_n}(\mathbb{C})$, $n = 1, 2, \ldots$, the total family of subalgebras mutually commuting and generating a norm-dense subalgebra of $\mathfrak{A}$. It is also clear, from our discussion, that if $\mathscr{B}$ is another uniformly matricial C^*-algebra with the same invariant $\{a_n\}$, then $\mathfrak{A} \cong \mathscr{B}$, even though $\mathfrak{A}$ and $\mathscr{B}$ may appear with distinct types $\{n_j\}$, $\{n_j'\}$ as first presented. What is not clear, at this point, is the uniqueness of the "prime power" decomposition of $\mathfrak{A}$. In other words, if $\mathfrak{A}$ and $\mathscr{B}$ have distinct invariants, are $\mathfrak{A}$ and $\mathscr{B}$ non-isomorphic? We shall see that this is the case (Theorem 12.1.1).

A more primitive question of this sort, at the heart of our considerations, is the following. If $\mathfrak{A}$ is of type $\{n_j\}$, is there a C^*-subalgebra (containing I) * isomorphic to $M_p(\mathbb{C})$, where p is a prime that divides none of the n_j? Specifically, does the CAR algebra (of type $\{2, 4, 8, \ldots\}$) have a C^*-subalgebra (containing I) * isomorphic to $M_3(\mathbb{C})$? Lemma 12.1.5 (and the considerations leading to it) will answer these questions negatively.

Suppose that $\mathfrak{A}$ and $\mathscr{B}$ have invariants $\{a_n\}$ and $\{b_n\}$, respectively, that $\mathfrak{A} \cong \mathscr{B}$, and that $a_k < b_k$ for some k. The isomorphism of $\mathscr{B}$ with $\mathfrak{A}$ will carry a subfactor of $\mathscr{B}$ of type $\mathrm{I}_{p_k^{b_0}}$, where $a_k < b_0 < \infty$, onto such a subfactor, $\mathfrak{A}_0$, of $\mathfrak{A}$. With $\{\mathfrak{A}_j\}$ a generating nest for $\mathfrak{A}$, p_k^{ko} divides no n_j, by definition of a_k. Nonetheless, with ε positive and $\{E_{rs}\}$ a self-adjoint system of $p_k^{bo} \times p_k^{bo}$ matrix units for $\mathfrak{A}_0$, it is possible to choose j so large that $\|E_{rs} - A_{rs}\| < \varepsilon$ for $r, s = 1, \ldots, p_k^{bo}$, where A_{rs} is an operator in the unit ball of $\mathfrak{A}_j$. We shall see, in Lemma 12.1.5, that if such an approximation is possible for a single (minimal) projection (say, E_{11}) in $\mathfrak{A}_0$ with $\frac{1}{8}$ for ε, then p_k^{bo} divides n_j. This contradiction will establish the following theorem.

12.1.1. THEOREM. *If $\mathfrak{A}$ and $\mathscr{B}$ are matricial C^*-algebras with invariants $\{a_n\}$ and $\{b_n\}$, respectively, then $\mathfrak{A}$ is * isomorphic to $\mathscr{B}$ if and only if $a_n = b_n$ for all n.*

A special property of finite representations of matricial C^*-algebras will be useful to us both in the present study and when we investigate matricial von Neumann algebras (see Section 12.2). If $\mathfrak{A}$ is a matricial C^*-algebra, it possesses a unique normalized trace (which restricts to the normalized trace on each matrix algebra of a generating nest for $\mathfrak{A}$).

12.1.2. PROPOSITION. *If the C^*-algebra $\mathfrak{A}$ acting on the Hilbert space $\mathscr{H}$ admits at most one trace and its weak-operator closure $\mathfrak{A}^-$ is finite, then $\mathfrak{A}^-$ is a factor.*

Proof. If τ is the normalized center-valued trace on $\mathfrak{A}^-$, P is a non-zero central projection in $\mathfrak{A}^-$, x is a unit vector in the range of P, and y is a unit vector in $\mathscr{H}$, then $\omega_x \circ \tau$ and $\omega_y \circ \tau$ restrict to the unique trace on $\mathfrak{A}$. From ultraweak continuity of τ on $\mathfrak{A}^-$ (see Theorem 8.2.8(vi)), $\omega_x \circ \tau$ and $\omega_y \circ \tau$ agree on $\mathfrak{A}^-$. Thus

$$(\omega_x \circ \tau)(P) = \langle \tau(P)x, x\rangle = \langle Px, x\rangle = 1 = \langle Py, y\rangle = (\omega_y \circ \tau)(P).$$

Hence $P = I$, and $\mathfrak{A}^-$ is a factor. ■

12.1.3. COROLLARY. *If $\mathscr{M}$ is a finite von Neumann algebra containing an ultraweakly dense matricial C^*-algebra, then $\mathscr{M}$ is a factor (of type II_1).*

By applying the GNS construction to the trace τ on a matricial C^*-algebra $\mathfrak{A}$, we produce a representation of $\mathfrak{A}$, necessarily faithful, since $\mathfrak{A}$ is simple (Proposition 10.4.18), with a generating unit trace vector for the image and, hence, for its ultraweak closure. It follows, from Theorem 7.2.15, that the ultraweak closure of the image is finite, and, from Corollary 12.1.3, that it is a factor of type II_1.

12.1.4. PROPOSITION. *Each matricial C^*-algebra $\mathfrak{A}$ has a faithful representation as an ultraweakly dense C^*-subalgebra of a factor of type II_1*

with a generating trace vector. Two such representations of $\mathfrak{A}$ are unitarily equivalent.

Proof. The existence of one such representation is established in the preceding discussion by applying the GNS construction to the unique trace τ on $\mathfrak{A}$. If φ and ψ are representations of $\mathfrak{A}$ on $\mathscr{H}$ and $\mathscr{K}$ with generating unit trace vectors x_0 and y_0, respectively, then the mapping $\varphi(A)x_0 \to \psi(A)y_0$ extends to a unitary transformation U of $\mathscr{H}$ onto $\mathscr{K}$ such that $U\varphi(A)U^{-1} = \psi(A)$ for all A in $\mathfrak{A}$, since $\tau = \omega_{x_0} \circ \varphi = \omega_{y_0} \circ \psi$ (see the proof of Proposition 4.5.3). ■

12.1.5. LEMMA. *If $\mathscr{R}$ is a finite von Neumann algebra containing subfactors $\mathscr{M}$ and $\mathscr{N}$ of types I_m and I_n, respectively, and E is a minimal projection in $\mathscr{M}$ such that $\|E - A\| < \frac{1}{8}$ for some A in the unit ball of $\mathscr{N}$, then m divides n.*

Proof. Since

$$\|E - \tfrac{1}{2}(A + A^*)\| \leq \tfrac{1}{2}(\|E - A\| + \|(E - A)^*\|) < \tfrac{1}{8},$$

we may assume that A is self-adjoint in $\mathscr{N}$. Thus $A = \lambda_1 F_1 + \cdots + \lambda_k F_k$, where $F_1, \ldots, F_k$ are orthogonal (spectral) projections (for A) in $\mathscr{N}$ and $-1 \leq \lambda_1 < \lambda_2 < \cdots < \lambda_k \leq 1$. Now

$$\|A - A^2\| \leq \|E - A\| \, \|A\| + \|I - E\| \, \|A - E\| < \tfrac{1}{4};$$

so that $|\lambda_j - \lambda_j^2| < \frac{1}{4}$. Since $|\lambda_j| \cdot |1 - \lambda_j| < \frac{1}{4}$, either $|\lambda_j| < \frac{1}{2}$ or $|1 - \lambda_j| < \frac{1}{2}$ —that is, λ_j is in either $(-\frac{1}{2}, \frac{1}{2})$ or $(\frac{1}{2}, 1]$. Suppose $\lambda_1, \ldots, \lambda_{j-1}$ are in $(-\frac{1}{2}, \frac{1}{2})$ and $\lambda_j, \ldots, \lambda_k$ are in $(\frac{1}{2}, 1]$. Let F be $F_j + \cdots + F_k$. Then $\|A - F\| < \frac{1}{2}$ so that $\|E - F\| < \frac{5}{8}$.

From Proposition 2.5.14, the range projection $R(EF)$ of EF is $E - (E \wedge (I - F))$. But $E \wedge (I - F) = 0$, for otherwise there is a unit vector x in the ranges of E and $I - F$ so that

$$1 = \|x\| = \|(E - F)x\| \leq \|E - F\| < \tfrac{5}{8}.$$

Thus $R(EF) = E$ and, symmetrically, $F = R(FE) \sim R((FE)^*)$. Hence E and F are equivalent in $\mathscr{R}$ (see Proposition 6.1.6). Since E is a minimal projection in $\mathscr{M}$, a subfactor of type I_m of $\mathscr{R}$, and F is a projection in $\mathscr{N}$, a subfactor of type I_n of $\mathscr{R}$, the (normalized) center-valued trace on $\mathscr{R}$ assigns $m^{-1}I$ to E and $qn^{-1}I$ to F, where q is some positive integer. Thus $nm^{-1} = q$ and m divides n. ■

As an immediate corollary of the preceding lemma (and Proposition 12.1.4), we have the following corollary.

12.1.6. COROLLARY. *If $\mathfrak{A}$ is a matricial C*-algebra of type $\{n_j\}$ and $\mathscr{N}$ is a subfactor of type I_n contained in $\mathfrak{A}$, then n divides some n_j.*

Corollary 12.1.6 and the discussion preceding the statement of Theorem 12.1.1 complete the proof of Theorem 12.1.1.

12.1.7. REMARK. It follows from Theorem 12.1.1 and the comments preceding it that there are an uncountable number of isomorphism classes of matricial C^*-algebras. For each possible invariant $\{a_n\}$, where a_n is either 0, or a positive integer, or ∞, and $\sum a_n = \infty$, there is a matricial C^*-algebra with that invariant. From tensor product considerations, each matricial C^*-algebra appears as a subalgebra of $\mathfrak{A}_\infty$ (type $\{n!\}$ with invariant $\{\infty, \infty, \ldots\}$). ■

12.1.8. REMARK. In Example 10.4.19 we constructed c inequivalent irreducible representations of the CAR algebra $\mathfrak{A}$ (the matricial C^*-algebra with invariant $\{\infty, 0, 0, \ldots\}$). If π_1 and π_2 are inequivalent irreducible representations of $\mathfrak{A}$ on $\mathscr{H}_{\pi_1}$ and $\mathscr{H}_{\pi_2}$ and $\pi_0 = \pi_1 \oplus \pi_2$ then $\pi_0(\mathfrak{A})^- = \mathscr{B}(\mathscr{H}_{\pi_1}) \oplus \mathscr{B}(\mathscr{H}_{\pi_2})$, from Corollary 10.3.9, where, as usual, $\pi(\mathfrak{A})^-$ denotes the weak-operator closure of $\pi(\mathfrak{A})$. It follows that $\pi_0(\mathfrak{A})^-$ is a von Neumann algebra of type I_∞ whose center is generated by two projections (and, in particular, is non-trivial). The assumption that $\mathfrak{A}^-$ is finite, in Proposition 12.1.2, is essential. ■

A matricial C^*-algebra $\mathfrak{A}$ is countably generated (equivalently, norm-separable) by the matrix units for the algebras in a generating nest. Moreover, each finite set of operators in $\mathfrak{A}$ can be approximated in norm as closely as we wish by operators in some finite type I subfactor of $\mathfrak{A}$—one of the algebras in a generating nest. The converse holds.

12.1.9. THEOREM. *If $\mathfrak{A}$ is a countably generated C^*-algebra, then $\mathfrak{A}$ is matricial if and only if for each positive ε and finite subset $\{A_1, \ldots, A_p\}$ of $\mathfrak{A}$ there is a finite type* I *subfactor $\mathscr{N}$ of $\mathfrak{A}$ and operators $B_1, \ldots, B_p$ in $\mathscr{N}$ such that $\|A_j - B_j\| < \varepsilon$, $j = 1, \ldots, p$.*

This result provides a useful criterion for determining when a C^*-algebra is matricial. It tells us, for example, that the norm closure of an ascending sequence of matricial C^*-algebras (each containing I) is, itself, a matricial C^*-algebra. Since its proof involves constructions entirely similar to those used in the proof of Theorem 12.2.2 (see Lemmas 12.2.3–12.2.6)—though, in the present case, involving the norm topology rather than the strong-operator topology—we shall give only a sketch of the proof.

Sketch of proof of Theorem 12.1.9. Let $\{A_1, A_2, \ldots\}$ be a countable family of operators generating $\mathfrak{A}$. Suppose we can show that, given a positive

ε, a finite subfactor $\mathcal{N}$ of type I in $\mathfrak{A}$ and a finite subset $\{A_1, \ldots, A_p\}$ of $\{A_1, A_2, \ldots\}$, there is a subfactor $\mathcal{M}$ of $\mathfrak{A}$, finite and of type I containing $\mathcal{N}$ and containing operators $B_1, \ldots, B_p$ such that $\|A_j - B_j\| < \varepsilon$ for j in $\{1, \ldots, p\}$. In this case, we choose $\mathcal{N}_1$ a finite type I subfactor of $\mathfrak{A}$ containing an operator B_{11} such that $\|A_1 - B_{11}\| < 1$. At the second stage, we choose $\mathcal{N}_2$ a finite type I subfactor of $\mathfrak{A}$ containing $\mathcal{N}_1$ and operators B_{12}, B_{22} such that $\|A_1 - B_{12}\| < \frac{1}{2}$ and $\|A_2 - B_{22}\| < \frac{1}{2}$. At the kth stage, we choose $\mathcal{N}_k$ a finite type I subfactor of $\mathfrak{A}$ containing $\mathcal{N}_{k-1}$ and operators $B_{1k}, \ldots, B_{kk}$ such that

$$\|A_1 - B_{1k}\| < k^{-1}, \qquad \ldots, \qquad \|A_k - B_{kk}\| < k^{-1}.$$

Then each A_j is in the norm closure of $\bigcup_{k=1}^{\infty} \mathcal{N}_k$. Thus this norm closure coincides with $\mathfrak{A}$; and $\mathfrak{A}$ is matricial.

To construct $\mathcal{M}$, given $\mathcal{N}$, $\{A_1, \ldots, A_p\}$, and ε as indicated, we proceed as follows. Let $\{E_{rs}\}$ be a self-adjoint $n \times n$ system of matrix units for the type I_n factor $\mathcal{N}$. By hypothesis we can choose $\mathcal{N}_0$, a finite subfactor of type I in $\mathfrak{A}$ containing operators A_{rs} in its unit ball and operators $C_1, \ldots, C_p$ such that $\|A_{rs} - E_{rs}\| < \varepsilon_1$ and $\|A_j - C_j\| < \varepsilon_1$, where ε_1 is a preassigned positive number. In the style of Lemma 12.1.5 (and, more closely, in the style of Lemmas 12.2.3 and 12.2.4), we can replace each A_{rs} by F_{rs} so that $\{F_{rs}\}$ is a self-adjoint system of $n \times n$ matrix units in $\mathcal{N}_0$ and $\|E_{rs} - F_{rs}\| < f(\varepsilon_1, n)$, where, for fixed n, $f(\varepsilon_1, n) \to 0$ as $\varepsilon_1 \to 0$. (Since $\sum_{j=1}^{n} E_{jj} = I$, $\|I - \sum_{j=1}^{n} F_{jj}\|$ is small and $\sum_{j=1}^{n} F_{jj} = I$.) In this case $\|E_{11} - E_{11}F_{11}E_{11}\|$ is small; so that $E_{11}F_{11}E_{11}$ and $(E_{11}F_{11}E_{11})^{1/2}$ are invertible in the Banach algebra $E_{11}\mathfrak{A}E_{11}$. Thus if $U_1(E_{11}F_{11}E_{11})^{1/2}$ is the polar decomposition of $F_{11}E_{11}$, then U_1 is a partial isometry in $\mathfrak{A}$ with initial projection E_{11} and final projection F_{11}. Moreover, $\|U_1 - E_{11}\|$ is small (depending on ε_1 and n). Let U_r be $F_{r1}U_1E_{1r}$. Then U_r is a partial isometry in $\mathfrak{A}$ with initial projection E_{rr} and final projection F_{rr}, and $\|U_r - E_{rr}\|$ is small. If $U = \sum_{r=1}^{n} U_r$, then U is a unitary operator in $\mathfrak{A}$, $\|I - U\|$ is small, and

$$\begin{aligned} UE_{rs}U^* &= (\textstyle\sum U_j)E_{rs}(\sum U_k^*) = (\sum F_{j1}U_1E_{1j})E_{rs}(\sum E_{k1}U_1^*F_{1k}) \\ &= F_{r1}U_1E_{1r}E_{rs}E_{s1}U_1^*F_{1s} = F_{r1}U_1E_{11}U_1^*F_{1s} = F_{r1}F_{11}F_{1s} = F_{rs}. \end{aligned}$$

Thus $U^*\mathcal{N}_0U$ is a finite subfactor $\mathcal{M}$ of type I in $\mathfrak{A}$ containing $\mathcal{N}$ and operators B_1 $(=U^*C_1U), \ldots, B_p$ $(=U^*C_pU)$ such that $\|A_j - B_j\| < \varepsilon$, $j \in \{1, \ldots, p\}$ (where ε_1 is chosen small). ■

12.1.10. Remark. From the preceding argument, we see that each finite type I subfactor $\mathcal{N}$ of a uniformly matricial C^*-algebra $\mathfrak{A}$ is contained in some generating nest for $\mathfrak{A}$ and is therefore one of a countably infinite commuting family of such subfactors that generates $\mathfrak{A}$. From Lemma 11.4.14, the C^*-algebra generated by those subfactors in the commuting family other than $\mathcal{N}$ is precisely $\mathcal{N}^c$ (the commutant of $\mathcal{N}$ in $\mathfrak{A}$). Thus $\mathcal{N}^c$ is itself a uniformly matricial C^*-algebra. ■

Bibliography: [5, 29]

12.2. The finite matricial factor

In Section 12.1, we studied uniformly matricial C^*-algebras. We noted that the weak-operator closure of the image under the GNS representation corresponding to the (unique) tracial state is a factor of type II_1. We say that a von Neumann algebra is *matricial* when it is the weak-operator closure of a uniformly matricial C^*-algebra. The main result of this section (Theorem 12.2.1) states that all finite matricial von Neumann algebras are * isomorphic.

We shall make use of the trace norm on a factor $\mathcal{M}$ of type II_1. Recall, from Remark 8.5.9, that $\|A\|_2 = (\text{trace } A^*A)^{1/2}$. In case $\mathcal{M}$ has a trace vector x_0, $\|A\|_2 = \|Ax_0\|$, so that strong-operator convergence implies trace-norm convergence. In this same case, convergence in trace norm on bounded subsets of $\mathcal{M}$ implies strong-operator convergence; for if $\{\|T_a\|\}$ is bounded and $\|T_a - T\|_2 \to_a 0$, then

$$\|(T_a - T)A'x_0\| = \|A'(T_a - T)x_0\| \underset{a}{\to} 0 \qquad (A' \in \mathcal{M}').$$

But x_0 is generating for $\mathcal{M}'$ since it is separating for $\mathcal{M}$. Thus $\{T_a\}$ is strong-operator convergent to T. For most of the purposes of this section, we may assume that $\mathcal{M}$ has a trace vector (for example, by considering $\mathcal{M}$ in its universal normal representation—see the discussion following Corollary 7.1.5).

Suppose that $\mathcal{M}$ and $\mathcal{N}$ are finite matricial von Neumann algebras (hence, from Corollary 12.1.3, factors) and that $\mathfrak{A}$ and $\mathcal{B}$ are * isomorphic weak-operator-dense uniformly matricial C^*-subalgebras of $\mathcal{M}$ and $\mathcal{N}$, respectively. In the manner just described, we may assume that the normalized tracial states on $\mathcal{M}$ and $\mathcal{N}$ have the forms $\omega_{x_0} | \mathcal{M}$ and $\omega_{y_0} | \mathcal{N}$. The factors resulting from restricting to the cyclic projections generated by x_0 and y_0 are * isomorphic to the original factors (see Proposition 5.5.5) and may be viewed as the weak-operator closures of the images of the (abstract) C^*-algebra to which $\mathfrak{A}$ and $\mathcal{B}$ are * isomorphic under the GNS representations corresponding to tracial states. From Proposition 12.1.4, these GNS representations are unitarily equivalent and, in particular, $\mathcal{M}$ and $\mathcal{N}$ are * isomorphic.

To prove that all finite matricial von Neumann algebras are * isomorphic, it will suffice, then, to show that each such contains, as a weak-operator-dense subalgebra, a specific uniformly matricial C^*-algebra. We shall show that this is the case for the uniformly matricial C^*-algebra $\mathfrak{A}_\infty$ (see Remark 12.1.7—$\{a_n\}$ has ∞ for each a_n).

Our program is to show (Lemma 12.2.6) that if $\mathscr{N}_0$ is a subfactor of type I_n of a finite matricial factor $\mathscr{M}$, ε is a positive number, m is a positive integer, and $A_1, \ldots, A_p$ are operators in the unit ball of $\mathscr{M}$, then some subfactor of type I_{mq} contains $\mathscr{N}_0$ and operators $B_1, \ldots, B_p$ in its unit ball such that $\|A_j - B_j\|_2 < \varepsilon$. If we have this result, we can approximate compatible matrix unit systems of a generating nest for $\mathscr{M}$ more and more closely in trace norm by subfactors of $\mathscr{M}$ finite and of type I, where the orders of the subfactors contain given prime-power divisors. At each stage we absorb the subfactor of the preceding stage and introduce a divisor consisting of a suitable product of primes (for example, at the nth stage we introduce $p_1^n \cdots p_n^n$ as a divisor of the order). In this way, we construct a uniformly matricial C^*-subalgebra $\mathfrak{A}$ in $\mathscr{M}$ with invariant $\{\infty, \infty, \ldots\}$ such that each matrix unit T of the compatible system for the original generating nest is a limit in trace norm of a sequence of operators in the unit ball of $\mathfrak{A}$. Hence $T \in \mathfrak{A}^-$ and $\mathfrak{A}^- = \mathscr{M}$.

From this discussion, once we have proved Lemma 12.2.6, we have the following theorem.

12.2.1. Theorem. *All finite matricial von Neumann algebras are * isomorphic.*

The proof of Lemma 12.2.6 is effected with the aid of some preliminary lemmas. Our plan is to approximate the $n \times n$ matrix units of the factor $\mathscr{N}_0$ of type I_n, to be absorbed, as well as the operators $A_1, \ldots, A_p$, very closely in trace norm by operators in the unit ball of some factor $\mathscr{M}_1$ in the generating nest. Working in $\mathscr{M}_1' \cap \mathscr{M}$ (a factor of type II_1), a factor of type I_n can be chosen that generates with $\mathscr{M}_1$ a subfactor $\mathscr{M}_2$ of $\mathscr{M}$ of type I_q, where q is divisible by mn. The original approximation is still valid for $\mathscr{M}_2$ (since it contains $\mathscr{M}_1$). In the next step, we adjust the operators in $\mathscr{M}_2$ approximating the $n \times n$ matrix unit system in $\mathscr{N}_0$ so that the adjusted operators form a self-adjoint system of $n \times n$ matrix units in $\mathscr{M}_2$ for which the sum of the projections (principal units) is I (see Lemma 12.2.4). For the principal units to have sum I, it is necessary to arrange, as outlined, that $\mathscr{M}_2$ have order divisible by n. The basic adjustment of operators approximated in trace norm by matrix units so that the adjusted operators are matrix units is carried out in Lemma 12.2.3. Since $\mathscr{N}_0$ and $\mathscr{M}_2$ are known, now, to have $n \times n$ matrix unit systems close to one another in trace norm, there is a unitary operator in $\mathscr{M}$ that maps the matrix unit system in $\mathscr{M}_2$ onto that of $\mathscr{N}_0$ and is close to I in trace norm. The transform $\mathscr{M}_0$ of $\mathscr{M}_2$ contains $\mathscr{N}_0$, since $\mathscr{N}_0$ is generated by its $n \times n$ matrix unit system. The approximation to $A_1, \ldots, A_p$ in trace norm is still possible with operators in $\mathscr{M}_0$, since the unitary operator is close to I in trace norm. Construction of the unitary operator is carried out in Lemma 12.2.5.

The preceding discussion and Lemmas 12.2.3–12.2.6 establish the following important condition for a finite factor to be matricial.

12.2.2. THEOREM. *If $\mathscr{M}$ is a countably generated factor of type* II_1 *such that for each positive ε and each finite set $\{A_1, \ldots, A_p\}$ of operators in the unit ball of $\mathscr{M}$ there is a finite type* I *subfactor $\mathscr{N}$ of $\mathscr{M}$ and operators $B_1, \ldots, B_p$ in the unit ball of $\mathscr{N}$ such that $\|A_j - B_j\|_2 < \varepsilon$ when $j \in \{1, \ldots, p\}$, then $\mathscr{M}$ is matricial.*

The assumption that $\mathscr{M}$ be countably generated is automatically satisfied if the underlying Hilbert space is separable. If $\{A_n\}$ is a countable set of operators in $\mathscr{M}$ generating it as a von Neumann algebra, it replaces the compatible matrix unit system for a weak-operator-dense matricial C^*-algebra in $\mathscr{M}$ in the discussion preceding the statement of Theorem 12.2.1. In that discussion, the generating nest serves only to provide us with the finite type I subfactor $\mathscr{N}$ appearing in the statement of Theorem 12.2.2. In either case, Lemmas 12.2.3–12.2.6, which follow, complete the proof.

The criterion obtained from Theorem 12.2.2 by omitting reference to the unit ball in $\mathscr{M}$ and $\mathscr{N}$ is also valid (see Exercise 12.4.32)—though the stated criterion suffices for most purposes. The result for C^*-algebras corresponding to Theorem 12.2.2 is Theorem 12.1.9. As in the C^*-algebra case, this criterion tells us that a finite factor appearing as the closure of an ascending sequence of matricial C^*-algebras (or von Neumann algebras) is a matricial factor.

12.2.3. LEMMA. *If E and F are equivalent orthogonal projections in a finite factor $\mathscr{M}$, V is a partial isometry in $\mathscr{M}$ with initial projection E and final projection F, $\mathscr{N}$ is a type I_n subfactor of $\mathscr{M}$, and A, B, T are operators in the unit ball of $\mathscr{N}$ such that $\|A - E\|_2 < b\,(<10^{-12})$, $\|B - F\|_2 < b$, and $\|T - V\|_2 < b$, then there are equivalent orthogonal projections M and N and a partial isometry W in $\mathscr{N}$ such that $W^*W = M$, $WW^* = N$, and*

$$\|M - E\|_2 < 56b^{1/128}, \qquad \|N - F\|_2 < 56b^{1/128}, \qquad \|W - V\|_2 < 56b^{1/128}.$$

If A is itself a projection, we can choose M so that $M \le A$.

Proof. Replacing A and B by $\frac{1}{2}(A + A^*)$ and $\frac{1}{2}(B + B^*)$, we may assume that A and B are self-adjoint. Using Remark 8.5.9, we have

$$\|A - A^2\|_2 \le \|A(E - A)\|_2 + \|(A - E)(I - E)\|_2 < 2b.$$

Since A is a self-adjoint element in the unit ball of $\mathscr{N}$, $A = \sum_{j=1}^{m} \lambda_j G_j$, where $|\lambda_j| \le 1$ and $\{G_j\}$ is an orthogonal family of projections in $\mathscr{N}$. If $d = b^{1/4}$, $X_0 = \{j : \lambda_j \notin [-d, d] \cup [1 - d, 1]\}$, and $j \in X_0$, then $d^2 < |\lambda_j| \cdot |1 - \lambda_j|$. Thus

$$d^4 \sum_{j \in X_0} \tau(G_j) < \|A - A^2\|_2^2 < 4b^2,$$

where τ is the normalized trace on $\mathscr{M}$. Let M_0 be the sum of those G_j for which $1 - d \le \lambda_j$. Then

$$\|A - M_0\|_2^2 \le d^2 \sum_{j \notin X_0} \tau(G_j) + \sum_{j \in X_0} \tau(G_j) < b^{1/2} + 4b,$$

and, since $b < 10^{-2}$,

$$\|M_0 - E\|_2 < (b^{1/2} + 4b)^{1/2} + b < 2b^{1/4}.$$

In case A is a projection, let M_0 be A (indeed, the preceding discussion will yield A as M_0). As $b < 10^{-2}$, we have

$$\|AB\|_2 \le \|(A - E)B\|_2 + \|E(B - F)\|_2 < 2b$$

(and $\|BA\|_2 < 2b$); so that

$$\begin{aligned} \|B - (I - M_0)B(I - M_0)\|_2 &\le 2\|BM_0\|_2 + \|M_0 B\|_2 \\ &\le 2(\|BA\|_2 + \|B(M_0 - A)\|_2) + \|AB\|_2 \\ &\quad + \|(M_0 - A)B\|_2 \\ &\le 6b + 3(b^{1/2} + 4b)^{1/2} < 4b^{1/4}. \end{aligned}$$

Thus $\|(I - M_0)B(I - M_0) - F\|_2 < 5b^{1/4}$ $(<10^{-2})$. Applying the previous argument with $(I - M_0)B(I - M_0)$ and F in place of A and E, we can construct a projection N_0 in $(I - M_0)\mathscr{N}(I - M_0)$ such that $\|N_0 - F\|_2 < 2(5b^{1/4})^{1/4} < 5b^{1/16}$.

If $N_0 T M_0 = S = W(S^*S)^{1/2}$ $(=(SS^*)^{1/2}W)$, where W is a partial isometry in $\mathscr{N}$ such that $W^*W = M \le M_0$ and $WW^* = N \le N_0$, then

$$\|S - V\|_2 = \|N_0 T M_0 - FVE\|_2 \le 5b^{1/16} + b + 2b^{1/4} \le 6b^{1/16}.$$

Thus

$$\|S^*S - V^*V\|_2 = \|S^*S - E\|_2 \le 12b^{1/16}$$

and

$$\|S^*S - (S^*S)^2\|_2 < 24b^{1/16}.$$

Arguing, now, as at the beginning of this proof with S^*S in place of A, d_0 $(=(12b^{1/16})^{1/4})$ in place of d, and Y_0 $(=\{j : \lambda_j \notin [0, d_0] \cup [1 - d_0, 1]\})$

in place of X_0, where $S^*S = \sum_{j=1}^{m} \lambda_j G_j$ with $\{G_j\}$ an orthogonal family of (spectral) projections in $\mathscr{N}$ for S^*S; we have $d_0^2 < |\lambda_j| \cdot |1 - \lambda_j|$ if $j \in Y_0$, so that $\sum_{j \in Y_0} \tau(G_j) < 48b^{1/16}$. Thus

$$\|(S^*S)^{1/2} - S^*S\|_2^2 = \sum_{j \notin Y_0} (\lambda_j^{1/2} - \lambda_j)^2 \tau(G_j) + \sum_{j \in Y_0} (\lambda_j^{1/2} - \lambda_j)^2 \tau(G_j)$$
$$\leq d_0 + 48b^{1/16} < 2b^{1/64} + 48b^{1/16} < 50b^{1/64}$$

and (since $W = WM = WM_0$ and $S = W(S^*S)^{1/2}$)

$$\begin{aligned}\|W - V\|_2 &\leq \|W(M_0 - E)\|_2 + \|W(E - S^*S)\|_2 \\ &\quad + \|W(S^*S - (S^*S)^{1/2})\|_2 + \|S - V\|_2 \\ &\leq 2b^{1/4} + 12b^{1/16} + 8b^{1/128} + 6b^{1/16} < 28b^{1/128}.\end{aligned}$$

Hence

$$\|W^*W - V^*V\|_2 = \|M - E\|_2 < 56b^{1/128}$$

and

$$\|WW^* - VV^*\|_2 = \|N - F\|_2 < 56b^{1/128}. \quad ■$$

12.2.4. LEMMA. *There is a positive integer k_n and a positive real-valued function f defined for pairs of positive integers (k, n) for which $k_n \leq k$ such that, for each n, $f(k, n)$ tends to 0 as k tends to ∞ and such that if $\mathscr{M}$ is a factor of type II_1, $\mathscr{N}_0$ is a subfactor of type I_n, $\mathscr{M}_0$ a subfactor of type I_{np}, $\{E_{rs}\}$ a self-adjoint system of $n \times n$ matrix units for $\mathscr{N}_0$ and $\{A_{rs}\}$ a set of n^2 operators in the unit ball of $\mathscr{M}_0$ for which $\|E_{rs} - A_{rs}\|_2 < k^{-1}$, where $k_n \leq k$, then there is a self-adjoint system $\{F_{rs}\}$ of $n \times n$ matrix units in $\mathscr{M}_0$ such that $\sum_{r=1}^{n} F_{rr} = I$ and $\|E_{rs} - F_{rs}\|_2 < f(k, n)$.*

Proof. From Lemma 12.2.3, if $k^{-1} < 10^{-12}$, we can find orthogonal projections M_{11}^1, M_{22}^1 and a partial isometry M_{21}^1 in $\mathscr{M}_0$ such that $M_{21}^1 M_{12}^1 = M_{22}^1$, $M_{12}^1 M_{21}^1 = M_{11}^1$, where $M_{12}^1 = M_{21}^{1*}$, and

$$\|E_{jh} - M_{jh}^1\|_2 < 56k^{-1/128} \qquad (j, h = 1, 2).$$

Since

$$\begin{aligned}&\|E_{j3} - (I - M_{11}^1)A_{j3}(I - M_{11}^1)\|_2 \\ &\quad = \|(I - E_{11})E_{j3}(I - E_{11}) - (I - M_{11}^1)A_{j3}(I - M_{11}^1)\|_2 \\ &\quad \leq 112k^{-1/128} + k^{-1},\end{aligned}$$

where $j = 2, 3$, we can apply Lemma 12.2.3, again, to the algebra $(I - M_{11}^1)\mathscr{M}_0(I - M_{11}^1)$ with E_{22}, E_{33}, E_{32} in place of E, F, V of that lemma, M_{22}^1, $(I - M_{11}^1)A_{33}(I - M_{11}^1)$, $(I - M_{11}^1)A_{32}(I - M_{11}^1)$ in place of A, B, T

and $112k^{-1/128} + k^{-1}$ in place of b. This application yields orthogonal projections M_{22}^2, M_{33}^2 and a partial isometry M_{32}^2 (all in $(I - M_{11}^1)\mathscr{M}_0(I - M_{11}^1)$) such that $M_{32}^2 M_{23}^2 = M_{33}^2$ (where $M_{23}^2 = M_{32}^{2*}$), $M_{23}^2 M_{32}^2 = M_{22}^2 \leq M_{22}^1$ (since A is the projection M_{22}^1, in the present case) and such that

$$\|E_{jh} - M_{jh}^2\|_2 \leq 56(112k^{-1/128} + k^{-1})^{1/128} \qquad (j, h = 2, 3).$$

Let M_{12}^2 be $M_{12}^1 M_{22}^2$, M_{21}^2 be M_{12}^{2*} $(=M_{22}^2 M_{21}^1)$, and M_{11}^2 be $M_{12}^2 M_{21}^2$. Then, since

$$\begin{aligned}\|M_{22}^1 - M_{22}^2\|_2 &\leq \|M_{22}^1 - E_{22}\|_2 + \|E_{22} - M_{22}^2\|_2 \\ &\leq 56(112k^{-1/128} + k^{-1})^{1/128} + 56k^{-1/128},\end{aligned}$$

we have

$$\begin{aligned}\|M_{12}^2 - E_{12}\|_2 \leq{} & \|M_{12}^1 M_{22}^2 - M_{12}^1 M_{22}^1\|_2 \\ & + \|M_{12}^1 M_{22}^1 - M_{12}^1 E_{22}\|_2 + \|M_{12}^1 E_{22} - E_{12}E_{22}\|_2 \\ \leq{} & 57(112k^{-1/128} + k^{-1})^{1/128}\end{aligned}$$

and

$$\|M_{11}^2 - E_{11}\|_2 \leq \|M_{12}^2 M_{21}^2 - E_{12}E_{21}\|_2 \leq 114(112k^{-1/128} + k^{-1})^{1/128}.$$

Thus

$$\|M_{jh}^2 - E_{jh}\|_2 \leq 114(112k^{-1/128} + k^{-1})^{1/128} \qquad (j, h = 1, 2, 3, |j - h| \leq 1).$$

Continuing in this way (at the next stage, we construct M_{33}^3, M_{44}^3, and M_{43}^3 such that $M_{43}^3 M_{34}^3 = M_{44}^3$ and $M_{34}^3 M_{43}^3 = M_{33}^3 \leq M_{33}^2$; and then we replace M_{32}^2 by $M_{33}^3 M_{32}^2$ labeled as M_{32}^3, then M_{22}^2 by $M_{23}^3 M_{32}^3$, and so forth), we construct mutually orthogonal projections $M_{11}, \ldots, M_{nn}$ and partial isometries M_{j+1j} in $\mathscr{M}_0$ such that $M_{j+1j}M_{jj+1} = M_{j+1j+1}$, $M_{jj+1}M_{j+1j} = M_{jj}$, where $M_{jj+1} = M_{j+1j}^*$, and $\|E_{jj} - M_{jj}\|_2 < f_0(k, n)$, $\|E_{j+1j} - M_{j+1j}\|_2 < f_0(k, n)$, where, for each n, $f_0(k, n)$ tends to 0 as k tends to infinity. To construct M_{33}^3, M_{44}^3, and M_{43}^3, k^{-1} must have been chosen so small, at the outset, that $457(112k^{-1/128} + k^{-1})^{1/128} < 10^{-12}$. If $1 \leq s < r \leq n$, define M_{rs} to be $M_{rr-1} \cdots M_{s+1s}$ and M_{sr} to be M_{rs}^*. Then $\{M_{rs}\}$ is a self-adjoint system of $n \times n$ matrix units in $\mathscr{M}_0$, and $\|E_{rs} - M_{rs}\|_2 < nf_0(k, n)$. Now

$$\|I - \sum_{r=1}^{n} M_{rr}\|_2 \leq \sum_{r=1}^{n} \|E_{rr} - M_{rr}\|_2 < nf_0(k, n).$$

Since $M_{11}, \ldots, M_{nn}$ are orthogonal equivalent projections in $\mathscr{M}_0$ and $\mathscr{M}_0$ is of type I_{np}; $I - \sum_{r=1}^{n} M_{rr}$ is the sum of n orthogonal equivalent projections $N_{11}, \ldots, N_{nn}$ in $\mathscr{M}_0$. Let $\{N_{rs}\}$ be a self-adjoint system of $n \times n$ matrix units in $\mathscr{M}_0$ (formed on $\{N_{rr}\}$); and let F_{rs} be $M_{rs} + N_{rs}$. Then $\{F_{rs}\}$ is a self-adjoint system of $n \times n$ matrix units in $\mathscr{M}_0$ and $\sum_{r=1}^{n} F_{rr} = I$. Since

$$\|N_{rs}\|_2^2 = \|N_{ss}\|_2^2 < nf_0(k, n)^2,$$

we have

$$\|E_{rs} - F_{rs}\|_2 \leq \|E_{rs} - M_{rs}\|_2 + \|N_{rs}\|_2 \leq (n + n^{1/2})f_0(k, n).$$

Let $f(k, n)$ be $(n + n^{1/2})f_0(k, n)$; and let k_n be chosen so large that if $k^{-1} < k_n^{-1}$ at the outset, then, at the last stage of the construction, Lemma 12.2.3 applies to allow us to find M_{nn} and M_{nn-1}. ■

12.2.5. Lemma. *If $\{E_{rt}\}$ and $\{F_{rt}\}$ are self-adjoint systems of $n \times n$ matrix units in a finite factor $\mathscr{M}$, $I = \sum_{r=1}^{n} E_{rr} = \sum_{r=1}^{n} F_{rr}$, and $\|E_{rt} - F_{rt}\|_2 < a$, then there is a unitary operator U in $\mathscr{M}$ such that $UE_{rt}U^* = F_{rt}$, for r, t in $\{1, \ldots, n\}$, and $\|U - I\|_2 < (7n - 1)a$.*

Proof. Let $V_{11}(E_{11}F_{11}E_{11})^{1/2}$ be the polar decomposition of $F_{11}E_{11}$; so that V_{11} is a partial isometry in $\mathscr{M}$ with initial projection E, the range projection of $E_{11}F_{11}$, and final projection F, the range projection of $F_{11}E_{11}$. Using Remark 8.5.9, we have

$$\begin{aligned}&[(E_{11}F_{11}E_{11})^{1/2} - E_{11}]^2\\ &\quad = [(E_{11}F_{11}E_{11})^{1/2} - E_{11}]^2E_{11}\\ &\quad \leq [(E_{11}F_{11}E_{11})^{1/2} - E_{11}]^2[(E_{11}F_{11}E_{11})^{1/2} + E_{11}]^2\\ &\quad = [E_{11}F_{11}E_{11} - E_{11}]^2 = [E_{11}(F_{11} - E_{11})E_{11}]^2\end{aligned}$$

and

$$\|(E_{11}F_{11}E_{11})^{1/2} - E_{11}\|_2 \leq \|E_{11}(F_{11} - E_{11})E_{11}\|_2 \leq \|F_{11} - E_{11}\|_2 < a.$$

Thus

$$\begin{aligned}\|V_{11} - E_{11}\|_2 &= \|V_{11}E_{11} - E_{11}\|_2\\ &\leq \|V_{11}E_{11} - V_{11}(E_{11}F_{11}E_{11})^{1/2}\|_2 + \|F_{11}E_{11} - E_{11}\|_2 < 2a;\end{aligned}$$

and

$$\|E - E_{11}\|_2 = \|V_{11}^*V_{11} - E_{11}^2\|_2 < 4a.$$

Since $E \sim F$ and $E_{11} \sim F_{11}$ in $\mathscr{M}$, $E_{11} - E \sim F_{11} - F$. Let W_{11} be a partial isometry in $\mathscr{M}$ with initial projection $E_{11} - E$ and final projection $F_{11} - F$. If $U_1 = W_{11} + V_{11}$, then U_1 is a partial isometry in $\mathscr{M}$ with initial projection E_{11} and final projection F_{11}. Moreover,

$$\|U_1 - E_{11}\|_2 \leq \|W_{11}\|_2 + \|V_{11} - E_{11}\|_2 = \|E - E_{11}\|_2 + \|V_{11} - E_{11}\|_2 < 6a.$$

Let U_j be $F_{j1}U_1E_{1j}$. Then $U_j^*U_j = E_{jj}$, $U_jU_j^* = F_{jj}$ and, when $j \in \{2, 3, \ldots, n\}$,

$$\|U_j - E_{jj}\|_2 = \|F_{j1}U_1E_{1j} - E_{j1}E_{11}E_{1j}\|_2 < 7a.$$

Let U be $\sum_{j=1}^n U_j$. Then U is a unitary operator in $\mathscr{M}$ and

$$\|U - I\|_2 \le (7n - 1)a.$$

We have

$$UE_{rt}U^* = F_{r1}U_1E_{1r}E_{rt}E_{t1}U_1^*F_{1t} = F_{rt}. \quad \blacksquare$$

12.2.6. LEMMA. *If $\mathscr{M}$ is a matricial factor of type* II_1, *$\mathscr{N}_0$ is a subfactor of type* I_n, *$A_1, \ldots, A_p$ are operators in the unit ball of $\mathscr{M}$, ε is a positive number, and m is a positive integer, there is a subfactor $\mathscr{M}_0$ of $\mathscr{M}$ of type* I_{mr} *containing $\mathscr{N}_0$ and operators $B_1, \ldots, B_p$ in the unit ball of $\mathscr{M}_0$ such that $\|A_j - B_j\|_2 < \varepsilon$.*

Proof. Let $\mathscr{M}_1$ be a subfactor of type I_t of $\mathscr{M}$ (from a generating nest for $\mathscr{M}$) such that $\|E_{rs} - A_{rs}\|_2 < k^{-1}$ and $\|A_j - B_j'\|_2 < \varepsilon/2$, where A_{rs} and B_j' are in the unit ball of $\mathscr{M}_1$, $\{E_{rs}\}$ is a self-adjoint system of $n \times n$ matrix units for $\mathscr{N}_0$ and k is so large that $f(k, n) < \varepsilon[6(7n - 1)]^{-1}$. Then $\mathscr{M}_1' \cap \mathscr{M}$ is a subfactor of $\mathscr{M}$ of type II_1. Let $\mathscr{M}_2$ be a subfactor of $\mathscr{M}_1' \cap \mathscr{M}$ of type I_{mn}. Then $\mathscr{M}_1$ and $\mathscr{M}_2$ generate a subfactor $\mathscr{M}_3$ of type I_{mnt} (see Lemma 11.4.8). From Lemma 12.2.4, there is a self-adjoint system of $n \times n$ matrix units $\{F_{rs}\}$ in $\mathscr{M}_3$ such that $\sum_{r=1}^n F_{rr} = I$ and $\|E_{rs} - F_{rs}\|_2 < f(k, n)$ (for this, k is chosen greater than k_n). Applying Lemma 12.2.5, we find a unitary operator U in $\mathscr{M}$ such that $UF_{rs}U^* = E_{rs}$ and $\|U - I\|_2 < \varepsilon/6$. Then $U\mathscr{M}_3U^*$ is a subfactor $\mathscr{M}_0$ of type I_{mnt} containing $\mathscr{N}_0$. Moreover, if $B_j = UB_j'U^*$, then $\|A_j - B_j\|_2 \le \|A_j - B_j'\|_2 + \|B_j' - UB_j'U^*\|_2 < \varepsilon$. $\blacksquare$

In Example 6.7.7 we considered the group Π of permutations that leave fixed all but a finite set of integers. We noted that Π is an i.c.c. group and, by virtue of Theorem 6.7.5, that its left and right von Neumann group algebras $\mathscr{L}_\Pi$ and $\mathscr{R}_\Pi$ are factors of type II_1. If $\{g_1, \ldots, g_n\}$ is a finite subset of Π, then, for some integer m, the permutations $g_1, \ldots, g_n$ leave fixed all integers not in $\{-m, \ldots, m\}$. It follows that $\{g_1, \ldots, g_n\}$ lies in the finite subgroup of Π consisting of those permutations that move only the integers $-m, \ldots, m$. Each finite subset of Π generates a finite subgroup. We say that a group in which each finite subset generates a finite group is *locally finite.* Equivalently G is locally finite if and only if its set of finite subgroups is directed by inclusion and G is the union of its finite subgroups. If G is countable, then G is locally finite if and only if it is the union of an ascending sequence of finite subgroups.

If G_0 is a finite subgroup of G, then $\{L_g : g \in G_0\}$ generates a finite-dimensional C^*-subalgebra of $\mathscr{L}_G$. It follows that if G is a countable locally finite group, $\mathscr{L}_G$ is the strong-operator closure of an ascending sequence of finite-dimensional C^*-algebras. We conclude from this discussion (by applying a slightly extended form of Theorem 12.2.2—see Remark 12.2.8 below), that $\mathscr{L}_G$ is the finite matricial factor whenever G is a locally finite i.c.c. group—in particular, $\mathscr{L}_\Pi$ is the finite matricial factor.

12.2.7. PROPOSITION. *If G is a locally finite* i.c.c. *group, $\mathscr{L}_G$ is the finite matricial factor.*

In Example 6.7.6 we noted that $\mathscr{F}_n$, the free group on n generators, is an i.c.c. group when $n \geq 2$. It is proved in Theorem 6.7.8 that $\mathscr{L}_\Pi$ and $\mathscr{L}_{\mathscr{F}_2}$ are not * isomorphic. Since $\mathscr{L}_\Pi$ is the finite matricial factor it follows that $\mathscr{L}_{\mathscr{F}_2}$ is not matricial.

12.2.8. REMARK. The condition that $\{A_1, \ldots, A_p\}$ be approximable by operators in a finite type I subfactor of $\mathscr{M}$, in Theorem 12.2.2, can be weakened to approximation by operators in a finite-dimensional C^*-subalgebra $\mathscr{R}$ of $\mathscr{M}$ since each finite set in such a subalgebra can be approximated by a finite set in a finite type I subfactor. To see this, we suppose $A_1, \ldots, A_p$ are in $\mathscr{R}$. Note that $\mathscr{R}$ is a finite type I von Neumann algebra. From Proposition 6.6.6 there are mutually orthogonal projections $E_1, \ldots, E_n$ central in $\mathscr{R}$ (but not in $\mathscr{M}$) such that $\mathscr{R}E_j$ is a finite type I factor and $\sum E_j$ is the unit of $\mathscr{R}$. Let $\{E_{rs}^{(j)}\}$ be a system of matrix units for $\mathscr{R}E_j$. We can assume that $\{A_1, \ldots, A_p\}$ is the set of all $E_{rs}^{(j)}$. If $F_{11}^{(j)}$ is a projection in $\mathscr{M}$ slightly smaller (in operator order and trace norm) than $E_{11}^{(j)}$ and $F_{rs}^{(j)}$ is $E_{r1}^{(j)}F_{11}^{(j)}E_{1s}^{(j)}$, then $F_{rs}^{(j)}$ is near $E_{rs}^{(j)}$ and $\{F_{rs}^{(j)}\}$ is a system of matrix units for a finite type I factor. In this process, we can arrange that the dimension of the $F_{11}^{(j)}$ are mutually commensurable, so that each is the sum of a finite set of projections orthogonal and equivalent to each other for all j. Using this subdivision of the $F_{11}^{(j)}$, we can now construct a system of matrix units consistent with all the $F_{rs}^{(j)}$, generating a finite type I subfactor of $\mathscr{M}$ containing all $\{F_{rs}^{(j)}\}$. ■

12.2.9. REMARK. The criterion of Remark 12.2.8 allows us to conclude that $E\mathscr{M}E$ is the finite matricial factor, when $\mathscr{M}$ is the finite matricial factor and E is a projection in $\mathscr{M}$. To see this, let E have dimension b relative to $\mathscr{M}$, let a positive ε be given, and let F be a subprojection of E in $\mathscr{M}$ of rational dimension m/n, where $b - (\varepsilon/3)^2 < m/n < b$. From the proof Theorem 12.2.1, $\mathscr{M}$ contains the uniformly matricial C^*-algebra $\mathfrak{A}_\infty$ as an ultraweakly dense subalgebra. Hence there is a generating nest $\{\mathscr{M}_j\}$ of finite type I factors for $\mathscr{M}$ containing projections of every rational dimension. In particular, there is a projection F_0 of dimension m/n in some $\mathscr{M}_j$. There is a unitary operator U in $\mathscr{M}$ such that $UF_0U^* = F$. Upon replacing $\{\mathscr{M}_j\}$ by $\{U\mathscr{M}_jU^*\}$, we may assume that $F \in \mathcal{M}_j$. With $A_1, \ldots, A_p$ in $(E\mathcal{M}E)_1$, there are $B_1, \ldots, B_p$ in $(\mathcal{M}_k)_1$, for some k greater than j, such that $\|A_j - B_j\|_2 < \varepsilon/3$ when $j = 1, \ldots, p$. In this case, $FB_1F, \ldots, FB_pF$ are in $F\mathcal{M}_kF$; and

$$\|A_j - FB_jF\|_2 = \|EA_jE - FB_jF\|_2 < \varepsilon. \quad \blacksquare$$

Bibliography: [13, 58]

12.3. States and representations of matricial C^*-algebras

In Section 11.4 we studied C^*-algebras that appear as the tensor product of C^*-algebras. We introduced product states on a C^*-algebra $\mathfrak{A}$ generated by a commuting family of C^*-subalgebras. In Section 12.1 we noted that a uniformly matricial C^*-algebra is the tensor product of a countable family of finite type I factors (formed from a generating nest). In this section, we study product states of uniformly matricial C^*-algebras and their corresponding GNS representations.

We begin with some general results that give criteria for states to engender quasi-equivalent representations of a uniformly matricial C^*-algebra $\mathfrak{A}$ (Proposition 12.3.2). We prove that the automorphism group of $\mathfrak{A}$ acts transitively on the pure states of $\mathfrak{A}$ (Theorem 12.3.4). The key to this result is Lemma 12.3.5, which states that a uniformly matricial C^*-algebra acting on a separable Hilbert space and having the same ultraweak closure as a C^*-algebra * isomorphic to it can be transformed onto the isomorphic algebra by a unitary operator in that common closure.

The uniformly matricial C^*-algebra with invariant $\{\infty, 0, 0, 0, \ldots\}$ plays an important role in physical applications. Its representations correspond to the representations of the canonical anticommutation relations (CAR); and the algebra is called the CAR algebra. The latter part of this section deals with that algebra exclusively. We study special product states parametrized by a number a in $[0, 1]$. In Theorem 12.3.8 we identify the type of the ultraweak closure of the image of the CAR algebra under the GNS representations corresponding to these states. We show that it is a factor of type III when a is not 0, 1, or $\frac{1}{2}$. In Theorem 12.3.14 we show that the two factors of type III corresponding to a and b are not * isomorphic when $0 < b < a < \frac{1}{2}$.

12.3.1. LEMMA. *If $\{\mathfrak{A}_n\}$ is a generating nest for the uniformly matricial C^*-algebra $\mathfrak{A}$ acting on the Hilbert space $\mathscr{H}$ and $\mathfrak{A}_n^c = \mathfrak{A}_n' \cap \mathfrak{A}$, then*

$$(\mathfrak{A}_n^c)^- = \mathfrak{A}_n' \cap \mathfrak{A}^- \tag{1}$$

and

$$\bigcap (\mathfrak{A}_n^c)^- = \mathfrak{A}' \cap \mathfrak{A}^-. \tag{2}$$

Proof. On replacing $\{\mathfrak{A}_n\}$ by the corresponding commuting family of finite type I factors (see Remark 12.1.10) and identifying $\mathfrak{A}_n^c$ in these terms, we can deduce (1) from Lemma 11.4.14. From (1),

$$\bigcap_n (\mathfrak{A}_n^c)^- = \bigcap_n (\mathfrak{A}_n' \cap \mathfrak{A}^-) = (\bigcap_n \mathfrak{A}_n') \cap \mathfrak{A}^- = \mathfrak{A}' \cap \mathfrak{A}^-,$$

since $\{\mathfrak{A}_n\}$ generates $\mathfrak{A}$.

Another argument for (1) proceeds as follows. Let G be the group of unitary operators in $\mathfrak{A}_n$ generated by the group of permutation unitary matrices and the group of self-adjoint, diagonal, unitary matrices relative to a given matrix representation of $\mathfrak{A}_n$. Then G is a finite group generating $\mathfrak{A}_n$. Let $\eta(T)$ be $m^{-1}\sum_{U\in G} UTU^*$ for T in $\mathscr{B}(\mathscr{H})$, where m is the order of G. The mapping η is strong-operator continuous, $\eta(\mathscr{B}(\mathscr{H})) \subseteq \mathfrak{A}_n'$, $\eta(\mathfrak{A}) = \mathfrak{A}_n^c$, and η is the identity mapping on $\mathfrak{A}_n'$. Thus

$$\mathfrak{A}_n' \cap \mathfrak{A}^- = \eta(\mathfrak{A}_n' \cap \mathfrak{A}^-) \subseteq \eta(\mathfrak{A}^-) \subseteq \eta(\mathfrak{A})^- = (\mathfrak{A}_n^c)^- \subseteq \mathfrak{A}_n' \cap \mathfrak{A}^-;$$

from which we conclude (1). ∎

12.3.2. PROPOSITION. *If $\mathfrak{A}$ is a uniformly matricial C^*-algebra, $\{\mathfrak{A}_n\}$ is a generating nest for $\mathfrak{A}$, and ρ_1, ρ_2 are primary states of $\mathfrak{A}$, then the GNS representations π_1, π_2 corresponding to ρ_1, ρ_2 are quasi-equivalent if there is some n such that $\|(\rho_1 - \rho_2) \mid \mathfrak{A}_n^c\| < 2$ and only if $\|(\rho_1 - \rho_2) \mid \mathfrak{A}_n^c\| \to_n 0$.*

Proof. Suppose, first, that π_1 and π_2 are not quasi-equivalent. Since π_1 and π_2 are primary, they are disjoint, by Proposition 10.3.12(ii). Hence, from Theorem 10.3.5,

$$(\pi_1 \oplus \pi_2)(\mathfrak{A})^- = \pi_1(\mathfrak{A})^- \oplus \pi_2(\mathfrak{A})^-.$$

If x_1 and x_2 are unit vectors in the representation spaces $\mathscr{H}_1$ and $\mathscr{H}_2$ for π_1 and π_2 such that $\rho_1(A) = \langle \pi_1(A)x_1, x_1\rangle$ and $\rho_2(A) = \langle \pi_2(A)x_2, x_2\rangle$ for all A in $\mathfrak{A}$, then $(\omega_{x_1} - \omega_{x_2})(P_1 - P_2) = 2$, where P_1 and P_2 are the orthogonal projections of $\mathscr{H}_1 \oplus \mathscr{H}_2$ onto $\mathscr{H}_1$ and $\mathscr{H}_2$, respectively. It follows that $\|(\omega_{x_1} - \omega_{x_2}) \mid \mathscr{C}\| = 2$, where $\mathscr{C}$ is the center of $(\pi_1 \oplus \pi_2)(\mathfrak{A})^-$. From Lemma 12.3.1, $\mathscr{C} = \bigcap_n ((\pi_1 \oplus \pi_2)(\mathfrak{A}_n)^c)^-$. Using the Kaplansky density theorem, we have

$$\begin{aligned} 2 = \|(\omega_{x_1} - \omega_{x_2})|((\pi_1 \oplus \pi_2)(\mathfrak{A}_n)^c)^-\| &= \|(\omega_{x_1} - \omega_{x_2})|(\pi_1 \oplus \pi_2)(\mathfrak{A}_n)^c\| \\ &= \|(\rho_1 - \rho_2)|\mathfrak{A}_n^c\|. \end{aligned}$$

Suppose, now, that π_1 and π_2 are quasi-equivalent. Let α be a * isomorphism of $\pi_1(\mathfrak{A})^-$ onto $\pi_2(\mathfrak{A})^-$ such that $\pi_2 = \alpha \circ \pi_1$. With x_1 and x_2 as above, $\omega_{x_2} \circ \alpha$ is a normal state of $\pi_1(\mathfrak{A})^-$. From Lemma 12.3.1, the center $\{\lambda I\}$ of $\pi_1(\mathfrak{A})^-$ is $\bigcap_n (\pi_1(\mathfrak{A}_n)^c)^-$. Thus, from Lemma 11.4.12,

$$\|(\omega_{x_1} - \omega_{x_2} \circ \alpha) \mid (\pi_1(\mathfrak{A}_n)^c)^-\| = \|(\rho_1 - \rho_2) \mid \mathfrak{A}_n^c\| \underset{n}{\to} 0. \quad ∎$$

If ρ_1 and ρ_2 are states of a C^*-algebra $\mathfrak{A}$, we say that ρ_1 and ρ_2 are *automorphic* when there is a * automorphism α of $\mathfrak{A}$ such that $\rho_2 = \rho_1 \circ \alpha$. This relation on states of $\mathfrak{A}$ is an equivalence relation. Let π_1 and π_2 be the GNS representations of $\mathfrak{A}$ corresponding to ρ_1 and ρ_2; and let x_1 and x_2 be unit generating vectors for $\pi_1(\mathfrak{A})$ and $\pi_2(\mathfrak{A})$ such that $\rho_1 = \omega_{x_1} \circ \pi_1$ and $\rho_2 = \omega_{x_2} \circ \pi_2$. If $\rho_2 = \rho_1 \circ \alpha$, then $\rho_2 = \omega_{x_1} \circ \pi_1 \circ \alpha$. It follows that the representation $\pi_1 \circ \alpha$ of $\mathfrak{A}$ (with image $\pi_1(\mathfrak{A})$) is unitarily equivalent to π_2. Thus two automorphic states engender representations φ_1 and φ_2 with the same image and having a unit generating vector x_0 such that the given states are $\omega_{x_0} \circ \varphi_1$ and $\omega_{x_0} \circ \varphi_2$. A partial converse to this last observation is valid.

12.3.3. PROPOSITION. *If the* GNS *representations π_1 and π_2 of the C^*-algebra $\mathfrak{A}$ corresponding to the states ρ_1 and ρ_2 are faithful, then ρ_1 and ρ_2 are automorphic if and only if these representations can be realized on the same Hilbert space with the same image for which there is a generating vector x_0 such that $\rho_1 = \omega_{x_0} \circ \pi_1$ and $\rho_2 = \omega_{x_0} \circ \pi_2$.*

If ρ_1 and ρ_2 are pure states of $\mathfrak{A}$, π_1 and π_2 are faithful, and $\pi_1(\mathfrak{A}) = U\pi_2(\mathfrak{A})U^{-1}$ for some unitary U, then ρ_1 and ρ_2 are automorphic.

Proof. The preceding discussion establishes one of the implications of the first assertion of this proposition. If π_1 and π_2 are faithful, the mapping $A \to \pi_1^{-1}(\pi_2(A))$ defines a * automorphism α of $\mathfrak{A}$. If $\rho_1 = \omega_{x_0} \circ \pi_1$ and $\rho_2 = \omega_{x_0} \circ \pi_2$, then, since $\pi_2 = \pi_1 \circ \alpha$, $\rho_2 = \omega_{x_0} \circ \pi_1 \circ \alpha = \rho_1 \circ \alpha$; and ρ_1 is automorphic to ρ_2.

Under the conditions of the second assertion of this proposition, with β the * automorphism of $\mathfrak{A}$ defined by

$$\beta(A) = \pi_1^{-1}(U\pi_2(A)U^{-1}) \qquad (A \in \mathfrak{A})$$

we have

$$\begin{aligned}(\rho_1 \circ \beta)(A) &= (\omega_{x_1} \circ \pi_1 \circ \pi_1^{-1})(U\pi_2(A)U^{-1}) \\ &= \langle \pi_2(A)U^{-1}x_1, U^{-1}x_1\rangle = (\omega_{y_1} \circ \pi_2)(A) \qquad (A \in \mathfrak{A}),\end{aligned}$$

where $y_1 = U^{-1}x_1$ and $\rho_1 = \omega_{x_1} \circ \pi_1$. Thus ρ_1 and $\omega_{y_1} \circ \pi_2$ are automorphic. From the (essential) uniqueness of the GNS representation, π_2 corresponds to $\omega_{y_1} \circ \pi_2$ as well as to ρ_2. Since π_2 is irreducible, it follows from Theorem 10.2.6 that there is a unitary element V in $\mathfrak{A}$ for which

$$\rho_2(VAV^*) = (\omega_{y_1} \circ \pi_2)(A) \qquad (A \in \mathfrak{A}).$$

In particular, ρ_2 is automorphic to $\omega_{y_1} \circ \pi_2$ and, hence, to ρ_1. ∎

The special properties of uniformly matricial C^*-algebras will allow us to prove the following theorem.

12.3.4. THEOREM. *All pure states of a uniformly matricial C^*-algebra are automorphic.*

By virtue of Proposition 12.3.3, it will suffice to show that if $\mathfrak{A}_1$ and $\mathfrak{A}_2$ are * isomorphic uniformly matricial C^*-algebras acting irreducibly on Hilbert spaces $\mathscr{H}_1$ and $\mathscr{H}_2$, there is a unitary transformation U of $\mathscr{H}_2$ onto $\mathscr{H}_1$ such that $\mathfrak{A}_1 = U\mathfrak{A}_2U^{-1}$. Since $\mathfrak{A}_1$ and $\mathfrak{A}_2$ are norm-separable, their cyclic representations occur on separable Hilbert spaces. Thus $\mathscr{H}_1$ and $\mathscr{H}_2$ may be identified by a unitary transformation. We may assume that $\mathfrak{A}_1$ and $\mathfrak{A}_2$ act on the same Hilbert space $\mathscr{H}$. In this case $\mathfrak{A}_1^- = \mathfrak{A}_2^- = \mathscr{B}(\mathscr{H})$. From these comments, Theorem 12.3.4 will be proved when we establish the following lemma.

12.3.5. LEMMA. *If two * isomorphic uniformly matricial C^*-algebras $\mathfrak{A}$ and $\mathscr{B}$ acting on the separable Hilbert space $\mathscr{H}$ have the same ultraweak closure $\mathscr{R}$, then there is a unitary operator U in $\mathscr{R}$ such that $U\mathfrak{A}U^{-1} = \mathscr{B}$.*

Proof. We note, first, that if $\mathscr{M}$ and $\mathscr{N}$ are factors contained in $\mathfrak{A}$ and $\mathscr{B}$, respectively, with both $\mathscr{M}$ and $\mathscr{N}$ of type I_n, then there is a unitary operator V in $\mathscr{R}$ such that $V\mathscr{M}V^{-1} = \mathscr{N}$. For this, let $\{E_{rs}\}$ and $\{F_{rs}\}$ be self-adjoint $n \times n$ systems of matrix units for $\mathscr{M}$ and $\mathscr{N}$. If V_0 is a partial isometry in $\mathscr{R}$ with initial projection E_{11} and final projection F_{11}, then $\sum_j F_{j1}V_0E_{1j}$ will serve as V.

Let $\mathscr{M}_0$ be a finite factor of type I_m in $\mathfrak{A}$ such that $\mathscr{M} \subseteq \mathscr{M}_0$ (where $\mathscr{M}$ and V are as above). Let a finite set of vectors $x_1, \ldots, x_r$ in $\mathscr{H}$ and a positive ε be given. We prove, next, that there is a unitary operator W in $\mathscr{R}$ such that $W\mathscr{M}_0W^{-1} \subseteq \mathscr{B}$, $WAW^{-1} = VAV^{-1}(A \in \mathscr{M})$, and $\|(V - W)x_j\| < \varepsilon$ for each j in $\{1, \ldots, r\}$. Since $V\mathscr{M}V^{-1}$ $(= \mathscr{N})$ is a finite type I factor, by Remark 12.1.10 its commutant relative to $V\mathfrak{A}V^{-1}$ is a uniformly matricial C^*-algebra and is * isomorphic to the relative commutant of $\mathscr{N}$ in $\mathscr{B}$ (that is, to $\mathscr{N}' \cap \mathscr{B}$). Now $V\mathfrak{A}V^{-1}$ is weak-operator dense in $\mathscr{R}$ so that the weak-operator closure of the first-mentioned relative commutant is $\mathscr{N}' \cap \mathscr{R}$, which is the weak-operator closure of the second-mentioned relative commutant, from Lemma 12.3.1. Thus, from the result proved in the first paragraph, there is a unitary operator V_1 in $\mathscr{N}' \cap \mathscr{R}$ such that

$$V_1(V\mathscr{M}_0V^{-1} \cap \mathscr{N}')V_1^{-1} \subseteq \mathscr{B}.$$

(Note, for this, that $\mathscr{N} = V\mathscr{M}V^{-1} \subseteq V\mathscr{M}_0V^{-1} \subseteq V\mathfrak{A}V^{-1}$.) Since V_1 is in the strong-operator closure $\mathscr{N}' \cap \mathscr{R}$ of $\mathscr{N}' \cap \mathscr{B}$, by Corollary 5.3.7, there is a unitary operator V_2^{-1} in $\mathscr{N}' \cap \mathscr{B}$ such that $\|(V_1 - V_2^{-1})Vx_j\| < \varepsilon$ when $j \in \{1, \ldots, r\}$. It follows that $\|(V_2V_1V - V)x_j\| < \varepsilon$, V_2V_1V may be chosen as W.

Let $\{A_j\}$ and $\{B_j\}$ be countable norm-dense subsets of $\mathfrak{A}$ and $\mathscr{B}$, respectively; and let $\{x_j\}$ be a countable dense subset of $\mathscr{H}$. We shall construct ascending sequences $\{\mathscr{M}_n\}$ and $\{\mathscr{N}_n\}$ of finite type I factors in $\mathfrak{A}$ and $\mathscr{B}$, respectively, and a sequence $\{U_n\}$ of unitary operators in $\mathscr{R}$ satisfying:

(i) there are operators $S_1, \ldots, S_n$ in $\mathscr{M}_n$ and $T_1, \ldots, T_n$ in $\mathscr{N}_n$ such that $\|A_j - S_j\| < 1/n$ and $\|B_j - T_j\| < 1/n$,

(ii) $\|(U_n - U_{n+1})x_j\| < 2^{-n}$, when $j \in \{1, \ldots, n\}$,

(iii) $U_n \mathscr{M}_n U_n^{-1} = \mathscr{N}_n$, $U_n A U_n^{-1} = U_{n-1} A U_{n-1}^{-1}$ for A in $\mathscr{M}_{n-1}$.

If we have $\{\mathscr{M}_n\}$, $\{\mathscr{N}_n\}$, and $\{U_n\}$, we complete the proof as follows. Since $\{x_j\}$ is dense in $\mathscr{H}$, each U_n is unitary, and $\{U_n x_j\}$ is Cauchy convergent for each j; $\{U_n\}$ is strong-operator convergent to an isometry U in $\mathscr{R}$. With B in $\mathscr{N}_n$,

$$\langle U^*BUx_j, x_k\rangle = \lim_m \langle BU_m x_j, U_m x_k\rangle = \langle U_n^{-1} BU_n x_j, x_k\rangle,$$

so that $U^*BU = U_n^{-1}BU_n$. It follows that $B \to U^*BU$ is a * isomorphism of $\bigcup_n \mathscr{N}_n$ *onto* $\bigcup_n \mathscr{M}_n$. Since $U^*A^*BU = U^*A^*UU^*BU$, for all x and y in $\mathscr{H}$ and A and B in $\bigcup_n \mathscr{N}_n$, $\langle BUx, AUy\rangle = \langle UU^*BUx, AUy\rangle$. If $E = UU^*$, then $\langle (I - E)BUx, AUy\rangle = 0$; and, taking strong-operator limits, this same equality holds for all A and B in $\mathscr{R}$. We replace A and B by U^* and conclude that $\langle (I - E)x, y\rangle = 0$ for all x and y in $\mathscr{H}$. Thus

$$UU^* = E = I = U^*U;$$

and U is the desired unitary operator.

It remains to construct $\{\mathscr{M}_n\}$, $\{\mathscr{N}_n\}$, and $\{U_n\}$ with properties (i), (ii), and (iii). For $\mathscr{M}_1$ and $\mathscr{N}_1$, choose algebras of the same type from generating nests for $\mathfrak{A}$ and $\mathscr{B}$ ensuring that there are operators S_1 (in $\mathscr{M}_1$) and T_1 (in $\mathscr{N}_1$) such that $\|A_1 - S_1\| < 1$ and $\|B_1 - T_1\| < 1$. For U_1, we choose a unitary operator in $\mathscr{R}$ mapping $\mathscr{M}_1$ onto $\mathscr{N}_1$ (by using the result of the first paragraph). Suppose that we have constructed $\{\mathscr{M}_1, \ldots, \mathscr{M}_n\}$, $\{\mathscr{N}_1, \ldots, \mathscr{N}_n\}$, and $\{U_1, \ldots, U_n\}$ satisfying conditions (i), (ii), (iii). By using the argument of Theorem 12.1.9, we can find a finite type I factor $\mathscr{M}_0$ containing $\mathscr{M}_n$ and operators $S_1, \ldots, S_{n+1}$ such that $\|A_j - S_j\| < 1/n + 1$ when $j \in \{1, \ldots, n + 1\}$. The result proved in the second paragraph allows us to find a unitary operator U_0 in $\mathscr{R}$ mapping $\mathscr{M}_0$ into $\mathscr{B}$ and such that $\|(U_0 - U_n)x_j\| < 2^{-(n+1)}$ when $j \in \{1, \ldots, n\}$ and $U_0 A U_0^{-1} = U_n A U_n^{-1}$ when $A \in \mathscr{M}_n$. Applying this same result in the argument of Theorem 12.1.9, we construct a finite type I factor $\mathscr{N}_{n+1}$ in $\mathscr{B}$ containing $U_0 \mathscr{M}_0 U_0^{-1}$ and operators $T_1, \ldots, T_{n+1}$ such that $\|B_j - T_j\| < 1/n + 1$ when $j \in \{1, \ldots, n + 1\}$. Again, from the result of the second paragraph, we can find a unitary operator U_{n+1}^{-1} in $\mathscr{R}$ mapping $\mathscr{N}_{n+1}$ into $\mathfrak{A}$ and such that $\|(U_{n+1}^{-1} - U_0^{-1})U_0 x_j\| < 2^{-(n+1)}$ when

$j \in \{1, \ldots, n\}$ and $U_{n+1}^{-1}U_0A_0U_0^{-1}U_{n+1} = A_0$ for each A_0 in $\mathscr{M}_0$. Then $\|(U_0 - U_{n+1})x_j\| < 2^{-(n+1)}$ and $\|(U_n - U_{n+1})x_j\| < 2^{-n}$ when $j \in \{1, \ldots, n\}$. If we let $\mathscr{M}_{n+1}$ be $U_{n+1}^{-1}\mathscr{N}_{n+1}U_{n+1}$, then $\mathscr{M}_{n+1}$ contains $U_0^{-1}U_0\mathscr{M}_0U_0^{-1}U_0$, which, in turn, contains $\mathscr{M}_n$ and the operators $S_1, \ldots, S_{n+1}$. In addition, with A in $\mathscr{M}_n$, A is in $\mathscr{M}_0$; so that $U_{n+1}^{-1}U_0AU_0^{-1}U_{n+1} = A$. Thus $U_{n+1}AU_{n+1}^{-1} = U_0AU_0^{-1} = U_nAU_n^{-1}$. ■

If ρ_1 and ρ_2 are pure states of the norm-separable C^*-algebra $\mathfrak{A}$ then $\pi_1(\mathfrak{A})^-$ is * isomorphic to $\pi_2(\mathfrak{A})^-$ as both are * isomorphic to $\mathscr{B}(\mathscr{H})$ with $\mathscr{H}$ a separable Hilbert space, where π_1 and π_2 are the GNS representations corresponding to ρ_1 and ρ_2. In general, we say that two states of a C^*-algebra are *algebraically equivalent* when their corresponding GNS representations have images with * isomorphic ultraweak closures. From the discussion preceding Proposition 12.3.3, automorphic states are algebraically equivalent. Lemma 12.3.5 allows us to prove the following extension of Theorem 12.3.4. (To see that it is an extension of Theorem 12.3.4, recall that, by Proposition 10.3.7(i), quasi-equivalent irreducible representations are equivalent and apply Theorem 10.2.6.)

12.3.6. THEOREM. *Two states ρ_1 and ρ_2 of a uniformly matricial C^*-algebra $\mathfrak{A}$ are algebraically equivalent if and only if there is a * automorphism α of $\mathfrak{A}$ such that the representations engendered by $\rho_1 \circ \alpha$ and ρ_2 are quasi-equivalent.*

Proof. Suppose there is an α as described. Let π_1, π_2, and π be the representations engendered by ρ_1, ρ_2, and $\rho_1 \circ \alpha$, respectively. By assumption there is a * isomorphism β of $\pi(\mathfrak{A})^-$ onto $\pi_2(\mathfrak{A})^-$ such that $\beta \circ \pi = \pi_2$. The discussion preceding Proposition 12.3.3 tells us that π is (unitarily equivalent to) $\pi_1 \circ \alpha$; so that $(\pi_1 \circ \alpha)(\mathfrak{A})$ and $\pi(\mathfrak{A})$ coincide. But $(\pi_1 \circ \alpha)(\mathfrak{A})$ is $\pi_1(\mathfrak{A})$. Hence β is a * isomorphism of $\pi_1(\mathfrak{A})^-$ onto $\pi_2(\mathfrak{A})^-$ and ρ_1 is algebraically equivalent to ρ_2.

The foregoing applies to all C^*-algebras. Assume that ρ_1 and ρ_2 are algebraically equivalent. We make specific use, now, of the assumption that $\mathfrak{A}$ is matricial so that Lemma 12.3.5 applies. Suppose that β is a * isomorphism of $\pi_1(\mathfrak{A})^-$ onto $\pi_2(\mathfrak{A})^-$, where π_1 and π_2 are the representations of $\mathfrak{A}$ engendered by ρ_1 and ρ_2. Then $\beta(\pi_1(\mathfrak{A}))$ and $\pi_2(\mathfrak{A})$ are * isomorphic matricial C^*-algebras with the same weak-operator closure $\pi_2(\mathfrak{A})^-$. From Lemma 12.3.5, there is a unitary operator U in $\pi_2(\mathfrak{A})^-$ such that

$$U\beta(\pi_1(\mathfrak{A}))U^{-1} = \pi_2(\mathfrak{A}).$$

Then $A \to \pi_2^{-1}(U\beta(\pi_1(A))U^{-1})$ is a * automorphism α of $\mathfrak{A}$. Now $\rho_2 \circ \alpha$ engenders the representation $\pi_2 \circ \alpha$; and $(\pi_2 \circ \alpha)(A) = U\beta(\pi_1(A))U^{-1}$. That is, $\rho_2 \circ \alpha$ engenders a representation unitarily equivalent to $\beta \circ \pi_1$. As β is a * isomorphism on $\pi_1(\mathfrak{A})^-$, $\beta \circ \pi_1$ is quasi-equivalent to π_1. Hence ρ_1 and $\rho_2 \circ \alpha$ engender quasi-equivalent representations of $\mathfrak{A}$. ■

12.3.7. DEFINITION. A state ρ of the CAR algebra $\mathfrak{A}$ is said to be of *product type a* when there is a mutually commuting family $\{\mathcal{N}_j\}$ of factors of type I_2 generating $\mathfrak{A}$ such that, for each j, $\rho(E_{11}) = a$ and $\rho(E_{12}) = 0$ for some self-adjoint set $\{E_{11}, E_{12}, E_{21}, E_{22}\}$ of matrix units for $\mathcal{N}_j$; and $\rho = \otimes \rho \mid \mathcal{N}_j$ (in the sense of Proposition 11.4.6 and the comment following it).

For the remainder of this section, we shall study the representation engendered by a state ρ of $\mathfrak{A}$ of product type a. It is clear that if ρ and η are both states of $\mathfrak{A}$ of product type a, then ρ and η are automorphic. If φ and ψ are the representations engendered by ρ and η and $\mathcal{M} = \varphi(\mathfrak{A})^-$, $\mathcal{N} = \psi(\mathfrak{A})^-$, it follows, in particular, that $\mathcal{M}$ and $\mathcal{N}$ are * isomorphic. In Theorem 12.3.8, we note that product states of type a are primary; so that $\mathcal{M}$ is a factor. We determine the type of $\mathcal{M}$ in terms of a. It is a simple observation that ρ is of product type a if it is of product type $1 - a$ (interchange the roles of the matrix units E_{11} and E_{22}); so that we need consider only those a in $[0, \frac{1}{2}]$. We note, in Theorem 12.3.8, that $\mathcal{M}$ is of type III when $a \in (0, \frac{1}{2})$.

The results following Theorem 12.3.8 will establish (Theorem 12.3.14) that, if $0 < b < a < \frac{1}{2}$, the two factors of type III associated with product states of types a and b are not * isomorphic.

In the discussion that follows, we draw on the construction studied in Section 8.6. When applied to the restricted direct sum G (all but a finite number of coordinates equal to the identity) of a countable family $\{G_j\}_{j=0,\pm 1,\pm 2,\ldots}$ of two-element groups G_j ($=\{0, 1\}$ with the group operation of addition modulo 2) and the space S formed from the topological product of the spaces G_j provided with a suitable (product probability) measure, it will allow us to display a representation unitarily equivalent to that engendered by a state of the CAR algebra of product type a ($\neq 0, 1$).

We denote by g_j the element of G whose coordinate in G_j is 1 and all other coordinates 0, by z_g the function on G that is 1 at g and 0 elsewhere, by p_j the (projection) mapping that assigns to a point in S its jth coordinate, by y_{j0} and y_{j1} the characteristic functions of the subsets S_{j0} and S_{j1} of S consisting of points s for which $p_j(s)$ is 0 and 1, respectively. Let m_j be the measure on G_j that assigns a and $1 - a$ to the singleton sets $\{0\}$ and $\{1\}$; and let m be the Borel measure on S formed as the product of the measures m_j. More in the spirit of product states, we may view $C(S)$ as the tensor product of the (C^*-) algebras $C(G_j)$, $j \in \{0, \pm 1, \pm 2, \ldots\}$, and m as the regular Borel measure on S corresponding to the product of the states ρ_j on $C(G_j)$ resulting from "integration" relative to m_j—that is, the state on the product that assigns $\Pi_{j=-n}^{n}(aa_{j0} + (1-a)a_{j1})$ to $\Pi_{j=-n}^{n}(a_{j0}y_{j0} + a_{j1}y_{j1})$ $(=f)$. For the identification of $C(S)$ with $\otimes C(G_j)$, we employ Proposition 11.4.3, after observing that the subalgebra of $C(S)$ generated by the canonical images of the C^*-algebras $C(G_j)$ is norm dense in $C(S)$. This subalgebra consists of all linear combinations of functions of the form f (above); and it is norm dense by a simple application of the Stone–Weierstrass theorem.

Since S is the unrestricted direct sum of the groups G_j as well as being a compact Hausdorff space, and G is a subgroup of this unrestricted direct sum, G acts on S by translation. We examine the Borel measure space (S, m) with this action of G in the context of the construction described in Section 8.6, noting first, that the conditions (A), (B), (C), and ergodicity (preceding Lemma 8.6.5) are satisfied. For (A), the family $\{S_{j0}, S_{j1} : j = 0, \pm 1, \ldots\}$ of Borel subsets of S suffices. For (B), the action of g on S is a homeomorphism and, so, preserves Borel sets. As for the preservation of null sets, the elements g_j of G generate G; so that it will suffice to show that each g_j preserves null sets. For this we note that the Radon–Nikodým derivative of $m \circ g_j$ with respect to m is $a^{-1}(1-a)y_{j0} + a(1-a)^{-1}y_{j1}$, a positive continuous function on S vanishing nowhere. For (C), the action of G on S is group translation; so that each element of G other than the identity acts in a fixed-point-free manner on S.

In the notation of Section 8.6, $\mathscr{H}$ is $L_2(S, m)$, M_y is the multiplication operator on $\mathscr{H}$ corresponding to the bounded Borel function y on S, and U_g is the unitary operator on $\mathscr{H}$ corresponding to translation by g. Let L_g be the (unitary) operator on $l_2(G)$ corresponding to (left) translation by g. The Hilbert space $\mathscr{K}$ $(=\sum_g \oplus \mathscr{H})$ used in Section 8.6 can be identified in the usual way with $\mathscr{H} \otimes l_2(G)$. The von Neumann algebra $\mathscr{R}$ generated by $\mathscr{A} \otimes \mathbb{C}I$ and $\{U_g \otimes L_g : g \in G\}$, where $\mathscr{A}$ is the multiplication algebra of $L_2(S, m)$, is then the one constructed in Section 8.6 (since $A \otimes I$ and $U_g \otimes L_g$ have operator matrices $[\delta_{p,q}A]$ and $[\delta_{p,gq}U_g]$, respectively, by the discussion preceding Proposition 2.6.13). Without ergodicity we do not know that $\mathscr{R}$ is a factor. We shall prove that $\mathscr{R}$ is a factor by identifying it with the weak-operator closure of the image of the CAR algebra $\mathfrak{A}$ under the representation engendered by a product state ρ of type a—a critical step in our program, in any event. For this we observe that there is a family $\{\mathscr{N}_j\}$ of mutually commuting factors of type I_2 generating $\mathfrak{A}$ such that $\rho = \otimes(\rho \mid \mathscr{N}_j)$. Remark 11.4.16 applies, and ρ is primary. When we know that $\mathscr{R}$ is a factor, it will follow that G acts ergodically on S (from the discussion following Proposition 8.6.9).

For this identification of $\mathscr{R}$, let $\mathscr{N}_j$ be the subalgebra of $\mathscr{R}$ generated by $M_{y_{j0}} \otimes I (= E_{11}^{(j)})$, $M_{y_{j1}} \otimes I (= E_{22}^{(j)})$, $(M_{y_{j0}}U_{g_j}) \otimes L_{g_j} (= E_{12}^{(j)})$, and $(E_{12}^{(j)})^*$ $(=E_{21}^{(j)})$. Let $\tilde{1}$ be the constant function 1 on S, and let x_0 be $\tilde{1} \otimes z_e$, where e is the identity element of G. Then $\{\mathcal{N}_j\}$ is a mutually commuting family of factors of type I_2 that generates $\mathcal{R}$. For this, we need note only that the set of linear combinations of functions of the form $\prod_{j=-n}^{n}(a_{j0}y_{j0}+a_{j1}y_{j1})$ is norm dense in $C(S)$ (from the Stone–Weierstrass theorem—as observed earlier) and that $\{g_j\}$ generates G. Thus the CAR algebra $\mathfrak{A}$ can be identified with the weak-operator-dense C^*-subalgebra of $\mathcal{R}$ generated by $\{\mathcal{N}_j\}$. If $\rho = \omega_{x_0}|\mathfrak{A}$, then since $\langle (M_y U_g \otimes L_g)(\tilde{1} \otimes z_e), \tilde{1} \otimes z_e \rangle = 0$ $(g \neq e, M_y \in \mathcal{A})$,

$$\rho\left(\prod_{j=-n}^{n}\left(\sum_{r,s=1}^{2} a_{rs}^{(j)} E_{rs}^{(j)}\right)\right) = \prod_{j=-n}^{n} (aa_{11}^{(j)} + (1-a)a_{22}^{(j)});$$

so that ρ is a product state of type a. Once we note that x_0 is generating for $\mathfrak{A}$, it follows that $\mathfrak{A}$ acting on $\mathcal{K}$ is (unitarily equivalent to) the representation of $\mathfrak{A}$ engendered by ρ. With y' in $C(S)$, we have that $(M_{y'} \otimes I)(\tilde{1} \otimes z_e) = y' \otimes z_e$; so that $y \otimes z_e$ is in the closure of $\mathfrak{A}x_0$ for each y in $\mathcal{H}$. With g in G, $g = g_{j(1)} \cdots g_{j(r)}$ for some $j(1), \ldots, j(r)$; and $U_g \otimes L_g = \prod_{k=1}^{r} U_{g_{j(k)}} \otimes L_{g_{j(k)}}$. Thus $(U_g \otimes L_g)((U_g^* y) \otimes z_e)$ $(= y \otimes z_g)$ is in the closure of $\mathfrak{A}x_0$ for all y in $\mathcal{H}$ and g in G. Since $\{z_g : g \in G\}$ is an orthonormal basis for $l_2(G)$; $y \otimes z$ is in the closure of $\mathfrak{A}x_0$ for all y in $\mathcal{H}$ and z in $l_2(G)$; so that this closure is $\mathcal{K}$. Thus $\mathcal{R}$ is a factor and G acts ergodically on S.

We complete this discussion by determining the type of $\mathcal{R}$ with the aid of the criteria described in Proposition 8.6.10. If a is 0 or 1, $\rho \mid \mathcal{N}_j$ is pure; so that $\rho(= \otimes\rho|\mathcal{N}_j)$ is a pure state of $\mathfrak{A}$, by Proposition 11.4.7, and $\pi_\rho(\mathfrak{A})^- = \mathcal{B}(\mathcal{H}_\rho)$. If $a = \frac{1}{2}$, $\rho \mid \mathcal{N}_j$ is the normalized trace on $\mathcal{N}_j$, and $\rho\,(= \otimes\rho \mid \mathcal{N}_j)$ is the (unique) normalized trace on $\mathfrak{A}$, by Proposition 11.4.7. Thus $\pi_\rho(\mathfrak{A})^-$ is the matricial factor of type II_1, in this case.

Suppose, now, that $a \in (0, 1)$ and $a \neq \frac{1}{2}$. If X_n is the (open) subset of S consisting of all points s such that $p_j(s) = p_j(s_0)$ for j in $\{-n, \ldots, n\}$, where s_0 is a given point of S, then $m(X_n) = a^k(1-a)^{2n+1-k} \to_n 0$ and $\bigcap_n X_n = \{s_0\}$, where $p_j(s_0) = 0$ for k values of j in $\{-n. \ldots, n\}$. Thus $m(\{s_0\}) = 0$ and, from Proposition 8.6.10(i), $\mathcal{R}$ is not of type I.

Suppose $\mathcal{R}$ is of type II_1 and τ is the normalized trace on $\mathcal{R}$. Then τ_0 and the product state ρ of type a engender quasi-equivalent representations of $\mathfrak{A}$, where $\tau_0 = \tau \circ \pi_\rho(\in \mathcal{N}(\pi_\rho))$, from Proposition 10.3.14. By Proposition 12.3.2, $\|(\tau_0 - \rho)|\mathfrak{A}_n^c\| \to_n 0$ and, in particular, $\|(\tau_0 - \rho)|\mathcal{N}_j\| \to_j 0$. But $|(\tau_0 - \rho)(E_{11}^{(j)})| = |a - \frac{1}{2}|$ for all j. Thus $\mathcal{R}$ is not of type II_1.

Suppose $\mathcal{R}$ is of type II_∞ and ρ_0 is a non-zero normal semi-finite tracial weight on $\mathcal{R}$. Then, as in the proofs of Lemma 8.6.3 and Proposition 8.6.10, there is a G-invariant σ-finite infinite Borel measure m_0 on S whose value at a Borel subset X of S is $\rho_0(M_x \otimes I)$, where x is the characteristic function of X. Since $\mathcal{R}$ is a factor, ρ_0 is faithful and m is absolutely continuous with respect to

m_0. On the other hand, by construction of m_0, m_0 is absolutely continuous with respect to m. Hence, the Radon–Nikodým derivative h of m with respect to m_0 is positive almost everywhere. Since $1 = m(S) = \int h(s)\,dm_0(s)$, $h \in L_1(S, m_0)$. The mapping that assigns $E_{rs}^{(j+1)}$ to $E_{rs}^{(j)}$ for each integer j and r, s in $\{1, 2\}$ determines an automorphism of $\mathfrak{A}$ that leaves ρ invariant. It follows that this automorphism is implemented by a unitary operator on $\mathscr{H}$ (that leaves x_0 fixed) and extends to a * automorphism α of $\mathscr{R}$. From Proposition 8.5.5, $\rho_0 \circ \alpha = c\rho_0$, for some positive scalar c. Replacing α by α^{-1}, if necessary, we may assume that $c \le 1$.

We show that $c = 1$. By construction α restricts to an automorphism of $\mathscr{A} \otimes \mathbb{C}I$ induced by the transformation of S that shifts coordinates by 1. Denote by α, again, this shift transformation on S. Let ε be a positive number, X_0 be a Borel subset of S such that $0 < m_0(X_0) < \infty$, and y_k be the characteristic function of $\alpha^k(X_0)$. Then, if we assume $c < 1$, $m_0(\alpha^k(X_0)) = c^k m_0(X_0) \to 0$ as $k \to \infty$. Thus

$$\begin{aligned} m(X_0) = m(\alpha^k(X_0)) &= \int_S y_k(s)\,dm(s) = \int y_k(s)h(s)\,dm_0(s) \\ &= \int_{\alpha^k(X_0)} h(s)\,dm_0(s) \underset{k}{\to} 0, \end{aligned}$$

by absolute continuity of the indefinite integral, $\int h\,dm_0$ (since $h \in L_1(S, m_0)$). From absolute continuity of m_0 with respect to m, it follows that $m_0(X_0) = 0$, contradicting the choice of X_0. Thus $c = 1$, and $m_0 \circ \alpha = m_0$.

Since $m_0(S) = \infty$, m_0 is not a scalar multiple of m; and h is not (almost everywhere) constant. Let y be the characteristic function of a subset X of S of the form $\{s : a_1 < h(s) < a_2\}$ such that both X and $S\backslash X$ have positive m-measure. With f a bounded measurable function on S,

$$\begin{aligned} \int f(\alpha(s))h(s)\,dm_0(s) &= \int f(\alpha(s))\,dm(s) = \int f(\alpha(s))\,dm(\alpha(s)) \\ &= \int f(s)\,dm(s) = \int f(s)h(s)\,dm_0(s) \\ &= \int f(\alpha(s))h(\alpha(s))\,dm_0(\alpha(s)) \\ &= \int f(\alpha(s))h(\alpha(s))\,dm_0(s); \end{aligned}$$

so that $h = h \circ \alpha$ almost everywhere. Hence $y = y \circ \alpha$ almost everywhere; and $\alpha(E) = E$, where E is the projection $M_y \otimes I$. With F the projection in $\mathscr{R}$

corresponding to $y_{j_1 0} \cdots y_{j_r 0} \cdot y_{k_1 1} \cdots y_{k_s 1}$ $(=y')$, $\{\alpha^n(F)\}$ has a weak-operator limiting point T that is an element of $\bigcap_n (\mathfrak{A}_n^c)^-$. From Lemma 12.3.1 and the fact that $\mathscr{R}$ is a factor, T is a scalar. Since $\rho(\alpha^n(F)) = \rho(F) = a^r(1-a)^s$ and ρ extends to a normal state of $\mathscr{R}$ (namely, $\omega_{x_0} | \mathscr{R}$), we see that $T = a^r(1-a)^s I$. Hence $\omega_{x_0}(E\alpha^n(F)) = \omega_{x_0}(\alpha^n(EF)) = \omega_{x_0}(EF)$ is near $\omega_{x_0}(E)\omega_{x_0}(F)$ for certain values of n; and $\omega_{x_0}(EF) = \omega_{x_0}(E)\omega_{x_0}(F)$. As the functions of the form y' generate a norm-dense subalgebra of $C(S)$ linearly and $\{M_{\tilde{y}} \otimes I : \tilde{y} \in C(S)\}$ is strong-operator dense in $\mathscr{A} \otimes \mathbb{C}I$, $\omega_{x_0}(EA) = \omega_{x_0}(E)\omega_{x_0}(A)$ for each A in $\mathscr{A}$. In particular, $0 = \omega_{x_0}(E(I-E)) = \omega_{x_0}(E)\omega_{x_0}(I-E)$; so that either $\omega_{x_0}(E) = m(X) = 0$ or $\omega_{x_0}(I-E) = m(S \setminus X) = 0$, contradicting the choice of X. Hence $\mathscr{R}$ is not of type II_∞. It follows that for a in $(0, 1)$ and a different from $\frac{1}{2}$, $\mathscr{R}$ is of type III.

We summarize the results of the preceding discussion in the theorem that follows.

12.3.8. THEOREM. *If ρ is a state of the CAR algebra $\mathfrak{A}$ of product type a, then the weak-operator closure $\mathscr{M}$ of the image of the representation of $\mathfrak{A}$ engendered by ρ is a factor. If a is 0 or 1, ρ is a pure state of $\mathfrak{A}$ and $\mathscr{M}$ is of type I_∞. If a is $\frac{1}{2}$, ρ is the normalized trace on $\mathfrak{A}$ and $\mathscr{M}$ is of type II_1. If $a \in (0, 1)$ and $a \neq \frac{1}{2}$, then $\mathscr{M}$ is of type* III.

12.3.9. REMARK. Let $\{E_{jk}\}$ be the standard system of matrix units for $M_n(\mathbb{C})$ and let ρ be a linear functional on $M_n(\mathbb{C})$. If $H = \sum_{j,k=1}^n \rho(E_{jk})E_{kj}$, then $\rho(A) = \mathrm{Tr}(HA)$ for each A in $M_n(\mathbb{C})$ (where $\mathrm{Tr}((a_{jk})) = \sum_{j=1}^n a_{jj}$). If ρ is hermitian, H is self-adjoint (and conversely); and ρ is positive precisely when H is positive. Since there is a *unique* H for which $\mathrm{Tr}(HA) = \rho(A)$, the set of eigenvalues $\{h_1, \ldots, h_n\}$ of H (listed in decreasing order) is an invariant of ρ. We refer to $\{h_1, \ldots, h_n\}$ as the *eigenvalue list* of ρ. If H is self-adjoint and U is a unitary matrix such that $UHU^* = \sum_{j=1}^n h_j E_{jj}$, then, when $\|A\| \le 1$ and $UAU^* = \sum_{j,k=1}^n b_{jk} E_{jk}$,

$$\begin{aligned} |\rho(A)| = |\mathrm{Tr}(HA)| = |\mathrm{Tr}(UHU^*UAU^*)| &= \Big|\sum_{j=1}^n h_j b_{jj}\Big| \\ &\le \sum_{j=1}^n |h_j| = \mathrm{Tr}(|H|). \end{aligned}$$

On the other hand, if $UAU^* = \sum_{j=1}^n s_j E_{jj}$ where $s_j h_j = |h_j|$ (and $s_j = 0$ when $h = 0$), then $\|A\| \le 1$ and $\rho(A) = \mathrm{Tr}(HA) = \mathrm{Tr}(|H|)$. Thus $\|\rho\| = \mathrm{Tr}(|H|)$.

If the linear functional η on $M_m(\mathbb{C})$ corresponds (similarly) to K, then $\rho \otimes \eta$ corresponds to $H \otimes K$, since the tracial state on $M_n(\mathbb{C}) \otimes M_m(\mathbb{C})$ is the (tensor) product of the tracial states on $M_n(\mathbb{C})$ and $M_m(\mathbb{C})$. Thus the eigenvalue list of $\rho \otimes \eta$ is a rearrangement of $\{h_1k_1, \ldots, h_nk_m\}$, where $\{k_1, \ldots, k_m\}$ is the eigenvalue list of K. ■

In the lemma that follows, we find a lower bound for the norm difference of two states of $M_n(\mathbb{C})$ in terms of their eigenvalue lists.

12.3.10. LEMMA. *If ρ and η are states of the factor $\mathcal{M}_n$ consisting of all operators on a Hilbert space of finite dimension n, then*

$$\sum_{j=1}^{n} |h_j - k_j| \le \|\rho - \eta\|,$$

where $\{h_1, \ldots, h_n\}$ and $\{k_1, \ldots, k_n\}$ are the eigenvalue lists of ρ and η.

Proof. From Remark 12.3.9, there are operators H and K in $\mathcal{M}_n$ such that $\rho(A) = \mathrm{Tr}(HA)$ and $\eta(A) = \mathrm{Tr}(KA)$ for all A in $\mathcal{M}_n$; and $\|\rho - \eta\| = \mathrm{Tr}(|H - K|)$. We note that if A and B are self-adjoint operators in $\mathcal{M}_n$ with eigenvalues $\{a_1, \ldots, a_n\}$ and $\{b_1, \ldots, b_n\}$, respectively, listed in decreasing order, and $B \le A$, then $b_j \le a_j$ for j in $\{1, \ldots, n\}$. This follows, at once, from the "minimax principle" characterizing a_j as

$$\min\{\max\{\langle Ax, x\rangle : \|x\| = 1, Ex = x\} : n - j + 1 \le \dim E\};$$

since $\langle Bx, x\rangle \le \langle Ax, x\rangle$ for all x. Once this is established, we complete the proof by noting that $\pm(H - K) \le |H - K|$ so that $\frac{1}{2}[H + K + |H - K|]$ is greater than both H and K. Thus $a_j \ge h_j$ and $a_j \ge k_j$, where $\{a_1, \ldots, a_n\}$, $\{h_1, \ldots, h_n\}$, and $\{k_1, \ldots, k_n\}$ are the eigenvalues of $\frac{1}{2}[H + K + |H - K|]$, H, and K, listed in decreasing order. Thus $a_j \ge \frac{1}{2}[h_j + k_j + |h_j - k_j|]$, when $j \in \{1, \ldots, n\}$; from which

$$\begin{aligned}\frac{1}{2}\left(\sum_{j=1}^{n} (h_j + k_j) + \sum_{j=1}^{n} |h_j - k_j|\right) \le \sum_{j=1}^{n} a_j &= \mathrm{Tr}(\tfrac{1}{2}[H + K + |H - K|]) \\ &= \tfrac{1}{2}\sum_{j=1}^{n} (h_j + k_j) + \tfrac{1}{2}\mathrm{Tr}(|H - K|),\end{aligned}$$

and

$$\sum_{j=1}^{n} |h_j - k_j| \le \mathrm{Tr}(|H - K|) = \|\rho - \eta\|.$$

To prove the minimax principle, observe that if $Ax_j = a_jx_j$, $\|x_j\| = 1$, $\{x_1, \ldots, x_n\}$ are mutually orthogonal, and E_j is the projection with range $[x_j, x_{j+1}, \ldots, x_n]$, then $\dim E_j = n - j + 1$ and

$$a_j = \langle Ax_j, x_j\rangle = \max\{\langle Ax, x\rangle : \|x\| = 1, E_jx = x\}.$$

Thus the "minimax" does not exceed a_j. On the other hand, if E is a projection in $\mathscr{M}_n$ of dimension $n - j + 1$, then there is a unit vector x_0 in the range of E and in the (j-dimensional) space generated by $\{x_1, \ldots, x_j\}$. Hence

$$a_j = \min\{\langle Ax, x\rangle : \|x\| = 1, x = r_1x_1 + \cdots + r_jx_j\} \le \langle Ax_0 x_0\rangle$$
$$\le \max\{\langle Ax, x\rangle : \|x\| = 1, Ex = x\}.$$

Thus the "minimax" is not less than a_j. ■

12.3.11. LEMMA. *If $\rho = (\rho \mid \mathscr{M}_1) \otimes (\rho \mid \mathscr{M}_2)$, ρ and η are states of $\mathscr{M}_1 \otimes \mathscr{M}_2$, $\mathscr{M}_1$ is a factor of type I_2 and $\mathscr{M}_2$ is a factor of type I_n, and*

(i) *$\rho \mid \mathscr{M}_1$, $\rho \mid \mathscr{M}_2$, ρ, and η have respective eigenvalue lists $\{1 - a, a\}$, $\{a_1, \ldots, a_n\}$, $\{c_1, \ldots, c_{2n}\}$, and $\{b_1, \ldots, b_{2n}\}$,*
(ii) *$b_j b_k^{-1} \le b(1 - b)^{-1}$ when $b_j < b_k$, where $0 \le b \le a \le \frac{1}{2}$;*

then

$$\min\left\{\frac{1}{2} - a, \frac{a - b}{2}\right\} \le \|\rho - \eta\|. \tag{3}$$

Proof. From Lemma 12.3.10, $\sum_{j=1}^{2n} |c_j - b_j| \le \|\rho - \eta\|$. From the given form of ρ, $\{aa_1, \ldots, aa_n, (1 - a)a_1, \ldots, (1 - a)a_n\}$ is a rearrangement of $\{c_1, \ldots, c_{2n}\}$. Thus

$$\sum_{j=1}^{n} (|aa_j - d_j| + |(1 - a)a_j - e_j|) = \sum_{j=1}^{2n} |c_j - b_j| \le \|\rho - \eta\|, \tag{4}$$

where $\{d_1, \ldots, d_n, e_1, \ldots, e_n\}$ is a rearrangement of $\{b_1, \ldots, b_{2n}\}$. Let m_j and M_j be, respectively, the smaller and larger of d_j and e_j. If $m_j < M_j$, then $m_j M_j^{-1} \le b(1 - b)^{-1} \le a(1 - a)^{-1}$, from (ii). Hence, if $M_j \ne 0 \ne a$,

$$\begin{aligned}
|aa_j - d_j| + |(1 - a)a_j - e_j| &= a\left|a_j - \frac{d_j}{a}\right| + (1 - a)\left|a_j - \frac{e_j}{1 - a}\right| \\
&\ge a\left|a_j - \frac{d_j}{a}\right| + a\left|a_j - \frac{e_j}{1 - a}\right| \\
&\ge a\left|\frac{e_j}{1 - a} - \frac{d_j}{a}\right| \\
&\ge a\left|\frac{M_j}{1 - a} - \frac{m_j}{a}\right| \\
&= M_j\left|\frac{a}{1 - a} - \frac{m_j}{M_j}\right| \\
&\ge M_j \min\left\{\frac{a}{1 - a} - \frac{b}{1 - b}, 1 - \frac{a}{1 - a}\right\} \\
&\ge M_j \min\{a - b, 1 - 2a\}.
\end{aligned}$$

Note that the inequality derived above,

$$(5)\qquad |aa_j - d_j| + |(1-a)a_j - e_j| \geq M_j \min\{a-b, 1-2a\}$$

is valid when M_j or a is 0 as well. Since $m_j \leq M_j$, we conclude from (5) and (4) that

$$\begin{aligned}\min\left\{\frac{a-b}{2}, \frac{1}{2} - a\right\} &= \frac{1}{2}\min\{a-b, 1-2a\}\sum_{j=1}^{2n} b_j\\ &= \frac{1}{2}\min\{a-b, 1-2a\}\sum_{j=1}^{n}(M_j + m_j)\\ &\leq \sum_{j=1}^{n} M_j \min\{a-b, 1-2a\}\\ &\leq \sum_{j=1}^{n}(|aa_j - d_j| + |(1-a)a_j - e_j|) \leq \|\rho - \eta\|,\end{aligned}$$

which establishes (3). ■

12.3.12. LEMMA. *Let $\mathfrak{A}$ be a uniformly matricial C^*-algebra with $\{\mathfrak{A}_j\}$ and $\{\mathscr{B}_j\}$ two sets of mutually commuting C^*-subalgebras of $\mathfrak{A}$, each set generating $\mathfrak{A}$ and each $\mathfrak{A}_j$ and $\mathscr{B}_j$ containing I and * isomorphic to a finite type* I *factor. Let $\{E_{rs}^j\}$ be a self-adjoint system of matrix units for $\mathfrak{A}_j$. Given a positive ε and a positive integer n, there are positive integers $m(>n)$ and n_0 and a self-adjoint system of matrix units $\{F_{rs}\}$ in the algebra generated by $\{\mathscr{B}_{n+1}, \ldots, \mathscr{B}_m\}$ such that, $\|F_{rs} - E_{rs}^{n_0}\| < \varepsilon$, for all r and s.*

Proof. If $\tilde{\mathscr{B}}_n$ is the C^*-subalgebra of $\mathfrak{A}$ generated by $\{\mathscr{B}_1, \ldots, \mathscr{B}_n\}$, it will suffice to find n_0 and a self-adjoint system of matrix units $\{F_{rs}^0\}$ in $\tilde{\mathscr{B}}_n^c$ such that $\|F_{rs}^0 - E_{rs}^{n_0}\| < \varepsilon/2$, for all r and s; for $\tilde{\mathscr{B}}_n^c$ is the C^*-subalgebra of $\mathfrak{A}$ generated by $\{\mathscr{B}_{n+1}, \mathscr{B}_{n+2}, \ldots\}$ and we can find (as in Lemma 12.2.4) some m (exceeding n) and a self-adjoint system of matrix units $\{F_{rs}\}$ in the algebra generated by $\{\mathscr{B}_{n+1}, \ldots, \mathscr{B}_m\}$ such that $\|F_{rs} - F_{rs}^0\| < \varepsilon/2$.

To find $\{F_{rs}^0\}$ and n_0, approximate a self-adjoint system of matrix units for $\tilde{\mathscr{B}}_n$ very closely (as in Lemma 12.2.4) by a system of matrix units in $\tilde{\mathfrak{A}}_p$, the C^*-subalgebra of $\mathfrak{A}$ generated by $\{\mathfrak{A}_1, \ldots, \mathfrak{A}_p\}$, and let $\tilde{\mathfrak{A}}_0$ be the C^*-subalgebra of $\tilde{\mathfrak{A}}_p$ generated by these approximating matrix units. As in Lemma 12.2.5, there is a unitary operator U in $\mathfrak{A}$ close in norm (depending on the given n in the sense that this n determines the order of $\tilde{\mathscr{B}}_n$ with the generating nest $\{\tilde{\mathscr{B}}_j\}$ given) to I such that $U\tilde{\mathfrak{A}}_0 U^{-1} = \tilde{\mathscr{B}}_n$. If $n_0 > p$, then $E_{rs}^{n_0} \in \tilde{\mathfrak{A}}_0^c$; so that $\{UE_{rs}^{n_0}U^{-1}\}$ is the desired system $\{F_{rs}^0\}$ in $\tilde{\mathscr{B}}_n^c$. ■

12.3.13. LEMMA. *If ρ and η are product states of types a and b, respectively, where $0 < b < a < \frac{1}{2}$, of the CAR algebra $\mathfrak{A}$, then the* GNS *representations corresponding to ρ and η are not quasi-equivalent.*

Proof. Let $\{\mathfrak{A}_n\}$ and $\{\mathscr{B}_n\}$ be sets of mutually commuting C^*-subalgebras of $\mathfrak{A}$, each set generating $\mathfrak{A}$ (in norm) and each $\mathfrak{A}_n$ and $\mathscr{B}_n$ containing I and $*$ isomorphic to $M_2(\mathbb{C})$, such that $\rho = \otimes(\rho|\mathfrak{A}_n)$, $\eta = \otimes(\eta|\mathscr{B}_n)$, and $\mathfrak{A}_n$ and $\mathscr{B}_n$ have self-adjoint systems of 2×2 matrix units $\{E^n_{rs}\}$ and $\{F^n_{rs}\}$ for which

$$\rho(E^n_{11}) = a, \qquad \rho(E^n_{22}) = 1 - a, \qquad \rho(E^n_{12}) = 0 = \eta(F^n_{12}), \tag{6}$$
$$\eta(F^n_{11}) = b, \qquad \eta(F^n_{22}) = 1 - b.$$

We assume, now, that ρ and η correspond to quasi-equivalent representations and derive a contradiction from this assumption. By Theorem 12.3.8 ρ and η are primary states; Proposition 12.3.2 applies and $\|(\rho - \eta)|\tilde{\mathscr{B}}^c_j\| \to_j 0$, where $\tilde{\mathscr{B}}_j$ is the C^*-subalgebra of $\mathfrak{A}$ generated by $\{\mathscr{B}_1, \ldots, \mathscr{B}_j\}$. Let c be $\min\{\frac{1}{2} - a, (a - b)/2\}$. Choose n so that $\|(\rho - \eta) | \tilde{\mathscr{B}}^c_n\| < c/9$. From Lemma 12.3.12 there is an m (exceeding n) and a self-adjoint system of 2×2 matrix units $\{F_{rs}\}$ in $\mathscr{B}_0$ for which $F_{11} + F_{22} = I$, where $\mathscr{B}_0$ is the C^*-subalgebra of $\mathfrak{A}$ generated by $\{\mathscr{B}_{n+1}, \mathscr{B}_{n+2}, \ldots, \mathscr{B}_m\}$; and there is an n_0 such that

$$\|E^{n_0}_{rs} - F_{rs}\| < \frac{ca^2}{18}. \tag{7}$$

(Henceforth, we write E_{rs} in place of $E^{n_0}_{rs}$.) Since $0 < c < a < \frac{1}{2}$ and

$$|a - \rho(F_{11})| = |\rho(E_{11}) - \rho(F_{11})| \le \|E_{11} - F_{11}\| < \frac{ca^2}{18},$$

we have

$$\tfrac{8}{9}a \le \rho(F_{11}) \qquad \text{and} \qquad |a^{-1} - \rho(F_{11})^{-1}| \le \frac{c}{16}. \tag{8}$$

Let $\rho_0(A)$ be $\rho(F_{11})^{-1}[a\rho(F_{11}AF_{11}) + (1 - a)\rho(F_{12}AF_{21})]$ for each A in $\mathfrak{A}$. Clearly ρ_0 is a state of $\mathfrak{A}$. Note that $\rho_0 = (\rho_0 \mid \mathscr{B}) \otimes (\rho_0 \mid \mathscr{B}^c)$, where $\mathscr{B}$ is the algebra generated by $\{F_{rs}\}$; for with B in $\mathscr{B}^c$, $\rho_0(B) = \rho(F_{11})^{-1}\rho(F_{11}B)$, so that

$$\rho_0(F_{11}B) = \rho_0(F_{11})\rho_0(B)$$

and

$$\rho_0(F_{12}B) = \rho_0(BF_{21}) = 0 = \rho_0(F_{12})\rho_0(B).$$

Note, too, that $\rho_0(F_{11}) = a$, $\rho_0(F_{22}) = 1 - a$ and $\rho_0(F_{12}) = \rho_0(F_{21}) = 0$; so that $\rho_0 \mid \mathscr{B}$ has $\{1 - a, a\}$ as its eigenvalue list.

We show, next, that

$$\rho(E_{11}AE_{22}) = \rho(E_{22}AE_{11}) = 0 \tag{9}$$

and that

(10)

$$\rho(E_{22}AE_{22}) = (1-a)\rho(E_{11})^{-1}\rho(E_{12}AE_{21}) = (1-a)a^{-1}\rho(E_{12}AE_{21})$$

for each A in $\mathfrak{A}$. By linearity and norm continuity (in A) of each term of (9) and (10), it suffices to establish (9) and (10) when $A = A_1A_2$, where $A_1 \in \mathfrak{A}_{n_0}$ and $A_2 \in \mathfrak{A}_{n_0}^c$. Since $\rho = (\rho \mid \mathfrak{A}_{n_0}) \otimes (\rho \mid \mathfrak{A}_{n_0}^c)$, from (6) we have that

$$\rho(E_{11}A_1A_2E_{22}) = \rho(E_{11}A_1E_{22})\rho(A_2) = 0.$$

As $A_1 = \sum_{r,s=1}^{2} a_{rs}E_{rs}$ for some scalars a_{rs}, we have

$$\rho(E_{22}A_1A_2E_{22}) = \rho(E_{22}A_1E_{22})\rho(A_2) = (1-a)a_{22}\rho(A_2)$$

and

$$\rho(E_{12}A_1A_2E_{21}) = aa_{22}\rho(A_2);$$

from which (10) follows.

If $A \in (\mathfrak{A})_1$, applying (9), (7), (10), and (8),

$$\begin{aligned}
&|\rho(F_{11}AF_{22}) - \rho_0(F_{11}AF_{22})| = |\rho(F_{11}AF_{22}) - \rho(E_{11}AE_{22})| < \frac{ca^2}{9},\\
&|\rho(F_{22}AF_{22}) - \rho_0(F_{22}AF_{22})|\\
&\quad = |\rho(F_{22}AF_{22}) - (1-a)\rho(F_{11})^{-1}\rho(F_{12}AF_{21})|\\
&\quad \le |\rho(F_{22}AF_{22}) - \rho(E_{22}AE_{22})|\\
&\qquad + |(1-a)[a^{-1}\rho(E_{12}AE_{21}) - \rho(F_{11})^{-1}\rho(F_{12}AF_{21})| < \frac{4c}{9},
\end{aligned}$$

and

$$|\rho(F_{11}AF_{11}) - \rho_0(F_{11}AF_{11})| = |\rho(F_{11}AF_{11})(1 - a\rho(F_{11})^{-1})| < \frac{ca}{16}.$$

Thus

$$|(\rho - \rho_0)(A)| = \left|(\rho - \rho_0)\left(\sum_{r,s=1}^{2} F_{rr}AF_{ss}\right)\right| < \frac{7c}{9},$$

and $\|\rho - \rho_0\| < 7c/9$. Hence

(11)
$$\begin{aligned}
\|(\eta - \rho_0) \mid \mathscr{B}_0\| &\le \|(\eta - \rho)\mid\mathscr{B}_0\| + \|(\rho - \rho_0)\mid\mathscr{B}_0\|\\
&< \|(\eta - \rho)\mid\mathscr{B}_n^c\| + \tfrac{7}{9}c < c.
\end{aligned}$$

The eigenvalues of $\eta|\mathscr{B}_0$ are $(1-b)^{m-n-k}b^k$, $k \in \{0, 1, \ldots, m-n\}$ (with repetitions). The quotient of the smaller by the larger of two of these eigenvalues is $b^r(1-b)^{-r}$ $(\leq b(1-b)^{-1})$ for some positive integer r. The hypotheses of Lemma 12.3.11 apply to $\eta|\mathscr{B}_0$ and $\rho_0|\mathscr{B}_0$; so that

$$c \leq \|(\eta - \rho_0)|\mathscr{B}_0\|,$$

contradicting (11). ∎

12.3.14. THEOREM. *If ρ and η are states of the CAR algebra $\mathfrak{A}$, ρ is of product type a, η is of product type b, and $0 \leq b < a \leq \frac{1}{2}$, then the ultraweak closures $\mathscr{M}_1$ and $\mathscr{M}_2$ of the images of $\mathfrak{A}$ under the* GNS *representations corresponding to ρ and η are (matricial) factors that are not * isomorphic. If $0 < b < a < \frac{1}{2}$, then $\mathscr{M}_1$ and $\mathscr{M}_2$ are of type* III.

Proof. As noted in Theorem 12.3.8, the product states of $\mathfrak{A}$ are primary, so that $\mathscr{M}_1$ and $\mathscr{M}_2$ are factors. If $\mathscr{M}_1$ and $\mathscr{M}_2$ are * isomorphic, that is, if ρ and η are algebraically equivalent there is, from Theorem 12.3.6, a * automorphism α of $\mathfrak{A}$ such that $\rho \circ \alpha$ and η correspond to quasi-equivalent representations of $\mathfrak{A}$. But Lemma 12.2.13 rules this out when $0 < b < a < \frac{1}{2}$, since $\rho \circ \alpha$ is, again, of product type a and η is of product type b.

An application of Theorem 12.3.8 complete the proof. If $b = 0$, $\mathscr{M}_2$ is of type I_∞. If $a = \frac{1}{2}$, $\mathscr{M}_1$ is of type II_1. ∎

Bibliography: [4, 12, 13, 72]

12.4. Exercises

12.4.1. Let E and F be projections in a von Neumann algebra $\mathscr{R}$ such that $\|E - F\| < 1$. Show that

(i) $E \sim F$ [*Hint*. See the proof of Lemma 12.1.5.];
(ii) there is a unitary operator U in $\mathscr{R}$ such that $UEU^* = F$.

12.4.2. Let E and $\mathfrak{A}$ be a projection and a C^*-algebra, respectively, acting on a Hilbert space $\mathscr{H}$. Suppose $0 < a \leq \frac{1}{8}$ and $\|E - A\| < a$ for some A in $(\mathfrak{A})_1$. Show that

(i) we can replace A by a self-adjoint operator in $(\mathfrak{A})_1$ [*Hint*. Consider $(A + A^*)/2$.];
(ii) $\|A^2 - A\| < 2a$ [*Hint*. Note that $A - A^2 = A(I - A) - A(I - E) + A(I - E) - E(I - E)$.];

(iii) each s in sp A is such that $0 < s^2 - s + 2a$, $0 < s - s^2 + 2a$, that s lies in the interval $[\frac{1}{2} - (\frac{1}{4} + 2a)^{1/2}, \frac{1}{2} + (\frac{1}{4} + 2a)^{1/2}]$, and s is not in

$$[\tfrac{1}{2} - (\tfrac{1}{4} - 2a)^{1/2}, \tfrac{1}{2} + (\tfrac{1}{4} - 2a)^{1/2}],$$

(iv) there is a projection F in $\mathfrak{A}$ such that

$$\|E - F\| < a + \tfrac{1}{2} - (\tfrac{1}{4} - 2a)^{1/2}.$$

[*Hint*. Consider $f(A)$, where f is 0 to the left and 1 to the right of the second interval described in (iii).]

12.4.3. Let $\mathfrak{A}$ be a C^*-algebra and $\mathscr{S}$ be a subset that generates a dense linear subspace of $\mathfrak{A}$. Suppose that for each finite set of elements $A_1, \ldots, A_n$ in $\mathscr{S}$ there is a state ρ of $\mathfrak{A}$ satisfying

$$\rho(A_j A_k) = \rho(A_k A_j) \qquad (j, k \in \{1, \ldots, n\}).$$

Show that $\mathfrak{A}$ has a tracial state.

12.4.4. Let $\mathfrak{A}$ be a C^*-algebra and $\{\mathfrak{A}_a : a \in \mathbb{A}\}$ be a family of C^*-subalgebras of $\mathfrak{A}$ totally ordered by inclusion and such that $\bigcup_{a \in \mathbb{A}} \mathfrak{A}_a$ is norm dense in $\mathfrak{A}$. Show that

(i) $\mathfrak{A}$ is a simple C^*-algebra if each $\mathfrak{A}_a$ is simple [*Hint*. Consider the quotient mapping modulo an ideal and recall that * isomorphisms are isometric.];
(ii) $\mathfrak{A}$ admits a trace if each $\mathfrak{A}_a$ admits a trace [*Hint*. Use Exercise 12.4.3.];
(iii) $\mathfrak{A}$ has a unique trace if each $\mathfrak{A}_a$ has a unique trace;
(iv) $\mathfrak{A}$ is simple and has a unique trace if each $\mathfrak{A}_a$ is a factor of type II_1;
(v) $\mathfrak{A}$ is not uniformly matricial if each $\mathfrak{A}_a$ is a factor of type II_1.

12.4.5. Let $\mathscr{M}$ be a factor of type II_1, E a projection in $\mathscr{M}$, $\mathscr{N}$ a finite type I subfactor of $\mathscr{M}$, and A an operator in $(\mathscr{N})_1$ such that

$$\|A - E\|_2 < b < 10^{-2}.$$

Show that there is a projection M in $\mathscr{N}$ such that

$$\|M - E\|_2 < 2b^{1/4}.$$

[*Hint*. See the proof of Lemma 12.2.3.]

12.4.6. Let $\mathscr{M}$ be a factor acting on a Hilbert space $\mathscr{H}$.

(i) Show that a finite orthogonal family of projection $\{E_1, \ldots, E_m\}$ with sum I lies in a finite type I subfactor of $\mathcal{M}$ if and only if each E_j is a finite sum of monic projections in $\mathcal{M}$. (See the introduction to Section 8.2. We say that E_j is a *rational projection* in $\mathcal{M}$.) [*Hint*. When $\mathcal{M}$ is infinite, use Exercise 6.9.4 to show that each $E_j \sim I$. When $\mathcal{M}$ is of type II_1, use Lemma 6.5.6.]

(ii) Suppose $\mathcal{N}$ is a finite type I subfactor of $\mathcal{M}$, E is a projection in $\mathcal{M}$, and B is an operator in $\mathcal{N}$ such that $\|E - B\| < 1/8$. Show that E is rational. [*Hint*. Use (i) and Exercises 12.4.1 and 12.4.2.]

(iii) Show that a normal operator in $\mathcal{M}$ lies in a finite type I subfactor of $\mathcal{M}$ if and only if it is a finite linear combination of orthogonal rational projections with sum I. (We call such an operator *rational*.)

12.4.7. Let $\mathcal{M}$ be a factor and $\{E_1, \ldots, E_m\}$ be an orthogonal family of rational projections (see Exercise 12.4.6) in $\mathcal{M}$ with sum I. Let

$$\{E(j, k, r): j, k \in \{1, \ldots, n(r)\}\} \qquad (r \in \{1, \ldots, m\})$$

be a self-adjoint system of $n(r) \times n(r)$ matrix units in $E_r \mathcal{M} E_r$ such that $\sum_{j=1}^{n(r)} E(j, j, r) = E_r$.

(i) Suppose $\mathcal{M}$ is infinite. Show that each $E(j, j, r) \sim I$ and that there is a finite type I subfactor of $\mathcal{M}$ containing all $E(j, k, r)$. [*Hint*. Use Exercise 6.9.4 and Lemma 6.6.4.]

(ii) Suppose $\mathcal{M}$ is finite. Show that there is a finite type I subfactor of $\mathcal{M}$ containing all $E(j, k, r)$. [*Hint*. When $\mathcal{M}$ is of type II_1, use Lemma 6.5.6 to express $E(1, 1, r)$ as a sum of equivalent orthogonal projections such that all are equivalent as r varies. Use Lemma 6.6.4 to replace $\{E(j, k, r)\}$ by a system of matrix units in $E_r \mathcal{M} E_r$ containing the subprojections of $E(1, 1, r)$ for each r. Now use the argument for (i) with the new matrix unit systems.]

12.4.8. Let $E_1, \ldots, E_n$ be a commuting family of projections in a factor $\mathcal{M}$ of type II_1 acting on a Hilbert space $\mathcal{H}$. Let $x_1, \ldots, x_m$ be vectors in $\mathcal{H}$ and ε be a positive number. Show that

(i) there is an orthogonal family $\{F_1, \ldots F_r\}$ of projections in $\mathcal{M}$ such that each E_j is a finite sum of projections in $\{F_1, \ldots, F_r\}$;

(ii) with $\{F_1, \ldots, F_r\}$ as in (i), there is a family of rational projections (see Exercise 12.4.6) $\{G_1, \ldots, G_r\}$ in $\mathcal{M}$ such that $G_j \le F_j$ and

$$\|(F_j - G_j)x_k\| < \varepsilon/r \qquad (j \in \{1, \ldots, r\}, \quad k \in \{1, \ldots, m\});$$

(iii) there is a finite type I subfactor $\mathcal{N}$ of $\mathcal{M}$ and projections $N_1, \ldots, N_n$ in $\mathcal{N}$ such that $N_j \le E_j$ and

$$\|(E_j - N_j)x_k\| < \varepsilon \qquad (j \in \{1, \ldots, n\}, \quad k \in \{1, \ldots, m\}).$$

12.4.9. Let $\mathcal{M}$ be a factor of type II_1 acting on a Hilbert space $\mathcal{H}$. Let $\{E_1, \ldots, E_n\}$ be an orthogonal family of projections in $\mathcal{M}$ and

$$\{E(j, k, r): j, k \in \{1, \ldots, n(r)\}\}$$

be a self-adjoint system of $n(r) \times n(r)$ matrix units in $E_r \mathcal{M} E_r$ such that $\sum_{j=1}^{n(r)} E(j, j, r) = E_r$. With ε a positive number and $x_1, \ldots, x_m$ vectors in $\mathcal{H}$, show that there are a finite type I subfactor $\mathcal{N}$ of $\mathcal{M}$ and operators $F(j, k, r)$ in $\mathcal{N}$ such that

$$\|[E(j, k, r) - F(j, k, r)]x_h\| < \varepsilon$$

for all j, k in $\{1, \ldots, n(r)\}$, h in $\{1, \ldots, m\}$, and r in $\{1, \ldots, n\}$, and

$$\{F(j, k, r): j, k \in \{1, \ldots, n(r)\}\}$$

is a self-adjoint system of $n(r) \times n(r)$ matrix units in $F_r \mathcal{M} F_r$, where $F_r = \sum_{j=1}^{n(r)} F(j, j, r)$ and $\{F_r : r \in \{1, \ldots, n\}\}$ is an orthogonal family of projections in $\mathcal{N}$. [*Hint*. Use Exercise 12.4.8 to find rational subprojections $F(1, 1, r)$ of $E(1, 1, r)$. Construct $F(j, k, r)$ from $E(j, 1, r)$ and $F(1, 1, r)$. Use Exercise 12.4.7 to find $\mathcal{N}$.]

12.4.10. Let $\mathcal{R}$ be a type I_n von Neumann subalgebra of a factor $\mathcal{M}$ of type II_1, where n is a finite cardinal, and let ε be a positive real number.

(i) With $\{A_1, \ldots, A_m\}$ a finite set of operators in $(\mathcal{R})_1$, show that there are a finite type I subfactor $\mathcal{N}$ of $\mathcal{M}$ and operators $B_1, \ldots, B_m$ in $(\mathcal{N})_1$ such that

$$\|A_h - B_h\|_2 < \varepsilon \qquad (h \in \{1, \ldots, m\}).$$

[*Hint*. Use the (unique) tracial state on $\mathcal{M}$ to represent $\mathcal{M}$ on a Hilbert space $\mathcal{H}$ so that for some unit (trace) vector x_0 in $\mathcal{H}$, $\|A\|_2 = \|Ax_0\|$ when $A \in \mathcal{M}$. Choose matrix units $\{E_{jk}\}$ for $\mathcal{R}$ such that E_{11} is an abelian projection in $\mathcal{R}$. With A_h expressed as $\sum_{j,k=1}^{n} C(h, j, k)E_{jk}$, where $C(h, j, k)$ is in the center of $\mathcal{R}$, use the spectral theorem to find a finite family $\{Q_1, \ldots, Q_r\}$ of central projections in $\mathcal{R}$ and complex scalars $\lambda(h, j, k, t)$ such that

$$\|C(h, j, k)Q_t - \lambda(h, j, k, t)Q_t\| < \varepsilon/2n^2.$$

Apply the result of Exercise 12.4.9.]

(ii) With A a normal operator in $\mathcal{M}$, show that there are a finite type I subfactor $\mathcal{N}$ of $\mathcal{M}$ and a B in $\mathcal{N}$ such that $\|A - B\|_2 < \varepsilon$.

(iii) If A_1 and A_2 are normal operators in $\mathcal{M}$, are there a finite type I subfactor $\mathcal{N}$ of $\mathcal{M}$ and operators B_1 and B_2 in $\mathcal{N}$ such that $\|A_j - B_j\|_2 < \varepsilon$ $(j \in \{1, 2\})$? Proof? Counterexample? [*Hint*. Consider the factor $\mathcal{L}_{\mathcal{F}_2}$ studied in Theorem 6.7.8, and note that the set of operators in $\mathcal{M}$ commuting with a given finite type I subfactor of $\mathcal{M}$ is a subfactor of $\mathcal{M}$ of type II_1.]

12.4.11. Let E and F be projections on a Hilbert space $\mathscr{H}$, and let $\mathscr{R}$ be the von Neumann algebra generated by E, F, and I. Show that

(i) $(E - F)^2$ commutes with E and F;

(ii) $E\mathscr{R}E = \mathscr{C}E$, where $\mathscr{C}$ is the center of $\mathscr{R}$ [*Hint.* Use (i) and consider $(E - F)^2E$.];

(iii) E and $I - E$ are abelian projections in $\mathscr{R}$;

(iv) $\mathscr{R}$ is either abelian, of type I_2, or the direct sum of an abelian von Neumann algebra and one of type I_2.

12.4.12. Suppose $\mathfrak{A}$ is generated as a C^*-algebra by two projections and I. Let π be an irreducible representation of $\mathfrak{A}$ on a Hilbert space $\mathscr{H}$. Show that $\mathscr{H}$ has dimension at most 2. [*Hint.* Use Exercise 12.4.11.]

12.4.13. Let $\mathscr{R}$ be a von Neumann algebra that has no minimal projections. With E a projection in $\mathscr{R}$ and ω a normal state of $\mathscr{R}$, show that there is a family $\{E_\lambda : 0 \le \lambda \le \omega(E)\}$ of projections E_λ in $\mathscr{R}$ such that $E_0 = 0$, $E_{\omega(E)} = E$, $\omega(E_\lambda) = \lambda$ for each λ in $[0, \omega(E)]$, and $E_\lambda \le E_{\lambda'}$ when $\lambda \le \lambda'$. [*Hint.* Let $\omega(E)$ be a and ω' be $a^{-1}\omega | E\mathscr{R}E$. Use a maximality argument on orthogonal families of projections in $E\mathscr{R}E$ such that the values of ω' at their unions does not exceed $\frac{1}{2}$ to produce a projection $E_{a/2}$ in $E\mathscr{R}E$ for which $\omega'(E_{a/2}) = \frac{1}{2}$. Now find E_{ar} such that $\omega'(E_{ar}) = r$ for each dyadic rational r in $[0, 1]$.]

12.4.14. In solving (i), (ii), and (iii), make the following (inductive) assumption:

$(*)$ If $\mathscr{R}$ is a von Neumann algebra that has no minimal projections, then for each set of n normal states $\omega_1, \ldots, \omega_n$ of $\mathscr{R}$ such that ω_1 has support I, there is a projection E in $\mathscr{R}$ such that $\omega_1(E) = \cdots = \omega_n(E) = \frac{1}{2}$.

(i) Show that if F is a projection in $\mathscr{R}$ such that $\omega_1(F) = \cdots = \omega_n(F) (= a)$, where $\omega_1, \ldots, \omega_n$ are as described in $(*)$, then there is a family $\{F_\lambda : 0 \le \lambda \le a\}$ of projections F_λ in $\mathscr{R}$ such that $F_0 = 0$, $F_a = F$, $\omega_1(F_\lambda) = \cdots = \omega_n(F_\lambda) = \lambda$ for each λ in $[0, a]$, and $F_\lambda \le F_{\lambda'}$ when $\lambda \le \lambda'$. [*Hint.* Use $(*)$ to construct F_{ar} for each dyadic rational r in $[0, 1]$ by considering the restrictions of $\omega_1, \ldots, \omega_n$ to $F\mathscr{R}F$.]

(ii) With the notation of (i), show that $\lambda \to \omega(F_\lambda)$ is continuous on $[0, a]$ for each normal state ω of $\mathscr{R}$. [*Hint.* Consider the faithful representation of $\mathscr{R}$ corresponding to ω_1, in which $\omega_1 = \omega_x | \mathscr{R}$ with x a generating and separating vector for $\mathscr{R}$. Note that $\omega = \omega_y | \mathscr{R}$ and that $y = \lim A'_n x$ for some sequence $\{A'_n\}$ in $\mathscr{R}'$.]

(iii) With $\omega_1, \ldots, \omega_n$, ω normal states of $\mathscr{R}$ such that ω_1 has support I, show that there is a projection M in $\mathscr{R}$ such that $\omega_1(M) = \cdots = \omega_n(M) =$

$\omega(M) = \frac{1}{2}$. [*Hint.* With E and $I - E$ in place of F, and $\{E_\lambda\}$ and $\{F_\lambda\}$, respectively, the families of projections in $\mathscr{R}$ described in (i), note that the function $\lambda \to \omega(E_{\lambda/2} + F_{(1-\lambda)/2})$ on $[0, 1]$ is continuous and takes the value $\frac{1}{2}$.]

(iv) Conclude that $(*)$ holds for all n in $\mathbb{N}$.

(v) Show that if $\omega_1, \ldots, \omega_n$ are normal states of $\mathscr{R}$ (do not assume that ω_1 has support I), there is a family $\{E_\lambda : \lambda \in [0, 1]\}$ of projections in $\mathscr{R}$ such that $\omega_1(E_\lambda) = \cdots = \omega_n(E_\lambda) = \lambda$ for each λ in $[0, 1]$ and $E_\lambda \le E_{\lambda'}$ when $\lambda \le \lambda'$.

12.4.15. (i) Let $\mathscr{R}$ be a von Neumann algebra that has no minimal projections and $\omega_1, \ldots, \omega_n$ be normal states of $\mathscr{R}$. Show that if E_1 is a projection in $\mathscr{R}$ and λ_1 is in $[0, 1]$, then there is a projection E in $E_1\mathscr{R}E_1$ such that $\omega_j(E) = \omega_j(\lambda_1 E_1)$ for each j in $\{1, \ldots, n\}$. [*Hint.* Restrict each ω_j to $E_1\mathscr{R}E_1$ and use Exercise 12.4.14(v).]

(ii) Assume (inductively) that,

$(**)$ if $\{E_1, \ldots, E_k\}$ is a family of k mutually orthogonal projections in a von Neumann algebra $\mathscr{R}$ that has no minimal projections, $\{\omega_1, \ldots, \omega_n\}$ is a finite set of normal states of $\mathscr{R}$, and $\lambda_1, \ldots, \lambda_k$ are in $[0, 1]$, then there is a projection E in $F\mathscr{R}F$ such that $\omega_j(E) = \omega_j(A)$ for each j in $\{1, \ldots, n\}$, where $A = \lambda_1 E_1 + \cdots + \lambda_k E_k$ and $F = E_1 + \cdots + E_k$.

Show that if $\{E_1, \ldots, E_{k+1}\}$ is a family of $k + 1$ mutually orthogonal projections in $\mathscr{R}$, $\{\omega_1, \ldots, \omega_n\}$ is a finite set of normal states of $\mathscr{R}$, and $\lambda_1, \ldots, \lambda_{k+1}$ are in $[0, 1]$, then there is a projection E in $G\mathscr{R}G$ such that $\omega_j(E) = \omega_j(B)$ for each j in $\{1, \ldots, n\}$, where $B = \lambda_1 E_1 + \cdots + \lambda_{k+1} E_{k+1}$ and $G = E_1 + \cdots + E_{k+1}$. [*Hint.* Restrict ω_j to each of $F\mathscr{R}F$ and $E_{k+1}\mathscr{R}E_{k+1}$, where $F = E_1 + \cdots + E_k$ and apply $(**)$ and (i).]

(iii) Conclude that $(**)$ is valid for each k in $\mathbb{N}$.

(iv) Show that the set $\mathscr{P}$ of projections in a von Neumann algebra $\mathscr{R}$ that has no minimal projections has weak-operator closure $(\mathscr{R})_1^+$. (Compare Exercise 5.7.8(i).)

12.4.16. Let $\mathscr{R}$ be a type I von Neumann subalgebra of a factor $\mathscr{M}$ of type II_1 and let $\{P_n\}$ be the family of central projections in $\mathscr{R}$ such that $\mathscr{R}P_n$ is of type I_n or $P_n = 0$ and $\sum_n P_n = I$. Suppose $P_n = 0$ when $n > m$ for some finite cardinal m.

(i) With $\{A_1, \ldots, A_n\}$ a finite set of operators in $(\mathscr{R})_1$ and ε a positive number, show that there are a finite type I subfactor $\mathscr{N}$ of $\mathscr{M}$ and operators $B_1, \ldots, B_n$ *in* $(\mathscr{N})_1$ such that

$$\|A_h - B_h\|_2 < \varepsilon \qquad (h \in \{1, \ldots, n\}).$$

[*Hint.* Use Exercises 12.4.10 and 12.4.9.]

(ii) With E and F projections in $\mathscr{M}$ and ε a positive number, show that there are a finite type I factor $\mathscr{N}$ and operators A and B in $(\mathscr{N})_1$ such that

$$\|A - E\|_2 < \varepsilon, \qquad \|B - F\|_2 < \varepsilon.$$

[*Hint*, Use (i) and Exercise 12.4.11.]

(iii) With the notation of (ii), show that A and B can be chosen to be projections. [*Hint*. Use Exercise 12.4.5.]

(iv) With H_1 and H_2 in $(\mathscr{M})_1^+$ and $\mathscr{U}_1$ and $\mathscr{U}_2$ ultraweakly open sets in $(\mathscr{M})_1$ containing H_1 and H_2, respectively, show that there are a finite type I subfactor $\mathscr{N}$ of $\mathscr{M}$ and projections E_1 and E_2 in $\mathscr{N}$ such that $E_1 \in \mathscr{U}_1$ and $E_2 \in \mathscr{U}_2$. [*Hint*. Use (iii) and Exercise 12.4.15(iv).]

12.4.17. Let α be a * automorphism of a von Neumann algebra $\mathscr{R}$. Suppose that there is an A in $\mathscr{R}$ such that $C_A = I$ and $A\alpha(B) = BA$ for each B in $\mathscr{R}$. With VH the polar decomposition of A, show that

(i) H is in the center of $\mathscr{R}$ [*Hint*. Prove that $H^2\alpha(B) = \alpha(B)H^2$ for each B in $\mathscr{R}$. Use the fact that $\alpha(B^*) = \alpha(B)^*$.];

(ii) $C_H = I$;

(iii) $\alpha(B) = V^*BV$ for each B in $\mathscr{R}$ [*Hint*. Note that $H(\alpha(B) - V^*BV) = 0$ and use Theorem 5.5.4.];

(iv) V is a unitary operator in $\mathscr{R}$; conclude that α is inner. [*Hint*. Consider $I - R(V)$.]

12.4.18. Let α be a * automorphism of a von Neumann algebra $\mathscr{R}$. When there is no element A in $\mathscr{R}$, other than 0, such that $A\alpha(B) = BA$ for each B in $\mathscr{R}$, we say that α *acts freely* (*on* $\mathscr{R}$).

(i) Suppose $A \in \mathscr{R}$ and $A\alpha(B) = BA$ for each B in $\mathscr{R}$. Show that $\alpha(C_A) = C_A$. [*Hint*. Consider C_A and $\alpha^{-1}(C_A)$ in place of B.]

(ii) Show that either α acts freely on $\mathscr{R}$ or there is a non-zero central projection Q in $\mathscr{R}$ such that $\alpha(Q) = Q$ and $\alpha|\mathscr{R}Q$ is inner. [Hint. Use Exercise 12.4.17.]

(iii) Show that there is a central projection P in $\mathscr{R}$ uniquely defined by the conditions: $\alpha(P) = P$, P is 0 or $\alpha|\mathscr{R}P$ is inner, and $\alpha|\mathscr{R}(I - P)$ acts freely. [*Hint*. Use (ii) and a maximality argument.]

12.4.19. With the notation of Exercise 11.5.25, suppose that the automorphism α of (iii) of that exercise is inner. Show that $\mathscr{R}$ is a factor. [*Hint*. Assume that the center of $\mathscr{R}$ contains an operator that is not a scalar and use Proposition 11.1.8.]

12.4.20. With the notation of Exercise 12.4.19, let β be a * automorphism of $\mathscr{R}$ and suppose that α is inner. Show that

(i) $\beta \bar{\otimes} \beta^{-1}$ is inner [*Hint*. Consider $(\beta \bar{\otimes} \iota)\alpha(\beta \bar{\otimes} \iota)^{-1}\alpha$, where ι is the identity automorphism of $\mathscr{R}$, and note that the set of inner * automorphisms of $\mathscr{R}$ is a normal subgroup of the set of all * automorphisms.];

(ii) $\Phi_z(U_0)(\beta(A) \otimes I) = (A \otimes I)\Phi_z(U_0)$ for each A in $\mathscr{R}$ and each unit vector z in $\mathscr{H}$, where U_0 is a unitary operator that lies in $\mathscr{R} \bar{\otimes} \mathscr{R}$ and implements $\beta \bar{\otimes} \beta^{-1}$, and Φ_z is as in Proposition 11.2.24;

(iii) β is inner. [*Hint*. Use (ii), Exercise 12.4.17, and Proposition 11.2.24 extended so that the condition that T be positive is removed.]

12.4.21. Let $\mathscr{N}$ be a subfactor of a factor $\mathscr{M}$ of type II_1 and α be the * automorphism of $\mathscr{M} \bar{\otimes} \mathscr{M}$ (described in Exercise 11.5.25) that assigns $B \otimes A$ to $A \otimes B$ for all A and B in $\mathscr{M}$. Let τ be the (unique) tracial state on $\mathscr{M} \bar{\otimes} \mathscr{M}$ and φ be the conditional expectation described in Exercise 8.7.28, mapping $\mathscr{M} \bar{\otimes} \mathscr{M}$ onto $\mathscr{N} \bar{\otimes} \mathscr{N}$ in this case. Suppose U is a unitary operator in $\mathscr{M} \bar{\otimes} \mathscr{M}$ that implements $\alpha | \mathscr{N} \bar{\otimes} \mathscr{N}$.

(i) Show that $\varphi(U)\alpha(T) = T\varphi(U)$ for each T in $\mathscr{N} \bar{\otimes} \mathscr{N}$.

Suppose $\mathscr{N}$ admits an outer * automorphism.

(ii) Show that $\varphi(U) = 0$. [*Hint*. Use Exercises 12.4.20(iii) and 12.4.17.]

(iii) Show that $\tau(UT) = 0$ $(T \in \mathscr{N} \bar{\otimes} \mathscr{N})$.

12.4.22. Let $\mathscr{M}$ be the factor of type II_1 (described in Example 8.6.12) constructed from the interval $[0, 1)$ $(= S)$ with Lebesgue measure m and the group G of translations modulo 1 by rationals. With E_j the projection in $\mathscr{M}$ corresponding to the characteristic function of the interval $[(j-1)n^{-1}, jn^{-1})$ for j in $\{1, \ldots, n\}$ and V_n the unitary operator in $\mathscr{M}$ corresponding to translation by n^{-1}, let $\mathscr{M}_n$ be the von Neumann subalgebra of $\mathscr{M}$ generated by $\{E_j\}$ and V_n. Show that

(i) $\mathscr{M}_n$ is a factor of type I_n [*Hint*. Consider $V_n^{-j}E_1$ and use Lemma 6.6.4.];

(ii) $\mathscr{M}_n \subseteq \mathscr{M}_m$ when m is divisible by n;

(iii) $\bigcup_{n=1}^{\infty} \mathscr{M}_{n!}$ is strong-operator dense in $\mathscr{M}$ [*Hint*. Consider subintervals of $[0, 1)$ with rational endpoints.];

(iv) $\mathscr{M}$ is the matricial factor of type II_1.

12.4.23. Adopt the notation of Exercise 12.4.22, and let G_0 be the set of translations modulo 1 by dyadic rationals. Show that

(i) G_0 is a subgroup of G;

(ii) G_0 acts ergodically on S [*Hint*. Modify the argument of the last paragraph of Example 8.6.12.];

(iii) the von Neumann subalgebra $\mathscr{M}_0$ of $\mathscr{M}$ generated by the operators corresponding to the multiplication algebra $\mathscr{A}$ of (S, m) and the unitary representation of G restricted to G_0 is a factor of type II_1 distinct from $\mathscr{M}$;
(iv) $\mathscr{M}_0' \cap \mathscr{M} = \mathbb{C}I$. [*Hint*. Note that a maximal abelian subalgebra of $\mathscr{M}$ is contained in $\mathscr{M}_0$.]

12.4.24. With the notation of Exercise 12.4.23, show that

(i) $\mathscr{M}_0$ is matricial and is, therefore, *the* matricial factor of type II_1 [*Hint*. Use Exercise 12.4.22.];
(ii) $V^*\mathscr{M}_0 V = \mathscr{M}_0$, where V is the unitary operator in $\mathscr{M}$ corresponding to translation (modulo 1) by $\frac{1}{3}$ on $[0, 1)$ [*Hint*. Note that G is abelian and use 8.6(1) and 8.6(2).];
(iii) the mapping $T \to V^*TV (T \in \mathscr{M}_0)$, with V as in (ii), is an outer * automorphism of $\mathscr{M}_0$ [*Hint*. Use Exercise 12.4.23.];
(iv) each matricial factor of type II_1 admits an outer * automorphism.

(The fact that $\mathscr{M}_0$ is matricial, noted in (i), is also a consequence of a general result [13: Corollary 2]: Each subfactor of type II_1 of the matricial factor of type II_1 is matricial.)

12.4.25. Let $\mathscr{R}$ be a von Neumann algebra of type II_1.

(i) Suppose $\{\mathscr{N}_j\}$ is an ascending sequence of distinct type I subfactors of $\mathscr{R}$ (each containing I). Show that the ultraweak closure of $\bigcup_j \mathscr{N}_j$ in $\mathscr{R}$ is a matricial type II_1 subfactor of $\mathscr{R}$.
(ii) Show that each type I subfactor of $\mathscr{R}$ is contained in a matricial type II_1 subfactor of $\mathscr{R}$.

12.4.26. Let $\mathscr{R}_1$ and $\mathscr{R}_2$ be von Neumann algebras acting on Hilbert spaces $\mathscr{H}_1$ and $\mathscr{H}_2$, and let $\mathscr{S}_1$ and $\mathscr{S}_2$ be subsets of $(\mathscr{R}_1)_1$ and $(\mathscr{R}_2)_1$, respectively. Suppose A_1 and A_2 are in the ultraweak closures of $\mathscr{S}_1$ and $\mathscr{S}_2$, respectively. Show that $A_1 \otimes A_2$ is in the ultraweak closure of $\mathscr{S}_1 \otimes \mathscr{S}_2$, where

$$\mathscr{S}_1 \otimes \mathscr{S}_2 = \{S_1 \otimes S_2 : S_1 \in \mathscr{S}_1, S_2 \in \mathscr{S}_2\}.$$

[*Hint*. Use Proposition 11.2.8.]

12.4.27. Let $\mathscr{M}$ be a factor of type II_1 and τ be the (unique) tracial state on $\mathscr{M} \mathbin{\overline{\otimes}} \mathscr{M}$. Recall from Exercise 11.5.25 that the mapping that assigns $B \otimes A$ to $A \otimes B$, for A and B in $\mathscr{M}$, extends to a * automorphism α of $\mathscr{M} \mathbin{\overline{\otimes}} \mathscr{M}$.

(i) Suppose U is a unitary operator in $\mathscr{M}\bar{\otimes}\mathscr{M}$ that implements α. Show that $\tau(U(E\otimes F))=0$ when E and F are projections in $\mathscr{M}$. [*Hint.* Use Exercises 12.4.16 (iii), 12.4.25(ii), 12.4.24(iv), and 12.4.21(iii).]

(ii) With A and B in $(\mathscr{M})_1^+$, show that $A\otimes B$ is in the weak-operator closure of $\mathscr{P}\otimes\mathscr{P}$, where

$$\mathscr{P}\otimes\mathscr{P}=\{E\otimes F: E, F\in\mathscr{P}\}$$

and $\mathscr{P}$ is the set of projections in $\mathscr{M}$. [*Hint.* Use Exercises 12.4.15 and 12.4.26.]

(iii) With A and B in $(\mathscr{M})_1^+$ and U as in (i), show that $\tau(U(A\otimes B))=0$.

(iv) Show that α is not inner.

12.4.28. Suppose $\mathscr{R}$ is a countably generated, countably decomposable von Neumann algebra. Show that

(i) $(\mathscr{R})_1$ has a countable, strong-operator-dense subset [*Hint.* Use the Kaplansky density theorem.];

(ii) each von Neumann subalgebra of $\mathscr{R}$ is countably generated. [*Hint.* Use Exercise 5.7.46 to find a metric on $(\mathscr{R})_1$ whose associated metric topology is the strong-operator topology. Then use separability arguments in conjunction with (i).]

12.4.29. Let $\mathscr{R}$ be a von Neumann algebra of type II_1.

(i) Show that the ultraweak closure of the union of a family of subfactors of $\mathscr{R}$ of type II_1 totally ordered by inclusion is a subfactor of type II_1. [*Hint.* Use Exercise 12.4.4 and Proposition 12.1.2.]

Suppose $\mathscr{R}$ is countably generated (as a von Neumann algebra) with countably decomposable center (for example—$\mathscr{R}$ acts on a separable Hilbert space). Show that

(ii) each matricial subfactor of $\mathscr{R}$ is contained in a maximal matricial subfactor of $\mathscr{R}$ [*Hint.* Use Corollary 8.2.9 and Exercise 12.4.28 to show that each von Neumann subalgebra of $\mathscr{R}$ is countably generated. Use Theorem 12.2.2.];

(iii) each finite type I subfactor of $\mathscr{R}$ is contained in a maximal matricial subfactor. [*Hint.* Use Exercise 12.4.25 and (ii).]

12.4.30. Let $\mathscr{R}$ be a von Neumann algebra of type II_1, and let $\mathscr{M}$ be a maximal matricial von Neumann subalgebra of $\mathscr{R}$. Show that

(i) $\mathscr{M}$ is a factor of type II_1;

(ii) $\mathscr{M}'\cap\mathscr{R}$ is a factor if $(\mathscr{M}'\cap\mathscr{R})'\cap\mathscr{R}=\mathscr{M}$;

(iii) $(\mathscr{M}'\cap\mathscr{R})'\cap\mathscr{R}\neq\mathscr{M}$ unless $\mathscr{R}$ is a matricial factor.

12.4.31. We say that a von Neumann algebra $\mathscr{R}$ is *normal* when each von Neumann subalgebra $\mathscr{S}$ of $\mathscr{R}$ coincides with its own *relative double commutant* (that is, $(\mathscr{S}' \cap \mathscr{R})' \cap \mathscr{R} = \mathscr{S}$). Show that

(i) $\mathscr{R}$ is a factor if $\mathscr{R}$ is normal;
(ii) $\mathscr{R}$ is normal if $\mathscr{R}$ is a type I factor;
(iii) $\mathscr{R}$ is not normal if $\mathscr{R}$ is a type II_1 factor [*Hint.* Use Exercises 12.4.23 and 12.4.30.];
(iv) $\mathscr{R}$ is not normal if $\mathscr{R}$ is a factor of type II_∞. [*Hint.* Use Theorem 6.7.10.]

12.4.32. Let $\mathscr{M}$ be a factor of type II_1.

(i) Show that $\|U(H) - U(H_0)\|_2 \le 2\|H - H_0\|_2$ for each pair of self-adjoint operators H and H_0 in $\mathscr{M}$, where $U(H)$ is the Cayley transform appearing in Lemma 5.3.3. [*Hint.* Use the trace representation and Remark 8.5.9.]

(ii) Suppose h is a continuous real-valued function vanishing at ∞ on $\mathbb{R}$. Let $\{H_n\}$ be a sequence of self-adjoint operators in $\mathscr{M}$ such that $\{\|H_n - H_0\|_2\}$ tends to 0 as n tends to ∞. Show that $\{h(H_n)\}$ is strong-operator convergent to $h(H_0)$. [*Hint.* Use Exercise 8.7.3, the proof of Theorem 5.3.4, and (i).]

(iii) With H_0 a self-adjoint operator in $\mathscr{M}$, h as in (ii), and ε a positive number, find a positive δ such that $\|h(H) - h(H_0)\|_2 < \varepsilon$ provided that $\|H - H_0\|_2 < \delta$ and $H \in \mathscr{M}$.

(iv) Suppose $\mathscr{M}$ is countably generated and for each finite subset $\{A_1, \ldots, A_p\}$ of $\mathscr{M}$ and positive ε there are a finite type I subfactor $\mathscr{N}$ of $\mathscr{M}$ and operators $B_1, \ldots, B_p$ in $\mathscr{N}$ such that

$$\|A_j - B_j\|_2 < \varepsilon \qquad (j \in \{1, \ldots, p\}).$$

Show that $\mathscr{M}$ is matricial. [*Hint.* Use (iii) and Theorem 12.2.2.]

12.4.33. Let $\mathscr{M}$ be a countably generated factor of type II_1, and let $\mathscr{S}$ be a self-adjoint subset of $(\mathscr{M})_1$ that generates $\mathscr{M}$ as a von Neumann algebra. Suppose that for each positive ε and each finite subset $\{S_1, \ldots, S_n\}$ of $\mathscr{S}$, there are a finite type I subfactor $\mathscr{N}$ of $\mathscr{M}$ and operators $T_1, \ldots, T_n$ in $(\mathscr{N})_1$ such that $\|S_j - T_j\|_2 < \varepsilon$ for each j in $\{1, \ldots, n\}$. Show that $\mathscr{M}$ is matricial.

12.4.34. Let $\mathscr{N}$ be a type I subfactor of a factor $\mathscr{M}$, E be a minimal projection in $\mathscr{N}$, and $\mathscr{M}_0$ be the set of elements in $\mathscr{M}$ that commute with all elements of $\mathscr{N}$. Show that

(i) $E\mathscr{M}E = \mathscr{M}_0 E$ [*Hint.* Use Lemma 6.6.3.];

(ii) $(E \otimes E)(\mathcal{M} \mathbin{\overline{\otimes}} \mathcal{M})(E \otimes E) = (\mathcal{M}_0 \mathbin{\overline{\otimes}} \mathcal{M}_0)(E \otimes E)$;

(iii) $\tilde{\mathcal{N}} = \mathcal{M}_0 \mathbin{\overline{\otimes}} \mathcal{M}_0$, where $\tilde{\mathcal{N}}$ is the set of elements in $\mathcal{M} \mathbin{\overline{\otimes}} \mathcal{M}$ that commute with $\mathcal{N} \mathbin{\overline{\otimes}} \mathcal{N}$. [*Hint.* Consider the mapping $\tilde{T} \to \tilde{T}(E \otimes E)$ $(\tilde{T} \in \tilde{\mathcal{N}})$.]

12.4.35. Suppose $\mathcal{M}$ is a von Neumann algebra of type II_∞ and α is the * automorphism of $\mathcal{M} \mathbin{\overline{\otimes}} \mathcal{M}$ (described in Exercise 11.5.25) that assigns $B \otimes A$ to $A \otimes B$ for each A and B in $\mathcal{M}$.

(i) Suppose α is inner. Show that $\mathcal{M}$ has the form $n \otimes \mathcal{M}_0$, where n is an infinite cardinal and $\mathcal{M}_0$ is a factor of type II_1. [*Hint.* Use Exercise 12.4.19 and Theorem 6.7.10.]

(ii) Show that if α is inner, then α restricted to $(\mathcal{M}_0 \otimes I_n) \mathbin{\overline{\otimes}} (\mathcal{M}_0 \otimes I_n)$ is inner. [*Hint.* Use (i), Corollary 9.3.5, and Exercise 12.4.34.]

(iii) Conclude that α is outer (that is, *not inner*). [*Hint.* Use the result of Exercise 12.4.27.]

12.4.36. Let $\mathcal{R}$ and $\mathcal{S}$ be von Neumann algebras and ρ and σ be non-zero elements of $\mathcal{R}_\sharp$ and $\mathcal{S}_\sharp$, respectively. Show that

(i) there is a unique element $\rho \otimes \sigma$ of $(\mathcal{R} \mathbin{\overline{\otimes}} \mathcal{S})_\sharp$ such that $(\rho \otimes \sigma)(R \otimes S) = \rho(R)\sigma(S)$ for each R in $\mathcal{R}$ and S in $\mathcal{S}$ and that $\|\rho \otimes \sigma\| = \|\rho\| \, \|\sigma\|$ [*Hint.* Use Theorem 11.2.10 to consider $\mathcal{R}$ and $\mathcal{S}$ in their universal normal representations. Then apply Corollary 7.3.3.];

(ii) there are unique operators $\Phi_\sigma(\tilde{T})$ and $\Psi_\rho(\tilde{T})$ in $\mathcal{R}$ and $\mathcal{S}$, respectively, corresponding to each $\tilde{T}$ in $\mathcal{R} \mathbin{\overline{\otimes}} \mathcal{S}$, satisfying

$$\rho'(\Phi_\sigma(\tilde{T})) = (\rho' \otimes \sigma)(\tilde{T}), \qquad \sigma'(\Psi_\rho(\tilde{T})) = (\rho \otimes \sigma')(\tilde{T})$$

for each ρ' in $\mathcal{R}_\sharp$ and each σ' in $\mathcal{S}_\sharp$ [*Hint.* Recall that $\mathcal{R}$ and $\mathcal{S}$ are the norm duals of $\mathcal{R}_\sharp$ and $\mathcal{S}_\sharp$.];

(iii) Φ_σ and Ψ_ρ (as defined by (ii)) are ultraweakly continuous linear mappings of $\mathcal{R} \mathbin{\overline{\otimes}} \mathcal{S}$ onto $\mathcal{R}$ and $\mathcal{S}$, respectively, satisfying

$$\Phi_\sigma((A \otimes I)\tilde{T}(B \otimes I)) = A\Phi_\sigma(\tilde{T})B,$$

$$\Psi_\rho((I \otimes C)\tilde{T}(I \otimes D)) = C\Psi_\rho(\tilde{T})D$$

for each $\tilde{T}$ in $\mathcal{R} \mathbin{\overline{\otimes}} \mathcal{S}$, A, B in $\mathcal{R}$, and C, D in $\mathcal{S}$, and that

$$\Phi_\sigma(R \otimes S) = \sigma(S)R, \qquad \Psi_\rho(R \otimes S) = \rho(R)S$$

when $R \in \mathcal{R}$ and $S \in \mathcal{S}$;

(iv) $\Phi_\sigma(\tilde{T}) \in \mathcal{R}_0$ and $\Psi_\rho(\tilde{T}) \in \mathcal{S}_0$ if $\tilde{T} \in \mathcal{R}_0 \mathbin{\overline{\otimes}} \mathcal{S}_0$, where $\mathcal{R}_0$ and $\mathcal{S}_0$ are von Neumann subalgebras of $\mathcal{R}$ and $\mathcal{S}$, respectively [*Hint.* Consider, first, the case where $\tilde{T} = R_0 \otimes S_0$ with R_0 in $\mathcal{R}_0$ and S_0 in $\mathcal{S}_0$. Then use (iii).];

(v) $\tilde{T} \in \mathscr{R}_0 \bar{\otimes} \mathscr{S}_0$ if $\Phi_{\sigma'}(\tilde{T}) \in \mathscr{R}_0$ and $\Psi_{\rho'}(\tilde{T}) \in \mathscr{S}_0$ for each σ' in $\mathscr{S}_\sharp$ and each ρ' in $\mathscr{R}_\sharp$. [*Hint*. With A' in $\mathscr{R}_0'$, show that

$$\langle (A' \otimes I)\tilde{T}(x \otimes y), u \otimes v\rangle = \langle \tilde{T}(A' \otimes I)(x \otimes y), u \otimes v\rangle$$

for all x and u in $\mathscr{H}$ and y and v in $\mathscr{K}$. Let σ' be $\omega_{y,v}|\mathscr{S}$, ρ' be $\omega_{x, A'^*u}|\mathscr{R}$, and ρ'' be $\omega_{A'x,u}|\mathscr{R}$, for this. Use Theorem 11.2.16.]

12.4.37. Let $\mathscr{R}$ and $\mathscr{S}$ be von Neumann algebras acting on Hilbert spaces $\mathscr{H}$ and $\mathscr{K}$, respectively. Suppose $\mathscr{R}_0$ and $\mathscr{S}_0$ are von Neumann subalgebras of $\mathscr{R}$ and $\mathscr{S}$, respectively. Show that

(i) $(\mathscr{R}'_0 \cap \mathscr{R})\bar{\otimes}(\mathscr{S}'_0 \cap \mathscr{S}) = (\mathscr{R}_0 \bar{\otimes} \mathscr{S}_0)' \cap (\mathscr{R} \bar{\otimes} \mathscr{S})$ [*Hint*. Use Exercise 12.4.36(v).]; (The result of Exercise 12.4.34(iii) is a special case of this formula.)

(ii) $\mathscr{A} \bar{\otimes} \mathscr{B}$ is a maximal abelian subalgebra of $\mathscr{R} \bar{\otimes} \mathscr{S}$ when $\mathscr{A}$ and $\mathscr{B}$ are maximal abelian subalgebras of $\mathscr{R}$ and $\mathscr{S}$, respectively;

(iii) $\mathscr{C} \bar{\otimes} \mathscr{D}$ is the center of $\mathscr{R} \bar{\otimes} \mathscr{S}$ when $\mathscr{C}$ is the center of $\mathscr{R}$ and $\mathscr{D}$ is the center of $\mathscr{S}$ by using (i);

(iv) $\mathscr{R}'_0 \cap \mathscr{R} = \mathbb{C}I$ and $\mathscr{S}'_0 \cap \mathscr{S} = \mathbb{C}I$ if and only if we have that $(\mathscr{R}_0 \bar{\otimes} \mathscr{S}_0)' \cap (\mathscr{R} \bar{\otimes} \mathscr{S}) = \mathbb{C}I$.

12.4.38. Let $\mathscr{R}$ be a matricial von Neumann algebra, $\{\mathscr{M}_n\}$ a generating nest of finite type I factors for $\mathscr{R}$, and $\mathfrak{X}^\#$ a dual normal $\mathscr{R}$-module (as described in Exercise 10.5.13). Choose a compatible self-adjoint system of matrix units for $\mathscr{M}_n$, and let $\mathscr{G}_n$ be the set (finite group) of unitary operators in $\mathscr{M}_n$ whose matrix representations relative to these matrix units have only 1 and -1 as non-zero entries. Let $\mathscr{G}$ be the locally finite group $\bigcup_n \mathscr{G}_n$ and μ an invariant mean on $\mathscr{G}$ (as described in Exercise 3.5.7).

(i) Show that the linear span of $\mathscr{G}$ is $\bigcup_n \mathscr{M}_n$; conclude that this linear span is ultraweakly dense in $\mathscr{R}$.

(ii) Let $\mathfrak{X}$ be the predual of $\mathfrak{X}^\#$ and $f_x(U)$ be $[U^*\delta(U)](x)$ for each U in $\mathscr{G}$ and x in $\mathfrak{X}$, with δ a derivation of $\mathscr{R}$ into $\mathfrak{X}^\#$. Show that f_x is bounded on $\mathscr{G}$ and that $x \to \mu(f_x)$ is an element ρ_0 of $\mathfrak{X}^\#$. [*Hint*. Use Exercise 4.6.66.]

(iii) With the notation of (ii), show that

$$\delta(V) = V\rho_0 - \rho_0 V \qquad (V \in \mathscr{G}).$$

[*Hint*. Define $g_x(U)$ to be $[V^*U^*\delta(U)V](x)$ and $f_x^V(U)$ to be $f_x(UV)$. Use the fact that δ is a derivation and μ is an invariant mean to establish the desired relation.]

(iv) Show that $\delta(A) = A\rho_0 - \rho_0 A$ for each A in $\mathscr{R}$. [*Hint*. Use (iii) to establish this for each A in the linear span of $\mathscr{G}$. Then use (i) and Exercise 10.5.13.]

(v) Note that the result of (iv) is valid when $\mathscr{R}$ is assumed just to have an *amenable* group (one having an invariant mean) of unitary operators whose linear span is ultraweakly dense in $\mathscr{R}$. Show that $\mathscr{R}$ has such a group when it is the ultraweak closure of an ascending family of finite-dimensional self-adjoint subalgebras, and in particular, when $\mathscr{R}$ is abelian and acts on a separable Hilbert space.

12.4.39. Let $\mathscr{H}$ be a separable Hilbert space and $\mathscr{H}_n$ be the n-fold tensor product $\mathscr{H} \otimes \cdots \otimes \mathscr{H}$ of $\mathscr{H}$ with itself. Show that

(i) there are operators U_σ and S_n^- on $\mathscr{H}_n$ such that

$$U_\sigma(x_1 \otimes \cdots \otimes x_n) = x_{\sigma(1)} \otimes \cdots \otimes x_{\sigma(n)} \quad , \quad S_n^- = (n!)^{-1} \sum_\sigma \chi(\sigma) U_\sigma,$$

for all $x_1, \ldots, x_n$ in $\mathscr{H}$, where σ is a permutation of $\{1, \ldots, n\}$ and $\chi(\sigma)$ is its sign ($\chi(\sigma) = 1$ if σ is even and $\chi(\sigma) = -1$ if σ is odd), with U_σ a unitary operator and S_n^- a projection on $\mathscr{H}_n$ [*Hint*. Establish that $(x_1, \ldots, x_n) \to x_{\sigma(1)} \otimes \cdots \otimes x_{\sigma(n)}$ is a weak Hilbert–Schmidt mapping of $\mathscr{H} \times \cdots \times \mathscr{H}$ into $\mathscr{H}_n$ and apply Theorem 2.6.4. Note that $U_{\sigma'} U_\sigma = U_{\sigma\sigma'}$, $\chi(\sigma\sigma') = \chi(\sigma)\chi(\sigma')$, and $\chi(\sigma^{-1}) = \chi(\sigma)$.];

(ii)

$$\langle x_1 \wedge \cdots \wedge x_n, y_1 \wedge \cdots \wedge y_n \rangle = \det(\langle x_j, y_k \rangle),$$

where $x_1 \wedge \cdots \wedge x_n = (n!)^{1/2} S_n^-(x_1 \otimes \cdots \otimes x_n)$; (The vector $x_1 \wedge \cdots \wedge x_n$ in $\mathscr{H}_n$ is referred to as the *exterior* or *wedge product* of $x_1, \ldots, x_n$.)

(iii) $x_1 \wedge \cdots \wedge x_n = 0$ if and only if $\{x_1, \ldots, x_n\}$ are linearly dependent and $\langle x_1 \wedge \cdots \wedge x_n, y_1 \wedge \cdots \wedge y_n \rangle = 0$ with $x_1 \wedge \cdots \wedge x_n$ a non-zero vector in $\mathscr{H}_n$ if and only if there is a non-zero vector in $[x_1, \ldots, x_n]$ orthogonal to $[y_1, \ldots, y_n]$ [*Hint*. Use (ii).];

(iv) $(x_1, \ldots, x_n) \to x_1 \wedge \cdots \wedge x_n$ is an alternating multilinear mapping Λ of $\mathscr{H} \times \cdots \times \mathscr{H}$ into the range $\mathscr{H}_n^{(a)}$ of S_n^- (that is, Λ is linear in each coordinate and for each permutation σ of $\{1, \ldots, n\}$

$$\Lambda(x_1, \ldots, x_n) = \chi(\sigma)\Lambda(x_{\sigma(1)}, \ldots, x_{\sigma(n)})),$$

and there is a (unique) bounded linear mapping $\hat{\alpha}$ of $\mathscr{H}_n^{(a)}$ into $\mathscr{K}$ such that $\alpha = \hat{\alpha} \circ \Lambda$ when α is a weak Hilbert–Schmidt, alternating, multilinear mapping of $\mathscr{H} \times \cdots \times \mathscr{H}$ into a Hilbert space $\mathscr{K}$ [*Hint*. Use Theorem 2.6.4 to express α as $\tilde{\alpha} \circ p$ with $\tilde{\alpha}$ a bounded linear mapping of $\mathscr{H}_n$ into $\mathscr{K}$. Show that $\tilde{\alpha} S_n^- = \tilde{\alpha}$, and let $\hat{\alpha}$ be $(n!)^{-1/2}\tilde{\alpha}$.];

(v) $\{e_{j(1)} \wedge \cdots \wedge e_{j(n)} : j(1) < \cdots < j(n)\}$ is an orthonormal basis for the range $\mathscr{H}_n^{(a)}$ of S_n^-, where $\{e_m\}$ is an orthonormal basis for $\mathscr{H}$ [*Hint*. Use (iii) and (iv).];

(vi) there is a unique bounded linear mapping $a_n(x)^*$ of $\mathscr{H}_n^{(a)}$ into $\mathscr{H}_{n+1}^{(a)}$ that assigns $x \wedge x_1 \wedge \cdots \wedge x_n$ to $x_1 \wedge \cdots \wedge x_n$ for all $x_1, \ldots, x_n$ in $\mathscr{H}$. [*Hint.* Let x be e_1, with $\{e_m\}$ an orthonormal basis for $\mathscr{H}$, and show that $(x_1, \ldots, x_n) \to x \wedge x_1 \wedge \cdots \wedge x_n$ is a weak Hilbert–Schmidt, alternating, multilinear mapping of $\mathscr{H} \times \cdots \times \mathscr{H}$ into $\mathscr{H}_{n+1}^{(a)}$. Use (iv).]

12.4.40. In the notation of Exercise 12.4.39, let $\mathscr{H}_{\mathscr{F}}^{(a)}$ be $\sum_{n=0}^{\infty} \oplus \mathscr{H}_n^{(a)}$, where $\mathscr{H}_0^{(a)}$ is a one-dimensional space generated by a unit vector x_0. (The space $\mathscr{H}_{\mathscr{F}}^{(a)}$ is referred to as *antisymmetric Fock space* and x_0 is the *Fock vacuum.*) Let $\{e_m\}$ be an orthonormal basis for $\mathscr{H}$. Show that

(i) $\sum_{n=0}^{\infty} \oplus a_n(e_1)^* (= a(e_1)^*)$ is a partial isometry with initial space $\mathscr{K}$ spanned by

$$\{x_0, e_{j(1)} \wedge \cdots \wedge e_{j(n)} : 1 < j(1) < \cdots < j(n),\ n = 1, 2, \ldots\}$$

and final space $\mathscr{H}_{\mathscr{F}}^{(a)} \ominus \mathscr{K}$, where $a_0(e_1)^*(cx_0) = ce_1$ [*Hint.* Use Exercise 12.4.39(iii) and (v).];

(ii) the mapping $x \to a(x)^*$ is linear, and

$$(*) \qquad \begin{aligned} a(x)a(x)^* + a(x)^*a(x) &= \langle x, x\rangle I, \\ a(y)a(x)^* + a(x)^*a(y) &= \langle x, y\rangle I, \\ a(x)a(y) + a(y)a(x) &= 0, \end{aligned}$$

for all x and y in $\mathscr{H}$, and conclude that $\{a(e_j)^*\}$ is a representation of the CAR in the sense of Exercise 10.5.89 [*Hint.* Use (i) for the first relation of $(*)$ and polarize $(*)$ for the second.]; (The representation of the CAR described here is referred to as the *Fock representation.*)

(iii)

$$a(x)(x_1 \wedge \cdots \wedge x_n) = \sum_{j=1}^{n} (-1)^{j+1} \langle x, x_j\rangle x_1 \wedge \cdots \wedge x_{j-1} \wedge x_{j+1} \wedge \cdots \wedge x_n$$

and $a(x)x_0 = 0$ for all $x, x_1, \ldots, x_n$ in $\mathscr{H}$ [*Hint.* Use Exercise 12.4.39(ii) and expand the determinant expression for $\langle x \wedge y_2 \wedge \cdots \wedge y_n, x_1 \wedge \cdots \wedge x_n\rangle$ in terms of its first row.]; (The operators $a(x)$ and $a(x)^*$ are referred to as *annihilators* and *creators*, respectively.)

(iv) $\{a(x), a(x)^* : x \in \mathscr{H}\}$ generates the CAR algebra $\mathfrak{A}$ (on $\mathscr{H}_{\mathscr{F}}^{(a)}$) [*Hint.* Use Exercise 10.5.89.];

(v) the self-adjoint operator algebra $\mathfrak{A}_0$ that is generated by $\{a(x), a(x)^* : x \in \mathscr{H}\}$ consists of linear combinations of I and products $a(x_1)^* \cdots a(x_n)^* a(y_1) \cdots a(y_m)$ with all creators to the left and all annihilators to the right (*Wick-ordered monomials*) and is norm dense in the CAR algebra $\mathfrak{A}$; each annihilator is in the left kernel of $\omega_0 (= \omega_{x_0} | \mathfrak{A})$ and the null space of

$\omega_0|\mathfrak{A}_0$ is the linear span of the Wick-ordered monomials (other than scalar multiples of I) [*Hint.* Use (iii) and (iv).]; (The state ω_0 is referred to as the *Fock vacuum state.*)

(vi) if ρ is a state of $\mathfrak{A}$ such that $\rho \le t\omega_0$ for some positive real t, then $\rho = \omega_0$; and conclude that ω_0 is pure and the Fock representation is irreducible. [*Hint.* Note that each annihilator is in the left kernel of ρ and that the restrictions of ρ and ω_0 to $\mathfrak{A}_0$ have the same null space. Show that x_0 is generating for $\mathfrak{A}_0$.]

CHAPTER 13

CROSSED PRODUCTS

This chapter is concerned with a construction by which, given a representation of a group by * automorphisms of a von Neumann algebra, another von Neumann algebra, the "crossed product" is produced. Two forms of the theory are developed, one dealing with crossed products by automorphic representations of *discrete* groups, the other based on *continuous* automorphic representations of the additive group $\mathbb{R}$ of real numbers.

Discrete crossed products are studied in Section 13.1. They have previously been encountered (though not by name) in Section 8.6; for each of the factors described there (and hence, also, each of the continuum of type III factors occurring in Section 12.3) is the crossed product of a maximal abelian von Neumann algebra by a discrete group of automorphisms. It is proved below that the crossed product of a factor by a discrete group of outer automorphisms is again a factor; and by determining suitable modular automorphism groups, certain discrete crossed products of factors of type II_{∞} are identified as being type III factors. From this, and by means of an invariant derived from modular automorphism groups, it is shown that a continuum of non-isomorphic type III factors can be obtained as discrete crossed products of a single factor of type II_{∞}. In addition, a method is given for calculating the invariant just mentioned, for certain matricial von Neumann algebras.

Section 13.2 is concerned with continuous crossed products, the main result being a duality theorem. Given a continuous automorphic representation of $\mathbb{R}$ on a von Neumann algebra, one can construct both the crossed-product von Neumann algebra and also a "dual" representation of $\mathbb{R}$ by automorphisms of the latter algebra. This permits the formation of the "second crossed product"; and the duality theorem asserts that it is * isomorphic to the tensor product of the original von Neumann algebra with a factor of type I_{∞}.

Modular automorphism groups can be viewed as continuous automorphic representations of $\mathbb{R}$, and the corresponding crossed products are studied in Section 13.3. It is shown that, for a given von Neumann algebra, all its modular automorphism groups yield the same crossed-product algebra (up to unitary equivalence). The latter algebra is semi-finite, and is of type II_{∞} when the original algebra is type III. From this last result, together with the duality theorem, every type III von Neumann algebra is shown to be * isomorphic to the crossed product of an algebra of type II_∞ by a continuous automorphic representation of $\mathbb{R}$.

13.1. Discrete crossed products

Suppose that $\mathcal{M}$ is a von Neumann algebra acting on a Hilbert space $\mathcal{H}$, and G is a discrete group (with unit e). By an *automorphic representation* of G on $\mathcal{M}$, we mean a homomorphism $\alpha: g \to \alpha_g$ from G into the group of * automorphisms of $\mathcal{M}$. Such a representation α is said to be *unitarily implemented* if there is a unitary representation $g \to U(g)$ of G on $\mathcal{H}$, such that $\alpha_g(A) = U(g)AU(g)^*$ for all A in $\mathcal{M}$ and g in G; in this case, $U(g)\mathcal{M}U(g)^* = \mathcal{M}$, and therefore $U(g)\mathcal{M}'U(g)^* = \mathcal{M}'$ for each g in G.

We shall give two definitions of the "crossed product" $\mathcal{R}(\mathcal{M}, \alpha)$, a von Neumann algebra acting on $\mathcal{H} \otimes l_2(G)$, when α is an automorphic representation of G on $\mathcal{M}$ $(\subseteq \mathcal{B}(\mathcal{H}))$. The first (abstract) form of the definition applies in general. The second (implemented) form, which is more convenient for certain computational purposes, is applicable only when α is unitarily implemented.

Crossed products are used mainly for studying properties of von Neumann algebras that are invariant under * isomorphism; and for this purpose, the distinction between the abstract and implemented forms is not important. If α is an automorphic representation of G on $\mathcal{M}$, and θ is a * isomorphism from $\mathcal{M}$ onto another von Neumann algebra $\mathcal{N}$, we can form the automorphic representation $\beta: g \to \beta_g = \theta \circ \alpha_g \circ \theta^{-1}$ of G on $\mathcal{N}$. It turns out that $\mathcal{R}(\mathcal{N}, \beta)$ is * isomorphic to $\mathcal{R}(\mathcal{M}, \alpha)$; and for suitably chosen $\mathcal{N}$ and θ, β is unitarily implemented (Proposition 13.1.2). Moreover, the abstract and implemented crossed products are unitarily equivalent, when the automorphic representation is unitarily implemented. For these reasons, it is usually possible to work with the more convenient, implemented, form.

Suppose, then, that $\mathcal{H}$ is a Hilbert space, G a discrete group. We shall write the elements of $l_2(G)$ as complex-valued functions y on G (with $\sum |y(g)|^2 < \infty$). We denote by $\{y_g: g \in G\}$ the orthonormal basis of $l_2(G)$ determined by $y_g(h) = \delta_{g,h}$, write E_g for the one-dimensional projection whose range contains y_g, and define the left-translation unitary operator l_g on $l_2(G)$ by $(l_g y)(h) = y(g^{-1}h)$. Simple calculations show that $l_{gh} = l_g l_h$, that $l_g y_h = y_{gh}$, and that $l_g E_h l_g^* = E_{gh}$. Elements of the Hilbert space $\sum_{g \in G} \oplus \mathcal{H}$ may be written as functions $x: G \to \mathcal{H}$ (with $\sum \|x(g)\|^2 < \infty$), and the equation

$$(1) \qquad Wx = \sum_{g \in G} x(g) \otimes y_g$$

defines a unitary operator W from $\sum_g \oplus \mathscr{H}$ onto $\mathscr{H} \otimes l_2(G)$. Operators S acting on $\sum_g \oplus \mathscr{H}$ will be represented in the usual way by matrices $[S(p, q)]$, in which p, q run through G and $S(p, q) \in \mathscr{B}(\mathscr{H})$; while an operator T acting on $\mathscr{H} \otimes l_2(G)$ will be represented by the matrix of W^*TW.

Now suppose, further, that $\mathscr{M}$ is a von Neumann algebra acting on $\mathscr{H}$, and $\alpha: g \to \alpha_g$ is an automorphic representation of G on $\mathscr{M}$. When $A \in \mathscr{M}$ and $g \in G$, define operators $\Psi(A)$ and L_g, acting on $\mathscr{H} \otimes l_2(G)$, by

(2) $$\Psi(A) = \sum_{g \in G} \alpha_g^{-1}(A) \otimes E_g, \qquad L_g = I \otimes l_g.$$

Straightforward calculation, based on (1) and (2), shows that

(3) $$W^*\Psi(A)W = \sum_{g \in G} \oplus \alpha_g^{-1}(A) \qquad (A \in \mathscr{M});$$

and, from this, $\Psi(A)$ has matrix $[\delta_{p,q}\alpha_p^{-1}(A)]$. Since $l_{gh} = l_g l_h$, while l_g has (numerical) matrix $[\delta_{p,gq}]$ with respect to the orthonormal basis $\{y_p : p \in G\}$ of $l_g(G)$, it follows that

(4) $$L_{gh} = L_g L_h,$$

and that L_g has matrix $[\delta_{p,gq}I]$. From (2) and the fact that $l_g E_h l_g^* = E_{gh}$ (or by the appropriate matrix calculations), we deduce that

(5) $$L_g \Psi(A) L_g^* = \Psi(\alpha_g(A)).$$

From (2), (3), (4), and (5), Ψ is a * isomorphism from $\mathscr{M}$ onto a von Neumann subalgebra $\Psi(\mathscr{M})$ of $\mathscr{B}(\mathscr{H} \otimes l_2(G))$, and $g \to L_g$ is a unitary representation of G that implements the automorphic representation $g \to \Psi \circ \alpha_g \circ \Psi^{-1}$ of G on $\Psi(\mathscr{M})$. Moreover, the set $\mathscr{R}_0$, consisting of all finite sums of operators of the form $L_g \Psi(A)$, is a * subalgebra of $\mathscr{M} \mathbin{\overline{\otimes}} \mathscr{B}(l_2(G))$; and $\mathscr{R}_0^-$ is the von Neumann subalgebra of $\mathscr{M} \mathbin{\overline{\otimes}} \mathscr{B}(l_2(G))$ generated by $\{\Psi(A), L_g : A \in \mathscr{M}, g \in G\}$.

13.1.1. Definition (Abstract crossed product). If $\mathscr{M}$ is a von Neumann algebra acting on a Hilbert space $\mathscr{H}$, G is a discrete group, and $\alpha: g \to \alpha_g$ is an automorphic representation of G on $\mathscr{M}$, the (*abstract*) *crossed product* of $\mathscr{M}$ by α is the von Neumann algebra $\mathscr{R}(\mathscr{M}, \alpha)$, acting on $\mathscr{H} \otimes l_2(G)$, generated by the operators

$$\Psi(A) = \sum_{g \in G} \alpha_g^{-1}(A) \otimes E_g, \qquad L_g = I \otimes l_g \qquad (A \in \mathscr{M}, \quad g \in G). \quad \blacksquare$$

13.1.2. Proposition. *Suppose that $\mathscr{M}$ is a von Neumann algebra, G is a discrete group, and $\alpha: g \to \alpha_g$ is an automorphic representation of G on $\mathscr{M}$.*

(i) *If θ is a * isomorphism from $\mathscr{M}$ onto a von Neumann algebra $\mathscr{N}$, and β is the automorphic representation $g \to \beta_g = \theta \circ \alpha_g \circ \theta^{-1}$ of G on $\mathscr{N}$, then we have that $\mathscr{R}(\mathscr{N}, \beta)$ is * isomorphic to $\mathscr{R}(\mathscr{M}, \alpha)$.*

(ii) *There is a * isomorphism θ, from $\mathscr{M}$ onto a von Neumann algebra $\mathscr{N}$, such that the automorphic representation $g \to \theta \circ \alpha_g \circ \theta^{-1}$ of G on $\mathscr{N}$ is unitarily implemented.*

Proof. (i) If ι is the identity automorphism of $\mathscr{B}(l_2(G))$, and φ is the * isomorphism $\theta \mathbin{\overline{\otimes}} \iota$ from $\mathscr{M} \mathbin{\overline{\otimes}} \mathscr{B}(l_2(G))$ onto $\mathscr{N} \mathbin{\overline{\otimes}} \mathscr{B}(l_2(G))$, then $\varphi(\mathscr{R}(\mathscr{M}, \alpha))$ is the von Neumann algebra generated by the operators $\varphi(\Psi(A))$ $(A \in \mathscr{M})$ and $\varphi(L_g)$ $(g \in G)$. Now

$$\varphi(\Psi(A)) = \sum_{g \in G} \theta(\alpha_g^{-1}(A)) \otimes E_g = \sum_{g \in G} \beta_g^{-1}(\theta(A)) \otimes E_g,$$

$\varphi(L_g) = I \otimes l_g$, and these operators generate $\mathscr{R}(\mathscr{N}, \beta)$; so $\varphi(\mathscr{R}(\mathscr{M}, \alpha)) = \mathscr{R}(\mathscr{N}, \beta)$.

(ii) We have already noted, in the discussion preceding Definition 13.1.1, that the automorphic representation $g \to \Psi \circ \alpha_g \circ \Psi^{-1}$ of G on $\Psi(\mathscr{M})$ is unitarily implemented. ■

We now consider implemented crossed products. Suppose that $\alpha\colon g \to \alpha_g$ is an automorphic representation of G on $\mathscr{M}$ $(\subseteq \mathscr{B}(\mathscr{H}))$ that is implemented by a unitary representation $g \to U(g)$ of G on $\mathscr{H}$. The equation

$$U = \sum_{g \in G} U(g) \otimes E_g \tag{6}$$

defines a unitary operator U on $\mathscr{H} \otimes l_2(G)$, and U has matrix $[\delta_{p,q} U(p)]$. If we write $\Phi(A)$ and $V(g)$ for the "transforms" under U of $\Psi(A)$ and L_g, then straightforward matrix calculations show that

$$\Phi(A) = U\Psi(A)U^* = A \otimes I, \qquad V(g) = UL_gU^* = U(g) \otimes l_g, \tag{7}$$

and that $\Phi(A)$, $V(g)$ have matrices $[\delta_{p,q} A]$, $[\delta_{p,gq} U(g)]$, respectively. From Definition 13.1.1, $U\mathscr{R}(\mathscr{M}, \alpha)U^*$ is the von Neumann algebra generated by $\{A \otimes I,\ U(g) \otimes l_g : A \in \mathscr{M},\ g \in G\}$; and it is reasonable to introduce implemented crossed products (allowing slightly inconsistent use of the symbol $\mathscr{R}(\mathscr{M}, \alpha)$) as follows.

13.1.3. Definition (Implemented crossed product). If $\mathscr{M}$ is a von Neumann algebra acting on a Hilbert space $\mathscr{H}$, G is a discrete group, and $\alpha\colon g \to \alpha_g$ is an automorphic representation of G on $\mathscr{M}$ that is implemented by a unitary representation $g \to U(g)$ of G on $\mathscr{H}$, then the (*implemented*) *crossed product* of $\mathscr{M}$ by α is the von Neumann algebra $\mathscr{R}(\mathscr{M}, \alpha)$, acting on $\mathscr{H} \otimes l_2(G)$, that is generated by the operators

$$\Phi(A) = A \otimes I, \qquad V(g) = U(g) \otimes l_g \qquad (A \in \mathscr{M}, \quad g \in G). \quad ■$$

The situation under consideration in Definition 13.1.3 is one already encountered (with some additional restrictions) in Section 8.6, *An operator-theoretic construction*. There, $\mathscr{A}$ (in place of $\mathscr{M}$) was a von Neumann algebra acting on a Hilbert space $\mathscr{H}$, G was a discrete group, and $g \to U(g)$ was a unitary representation of G on $\mathscr{H}$, such that $U(g)\mathscr{A}U(g)^* = \mathscr{A}$ for each g in G. Additional assumptions, in force in Section 8.6 (but dropped in the present section), were that $\mathscr{A}$ was maximal abelian and that G acted freely on $\mathscr{A}$. The von Neumann algebra $\mathscr{R}$ considered in Proposition 8.6.1 (and shown, in certain circumstances, to be a factor) acted on the Hilbert space $\sum_{g \in G} \oplus \mathscr{H}$ $(=\mathscr{K})$, and was generated by the operators $\Phi(A)$, $V(g)(A \in \mathscr{A}, g \in G)$ having matrices $[\delta_{p,q}A]$, $[\delta_{p,gq}U(g)]$, respectively. If $\mathscr{K}$ is identified with $\mathscr{H} \otimes l_2(G)$, by means of the unitary transformation W in (1), the operators $\Phi(A)$ and $V(g)$ of Section 8.6 are precisely the same as those occurring in (7) and in Definition 13.1.3. Accordingly, the von Neumann algebra $\mathscr{R}$ of Section 8.6 is (unitarily equivalent to) the (implemented) crossed product $\mathscr{R}(\mathscr{A}, \alpha)$, where α is the automorphic representation of G on $\mathscr{A}$ defined by $\alpha_g(A) = U(g)AU(g)^*$.

In the more general situation considered in Definition 13.1.3, the mapping $\Phi\colon A \to A \otimes I$ is a * isomorphism from $\mathscr{M}$ onto a von Neumann subalgebra of the implemented crossed product $\mathscr{R}(\mathscr{M}, \alpha)$. It follows from (4) and (7) that $g \to V(g)$ is a unitary representation of G on $\mathscr{H} \otimes l_2(G)$, that

$$(8) \quad V(g)\Phi(A)V(g)^* = \Phi(U(g)AU(g)^*) = \Phi(\alpha_g(A)) \qquad (A \in \mathscr{M}, \quad g \in G),$$

and hence that $V(g)\Phi(\mathscr{M})V(g)^* = \Phi(\mathscr{M})$. From this, the set of all finite sums of operators of the form $V(g)\Phi(A)$ is a self-adjoint algebra, whose strong-operator closure is $\mathscr{R}(\mathscr{M}, \alpha)$. Straightforward matrix calculations, described below, show that $\mathscr{R}(\mathscr{M}, \alpha)$ consists of all elements of $\mathscr{B}(\mathscr{H} \otimes l_2(G))$ having matrix of the form $[U(pq^{-1})A(pq^{-1})]$, for some mapping $g \to A(g) : G \to \mathscr{M}$; while the commutant $\mathscr{R}(\mathscr{M}, \alpha)'$ consists of all operators with matrix of the form $[U(p)A'(q^{-1}p)U(p)^*]$, for some mapping $g \to A'(g) : G \to \mathscr{M}'$. Observe, from this, that the matrix of an operator in $\mathscr{R}(\mathscr{M}, \alpha)$ has the same operator, an element $A(e)$ of $\mathscr{M}$, at each diagonal position; moreover, since $U(g)$ implements an automorphism of $\mathscr{M}'$ (as well as an automorphism of $\mathscr{M}$), the matrix of an operator in $\mathscr{R}(\mathscr{M}, \alpha)'$ has all its entries in $\mathscr{M}'$. The matrix calculations needed to establish the results just stated are almost identical with those occurring in the first five paragraphs of the proof of Proposition 8.6.1. The only significant difference occurs almost at the outset, and is due to the fact that $\mathscr{M}$ (unlike $\mathscr{A}$ in Section 8.6) is not assumed to be maximal abelian. Thus $\Phi(\mathscr{M})$, corresponding to the algebra $\mathscr{M} \otimes I_n$ of all diagonal matrices with the same element of $\mathscr{M}$ at each diagonal position, has commutant $\Phi(\mathscr{M})'$, corresponding to the algebra $n \otimes \mathscr{M}'$ of (bounded) matrices with all entries in $\mathscr{M}'$ (see Lemma 6.6.2); but, in contrast with the proof of Proposition 8.6.1, in the present case we must distinguish $n \otimes \mathscr{M}'$ from $n \otimes \mathscr{M}$. With this modification, the earlier argument applies in the present context.

In the remainder of this section, we shall be concerned with properties of von Neumann algebras that are invariant under * isomorphism. Throughout, α is an automorphic representation of G on $\mathscr{M}$ ($\subseteq \mathscr{B}(\mathscr{H})$). In *proving* results, we assume also that α is implemented by a unitary representation $g \to U(g)$ of G on $\mathscr{H}$; and we use without further comment the notation and information set out in the preceding paragraph. By Proposition 13.1.2, the results remain valid for abstract crossed products as well.

13.1.4. PROPOSITION. *If ρ is a faithful normal state of $\mathscr{M}$, there is a faithful normal state ω of $\mathscr{R}(\mathscr{M}, \alpha)$, defined by the equation*

$$\omega(R) = \rho(R(e, e)) \qquad (R \in \mathscr{R}(\mathscr{M}, \alpha)),$$

where R has matrix $[R(p, q)]$.

Proof. When $R \in \mathscr{R}(\mathscr{M}, \alpha)$, its matrix $[R(p, q)]$ has the form $[U(pq^{-1})A(pq^{-1})]$ for some mapping $g \to A(g): G \to \mathscr{M}$. Thus $R(e, e) = A(e) \in \mathscr{M}$ and ω, as defined, is a normal state of $\mathscr{R}(\mathscr{M}, \alpha)$. Moreover,

$$\omega(R^*R) = \rho\left(\sum_{g \in G} R(g, e)^*R(g, e)\right) = \rho\left(\sum_{g \in G} A(g)^*A(g)\right).$$

If $\omega(R^*R) = 0$, then $\sum A(g)^*A(g) = 0$, since ρ is a faithful state of $\mathscr{M}$; so $A(g) = 0$ for all g, $R(p, q) = 0$ for all p and q, and thus $R = 0$. Hence ω is faithful. ∎

13.1.5. PROPOSITION. *Suppose that $\mathscr{M}$ is a factor.*

(i) *If U is a unitary operator acting on $\mathscr{H}$, $U\mathscr{M}U^* = \mathscr{M}$, $A \in \mathscr{M}$, $A' \in \mathscr{M}'$, and $A' = AU \neq 0$, then U implements an inner automorphism of $\mathscr{M}$.*

(ii) *If α_g is an outer automorphism of $\mathscr{M}$, for all $g(\neq e)$ in G, then $\mathscr{R}(\mathscr{M}, \alpha)$ is a factor.*

Proof. (i) Since $A' = AU \neq 0$, $A'A'^* = AA^* \in \mathscr{M} \cap \mathscr{M}'$; so $A'A'^* = AA^* = a^{-2}I$, for some positive real number a. We can define partial isometries, V in $\mathscr{M}$ and V' in $\mathscr{M}'$, by $V = aA$, $V' = aA'$; and $V' = VU$. Since

$$I = a^2A'A'^* = V'V'^* = VUV'^* = VW'U,$$

where $W' = UV'^*U^* \in U\mathscr{M}'U^* = \mathscr{M}'$, it follows that $VW' = U^*$ (and V, W' commute). From this, the partial isometries V, W', and hence also V' ($= VU$), are invertible, and are therefore unitary operators. Thus $U = V^*V'$, and U implements the same inner automorphism of $\mathcal{M}$ as does V^*.

(ii) If $R \in \mathcal{R}(\mathcal{M}, \alpha) \cap \mathcal{R}(\mathcal{M}, \alpha)'$, the matrix of R can be written in both the forms $[U(pq^{-1})A(pq^{-1})]$ and $[U(p)A'(q^{-1}p)U(p)^*]$, for suitable mappings

$$g \to A(g) : G \to \mathcal{M} \qquad \text{and} \qquad g \to A'(g) : G \to \mathcal{M}';$$

and $U(pq^{-1})A(pq^{-1}) = U(p)A'(q^{-1}p)U(p)^*$. With $q = e$, we obtain $A'(p) = A(p)U(p)$. Since $U(p)$ implements α_p, an outer automorphism of $\mathcal{M}$ when $p \neq e$, it follows from (i) that $A(p) = 0$ $(p \neq e)$. Also, $A'(e) = A(e) \in \mathcal{M} \cap \mathcal{M}'$, whence $A(e) = aI$ for some scalar a. Thus R has matrix $[\delta_{p,q}aI]$, and $R = aI$. ■

We now show that, in certain cases, it is possible to determine the modular automorphism group of $\mathcal{R}(\mathcal{M}, \alpha)$, corresponding to a state ω of the type occurring in Proposition 13.1.4.

13.1.6. PROPOSITION. *Suppose that $\mathcal{M}$ is a factor, τ is a faithful normal semi-finite tracial weight on $\mathcal{M}$, ρ is a faithful normal state of $\mathcal{M}$, and $\{\sigma_t\}$ is the modular automorphism group of $\mathcal{R}(\mathcal{M}, \alpha)$, corresponding to the faithful normal state ω constructed from ρ as in Proposition* 13.1.4. *Then there is a homomorphism $g \to a_g$ from G into the multiplicative group of positive real numbers, and a positive self-adjoint invertible operator H affiliated with $\mathcal{M}$, such that*

$$\tau \circ \alpha_g = a_g \tau, \quad \sigma_t(R) = W_t R W_t^* \qquad (g \in G, \quad t \in \mathbb{R}, \quad R \in \mathcal{R}(\mathcal{M}, \alpha)),$$

where W_t is the unitary operator with matrix $[\delta_{p,q} a_p^{it} H^{it}]$.

Proof. From Theorem 8.5.7, every faithful normal semi-finite tracial weight on $\mathcal{M}$ is a multiple of τ. Hence $\tau \circ \alpha_g = a_g \tau$ for some positive real number a_g; and $a_{gh} = a_g a_h$ since

$$a_{gh}\tau = \tau \circ \alpha_{gh} = (\tau \circ \alpha_g) \circ \alpha_h = a_g \tau \circ \alpha_h = a_g a_h \tau.$$

For all real t, let X_t be the unitary operator, acting on $\mathcal{H} \otimes l_2(G)$, whose matrix is $[\delta_{p,q} a_p^{it} I]$. Simple matrix calculations show that $X_t \Phi(A) X_t^* = \Phi(A)$ $(A \in \mathcal{M})$ and $X_t V(g) X_t^* = a_g^{it} V(g)$ $(g \in G)$; so $X_t \mathcal{R}(\mathcal{M}, \alpha) X_t^* = \mathcal{R}(\mathcal{M}, \alpha)$.

By Lemma 9.2.19 there is a positive element K in the unit ball of $\mathcal{M}$, such that $\tau(I - K) < \infty$, K and $I - K$ are both one-to-one mappings, and $\tau((I - K)A) = \rho(KA) = \rho(AK)$ $(A \in \mathcal{M})$. Thus $K^{-1}(I - K)$ is a positive self-adjoint invertible operator H affiliated with $\mathcal{M}$. For all real t, H^{it} is a unitary operator in $\mathscr{M}$; so there is a unitary element Y_t of $\mathscr{R}(\mathscr{M}, \alpha)$ that has matrix $[\delta_{p,q}H^{it}]$. By matrix multiplication, $W_t = X_t Y_t$; so from the preceding paragraph, W_t implements a * isomorphism β_t of $\mathscr{R}(\mathscr{M}, \alpha)$.

In order to show that $\{\beta_t\}$ is the modular automorphism group corresponding to ω, it now suffices to verify that it satisfies the weakened form of the modular condition required in Lemma 9.2.17. To this end, we first describe a suitable everywhere-dense * subalgebra $\mathfrak{A}$ of $\mathscr{R}(\mathscr{M}, \alpha)$. For each $n = 3, 4, \ldots$, let E_n be the spectral projection for K, corresponding to the interval $[n^{-1}, 1 - n^{-1}]$. Then $\{E_n\}$ is an increasing sequence of projections in $\mathscr{M}$; and $\lim E_n = I$, since $0 \le K \le I$ and 0, 1 are not eigenvalues of K. Simple calculations (involving (8)) show that, for a fixed value of n, products and adjoints of operators of the form $\Phi(E_n)V(g)\Phi(AE_n)$ (with A in $\mathscr{M}$ and g in G) are again of this form. The linear space generated by all such operators is therefore a * subalgebra $\mathfrak{A}_n$ of $\mathscr{R}(\mathscr{M}, \alpha)$; and then, since $\Phi(E_n) = \Phi(E_{n+1})V(e)\Phi(E_n E_{n+1}) \in \mathfrak{A}_{n+1}$, we have $\mathfrak{A}_n = \Phi(E_n)\mathfrak{A}_{n+1}\Phi(E_n) \subseteq \mathfrak{A}_{n+1}$. Hence $\bigcup \mathfrak{A}_n$ is a * subalgebra $\mathfrak{A}$ of $\mathscr{R}(\mathscr{M}, \alpha)$; and $\mathfrak{A}^- = \mathscr{R}(\mathscr{M}, \alpha)$, since $\lim E_n = I$.

We now have to show that, given any R and S in $\mathfrak{A}$, there is a complex-valued function f, bounded and continuous on $\{z \in \mathbb{C} : 0 \le \operatorname{Im} z \le 1\}$, and analytic on the interior of that strip, with boundary values

$$f(t) = \omega(\beta_t(R)S), \qquad f(t + i) = \omega(S\beta_t(R)) \qquad (t \in \mathbb{R}).$$

By linearity, it suffices to consider the case in which $R = \Phi(E_n)V(g)\Phi(AE_n)$ for some A in $\mathscr{M}$, g in G, and $n\ (\ge 3)$. Thus R has matrix $[\delta_{p,gq}E_n U(g)AE_n]$, while S has matrix $[U(pq^{-1})A(pq^{-1})]$ for some mapping $g \to A(g)$ from G into $\mathscr{M}$. By calculating the (e, e) component, in the matrices of $\beta_t(R)S$ and $S\beta_t(R)$, we obtain

$$\omega(\beta_t(R)S) = a_g^{it}\rho(H^{it}E_n U(g)AE_n H^{-it}U(g)^*A(g^{-1})),$$

$$\omega(S\beta_t(R)) = a_g^{it}\rho(U(g)^*A(g^{-1})H^{it}E_n U(g)AE_n H^{-it}).$$

Since $K \in \mathscr{M}$, and by virtue of our choice of E_n, we can find elements C_n and D_n of $E_n\mathscr{M}E_n$ such that

$$E_n = (I - K)C_n = D_n K = KD_n.$$

Indeed, the restrictions to $E_n(\mathscr{H})$, of the operators K, $I - K$, H, all have bounded inverses; and $H^{-1}(I - K)C_n = KC_n$, $KHE_n = (I - K)E_n$, since $H = K^{-1}(I - K)$. The equation

$$f(z) = a_g^{iz}\rho(H^{iz}E_n U(g)AH^{-iz}E_n U(g)^*A(g^{-1}))$$

defines an entire function f that is bounded on each strip of finite width parallel to the real axis. Moreover, $f(t) = \omega(\beta_t(R)S)$ for all real t, since E_n commutes with H^{-it}; while

$$\begin{aligned}
f(t+i) &= a_g^{it-1}\rho(H^{it}H^{-1}E_nU(g)AHH^{-it}E_nU(g)^*A(g^{-1}))\\
&= a_g^{it-1}\rho(H^{it}H^{-1}(I-K)C_nU(g)AE_nHE_nH^{-it}U(g)^*A(g^{-1}))\\
&= a_g^{it-1}\rho(H^{it}KC_nU(g)AD_nKHE_nH^{-it}U(g)^*A(g^{-1}))\\
&= a_g^{it-1}\rho(KH^{it}C_nU(g)AD_n(I-K)E_nH^{-it}U(g)^*A(g^{-1}))\\
&= a_g^{it-1}\tau((I-K)H^{it}C_nU(g)AD_nE_nH^{-it}(I-K)U(g)^*A(g^{-1}))\\
&= a_g^{it-1}\tau(A(g^{-1})(I-K)H^{it}C_nU(g)AD_nE_nH^{-it}(I-K)U(g)^*)\\
&= a_g^{it-1}\tau(\alpha_g(U(g)^*A(g^{-1})H^{it}(I-K)C_nU(g)AD_nE_nH^{-it}(I-K)))\\
&= a_g^{it}\tau(U(g)^*A(g^{-1})H^{it}E_nU(g)AD_nE_nH^{-it}(I-K))\\
&= a_g^{it}\rho(U(g)^*A(g^{-1})H^{it}E_nU(g)AD_nE_nH^{-it}K)\\
&= a_g^{it}\rho(U(g)^*A(g^{-1})H^{it}E_nU(g)AD_nKE_nH^{-it})\\
&= a_g^{it}\rho(U(g)^*A(g^{-1})H^{it}E_nU(g)AE_nH^{-it})\\
&= \omega(S\beta_t(R)). \quad \blacksquare
\end{aligned}$$

There is a more natural version of Proposition 13.1.6, obtained by using τ itself, and the corresponding weight on $\mathscr{R}(\mathscr{M},\alpha)$, in place of ρ and ω, respectively. This avoids the assumption (implicit in any reference to faithful normal states) that the algebras in question are countably decomposable. Moreover, both the statement and the proof of the proposition are simplified, in that H is replaced, throughout, by I. Of course, this version relies on modular theory, in the context of weights rather than states, as developed at the end of Section 9.2.

13.1.7. PROPOSITION. *Suppose that $\mathscr{M}$ is a semi-finite factor, α_g is an outer automorphism of $\mathscr{M}$ for all g ($\neq e$) in G, ρ is a faithful normal state of $\mathscr{M}$, and $\{\sigma_t\}$ is the modular automorphism group of $\mathscr{R}(\mathscr{M},\alpha)$, corresponding to the faithful normal state ω constructed as in Proposition 13.1.4. Let a_g ($g \in G$) be the positive scalars determined by the condition $\tau\circ\alpha_g = a_g\tau$, where τ is a faithful normal semi-finite tracial weight on $\mathscr{M}$. Then for each real number t, σ_t is an inner automorphism of $\mathscr{R}(\mathscr{M},\alpha)$ if and only if $a_g^{it} = 1$ for each g in G.*

Proof. By Proposition 13.1.6, $\sigma_t(R) = W_tRW_t^*$, where the matrix of W_t is $[\delta_{p,q}a_p^{it}U_t]$ for some unitary operator U_t in $\mathscr{M}$. Thus $W_t = X_tY_t$, where $Y_t = \Phi(U_t) \in \mathscr{R}(\mathscr{M},\alpha)$, and X_t is the unitary operator with matrix $[\delta_{p,q}a_p^{it}I]$; and σ_t is inner if and only if X_t implements an inner automorphism of $\mathscr{R}(\mathscr{M},\alpha)$. Clearly, then, σ_t is inner if $a_g^{it} = 1$ for each g in G.

Conversely, suppose that σ_t is inner. Since X_t implements an inner automorphism of $\mathscr{R}(\mathscr{M},\alpha)$, it can be expressed as $V'V^*$, where V ($\in \mathscr{R}(\mathscr{M},\alpha)$)

and V' ($\in \mathscr{R}(\mathscr{M}, \alpha)'$) are unitary operators. The matrices of V and V' have the form $[U(pq^{-1})A(pq^{-1})]$ and $[U(p)A'(q^{-1}p)U(p)^*]$, respectively. Since $X_t V = V'$,

$$a_p^{it}U(pq^{-1})A(pq^{-1}) = U(p)A'(q^{-1}p)U(p)^* \qquad (p, q \in G). \tag{9}$$

When $q = e \neq p$, we obtain $A'(p) = a_p^{it}A(p)U(p)$; and $A(p) = 0$, by Proposition 13.1.5(i), since α_p is an outer automorphism of $\mathscr{M}$. Since V is unitary and $A(p) = 0$ when $p \neq e$, it follows that $A(e) \neq 0$. From (9), with $p = q$,

$$a_p^{it}A(e) = U(p)A'(e)U(p)^* \qquad (p \in G);$$

so $A(e) \in \mathscr{M} \cap \mathscr{M}'$. Thus $A(e)$ (and therefore, also, $A'(e)$) is a scalar (non-zero) multiple of I, and $a_p^{it}A(e) = A'(e)$. Accordingly, a_p^{it} is independent of p, and $a_p^{it} = a_e^{it} = 1 (p \in G)$. ∎

13.1.8. Corollary. *Under the conditions set out in Proposition* 13.1.7, $\mathscr{R}(\mathscr{M}, \alpha)$ *is a semi-finite factor if* $a_g = 1$ *for all* g *in* G, *and is a type* III *factor if* $a_g \neq 1$ *for some* g *in* G.

Proof. If $a_g \neq 1$ for some g in G, $a_g^{it} \neq 1$ for suitable real numbers t. For such t, σ_t is an outer automorphism by Proposition 13.1.7. Thus $\mathscr{R}(\mathscr{M}, \alpha)$, is not semi-finite (Theorem 9.2.21) and is therefore type III, since it is a factor by Proposition 13.1.5(ii).

If $a_g = 1$ for all g in G then, from Proposition 13.1.6, there is a positive self-adjoint invertible operator H affiliated with $\mathscr{M}$ such that σ_t is implemented by the unitary operator W_t whose matrix is $[\delta_{p,q}H^{it}]$. From Remark 11.2.39, $H \otimes I$ is a positive self-adjoint invertible operator affiliated with $\mathscr{A} \otimes \mathbb{C}I$ ($\subseteq \mathscr{R}(\mathscr{M}, \alpha)$), where $\mathscr{A}$ is the (abelian) von Neumann algebra generated by H. The mapping $A \to A \otimes I$ defines a σ-normal isomorphism of $\mathscr{N}(\mathscr{A})$ onto $\mathscr{N}(\mathscr{A} \otimes \mathbb{C}I)$. From Proposition 5.6.30, the Borel function calculus on $\mathscr{N}(\mathscr{A})$ and that on $\mathscr{N}(\mathscr{A} \otimes \mathbb{C}I)$ "commute" with this isomorphism. Thus $(H \otimes I)^{it} = H^{it} \otimes I = W_t$. It now follows from Theorem 9.2.21 that $\mathscr{R}(\mathscr{M}, \alpha)$ is semi-finite.

It is of interest to consider an alternative proof that $\mathscr{R}(\mathscr{M}, \alpha)$ is semi-finite when $a_g = 1$ for all g in G. In this case, $\mathscr{R}(\mathscr{M}, \alpha)$ has a faithful normal semi-finite tracial weight τ_1. Indeed, if τ is such a weight on $\mathscr{M}$, it suffices to define $\tau_1(R) = \tau(R(e, e))$, where R (in $\mathscr{R}(\mathscr{M}, \alpha)^+$) has matrix $[R(p, q)]$. The straightforward argument showing that τ_1 has the stated properties is similar to the proof of the corresponding part of Lemma 8.6.3. ∎

Suppose that $\mathscr{M}$ is a semi-finite factor, and α_g is an outer automorphism of $\mathscr{M}$ for all g ($\neq e$) in G. Each faithful normal state ρ of $\mathscr{M}$ gives rise, as in Proposition 13.1.4, to a faithful normal state ω of the factor $\mathscr{R}(\mathscr{M}, \alpha)$, and to the corresponding modular automorphism group $\{\sigma_t^\omega\}$ of $\mathscr{R}(\mathscr{M}, \alpha)$. For states ω of this type, Proposition 13.1.7 identifies the set

$$\{t \in \mathbb{R} : \sigma_t^\omega \text{ is an inner automorphism of } \mathscr{R}(\mathscr{M}, \alpha)\}$$

as being $\{t \in \mathbb{R} : a_g^{it} = 1 \text{ for all } g \text{ in } G\}$, and therefore independent of the choice of ω (within the given class of states). We now prove a general result of this type, and use it to introduce an algebraic invariant for von Neumann algebras.

13.1.9. THEOREM. *Suppose that λ and μ are faithful normal states of a von Neumann algebra $\mathscr{R}$, and $\{\sigma_t^\lambda\}$, $\{\sigma_t^\mu\}$ are the corresponding modular automorphism groups of $\mathscr{R}$. Then there is a strong-operator-continuous mapping $t \to U_t$, from $\mathbb{R}$ into the unitary group of $\mathscr{R}$, such that*

$$\sigma_t^\mu(A) = U_t\sigma_t^\lambda(A)U_t^*, \qquad U_{s+t} = U_s\sigma_s^\lambda(U_t) \qquad (s, t \in \mathbb{R}; \quad A \in \mathscr{R}).$$

For each real number t, the automorphism σ_t^λ of $\mathscr{R}$ is inner if and only if σ_t^μ is inner.

Proof. Let $\mathscr{S}$ be the von Neumann algebra $2 \otimes \mathscr{R}$ of all 2×2 matrices with entries in $\mathscr{R}$; and for $j, k = 1, 2$, denote by E_{jk} $(\in \mathscr{S})$ the matrix that has I in the (j, k) position and zeros elsewhere. Define a faithful normal state ω of $\mathscr{S}$ by $\omega([A_{jk}]) = \lambda(A_{11}) + \mu(A_{22})$, and let $\{\sigma_t^\omega\}$ be the corresponding modular automorphism group of $\mathscr{S}$. From Proposition 9.2.14(iii), $\sigma_t^\omega(E_{jj}) = E_{jj}$ for all real t, since $\omega(E_{jj}S) = \omega(SE_{jj})$ $(S \in \mathscr{S})$; so $\sigma_t^\omega(E_{jj}\mathscr{S}E_{kk}) = E_{jj}\mathscr{S}E_{kk}$ for j, $k = 1, 2$. Accordingly, if $A \in \mathscr{R}$ and S is the 2×2 matrix with A in the (j, k) position and zeros elsewhere, then $\sigma_t^\omega(S)$ has an element $\sigma_t^{jk}(A)$ of $\mathscr{R}$ in the (j, k) position, and zeros elsewhere. In this way, for each real t, we obtain four mappings $\sigma_t^{jk}: \mathscr{R} \to \mathscr{R}$; and $\sigma_t^{jk}(A)$ is strong-operator continuous as a function of t for each A in $\mathscr{R}$.

Since $\{\sigma_t^\omega\}$ is a one-parameter group of * automorphisms of $\mathscr{S}$, satisfying the modular condition relative to ω, it follows by restriction to $E_{11}\mathscr{S}E_{11}$ that $\{\sigma_t^{11}\}$ is a one-parameter group of * automorphisms of $\mathscr{R}$, and satisfies the modular condition relative to λ. Thus $\sigma_t^{11} = \sigma_t^\lambda$; and similarly $\sigma_t^{22} = \sigma_t^\mu$. Since σ_t^ω preserves adjoints, $\sigma_t^{21}(A)^* = \sigma_t^{12}(A^*)$ for each A in $\mathscr{R}$.

Let U_t be $\sigma_t^{21}(I)$, so that $U_t^* = \sigma_t^{12}(I)$. By applying σ_t^ω to the relations

$$\begin{bmatrix} 0 & 0 \\ 0 & A \end{bmatrix} = \begin{bmatrix} 0 & 0 \\ I & 0 \end{bmatrix}\begin{bmatrix} A & 0 \\ 0 & 0 \end{bmatrix}\begin{bmatrix} 0 & I \\ 0 & 0 \end{bmatrix}, \qquad \begin{bmatrix} I & 0 \\ 0 & 0 \end{bmatrix} = \begin{bmatrix} 0 & I \\ 0 & 0 \end{bmatrix}\begin{bmatrix} 0 & 0 \\ I & 0 \end{bmatrix},$$

we obtain $\sigma_t^\mu(A) = U_t\sigma_t^\lambda(A)U_t^*$ (in particular, $I = U_tU_t^*$) and $I = U_t^*U_t$. By applying σ_s^ω to the relation

$$\sigma_t^\omega\left(\begin{bmatrix} 0 & 0 \\ I & 0 \end{bmatrix}\right) = \begin{bmatrix} 0 & 0 \\ I & 0 \end{bmatrix}\begin{bmatrix} U_t & 0 \\ 0 & 0 \end{bmatrix},$$

we obtain $U_{s+t} = U_s\sigma_s^\lambda(U_t)$.

Since $\sigma_t^\mu = \alpha_t \circ \sigma_t^\lambda$, where α_t is the inner automorphism of $\mathscr{R}$ implemented by U_t, it is apparent that σ_t^μ is inner if and only if σ_t^λ is inner. ∎

When $\mathscr{R}$ is a countably decomposable von Neumann algebra, let $\{\sigma_t\}$ be the modular automorphism group corresponding to a faithful normal state ω of $\mathscr{R}$; and define

$$T(\mathscr{R}) = \{t \in \mathbb{R} : \sigma_t \text{ is an inner automorphism of } \mathscr{R}\}.$$

It is apparent that $T(\mathscr{R})$ is an additive subgroup of $\mathbb{R}$; by Theorem 13.1.9, it does not depend on the choice of ω, and so constitutes an algebraic invariant for $\mathscr{R}$. This invariant does not distinguish between different semi-finite von Neumann algebras, since (Theorem 9.2.21) $T(\mathscr{R}) = \mathbb{R}$ when $\mathscr{R}$ is semi-finite. However, in Theorem 13.1.11 we shall use crossed products to show that, given any positive real number a, there are type III factors $\mathscr{R}$ for which $T(\mathscr{R}) = \{0, \pm a, \pm 2a, \ldots\}$.

Theorem 13.1.9 remains valid, with semi-finite weights in place of states. Accordingly, for countably decomposable von Neumann algebras, it makes no difference if $T(\mathscr{R})$ is defined in terms of the modular automorphism group corresponding to a faithful normal semi-finite weight on $\mathscr{R}$. In this way, the invariant can be defined also for algebras that are not countably decomposable.

13.1.10. Proposition. *If $0 < \lambda < 1$, $\mathscr{R}$ is a finite matricial factor, $\mathscr{K}$ is a separable infinite-dimensional Hilbert space, $\mathscr{M}$ is the factor $\mathscr{R} \mathbin{\overline{\otimes}} \mathscr{B}(\mathscr{K})$ of type II_∞, and τ is a faithful normal semi-finite tracial weight on $\mathscr{M}$, there is a * automorphism θ of $\mathscr{M}$ for which $\tau \circ \theta = \lambda\tau$.*

Proof. There is a system $\{F_{jk} : j, k = 1, 2, \ldots\}$ of matrix units in $\mathscr{M}$, such that $\sum F_{jj} = I$ and $F_{11}\mathscr{M}F_{11}$ is * isomorphic to $\mathscr{R}$. Thus F_{11} is a finite projection in $\mathscr{M}$; and we may assume that τ has been normalized so that $\tau(F_{11}) = 1$, whence $\tau | F_{11}\mathscr{M}F_{11}$ is the faithful normal tracial state on $F_{11}\mathscr{M}F_{11}$. Since $F_{11}\mathscr{M}F_{11}$ is a factor of type II_1, it contains a projection E_{11} for which $\tau(E_{11}) = \lambda$. By Proposition 6.3.12, there is an orthogonal sequence $\{E_{jj} : j = 1, 2, \ldots\}$ of projections in $\mathscr{M}$, with sum I, each equivalent to E_{11}; and these may be augmented by suitable partial isometries in $\mathscr{M}$, to form a system $\{E_{jk} : j, k = 1, 2, \ldots\}$ of matrix units.

From Remark 12.2.9, there is a * isomorphism φ from the matricial factor $F_{11}\mathscr{M}F_{11}$ onto $E_{11}\mathscr{M}E_{11}$; so there is the corresponding * isomorphism $\aleph_0 \otimes \varphi$ between the algebras $\aleph_0 \otimes F_{11}\mathscr{M}F_{11}$ and $\aleph_0 \otimes E_{11}\mathscr{M}E_{11}$ of (countably) infinite matrices. We can consider these matrices as being indexed by the positive integers; and if G is the element of $\aleph_0 \otimes F_{11}\mathscr{M}F_{11}$ that has the unit in the (1, 1) position and zeros elsewhere, then $(\aleph_0 \otimes \varphi)(G)$

is the corresponding element of $\aleph_0 \otimes E_{11}\mathscr{M}E_{11}$. Moreover, there are * isomorphisms

$$\varphi_1 : \mathscr{M} \to \aleph_0 \otimes F_{11}\mathscr{M}F_{11}, \qquad \varphi_2 : \mathscr{M} \to \aleph_0 \otimes E_{11}\mathscr{M}E_{11},$$

such that $\varphi_1(F_{11}) = G$ and $\varphi_2(E_{11}) = (\aleph_0 \otimes \varphi)(G)$. Accordingly, $\theta(F_{11}) = E_{11}$, where θ is the * automorphism $\varphi_2^{-1} \circ (\aleph_0 \otimes \varphi) \circ \varphi_1$ of $\mathscr{M}$. Since $\tau \circ \theta$ is a multiple of τ, and $\tau(\theta(F_{11})) = \tau(E_{11}) = \lambda = \lambda\tau(F_{11})$, it follows that $\tau \circ \theta = \lambda\tau$. ■

We now show how automorphisms of the type considered in Proposition 13.1.10 give rise to factors of type III.

13.1.11. THEOREM. *Suppose that* $0 < \lambda < 1$, $\mathscr{M}$ *is a countably decomposable factor of type* II_∞, τ *is a faithful normal semi-finite tracial weight on* $\mathscr{M}$, *and* θ *is a * automorphism of* $\mathscr{M}$ *for which* $\tau \circ \theta = \lambda\tau$. *If* $\mathbb{Z}$ *is the additive group of integers, and* α *is the automorphic representation* $n \to \alpha_n = \theta^n$ *of* $\mathbb{Z}$ *on* $\mathscr{M}$, *then* $\mathscr{R}(\mathscr{M}, \alpha)$ *is a type* III *factor, and*

$$T(\mathscr{R}(\mathscr{M}, \alpha)) = \{0, \pm a, \pm 2a, \ldots\},$$

where $a = 2\pi/|\log \lambda|$.

Proof. For each n in $\mathbb{Z}$, $\tau \circ \alpha_n = \lambda^n\tau$; in the notation of Proposition 13.1.7, $a_n = \lambda^n$. When $n \neq 0$, α_n is an outer automorphism, since $\tau \circ \alpha_n \neq \tau$; so $\mathscr{R}(\mathscr{M}, \alpha)$ is a type III factor by Corollary 13.1.8. Upon calculating $T(\mathscr{R}(\mathscr{M}, \alpha))$ by using the modular automorphism group of a state ω of the type considered in Proposition 13.1.7, we obtain

$$\begin{aligned} T(\mathscr{R}(\mathscr{M}, \alpha)) &= \{t \in \mathbb{R} : a_n^{it} = 1 \ (n \in \mathbb{Z})\} \\ &= \{t \in \mathbb{R} : \lambda^{it} = 1\} \\ &= \{0, \pm a, \pm 2a, \ldots\}, \end{aligned}$$

where $a = 2\pi/|\log \lambda|$. ■

Theorem 13.1.11 remains true without the assumption that $\mathscr{M}$ is countably decomposable. In this broader context, of course, faithful normal *states* are no longer available. We have already noted, however, that there is a (more natural) variant of Proposition 13.1.6, which uses the trace τ itself, and the corresponding *weight* on $\mathscr{R}(\mathscr{M}, \alpha)$; and the same applies to Proposition 13.1.7 and Corollary 13.1.8. From these results, and by using the modular automorphism group associated with this weight to determine $T(\mathscr{R}(\mathscr{M}, \alpha))$, one obtains the more general version of Theorem 13.1.11.

When $0 < \lambda < 1$, we may apply Theorem 13.1.11, with the factor $\mathscr{M}$ and the * automorphism θ described in Proposition 13.1.10, to prove the existence of a type III factor $\mathscr{R}(\mathscr{M}, \alpha)$ such that

$$T(\mathscr{R}(\mathscr{M}, \alpha)) = \{0, \pm a, \pm 2a, \ldots\},$$

where $a = 2\pi/|\log \lambda|$. By varying λ, we obtain a continuum of type III factors, no two of which are * isomorphic. Although it is by no means obvious, it can be shown that this family of factors is the same as the one constructed in Section 12.3 [13].

The remainder of the present section is concerned with further properties of the invariant T. We show, in particular, how it can be determined for a class of factors that properly includes those considered in Section 12.3. For this purpose, we shall require a further identification of the modular structure of a tensor product of von Neumann algebras. From Remark 11.2.36, we have that $\Delta_1 \otimes \Delta_2$ is the modular operator for $\mathscr{R}_1 \bar{\otimes} \mathscr{R}_2$, where Δ_1 and Δ_2 are the modular operators for $\mathscr{R}_1$ and $\mathscr{R}_2$, respectively. In the proposition that follows, we show that the modular automorphism group of $\mathscr{R}_1 \bar{\otimes} \mathscr{R}_2$ is $\{\sigma_t^{(1)} \bar{\otimes} \sigma_t^{(2)}\}$, where $\{\sigma_t^{(1)}\}$ and $\{\sigma_t^{(2)}\}$ are the modular automorphisms groups of $\mathscr{R}_1$ and $\mathscr{R}_2$, respectively. This can be proved by establishing the formula

$$(\Delta_1 \otimes \Delta_2)^{it} = \Delta_1^{it} \otimes \Delta_2^{it}.$$

We shall proceed, however, by using the characterization of the modular groups obtained in Lemma 9.2.17.

13.1.12. PROPOSITION. *If ω_j is a faithful normal state of a von Neumann algebra $\mathscr{R}_j$, and $\{\sigma_t^{(j)}\}$ is the corresponding modular automorphism group $(j = 1, 2)$, then $\omega_1 \bar{\otimes} \omega_2$ is a faithful normal state of $\mathscr{R}_1 \bar{\otimes} \mathscr{R}_2$, and its modular automorphism group is $\{\sigma_t^{(1)} \bar{\otimes} \sigma_t^{(2)}\}$.*

Proof. Upon replacing $\mathscr{R}_j$ by a suitable * isomorphic von Neumann algebra (for example, its universal normal representation) we may suppose that $\omega_j = \omega_{x(j)}|\mathscr{R}_j$, where $x(j)$ is a separating vector for $\mathscr{R}_j$; and $\omega_1 \bar{\otimes} \omega_2 = \omega_{x(1)\otimes x(2)}|\mathscr{R}_1 \bar{\otimes} \mathscr{R}_2$. Since $x(j)$ is generating for $\mathscr{R}_j'$, $x(1) \otimes x(2)$ is generating for $(\mathscr{R}_1 \bar{\otimes} \mathscr{R}_2)'$ and hence separating for $\mathscr{R}_1 \bar{\otimes} \mathscr{R}_2$; so $\omega_1 \bar{\otimes} \omega_2$ is faithful.

It is apparent that $\{\sigma_t^{(1)} \bar{\otimes} \sigma_t^{(2)}\}$ is a one-parameter group of * automorphisms of $\mathscr{R}_1 \bar{\otimes} \mathscr{R}_2$. In order to identify it as the modular automorphism group corresponding to $\omega_1 \bar{\otimes} \omega_2$, it suffices to verify the weakened form of the modular condition required in Lemma 9.2.17. For this, let $\mathfrak{A}$ be the strong-operator-dense * subalgebra of $\mathscr{R}_1 \bar{\otimes} \mathscr{R}_2$ consisting of all finite sums of simple tensors. We have to show that, given any A and B in $\mathfrak{A}$, there is a complex-valued function f, bounded and continuous on the strip $\{z \in \mathbb{C} : 0 \le \operatorname{Im} z \le 1\}$, analytic on the interior of that strip, and with boundary values

$$f(t) = (\omega_1 \bar{\otimes} \omega_2)((\sigma_t^{(1)} \bar{\otimes} \sigma_t^{(2)})(A)B),$$
$$f(t + i) = (\omega_1 \bar{\otimes} \omega_2)(B(\sigma_t^{(1)} \bar{\otimes} \sigma_t^{(2)})(A)).$$

By linearity, we may suppose that $A = A_1 \otimes A_2$, $B = B_1 \otimes B_2$; then

$$(\omega_1 \bar{\otimes} \omega_2)((\sigma_t^{(1)} \bar{\otimes} \sigma_t^{(2)})(A)B) = \omega_1(\sigma_t^{(1)}(A_1)B_1)\omega_2(\sigma_t^{(2)}(A_2)B_2),$$
$$(\omega_1 \bar{\otimes} \omega_2)(B(\sigma_t^{(1)} \bar{\otimes} \sigma_t^{(2)}(A)) = \omega_1(B_1\sigma_t^{(1)}(A_1))\omega_2(B_2\sigma_t^{(2)}(A_2)),$$

and the existence of a suitable function f is apparent from the fact that $\{\sigma_t^{(j)}\}$ satisfies the modular condition relative to ω_j. ∎

For infinite tensor products of finite type I factors, we have the following analogue of Proposition 13.1.12.

13.1.13. THEOREM. *Suppose that $\{\mathfrak{A}_j : j = 1, 2, \ldots\}$ is a sequence of mutually commuting finite type* I *factors acting on a Hilbert space $\mathscr{H}$ (and each containing the unit of $\mathscr{B}(\mathscr{H})$), $\mathfrak{A}$ is the C^*-algebra generated by $\bigcup \mathfrak{A}_j$, x is a unit cyclic vector for $\mathfrak{A}$, and $\omega_x|\mathfrak{A}$ is a product state $\otimes\rho_j$, where ρ_j is a faithful state of $\mathfrak{A}_j$ $(j = 1, 2, \ldots)$. Then $\omega_x|\mathfrak{A}^-$ is a faithful normal state of $\mathfrak{A}^-$; the corresponding modular automorphism group $\{\sigma_t\}$ of $\mathfrak{A}^-$ leaves each $\mathfrak{A}_j$ invariant, and $\{\sigma_t|\mathfrak{A}_j\}$ is the modular automorphism group of $\mathfrak{A}_j$ corresponding to ρ_j.*

Proof. For $n = 1, 2, \ldots$, let $\mathscr{M}_n$ and $\mathscr{M}_n^c$ be the C^*-algebras generated by $\bigcup\{\mathfrak{A}_j : j \le n\}$ and $\bigcup\{\mathfrak{A}_j : j > n\}$ respectively. From Proposition 13.1.12, together with the associativity of $\otimes$, $\rho_1 \otimes \cdots \otimes \rho_n$ is a faithful state of the spatial tensor product $\mathfrak{A}_1 \otimes \cdots \otimes \mathfrak{A}_n$ (of course, $\otimes$ and $\bar{\otimes}$ coincide, in this finite-dimensional situation). Since $\mathfrak{A}$ is (* isomorphic to) $\otimes\mathfrak{A}_j$, by Proposition 11.4.9, there is a * isomorphism ψ_n from $\mathfrak{A}_1 \otimes \cdots \otimes \mathfrak{A}_n$ onto $\mathscr{M}_n$, such that $\psi_n(A_1 \otimes \cdots \otimes A_n) = A_1A_2 \cdots A_n$. Moreover, $\rho_1 \otimes \cdots \otimes \rho_n = \omega_x \circ \psi_n$, since ω_x is a product state of $\mathfrak{A}$ and $\omega_x|\mathfrak{A}_j = \rho_j$; so $\omega_x|\mathscr{M}_n$ is a faithful state of the finite type I factor $\mathscr{M}_n$.

We assert that x is a generating vector for $\mathfrak{A}'$. For this, it suffices to show that $\mathscr{M}_n x \subseteq \mathfrak{A}'x$ $(n = 1, 2, \ldots)$, since x is generating for $\mathfrak{A}$ $(=(\bigcup \mathscr{M}_n)^=)$. From Lemma 11.4.8, there is a * isomorphism φ from $\mathfrak{A}$ onto $\mathscr{M}_n \otimes \mathscr{M}_n^c$, such that $\varphi(MB) = M \otimes B$ whenever $M \in \mathscr{M}_n$ and $B \in \mathscr{M}_n^c$. By restriction of the identity mapping on $\mathfrak{A}$, we obtain representations $\varphi_1 : M \to M|[\mathscr{M}_n x]$ of $\mathscr{M}_n$ on $[\mathscr{M}_n x]$, $\varphi_2 : B \to B|[\mathscr{M}_n^c x]$ of $\mathscr{M}_n^c$ on $[\mathscr{M}_n^c x]$. Both $\varphi_1(\mathscr{M}_n)$ and $\varphi_2(\mathscr{M}_n^c)$ have x as a generating vector; in the case of $\varphi_1(\mathscr{M}_n)$, it is also separating, since $\omega_x|\mathscr{M}_n$ is faithful. Thus $(\varphi_1 \otimes \varphi_2) \circ \varphi$ is a representation ψ of $\mathfrak{A}$ on the space $[\mathscr{M}_n x] \otimes [\mathscr{M}_n^c x]$, with a cyclic vector $x \otimes x$. Since $\omega_x|\mathfrak{A}$ is a product state,

$$\begin{aligned}\langle \psi(MB)(x \otimes x), x \otimes x \rangle &= \langle (\varphi_1 \otimes \varphi_2)(M \otimes B)(x \otimes x), x \otimes x \rangle \\ &= \langle Mx, x \rangle \langle Bx, x \rangle = \langle MBx, x \rangle,\end{aligned}$$

when $M \in \mathscr{M}_n$ and $B \in \mathscr{M}_n^c$; so

$$\langle \psi(A)(x \otimes x), x \otimes x \rangle = \langle Ax, x \rangle \qquad (A \in \mathfrak{A}).$$

Accordingly, there is a unitary operator U, from $\mathscr{H}$ onto $[\mathscr{M}_n x] \otimes [\mathscr{M}_n^c x]$, such that $Ux = x \otimes x$ and $\psi(A) = UAU^*$ $(A \in \mathfrak{A})$.

The proof is now completed by transferring all questions from $\mathfrak{A}$ to $U\mathfrak{A}U^*$. When $M \in \mathscr{M}_n$,

$$\begin{aligned} Mx = U^*UMU^*Ux &= U^*\psi(M)(x \otimes x) \\ &= U^*(\varphi_1(M) \otimes I)(x \otimes x) = U^*(\varphi_1(M)x \otimes x). \end{aligned}$$

Since $\varphi_1(\mathscr{M}_n)$ acts on a finite-dimensional Hilbert space, and has x as a separating vector, $\varphi_1(M)x = S'x$ for some S' in $\varphi_1(\mathscr{M}_n)'$. Since $S' \otimes I$ commutes with $\varphi_1(\mathscr{M}_n) \otimes \varphi_2(\mathscr{M}_n^c)$ $(= U\mathfrak{A}U^*)$, $U^*(S' \otimes I)U$ is an element A' of $\mathfrak{A}'$, and

$$\begin{aligned} A'x = U^*(S' \otimes I)(x \otimes x) &= U^*(S'x \otimes x) \\ &= U^*(\varphi_1(M)x \otimes x) = Mx. \end{aligned}$$

This shows that $\mathscr{M}_n x \subseteq \mathfrak{A}'x$; whence x is a generating vector for $\mathfrak{A}'$, and $\omega_x | \mathfrak{A}^-$ is a faithful normal state of $\mathfrak{A}^-$.

It remains to prove the stated results concerning the modular automorphism group $\{\sigma_t\}$ corresponding to $\omega_x | \mathfrak{A}^-$; and since we can renumber the sequence $\{\mathfrak{A}_j\}$, it suffices to consider only $\mathfrak{A}_1$. Observe that

$$U\mathfrak{A}^-U^* = [\varphi_1(\mathscr{M}_n) \otimes \varphi_2(\mathscr{M}_n^c)]^- = \varphi_1(\mathscr{M}_n) \mathbin{\overline{\otimes}} \varphi_2(\mathscr{M}_n^c)^-.$$

From the preceding paragraph, the normal product state $\omega_{Ux} | U\mathfrak{A}^-U^*$ $(= (\omega_x \mathbin{\overline{\otimes}} \omega_x) | U\mathfrak{A}U^*)$ is faithful. By Proposition 13.1.12, the corresponding modular automorphism group $\{\alpha_t\}$ of $U\mathfrak{A}^-U^*$ leaves the algebra $U\mathscr{M}_nU^*$ $(= \{\varphi_1(M) \otimes I : M \in \mathscr{M}_n\})$ invariant; and $\{\alpha_t | U\mathscr{M}_nU^*\}$ is the modular automorphism group corresponding to $\omega_{Ux} | U\mathscr{M}_nU^*$. Accordingly, $\{\sigma_t\}$ leaves $\mathscr{M}_n$ invariant, and $\{\sigma_t | \mathscr{M}_n\}$ is the modular automorphism group corresponding to $\omega_x | \mathscr{M}_n$. By taking $n = 1$, we obtain the required results concerning $\mathfrak{A}_1$, since $\mathscr{M}_1 = \mathfrak{A}_1$ and $\omega_x | \mathfrak{A}_1 = \rho_1$. ■

We can rephrase the statement about $\{\sigma_t\}$, in Theorem 13.1.13, as the assertion $\sigma_t | \mathfrak{A} = \otimes \sigma_t^{(j)}$, where $\{\sigma_t^{(j)}\}$ is the modular automorphism group of $\mathfrak{A}_j$, corresponding to ρ_j. In order to compute the invariant $T(\mathfrak{A}^-)$, we need the following criterion to determine whether σ_t is inner.

13.1.14. THEOREM. *Suppose that $\{\mathfrak{A}_j : j = 1, 2, \ldots\}$ is a sequence of mutually commuting finite type* I *factors acting on a Hilbert space $\mathscr{H}$ (and each containing the unit of $\mathscr{B}(\mathscr{H})$), $\mathfrak{A}$ is the C*-algebra generated by $\bigcup \mathfrak{A}_j$, x is a unit cyclic vector for $\mathfrak{A}$, and $\omega_x | \mathfrak{A}$ is a product state $\otimes \rho_j$, where ρ_j is a faithful state of $\mathfrak{A}_j$ $(j = 1, 2, \ldots)$. For each j, let U_j be a unitary operator in $\mathfrak{A}_j$ in the centralizer of ρ_j, and let θ_j be the * automorphism it induces on $\mathfrak{A}_j$. Then the ** *automorphism* $\theta = \otimes\theta_j$ *extends to a* * *automorphism* $\bar{\theta}$ *of* $\mathfrak{A}^-$; *and* $\bar{\theta}$ *is inner if and only if* $\sum\{1 - |\rho_j(U_j|\} < \infty$.

Proof. For each j, $\rho_j \circ \theta_j = \rho_j$, since U_j lies in the centralizer of ρ_j; so $\omega_x \circ \theta = \omega_x | \mathfrak{A}$. The equation $WAx = \theta(A)x$ $(A \in \mathfrak{A})$ therefore determines a unitary operator W that implements θ; moreover, W implements an automorphism $\bar{\theta}$ of $\mathfrak{A}^-$ that extends θ.

Upon replacing U_j by $a_j U_j$, where a_j is a suitable scalar, with $|a_j| = 1$, we may assume that $\rho_j(U_j) \geq 0$.

Suppose that $\sum\{1 - \rho_j(U_j)\} < \infty$. Let V_n be the unitary $U_1 U_2 \cdots U_n$ in $\mathfrak{A}$, and note that

$$V_n A V_n^* = \theta_j(A) \qquad (A \in \mathfrak{A}_j \ , \quad 1 \leq j \leq n),$$

since $U_k \in \mathfrak{A}_j'$ when $k \neq j$. We shall prove that the sequence $\{V_n\}$ is strong-operator convergent to a unitary operator V in $\mathfrak{A}^-$, and $\{V_n^*\}$ is strong-operator convergent to V^*. Once this is done, we have

$$VAV^* = \theta_j(A) = \theta(A) = \bar{\theta}(A) \qquad (A \in \mathfrak{A}_j);$$

whence $\bar{\theta}(A) = VAV^*$ for all A in $\mathfrak{A}^-$, and $\bar{\theta}$ is inner, since $\bigcup \mathfrak{A}_j$ generates $\mathfrak{A}^-$ as a von Neumann algebra. To establish the strong-operator convergence of $\{V_n\}$, note that if $1 \leq m < n$,

$$\begin{aligned}
\|V_{m-1}x - V_n x\|^2 &= \|V_{m-1}(I - U_m U_{m+1} U_{m+2} \cdots U_n)x\|^2 \\
&= \|(I - U_m U_{m+1} \cdots U_n)x\|^2 \\
&= \omega_x((I - U_m U_{m+1} \cdots U_n)^*(I - U_m U_{m+1} \cdots U_n)) \\
&= 2 - 2 \operatorname{Re} \omega_x(U_m U_{m+1} \cdots U_n) \\
&= 2 - 2\rho_m(U_m)\rho_{m+1}(U_{m+1}) \cdots \rho_n(U_n).
\end{aligned}$$

The last quantity tends to zero as $m = \min(m, n) \to \infty$, since the conditions $\rho_j(U_j) \geq 0$, $\sum\{1 - \rho_j(U_j)\} < \infty$ ensure the convergence of the infinite product $\prod_{j=k}^{\infty} \rho_j(U_j)$ (with non-zero limit), for sufficiently large k. Hence the sequence $\{V_n x\}$ converges, in norm, to an element of $\mathscr{H}$; and the same is true of $\{V_n A'x\}$ $(=\{A'V_n x\})$ for each A' in $\mathfrak{A}'$. Since x is a generating vector for $\mathfrak{A}'$, by Theorem 13.1.13, it now follows that the (bounded) sequence $\{V_n\}$ is strong-operator convergent (and thus, also, weak-operator convergent) to an element V of $\mathfrak{A}^-$. The same argument can be used to prove that $\{V_n^*\}$ converges in these topologies; and its limit is V^*, since the adjoint operation is weak-operator continuous. Finally, $VV^* = V^*V = I$, since operator multiplication is strong-operator continuous on bounded sets.

Conversely, suppose that $\bar{\theta}$ is implemented by a unitary operator V in $\mathfrak{A}^-$. For $n = 1, 2, \ldots$, let $\mathscr{M}_n$ and $\mathscr{M}_n^c$ be the C^*-algebras generated by

$\bigcup\{\mathfrak{A}_j : j \le n\}$ and $\bigcup\{\mathfrak{A}_j : j > n\}$, respectively. Since $\omega_x|\mathfrak{A}$ is the product state $\otimes\rho_j$,

$$\omega_x(A_1A_2\cdots A_nB) = \rho_1(A_1)\rho_2(A_2)\cdots\rho_n(A_n)\omega_x(B)$$

when $A_1\in\mathfrak{A}_1,\ldots, A_n\in\mathfrak{A}_n$, $B\in\mathscr{M}_n^c$, by Remark 11.4.16. By ultraweak continuity, this remains true for all B in $(\mathscr{M}_n^c)^-$.

Since $\omega_x(V^*V) = 1$ and $V^*\in(\bigcup \mathscr{M}_n)^-$, we can choose an integer k and elements A_j of $(\mathfrak{A}_j)_1$ $(j = 1,\ldots, k)$, for which

$$\omega_x(A_1A_2\cdots A_kV) \ne 0.$$

When $n \ge k$, $U_1U_2\cdots U_n$ implements the same automorphism, $\theta|\mathscr{M}_n$ of $\mathscr{M}_n$, as does V; from this, and since $\mathscr{M}'_n\cap\mathfrak{A}^- = (\mathscr{M}_n^c)^-$ by Lemma 11.4.14, $V = U_1U_2\cdots U_nV_n$, for some unitary operator V_n in $(\mathscr{M}_n^c)^-$. Hence

$$\begin{aligned}
0 &\ne |\omega_x(A_1A_2\cdots A_kV)|\\
&= |\omega_x(A_1A_2\cdots A_kU_1U_2\cdots U_nV_n)|\\
&= |\omega_x(A_1U_1A_2U_2\cdots A_kU_kU_{k+1}\cdots U_nV_n)|\\
&= |\rho_1(A_1U_1)\rho_2(A_2U_2)\cdots\rho_k(A_kU_k)\rho_{k+1}(U_{k+1})\cdots\rho_n(U_n)\omega_x(V_n)|\\
&\le \rho_{k+1}(U_{k+1})\rho_{k+2}(U_{k+2})\cdots\rho_n(U_n).
\end{aligned}$$

Since $0 \le \rho_j(U_j) \le 1$, it now follows that the infinite product $\prod_{j=k}^{\infty}\rho_j(U_j)$ converges (with non-zero limit), whence $\sum\{1-\rho_j(U_j)\} < \infty$. ■

We now use the two preceding theorems to study certain representations of the CAR algebra.

13.1.15. THEOREM. *Suppose that $\{\mathfrak{A}_r : r = 1, 2,\ldots\}$ is a sequence of factors of type I_2, $\{E_{jk}^{(r)} : j, k = 1, 2\}$ is a system of matrix units for $\mathfrak{A}_r$, $\{a_r\}$ is a sequence of real numbers satisfying $0 < a_r \le \frac{1}{2}$, ρ_r is the faithful state of $\mathfrak{A}_r$ defined by*

$$\rho_r\left(\sum_{j,k=1}^{2} c_{jk}E_{jk}^{(r)}\right) = a_rc_{11} + (1-a_r)c_{22},$$

and $b_r = \log(a_r^{-1}(1-a_r))$. If $\mathfrak{A} = \otimes\mathfrak{A}_r$, $\rho = \otimes\rho_r$, and π_ρ is the representation engendered by ρ, then $\pi_\rho(\mathfrak{A})^-$ is a factor and

$$\begin{aligned}
T(\pi_\rho(\mathfrak{A})^-) &= \left\{t\in\mathbb{R} : \sum_{r=1}^{\infty}[1-|a_r^{1+it} + (1-a_r)^{1+it}|] < \infty\right\}\\
&= \left\{t\in\mathbb{R} : \sum_{r=1}^{\infty} e^{-b_r}\sin^2(\tfrac{1}{2}b_rt) < \infty\right\}.
\end{aligned}$$

Proof. By Theorem 11.4.15 and Remark 11.4.16, $\mathfrak{A}$ is simple and ρ is a primary state; so π_ρ is faithful, and $\pi_\rho(\mathfrak{A})^-$ is a factor. Upon identifying $\mathfrak{A}$ with $\pi_\rho(\mathfrak{A})$, ρ takes the form $\omega_x|\mathfrak{A}$ for some cyclic unit vector x, and we reduce to the type of situation considered in Theorems 13.1.13 and 13.1.14.

When $A \in \mathfrak{A}_r$, $\rho_r(A) = 2\tau_r(H_r A)$, where τ_r is the (normalized) tracial state of $\mathfrak{A}_r$ and $H_r = a_r E_{11}^{(r)} + (1 - a_r)E_{22}^{(r)}$. Given A and B in $\mathfrak{A}_r$, the entire function

$$f(z) = 2\tau_r(H_r^{1+iz} A H_r^{-iz} B)$$

satisfies

$$f(t) = 2\tau_r(H_r H_r^{it} A H_r^{-it} B) = \rho_r(H_r^{it} A H_r^{-it} B),$$
$$f(t+i) = 2\tau_r(H_r^{it} A H_r^{-it} H_r B) = 2\tau_r(H_r B H_r^{it} A H_r^{-it}) = \rho_r(B H_r^{it} A H_r^{-it}).$$

Accordingly, the modular automorphism group $\{\sigma_t^{(r)}\}$ of $\mathfrak{A}_r$ corresponding to ρ_r, is given by $\sigma_t^{(r)}(A) = H_r^{it} A H_r^{-it}$; for the preceding calculation shows that the modular condition is satisfied. Moreover, H_r (and, hence, also H_r^{it}) lies in the centralizer of ρ_r, since $\rho_r(H_r A) = \rho_r(A H_r) = 2\tau_r(H_r A H_r)$ for all A in $\mathfrak{A}_r$.

From Theorem 13.1.13, $\omega_x|\mathfrak{A}^-$ is a faithful normal state; so we can use the corresponding modular automorphism group $\{\sigma_t\}$ of $\mathfrak{A}^-$ to determine $T(\mathfrak{A}^-)$. By the same theorem, σ_t is the (unique) extension of $\otimes\sigma_t^{(r)}$ to an automorphism of $\mathfrak{A}^-$. From Theorem 13.1.14, σ_t is inner if and only if $\sum\{1 - |\rho_r(H_r^{it})|\} < \infty$. Now

$$\rho_r(H_r^{it}) = \rho_r(a_r^{it} E_{11}^{(r)} + (1 - a_r)^{it} E_{22}^{(r)}) = a_r^{1+it} + (1 - a_r)^{1+it}.$$

Thus, σ_t is inner if and only if $\sum\{1 - |a_r^{1+it} + (1 - a_r)^{1+it}|\} < \infty$.

Since $1 - a_r = a_r \exp b_r$,

$$|a_r^{1+it} + (1 - a_r)^{1+it}| = |a_r^{1+it}(1 + e^{b_r(1+it)})|$$
$$= a_r[1 + 2e^{b_r}\cos(b_r t) + e^{2b_r}]^{1/2} = c_r^{1/2},$$

where (since $a_r = (1 + \exp b_r)^{-1}$),

$$c_r = \frac{1 + 2e^{b_r}\cos(b_r t) + e^{2b_r}}{1 + 2e^{b_r} + e^{2b_r}}.$$

Thus

$$1 - |a_r^{1+it} + (1 - a_r)^{1+it}| = 1 - c_r^{1/2} = (1 - c_r)/(1 + c_r^{1/2})$$
$$= \frac{4e^{-b_r}\sin^2(\tfrac{1}{2}b_r t)}{(1 + c_r^{1/2})(1 + 2e^{-b_r} + e^{-2b_r})}$$

Since $0 \le c_r \le 1$ and $b_r \ge 0$,

$$\tfrac{1}{2}e^{-b_r}\sin^2(\tfrac{1}{2}b_r t) \le 1 - |a_r^{1+it} + (1 - a_r)^{1+it}| \le 4e^{-b_r}\sin^2(\tfrac{1}{2}b_r t);$$

so $\sum\{1 - |a_r^{1+it} + (1 - a_r)^{1+it}|\} < \infty$ if and only if $\sum e^{-b_r}\sin^2(\tfrac{1}{2}b_r t) < \infty$. ■

We consider some examples of representations π_ρ of the CAR algebra $\mathfrak{A}$, obtained by appropriate choice of the sequence $\{a_r\}$ in Theorem 13.1.15.

If $0 < a < \frac{1}{2}$ and each a_r is a, ρ is a product state of type a, and the corresponding factor $\pi_\rho(\mathfrak{A})^-$ is one of those considered in Section 12.3. Each b_r is $\log(a^{-1}(1-a))$ ($=b$), so the series $\sum e^{-b_r}\sin^2(\frac{1}{2}b_r t)$ converges if and only if $\sin(\frac{1}{2}bt) = 0$. Hence

$$T(\pi_\rho(\mathfrak{A})^-) = \{0, \pm c, \pm 2c, \ldots\},$$

where $c = 2\pi/b = 2\pi/\log(a^{-1}(1-a))$; and $\pi_\rho(\mathfrak{A})^-$ is a type III factor. Different choices of a in $(0, \frac{1}{2})$ give different values of c, and so lead to non-isomorphic factors. This provides an alternative approach to some of the results obtained in Section 12.3.

If $a_r \to \frac{1}{2}$ as $r \to \infty$, then $b_r \to 0$ and $e^{-b_r}\sin^2(\frac{1}{2}b_r t)/(\frac{1}{2}b_r t)^2 \to 1$. If, further, $\sum b_r^2$ diverges, then $\sum e^{-b_r}\sin^2(\frac{1}{2}b_r t)$ diverges for all non-zero t; so $T(\pi_\rho(\mathfrak{A})^-) = \{0\}$, and $\pi_\rho(\mathfrak{A})^-$ is a type III factor. If $\sum b_r^2$ converges, then $\sum e^{-b_r}\sin^2(\frac{1}{2}b_r t)$ converges for all real t, whence $T(\pi_\rho(\mathfrak{A})^-) = \mathbb{R}$; in fact, in this case, $\pi_\rho(\mathfrak{A})^-$ is the finite matricial factor (see Exercise 13.4.9).

If $a_r \to 0$ as $r \to \infty$, then $b_r \to \infty$. If $\sum e^{-b_r} < \infty$, then $\sum e^{-b_r}\sin^2(\frac{1}{2}b_r t) < \infty$ for all real t, and $T(\pi_\rho(\mathfrak{A})^-) = \mathbb{R}$; in fact, $\pi_\rho(\mathfrak{A})^-$ is a type I_∞ factor in this case (see Exercise 13.4.10). When $a_r \to 0$ and $\sum e^{-b_r}$ diverges, $\pi_\rho(\mathfrak{A})^-$ is a type III factor, and there are various possibilities for $T(\pi_\rho(\mathfrak{A})^-)$, depending on the sequence $\{b_r\}$ (see Exercises 13.4.11, 13.4.12, 13.4.13, and 13.4.14).

If $a_{2r} \to \frac{1}{2}$ and $\sum b_{2r}^2 < \infty$, while $a_{2r+1} \to 0$ and $\sum e^{-b_{2r+1}} < \infty$, the ultraweak closures of the representations engendered by the states $\rho | \bigotimes \mathfrak{A}_{2r}$ and $\rho | \bigotimes \mathfrak{A}_{2r+1}$ are factors of types II_1 and I_∞, respectively. It is not difficult to show that $\pi_\rho(\mathfrak{A})^-$ is unitarily equivalent to the (von Neumann algebra) tensor product of these two factors, and is therefore a factor of type II_∞ (see Exercise 13.4.15).

In using Proposition 13.1.12 to calculate $T(\mathscr{R}_1 \bar{\otimes} \mathscr{R}_2)$ from $T(\mathscr{R}_1)$ and $T(\mathscr{R}_2)$, we require the following result.

13.1.16. THEOREM. *If α_j is a * automorphism of a von Neumann algebra $\mathscr{R}_j$ $(j = 1, 2)$, then the * automorphism $\alpha_1 \bar{\otimes} \alpha_2$ of $\mathscr{R}_1 \bar{\otimes} \mathscr{R}_2$ is inner if and only if both α_1 and α_2 are inner.*

Proof. If α_j is implemented by a unitary operator U_j $(\in \mathscr{R}_j)$ for $j = 1, 2$, then $\alpha_1 \bar{\otimes} \alpha_2$ is implemented by $U_1 \otimes U_2$ $(\in \mathscr{R}_1 \bar{\otimes} \mathscr{R}_2)$.

Conversely, suppose that $\alpha_1 \bar{\otimes} \alpha_2$ is the inner automorphism associated with a unitary element W of $\mathscr{R}_1 \bar{\otimes} \mathscr{R}_2$. For each A in $\mathscr{R}_1$, $(\alpha_1 \bar{\otimes} \alpha_2)(A \otimes I) = \alpha_1(A) \otimes I$; so

$$W(A \otimes I) = (\alpha_1(A) \otimes I)W \qquad (A \in \mathscr{R}_1).$$

Now $W \in \mathscr{R}_1 \bar{\otimes} \mathscr{B}(\mathscr{H}_2)$, where $\mathscr{H}_2$ is the Hilbert space on which $\mathscr{R}_2$ acts. If we represent tensor product operators in the usual way by matrices (indexed by some set $\mathbb{A}$), then W corresponds to a matrix $[W_{ab}]$ with entries in $\mathscr{R}_1$, while $A \otimes I$ and $\alpha_1(A) \otimes I$ correspond to $[\delta_{ab} A]$ and $[\delta_{ab}\alpha_1(A)]$, respectively. The last equation thus entails

$$(10) \qquad W_{ab}A = \alpha_1(A)W_{ab} \qquad (A \in \mathscr{R}_1; \quad a, b \in \mathbb{A}).$$

Upon replacing A by a unitary operator U in $\mathscr{R}_1$, it follows from (10) that

$$W_{ab}W_{ab}^* = \alpha_1(U)W_{ab}W_{ab}^*\alpha_1(U)^*, \qquad U^*W_{ab}^*W_{ab}U = W_{ab}^*W_{ab};$$

so both $W_{ab}W_{ab}^*$ and $W_{ab}^*W_{ab}$ lie in the center $\mathscr{C}$ of $\mathscr{R}_1$. Hence their range projections (that is, the range projections of W_{ab} and W_{ab}^*) lie in $\mathscr{C}$. From this, and since W_{ab} and W_{ab}^* have the same central carrier P_{ab}, it follows that W_{ab} and W_{ab}^* both have range projection P_{ab}. Accordingly, in the polar decomposition $V_{ab}H_{ab}$ of W_{ab}, V_{ab} is a partial isometry with initial and final projections both equal to P_{ab}, while $H_{ab} = [W_{ab}^*W_{ab}]^{1/2} \in \mathscr{C}$. From (10)

$$V_{ab}AH_{ab} = V_{ab}H_{ab}A = W_{ab}A = \alpha_1(A)W_{ab} = \alpha_1(A)V_{ab}H_{ab}$$

for each A in $\mathscr{R}_1$. Since H_{ab} has range projection P_{ab} $(\in \mathscr{C})$, the initial projection of V_{ab}, it now follows that

$$V_{ab}A = \alpha_1(A)V_{ab} \qquad (A \in \mathscr{R}_1; \quad a, b \in \mathbb{A}).$$

Note also that $\bigvee P_{ab} = I$, since $[W_{ab}]$ is the matrix of a unitary operator in $\mathscr{R}_1 \bar{\otimes} \mathscr{B}(\mathscr{H}_2)$.

Let $\{Q_c\}$ be an orthogonal family of non-zero central projections in $\mathscr{R}_1$, maximal subject to the following condition: for each c, there is a partial isometry V_c in $\mathscr{R}_1$, with both initial and final projection equal to Q_c, such that $V_cA = \alpha_1(A)V_c$ for each A in $\mathscr{R}_1$. Then $I - \sum Q_c = 0$; for otherwise, one can add to the family $\{Q_c\}$ a projection $Q(\neq 0)$ of the form $(I - \sum Q_c)P_{ab}$ (with QV_{ab} as the corresponding partial isometry), contrary to the maximality assumption. Since $\sum Q_c = I$ and $V_cA = \alpha_1(A)V_c$, it follows that $\sum V_c$ is a unitary operator V in $\mathscr{R}_1$, and $VA = \alpha_1(A)V$ $(A \in \mathscr{R}_1)$. Thus α_1 (and, similarly, α_2) is inner. ■

13.1.17. COROLLARY. $T(\mathscr{R}_1 \bar{\otimes} \mathscr{R}_2) = T(\mathscr{R}_1) \cap T(\mathscr{R}_2)$.

Proof. This follows at once from Theorem 13.1.16 and Proposition 13.1.12 (except that the latter results must be replaced by its analogue in terms of semi-finite weights, if either $\mathscr{R}_1$ or $\mathscr{R}_2$ is not countably decomposable). ■

Bibliography: [12, 56, 101]

13.2. Continuous crossed products

In this section we consider the "continuous crossed product" $\mathscr{R}(\mathscr{M}, \alpha)$ of a von Neumann algebra $\mathscr{M}$ by a continuous one-parameter group $\{\alpha_t\}$ of * automorphisms of $\mathscr{M}$. As in Section 13.1, there are both "abstract" and "implemented" crossed products; and the relations between them are analogous to those already established in the discrete case. The main result is a duality theorem (13.2.9): given $\mathscr{M}$ and $\{\alpha_t\}$ as above, there is a dual one-parameter group $\{\hat{\alpha}_t\}$ of * automorphisms of $\mathscr{R}(\mathscr{M}, \alpha)$; and the second crossed product, $\mathscr{R}(\mathscr{R}(\mathscr{M}, \alpha), \hat{\alpha})$, is * isomorphic to $\mathscr{M} \mathbin{\overline{\otimes}} \mathscr{B}(L_2(\mathbb{R}))$, where Lebesgue measure is used on $\mathbb{R}$. In Section 13.3, the duality theorem is applied in studying the structure of type III von Neumann algebras.

Suppose that G is a topological group, $\mathscr{M}$ is a von Neumann algebra acting on a Hilbert space $\mathscr{H}$, and $\alpha : g \to \alpha_g$ is an automorphic representation of G on $\mathscr{M}$. We say that α is *continuous* if, for each A in $\mathscr{M}$, the mapping $g \to \alpha_g(A) : G \to \mathscr{M}$ is continuous relative to the weak-operator topology on $\mathscr{M}$. We say that α is *unitarily implemented* if there is a continuous unitary representation $g \to U(g)$ of G on $\mathscr{H}$, such that $\alpha_g(A) = U(g)AU(g)^*$ for each A in $\mathscr{M}$ and g in G. Continuity of α is equivalent to the (apparently more restrictive) condition that $g \to \alpha_g(A)$ is continuous with respect to the strong-operator topology on $\mathscr{M}$. To see this, it suffices to consider the case in which A is replaced by a unitary element U of $\mathscr{M}$. If α is continuous, then, for each x in $\mathscr{H}$, the mapping $g \to \alpha_g(U)x : G \to \mathscr{H}$ is continuous relative to the weak topology on $\mathscr{H}$. Since $\|\alpha_g(U)x\| = \|x\|$ for all g in G, it follows from Proposition 2.3.5 that $g \to \alpha_g(U)x$ is continuous with respect to the norm topology on $\mathscr{H}$, whence $g \to \alpha_g(U)$ is strong-operator continuous.

There is a theory of continuous crossed products $\mathscr{R}(\mathscr{M}, \alpha)$, in which $\alpha: g \to \alpha_g$ is a continuous automorphic representation of a locally compact abelian group G on a von Neumann algebra $\mathscr{M}$. In this context, there is a continuous automorphic representation $\hat{\alpha}: p \to \hat{\alpha}_p$ of the dual group $\hat{G}$ on $\mathscr{R}(\mathscr{M}, \alpha)$; and $\mathscr{R}(\mathscr{R}(\mathscr{M}, \alpha), \hat{\alpha})$ is * isomorphic to $\mathscr{M} \mathbin{\overline{\otimes}} \mathscr{B}(L_2(G))$, where Haar measure is used on G [97]. We shall consider only the case in which G is the additive group $\mathbb{R}$ (whence $\hat{G}$ can be identified with $\mathbb{R}$, and $\{\alpha_g\}$, $\{\hat{\alpha}_g\}$ are one-parameter groups). The reader who is familiar with Fourier analysis on locally compact abelian groups will see that the theory generalizes easily to the broader context.

Throughout the remainder of this section, we shall be concerned with a von Neumann algebra $\mathscr{M}$ acting on a Hilbert space $\mathscr{H}$, and a continuous automorphic representation $\alpha: t \to \alpha_t$ of $\mathbb{R}$ on $\mathscr{M}$. In references to measure, measurability, etc., it is Lebesgue measure (on the σ-algebra of Lebesgue measurable subsets of $\mathbb{R}$) that is intended.

We begin by describing the Hilbert space (which we shall then identify with $\mathscr{H} \otimes L_2(\mathbb{R})$) on which the crossed product von Neumann algebra will act. A mapping $x\colon \mathbb{R} \to \mathscr{H}$ is said to be measurable if the complex-valued function $\langle x(s), u\rangle$ of s is measurable on $\mathbb{R}$ for each u in $\mathscr{H}$, and the closed subspace $\mathscr{H}_x$ of $\mathscr{H}$ generated by $\{x(s) : s \in \mathbb{R}\}$ is separable. Given two such measurable mappings x and y, from $\mathbb{R}$ into $\mathscr{H}$, the complex-valued function $\langle x(s), y(s)\rangle$ of s is measurable on $\mathbb{R}$; for if $\{u_1, u_2, \ldots\}$ is an orthonormal basis of $\mathscr{H}_x$ and E is the projection from $\mathscr{H}$ onto $\mathscr{H}_x$,

$$\begin{aligned}\langle x(s), y(s)\rangle &= \langle x(s), Ey(s)\rangle \\ &= \sum_{j=1}^{\infty} \langle x(s), u_j\rangle\langle u_j, Ey(s)\rangle = \sum_{j=1}^{\infty} \langle x(s), u_j\rangle\langle u_j, y(s)\rangle.\end{aligned}$$

In particular, the function $\|x(s)\|^2$ of s is measurable.

Let $\mathscr{V}$ be the set of all measurable mappings $x\colon \mathbb{R} \to \mathscr{H}$ for which $\int_{\mathbb{R}} \|x(s)\|^2\, ds < \infty$. It is apparent that $\mathscr{V}$ is a complex-vector space, with a (positive semi-definite) inner product defined by $\langle x, y\rangle = \int_{\mathbb{R}} \langle x(s), y(s)\rangle\, ds$. The subspace $\mathscr{N} = \{x \in \mathscr{V} : \langle x, x\rangle = 0\}$ consists of those x in $\mathscr{V}$ (the null functions) for which $x(s) = 0$ almost everywhere. We denote by $L_2(\mathbb{R}, \mathscr{H})$ the quotient space $\mathscr{V}/\mathscr{N}$, with the induced (positive definite) inner product; and we adopt the usual convention of referring to elements of $L_2(\mathbb{R}, \mathscr{H})$ as functions, although strictly speaking they are equivalence classes modulo null functions.

13.2.1. PROPOSITION. *$L_2(\mathbb{R}, \mathscr{H})$ is a Hilbert space, and there is a unitary operator W from $\mathscr{H} \otimes L_2(\mathbb{R})$ onto $L_2(\mathbb{R}, \mathscr{H})$, such that*

$$(W(u \otimes f))(s) = f(s)u \qquad (u \in \mathscr{H},\quad f \in L_2(\mathbb{R}),\quad s \in \mathbb{R}).$$

Proof. To show that $L_2(\mathbb{R}, \mathscr{H})$ is a Hilbert space, it remains to prove only that it is complete. For this, suppose that $\{x_n\}$ is a Cauchy sequence in $L_2(\mathbb{R}, \mathscr{H})$, let $\{n(1), n(2), \ldots\}$ be a strictly increasing sequence of positive integers such that $\|x_m - x_n\| < 2^{-j}$ whenever $m, n \geq n(j)$, and let $y_j = x_{n(j)}$. The non-negative function $g_j(s) = \|y_j(s) - y_{j+1}(s)\|$ is an element of $L_2(\mathbb{R})$, and

$$\|g_j\| = \|y_j - y_{j+1}\| = \|x_{n(j)} - x_{n(j+1)}\| < 2^{-j}.$$

For $k = 1, 2, \ldots$,

$$\begin{aligned}\int_{\mathbb{R}} \left[\sum_{j=1}^{k} \|y_j(s) - y_{j+1}(s)\|\right]^2 ds &= \int_{\mathbb{R}} \left[\sum_{j=1}^{k} g_j(s)\right]^2 ds \\ &= \|g_1 + \cdots + g_k\|^2 \leq [\|g_1\| + \cdots + \|g_k\|]^2 < 1;\end{aligned}$$

so

$$\lim_{k\to\infty}\int_{\mathbb{R}}\left[\sum_{j=1}^{k}\|y_j(s)-y_{j+1}(s)\|\right]^2 ds \le 1.$$

It now follows, by applying the monotone convergence theorem, that $\sum_1^\infty \|y_j(s)-y_{j+1}(s)\| < \infty$ for almost all s in $\mathbb{R}$. If Z denotes the exceptional null set, then $\{y_j(s)\}$ converges (in the norm topology on $\mathscr{H}$) whenever $s \in \mathbb{R}\backslash Z$; and we may define a mapping $x\colon \mathbb{R}\to\mathscr{H}$ by

$$x(s) = \begin{cases} \lim y_j(s) = \lim x_{n(j)}(s) & (s\in\mathbb{R}\backslash Z) \\ 0 & (s\in Z) \end{cases}.$$

When $n \ge n(k)$ and $j \ge k$,

$$\begin{aligned}\int_{\mathbb{R}}\|x_n(s)-y_j(s)\|^2\,ds &= \|x_n - y_j\|^2\\ &= \|x_n - x_{n(j)}\|^2 < 2^{-2k};\end{aligned}$$

and $\|x_n(s)-y_j(s)\| \to \|x_n(s)-x(s)\|$ as $j\to\infty$, when $s\in\mathbb{R}\backslash Z$. From Fatou's lemma,

$$\int_{\mathbb{R}}\|x_n(s)-x(s)\|^2\,ds \le 2^{-2k} \qquad (n\ge n(k)).$$

For $n = 1, 2, \ldots$, there is a separable subspace $\mathscr{H}_n$ of $\mathscr{H}$ that contains all values of x_n. The separable subspace $\bigvee \mathscr{H}_n$ contains $x_n(s)$ for all n and s, and so contains $x(s)$ (which is either 0 or $\lim x_{n(j)}(s)$) and $x_n(s)-x(s)$. When $u\in\mathscr{H}$, the function $\langle x_n(s)-x(s), u\rangle$ of s is measurable on $\mathbb{R}$, since it is the limit almost everywhere (as $j\to\infty$) of $\langle x_n(s)-x_{n(j)}(s), u\rangle$. Accordingly,

$$x_n - x \in L_2(\mathbb{R},\mathscr{H}), \qquad \|x_n - x\| \le 2^{-k} \qquad (n\ge n(k));$$

so $\{x_n\}$ converges to x ($\in L_2(\mathbb{R},\mathscr{H})$), whence $L_2(\mathbb{R},\mathscr{H})$ is complete.

Suppose that $f_1,\ldots,f_n\in L_2(\mathbb{R})$, $u_1,\ldots,u_n\in\mathscr{H}$. The equation $x(s) = \sum f_j(s)u_j$ defines an element x of $L_2(\mathbb{R},\mathscr{H})$, and

$$\begin{aligned}\|x\|^2 &= \int_{\mathbb{R}}\langle x(s), x(s)\rangle\,ds\\ &= \sum_{j,k=1}^{n}\langle u_j, u_k\rangle\int_{\mathbb{R}} f_j(s)\overline{f_k(s)}\,ds\\ &= \sum_{j,k=1}^{n}\langle u_j\otimes f_j, u_k\otimes f_k\rangle = \Big\|\sum_{j=1}^{n} u_j\otimes f_j\Big\|^2.\end{aligned}$$

Accordingly, there is an isometric linear mapping W, from $\mathscr{H} \otimes L_2(\mathbb{R})$ into $L_2(\mathbb{R}, \mathscr{H})$, such that

$$(W(u \otimes f))(s) = f(s)u \qquad (u \in \mathscr{H}, \quad f \in L_2(\mathbb{R}), \quad s \in \mathbb{R}).$$

The range of W is a closed subspace, and it remains to prove that it is the whole of $L_2(\mathbb{R}, \mathscr{H})$. For this, suppose that $x \in L_2(\mathbb{R}, \mathscr{H})$ and x is orthogonal to the range of W; we have to show that $x = 0$. Let $\mathscr{H}_x$ be a separable subspace of $\mathscr{H}$ that contains all the values of x, and let $\{u_1, u_2, \ldots\}$ be an orthonormal basis of $\mathscr{H}_x$. For $j = 1, 2, \ldots,$

$$\int_{\mathbb{R}} f(s)\langle u_j, x(s)\rangle \, ds = \langle W(u_j \otimes f), x\rangle = 0$$

for every f in $L_2(\mathbb{R})$. Thus $\langle u_j, x(s)\rangle = 0$ for almost all s, and $x(s) = 0$ almost everywhere. It follows that $x = 0$. ■

We shall frequently identify $\mathscr{H} \otimes L_2(\mathbb{R})$ with $L_2(\mathbb{R}, \mathscr{H})$ by means of the unitary operator W occurring in Proposition 13.2.1, thus interpreting $u \otimes f$ as the function $s \to f(s)u : \mathbb{R} \to \mathscr{H}$. When $g \in L_\infty(\mathbb{R})$, the equation $(M_g x)(s) = g(s)x(s)$ defines a bounded linear operator M_g acting on $L_2(\mathbb{R}, \mathscr{H})$. Since

$$\begin{aligned}(M_g(u \otimes f))(s) &= g(s)(u \otimes f)(s) \\ &= g(s)f(s)u = (u \otimes gf)(s),\end{aligned}$$

it follows that $M_g = I \otimes m_g$, where m_g is the operator (acting on $L_2(\mathbb{R})$) of multiplication by g.

13.2.2. PROPOSITION. *Suppose that $a: s \to a(s)$ is a weak-operator continuous mapping from $\mathbb{R}$ into a bounded subset of $\mathscr{B}(\mathscr{H})$.*

(i) *The equation*

$$(Ax)(s) = a(s)x(s) \qquad (x \in L_2(\mathbb{R}, \mathscr{H}), \quad s \in \mathbb{R})$$

defines a bounded linear operator A acting on $L_2(\mathbb{R}, \mathscr{H})$. If $a(s) = g(s)A_0$, where $A_0 \in \mathscr{B}(\mathscr{H})$ and g is a bounded continuous complex-valued function on $\mathbb{R}$, then $A = A_0 \otimes m_g$, where m_g is the operator (acting on $L_2(\mathbb{R})$) of multiplication by g.

(ii) *Suppose that $\mathscr{M}$ is a von Neumann algebra acting on $\mathscr{H}$, and $a(s) \in \mathscr{M}$ for all real s. Then $A \in \mathscr{M} \mathbin{\bar{\otimes}} \mathscr{A}$ $(\subseteq \mathscr{M} \mathbin{\bar{\otimes}} \mathscr{B}(L_2(\mathbb{R})))$, where $\mathscr{A}$ is the multiplication algebra $\{m_g : g \in L_\infty(\mathbb{R})\}$ acting on $L_2(\mathbb{R})$. If $\mathscr{N}$ is a von Neumann algebra acting on a Hilbert space $\mathscr{K}$, and θ is a * isomorphism from $\mathscr{M}$ onto $\mathscr{N}$, the equation $b(s) = \theta(a(s))$ defines a weak-operator continuous mapping b from $\mathbb{R}$ into a bounded subset of $\mathscr{N}$. If B is the corresponding operator acting on $L_2(\mathbb{R}, \mathscr{K})$, defined by $(By)(s) = b(s)y(s)$, then $B = (\theta \mathbin{\bar{\otimes}} \iota)(A)$, where ι denotes the identity automorphism of $\mathscr{B}(L_2(\mathbb{R}))$.*

Proof. (i) When $x \in L_2(\mathbb{R}, \mathscr{H})$, let $\mathscr{H}_x$ be a separable subspace of $\mathscr{H}$ that contains all the values of x, and let $\mathscr{H}_0$ be the separable subspace generated by $\bigcup\{a(s)(\mathscr{H}_x) : s \text{ rational}\}$. The equation $(Ax)(s) = a(s)x(s)$ defines a mapping $Ax: \mathbb{R} \to \mathscr{H}$. When $u \in \mathscr{H}_0^\perp$ and $t \in \mathbb{R}$, $\langle a(s)x(t), u\rangle$ is a continuous function of s that vanishes at all rational s and is therefore zero throughout $\mathbb{R}$. Thus $a(s)x(t) \in \mathscr{H}_0^{\perp\perp} = \mathscr{H}_0$ for real s and t; in particular, $(Ax)(s) = a(s)x(s) \in \mathscr{H}_0$.

Suppose that $\{u_1, u_2, \ldots\}$ is an orthonormal basis of $\mathscr{H}_x$, and E is the projection from $\mathscr{H}$ onto $\mathscr{H}_x$. When $u \in \mathscr{H}$,

$$\begin{aligned}\langle (Ax)(s), u\rangle &= \langle a(s)x(s), u\rangle = \langle x(s), Ea(s)^*u\rangle \\ &= \sum_j \langle x(s), u_j\rangle\langle u_j, Ea(s)^*u\rangle \\ &= \sum_j \langle x(s), u_j\rangle\langle a(s)u_j, u\rangle;\end{aligned}$$

and the right-hand side is measurable on $\mathbb{R}$, as a function of s, since x is measurable and a is weak-operator continuous.

The preceding argument shows that Ax is a measurable mapping from $\mathbb{R}$ into $\mathscr{H}$. If $M = \sup\|a(s)\|$ $(<\infty)$, then $\|(Ax)(s)\| = \|a(s)x(s)\| \le M\|x(s)\|$, and thus

$$\int_{\mathbb{R}} \|(Ax)(s)\|^2\, ds \le M^2 \int_{\mathbb{R}} \|x(s)\|^2\, ds < \infty.$$

Accordingly, A is a bounded linear operator acting on $L_2(\mathbb{R}, \mathscr{H})$, and $\|A\| \le M$.

Suppose next that $a(s) = g(s)A_0$, where $A_0 \in \mathscr{B}(\mathscr{H})$ and g is a bounded continuous complex-valued function on $\mathbb{R}$. Given u in $\mathscr{H}$ and f in $L_2(\mathbb{R})$,

$$\begin{aligned}(A(u \otimes f))(s) &= a(s)(u \otimes f)(s) \\ &= g(s)A_0(f(s)u) \\ &= g(s)f(s)A_0u \\ &= (A_0u \otimes gf)(s) = (A_0u \otimes m_g f)(s).\end{aligned}$$

Thus $A(u \otimes f) = (A_0 \otimes m_g)(u \otimes f)$, whence $A = A_0 \otimes m_g$.

(ii) Suppose now that $a(s) \in \mathscr{M}$ for all real s; and again let $M = \sup\|a(s)\|$. We shall prove below that there is a sequence $\{a_n\}$ of mappings from $\mathbb{R}$ into $\mathscr{M}$, each having the form

$$(1) \qquad a_n(s) = \sum_{j=1}^{k(n)} g_{nj}(s)A_{nj}$$

for suitable bounded continuous real-valued functions g_{nj} on $\mathbb{R}$ and operators A_{nj} in $\mathscr{M}$ $(1 \le j \le k(n))$, such that $\|a_n(s)\| \le M$ and $\{a_n(s)\}$ is weak-operator convergent to $a(s)$ for all real s.

Before proving the existence of such a sequence $\{a_n\}$, we observe that this suffices to complete the proof of the proposition. Indeed, it is apparent that $a_n(s)$ is weak-operator continuous as a function of s (as well as bounded). By (i), the corresponding operator A_n acting on $L_2(\mathbb{R}, \mathscr{H})$ is given by

$$A_n = \sum_{j=1}^{k(n)} A_{nj} \otimes m_{g_{nj}};$$

so $A_n \in \mathscr{M} \mathbin{\overline{\otimes}} \mathscr{A}$. Given x and y in $L_2(\mathbb{R}, \mathscr{H})$,

$$\langle a(s)x(s), y(s)\rangle = \lim_{n\to\infty} \langle a_n(s)x(s), y(s)\rangle,$$

$$|\langle a_n(s)x(s), y(s)\rangle| \leq M\|x(s)\|\ \|y(s)\|,$$

and the function $M\|x(s)\|\ \|y(s)\|$ of s lies in $L_1(\mathbb{R})$. By the dominated convergence theorem

$$\int_{\mathbb{R}} \langle a(s)x(s), y(s)\rangle\, ds = \lim_{n\to\infty} \int_{\mathbb{R}} \langle a_n(s)x(s), y(s)\rangle\, ds;$$

that is, $\langle Ax, y\rangle = \lim\langle A_n x, y\rangle$. Thus A is the weak-operator limit of the sequence $\{A_n\}$, whence $A \in \mathscr{M} \mathbin{\overline{\otimes}} \mathscr{A}$ $(\subseteq \mathscr{M} \mathbin{\overline{\otimes}} \mathscr{B}(L_2(\mathbb{R})))$.

If θ is a * isomorphism from $\mathscr{M}$ onto a von Neumann algebra $\mathscr{N}$ $(\subseteq \mathscr{B}(\mathscr{K}))$, then θ is weak-operator continuous on bounded sets, as is the * isomorphism $\theta \mathbin{\overline{\otimes}} \iota$ from $\mathscr{M} \mathbin{\overline{\otimes}} \mathscr{B}(L_2(\mathbb{R}))$ onto $\mathscr{N} \mathbin{\overline{\otimes}} \mathscr{B}(L_2(\mathbb{R}))$. The equations

$$b(s) = \theta(a(s)), \qquad b_n(s) = \theta(a_n(s))$$

define weak-operator continuous mappings b and b_n, from $\mathbb{R}$ into $\mathscr{N}$, and

$$\|b(s)\| \leq M, \qquad \|b_n(s)\| \leq M, \quad b(s) = \lim b_n(s) \qquad (s \in \mathbb{R}),$$

where the limit is in the weak-operator topology. There are corresponding bounded linear operators B and B_n, acting on $L_2(\mathbb{R}, \mathscr{K})$, defined by

$$(By)(s) = b(s)y(s), \qquad (B_n y)(s) = b_n(s)y(s) \qquad (y \in L_2(\mathbb{R}, \mathscr{K}), \quad s \in \mathbb{R}),$$

and satisfying $\|B\| \leq M$, $\|B_n\| \leq M$. The argument used above to prove that $A = \lim A_n$ can now be applied again to show that $B = \lim B_n$. Since

$$b_n(s) = \theta(a_n(s)) = \sum_{j=1}^{k(n)} g_{nj}(s)\theta(A_{nj}),$$

it follows that

$$B_n = \sum_{j=1}^{k(n)} \theta(A_{nj}) \otimes m_{g_{nj}} = (\theta \mathbin{\overline{\otimes}} \iota)\left(\sum_{j=1}^{k(n)} A_{nj} \otimes m_{g_{nj}}\right) = (\theta \mathbin{\overline{\otimes}} \iota)(A_n);$$

and when $n \to \infty$, we obtain $B = (\theta \mathbin{\overline{\otimes}} \iota)(A)$.

It remains to prove the existence of a sequence $\{a_n\}$ with the stated properties. For each $n = 1, 2, \ldots$, let $\{G_{nj} : j = 1, \ldots, k(n)\}$ be a covering of the closed interval $[-n, n]$ by open intervals of length not exceeding n^{-1}. Let s_{nj} be a point in G_{nj}, and let $A_{nj} = a(s_{nj})$. We can now define $a_n : \mathbb{R} \to \mathscr{M}$ as in (1), choosing the continuous functions g_{nj} so that

$$g_{nj}(s) \geq 0, \qquad \sum_{j=1}^{k(n)} g_{nj}(s) \leq 1 \quad (s \in \mathbb{R}), \qquad \sum_{j=1}^{k(n)} g_{nj}(s) = 1 \quad (s \in [-n, n]),$$

and g_{nj} vanishes outside G_{nj}. Since $\|A_{nj}\| \leq M$, it is apparent that $\|a_n(s)\| \leq M$. When $u, v \in \mathscr{H}$ and $t \in \mathbb{R}$, we have $t \in [-n, n]$, and thus $\sum_j g_{nj}(t) = 1$ for all sufficiently large n. Thus

$$\begin{aligned}
&|\langle a(t)u, v\rangle - \langle a_n(t)u, v\rangle| \\
&\quad = |\langle [a(t) - \sum_{j=1}^{k(n)} g_{nj}(t)a(s_{nj})]u, v\rangle| \\
&\quad = |\langle \sum_{j=1}^{k(n)} g_{nj}(t)[a(t) - a(s_{nj})]u, v\rangle| \\
&\quad \leq \sum_{j=1}^{k(n)} g_{nj}(t)|\langle a(t)u, v\rangle - \langle a(s_{nj})u, v\rangle| \\
&\quad = \sum_{\substack{1 \leq j \leq k(n) \\ t \in G_{nj}}} g_{nj}(t)|\langle a(t)u, v\rangle - \langle a(s_{nj})u, v\rangle| \\
&\quad \leq \max\{|\langle a(t)u, v\rangle - \langle a(s_{nj})u, v\rangle| : 1 \leq j \leq k(n), t \in G_{nj}\} \\
&\quad \leq \max\{|\langle a(t)u, v\rangle - \langle a(s)u, v\rangle| : s \in [t - n^{-1}, t + n^{-1}]\}.
\end{aligned}$$

The right-hand side has limit zero as $n \to \infty$, since the function $\langle a(s)u, v\rangle$ of s is continuous; so $a(t)$ is the weak-operator limit of $\{a_n(t)\}$. ■

We now introduce the operators that will be used to define the (abstract, continuous) crossed product $\mathscr{R}(\mathscr{M}, \alpha)$ and the dual automorphic representation $\hat{\alpha}$ of $\mathbb{R}$ on $\mathscr{R}(\mathscr{M}, \alpha)$, when α is a continuous automorphic representation of $\mathbb{R}$ on $\mathscr{M}$ $(\subseteq \mathscr{B}(\mathscr{H}))$. For each A in $\mathscr{M}$, the mapping $s \to \alpha_s^{-1}(A) : \mathbb{R} \to \mathscr{M}$ is bounded and weak-operator continuous. By Proposition 13.2.2, the equation

$$(2) \qquad (\Psi(A)x)(s) = \alpha_s^{-1}(A)x(s) \qquad (x \in L_2(\mathbb{R}, \mathscr{H}), \quad s \in \mathbb{R})$$

defines a bounded linear operator $\Psi(A)$ acting on $L_2(\mathbb{R}, \mathscr{H})$; and $\Psi(A) \in \mathscr{M} \bar{\otimes} \mathscr{A}$, where $\mathscr{A}$ is the multiplication algebra $\{m_g : g \in L_\infty(\mathbb{R})\}$. For all real t and p, the equations

$$(3) \qquad (l_t f)(s) = f(s - t), \qquad (w_p f)(s) = e^{-isp} f(s),$$

$$(4) \qquad (L_t x)(s) = x(s - t), \qquad (W_p x)(s) = e^{-isp} x(s)$$

($f \in L_2(\mathbb{R})$, $x \in L_2(\mathbb{R}, \mathscr{H})$, $s \in \mathbb{R}$) define unitary operators, l_t and w_p acting on $L_2(\mathbb{R})$, L_t and W_p acting on $L_2(\mathbb{R}, \mathscr{H})$; and $w_p \in \mathscr{A}$.

If we were developing the theory in the more general context of a continuous automorphic representation of a locally compact abelian group G, these unitary operators would be defined for all t in G and all p in the dual group $\hat{G}$. The present treatment deals only with the case in which $G = \mathbb{R}$, however; and $\hat{\mathbb{R}}$ is identified with $\mathbb{R}$ by means of the duality $(t, p) \to \exp itp$.

13.2.3. Proposition. (i) *The mapping Ψ, defined by* (2), *is a * isomorphism from $\mathscr{M}$ onto a von Neumann subalgebra of $\mathscr{M} \bar{\otimes} \mathscr{A}$, where $\mathscr{A}$ is the multiplication algebra $\{m_g : g \in L_\infty(\mathbb{R})\}$ acting on $L_2(\mathbb{R})$.*

(ii) *Each of the mappings $t \to l_t$, $t \to L_t$, $p \to w_p$, $p \to W_p$ is a continuous unitary representation of $\mathbb{R}$, and $L_t = I \otimes l_t$, $W_p = I \otimes w_p$.*

(iii) *When $A \in \mathscr{M}$ and $t, p \in \mathbb{R}$,*

$$w_p l_t w_p^* = e^{-itp} l_t, \qquad L_t \Psi(A) L_t^* = \Psi(\alpha_t(A)),$$

$$W_p \Psi(A) W_p^* = \Psi(A), \qquad W_p L_t W_p^* = e^{-itp} L_t.$$

Proof. (i) It is apparent that Ψ is a * homomorphism, and we have already noted that $\Psi(\mathscr{M}) \subseteq \mathscr{M} \bar{\otimes} \mathscr{A}$. If $A \in \mathscr{M}$ and $\Psi(A) = 0$, then

$$\int_{\mathbb{R}} |f(s)|^2 \|\alpha_s^{-1}(A)u\|^2 \, ds = \|\Psi(A)(u \otimes f)\|^2 = 0$$

for all u in $\mathscr{H}$ and f in $L_2(\mathbb{R})$. It follows that the (continuous) function $\|\alpha_s^{-1}(A)u\|$ of s vanishes throughout $\mathbb{R}$; with $s = 0$, we obtain $Au = 0$ ($u \in \mathscr{H}$), so $A = 0$. Hence Ψ is a * isomorphism from $\mathscr{M}$ into $\mathscr{M} \bar{\otimes} \mathscr{A}$.

In order to prove that $\Psi(\mathscr{M})$ is a von Neumann subalgebra of $\mathscr{M} \bar{\otimes} \mathscr{A}$, it suffices to verify that its unit ball $\Psi((\mathscr{M})_1)$ is weak-operator compact; and this will follow if we show that $\Psi | (\mathscr{M})_1$ is weak-operator continuous. Vectors of the type $x = \sum_1^n u_j \otimes f_j$, where $u_1, \ldots, u_n \in \mathscr{H}$ and each f_j ($\in L_2(\mathbb{R})$) is continuous and has compact support, form an everywhere-dense subset of $L_2(\mathbb{R}, \mathscr{H})$ ($= \mathscr{H} \otimes L_2(\mathbb{R})$). Accordingly, the corresponding vector states determine the weak-operator topology on $\Psi((\mathscr{M})_1)$; and it suffices to show, for such x, that the state $\omega_x \circ \Psi$ of $\mathscr{M}$ is weak-operator continuous on $(\mathscr{M})_1$.

If $\{A_a\}$ is a bounded increasing net of self-adjoint elements of $\mathscr{M}$, with limit A, then

$$\begin{aligned}\langle(\Psi(A_a)x)(s), x(s)\rangle &= \langle\alpha_s^{-1}(A_a)x(s), x(s)\rangle \\ &= \sum_{j,k=1}^{n} f_j(s)\overline{f_k(s)}\langle\alpha_s^{-1}(A_a)u_j, u_k\rangle\end{aligned}$$

for all real s; and there is a similar equation with A in place of A_a. From this, and since the * automorphisms α_s^{-1} of $\mathscr{M}$ are weak-operator continuous on bounded sets, it follows that the functions $\langle(\Psi(A_a)x)(s), x(s)\rangle$ of s are continuous, have supports inside a fixed compact subset of $\mathbb{R}$, and form an increasing net that converges to the continuous function $\langle(\Psi(A)x)(s), x(s)\rangle$. By Dini's theorem, the convergence is uniform, and thus

$$\int_{\mathbb{R}}\langle(\Psi(A_a)x)(s), x(s)\rangle\, ds \underset{a}{\rightarrow} \int_{\mathbb{R}}\langle(\Psi(A)x)(s), x(s)\rangle\, ds;$$

that is, $\langle\Psi(A_a)x, x\rangle \underset{a}{\rightarrow} \langle\Psi(A)x, x\rangle$. This shows that the state $\omega_x \circ \Psi$ of $\mathscr{M}$ is normal, and therefore weak-operator continuous on $(\mathscr{M})_1$.

(ii) Straightforward calculations show that the mappings $t \rightarrow l_t$, $p \rightarrow w_p$ are unitary representations of $\mathbb{R}$ on $L_2(\mathbb{R})$. When $u \in \mathscr{H}$ and $f \in L_2(\mathbb{R})$,

$$\begin{aligned}(L_t(u \otimes f))(s) &= (u \otimes f)(s - t) = f(s - t)u \\ &= (l_t f)(s)u = (u \otimes l_t f)(s);\end{aligned}$$

so $L_t(u \otimes f) = u \otimes l_t f$, and $L_t = I \otimes l_t$. In the discussion preceding Proposition 13.2.2, we have noted that $M_g = I \otimes m_g$ $(g \in L_\infty(\mathbb{R}))$; and by taking $g(s) = \exp(-isp)$, we obtain $W_p = I \otimes w_p$.

It now suffices to show that the unitary representations $t \rightarrow l_t$, $p \rightarrow w_p$ are continuous. Since continuous functions f with compact support form an everywhere-dense subset of $L_2(\mathbb{R})$, they determine the strong-operator topology on bounded sets in $\mathscr{B}(L_2(\mathbb{R}))$. For such f,

$$\lim_{t\rightarrow 0}\|l_t f - f\|^2 = \lim_{t\rightarrow 0}\int_{\mathbb{R}} |f(s - t) - f(s)|^2\, ds = 0,$$

$$\lim_{p\rightarrow 0}\|w_p f - f\|^2 = \lim_{p\rightarrow 0}\int_{\mathbb{R}} |e^{isp} - 1|^2 |f(s)|^2\, ds = 0,$$

since (after confining attention to small values of t in the first equation) the integrand in both cases vanishes outside a fixed compact set and converges uniformly to 0.

(iii) The required relations may all be verified by straightforward calculation. ■

By Proposition 13.2.3(i) and (ii), the operators $\Psi(A)$ $(A \in \mathcal{M})$ and L_t $(t \in \mathbb{R})$ all lie in $\mathcal{M} \,\overline{\otimes}\, \mathcal{B}(L_2(\mathbb{R}))$. From this, together with part (iii) of that proposition, it is easily verified that the set of all finite sums of operators of the form $L_t \Psi(A)$ is a * subalgebra $\mathcal{R}_0$ of $\mathcal{M} \,\overline{\otimes}\, \mathcal{B}(L_2(\mathbb{R}))$, and $W_p \mathcal{R}_0 W_p^* = \mathcal{R}_0$ for all real p. Accordingly, the von Neumann algebra generated by the set $\{\Psi(A), L_t : A \in \mathcal{M}, t \in \mathbb{R}\}$ is the * subalgebra $\mathcal{R}_0^-$ of $\mathcal{M} \,\overline{\otimes}\, \mathcal{B}(L_2(\mathbb{R}))$; and W_p implements a * automorphism $\hat{\alpha}_p$ of $\mathcal{R}_0^-$. Since $p \to W_p$ is a continuous unitary representation of $\mathbb{R}$, $\hat{\alpha} : p \to \hat{\alpha}_p$ is a continuous automorphic representation.

13.2.4. Definition (Abstract continuous crossed product). If $\mathcal{M}$ is a von Neumann algebra acting on a Hilbert space $\mathcal{H}$, and α is a continuous automorphic representation of $\mathbb{R}$ on $\mathcal{M}$, the (*abstract, continuous*) *crossed product* of $\mathcal{M}$ by α is the von Neumann algebra $\mathcal{R}(\mathcal{M}, \alpha)$, acting on $L_2(\mathbb{R}, \mathcal{H})$ $(= \mathcal{H} \otimes L_2(\mathbb{R}))$, that is generated by the operators $\Psi(A)$, L_t $(A \in \mathcal{M}, t \in \mathbb{R})$ defined in (2) and (4). The *dual representation* $\hat{\alpha}$ is the continuous automorphic representation $p \to \hat{\alpha}_p$ of $\mathbb{R}$, where $\hat{\alpha}_p$ is the * automorphism of $\mathcal{R}(\mathcal{M}, \alpha)$ that is implemented by the unitary operator W_p defined in (4). ■

13.2.5. Proposition. *Suppose that $\alpha: t \to \alpha_t$ is a continuous automorphic representation of $\mathbb{R}$ on a von Neumann algebra $\mathcal{M}$, and $\hat{\alpha}: p \to \hat{\alpha}_p$ is the dual representation of $\mathbb{R}$ on $\mathcal{R}(\mathcal{M}, \alpha)$.*

(i) *If θ is a * isomorphism from $\mathcal{M}$ onto a von Neumann algebra $\mathcal{N}$, the equation $\beta_t = \theta \circ \alpha_t \circ \theta^{-1}$ defines a continuous automorphic representation $\beta: t \to \beta_t$ of $\mathbb{R}$ on $\mathcal{N}$. There is a * isomorphism φ, from $\mathcal{R}(\mathcal{M}, \alpha)$ onto $\mathcal{R}(\mathcal{N}, \beta)$, such that the dual representation $\hat{\beta}$ of $\mathbb{R}$ on $\mathcal{R}(\mathcal{N}, \beta)$ is given by $\hat{\beta}_p = \varphi \circ \hat{\alpha}_p \circ \varphi^{-1}$; and if θ is unitarily implemented, then so is φ.*

(ii) *Under the conditions set out in* (i), *the von Neumann algebras $\mathcal{R}(\mathcal{R}(\mathcal{M}, \alpha), \hat{\alpha})$ and $\mathcal{R}(\mathcal{R}(\mathcal{N}, \beta), \hat{\beta})$ are * isomorphic; and they are unitarily equivalent if θ is unitarily implemented.*

(iii) *There is a * isomorphism θ, from $\mathcal{M}$ onto a von Neumann algebra $\mathcal{N}$, such that the continuous automorphic representation $t \to \theta \circ \alpha_t \circ \theta^{-1}$ of $\mathbb{R}$ on $\mathcal{N}$ is unitarily implemented.*

Proof. (i) Since θ is weak-operator continuous on bounded sets in $\mathcal{M}$, the automorphic representation $\beta: t \to \beta_t = \theta \circ \alpha_t \circ \theta^{-1}$ of $\mathbb{R}$ on $\mathcal{N}$ is continuous. Corresponding to the operators $\Psi(A)$, L_t, W_p defined in (2) and (4), there are similar operators $\Psi^{(\beta)}(B)$ $(B \in \mathcal{N})$, $L_t^{(\beta)}$, $W_p^{(\beta)}$ constructed from the representation β. Since $\theta(\alpha_s^{-1}(A)) = \beta_s^{-1}(\theta(A))$, it follows from Proposition 13.2.2(ii) that $(\theta \mathbin{\overline{\otimes}} \iota)(\Psi(A)) = \Psi^{(\beta)}(\theta(A))$ for each A in $\mathscr{M}$, where ι is the identity automorphism on $\mathscr{B}(L_2(\mathbb{R}))$. Moreover, $(\theta \mathbin{\overline{\otimes}} \iota)(L_t) = L_t^{(\beta)}$ and $(\theta \mathbin{\overline{\otimes}} \iota)(W_p) = W_p^{(\beta)}$, since $L_t = I \otimes l_t$, $W_p = I \otimes w_p$ (and there are similar equations for $L_t^{(\beta)}$, $W_p^{(\beta)}$). Accordingly, $\theta \mathbin{\overline{\otimes}} \iota$ carries $\mathscr{R}(\mathscr{M}, \alpha)$ onto $\mathscr{R}(\mathscr{N}, \beta)$, and

$$\begin{aligned}(\theta \mathbin{\overline{\otimes}} \iota)(\hat{\alpha}_p(R)) &= (\theta \mathbin{\overline{\otimes}} \iota)(W_p R W_p^*)\\ &= W_p^{(\beta)}(\theta \mathbin{\overline{\otimes}} \iota)(R) W_p^{(\beta)*} = \hat{\beta}_p((\theta \mathbin{\overline{\otimes}} \iota)(R))\end{aligned}$$

for each R in $\mathscr{R}(\mathscr{M}, \alpha)$. By restriction, $\theta \mathbin{\overline{\otimes}} \iota$ induces a * isomorphism φ from $\mathscr{R}(\mathscr{M}, \alpha)$ onto $\mathscr{R}(\mathscr{N}, \beta)$, and $\hat{\beta}_p = \varphi \circ \hat{\alpha}_p \circ \varphi^{-1}$; and if θ is implemented by a unitary operator U, then φ is implemented by $U \otimes I$.

(ii) Upon replacing $\mathscr{M}$, $\mathscr{N}$, α, and θ by $\mathscr{R}(\mathscr{M}, \alpha)$, $\mathscr{R}(\mathscr{N}, \beta)$, $\hat{\alpha}$, and φ, respectively, (ii) is an immediate consequence of (i).

(iii) From Proposition 13.2.3, the continuous unitary representation $t \to L_t$ of $\mathbb{R}$ implements the automorphic representation $t \to \Psi \circ \alpha_t \circ \Psi^{-1}$ of $\mathbb{R}$ on the von Neumann algebra $\Psi(\mathscr{M})$. ■

Suppose that the continuous automorphic representation $\alpha: t \to \alpha_t$ of $\mathbb{R}$ on $\mathscr{M}$ $(\subseteq \mathscr{B}(\mathscr{H}))$ is implemented by a continuous unitary representation $t \to U(t)$ of $\mathbb{R}$ on $\mathscr{H}$. From Proposition 13.2.2(i), the equation

$$(Ux)(s) = U(s)x(s) \tag{5}$$

defines a bounded linear operator U acting on $L_2(\mathbb{R}, \mathscr{H})$; and it is apparent that U is unitary. When $x \in L_2(\mathbb{R}, \mathscr{H})$,

$$\begin{aligned}(U\Psi(A)U^*x)(s) &= U(s)(\Psi(A)U^*x)(s) = U(s)\alpha_s^{-1}(A)(U^*x)(s)\\ &= U(s)\alpha_s^{-1}(A)U(s)^*x(s) = Ax(s) = ((A \otimes I)x)(s),\\ (UL_tU^*x)(s) &= U(s)(L_tU^*x)(s) = U(s)(U^*x)(s-t)\\ &= U(s)U(s-t)^*x(s-t) = U(t)(L_tx)(s)\\ &= ((U(t) \otimes I)L_tx)(s) = ((U(t) \otimes I)(I \otimes l_t)x)(s)\\ &= ((U(t) \otimes l_t)x)(s)\end{aligned}$$

When $g \in L_\infty(\mathbb{R})$ and M_g $(= I \otimes m_g)$ is defined as in the discussion preceding Proposition 13.2.2,

$$\begin{aligned}(UM_gU^*x)(s) &= U(s)(M_gU^*x)(s) = g(s)U(s)(U^*x)(s)\\ &= g(s)x(s) = (M_gx)(s).\end{aligned}$$

Hence

(6) $U\Psi(A)U^* = A \otimes I, \quad UL_tU^* = U(t) \otimes l_t, \quad U(I \otimes m_g)U^* = I \otimes m_g,$

whenever $A \in \mathscr{M}, t \in \mathbb{R}$, and $g \in L_\infty(\mathbb{R})$. With g defined by $g(s) = \exp(-isp)$, we obtain $UW_pU^* = W_p$.

From (6), $U\mathscr{R}(\mathscr{M}, \alpha)U^*$ is the von Neumann algebra generated by the operators $A \otimes I$, $U(t) \otimes l_t$ $(A \in \mathscr{M}, t \in \mathbb{R})$. Moreover, if φ denotes the * isomorphism $R \to URU^*$ from $\mathscr{R}(\mathscr{M}, \alpha)$ onto $U\mathscr{R}(\mathscr{M}, \alpha)U^*$, then

$$\varphi(\hat{\alpha}_p(R)) = UW_pRW_p^*U^* = W_pURU^*W_p^* = W_p\varphi(R)W_p^*$$

for each R in $\mathscr{R}(\mathscr{M}, \alpha)$. It follows that the continuous unitary representation $p \to W_p$ of $\mathbb{R}$ implements the continuous automorphic representation $\beta: p \to \varphi \circ \hat{\alpha}_p \circ \varphi^{-1}$ of $\mathbb{R}$ on $U\mathscr{R}(\mathscr{M}, \alpha)U^*$. By Proposition 13.2.5(i), with $\mathscr{M}, \mathscr{N}, \alpha, \theta$ replaced by $\mathscr{R}(\mathscr{M}, \alpha)$, $U\mathscr{R}(\mathscr{M}, \alpha)U^*$, $\hat{\alpha}$, φ, respectively, the crossed product $\mathscr{R}(U\mathscr{R}(\mathscr{M}, \alpha)U^*, \beta)$ is unitarily equivalent to $\mathscr{R}(\mathscr{R}(\mathscr{M}, \alpha), \hat{\alpha})$.

In view of the preceding discussion, it is not unreasonable to view $U\mathscr{R}(\mathscr{M}, \alpha)U^*$ as the "implemented" crossed product of $\mathscr{M}$ by α, with β as the corresponding dual representation, since these are obtained by unitary transformation from the "abstract" version. The following definition (which involves slight ambiguity of notation) effects this.

13.2.6. Definition (Implemented continuous crossed product). If $\mathscr{M}$ is a von Neumann algebra acting on a Hilbert space $\mathscr{H}$, α is a continuous automorphic representation of $\mathbb{R}$ on $\mathscr{M}$, and α is implemented by a continuous unitary representation $t \to U(t)$ of $\mathbb{R}$ on $\mathscr{H}$, the *(implemented, continuous) crossed product* of $\mathscr{M}$ by α is the von Neumann algebra $\mathscr{R}(\mathscr{M}, \alpha)$, acting on $L_2(\mathbb{R}, \mathscr{H})\,(= \mathscr{H} \otimes L_2(\mathbb{R}))$, that is generated by the operators $A \otimes I$, $U(t) \otimes l_t\,(A \in \mathscr{M},\ t \in \mathbb{R})$. The *dual representation* $\hat{\alpha}$ is the continuous automorphic representation $p \to \hat{\alpha}_p$ of $\mathbb{R}$, where $\hat{\alpha}_p$ is the * automorphism of $\mathscr{R}(\mathscr{M}, \alpha)$ that is implemented by W_p. ■

We have four candidates for the second crossed product $\mathscr{R}(\mathscr{R}(\mathscr{M}, \alpha), \hat{\alpha})$, when α is unitarily implemented; for we must first decide whether to take $\mathscr{R}(\mathscr{M}, \alpha)$ (together with the dual action $\hat{\alpha}$) in the abstract or implemented sense, and in each case we have the same choice in forming the crossed product of $\mathscr{R}(\mathscr{M}, \alpha)$ by $\hat{\alpha}$. However, from the discussion preceding Definition 13.2.6, it is apparent that all four algebras so obtained are unitarily equivalent. When α is not unitarily implemented, $\mathscr{R}(\mathscr{M}, \alpha)$ and $\hat{\alpha}$ must be taken in the abstract version, but the crossed product of $\mathscr{R}(\mathscr{M}, \alpha)$ by $\hat{\alpha}$ can be taken in either sense; once again, the algebras so obtained are unitarily equivalent.

In proving the duality theorem, we make use of some results of the theory of Fourier transforms proved in Section 3.2, *The Banach algebra* $L_1(\mathbb{R})$ *and Fourier analysis*. We prove a proposition concerning the operators l_t and w_p, defined in (3), with the aid of Corollary 3.2.28(ii) (the "uniqueness theorem").

13.2.7. PROPOSITION. (i) *The weak-operator-closed linear subspace of* $\mathscr{B}(L_2(\mathbb{R}))$ *generated by the operators* w_p $(p \in \mathbb{R})$ *is the multiplication algebra* $\mathscr{A} = \{m_g : g \in L_\infty(\mathbb{R})\}$.

(ii) *The von Neumann algebra generated by the operators* w_p, l_t $(p, t \in \mathbb{R})$ *is* $\mathscr{B}(L_2(\mathbb{R}))$.

Proof. (i) Since $\mathscr{A}$ is weak-operator closed and $w_p \in \mathscr{A}$, it suffices to show that the condition $\omega(w_p) = 0$ $(p \in \mathbb{R})$, for a weak-operator continuous linear functional ω on $\mathscr{A}$, entails $\omega = 0$. Given such an ω, we may choose functions $f_1, \ldots, f_n, g_1, \ldots, g_n$ in $L_2(\mathbb{R})$, so that $\omega(A) = \sum \langle Af_j, g_j \rangle$ for each A in $\mathscr{A}$. Now every element of $\mathscr{A}$ has the form m_h, for some h in $L_\infty(\mathbb{R})$, and

$$\omega(m_h) = \sum_{j=1}^{n} \int_{\mathbb{R}} h(s) f_j(s) \overline{g_j(s)}\, ds = \int_{\mathbb{R}} h(s) k(s)\, ds$$

where k is the L_1 function defined by $k(s) = \sum f_j(s)\overline{g_j(s)}$. When $h(s) = \exp isp$, m_h is w_{-p}; and we obtain

$$\hat{k}(p) = (2\pi)^{-1/2} \int_{\mathbb{R}} e^{isp} k(s)\, ds = (2\pi)^{-1/2}\omega(w_{-p}) = 0$$

for all real p. By the uniqueness theorem, k is a null function; so $\omega(m_h) = \int_{\mathbb{R}} h(s)k(s)\, ds = 0$ for all h in $L_\infty(\mathbb{R})$, and thus $\omega = 0$.

(ii) Let $\mathscr{R}$ be the von Neumann algebra generated by the operators w_p, l_t $(p, t \in \mathbb{R})$. By (i), $\mathscr{R}$ contains the maximal abelian von Neumann algebra $\mathscr{A}$; so $\mathscr{R}' \subseteq \mathscr{A}' = \mathscr{A}$. If E' is a projection in $\mathscr{R}'$, then $E' = m_h$, where h is the characteristic function of a Borel subset X of $\mathbb{R}$. Since $E' = l_t E' l_t^* = l_t m_h l_t^*$, and $l_t m_h l_t^*$ is the operator of multiplication by the characteristic function of the set $X + t = \{s + t : s \in X\}$, the symmetric difference of the sets X and $X + t$ has Lebesgue measure zero, for all real t. We have already noted, in Example 8.6.13, that the group of (rational) translations of $\mathbb{R}$ acts ergodically, for Lebesgue measure. Thus either X or $\mathbb{R}\backslash X$ is a null set, and E' is either 0 or I. Accordingly, $\mathscr{R}'$ consists of scalars, and $\mathscr{R} = \mathscr{B}(L_2(\mathbb{R}))$. ■

We recall the discussion and notation of Theorem 3.2.31 and note that, if f is a continuous complex-valued function on $\mathbb{R}$ with support in a finite interval and p, q, s, t are real, then

$$\begin{aligned}(w_tTf)(q) &= e^{-itq}(Tf)(q) = e^{-itq}\hat{f}(q)\\ &= (2\pi)^{-1/2}\int_{\mathbb{R}} e^{i(s-t)q}f(s)\,ds = (2\pi)^{-1/2}\int_{\mathbb{R}} e^{isq}f(s+t)\,ds\\ &= (2\pi)^{-1/2}\int_{\mathbb{R}} e^{isq}(l_{-t}f)(s)\,ds = (Tl_{-t}f)(q),\\ (l_pTf)(q) &= (Tf)(q-p) = \hat{f}(q-p)\\ &= (2\pi)^{-1/2}\int_{\mathbb{R}} e^{iqs}e^{-ips}f(s)\,ds\\ &= (2\pi)^{-1/2}\int_{\mathbb{R}} e^{isq}(w_pf)(s)\,ds = (Tw_pf)(q).\end{aligned}$$

Hence,

$$(7) \qquad w_tT = Tl_{-t}\ , \qquad l_pT = Tw_p.$$

We summarize, as a theorem, the results of the preceding discussion.

13.2.8. THEOREM. *There is a unitary operator T, from $L_2(\mathbb{R})$ onto $L_2(\mathbb{R})$, such that*

$$(Tf)(p) = (2\pi)^{-1/2}\int_{\mathbb{R}} e^{isp}f(s)\,ds \qquad (p\in\mathbb{R})$$

when f is continuous and has compact support. Moreover,

$$w_t = Tl_{-t}T^*, \qquad l_p = Tw_pT^* \qquad (t,\ p\in\mathbb{R}).$$

We can now state and prove the duality theorem.

13.2.9. THEOREM. *If $\mathscr{M}$ is a von Neumann algebra acting on a Hilbert space $\mathscr{H}$, $\alpha\colon t\to\alpha_t$ is a continuous automorphic representation of $\mathbb{R}$ on $\mathscr{M}$, and $\hat{\alpha}\colon p\to\hat{\alpha}_p$ is the dual representation of $\mathbb{R}$ on $\mathscr{R}(\mathscr{M},\alpha)$, then the second (continuous) crossed product $\mathscr{R}(\mathscr{R}(\mathscr{M},\alpha),\hat{\alpha})$ is * isomorphic to $\mathscr{M}\,\overline{\otimes}\,\mathscr{B}(L_2(\mathbb{R}))$.*

Proof. By the discussion following Definition 13.2.6, the various possible interpretations of the second crossed product yield unitarily equivalent von Neumann algebras; so it does not matter which of them we use. From Proposition 13.2.5, there is a von Neumann algebra $\mathscr{N}$, * isomorphic to $\mathscr{M}$, and a unitarily implemented continuous automorphic representation β of $\mathbb{R}$ on $\mathscr{N}$, such that $\mathscr{R}(\mathscr{R}(\mathscr{M},\alpha),\hat{\alpha})$ is * isomorphic to $\mathscr{R}(\mathscr{R}(\mathscr{N},\beta),\hat{\beta})$. Of course,

$\mathscr{M} \mathbin{\overline{\otimes}} \mathscr{B}(L_2(\mathbb{R}))$ is * isomorphic to $\mathscr{N} \mathbin{\overline{\otimes}} \mathscr{B}(L_2(\mathbb{R}))$; so upon replacing $\mathscr{M}$ and α by $\mathscr{N}$ and β, we reduce to the case in which α is implemented by a continuous unitary representation $t \to U(t)$ of $\mathbb{R}$ on $\mathscr{H}$.

We adopt the "implemented" interpretation, for both crossed products (and hence, also, for the dual representation $\hat{\alpha}$); we write $\mathscr{R}_0$ for $\mathscr{R}(\mathscr{R}(\mathscr{M}, \alpha), \hat{\alpha})$, and use the notation introduced in (2), (3), and (4). Thus $\mathscr{R}(\mathscr{M}, \alpha)$ acts on $\mathscr{H} \otimes L_2(\mathbb{R})$ $(= L_2(\mathbb{R}, \mathscr{H}))$ and is generated by the operators $A \otimes I$, $U(t) \otimes l_t$ $(A \in \mathscr{M}, t \in \mathbb{R})$, while the automorphism $\hat{\alpha}_p$ is implemented by W_p $(= I \otimes w_p)$. The second crossed product $\mathscr{R}_0$ acts on $L_2(\mathbb{R}, \mathscr{H}) \otimes L_2(\mathbb{R})$, and is generated by the operators $R \otimes I$, $W_p \otimes l_p$ $(R \in \mathscr{R}(\mathscr{M}, \alpha), p \in \mathbb{R})$. If we again identify $L_2(\mathbb{R}, \mathscr{H})$ with $\mathscr{H} \otimes L_2(\mathbb{R})$, it follows that $\mathscr{R}_0$ acts on the space $\mathscr{H} \otimes L_2(\mathbb{R}) \otimes L_2(\mathbb{R})$, and is generated by the operators

$$A \otimes I \otimes I, \qquad U(t) \otimes l_t \otimes I, \qquad I \otimes w_p \otimes l_p \qquad (A \in \mathscr{M}; \quad t, p \in \mathbb{R}).$$

Let T be the Fourier transform, considered as a unitary operator acting on $L_2(\mathbb{R})$; and let V_1 be $I \otimes I \otimes T$, acting on $\mathscr{H} \otimes L_2(\mathbb{R}) \otimes L_2(\mathbb{R})$. Since $T^* l_p T = w_p$, $V_1^* \mathscr{R}_0 V_1$ is the von Neumann algebra $\mathscr{R}_1$ generated by the operators

$$A \otimes I \otimes I, \qquad U(t) \otimes l_t \otimes I, \qquad I \otimes w_p \otimes w_p \qquad (A \in \mathscr{M}; \quad t, p \in \mathbb{R}).$$

In accordance with Proposition 13.2.1, we now identify $L_2(\mathbb{R}) \otimes L_2(\mathbb{R})$ with $L_2(\mathbb{R}, L_2(\mathbb{R}))$ in such a way that, when $f, g \in L_2(\mathbb{R})$, the element $f \otimes g$ of $L_2(\mathbb{R}, L_2(\mathbb{R}))$ is given by $(f \otimes g)(s) = g(s)f$. By Proposition 13.2.2(i), when $x \in L_2(\mathbb{R}, L_2(\mathbb{R}))$ we have

$$((w_p \otimes I)x)(s) = w_p x(s),$$
$$((l_r \otimes I)x)(s) = l_r x(s),$$
$$((w_p \otimes w_p)x)(s) = e^{-isp} w_p x(s)$$

(for the third equation, recall that w_p is the operator on $L_2(\mathbb{R})$ of multiplication by the bounded continuous function e^{-isp}). Again by Proposition 13.2.2(i), the equation

$$(Vx)(s) = l_s x(s)$$

defines a unitary operator V on $L_2(\mathbb{R}, L_2(\mathbb{R}))$. Since $w_p l_s w_p^* = e^{-isp} l_s$ by Proposition 13.2.3(iii), we have

$$\begin{aligned} (V(w_p \otimes w_p)x)(s) &= l_s((w_p \otimes w_p)x)(s) = e^{-isp} l_s w_p x(s) \\ &= w_p l_s x(s) = w_p (Vx)(s) = ((w_p \otimes I)Vx)(s), \end{aligned}$$

$$\begin{aligned} (V(l_r \otimes I)x)(s) &= l_s((l_r \otimes I)x)(s) = l_s l_r x(s) \\ &= l_r l_s x(s) = l_r (Vx)(s) = ((l_r \otimes I)Vx)(s). \end{aligned}$$

It follows that $V(w_p \otimes w_p)V^* = w_p \otimes I$ and $V(l_r \otimes I)V^* = l_r \otimes I$. Accordingly, if V_2 is the unitary operator $I \otimes V$ acting on $\mathscr{H} \otimes L_2(\mathbb{R}, L_2(\mathbb{R}))$ $(= \mathscr{H} \otimes L_2(\mathbb{R}) \otimes L_2(\mathbb{R}))$, the von Neumann algebra $V_2 \mathscr{R}_1 V_2^*$ $(= V_2 V_1^* \mathscr{R}_0 V_1 V_2^*)$ is generated by the operators

$$A \otimes I \otimes I, \qquad U(t) \otimes l_t \otimes I, \qquad I \otimes w_p \otimes I \qquad (A \in \mathscr{M}; \quad t, p \in \mathbb{R}).$$

Hence $\mathscr{R}_0$ is * isomorphic to the von Neumann algebra $\mathscr{R}_2$, acting on $\mathscr{H} \otimes L_2(\mathbb{R})$, that is generated by the operators

$$A \otimes I, \qquad U(t) \otimes l_t, \qquad I \otimes w_p \qquad (A \in \mathscr{M}; \quad t, p \in \mathbb{R}).$$

Equation (5) defines a unitary operator U acting on $L_2(\mathbb{R}, \mathscr{H})$ $(= \mathscr{H} \otimes L_2(\mathbb{R}))$, and $\mathscr{R}_0$ is * isomorphic to $U^* \mathscr{R}_2 U$ $(= \mathscr{R}_3)$; so it now suffices to show that $\mathscr{R}_3 = \mathscr{M} \mathbin{\overline{\otimes}} \mathscr{B}(L_2(\mathbb{R}))$. Since $L_t = I \otimes l_t$, while w_p is a multiplication operator, it follows from (6) that $\mathscr{R}_3$ is the von Neumann algebra generated by the operators

$$\Psi(A), \qquad I \otimes l_t, \qquad I \otimes w_p \qquad (A \in \mathscr{M}; \quad t, p \in \mathbb{R}).$$

Now $\Psi(A) \in \mathscr{M} \mathbin{\overline{\otimes}} \mathscr{B}(L_2(\mathbb{R}))$, and the operators $I \otimes l_t$, $I \otimes w_p$ generate $\mathbb{C}_{\mathscr{H}} \mathbin{\overline{\otimes}} \mathscr{B}(L_2(\mathbb{R}))$, by Propositions 13.2.3(i), 13.2.7(ii); so

$$\mathbb{C}_{\mathscr{H}} \mathbin{\overline{\otimes}} \mathscr{B}(L_2(\mathbb{R})) \subseteq \mathscr{R}_3 \subseteq \mathscr{M} \mathbin{\overline{\otimes}} \mathscr{B}(L_2(\mathbb{R})).$$

If $S \in \mathscr{R}_3'$, then S commutes with $\mathbb{C}_{\mathscr{H}} \mathbin{\overline{\otimes}} \mathscr{B}(L_2(\mathbb{R}))$; so $S = B \otimes I$ for some B in $\mathscr{B}(\mathscr{H})$, and thus $(Sx)(s) = Bx(s)$ $(x \in L_2(\mathbb{R}, \mathscr{H}),\ s \in \mathbb{R})$. Given u in $\mathscr{H}$, let $x = u \otimes f$, where f is a continuous function in $L_2(\mathbb{R})$, and $f(0) = 1$. Since S commutes with $\Psi(A)$ for each A in $\mathscr{M}$, while

$$(S\Psi(A)x)(s) = B\alpha_s^{-1}(A)x(s) = f(s)B\alpha_s^{-1}(A)u,$$

$$(\Psi(A)Sx)(s) = \alpha_s^{-1}(A)Bx(s) = f(s)\alpha_s^{-1}(A)Bu,$$

it follows that $f(s)B\alpha_s^{-1}(A)u = f(s)\alpha_s^{-1}(A)Bu$ for almost all s in $\mathbb{R}$. Since the functions of s occurring in this equation are continuous, equality occurs for all real s; and when $s = 0$, we obtain $BAu = ABu$ $(u \in \mathscr{H}, A \in \mathscr{M})$. Hence $B \in \mathscr{M}'$, $S = B \otimes I \in (\mathscr{M} \mathbin{\overline{\otimes}} \mathscr{B}(L_2(\mathbb{R})))'$, and $\mathscr{R}_3' \subseteq (\mathscr{M} \mathbin{\overline{\otimes}} \mathscr{B}(L_2(\mathbb{R})))'$. Thus $\mathscr{R}_3 \supseteq \mathscr{M} \mathbin{\overline{\otimes}} \mathscr{B}(L_2(\mathbb{R}))$, and the reverse inclusion has already been proved. ■

13.2.10. COROLLARY. *If α is a continuous automorphic representation of $\mathbb{R}$ on a properly infinite von Neumann algebra $\mathscr{M}$, and $\hat{\alpha}$ is the dual representation of $\mathbb{R}$ on $\mathscr{R}(\mathscr{M}, \alpha)$, then the second (continuous) crossed product $\mathscr{R}(\mathscr{R}(\mathscr{M}, \alpha), \hat{\alpha})$ is * isomorphic to $\mathscr{M}$.*

Proof. From the duality theorem, it suffices to show that $\mathscr{M}$ is * isomorphic to $\mathscr{M} \mathbin{\overline{\otimes}} \mathscr{B}(L_2(\mathbb{R}))$. By repeated application of the halving lemma

(6.3.3), as in the proof of Theorem 6.3.4, we can find an orthogonal sequence $\{G_j\}$ of projections in $\mathscr{M}$, with sum I and each equivalent to I.

Since $G_1 \sim I$, the von Neumann algebras $\mathscr{M}$ and $G_1\mathscr{M}G_1$ are unitarily equivalent. Now $\mathscr{M}$ is * isomorphic to the algebra $\aleph_0 \otimes G_1\mathscr{M}G_1$ of bounded operators having (countably infinite) matrices with all entries in $G_1\mathscr{M}G_1$. Thus $\mathscr{M}$ is * isomorphic to $\aleph_0 \otimes \mathscr{M}$; and this, in turn, is unitarily equivalent to $\mathscr{M} \bar{\otimes} \mathscr{B}(L_2(\mathbb{R}))$, since the Hilbert space $L_2(\mathbb{R})$ has dimension $\aleph_0$. ∎

In the next section, we shall need the following result.

13.2.11. Lemma. *If α is a continuous automorphic representation of $\mathbb{R}$ on a von Neumann algebra $\mathscr{M}$, $\hat{\alpha}$ is the dual representation of $\mathbb{R}$ on the (abstract, continuous) crossed product $\mathscr{R}(\mathscr{M}, \alpha)$, and $\Psi\colon \mathscr{M} \to \mathscr{R}(\mathscr{M}, \alpha)$ is the * isomorphism defined by (2), then*

$$\Psi(\mathscr{M}) = \{R \in \mathscr{R}(\mathscr{M}, \alpha) : \hat{\alpha}_p(R) = R \ (p \in \mathbb{R})\}.$$

Proof. By Proposition 13.2.3(iii),

$$\hat{\alpha}_p(\Psi(A)) = W_p\Psi(A)W_p^* = \Psi(A) \qquad (A \in \mathscr{M}, \quad p \in \mathbb{R}).$$

Hence

$$\Psi(\mathscr{M}) \subseteq \{R \in \mathscr{R}(\mathscr{M}, \alpha) : \hat{\alpha}_p(R) = R \ (p \in \mathbb{R})\}.$$

and it remains to prove the reverse inclusion.

By Proposition 13.2.5(iii) there is a * isomorphism θ, from $\mathscr{M}$ onto a von Neumann algebra $\mathscr{N}$, such that the continuous automorphic representation $\beta\colon t \to \beta_t = \theta \circ \alpha_t \circ \theta^{-1}$ of $\mathbb{R}$ on $\mathscr{N}$ is unitarily implemented. Let $\Psi^{(\beta)}\colon \mathscr{N} \to \mathscr{R}(\mathscr{N}, \beta)$ be the * isomorphism analogous to $\Psi\colon \mathscr{M} \to \mathscr{R}(\mathscr{M}, \alpha)$; and let $\hat{\beta}$ be the dual representation of $\mathbb{R}$ on $\mathscr{R}(\mathscr{N}, \beta)$. By Proposition 13.2.5(i), there is a * isomorphism φ, from $\mathscr{R}(\mathscr{M}, \alpha)$ onto $\mathscr{R}(\mathscr{N}, \beta)$, such that $\hat{\beta}_p = \varphi \circ \hat{\alpha}_p \circ \varphi^{-1}$ for all real p; and the proof of that result shows also that $\varphi(\Psi(A)) = (\theta \bar{\otimes} \iota)(\Psi(A)) = \Psi^{(\beta)}(\theta(A))$, when $A \in \mathscr{M}$. Thus φ carries $\Psi(\mathscr{M})$ onto $\Psi^{(\beta)}(\mathscr{N})$, and carries the fixed point algebra $\{R \in \mathscr{R}(\mathscr{M}, \alpha) : \hat{\alpha}_p(R) = R \ (p \in \mathbb{R})\}$ onto the corresponding subalgebra of $\mathscr{R}(\mathscr{N}, \beta)$. Accordingly, upon replacing $\mathscr{M}$, α by $\mathscr{N}$, β, respectively, we may suppose that α is implemented by a continuous unitary representation $t \to U(t)$ of $\mathbb{R}$ on $\mathscr{H}$ (the Hilbert space on which $\mathscr{M}$ acts).

Let U be the unitary operator defined in (5); and recall, from the discussion following (6), that U commutes with W_p $(= I \otimes w_p)$. Then $\mathscr{R}(\mathscr{M}, \alpha) \subseteq \mathscr{M} \bar{\otimes} \mathscr{B}(L_2(\mathbb{R}))$, while the implemented crossed product $U\mathscr{R}(\mathscr{M}, \alpha)U^*$ is generated by the operators $A \otimes I$, $U(t) \otimes l_t$ $(A \in \mathscr{M}, t \in \mathbb{R})$, and so commutes with $I \otimes l_t$.

Suppose that $R \in \mathscr{R}(\mathscr{M}, \alpha)$ and $\hat{\alpha}_p(R) = R$ for all real p. Then R commutes with $I \otimes w_p$ (since $\hat{\alpha}_p$ is implemented by $I \otimes w_p$). From the preceding paragraph, it now follows that URU^* commutes with both $I \otimes w_p$ and $I \otimes l_t$ for all real p and t. By Proposition 13.2.7(ii), URU^* commutes with $\mathbb{C}_{\mathscr{H}} \mathbin{\overline{\otimes}} \mathscr{B}(L_2(\mathbb{R}))$, and so has the form $A \otimes I$ for some A in $\mathscr{B}(\mathscr{H})$. For each x in $L_2(\mathbb{R}, \mathscr{H})$,

$$\begin{aligned}(Rx)(s) &= (U^*(A \otimes I)Ux)(s) = U(-s)((A \otimes I)Ux)(s)\\ &= U(-s)A(Ux)(s) = U(-s)AU(s)x(s).\end{aligned}$$

If we show that $A \in \mathscr{M}$, then $U(-s)AU(s) = \alpha_s^{-1}(A)$ and, from (2), $R = \Psi(A) \in \Psi(\mathscr{M})$.

Since $R \in \mathscr{M} \mathbin{\overline{\otimes}} \mathscr{B}(L_2(\mathbb{R}))$, R commutes with $A' \otimes I$ for each A' in $\mathscr{M}'$. When $x \in L_2(\mathbb{R}, \mathscr{H})$,

$$U(-s)AU(s)A'x(s) - A'U(-s)AU(s)x(s) = ((R(A' \otimes I) - (A' \otimes I)R)x)(s),$$

and the right-hand side vanishes for almost all s in $\mathbb{R}$. When x has the form $u \otimes f$, with u in $\mathscr{H}$ and f a continuous function in $L_2(\mathbb{R})$ satisfying $f(0) = 1$, the left-hand side is continuous in the space $\mathscr{H}$ as a function of s, and so vanishes for all s. When $s = 0$, we obtain $(AA' - A'A)u = 0$ ($u \in \mathscr{H}$, $A' \in \mathscr{M}'$); so $A \in \mathscr{M}'' = \mathscr{M}$, and $R = \Psi(A)$. ∎

Bibliography: [16, 97]

13.3. Crossed products by modular automorphism groups

If $\mathscr{M}$ is a countably decomposable von Neumann algebra acting on a Hilbert space $\mathscr{H}$, and $\{\sigma_t\}$ is the modular automorphism group corresponding to a faithful normal state ω of $\mathscr{M}$, the mapping $\sigma: t \to \sigma_t$ is a continuous automorphic representation of $\mathbb{R}$ on $\mathscr{M}$. Thus we can form the (abstract, continuous) crossed product von Neumann algebra $\mathscr{R}(\mathscr{M}, \sigma)$, acting on the Hilbert space $L_2(\mathbb{R}, \mathscr{H})$ ($= \mathscr{H} \otimes L_2(\mathbb{R})$), and the dual representation $\hat{\sigma}$ of $\mathbb{R}$ on $\mathscr{R}(\mathscr{M}, \sigma)$. We show first that, up to unitary equivalence, these constructs are independent of the choice of the faithful normal state ω. After that, we use such crossed products in studying the structure of countably decomposable type III von Neumann algebras. By using weights in place of states, and by use of modular theory in that more general context, the main results obtained in this section can be extended, so as to apply without countable decomposability restrictions [97].

Throughout, we shall continue to use the notation introduced in Section 13.2.

13.3.1. THEOREM. *Suppose that $\mathscr{M}$ is a countably decomposable von Neumann algebra acting on a Hilbert space $\mathscr{H}$, λ and μ are faithful normal states of $\mathscr{M}$, $\{\sigma_t^\lambda\}$ and $\{\sigma_t^\mu\}$ are the corresponding modular automorphism groups, and $\hat{\sigma}^\lambda$, $\hat{\sigma}^\mu$ are the dual representations of $\mathbb{R}$ on the crossed-product von Neumann algebras $\mathscr{R}(\mathscr{M}, \sigma^\lambda)$, $\mathscr{R}(\mathscr{M}, \sigma^\mu)$, respectively. Then there is a unitary operator U, acting on the Hilbert space $L_2(\mathbb{R}, \mathscr{H})$, such that $U\mathscr{R}(\mathscr{M}, \sigma^\lambda)U^* = \mathscr{R}(\mathscr{M}, \sigma^\mu)$ and $\hat{\sigma}_p^\mu = \varphi \circ \hat{\sigma}_p^\lambda \circ \varphi^{-1}$ $(p \in \mathbb{R})$, where φ is the * isomorphism $R \to URU^*$ from $\mathscr{R}(\mathscr{M}, \sigma^\lambda)$* onto *$\mathscr{R}(\mathscr{M}, \sigma^\mu)$. The von Neumann algebras $\mathscr{R}(\mathscr{R}(\mathscr{M}, \sigma^\lambda), \hat{\sigma}^\lambda)$ and $\mathscr{R}(\mathscr{R}(\mathscr{M}, \sigma^\mu), \hat{\sigma}^\mu)$ are unitarily equivalent.*

Proof. In constructing $\mathscr{R}(\mathscr{M}, \sigma^\lambda)$ and the dual representation $\hat{\sigma}^\lambda$, we use operators $\Psi_\lambda(A)$, L_t, W_p, defined by (the analogue of) 13.2(2) and 13.2(4); and we require similar notation in relation to μ.

By Theorem 13.1.9, there is a strong-operator continuous mapping $t \to U_t$, from $\mathbb{R}$ into the unitary group of $\mathscr{M}$, such that

$$\sigma_t^\mu(A) = U_t \sigma_t^\lambda(A) U_t^*, \qquad U_{s+t} = U_s \sigma_s^\lambda(U_t) \qquad (s, t \in \mathbb{R}; \quad A \in \mathscr{M}).$$

It follows easily from Proposition 13.2.2(i) that the equation

$$(Ux)(s) = U_{-s}x(s) \qquad (x \in L_2(\mathbb{R}, \mathscr{H}), \quad s \in \mathbb{R})$$

defines a unitary operator U acting on $L_2(\mathbb{R}, \mathscr{H})$, and $(U^*x)(s) = U_{-s}^*x(s)$.

Straightforward calculations show that U commutes with W_p for all real p, and that

$$U\Psi_\lambda(A)U^* = \Psi_\mu(A) \qquad (A \in \mathscr{M}).$$

Since $U_{t-s} = U_{-s}\sigma_{-s}^\lambda(U_t) = \sigma_{-s}^\mu(U_t)U_{-s}$, we have $U_{-s} = \sigma_{-s}^\mu(U_t^*)U_{t-s}$; so when $x \in L_2(\mathbb{R}, \mathscr{H})$

$$\begin{aligned}(UL_tx)(s) &= U_{-s}(L_tx)(s) = \sigma_{-s}^\mu(U_t^*)U_{t-s}x(s-t)\\ &= \sigma_{-s}^\mu(U_t^*)(Ux)(s-t) = \sigma_{-s}^\mu(U_t^*)(L_tUx)(s) = (\Psi_\mu(U_t^*)L_tUx)(s).\end{aligned}$$

Thus

$$UL_tU^* = \Psi_\mu(U_t^*)L_t \qquad (t \in \mathbb{R});$$

and from this, together with the previous displayed equation, we obtain

$$U\Psi_\lambda(U_t)L_tU^* = L_t \qquad (t \in \mathbb{R}).$$

It follows from the preceding paragraph that U implements a * isomorphism φ from $\mathscr{R}(\mathscr{M}, \sigma^\lambda)$ onto $\mathscr{R}(\mathscr{M}, \sigma^\mu)$. Moreover

$$\begin{aligned}\varphi(\hat{\sigma}_p^\lambda(R)) &= UW_pRW_p^*U^*\\ &= W_pURU^*W_p^* = \hat{\sigma}_p^\mu(\varphi(R)) \quad (R \in \mathscr{R}(\mathscr{M}, \sigma^\lambda)),\end{aligned}$$

and thus $\hat{\sigma}_p^\mu = \varphi \circ \hat{\sigma}_p^\lambda \circ \varphi^{-1}$ for all real p.

With $\mathcal{M}$, $\mathcal{N}$, θ, and α replaced by $\mathcal{R}(\mathcal{M}, \sigma^{\lambda})$, $\mathcal{R}(\mathcal{M}, \sigma^{\mu})$, φ, and $\hat{\sigma}^{\lambda}$, respectively, it now follows from Proposition 13.2.5(i) that $\mathcal{R}(\mathcal{R}(\mathcal{M}, \sigma^{\lambda}), \hat{\sigma}^{\lambda})$ and $\mathcal{R}(\mathcal{R}(\mathcal{M}, \sigma^{\mu}), \hat{\sigma}^{\mu})$ are unitarily equivalent. ■

Suppose that $\mathcal{M}$ is a countably decomposable von Neumann algebra, and $\{\sigma_t\}$ is the modular automorphism group corresponding to a faithful normal state ω of $\mathcal{M}$. The theorem just proved shows that, up to unitary equivalence, the crossed product $\mathcal{M}_0 = \mathcal{R}(\mathcal{M}, \sigma)$ and the dual representation $\hat{\sigma}$ of $\mathbb{R}$ on $\mathcal{M}_0$ are independent of the choice of ω. After a series of preparatory results, we shall show (Theorem 13.3.6) that $\mathcal{M}_0$ is semi-finite. When the original algebra $\mathcal{M}$ is itself semi-finite, the relation between $\mathcal{M}$ and $\mathcal{M}_0$ is simple (see Exercise 13.4.16) but not useful. In contrast, when $\mathcal{M}$ is type III, we shall see (Theorem 13.3.7) that $\mathcal{M}_0$ is of type II_{∞}; and by Corollary 13.2.10, $\mathcal{M}$ is * isomorphic to $\mathcal{R}(\mathcal{M}_0, \hat{\sigma})$. Thus each countably decomposable type III von Neumann algebra can be expressed, in a canonical way, as the crossed product of a type II_{∞} von Neumann algebra $\mathcal{M}_0$ by a continuous automorphic representation $\hat{\sigma}$ of $\mathbb{R}$ on $\mathcal{M}_0$.

We shall prove that $\mathcal{M}_0$ is semi-finite by giving an explicit construction of a faithful normal semi-finite tracial weight τ. In fact, τ has the additional property, described as *relative invariance* under $\hat{\sigma}$, that $\tau \circ \hat{\sigma}_p = e^{-p}\tau$ for all real p.

Under the conditions set out above, there is a * isomorphism θ, from $\mathcal{M}$ onto a von Neumann algebra $\mathcal{N}$, such that the faithful normal state $\lambda = \omega \circ \theta^{-1}$ of $\mathcal{N}$ is a vector state, arising from a separating and generating vector for $\mathcal{N}$. For example, in view of Corollary 7.1.7, it suffices to take, for θ, the (faithful) representation of $\mathcal{M}$ engendered by ω. The modular automorphism group $\{\sigma_t^{\lambda}\}$ of $\mathcal{N}$ is given by $\sigma_t^{\lambda} = \theta \circ \sigma_t \circ \theta^{-1}$, since it is immediately verified that $\{\theta \circ \sigma_t \circ \theta^{-1}\}$ satisfies the modular condition relative to λ. From Proposition 13.2.5, there is a * isomorphism φ, from $\mathcal{R}(\mathcal{M}, \sigma)$ onto $\mathcal{R}(\mathcal{N}, \sigma^{\lambda})$, such that $\hat{\sigma}_p^{\lambda} = \varphi \circ \hat{\sigma}_p \circ \varphi^{-1}$. Accordingly, in proving the results described in the preceding two paragraphs, we may work with $\mathcal{N}$ and λ in place of $\mathcal{M}$ and ω, thus reducing to the case in which ω arises from a separating and generating vector.

We now review some notation that will be used in the remainder of this section. Throughout, $\mathcal{M}$ denotes a countably decomposable von Neumann algebra acting on a Hilbert space $\mathcal{H}$, v is a separating and generating vector for $\mathcal{M}$, and $\{\sigma_t\}$ is the modular automorphism group corresponding to the faithful normal state $\omega_v|\mathcal{M}$. We denote by T the unitary operator, acting on $L_2(\mathbb{R})$, derived as in Theorem 13.2.8 from the Fourier transform. When $B \in \mathcal{B}(L_2(\mathbb{R}))$, we write $\hat{B}$ for T^*BT, so that $w_p = \hat{l}_p$ and $l_t = \hat{w}_{-t}$, and the mapping $B \to \hat{B}$ is a * automorphism of $\mathcal{B}(L_2(\mathbb{R}))$. We write $\mathcal{A}$ for the multiplication algebra $\{m_g : g \in L_{\infty}(\mathbb{R})\}$ acting on $L_2(\mathbb{R})$, and $\hat{\mathcal{A}}$ for $T^*\mathcal{A}T$

$(=\{\hat{m}_g : g \in L_\infty(\mathbb{R})\})$. Since $\mathscr{A}$ is maximal abelian, so is $\hat{\mathscr{A}}$. The linear span of the operators w_t $(t \in \mathbb{R})$ is a * subalgebra of $\mathscr{A}$ that is weak-operator (equivalently, ultraweakly) dense in $\mathscr{A}$ by Proposition 13.2.7(i); so the linear span of $\{l_t : t \in \mathbb{R}\}$ is a weak-operator-dense * subalgebra of $\hat{\mathscr{A}}$. From this, and since $L_t = I \mathbin{\overline{\otimes}} l_t$, $\mathbb{C}_{\mathscr{H}} \mathbin{\overline{\otimes}} \hat{\mathscr{A}}$ is the weak-operator-closed linear span of $\{L_t : t \in \mathbb{R}\}$. Thus $\mathscr{R}(\mathscr{M}, \sigma)$, which is the weak-operator-closed subspace generated by $\{L_t \Psi(A) : A \in \mathscr{M},\ t \in \mathbb{R}\}$, contains $\mathbb{C}_{\mathscr{H}} \mathbin{\overline{\otimes}} \hat{\mathscr{A}}$, and is also the weak-operator-closed linear span of $\{(I \otimes \hat{m}_g)\Psi(A) : A \in \mathscr{M},\ g \in L_\infty(\mathbb{R})\}$.

The symbols X and Y will always denote bounded measurable subsets of $\mathbb{R}$. We write e_X $(\in \mathscr{A})$ for the projection of multiplication by the characteristic function of X, and define E_X in $\mathbb{C}_{\mathscr{H}} \mathbin{\overline{\otimes}} \hat{\mathscr{A}}$ $(\subseteq \mathscr{R}(\mathscr{M}, \sigma))$ by $E_X = I \otimes \hat{e}_X$. Since $\{e_{[-n,n]}\}$ is an increasing sequence of projections with limit I, the same is true of $\{E_{[-n,n]}\}$. The equation

$$f_X(q) = \begin{cases} e^{-(1/2)q} & (q \in X) \\ 0 & (q \in \mathbb{R}\setminus X), \end{cases}$$

defines an element f_X of $L_2(\mathbb{R})$, and $e_X f_Y = f_X$ when $X \subseteq Y$. Let τ_X be the vector state of $\mathscr{R}(\mathscr{M}, \sigma)$, corresponding to the vector $v \otimes T^* f_X$. In the sequence of lemmas that follow, we shall show that $\tau_X | E_X \mathscr{R}(\mathscr{M}, \sigma) E_X$ is a tracial state, and that the equation $\tau(R) = \sup \tau_X(R)$ $(R \in \mathscr{R}(\mathscr{M}, \sigma)^+)$ defines a faithful normal semi-finite tracial weight τ that is relatively invariant under $\hat{\sigma}$. (We use "state" loosely—not insisting on normalization at I.)

The automorphism $\hat{\sigma}_p$ of $\mathscr{R}(\mathscr{M}, \sigma)$ is implemented by the unitary operator W_p $(= I \otimes w_p = I \otimes \hat{l}_p)$, which commutes with $\Psi(A)$ for each A in $\mathscr{M}$. A straightforward calculation shows that, when $g \in L_\infty(\mathbb{R})$, $l_p m_g l_p^* = m_h$, where h $(\in L_\infty(\mathbb{R}))$ is defined by $h(q) = g(q - p)$. Thus

$$\begin{aligned} \hat{\sigma}_p((I \otimes \hat{m}_g)\Psi(A)) &= \hat{\sigma}_p(I \otimes \hat{m}_g)\Psi(A) \\ &= (I \otimes \hat{l}_p \hat{m}_g \hat{l}_p^*)\Psi(A) = (I \otimes \hat{m}_h)\Psi(A). \end{aligned}$$

When $A = I$ and g is the characteristic function of X, we obtain $\hat{\sigma}_p(E_X) = E_{X+p}$, where $X + p = \{q + p : q \in X\}$.

13.3.2. LEMMA. (i) *If* $A \in \mathscr{M}$, $B \in \mathscr{B}(L_2(\mathbb{R}))$, and $f \in L_2(\mathbb{R})$,

$$\omega_{v \otimes f}((I \otimes B)\Psi(A)) = \omega_v(A)\omega_f(B).$$

(ii) *If X is a bounded measurable subset of* $\mathbb{R}$, $A \in \mathscr{M}$, *and* $g \in L_\infty(\mathscr{R})$,

$$\tau_X((I \otimes \hat{m}_g)\Psi(A)) = \omega_v(A) \int_X g(q) e^{-q}\, dq.$$

Proof. (i) We recall that $v \otimes f$ is identified as an element of $L_2(\mathbb{R}, \mathscr{H})$, defined by $(v \otimes f)(s) = f(s)v$. Since $\omega_v | \mathscr{M}$ is invariant under $\{\sigma_t\}$,

$$\begin{aligned}
\omega_{v\otimes f}((I \otimes B)\Psi(A)) &= \langle \Psi(A)(v \otimes f), v \otimes B^*f \rangle \\
&= \int_{\mathbb{R}} \langle (\Psi(A)(v \otimes f))(s), (v \otimes B^*f)(s) \rangle \, ds \\
&= \int_{\mathbb{R}} \langle \sigma_s^{-1}(A)(f(s)v), (B^*f)(s)v \rangle \, ds \\
&= \int_{\mathbb{R}} \langle \sigma_s^{-1}(A)v, v \rangle f(s)\overline{(B^*f)(s)} \, ds \\
&= \omega_v(A) \int_{\mathbb{R}} f(s)\overline{(B^*f)(s)} \, ds \\
&= \omega_v(A)\langle f, B^*f \rangle = \omega_v(A)\omega_f(B).
\end{aligned}$$

(ii) By taking $f = T^*f_X$ in (i), we obtain

$$\begin{aligned}
\tau_X((I \otimes \hat{m}_g)\Psi(A)) &= \omega_v(A)\langle \hat{m}_g T^*f_X, T^*f_X \rangle \\
&= \omega_v(A)\langle T\hat{m}_g T^*f_X, f_X \rangle = \omega_v(A)\langle m_g f_X, f_X \rangle \\
&= \omega_v(A) \int_{\mathbb{R}} g(q)|f_X(q)|^2 \, dq \\
&= \omega_v(A) \int_X g(q)e^{-q} \, dq. \quad \blacksquare
\end{aligned}$$

13.3.3. LEMMA. *Suppose that X and Y are bounded measurable subsets of $\mathbb{R}$, and $p \in \mathbb{R}$.*

(i) *If $X \subseteq Y$, $\tau_Y(E_X R E_X) = \tau_X(R)$ for all R in $\mathscr{R}(\mathscr{M}, \sigma)$.*
(ii) *If $X \cap Y$ is a null set, $\tau_{X\cup Y} = \tau_X + \tau_Y$.*
(iii) *$\tau_X \circ \hat{\sigma}_p = e^{-p}\tau_{X-p}$, where $X - p = \{q - p : q \in X\}$.*

Proof. (i) If $X \subseteq Y$, we have $e_X f_Y = f_X$, and therefore

$$\begin{aligned}
E_X(v \otimes T^*f_Y) &= (I \otimes \hat{e}_X)(v \otimes T^*f_Y) \\
&= v \otimes (T^*e_X T T^*f_Y) = v \otimes T^*f_X.
\end{aligned}$$

Thus

$$\begin{aligned}
\tau_Y(E_X R E_X) &= \langle E_X R E_X(v \otimes T^*f_Y), v \otimes T^*f_Y \rangle \\
&= \langle R(v \otimes T^*f_X), v \otimes T^*f_X \rangle = \tau_X(R).
\end{aligned}$$

(ii) If $X \cap Y$ is a null set, it is apparent from Lemma 13.3.2(ii) that $\tau_{X\cup Y}(R) = \tau_X(R) + \tau_Y(R)$, when R has the form $(I \otimes \hat{m}_g)\Psi(A)$, with A in $\mathscr{M}$

and g in $L_\infty(\mathbb{R})$. Since $\mathcal{R}(\mathcal{M}, \sigma)$ is the weak-operator closed linear span of such operators R, while τ_X, τ_Y, and $\tau_{X\cup Y}$ are vector states, it follows that $\tau_{X\cup Y} = \tau_X + \tau_Y$.

(iii) It suffices to verify that $\tau_X(\hat{\sigma}_p(R)) = e^{-p}\tau_{X-p}(R)$ when R has the form $(I \otimes \hat{m}_g)\Psi(A)$, with A in $\mathcal{M}$ and g in $L_\infty(\mathbb{R})$. In this case, $\hat{\sigma}_p(R) = (I \otimes \hat{m}_h)\Psi(A)$, where $h(q) = g(q - p)$; and from Lemma 13.3.2(ii)

$$\begin{aligned}\tau_X(\hat{\sigma}_p(R)) &= \tau_X((I \otimes \hat{m}_h)\Psi(A)) = \omega_v(A)\int_X h(q)e^{-q}\,dq \\ &= \omega_v(A)e^{-p}\int_X g(q-p)e^{-(q-p)}\,dq = e^{-p}\omega_v(A)\int_{X-p} g(q)e^{-q}\,dq \\ &= e^{-p}\tau_{X-p}((I \otimes \hat{m}_g)\Psi(A)) = e^{-p}\tau_{X-p}(R). \quad \blacksquare\end{aligned}$$

13.3.4. LEMMA. *The equation*

$$\tau(R) = \sup \tau_X(R) \qquad (R \in \mathcal{R}(\mathcal{M}, \sigma)^+),$$

in which the supremum is taken over all bounded measurable subsets X of $\mathbb{R}$, defines a faithful normal semi-finite weight τ on $\mathcal{R}(\mathcal{M}, \sigma)$. Moreover,

$$\tau(E_X R E_X) = \tau_X(E_X R E_X) = \tau_X(R),$$

$$\tau(R) = \lim_{n\to\infty} \tau_{[-n,n]}(R)$$

for all R in $\mathcal{R}(\mathcal{M}, \sigma)^+$; and $\tau \circ \hat{\sigma}_p = e^{-p}\tau$ for all real p.

Proof. The equation $\tau(R) = \sup \tau_X(R)$ defines a mapping τ from $\mathcal{R}(\mathcal{M}, \sigma)^+$ into $[0, \infty]$, and it is apparent from Lemma 13.3.3(iii) that $\tau \circ \hat{\sigma}_p = e^{-p}\tau$. From parts (i) and (ii) of that lemma,

$$\tau_Y(R) \geq \tau_X(R) = \tau_Y(E_X R E_X)$$

when $R \in \mathcal{R}(\mathcal{M}, \sigma)^+$ and $X \subseteq Y$. From this, and since $X \subseteq [-n, n]$ for all sufficiently large integers n, it follows that

$$\tau_X(R) \leq \lim_{n\to\infty} \tau_{[-n,n]}(R), \qquad \tau_X(R) = \lim_{n\to\infty} \tau_{[-n,n]}(E_X R E_X),$$

each of these limits being that of an increasing sequence. Thus

$$\tau(R) = \sup \tau_X(R) = \lim_{n\to\infty} \tau_{[-n,n]}(R),$$

and therefore

$$\tau(E_X R E_X) = \lim_{n\to\infty} \tau_{[-n,n]}(E_X R E_X) = \tau_X(R).$$

From the last equation, τ takes finite values on the positive part of the weak-operator dense * subalgebra $\bigcup E_X\mathscr{R}(\mathscr{M},\sigma)E_X$ of $\mathscr{R}(\mathscr{M},\sigma)$; and upon replacing R by E_XRE_X, we obtain $\tau(E_XRE_X) = \tau_X(E_XRE_X)$. Moreover, by Lemma 13.3.3(ii)

$$\tau(R) = \lim_{n\to\infty} \tau_{[-n,n]}(R)$$

$$= \lim_{n\to\infty} \sum_{j=-n}^{n-1} \tau_{[j,j+1]}(R) = \sum_{j=-\infty}^{\infty} \tau_{[j,j+1]}(R);$$

so τ is the sum of the vector states $\tau_{[j,j+1]}$ $(j = 0, \pm 1, \pm 2, \ldots)$.

The preceding argument shows that τ is a normal semi-finite weight on $\mathscr{R}(\mathscr{M},\sigma)$, and it remains to show that τ is faithful. For this, suppose that $R\in\mathscr{R}(\mathscr{M},\sigma)$ and $\tau(R^*R) = 0$; we have to show that $R = 0$. For every bounded measurable subset X of $\mathbb{R}$,

$$0 = \tau(R^*R) \geq \tau_X(R^*R) = \|R(v\otimes T^*f_X)\|^2,$$

and thus $R(v\otimes T^*f_X) = 0$. It now suffices to show that the set of all vectors of the form $v\otimes T^*f_X$ is separating for $\mathscr{R}(\mathscr{M},\sigma)$; equivalently, that it is generating for $\mathscr{R}(\mathscr{M},\sigma)'$.

If $f\in L_2(\mathbb{R})$, and $\langle f, T^*f_X\rangle = 0$ for every bounded measurable subset X of $\mathbb{R}$, then

$$0 = \langle Tf, f_X\rangle = \int_X (Tf)(q)e^{-(1/2)q}\, dq$$

for each such X. Thus $(Tf)(q) = 0$ for almost all q, whence $Tf = 0$ and therefore $f = 0$. This shows that the vectors T^*f_X generate an everywhere-dense subspace of $L_2(\mathbb{R})$; and $\mathscr{M}'v$ is everywhere dense in $\mathscr{H}$, since v is a separating vector for $\mathscr{M}$. Accordingly, vectors of the form $A'v\otimes T^*f_X$ (with A' in $\mathscr{M}'$) generate an everywhere-dense subspace of $\mathscr{H}\otimes L_2(\mathbb{R})$. Since $\mathscr{R}(\mathscr{M},\sigma)\subseteq \mathscr{M}\,\overline{\otimes}\,\mathscr{B}(L_2(\mathbb{R}))$, $A'\otimes I\in\mathscr{R}(\mathscr{M},\sigma)'$ when $A'\in\mathscr{M}'$; so the set of all vectors of the form $v\otimes T^*f_X$ is generating for $\mathscr{R}(\mathscr{M},\sigma)'$. ∎

In order to show that τ is a tracial weight, we require one more lemma, the proof of which makes further use of Fourier transforms. Suppose that g is a complex-valued function, defined and having a continuous second derivative g'' throughout $\mathbb{R}$, and vanishing outside a compact interval $[-c, c]$. Then $g\in L_1(\mathbb{R})\cap L_\infty(\mathbb{R})$. The Fourier transform $\hat{g}$ can be defined, for all *complex* z, by

$$\hat{g}(z) = (2\pi)^{-1/2}\int_{\mathbb{R}} e^{iqz}g(q)\, dq = (2\pi)^{-1/2}\int_{-c}^{c} e^{iqz}g(q)\, dq.$$

By differentiation under the integral sign, it follows that $\hat{g}$ is an entire function. Moreover,

$$\hat{g}(s + ia) = (2\pi)^{-1/2} \int_{-c}^{c} e^{iqs} e^{-qa} g(q)\, dq \qquad (s, a \in \mathbb{R});$$

so

$$|\hat{g}(s + ia)| \leq (2\pi)^{-1/2} e^{c|a|} \int_{-c}^{c} |g(q)|\, dq,$$

and $\hat{g}$ is bounded on every strip of finite width parallel to the real axis. Upon integration by parts twice, we obtain

$$\hat{g}(s) = -(2\pi)^{-1/2} s^{-2} \int_{-c}^{c} e^{iqs} g''(q)\, dq \qquad (s \neq 0);$$

so $|\hat{g}(s)| \leq K|s|^{-2}$, where $K = (2\pi)^{-1/2} \int |g''(q)|\, dq$. From this, and since $\hat{g}$ is bounded and continuous on $\mathbb{R}$, it follows that $\hat{g}|\mathbb{R} \in L_1(\mathbb{R})$. By applying this integration-by-parts argument, with $e^{-qa}g(q)$ in place of $g(q)$ (or by using the Riemann–Lebesgue lemma for this function), we deduce also that $\hat{g}(s + ia) \to 0$ when $|s| \to \infty$ for each fixed real number a.

13.3.5. LEMMA. *Suppose that $A, B \in \mathcal{M}$, X is a compact interval $[-c, c]$, and g, h are complex-valued functions, defined and having continuous second derivatives on $\mathbb{R}$, and vanishing outside X. Then*

$$\tau_X((I \otimes \hat{m}_g)\Psi(A)(I \otimes \hat{m}_h)\Psi(B)) = \tau_X((I \otimes \hat{m}_h)\Psi(B)(I \otimes \hat{m}_g)\Psi(A)).$$

Proof. From the discussion preceding the statement of the lemma, g, $h \in L_1(\mathbb{R}) \cap L_\infty(\mathbb{R})$ and $\hat{g}, \hat{h} \in L_1(\mathbb{R})$. When $x_1, x_2 \in L_2(\mathbb{R}, \mathscr{H})$ $(= \mathscr{H} \otimes L_2(\mathbb{R}))$,

$$((I \otimes m_h)x_1)(q) = h(q)x_1(q), \qquad ((I \otimes w_s)x_1)(q) = e^{-isq}x_1(q).$$

From the inversion theorem for Fourier transforms, together with Fubini's theorem,

$$\begin{aligned}
\langle (I \otimes m_h)x_1, x_2 \rangle &= \int_{\mathbb{R}} h(q)\langle x_1(q), x_2(q)\rangle\, dq \\
&= (2\pi)^{-1/2} \int_{\mathbb{R}} \left(\int_{\mathbb{R}} e^{-isq} \hat{h}(s)\, ds \right) \langle x_1(q), x_2(q)\rangle\, dq \\
&= (2\pi)^{-1/2} \int_{\mathbb{R}} \hat{h}(s) \left(\int_{\mathbb{R}} \langle e^{-isq} x_1(q), x_2(q)\rangle\, dq \right) ds \\
&= (2\pi)^{-1/2} \int_{\mathbb{R}} \hat{h}(s) \langle (I \otimes w_s)x_1, x_2\rangle\, ds.
\end{aligned}$$

Upon replacing x_j by $(I \otimes T)x_j$, and recalling that $T^* m_h T = \hat{m}_h$, $T^* w_s T = l_{-s}$, we obtain

$$\langle (I \otimes \hat{m}_h)x_1, x_2 \rangle = (2\pi)^{-1/2} \int_{\mathbb{R}} \hat{h}(s) \langle (I \otimes l_{-s})x_1, x_2 \rangle \, ds.$$

In this last equation, we may take $x_1 = \Psi(B)(v \otimes T^* f_X)$ and $x_2 = [(I \otimes \hat{m}_g)\Psi(A)]^*(v \otimes T^* f_X)$; and since τ_X is the vector state of $\mathscr{R}(\mathscr{M}, \sigma)$ corresponding to $v \otimes T^* f_X$, it follows that

$$\begin{aligned} &\tau_X((I \otimes \hat{m}_g)\Psi(A)(I \otimes \hat{m}_h)\Psi(B)) \\ &\qquad = (2\pi)^{-1/2} \int_{\mathbb{R}} \hat{h}(s)\tau_X((I \otimes \hat{m}_g)\Psi(A)(I \otimes l_{-s})\Psi(B)) \, ds. \end{aligned}$$

Since $I \otimes l_t = L_t$ and $L_t \Psi(A) L_t^* = \Psi(\sigma_t(A))$,

$$\tau_X((I \otimes \hat{m}_g)\Psi(A)(I \otimes l_{-s})\Psi(B)) = \tau_X((I \otimes \hat{m}_g l_{-s})\Psi(\sigma_s(A)B)).$$

Now $\hat{m}_g l_{-s} = \hat{m}_g \hat{w}_s = \hat{m}_k$, where k $(\in L_\infty(\mathbb{R}))$ is defined by $k(q) = \exp(-isq)g(q)$; so from Lemma 13.3.2(ii), and since g vanishes outside X,

$$\begin{aligned} \tau_X((I \otimes \hat{m}_g)\Psi(A)(I \otimes l_{-s})\Psi(B)) &= \omega_v(\sigma_s(A)B) \int_X k(q)e^{-q} \, dq \\ &= \omega_v(\sigma_s(A)B) \int_{\mathbb{R}} e^{iq(i-s)} g(q) \, dq \\ &= (2\pi)^{1/2} \omega_v(\sigma_s(A)B)\hat{g}(i - s). \end{aligned}$$

Thus

$$\tau_X((I \otimes \hat{m}_g)\Psi(A)(I \otimes \hat{m}_h)\Psi(B)) = \int_{\mathbb{R}} \omega_v(\sigma_s(A)B)\hat{h}(s)\hat{g}(i - s) \, ds;$$

and by symmetry,

$$\tau_X((I \otimes \hat{m}_h)\Psi(B)(I \otimes \hat{m}_g)\Psi(A)) = \int_{\mathbb{R}} \omega_v(\sigma_s(B)A)\hat{g}(s)\hat{h}(i - s) \, ds.$$

We have to show that the integrals occurring in the last two equations are equal.

Since $\{\sigma_s\}$ satisfies the modular condition, relative to the state $\omega_v | \mathscr{M}$, there is a complex-valued function F, bounded and continuous on the strip $\Omega = \{z \in \mathbb{C} : 0 \le \operatorname{Im} z \le 1\}$, analytic on the interior of Ω, and with boundary values

$$F(s) = \omega_v(\sigma_s(A)B), \qquad F(s + i) = \omega_v(B\sigma_s(A)) = \omega_v(\sigma_{-s}(B)A).$$

The function $F(z)\hat{h}(z)\hat{g}(i-z)$ is bounded and continuous on Ω, analytic inside Ω, and tends to 0 when $|\text{Re}\, z| \to \infty$. By integrating this function round the rectangle with vertices $\pm a$, $\pm a + i$, and letting $a \to \infty$, it follows from Cauchy's theorem and the dominated convergence theorem that

$$\begin{aligned}\int_{\mathbb{R}} F(s)\hat{h}(s)\hat{g}(i-s)\,ds &= \int_{\mathbb{R}} F(s+i)\hat{h}(s+i)\hat{g}(-s)\,ds \\ &= \int_{\mathbb{R}} F(i-s)\hat{g}(s)\hat{h}(i-s)\,ds;\end{aligned}$$

that is

$$\int_{\mathbb{R}} \omega_v(\sigma_s(A)B)\hat{h}(s)\hat{g}(i-s)\,ds = \int_{\mathbb{R}} \omega_v(\sigma_s(B)A)\hat{g}(s)\hat{h}(i-s)\,ds. \quad \blacksquare$$

13.3.6. THEOREM. *Suppose that $\mathscr{M}$ is a countably decomposable von Neumann algebra acting on a Hilbert space $\mathscr{H}$, $\{\sigma_t\}$ is the modular automorphism group corresponding to a faithful normal state ω of $\mathscr{M}$, and $\hat{\sigma}: p \to \hat{\sigma}_p$ is the dual representation of $\mathbb{R}$ on $\mathscr{R}(\mathscr{M}, \sigma)$. Then $\mathscr{R}(\mathscr{M}, \sigma)$ is semi-finite, and has a faithful normal semi-finite tracial weight τ satisfying $\tau \circ \hat{\sigma}_p = e^{-p}\tau$ $(p \in \mathbb{R})$.*

Proof. We have already noted that it is sufficient to consider the case in which $\omega = \omega_v|\mathscr{M}$, where v is a separating and generating vector for $\mathscr{M}$. In that case, it remains only to prove that the weight τ, described in Lemma 13.3.4, is tracial; so we have to show that $\tau(R^*R) = \tau(RR^*)$ for each R in $\mathscr{R}(\mathscr{M}, \sigma)$.

The projections $E_{[-n,n]}$ form an increasing sequence with limit I. When $1 \le m \le n$, it follows from Lemma 13.3.4 that

$$\begin{aligned}\tau_{[-n,n]}(E_{[-n,n]}R^*E_{[-n,n]}RE_{[-n,n]}) &= \tau_{[-n,n]}(R^*E_{[-n,n]}R) \\ &\ge \tau_{[-n,n]}(R^*E_{[-m,m]}R);\end{aligned}$$

so

$$\begin{aligned}&\liminf_{n\to\infty} \tau_{[-n,n]}(E_{[-n,n]}R^*E_{[-n,n]}RE_{[-n,n]}) \\ &\qquad \ge \lim_{n\to\infty} \tau_{[-n,n]}(R^*E_{[-m,m]}R) = \tau(R^*E_{[-m,m]}R).\end{aligned}$$

When $m \to \infty$, it results from the normality of τ that

$$\liminf_{n\to\infty} \tau_{[-n,n]}(E_{[-n,n]}R^*E_{[-n,n]}RE_{[-n,n]}) \ge \tau(R^*R).$$

However,

$$\tau(R^*R) \geq \tau_X(R^*R) \geq \tau_X(R^*E_X R) = \tau_X(E_X R^* E_X R E_X)$$

for every bounded measurable subset X of $\mathbb{R}$. Thus

$$\tau(R^*R) = \lim_{n\to\infty} \tau_{[-n,n]}(E_{[-n,n]}R^*E_{[-n,n]}RE_{[-n,n]});$$

and upon replacing R by R^*, we obtain the corresponding formula for $\tau(RR^*)$.

In order to show that $\tau(R^*R) = \tau(RR^*)$, it now suffices to prove that $\tau_X | E_X\mathscr{R}(\mathscr{M}, \sigma)E_X$ is a tracial state, when X is a compact interval $[-c, c]$. Now $E_X = I \otimes \hat{m}_k$, where k is the characteristic function of X. We can choose a sequence of functions $k_n\colon \mathbb{R} \to [0, 1]$, each having a continuous second derivative and vanishing outside X, such that $k(s) = \lim k_n(s)$ except at the endpoints $\pm c$. Suppose that $A, B \in \mathscr{M}$ and $g, h \in L_\infty(\mathbb{R})$. There are sequences $\{g_n\}$, $\{h_n\}$ of continuous complex-valued functions on X, such that $|g_n(s)| \leq \|g\|$ and $|h_n(s)| \leq \|h\|$ throughout X, while $g_n(s) \to g(s)$, $h_n(s) \to h(s)$ almost everywhere on X. We may suppose further that g_n, h_n are polynomials (and so are defined throughout $\mathbb{R}$); because, for example, g_n can be replaced by a polynomial that approximates $(1 - n^{-1})g_n$ within $n^{-1}\|g\|$, uniformly on X. The functions $k_n g_n$, $k_n h_n$ have continuous second derivatives throughout $\mathbb{R}$, and vanish outside X; so by Lemma 13.3.5,

$$\begin{aligned}(1)\qquad &\tau_X((I \otimes \hat{m}_{k_n g_n})\Psi(A)(I \otimes \hat{m}_{k_n h_n})\Psi(B)) \\ &\quad = \tau_X((I \otimes \hat{m}_{k_n h_n})\Psi(B)(I \otimes \hat{m}_{k_n g_n})\Psi(A)).\end{aligned}$$

Since $|k_n(s)g_n(s)| \leq \|g\|$ for all real s, and $k_n(s)g_n(s) \to k(s)g(s)$ almost everywhere, it follows from the dominated convergence theorem that

$$\lim_{n\to\infty} \int_{\mathbb{R}} |[k_n(s)g_n(s) - k(s)g(s)]f(s)|^2\, ds = 0$$

for every f in $L_2(\mathbb{R})$. Hence

$$\|m_{k_n g_n}\| \leq \|g\|, \qquad m_{k_n g_n} \to m_{kg}(= m_k m_g)$$

in the strong-operator topology, and therefore

$$\begin{aligned} I \otimes \hat{m}_{k_n g_n} &= I \otimes T^* m_{k_n g_n} T \\ &\to I \otimes T^* m_k m_g T = I \otimes \hat{m}_k \hat{m}_g = E_X(I \otimes \hat{m}_g)\end{aligned}$$

(again, in the strong-operator topology). A similar argument applies to the sequence $k_n h_n$; and by taking limits, as $n \to \infty$, in (1) we obtain

$$\tau_X(E_X(I \otimes \hat{m}_g)\Psi(A)E_X(I \otimes \hat{m}_h)\Psi(B)) = \tau_X(E_X(I \otimes \hat{m}_h)\Psi(B)E_X(I \otimes \hat{m}_g)\Psi(A)).$$

Since $\mathscr{R}(\mathscr{M}, \sigma)$ is the weak-operator closed linear span of elements of the form $(I \otimes \hat{m}_g)\Psi(A)$, it now follows that

$$\tau_X(E_X R_1 E_X R_2) = \tau_X(E_X R_2 E_X R_1) \qquad (R_1, R_2 \in \mathscr{R}(\mathscr{M}, \sigma));$$

and $\tau_X | E_X \mathscr{R}(\mathscr{M}, \sigma) E_X$ is a tracial state. ■

13.3.7. THEOREM. *If $\mathscr{M}$ is a countably decomposable type* III *von Neumann algebra, and $\{\sigma_t\}$ is the modular automorphism group corresponding to a faithful normal state ω of $\mathscr{M}$, then $\mathscr{R}(\mathscr{M}, \sigma)$ is of type* II_∞.

Proof. We show first that $\mathscr{R}(\mathscr{M}, \sigma)$ is properly infinite. For this, suppose that E is a finite projection in the center of $\mathscr{R}(\mathscr{M}, \sigma)$; we have to prove that $E = 0$. The mapping $A \to \Psi(A)E$ is an ultraweakly continuous * homomorphism φ, from $\mathscr{M}$ into the finite von Neumann algebra $\mathscr{R}(\mathscr{M}, \sigma)E$. The kernel of φ is an ultraweakly closed two-sided ideal in $\mathscr{M}$, and, from Theorem 6.8.8, has the form $\mathscr{M}(I - F)$ for some projection F in the center of $\mathscr{M}$. The restriction $\varphi | \mathscr{M}F$ is a * isomorphism from $\mathscr{M}F$ into $\mathscr{R}(\mathscr{M}, \sigma)E$, and $\varphi(F) = \varphi(F + I - F) = E$. Since E is finite in $\mathscr{R}(\mathscr{M}, \sigma)E$, F is finite in $\mathscr{M}F$ and hence, also, in $\mathscr{M}$. Thus $F = 0$, and $E = \varphi(F) = 0$, since $\mathscr{M}$ is type III.

Since $\mathscr{R}(\mathscr{M}, \sigma)$ is properly infinite by the preceding paragraph and semifinite by Theorem 13.3.6, it now remains to show that it has no central portion of type I. We assume the contrary, and in due course arrive at a contradiction. Let P be the largest central projection in $\mathscr{R}(\mathscr{M}, \sigma)$ for which $\mathscr{R}(\mathscr{M}, \sigma)P$ is type I. Then P is non-zero (by our assumption), and is invariant under every * automorphism of $\mathscr{R}(\mathscr{M}, \sigma)$; in particular, $\hat{\sigma}_p(P) = P$ for all real p.

We assert that there is a projection Q in the center of $\mathscr{R}(\mathscr{M}, \sigma)$, such that $\hat{\sigma}_p(Q)$ is a decreasing function of p, and

$$\lim_{p \to \infty} \hat{\sigma}_p(Q) = 0, \qquad \lim_{p \to -\infty} \hat{\sigma}_p(Q) = P.$$

To this end, we introduce projections e_r (acting on $L_2(\mathbb{R})$) and E_r ($\in \mathscr{R}(\mathscr{M}, \sigma)$) for all real r, as follows: e_r is the operator of multiplication by the characteristic function g_r of the interval $[r, \infty)$, and $E_r = I \otimes \hat{e}_r$. Thus E_r is a decreasing function of r, with limits 0 at ∞ and I at $-\infty$. From the discussion immediately preceding Lemma 13.3.2, with g_r in place of g, we obtain $\hat{\sigma}_p(E_r) = E_{p+r}$. From Lemmas 13.3.4 and 13.3.2(ii)

$$\begin{aligned}
\tau(E_r) &= \lim_{n \to \infty} \tau_{[-n, n]}(E_r) \\
&= \lim_{n \to \infty} \int_{-n}^{n} g_r(q) e^{-q}\, dq = \int_r^\infty e^{-q}\, dq = e^{-r}.
\end{aligned}$$

Let Q be the central carrier of PE_0 relative to $\mathscr{R}(\mathscr{M}, \sigma)$. Since $\hat{\sigma}_p(Q)$ is the central carrier of PE_p $(=\hat{\sigma}_p(PE_0))$, it follows that $\hat{\sigma}_p(Q)$ increases when p decreases, and $PE_p \leq \hat{\sigma}_p(Q) \leq P$. Since PE_p has limit P when $p \to -\infty$, the same is true of $\hat{\sigma}_p(Q)$. When p increases to ∞, $\hat{\sigma}_p(Q)$ decreases to a projection $Q_\infty(\leq P)$ in the center of $\mathscr{R}(\mathscr{M}, \sigma)$. In order to show that $Q_\infty = 0$, we use the fact that $\mathscr{R}(\mathscr{M}, \sigma)P$ is type I, so that P is the central carrier of an abelian projection F in $\mathscr{R}(\mathscr{M}, \sigma)$. The abelian projection $F\hat{\sigma}_p(Q)$ has the same central carrier, $\hat{\sigma}_p(Q)$, as does PE_p; so $PE_p \precsim F\hat{\sigma}_p(Q)$. Thus

$$e^{-p} = \tau(E_p) \geq \tau(PE_p) \geq \tau(F\hat{\sigma}_p(Q)) \geq \tau(FQ_\infty) \qquad (p \in \mathbb{R}),$$

and therefore $\tau(FQ_\infty) = 0$. Since τ is faithful and $Q_\infty \leq P = C_F$, it now follows that $Q_\infty = 0$.

By using the properties of Q, as set out at the beginning of the preceding paragraph, we shall prove that there is a non-zero finite projection G in $\mathscr{R}(\mathscr{M}, \sigma)$ that is invariant under $\{\hat{\sigma}_p\}$. Once this has been done, it follows from Lemma 13.2.11 that $G \in \Psi(\mathscr{M})$; and G is finite relative to $\Psi(\mathscr{M})$ since it is finite in $\mathscr{R}(\mathscr{M}, \sigma)$. Thus $\Psi(\mathscr{M})$ is * isomorphic to the type III von Neumann algebra $\mathscr{M}$, yet contains a non-zero finite projection; and we obtain the contradiction required to complete the proof of the theorem. Accordingly, it suffices to prove the existence of a projection G that has the stated properties.

Since $\hat{\sigma}_p(Q)\hat{\sigma}_q(Q) = \hat{\sigma}_q(Q)\hat{\sigma}_p(Q) = \hat{\sigma}_q(Q)$ when $p \leq q$, the central projection $F = Q - \hat{\sigma}_1(Q)$ in $\mathscr{R}(\mathscr{M}, \sigma)$ satisfies $\hat{\sigma}_p(F)\hat{\sigma}_q(F) = 0$ when $|p - q| \geq 1$. Moreover,

$$\sum_{j=-n}^{n-1} \hat{\sigma}_j(F) = \sum_{j=-n}^{n-1} \{\hat{\sigma}_j(Q) - \hat{\sigma}_{j+1}(Q)\} = \hat{\sigma}_{-n}(Q) - \hat{\sigma}_n(Q),$$

and thus

$$\sum_{j=-\infty}^{\infty} \hat{\sigma}_j(F) = P, \tag{2}$$

the series converging in the strong-operator topology.

Since F is a non-zero central projection in $\mathscr{R}(\mathscr{M}, \sigma)$, while the weight τ is faithful and semi-finite, we can choose A in $\mathscr{R}(\mathscr{M}, \sigma)$ so that $0 < A \leq F(\leq P)$ and $0 < \tau(A) < \infty$. Let c be a real number such that $0 < c < \tau(A)$. From the definition of τ, in Lemma 13.3.4, there is a bounded measurable subset X of $\mathbb{R}$ for which $c < \tau_X(A)$. Since τ_X is a normal state, while $A = AF$, the function $\tau_X(\hat{\sigma}_p(A)F)$ of p is continuous throughout $\mathbb{R}$, and takes the value $\tau_X(A)(>c)$ when $p = 0$. Hence there is a positive integer k such that

$$\tau_X(\hat{\sigma}_p(A)F) > c \qquad (|p| < 2^{-k}).$$

The reasoning that now follows can be viewed, intuitively, as being concerned with the existence and properties of an element, $\int_{\mathbb{R}} \hat{\sigma}_p(A)\, dp$, of $\mathscr{R}(\mathscr{M}, \sigma)$; the convergence of this integral can be deduced from the properties of the projection F. In the formal argument, however, this integral is replaced by certain approximating sums. We assert that, for each positive integer n, the equation

$$R_n = 2^{-n} \sum_{r=-\infty}^{\infty} \hat{\sigma}_{r/2^n}(A) \tag{3}$$

defines an element R_n of $\mathscr{R}(\mathscr{M}, \sigma)$, the series being strong-operator convergent; and we claim also that

$$0 \le R_n \le 2P, \tag{4}$$

$$2^{-k}c \le \tau_X(R_nF) \le \tau(R_nF) \le 2e\tau(A) \qquad (n \ge k). \tag{5}$$

Since the series occurring in (3) consists of positive operators, both the strong-operator convergence of this series and also the inequality (4) follow at once, if we show that all the finite sums

$$R(m, n) = 2^{-n} \sum_{r=-m}^{m-1} \hat{\sigma}_{r/2^n}(A) \qquad (m = 1, 2, \ldots)$$

are dominated by $2P$. Now $A = AF = AP$, hence $\hat{\sigma}_p(A) = \hat{\sigma}_p(A)P$ for all real p, and therefore $R(m, n) = R(m, n)P$. Moreover,

$$\hat{\sigma}_p(A)\hat{\sigma}_q(F) = \hat{\sigma}_p(A)\hat{\sigma}_p(F)\hat{\sigma}_q(F) = 0 \qquad (|p - q| \ge 1)$$

(in particular, $\hat{\sigma}_p(A)F = 0$ when $|p| \ge 1$). For each positive integer j,

$$R(m, n)\hat{\sigma}_j(F) = 2^{-n} \sum_{r=-m}^{m-1} \hat{\sigma}_{r/2^n}(A)\hat{\sigma}_j(F),$$

and there are at most $2^{n+1} - 1$ non-zero terms on the right-hand side (those in which $|j - r/2^n| < 1$). Since each of these terms is dominated by $\hat{\sigma}_j(F)$, it follows that $R(m, n)\hat{\sigma}_j(F) \le 2\hat{\sigma}_j(F)$. Summation over all integers j now yields (with the aid of (2)) the required inequality $R(m, n) \le 2P$.

The preceding paragraph proves the existence of an element R_n of $\mathscr{R}(\mathscr{M}, \sigma)$, defined by (3) and satisfying (4). Since τ is normal and

$$\tau \circ \hat{\sigma}_p = e^{-p}\tau \quad (p \in \mathbb{R}), \qquad \hat{\sigma}_p(A)F = 0 \quad (|p| \ge 1),$$

we have

$$\begin{aligned} \tau(R_nF) &= 2^{-n} \sum_{r=-\infty}^{\infty} \tau(\hat{\sigma}_{r/2^n}(A)F) \\ &= 2^{-n} \sum_{|r|<2^n} \tau(\hat{\sigma}_{r/2^n}(A)F) \\ &\le 2^{-n} \sum_{|r|<2^n} \tau(\hat{\sigma}_{r/2^n}(A)) \\ &\le 2^{-n}(2^{n+1} - 1)e\tau(A) < 2e\tau(A). \end{aligned}$$

Also,

$$\tau(R_nF) \geq \tau_X(R_nF)$$
$$= 2^{-n} \sum_{r=-\infty}^{\infty} \tau_X(\hat{\sigma}_{r/2^n}(A)F) \geq 2^{-k}c$$

when $n \geq k$, since

$$\tau_X(\hat{\sigma}_{r/2^n}(A)F) > c \qquad (r = 0, 1, \ldots, 2^{n-k} - 1).$$

This completes the proof of (5).

Let $\mathscr{S}_n$ $(n \geq k)$ be the set of all elements R in $\mathscr{R}(\mathscr{M}, \sigma)$ for which

$$0 \leq R \leq 2P, \qquad 2^{-k}c \leq \tau_X(RF), \qquad \tau(RF) \leq 2e\tau(A),$$
$$\hat{\sigma}_{r/2^n}(R) = R \qquad (r = 0, \pm 1, \pm 2, \ldots).$$

Since the condition $\tau(RF) \leq 2e\tau(A)$ is equivalent to the requirement that $\tau_Y(RF) \leq 2e\tau(A)$ for every bounded measurable subset Y of $\mathbb{R}$, and each τ_Y is a normal state, it is apparent that $\mathscr{S}_n$ is weak-operator compact. Furthermore, $0 \notin \mathscr{S}_n$, $\mathscr{S}_{n+1} \subseteq \mathscr{S}_n$, and $\mathscr{S}_n$ is non-empty since it contains the operator R_n defined by (3). Accordingly there is at least one element R in $\bigcap \mathscr{S}_n$. It is apparent that $0 < R \leq 2P$ and $\tau(RF) < \infty$; moreover, $\hat{\sigma}_p(R) = R$ for all dyadic rational numbers p, and by continuity this remains true for all real p.

The fixed-point algebra $\{S \in \mathscr{R}(\mathscr{M}, \sigma) : \hat{\sigma}_p(S) = S \ (p \in \mathbb{R})\}$ is a von Neumann subalgebra of $\mathscr{R}(\mathscr{M}, \sigma)$ (in fact, it is $\Psi(\mathscr{M})$) that contains R and so contains each spectral projection G of R. For a suitable choice of G, there is a positive scalar a such that $0 < G \leq aR$; and then

$$0 < G \leq P, \qquad \tau(GF) \leq a\tau(RF) < \infty.$$

Since τ is a faithful tracial weight, the projection GF in $\mathscr{R}(\mathscr{M}, \sigma)$ is finite, and hence so is $G\hat{\sigma}_p(F)$ $(= \hat{\sigma}_p(GF))$. Now

$$G = GP = \sum_{j=-\infty}^{\infty} G\hat{\sigma}_j(F);$$

and since the finite projections $G\hat{\sigma}_j(F)$ in $\mathscr{R}(\mathscr{M}, \sigma)$ have pairwise-orthogonal central carriers, it follows that G is finite.

From our assumption that $\mathscr{R}(\mathscr{M}, \sigma)$ has a central portion of type I, we have now deduced the existence of a non-zero finite projection G that is invariant under $\{\hat{\sigma}_p\}$. As already noted, this leads to a contradiction. ■

By taking $\mathscr{M}_0 = \mathscr{R}(\mathscr{M}, \sigma)$, and appealing to Corollary 13.2.10 and Theorems 13.3.6 and 13.3.7, we obtain the following result on the structure of type III von Neumann algebras.

13.3.8. COROLLARY. *If $\mathcal{M}$ is a countably decomposable type* III *von Neumann algebra, there is a continuous automorphic representation $\hat{\sigma}$ of $\mathbb{R}$ on a type* II_∞ *von Neumann algebra $\mathcal{M}_0$, and a faithful normal semi-finite tracial weight τ* on *$\mathcal{M}_0$, such that $\tau \circ \hat{\sigma}_p = e^{-p}\tau$ for all real p, and $\mathcal{M}$ is * isomorphic to $\mathcal{R}(\mathcal{M}_0, \hat{\sigma})$.*

Bibliography: [16, 97]

13.4. Exercises

13.4.1. Suppose that $\mathcal{M}$ is a von Neumann algebra acting on a Hilbert space $\mathcal{H}$, G is a discrete group with unit e, $\alpha: g \to \alpha_g$ is an automorphic representation of G on $\mathcal{M}$ that is implemented by a unitary representation $g \to U(g)$ of G on $\mathcal{H}$, and $\mathcal{R}(\mathcal{M}, \alpha)$ is the (implemented) crossed product of $\mathcal{M}$ by α. Recall, from Definition 13.1.3 and the discussion following it, that $\mathcal{R}(\mathcal{M}, \alpha)$ acts on the Hilbert space $\mathcal{H} \otimes \mathcal{K}'$, where $\mathcal{K}'$ is $l_2(G)$, and that an element R of $\mathcal{B}(\mathcal{H} \otimes \mathcal{K}')$ lies in $\mathcal{R}(\mathcal{M}, \alpha)$ if and only if R is represented by a matrix of the form $[U(pq^{-1})A(pq^{-1})]_{p,q \in G}$ for some mapping $g \to A(g)$ from G into $\mathcal{M}$. Given such an element R of $\mathcal{R}(\mathcal{M}, \alpha)$, let $\Phi'(R)$ be the element $A(e) \otimes I$ of $\mathcal{M} \mathbin{\bar{\otimes}} \mathbb{C}_{\mathcal{K}'}$. Show that

(i) Φ' is a conditional expectation from $\mathcal{R}(\mathcal{M}, \alpha)$ onto $\mathcal{M} \mathbin{\bar{\otimes}} \mathbb{C}_{\mathcal{K}'}$ (see Exercise 8.7.23);
(ii) Φ' is *faithful*, in the sense that $\Phi'(R) \neq 0$ when $0 \neq R \in \mathcal{R}(\mathcal{M}, \alpha)^+$;
(iii) Φ' is weak-operator continuous on the unit ball of $\mathcal{R}(\mathcal{M}, \alpha)$.

13.4.2. Suppose that $\mathcal{M}$ is a type III von Neumann algebra, G is a discrete group, and $\alpha: g \to \alpha_g$ is an automorphic representation of G on $\mathcal{M}$. Show that the crossed product von Neumann algebra $\mathcal{R}(\mathcal{M}, \alpha)$ is of type III. (Note that this is in sharp contrast with the situation for *continuous* crossed products—see, for example, Theorem 13.3.7.) [*Hint*: Use Exercise 13.4.1 and Proposition 11.2.25.]

13.4.3. Suppose that G is a countable locally finite i.c.c. group, $\mathcal{L}_G$ is the factor of type II_1 acting on $l_2(G)$ described in Section 6.7, $\mathcal{K}$ is a separable infinite-dimensional Hilbert space, and $\mathcal{M}$ is the von Neumann algebra $\mathcal{L}_G \mathbin{\bar{\otimes}} \mathcal{B}(\mathcal{K})$. Show that

(i) $\mathcal{M}$ is a factor of type II_∞ with commutant $\mathcal{M}'$ of type II_1;
(ii) there is a * automorphism of $\mathcal{M}$ that is not unitarily implemented. [*Hint*. Use Propositions 12.2.7, 13.1.10, and Exercise 9.6.33.]

13.4.4. Suppose that $\mathscr{R}$ is a factor of type II_1, $\mathscr{K}$ is a separable infinite-dimensional Hilbert space, $\mathscr{M}$ is the factor $\mathscr{R} \bar{\otimes} \mathscr{B}(\mathscr{K})$ of type II_∞, τ_1 is the unique tracial state on $\mathscr{R}$, and τ_∞ is a faithful normal semi-finite tracial weight on $\mathscr{M}$. Let c be a real number such that $0 < c \leq 1$, and let E be a projection in $\mathscr{R}$ such that $\tau_1(E) = c$ (see Proposition 8.5.3). Show that $\mathscr{R}$ is * isomorphic to $E\mathscr{R}E$ if and only if there is a * automorphism θ of $\mathscr{M}$ such that $\tau_\infty \circ \theta = c\tau_\infty$. [*Hint*. Most of the necessary ideas can be found in the proof of Proposition 13.1.10.]

13.4.5. Suppose that $\mathscr{R}$ is a factor of type II_1, $\mathscr{K}$ is a separable infinite-dimensional Hilbert space, $\mathscr{M}$ is the factor $\mathscr{R} \bar{\otimes} \mathscr{B}(\mathscr{K})$ of type II_∞, and τ_∞ is a faithful normal semi-finite tracial weight on $\mathscr{M}$. Suppose also that c is a positive real number, $n(\geq c)$ is a positive integer, and E is a projection in the type II_1 factor $n \otimes \mathscr{R}$ such that $\tau_n(E) = c/n$, where τ_n is the unique tracial state on $n \otimes \mathscr{R}$. Show that $\mathscr{R}$ is * isomorphic to $E(n \otimes \mathscr{R})E$ if and only if there is a * automorphism θ of $\mathscr{M}$ such that $\tau_\infty \circ \theta = c\tau_\infty$. [*Hint*. Adapt the argument required in solving Exercise 13.4.4.]

13.4.6. Suppose that $\mathscr{R}$ is a factor of type II_1, $\mathscr{K}$ is a separable infinite-dimensional Hilbert space, $\mathscr{M}$ is the factor $\mathscr{R} \bar{\otimes} B(K)$ of type II_∞, and τ_∞ is a faithful normal semi-finite tracial weight on $\mathscr{M}$. Let $\mathscr{F}\mathscr{G}(\mathscr{R})$ be the set of all positive real numbers c with the following property: there is * automorphism θ of $\mathscr{M}$ such that $\tau_\infty \circ \theta = c\tau_\infty$. (Note that Exercise 13.4.5 provides an alternative characterization of the elements of $\mathscr{F}\mathscr{G}(\mathscr{R})$.) Show that

(i) $\mathscr{F}\mathscr{G}(\mathscr{R})$ is a subgroup of the multiplicative group of positive real numbers (we refer to $\mathscr{F}\mathscr{G}(\mathscr{R})$ as the *fundamental group* of $\mathscr{R}$);

(ii) $\mathscr{F}\mathscr{G}(\mathscr{R}) = \mathscr{F}\mathscr{G}(E\mathscr{R}E) = \mathscr{F}\mathscr{G}(n \otimes \mathscr{R})$ when E is a non-zero projection in $\mathscr{R}$ and n is a positive integer.

13.4.7. Suppose $\mathscr{M}$ is a factor of type II_∞ with commutant $\mathscr{M}'$ of type II_1 acting on a separable Hilbert space $\mathscr{H}$. Let τ_∞ be a faithful normal semi-finite tracial weight on $\mathscr{M}$, E be a projection in $\mathscr{M}$ such that $\tau_\infty(E) = 1$, and $\mathscr{M}_0$ be $E\mathscr{M}E$. Show that

(i) the mapping, $f\colon \theta \to \tau_\infty(\theta(E))$, is a homomorphism of the group $\mathrm{aut}(\mathscr{M})$ of * automorphisms of $\mathscr{M}$ onto $\mathscr{F}\mathscr{G}(\mathscr{M}_0)$ (defined in Exercise 13.4.6). [*Hint*. Use Proposition 8.5.5.];

(ii) the kernel of the mapping f (defined in (i)) is the subgroup $\mathrm{aut}_s(\mathscr{M})$ of $\mathrm{aut}(\mathscr{M})$ consisting of those automorphisms of $\mathscr{M}$ that are implemented by unitary operators on $\mathscr{H}$ [*Hint*. Use the result of Exercise 9.6.33.];

(iii) the mapping, $\tilde{f} : \theta\, \mathrm{aut}_s(\mathscr{M}) \to f(\theta)$, is an isomorphism of the quotient group $\mathrm{aut}(\mathscr{M})/\mathrm{aut}_s(\mathscr{M})$ onto $\mathscr{F}\mathscr{G}(\mathscr{M}_0)$.

13.4.8. With the notation of Exercise 13.4.6, determine $\mathscr{F}\mathscr{G}(\mathscr{R})$ when $\mathscr{R}$ is the finite matricial factor.

13.4.9. With the notation of Theorem 13.1.15, suppose $a_r \to \frac{1}{2}$ as $r \to \infty$ and $\sum_{r=1}^{\infty} b_r^2 < \infty$. Identify $\mathfrak{A}$ with $\pi_\rho(\mathfrak{A})$, as in the proof of Theorem 13.1.15, so that $\mathfrak{A}^-$ is a factor and $\rho = \omega_x|\mathfrak{A}$ for some cyclic unit vector x for $\mathfrak{A}$. Suppose

$$H'_r = \tfrac{1}{2}[a_r^{-1}E_{11}^{(r)} + (1 - a_r)^{-1}E_{22}^{(r)}] \qquad (r = 1, 2, \ldots),$$

$$K_n = H'_1 H'_2 \cdots H'_n \qquad (n = 1, 2, \ldots),$$

$$\tau'_n(A) = \omega_x(AK_n) \qquad (A \in \mathfrak{A}^-).$$

Show that

(i) K_n is in the centralizer of $\omega_x|\mathfrak{A}^-$;
(ii) τ'_n is a vector state of $\mathfrak{A}^-$;
(iii) $\|\tau'_m - \tau'_n\|^2 \le \rho((K_m - K_n)^2)$;
(iv) the sequence $\{\tau'_n\}$ converges in norm to a normal state τ' of $\mathfrak{A}^-$;
(v) τ' is a tracial state of $\mathfrak{A}^-$;
(vi) $\mathfrak{A}^-$ is a (by Theorem 12.2.1, is the unique) finite matricial factor.

13.4.10. With the notation of Theorem 13.1.15, suppose $\sum_{r=1}^{\infty} a_r < \infty$ (this is equivalent to the assumption that $\sum_{r=1}^{\infty} e^{-b_r} < \infty$, since $0 < a_r = (1 + e^{b_r})^{-1}$). Identify $\mathfrak{A}$ with $\pi_\rho(\mathfrak{A})$, as in the proof of Theorem 13.1.15, so that $\mathfrak{A}^-$ is a factor and $\rho = \omega_x|\mathfrak{A}$ for some cyclic unit vector x for $\mathfrak{A}$. For $r = 1, 2, \ldots$, let ω_r be the pure state of $\mathfrak{A}_r$ given by

$$\omega_r\left(\sum_{j,k=1}^{2} c_{jk} E_{jk}^{(r)}\right) = c_{22},$$

and let H_r be the element $(1 - a_r)^{-1/2} E_{22}^{(r)}$ of $\mathfrak{A}_r$. Define a sequence $\{\sigma_n\}$ of states of $\mathfrak{A}$ by

$$\sigma_n = \omega_1 \otimes \omega_2 \otimes \cdots \otimes \omega_n \otimes \rho_{n+1} \otimes \rho_{n+2} \otimes \cdots.$$

Show that

(i) $\sigma_n = \omega_{K_n x}|\mathfrak{A}$, where $K_n = H_1 H_2 \cdots H_n$;
(ii) $\|K_n x - K_m x\|^2 = 2[1 - \prod_{r=m+1}^{n}(1 - a_r)^{1/2}] \quad (1 \le m < n)$;
(iii) the sequence $\{K_n x\}$ converges in norm to a unit vector y;
(iv) $\omega_y|\mathfrak{A}$ is a pure state of $\mathfrak{A}$;
(v) $\mathfrak{A}^-$ is a type I_∞ factor.

13.4.11. With the notation of Theorem 13.1.15, suppose that $a_r \to 0$ and $\sum_{r=1}^{\infty} a_r$ diverges.

(i) Show that $b_r \to \infty$ and $\sum_{r=1}^{\infty} e^{-b_r}$ diverges.

(ii) Suppose that $\{c_r\}$ and $\{d_r\}$ are sequences of positive real numbers such that $\sum_{r=1}^{\infty} c_r$ diverges and $d_r \to \infty$, and a, b are real numbers with $a < b$. Show that the set

$$\left\{t \in \mathbb{R} : a < t < b,\ \sum_{r=1}^{\infty} c_r \sin^2(\tfrac{1}{2}d_r t) \text{ diverges}\right\}$$

has positive Lebesgue measure. [*Hint*. Suppose the contrary. Upon deleting a finite number of terms from the sequences $\{c_r\}$, $\{d_r\}$, reduce to the case in which $4d_r^{-1} < b - a$ for $r = 1, 2, \ldots$. Consider

$$\int_a^b f_n(t)\, dt,$$

where

$$f_n(t) = (c_1 + c_2 + \cdots + c_n)^{-1}(c_1 \sin^2(\tfrac{1}{2}d_1 t) + \cdots + c_n \sin^2(\tfrac{1}{2}d_n t)).]$$

(iii) Deduce from (i), (ii), and Theorem 13.1.15 that $\pi_\rho(\mathfrak{A})^-$ is a type III factor, and the complement in $\mathbb{R}$ of the set $T(\pi_\rho(\mathfrak{A})^-)$ meets each non-empty open subset of $\mathbb{R}$ in a set of positive Lebesgue measure.

13.4.12. With the notation of Theorem 13.1.15, show that $T(\pi_\rho(\mathfrak{A})^-) = \{0\}$ if

$$a_n = \frac{1}{n+1} \qquad (n = 1, 2, \ldots).$$

[*Hint*. Note that $b_n = \log n$, and that it is necessary to show that the series $\sum_{n=1}^{\infty} n^{-1} \sin^2(\tfrac{1}{2}t \log n)$ diverges for all non-zero real numbers t. When $t > 0$, define $c(>1)$ by the condition $t \log c = \pi$, and for $k = 1, 2, \ldots$, let n_k be the largest positive integer such that

$$\tfrac{1}{2}t \log n_k < 2k\pi + (\pi/4).$$

Consider the sum

$$\sum_{n=n_k+1}^{[cn_k]} \frac{1}{n} \sin^2(\tfrac{1}{2}\, t \log n),$$

where $[cn_k]$ denotes the largest integer not exceeding cn_k.]

13.4.13. With the notation of Theorem 13.1.15, show that

$$T(\pi_\rho(\mathfrak{A})^-) = \{0, \pm 2\pi, \pm 4\pi, \ldots\}$$

if $b_n = [\log n]$ $(n = 3, 4, \ldots)$, where $[x]$ denotes the largest integer not exceeding x. [*Hint*. When t is a multiple of 2π, each term of the series $\sum_{n=1}^{\infty} e^{-b_n} \sin^2(\frac{1}{2}b_n t)$ is 0, since b_n is an integer. In view of this, and from Theorem 13.1.15, it suffices to prove that the series diverges when t is not a multiple of 2π. For such t, by considering the cases in which $t/2\pi$ is rational and irrational, show that there is a strictly increasing sequence $\{m_k\}$ of positive integers and a positive real number b, such that

$$|\sin(\tfrac{1}{2}m_k t)| \geq b \qquad (k = 1, 2, \ldots).$$

Let n_k be the largest integer for which $\log n_k < m_k$. Show that

$$b_n = m_k \qquad (n_k + 1 \leq n \leq 2n_k),$$

and consider

$$\sum_{n=n_k+1}^{2n_k} e^{-b_n} \sin^2(\tfrac{1}{2}b_n t).]$$

13.4.14. With the notation of Theorem 13.1.15, suppose that, for $n = 1, 2, \ldots,$

$$b_r = n! \qquad \text{when} \quad [e^{n!}] < r \leq [e^{(n+1)!}],$$

where $[x]$ denotes the largest integer not exceeding x. Show that $T(\pi_\rho(\mathfrak{A})^-)$ contains each rational multiple of 2π but is not the whole of $\mathbb{R}$.

13.4.15. With the notation of Theorem 13.1.15, suppose that

$$a_{2r-1} \to 0, \qquad a_{2r} \to \tfrac{1}{2}, \qquad \text{as} \quad r \to \infty,$$

$$\sum_{r=1}^{\infty} a_{2r-1} < \infty, \qquad \sum_{r=1}^{\infty} b_{2r}^2 < \infty.$$

Show that the factor $\pi_\rho(\mathfrak{A})^-$ is of type II_∞. [*Hint*. Let

$$\mathfrak{A}_o = \bigotimes_{r=1}^{\infty} \mathfrak{A}_{2r-1}, \qquad \rho_o = \bigotimes_{r=1}^{\infty} \rho_{2r-1},$$

$$\mathfrak{A}_e = \bigotimes_{r=1}^{\infty} \mathfrak{A}_{2r}, \qquad \rho_e = \bigotimes_{r=1}^{\infty} \rho_{2r}.$$

Identify $\mathfrak{A}_o$ with $\pi_{\rho_o}(\mathfrak{A}_o)$ and $\mathfrak{A}_e$ with $\pi_{\rho_e}(\mathfrak{A}_e)$ (see the first paragraph of the proof of Theorem 13.1.15) so that

$$\rho_o = \omega_{x_o}|\mathfrak{A}_o \quad ; \quad \rho_e = \omega_{x_e}|\mathfrak{A}_e$$

for some cyclic unit vectors, x_o for $\mathfrak{A}_o$ and x_e for $\mathfrak{A}_e$, and $\mathfrak{A}_o^-$, $\mathfrak{A}_e^-$ are factors of types I_∞, II_1, respectively, by Exercises 13.4.10 and 13.4.9. Show that there is a * isomorphism φ from $\mathfrak{A}$ onto $\mathfrak{A}_o \otimes \mathfrak{A}_e$ such that

$$\rho = \omega_{x_o \otimes x_e} \circ \varphi.$$

Deduce that the C^*-algebra $\pi_\rho(\mathfrak{A})$ is unitarily equivalent to $\mathfrak{A}_o \otimes \mathfrak{A}_e$.]

13.4.16. Suppose that $\mathscr{M}$ is a countably decomposable semi-finite von Neumann algebra and $\{\sigma_t\}$ is the modular automorphism group corresponding to a faithful normal state ω of $\mathscr{M}$. Show that the continuous crossed product $\mathscr{R}(\mathscr{M}, \sigma)$ is unitarily equivalent to $\mathscr{M} \mathbin{\bar{\otimes}} \mathscr{A}$, where $\mathscr{A}$ is the multiplication algebra corresponding to Lebesgue measure on $\mathbb{R}$. [*Hint*. By Theorem 9.2.21, there is a continuous unitary representation $t \to U(t)$: $\mathbb{R} \to \mathscr{M}$ such that $U(t)$ implements σ_t. Interpret $\mathscr{R}(\mathscr{M}, \sigma)$ as an implemented continuous crossed product (Definition 13.2.6) and use Theorem 13.2.8 and Proposition 13.2.7(i).]

13.4.17. Let $\mathscr{R}$ be a von Neumann algebra acting on a Hilbert space $\mathscr{H}$, $t \to \alpha_t$ be an automorphic representation of $\mathbb{R}$ on $\mathscr{R}$ implemented by a strong-operator-continuous one-parameter unitary group $t \to U_t$ with U_t in $\mathscr{R}$ for each real t, and $\mathscr{S}$ be the crossed product of $\mathscr{R}$ by α. Show that

(i) $\mathscr{S}$ is * isomorphic to $\mathscr{R} \mathbin{\bar{\otimes}} \mathscr{A}$ where $\mathscr{A}$ is the multiplication algebra of $L_2(\mathbb{R})$;
(ii) $\mathscr{S}$ is of type I, II, or III when $\mathscr{R}$ is of type I, II, or III, respectively.

13.4.18. Let $\mathscr{R}$ be a von Neumann algebra acting on a separable Hilbert space $\mathscr{H}$, u be a separating and generating unit vector for $\mathscr{R}$, and J be the modular conjugation corresponding to $(\mathscr{R}, u)$.

(i) Find a norm-continuous one-parameter unitary group $t \to U_t$ on $\mathscr{H}$ that generates a maximal abelian subalgebra of $\mathscr{R}$. [*Hint*. Use Exercise 9.6.41.]
(ii) Show that U_tJU_tJ $(=V_t)$ is a norm-continuous one-parameter unitary group and that $V_tAV_t^* = U_tAU_t^*$ $(=\alpha_t(A))$ for each A in $\mathscr{R}$ and t in $\mathbb{R}$.
(iii) Show that $\mathscr{R}$ and $\{V_t : t \in \mathbb{R}\}$ generate a type I von Neumann algebra $\mathscr{T}$. [*Hint*. Note that $\{JU_tJ\}$ generates a maximal abelian subalgebra of $\mathscr{R}'$ and use Exercise 9.6.1.]
(iv) With α_t as in (ii) and $\mathscr{T}$ as in (iii), show that the crossed product of $\mathscr{R}$ by α is not isomorphic to $\mathscr{T}$ when $\mathscr{R}$ is of type II or III. [*Hint*. Use Exercise 13.4.17.]

13.4.19. Let $\mathscr{R}$ be a von Neumann algebra acting on a Hilbert space $\mathscr{H}$, aut($\mathscr{R}$) be the group of * automorphisms of $\mathscr{R}$, and i($\mathscr{R}_{\#}$) be the group of isometries of the predual $\mathscr{R}_{\#}$ of $\mathscr{R}$ onto itself. The *strong topology* on i($\mathscr{R}_{\#}$) has a basic open neighborhood of a_0 determined by a finite set $\{\omega_1, \ldots, \omega_n\}$ of elements of $\mathscr{R}_{\#}$ and consists of those a in i($\mathscr{R}_{\#}$) such that $\|a(\omega_j) - a_0(\omega_j)\| < 1$ for each j in $\{1, \ldots, n\}$.

(i) Show that each α in aut($\mathscr{R}$) is the (Banach space) adjoint of a (unique) $\alpha_{\#}$ in i($\mathscr{R}_{\#}$) and that the mapping $\alpha \to \alpha_{\#}$ is an anti-isomorphism of the group aut($\mathscr{R}$) with its image in i($\mathscr{R}_{\#}$).

The mapping of (i) transfers the strong topology on i($\mathscr{R}_{\#}$) to a topology on aut($\mathscr{R}$) we shall call the *bounded weak-operator* (bw-) *topology*.

(ii) Show that i($\mathscr{R}_{\#}$) is a topological group in the strong topology, and conclude that aut($\mathscr{R}$) is a topological group in the bw-topology.

(iii) Suppose $\mathscr{R}$ has a generating and separating vector. Show that there is a (strong-operator-) continuous unitary representation of aut($\mathscr{R}$) (provided with its bw-topology) on $\mathscr{H}$ that implements (the identity representation of) aut($\mathscr{R}$). [*Hint*. Use Exercises 9.6.65 and 9.6.60(iv).]

13.4.20. With the notation of Exercise 13.4.19, suppose $g \to U_g$ is a strong-operator continuous unitary representation of a topological group G on $\mathscr{H}$ such that $U_g A U_g^* \,(= \alpha_g(A)) \in \mathscr{R}$ for each g in G and A in $\mathscr{R}$. Show that $g \to \alpha_g$ is a continuous homomorphism of G into aut($\mathscr{R}$) provided with its bw-topology.

13.4.21. Let $\mathscr{R}$ be a von Neumann algebra and $\mathscr{B}(\mathscr{R})$ be the linear space of bounded linear transformations of $\mathscr{R}$ into itself equipped with the weak topology induced by the family.

$$\{\beta \to \omega(\beta(A)) \; (\beta \in \mathscr{B}(\mathscr{R})) : \omega \in \mathscr{R}_{\#},\ A \in \mathscr{R}\}$$

of linear functionals on $\mathscr{B}(\mathscr{R})$. Show that

(i) an automorphic representation α of a topological group G on $\mathscr{R}$ is continuous in the sense of Section 13.2 if and only if the mapping α is continuous from G into $\mathscr{B}(\mathscr{R})$;

(ii) the mapping that assigns to each unitary operator U in $\mathscr{R}$ the automorphism $A \to UAU^* (A \in \mathscr{R})$ is a continuous mapping from $\mathscr{U}(\mathscr{R})$, the unitary group of $\mathscr{R}$ equipped with the weak-operator topology, into $\mathscr{B}(\mathscr{R})$. [*Hint*. Use Exercise 5.7.5.]

13.4.22. Let $\mathscr{R}$ be a von Neumann algebra and α be an automorphic representation of $\mathbb{R}$ on $\mathscr{R}$ (not assumed continuous). Suppose that each $\alpha(t)$ is inner and G is the family of unitary operators in $\mathscr{R}$ that implement some $\alpha(t)$.

(i) Show that G is a subgroup of the group of unitary operators in $\mathscr{R}$ and that U and V in G implement the same automorphism of $\mathscr{R}$ if and only if $V = CU$ for some unitary operator C in the center of $\mathscr{R}$.

(ii) Suppose U and V in G implement $\alpha(t)$ and $\alpha(s)$, respectively. Show that $UV = VU$ if t/s is rational [*Hint*. Suppose $t/s = n/m$ with n and m integers. Choose W in G implementing $\alpha(t/n)$ and note that W^n implements $\alpha(t)$. Conclude that $U = CW^n$ for some central unitary operator C, and that $WU = UW$. Then note that $W^m U = UW^m$.]

(iii) Suppose α is continuous. Show that G is abelian.

13.4.23. Let α be a continuous automorphic representation of $\mathbb{R}$ on a factor $\mathscr{M}$ by inner automorphisms, G be the family of unitary operators in $\mathscr{M}$ that implement some $\alpha(t)$, $\mathfrak{A}$ be the C^*-subalgebra of $\mathscr{M}$ generated by G and ρ be a pure state of $\mathfrak{A}$. Show that

(i) ρ restricts to a character ξ of G (norm-, but not necessarily, strong-operator-) continuous on G [*Hint*. Use Exercise 13.4.22.];

(ii) the kernel H of ξ and $\{cI : |c| = 1\}$ $(=C)$ generate G, and $H \cap C = \{I\}$;

(iii) the mapping that assigns to each U in H the automorphism that U implements on $\mathscr{M}$ is an isomorphism of H onto $\{\alpha(t) : t \in \mathbb{R}\}$, conclude that there is a homomorphism of $\mathbb{R}$ into the unitary group of $\mathscr{M}$ that implements α. (In general, this homomorphism will not be strong-operator continuous.)

13.4.24. Let $\mathscr{R}$ be a von Neumann algebra acting on a Hilbert space $\mathscr{H}$, u be a separating and generating unit vector for $\mathscr{R}$, (J, Δ) be the modular structure corresponding to $(\mathscr{R}, u)$, and $t \to \sigma_t$ be the modular automorphism group of $\mathscr{R}$ implemented by $t \to \Delta^{it}$. Suppose that each σ_t is an inner automorphism of $\mathscr{R}$. Show that

(i) there is a maximal abelian subalgebra $\mathscr{A}$ of $\mathscr{R}$ such that $\sigma_t(A) = A$ for each A in $\mathscr{A}$ [*Hint*. Use Exercise 13.4.22.];

(ii) with $\mathscr{A}$ as in (i), $\mathscr{A}$ is contained in the centralizer of $\omega_u|\mathscr{R}$;

(iii) there is an ultraweakly continuous conditional expectation Φ_u (see Exercise 8.7.23) mapping $\mathscr{R}$ onto $\mathscr{A}$ such that $\Phi_u \circ \sigma_t = \Phi_u$ for all t in $\mathbb{R}$. [*Hint*. With H in $(\mathscr{R}^+)_1$, define $\omega_0(A)$ to be $\omega_u(HA)$ for A in $\mathscr{A}$. Show that $0 \le \omega_0 \le \omega_u|\mathscr{A}$ and use Theorem 7.3.13 to construct Φ_u. Show that $\omega_{Au} \circ \Phi_u$ is strong-operator continuous at 0 on $\mathscr{R}$ and use Lemma 7.1.3 to prove that Φ_u is ultraweakly continuous.];

(iv) $\omega_u \circ \Phi_u = \omega_u|\mathscr{R}$, Φ_u is uniquely defined by this equality, and Φ_u is faithful (see Exercise 8.7.28(iii)).

13.4.25. Let $\mathcal{M}$ be the (left) von Neumann algebra $\mathcal{L}_{\mathcal{F}_2}$ corresponding to the free group $\mathcal{F}_2$ on two generators a and b (see Example 6.7.6) and let u be the unit trace vector for $\mathcal{M}$ corresponding to the function that takes the value 1 at the unit element of $\mathcal{F}_2$ and 0 at all other elements.

(i) Identify the modular group of $\mathcal{M}$ corresponding to $(\mathcal{M}, u)$.

(ii) Identify the group of unitary elements in $\mathcal{M}$ that implement some element of the modular group found in (i).

(iii) Let $\mathcal{A}$ be the maximal abelian subalgebra of $\mathcal{M}$ corresponding to the generator a. (See Exercise 6.9.42.) Note that $\mathcal{A}$ is contained in the centralizer of $\omega_u|\mathcal{M}$, and let Φ be the conditional expectation of $\mathcal{M}$ onto $\mathcal{A}$ described in Exercise 13.4.24(iii). How does Φ compare to φ of Exercise 8.7.28(iii) (where $\mathcal{A}$ and $\mathcal{M}$ replace $\mathcal{R}$ and $\mathcal{S}$, respectively)?

(iv) With L_x in $\mathcal{M}$, show that $\Phi(L_x) = L_{x'}$, where $x'(a^n) = x(a^n)$ for each n in $\mathbb{Z}$ and $x'(c) = 0$ when $c \notin \{a^n : n \in \mathbb{Z}\}$. [*Hint*. Recall from Exercise 13.4.24, that $\omega_u \circ \Phi = \omega_u|\mathcal{M}$.]

CHAPTER 14

DIRECT INTEGRALS AND DECOMPOSITIONS

In Section 2.6, *Direct sums*, we studied direct sums of Hilbert spaces. In Chapter 5 (following Corollary 5.5.7), we considered direct sums of von Neumann algebras. The present chapter deals with a useful generalization of the concept of "direct sum" as it applies to Hilbert-space constructs. In this generalization the "discrete" index set X of the sum is replaced by a (suitably restricted) measure space (X, μ). In the simplest case, with one-dimensional component Hilbert spaces of complex numbers, the generalization amounts to passing from $l_2(X)$ to $L_2(X, \mu)$. In the case of direct sums, we assign Hilbert-space constructs, for example, operators, to each point of X and "add" them. In the theory of direct integrals, we assign such constructs to each point of the measure space (X, μ) and "integrate" them. For the case of direct sums, we may have to impose a convergence condition (especially when X is infinite). For the case of direct integrals, we must impose both measurability restrictions (on the assignment of constructs to points) and convergence (that is, integrability) restrictions.

To avoid the possible pitfalls inherent in the consideration of measure spaces of a very general nature, we shall assume, throughout this chapter, that our measure space (X, μ) consists of a locally compact σ-compact space X (that is, X is the countable union of compact sets) and μ is a positive Borel measure on X (taking finite values on compact sets). At the same time, many of the measure-theoretic arguments we give will involve eliminating collections of subsets of X of measure 0 ("μ-null sets," or simply, "null sets," when the context makes clear what is intended). Of course, these collections must be countable for such an argument to be effective. The possibility of keeping these collections countable relies on an assumption of separability of the Hilbert spaces that enter our discussion. This assumption applies throughout the chapter. At a certain stage (following Theorem 14.1.21), we shall want to assume that our measure space can be given a metric in which it is complete and separable. The reader who finds this assumption reassuring is urged to consider it in force throughout the chapter. There is no serious loss of generality if we think of X as the unit interval plus at most a countable number of atoms and μ as Lebesgue measure on the unit interval.

The chapter is divided into three sections. The first, and longest section, describes Hilbert spaces that are direct integrals and develops their theory. In particular, the operators and von Neumann algebras that are decomposable relative to such a direct integral are studied. Section 14.2 deals with the possibility of decomposing a given Hilbert space as a direct integral of Hilbert spaces relative to a given abelian von Neumann algebra on it. Section 14.3 is an appendix composed of those less standard measure-theoretic results needed in the earlier sections of this chapter.

14.1. Direct integrals

In this section, we define direct-integral decompositions of Hilbert spaces, operators that are decomposable and diagonalizable relative to such a decomposition, and von Neumann algebras that are decomposable relative to such a decomposition. We study the basic properties of these constructs.

If we follow this development in the familiar special case of direct sums of Hilbert spaces (the case of direct integral decompositions over discrete measure spaces), the point of view we adopt is that each vector of the direct sum is a function on the index set to the various Hilbert spaces (subspaces) that make up the direct sum. To guarantee that we have the full direct sum rather than a proper subspace, we make the technical assumption embodied in Definition 14.1.1(ii). The diagonalizable operators are those that are scalars on each of the spaces; and the decomposable operators are those that transform the subspaces of the direct sum into themselves (see Definition 14.1.6). While it is relatively easy to show that the bound of a decomposable operator is the supremum of the bounds of its various components, the corresponding result (Proposition 14.1.9) for direct integrals requires some more effort and care. As one might suspect from the case of direct sums, the families of decomposable operators and diagonalizable operators form von Neumann algebras with the latter the center of the former (Theorem 14.1.10). Direct-integral decompositions of representations of C^*-algebras and states appear (Definition 14.1.12) in a manner analogous to their direct-sum decompositions.

Defining direct integrals of von Neumann algebras requires a more circumspect approach than is needed for their direct sums. The countability demands of the measure-theoretic situation require us to operate from some countable "staging area." A norm-separable C^*-subalgebra and the components of its identity representation are used for this. (See Definition 14.1.14.) Fine points of normality of components of normal states and the nature of the components of projections with special properties (for example, abelian, finite, etc.) take on greater significance in the context of direct integrals (see

Lemmas 14.1.19 and 14.1.20) and allow us to identify the types of the components of the von Neumann algebra. (See Theorem 14.1.21.) The type III situation presents some special problems that have been avoided to that point. To illustrate these difficulties, note that if we form the direct sum, $\sum \oplus \mathscr{H}_a$, of Hilbert spaces $\mathscr{H}_a$ $(a \in \mathbb{A})$ and have assigned to each index a some collection $\mathscr{S}_a$ of bounded operators on $\mathscr{H}_a$, there is no problem in selecting an operator T_a from each $\mathscr{S}_a$ and forming the direct sum operator, $\sum \oplus T_a$ (provided $\{\|T_a\| : a \in \mathbb{A}\}$ is bounded). In the case of direct integrals, where the index family $\mathbb{A}$ must be replaced by the measure space (X, μ), we have the added requirement that the selection must be made in a "measurable manner." The techniques of Borel structures and analytic sets used in establishing the measurable selection principle needed for this appear in the appendix (Section 14.3). The results that draw on this principle appear at the end of this section—notably, the result that the components of the commutant are the commutants of the components (Proposition 14.1.24) and the proof that the components of a type III von Neumann algebra are of type III.

14.1.1. Definition. If X is a σ-compact locally compact (Borel measure) space, μ is the completion of a Borel measure on X, and $\{\mathscr{H}_p\}$ is a family of separable Hilbert spaces indexed by the points p of X, we say that a separable Hilbert space $\mathscr{H}$ is the *direct integral* of $\{\mathscr{H}_p\}$ over (X, μ) (we write: $\mathscr{H} = \int_X \oplus \mathscr{H}_p \, d\mu(p)$) when, to each x in $\mathscr{H}$, there corresponds a function $p \to x(p)$ on X such that $x(p) \in \mathscr{H}_p$ for each p and

(i) $p \to \langle x(p), y(p) \rangle$ is μ-integrable, when $x, y \in \mathscr{H}$, and $\langle x, y \rangle = \int_X \langle x(p), y(p) \rangle \, d\mu(p)$

(ii) if $u_p \in \mathscr{H}_p$ for all p in X and $p \to \langle u_p, y(p) \rangle$ is integrable for each y in $\mathscr{H}$, then there is a u in $\mathscr{H}$ such that $u(p) = u_p$ for almost every p. We say that $\int_X \oplus \mathscr{H}_p \, d\mu(p)$ and $p \to x(p)$ are the (*direct integral*) *decompositions* of $\mathscr{H}$ and x, respectively. ■

14.1.2. Remark. From (ii) of the preceding definition, with x and y in $\mathscr{H}$ there is a z in $\mathscr{H}$ such that $ax(p) + y(p) = z(p)$ for almost every p. Since

$$\langle ax + y - z, u \rangle = \int_X \langle ax(p) + y(p) - z(p), u(p) \rangle \, d\mu(p) = 0$$

for all u in $\mathscr{H}$, it follows that $z = ax + y$. That is, the function corresponding to $ax + y$ agrees with $p \to ax(p) + y(p)$ almost everywhere. It follows that if $x(p) = y(p)$ almost everywhere, then $x = y$; for then $(x - y)(p) = 0$ almost everywhere and, from (i) of Definition 14.1.1, $\|x - y\|^2 = 0$. ■

It follows, as well, from (i) and (ii) that the span of $\{x(p) : x \in \mathscr{H}\}$ is $\mathscr{H}_p$ for almost all p. In the lemma that follows, we prove an expanded form of this fact that will be useful to us.

14.1.3. LEMMA. *If $\{x_a\}$ is a set spanning $\mathscr{H}$, then $\mathscr{H}_p^0 = \mathscr{H}_p$ for almost every p, where $\mathscr{H}_p^0$ is the closed subspace of $\mathscr{H}_p$ spanned by $\{x_a(p)\}$.*

Proof. If $X_0 = \{p : p \in X, \mathscr{H}_p^0 \neq \mathscr{H}_p\}$ and u_p is a unit vector in $\mathscr{H}_p \ominus \mathscr{H}_p^0$ or 0 as $p \in X_0$ or $p \notin X_0$, then $0 = \langle u_p, x_a(p) \rangle$ for all p. With y in $\mathscr{H}$, let $\{y_n\}$ be a sequence of finite linear combinations of elements in $\{x_a\}$ such that $\|y - y_n\| \to 0$. If $y_j = b_1 x_{a_1} + \cdots + b_n x_{a_n}$, then $y_j(p) = b_1 x_{a_1}(p) + \cdots + b_n x_{a_n}(p)$ except for p in a null set N_j. Thus $0 = \langle u_p, y_j(p) \rangle$ for p in $X \backslash N_j$. Since

$$\|y - y_n\|^2 = \int_X \|y(p) - y_n(p)\|^2 \, d\mu(p) \to 0,$$

some subsequence $\{\|y(p) - y_{n_k}(p)\|\}$ tends to 0 except for p in a null set N_0. For p not in the null set $\bigcup_{j=0}^{\infty} N_j$, then, $\langle u_p, y(p) \rangle = 0$. In particular, $p \to \langle u_p, y(p) \rangle$ is integrable for each y in $\mathscr{H}$. From Definition 14.1.1(ii), there is a u in $\mathscr{H}$ such that $u_p = u(p)$ almost everywhere. But

$$0 = \langle u_p, u(p) \rangle = \langle u_p, u_p \rangle$$

almost everywhere. As u_p is a unit vector when p is in X_0, X_0 is a null set. ∎

14.1.4. EXAMPLES. (a) The space $L_2(X, \mu)$ is itself the direct integral of one-dimensional Hilbert spaces $\{C_p\}$ (each identified with the complex numbers). To see this, select from each equivalence class of functions in $L_2(X, \mu)$ a representative f. Then (i) of Definition 14.1.1 is a consequence of the definition of $L_2(X, \mu)$. For (ii) of that definition, we note that if f is a complex-valued function on X such that $f \cdot g \in L_1(X, \mu)$ for each g in $L_2(X, \mu)$, then $f \in L_2(X, \mu)$. (Compare Exercise 1.9.30.)

(b) The (discrete) direct sum of a countable family of Hilbert spaces $\{\mathscr{H}_n\}$ may be viewed as the direct integral of $\{\mathscr{H}_n\}$ over the space of natural numbers provided with the measure that assigns to each subset the number of elements it contains. Each element of the direct sum is a function $n \to x(n)$ with domain $\mathbb{N}$, where $x(n) \in \mathscr{H}_n$. If y is another element with corresponding function $n \to y(n)$, then

$$\langle x, y \rangle = \sum_{n=1}^{\infty} \langle x(n), y(n) \rangle,$$

by definition of the inner product on the direct sum. But the sum in this last equality is the integral relative to the ("counting") measure on $\mathbb{N}$ just described; and (i) of Definition 14.1.1 is fulfilled.

To verify (ii) of that definition, suppose $u_n \in \mathscr{H}$ for each n in $\mathbb{N}$; and suppose that $\sum_{n=1}^{\infty} |\langle u_n, y(n)\rangle| < \infty$ for each y in the direct sum. Let f be a function in $l_2(\mathbb{N})$, and let $y(n)$ be $\|u_n\|^{-1}\overline{f(n)}u_n$ if $u_n \neq 0$ and 0 if $u_n = 0$. Then $\sum_{n=1}^{\infty}\langle u_n, y(n)\rangle = \sum_{n=1}^{\infty}\|u_n\| \cdot f(n) < \infty$. It follows that $n \to \|u_n\|$ is in $l_2(\mathbb{N})$. (See Exercise 1.9.30.) Thus u is in the direct sum, where $u(n) = u_n$. ■

14.1.5. Remark. If $\mathscr{H}$ is the direct integral of $\{\mathscr{H}_p\}$ over (X, μ), it may occur that the spaces $\mathscr{H}_p$ have varying dimensions (finite as well as countably infinite under our separability assumption). We note that the set X_n of points p in X at which $\mathscr{H}_p$ has dimension n is measurable. To see this, let $\{x_j\}$ be an orthonormal basis for $\mathscr{H}$. Let $r_1, r_2, \ldots$ be an enumeration of the (complex) rationals, where $r_1 = 1$. With $j_1, \ldots, j_n, k_1, \ldots, k_n$, and m positive integers, let $X_{j,k,m}$ be $\{p : \|r_{j_1}x_{k_1}(p) + \cdots + r_{j_n}x_{k_n}(p)\| < m^{-1}\}$ (where j and k denote the ordered n-tuples $\langle j_1, \ldots, j_n\rangle$ and $\langle k_1, \ldots, k_n\rangle$, respectively, some $j_h = 1$, and $\{k_1, \ldots, k_n\}$ are distinct). From Lemma 14.1.3, with the exception of points p in a null set X_0, $\{x_j(p)\}$ generates $\mathscr{H}_p$. For p not in X_0, $\mathscr{H}_p$ has dimension less than n precisely when p lies in $\bigcap_{k,m}\bigcup_j X_{j,k,m}$. Thus the set of points at which $\mathscr{H}_p$ has dimension less than n is measurable; and each X_n is measurable. ■

14.1.6. Definition. If $\mathscr{H}$ is the direct integral of $\{\mathscr{H}_p\}$ over $\{X, \mu\}$, an operator T in $\mathscr{B}(\mathscr{H})$ is said to be *decomposable* when there is a function $p \to T(p)$ on X such that $T(p) \in \mathscr{B}(\mathscr{H}_p)$ and, for each x in $\mathscr{H}$, $T(p)x(p) = (Tx)(p)$ for almost every p. If, in addition, $T(p) = f(p)I_p$, where I_p is the identity operator on $\mathscr{H}_p$, we say that T is *diagonalizable.* ■

14.1.7. Remark. If $p \to T(p)$ and $p \to T'(p)$ are decompositions of T, then $T(p) = T'(p)$ almost everywhere. For this, let $\{x_j\}$ be a denumerable set spanning $\mathscr{H}$. From Lemma 14.1.3, there is a null set N_0 such that $\{x_j(p)\}$ spans $\mathscr{H}_p$ for p in $X\backslash N_0$. At the same time,

$$T(p)x_j(p) = (Tx_j)(p) = T'(p)x_j(p),$$

except for p in a null set N_j. It follows that the (bounded) operators $T(p)$ and $T'(p)$ coincide on $X\backslash N$, where $N = \bigcup_{j=0}^{\infty} N_j$.

Conversely, if T and S are decomposable and $T(p) = S(p)$ almost everywhere, then $T = S$; for, then,

$$\begin{aligned}\langle Tx, y\rangle &= \int_X \langle (Tx)(p), y(p)\rangle\, d\mu(p) = \int_X \langle T(p)x(p), y(p)\rangle\, d\mu(p)\\ &= \int_X \langle S(p)x(p), y(p)\rangle\, d\mu(p) = \langle Sx, y\rangle\end{aligned}$$

for all x and y in $\mathscr{H}$.

If f is a bounded measurable function on X, then $p \to \langle f(p)x(p), y(p)\rangle$ is integrable for all x and y in $\mathscr{H}$. From Definition 14.1.1(ii), there is a z in $\mathscr{H}$ such that $f(p)x(p) = z(p)$ almost everywhere. Defining $M_f x$ to be z, we have that M_f is a diagonalizable operator with decomposition $p \to f(p)I_p$. In particular, if f is the characteristic function of some measurable set X_0, then M_f is a projection—*the diagonalizable projection corresponding to* X_0. If H is a diagonalizable positive operator, it will follow from Proposition 14.1.9 that H has the form M_f with f measurable and essentially bounded. From this we can conclude the same for each diagonalizable operator. ∎

14.1.8. PROPOSITION. *If $\mathscr{H}$ is the direct integral of $\{\mathscr{H}_p\}$ over (X, μ) and T_1, T_2 are decomposable operators in $\mathscr{B}(\mathscr{H})$, then $aT_1 + T_2$, T_1T_2, T_1^*, and I are decomposable and the following relations hold for almost every p:*

(i) $(aT_1 + T_2)(p) = aT_1(p) + T_2(p)$;
(ii) $(T_1T_2)(p) = T_1(p)T_2(p)$;
(iii) $T_1^*(p) = T_1(p)^*$;
(iv) $I(p) = I_p$.

Moreover,

(v) *if $T_1(p) \le T_2(p)$ almost everywhere then $T_1 \le T_2$.*

Proof. For (i) note that, given x in $\mathscr{H}$, and defining $(aT_1 + T_2)(p)$ to be $aT_1(p) + T_2(p)$, we have

$$\begin{aligned}(aT_1 + T_2)(p)x(p) &= aT_1(p)x(p) + T_2(p)x(p) = (aT_1x)(p) + (T_2x)(p)\\ &= (aT_1x + T_2x)(p) = ((aT_1 + T_2)x)(p)\end{aligned}$$

for almost every p, from Definition 14.1.6 and Remark 14.1.2. Thus $aT_1 + T_2$ is decomposable with decomposition $p \to aT_1(p) + T_2(p)$.

Similarly, defining $(T_1T_2)(p)$ to be $T_1(p)T_2(p)$, we have

$$(T_1T_2)(p)x(p) = T_1(p)(T_2(p)x(p)) = T_1(p)((T_2x)(p)) = (T_1T_2x)(p),$$

almost everywhere, for each x in $\mathscr{H}$. Thus T_1T_2 is decomposable with decomposition $p \to T_1(p)T_2(p)$.

Defining $T^*(p)$ to be $T(p)^*$, we have

$$\langle T^*(p)x(p), y(p)\rangle = \langle x(p), T(p)y(p)\rangle = \langle x(p), (Ty)(p)\rangle$$

almost everywhere; and $p \to \langle x(p), (Ty)(p)\rangle$ is integrable. From Definition 14.1.1.(ii), there is a z in $\mathscr{H}$ such that $T^*(p)x(p) = z(p)$ almost everywhere. Since

$$\begin{aligned}\langle T^*x - z, y\rangle &= \langle x, Ty\rangle - \langle z, y\rangle\\ &= \int_X \langle x(p), T(p)y(p)\rangle\, d\mu(p) - \int_X \langle T^*(p)x(p), y(p)\rangle\, d\mu(p) = 0\end{aligned}$$

for each y in $\mathscr{H}$, $T^*x - z = 0$. Thus $(T^*x)(p) = z(p) = T(p)^*x(p)$ almost everywhere, and T^* is decomposable with decomposition $p \to T(p)^*$.

Defining $I(p)$ to be I_p, we have

$$I(p)x(p) = I_p x(p) = x(p) = (Ix)(p),$$

so that I is decomposable with decomposition $p \to I_p$.

If $T_1(p) \le T_2(p)$ almost everywhere and $x \in \mathscr{H}$,

$$\langle T_1 x, x\rangle = \int_X \langle (T_1 x)(p), x(p)\rangle \, d\mu(p) = \int_X \langle T_1(p)x(p), x(p)\rangle \, d\mu(p)$$

$$\le \int_X \langle T_2(p)x(p), x(p)\rangle \, d\mu(p) = \langle T_2 x, x\rangle;$$

so that $T_1 \le T_2$. ∎

The converse to (*v*) of Proposition 14.1.8 is valid and allows us to show that $p \to \|T(p)\|$ is essentially bounded with essential bound $\|T\|$ for a decomposable operator T.

14.1.9. PROPOSITION. *If $\mathscr{H}$ is the direct integral of $\{\mathscr{H}_p\}$ over (X, μ) and A_1, A_2 are decomposable, self-adjoint operators on $\mathscr{H}$ such that $A_1 \le A_2$, then $A_1(p) \le A_2(p)$ almost everywhere. If T is decomposable, then $p \to \|T(p)\|$ is an essentially bounded measurable function with essential bound $\|T\|$.*

Proof. From Proposition 14.1.8(i), $A_2 - A_1$ is a positive, decomposable operator with decomposition $A_2(p) - A_1(p)$. Thus it will suffice to show that, if $0 \le H$ and H is decomposable, then $0 \le H(p)$ almost everywhere. Choosing a dense denumerable subset of $\mathscr{H}$ and forming finite linear combinations of its elements with rational coefficients, we construct a dense denumerable subset $\{x_j\}$ of $\mathscr{H}$ that is a linear space over the rationals. From Lemma 14.1.3, $\{x_j(p)\}$ spans $\mathscr{H}_p$ for p not in some null set N_0. For each finite rational-linear combination, $r_1x_1 + \cdots + r_nx_n$, there is an x_j equal to it; and, from Remark 14.1.2, there is a null set outside of which $r_1x_1(p) + \cdots + r_nx_n(p) = x_j(p)$. If $N_1, N_2, \ldots$ are these null sets (corresponding to an enumeration of the rational-linear combinations), then $\{x_j(p)\}$ is a rational-linear space spanning $\mathscr{H}_p$ for p not in $\bigcup_{j=0}^{\infty} N_j$, a null set N. With p not in N, then, $\{x_j(p)\}$ is dense in $\mathscr{H}_p$.

If $0 \le H$, then $0 \le \langle Hx_j, x_j\rangle = \int_X \langle H(p)x_j(p), x_j(p)\rangle \, d\mu(p)$. Suppose $H(p)x_j(p), x_j(p)\rangle < a < 0$ for p in some subset X_0 of X of finite positive measure. With f the characteristic function of X_0, $p \to \langle f(p)x_j(p), y(p)\rangle$ is integrable for each y in $\mathscr{H}$; so that (just as in Remark 14.1.7), for some z in $\mathscr{H}$, $z(p) = f(p)x_j(p)$ almost everywhere. In this case

$$\langle Hz, z\rangle = \int_X \langle H(p)f(p)x_j(p), f(p)x_j(p)\rangle \, d\mu(p)$$

$$= \int_{X_0} \langle H(p)x_j(p), x_j(p)\rangle \, d\mu(p) \leq a\mu(X_0) < 0,$$

contradicting the assumption that $0 \leq H$. Therefore $0 \leq \langle H(p)x_j(p), x_j(p)\rangle$ except for p in a null set M_j. If $M = \bigcup_{j=1}^{\infty} M_j$ and $p \notin N \cup M$, then $0 \leq \langle H(p)x_j(p), x_j(p)\rangle$ with $\{x_j(p)\}$ a dense subset of $\mathscr{H}_p$. It follows that $0 \leq H(p)$ for p not in $N \cup M$.

With T decomposable, T^* and T^*T are decomposable with decompositions $T^*(p)$ and $T^*(p)T(p)$, respectively, from Proposition 14.1.8. Since $\|T(p)\|^2 = \|T^*(p)T(p)\|$; to show that $p \to \|T(p)\|$ is measurable and essentially bounded with essential bound $\|T\|$, it will suffice to deal with a positive decomposable operator H on $\mathscr{H}$. Now $0 \leq H \leq \|H\|I$ so that, from what we have just proved, $0 \leq H(p) \leq \|H\|I_p$ almost everywhere. Conversely, from Proposition 14.1.8, if $0 \leq H(p) \leq aI_p$ almost everywhere, then $0 \leq H \leq aI$ and $\|H\| \leq a$. It follows that the essential bound of $p \to \|H(p)\|$ is $\|H\|$.

To establish that $p \to \|H(p)\|$ is measurable, we make use of $\{x_j\}$ and N, introduced in the first paragraph of this proof. If s (>0) is rational then the set X_s of points p not in N where $H(p) \leq sI_p$ is

$$\bigcap_{j=1}^{\infty} \{p : \langle H(p)x_j(p), x_j(p)\rangle \leq s\|x_j(p)\|^2, p \notin N\}.$$

Now $\|H(p)\|$ lies in an open interval if and only if there are rationals r and s in that interval such that $H(p) \nleq rI_p$ and $H(p) \leq sI_p$; so that the set of such p in $X \setminus N$ is a countable union of the sets $X_s \setminus X_r$. Thus $p \to \|H(p)\|$ is measurable. ∎

14.1.10. THEOREM. *If $\mathscr{H}$ is the direct integral of $\{\mathscr{H}_p\}$ over (X, μ), the set $\mathscr{R}$ of decomposable operators is a von Neumann algebra with abelian commutant $\mathscr{R}'$ coinciding with the family $\mathscr{C}$ of diagonalizable operators.*

Proof. From Proposition 14.1.8, $\mathscr{R}$ is a self-adjoint algebra of operators on $\mathscr{H}$ containing I. It remains to show that $\mathscr{R}$ is strong-operator closed. Let A be an operator of norm 1 in the strong-operator closure of $\mathscr{R}$, and let $(x_j\}$ be a denumerable dense subset of $\mathscr{H}$. Using the Kaplansky density theorem, there is a sequence $\{T_n\}$ of operators in the unit ball of $\mathscr{R}$ such that $T_n x_j \to Ax_j$ for all j. Then for all j,

$$\|T_n x_j - Ax_j\|^2 = \int_X \|T_n(p)x_j(p) - (Ax_j)(p)\|^2 \, d\mu(p) \to 0.$$

There is a subsequence $\{T_{n1}\}$ of $\{T_n\}$ such that $\|T_{n1}(p)x_1(p) - (Ax_1)(p)\|$ tends to 0 almost everywhere. Again, there is a subsequence $\{T_{n2}\}$ of $\{T_{n1}\}$ such that $\|T_{n2}(p)x_2(p) - (Ax_2)(p)\|$ tends to 0 almost everywhere. With $\{T_{nn}\}$ the "diagonal" (that is T_{11}, $T_{22}, \ldots$) of these subsequences, we have that $\|T_{nn}(p)x_j(p) - (Ax_j)(p)\| \to_n 0$ almost everywhere. Using Lemma 14.1.3 and Proposition 14.1.9, there is a null set N such that $\{x_j(p)\}$ spans $\mathscr{H}_p$, $\|T_{nn}(p)\| \le 1$ for all n, and $T_{nn}(p)x_j(p) \to_n (Ax_j)(p)$ for all j, when $p \in X \backslash N$. It follows that, for p in $X \backslash N$, there is an operator $A(p)$ in the unit ball of $\mathscr{B}(\mathscr{H}_p)$ such that $A(p)x_j(p) = (Ax_j)(p)$ for all j.

With x in $\mathscr{H}$, let $(x_{j'}\}$ be a sequence chosen from $\{x_j\}$ tending to x. Using the L_2-subsequence argument of the preceding paragraph, we can choose a subsequence $\{x_{j''}\}$ of $\{x_{j'}\}$ such that $x_{j''}(p) \to x(p)$ and $(Ax_{j''})(p) \to (Ax)(p)$ for p not in some null set M. Then for p not in $N \cup M$, $A(p)x(p) = (Ax)(p)$. Thus A is decomposable with decomposition $p \to A(p)$, $\mathscr{R}$ is strong-operator closed, and $\mathscr{R}$ is a von Neumann algebra. With $\mathscr{C}$ in place of $\mathscr{R}$, this same argument shows that $\mathscr{C}$ is a von Neumann algebra.

If A is diagonalizable with decomposition $f(p)I_p$ and T is decomposable, then AT and TA are decomposable with decompositions $f(p)I_pT(p)$ and $T(p)f(p)I_p$, respectively, from Proposition 14.1.8(ii). Since AT and TA have the same decompositions, $AT = TA$ (from Remark 14.1.7), and $A \in \mathscr{R}'$. We show that $\mathscr{R} = \mathscr{C}'$, and since, as just noted, $\mathscr{C}$ is a von Neumann algebra, $\mathscr{R}' = \mathscr{C}'' = \mathscr{C}$. As $\mathscr{R} \subseteq \mathscr{C}'$ and $\mathscr{R}$ and $\mathscr{C}'$ are von Neumann algebras, in order to show that $\mathscr{R} = \mathscr{C}'$, it will suffice to show that each projection E in $\mathscr{C}'$ is in $\mathscr{R}$. For this, let $\{u_j\}$ and $\{v_j\}$ be orthonormal bases for $E(\mathscr{H})$ and $(I - E)(\mathscr{H})$, respectively; and let $\{x_j\}$ be an enumeration of the set of finite rational-linear combinations of elements in $\{u_j, v_j\}$. As in the first paragraph of the proof of Proposition 14.1.9, there is a null set N such that, if $r_1x_1 + \cdots + r_nx_n = x_j$, then $r_1x_1(p) + \cdots + r_nx_n(p) = x_j(p)$, for rationals $r_1, \ldots, r_n$, and $\{x_j(p)\}$ is dense in $\mathscr{H}_p$, when $p \notin N$. For p not in N, let $E(p)$ be the projection with range spanned by $\{u_j(p)\}$. If u is a finite rational-linear combination of elements in $\{u_j\}$ and $p \notin N$, $(Eu)(p) = u(p) = E(p)u(p)$. Let v be a finite rational-linear combination of $\{v_j\}$. Suppose, for the moment, that we know $\langle u_j(p), v(p)\rangle = 0$ if $p \notin M$ for some null set M. Then $0 = E(p)v(p) = (Ev)(p)$ for p not in $N \cup M$. Hence, if $p \notin N \cup M$, $(Ex_j)(p) = E(p)x_j(p)$. With x in $\mathscr{H}$, there is a sequence $\{x_{j'}\}$ of elements in $\{x_j\}$ tending to x. As in the preceding paragraph of this proof, there is a null set N_0 such that $(Ex)(p) = E(p)x(p)$ if $p \notin N_0 \cup N \cup M$. Thus $E \in \mathscr{R}$.

It remains to prove that $\langle u(p), v(p)\rangle = 0$ almost everywhere when $Eu = u$ and $Ev = 0$. At this point, we use the assumption that E commutes with $\mathscr{C}$. Let P be the diagonalizable projection corresponding to the measurable subset X_0 of X. (See Remark 14.1.7.) Then

$$0 = \langle Pu, Ev\rangle = \langle EPu, v\rangle = \langle PEu, v\rangle$$
$$= \langle Pu, v\rangle = \int_{X_0} \langle u(p), v(p)\rangle \, d\mu(p).$$

Since this holds for each measurable subset X_0 of X, $\langle u(p), v(p)\rangle = 0$ almost everywhere. ■

Note that the first two paragraphs of the preceding proof establish that if $\{T_n\}$ is a (bounded) sequence of decomposable operators converging to A in the strong-operator topology, then A is decomposable and some subsequence $\{T_{n'}\}$ of $\{T_n\}$ is such that $\{T_{n'}(p)\}$ converges to $\{A(p)\}$ almost everywhere. If $\{T_n\}$ is monotone, then $\{T_n(p)\}$ is monotone for almost all p, from Proposition 14.1.9, and the sequence $\{T_n(p)\}$, itself, is strong-operator convergent to $\{A(p)\}$ almost everywhere. In particular, if $\{E_n\}$ is an orthogonal family of projections with sum E, then $\sum E_n(p) = E(p)$ almost everywhere.

14.1.11. EXAMPLES. (a) With reference to Example 14.1.4(a), the algebra of decomposable operators on $L_2(X, \mu)$ (considered as a direct integral of one-dimensional spaces) coincides with the algebra of diagonalizable operators. It is the (maximal abelian) multiplication algebra of $L_2(X, \mu)$ see Example 5.1.6).

(b) In case $\mathscr{H}$ is the discrete direct sum of a countable family $\{\mathscr{H}_n\}$ of Hilbert spaces (see Example 14.1.4(b)) each decomposable operator T is the direct sum (see Section 2.6, *Direct sums*) of a family $\{T_n\}$ of operators T_n on $\mathscr{H}_n$ (so that $T\{x_n\} = \{T_n x_n\}$ and $\|T\| = \sup\{\|T_n\|\}$). In case T is diagonalizable, each T_n is a scalar. ■

In the definition that follows, we refer to representations of general C^*-algebras rather than norm-separable algebras simply because norm-separability plays no role in the definition. In practice, however, we shall have to assume that our algebra is norm-separable in order to prove the results that interest us.

14.1.12. DEFINITION. If $\mathscr{H}$ is the direct integral of Hilbert spaces $\{\mathscr{H}_p\}$ over (X, μ), a representation φ of a C^*-algebra $\mathfrak{A}$ on $\mathscr{H}$ is said to be *decomposable* over (X, μ) when there is representation φ_p of $\mathfrak{A}$ on $\mathscr{H}_p$ such that $\varphi(A)$ is decomposable for each A in $\mathfrak{A}$ and $\varphi(A)(p) = \varphi_p(A)$ almost everywhere. If $\varphi(A)$ is diagonalizable as well, for each A in $\mathfrak{A}$, we say that φ is *diagonalizable*. The mapping $p \to \varphi_p$ is said to be a *decomposition* (or *diagonalization*) of φ. A state ρ of $\mathfrak{A}$ is said to be *decomposable* with decomposition $p \to \rho_p$ when ρ_p is a positive linear functional on $\mathfrak{A}$ for each p, such that $\rho_p(A) = 0$ when $\varphi_p(A) = 0$, $p \to \rho_p(A)$ is integrable for each A in $\mathfrak{A}$, and $\rho(A) = \int_X \rho_p(A)\, d\mu(p)$. ■

As in Remark 14.1.7, which applies to decompositions of operators, one can show that if $p \to \varphi_p$ and $p \to \varphi'_p$ are decompositions of φ, a representation of the norm-separable C^*-algebra $\mathfrak{A}$, then $\varphi_p = \varphi'_p$ almost everywhere (see Exercise 14.4.2); and that if φ and φ' are decomposable representations of $\mathfrak{A}$ whose decompositions are equal almost everywhere, then $\varphi = \varphi'$ (see Exercise 14.4.1). The condition that $\rho_p(A) = 0$ when $\varphi_p(A) = 0$ guarantees that there is a positive linear functional ρ'_p of $\varphi_p(\mathfrak{A})$ such that $\rho'_p(\varphi_p(A)) = \rho_p(A)$ for all A in $\mathfrak{A}$. In application, it will suffice to have ρ_p defined on the complement of a null set N; for with p in N, we can let ρ_p be $\eta \circ \varphi_p$, where η is an arbitary state of $\mathfrak{A}_p$. The resulting mapping $p \to \rho_p$, defined for all p in X, will satisfy the conditions of Definition 14.1.12.

14.1.13. THEOREM. *If $\mathscr{H}$ is a direct integral of Hilbert spaces $\{\mathscr{H}_p\}$ over (X, μ) and φ is a representation of the norm-separable C^*-algebra $\mathfrak{A}$ in the algebra of decomposable operators, then there is a null set N and a representation φ_p for each p in $X \setminus N$ such that $p \to \varphi_p$ is the decomposition of φ. If ρ is a state of $\mathfrak{A}$ and x_0 is a vector in $\mathscr{H}$ such that $\rho(A) = \langle \varphi(A)x_0, x_0 \rangle$ for each A in $\mathfrak{A}$, then ρ is decomposable with decomposition $p \to \rho_p$, where $\rho_p(A) = \langle \varphi_p(A)x_0(p), x_0(p) \rangle$.*

Proof. Let $\mathfrak{A}_0$ be the self-adjoint algebra over the rationals consisting of finite rational-linear combinations of finite products from a self-adjoint denumerable generating set for $\mathfrak{A}$. Let $\mathscr{H}_0$ be a dense denumerable rational-linear space in $\mathscr{H}$. With A_1, A_2 in $\mathfrak{A}_0$ and r_1, r_2 rationals,

$$(r_1\varphi(A_1) + r_2\varphi(A_2))(p) = r_1\varphi(A_1)(p) + r_2\varphi(A_2)(p),$$

$$(\varphi(A_1)\varphi(A_2))(p) = \varphi(A_1)(p)\varphi(A_2)(p),$$

and

$$\varphi(A_1)^*(p) = \varphi(A_1)(p)^*$$

for almost every p, from Proposition 14.1.8. There is a countable union N_0 of null sets such that these relations hold on $\mathfrak{A}_0$ for all p in $X \setminus N_0$, that is $p \to \varphi(A)(p)$ is a representation φ_p^0 of $\mathfrak{A}_0$ in $\mathscr{B}(\mathscr{H}_p)$ for p in $X \setminus N_0$.

Using Proposition 14.1.9 in this same way, we can locate a null set N_1 such that $-I_p \le \varphi(A)(p) \le I_p$ for each self-adjoint A in the unit ball of $\mathfrak{A}_0$

and all p in $X \setminus N_1$. Thus φ_p^0 is bounded on $\mathfrak{A}_0$ and extends (uniquely) to a representation φ_p of $\mathfrak{A}$ on $\mathscr{H}_p$, for p not in $N_0 \cup N_1$.

To see that $\varphi_p(A) = \varphi(A)(p)$ almost everywhere, let $\{A_n\}$ be a sequence in $\mathfrak{A}_0$ such that $\|A_n - A\| \to 0$. Then if $p \notin N_0 \cup N_1$, $\|\varphi_p(A_n) - \varphi_p(A)\| \to 0$. Employing $\mathscr{H}_0$ together with the L_2-subsequence, "diagonal" argument of (the first paragraph of) the proof of Theorem 14.1.10, since we have that $\|\varphi(A_n) - \varphi(A)\| \to 0$, there is a null set N_2 such that $\{\varphi_p(A_n)\}$ $(=\{\varphi(A_n)(p)\})$ is strong-operator convergent to $\varphi(A)(p)$ if $p \notin N_0 \cup N_1 \cup N_2$. Thus if $p \notin N_0 \cup N_1 \cup N_2$, then $\varphi_p(A) = \varphi(A)(p)$, φ is decomposable, and $p \to \varphi_p$ is its decomposition.

If we define $\rho_p(A)$ for p in $X \setminus (N_0 \cup N_1)$ to be $\langle \varphi_p(A)x_0(p), x_0(p) \rangle$, then, with A and N_2 as above,

$$\langle \varphi(A)(p)x_0(p), x_0(p) \rangle = \langle \varphi_p(A)x_0(p), x_0(p) \rangle$$

for p in $X \setminus (N_0 \cup N_1 \cup N_2)$; and

$$\begin{aligned} \rho(A) = \langle \varphi(A)x_0, x_0 \rangle &= \int_X \langle \varphi(A)(p)x_0(p), x_0(p) \rangle \, d\mu(p) \\ &= \int_X \rho_p(A) \, d\mu(p). \end{aligned}$$

Thus ρ is decomposable with decomposition $p \to \rho_p$. ■

14.1.14. Definition. If $\mathscr{H}$ is the direct integral of Hilbert spaces $\{\mathscr{H}_p\}$ over (X, μ), a von Neumann algebra $\mathscr{R}$ on $\mathscr{H}$ is said to be *decomposable* with decomposition $p \to \mathscr{R}_p$ when $\mathscr{R}$ contains a norm-separable strong-operator-dense C^*-subalgebra $\mathfrak{A}$ for which the identity representation ι is decomposable and such that $\iota_p(\mathfrak{A})$ is strong-operator dense in $\mathscr{R}_p$ almost everywhere. ■

The lemma that follows establishes that the decomposition $p \to \mathscr{R}_p$ of $\mathscr{R}$ is independent of the C^*-subalgebra $\mathfrak{A}$ giving rise to the decomposition. Before proceeding to that lemma, however, we remark on some consequences of the preceding definition. Since ι is decomposable, $\mathfrak{A}$ consists of decomposable operators (from Definition 14.1.12). Thus each operator A in $\mathscr{R}$ is decomposable, from Theorem 14.1.10. At the same time, the argument of the first paragraph of that theorem assures us that $A(p) \in \mathscr{R}_p$ for almost all p.

14.1.15. Lemma. *If $\mathscr{H}$ is the direct integral of Hilbert spaces $\{\mathscr{H}_p\}$ over (X, μ), $\mathfrak{A}$ and $\mathscr{B}$ are norm-separable C^*-subalgebras of the algebra of decomposable operators, and $\mathfrak{A}^- = \mathscr{B}^-$, then $\mathfrak{A}_p^- = \mathscr{B}_p^-$ almost everywhere, where $\mathfrak{A}_p$ and $\mathscr{B}_p$ are the images in $\mathscr{B}(\mathscr{H}_p)$ of the decomposition of the identity representations of $\mathfrak{A}$ and $\mathscr{B}$.*

Proof. From Theorem 14.1.13, there is a null set N_0 such that $A \to A(p)$ and $B \to B(p)$ are representations of $\mathfrak{A}$ and $\mathscr{B}$ on $\mathscr{H}_p$, when $p \notin N_0$. Let $\mathfrak{A}_0$ and $\mathscr{B}_0$ be (norm-)dense denumerable subsets of $\mathfrak{A}$ and $\mathscr{B}$; and let $\mathscr{H}_0$ be a dense denumerable rational-linear space in $\mathscr{H}$. If $B \in \mathscr{B}_0$, since $\mathfrak{A}^- = \mathscr{B}^-$, there is a sequence $\{A_n\}$ in the ball of radius $\|B\|$ in $\mathfrak{A}_0$ such that $A_n x \to Bx$ for each x in $\mathscr{H}_0$. Again, using an L_2 – subsequence "diagonal" argument (as in the proof of Theorem 14.1.10), there is a subsequence $\{A_{n'}\}$ of $\{A_n\}$ and a null set N such that $A_{n'}(p)x(p) \to B(p)x(p)$ for all x in $\mathscr{H}$, when $p \notin N$. Let N_1 be the (countable) union of null sets formed by applying this process to each B in $\mathscr{B}_0$. Then $B(p) \in \mathfrak{A}_p^-$ for all B in $\mathscr{B}_0$; hence, since such $B(p)$ form a norm-dense subset of $\mathscr{B}_p$, $\mathscr{B}_p^- \subseteq \mathfrak{A}_p^-$, when $p \notin N_0 \cup N_1$. Similarly, there are null sets M_0 and M_1 such that $\mathfrak{A}_p^- \subseteq \mathscr{B}_p^-$ when $p \notin M_0 \cup M_1$. Thus $\mathfrak{A}_p^- = \mathscr{B}_p^-$ almost everywhere. ∎

Once we establish that von Neumann algebras on separable Hilbert spaces have strong-operator-dense C^*-subalgebras that are norm separable, combining Theorem 14.1.13 with 14.1.15, we have the following theorem.

14.1.16. THEOREM. *If $\mathscr{H}$ is the direct integral of Hilbert spaces $\{\mathscr{H}_p\}$ over (X, μ) and $\mathscr{R}$ is a von Neumann subalgebra of the algebra of decomposable operators, then $\mathscr{R}$ is decomposable with unique decomposition $p \to \mathscr{R}_p$.*

Although it could be proved by introducing an appropriate metric and quoting some elementary results from the topology of separable metric spaces, we make use of operator-theoretic techniques instead to prove the following lemma.

14.1.17. LEMMA. *Each von Neumann algebra $\mathscr{R}$ acting on a separable Hilbert space $\mathscr{H}$ contains a strong-operator-dense norm-separable C^*-subalgebra.*

Proof. With $\{x_j\}$ a dense denumerable subset of $\mathscr{H}$, the vector $(\|x_1\|^{-1}x_1, (2\|x_2\|)^{-1}x_2, \ldots)$ is separating for $(\iota \oplus \iota \oplus \cdots)(\mathscr{R})$, where ι is the identity representation of $\mathscr{R}$. Thus we may assume that $\mathscr{R}$ acting on $\mathscr{H}$ has a separating vector; so that each normal linear functional on $\mathscr{R}$ has the form $\omega_{x,y}|\mathscr{R}$. Choose A_{jk} in $(\mathscr{R})_1$ so that $\omega_{x_j,x_k}(A_{jk}) \geq \|\omega_{x_j,x_k}|\mathscr{R}\| - \frac{1}{4}$; and let $\mathfrak{A}$ be the (norm-separable) C^*-algebra generated by $\{A_{jk}\}$. If $\|\omega_{x,y}|\mathfrak{A}\| = 0$, $\|\omega_{x,y}|\mathscr{R}\| = 1$ and $\|x_j - x\|$, $\|x_k - y\|$ are small, then $\|\omega_{x,y}|\mathscr{R} - \omega_{x_j,x_k}|\mathscr{R}\|$ is small. But

$$\begin{aligned}\|\omega_{x,y}|\mathscr{R} - \omega_{x_j,x_k}|\mathscr{R}\| &\geq |(\omega_{x,y} - \omega_{x_j,x_k})(A_{jk})| = \omega_{x_j,x_k}(A_{jk}) \\ &\geq \|\omega_{x_j,x_k}|\mathscr{R}\| - \tfrac{1}{4} \geq \|\omega_{x,y}|\mathscr{R}\| - \tfrac{1}{2} = \tfrac{1}{2}.\end{aligned}$$

Hence each normal linear functional annihilating $\mathfrak{A}$ annihilates $\mathscr{R}$; and $\mathfrak{A}^{-} = \mathscr{R}$. ■

14.1.18. PROPOSITION. *If $\mathscr{H}$ is the direct integral of Hilbert spaces $\{\mathscr{H}_p\}$ over (X, μ), $\mathscr{R}$ and $\mathscr{S}$ are decomposable von Neumann algebras on $\mathscr{H}$, each containing the algebra $\mathscr{C}$ of diagonalizable operators, with decompositions $p \to \mathscr{R}_p$, $p \to \mathscr{S}_p$, and A is a decomposable operator on $\mathscr{H}$, then $A \in \mathscr{S}$ if and only if $A(p) \in \mathscr{S}_p$ almost everywhere. If $\mathscr{R}_p = \mathscr{S}_p$ almost everywhere, then $\mathscr{R} = \mathscr{S}$.*

Proof. Suppose, first, that $A \geq 0$, $A(p) \in \mathscr{S}_p$ almost everywhere, and A is not in $\mathscr{S}$. The Hahn–Banach theorem ("separation" form) applied to $\mathscr{B}(\mathscr{H})$ with its (locally convex) weak-operator topology, provides us with a hermitian normal linear functional ρ on $\mathscr{B}(\mathscr{H})$ annihilating $\mathscr{S}$ but not A. From Theorem 7.4.7, there are positive normal functionals ρ^+ and ρ^- such that $\rho = \rho^+ - \rho^-$. Since $\rho|\mathscr{S} = 0$, $\rho^+|\mathscr{S} = \rho^-|\mathscr{S}$. From Theorem 7.1.12, there are countable families $\{x_n\}$, $\{y_n\}$ of vectors in $\mathscr{H}$ with $\sum \|x_n\|^2$ and $\sum \|y_n\|^2$ finite such that $\rho^+ = \sum \omega_{x_n}$ and $\rho^- = \sum \omega_{y_n}$. If H is a positive operator in $\mathscr{S}$,

$$\begin{aligned}\rho^+(H) &= \sum_{n=1}^{\infty} \langle Hx_n, x_n\rangle = \sum_{n=1}^{\infty} \int_X \langle H(p)x_n(p), x_n(p)\rangle \, d\mu(p)\\ &= \int_X \left(\sum_{n=1}^{\infty} \langle H(p)x_n(p), x_n(p)\rangle \right) d\mu(p)\\ &= \rho^-(H) = \int_X \left(\sum_{n=1}^{\infty} \langle H(p)y_n(p), y_n(p)\rangle \right) d\mu(p),\end{aligned}$$

where $f\colon p \to \sum_{n=1}^{\infty}\langle H(p)x_n(p), x_n(p)\rangle$ and $g\colon p \to \sum_{n=1}^{\infty}\langle H(p)y_n(p), y_n(p)\rangle$ are understood as the L_1-limits of the finite partial sums (which form Cauchy sequences of positive, integrable functions on X). At the same time, these sums converge almost everywhere (since a subsequence of each of these sequences does and each sequence is monotone increasing). As $\mathscr{S}$ contains $\mathscr{C}$, we can replace H by its product with the diagonalizable projection corresponding to a measurable subset X_0 of X (see Remark 14.1.7). In this case, we have $\int_{X_0} f(p)\, d\mu(p) = \int_{X_0} g(p)\, d\mu(p)$. Thus $f = g$ almost everywhere. Since the null set involved in this last equality varies with H, we cannot assert that $(\sum_{n=1}^{\infty} \omega_{x_n(p)})|\mathscr{S}_p = (\sum_{n=1}^{\infty} \omega_{y_n(p)})|\mathscr{S}_p$ at this point. However, since $\mathscr{S}$ is decomposable, it contains a norm-separable C^*-subalgebra $\mathfrak{A}$ such that $B \to B(p)$ is a representation of $\mathfrak{A}$ on $\mathscr{H}_p$ with range $\mathfrak{A}_p$ strong-operator dense in $\mathscr{S}_p$ for almost all p. If $\{H_j\}$ is the denumerable set of positive operators in $\mathfrak{A}$ obtained by expressing each operator in a (norm-)dense denumerable subset of $\mathfrak{A}$ as a linear combination of (four) positive operators, then, except for p in some null set N,

$$\sum_{n=1}^{\infty} \langle H_j(p)x_n(p), x_n(p)\rangle = \sum_{n=1}^{\infty} \langle H_j(p)y_n(p), y_n(p)\rangle$$

for all j. Since the functionals on $\mathscr{B}(\mathscr{H}_p)$ involved in this last equality are normal, $(\sum \omega_{x_n(p)})|\mathscr{S}_p = (\sum \omega_{y_n(p)})|\mathscr{S}_p$ for p not in N. Now $A(p)\in\mathscr{S}_p$ almost everywhere, by assumption, whence we have that $\sum_{n=1}^{\infty}\langle A(p)x_n(p), x_n(p)\rangle = \sum_{n=1}^{\infty}\langle A(p)y_n(p), y_n(p)\rangle$ almost everywhere. Integrating over X, this yields $\rho^+(A) = \sum \omega_{x_n}(A) = \sum \omega_{y_n}(A) = \rho^-(A)$; so that $\rho(A) = 0$—contradicting the choice of ρ. Thus $A\in\mathscr{S}$. The last assertion of the statement follows from this.

If A is a self-adjoint decomposable operator such that $A(p)\in\mathscr{S}_p$ almost everywhere, we have that $A = \|A\|I - (\|A\|I - A)$ and that $A(p) = \|A\|I_p - (\|A\|I_p - A(p))$ almost everywhere. Hence, from the preceding argument, $\|A\|I - A\in\mathscr{S}$. Thus $A\in\mathscr{S}$. ■

14.1.19. Lemma. *If $\mathscr{H}$ is the direct integral of Hilbert spaces $\{\mathscr{H}_p\}$ over (X, μ), $\mathscr{R}$ is a decomposable von Neumann algebra on $\mathscr{H}$, and ω is a normal state of $\mathscr{R}$, then there is a mapping, $p \to \omega_p$, where ω_p is a positive normal linear functional on $\mathscr{R}_p$, and $\omega(A) = \int_X \omega_p(A(p))\, d\mu(p)$ for each A in $\mathscr{R}$. If $\mathscr{R}$ contains the algebra $\mathscr{C}$ of diagonalizable operators and $\omega|E\mathscr{R}E$ is faithful or tracial, for some projection E in $\mathscr{R}$, then $\omega_p|E(p)\mathscr{R}_pE(p)$ is, accordingly, faithful or tracial almost everywhere.*

Proof. From Theorem 7.1.12, there is a countable set of vectors $\{y_n\}$ in $\mathscr{H}$ such that $\omega(A) = \sum \langle Ay_n, y_n\rangle$ for A in $\mathscr{R}$ and

$$1 = \sum_{n=1}^{\infty} \|y_n\|^2 = \sum_{n=1}^{\infty} \int_X \langle y_n(p), y_n(p)\rangle\, d\mu(p).$$

It follows that $\sum_{n=1}^{\infty}\langle y_n(p), y_n(p)\rangle$ is finite almost everywhere; so that $A_p \to \sum_{n=1}^{\infty}\langle A_p y_n(p), y_n(p)\rangle$ defines a positive normal linear functional on $\mathscr{B}(\mathscr{H}_p)$. We denote the restriction of this functional to $\mathscr{R}_p$ by ω_p. As in the proof of Proposition 14.1.18, it follows, now, that

$$\begin{aligned}\int_X \omega_p(A(p))\, d\mu(p) &= \int_X \left(\sum_{n=1}^{\infty} \langle A(p)y_n(p), y_n(p)\rangle\right) d\mu(p)\\ &= \sum_{n=1}^{\infty} \int_X \langle A(p)y_n(p), y_n(p)\rangle\, d\mu(p) = \sum_{n=1}^{\infty} \langle Ay_n, y_n\rangle = \omega(A)\end{aligned}$$

for each positive A in $\mathscr{R}$. By expressing an arbitrary operator in $\mathscr{R}$ as a linear combination of four positive operators in $\mathscr{R}$, we have the same equality, now, for each A in $\mathscr{R}$.

If $\mathscr{C} \subseteq \mathscr{R}$, then $\mathscr{R}' \subseteq \mathscr{C}'$; so that $\mathscr{R}'$ is a decomposable von Neumann algebra. (See Theorem 14.1.10.) Let $\{A'_m\}$ be a denumerable strong-operator-dense subset of $\mathscr{R}'$. Let $\{E, A_n\}$ be a norm-dense subset of a strong-operator-dense norm-separable C^*-subalgebra $\mathfrak{A}$ of $\mathscr{R}$. If $\omega | E\mathscr{R}E$ is faithful, then $\{A'_m E y_n\}$ spans $E(\mathscr{H})$, for ω annihilates the projection (in $E\mathscr{R}E$) on the orthogonal complement of this span in $E(\mathscr{H})$. Now $A'_m(p)$ commutes with $\mathfrak{A}_p$ for almost every p; so that $\{A'_m(p)\}$ and $\mathscr{R}_p$ commute almost everywhere. At the same time, $\{A'_m(p)E(p)y_n(p)\}$ spans $E(p)(\mathscr{H}_p)$ almost everywhere. To see this, choose a denumerable set $\{x_k\}$ spanning $(I - E)(\mathscr{H})$; so that $\{x_k, A'_m E y_n\}$ spans $\mathscr{H}$. From Lemma 14.1.3, $\{x_k(p), A'_m(p)E(p)y_n(p)\}$ spans $\mathscr{H}_p$ almost everywhere. Since $Ex_k = 0$ for all k, $E(p)x_k(p) = 0$ for all k, almost everywhere. Thus $\{A'_m(p)E(p)y_n(p)\}$ spans $E(p)(\mathscr{H}_p)$ almost everywhere. Hence there is a null set N such that $\{\mathscr{R}'_p E(p)y_n(p),\ n = 1, 2, \ldots\}$ spans $E(p)(\mathscr{H}_p\}$ when $p \notin N$. If H_p is a positive operator in $E(p)\mathscr{R}_p E(p)$ such that $\omega_p(H_p) = 0$ for some p not in N, then $0 = H_p y_n(p) = H_p E(p) y_n(p)$ for all n, so that $H_p \mathscr{R}'_p E(p) y_n(p) = 0$ and $0 = H_p E(p) = H_p$. Thus $\omega_p | E(p)\mathscr{R}_p E(p)$ is faithful when $p \notin N$.

With $\omega | E\mathscr{R}E$ tracial and P the diagonalizable projection corresponding to a measurable subset X_0 of X (see Remark 14.1.7) we have

$$\begin{aligned}\int_{X_0} &\omega_p(E(p)A_n(p)E(p)A_m(p)E(p))\, d\mu(p)\\ &= \omega(PEA_nEA_mE) = \omega(PEA_mEA_nE)\\ &= \int_{X_0} \omega_p(E(p)A_m(p)E(p)A_n(p)E(p))\, d\mu(p).\end{aligned}$$

Thus $\omega_p(E(p)A_n(p)E(p)A_m(p)E(p)) = \omega_p(E(p)A_m(p)E(p)A_n(p)E(p))$ almost everywhere—first for the given n and m, then for all n and m. Since ω_p is a normal state and $E(p)\mathfrak{A}_p E(p)$ is strong-operator dense in $E(p)\mathscr{R}_p E(p)$, $\omega_p | E(p)\mathscr{R}_p E(p)$ is tracial almost everywhere. ■

The mapping $p \to \omega_p$, of the preceding theorem, gives rise to a mapping $p \to \omega'_p$, where $\omega'_p = \omega_p \circ \iota_p$ and ι is the identity representation of $\mathscr{R}$ on $\mathscr{H}$. Although decompositions of states are defined for norm-separable C^*-algebras (Definition 14.1.12), the mapping $p \to \omega'_p$ is, in effect, a decomposition of the normal state ω into positive normal linear functionals ω'_p.

14.1.20. LEMMA. *If $\mathscr{H}$ is the direct integral of Hilbert spaces $\{\mathscr{H}_p\}$ over (X, μ), $\mathscr{R}$ is a decomposable von Neumann algebra on $\mathscr{H}$, and E is a projection in $\mathscr{R}$, then the following assertions hold almost everywhere*:

(i) *$E(p)$ is a projection in $\mathscr{R}_p$;*
(ii) *if E is in the center of $\mathscr{R}$, then $E(p)$ is in the center of $\mathscr{R}_p$;*
(iii) *if $E \sim F$ in $\mathscr{R}$, then $E(p) \sim F(p)$ in $\mathscr{R}_p$;*
(iv) *if E is abelian in $\mathscr{R}$, then $E(p)$ is abelian in $\mathscr{R}_p$;*
(v) *$C_E(p) = C_{E(p)}$;*
(vi) *if E is properly infinite in $\mathscr{R}$ and $C_E = I$, then $E(p)$ is properly infinite in $\mathscr{R}_p$;*
(vii) *if E is finite in $\mathscr{R}$ and $\mathscr{R}$ contains the diagonalizable operators, then $E(p)$ is finite in $\mathscr{R}_p$.*

Proof. (i) Since $E^2 = E = E^*$, we have $E(p)^2 = E(p) = E(p)^*$ almost everywhere, from Proposition 14.1.8; so that $E(p)$ is a projection in $\mathscr{R}_p$ almost everywhere.

(ii) If $ET = TE$ with T in $\mathscr{R}$, then $E(p)T(p) = T(p)E(p)$ almost everywhere, again, from Proposition 14.1.8. Allowing T to take on a denumerable dense set of values in a strong-operator-dense norm-separable C^*-subalgebra $\mathfrak{A}$ of $\mathscr{R}$, we have that $E(p)$ commutes with $\mathfrak{A}_p$ almost everywhere. Thus $E(p)$ is in the center of $\mathscr{R}_p$ almost everywhere, if E is in the center of $\mathscr{R}$.

(iii) If there is a partial isometry V in $\mathscr{R}$ such that $V^*V = E$ and $VV^* = F$, then $V^*(p)V(p) = E(p)$ and $V(p)V^*(p) = F(p)$ almost everywhere, from Proposition 14.1.8. Thus $E(p) \sim F(p)$ in $\mathscr{R}_p$ almost everywhere.

(iv) If E is abelian in $\mathscr{R}$ then $EAEBE = EBEAE$ for each A and B in $\mathscr{R}$. Thus $E(p)A(p)E(p)B(p)E(p) = E(p)B(p)E(p)A(p)E(p)$ almost everywhere. Letting A and B take on a denumerable (norm-)dense set of values in the strong-operator-dense norm-separable C^*-subalgebra $\mathfrak{A}$ of $\mathscr{R}$, we conclude that $E(p)\mathfrak{A}_pE(p)$ is abelian almost everywhere. Hence $E(p)\mathscr{R}_pE(p)$ is abelian almost everywhere; and $E(p)$ is an abelian projection in $\mathscr{R}_p$ almost everywhere.

(v) From Proposition 5.5.2, the range of C_E is $[\mathscr{R}E(\mathscr{H})]$. With $\{x_j\}$ a denumerable set spanning $E(\mathscr{H})$ and $\{A_n\}$ a denumerable strong-operator-dense subset of $\mathscr{R}$, $\{A_nx_j\}$ spans $C_E(\mathscr{H})$. Using Lemma 4.1.3 and arguing as in the second paragraph of the proof of Lemma 14.1.19, we have that $\{x_j(p)\}$ spans $E(p)(\mathscr{H}_p)$ and $\{A_n(p)x_j(p)\}$ spans $C_E(p)(\mathscr{H}_p)$, almost everywhere. If $\mathfrak{A}$ is the norm-separable C^*-subalgebra of $\mathscr{R}$ generated by $\{A_n\}$, then $\{A_n(p)\}$ generates $\mathfrak{A}_p$ and $\mathfrak{A}_p^- = \mathscr{R}_p$ almost everywhere. Thus $\{A_n(p)x_j(p)\}$ spans $C_{E(p)}(\mathscr{H}_p)$ almost everywhere; and $C_E(p) = C_{E(p)}$ almost everywhere.

(vi) If E is properly infinite in $\mathscr{R}$ and $C_E = I$, there is a subprojection F of E such that $E \sim F \sim E - F$ in $\mathscr{R}$, from Lemma 6.3.3. Thus $E(p) \sim F(p) \sim E(p) - F(p)$ in $\mathscr{R}_p$, almost everywhere, (from (iii)) and either $E(p) = 0$ or $E(p)$

is properly infinite, almost everywhere. But $C_E = I$, so that $I_p = C_E(p) = C_{E(p)}$ almost everywhere (from (v)). Thus $E(p)$ is properly infinite almost everywhere.

(vii) If E is finite in $\mathscr{R}$, then $E\mathscr{R}E$ admits a faithful normal tracial state. Composing this state with the mapping $A \to EAE$ yields a normal state ω of $\mathscr{R}$ that is both tracial and faithful on $E\mathscr{R}E$. If $\mathscr{R}$ contains the diagonalizable operators, Lemma 14.1.19 applies, and $\omega_p | E(p)\mathscr{R}_p E(p)$ is a faithful tracial state for almost all p. Thus $E(p)$ is finite in $\mathscr{R}_p$ for almost all p. ∎

In the next theorem, we identify the types of the components in the decomposition of von Neumann algebras of various types. For unity of statement, we have included the algebras of type III, although the proof that we give that the components are of type III, in this case, requires a special technique whose development we postpone to the discussion after the theorem.

For the proof of the following theorem, we shall make use of some simple observations concerning decompositions of subspaces. If $\mathscr{H}$ is the direct integral of $\{\mathscr{H}_p\}$ over (X, μ), $\mathscr{R}$ is a decomposable von Neumann algebra on $\mathscr{H}$ with decomposition $p \to \mathscr{R}_p$, and E is a decomposable projection, then $E(\mathscr{H})$ has a direct integral decomposition and $E\mathscr{R}E$ is decomposable relative to it with decomposition $p \to E(p)\mathscr{R}_p E(p)$. To see this, note that $E \in \mathscr{C}'$, where $\mathscr{C}$ is the algebra of diagonalizable operators on $\mathscr{H}$; so that E has a central carrier relative to $\mathscr{C}'$ in $\mathscr{C}$. This central carrier is the diagonalizable projection corresponding to a measurable subset X_0 of X; and $E(\mathscr{H})$ is the direct integral of $\{E(p)(\mathscr{H}_p)\}$ over (X_0, μ). (Although X_0 is a measurable subset of X and not itself locally compact, our results apply without change to this situation; and we can speak of direct integral decompositions over (X_0, μ). Indeed this direct integral is essentially the same as that over (X, μ_0) where $\mu_0(Y) = \mu(Y \cap X_0)$ for a measurable subset Y of X.) If T is decomposable on $\mathscr{H}$, then ETE is decomposable on $E(\mathscr{H})$ (with decomposition $p \to E(p)T(p)E(p)$). If S_0 is decomposable on $E(\mathscr{H})$, then S on $\mathscr{H}$, agreeing with S_0 on $E(\mathscr{H})$ and 0 on $(I - E)(\mathscr{H})$ is decomposable (with decomposition $S(p) = S_0(p)$ for p in X_0 and $S(p) = 0$ for p in $X \setminus X_0$) and $ESE = S$. Thus $E\mathscr{C}'E$ is the algebra of decomposable operators on $E(\mathscr{H})$ and $\mathscr{C}E$ is the algebra of diagonalizable operators. Moreover, the decomposition of $E\mathscr{R}E$ is $p \to E(p)\mathscr{R}_p E(p)$. If $\mathscr{C} \subseteq \mathscr{R}$, then $\mathscr{C}E \subseteq E\mathscr{R}E$.

14.1.21. Theorem. *If $\mathscr{H}$ is the direct integral of Hilbert spaces $\{\mathscr{H}_p\}$ over (X, μ) and $\mathscr{R}$ is a decomposable von Neumann algebra containing the diagonalizable operators, then $\mathscr{R}$ is of type* I_n, II_1, II_∞, *or* III *if and only if, correspondingly, $\mathscr{R}_p$ is of type* I_n, II_1, II_∞, *or* III *almost everywhere.*

Proof. If $\mathscr{R}$ is of type I_n (n possibly infinite), there are n orthogonal equivalent abelian projections $E_1, \ldots, E_n$ in $\mathscr{R}$ with sum I. From Lemma 14.1.20 and Proposition 14.1.8, $E_j(p)$ is abelian in $\mathscr{R}_p$, $C_{E_j(p)} = I_p$, $E_j(p) \sim E_k(p)$, and $\sum_{j=1}^{\infty} E_j(p) = I_p$ almost everywhere. It follows that $\mathscr{R}_p$ is of type I_n almost everywhere.

If $\mathscr{R}$ is of type II_1, then from Lemma 14.1.20(vii), I_p is finite in $\mathscr{R}_p$ almost everywhere. An infinite orthogonal family of projections in $\mathscr{R}$, each with central carrier I, gives rise to an infinite orthogonal family each with central carrier I_p in $\mathscr{R}_p$ almost everywhere. Thus $\mathscr{R}_p$ is of type II_1 almost everywhere.

If $\mathscr{R}$ is of type II_∞, then from Lemma 14.1.20(vi), I_p is properly infinite in $\mathscr{R}_p$ almost everywhere. If E is a finite projection in $\mathscr{R}$ with central carrier I, E has an infinite orthogonal family of subprojections each with central carrier I. Thus $E(p)$ is finite in $\mathscr{R}_p$ and admits an infinite, orthogonal family of subprojections with central carrier I_p almost everywhere. Thus $\mathscr{R}_p$ is of type II_∞ almost everywhere.

Assuming, as we shall at this point (see the discussion following the proof of Proposition 14.1.24), that $\mathscr{R}_p$ is of type III almost everywhere when $\mathscr{R}$ is of type III, there is no difficulty in showing that the types of $\mathscr{R}_p$ determine that of $\mathscr{R}$. If $\mathscr{R}_p$ is of type either I_n, or II_1, or II_∞, or III, almost everywhere and $\mathscr{R}$ has a central projection Q such that $\mathscr{R}Q$ is not of the corresponding type, then, from the discussion preceding this theorem $(\mathscr{R}Q)_p$ is $Q(p)\mathscr{R}_pQ(p)$. From (ii) of Lemma 14.1.20, $Q(p)$ is in the center of $\mathscr{R}_p$ for almost all p; so that $Q(p)\mathscr{R}_pQ(p) = \mathscr{R}_pQ(p)$ almost everywhere. Thus $\mathscr{R}Q$ acting on $Q(\mathscr{H})$ is a decomposable algebra of one type, containing the diagonalizable operators, whose components $\mathscr{R}_pQ(p)$ are of a different type almost everywhere. But this contradicts our assumption in the type III case and what we have established in the other cases. Thus $\mathscr{R}$ has the same type as its components. ∎

We have deferred the proof that a type III von Neumann algebra containing the diagonalizable operators has type III components in its decomposition until after the proof of Proposition 14.1.24. For its proof, we use a special type of argument that must wait until some preliminary results have been developed. The argument involves a measurable "selection" or "cross-section" principle. This principle entails an excursion into an area we have not encountered thus far and requires methods for its proof foreign to those we have been using. For this reason, we have placed the discussion of this principle in an appendix to this chapter (Section 14.3).

The measurable-selection principle makes it possible for us to use a very natural (and powerful) strategy of proof in decomposition theory (that we have, nevertheless, avoided until now). This strategy is best described with an illustration. Suppose $\{\mathscr{H}_p\}$ is the decomposition of $\mathscr{H}$ over (X, μ). If $\mathscr{R}$ is a decomposable von Neumann algebra containing the algebra of diagonalizable operators, the same is true of $\mathscr{R}'$. Is $(\mathscr{R}')_p$ equal to $(\mathscr{R}_p)'$ almost everywhere? The affirmative answer to this question is the substance of Proposition 14.1.24. Let us consider how we might prove it. If $\{A_j\}$ and $\{A'_j\}$ are denumerable families of operators, strong-operator dense in $\mathscr{R}$ and $\mathscr{R}'$, respectively, then, except for p in a null set N, all $A_k(p)$ commute with all $A'_j(p)$. Thus $(\mathscr{R}')_p \subseteq (\mathscr{R}_p)'$ for p not in N. (This argument has appeared in the proof of Lemma 14.1.19.) If $(\mathscr{R}_p)' = (\mathscr{R}')_p$, let A'_p be 0; otherwise choose A'_p in $(\mathscr{R}_p)' \backslash (\mathscr{R}')_p$. Suppose that we can make our choice of A'_p in $(\mathscr{R}_p)' \backslash (\mathscr{R}')_p$ in a measurable manner—specifically, so that $A'_p = A'(p)$ almost everywhere for some decomposable A'. Then $A'(p)$ commutes with all $A_j(p)$ almost everywhere; so that A' commutes with all A_j and $A' \in \mathscr{R}'$. But, then, $A'(p) = A'_p \in (\mathscr{R}')_p$ almost everywhere; and $(\mathscr{R}_p)' = (\mathscr{R}')_p$ almost everywhere.

The question of the possibility of making a "measurable selection" of A'_p in $(\mathscr{R}_p)' \backslash (\mathscr{R}')_p$ masks an additional problem. Is the set at which $(\mathscr{R}_p)' = (\mathscr{R}')_p$ a measurable set? The application of the measurable-selection principle entails establishing that this set is measurable. The topics presented in Section 14.3 supply the techniques for proving this measurability as well as the measurable-selection principle. What is needed is the fact that an "analytic subset" of X (the continuous image of a complete separable metric space) is μ-measurable. The selection principle amounts to the fact that, if we associate with each p in a subset X_0 of X a non-null subset $\mathscr{S}_p$ of a complete separable metric (csm or Polish) space (one with a denumerable base admitting a complete metric)—usually the unit ball in $\mathscr{B}(\mathscr{K})$—and the set $\mathscr{S}$ of pairs (p, A) with p in X, A in $\mathscr{S}_p$ is an analytic set, then there is a measurable mapping, $p \to A_p$, from a measurable subset of X containing X_0 into the csm space such that $A_p \in \mathscr{S}_p$. (The mapping "selects" A_p from the non-null sets $\mathscr{S}_p$ in a measurable manner.) The precise details and proofs appear in Section 14.3.

In order to apply these measurability and measurable-selection techniques to our direct-integral problems, we must transfer our decomposition $\{\mathscr{H}_p\}$ over (X, μ) to a single Hilbert space $\mathscr{K}$ in a measurable manner and develop the notion of measurable mappings from X into $\mathscr{K}$ and $\mathscr{B}(\mathscr{K})$. We carry out this program in the following few paragraphs. The lemma that follows prepares us for the transfer to a single space $\mathscr{K}$.

14.1.22. LEMMA. *If $\mathscr{H}$ is the direct integral of Hilbert spaces $\{\mathscr{H}_p\}$ over (X, μ), then the algebra $\mathscr{R}$ of decomposable operators is of type* I_n *if and only if $\mathscr{H}_p$ is n-dimensional almost everywhere. If $\mathscr{R}$ is of type* I_n *and $\{E_j : j = 1, \ldots, n\}$ is an orthogonal family of abelian projections with central carriers I and sum I, then, with x_j a generating vector for E_j under $\mathscr{R}'$ $(=\mathscr{C})$, $x_j(p) \neq 0$ and $\{x_j(p)\}$ is an orthogonal family of vectors that span $\mathscr{H}_p$ for almost all p.*

Proof. If $\mathscr{R}$ is of type I_n and the family $\{E_j : j = 1, \ldots, n\}$ is as in the statement of this lemma, then $\{E_j(p)\}$ is an orthogonal family of abelian projections with central carriers I_p and sum I_p for almost all p. In particular, $\mathscr{H}_p$ has dimension not less than n for almost all p. We show that the dimension of $\mathscr{H}_p$ does not exceed n, for almost all p, by establishing the last assertion of this lemma.

With j distinct from k, let X_0 be the set of p such that $\langle x_j(p), x_k(p)\rangle \neq 0$. If Q is the diagonalizable projection corresponding to a measurable subset Y of X_0, then

$$0 = \langle QE_jx_j, E_kx_k\rangle = \langle Qx_j, x_k\rangle = \int_Y \langle x_j(p), x_k(p)\rangle \, d\mu(p).$$

Since this holds for every measurable subset Y of X_0, $\langle x_j(p), x_k(p)\rangle = 0$ almost everywhere; and X_0 is a null set. Thus $\{x_j(p)\}$ is an orthogonal family for almost all p.

Let $\{C_j\}$ be a denumerable strong-operator-dense subset of $\mathscr{C}$. Since x_k generates E_k under $\mathscr{C}$, $\{C_jx_k : j = 1, 2, \ldots\}$ spans $E_k(\mathscr{H})$. As $\{E_k\}$ has sum I, $\{C_jx_k : j, k = 1, 2, \ldots\}$ spans $\mathscr{H}$. Thus $\{C_j(p)x_k(p) : j, k = 1, 2, \ldots\}$ spans $\mathscr{H}_p$ for almost all p, from Lemma 14.1.3. Except for p in some null set $C_j(p)$ is a scalar multiple of I_p for all j (since C_j is diagonalizable). Thus $\{x_j(p)\}$ spans $\mathscr{H}_p$ for almost all p; and $\mathscr{H}_p$ has dimension n.

Since x_j generates E_j and $C_{E_j} = I$, the range, $[\mathscr{R}\mathscr{C}x_j]$ $(=[\mathscr{R}x_j])$, of C_{E_j} is $\mathscr{H}$ (see Proposition 5.5.2). Thus x_j is separating for $\mathscr{C}$ (see Proposition 5.5.11 in conjunction with Theorem 14.1.10). It follows (from Lemma 14.1.19, for example) that $x_j(p) \neq 0$ almost everywhere.

Suppose, now, that $\mathscr{H}_p$ has dimension n for almost all p. Since $\mathscr{R}' = \mathscr{C}$ and $\mathscr{C}$ is abelian, $\mathscr{R}$ is of type I (see Theorem 9.1.3) and there are central projections P_m such that $\mathscr{R}P_m$ is of type I_m or $P_m = 0$ (see Theorem 6.5.2). Assume $P_m \neq 0$ for some m. Since $P_m \in \mathscr{C}$, P_m is the diagonalizable projection corresponding to some measurable subset X_0 of X of positive measure (see Remark 14.1.7). In this case, $P_m(\mathscr{H})$ is the direct integral of $\{\mathscr{H}_p\}$ over (X_0, μ), $\mathscr{C}P_m$ is the algebra of diagonalizable operators, and $\mathscr{R}P_m$ is the algebra of decomposable operators, relative to this decomposition of $P_m(\mathscr{H})$. Since $\mathscr{R}P_m$ is of type I_m, from what we have just proved, $\mathscr{H}_p$ is m dimensional for almost all p in X_0. By assumption, $m = n$. Thus $P_m = 0$ unless $m = n$; and $\mathscr{R}$ is of type I_n. ■

With the notation of Lemma 14.1.22, we carry out the transfer of the decomposition $\{\mathscr{H}_p\}$ to a single Hilbert space $\mathscr{K}$ of dimension n. Let $\{y_j\}$ be an orthonormal basis for $\mathscr{K}$, and let N be a null set X such that $x_j(p) \neq 0$ for all j and $\{x_j(p)\}$ is an orthogonal family of vectors spanning $\mathscr{H}_p$ for each p in

$X \backslash N$. There is a unique unitary transformation U_p of $\mathscr{H}_p$ onto $\mathscr{K}$, for p not in N, such that $U_p x_j(p) = \|x_j(p)\| y_j$ for all j. With x in $\mathscr{H}$,

$$x(p) = \sum_{j=1}^{n} \langle x(p), x_j(p)\rangle \|x_j(p)\|^{-2} x_j(p)$$

and

$$U_p x(p) = \sum_{j=1}^{n} \langle x(p), x_j(p)\rangle \|x_j(p)\|^{-1} y_j \quad \left(= \sum_{j=1}^{n} a_j(p) y_j \right).$$

If y is a vector in $\mathscr{K}$, then $\langle U_p x(p), y\rangle = \sum_{j=1}^{n} \langle y_j, y\rangle a_j(p)$ for each p not in N. Each function $p \to \langle y_j, y\rangle a_j(p)$ is measurable, so that $p \to \langle U_p x(p), y\rangle$ is the (finite) limit almost everywhere of measurable functions. Thus $p \to \langle U_p x(p), y\rangle$ is measurable. We say that the mapping $p \to U_p x(p)$ of X into $\mathscr{K}$ is *weakly measurable* in this case (that is, when the functions $p \to \langle U_p x(p), y\rangle$ are measurable for all y in $\mathscr{K}$). At the same time, $p \to U_p A(p) U_p^{-1}$ is a weakly measurable mapping of X into $\mathscr{B}(\mathscr{K})$ for each decomposable A on $\mathscr{H}$ (that is, $p \to U_p A(p) U_p^{-1} y$ is weakly measurable for each y in $\mathscr{K}$). To see this, note that it suffices, for weak-measurability of a mapping $p \to z(p)$ of X into $\mathscr{K}$, to verify that $p \to \langle z(p), y_j\rangle$ is measurable for each y_j in the orthonormal basis $\{y_j\}$; for, as above, $p \to \langle z(p), y\rangle$ is, then, the (finite) limit almost everywhere of measurable functions on X. In the same way, it suffices, for weak-measurability of a mapping $p \to B_p$ of X into $\mathscr{B}(\mathscr{K})$, to verify weak-measurability of the functions $p \to B_p y_j$ for each j, and, hence, of $\langle B_p y_j, y_k\rangle$ for each pair j, k. In the present case,

$$\langle U_p A(p) U_p^{-1} y_j, y_k\rangle = (\|x_j(p)\| \cdot \|x_k(p)\|)^{-1} \langle A(p) x_j(p), x_k(p)\rangle;$$

so that $p \to U_p A(p) U_p^{-1}$ is weakly measurable.

It will be useful for us to note, in passing, that $p \to B_p^*$ is weakly measurable if $p \to B_p$ is (clear, from the definition) and that $p \to T_p S_p$ is weakly measurable if each of $p \to T_p$ and $p \to S_p$ is. For this, we observe that we can recast the criterion for weak-measurability of $p \to B_p$. Let $B_p y_j$ be $\sum_{k=1}^{n} b_k(p) y_k$. Then $p \to \langle B_p y_j, y_k\rangle$ $(= b_k(p))$ is measurable; so that $p \to B_p$ (and, by the same argument, $p \to z(p)$) is weakly measurable if and only if each of the functions b_k is measurable (the functions $a_k(p)$, where $z(p) = \sum_{k=1}^{n} a_k(p) y_k$, in the case of $p \to z(p)$). It follows that $p \to z(p)$ and $p \to B_p$ are *strongly measurable* (that is, measurable as mappings into $\mathscr{K}$ in its metric and $\mathscr{B}(\mathscr{K})$ in its strong-operator topology) if and only if they are weakly measurable. We say, henceforth, that such mappings are *measurable.* Note, too, that $p \to \langle u(p), v(p)\rangle$ is measurable if $p \to u(p)$ and $p \to v(p)$ are; for it can be factored as $p \to (u(p), v(p)) \to \langle u(p), v(p)\rangle$, a measurable mapping of X into $\mathscr{K} \times \mathscr{K}$ (since $\mathscr{K}$ is separable) followed by a continuous mapping of $\mathscr{K} \times \mathscr{K}$ into $\mathbb{C}$. Returning to the mappings $p \to T_p$ and $p \to S_p$, we see that if $S_p y_j = \sum_{m=1}^{n} s_{jm}(p) y_m$ and $T_p^* y_k = \sum_{m=1}^{n} t_{km}^*(p) y_m$, then

$$\langle T_p S_p y_j, y_k\rangle = \sum_{m=1}^{n} s_{jm}(p)\overline{t_{km}^*(p)}.$$

Since $p \to \sum_{m=1}^{n} s_{jm}(p)\overline{t_{km}^*(p)}$ is a measurable function (finite almost everywhere) $p \to S_p T_p$ is measurable.

In the case of a general direct-integral decomposition $\{\mathscr{H}_p\}$ of $\mathscr{H}$ over (X, μ), the set X_n of points p such that $\mathscr{H}_p$ has dimension n is measurable (see Remark 14.1.5) and corresponds to a diagonalizable projection P_n such that $\mathscr{R}P_n$ is of type I_n (if X_n is not a null set) and is the algebra of decomposable operators arising from the decomposition of $P_n(\mathscr{H})$ as $\{\mathscr{H}_p\}$ over (X_n, μ). Associated with this decomposition is the family of unitary transformations U_{np}, p in X_n, of $\mathscr{H}_p$ onto a fixed Hilbert space $\mathscr{K}_n$ of dimension n. If we form $\mathscr{K}$, the direct sum of $\mathscr{K}_n$, $n = 1, 2, \ldots, \aleph_0$ (where $\mathscr{K}_n$ is absent if X_n is a null set), then the total family $\{U_{np} : n = 1, 2, \ldots, \aleph_0;\ p \in X\}$ transfers the constructs of this general decomposition of $\mathscr{H}$ onto the one space $\mathscr{K}$ in a "measurable" manner. We summarize this discussion in the lemma that follows.

14.1.23. Lemma. *If $\mathscr{H}$ is the direct integral of the Hilbert spaces $\{\mathscr{H}_p\}$ over (X, μ) and X_n is the set of points p in X at which $\mathscr{H}_p$ has dimension n, then, if X_n is not a null set for μ, the diagonalizable projection P_n corresponding to X_n is the maximal central projection in $\mathscr{R}$, the algebra of decomposable operators, such that $\mathscr{R}P_n$ is of type I_n. In this case, there is a family $\{U_{np} : p \in X_n\}$ such that U_{np} is a unitary transformation of $\mathscr{H}_p$ onto a fixed Hilbert space $\mathscr{K}_n$ of dimension n. If $\mathscr{K}$ is the direct sum of those $\mathscr{K}_n$ such that X_n is not a μ-null set and U_p is U_{np} when $p \in X_n$, then $\{U_p\}$ is a family of unitary transformations such that U_p maps $\mathscr{H}_p$ into $\mathscr{K}$, $p \to U_p x(p)$ is measurable for each x in $\mathscr{H}$, and $p \to U_p A(p) U_p^*$ is measurable for each A in $\mathscr{R}$.*

We illustrate the actual technical use of the measurable-selection principle by giving the full argument for the result on decomposition of commutants whose proof was sketched prior to Lemma 14.1.22. For the remainder of this chapter, the assumption that X is metrizable as a csm space is in force.

14.1.24. Proposition. *If $\mathscr{H}$ is the direct integral of $\{\mathscr{H}_p\}$ over (X, μ) and $\mathscr{R}$ is a decomposable von Neumann algebra on $\mathscr{H}$ containing the algebra $\mathscr{C}$ of* *diagonalizable operators, then* $\mathscr{R}'$ *is decomposable and* $(\mathscr{R}')_p = (\mathscr{R}_p)'$ *almost everywhere.*

Proof. Let $\{A_j\}$ and $\{A'_j\}$ be denumerable strong-operator-dense subsets of the unit ball in each of $\mathscr{R}$ and $\mathscr{R}'$, respectively. Let $\{y_j\}$ be a denumerable dense subset of $\mathscr{H}$. To employ the techniques of Section 14.3, we introduce the Hilbert space $\mathscr{K}$ of Lemma 14.1.23 and the family of unitary transformations described there. We use the unit ball $\mathscr{B}_1$ of $\mathscr{B}(\mathscr{K})$ provided with its strong-operator topology and the translationally invariant metric, $d(S, T) = \sum_{j=1}^{\infty} 2^{-j}\|(S - T)e_j\|$, where $\{e_j\}$ is an orthonormal basis for $\mathscr{K}$, which is (uniformly) compatible with this topology, as our csm space. Since $\mathscr{C} \subseteq \mathscr{R}$, $\mathscr{R}'$ is decomposable. Let F_n be the orthogonal projection of $\mathscr{K}$ onto $\mathscr{K}_n$. We consider the set of pairs (p, A) in $X \times \mathscr{B}_1$ satisfying the following conditions:

(i) $AU_pA_jU_p^* = U_pA_j(p)U_p^*A$, $j = 1, 2, \ldots$, $p \in X_n$, and $F_nAF_n = A$;
(ii) there are positive integers m and h such that, for each j, there is a positive integer k not exceeding h for which

$$(*) \quad \|(A - U_pA'_j(p)U_p^*)U_py_k(p)\| \geq 1/m, \qquad p \in X_n, \qquad \text{and} \qquad F_nAF_n = A.$$

Except for p in a Borel μ-null set N_0 in X, a pair (p, A) satisfies (i) if and only if $p \in X_n$ and $A \in U_p(\mathscr{R}_p)'U_p^*$ and satisfies (ii) if and only if $F_nAF_n = A$, $p \in X_n$, and $A \notin U_p(\mathscr{R}')_pU_p^*$. We may also assume that $(X\backslash N_0) \cap X_n$ are disjoint for different n and have union $X\backslash N_0$. Since $p \to U_pA_j(p)U_p^*$, $p \to U_pA'_j(p)U_p^*$, and $p \to U_py_j(p)$ are measurable for all j; there is a Borel μ-null set N_1 such that, restricted to $X\backslash N_1$, these mappings into $\mathscr{B}_1$ and $\mathscr{K}$ are Borel for all j (see Lemma 14.3.1). Let X_0 be $X\backslash(N_0 \cup N_1)$. It follows that the mappings $(p, A) \to U_pA_j(p)U_p^*A$ and $(p, A) \to AU_pA_j(p)U_p^*$, from $X_0 \times \mathscr{B}_1$, with its (product) topological Borel structure, to $\mathscr{B}_1$, are Borel mappings. Indeed, they can be factored as the composition of the strong-operator continuous mappings $(U_pA_j(p)U_p^*, A) \to U_pA_j(p)U_p^*A$ and $(U_pA_j(p)U_p^*, A) \to AU_pA_j(p)U_p^*$ of $\mathscr{B}_1 \times \mathscr{B}_1$ into $\mathscr{B}_1$ with the mapping $(p, A) \to (U_pA_j(p)U_p^*, A)$ from $X_0 \times \mathscr{B}_1$ into $\mathscr{B}_1 \times \mathscr{B}_1$; and this last mapping is Borel relative to the (product) topological Borel structures since the strong-operator topology on $\mathscr{B}_1$ has a countable base by virtue of the separability of $\mathscr{K}$. The subset $\mathscr{S}_{jn}$ of $(X_n \cap X_0) \times (F_n\mathscr{B}_1F_n)$ where these mappings agree, for a given j, is a Borel set, as is $\bigcap_{j=1}^{\infty} \mathscr{S}_{jn}$ $(= \mathscr{S}'_n)$, the subset of $X_0 \times \mathscr{B}_1$ fulfilling condition (i). Again, $\bigcup_{n=1}^{\infty} \mathscr{S}'_n$ $(= \mathscr{S}')$ is the Borel subset of points (p, A) in $X_0 \times \mathscr{B}_1$ such that $A \in U_p(\mathscr{R}_p)'U_p^*$.

Similarly, the subset $\mathscr{S}_{jkmn}$ of $(X_n \cap X_0) \times (F_n\mathscr{B}_1F_n)$ satisfying $(*)$, for given j, k, m, and n, is a Borel set, since $(p, A) \to (U_pA'_j(p)U_p^*, A, U_py_k(p))$ is a Borel mapping of $X_0 \times \mathscr{B}_1$ into $\mathscr{B}_1 \times \mathscr{B}_1 \times \mathscr{K}$ and the mapping

$(U_pA'_j(p)U_p^*, A, U_py_k(p)) \to \|(A - U_pA'_j(p)U_p^*)U_py_k(p)\|$ from $\mathscr{B}_1 \times \mathscr{B}_1 \times \mathscr{K}$ into $\mathbb{R}$ is continuous. Now

$$\bigcup_{h,n,m=1}^{\infty} \bigcap_{j=1}^{\infty} \bigcup_{k=1}^{h} \mathscr{S}_{jkmn}$$

is the subset $\mathscr{S}''$ of pairs (p, A) in $X_0 \times \mathscr{B}_1$ such that, for some n, $F_nAF_n = A$, $p \in X_n$ and $A \notin U_p(\mathscr{R}')_pU_p^*$. Thus $\mathscr{S}' \cap \mathscr{S}''$ $(=\mathscr{S})$ is the subset of pairs (p, A) in $X_0 \times \mathscr{B}_1$ such that $A \in (U_p(\mathscr{R}_p)'U_p^*)\backslash(U_p(\mathscr{R}')_pU_p^*$ and is Borel subset of $X \times \mathscr{B}_1$.

From Theorem 14.3.5, $X \times \mathscr{B}_1$ is a csm space and $\mathscr{S}$ is an analytic subset of it. The image X_1 of $\mathscr{S}$ under the projection mapping $X \times \mathscr{B}_1$ onto its X-coordinate is (an analytic, hence, measurable subset of X and is) precisely the set of points p in X_0 for which $(\mathscr{F}_p =)$ $U_p(\mathscr{R}_p)'U_p^*\backslash U_p(\mathscr{R}')_pU_p^* \neq \varnothing$. From Theorem 14.3.6 there is a measurable mapping $p \to U_pA'_pU_p^*$ from X_1 into $\mathscr{B}_1$ such that $U_pA'_pU_p^* \in \mathscr{F}_p$. Then A'_p $(=U_p^*U_pA'_pU_p^*U_p) \in (\mathscr{R}_p)'\backslash(\mathscr{R}')_p$ for p in X_1. Defining A'_p to be 0 for p in $X\backslash X_1$, with x and y in $\mathscr{H}$, we have

$$\langle A'_px(p), y(p)\rangle = \langle U_pA'_pU_p^*U_px(p), U_py(p)\rangle;$$

so that $p \to \langle A'_px(p), y(p)\rangle$ is measurable. Since

$$|\langle U_pA'_pU_p^*U_px(p), U_py(p)\rangle| \leq \|x(p)\| \cdot \|y(p)\|,$$

$p \to \langle A'_px(p), y(p)\rangle$ is integrable. It follows that $A'_px(p) = (A'x)(p)$ almost everywhere, for some $A'x$ in $\mathscr{H}$, from Definition 14.1.1. Moreover, A' is linear, since each A'_p is linear; and

$$\begin{aligned} |\langle A'x, y\rangle| &= \left|\int_X \langle (A'x)(p), y(p)\rangle \, d\mu(p)\right| \leq \int_X \|x(p)\| \cdot \|y(p)\| \, d\mu(p) \\ &\leq \left(\int_X \|x(p)\|^2 \, d\mu(p)\right)^{1/2}\left(\int_X \|y(p)\|^2 \, d\mu(p)\right)^{1/2} = \|x\| \cdot \|y\|, \end{aligned}$$

so that $\|A'\| \leq 1$. It follows that A' is a decomposable operator on $\mathscr{H}$ such that $A'(p) \in (\mathscr{R}_p)'\backslash(\mathscr{R}')_p$ for all p in X_1. But, in this case, $A'(p)A_j(p) = A_j(p)A'(p)$ almost everywhere for all j. (Recall that $A'(p) = 0$ for p not in X_1). Thus $A'A_j = A_jA'$ for all j, and $A' \in \mathscr{R}'$. From the discussion following Definition 14.1.14, $A'(p) \in (\mathscr{R}')_p$ for almost all p. Hence X_1 is a μ-null set, and $(\mathscr{R}_p)' = (\mathscr{R}')_p$ almost everywhere. ■

We complete the argument of Theorem 14.1.21 by showing that if a decomposable von Neumann algebra $\mathscr{R}$ of type III contains the algebra $\mathscr{C}$ of diagonalizable operators on a Hilbert space $\mathscr{H}$, the direct integral of $\{\mathscr{H}_p\}$

over (X, μ), then $\mathscr{R}_p$ is of type III almost everywhere. Since $\mathscr{C} \subseteq \mathscr{R}$ and $\mathscr{R} \subseteq \mathscr{C}'$ ($\mathscr{R}$ is decomposable), $\mathscr{C}$ is contained in the center of $\mathscr{R}$. Since $\mathscr{R}$ is of type III, the algebra $\mathscr{C}'$ of decomposable operators on $\mathscr{H}$ must be of type I_∞; for a non-zero, central projection P in $\mathscr{C}'$ lies in $\mathscr{C}$ (the center of $\mathscr{C}'$). Hence P lies in the center of $\mathscr{R}$. Thus P is infinite in $\mathscr{R}$, and, therefore, in $\mathscr{C}'$.

Employing the transfer process from $\{\mathscr{H}_p\}$ to a Hilbert space $\mathscr{K}$ (Lemma 14.1.23), since $\mathscr{C}'$ is of type I_∞, we can use a single infinite-dimensional $\mathscr{K}$ rather than a direct sum. Let $\{U_p : p \text{ in } X\}$ be the family of unitary transformations U_p of $\mathscr{H}_p$ onto $\mathscr{K}$ with the properties noted in Lemma 14.1.23. Let $\mathscr{B}_0$ be the unit ball in the set of self-adjoint operators in $\mathscr{B}(\mathscr{K})$ provided with its strong-operator topology (in which it is a complete separable metric space).

Since $\mathscr{R}$ is of type III and $\mathscr{H}$ is separable, there is a unit separating (and generating) vector y_0 for $\mathscr{R}$ (see Proposition 9.1.6). It follows (from Lemma 14.1.19, for example) that $y_0(p)$ is separating for $\mathscr{R}_p$, except for p in a null set N_0. Let $\{A_j\}$ and $\{A'_j\}$ be denumerable strong-operator-dense subsets of the intersection of the unit ball in $\mathscr{B}(\mathscr{H})$ with the self-adjoint operators in $\mathscr{R}$ and $\mathscr{R}'$, respectively. Then $\{A_j(p)\}$ and $\{A'_j(p)\}$ generate $\mathscr{R}_p$ and, from Proposition 14.1.24, $\mathscr{R}'_p$, respectively, except for p in a null set N_1. Let $\{y_j\}$ be a denumerable dense rational-linear subspace of $\mathscr{H}$, so that $\{y_j(p)\}$ is such a subspace of $\mathscr{H}_p$, except for p in some null set N_2. We consider the pairs (p, E) in $X \times \mathscr{B}_0$ satisfying the following three conditions:

(i) $E = E^2 \neq 0$.

(ii) $EU_pA'_j(p)U_p^* = U_pA'_j(p)U_p^*E,\ j = 1, 2, \ldots$.

(iii) For each positive integer n, there is a positive integer m such that

$$\begin{aligned}|\langle EU_pA_j(p)U_p^*EU_py_m(p), EU_pA_k(p)U_p^*EU_py_m(p)\rangle \\ -\langle EU_pA_k(p)U_p^*EU_py_m(p), EU_pA_j(p)U_p^*EU_py_m(p)\rangle| \\ < \frac{\|EU_py_m(p)\|^2}{n} \qquad \text{for} \quad j, k = 1, 2, \ldots.\end{aligned}$$

If p in $X\backslash N$, where $N = N_0 \cup N_1 \cup N_2$, and E in $\mathscr{B}_0$ satisfy (i), (ii), and (iii), then E is a projection, from (i) (since $\mathscr{B}_0$ consists of self-adjoint operators), $E \in U_p\mathscr{R}_pU_p^*$, from (ii) (since $\{A'_j(p)\}$ generates $\mathscr{R}'_p$), and $EU_p\mathscr{R}_pU_p^*E$ is not properly infinite, from (iii). To see this last, note that, from (iii), the vector state of $EU_p\mathscr{R}_pU_p^*E$ associated with $EU_py_m(p)$ approximates a trace to within $1/n$ since $\{EU_pA_j(p)U_p^*E\}$ is strong-operator dense in the set of self-adjoint elements in the unit ball of $EU_p\mathscr{R}_pU_p^*E$. Passing to a limit (of some subnet of these states in the weak* compact unit ball of the dual of $EU_p\mathscr{R}_pU_p^*E$), we see that $EU_p\mathscr{R}_pU_p^*E$ possesses a tracial state ρ and E is not properly infinite.

Conversely, if E is a non-zero projection in $U_p\mathscr{R}_pU_p^*$ that is not properly infinite, there is a normal tracial state of $EU_p\mathscr{R}_pU_p^*E$ (compose the center-valued trace of Theorem 8.2.8 on a finite summand of $EU_p\mathscr{R}_pU_p^*E$ with a vector state, for this), which must be a vector state, from Theorem 7.2.3, since $EU_py_0(p)$ is separating for $EU_p\mathscr{R}_pU_p^*E$, when $p \notin N_0$, If we choose $\{y_m(p)\}$, a sequence of vectors in $\{y_j(p)\}$ such that $\{U_py_m(p)\}$ tends to a vector corresponding to that vector state (this is possible when $p \notin N_2$), we see that p (in $X\backslash N$) and E satisfy (i), (ii), and (iii). Thus (i), (ii), and (iii) determine precisely the set Y_0 of points p in $X\backslash N$ at which $\mathscr{R}_p$ contains a non-zero projection that is not properly infinite, that is, at which $\mathscr{R}_p$ is not of type III—equivalently, at which the set of E in $\mathscr{B}_0$, such that p and E satisfy (i), (ii), and (iii), is non-empty. To apply the measurable-selection principle, we shall locate a Borel (hence, analytic) subset $\mathscr{S}$ of $X \times \mathscr{B}_0$ consisting of pairs (p, E) satisfying (i), (ii), and (iii), such that the image of $\mathscr{S}$ under the X-coordinate projection is an analytic set X_1 differing from Y_0 by a μ-null set. Suppose, for the moment, that we have found $\mathscr{S}$. There is, then, a measurable mapping $p \to U_pE_pU_p^*$ from X_1 into $\mathscr{B}_0$ such that p and $U_pE_pU_p^*$ satisfy (i), (ii), and (iii). Defining E_p to be 0 for p in $X\backslash X_1$, we have, precisely as at the end of the proof of Proposition 14.1.24, that $E_p = E(p)$ almost everywhere for some decomposable operator E on $\mathscr{H}$. Since $E(p)$ is a projection in $\mathscr{R}_p$ almost everywhere, E is a projection in $\mathscr{R}$ (from Proposition 14.1.18, by forming the von Neumann algebra generated by $\mathscr{R}$ and E). Since $\mathscr{R}$ is of type III, either $E = 0$ or E is properly infinite. If $E = 0$, $E(p) = 0$ almost everywhere, X_1 is, therefore, a μ-null set, and $\mathscr{R}_p$ is of type III almost everywhere. If E is properly infinite, there is a projection F in $\mathscr{R}$ such that $F < E$ and $E \sim F \sim E - F$, from which, $F(p) \le E(p)$ and $E(p) \sim F(p) \sim E(p) - F(p)$ (see Proposition 14.1.9 and Lemma 14.1.20). In this case, $E(p)$ is either 0 or properly infinite almost everywhere. In either event, X_1 is a μ-null set, and $\mathscr{R}_p$ is of type III almost everywhere.

It remains to locate a Borel subset $\mathscr{S}$ of $X \times \mathscr{B}_0$ with the properties noted. Using Lemma 14.3.1 let X_0 be a Borel subset of X such that the measurable mappings $p \to U_pA_j(p)U_p^*$, $p \to U_pA_j'(p)U_p^*$, and $p \to U_py_j(p)$ are Borel mappings on X_0 for all j and such that $X\backslash X_0$ is a μ-null set containing N. Since $(p, E) \to E$ and $(p, E) \to E^2$ are continuous mappings of $X_0 \times \mathscr{B}_0$ into $\mathscr{B}_0$, the points at which they agree form a Borel subset $\mathscr{S}_0'$ of $X_0 \times \mathscr{B}_0$; and $\mathscr{S}_0'\backslash(X_0 \times \{0\})$ is a Borel subset $\mathscr{S}_0$ of $X_0 \times \mathscr{B}_0$. The pairs in $\mathscr{S}_0$ are those at which (i) is satisfied. Just as in the proof of Proposition 14.1.24, the set of pairs $\mathscr{S}_1$ in $X_0 \times \mathscr{B}_0$ satisfying (ii) is a Borel set. Again, the inequality of (iii) determines, for each j, k, n, and m, a Borel subset $\mathscr{S}_{jknm}$, and

$$\bigcap_{n=1}^{\infty} \bigcup_{m=1}^{\infty} \bigcap_{j,k=1}^{\infty} \mathscr{S}_{jknm}$$

is the Borel set $\mathscr{S}_2$ of points of $X_0 \times \mathscr{B}_0$ satisfying (iii). Let $\mathscr{S}$ be $\mathscr{S}_0 \cap \mathscr{S}_1 \cap \mathscr{S}_2$. Then $\mathscr{S}$ is precisely the set of points in $X_0 \times \mathscr{B}_0$ satisfying (i), (ii), and (iii). The image X_1 of $\mathscr{S}$ under the X-coordinate projection is an analytic set differing from Y_0 by a μ-null set. ■

Bibliography: [21, 54, 67]

14.2. Decompositions relative to abelian algebras

In Section 14.1, we developed the basic theory of Hilbert spaces that are direct integrals (as characterized in Definition 14.1.1) and studied the resulting decompositions of decomposable operators (Definition 14.1.6) and decomposable von Neumann algebras (Definition 14.1.14). In the present section, we discuss the possibility of recognizing a given separable Hilbert space as a direct integral and a given von Neumann algebra on it as decomposable. More precisely, we shall begin by asking when $\mathscr{H}$ is the direct integral of spaces $\{\mathscr{H}_p\}$ in such a way that a given abelian von Neumann algebra $\mathscr{A}$ on it appears as the algebra of diagonalizable operators. We shall see that this is always the case (Theorem 14.2.1) and note the details of the construction expressing $\mathscr{H}$ as such a direct integral. Theorem 14.1.10 tells us, then, that $\mathscr{A}'$ is the (von Neumann) algebra of all decomposable operators on $\mathscr{H}$ relative to this direct integral decomposition; and Theorem 14.1.16 tells us that a von Neumann algebra is decomposable if and only if it is a subalgebra of $\mathscr{A}'$. In this perpsective, given an abelian von Neumann algebra $\mathscr{A}$ on $\mathscr{H}$ and a von Neumann subalgebra $\mathscr{R}$ of $\mathscr{A}'$ (equivalently, $\mathscr{R}'$ containing $\mathscr{A}$), we may ask ourselves about the effect of a special relation between $\mathscr{A}$ and $\mathscr{R}$ on the components $\mathscr{R}_p$ of the decomposition of $\mathscr{R}$. For example, if $\mathscr{A}$ is the center of $\mathscr{R}$, or $\mathscr{A}$ is a maximal abelian subalgebra of $\mathscr{R}'$, can anything specific be said about $\mathscr{R}_p$? We shall see (Theorems 14.2.2 and 14.2.4) that $\mathscr{A}$ is the center of $\mathscr{R}$ if and only if $\mathscr{R}_p$ is a factor almost everywhere, and $\mathscr{A}$ is maximal abelian in $\mathscr{R}'$ if and only if $\mathscr{R}_p$ is $\mathscr{B}(\mathscr{H}_p)$ almost everywhere.

The problem of expressing $\mathscr{H}$ as a direct integral of Hilbert spaces is largely one of identification. The serious work was done in Section 9.3 when we described the spatial action of type I von Neumann algebras and in Section 9.4 when we described the maximal abelian algebras on separable Hilbert spaces. To illustrate this, we begin by considering the simplest instance. Suppose $\mathscr{A}$ is a maximal abelian subalgebra of $\mathscr{B}(\mathscr{H})$. From Theorem 9.4.1, $\mathscr{A}$ is unitarily equivalent to exactly one of the multiplication algebras $\mathscr{A}_c$, $\mathscr{A}_j$ $(1 \le j \le \aleph_0)$, or $\mathscr{A}_c \oplus \mathscr{A}_k$ $(1 \le k \le \aleph_0)$. If we denote by (X, μ) the appropriate measure space associated with $\mathscr{A}$ ([0, 1] with Lebesgue measure in the case of $\mathscr{A}_c$, and S_j in the case of $\mathscr{A}_j$—see the discussion preceding Theorem 9.4.1 for this notation), then there is a unitary transformation of $\mathscr{H}$ onto $L_2(X, \mu)$ that carries $\mathscr{A}$ onto the multiplication algebra of $L_2(X, \mu)$. Thus each vector in $\mathscr{H}$ corresponds to a function f in $L_2(X, \mu)$. We have noted in Examples 14.1.4(a) and 14.1.11(a) that $L_2(X, \mu)$ is a direct integral of one-dimensional Hilbert spaces and that the algebra of diagonalizable operators is the multiplication algebra of $L_2(X, \mu)$. Thus we have our desired (unitary equivalence of $\mathscr{H}$ with a) direct integral decomposition of $\mathscr{H}$ in which $\mathscr{A}$ is (unitarily equivalent to) the algebra of diagonalizable operators.

The next level of complexity occurs when $\mathscr{A}'$ is of type I_2. In this case, Theorem 9.3.2 tells us that $\mathscr{A}$ is unitarily equivalent to $1 \otimes (\mathscr{A}_0 \otimes I_2)$, where $\mathscr{A}_0$ is a maximal abelian algebra acting on a (separable) Hilbert space $\mathscr{K}$ (see the discussion following Theorem 6.6.1 for this notation). In other words, there is a unitary transformation of $\mathscr{H}$ onto $\mathscr{K} \oplus \mathscr{K}$ that carries $\mathscr{A}$ onto the set of operators $\{A \oplus A : A \in \mathscr{A}_0\}$. If we view $\mathscr{K}$ as $L_2(X, \mu)$ and $\mathscr{A}_0$ as the multiplication algebra on it (so that $\mathscr{A}$ is isomorphic to $\mathscr{A}_0$, though not unitarily equivalent to it), an element of $\mathscr{H}$ is transformed onto a pair (f, g) of functions in $L_2(X, \mu)$. To each p in X, there corresponds $(f(p), g(p))$, a vector of a two-dimensional Hilbert space $\mathscr{H}_p$. If an operator A in $\mathscr{A}_0$ corresponds to multiplication by h, then, since $A \oplus A$ transforms (f, g) onto (hf, hg) which has component $(h(p)f(p), h(p)g(p))$ in $\mathscr{H}_p$, $A \oplus A$ on $\mathscr{K} \oplus \mathscr{K}$ has component $h(p)I_p$ on $\mathscr{H}_p$. It is readily verified that $\mathscr{K} \oplus \mathscr{K}$ is the direct integral of $\{\mathscr{H}_p\}$ and that $\mathscr{A}_0 \otimes I_2$ (which corresponds to $\mathscr{A}$) is the algebra of diagonalizable operators relative to this decomposition.

If $\mathscr{A}'$ is of type I_n, with n finite, then $\mathscr{A}$ is unitarily equivalent to $\mathscr{A}_0 \otimes I_n$, and the preceding discussion is altered only in replacing pairs by n-tuples. In a formal sense, the same is true when $\mathscr{A}'$ is of type I_∞. In this case, the Hilbert space to which $\mathscr{H}$ is unitarily equivalent consists of sequences $\{f_j\}$ of functions f_j in $L_2(X, \mu)$ such that $\sum_{j=1}^{\infty} \|f_j\|_2^2$ is finite. It follows that $\sum_{j=1}^{\infty} |f_j(p)|^2$ is finite for almost all p. If we start with an orthonormal basis $\{x_n\}$ in $\mathscr{H}$ and the corresponding sequences $\{f_{nj}\}$ in $\mathscr{K} \oplus \mathscr{K} \otimes \cdots$, except for p in a μ-null set, we associate with each x_n the vector $\{f_{n1}(p), f_{n2}(p), \ldots\}$ in l_2 $(= \mathscr{H}_p)$. Again, $\mathscr{H}$ is (unitarily equivalent to) the direct integral of $\{\mathscr{H}_p\}$ and $\mathscr{A}$ is (unitarily equivalent to) the algebra of diagonalizable operators relative to this direct integral decomposition.

In the most general situation, $\mathscr{A}'$ is the direct sum $\sum \oplus \mathscr{A}'P_n$, $n = 1, 2, \ldots, \aleph_0$, where P_n is a projection in $\mathscr{A}$ and $\mathscr{A}'P_n$ is of type I_n or $P_n = 0$. In this case (with $P_n \neq 0$), we apply the preceding considerations to the abelian von Neumann algebra $\mathscr{A}P_n$ acting on $P_n(\mathscr{H})$ to construct the measure space (X_n, μ_n) such that $P_n(\mathscr{H})$ is unitarily equivalent to the n-fold direct sum $\mathscr{K}_n$ of $L_2(X_n, \mu_n)$ with itself and such that $\mathscr{A}P_n$ is carried onto the algebra of diagonalizable operators on $\mathscr{K}_n$ relative to the direct integral decomposition

$\{\mathscr{H}_p\}$ of $\mathscr{K}_n$ over (X_n, μ_n) we described. Let (X, μ) be the direct sum of these measure spaces. Then the total family $\{\mathscr{H}_p\}$ constitutes a direct integral decomposition of $\sum \oplus \mathscr{K}_n$ $(=\mathscr{K})$, and the direct sum of the unitary transformations of $P_n(\mathscr{H})$ onto $\mathscr{K}_n$ is a unitary transformation of $\mathscr{H}$ onto $\mathscr{K}$ that carries $\mathscr{A}$ onto the algebra of diagonalizable operators relative to this decomposition. We summarize this discussion in the theorem that follows.

14.2.1 THEOREM. *If $\mathscr{A}$ is an abelian von Neumann algebra on the separable Hilbert space $\mathscr{H}$ there is a (locally compact complete separable metric) measure space (X, μ) such that $\mathscr{H}$ is (unitarily equivalent to) the direct integral of Hilbert spaces $\{\mathscr{H}_p\}$ over (X, μ) and $\mathscr{A}$ is (unitarily equivalent to) the algebra of diagonalizable operators relative to this decomposition.*

With Theorem 14.2.1 in mind, we may speak of the (direct integral) decomposition of a (separable) Hilbert space $\mathscr{H}$ relative to an abelian von Neumann algebra $\mathscr{A}$ on $\mathscr{H}$, as well as the decomposition of a von Neumann subalgebra $\mathscr{R}$ of $\mathscr{A}'$ relative to $\mathscr{A}$. We study, now, the effect of special assumptions about the relation between $\mathscr{R}$ and $\mathscr{A}$ on the components $\mathscr{R}_p$.

14.2.2. THEOREM. *If $\mathscr{A}$ is an abelian von Neumann subalgebra of the center $\mathscr{C}$ of a von Neumann algebra $\mathscr{R}$ on a separable Hilbert space $\mathscr{H}$ and $\{\mathscr{H}_p\}$ is the direct integral decomposition of $\mathscr{H}$ relative to $\mathscr{A}$, then $\mathscr{C}_p$ is the center of $\mathscr{R}_p$ almost everywhere. In particular, $\mathscr{R}_p$ is a factor almost everywhere if and only if $\mathscr{A} = \mathscr{C}$.*

Proof. Since $\mathscr{A} \subseteq \mathscr{C} \subseteq \mathscr{R}$, $\mathscr{R}' \subseteq \mathscr{C}' \subseteq \mathscr{A}'$. As $\mathscr{C} \subseteq \mathscr{R}'$, $\mathscr{C} \subseteq \mathscr{R} \subseteq \mathscr{C}' \subseteq \mathscr{A}'$, and each of $\mathscr{R}$, $\mathscr{R}'$, $\mathscr{C}$, and $\mathscr{C}'$ is decomposable (relative to $\mathscr{A}$). Let $\mathfrak{A}_1$ and $\mathfrak{A}_2$ be norm-separable strong-operator-dense C^*-subalgebras of $\mathscr{R}$ and $\mathscr{R}'$, respectively; and let $\mathfrak{A}$ be the C^*-algebra they generate. Then $\mathfrak{A}$ is a norm-separable strong-operator-dense C^*-subalgebra of the von Neumann algebra $\mathscr{C}'$ generated by $\mathscr{R}$ and $\mathscr{R}'$. It follows that $\mathscr{R}_p$, $(\mathscr{R}')_p$, and $(\mathscr{C}')_p$ are the strong-operator closures of $(\mathfrak{A}_1)_p$, $(\mathfrak{A}_2)_p$, and $\mathfrak{A}_p$, respectively. From Proposition 14.1.24, $(\mathscr{R}')_p = (\mathscr{R}_p)'$ and $(\mathscr{C}')_p = (\mathscr{C}_p)'$ almost everywhere; so that $(\mathfrak{A}_1)_p$ and $(\mathfrak{A}_2)_p$ generate the commutant of the center of $\mathscr{R}_p$ almost everywhere. Since $(\mathfrak{A}_1)_p$ and $(\mathfrak{A}_2)_p$ generate $\mathfrak{A}_p$, $(\mathscr{C}_p)'$ is the commutant of the center of $\mathscr{R}_p$; and $\mathscr{C}_p$ is the center of $\mathscr{R}_p$, almost everywhere.

If $\mathscr{C} = \mathscr{A}$, then $\mathscr{C}_p$ is the algebra of scalars, since $\mathscr{A}$ is the algebra of diagonalizable operators; and $\mathscr{R}_p$ is a factor almost everywhere. Conversely, if $\mathscr{C}_p$ is the algebra of scalars (that is, if $\mathscr{R}_p$ is a factor) almost everywhere, then $\mathscr{C}_p = \mathscr{A}_p$ almost everywhere and $\mathscr{C} = \mathscr{A}$, from Proposition 14.1.18. ■

Combining Theorems 14.1.21 and 14.2.2, we have the following corollary.

14.2.3. COROLLARY. *If $\mathscr{R}$ is a von Neumann algebra of type* I_n, II_1, II_∞, *or* III *acting on a separable Hilbert space $\mathscr{H}$, the components $\mathscr{R}_p$ of $\mathscr{R}$ in its direct integral decomposition relative to its center are, almost everywhere, factors of type* I_n, II_1, II_∞, *or* III, *respectively.*

The decomposition of a von Neumann algebra $\mathscr{R}$ relative to its center is referred to as "the central decomposition of $\mathscr{R}$ (into factors)." To what extent is this central decomposition of $\mathscr{R}$ into factors unique? In the most primitive case, when the center $\mathscr{C}$ of $\mathscr{R}$ contains two minimal projections Q_1 and Q_2 with sum I, the measure space can be taken to consist of two points p_1 and p_2, each with positive measure, and $\mathscr{R}_{p_1}$ is (unitarily equivalent to) $\mathscr{R}Q_1$ acting on $Q_1(\mathscr{H})$ while $\mathscr{R}_{p_2}$ is (unitarily equivalent to) $\mathscr{R}Q_2$ acting on $Q_2(\mathscr{H})$. In this case, $\mathscr{H}$ is the direct integral of $\{\mathscr{H}_{p_1}, \mathscr{H}_{p_2}\}$, where $\mathscr{H}_{p_1} = Q_1(\mathscr{H})$ and $\mathscr{H}_{p_2} = Q_2(\mathscr{H})$.

If $\{\mathscr{H}_p\}$ is another direct integral decomposition of $\mathscr{H}$ over (X, μ) relative to which $\mathscr{R}$ is a decomposable algebra containing the diagonalizable operators and $\mathscr{R}_p$ is a factor almost everywhere, Theorem 14.2.2 tells us that $\mathscr{C}$ is the algebra of diagonalizable operators. We know that $\mathscr{C}$ is * isomorphic to the multiplication algebra of $L_2(X, \mu)$; so that, measure-theoretic minutiae aside, X consists of two points q_1 and q_2 to each of which μ assigns positive measure. With Q_1', Q_2' the corresponding minimal projections in $\mathscr{C}$, $\mathscr{R}_{q_1}$ and $\mathscr{R}_{q_2}$ are unitarily equivalent to $\mathscr{R}Q_1'$ and $\mathscr{R}Q_2'$, respectively. Of course the decomposition of $\mathscr{R}$ into factors is "unique" in this case. We have, simply, to discover which of Q_1 or Q_2 the projections Q_1' and Q_2' are and "match" the corresponding factors.

The essence of the uniqueness is the "converse" part of Theorem 14.2.2 stating that each decomposition of an algebra $\mathscr{R}$ containing the diagonalizable operators into factors is such that the center of $\mathscr{R}$ coincides with the algebra of diagonalizable operators. In the case of more general direct-integral decompositions into factors, the basic ingredients of the preceding discussion still apply. The center $\mathscr{C}$ of $\mathscr{R}$ is the algebra of diagonalizable operators and is * isomorphic to the multiplication algebra $\mathscr{A}_1$ of the measure space (X_1, μ_1) of the decomposition. If $\mathscr{A}_2$ is the multiplication algebra of (X_2, μ_2) and $\mathscr{C}$ is the algebra of diagonalizable operators in a decomposition of $\mathscr{R}$ over (X_2, μ_2), the isomorphisms of $\mathscr{C}$ with $\mathscr{A}_1$ and $\mathscr{A}_2$ provide us with an isomorphism φ of $\mathscr{A}_1$ onto $\mathscr{A}_2$. Since $\mathscr{A}_1$ and $\mathscr{A}_2$ are maximal abelian algebras, Theorem 9.3.1 applies and assures us that φ is implemented by a unitary transformation. Following the pattern of the argument in the case of the two-point space, we should hope to map X_1 onto X_2 in a manner that implements φ. It is too much to expect that we can find a one-to-one mapping of X_1 onto X_2 that preserves measurable sets and measure 0 sets; for X_1 and X_2 may not even have the same cardinality. For example, nothing prevents us from adding a third point to the two-point space and assigning it 0 measure, in the framework of the general direct integral theory. But we may hope to exclude Borel subsets of measure 0 from each of X_1 and X_2 and map the remaining portions of the space onto one another by a one-to-one mapping such that both the mapping and its inverse preserve measurable sets and measure 0 sets. That there is such a mapping (in the case of our restricted measure spaces) is the substance of a theorem due to von Neumann [63], whose proof is a measure-theoretic construction. (See Theorem 14.3.4.) Suppose that Y_1 and Y_2 are Borel subsets of X_1 and X_2 such that $\mu_1(X_1 \setminus Y_1) = \mu_2(X_2 \setminus Y_2) = 0$, and η is a one-to-one mapping of Y_1 onto Y_2 that, together with η^{-1}, preserves measurable sets and measure 0 sets and for which $\varphi(M_{f_1}) = M_{f_2}$ where $f_1 = f_2 \circ \eta$ almost everywhere. Let $\mu_2 \circ \eta$ denote the measure on X_1 (equivalent to μ_1) that assigns the measure $\mu_2(\eta(Y))$ to a measurable subset Y of Y_1. There is a measurable function f, finite and positive almost everywhere on X_1, such that for each $\mu_2 \circ \eta$-integrable function g on X_1,

$$\int_{X_1} g(p)(f(p))^2\, d\mu_1(p) = \int_{X_1} g(p)\, d\mu_2 \circ \eta(p).$$

Let $\{x_j\}$ be a dense denumerable rational-linear subspace of $\mathscr{H}$ so that $\{x_j(p)\}$ and $\{x_j(q)\}$ are such subspaces of $\mathscr{H}_p$ and $\mathscr{H}_q$ for almost all p in X_1 and almost all q in X_2. Let $U_p x_j(p)$ be $f(p)x_j(\eta(p))$. We note that U_p extends to a unitary transformation of $\mathscr{H}_p$ onto $\mathscr{H}_{\eta(p)}$. Let Z_1 and Z_2 be measurable subsets of Y_1 and Y_2 such that $\eta(Z_1) = Z_2$. From our assumption that η implements φ and that φ is engendered by the isomorphisms of $\mathscr{C}$ with $\mathscr{A}_1$ and $\mathscr{A}_2$; Z_1 and Z_2 correspond to the same diagonalizable projection P in $\mathscr{C}$. Since $\langle x_j(\eta(p)), x_k(\eta(p))\rangle$ is $\mu_2 \circ \eta$-integrable on X_1,

$$\begin{aligned}
\int_{Z_1} \langle U_p x_j(p), U_p x_k(p)\rangle\, d\mu_1(p) &= \int_{Z_1} \langle x_j(\eta(p)), x_k(\eta(p))\rangle (f(p))^2\, d\mu_1(p) \\
&= \int_{Z_1} \langle x_j(\eta(p)), x_k(\eta(p))\rangle\, d\mu_2 \circ \eta(p) \\
&= \int_{Z_2} \langle x_j(q), x_k(q)\rangle\, d\mu_2(q) \\
&= \langle x_j, P x_k\rangle = \int_{Z_1} \langle x_j(p), x_k(p)\rangle\, d\mu_1(p).
\end{aligned}$$

As this holds for all measurable subsets Z_1 of Y_1, $p \to \langle U_p x_j(p), U_p x_k(p)\rangle$ is μ_1-integrable over X_1 and $\langle U_p x_j(p), U_p x_k(p)\rangle = \langle x_j(p), x_k(p)\rangle$ for all j and k except for a μ_1-null set. It follows, since $\{x_j(p)\}$ and $\{x_j(q)\}$ are dense rational-linear subspaces of $\mathscr{H}_p$ and $\mathscr{H}_q$ almost everywhere, that U_p extends to a unitary transformation of $\mathscr{H}_p$ onto $\mathscr{H}_{\eta(p)}$, for almost all p.

If A is a decomposable operator on $\mathscr{H}$ relative to (X_1, μ_1) then A commutes with $\mathscr{C}$; so that A is decomposable relative to (X_2, μ_2). Let $\{A_k\}$ be a denumerable self-adjoint strong-operator-dense subalgebra of $\mathscr{R}$ over the rationals; and let $\mathfrak{A}$ be its norm closure (a norm-separable C^*-subalgebra of $\mathscr{R}$). If $\{x_{j'}\}$ is a sequence of vectors in $\{x_j\}$ tending to $A_k x_j$, an L_2-subsequence argument tells us that, for some subsequence, $\{x_{j''}(p)\}$ tends to $A_k(p)x_j(p)$ for almost all p and $\{x_{j''}(q)\}$ tends to $A_k(q)x_j(q)$ for almost all q. Thus $U_p x_{j''}(p)$ $(=f(p)x_{j''}(\eta(p)))$ tends to $f(p)A_k(\eta(p))x_j(\eta(p))$ for almost all p. Since U_p is continuous, $U_p A_k(p)x_j(p) = f(p)A_k(\eta(p))x_j(\eta(p)) = A_k(\eta(p))U_p x_j(p)$ for almost all p and all j and k. Now $U_p A_k(p) - A_k(\eta(p))U_p$ is continuous and $\{x_j(p)\}$ is dense in $\mathscr{H}_p$; so that $U_p A_k(p)U_p^{-1} = A_k(\eta(p))$ for all k and almost all p. As $\{A_k(p)\}$ generates $\mathfrak{A}_p$ and $\{A_k(\eta(p))\}$ generates $\mathfrak{A}_{\eta(p)}$; and $\mathscr{R}_p = \mathfrak{A}_p^-$, $\mathscr{R}_{\eta(p)} = \mathfrak{A}_{\eta(p)}^-$, it follows that $U_p \mathscr{R}_p U_p^{-1} = \mathscr{R}_{\eta(p)}$ for almost all p. This establishes the uniqueness of the decomposition of $\mathscr{R}$ into factors.

14.2.4. THEOREM. *If $\mathscr{R}$ is a von Neumann algebra acting on a separable Hilbert space $\mathscr{H}$ and $\mathscr{H}$ is the direct integral of $\{\mathscr{H}_p\}$ in a decomposition relative to an abelian von Neumann subalgebra $\mathscr{A}$ of $\mathscr{R}'$, then $\mathscr{R}_p = \mathscr{B}(\mathscr{H}_p)$ almost everywhere if and only if $\mathscr{A}$ is a maximal abelian subalgebra of $\mathscr{R}'$.*

Proof. Since $\mathscr{A} \subseteq \mathscr{R}'$; $\mathscr{R} \subseteq \mathscr{A}'$, and $\mathscr{R}$ is decomposable. Let $\{B_j\}$ be a denumerable strong-operator-dense self-adjoint subalgebra (over the rationals) of $\mathscr{R}$, and let $\{A_j\}$ be such a subalgebra of $\mathscr{A}$. Let $\mathfrak{A}_1$ and $\mathfrak{A}_2$ be their norm closures (the C^*-algebras they generate). If $\mathfrak{A}$ is the C^*-algebra generated by $\mathfrak{A}_1$ and $\mathfrak{A}_2$, then $\mathfrak{A}$ is the norm closure of the set of finite sums of $B_j A_k$ (an algebra over the rationals, since $B_j A_k = A_k B_j$), and $\mathfrak{A}$ is strong-operator dense in the von Neumann algebra $\mathscr{S}$ generated by $\mathscr{R}$ and $\mathscr{A}$. Thus $\mathscr{R}_p$ and $\mathscr{S}_p$ are the strong-operator closures of $(\mathfrak{A}_1)_p$ and $\mathfrak{A}_p$, respectively. But $\mathfrak{A}_p = (\mathfrak{A}_1)_p$, and $\mathscr{R}_p = \mathscr{S}_p$ almost everywhere, since $(B_j A_k)(p) = B_j(p)a_k(p)I_p = a_k(p)B_j(p)$, where $a_k(p)$ is a complex number. (Recall that A_k is a diagonalizable operator and that $\mathscr{R}$ is not assumed to contain $\mathscr{A}$; so that the conclusion $\mathscr{R}_p = \mathscr{S}_p$ does not imply that $\mathscr{R} = \mathscr{S}$ nor does it conflict with Proposition 14.1.18.)

If $\mathscr{R}_p = \mathscr{B}(\mathscr{H}_p)$ almost everywhere, then $\mathscr{S}_p = \mathscr{B}(\mathscr{H}_p)$ almost everywhere and $\mathscr{S}$ contains $\mathscr{A}$, the algebra of diagonalizable operators. On the other hand $\mathscr{A}'$ has center $\mathscr{A}$, so that $(\mathscr{A}')_p = (\mathscr{A}_p)'$, from Proposition 14.1.24, and $\mathscr{A}_p$ is the algebra of scalars, since $\mathscr{A}$ is the algebra of diagonalizable operators. Thus $(\mathscr{A}')_p = \mathscr{B}(\mathscr{H}_p) = \mathscr{S}_p$ almost everywhere. Applying Proposition 14.1.18, $\mathscr{A}' = \mathscr{S}$; and $\mathscr{A}$ is $\mathscr{S}'$. Since an operator commutes with the (von Neumann) algebra generated by $\mathscr{R}$ and $\mathscr{A}$ if and only if it commutes with both $\mathscr{R}$ and $\mathscr{A}$, $\mathscr{R}' \cap \mathscr{A}' = \mathscr{S}' = \mathscr{A}$. Thus $\mathscr{A}$ is maximal abelian in $\mathscr{R}'$ when $\mathscr{R}_p = \mathscr{B}(\mathscr{H}_p)$ almost everywhere.

Suppose, now, that $\mathscr{A}$ is maximal abelian in $\mathscr{R}'$; so that $\mathscr{S}' = \mathscr{A}$. Then $\mathscr{A}_p = \{aI_p\} = (\mathscr{S}')_p = (\mathscr{S}_p)'$ almost everywhere, from Proposition 14.1.24; and $\mathscr{R}_p = \mathscr{S}_p = \mathscr{B}(\mathscr{H}_p)$ almost everywhere. ■

14.2.5. COROLLARY. *If φ is a representation of a C^*-algebra $\mathfrak{A}$ on a separable Hilbert space $\mathscr{H}$, the decomposition of φ relative to an abelian von Neumann subalgebra $\mathscr{A}$ of $\varphi(\mathfrak{A})'$ has components that are irreducible almost everywhere if and only if $\mathscr{A}$ is maximal abelian in $\varphi(\mathfrak{A})'$.*

Bibliography: [21, 54, 67]

14.3. Appendix—Borel mappings and analytic sets

A Borel structure on a set X is a family $\mathscr{B}$ of subsets (the *Borel sets*) containing $\varnothing$, the complement of each set in $\mathscr{B}$, and the union of each countable subfamily of $\mathscr{B}$. Since the intersection of Borel structures on X is a Borel structure and the set of all subsets of X is a Borel structure on X, each family $\mathscr{F}$ of subsets of X is contained in a smallest Borel structure, *the Borel structure generated by* $\mathscr{F}$. If X is a topological space the Borel structure generated by the open sets (equivalently, the closed sets) is called *the topological* Borel structure. For our purposes, we may restrict our attention to topological spaces and their topological Borel structures, although some of the results that follow are valid for the more general Borel structures.

If $\mathscr{B}$ is a Borel structure on X and μ is a (positive) measure defined on the sets of $\mathscr{B}$, then the family $\bar{\mathscr{B}}$ of sets of the form $X_0 \cup N_0$, where $X_0 \in \mathscr{B}$ and N_0 is a subset of a μ-null set, is a Borel structure on X. Defining $\bar{\mu}(X_0 \cup N_0)$ as $\mu(X_0)$, $\bar{\mu}$ is a measure defined on the sets of $\bar{\mathscr{B}}$ and each subset of a $\bar{\mu}$-null set is in $\bar{\mathscr{B}}$. We say that a measure space $(X, \bar{\mu})$ with this property (each subset of a null set is measurable) is *complete*. In the present case $(X, \bar{\mu})$ is the *completion* of (X, μ). If X is a topological space and $\mathscr{B}$ is its topological Borel structure, we refer to either (X, μ) or $(X, \bar{\mu})$ as a *Borel measure space*.

We say that a Borel measure space (X, μ) is *regular* when (as agreed in Remark 1.7.6) $\mu(C) < \infty$ for each compact set C and each measurable subset X_0 differs from both the intersection of a descending sequence $\{\mathscr{O}_n\}$ of open sets containing it and the union of an ascending sequence $\{C_n\}$ of closed sets contained in it by μ-null sets. In this case, if some $\mu(\mathscr{O}_n) < \infty$ when $\mu(X_0) < \infty$, then $\lim \mu(\mathscr{O}_n) = \mu(X_0) = \lim \mu(C_n)$. Either the condition on open sets or the condition on closed sets implies the other. For example, knowing the condition on open sets and given a measurable set X_0, let $\{\mathscr{O}_n\}$ be a descending sequence of open sets containing $X\backslash X_0$ such that $\mu((\bigcap \mathscr{O}_n)\backslash(X\backslash X_0)) = 0$. But $(\bigcap \mathscr{O}_n)\backslash(X\backslash X_0) = X_0\backslash(\bigcup(X\backslash\mathscr{O}_n))$; $X\backslash\mathscr{O}_n$

$(=C_n)$ is a closed subset of X_0 (since $X \setminus X_0 \subseteq \mathcal{O}_n$), and $\{C_n\}$ is ascending since $\{\mathcal{O}_n\}$ is descending.

If $\mathcal{B}_1$ is a Borel structure on X_1 and $\mathcal{B}_2$ is a Borel structure on X_2, a mapping f of X_1 into X_2 is said to be a Borel mapping when $f^{-1}(X_2') \in \mathcal{B}_1$ for each X_2' in $\mathcal{B}_2$. If $\{X_b\}$ is a subfamily of $\mathcal{B}_2$ generating the Borel structure $\mathcal{B}_2$ and f is a mapping of X_1 into X_2 such that $f^{-1}(X_b) \in \mathcal{B}_1$ for each b, then f is a Borel mapping. To see this, note that, since f^{-1} preserves countable unions and complements and since $\mathcal{B}_1$ is a Borel structure, the family $\mathcal{F}$ of those sets X_2' in $\mathcal{B}_2$ such that $f^{-1}(X_2') \in \mathcal{B}_1$ is a Borel structure. By assumption $\{X_b\} \subseteq \mathcal{F}$. Since $\{X_b\}$ generates $\mathcal{B}_2$, $\mathcal{F} = \mathcal{B}_2$, and f is a Borel mapping.

If (X, μ) is a complete Borel measure space, a mapping f of X into a space Y with a Borel structure is said to be *measurable* when $f^{-1}(Y_0)$ is μ-measurable for each Borel subset Y_0 of Y. Since μ is complete, $f^{-1}(Y_0)$ may not be a (topological) Borel set in X (and f may not be a Borel mapping). In the lemma that follows, we note the possibility of finding a Borel μ-null set on the complement of which f is a Borel mapping when the Borel structure in Y is countably generated.

14.3.1. LEMMA. *If (X, μ) is a complete Borel measure space and f is a measurable mapping of X into space Y with a countably generated Borel structure, then there is a Borel μ-null set N of X such that $f|(X \setminus N)$ is a Borel mapping.*

Proof. Let $\{Y_j\}$ be a denumerable family of Borel subsets of Y that generates the Borel structure. Then $f^{-1}(Y_j)$ is a measurable subset of X and, hence, is the union of a Borel subset X_j of X and a subset of a Borel μ-null set N_j. Let N be $\bigcup_{j=1}^{\infty} N_j$. Then

$$f^{-1}(Y_j) \cap (X \setminus N) = X_j \cap (X \setminus N),$$

and $X_j \cap (X \setminus N)$ is a Borel subset of $X \setminus N$. Thus $f|(X \setminus N)$ is a Borel mapping. ∎

14.3.2. LEMMA. *If (X, μ) is a regular Borel measure space, Y is a completely regular space such that $C(Y)$ is countably generated, and η is a mapping of X into Y that is Borel and one-to-one on a compact subset $\mathcal{K}$, then there is a Borel subset $\mathcal{K}_0$ of $\mathcal{K}$ such that η is a Borel isomorphism on $\mathcal{K}_0$ and $\mu(\mathcal{K} \setminus \mathcal{K}_0) = 0$.*

Proof. Let $\mathcal{F}_0$ be the (denumerable) subalgebra over the rationals generated by a countable generating family of functions in $C(Y)$ (so that $\mathcal{F}_0$ is norm, dense in the algebra $C(Y)$ of continuous bounded functions on Y). Since η is a Borel mapping on $\mathcal{K}$, $f \circ \eta$ is a Borel function on $\mathcal{K}$ for each f in $C(Y)$. Let $f_1, f_2, \ldots$ be an enumeration of the functions in $\mathcal{F}_0$. By Lusin's theorem, given a positive ε, there is a closed subset $\mathscr{K}_1$ of $\mathscr{K}$ such that $\mu(\mathscr{K}\backslash\mathscr{K}_1) < \varepsilon/2$ and $f_1 \circ \eta$ is continuous on $\mathscr{K}_1$. Again, there is a closed subset $\mathscr{K}_2$ of $\mathscr{K}_1$ such that $f_2 \circ \eta$ is continuous on $\mathscr{K}_2$ and $\mu(\mathscr{K}_1\backslash\mathscr{K}_2) < \varepsilon/4$. Inductively, there is a closed subset $\mathscr{K}_n$ of $\mathscr{K}_{n-1}$ such that $f_n \circ \eta$ is continuous on $\mathscr{K}_n$ and $\mu(\mathscr{K}_{n-1}\backslash\mathscr{K}_n) < \varepsilon/2^n$. Let $\mathscr{K}_\varepsilon$ be $\bigcap_{n=1}^{\infty} \mathscr{K}_n$. Then $\mu(\mathscr{K}\backslash\mathscr{K}_\varepsilon) < \varepsilon$ and $\mathscr{K}_\varepsilon$ is a closed set on which all $f_n \circ \eta$ are continuous. Since $\{f_n\}$ determines the topology of Y, η is a continuous mapping of $\mathscr{K}_\varepsilon$ into Y. As η is one-to-one on $\mathscr{K}_\varepsilon$, Y is Hausdorff, and $\mathscr{K}_\varepsilon$ is compact, η is a homeomorphism on $\mathscr{K}_\varepsilon$. It follows that $\eta(\mathscr{K}_\varepsilon)$ is compact and that the image of each Borel set in $\mathscr{K}_\varepsilon$ under η is a Borel set in $\eta(\mathscr{K}_\varepsilon)$ and hence in Y.

Let $\mathscr{K}_0$ be $\bigcup_{n=1}^{\infty} \mathscr{K}_{1/n}$. Then $\mathscr{K}_0$ is a Borel subset of X, $\mathscr{K}_0 \subseteq \mathscr{K}$, and $\mu(\mathscr{K}\backslash\mathscr{K}_0) = 0$. If X_0 is a Borel subset of $\mathscr{K}_0$, then $X_0 \cap \mathscr{K}_{1/n}$ is a Borel subset of $\mathscr{K}_{1/n}$ for each n; and $\eta(X_0) = \bigcup_{n=1}^{\infty} \eta(X_0 \cap \mathscr{K}_{1/n})$. Thus $\eta(X_0)$ is a Borel set in Y and η is a Borel isomorphism on $\mathscr{K}_0$. ■

14.3.3. PROPOSITION. *If (X, μ) is a σ-compact regular Borel measure space, $\mathscr{S}$ is a subset of X of finite measure, and η is a one-to-one Borel mapping of $\mathscr{S}$ into a completely regular space Y for which $C(Y)$ is countably generated, then there is a Borel subset X_0 of $\mathscr{S}$ such that $\mu(\mathscr{S}\backslash X_0) = 0$, $\eta(X_0)$ is a Borel subset of Y, and η is a Borel isomorphism of X_0 onto $\eta(X_0)$.*

Proof. Since X is regular, there is an ascending sequence $\{C_n\}$ of closed subsets of $\mathscr{S}$ such that $\mu(\mathscr{S}\backslash\bigcup_{n=1}^{\infty} C_n) = 0$. As X is σ-compact, there is an ascending sequence $\{\mathscr{K}'_n\}$ of compact subsets of X such that $X = \bigcup_{n=1}^{\infty} \mathscr{K}'_n$. Then $\mathscr{K}'_n \cap C_n\ (= \mathscr{K}_n)$ is compact, $\{\mathscr{K}_n\}$ is an ascending sequence of compact subsets of $\mathscr{S}$, and $\mu(\mathscr{S}\backslash\bigcup_{n=1}^{\infty} \mathscr{K}_n) = 0$ (for, with $\{\mathscr{K}'_n\}$ and $\{C_n\}$ ascending, $(\bigcup_{n=1}^{\infty} \mathscr{K}'_n) \cap (\bigcup_{n=1}^{\infty} C_n) = \bigcup_{n=1}^{\infty} \mathscr{K}_n$). From Lemma 14.3.2, each $\mathscr{K}_n$ contains a Borel subset B_n such that $\mu(\mathscr{K}_n\backslash B_n) = 0$, $\eta(B_n)$ is a Borel set, and η is a Borel isomorphism of B_n onto $\eta(B_n)$. By taking $\bigcup_{n=1}^{\infty} B_n$ as X_0, our result follows. ■

14.3.4. THEOREM. *If X_1 and X_2 are σ-compact separable metric spaces, μ_1 and μ_2 are complete regular Borel measures on X_1 and X_2, and φ is an isomorphism of $L_\infty(X_1, \mu_1)$ onto $L_\infty(X_2, \mu_2)$, then there are Borel subsets X'_1, X'_2 of X_1, X_2 and a Borel isomorphism η of X'_1 onto X'_2 that identifies μ_1-null sets with μ_2-null sets such that $\mu_1(X_1\backslash X'_1) = \mu_2(X_2\backslash X'_2) = 0$ and $f|X'_1 = \varphi(f) \circ \eta$ for each f in $L_\infty(X_1, \mu_1)$.*

Proof. Throughout this proof, the subscript j takes the values 1 and 2. Since X_j is separable and μ_j is regular, $L_2(X_j, \mu_j)$ is a separable Hilbert space. Its multiplication algebra $\tilde{\mathscr{A}}_j$ admits a separating (and generating) unit vector x_j, by Corollary 5.5.17. In fact, since the state $\omega_{x_2} \circ \tilde{\varphi}$ is normal, where $\tilde{\varphi}(M_f) = M_{\varphi(f)}$; by Theorem 7.2.3, it is a vector state of $\tilde{\mathscr{A}}_1$, and we may choose x_1 so that $\omega_{x_1}|\tilde{\mathscr{A}}_1 = \omega_{x_2} \circ \tilde{\varphi}$. If we assign the value $\langle M_{f_j} x_j, x_j \rangle$ to a measurable subset $\mathscr{S}_j$ of X_j, where f_j is the characteristic function of $\mathscr{S}_j$, the resulting set function μ_j' is a positive regular Borel measure on X_j, equivalent to μ_j, with the following properties:

(i) $\mu_j'(X_j) = 1$;

(ii) $\displaystyle\int_{X_1} f(p_1)\, d\mu_1'(p_1) = \int_{X_2} \varphi(f)(p_2)\, d\mu_2'(p_2) \qquad \text{for } f \text{ in } L_\infty(X_1, \mu_1).$

If we can establish the assertion of this theorem for the measure spaces (X_1, μ_1') and (X_2, μ_2'), then it follows for (X_1, μ_1) and (X_2, μ_2). We assume, henceforth, that μ_j satisfies (i) and (ii).

Let $\tilde{\mathscr{B}}_j$ be the denumerable strong-operator-dense subalgebra over the rationals of $\tilde{\mathscr{A}}_j$ generated by the characteristic functions of the open sets in a denumerable basis for the topology of X_j. Let $\tilde{\mathscr{C}}_1$ and $\tilde{\mathscr{C}}_2$ be the algebras over the rationals generated by $\{\tilde{\mathscr{B}}_1, \tilde{\varphi}^{-1}(\tilde{\mathscr{B}}_2)\}$ and $\{\tilde{\varphi}(\tilde{\mathscr{B}}_1), \tilde{\mathscr{B}}_2\}$, respectively. Let $\mathscr{B}_j$ and $\mathscr{C}_j$ be the algebras of equivalence classes of functions in $L_\infty(X_j, \mu_j)$ corresponding to $\tilde{\mathscr{B}}_j$ and $\tilde{\mathscr{C}}_j$. Choosing a representative f_j for each (real) class in $\mathscr{C}_j$, we can find a μ_j-null set N_j such that f_j is a Borel function on $X_j \backslash N_j$, $f_j(X_j \backslash N_j) \subseteq [-\|f_j\|_\infty, \|f_j\|_\infty]$, and the algebraic operations in $\mathscr{C}_j$ correspond to pointwise operations on $X_j \backslash N_j$. We may treat $\mathscr{C}_j$ (and $\mathscr{B}_j$), now, as algebras of functions. It will be convenient to assume, as we may, that N_j contains the complement of the support of μ_j. (See Remark 3.4.13.) It follows that each point p_j in $X_j \backslash N_j$ corresponds to a bounded multiplicative linear functional ρ_{p_j} (hence, from Theorem 3.4.7, pure state) on $\tilde{\mathscr{D}}_j$, the norm closure of $\tilde{\mathscr{C}}_j$. Since $\tilde{\mathscr{D}}_2$ is an abelian C^*-algebra, there is a * isomorphism $\tilde{\psi}_2$ of $\tilde{\mathscr{D}}_2$ onto $C(Y)$, where Y is the compact Hausdorff space of pure states of $\tilde{\mathscr{D}}_2$ with its weak* topology (see Theorem 4.4.3). Let $\tilde{\psi}_1$ be $\tilde{\psi}_2 \circ (\tilde{\varphi}|\tilde{\mathscr{D}}_1)$; and let ψ_j be the corresponding isomorphism of $\mathscr{D}_j$, the algebra of functions in $L_\infty(X_j, \mu_j)$ associated with $\tilde{\mathscr{D}}_j$. Then $\rho_{p_j} \circ \tilde{\psi}_j^{-1}$ is a multiplicative linear functional on $C(Y)$, and corresponds to a point $\xi_j(p_j)$ of Y, by Corollary 3.4.2. By construction $\psi_j(f_j)(\xi_j(p_j)) = f_j(p_j)$ for f_j in $\mathscr{C}_j$ and p_j in $X_j \backslash N_j$.

The mapping $\psi_j(f_j) \to \int_{X_j} f_j(p_j)\, d\mu_j(p_j)$ $(f_j \in \mathscr{D}_j)$ defines a state of $C(Y)$ and, hence, a regular Borel measure ν_j on Y such that $\nu_j(Y) = 1$. Since $\psi_2(\varphi(f_1)) = \psi_1(f_1)$ for f_1 in $\mathscr{D}_1$; from (ii), the two states of $C(Y)$ determining the measures ν_1 and ν_2 are identical. Thus $\nu_1 = \nu_2$ $(=\nu)$. By construction

$$\int_{X_j} f_j(p_j)\, d\mu_j(p_j) = \int_Y \psi_j(f_j)(y)\, d\nu(y)$$

for f_j in $\mathscr{D}_j$. Since $\tilde{\mathscr{C}}_j$ is norm dense in $\tilde{\mathscr{D}}_j$, the functions $\psi_j(\mathscr{C}_j)$ determine the topology of Y. As $f_j|(X_j \backslash N_j)$ is a Borel function and $\psi_j(f_j) \circ \xi_j = f_j|(X_j \backslash N_j)$ for f_j in $\mathscr{C}_j$, ξ_j is a Borel mapping of $X_j \backslash N_j$ into Y. Moreover, ξ_j is one-to-one, since the (representing) functions in $\mathscr{B}_j$ separate the points of $X_j \backslash N_j$. From Proposition 14.3.3, there is a (σ-compact) Borel set X_j'' contained in $X_j \backslash N_j$ such that $\mu_j(X_j'') = 1$, $\xi_j(X_j'')$ $(= Y_j)$ is a Borel set in Y, and ξ_j is a Borel isomorphism of X_j'' onto Y_j.

Having just noted that μ_j and ν induce the same integration when functions in $\mathscr{D}_j$ are "transported" by ψ_j and that ξ_j^{-1} "induces" ψ_j on $\mathscr{C}_j$, a dense subset of $\mathscr{D}_j$, we should be led to feel that $\nu(Y_0) = \mu_j(\xi_j^{-1}(Y_0))$ for each measurable subset Y_0 of Y—equivalently, that

$$(1) \qquad \int_Y g_j(y)\, d\nu(y) = \int_{X_j''} g_j(\xi_j(p_j))\, d\mu_j(p_j)$$

for each characteristic function g_j of a measurable subset Y_0 of Y. We prove (1) in the argument that follows, noting in the next computation that it holds when g_j is $\psi_j(f_j)$ and f_j is the selected representative of an equivalence class in $\mathscr{C}_j$ $(\subseteq \mathscr{D}_j)$. For such an f_j,

$$\begin{aligned}
\int_Y \psi_j(f_j)(y)\, d\nu(y) &= \int_{X_j} f_j(p_j)\, d\mu_j(p_j) \\
&= \int_{X_j \backslash N_j} f_j(p_j)\, d\mu_j(p_j) \\
&= \int_{X_j \backslash N_j} \psi_j(f_j)(\xi_j(p_j))\, d\mu_j(p_j) \\
&= \int_{X_j''} \psi_j(f_j)(\xi_j(p_j))\, d\mu_j(p_j).
\end{aligned}$$

Let $\mathscr{F}_j$ be the class of all bounded complex-valued Borel functions g_j on Y for which (1) holds. Then $\mathscr{F}_j$ is linear, norm closed, and contains a norm-dense subset of $C(Y)$. Thus $\mathscr{F}_j$ contains $C(Y)$. From the monotone convergence theorem, if a bounded function g_j is the pointwise limit, *everywhere* on Y, of an increasing sequence of (real-valued) functions in $\mathscr{F}_j$, then $g_j \in \mathscr{F}_j$. We show, next, that $\mathscr{F}_j$ contains the characteristic function of each open set in Y. For this note that, since Y has a countable generating family of continuous functions, it is homeomorphic to a subset of the product of a countable number of unit intervals. Hence Y admits a metric (compatible with its topology). If $\mathscr{O}$ is an open subset of Y and $f_n(y)$ is $\min\{n \operatorname{dist}(y, Y \backslash \mathscr{O}), 1\}$, then $\{f_n\}$ is an increasing sequence of functions in $C(Y)$ tending pointwise to the characteristic function of $\mathscr{O}$; and this function is in $\mathscr{F}_j$.

Let $\mathscr{S}_j$ be the class of all Borel subsets of Y whose characteristic functions satisfy (1). From the results of the preceding paragraph, $\mathscr{S}_j$ contains all the open sets in Y. Since $\mathscr{F}_j$ is linear and contains 1, the complement of a set in $\mathscr{S}_j$ lies in $\mathscr{S}_j$. Moreover, the countable union of sets in $\mathscr{S}_j$ is in $\mathscr{S}_j$ (from the "monotone closure" property of $\mathscr{F}_j$). Hence $\mathscr{S}_j$ contains all the Borel sets in Y. Applying (1) to the characteristic function g_j of Y_j, we have $\nu(Y_j) = \mu_j(X_j'') = 1$. Hence $\nu(Y_1 \cap Y_2) = 1$; and applying (1) to the characteristic function of $Y_1 \cap Y_2$, we have $\mu_j(X_j') = 1$, where X_j' is the Borel subset $\xi_j^{-1}(Y_1 \cap Y_2)$ of X_j''.

For p in X_1', let $\eta(p)$ be $\xi_2^{-1}(\xi_1(p))$. Then η is a one-to-one, measure preserving, Borel isomorphism of X_1' onto X_2', and $\mu_j(X_j \backslash X_j') = 0$. If $p \in X_1'$ and $f \in \mathscr{C}_1$, then

$$\begin{aligned}(\varphi(f) \circ \eta)(p) &= (\varphi(f) \circ \xi_2^{-1})(\xi_1(p)) = (\psi_2(\varphi(f)))(\xi_1(p)) \\ &= (\psi_1(f))(\xi_1(p)) = f(p).\end{aligned}$$

But the mapping $M_g \to M_{g \circ \eta}$ $(g \in L_\infty(X_2, \mu_2))$ is a * isomorphism $\tilde{\varphi}'$ of $\tilde{\mathscr{A}}_2$ onto $\tilde{\mathscr{A}}_1$. Since $\tilde{\varphi}^{-1}$ and $\tilde{\varphi}'$ agree on the self-adjoint strong-operator-dense subalgebra $\tilde{\mathscr{C}}_2$, they agree on $\tilde{\mathscr{A}}_2$. Thus η is a Borel isomorphism of X_1' onto X_2' such that $f|X_1' = \varphi(f) \circ \eta$ when $f \in L_\infty(X_1, \mu_1)$, and $\mu_j(X_j \backslash X_j') = 0$. ■

A topological space X is said to be a csm (complete separable metrizable) space if it has a countable base for its open sets and admits a metric in which it is complete. The set consisting of a point chosen in each set of a countable base is dense in X so that, with its metric, X is a complete separable metric space. Let d be a metric on X in which it is complete. Defining $d'(p, q)$ to be $d(p, q)[1 + d(p, q)]^{-1}$, we have that d' is a metric on X, $d'(p, q) \leq \min(d(p, q), 1)$, and $d(p, q) \leq 2d'(p, q)$ when $d'(p, q) \leq \frac{1}{2}$ (since $d(p, q) = d'(p, q)[1 - d'(p, q)]^{-1}$). It follows that d and d' determine the same topology on X and that X is complete relative to d'. There is no loss of generality in assuming, therefore, that our csm space X is equipped with a metric bounded by 1 (that is, $d(p, q) \leq 1$ for each pair p, q in X).

If (X_n, d_n), $n = 1, 2, \ldots$ are csm spaces (d_n a metric bounded by 1), their topological product X is a csm space. To see this, let $d(\{p_n\}, \{q_n\})$ be $\sum 2^{-n} d_n(p_n, q_n)$. Then d is a metric on X, compatible with the product topology, relative to which X is complete. Choose a point p_n in each X_n; and let $\{p_{nj} : j = 1, 2, \ldots\}$ be a denumerable dense subset of X_n. Then $\{(p_{1j_1}, p_{2j_2}, \ldots, p_{nj_n}, p_{n+1}, p_{n+2}, \ldots)\}$ is a denumerable dense subset of X. If the spaces X_n are disjoint and X_0 is their (topological) sum, then, defining $d_0(p, q)$ to be 1 when $p \in X_n$, $q \in X_m$, and $n \neq m$, and $d_0(p_n, q_n)$ to be $d_n(p_n, q_n)$ when p_n, q_n are in X_n, d_0 is a metric on X_0 compatible with its sum topology relative to which X_0 is complete. Thus X is a csm space.

Of course each closed subset of a csm space X is a csm space. The same is true of an open subset $\mathcal{O}$. With p, q in $\mathcal{O}$, let $d'(p, q)$ be

$$|(\inf\{d(p, p') : p' \in X \backslash \mathcal{O}\})^{-1} - (\inf\{d(q, q') : q' \in X \backslash \mathcal{O}\})^{-1}| + d(p, q).$$

Then d' is a metric on $\mathcal{O}$ compatible with its (induced) topology relative to which it is complete. To see that $\mathcal{O}$ is complete relative to d', note that if $d'(p_n, p_m) \to 0$, then $d(p_n, p_m) \to 0$; so that there is a p in X such that $d(p_n, p) \to 0$. If $p \in X \backslash \mathcal{O}$, $[\inf\{d(p_n, p') : p' \in X \backslash \mathcal{O}\}]^{-1} \to \infty$, contradicting the fact that $d'(p_n, p_m) \to 0$ as n, $m \to \infty$. Thus $p \in \mathcal{O}$ and (by compatibility) $d'(p_n, p) \to 0$. Hence $(\mathcal{O}, d')$ is a csm space.

It follows from the preceding discussion that the space $\mathbb{Z}_+$ of positive integers (and 0) is complete relative to the metric d, where $d(m, n) = |m - n|(1 + |m - n|)^{-1}$ and that $\mathbb{S}$, the countable product of $\mathbb{Z}_+$ with itself (that is, the space of sequences of positive integers) is complete relative to d', where

$$d'(\{m_j\}, \{n_j\}) = \sum_{j=1}^{\infty} 2^{-j}|m_j - n_j|(1 + |m_j - n_j|)^{-1},$$

and d' is compatible with the product topology on $\mathbb{S}$. Thus $(\mathbb{S}, d')$ is a csm space. We prove that each (non-empty) csm space (X, d) is the continuous image of $\mathbb{S}$. The closed balls with radius $\frac{1}{2}\varepsilon$ and centers a dense denumerable subset of X form a denumerable covering of X by non-null closed sets with diameter not exceeding ε. Let $X_1, X_2, \ldots$ be such a covering for diameter $\frac{1}{2}$. Let $X_{n1}, X_{n2}, \ldots$ be such a covering of X_n for diameter $\frac{1}{4}$. Let $X_{nm1}, X_{nm2} \ldots$ be such a covering of X_{nm} for diameter $\frac{1}{8}$. Continuing in this way, we define closed non-null sets $X_{n_1 n_2 \cdots n_k}$ with diameter not exceeding 2^{-k} such that $X_{n_1 \cdots n_k} \subseteq X_{n_1 \cdots n_{k-1}}$, for each finite sequence $(n_1, \ldots, n_k)$ of positive integers. With $(n_1, n_2, \ldots)$ in $\mathbb{S}$, choose a point p_k in $X_{n_1 \cdots n_k}$ for each k. Then $\{p_k\}$ is Cauchy convergent in X and tends to a limit $f(n_1, n_2, \ldots)$ in X. If $\{p'_k\}$ is an alternative choice of defining sequence, $d(p_k, p'_k) \le 2^{-k}$, since both p_k and p'_k lie in $X_{n_1 \cdots n_k}$. Thus $\{p'_k\}$ tends to $f(n_1, \ldots, n_k)$; and f is a mapping of $\mathbb{S}$ into X. If $(n_1, n_2, \ldots)$ and $(m_1, m_2, \ldots)$ are suitably close in $\mathbb{S}$, then $n_1 = m_1, \ldots, n_k = m_k$ for some large k. Hence $d(f(n_1, n_2, \ldots), f(m_1, m_2, \ldots)) \le 2^{-k}$ (since both $f(n_1, n_2, \ldots)$ and $f(m_1, m_2, \ldots)$ lie in $X_{n_1 \cdots n_k}$); and f is continuous. If $p \in X$, p lies in some X_{n_1} (since $X_1, X_2, \ldots$ is a covering of X) and p is in some $X_{n_1 n_2}$ (since $X_{n_1 1}, X_{n_1 2}, \ldots$ is a covering of X_{n_1}). Continuing in this way, we construct a sequence $\{n_1, n_2, \ldots\}$ such that $f(n_1, n_2, \ldots) = p$. Thus f is a continuous mapping of $\mathbb{S}$ onto X.

A subset X_0 of a csm space X is said to be *analytic* if it is the continuous image of a csm space. Of course each csm space is an analytic subset of itself. Composing mappings, we see that the image of an analytic set in a csm space under a continuous mapping is an analytic set. If X_n is a csm space, $n = 1$, $2, \ldots$ and f_n is a continuous mapping of X_n into the csm space X (so that $f_n(X_n)$ is an analytic subset of X) then, defining f to be f_n on X_n, as a subset of the sum space X_0, f is a continuous mapping of X_0, a csm space, onto $\bigcup f(X_n)$. Thus the countable union of analytic subsets of a space is analytic.

Again, the product Y of the spaces X_n is a csm space. If π_j is the jth coordinate mapping of Y onto X_j, the set Y_{nm} of points of Y at which the two (continuous) mappings $f_n \circ \pi_n$ and $f_m \circ \pi_m$ agree is closed. Thus $\bigcap_{n,m} Y_{nm}$ $(= Y_0)$ is a closed subset of Y. Hence Y_0 is a csm space; and all the mappings $f_n \circ \pi_n$ coincide on Y_0 to determine a single continuous mapping of Y_0 onto $\bigcap f_n(X_n)$. Therefore a countable intersection of analytic sets is analytic. It follows that the family of analytic subsets of a csm space X whose complements are also analytic subsets of X is a Borel structure on X containing the closed sets. (Note, for this, that, with cautious and strict interpretation, $\varnothing$ is a csm space and is the continuous image of itself under the identity mapping on it.) Hence each Borel subset of a csm space is analytic.

Our next goal is to prove a result approximating a converse to the preceding conclusion: If (X, μ) is a complete σ-compact csm Borel measure space and A is an analytic subset of X, then A is measurable. We shall prove this by showing that there is a compact csm space Y, a descending sequence $\{Y_n\}$ of σ-compact subsets of Y with intersection B, and a continuous mapping f of Y into X, such that $f(B) = A$. We first show that, under these circumstances, A is measurable after a few easy measure-theoretic observations that will be needed for the argument.

If A' is a subset of the finite measure space (X, μ), we define $\mu^*(A')$ as $\inf\{\mu(S) : A' \subseteq S,\ S \text{ measurable}\}$. Of course $\mu^*(A') \le \mu^*(B')$ if $A' \subseteq B'$. If S_n is a measurable set containing A' such that $\mu(S_n) \le \mu^*(A') + 1/n$, then $\bigcap S_n$ is a measurable set containing A' with measure not exceeding $\mu^*(A')$. Thus $\mu(\bigcap S_n) = \mu^*(A')$. If $\{A_n\}$ is an ascending sequence of sets with union A', S_n is a measurable set containing A_n such that $\mu(S_n) = \mu^*(A_n)$, and $T_n = S_n \cap S_{n+1} \cap \cdots$, then $\mu(T_n) = \mu(S_n) = \mu^*(A_n)$ since T_n is a measurable set and $A_n \subseteq T_n \subseteq S_n$. Moreover, $\{T_n\}$ is an ascending sequence of measurable sets with (measurable) union T containing A'; and $\mu(T) = \lim \mu(T_n) = \lim \mu^*(A_n)$. Thus $\mu^*(A') \le \lim \mu^*(A_n)$. Since $A_n \subseteq A'$ for each n, $\lim \mu^*(A_n) \le \mu^*(A')$; and $\mu^*(A') = \lim \mu^*(A_n)$. If the measure space (X, μ) is complete and, for each positive m, there is a measurable subset C_m of A' such that $\mu^*(A') - 1/m \le \mu(C_m)$, then A' is measurable. To see this, choose S measurable such that $\mu(S) = \mu^*(A')$ and $A' \subseteq S$. Then $\bigcup C_m$ $(= C)$ is measurable and $\mu^*(A') \le \mu(C) \le \mu(S) = \mu^*(A')$ while $C \subseteq A' \subseteq S$. Thus $\mu(S \backslash C) = 0$ and A' is the union of C and the μ-null set $A' \backslash C$.

We are assuming that $X = \bigcup K_n$ with K_n compact and $\mu(K_n)$ finite. We want to show that the analytic subset A of X is measurable. Since $A \cap K_n$ is analytic and $A = \bigcup(A \cap K_n)$, we may assume that $\mu(X)$ is finite and that X is compact. With these assumptions in force and with the notation adopted earlier ($A = f(B)$ and $B = \bigcap Y_n$), we use the discussion of the preceding paragraph to show that A is measurable. There is an ascending sequence $\{K_{nj}\}$ of compact sets with union Y_n. Since $B \subseteq Y_1$, $f(B) = \bigcup_j f(B \cap K_{1j})$.

Thus $\mu^*(f(B)) = \lim \mu^*(f(B \cap K_{1j}))$; and there is a j_1 such that

$$\mu^*(f(B)) - 1/m < \mu^*(f(B \cap K_{1j_1})) \leq \mu(f(K_{1j_1})).$$

Since $B \cap K_{1j_1} \subseteq Y_2$, $B \cap K_{1j_1} = \bigcup_j (B \cap K_{1j_1} \cap K_{2j})$ and, again, there is a j_2 such that

$$\mu^*(f(B)) - 1/m < \mu^*(f(B \cap K_{1j_1} \cap K_{2j_2})) \leq \mu(f(K_{1j_1} \cap K_{2j_2})).$$

Continuing, we find j_n such that

$$\mu^*(f(B)) - 1/m < \mu(f(K_{1j_1} \cap \cdots \cap K_{nj_n})).$$

Now $\{K_{1j_1} \cap \cdots \cap K_{nj_n}\}$ is a descending sequence of compact sets with (compact) intersection K so that $\{f(K_{1j_1} \cap \cdots \cap K_{nj_n})\}$ is a descending sequence of compact sets with (compact) intersection S containing $f(K)$. If $p \in S$, there is a q_n in $K_{1j_1} \cap \cdots \cap K_{nj_n}$ such that $f(q_n) = p$. The closures of the sets $\{q_j : j = n, n+1, \ldots\}$, $n = 1, 2, \ldots$ form a descending sequence of non-null compact subsets of Y and their intersection is non-null. With q in this intersection, $q \in K$ and $f(q) = p$. Thus $S = f(K)$. Hence $\mu^*(f(B)) - 1/m \leq \mu(f(K))$. But $K \subseteq K_{nj_n} \subseteq Y_n$ for all n; so that $K \subseteq B$ and $f(K) \subseteq f(B)$. From the discussion of the preceding paragraph, $f(B)(=A)$ is measurable. It remains to find Y, $\{Y_n\}$, and f.

We have seen that there is a continuous mapping g of $\mathbb{S}$ into X such that $g(\mathbb{S}) = A$. Let $\bar{\mathbb{R}}_n$ be a copy of the one-point compactification of $\mathbb{R}$ for $n = 1, 2, \ldots$; and let W be the product of the $\bar{\mathbb{R}}_n$. If $Y = W \times X$, then Y is a compact (csm) space and $\mathbb{S}$ can be viewed as a subset of W. Let B be the graph of g (in Y) and f the projection of Y onto X. Then $f(B) = A$ and f is continuous. Let F_{nm} be the union of all closed intervals of lengths $1/m$ with centers 0 or a positive integer in $\bar{\mathbb{R}}_n$. If

$$F_n = (F_{1n} \times F_{2n} \times \cdots \times F_{nn} \times \bar{\mathbb{R}}_{n+1} \times \bar{\mathbb{R}}_{n+2} \times \cdots) \times X,$$

then $\{F_n\}$ is a descending sequence of σ-compact sets in Y with intersection $\mathbb{S} \times X$. Since g is a continuous mapping of $\mathbb{S}$ into X its graph B is closed in $\mathbb{S} \times X$; so that $B^- \cap (\mathbb{S} \times X) = B$, where B^- is the closure of B in Y. Thus $B = B^- \cap F_1 \cap F_2 \cap \cdots$. Since B^- is compact (closed in Y), it follows that $\{B^- \cap F_n\}$ is a descending sequence of σ-compact sets in Y with intersection B. (In the notation of the preceding paragraph, $Y_n = B^- \cap F_n$ and $K_{nj} = B^- \cap C_{nj}$, where $\{C_{nj}\}$ is an ascending sequence of compact subsets of Y with union F_n.)

We summarize the information of the preceding discussion, needed for reference, in the theorem that follows.

14.3.5. THEOREM. *If X is a separable topological space metrizable as a complete metric space (X is a* csm *space), then each Borel set is analytic. If in addition (X, μ) is a complete, Borel measure space, $X = \bigcup K_n$ with K_n compact and $\mu(K_n)$ finite, then each analytic set is measurable.*

We are now in a position to prove the measurable-selection principle used in establishing the last results of Section 14.1. For this purpose, we introduce the total (*lexicographical*) order on $\mathbb{S}$ in which $(n_1, n_2, \ldots) < (m_1, m_2, \ldots)$ when, for some k, $n_1 = m_1, \ldots, n_k = m_k$ and $n_{k+1} < m_{k+1}$. Each closed set C in $\mathbb{S}$ has a smallest element in this ordering. To see this, let c_1 be an element of C with smallest first coordinate. Among the elements of C with the same first coordinate as c_1 choose one, c_2, with smallest second coordinate. Continuing in this way, we construct a sequence $\{c_n\}$ of elements of C tending to the element c formed from these smallest coordinates. Since C is closed, $c \in C$ and c is the smallest element in C.

If (X, μ) is a complete Borel measure space, $X = \bigcup K_n$ with K_n compact and $\mu(K_n)$ finite, and X is a csm space, then each analytic subset A of X is μ-measurable (from Theorem 14.3.5). There is a continuous mapping f of $\mathbb{S}$ into X with image A. A prototype (and test) of our measurable-selection principle might involve finding a measurable "cross-section" for f; that is, a measurable mapping g of A into $\mathbb{S}$ such that $f \circ g$ is the identity on A. (In effect, we want to select a point $g(p)$ in $f^{-1}(p)$ for each p in A in such a way that g is measurable.) Since f is continuous, $f^{-1}(p)$ is closed in $\mathbb{S}$. Let $g(p)$ be its smallest element (in the lexicographical ordering). We show that g is measurable by proving that the inverse image under g of each set in a base for the open sets of $\mathbb{S}$ is measurable. As our base, we use sets of the form $\{n_1\} \times \{n_2\} \times \cdots \times \{n_k\} \times \mathbb{Z}_+ \times \mathbb{Z}_+ \times \cdots$. If s and s' are, respectively, the points $(n_1, n_2, \ldots, n_k, 0, 0, \ldots)$ and $(n_1, n_2, \ldots, n_k + 1, 0, 0, \ldots)$, this set is $\{s_0 : s_0 \in \mathbb{S},\ s \le s_0 < s'\}$ $(= \langle s \rangle)$. (It is closed as well as open.) Now $p \in g^{-1}(\langle s \rangle)$ if and only if $f^{-1}(p)$ has its smallest element in $\langle s \rangle$. Thus $g^{-1}(\langle s \rangle) = f(\mathbb{S}_{s'}) \backslash f(\mathbb{S}_s)$ where $\mathbb{S}_s = \{s_0 : s_0 \in \mathbb{S}, s_0 < s\}$. Since $\mathbb{S}_s$ and $\mathbb{S}_{s'}$ are open in $\mathbb{S}$, they are analytic subsets of $\mathbb{S}$. Hence $f(\mathbb{S}_{s'})$ and $f(\mathbb{S}_s)$ are analytic subsets of X. It follows that $g^{-1}(\langle s \rangle)$ is measurable, and g is a measurable mapping.

If Y is a csm space and $\mathscr{S}$ is an analytic subset of $X \times Y$, the image A of $\mathscr{S}$ under the projection mapping π_0 of $X \times Y$ onto X is an analytic subset of X. The set A consists of precisely those points p of X such that the set Y_p of points q in Y with (p, q) in $\mathscr{S}$ is non-empty. The measurable-selection principle asserts that, in these circumstances, we can select a point from each Y_p in a measurable manner—that is, there is a measurable mapping η of A into Y such that $\eta(p) \in Y_p$ for each p. We know that there is a continuous mapping f_0 of $\mathbb{S}$ into $X \times Y$ with image $\mathscr{S}$. Thus $\pi_0 \circ f_0$ is a continuous mapping f of $\mathbb{S}$ onto A. From the discussion of the preceding paragraph, there is a measurable mapping g from A into $\mathbb{S}$ such that $f \circ g$ is the identity on A. Let π_1 be the projection of $X \times Y$ onto Y. Then $\pi_1 \circ f_0 \circ g$ is a measurable mapping η of A into Y; and, for each p in A, $f_0(g(p))$ $(=(p', q)) \in \mathscr{S}$. Now $p' = \pi_0 f_0(g(p)) = f(g(p)) = p$, so that $\eta(p) = \pi_1(f_0(g(p))) = \pi_1(p, q) = q \in Y_p$; and η is the desired measurable mapping of A into Y.

We summarize this discussion in our measurable-selection principle.

14.3.6. THEOREM. *If (X, μ) is a complete Borel measure space, $X = \bigcup K_n$ with K_n compact and $\mu(K_n)$ finite, X and Y are* csm *spaces, π is the projection of $X \times Y$ onto X, $\mathscr{S}$ is an analytic subset of $X \times Y$, and $A = \pi(\mathscr{S})$, then there is a measurable mapping η of A into Y such that $(p, \eta(p)) \in \mathscr{S}$ for each p in A.*

14.4. Exercises

14.4.1. If $\mathscr{H}$ is the direct integral of the Hilbert spaces $\{\mathscr{H}_p\}$ over (X, μ) and φ, φ' are two decomposable representations of a C^*-algebra $\mathfrak{A}$ on $\mathscr{H}$, show that $\varphi = \varphi'$ if $\varphi_p = \varphi'_p$ almost everywhere.

14.4.2. If $\mathscr{H}$ is the direct integral of the Hilbert spaces $\{\mathscr{H}_p\}$ over (X, μ) and $p \to \varphi_p$, $p \to \varphi'_p$ are decompositions of the representation φ of the norm-separable C^*-algebra $\mathfrak{A}$, prove that $\varphi_p = \varphi'_p$ almost everywhere.

In the five exercises that follow, we outline some basic results of the theory of locally compact, abelian groups. Our goal, attained in Exercise 14.4.10, is a strengthening of the result of Exercise 13.4.23(iii). Our starting point is Haar measure on such groups [H : pp. 250–263]. We indicate the group operations additively (the group "product" of s and t is $s + t$ and the group "inverse" to t is $-t$), the group identity by 0, and the integral of f (in $L_1(G)$) relative to Haar measure by $\int f(t)\, dt$.

14.4.3. Let G be a σ-compact, locally compact, abelian group.

(i) Examine the discussion beginning with Definition 3.2.21 and through to the statement of Proposition 3.2.23, with G in place of $\mathbb{R}$, and conclude that it remains valid.

(ii) Show that the non-zero linear functional ρ on $L_1(G)$ is multiplicative if and only if there is a character χ of G (that is, a continuous homomorphism of G into $\mathbb{T}_1$) such that, for each f in $L_1(G)$,

$$\rho(f) = \int f(t)\chi(t)\, dt = \hat{f}(\chi).$$

[*Hint*. See the proof of Theorem 3.2.26.]

(iii) Let $\{\mathcal{U}_a : a \in \mathbb{A}\}$ be a neighborhood base of 0 in G such that each $\mathcal{U}_a$ has compact closure, and let u_a be a positive function in $L_1(G)$ such that $\|u_a\|_1 = 1$ and $u_a(t) = 0$ when $t \notin \mathcal{U}_a$. Show that $\|f * u_a - f\|_p \to 0$ over the net $\mathbb{A}$ (directed by inclusion of the neighborhoods $\mathcal{U}_a$) for each f in $L_p(G)$. [*Hint*. See Lemma 3.2.24.]

(iv) Show that, for each r in G, different from 0, there is a character χ of G such that $\chi(r) \neq 1$. [*Hint*. Use the pattern of the construction in the comments following the proof of Lemma 3.2.24, in conjunction with (iii), to obtain an abelian C^*-algebra $\mathfrak{A}_0(G)$ (acting on $L_2(G)$). The pure states of $\mathfrak{A}_0(G)$ give rise to characters of G through the construction indicated in the proof of Theorem 3.2.26.]

14.4.4. Let G be a σ-compact, locally compact, abelian group. Show that

(i) the mapping $t \to f_t$ ($t \in G, f \in L_p(G)$) is a continuous mapping from G into the normed space $L_p(G)$ [*Hint*. Establish this first, when f is continuous and vanishes outside a compact set.];

(ii) $f * g$ is continuous when $f \in L_1(G)$ and $g \in L_\infty(G)$;

(iii) $S - S$ $(=\{s - s' : s, s' \in S\})$ contains an (open) neighborhood of 0 when S is a measurable subset of G having positive measure. [*Hint*. Use (ii) with f and g replaced by the characteristic functions of S and $-S$.]

14.4.5. A mapping φ from a measure space into a topological space X is said to be measurable when $\varphi^{-1}(\mathcal{O})$ is a measurable set for each open subset $\mathcal{O}$ of X. Let G be a σ-compact, locally compact, abelian group, and let φ be a measurable homomorphism of G into a topological group H that contains a countable dense subset. Show that φ is continuous. [*Hint*. Use Exercise 14.4.4(iii).]

14.4.6. Let E be a topological group, C be a closed normal subgroup of E, χ be a continuous, idempotent ($\chi(\chi(g)) = \chi(g)$) homomorphism of E onto C, H be the kernel of χ, and φ be the quotient mapping of E onto E/C. Show that

(i) $\chi(c) = c$ for each c in C;

(ii) $s = hc$ for each s in E, where

$$h = s\chi(s)^{-1}(=\xi(s)) \in H, \qquad c = \chi(s) \in C;$$

(iii) $\varphi(\mathscr{U}) = \varphi(\mathcal{O})$ for each (relative open) subset $\mathscr{U}$ of H, where $\mathcal{O} = \xi^{-1}(\mathscr{U})$;

(iv) φ restricts to a topological group isomorphism (homeomorphism and group isomorphism) of H onto E/C provided with the quotient topology (the open subsets of E/C are the images under φ of the open subsets of E).

14.4.7. Let E be a topological group, C be a compact normal subgroup of E, and φ be the quotient mapping of E onto E/C. Suppose E/C is locally compact.

(i) With V a compact subset of E/C containing the unit C of E/C and $\{\mathcal{O}_a : a \in \mathbb{A}\}$ an open covering of $\varphi^{-1}(V)$ $(=U)$, let $\{\mathcal{O}_1(g), \dots \mathcal{O}_n(g)\}$ be a finite subcovering of gC for g in U. Let $\mathcal{O}$ be $\bigcup_{j=1}^{n} \mathcal{O}_j(g)$. Find an open neighborhood $\mathcal{O}_g$ of the unit e of E such that $gC\mathcal{O}_g \subseteq \mathcal{O}$. [*Hint*. For each h in gC, find an open neighborhood V_h of e such that $hV_hV_h \subseteq \mathcal{O}$.]

(ii) With the notation of (i), show that $\{\varphi(g\mathcal{O}_g) : g \in U\}$ is an open covering of V.

(iii) With the notation of (i) and (ii), let $\{\varphi(g(1)\mathcal{O}_{g(1)}), \dots, \varphi(g(m)\mathcal{O}_{g(m)})\}$ be a finite subcovering of V. Show that

$$\{\mathcal{O}_1(g(1)), \dots, \mathcal{O}_{n(1)}(g(1)), \dots, \mathcal{O}_1(g(m)), \dots, \mathcal{O}_{n(m)}(g(m))\}$$

is a covering of U and a finite subcovering of $\{\mathcal{O}_a : a \in \mathbb{A}\}$.

(iv) Conclude that U is compact and that E is locally compact.

14.4.8. Let $\mathcal{M}$ be a factor, α be a continuous automorphic representation of $\mathbb{R}$ on $\mathcal{M}$ by inner automorphisms, G be the group of unitary operators in $\mathcal{M}$ that implement the automorphisms $\alpha(t)$ $(t \in \mathbb{R})$, C be $\{cI : c \in \mathbb{T}_1\}$, and $\eta(t)$ be UC, where U in G implements $\alpha(t)$.

(i) Show that η is well defined and is a homomorphism of $\mathbb{R}$ onto G/C.

Suppose η in (i) is continuous, where G is provided with its weak-operator topology and G/C with its quotient topology (see Exercise 14.4.6(iv)). Let E be the topological group

$$\{(U, t) : U \in G, t \in \mathbb{R}, U \text{ implements } \alpha(t)\}$$

as a subgroup of $G \oplus \mathbb{R}$ with the product topology on $G \times \mathbb{R}$, C_0 be the (closed) subgroup $\{(cI, 0) : c \in \mathbb{T}_1\}$ of E, and $\pi(U, t)$ be t for each (U, t) in E. Show that

(ii) π is a continuous, open homomorphism of E onto $\mathbb{R}$ with kernel C_0, and conclude that E/C_0 is isomorphic and homeomorphic to $\mathbb{R}$ [*Hint* To show that π is open, note that a basis for the open sets of E consists of sets of the form

$$\{(U, t) : a < t < b, \eta(t) = UC, U \in \mathcal{O}, \mathcal{O} \text{ open in } G\}$$

and use the assumption that η is continuous.];

(iii) E is a σ-compact, locally compact, abelian group [*Hint.* Use Exercises 13.4.22 and 14.4.7.];

(iv) each character of C_0 is the restriction of a character of E [*Hint.* Use Exercise 3.5.38 to show that the group of characters of C_0 is $\mathbb{Z}$. With the aid of Exercise 14.4.3, note that the subgroup of $\mathbb{Z}$ consisting of restrictions of characters of E "separates" points of C_0. Conclude that this subgroup is $\mathbb{Z}$.];

(v) there is a continuous, idempotent homomorphism χ of E onto C_0 and a closed subgroup H of E that is homeomorphic and isomorphic to $\mathbb{R}$ [*Hint.* Use (iv) to extend the identity mapping on C_0 to a homomorphism χ with the desired properties. Use Exercise 14.4.6.];

(vi) there is a (continuous) one-parameter unitary group $t \to U_t$ that implements α such that U_t is in $\mathcal{M}$ for each t in $\mathbb{R}$.

14.4.9. Let $\mathcal{R}$ be a von Neumann algebra acting on a separable Hilbert space $\mathcal{H}$. Show that

(i) $(\mathcal{R})_1$ is a csm space when $(\mathcal{R})_1$ is endowed with its strong-operator * topology (determined by the semi-norms $T \to \|Tx\| + \|T^*x\|$) [*Hint.* Introduce a metric of the type described in Exercise 2.8.35. Use the argument of Proposition 2.5.11.];

(ii) the restriction of the strong-operator * topology to the unitary group $\mathcal{U}(\mathcal{R})$ of $\mathcal{R}$ coincides with the strong- (and weak-)operator topology on $\mathcal{U}(\mathcal{R})$ [*Hint.* Use Remark 2.5.10 and Exercise 5.7.5.];

(iii) $\mathcal{U}(\mathcal{R})$ is a closed subset of $(\mathcal{R})_1$ and conclude that $\mathcal{U}(\mathcal{R})$ is a csm space when $\mathcal{U}(\mathcal{R})$ is provided with its strong- (or weak-)operator topology.

14.4.10. Let $\mathcal{M}$ be a factor acting on a separable Hilbert space $\mathcal{H}$ and α be a continuous automorphic representation of $\mathbb{R}$ by inner automorphisms of $\mathcal{M}$.

(i) With E as in Exercise 14.4.8, show that E is a closed subset of $\mathcal{U}(\mathcal{M}) \times \mathbb{R}$, where $\mathcal{U}(\mathcal{M})$, the unitary group of $\mathcal{M}$, is endowed with its strong-operator topology.

(ii) Conclude from (i) and Exercise 14.4.9 that $\mathcal{U}(\mathcal{M}) \times \mathbb{R}$ is a csm space and E is an analytic subset of it.

(iii) Show that there is a measurable mapping $t \to V_t$ of $\mathbb{R}$ into $\mathcal{U}(\mathcal{M})$ such that $(V_t, t) \in E$ for each t in $\mathbb{R}$. [*Hint.* Use Theorem 14.3.6.]

(iv) With η as in Exercise 14.4.8, show that η is continuous. [*Hint.* Use Exercise 14.4.5.]

(v) Show that there is a (continuous) one-parameter unitary group $t \to U_t$ that implements α and such that each U_t is in $\mathcal{M}$.

14.4.11. Let $\mathcal{M}$ be a factor acting on a separable Hilbert space $\mathcal{H}$ and u be a separating and generating unit vector for $\mathcal{M}$. Let $t \to \sigma_t$ be the modular automorphism group corresponding to $(\mathcal{M}, u)$. Suppose each σ_t is inner. Show that

(i) $\mathcal{M}$ is semi-finite [*Hint*. Use Theorem 9.2.21 and Exercise 14.4.10.];
(ii) $\mathcal{R}(\mathcal{M}, \sigma)$ is * isomorphic to $\mathcal{M} \mathbin{\overline{\otimes}} \mathcal{A}$, where $\mathcal{A}$ is the multiplication algebra corresponding to Lebesgue measure on $\mathbb{R}$ [*Hint*. Use Exercise 13.4.17.];
(iii) $\mathcal{R}(\mathcal{M}, \sigma)$ is semi-finite.

14.4.12. Let $\mathcal{M}$ be a factor of type III acting on a separable Hilbert space. Show that

(i) $\mathcal{M}$ has a separating and generating vector [*Hint*. Use Proposition 9.1.6.];
(ii) $\mathcal{M}$ admits an outer automorphism. [*Hint*. Use Exercise 14.4.11.]

14.4.13. Let $\mathcal{R}$ be a von Neumann algebra of type III acting on a separable Hilbert space and let α be the * automorphism of $\mathcal{R} \mathbin{\overline{\otimes}} \mathcal{R}$ described in Exercise 11.5.25(iii). Show that α is outer. [*Hint*. Use Exercises 12.4.19, 12.4.20, and 14.4.12.]

14.4.14. Let X and Y be csm spaces, f be a continuous mapping of X into Y, and A be an analytic subset of Y. Show that $f^{-1}(A)$ is an analytic subset of X. [*Hint*. Choose V a csm space and g a continuous mapping of V onto A. Study $\pi(B)$, where B is the inverse image of the diagonal in $Y \times Y$ under the mapping $(x, v) \to (f(x), g(v))$ of $X \times V$ into $Y \times Y$ and π is the projection of $X \times V$ onto X.]

14.4.15. Let $\mathcal{R}$ be a von Neumann algebra acting on a separable Hilbert space $\mathcal{H}$. Show that

(i) $\{UU' : U \in \mathcal{U}(\mathcal{R}), U' \in \mathcal{U}(\mathcal{R}')\}$ $(= \mathcal{U}_{\mathrm{i}}(\mathcal{R}))$ is an analytic subset of the (csm) space $\mathcal{U}(\mathcal{H})$, where $\mathcal{U}(\mathcal{R})$, $\mathcal{U}(\mathcal{R}')$, and $\mathcal{U}(\mathcal{H})$ are the groups of unitary operators in $\mathcal{R}$, $\mathcal{R}'$, and $\mathcal{B}(\mathcal{H})$, respectively, each provided with its strong-operator topology [*Hint*. Use Exercise 14.4.9, and consider the mapping $(U, U') \to UU'$ of $\mathcal{U}(\mathcal{R}) \times \mathcal{U}(\mathcal{R}')$ into $\mathcal{U}(\mathcal{H})$.];

(ii) an automorphism of $\mathcal{R}$ implemented by a unitary operator V on $\mathcal{H}$ is inner if and only if $V \in \mathcal{U}_{\mathrm{i}}(\mathcal{R})$;

(iii) $T(\mathscr{R})$ is an analytic subset of $\mathbb{R}$. [*Hint*. Use the result of Exercise 14.4.14 in conjunction with (i) and (ii).]

14.4.16. Let $\mathscr{M}$ be a factor of type III acting on a separable Hilbert space $\mathscr{H}$. Show that

(i) $T(\mathscr{M})$ is a subset of $\mathbb{R}$ having Lebesgue measure 0 [*Hint*. Use Exercise 14.4.15(iii), Theorem 14.3.5, Exercise 14.4.4(iii), and Exercise 14.4.11(i).];

(ii) $\mathscr{M}$ admits an automorphism α such that α^n is outer for all positive integers n. [*Hint*. Use (i) and note that $\bigcup_{n=1}^{\infty}[n^{-1}\cdot T(\mathscr{M})]$ has Lebsegue measure 0.]

14.4.17. Let $\mathscr{M}$ be a factor of type III acting on a separable Hilbert space $\mathscr{H}$, and let α be a * automorphism of $\mathscr{M}$ implemented by a unitary operator U on $\mathscr{H}$. Suppose that α^n $(=\alpha_n)$ is outer for each non-zero integer n. Show that

(i) $\mathscr{R}(\mathscr{M}, \alpha)$ is a factor of type III [*Hint*. Use Proposition 13.1.5 and Exercise 13.4.2.];

(ii) $\Phi(\mathscr{M})' \cap \mathscr{R}(\mathscr{M}, \alpha) = \{cI : c \in \mathbb{C}\}$ [*Hint*. Recall the matrix descriptions of the elements of $\Phi(\mathscr{M})$ and of $\mathscr{R}(\mathscr{M}, \alpha)$. Use Exercise 12.4.17(iv).];

(iii) $\Phi(\mathscr{M})'$ is * isomorphic to $\mathscr{M}'$ [*Hint*. Use Proposition 9.1.6 and Theorem 7.2.9.];

(iv) $\mathscr{R}(\mathscr{M}, \alpha)'$ is a proper subset of $\Phi(\mathscr{M})'$ [*Hint*. Show that $U \otimes l_1$ is in $\mathscr{R}(\mathscr{M}, \alpha)$ and not in $\Phi(\mathscr{M})$.];

(v) $\Phi(\mathscr{M})'$ is not normal in the sense of Exercise 12.4.31.

14.4.18. Show that a von Neumann algebra acting on a separable Hilbert space is normal (in the sense of Exercise 12.4.31) if and only if it is a factor of type I. [*Hint*. Use Exercises 12.4.31, 14.4.16, and 14.4.17.]

14.4.19. Let $\mathscr{R}$ acting on a Hilbert space $\mathscr{H}$ be a von Neumann algebra, $\mathscr{S}$ be a von Neumann subalgebra of $\mathscr{R}$, and $t \to V_t$ be a (continuous) one-parameter unitary group on $\mathscr{H}$ that implements one-parameter groups $t \to \sigma'_t$ and $t \to \sigma_t$ of automorphisms of $\mathscr{R}$ and $\mathscr{S}$, respectively, where $t \to \sigma_t$ is the modular automorphism group of $\mathscr{S}$ corresponding to a faithful normal state ω of $\mathscr{S}$. Let Φ' be a faithful, ultraweakly continuous conditional expectation of $\mathscr{R}$ onto $\mathscr{S}$ and $\mathscr{G}$ be a group of unitary operators in $\mathscr{R}$ such that $\mathscr{G}$ and $\mathscr{S}$ generate $\mathscr{R}$ as a von Neumann algebra. Suppose that $V\mathscr{S}V^* = \mathscr{S}$, $\omega(VAV^*) = \omega(A)$ $(A \in \mathscr{S})$, $\sigma'_t(V) = V$, and $\Phi'(V) = 0$, for each V $(\neq I)$ in $\mathscr{G}$. Show that

(i) $\{V'_1A_1 + \cdots + V'_nA_n : V'_j \in \mathscr{G}, A_j \in \mathscr{S}\}$ $(=\mathfrak{A})$ is a self-adjoint subalgebra of $\mathscr{R}$ and $\mathfrak{A}^- = \mathscr{R}$;

(ii) $t \to \sigma'_t$ is the modular automorphism group of $\mathscr{R}$ corresponding to $\omega \circ \Phi'$. [*Hint*. Use (i) and Lemma 9.2.17.]

14.4.20. Let $\mathscr{M}$, acting on the Hilbert space $\mathscr{H}$, be the factor of type III constructed in Exercise 13.4.12, ω_0 be a faithful normal state of $\mathscr{M}$, (J, Δ) be the modular structure and $t \to \sigma_t$ be the modular automorphism group of $\mathscr{M}$ corresponding to ω_0. Thus $t \to \Delta^{it}$ $(=U(t))$ implements $t \to \sigma_t$. Let $\mathscr{R}(\mathscr{M}, \sigma)$ be the (implemented) crossed product of $\mathscr{M}$ by σ considered as an automorphic representation of $\mathbb{R}$ (as a *discrete* group) on $\mathscr{M}$. With the notation of Definition 13.1.3, let $\mathscr{R}$ be the von Neumann algebra (which acts on $\mathscr{H} \otimes l_2(\mathbb{R})$) generated by $(\mathscr{S} =)\Phi(\mathscr{M})$ and $\{V(t) : t \in G\}$, where G is a given subgroup of $\mathbb{R}$ (as a discrete group). Show that

(i) $\mathscr{R}$ is * isomorphic to $\mathscr{R}(\mathscr{M}, \alpha)$, where $\alpha = \sigma | G$ [*Hint*. Show that the projection of $\sum_{t \in \mathbb{R}} \oplus \mathscr{H}$ onto $\sum_{g \in G} \oplus \mathscr{H}$ commutes with $\mathscr{R}$, and the restriction of $\mathscr{R}$ to $\sum_{g \in G} \oplus \mathscr{H}$ is a * isomorphism of $\mathscr{R}$ onto $\mathscr{R}(\mathscr{M}, \alpha)$. Consider generators and matrix representations.];

(ii) $\mathscr{R}$ is a factor of type III. [*Hint*. Use Exercises 13.4.2, 13.4.12, and Proposition 13.1.5(ii).]

Let $\sigma'_t(T)$ be $V(t)TV(t)^*$ $(t \in \mathbb{R}, T \in \mathscr{R})$. With ω the faithful normal state of $\mathscr{S}$ such that $\omega \circ \Phi = \omega_0$, and Φ' the conditional expectation of $\mathscr{R}(\mathscr{M}, \sigma)$ onto $\mathscr{S}$ described in Exercise 13.4.1, let ω' be $(\omega \circ \Phi')|\mathscr{R}$. Show that

(iii) ω' is a faithful normal state of $\mathscr{R}$ and $t \to \sigma'_t$ is the modular automorphism group of $\mathscr{R}$ corresponding to ω' [*Hint*. Use the Exercises 13.4.1 and 14.4.19.];

(iv) $\mathscr{S}' \cap \mathscr{R}(\mathscr{M}, \sigma) = \{cI : c \in \mathbb{C}\}$ [*Hint*. Use Exercise 12.4.17(iv).];

(v) $G = T(\mathscr{R})$ [*Hint*. Use (iii) and (iv). Study the matrix representation of $V(t)$.];

(vi) there is a countably decomposable factor of type III for which the modular automorphism group consists of inner automorphisms. (Compare the result of Exercise 14.4.16(i).)

BIBLIOGRAPHY

General references

[H] P. R. Halmos, "Measure Theory." D. Van Nostrand, Princeton, New Jersey, 1950; reprinted, Springer-Verlag, New York, 1974.

[K] J. L. Kelley, "General Topology." D. Van Nostrand, Princeton, New Jersey, 1955; reprinted, Springer-Verlag, New York, 1975.

[R] W. Rudin, "Real and Complex Analysis," 2nd ed. McGraw-Hill, New York, 1974.

References

[1] W. Ambrose, Spectral resolution of groups of unitary operators, *Duke Math. J.* **11** (1944), 589–595.

[2] J. Anderson, Extreme points in sets of positive linear mappings on $\mathscr{B}(\mathscr{H})$, *J. Fnal. Anal.* **31** (1979), 195–217.

[3] J. Anderson and J. W. Bunce, A type II_{∞} factor representation of the Calkin algebra, *Amer. J. Math.* **99** (1977), 515–521.

[4] H. Araki and E. J. Woods, A classification of factors, *Publ. Res. Inst. Math. Sci. Kyoto* **4** (1968), 51–130.

[5] O. Bratteli, Inductive limits of finite dimensional C^*-algebras, *Trans. Amer. Math. Soc.* **171** (1972), 195–234.

[6] D. J. C. Bures, Certain factors constructed as infinite tensor products, *Compositio Math.* **15** (1963), 169–191.

[7] J. W. Calkin, Two-sided ideals and congruences in the ring of bounded operators in Hilbert space, *Ann. of Math.* **42** (1941), 839–873.

[8] M-D. Choi and E. G. Effros, Separable nuclear C^*-algebras and injectivity, *Duke Math. J.* **43** (1976), 309–322.

[9] M-D. Choi and E. G. Effros, Nuclear C^*-algebras and injectivity: the general case, *Indiana Univ. Math. J.* **26** (1977), 443–446.

[10] F. Combes, Poids sur une C^*-algèbre, *J. Math. Pures Appl.* **47** (1968), 57–100.

[11] F. Combes, Poids associé à une algèbre hilbertienne à gauche, *Compositio Math.* **23** (1971), 49–77.

[12] A. Connes, Une classification des facteurs de type III, *Ann. Sci. École Norm. Sup. Paris* **6** (1973), 133–252.

[13] A. Connes, Classification of injective factors, Cases II_1, II_{∞}, III_{λ}, $\lambda \neq 1$, *Ann. of Math.* **104** (1976), 73–115.

[14] A. Van Daele, The Tomita–Takesaki theory for von Neumann algebras with a separating and cyclic vector, *in* "C^*-Algebras and Their Applications to Statistical Mechanics and Quantum Field Theory" (*Proc. Internat. School of Physics "Enrico Fermi," Course LX, Varenna*, D. Kastler, ed., 1973), pp. 19–28. North-Holland Publ., Amsterdam, 1976.

[15] A. Van Daele, A new approach to the Tomita–Takesaki theory of generalized Hilbert algebras, *J. Fnal. Anal.* **15** (1974), 378–393.

[16] A. Van Daele, "Continuous Crossed Products and Type III von Neumann Algebras," London Math. Soc. Lecture Note Series 31. Cambridge University Press, London, 1978.

[17] J. Dixmier, Les anneaux d'opérateurs de classe finie, *Ann. Sci. École. Norm. Sup. Paris* **66** (1949), 209–261.

[18] J. Dixmier, Les fonctionnelles linéaires sur l'ensemble des opérateurs bornés d'un espace de Hilbert, *Ann. of Math.* **51** (1950), 387–408.

[19] J. Dixmier, Formes linéaires sur un anneau d'opérateurs, *Bull. Soc. Math. France* **81** (1953), 9–39.

[20] J. Dixmier, Sur les anneaux d'opérateurs dans les espaces hilbertiens, *C. R. Acad. Sci. Paris* **238** (1954), 439–441.

[21] J. Dixmier, "Les Algèbres d'Opérateurs dans l'Espace Hilbertien." Gauthier-Villars, Paris, 1957; 2nd ed., 1969.

[22] J. Dixmier, "Les C^*-Algèbres et leurs Représentations." Gauthier-Villars Paris, 1964. [*English translation*; C^*-Algebras. North-Holland Mathematical Library, Vol. 15. North-Holland Pub., Amsterdam, 1977.]

[23] J. Dixmier, Existence de traces non normales, *C. R. Acad. Sci. Paris* **262** (1966), 1107–1108.

[24] H. A. Dye, The Radon–Nikodým theorem for finite rings of operators, *Trans. Amer. Math. Soc.* **72** (1952), 243–280.

[25] E. G. Effros and E. C. Lance, Tensor products of operator algebras, *Adv. in Math.* **25** (1977), 1–34.

[26] J. M. G. Fell and J. L. Kelley, An algebra of unbounded operators, *Proc. Nat. Acad. Sci. U.S.A.* **38** (1952), 592–598.

[27] L. T. Gardner, On isomorphisms of C^*-algebras, *Amer. J. Math.* **87** (1965), 384–396.

[28] I. M. Gelfand and M. A. Neumark, On the imbedding of normed rings into the ring of operators in Hilbert space, *Mat. Sb.* **12** (1943), 197–213.

[29] J. G. Glimm, On a certain class of operator algebras, *Trans. Amer. Math. Soc.* **95** (1960), 318–340.

[30] J. G. Glimm and R. V. Kadison, Unitary operators in C^*-algebras, *Pacific J. Math.* **10** (1960), 547–556.

[31] E. L. Griffin, Some contributions to the theory of rings of operators, *Trans. Amer. Math. Soc.* **75** (1953), 471–504.

[32] E. L. Griffin, Some contributions to the theory or rings of operators. II, *Trans. Amer. Math. Soc.* **79** (1955), 389–400.

[33] U. Haagerup, Tomita's theory for von Neumann algebras with a cyclic and separating vector (private circulation, June 1973).

[34] U. Haagerup, All nuclear C^*-algebras are amenable, *Invent. Math.* **74** (1983), 305–319.

[35] H. Hahn, Über die Integrale des Herrn Hellinger und die Orthogonal invarianten der quadratischen Formen von unendlich vielen Veränderlichen, *Monatshefte für Mathematik und Physik* **23** (1912), 161–224.

[36] P. R. Halmos, "Introduction to Hilbert Space and the Theory of Spectral Multiplicity." Chelsea Publ., New York, 1951.

[37] F. Hansen and G. K. Pedersen, Jensen's inequality for operators and Löwner's theorem, *Math. Ann.* **258** (1982), 229–241.

[38] E. Hellinger, Neue Begründung der Theorie quadratischer Formen von unendlichvielen Veränderlichen, *J. für Math.* **136** (1909), 210–271.

[39] D. Hilbert, Grundzüge einer allgemeine Theorie der linearen Integralgleichungen IV, *Nachr. Akad. Wiss. Göttingen. Math.-Phys. Kl.* 1904, 49–91.

[40] R. V. Kadison, Isometries of operator algebras, *Ann. of Math.* **54** (1951), 325–338.

[41] R. V. Kadison, On the additivity of the trace in finite factors, *Proc. Nat. Acad. Sci. U.S.A.* **41** (1955), 385–387.
[42] R. V. Kadison, Irreducible operator algebras, *Proc. Nat. Acad. Sci. U.S.A.* **43** (1957), 273–276.
[43] R. V. Kadison, Unitary invariants for representations of operator algebras, *Ann. of Math.* **66** (1957), 304–379.
[44] R. V. Kadison, The trace in finite operator algebras, *Proc. Amer. Math. Soc.* **12** (1961), 973–977.
[45] R. V. Kadison, Similarity of operator algebras, *Acta Math.* **141** (1978), 147–163.
[46] I. Kaplansky, A theorem on rings of operators, *Pacific J. Math.* **1** (1951), 227–232.
[47] I. Kaplansky, Projections in Banach algebras, *Ann. of Math.* **53** (1951), 235–249.
[48] I. Kaplansky, Algebras of type I, *Ann. of Math.* **56** (1952), 460–472.
[49] I. Kaplansky, Representations of separable algebras, *Duke Math. J.* **19** (1952), 219–222.
[50] J. L. Kelley, Commutative operator algebras, *Proc. Nat. Acad. Sci. U.S.A.* **38** (1952), 598–605.
[51] E. C. Lance, On nuclear C^*-algebras, *J. Fnal. Anal.* **12** (1973), 157–176.
[52] E. C. Lance, Tensor products of C^*-algebras, *in* "C^*-algebras and Their Applications to Statistical Mechanics and Quantum Field Theory" (*Proc. Internat. School of Physics "Enrico Fermi," Course LX, Varenna*, D. Kastler, ed., 1973), pp. 154–166. North-Holland Publ., Amsterdam, 1976.
[53] G. W. Mackey, Induced representations of locally compact groups. II. The Frobenius reciprocity theorem, *Ann. of Math.* **58** (1953), 193–221.
[54] F. I. Mautner, Unitary representations of locally compact groups I, *Ann. of Math.* **51** (1950), 1–25.
[55] Y. Misonou, On the direct product of W^*-algebras, *Tôhoku. Math. J.* **6** (1954), 189–204.
[56] F. J. Murray and J. von Neumann, On rings of operators, *Ann. of Math.* **37** (1936), 116–229.
[57] F. J. Murray and J. von Neumann, On rings of operators, II, *Trans. Amer. Math. Soc.* **41** (1937), 208–248.
[58] F. J. Murray and J. von Neumann, On rings of operators. IV, *Ann. of Math.* **44** (1943), 716–808.
[59] H. Nakano, Unitärinvarianten im allgemeinen Euklidischen Raum, *Math. Ann.* **118** (1941), 112–133.
[60] H. Nakano, Unitärinvariante hypermaximale normale Operatoren, *Ann. of Math.* **42** (1941), 657–664.
[61] J. von Neumann, Zur Algebra der Funktionaloperationen und Theorie der normalen Operatoren, *Math. Ann.* **102** (1930), 370–427.
[62] J. von Neumann, Allgemeine Eigenwerttheorie Hermitescher Funktionaloperatoren, *Math. Ann.* **102** (1930), 49–131.
[63] J. von Neumann, Über Funktionen von Funktionaloperatoren, *Ann. of Math.* **32** (1931), 191–226.
[64] J. von Neumann, Über adjungierte Funktionaloperatoren, *Ann. of Math.* **33** (1932), 294–310.
[65] J. von Neumann, On infinite direct products, *Compositio Math.* **6** (1938), 1–77.
[66] J. von Neumann, On rings of operators. III, *Ann. of Math.* **41** (1940), 94–161.
[67] J. von Neumann, On rings of operators. Reduction theory, *Ann. of Math.* **50** (1949), 401–485.
[68] M. Neumark, Positive definite operator functions on a commutative group (in Russian, English summary), *Bull, Acad. Sci. URSS Ser. Math.* [*Izv. Akad. Nauk SSSR*] **7** (1943), 237–244.
[69] G. K. Pedersen, Measure theory for C^*-algebras, *Math. Scand.* **19** (1966), 131–145.

[70] G. K. Pedersen, "C^*-Algebras and Their Automorphism Groups," London Mathematical Society Monographs, Vol. 14. Academic Press, London, 1979.

[71] A. Plessner and V. Rohlin, Spectral theory of linear operators, II (in Russian), *Uspehi Mat. Nauk N.S.* **1** (1946), 71–191.

[72] R. T. Powers, Representations of uniformly hyperfinite algebras and their associated von Neumann rings, *Ann. of Math.* **86** (1967), 138–171.

[73] G. A. Reid, On the Calkin representations, *Proc. London Math. Soc.* **23** (1971), 547–564.

[74] M. Rieffel and A. Van Daele, The commutation theorem for tensor products of von Neumann algebras, *Bull. London Math. Soc.* **7** (1975), 257–260.

[75] M. Rieffel and A. Van Daele, A bounded operator approach to the Tomita–Takesaki theory, *Pacific J. Math.* **69** (1977), 187–221.

[76] F. Riesz, "Les Systèmes d'Équations Linéaires à une Infinité d'Inconnues." Gauthier-Villars, Paris, 1913.

[77] F. Riesz, Über die linearen Transformationen des komplexen Hilbertschen Raumes, *Acta Sci. Math. (Szeged)* **5** (1930–32), 23–54.

[78] S. Sakai, On topological properties of W^*-algebras, *Proc. Japan Acad.* **33** (1957), 439–444.

[79] S. Sakai, On linear functionals of W^*-algebras, *Proc. Japan Acad.* **34** (1958), 571–574.

[80] S. Sakai, A Radon–Nikodym theorem in W^*-algebras, *Bull. Amer. Math. Soc.* **71** (1965), 149–151.

[81] S. Sakai, On a problem of Calkin, *Amer. J. Math.* **88** (1966), 935–941.

[82] S. Sakai, "C^*-Algebras and W^*-Algebras," Ergebnisse der Mathematik und ihrer Grenzgebiete, 60. Springer-Verlag, Heidelberg, 1971.

[83] J. T. Schwartz, Two finite, non-hyperfinite, non-isomorphic factors, *Comm. Pure Appl. Math.* **16** (1963), 19–26.

[84] I. E. Segal, Irreducible representations of operator algebras, *Bull. Amer. Math. Soc.* **53** (1947), 73–88.

[85] I. E. Segal, Two-sided ideals in operator algebras, *Ann. of Math.* **50** (1949), 856–865.

[86] S. Sherman, The second adjoint of a C^*-algebra, *Proc. Int. Congress of Mathematicians, Cambridge, 1950*, Vol. 1. p. 470.

[87] M. H. Stone, On one-parameter unitary groups in Hilbert space, *Ann. of Math.* **33** (1932), 643–648.

[88] M. H. Stone, "Linear Transformations in Hilbert Space and Their Applications to Analysis," American Mathematical Society Colloquium Publications, Vol. 15. Amer. Math. Soc., New York, 1932.

[89] M. H. Stone, The generalized Weierstrass approximation theorem, *Math. Mag.* **21** (1948), 167–183, 237–254.

[90] M. H. Stone, Boundedness properties in function-lattices, *Canad. J. Math.* **1** (1949), 176–186.

[91] S. Strătilă and L. Zsidó, "Lectures on von Neumann Algebras." Abacus Press, Tunbridge Wells, 1979.

[92] Z. Takeda, Conjugate spaces of operator algebras, *Proc. Japan Acad.* **30** (1954), 90–95.

[93] M. Takesaki, On the conjugate space of operator algebra, *Tôhoku Math. J.* **10** (1958), 194–203.

[94] M. Takesaki, On the cross-norm of the direct product of C^*-algebras, *Tôhoku Math. J.* **16** (1964), 111–122.

[95] M. Takesaki, "Tomita's Theory of Modular Hilbert Algebras and Its Applications," Lecture Notes in Mathematics, Vol. 128. Springer-Verlag, Heidelberg, 1970.

[96] M. Takesaki, A short proof for the commutation theorem $(\mathcal{M}_1 \mathbin{\overline{\otimes}} \mathcal{M}_2)' = \mathcal{M}'_1 \mathbin{\overline{\otimes}} \mathcal{M}'_2$. "Lectures on Operator Algebras," Lecture Notes in Mathematics, Vol. 247, pp. 780–786. Springer-Verlag, Heidelberg, 1972.

[97] M. Takesaki, Duality for crossed products and the structure of von Neumann algebras of type III, *Acta Math.* **131** (1973), 249–310.

[98] M. Takesaki, "Theory of Operator Algebras I," Springer-Verlag, Heidelberg, 1979.

[99] M. Tomita, Standard forms of von Neumann algebras, *Fifth Functional Analysis Symposium of the Math. Soc. of Japan, Sendai, 1967.*

[100] J. Tomiyama, On the projection of norm one in W^*-algebras, *Proc. Japan Acad.* **33** (1957), 608–612.

[101] T. Turumaru, Crossed product of operator algebras, *Tôhoku Math. J.* **10** (1958), 355–365.

[102] F. Wecken, Unitärinvarianten selbstadjungierter Operatoren, *Math. Ann.* **116** (1939), 422–455.

[103] S. L. Woronowicz, "Operator Systems and Their Applications to the Tomita–Takesaki Theory," Lecture Note Series, Vol. 52. Aarhus Universitet, Matematisk Institut, Aarhus, 1979.

[104] F. J. Yeadon, A new proof of the existence of a trace in a finite von Neumann algebra, *Bull. Amer. Math. Soc.* **77** (1971), 257–260.

[105] L. Zsidó, A proof of Tomita's fundamental theorem in the theory of standard von Neumann algebras, *Rev. Roum. Math. Pure Appl.* **20** (1975), 609–619.

INDEX OF NOTATION

Algebras and related matters

Direct sums and integrals

Equivalences and orderings

Linear spaces

Linear topological spaces, Banach spaces, Hilbert spaces

Modular theory

Multiplicity theory

Sets and mappings

Special Banach spaces

States and weights

Tensor products and crossed products

INDEX

D

E

F

G

H

I

N

Q

R

S

T

V

W